Food Biochemistry and Food Processing

Food Biochemistry and Food Processing

Editor
Y. H. Hui

Associate Editors
Wai-Kit Nip
Leo M.L. Nollet
Gopinadhan Paliyath
Benjamin K. Simpson

Blackwell Publishing Professional
2121 State Avenue, Ames, Iowa 50014, USA

Orders: 1-800-862-6657
Office: 1-515-292-0140
Fax: 1-515-292-3348
Web site: www.blackwellprofessional.com

Blackwell Publishing Ltd
9600 Garsington Road, Oxford OX4 2DQ, UK
Tel.: 144 (0)1865 776868

Blackwell Publishing Asia
550 Swanston Street, Carlton, Victoria 3053, Australia
Tel.: 161 (0)3 8359 1011

First edition, 2006

Library of Congress Cataloging-in-Publication Data

Food biochemistry and food processing / editor, Y.H. Hui ; associate editors, Wai-Kit Nip . . . [et al.].— 1st ed.
p. cm.
Includes index.
ISBN-13: 978-0-8138-0378-4 (alk. paper)
ISBN-10: 0-8138-0378-0 (alk. paper)
1. Food industry and trade—Research. I. Hui, Y. H. (Yiu H.)
TP370.8.F66 2006
664—dc22

2005016405

The last digit is the print number: 9 8 7 6 5 4 3 2

Contents

Contributors

Harry Ako (Chapter 6)
Department of Molecular Biosciences and Bioengineering
University of Hawaii at Manoa
Honolulu, HI 96822, USA
Phone: 808-956-2012
Fax: 808-956-3542
Email: hako@hawaii.edu

I. A. Axarli (Chapter 8)
Enzyme Technology Laboratory
Department of Agricultural Biotechnology
Agricultural University of Athens
Iera Odos 75
11855 Athens, Greece

Dulal Borthakur (Chapter 3)
Department of Molecular Biosciences and Bioengineering
University of Hawaii at Manoa
Honolulu, Hawaii 96822, USA
Phone: 808-956-6600
Fax: 808-956-3542
Email: dulal@hawaii.edu

Terri D. Boylston (Chapter 26)
Food Science and Human Nutrition
Iowa State University
2547 Food Sciences Building
Ames, IA 50011, USA
Phone: 515-294-0077
Fax: 515-294-8181
Email: tboylsto@iastate.edu

Chung Chieh (Chapter 5)
Department of Chemistry
University of Waterloo
Waterloo, Ontario N2L 3G1, Canada
Phone (office): 519-888-4567 ext. 5816
Phone (home): 519-746-5133
Fax: 519-746-0435
Email: cchieh@uwaterloo.ca

Robin L.T. Churchill (Chapter 31)
Department of Environmental Biology
University of Guelph
Guelph, Ontario N1G 2W1, Canada

Nieves Corzo (Chapter 4)
Instituto de Fermentaciones Industriales (CSIC)
c/Juan de la Cierva, 3
28006 Madrid, Spain
Phone: 34 91 562 2900
Fax: 34 91 564 4853
Email: ifiv308@ifi.csic.es

C. M. Courtin (Chapter 25)
Department of Food and Microbial Technology
Faculty of Applied Bioscience and Engineering
Katholieke Universiteit Leuven
Kasteelpark Arenberg 20
B-3001 Leuven, Belgium
Phone: +32 16 321 634
Fax: +32 16 321 997
Email: christophe.courtin@agr.kuleuven.ac.be

N. Cross (Chapter 16)
Cross Associates
4461 North Keokuk Avenue
Apt. 1
Chicago, IL 60630, USA
Phone: 773-545-9289
Email: n.cross@sbcglobal.net

Maria Dolores del Castillo (Chapter 4)
Instituto de Fermentaciones Industriales (CSIC)
c/Juan de la Cierva, 3
28006 Madrid, Spain
Phone: 34 91 562 2900
Fax: 34 91 564 4853
Email: ifiv308@ifi.csic.es

J. A. Delcour (Chapter 25)
Department of Food and Microbial Technology
Faculty of Applied Bioscience and Engineering
Katholieke Universiteit Leuven
Kasteelpark Arenberg 20
B-3001 Leuven, Belgium
Phone: +32 16 321 634
Fax: +32 16 321 997
Email: jan.delcour@agr.kuleuven.ac.be

Juana Fernández-López (Chapter 15)
Departamento de Tecnología Agroalimentaria
Escuela Politécnica Superior de Orihuela
Universidad Miguel Hernández
Camino a Beniel s/n 03313 Desamparados
Orihuela (Alicante), Spain
Phone: +34 6 674 9656
Fax: +34 6 674 9609/674 9619
Email: j.fernandez@umh.es or
juana.fernaandez@accesosis.es

Alisdair R. Fernie (Chapter 11)
Max Planck Institute of Molecular Plant Physiology
Am Mühlenberg 1
14476 Golm, Germany

Patrick F. Fox (Chapter 19, 20)
Food Science and Technology
University College Cork
Cork, Ireland
Phone: 00 353 21 490 2362
Fax: 00 353 21 427 0001
Email: pff@ucc.ie

Peter Geigenberger (Chapter 11)
Max Planck Institute of Molecular Plant Physiology
Am Mühlenberg 1
14476 Golm, Germany
Email: geigenberger@mpimp-golm.mpg.de

Juliet A. Gerrard (Chapter 9)
School of Biological Sciences
University of Canterbury,
Christchurch, New Zealand
Phone: +64 03 364 2987
Fax: +64 03 364 2950
Email: juliet.gerrard@canterbury.ac.nz

J. Christopher Hall (Chapter 31)
Department of Environmental Biology
University of Guelph
Guelph, Ontario N1G 2W1, Canada
Phone: 519-824-4120 ext. 52740
Fax: 519-837-0442
Email: jchall@evb.uoguelph.ca

Dominic F. Houlihan (Chapter 18)
School of Biological Sciences
University of Aberdeen, Aberdeen, UK

Y. H. Hui (Editor, Chapter 16)
Science Technology System
P.O. Box 1374
West Sacramento, CA 95691, USA
Phone: 916-372-2655
Fax: 916-372-2690
Email: yhhui@aol.com

Chii-Ling Jeang (Chapter 7)
Department of Food Science
National Chung Hsing University
Taichung, Taiwan 40227, Republic of China
Phone: 886 4 228 62797
Fax: 886 4 228 76211
Email: cljeang@nchu.edu.tw

Sandro Jube (Chapter 3)
Department of Molecular Biosciences and Bioengineering
University of Hawaii at Manoa
Honolulu, HI 96822, USA
Phone: 808-956-8210
Fax: 808-956-3542
Email: sandro@hawaii.edu

Mary S. Kalamaki (Chapter 12)
Department of Pharmaceutical Sciences
Aristotle University of Thessaloniki
54124 Thessaloniki, Greece
Phone: +30 2310 412238
Fax: +30 2310 412238
Email: mskalamaki@panafonet.gr

Alan L. Kelly (Chapter 19, 20)
Food Science and Technology
University of Cork
Cork, Ireland
Phone: 00 353 21 490 3405
Fax: 00 353 21 427 0001
Email: a.kelly@ucc.ie

G. A. Kotzia (Chapter 8)
Enzyme Technology Laboratory
Department of Agricultural Biotechnology
Agricultural University of Athens
Iera Odos 75
11855 Athens, Greece

H. G. Kristinsson (Chapter 16)
University of Florida
Laboratory of Aquatic Food Biomolecular Research
Aquatic Food Products Program
Department of Food Science and Human Nutrition
Gainesville FL 32611, USA
Phone: 352-392-1991 ext. 500
Fax: 352-392-9467
Email: HGKristinsson@mail.ifas.ufl.edu

Michael Krogsgaard Nielsen (Chapter 17)
Food Biotechnology and Engineering Group
Food Biotechnology
BioCentrum-DTU
Technical University of Denmark
Phone: 45 45 25 25 92
Email: mkn@dfu.min.dk

N. E. Labrou (Chapter 8)
Enzyme Technology Laboratory
Department of Agricultural Biotechnology
Agricultural University of Athens
Iera Odos 75
11855 Athens, Greece
Phone and Fax: +30 210 529 4308
Email: Lambrou@aua.gr

Hung Lee (Chapter 31)
Department of Environmental Biology
University of Guelph
Guelph, Ontario N1G 2W1, Canada

M. H. Lim (Chapter 16)
Department of Food Science
University of Otago
Dunedin, New Zealand
Phone: 64 3 479 7953
Fax: 64 3 479 7567
miang.lim@stonebow.otago.ac.nz

Massimo Marcone (Chapter 2)
Department of Food Science
University of Guelph
Guelph, Ontario N1G 2W1, Canada
Phone: 519-824-4120, ext. 58334
Fax: 519-824-6631
Email: mmarcone@uoguelph.ca

Samuel A. M. Martin (Chapter 18)
School of Biological Sciences
University of Aberdeen, Aberdeen, UK

Vikram V. Mistry (Chapter 10)
Dairy Science Department
South Dakota State University
Brookings SD 57007, USA
Phone: 605-688-5731
Fax: 605-688-6276
Email: vikram_mistry@sdstate.edu

Dennis P. Murr (Chapter 21)
Department of Plant Agriculture
University of Guelph
Guelph, Ontario N1G 2W1, Canada
Phone: 519-824-4120 ext. 53578
Email: dmurr@uoguelph.ca

Judith A. Narvhus (Chapter 27)
Dept of Chemistry, Biotechnology, and Food Science
Norwegian University of Life Sciences
Box 5003
1432 Aas, Norway
Email: judith.narvhus@umb.no

Henrik Hauch Nielsen (Chapter 17)
Danish Institute for Fisheries Research
Department of Seafood Research
Søltofts Plads
Technical University of Denmark, Bldg. 221
DK-2800 Kgs. Lyngby, Denmark
Phone: +45 45 25 25 93
Fax: +45 45 88 47 74
Email: hhn@dfu.min.dk

Michael Krogsgaard Nielsen (Chapter 17)
Food Biotechnology and Engineering Group
Food Biotechnology
BioCentrum-DTU
Technical University of Denmark, Bldg. 221
DK-2800 Kgs. Lyngby, Denmark
Phone: +45 45 25 25 92
Fax: +45 45 88 47 74
Email: mkn@dfumin.dk

Wai-kit Nip (Associate Editor, Chapters 1, 6, 16)
Department of Molecular Biosciences and Bioengineering
University of Hawaii at Manoa
Honolulu, HI 96822, USA
Phone: 808-956-3852
Fax: 808-955-6942
Email: wknip@hawaii.edu

Leo M. L. Nollet (Associate Editor)
Hogeschool Gent
Department of Engineering Sciences
Schoonmeersstraat 52
B9000 Gent, Belgium
Phone: 00 329 242 4242
Fax: 00 329 243 8777
Email: leo.nollet@hogent.be

Joseph A. Odumeru (Chapter 30)
Laboratory Services Division
University of Guelph
95 Stone Road West
Guelph, Ontario N1H 8J7, Canada
Phone: 519-767-6243
Fax Number: 519-767-6240
Email: jodumeru@lsd.uoguelph.ca

Moustapha Oke (Chapters 22, 23)
Ontario Ministry of Agriculture and Food
1 Stone Road West, 2nd Floor SW
Guelph, Ontario N1G 4Y2, Canada
Email: Moustapha.oke@omaf.gov.on.ca

Gopinadhan Paliyath (Associate Editor, Chapters 21, 22, 23)
Department of Plant Agriculture
University of Guelph
Guelph, Ontario N1G 2W1, Canada
Phone: 519-824-4120, ext. 54856
Email: gpaliyat@uoguelph.ca

José Angel Pérez-Alvarez (Chapter 15)
Departamento de Tecnología Agroalimentaria
Escuela Politécnica Superior de Orihuela
Universidad Miguel Hernández
Camino a Beniel s/n 03313 Desamparados
Orihuela (Alicante), Spain
Phone: +34 06 674 9656
Fax: +34 06 674 9609/674 9619
Email: ja.perez@umh.es

D. Platis
Laboratory of Enzyme Technology
Department of Agricultural Biotechnology
Agricultural University of Athens
Iera Odos 75
11855 Athens, Greece

Bianca M. Poli (Chapter 18)
Department of Animal Production
University of Florence
Florence, Italy

Milagro Reig (Chapter 13)
Instituto de Agroquímica y Tecnología de Alimentos (CSIC)
P.O. Box 73
46100 Burjassot (Valencia), Spain

Douglas J.H. Shyu (Chapter 7)
Graduate Institute of Biotechnology
National Chung Hsing University
Taichung, Taiwan 40227, Republic of China
Phone: 886 4 228 840328
Fax: 886 4 228 53527
Email: dshyu@ms7.hinet.net

Benjamin K. Simpson (Associate Editor)
Department of Food Science
McGill University, MacDonald Campus
21111 Lakeshore Road
St. Anne Bellevue PQ H9X3V9, Canada
Phone: 514-398-7737
Fax: 514-398-7977
Email: simpson@macdonald.mcgill.ca

L. F. Siow (Chapter 16)
Department of Food Science
University of Otago
Dunedin, New Zealand
Phone: 64 3 479 7953
Fax: 64 3 479 7567

Terje Sørhaug (Chapter 27)
Department of Chemistry, Biotechnology, and Food Science
Norwegian University of Life Sciences
Box 5003
1432 Aas, Norway
Email: terje.sorhaug@umb.no

P. S. Stanfield (Chapter 16)
Dietetic Resources
794 Bolton St.
Twin Falls, ID 83301, USA
Phone: 208-733-8662
Email: PStanfld@PMT.org

Nikolaos G. Stoforos (Chapter 12)
Department of Chemical Engineering
Aristotle University of Thessaloniki
54124 Thessaloniki, Greece
Phone: +30 2310 996450
Fax: +30 2310 996259
Email: stoforos@cheng.auth.gr

Petros S. Taoukis (Chapter 12)
Laboratory of Food Chemistry and Technology
School of Chemical Engineering
National Technical University of Athens
Iroon Polytechniou 5
15780 Athens, Greece
Phone: +30 210 772 3171
Fax: +30 210 772 3163
Email: taoukis@chemeng.ntua.gr>

Fidel Toldrá (Chapters 13, 14, 28)
Instituto de Agroquímica y Tecnología
de Alimentos (CSIC)
Apt. 73
46100 Burjassot (Valencia), Spain
Phone: 34 96 390 0022
Fax: 34 96 363 6301
Email: ftoldra@iata.csic.es

Jason T.C. Tzen (Chapter 7)
Graduate Institute of Biotechnology
National Chung Hsing University
Taichung, Taiwan 40227, Republic of China
Phone: 886 4 228 840328
Fax: 886 4 228 53527
Email: tctzen@dragon.nchu.edu.tw

T. Verwimp (Chapter 25)
Department of Food and Microbial Technology
Faculty of Applied Bioscience and Engineering
Katholieke Universiteit Leuven
Kasteelpark Arenberg 20
B-3001 Leuven, Belgium
Phone: +32 16 321 634
Fax: +32 16 321 997
Email: tiny.verwimp@agr.kuleuven.ac.be

Oddur T. Vilhelmsson (Chapter 18)
Faculty of Natural Resource Sciences
University of Akureyri
Borgum, Room 243
IS-600 Akureyri, Iceland
Phone: +354 460 8503
Fax: +354 460 8999
Mobile: +354 697 4252
Email: oddurv@unak.is

Mar Villamiel (Chapters 4, 24)
Instituto de Fermentaciones Industriales (CSIC)
c/Juan de la Cierva, 3
28006 Madrid, Spain
Phone: 34 91 562 2900
Fax: 34 91 564 4853
Email: ifiv308@ifi.csic.es

Ronnie Willaert (Chapter 29)
Department of Ultrastructure
Flanders Interuniversity Institute for Biotechnology
Vrije Universiteit Brussel
Pleinlaan 2
B-1050 Brussels, Belgium
Phone: 32 2 629 18 46
Fax: 32 2 629 19 63
Email: Ronnie.Willaert@vub.ac.be

Preface

In the last 20 years, the role of food biochemistry has assumed increasing significance in all major disciplines within the categories of food science, food technology, food engineering, food processing, and food biotechnology. In the five categories mentioned, progress has advanced exponentially. As usual, dissemination of information on this progress is expressed in many media, both printed and electronic. Books are available for almost every specialty area within the five disciplines mentioned, numbering in the hundreds. As is well known, the two areas of food biochemistry and food processing are intimately related. However, books covering a joint discussion of these topics are not so common. This book attempts to fill this gap, using the following approaches:

- Principles of food biochemistry,
- Advances in selected areas of food biochemistry,
- Food biochemistry and the processing of muscle foods and milk,
- Food biochemistry and the processing of fruits, vegetables, and cereals,
- Food biochemistry and the processing of fermented foods, and
- Microbiology and food safety.

The above six topics are divided over 31 chapters. Subject matters discussed under each topic are briefly reviewed below.

- The principles of food biochemistry are explored in definitions, applications, and analysis and in advances in food biotechnology. Specific examples used include enzymes, protein cross-linking, chymosin in cheesemaking, starch synthesis in the potato tuber, pectinolytic enzymes in tomatoes, and food hydration chemistry and biochemistry.
- The chemistry and biochemistry of muscle foods and milk are covered under the color of muscle foods, raw meat and poultry, processed meat and poultry, seafood enzymes, seafood processing, proteomics and fish processing, milk constituents, and milk processing. The chemistry and biochemistry of fruits, vegetables, and cereals are covered in raw fruits, fruits processing, vegetable processing, rye flours, and nonenzymatic browning of cereal baking products. The chemistry and biochemistry of fermented foods touch on four groups of products: dairy products, bakery and cereal products, fermented meat, and beer.
- The topic of microbiology and food safety covers microbial safety and food processing, and emerging bacterial foodborne pathogens.

This reference and classroom text is a result of the combined effort of more than 50 professionals from industry, government, and academia. These professionals are from more than 15 countries and have diverse expertise and background in the discipline of food biochemistry and food processing. These experts were led by an international editorial team of five members from three countries. All these individuals, authors and editors, are responsible for assembling in one place the scientific topics of food biochemistry and food processing, in their immense complexity. In sum, the end product is unique, both in depth and breadth, and will serve as

- An essential reference on food biochemistry and food processing for professionals in the government, industry, and academia.

- A classroom text on food biochemistry and food processing in an undergraduate food science program.

The editorial team thanks all the contributors for sharing their experience in their fields of expertise. They are the people who made this book possible. We hope you enjoy and benefit from the fruits of their labor.

We know how hard it is to develop the content of a book. However, we believe that the production of a professional book of this nature is even more difficult. We thank the editorial and production teams at Blackwell Publishing for their time, effort, advice, and expertise. You are the best judge of the quality of this book.

Y. H. Hui
W. K. Nip
L. M. L. Nollet
G. Paliyath
B. K. Simpson

Part I
Principles

1
Food Biochemistry—An Introduction

W. K. Nip

INTRODUCTION

Food losses and food poisoning have been recognized for centuries, but the causes of these problems were not understood. Improvements in food products by proper handling and primitive processing were practiced without knowing the reasons. Food scientists and technologists started to investigate these problems about 60 years ago. Currently, some of these causes are understood, and others are still being investigated. These causes may be microbiological, physical (mechanical), and/or chemical (including biochemical). Food scientists and technologists also recognized long ago the importance of a background in biochemistry, in addition to the basic sciences (chemistry, physics, microbiology, and mathematics). This was demonstrated by a general biochemistry course requirement in the first Recommended Undergraduate Course Requirements of the Institute of Food Technologists (IFT) in the United States in the late 1960s. To date, food biochemistry is still not listed in the IFT recommended undergraduate course requirements. However, many universities in various countries now offer a graduate course in food biochemistry as an elective or have food biochemistry as a specialized area of expertise in their undergraduate and graduate programs. One of the reasons for not requiring such a course at the undergraduate level may be that a biochemistry course is often taken in the last two to three semesters before graduation, and there is no room for such a course in the last semesters. Also, the complexity of this area is very challenging and requires broader views of the students, such as those at the graduate level. However, the importance of food biochemistry is now recognized in the subdiscipline of food handling and processing, as many of these problems are biochemistry related. A content-specific journal, the Journal of Food Biochemistry, has also been available since 1977 for scholars to report their food biochemistry–related research results, even though they can also report their findings to other journals.

Understanding of food biochemistry followed by developments in food biotechnology in the past decades resulted in, besides better raw materials and products, improved human nutrition and food safety, and these developments are applied in the food industry. For example, milk-intolerant consumers in the past did not have the advantage of consuming dairy products. Now they can, with the availability of lactase (a biotechnological product) at the retail level in some developed countries. Lactose-free milk is also produced commercially in some developed industrial countries. The socially annoying problem of flatulence that results from consuming legumes can be overcome by taking "Beano™" (alpha-galactosidase preparation from food-grade *Aspigillus niger*) with meals. Shark meat is made more palatable by bleeding the shark properly right after catch to avoid the biochemical reaction of urease on urea, both naturally present in the shark's blood. Proper control of enzymatic activities also resulted in better products. Tomato juice production is improved by proper control of its pectic enzymes. Better color in potato chips is the result of control of the oxidative enzymes and removal of substrates from the cut potato slices. More tender beef is the result of proper aging of carcasses and sometimes the addition of protease(s) at the consumer level; although this result had been observed in the past, the reasons behind it were unknown. Ripening inhibition of bananas during transport is achieved by removal of the ripening hormone ethylene in the package to minimize the activities of the ripening enzymes, making bananas available worldwide all year round. Proper icing or seawater chilling of tuna after catch avoids/controls histamine production by inhibiting the activities of bacterial histamine producers, thus avoiding scombroid or histamine poisoning. These are just a few of the examples that will be discussed in more detail in this chapter and in the commodity chapters in this book.

Problems due to biochemical causes are numerous; some are simple, while others are fairly complex. These can be reviewed either by commodity group or by food component. This introductory chapter takes the latter approach by grouping the various food components and listing selected related enzymes and their biochemical reactions (without structural formulas) in tables and presenting brief discussions. This will give the readers another way of looking at food biochemistry, but as an introduction to the following material, effort is taken to avoid redundancy with the chapters on commodities that cover the related biochemical reactions in detail.

This chapter presents first selected biochemical changes in the macrocomponents of foods (carbohydrates, proteins, and amino acids), then lipids, then selected biochemical changes in flavors, plant pigments, and other compounds important in food handling and processing. Biotechnological developments as they relate to food handling and processing are introduced only briefly, as new advances are extensively reviewed in Chapter 3. As an example of complexity in the food biochemistry area, a diagram showing the relationship of similar biochemical reactions of selected food components (carbohydrates) in different commodities is presented. Examples of serial degradation of selected food components are also illustrated with two other diagrams.

It should be noted that the main purpose of this chapter is to present an overview of food biochemistry by covering some of the basic biochemical activities related to various food components and their relations with food handling and processing. A second purpose is to get more students interested in food biochemistry. Purposely, only essential references are cited in the text, to make it easier to read; more extensive listings of references are presented at the ends of tables and figures. Readers should refer to these references for details and also consult the individual commodity chapters in this book (and their references) for additional information.

BIOCHEMICAL CHANGES IN CARBOHYDRATES IN FOOD

CHANGES IN CARBOHYDRATES IN FOOD SYSTEMS

Carbohydrates are abundant in foods of plant origin, but are fairly limited in quantity in foods of animal origin. However, some of the biochemical changes and their effect(s) on food quality are common to all foods regardless of animal or plant origin, while others are specific to an individual food. Figure 1.1 shows the relationship between the enzymatic degradation of glycogen and starch (glycolysis) and lactic acid and alcohol formation, as well as the citric acid cycle. Even though glycogen and starch are glucose polymers of different origin, after they are converted to glucose by the appropriate respective enzymes, the glycolysis pathway is common to all foods. The conversions of glycogen in fish and mammalian muscles are now known to utilize different pathways, but they end up with the same glucose-6-phosphate. Lactic acid formation is an important phenomenon in rigor mortis and souring and curdling of milk, as well as in the manufacturing of sauerkraut and other fermented vegetables. Ethanol is an important end product in the production of alcoholic beverages, bread making, and to a much smaller extent in some overripe fruits. The citric acid cycle is also important in alcoholic fermentation, cheese maturation, and fruit ripening. In bread making, α-amylase, either added or from the flour itself, partially hydrolyzes the starch in flour to release glucose units as an energy source for yeast to grow and develop so that the dough can rise during the fermentation period before punching, proofing, and baking.

CHANGES IN CARBOHYDRATES DURING SEED GERMINATION

Table 1.1 lists some of the biochemical reactions related to germination of cereal grains and seeds, with their appropriate enzymes, in the production of glucose and glucose or fructose phosphates from their major carbohydrate reserve, starch. They are then converted to pyruvate through glycolysis, as outlined in Figure 1.1. From then on, the pyruvate is utilized in various biochemical reactions. The glucose and glucose/fructose phosphates are also used in the building of various plant structures. The latter two groups of reactions are beyond the scope of this chapter.

METABOLISM OF COMPLEX CARBOHYDRATES

Besides starch, plants also possess other subgroups of carbohydrates, such as cellulose, β-glucans, and pectins. Both cellulose and β-glucans are composed of glucose units but with different β-glycosidic linkages. They cannot be metabolized in the human body, but are important carbohydrate reserves in plants and can be metabolized into smaller molecules for utilization during seed germination. Pectic substances (pectins) are always considered as the "gluing compounds" in plants. They also are not metabolized in the human body. Together with cellulose and β-glucans, they are now classified in the dietary fiber or complex carbohydrate category.

Interest in pectin stems from the fact that in unripe (green) fruits, pectins exist in the propectin form, giving the fruit a firm/hard structure. Upon ripening, propectins are metabolized into smaller molecules, giving ripe fruits a soft texture. Proper control of the enzymatic changes in propectin is commercially important in fruits, such as tomatoes, apples, and persimmons. Tomato fruits usually don't ripen at the same time on the vines, but this can be achieved by genetically modifying their pectic enzymes (see below). Genetically modified tomatoes can now reach a similar stage of ripeness before consumption and processing without going through extensive manual sorting. Fuji apples can be kept in the refrigerator for a much longer time than other varieties of apples before getting to the soft grainy texture stage because the Fuji apple has lower pectic enzyme activity. Persimmons are hard in the unripe stage, but can be ripened to a very soft texture due to pectic enzyme activity as well as the degradation of its starches. Table 1.2 lists some of the enzymes and their reactions related to these complex carbohydrates.

METABOLISM OF LACTOSE AND ORGANIC ACIDS IN CHEESE PRODUCTION

Milk does not contain high molecular weight carbohydrates; instead its main carbohydrate is lactose. Lactose can be enzymatically degraded to glucose and galactose-6-phosphate by phospho-β-galactosidase (lactase) by lactic acid bacteria. Glucose and galactose-6-phosphate are then further metabolized to various smaller molecules through various biochemical reactions that are important in the flavor development of various cheeses. Table 1.3 lists some

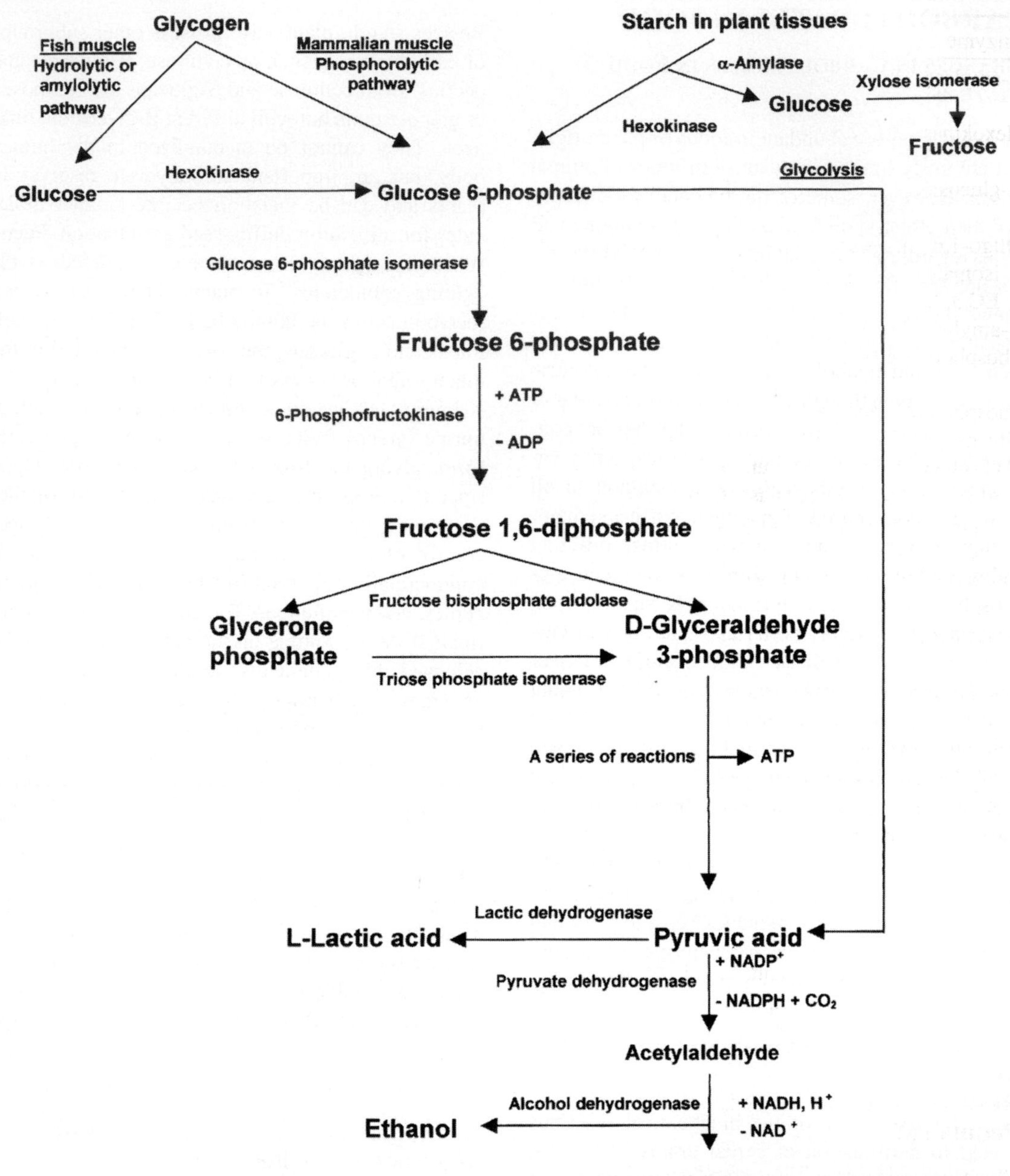

Figure 1.1. Degradation of glycogen and starch. α-Amylase (EC 3.2.1.1), Hexokinase (EC 2.7.1.1), Glucose-6-phosphate isomerase (EC 5.3.1.9), 6-Phosphofructokinase (EC 2.7.1.11), Fructose bisphosphate aldolase (EC 4.1.2.13), Triose phosphate isomerase (EC 5.3.1.1), L-lactic dehydrogenase (EC 1.1.1.27), Pyruvate dehydrogenase (NADP) (EC 1.2.1.51), Alcohol dehydrogenase (EC 1.1.1.1), Xylose isomerase (EC 5.3.15). [Eskin 1990, Lowrie 1992, Huff-Lonergan and Lonergan 1999, Cadwallader 2000, Gopakumar 2000, Simpson 2000, Greaser 2001, IUBMB-NC website (www.iumb.org)]

Table 1.1. Starch Degradation during Cereal Grain Germination

Enzyme	Reaction
α-amylase (EC 3.2.1.1)	Starch → glucose + maltose + maltotriose + α-limited dextrins + linear maltosaccharides
Hexokinase (EC 2.7.1.1)	D-hexose (glucose) + ATP → D-hexose (glucose)-6-phosphate + ADP
α-glucosidase (maltase, EC 3.2.1.20)	Hydrolysis of terminal, nonreducing 1,4-linked α-D-glucose residues with release of α-D-glucose
Oligo-1,6 glucosidase (limited dextrinase, isomaltase, sucrase isomerase, EC 3.2.1.10)	α-limited dextrin → linear maltosaccharides
β-amylase (EC 3.2.1.2)	Linear maltosaccharides → Maltose
Phosphorylase (EC 2.4.1.1)	Linear maltosaccharides + phosphate → α-D-glucose-1-phosphate
Phosphoglucomutase (EC 5.4.2.2)	α-D-glucose-1-phosphate → α-D-glucose-6-phosphate
Glucosephosphate isomerase (EC 5.3.1.9)	D-glucose-6-phosphate → D-fructose-6-phosphate
UTP-glucose 1-phosphate uridyl (UDP-glucose pyrophosphorylase, Glucose-1-phosphate uridyltransferase, EC 2.7.7.9)	UTP + α-D-glucose-1-phosphate → UDP-glucose + pyrophosphate transferase
Sucrose phosphate synthetase (EC 2.4.1.14)	UDP-glucose + D-fructose-6-phosphate → sucrose phosphate + UDP
Sugar phosphatase (EC 3.1.3.23)	Sugar phosphate (fructose-6-phosphate) → sugar (fructose) + inorganic phosphate
Sucrose phosphatase (EC 3.1.3.24)	Sucrose-6-F-phosphate → sucrose + inorganic phosphate
Sucrose synthetase (EC 2.4.1.13)	NDP-glucose + D-fructose → sucrose + NDP
β-fructose-furanosidase (invertase, succharase, EC 3.2.1.26)	Sucrose → glucose + fructose

Sources: Duffus 1987, Kruger and Lineback 1987, Kruger et al. 1987, Eskin 1990, Hoseney 1994, IUBMB-NC website (www.iubmb.org).

of these enzymatic reactions. This table also lists some enzymes and reaction products of organic acids present in very small amounts in milk. However, they are important flavor components (e.g., propionate, butyrate, acetaldehyde, diacetyl, and acetoine).

Removal of Glucose in Egg Powder Production

Glucose is present in very small quantities in egg albumen and egg yolk. However, in the production of dehydrated egg products, this small amount of glucose can undergo nonenzymatic reactions that lower the quality of the final products. This problem can be overcome by the glucose oxidase–catalase system. Glucose oxidase converts glucose to gluconic acid and hydrogen peroxide. The hydrogen peroxide is then decomposed into water and oxygen by the catalase. Application of this process is used almost exclusively for whole egg and other yolk-containing products. However, for dehydrated egg albumen, bacterial fermentation is applied to remove the glucose. Application of yeast fermentation to remove glucose is also possible. The exact processes of glucose removal in egg products are the proprietary information of individual processors (Hill 1986).

Production of Starch Sugars and Syrups

The hydrolysis of starch by means of enzymes (α- and β- amylases) and/or acid to produce glucose (dextrose) and maltose syrups has been practiced for many decades. Application of these biochemical reactions resulted in the availability of various starch (glucose and maltose) syrups, maltodextrins,

Table 1.2. Degradation of Complex Carbohydrates

Enzyme	Reaction
Cellulose degradation during seed germination[a]	
Cellulase (EC 3.2.1.4)	Endohydrolysis of 1,4-β-glucosidic linkages in cellulose and cereal β-D-glucans
Glucan 1,4-β-glucosidase (Exo-1,4-β-glucosidase, EC 3.2.1.74)	Hydrolysis of 1,4 linkages in 1,4-β-D-glucan so as to remove successive glucose units
Cellulose 1,4-β-cellubiosidase (EC3.2.1.91)	Hydrolysis of 1,4-β-D-glucosidic linkages in cellulose and cellotetraose releasing cellubiose from the nonreducing ends of the chains
β-galactosan degradation[a]	
β-galactosidase (EC 3.21.1.23)	β-(1→4)-linked galactan → D-galactose
β-glucan degradation[b]	
Glucan endo-1,6-β-glucosidase (EC 3.2.1.75)	Random hydrolysis of 1,6 linkages in 1,6-β-D-glucans
Glucan endo-1,4-β-glucosidase (EC 3.2.1.74)	Hydrolysis of 1,4 linkages in 1,4-β-D-glucans so as to remove successive glucose units
Glucan endo-1,3-β-D-glucanase (EC 3.2.1.58)	Successive hydrolysis of β-D-glucose units from the nonreducing ends of 1,3-β-D-glucans, releasing β-glucose
Glucan 1,3-β-glucosidase (EC 3.2.1.39)	1,3-β-D-glucans → α-D-glucose
Pectin degradation[b]	
Polygalacturonase (EC 3.2.1.15)	Random hydrolysis of 1,4-α-D-galactosiduronic linkages in pectate and other galacturonans
Galacturan 1,4-α-galacturonidase [Exopolygalacturonase, poly (galacturonate) hydrolase, EC 3.2.1.67)	(1,4-α-D-galacturoniside)$_n$ + H_2O → (1,4-α-D-galacturoniside)$_{n-1}$ + D-galacturonate
Pectate lyase (pectate transeliminase, EC 4.2.2.2)	Eliminative cleavage of pectate to give oligosaccharides with 4-deoxy-α-D-galact-4-enuronosyl groups at their nonreducing ends
Pectin lyase (EC 4.2.2.10)	Eliminative cleavage of pectin to give oligosaccharides with terminal 4-deoxy-6-methyl-α-D-galact-4-enduronosyl groups

Sources: [a]Duffus 1987, Kruger and Lineback 1987, Kruger et al. 1987, Smith 1999.
[b]Eskin 1990, IUBMB-NC website (www.iubmb.org).

maltose, and glucose for the food, pharmaceutical, and other industries. In the 1950s, researchers discovered that some xylose isomerase (D-xylose-keto-isomerase, EC 5.3.1.5) preparations possessed the ability to convert D-glucose to D-fructose. In the early 1970s, researchers succeeded in developing the immobilized enzyme technology for various applications. Because of the more intense sweetness of fructose as compared to glucose, selected xylose isomerase was successfully applied to this new technology with the production of high fructose syrup (called high fructose corn syrup in the United States). High fructose syrups have since replaced most of the glucose syrups in the soft drink industry. This is another example of the successful application of biochemical reactions in the food industry.

BIOCHEMICAL CHANGES OF PROTEINS AND AMINO ACIDS IN FOODS

Proteolysis in Animal Tissues

Animal tissues have similar structures even though there are slight differences between mammalian land animal tissues and aquatic (fish and shellfish) animal

Table 1.3. Changes in Carbohydrates in Cheese Manufacturing

Action, Enzyme or Enzyme System	Reaction
Formation of lactic acid	
Lactase (EC 3.2.1.108)	Lactose + H_2O → D-glucose + D-galactose
Tagatose pathway	Galactose-6-P → lactic acid
Embden-Meyerhoff pathway	Glucose → pyruvate → lactic acid
Formation of pyruvate from citric acid	
Citrate (*pro-3S*) lyase (EC 4.1.3.6)	Citrate → oxaloaceate
Oxaloacetate decarboxylase (EC 4.1.1.3)	Oxaloacetate → pyruvate + CO_2
Formation of propionic and acetic acids	
Propionate pathway	3 lactate → 2 propionate + 1 acetate + CO_2 + H_2O 3 alanine → propionic acid + 1 acetate + CO_2 + 3 ammonia
Formation of succinic acid	
Mixed acid pathway	Propionic acid + CO_2 → succinic acid
Formation of butyric acid	
Butyric acid pathway	2 lactate → 1 butyrate + CO_2 + $2H_2$
Formation of ethanol	
Phosphoketolase pathway	Glucose → acetylaldehyde → ethanol
Pyruvate decarboxylase (EC 4.1.1.1)	Pyruvate → acetylaldehyde + CO_2
Alcohol dehydrogenase (EC 1.1.1.1)	Acetylaldehyde + NAD + H^+ → ethanol + NAD^+
Formation of formic acid	
Pyruvate-formate lyase (EC 2.3.1.54)	Pyruvate + CoA → formic acid + acetyl CoA
Formation of diacetyl, acetoine, 2-3 butylene glycol	
Citrate fermentation pathway	Citrate → pyruvate → acetyl CoA → diacetyl → acetoine → 2-3 butylene glycol
Formation of acetic acid	
Pyruvate-formate lyase (EC 2.3.1.54)	Pyruvate + CoA → formic acid + acetyl CoA
Acetyl-CoA hydrolase (EC 3.1.2.1)	Acetyl CoA + H_2O → acetic acid + CoA

Sources: Schormuller 1968; Kilara and Shahani 1978; Law 1984a,b; Hutlins and Morris 1987; Kamaly and Marth 1989; Eskin 1990; Khalid and Marth 1990; Steele 1995; Walstra et al. 1999; IUBMB-NC website (www.iubmb.org).

tissues. The structure will break down slowly after the animal is dead. The desirable postmortem situation is meat tenderization, and the undesirable situation is tissue degradation/spoilage.

In order to understand these changes, it is important to understand the structure of animal tissues. Table 1.4 lists the location and major functions of myofibrillar proteins associated with contractile apparatus and cytoskeletal framework of animal tissues. Schematic drawings and pictures (microscopic, scanning, and transmission electronic microscopic images) of tissue macro- and microstructures are available in various textbooks and references. Chapter 13 in this book, Biochemistry of Raw Meat and Poultry, also shows a diagram of meat macro- and microstructures. To avoid redundancy, readers not familiar with meat structures are advised to refer to Figure 13.1 when reading the following two paragraphs that give a brief description of the muscle fiber structure and its degradation (Lowrie 1992, Huff-Lonergan and Lonergan 1999, Greaser 2001).

Individual muscle fibers are composed of myofibrils 1–2 μm thick and are the basic units of muscular contraction. The skeletal muscle of fish differs from that of mammals in that the fibers arranged between the sheets of connective tissue are much shorter. The connective tissue is present as short transverse sheets (myocommata) that divide the long fish muscles into segments (myotomes) corresponding in numbers to those of the vertebrae. A fine network of tubules, the sarcoplasmic reticulum separates the individual myofibrils. Within each fiber is a liquid matrix, referred to as the sarcoplasm, that contains mitochondria, enzymes, glycogen, adenosine triphosphate, creatine,

Table 1.4. Locations and Major Functions of Myofibrillar Proteins Associated with the Contractile Apparatus and Cytoskeletal Framework

Location	Protein	Major Function
Contractile apparatus		
A-band	Myosin	Muscle contraction
	c-protein	Binds myosin filaments
	F-, H-, I-proteins	Binds myosin filaments
M-line	M-protein	Binds myosin filaments
	Myomesin	Binds myosin filaments
	Creatine kinase	ATP synthesis
I-band	Actin	Muscle contraction
	Tropomyosin	Regulates muscle contraction
	Troponins T, I, C	Regulates muscle contraction
	β-, γ-actinins	Regulates actin filaments
Cytoskeletal framework		
GAP filaments	Connectin (titin)	Links myosin filaments to Z-line
N_2-Line	Nebulin	Unknown
By sarcolemma	Vinculin	Links myofibrils to sarcolemma
Z-line	α-actinin	Links actin filaments to Z-line
	Eu-actinin, filamin	Links actin filaments to Z-line
	Desmin, vimmentin	Peripheral structure to Z-line
	Synemin, Z-protein, Z-nin	Lattice structure of Z-line

Sources: Eskin 1990, Lowrie 1992, Huff-Lonergan and Lonergan 1999, Greaser 2001.

myoglobin, and other substances. Examination of myofibrils under a phase contrast light microscope shows them to be cross-striated due to the presence of alternating dark or A-bands and light or I-bands. These structures in the myofibril appear to be very similar in both fish and meat. A lighter band or H-zone transverses the A-band, while the I-band has a dark line in the middle known as the Z-line. A further dark line, the M-line, can also be observed at the center of the H-zone in some cases (not shown in Fig. 13.1). The basic unit of the myofibril is the sarcomere, defined as the unit between adjacent Z-lines. Examination of the sarcomere by electron microscope reveals two sets of filaments within the fibrils, a thick set consisting mainly of myosin and a thin set containing primarily of F-actin. In addition to the paracrystalline arrangement of the thick and thin set of filaments, there is a filamentous "cytoskeletal structure" composed of connectin and desmin.

Meat tenderization is the result of the synergetic effect of glycolysis and actions of proteases such as cathepsins and calpains. Meat tenderization is a very complex multifactorial process controlled by a number of endogenous proteases and some as yet poorly understood biological parameters. With currently available literature, the following explanation can be considered. At the initial postmortem stage, calpains, having optimal near neutral pH, attack certain proteins of the Z-line, such as desmin, filamin, nebulin, and to a lesser extent, connectin. With the progression of postmortem glycolysis, the pH drops to 5.5 to 6.5, which favors the action of cathepsins on myosin heavy chains, myosin light chains, α-actinin, tropnin C, and actin. This explanation does not rule out the roles played by other postmortem proteolytic systems that can contribute to this tenderization. (See Eskin 1990, Haard 1990, Huff-Lonergan and Lonergan 1999, Gopakumar 2000, Jiang 2000, Simpson 2000, Lowrie 1992, and Greaser 2001.)

Table 1.5 lists some of the more common enzymes used in meat tenderization. Papain, ficin, and bromelain are proteases of plant origin that can breakdown animal proteins. They have been applied in meat tenderization or in tenderizer formulations industrially or at the household or restaurant levels. Enzymes such as pepsins, trypsins, cathepsins, are well known in the degradation of animal tissues at various sites of the protein peptide chains. Enteropeptidase (enterokinase) is also known to activate trypsinogen by cleaving its peptide bond at Lys^6-Ile^7. Plasmin,

Table 1.5. Proteases in Animal Tissues and Their Degradation

Enzyme	Reaction
Acid/aspartyl proteases	
Pepsin A (Pepsin, EC 3.4.23.1)	Preferential cleavage, hydrophobic, preferably aromatic, residues in P1 and P′1 positions
Gastricsin (pepsin C, EC 3.4.23.3)	More restricted specificity than pepsin A; high preferential cleavage at Tyr bond
Cathepsin D (EC 3.4.23.5)	Specificity similar to, but narrower than that of pepsin A
Serine proteases	
Trypsin (α- and β-trypsin, EC 3.4.21.4)	Preferential cleavage: Arg-, Lys-
Chymotrypsin (Chymotrypsin A and B, EC 3.4.21.1)	Preferential cleavage: Tyr-, Trp-, Phe-, Leu-
Chymotrysin C (EC 3.4.21.2)	Preferential cleavage: Leu-, Tyr-, Phe-, Met-, Trp-, Gln-, Asn-
Pancreatic elastase (pancreato-peptidase E, pancreatic elastase I, EC 3.4.21.36)	Hydrolysis of proteins, including elastin. Preferential cleavage: Ala+
Plasmin (fibrinase, fibrinolysin, EC 3.4.21.7)	Preferential cleavage: Lys- > Arg-; higher selectivity than trypsin
Enteropeptidase (enterokinase, EC 3.4.21.9)	Activation of trypsinogen be selective cleavage of Lys^6-Ile^7 bond
Collagenase	General term for hydrolysis of collagen into smaller molecules
Thio/cysteine proteases	
Cathepsin B (cathepsin B1, EC 3.4.22.1)	Hydrolysis of proteins, with broad specificity for peptide bonds, preferentially cleaves -Arg-Arg- bonds in small molecule substrates
Papain (EC 3.4.22.2)	Hydrolysis of proteins, with broad specificity for peptide bonds, but preference for an amino acid bearing a large hydrophobic side chain at the P2 position. Does not accept Val in P1′
Fiacin (ficin, EC 3.4.23.3)	Similar to that of papain
Bromelain (3.4.22.4)	Broad specificity similar to that of pepsin A
γ-glutamyl hydrolase (EC 3.4.22.12 changed to 3.4.1.99)	Hydrolyzes γ-glutamyl bonds
Cathepsin H (EC 3.4.22.16)	Hydrolysis of proteins; acts also as an aminopeptidase (notably, cleaving Arg bond) as well as an endopeptidase
Calpain-1 (EC 3.4.22.17 changed to 3.4.22.50)	Limited cleavage of tropinin I, tropomyosin, and C-protein from myofibrils and various cytoskeletal proteins from other tissues. Activates phosphorylase, kinase, and cyclic-nucleotide-dependent protein kinase
Metalloproteases	
Procollagen N-proteinase (EC 3.4.24.14)	Cleaves N-propeptide of procollagen chain α1(I) at Pro+Gln and α1(II) and α2(I) at Ala+Gln

Sources: Eskin 1990, Haard 1990, Lowrie 1992, Huff-Lonergan and Lonergan 1999, Gopakumar 2000, Jiang 2000, Simpson 2000, Greaser 2001, IUBMB-NC website (www.iubmb.org).

pancreatic elastase and collagenase are responsible for the breakdown of animal connective tissues.

Transglutaminase Activity in Seafood Processing

Transglutaminase (TGase, EC 2.3.2.13) has the systematic name of protein-glutamine γ-glutamyltransferase. It catalyzes the acyl transfer reaction between γ-carboxyamide groups of glutamine residues in proteins, peptides, and various primary amines. When the ε-amino group of lysine acts as acyl acceptor, it results in polymerization and inter- or intramolecular cross-linking of proteins via formation of ε-(γ-glutamyl) lysine linkages. This occurs through exchange of the ε-amino group of the lysine residue for ammonia at the carboxyamide group of a glutamine residue in the protein molecule(s). Formation of covalent cross-links between proteins is the basis for TGase to modify the physical properties of protein foods. The addition of microbial TGase to surimi significantly increases its gel strength, particularly when the surimi has lower natural setting abilities (presumably due to lower endogenous TGase activity). Thus far, the primary applications of TGase in seafood processing have been for cold restructuring, cold gelation of pastes, or gel-strength enhancement through myosin cross-linking. In the absence of primary amines, water may act as the acyl acceptor, resulting in deamination of γ-carboxyamide groups of glutamine to form glutamic acid (Ashie and Lanier 2000).

Proteolysis during Cheese Fermentation

Chymosin (rennin) is an enzyme present in the calf stomach. In cheese making, lactic acid bacteria (starter) gradually lower the milk pH to the 4.7 that is optimal for coagulation by chymosin. Most lactic acid starters have limited proteolytic activities. However, other added lactic acid bacteria have much stronger proteolytic activities. These proteases and peptidases break down the milk caseins to smaller protein molecules and, together with the milk fat, provide the structure of various cheeses. Other enzymes such as decarboxylases, deaminases, and transaminase are responsible for the degradations of amino acids into secondary amines, indole, α-keto acids, and other compounds that give the typical flavor of cheeses. Table 1.6 lists some of these enzymes and their reactions.

Proteolysis in Germinating Seeds

Proteolytic activities are much lower in germinating seeds. Only aminopeptidase and carboxypeptidase A are better known enzymes (Table 1.7). They produce peptides and amino acids that are needed in the growth of the plant.

Proteases for Chill-Haze Reduction in Beer Production

In beer production, a small amount of protein is dissolved from the wheat and malt into the wort. During extraction of green beer from the wort, this protein fraction is also carried over to the beer. Because of its limited solubility in beer at lower temperatures, it precipitates out and causes hazing in the final product. Proteases of plant origin such as papain, ficin, and bromelain, and possibly other microbial proteases, can break down these proteins. Addition of one or more of these enzymes is commonly practiced in the brewing industry to reduce this chill-haze problem.

BIOCHEMICAL CHANGES OF LIPIDS IN FOODS

Changes in Lipids in Food Systems

Research reports on enzyme-induced changes in lipids in foods are abundant. In general, they are concentrated on changes in the unsaturated fatty acids or the unsaturated fatty moieties in acylglycerols (triglycerides). The most studied are linoleate (linoleic acid) and arachidonate (arachidonic acid) as they are quite common in many food systems (Table 1.8). Because of the number of double bonds in arachidonic acid, enzymatic oxidation can occur at various sites, and the responsible lipoxygenases are labeled according to these sites (Table 1.8).

Changes in Lipids during Cheese Fermentation

Milk contains a considerable amount of lipids and these milk lipids are subjected to enzymatic oxidation during cheese ripening. Under proper cheese maturation conditions, these enzymatic reactions starting from milk lipids create the desirable flavor compounds for these cheeses. These reactions are numerous and not completely understood, so only

Table 1.6. Proteolytic Changes in Cheese Manufacturing

Action and Enzymes	Reaction
Coagulation	
Chymosin (rennin, EC 3.4.23.4)	κ-Casein → Para-κ-casein + glycopeptide, similar to pepsin A
Proteolysis	
Proteases	Proteins → high molecular weight peptides + amino acids
Amino peptidases, dipeptidases, tripeptidases	Low molecular weight peptides → amino acids
Proteases, endopeptidases, aminopeptidases	High molecular weight peptides → low molecular weight peptides
Decomposition of amino acids	
Aspartate transaminase (EC 2.6.1.1)	L-Asparate + 2-oxoglutarate → oxaloacetate + L-glutamate
Methionine γ-lyase (EC 4.4.1.11)	L-methionine → methanethiol + NH_3 + 2-oxobutanolate
Tryptophanase (EC 4.1.99.1)	L-tryptophan + H_2O → indole + pyruvate + NH_3
Decarboxylases	Lysine → cadaverine
	Glutamate → aminobutyric acid
	Tyrosine → tyramine
	Tryptophan → tryptamine
	Arginine → putrescine
	Histidine → histamine
Deaminases	Alanine → pyruvate
	Tryptophan → indole
	Glutamate → α-ketoglutarate
	Serine → pyruvate
	Threonine → α-ketobutyrate

Sources: Schormuller 1968; Kilara and Shahani 1978; Law 1984a,b; Grappin et al. 1985; Gripon 1987; Kamaly and Marth 1989; Khalid and Marth 1990; Steele 1995; Walstra et al. 1999; IUBMB-NC website (www.iubmb.org).

Table 1.7. Protein Degradation in Germinating Seeds

Enzyme	Reaction
Aminopeptidase (EC 3.4.11.11* deleted in 1992, referred to corresponding aminopeptidase)	Neutral or aromatic aminoacyl-peptide + H_2O → neutral or aromatic amino acids + peptide
Carboxypeptidase A (EC 3.4.17.1)	Release of a C-terminal amino acid, but little or no action with -Asp, -Glu, -Arg, -Lys, or -Pro

Sources: Stauffer 1987a,b; Bewley and Black 1994; IUBMB-NC website (www.iubmb.org).

general reactions are provided (Table 1.9). Readers should refer to chapters 19, 20, and 26 in this book for a detailed discussion.

Lipid Degradation in Seed Germination

During seed germination, the lipids are degraded enzymatically to serve as energy source for plant growth and development. Because of the presence of a considerable amount of seed lipids in oilseeds, they have attracted the most attention, and various pathways in the conversion of fatty acids have been reported (Table 1.10). The fatty acids hydrolyzed from the oilseed glycerides are further metabolized into acyl-CoA. From acyl-CoA, it is converted to acetyl-CoA and eventually used to produce energy. It is reasonable to believe that similar patterns also exist in other nonoily seeds. Seed germination is important in production of malted barley flour for bread making and brewing. However, the changes of

Table 1.8. Enzymatic Lipid Oxidation in Food Systems

Enzyme	Reaction
Arachidonate-5-lipoxygenase (5-lipoxygenase, EC 1.13.11.34)	Arachidonate + O_2 → (6*E*, 8*Z*, 11*Z*, 14*Z*)-(5*S*)-5-hydroperoxyicosa-6-8-11,14-tetraenoate
Arachidonate-8-lipoxygenase (8-lipoxygeanse, EC 1.13.11.40)	Arachidonate + O_2 → (5Z, 9E, 11Z, 14Z)-(8R)-8-hydroperoxyicosa-5,9,11,14-tetraenoate
Arachidonate 12-lipoxygenase (12-lipoxygenase, EC 1.13.11.31)	Arachidonate + O_2 → (5Z, 8Z, 10E, 14Z)-(12S)-12-hydroperoxyicosa-5,8,10,14-tetraenoate
Arachidonate 15-lipoxygenase (15-lipoxygenase, EC 1.13.11.33)	Arachidonate + O_2 → (5*Z*, 8*Z*, 11*Z*, 13*E*)-(15*S*)-15-hydroperoxyicosa-5,8,11,13-tetraenoate
Lipoxygenase (EC 1.13.11.12)	Linoleate + O_2 → (9*Z*, 11*E*)-(13*S*)-13-hydroperoxyoctadeca-9, 11-dienoate

Sources: Lopez-Amaya and Marangoni 2000a,b; Pan and Kuo 2000; Kolakowska 2003; IUBMB-NC website (www.iubmb.org).

Table 1.9. Changes in Lipids in Cheese Manufacturing

Enzyme or Actions	Reaction
Lipolysis	
Lipases, esterases	Triglycerides → β-keto acids, acetoacetate, fatty acids
Acetoacetate decarboxylase (EC 4.1.1.4)	Acetoacetate + H^+ → acetone + CO_2
Acetoacetate-CoA ligase (EC 6.2.1.16)	Acetoacetate + ATP + CoA → acetyl CoA + AMP + diphosphate
Esterases	Fatty acids → esters
Conversion of fatty acids	
β-oxidation and decarboxylation	β-Keto acids → methyl ketones

Sources: Schormuller 1968, Kilara and Shahani 1978, IUBMB-NC website (www.iubmb.org).

Table 1.10. Lipid Degradation in Seed Germination

Enzyme or Enzyme System	Reaction
Lipase (oil body)	Triacylglycerol → diacylglycerol + fatty acid
	Triacylglycerol → monoacylglycerol + fatty acids
	Diacylglycerol → monoacylglycerol + fatty acid
	Fatty acid + CoA → acyl CoA
β-oxidation (glyoxysome)	Acyl CoA → acetyl CoA
Glyoxylate cycle (glyoxysome)	Acetyl CoA → succinate
Mitochondrion	Succinate → phosphoenol pyruvate
Reverse glycolysis (Cytosol)	Phosphoenol pyruvate → hexoses → sucrose

Sources: Bewley and Black 1994, Murphy 1999.

lipids in these seeds are of less importance than the activity of α-amylase.

BIOGENERATION OF FRESH-FISH ODOR

The main enzymes involved in biogeneration of the aroma in fresh fish have been reported as the 12- and 15-lipoxygenases (Table 1.8) and hydroperoxide lyase. The 12-lipoxygenase acts on specific polyunsaturated fatty acids and produces n-9-hydroperoxides. Hydrolysis of the 9-hydroperoxide of eicosapentenoic acid by specific hydroperoxide lyases leads to the formation of mainly (Z,Z)-3-6-nonadienal, which can undergo spontaneous or enzyme-catalyzed isomerization to (E,Z)-2,6-nonadienal. These aldehydes may undergo reduction to their corresponding alcohols. This conversion is a significant step in the general decline of the aroma intensity due to the fact that alcohols have somewhat higher odor detection thresholds than the aldehydes (Johnson and Linsay 1986, German et al. 1992).

BIOCHEMICALLY INDUCED FOOD FLAVORS

Many fruits and vegetables produce flavors that are significant in their acceptance and handling. There are a few well-known examples (Table 1.11). Garlic is well known for its pungent odor due to the enzymatic breakdown of its alliin to the thiosulfonate allicin, with the characteristic garlic odor. Strawberries have a very typical pleasant odor when they ripen. Biochemical production of the key compound responsible for strawberry flavor [2,5-dimethyl-4-hydroxy-2H-furan-3-one (DMHF)] is now known. It is the result of hydrolysis of terminal nonreducing β-D-glucose residues from DMHF-glucoside with release of β-D-glucose and DMHF. Lemon and orange seeds contain limonin, a bitter substance that can be hydrolyzed to limonate, which creates a less bitter taste sensation. Many cruciferous vegetables such as cabbage and broccoli have a sulfurous odor due to the production of a thiol after enzymatic hydrolysis of its glucoside. These are just some examples of biochemically induced fruit and vegetable flavors. Brewed tea darkens after it is exposed to air due to enzymatic oxidation. Flavors from cheese fermentation and fresh-fish odor have already been described earlier. Formation of fishy odor will be described later (see below). Readers interested in this subject should consult Wong (1989) for earlier findings of chemical reactions. Huang's review on biosynthesis of natural aroma compounds derived from amino acids, carbohydrates, and lipids should also be consulted (2005).

BIOCHEMICAL DEGRADATION AND BIOSYNTHESIS OF PLANT PIGMENTS

DEGRADATION OF CHLOROPHYL IN FRUIT MATURATION

Green fruits are rich in chlorophylls that are gradually degraded during ripening. Table 1.12 shows

Table 1.11. Selected Enzyme-Induced Flavor Reactions

Enzyme	Reaction
Alliin lyase (EC 4.4.1.4), (garlic, onion)	An *S*-alkyl-L-cysteine *S*-oxide → an alkyl sufenate + 2-aminoacrylate
β-glucosidase (EC 3.2.1.21) (strawberry)	Hydrolysis of terminal nonreducing β-D-glucose residues with release of β-D-glucose [2,5-Dimethyl-4-hydroxy-2H-furan-3-one (DMHF)-glucoside → DMHF]
Catechol oxidase (EC 1.10.3.1), (tea)	2 Catechol + O_2 → 2 1,2-benzoquinone + 2 H_2O
Limonin-D-ring-lactonase (EC 3.1.1.36) (lemon and orange seeds)	Limonoate-D-ring-lactone + H_2O → limonate
Thioglucosidase (EC 3.2.1.147) (cruciferous vegetables)	A thioglucoside + H_2O → A thiol + a sugar

Sources: Wong 1989, Eskin 1990, Chin and Lindsay 1994, Orruno et al. 2001, IUBMB-NC website (www.iubmb.org).

Table 1.12. Degradation of Chlorophyll

Enzyme	Reaction
Chlorophyllase (EC 3.1.1.4)	Chlorophyll → chlorophyllide + phytol
Magnesium dechelatase (EC not available)	Chlorophyllide a → phyeophorbide *a* + Mg^{2+}
Phyeophorbide *a* oxygenase (EC not available)	Phyeophorbide *a* + O_2 → red chlorophyll catabolite (RCC)
RCC reductase (EC not available)	RCC → fluorescent chlorophyll catabolite (FCC)
Various enzymes	FCC → nonfluorescent chlorophyll catabolites (NCC)

Sources: Eskin 1990, Dangl et al. 2000, IUBMB-NC website (www.iubmb.org).

some of the enzymatic reactions proposed in the degradation of chlorophyll *a* (Table 1.12).

Mevalonate and Isopentyl Diphosphate Biosynthesis prior to Formation of Carotenoids

Table 1.13 lists the sequence of reactions in the formation of (*R*)-mevalonate from acetyl-CoA, and from (*R*)-mevalonate to isopentyl diphosphate. Isopentyl diphosphate is a key building block for carotenoids (Croteau et al. 2000). Carotenoids are the group of fat-soluble pigments that provides the yellow to red colors of many common fruits such as yellow peaches, papayas, and mangoes. During postharvest maturation, these fruits show intense yellow to yellowish orange colors due to synthesis of carotenoids from its precursor isopentyl diphosphate, which is derived from (*R*)-mevalonate. Biosyntheses of carotenoids and terpenoids have a common precursor, (*R*)-mevalonate derived from acetyl-CoA (Table 1.13). (*R*)-mevalonate is also a building block for terpenoid biosynthesis (Croteau et al. 2000; IUBMB website).

Naringenin Chalcone Biosynthesis

Flavonoids are a group of interesting compounds that not only give fruits and vegetables various red, blue, or violet colors, but also are related to the group of bioactive compounds called stilbenes. They have a common precursor of *trans*-cinnamate branching out into two routes, one leading to the flavonoids, and the other leading to stilbenes (Table 1.14; IUBMB website). Considerable interest has been given to the stilbene trans 3,5,4'-trihydroxystilbene (commonly called reveratrol or resveratrol) in red grapes and red wine that may have potent antitumor properties and to another stilbene, combretastatin, with potential antineoplastic activity (Croteau

Table 1.13. Mevalonate and Isopentyl Diphosphate Biosyntheses

Enzyme	Reaction
Acetyl-CoA *C*-acetyltransferase (EC 2.3.1.9)	2 acetyl-CoA → acetoacetyl-Co-A + CoA
Hydroxymethylglutaryl-CoA-synthase (EC 2.3.3.10)	Acetoacetyl-CoA + acetyl-CoA + H_2O → (*S*)-3-hydroxy-3-methylglutaryl CoA + CoA
Hydroxymethylglutaryl-CoA reductase ($NADPH_2$) (EC 1.1.1.34)	(*S*)-3-hydroxy-3-methylglutaryl-CoA + 2 $NADPH_2$ → (*R*)-mevalonate + CoA + 2 NADP
Mevaldate reductase (EC 1.1.1.32)	(*R*)-mevalonate + NAD → mevaldate + $NADH_2$
Mevalonate kinase (EC 2.7.1.36)	(*R*)-mevalonate + ATP → (*R*)-5-phosphomevalonate +ADP
Phosphomevalonate kinase (EC 2.7.4.2)	(*R*)-5-phosphomevalonate + ATP → (*R*)-5-diphosphomevalonate + ADP
Diphosphomevalonate decarboxylase (EC 4.1.1.33)	(*R*)-5-diphosphomevalonate + ATP → isopentyl diphosphate + ADP + phosphate + CO_2

Sources: Croteau et al. 2000, IUBMB-NC Enzyme website (www.iubmb.org).

Table 1.14. Naringenin Chalcone Biosynthesis

Enzyme	Reaction
Phenylalanine ammonia-lyase (EC 4.3.1.5)	L-phenylalanine → *trans*-cinnamate + NH_3
Trans-cinnamate 4-monoxygenase (EC 1.14.13.11)	*Trans*-cinnamate + $NADPH_2$ + O_2 → 4-hydroxycinnamate + NADP + H_2O
4-Coumarate-CoA ligase (EC 6.2.1.12)	4-hydroxycinnamate (4-coumarate) + ATP + CoA → 4-coumaroyl-CoA + AMP + diphosphate
Naringinin-chalcone synthase (EC 2.3.1.74)	4-coumaroyl-CoA + **3** malonyl-CoA → naringinin chalcone + **4** CoA + **3** CO_2

Sources: Eskin 1990, Croteau et al. 2000, IUBMB-NC Enzyme website (www.iubmb.org).

et al. 2000). Table 1.14 gives the series of reactions in the biosynthesis of naringenin chalcone. Naringenin chalcone is the building block for flavonoid biosynthesis. The pathway for the biosynthesis of a stilbene pinosylvin and 3,4'5'-trihydroxystilbene has been postulated (IUBMB website).

SELECTED BIOCHEMICAL CHANGES IMPORTANT IN THE HANDLING AND PROCESSING OF FOODS

Production of Ammonia and Formaldehyde from Trimethylamine and Its N-oxide

Trimethylamine and its N-oxide have long been used as indices for freshness in fishery products. Degradation of trimethylamine and its N-oxide leads to the formation of ammonia and formaldehyde with undesirable odors. The pathway on the production of formaldehyde and ammonia from trimethylamine and its N-oxide is shown in Figure 1.2.

Production of Biogenic Amines

Most live pelagic and scombroid fish (e.g., tunas, sardines, and mackerel) contain an appreciable amount of histidine in the free state. In postmortem scombroid fish, the free histidine is converted by the bacterial enzyme histidine decarboxylase into free histamine. Histamine is produced in fish caught 40–50 hours after death when fish are not properly chilled. Improper handling of tuna and mackerel after harvest can produce enough histamine to cause food poisoning (called scombroid or histamine poisoning). The common symptoms of this kind of food poisoning are facial flushing, rashes, headache, and gastrointestinal disorder. These disorders seem to be strongly influenced by other related biogenic amines, such as putrescine and cadaverine, produced by similar enzymatic decarboxylation (Table 1.15). The

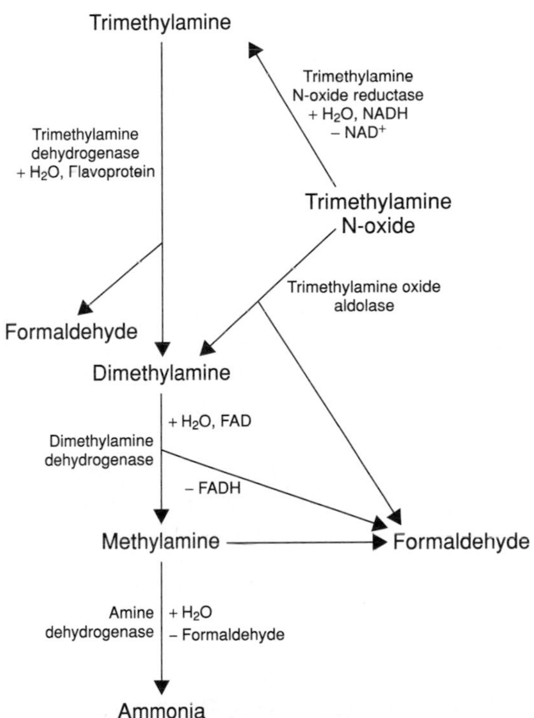

Figure 1.2. Degradation of trimethylamine and its N-oxide. Trimethylamine N-oxide reductase (EC 1.6.6.9), Trimethylamine dehydrogenase (EC 1.5.8.2), Dimethylamne dehydrogenase (EC 1.5.8.1), Amine dehydrogenase (EC 1.4.99.3). [Haard et al. 1982, Gopakumar 2000, Stoleo and Rehbein 2000, IUBMB-NC website (www.iubmb.org)]

Table 1.15. Secondary Amine Production in Seafoods

Enzyme	Reaction
Histidine decarboxyalse (EC 4.1.1.22)	L-Histidine → histamine + CO_2
Lysine decarboxylase (EC 4.1.1.18)	L-Lysine → cadaverine + CO_2
Ornithine decarboxylase (EC 4.1.1.17)	L-Ornithine → putrescine + CO_2

Sources: Gopakumar 2000, IUBMB-NC website (www.iubmb.org).

presence of putrescine and cadaverine is more significant in shellfish, such as shrimp. The detection and quantification of histamine is fairly simple and inexpensive. However, the detection and quantification of putrescine and cadaverine are more complicated and expensive. It is suspected that histamine may not be the real and main cause of poisoning, as histamine is not stable under strong acidic conditions such as pH 1 in the stomach. However, the U.S. Food and Drug Administration (FDA) has strict regulations governing the amount of histamine permissible in canned tuna, as an index of freshness of the raw materials, because of the simplicity of histamine analysis (Gopakumar 2000).

Production of Ammonia from Urea

Urea is hydrolyzed by urease (EC 3.5.1.5) to ammonia, which is one of the components of total volatile base (TVB). TVB nitrogen has been used as a quality index of seafood acceptability by various agencies (Johnson and Linsay 1986, Cadwallader 2000, Gopakumar 2000). A good example is shark, which contains fairly high amounts of urea in the live fish. Under improper handling, urea is converted to ammonia by urease, giving shark meat an ammonia odor that is not well accepted by consumers. To overcome this problem, the current practice of bleeding the shark near its tail right after harvest is very promising.

Adenosine Triphosphate Degradation

Adenosine triphosphate (ATP) is present in all biological systems. Its degradation in seafood has often been reported (Fig. 1.3) (Gill 2000, Gopakumar 2000). The degradation products, such as inosine and hypoxanthine, have been used individually or in combination as indices of freshness for many years.

Polyphenol Oxidase Browning

Polyphenol oxidase (PPO, EC 1.10.3.1, systematic name 1,2 benzenediol:oxygen oxidoreductase) is also labeled as phenoloxidase, phenolase, monophenol and diphenol oxidase, and tyrosinase. This enzyme catalyzes one of the most important color reactions that affects many fruits, vegetables, and seafood, especially crustaceans. This postmortem discoloration in crustacean species such as lobster, shrimp, and crab is also called melanosis or black spot. It connotes spoilage, is unacceptable to consumers, and thus reduces the market value of these products.

Polyphenol oxidase is responsible for catalyzing two basic reactions. In the first reaction, it catalyzes the hydroxylation of phenols with oxygen, to the *o*-position adjacent to an existing hydroxyl group. For example, tyrosine, a monohydroxy phenol, is present naturally in crustaceans. PPOs from shrimp and lobster are activated by trypsin or by a trypsin-like enzyme in the tissues to hydroxylate tyrosine with the formation of dihydroxylphenylamine (DOPA). The second reaction is the oxidation of the diphenol to *o*-benzoquinones, which are further oxidized to melanins (brown to dark products), usually by nonenzymatic mechanisms.

The major effect of reducing agents or antioxidants in the prevention of browning is their ability to reduce the *o*-quinones to the colorless diphenols, or to react irreversibly with the *o*-quinones to form stable colorless products. The use of reducing compounds is the most effective control method for PPO browning. The most widespread antibrowning treatment used by the food industry was the addition of sulfiting agents. However, because of safety concerns, other methods have been developed, including the use of other reducing agents (such as ascorbic acid and analogs, cysteine and glutathione), chelating agents (phosphates, EDTA), acidulants (citric

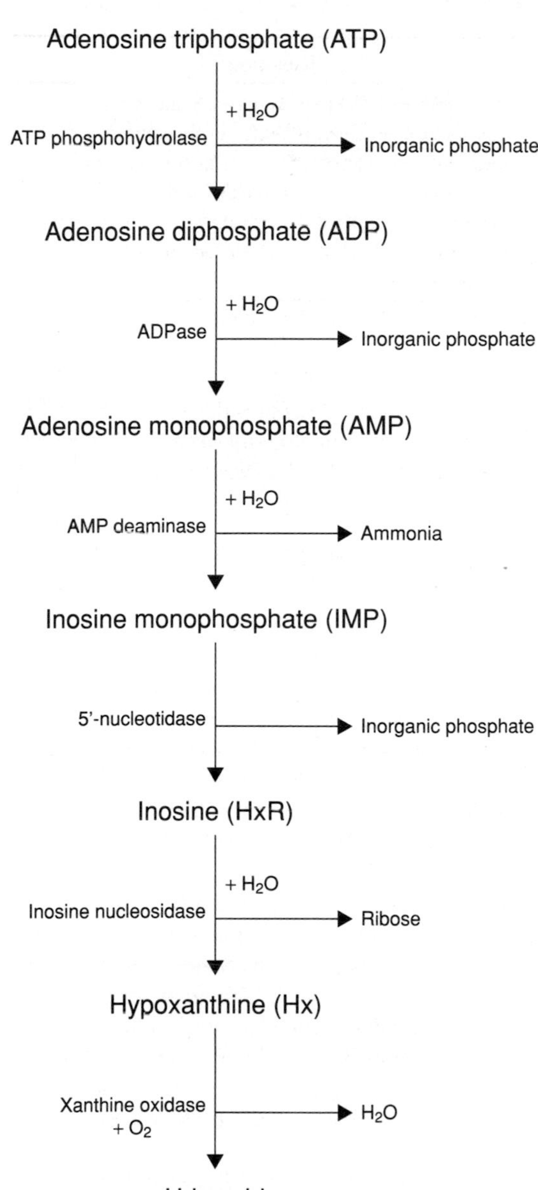

Figure 1.3. Degradation of adenosine triphosphate (ATP) in seafoods. ATP phosphohydrolase (EC 3.6.3.15), ADPase (EC 3.6.1.6), AMP deaminase (EC 3.5.4.6), 5'-nucleotidase (EC 3.1.3.5), Inosine nucleotidase (EC 3.1.3.5), Xanthine oxidase (EC 1.1.3.22). [Gill 2000, Gopakumar 2000, IUBMB-NC website (www.iubmb.org)]

acid, phosphoric acid), enzyme inhibitors, enzyme treatment, and complexing agents. Application of these inhibitors of enzymatic browning is strictly regulated in different countries (Eskin 1990, Gopakumar 2000, Kim et al. 2000).

Ethylene Production in Fruit Ripening

Ethylene acts as one of the initiators in fruit ripening. Its concentration is very low in green fruits but can accumulate inside the fruit and subsequently activates its own production. Table 1.16 lists the enzymes in the production of ethylene starting from methionine. The effect of ethylene is commonly observed in the shipping of bananas. The banana is a climacteric fruit with a fast ripening process. During shipping of green bananas, ethylene is removed through absorption by potassium permanganate to render a longer shelf life.

Reduction of Phytate in Cereals

Phytic acid (*myo*-inositol hexaphosphate) is the major phosphate reserve in many seeds. Since it exists as a mixed salt with elements such as potassium, magnesium, and calcium (and as such is called phytin or phytate), it is also a major source of these macronutrient elements in the seed. However, this salt form of macronutrients renders them unusable by the human body. During seed germination, phytase (4-phytase, phytate-6-phosphatase, EC 3.1.3.26) hydrolyzes the phytic acid to release phosphate, its associated phosphate cation, and 1-D-*myo*-inositol 1,2,3,4,5-pentakisphosphate. Breakdown of phytate is rapid and complete (Stauffer 1987a,b; Berger 1994; Bewley 1997). This enzymatic reaction releases the macronutrients from their bound forms so they are more easily utilized by the human body. This explains why breads utilizing flour from germinated wheat are more nutritious than those made from regular wheat flour.

BIOTECHNOLOGY IN FOOD PRODUCTION, HANDLING, AND PROCESSING

Biotechnology-Derived Food Enzymes

With the advancement of biotechnology, the food industry was not slow in jumping on the wagon for

Table 1.16. Ethylene Biosynthesis

Enzyme	Reaction
Methionine adenosyltransferase (adenosylmethionine synthase, EC 2.5.1.6)	L-methionine + ATP + H_2O → S-adenosyl-γ-methionine + diphosphate + phosphate
Aminocyclopropane carboxylate synthetase (EC 4.4.1.14)	S-adenosyl-γ-methionine → 1-aminocyclopropane-1-carboxylate + 5′-methylthio-adenosine
Aminocyclopropane carboxylate oxidase (EC 4.14.17.4)	1-aminocyclopropane-1-carboxylate + ascorbate + ½ O_2 → ethylene + dedroascorbate + CO_2 + HCN + H_2O

Sources: Eskin 1990, Bryce and Hill 1999, Crozier et al. 2000, Dangl et al. 2000, IUBMB-NC website (www.iubmb.org).

better processing aids. At least six biotechnology-derived enzymes have been developed: acetolactate decarboxylase, α-amylase, amylo-1,6-glucosidase, chymosin, lactase, and maltogenic α-amylase (Table 1.17). Chymosin has now been well adopted by the cheese industry because of reliable supply and reasonable cost. Lactase is also well accepted by the dairy industry for the production of lactose-free milk and as a dietary supplement for lactose-intolerant consumers. Amylases are also being used for the production of high fructose corn syrup and as an anti-staling agent for bread. The application of pectic enzymes in genetically modified tomatoes was mentioned earlier. It should be noted that each country has its own regulations governing the use of these biotechnology-derived enzymes.

Genetically Modified Microorganisms Useful in Food Processing

Like the biotechnology-derived food enzymes, genetically modified microorganisms are being developed for specific needs. Lactic acid bacteria and yeast have been developed to solve problems in the dairy, baking, and brewing industries (Table 1.18). As with the biotechnology-derived food enzymes, their use is governed by the regulations of individual countries.

CONCLUSION

The Institute of Food Technologists (IFT), formed in the United States in 1939, was the world's first such organization to pull together those working in food processing, chemistry, engineering, microbiology, and other subdisciplines who were trying to better understand food and help solve some of its related problems. Now, most countries have similar organizations, and the IFT has developed into a world organization and the leader in this field.

When we look back into the history of food science as a discipline, we see that it started out with a few universities in the United States, mainly in commodity departments, such as animal science, dairy science, horticulture, cereal science, poultry science, and fishery. Now, in the United States and Canada, most of these programs (about 50 in total) have evolved into a food science or food science and

Table 1.17. Selected Commercial Biotechnology-Derived Food Enzymes

Enzyme	Application
Acetolactate decarboxylase (EC 4.1.1.5)	Beer aging and diacetyl reduction
α-amylase (EC 3.2.1.1)	High fructose corn syrup (HFCS) production
Amylo-1,6-glucosidase (EC 3.2.1.33)	High fructose corn syrup (HFCS) production
Chymosin (EC 3.4.23.4)	Milk clotting in cheese manufacturing
Lactase (EC 3.2.1.108)	Lactose hydrolysis
Glucan 1,4-α-maltohydrolase (maltogenic α-amylase, EC 3.2.1.133)	Anti-staling in bread

Sources: Roller and Goodenough 1999, Anonymous 2000, IUBMB-NC website (www.iubmb.org).

Table 1.18. Selected Genetically Modified Microorganisms Useful in Food Processing

Microorganisms	Application
Lactobacillus lactis	Phage resistance, lactose metabolism, proteolytic activity, bacteriocin production
Saccharomyces (Baker's yeast)	Gas (carbon dioxide) production in sweet, high-sugar dough
Saccharomyces cervisiae (Brewer's yeast)	Manufacture of low-calorie beer (starch degradation)

Sources: Hill and Ross 1999, Roller and Goodenough 1999, Anonymous 2000.

human nutrition department. The IFT has played an important role in these developments. There are also programs in other countries where food science is grouped under other traditional disciplines such as biology or chemistry. However, some universities in a few countries put more emphasis on food science and form a school or a college. Many food science departments with a food biochemistry emphasis are now available all over the world, and they promote their programs through the Internet. These departments place their emphases on one or more commodities.

Research reports on various topics of food science and food technology have been published in various journals including the Journal of Food Science, Food Technology, and others. Food-related biochemical studies were published in various journals until 1977, when the first issue of Journal of Food Biochemistry was published. Although food biochemistry–related reports are still published in other journals, establishment of this journal is a milestone for this subdiscipline of food science. A few books with emphasis on food biochemistry in general and on specific commodities/components have also become available in the past 40 years.

Over the past several decades, many food biochemistry–related problems have been resolved, and these solutions have resulted in industry applications. Examples of such achievements as lactase, lactose-free milk, "Beano™," transgenic tomatoes with easier ripening control, application of transglutaminase to control seafood protein restructuring, proteases for meat tenderization, production of high fructose syrups, and others have been discussed earlier. With the recent interest and development in biotechnology, food biochemists are trying to apply this new technique to help solve many food-related biochemical problems. These may include but not be limited to those in food safety, improved nutrient content, delayed food spoilage, better raw materials for processing and product development, better processing technology, and less expensive flavoring materials. In the near future, we should not be surprised when researchers report breakthroughs that are food biochemistry related. In fact, this is expected, as we now have better trained researchers and more advanced research tools. Although its study requires a diversified background, food biochemistry is gaining more interest in the food science discipline. It is an area that will attract more students, especially with the current interest in biotechnology.

ACKNOWLEDGEMENTS

I would like to thank Prof. C. S. Tang, Department of Molecular Biosciences and Bioengineering, University of Hawaii at Manoa, Honolulu, Hawaii, and Prof. Mike Morgan, Proctor, Department of Food Science, University of Leeds, United Kingdom, for their constructive suggestions on the chapter outline and critical comments in the preparation of this chapter.

REFERENCES

Anonymous. 2000. IFT Expert Report on Biotechnology and Foods. Food Technology 54(10):37–56.

Ashie IA, Lanier TC. 2000. Transglutaminases in seafood processing. In: NF Haard, BK Simpson, editors, Seafood Enzymes. New York: Marcel Dekker, Inc. Pp. 147–166.

Berger M. 1994. Flour aging. In: B Godon, C Willm, editors. Primary Cereal Processing. New York: VCH Publishers, Inc. Pp. 439–452.

Bewley JD. 1997. Seed germination and dormancy. Plant Cell 9:1055–1066.

Bewley JD, Black M. 1994. Physiology of Development and Germination, 2nd ed. New York: Plenum Press. Pp. 293–344.

Bryce JH, Hill SA. 1999. Energy production and plant cells. In: PJ Lea, RC Leegood, editors, Plant Biochemistry and Molecular Biology Chichester: John Wiley and Sons. Pp. 1–28.

Cadwallader KR. 2000. Enzymes and flavor biogenesis. In: NF Haard, BK Simpson, editors, Seafood Enzymes. New York: Marcel Dekker, Inc. Pp. 365–383.

Chin HW, Lindsay RC. 1994. Modulation of volatile surfur compounds in cruciferous vegtebles. In: CJ Mussinan, ME Keelan, editors, Sulfur Compounds in Foods. Washington, DC: American Chemical Society. Pp. 90–104.

Croteau R, Kutchan TM, Lewis NG. 2000. Natural products (secondary metabolites). In: BB Buchenan, W Grussem, RL Jones, editors, Biochemistry and Molecular Biology of Plants. Rockwell, Maryland: American Society of Plant Physiologists. Pp. 1250–1318.

Crozier A, Kamiya Y, Bishop G, Yokota T. 2000. Biosynthesis of hormone and elicitor molecules. In: Buchenan BB, Grussem W, Jones RL, editors. Biochemistry and Molecular Biology of Plants. Rockwell, Maryland: American Society of Plant Physiologists. Pp. 850–929.

Dangl JL, Dietrich RA, Thomas H. 2000. Senescence and programmed cell death. In: BB Buchenan, W Grussem, RL Jones, editors, Biochemistry and Molecular Biology of Plants. Rockwell, Maryland: American Society of Plant Physiologists. Pp. 1044–1100.

Duffus CM. 1987. Physiological aspects of enzymes during grain development and germination. In: Kruger JE, Lineback D, Stauffer CE, editors, Enzymes and Their Role in Cereal Technology. St. Paul, Minnesota: American Association of Cereal Chemists. Pp. 83–116.

Eskin NAM. 1990. Biochemistry of Foods. San Diego: Academic Press. Pp. 17–44 (muscle to meat), 45–54 (meat pigments), 70–145 (fruits and vegetables), 185–196 (seed germination), 321–328 (brewing), 347–349, 349–351 (baking), 369–374 (milk coagulation), 376–389 (cheese ripening), 402–425 (enzymatic browning), 434–457 (off-flavor in milk), 467–527 (biotechnology).

German JB, Zhang, H, Berger R. 1992. Role of lipoxygenases in lipid oxidation in foods. In: AJ St. Angelo, editor, Lipid Oxidation in Food. Washington, DC: American Chemical Society. Pp. 74–92.

Gill T. 2000. Nucleotide-degrading enzymes. In: NF Haard, BK Simpson, editors, Seafood Enzymes. New York: Marcel Dekker, Inc. Pp. 37–68.

Gopakumar K. 2000. Enzymes and enzyme products as quality indices. In: NF Haard, BK Simpson, editors, Seafood Enzymes. New York: Marcel Dekker, Inc. Pp. 337–363.

Grappin R, Rank TC, Olson NF. 1985. Primary proteolysis of cheese proteins during ripening. J. Dairy Science 68:531–540.

Greaser M. 2001. Postmortem muscle Chemistry. In: YH Hui, WK Nip, RW Rogers, OA Young, editors, Meat Science and Applications. New York: Marcel Dekker, Inc. Pp. 21–37.

Gripon JC. 1987. Mould-ripened cheeses. In: PF Fox, editor, Cheese: Chemistry, Physics and Microbiology. London: Elsevier Applied Science. Pp. 121–149.

Haard CE, Flick GJ, Martin RE. 1982. Occurrence and significance of trimethylamine oxide and its derivatives in fish and shellfish. In: RE Martin, GJ Flick, CE Haard, DR Ward, editors, Chemistry and Biochemistry of Marine Food Products, Westport, Connecticut: AVI Publishing Company. Pp. 149–304.

Haard NF. 1990. Biochemical reactions in fish muscle during frozen storage. In: EG Bligh, editor, Seafood Science and Technology. London: Fishing News Books (Blackwell Scientific Publications, Ltd.) Pp. 176–209.

Haung TC. 2005. Biosynthesis of natural aroma compounds. In: YH Hui, JD Culbertson, S Duncan, ECY Li-Chan, CY Ma, CH Manley, TA McMeekin, WK Nip, LML Nollet, MS Rahman, YL Xiong, Handbook of Food Science. New York: Marcel Dekker, Inc./ Boca Raton, Florida: CRC Press. Forthcoming.

Hill C, Ross RP. 1999. Starter cultures for the dairy industry. In: S Roller, S Harlander, editors, Genetic Modification in the Food Industry. London: Blackie and Academic Professional. Pp. 174–192.

Hill WM. 1986. Desugarization of egg products. In: WJ Stadelman, OJ Cotterill, editors, Egg Science and Technology. Westport, Connecticut: AVI Publishing Co., Inc. Pp. 273–283.

Hoseney RC. 1994. Principles of Cereal Science and Technology. St. Paul, Minnesota: American Association of Cereal Chemists. Pp. 177–195 (malting and brewing), 229–273 (yeast-leavened products).

Huff-Lonergan E, Lonergan SM. 1999. Postmortem mechanisms of meat tenderization. In: YL Xiong, CT Ho, F Shahidi, editors. Quality Attributes of Muscle Foods. New York: Kluwer Academic/Plenum Publishers. Pp. 229–251.

Hutlins RW, Morris HA. 1987. Carbohydrate metabolism by *Streptococcus thermophilus*: A Review. Journal of Food Protection 50(10):876–884.

International Union of Biochemistry and Molecular Biology- Nomenclature Committee (IUBMB-NC). Http://www/iubmb.org.

Jiang ST. 2000. Enzymes and their effects on seafood texture. In: Haard NF, Simpson BK, editors. Seafood Enzymes. New York: Marcel Dekker, Inc. Pp. 411–450.

Johnson DB, Linsay RC. 1986. Enzymatic generation of volatile aroma compounds from fresh fish. In: TH Parliament, R Croteau, editors, Biogeneration of Aromas. Washington, DC: American Chemical Society. Pp. 201–219.

Kamaly KM, and Marth EH. 1989. Enzyme activities of *Lactic streptococci* and their role in maturation of cheese: A review. Journal of Dairy Science 72:1945–1966.

Khalid NM, Marth EH. 1990. Lactobacilli—Their enzymes and role in ripening and spoilage of cheese: A review. Journal of Dairy Science 73:2669–2684.

Kilara A, Shahani KM. 1978. Lactic fermentations of dairy foods and their biological significance. Journal of Dairy Science 61:1793–1800.

Kim J, Marshall MR, Wei CI. 2000. Polyphenoloxidase. In: NF Haard, BK Simpson, editors, Seafood Enzymes. New York: Marcel Dekker, Inc. Pp. 271–315.

Kolakowska A. 2003. Lipid oxidation in food systems. In: ZE Sikorski, A Kolakowska, editors, Chemical and Functional Properties of Food Lipids. Boca Raton: CRC Press. Pp. 133–166.

Kruger JE, Lineback DR. 1987. Carbohydrate-degrading enzymes in cereals. In: JE Kruger, D Lineback, CE Stauffer, editors, Enzymes and Their Role in Cereal Technology. St. Paul, Minnesota: American Association of Cereal Chemists. Pp. 117–139.

Kruger JE, Lineback D, Stauffer CE. 1987. Enzymes and Their Roles in Cereal Technology. St. Paul, Min nesota: American Association of Cereal Chemists.

Law BA. 1984a. Microorganisms and their enzymes in the maturation of cheeses. In: ME Bushell, editor, Progress in Industrial Microbiology, London: Elsevier. Pp. 245–283.

———. 1984b. Flavour Development in Cheeses. In: FL Davies, BA Law, editors, Advances in the Microbiology and Biochemistry of Cheese and Fermented Milks. London: Elsevier Applied Science Publishers. Pp. 187–208.

Lopez-Amaya C, Marangoni AG. 2000a. Lipases. In: NF Haard, BK Simpson, editors, Seafood Enzymes. New York: Marcel Dekker, Inc. Pp. 121–146.

———. 2000b. Phospholipases. In: NF Haard, BK Simpson, editors, Seafood Enzymes. New York: Marcel Dekker, Inc. Pp. 91–119.

Lowrie RA. 1992. Conversion of muscle into meat: Biochemistry. In: DA Ledward, DE Johnston, MK Knight, editors, The Chemistry of Muscle-based Foods. Cambridge: Royal Society of Chemistry. Pp. 43–61.

Murphy DJ. 1999. Plant lipids—Their metabolism, function, and utilization. In: PJ Lea, RC Leegood, editors, Plant Biochemistry and Molecular Biology. Chichester: John Wiley and Sons. Pp. 119–135.

Orruno E, Apenten RO, Zabetakis I. 2001. The role of beta-glucosidase in the biosynthesis of 2,5-diemthyle-4-hydroxy-3(H)-furanone in strawberry (*Fragaria* X *ananassa* cv. *Elsanta*). Flavour and Fragrance Journal 16(2): 81–84.

Pan BS, Kuo JM. 2000. Lipoxygenases. In: NF Haard, BK Simpson, editors, Seafood Enzymes. New York: Marcel Dekker, Inc. Pp. 317–336.

Roller S, Goodenough PW. 1999. Food enzymes. In: S Roller, S Harlander, editors, Genetic Modification in the Food Industry. London: Kluwer. Pp. 101–128.

Schormuller J. 1968. The chemistry and biochemistry of cheese ripening. Advances in Food Research. 16:231–334.

Simpson BK. 2000. Digestive proteinases from marine animals. In: NF Haard, BK Simpson, editors, Seafood Enzymes. New York: Marcel Dekker, Inc. Pp. 191–213.

Smith CJ. 1999. Carbohydrate biochemistry. In: PJ Lea, RC Leegood, editors, Plant Biochemistry and Molecular Biology. Chichester: John Wiley and Sons. Pp. 81–118.

Stauffer CE. 1987a. Proteases, peptidases, and inhibitors. In: JE Kruger, D Lineback, CE Stauffer, editors, Enzymes and Their Role in Cereal Technology. St. Paul, Minnesota: American Association of Cereal Chemists. Pp. 201–237.

———. 1987b. Ester hydrolases. In: Enzymes and Their Role in Cereal Technology. St. Paul, Minnesota: American Association of Cereal Chemists. Pp. 265–280.

Steele JL. 1995. Contribution of lactic acid bacteria to cheese ripening. In: EL Malin, MH Tunick, editors,

Chemistry of Structure-Function Relationships in Cheese. New York: Plenum Press. Pp. 209–220.

Stoleo CG, Rehbein H. 2000. TMAO-degrading enzymes. In: NF Haard, BK Simpson, editors, Seafood Enzymes. New York: Marcel Dekker, Inc. Pp. 167–190.

Walstra TJ, Geurts A, Jellema NA, van Boekel MAJS. 1999. Dairy Technology: Principles of Milk Properties and Processes. New York: Marcel Dekker, Inc. pp. 94–97 (enzymes), 325–362 (lactic fermentation), 541–553 (cheese making).

Wong, WSW. 1989. Mechanisms and Theory in Food Chemistry. New York: Van Nostrand Reinhold. Pp. 242–263.

2
Analytical Techniques in Food Biochemistry

M. Marcone

INTRODUCTION

Without question, food can be considered as a very complex and heterogeneous composition of hundreds, if not thousands, of different biochemical compounds. In the area of food biochemistry, the isolation and quantitative measurement of these chemical components has posed, and continues to pose, immense challenges to the analytical biochemist. Without the ability to measure both specifically and quantitatively those biochemical components in food matrices, further advancements in the understanding of how foods change during maturation or processing would not be possible.

Although it is impossible to address the quantitative analysis of all the different food components, the major techniques for the analysis of protein, lipids, carbohydrates, minerals, vitamins, and pigments will be addressed in detail in this chapter. The principles behind their analysis are the building blocks for other analytical determinations, including techniques such as gas chromatography, high performance liquid chromatography (HPLC), and spectroscopy, including infrared and mass spectroscopy.

PROTEIN ANALYSIS

Proteins are considered to be among the most abundant cell components and, except for storage proteins, are important for biological functions within the organism—plant or animal. Many food proteins have been purified and characterized over the years and found to range from approximately 5000 to more than a million Daltons. In general, they are all composed of various elements including carbon, hydrogen, nitrogen, oxygen, and sulfur. These elements are formed into twenty different amino acids, which are linked together by peptide bonds to form proteins. In general, nitrogen is the most distinguishing element in proteins, varying from approximately 13 to 19% due to variations in the specific amino acid composition of proteins (Chang 1998).

For the past several decades, protein analysis has been performed by determining the nitrogen content of the food product after complete acidic hydrolysis and digestion by the Kjeldahl method and subsequent conversion to protein content using various conversion factors (Chang 1998, Diercky and Huyghebaert 2000). As far back as the turn of the century, colorimetric protein determination methods such as the Biuret procedure (which exploited the development of the violet-purplish color that is produced when cupric ions complex with peptide bonds under alkaline conditions) became available. The color absorbance is measured at 540 nm, with the color intensity (absorbance) being proportional to the protein content (Chang 1998) with a sensitivity of 1–10 mg protein/mL. Over the years, further modifications were made with the development of the Lowry method (Lowry et al. 1951, Peterson 1979), which combines the Biuret reaction with the reduction of the Folin-Ciocalteau phenol reagent (phosphomolybdic-phosphotungstic acid) by tyrosine and tryptophan residues in the proteins. The resulting bluish color is read at 750 nm, which is highly sensitive for low protein concentrations (sensitivity 20–100 μg). Other methods exploit the tendency of proteins to absorb strongly in the ultraviolet spectrum (i.e., 280 nm), primarily due to tryptophan and tyrosine residues. Since the tryptophan and tyrosine contents in proteins are generally constant, the absorbance at 280 nm has been used to estimate the concentration of proteins using Beer's law. Because each protein has a unique aromatic amino acid composition, the extinction coefficient (E_{280}) must be determined for each individual protein for protein content estimation.

Although these methods are appropriate for quantitating the actual amounts of proteins available within a sample or commodity, they do not possess the ability to differentiate and quantitate the actual types of proteins within a mixture. The most currently used methods for detecting and/or quantitating specific protein components can be cataloged in the fields of spectrometry, chromatography, electrophoresis, or immunology or a combination of these (VanCamp and Huyghebaert 1996).

Electrophoresis is defined as the migration of charged molecules in a solution through an electrical field (Smith 1998). Although several forms of this technique exist, zonal electrophoresis (in which proteins are separated from a complex mixture into bands by migrating in aqueous buffers through a solid polymer matrix called a polyacrylamide gel) is perhaps the most common. In nondenaturing/native electrophoresis, proteins are separated based on their charge, size, and hydrodynamic shape. In denaturing polyacrylamide gel electrophoresis (PAGE), an anionic detergent, sodium dodecyl sulfate (SDS), is used to separate protein subunits by size (Smith 1998). Isoelectric focusing is a modification of electrophoresis in which proteins are separated by charge in an electrophoretic field on a gel matrix in which a pH gradient has been generated using ampholytes. Proteins will focus or migrate to the location in the pH gradient that equals the isoelectric point (pI) of the protein. Resolution is among the highest of any protein separation technique and can separate proteins with pI differences as small as 0.02 pH units (Smith 1998, Chang 1998). More recently, with the advent of capillary electrophoresis, proteins can be separated on the basis of charge or size in an electric field within a very short period of time. The primary difference between capillary electrophoresis and conventional electrophoresis (described above) is that a capillary tube is used in place of a polyacrylamide gel. Unlike a gel, which must be made and cast each time, the capillary tube can be reused over and over. Electrophoresis flow within the capillary also can influence separation of the proteins in capillary electrophoresis (Smith 1998).

High performance liquid chromatography (HPLC) is another extremely fast analytical technique that possesses excellent precision and specificity as well as the proven ability to separate protein mixtures into individual components. Many different kinds of HPLC techniques exist, depending on the nature of the column characteristics (chain length, porosity, etc.) and the elution characteristics (mobile phase, pH, organic modifiers). In principle, proteins can be analyzed based on the polarity, solubility, or size of their constituent components.

Reversed-phase chromatography was introduced in the 1950s (Howard and Martin 1950, Diercky and Huyghebaert 2000) and has become a widely applied HPLC method for the analysis of both proteins and a wide variety of other biological compounds. Reversed-phase chromatography is generally achieved on an inert column packing, typically covalently bonded with a high density of hydrophobic functional groups, such as linear hydrocarbons with 4, 8, or 18 residues in length, or the relatively more polar phenyl group. In fact, reversed-phase HPLC has proven itself useful and indispensable in the field of varietal identification. It has been shown that the processing quality of various grains depends on their physical and chemical characteristics, which are at least partially genetic in origin, and that a wide range of qualities exists within varieties of each species (Osborne 1996). The selection of the appropriate cultivar is therefore an important decision for a farmer, since it greatly influences the return he receives on his investment (Diercky and Huyghebaert 2000).

Size-exclusion chromatography separates protein molecules based on their size or, more precisely, their hydrodynamic volume, and it has become very popular in recent years. Size-exclusion chromatography utilizes uniform rigid particles whose uniform pores are sufficiently large for the protein molecules to enter. Large molecules do not enter the pores of the column particles and are therefore excluded, that is, they are eluted in the void volume of the column (i.e., elute first), whereas smaller molecules enter the column pores and therefore take longer to elute from the column. An application example of size-exclusion chromatography is the separation of soybean proteins (Oomah et al. 1994). In one particular study, nine peaks were eluted for soybean, corresponding to different protein size fractions; one peak showed a high variability for the relative peak area and could serve as a possible differentiation among different cultivars. Differences, qualitatively and quantitatively, in peanut seed protein composition were detected by size-exclusion chromatography and contributed to evaluation of genetic differences, processing conditions, and seed maturity. Basha (1990) found that size-exclusion chromatography was an excellent indicator of seed maturity. Basha (1990) discovered that the area of one particular component (peak) decreased with increasing maturity and remained unchanged towards later stages of seed maturity. The peak was present in all studied cultivars, all showing a "mature seed protein profile" with respect to this particular peak, which was therefore called "Maturin."

LIPID ANALYSIS

By definition, lipids are soluble in various organic solvents but insoluble in water. For this reason, lipid insolubility in water becomes an important distinguishing and analytical factor used in separating lipids from other cellar components such as carbohydrates and proteins (Min and Steenson 1998). Fats (solids at room temperature) and oils (liquid at room temperature) are composed primarily of tri-esters of glycerol with fatty acids and are commonly called triglycerides. Other major lipid types found in foods include free fatty acids, mono- and diacylglycerols, and phospholipids.

Fats and oils are widely distributed in nature and play many important biological roles, especially within cell membranes. In general, many naturally occurring lipids are composed of various numbers of fatty acids (one to three) with various chain lengths, usually greater than 12 carbons, although the vast majority of animal and vegetable fats are made up of fatty acid molecules of greater than 16 carbons.

The total lipid content of a food is commonly determined using various organic solvent extraction methods. Unfortunately, the wide range of relative hydrophobicity of different lipids makes the selection of a single universal solvent almost impossible for lipid extraction and quantitation (Min and Steenson 1998). In addition to various solvent extraction methods (using various solvents), there are nonsolvent wet extraction methods and other instrumental methods that utilize the chemical and physical properties of lipids for content determination.

Perhaps one of the most commonly used and easiest to perform methods is the Soxhlet method, a semicontinuous extraction method that allows for the sample in the extraction chamber to be completely submerged in solvents for 10 minutes or more before the extracted lipid and solvent are siphoned back into the boiling flask reservoir. The whole process is repeated numerous times until all the fat is removed. The fat content is determined either by measuring the weight loss of the sample or the weight of lipid removed.

Another excellent method for total fat determination includes supercritical fluid extraction. In this method, a compressed gas (usually CO_2) is brought to a specific pressure-temperature combination that allows it to attain supercritical solvent properties for the selective extraction of lipid from a matrix. In this way specific types of lipids can be selectively extracted while others remain in the matrix (Min and Steenson 1998). The dissolved fat is then separated from the compressed, liquified gas by a drop in pressure, and the precipitated lipid is then quantified as percent lipid by weight (Min and Steenson 1998).

Another method often used for total lipid quantitation is the infrared method, which is based on the absorption of infrared energy by fat at a wavelength of 5.73 μm (Min and Steenson 1998). In general, the more energy is absorbed at 5.73 μm, the higher the lipid content in the material. Near-infrared spectroscopy has been successfully used to measure the lipid content of various oilseeds, cereals, and meats; it has the added advantage of being nondisruptive to the sample, in contrast to other previously reviewed methods.

Although the above-cited methods are appropriate for quantitating the actual amounts of lipids within a given sample, they do not offer the ability to characterize the types of fatty acids within a mixture. Gas chromatography, however, does offer the ability to characterize these lipids in terms of their fatty acid composition (Pike 1998). First of all, mono-, di-, and triglycerides need to be isolated individually if a mixture exists, usually by simple adsorption chromatography on silica. The isolated glycerides can then be hydrolyzed to release individual fatty acids, which are subsequently converted to their ester form; that is, the glycerides are saponified and the fatty acids thus liberated are esterified to form fatty acid methyl esters. The fatty acids are now volatile and can be separated chromatographically using various packed and capillary columns using a variety of temperature-time gradients.

Separation of the actual mono-, di-, and triglycerides is usually much more problematic than determining their individual fatty acid constituents or building blocks. Although gas chromatography has also been used for this purpose, such methods give insufficient information to provide a complete triglyceride composition of a complex mixture. Such analyses are important for the edible oil industry for process and product quality control purposes as well as for the understanding of triglyceride biosynthesis and deposition in plant and animal cells (Marini 2000).

With HPLC analysis, Plattner et al. (1977) were able to establish that, under isocratic conditions, the logarithm of the elution volume of a triacylglycerol was directly proportional to the total number of carbon atoms (CN) and inversely proportional to the total number of double bonds (X) in the three fatty acyl chains (Marini 2000). The elution behavior is controlled by the equivalent carbon number (ECN) of a triacylglycerol, which may be defined as $ECN = CN - Xn$ where n is the factor for double bond contribution, normally close to 2.

The IUPAC Commission on Oils, Fats, and Derivatives undertook the development of a method for the determination of triglycerides in vegetable oils by liquid chromatography. Materials studied included soybean oil, almond oil, sunflower oil, olive oil, rapeseed oil, and blends of palm and sunflower oils and almond and sunflower oils (Marini 2000, Fireston 1994). AOAC International adopted this method for determination of triglycerides (by partition numbers) in vegetable oils by liquid chromatography as an IUPAC-AOC-AOAC method. In this method, triglycerides in vegetable oils are separated according to their equivalent carbon number by reversed-phase HPLC and detected by differential refractometry. Elution order is determined by calculating the equivalent carbon numbers, $ECN = s$ and $CN - 2n$, where CN is the carbon number and n is the number of double bonds (Marini 2000).

CARBOHYDRATE ANALYSIS

Carbohydrates play several important roles in foods, including among other things, imparting important physical properties to the foods as well as constituting a major source of energy in the human diet. In fact, it has been estimated that carbohydrates account for greater than 70% of the total daily caloric intake in many parts of the world (BeMiller and Low 1998).

Carbohydrates found in nature are almost exclusively of plant origin, with at least 90% of them occurring in the form of polysaccharides (BeMiller and Low 1998). Interestingly, although the most carbohydrates are in the form of polysaccharides, starches are about the only polysaccharide that is digestible by humans. The vast majority of polysaccharides are therefore nondigestible, and they have been divided into two classes, soluble and insoluble, which form what is commonly called dietary fiber.

For decades total carbohydrate was determined by exploiting the tendency of carbohydrates to condense with various phenolic-type compounds including phenol, orcinol, resorcinol, napthoresorcinol, and α-naphthol (BeMiller and Low 1998). The most widely used condensation was with phenol, which offered a rapid, simple, and specific determination for carbohydrates. Virtually all types of carbohydrates, mono-, di, oligo-, and polysaccharides, could be determined. After reaction with phenol in acid in the presence of heat, a stable color is produced that can be read spectrophotometrically. A standard curve is usually prepared with a carbohydrate similar to these being measured.

Although the above method was, and still is, used to quantitate the total amount of carbohydrate in a given sample, it does not offer the ability to determine the actual types and/or building blocks of individual carbohydrates. Earlier methods, which included paper chromatography, open column chromatography, and thin-layer chromatography, have largely been replaced by HPLC and/or gas chromatography (Peris-Tortajada 2000). Gas chromatography has been established as an important method in carbohydrate determinations since the early 1960s (Sweeley et al. 1963, Peris-Tortajada 2000), and several unique applications have since then been reported (El Rassi 1995).

For carbohydrates to be analyzed by gas chromatography, they must first be converted into volatile derivatives. Perhaps the most commonly used derivatizing agent is trimethylsilyl (TMS). In this procedure, the aldonic acid forms of carbohydrates are converted into their TMS ethers. The reaction mixtures are then injected directly into the chromatograph, and temperature programming is utilized to optimize the separation and identification of individual components. A flame ionization detector is still the detector of choice for carbohydrates. Unlike gas chromatography, HPLC analysis of carbohydrates requires no prior derivatization of carbohydrates and gives both qualitative (identification of peaks) and quantitative information for complex mixtures of carbohydrates. HPLC has been shown to be an excellent choice for the separation and analysis of a wide variety of carbohydrates, ranging from monosaccharides to oligosaccharides. For the analysis of larger polysaccharides, a hydrolysis step is required prior to chromatographic analysis. A variety of different columns can be used, with bonded amino phases used to separate carbohydrates with molecular weights up to about 2500, depending upon carbohydrate composition and, therefore, solubility properties (Peris-Tortajada 2000). The elution order on amine-bonded stationary phases is usually monosaccharide and sugar alcohols followed by disaccharides and oligosaccharides. Such columns have been successfully used to analyze carbohydrates in anything from fruits and vegetables all the way to processed foods such as cakes, confectionaries, beverages, and breakfast cereals (BeMiller and Low 1998). With larger polysaccharides, gel filtration becomes the preferred chromatographic technique, as found in the literature. Gel filtration media such as Sephadex® and Bio-Gel® have been successfully used to characterize polysaccharides according to molecular weight.

MINERAL ANALYSIS

Minerals are extremely important for the structural and physiological functioning of the body. It has been estimated that 98% of the calcium and 80% of the phosphorous in the human body are bound up within the skeleton (Hendricks 1998). Those minerals that are directly involved in physiological function (e.g., in muscle contraction) include sodium, calcium, potassium, and magnesium. Certain minerals (or macrominerals) are required in quantities of more than 100 mg per day; these include sodium, potassium, magnesium, phosphorous, calcium, chlorine, and sulfur. Another 10 minerals (trace minerals) are required in milligram quantities per day; these include silica, selenium, fluoride, molybdenum, manganese, chromium, copper, zinc, iodine, and iron (Hendricks 1998). Each of the macro- and trace minerals has a specific biochemical role in maintaining body function and is important to overall health and well-being.

Although minerals are naturally found in most food materials, some are added to foods during processing to accomplish certain objectives. An example of this is salt, which is added during processing to decrease water activity and to act as a preservative (e.g., pickles and cheddar cheese; Hendricks 1998). Iron is added to fortify white flour, and various other minerals such as calcium, iron, and zinc are added to various breakfast cereals. In fact, salt itself is fortified with iodine in North America in order to control goiter.

It should also be noted that food processing can decrease the mineral content (e.g., the milling of wheat removes the mineral-rich bran layer). During the actual washing and blanching of various foods, important minerals are often lost. It can therefore be concluded that accurate and specific methods for mineral determination are in fact important for nutritional purposes as well as for properly processing food products for both human and animal consumption.

In order to determine the total mineral content of a food material, the ashing procedure is usually performed. Ash refers to the inorganic residue that remains after ignition, or in some cases complete oxidation, of organic material (Harbers 1998). Ashing can be divided into three main types: (1) dry ashing (most commonly used); (2) wet ashing (oxidation), for samples with high fat content (such as meat products) or for preparation for elemental analysis; and (3) plasma ashing (low temperature) for when volatile elemental analysis is conducted.

In dry ashing, samples usually are incinerated in a muffle furnace at temperatures of 500–600°C. Most minerals are converted to oxides, phosphates, sulfates, chlorides, or silicates. Unfortunately, elements such as mercury, iron, selenium, and lead may be partially volatized using this procedure.

Wet ashing utilizes various acids to oxidize organic materials and minerals that are solubilized without their volatilization. Nitric and perchloric acids are often used, and reagent blanks are carried throughout the procedure and are subtracted from sample results.

In low-temperature plasma ashing, samples are treated in a similar way to those in dry ashing, but under a partial vacuum, with samples being oxidized by nascent oxygen formed by an electromagnetic field.

Although the above three methods have been proven to be appropriate for quantitating the total amount of mineral within a sample, they do not possess the ability to either differentiate or quantitate actual mineral elements within a mixture.

When atomic absorption spectrometers became widely used in the 1960s and 1970s, they paved the way for measuring trace amounts of mineral elements in various biological samples (Miller 1998). Essentially, atomic absorption spectroscopy is an analytical technique based on the absorption of ultraviolet or visible radiation by free atoms in the gaseous state. The sample must be first ashed and then diluted in weak acid. The solution is then atomized into a flame. According to Beer's Law, absorption is directly related to the concentration of a particular element in the sample.

Atomic emission spectroscopy differs from atomic absorption spectroscopy in that the source of radiation is, in fact, excited atoms or ions in the sample rather than from external source, has in part taken over. Atomic emission spectroscopy does have advantages with regard to sensitivity, interference, and multielement analysis (Miller 1998).

Recently, the use of ion-selective electrodes has made on-line testing of the mineral composition of samples a reality. In fact, many different electrodes have been developed for the direct measurement of various anions and cations such as calcium, bromide, fluoride, chloride, potassium, sulfide, and sodium (Hendricks 1998). Typically, levels down to 0.023 ppm can be measured. When working with ion-selective electrodes it is common procedure to measure a calibration curve.

VITAMIN ANALYSIS

By definition, vitamins are organic compounds of low molecular weight that must be obtained from external sources in the diet and are also essential for normal physiological and metabolic function (Russell 2000). Since the vast majority of vitamins cannot be synthesized by humans, they must be obtained from either food or dietary supplements. When vitamins are absent or at inadequate levels in the diet, deficiency disease commonly occurs, (e.g., scurvy and pellagra from a lack of ascorbic acid and niacin, respectively; Eitenmiller et al. 1998).

Analyses of vitamins in foods are performed for numerous reasons; for example, to check for regulatory compliance, to obtain data for nutrient labeling, or to study the changes in vitamin content attributable to food processing, packaging, and storage (Ball 2000). Therefore, numerous analytical methods have been developed to determine vitamin levels during processing and in the final product.

Vitamins have been divided into two distinct groups: (1) those that are water soluble (B vitamins, vitamin C) and (2) those that are fat soluble (vitamins A, D, E, and K).

The scientific literature contains numerous analytical methods for the quantitation of water-soluble vitamins including several bio-, calorimetric, and fluorescent assays that have been proven to be accurate, specific, and reproducible for both raw and processed food products (Eitenmiller et al. 1998). The scientific literature also contains an abundance of HPLC-based methodologies for the quantitation of water-soluble vitamins (Russell 2000). High performance liquid chromatography (HPLC) has become by far the most popular technique for the quantitation of these water-soluble vitamins. In general, it is the nonvolatile and hydrophilic nature of these vitamins that make them excellent candidates for reversed-phase HPLC analysis (Russell 2000). The ability to automate these analyses using autosamplers and robotics makes HPLC an increasingly popular technique. Since the vast majority of vitamins occur in food in trace amounts, detection sensitivity is paramount to their detection. Although ultraviolet absorbance is the most common detection method, both fluorescence and electrochemical detection are also used in specific cases. Refractive index detection is seldom used for vitamin detection due to its inherent lack of specificity and sensitivity.

During the 1960s, gas chromatography using packed columns was widely applied to the determination of various fat-soluble vitamins, especially vitamins D and E. Unfortunately, thin-layer chromatographic and open-column techniques were still necessary for preliminary separation of the vitamins, followed by derivatization to increase the vitamins' thermal stability and volatility (Ball 2000). More recently, the development of fused-silica, open tubular capillary columns has revived the use of gas chromatography, leading to a number of recent applications for the determination of fat-soluble vitamins, especially vitamin E (Marks 1988, Ulberth 1991, Kmostak and Kurtz 1993, Mariani and Bellan 1996). This being said, the method of choice for determining fat-soluble vitamins in foods is HPLC (Ball 2000). The interest in this chromatographic technique is due to the lack of need for derivatization and the greater separation and detection selectivity this technique offers. Various HPLC methods of analysis were introduced for the first time in the 1995 edition of the Official Methods of Analysis of AOAC International; these include vitamin A in milk (AOAC 992.04, 1995) and vitamins A (AOAC 992.26, 1995), E (AOAC 992.03, 1995), and K (AOAC 992.27, 1995) in various milk-based infant formulas (Ball 2000).

It should be noted that at present there is no universally recognized standard method for determining any of the fat-soluble vitamins that can be applied to all food types (Ball 2000).

PIGMENT ANALYSIS

Color is a very important characteristic of foods and is often one of the first quality attributes used to judge the quality or acceptability of a particular food (Schwartz 1998). There are a vast number of natural and synthetic pigments, both naturally occurring and added to foods, that contribute to food color. Of the naturally occurring pigments in foods, the vast majority can be divided into five major classes, four of which are distributed in plant tissues and one in animal tissues (Schwartz 1998). Of those found in plants, two types are lipid soluble (i.e., the chlorophylls and the carotenoids), and the other two are water soluble (i.e., anthocyanins and betalains). Carotenoids are found in animals but are not biosynthesized; that is, they are derived from plant sources (Schwartz 1998).

Several analytical methods have been developed for the analysis of chlorophylls in a wide variety of foods. Early spectrophotometric methods allowed for the quantitation of both chlorophyll a and chlorophyll b by measuring absorbance at the absorbance maxima of both chlorophyll types. Unfortunately, only fresh plant material could be assayed, as no pheophytin could be determined. This became the basis for the AOAC International spectrophotometric procedure (Method 992.04), which provides results for total chlorophyll content as well as for chlorophyll a and chlorophyll b quantitation.

Schwartz et al. (1981) described a simple reversed-phase HPLC method for the analysis of chlorophylls and their derivatives in fresh and processed plant tissues. This method simplified the determination of chemical alterations in chlorophyll during the processing of foods and allowed for the determination of pheophytins and pyropheophytins.

For carotenoid analysis, numerous HPLC methods have been developed, particularly for the specific separation of various carotenoids found in fruits and vegetables (Bureau and Bushway 1986). Both normal and reversed-phase methods have been used, with the reversed-phase methods predominating (Schwartz 1998). Reversed-phase chromatography on C-18 columns using isocratic elution procedures with mixtures of methanol and acetonitrile containing ethyl acetate, chloroform, or tetrahydrofuran have been found to be satisfactory (Schwartz 1998). Detection of carotenoids usually range from approximately 430 to 480 nm. Since β-carotene in hexane has an absorption maximum at 453 nm, many methods have detected a wide variety of carotenoids in this region (Schwartz 1998).

Measurements of anthocyanins have been performed by determining absorbance of diluted samples acidified to about pH 1.0 at wavelengths between 510 and 540 nm. Unfortunately, absorbance measurements of anthocyanin content provide only for a total quantification, and further information about the amounts of various individual anthocyanins must be obtained by other methods. Reversed-phase HPLC methods employing C-18 columns have been the methods of choice due to the water-soluble nature of anthocyanins. Mixtures of water and acetic, formic, or phosphoric acids are usually employed as part of the mobile phase.

Charged pigments such as betalains can be separated by electrophoresis, but HPLC has been found to provide for more rapid resolution and quantitation. Betalains have been separated on reversed-phase columns by using various ion-paring or ion-suppression techniques (Schwartz and Von Elbe 1980, Huang and Von Elbe 1985). This procedure allows for more interaction of the individual molecules of betaine with the stationary phase and better separation between individual components.

REFERENCES

AOAC International. 1995. Official Methods of Analysis, 16th edition. AOAC International, Gaithersburg, Maryland.

AOAC Official Method 992.03. 1995. Vitamin E activity (all-*rac*-α-tocopherol) in milk and milk-based infant formula. Liquid chromatographic method. In: Official Methods of Analysis of AOAC International, 16th edition, edited by MP Bueno, pp. 50-4–50-5. AOAC International, Arlington, Virginia.

AOAC Official Method 992.04. 1995. Vitamin A (Retinol isomers) in milk and milk-based infant formula. Liquid chromatographic method. In: Official Methods of Analysis of AOAC International, 16th edition, edited by MP Bueno, pp. 50-1–50-2. AOAC International, Arlington, Virginia.

AOAC Official Method 992.06. 1995. Vitamin A (Retinol isomers) in milk and milk-based infant formula. Liquid chromatographic method. In: Official Methods of Analysis of AOAC International, 16th edition, edited by MP Bueno, pp. 50-2–50-3. AOAC International, Arlington, Virginia.

AOAC Official Method 992.26. 1995. Vitamin D3 (Cholecalciferol) in ready-to-feed milk-based infant formula. Liquid chromatographic method. In: Official Methods of Analysis of AOAC International, 16th edition, edited by MP Bueno, pp. 50-5–50-6. AOAC International, Arlington, Virginia.

AOAC Official Method 992.27. 1995. *Trans*-vitamin K_1 (Phylloquinone) in ready-to-feed milk-based infant formula. Liquid chromatographic method. In: Official Methods of Analysis of AOAC International, 16th edition, edited by MP Bueno, pp. 50-6–50-8. AOAC International, Arlington, Virginia.

Ball GF. 2000. The fat-soluble vitamins. In: LML Nollet, editor, Food Analysis by HPLC, pp. 321–402. Marcel Dekker Inc., New York.

Basha SM. 1990. Protein as an indicator of peanut seed maturity. J. Agricultural and Food Chemistry 38: 373–376.

BeMiller JN, Low NH. 1998. Carbohydrate analysis. In: SS Nielsen, editor, Food Analysis, 2nd edition, pp. 167–187. Aspen Publication Inc., Gaithersburg, Maryland.

Bureau JL, Bushway RJ. 1986. HPLC Determination of carotenoids in fruits and vegetables in the United States. J Food Science 51:128–130.

Chang SKC. 1998. Protein analysis. In: SS Nielsen, editor, Food Analysis, 2nd edition, pp. 237–249. Aspen Publication Inc., Gaithersburg, Maryland.

Diercky S, Huyghebaert A. 2000. HPLC of food proteins. In: LML Nollet, editor, Food Analysis by HPLC, pp.127–167. Marcel Dekker Inc., New York.

Eitenmiller RR, Landen WO, Augustin J. 1998. Vitamin analysis. In: SS Nielsen, editor, Food Analysis, 2nd edition, pp. 281–291. Aspen Publication Inc., Gaithersburg, Maryland.

El Rassi Z. 1995. Carbohydrate Analysis High Performance Liquid Chromotography and Capillary Electrophoresis. Elsevier, Amsterdam.

Fireston D. 1994. J Assoc Off Anal Chem Int 77:954–961.

Harbers LH. 1998. Ash analysis. In: SS Nielsen, editor, Food Analysis, 2nd edition, pp. 141–149. Aspen Publication Inc., Gaithersburg, Maryland.

Hendricks DG. 1998. Mineral analysis. In: SS Nielsen, editor, Food Analysis, 2nd edition, pp. 151–165. Aspen Publication Inc., Gaithersburg, Maryland.

Howard GA, Martin AJP. 1950. Biochem J 46:532.

Huang HS, Von Elbe JH. 1985. Kinetics of degradation and regeneration of betanine. J Food Science 50:1115–1120, 1129.

Kmostak S, Kurtz DA. 1993. Rapid determination of supplemental vitamin E acetate in feed premixes by capillary gas chromatography. J Assoc Off Anal Chem Int 76:735–741.

Lowry OH, Rosebrough NJ, Farr K, Randall RJ. 1951. Protein determination. Anal. Biochem. 193:265–275.

Mariani C, Bellan G. 1996. Content of tocopherols, deidrotocopherols, tocodienols, tocotrienols in vegetable oils. Riv. Ital. Sostanze Grasse. 73:533–543 (In Italian).

Marini D. 2000. HPLC of lipids. In: LML Nollet, editor, Food Analysis by HPLC, pp. 169–249. Marcel Dekker Inc., New York.

Marks C. 1988. Determination of free tocopherols in deodorizer by capillary gas chromatography. J Am Oil Chem Soc 65:1936–1939.

Miller DD. 1998. Atomic absorption and emission spectroscopy. In: SS Nielsen, editor, Food Analysis, 2nd edition, pp. 425–442. Aspen Publication Inc., Gaithersburg, Maryland.

Min DB, Steenson DF. 1998. Crude fat analysis. In: SS Nielsen, editor, Food Analysis, 2nd edition, pp. 201–215. Aspen Publication Inc., Gaithersburg, Maryland.

Oomah BD, Voldeng H, Fregeau-Reid JA 1994. Characterization of soybean proteins by HPLC. Plant Foods for Human Nutrition. 45:25–263.

Osborne BG. 1996. Authenticity of cereals. In: PR Ashurst, MJ Dennis, editors, Food Authentication, pp. 171–197. Chapman and Hall, London, England.

Peris-Tortajada M. 2000. HPLC determination of carbohydrates in foods. In: LML Nollet, editor, Food Analysis by HPLC, pp. 287–302. Marcel Dekker Inc., New York.

Peterson GL. 1979. A new protein determination method. Anal. Biochem. 100:201–220.

Pike OA. 1998. Fat characterization. In: SS Nielsen, editor, Food Analysis, 2nd edition, pp. 217–235. Aspen Publication Inc., Gaithersburg, Maryland.

Plattner RD, Spencer GF, Kleiman R. 1977. J Am Oil Chem Soc 54:511–519.

Russell LF. 2000. Quantitative determination of water-soluble vitamins. In: LML Nollet, editor, Food Analysis by HPLC, pp. 403–476. Marcel Dekker Inc., New York.

Schwartz SJ. 1998. Pigment analysis. In: SS Nielsen, editor, Food Analysis, 2nd edition, pp. 293–304. Aspen Publication Inc., Gaithersburg, Maryland.

Schwartz SJ, Von Elbe JH. 1980. Quantitative determination of individual betacyanin pigments by high-performance liquid chromatography. J Agricultural and Food Chemistry. 28:540–543.

Schwartz SJ, Woo SL, Von Elbe JH. 1981. High performance liquid chromatography of chlorophylls and their derivatives in fresh and processed spinach. J Agricultural and Food Chemistry. 29:533–535.

Smith DM. 1998. Protein separation and characterization procedures. In: SS Nielsen, editor, Food Analysis, 2nd edition, pp. 251–263. Aspen Publication Inc., Gaithersburg, Maryland.

Sweeley CC, Bentley R, Makita M, Wells WW. 1963. J Am Chem Soc 85:2497–2502.

Ulberth F. 1991. Simultaneous determination of vitamin E isomers and cholesterol by GLC. J High Resolution Chromat 14:343–344.

VanCamp J, Huyghebaert A. 1996. Proteins. In: LML Nollet, editor, Handbook of Food Analysis. Vol. 1, Physical Characterization and Nutrient Analysis, pp. 277–310. Marcel Dekker, New York.

3

Recent Advances in Food Biotechnology Research

*S. Jube and D. Borthakur**

*Corresponding contributor.

INTRODUCTION

Modern biotechnology involves molecular techniques that use whole or parts of living organisms to produce or improve commercial products and processes. It is a relatively new and rapidly evolving branch of molecular biology, which started with the creation of the first recombinant gene 30 years ago. These techniques are, in many different ways, changing the way we live by improving the foods we eat, the beverages we drink, the clothes we wear, and the medicines we take. They also have enhanced other aspects of our lives through the development of new detection methods for early diagnosis of many diseases such as arteriosclerosis, cancer, diabetes, Parkinson's, and Alzheimer's. The application of biotechnology methods in the food and agricultural industry is one of the many aspects of biotechnology that has great impact on society. By the year 2050, it is expected that more than 10 billion people will be living on this planet, and it is also believed that there may not be enough resources to feed the world population (UNFPA 1995). Hunger and malnutrition already claim 24,000 lives a day in the developing countries of Asia, Africa, and Latin America (James 2003). Malnutrition, however, is not exclusive to developing nations. Many people in industrialized countries, although mostly well fed, still suffer from lack of proper nourishment.

Biotechnology is the scientific field that offers the greatest potential to stop hunger today and help avoid mass starvation in the future. Through biotechnology, scientists can enhance a crop's resistance to diseases and environmental stresses, allowing crops to be grown in relatively unproductive and unsuitable land. Recent developments in biotechnology will allow the production of more nutritious, safer, tastier, and healthier food. Advances in genetic engineering are revolutionizing the way we produce and consume food, and it is quite possible that in the next decade a large percentage of the food we eat will be bioengineered.

In this review, the recent advances, methods, and applications of biotechnology in the manufacture of food products from transgenic plants, animals, and microorganisms have been summarized. This article is not, by any means, a documentation of every application of biotechnology in the food industry, but a comprehensive review, which includes the most relevant examples. The results of important scientific research trying to improve the nutritional value of staple crops such as rice, potatoes, and soybeans will be discussed. This improvement can be achieved through the introduction of genes that encode for enzymes in the biosynthetic pathway of vitamins, essential amino acids, essential elements, and micronutrient binding proteins. Examples of genetic engineering of cattle, swine, poultry, and fish will be described, with the purpose of improving milk quality, decreasing fat content, increasing productivity/growth, and providing tolerance to freezing temperatures. The use of the mammary gland and egg as bioreactors for the production of important proteins will also be addressed. The third part of this article will focus on the role of microorganisms for the betterment of food products through the elimination of carcinogenic compounds from beverages, the inhibition of pathogenic bacteria from starter cultures, the production of healthier natural sweeteners, and the synthesis of beneficial compounds such as carotenoids. Finally, some examples on the use of biotechnology techniques in the detection of transgenic material and harmful pathogens in food products will be described. Other recent reviews of specific aspects of food biotechnology will complement the information provided in this article (Giddings et al. 2000, Kleter et al. 2001, Daniell and Dhingra 2002, Dove 2002, Sharma et al. 2002, Taylor and Hefle 2002, Vasil 2003).

BIOENGINEERED PLANTS

Genetic engineering methods have been extensively used to increase the quantity of different nutrients (vitamins, essential amino acids, minerals, and phytochemicals) and enhance their availability in plants. There are two main methods for transferring genes into plants for production of transgenic plants: *Agrobacterium*-mediated transformation and microprojectile bombardment. In the *Agrobacterium*-mediated transformation method, a genetically engineered strain of *Agrobacterium tumefaciens* is used to transfer the transgene into the plants. Some strains of *A. tumefaciens* have the natural ability to transfer a segment of their own DNA into plants for inducing crown-gall tumors. These crown gall–inducing wild-type strains of *A. tumefaciens* have a Ti (tumor inducing) plasmid that carries the genes for tumor induction. During the process of infection, *Agrobacterium* transfers a segment of Ti plasmid, known as T-DNA, to plant cells (Willmitzer et al. 1983). The Ti plasmid can be engineered into a two-plasmid (binary) system containing a "disarmed" Ti plasmid, in which the T-DNA has been deleted, and a small plasmid (referred to as a binary cloning vector) containing an "engineered" T-DNA segment. The disarmed Ti plasmid, which is maintained in an *A. tumefaciens* strain, serves as a helper, providing the transfer function for the engineered T-DNA, which contains a target gene and a plant selectable marker gene inserted between the T-DNA left and right borders. When the *A. tumefaciens* containing the disarmed Ti plasmid and the binary cloning vector is grown in the presence of acetosyringone, the *Agrobacterium vir* (virulence) gene proteins, which help transfer the engineered T-DNA region of the binary cloning vector to the plant cells, are produced (Zambryski 1988). The *Agrobacterium*-mediated transformation is the most commonly used method for genetic engineering of plants.

The microprojectile bombardment method, also known as the gene gun or biolistic transformation method, involves the delivery and expression of foreign DNA directly into individual plant cells (Klein et al. 1987). It has been proven to be a powerful method for transforming a large number of plant species, including monocots, which are often difficult to transform using *A. tumefaciens* (Vain et al. 1995). In this method, tungsten or gold spherical particles, approximately 4 μm in size, are coated with DNA and then accelerated to high speed and inserted into plant cells using a biolistic particle delivery system or a gene gun. Once the DNA gets inside a cell, it integrates into the plant DNA through some unknown process. It is not known whether integration of DNA into the chromosome requires the delivery of the microprojectiles into the plant nucleus. The microprojectile bombardment method has been used to transfer genes into various plant sections used in tissue culture regeneration, calli, cell suspensions, immature embryos, and pollens in a wide range of plant species. This method can also be used to transfer genes into chloroplasts and mitochondria, which cannot be accomplished by the *A. tumefaciens*-mediated gene transfer (Southgate et al. 1995).

Essential Vitamins

Vitamins play a crucial role in human health by controlling metabolism and assisting the biochemical processes that release energy from foods. They are important in the formation of hormones, blood cells, nervous-system chemicals, and genetic material. Vitamins combine with proteins to create metabolically active enzymes that are important in many chemical reactions. Of the 13 well-known vitamins, the body can only manufacture vitamin D; all others, such as vitamins A, C, and E, must be derived from the diet. Insufficient vitamin intake may cause a variety of health problems. Through biotechnology, scientists can increase the content of vitamins in certain crops, allowing a wider range of the world population to make use of their health benefits.

Vitamin A

Nearly two-thirds of the world population depends on rice as their major staple, and among them an estimated 300 million suffer from some degree of vitamin A deficiency (WHO 1997). This is a serious public health problem in a number of countries, including highly populated areas of Asia, Africa, and Latin America. The rice endosperm (the starchy interior part of the rice grain) does not contain any β-carotene, which is the precursor for vitamin A. Vitamin A is a component of the visual pigments of rod and cone cells in the retina, and its deficiency causes symptoms ranging from night blindness to total blindness. In Southeast Asia, it is estimated that a quarter of a million children go blind each year because of this nutritional deficiency. Plant foods such as carrots and many other vegetables contain β-carotene. Each β-carotene molecule is oxidatively cleaved in the intestine to yield two molecules of retinal, which can be then reduced to form retinol or vitamin A (Fig. 3.1).

Ingo Potrykus from the Swiss Federal Institute of Technology, Zurich, Switzerland, and Peter Beyer from the University of Freiburg recently developed transgenic rice, expressing genes for β-carotene biosynthesis in rice grains (Potrykus 2001). Rice endosperm naturally contains geranlylgeranyl pyrophosphate (GGPP), which is a precursor of the pathway for β-carotene biosynthesis. GGPP can be converted into β-carotene in four steps (Bartley et al. 1994) (Fig. 3.2). The bacterial enzyme phytoene desaturase (EC 1.14.99.30) encoded by the *crtI* gene can substitute the functions of both phytoene desaturase and ζ-carotene desaturase (EC 1.14.99.30) in plants (Armstrong 1994). To reduce the number of genes transformed into rice for the β-carotene pathway, the researchers used the *crtI* gene from the bacterium *Erwinia uredovora* (Ye et al. 2000). The *psy* gene encoding phytoene synthase (EC 2.5.1.32) and the *lcy* gene encoding lycopene β-cyclase used for transformation originated from the plant daffodil. The plant *psy* gene (cDNA) and the bacterial *crtI* gene were placed under the control of the endosperm-specific rice glutelin (Gt1) promoter and the 35S cauliflower mosaic virus (CaMV) promoter, respectively, and introduced in the binary plasmid pZPsC. Another plasmid, pZLcyH, was constructed by inserting the *lcy* gene from daffodil under the control of rice Gt1 promoter and the *aphIV* gene, for hygromycin resistance, under the control of 35S CaMV promoter. Plasmids pZPsC and pZLCyH were cotransformed into immature rice embryos by *Agrobacterium*-mediated transformation (Ye et al. 2000). All hygromycin-resistant transformants were screened for the presence of the *psy*, *crtI*, and *lcy* genes by Southern hybridization. A few of the transformed plants produced β-carotene in the endosperm, which caused the kernel to appear yellow. The selected line contained 1.6–4μg β-carotene per gram of endosperm and was established as "golden rice."

Vitamin C

Vitamin C or ascorbic acid, found in many plants, is an important component in human nutrition. It has antioxidant properties, improves immune cell and cardiovascular functions, prevents diseases linked to the connective tissue (Davey et al. 2000), and is required for iron utilization (Hallberg et al. 1989). Most animals and plants are able to synthesize ascorbic acid, but humans do not have the enzyme, L-gulono-1,4-lactone oxidoreductase (EC 1.1.3.8),

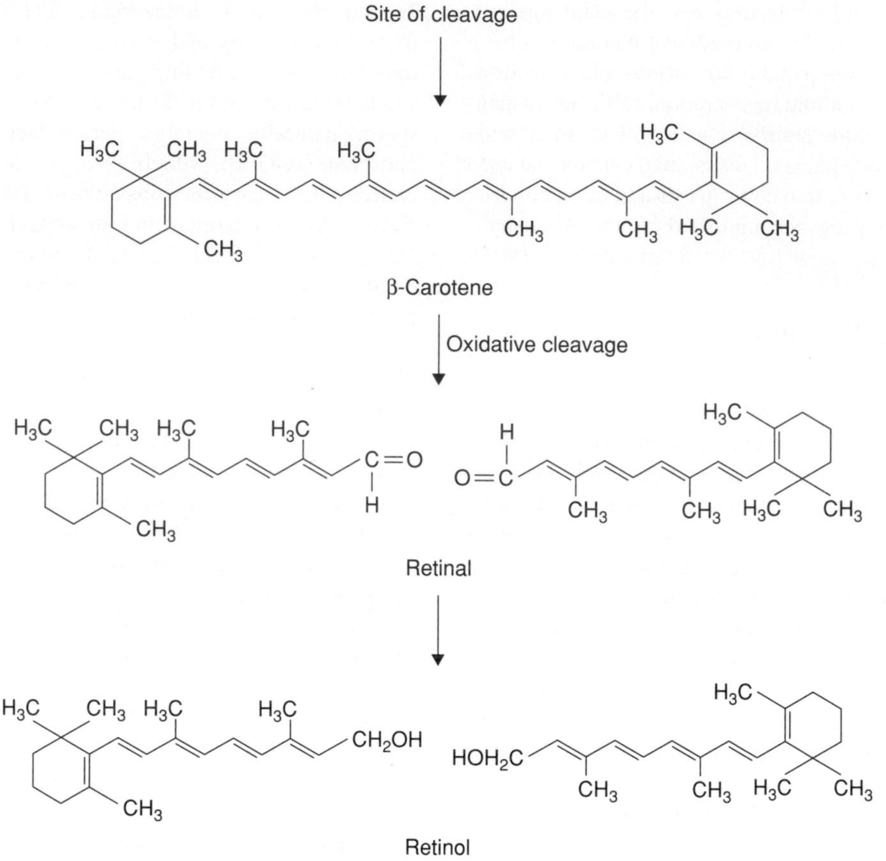

Site of cleavage

H$_3$C CH$_3$ H$_3$C H$_3$C H$_3$C CH$_3$ CH$_3$ CH$_3$ H$_3$C CH$_3$

β-Carotene

Oxidative cleavage

H$_3$C CH$_3$ H$_3$C H$_3$C C=O H O=C H H$_3$C CH$_3$ CH$_3$ CH$_3$ H$_3$C CH$_3$

Retinal

H$_3$C CH$_3$ H$_3$C H$_3$C CH$_2$OH HOH$_2$C H$_3$C CH$_3$ CH$_3$ CH$_3$ H$_3$C CH$_3$

Retinol

Figure 3.1. Formation of vitamin A from β-carotene.

necessary for the final step in ascorbic acid biosynthesis. For this reason, ascorbic acid needs to be consumed from dietary sources, especially from plants (Davey et al. 2000). The recent identification of the ascorbic acid pathway in plants opened the way to manipulating its biosynthesis and allowed the design of bioengineered plants that produce ascorbic acid at significantly higher levels. The biosynthetic pathway of ascorbic acid in animals differs from that in plants. In plants, vitamin C biosynthesis can be accomplished in two ways. First, D-galacturonic acid, which is released upon the hydrolysis of pectin (a major cell wall component), is converted into L-galactonic acid with the help of the enzyme D-galacturonic acid reductase (EC 2.7.1.44). L-galactonic acid is then readily converted into L-galactono-1,4-lactone, which is the immediate precursor of ascorbic acid (Fig. 3.3; Wheeler et al. 1998, Smirnoff et al. 2001). Researchers in Spain (Agius et al. 2003) isolated and characterized *GalUR*, a gene in strawberry that encodes the enzyme D-galacturonic acid reductase. The *GalUR* gene was amplified by polymerase chain reaction (PCR) as a 956 bp fragment and cloned into a binary vector behind a 35S CaMV promoter. The resulting plasmid was transformed into *E. coli* and delivered to *Agrobacterium* by triparental mating. Finally, the *GalUR* gene was introduced into *Arabidopsis thaliana* plants via *Agrobacterium*-mediated transformation. The expression of the strawberry *GalUR* gene in *A. thaliana* allowed the bioengineered plants to increase the biosynthesis of ascorbic acid by two to three times compared with the wild-type plants (Agius et al. 2003).

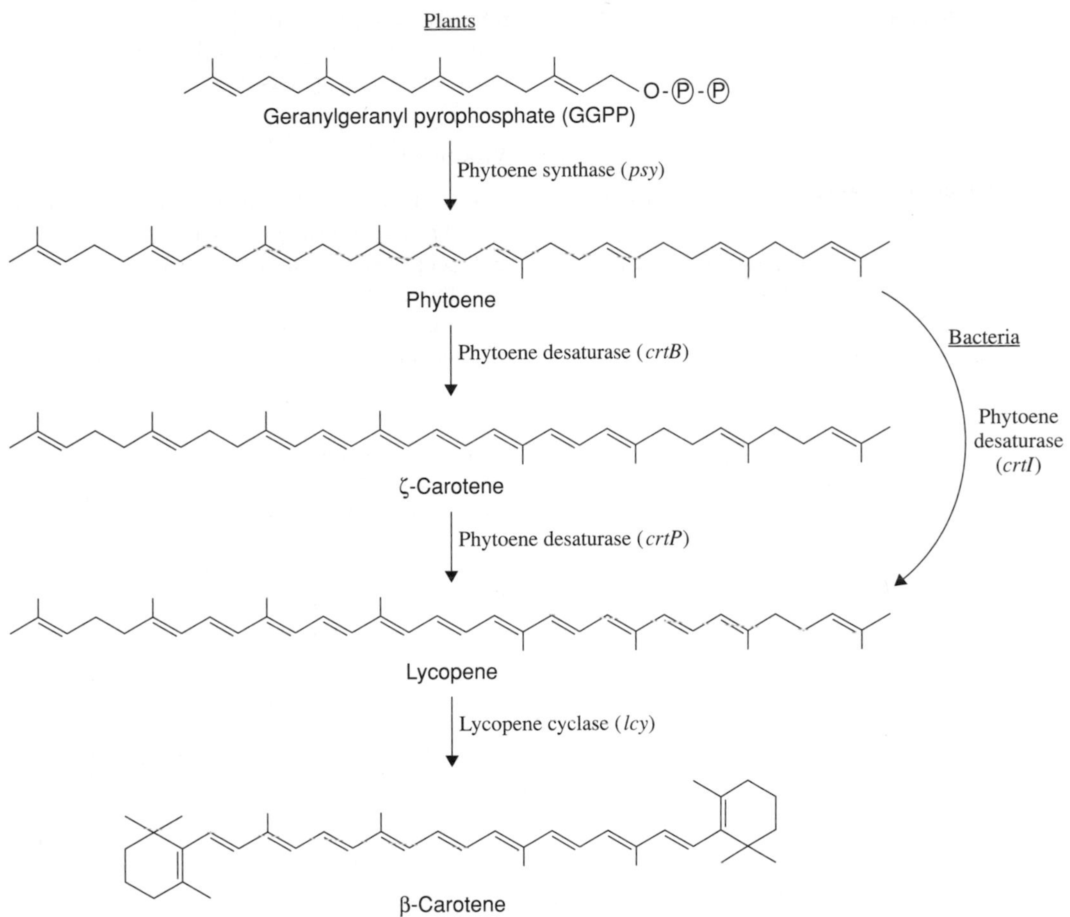

Figure 3.2. Biosynthetic pathway of β-carotene in plants and bacteria.

The second way by which plants synthesize vitamin C is through the recycling of used ascorbic acid (Fig. 3.4). During the first step of this recycling, ascorbic acid is oxidized, forming a radical called monodehydroascorbate (MDHA). Once MDHA is formed, it can be readily converted back into ascorbic acid by the enzyme monodehydroascorbate reductase (MDHAR) (EC 1.6.5.4) or further oxidized, forming dehydroascorbate (DHA). DHA can then undergo irreversible hydrolysis or be recycled to ascorbic acid by the enzyme dehydroascorbate reductase (DHAR) (EC 1.8.5.1), which uses the reductant glutathione (GSH) (Washko et al. 1992, Wheeler et al. 1998, Smirnoff et al. 2001). Researchers from the University of California, Riverside, hypothesized that by enhancing the expression of DHAR in plants, they could increase ascorbic acid synthesis, because a more efficient ascorbate recycling process would be achieved (Chen et al. 2003). To test their hypothesis, they isolated DHAR cDNA from wheat and expressed the gene in tobacco and maize plants. Tobacco plants were transformed by using *Agrobacterium*. A His tag was added to DHAR, which was then introduced in the binary vector pBI101, behind a 35S CaMV promoter. For maize, a DHAR without a His tag was placed under the control of the maize ubiquitin (Ub) promoter or the Shrunken 2 (Sh2) promoter in the pACH18 vector. Transgenic maize was generated by particle bombardment of the embryogenic callus.

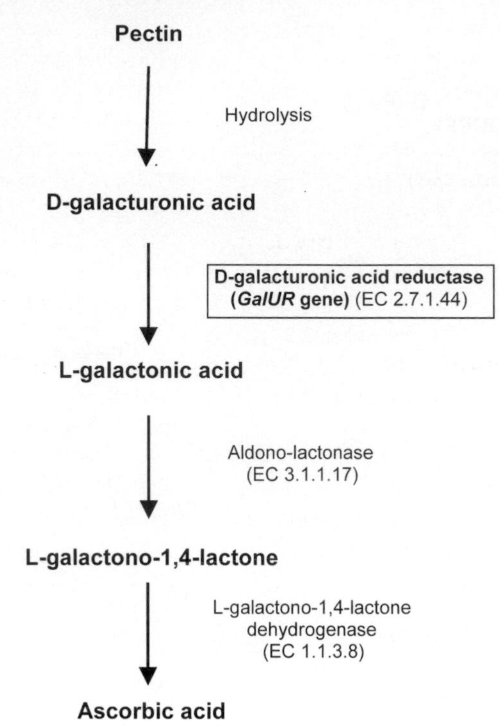

Figure 3.3. Biosynthetic pathway of vitamin C in plants.

DHAR expression was amplified up to 32 times in tobacco, and up to 100 times in maize, resulting in increased ascorbic acid levels of up to four-fold in the bioengineered plants (Chen et al. 2003).

Vitamin E

Vitamin E is a broad term used to describe a group of eight lipid-soluble antioxidants in the tocotrienol and tocopherol families that are synthesized by photosynthetic organisms, mainly plants (Hess 1993). Both the tocotrienol and tocopherol families can be distinguished into four different forms each (α, β, γ, δ), based on the number and position of methyl groups in the aromatic ring (Kamal-Eldin and Appelqvist 1996). Tocotrienols and tocopherols protect plants against oxidative stresses, and the antioxidant property of these molecules adds functional qualities to food products (Andlauer and Furst 1998). Vitamin E is an important component of mammalian diet, and excess intake has been shown to produce many beneficial therapeutic properties, including reduction of cholesterol levels, inhibition of breast cancer cell growth in vitro, decreased risk of cardiovascular diseases, and decreased incidence of many human degenerative disorders (Theriault et al. 1999).

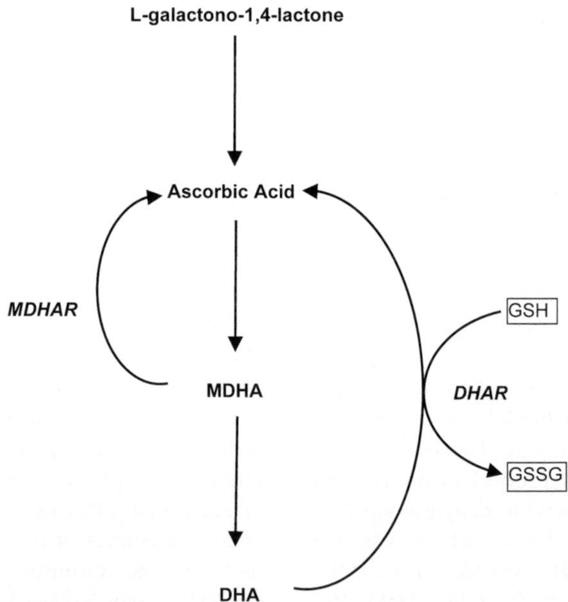

Figure 3.4. Oxidative pathway of vitamin C recycling.

Tocotrienols have more powerful antioxidant properties than tocopherols but are not absorbed as readily. The predominant forms of vitamin E in leaves and seeds are α-tocopherol and γ-tocotrienol, respectively (Munné-Bosch and Alegre 2002). While the biosynthesis of tocopherols and tocotrienols has been known for many years, the particular genes that encode for the different enzymes in the pathway have only recently been discovered. Researchers are trying to develop plants with increased vitamin E levels, and some positive results have already been achieved.

The first step in the pathway for the biosynthesis of both tocopherols and tocotrienols is the formation of homogentisic acid (HGA) from p-hydroxyphenyl-pyruvate, catalyzed by the enzyme p-hydroxphenyl-pyruvate dioxygenase (HPPD) (EC 1.13.11.27) (Fig. 3.5) (Grusack and DellaPenna 1999). Tocotrienol and tocopherol biosynthesis in plants originates from two different precursors. Tocotrienols are produced from the condensation of HGA and geranyl-geranyl diphosphate (GGDP), catalyzed by HGA geranylgeranyl transferase (HGGT) (EC 2.5.1.32), and tocopherols are formed from the condensation of HGA and phytyl diphosphate (PDP), catalyzed by HGA phytyl transferase (HPT) (EC 2.5.1.62) (Fig. 3.5) (Soll et al. 1980, Schultz et al. 1985, Collakova and DellaPenna 2001). Researchers from the Institute of Botany in Germany described the effects of constitutive expression of HPPD cDNA from barley *(Hordeum vulgare)* in tobacco plants. The HPPD gene was cloned into the pBinAR binary vector, in a *SmaI* cloning site located between the 35S CaMV promoter and the octopine synthase (EC 1.5.1.11) polyadenylation signal. The construct was then introduced into *Agrobacterium* GV3101, which was

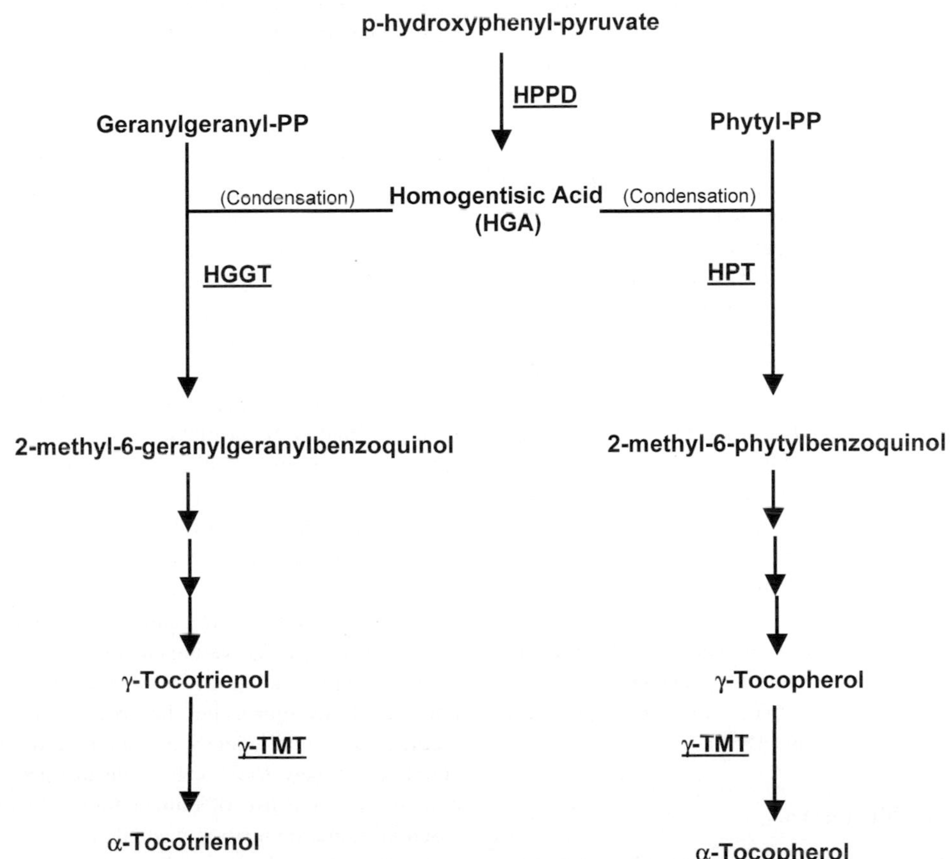

Figure 3.5. Biosynthetic pathway of vitamin E (α-tocotrienol and α-tocopherol). (Adapted from Cahoon et al. 2003.)

used to transform tobacco explants. The results showed that transgenic lines had a greater capacity for overall biosynthesis of HGA and produced a two-fold increase in vitamin E in the seeds. Vitamin E content in leaves was not affected (Falk et al. 2003).

In another approach towards vitamin E enhancement, Cahoon et al. (2003) reported the identification and isolation of a novel monocot gene that encodes HGGT, which is so far the only known enzyme specific for the synthesis of tocotrienols. These researchers found that the expression of the barley HGGT enhanced the tocotrienol synthesis by 10- to 15-fold in the leaves of *A. thaliana* and by six-fold in the seeds of corn. The barley HGGT cDNA was placed under the control of the 35S CaMV promoter and the nopaline synthase terminator. The construct was inserted into the binary vector pZS199 to generate plasmid pSH24. The plasmid was then introduced into *Agrobacterium* for transformation into tobacco and *A. thaliana* (Cahoon et al. 2003).

A third way by which vitamin E content in plants can be manipulated involves the last enzyme in the final step of the tocotrienol and tocopherol biosynthetic pathway, in which γ-tocotrienol and γ-tocopherol are converted to α-tocotrienol and α-tocopherol, respectively. This step is catalyzed by the enzyme γ-tocopherol methyltransferase (γ-TMT) (EC 2.1.1.95) (Fig. 3.5) (Shintani and DellaPenna 1998). α-tocopherol has the highest oxidative property among the members of the vitamin E family (Kamal-Eldin and Appelqvist 1996). Unfortunately, plant oils, which are the main dietary source of vitamin E, contain only a fractional amount of α-tocopherol but a high level of its precursor, γ-tocopherol. Shintani and DellaPenna overexpressed endogenous *A. thaliana* γ-TMT to enhance conversion of γ-tocopherol into α-tocopherol. They introduced the γ-TMT cDNA construct under the control of a 35S CaMV promoter in a binary vector into *A. thaliana* plants by *Agrobacterium*-mediated transformation. α-tocopherol content of bioengineered seeds was nine-fold greater than that of the wild-type seeds (Shintani and DellaPenna 1998).

Essential Minerals

To maintain a well functioning, healthy body, humans require 17 different essential minerals in their diet. Minerals are inorganic ions found in nature and cannot be made by living organisms. They can be divided into two classes: macronutrients and micronutrients. Macronutrients are the minerals that we need in large quantity, including calcium, phosphorus, sodium, magnesium, chlorine, sulfur, and silicon. Micronutrients, or trace minerals, are the minerals that are required in small amounts, of which iron is the most prevalent, followed by fluorine, zinc, copper, cobalt, iodine, selenium, manganese, molybdenum, and chromium. Although a balanced consumption of plant-based foods should naturally provide these nutrients, mineral deficiency, especially of iron, is widespread among the world population.

Iron

Even though iron is required in trace amounts, it is the most widespread nutrient deficiency worldwide. It is believed that about 30% of the world population suffers from serious nutritional problems caused by insufficient intake of iron (WHO 1992). Iron is an important constituent of hemoglobin, the oxygen-carrying component of the blood, and is also a part of myoglobin, which helps muscle cells to store oxygen. Low iron levels can cause the development of iron deficiency anemia. In an anemic person the blood contains a low level of oxygen, which result in many health problems including infant retardation (Walter et al. 1986), pregnancy complications (Murphy et al. 1986), low immune function (Murakawa et al. 1987), and tiredness (Basta et al. 1979). Iron is present in food in both inorganic (ferric and ferrous) and organic (heme and nonheme) forms. Heme iron, which is highly bioavailable, is derived primarily from the hemoglobin and myoglobin of flesh foods such as meats, fish, and poultry (Taylor et al. 1986). In humans, reduced iron (ferrous) is taken up more readily than oxidized (ferric) iron. Several approaches have been used in the fight against iron deficiency including nutraceutical supplementation, food fortification, and various methods of food preparation and processing (Maberly et al. 1994). So far, none of these approaches has been successful in eradicating iron deficiency, especially in developing countries. A new tool in the fight against nutrient deficiency is the use of biotechnology to improve essential mineral nutrition in staple crops.

At this time, there are basically two ways in which genetic engineering can be used for this purpose: (1) by increasing the concentration of the iron-binding protein ferritin and (2) by reducing the amount of iron-absorption inhibitor phytic acid. Although iron intake is important for human health, it can be toxic, so the ability to store and release iron in a controlled manner is crucial. The 450 kDa ferritin protein, found in animals, plants, and bacteria, can accumulate up to 4500 atoms of iron (Andrews et al. 1992). This protein consists of 24 subunits assembled into a hollow spherical structure within which iron is stored as a hydrous ferric oxide mineral core (Fig. 3.6). The two main functions of ferritin in living organisms are to supply iron for the synthesis of proteins such as ferredoxin and cytochromes and to prevent free radical damage to cells. Studies have shown that ferritin can be orally administrated and is effective for treatment of rat anemia (Beard et al. 1996), suggesting that increasing ferritin content of cereals may solve the problem of dietary iron deficiency in humans. Japanese researchers (Goto et al. 1999) introduced soybean ferritin cDNA into rice plants, under the control of a seed specific promoter, GluB-1, from the rice seed-storage protein gene encoding glutelin. The two advantages of this promoter are the accumulation of iron specifically in the rice grain endosperm, and its ability to induce ferritin at a high level. The ferritin cDNA was isolated from soybean cotyledons, inserted into the binary vector pGPTV-35S-bar, and transferred into rice using *Agrobacterium*. The iron content of the rice seed in the transgenic plants was three times greater than that of the untransformed wild-type plants.

Phytic acid, or phytate, is the major inhibitor of many essential minerals, including iron, zinc, and magnesium, and is believed to be directly responsible for the problem of iron deficiency (Ravindran et al. 1995). In cereal grains, phytic acid is the primary phosphate storage, and it is deposited in the aleurone storage vacuoles (Lott 1984). During seed germination, phytic acid is catalyzed into inorganic phosphorous, by the action of the hydrolytic enzyme phytase (EC 3.1.3.8) (Fig. 3.7). There is little or no phytase activity in the dry seeds or in the digestive tract of monogastric animals (Gibson and Ullah 1990, Lantzsch et al. 1992). In a recent study, it has been shown that phytase activity can be reestablished in mature dry seeds under optimum pH and temperature conditions (Brinch-Pedersen et al. 2002).

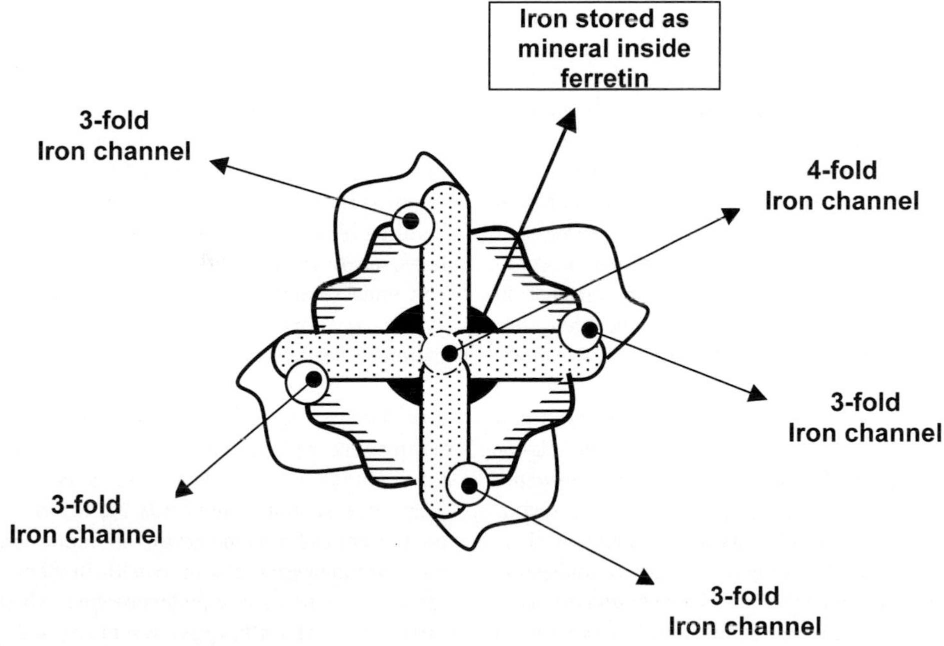

Figure 3.6. Iron binding protein ferritin.

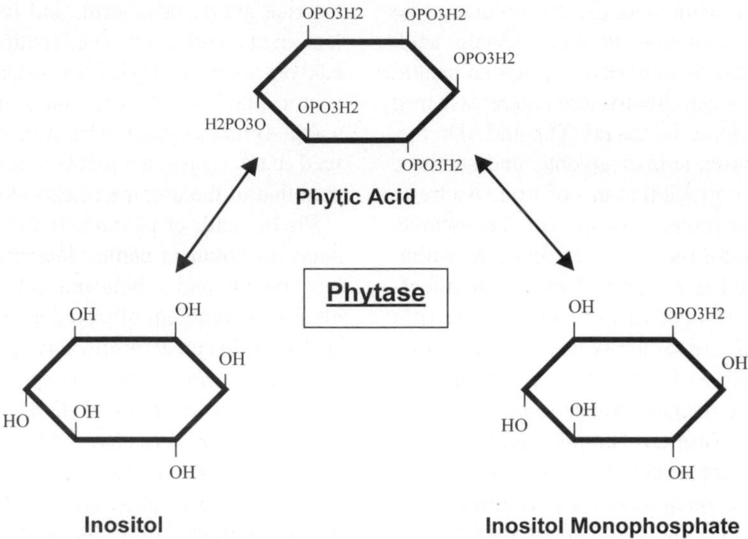

Figure 3.7. Phytic acid is degraded during seed germination by a specific enzyme called phytase [*myo*-inositol-(1,2,3,4,5,6)-hexa*kis*phosphate phosphohydrolase] (EC 3.1.3.8).

A reduction in the amount of phytic acid in staple foods is likely to result in a much greater bioavailability of iron and other essential minerals. Lucca et al. (2002) inserted a fungal *(Aspergillus fumigatus)* phytase cDNA into rice to increase the degradation of phytic acid. Rice suspension cells, derived from immature zygotic embryos, were used for biolistic transformation with the *A. fumigatus* phytase gene. Phytase from *A. fumigatus* was the enzyme of choice because it is heat stable and thus can refold into an active form after heat denaturation (Wyss et al. 1998). The main purpose of this research was to increase phytase activity during seed germination and to retain the enzyme activity in the seed after food processing and in the human digestive tract. Although the researchers achieved high expression levels of phytase in the rice endosperm, by placing it under the control of the strong tissue-specific globulin promoter, the thermotolerance of the transgenic rice was not as high as expected. It has been speculated that the reason for this unexpected low thermostability of the *A. fumigatus* phytase in transgenic rice is due to the interference of the cellular environment of the endosperm to maintain the enzyme in an active configuration (Holm et al. 2002). Further studies are needed to develop an endogenous phytase enzyme that is thermostable and maintains high activity in plant tissues.

Essential Amino Acids

Proteins are organic molecules formed by amino acids. The digestive system breaks down proteins into single amino acids so that they can enter into the bloodstream. Cells then use the amino acids as building blocks to form enzymes and structural proteins. There are two types of amino acids, essential and nonessential. Essential amino acids cannot be synthesized by animals, including humans, and therefore need to be acquired in the diet. The nine essential amino acids are histidine, isoleucine, leucine, lysine, methionine, phenylalanine, threonine, tryptophan, and valine. The body can synthesize nonessential amino acids as long as there is a proper intake of essential amino acids and calories. Proteins are present in foods in varying amounts; some foods have all nine essential amino acids in them, and they are referred to as complete proteins. Most animal products (meat, milk, eggs) provide a good source of complete proteins. Vegetables sources, on the other hand, are usually low on or missing certain essential amino acids. For instance, grains tend to lack lysine, while pulses are short in methionine (Miflin et al. 1999). In order to provide better nutrition from plant sources, it is essential to increase the content of essential amino acids in seed and tuber proteins. This is particularly important for countries where certain crops, such as rice, potatoes, and corn, are the main dietary source.

Lysine

Rice is one of the most important staple crops and is consumed by 65% of the world population on a daily basis (Lee et al. 2003). It is a good source of essential nutrients such as vitamins B1 (thiamin), B2 (riboflavin), and B3 (niacin), but it is low in the essential amino acids lysine and isoleucine (Fickler 1995). Adequate intake of lysine is essential because it serves many important functions in the body including aiding calcium absorption, collagen formation, and the production of antibodies, hormones, and enzymes. A deficiency in lysine may result in tiredness, inability to concentrate, irritability, bloodshot eyes, retarded growth, hair loss, anemia, and reproductive problems (Cooper 1996). Zheng et al. (1995) developed a transgenic rice with enhanced lysine content. They accomplished this by expressing the seed storage protein β-phaseolin from the common bean *(Phaseolus vulgaris)* in the grain of transgenic rice. The genomic and cDNA sequences of the β-phaseolin gene from *P. vulgaris* were placed under the control of either a rice seed–specific glutelin Gt1 promoter or the native β-phaseolin promoter. The vectors containing the β-phaseolin gene were transferred into the rice chromosome by protoplast-mediated transformation. Four percent of total endosperm protein in the transgenic rice was phaseolin, which resulted in a significant increase in the lysine content in rice (Zheng et al. 1995).

Methionine and Tyrosine

In terms of global food production, potato *(Solanum tuberosum)* is only behind rice, wheat, and corn on the list of the crop species that are most important for human nutrition worldwide (Chakraborty et al. 2000). There are four main purposes for the production of potatoes: for the fresh food market, for animal feed, for the food processing industry, and for nonfood industrial uses such as the manufacture of starch and alcohol (Chakraborty et al. 2000). Potato is a good source of potassium, iron, and vitamins C and B, but it is not a rich protein source. Potato proteins are limited in nutritive value because they lack the amino acids lysine, methionine, and tyrosine (Jaynes et al. 1986). A lack of methionine in a person's diet may result in an imbalanced uptake of other amino acids, as well as retardation in growth and development. Methionine is also the main supplier of sulfur, which prevents disorders of the hair, skin, and nails, helps lower cholesterol levels by increasing the liver's production of the phospholipid lecithin, and is a natural chelating agent for heavy metals (Cooper 1996).

Scientists from the National Center for Plant Genome Research in India isolated and cloned a gene that encodes for a seed-specific protein from *Amaranthus hypocondriacus* called amaranth seed albumin (AmA1) (Chakraborty et al. 2000). The advantages of using the AmA1 protein to improve crops' nutritional value are that (1) it is well balanced in the composition of all essential amino acids, (2) it is a nonallergenic protein, and (3) it is encoded by a single gene, *AmA1*. This gene was cloned into a binary vector, under the control of a constitutive 35S CaMV promoter (plasmid pSB8) and a tuber-specific, granule-bound starch synthase (EC 2.4.1.21) promoter (plasmid pSB8G). The *AmA1* gene constructs from these two binary plasmids were introduced into potato through *Agrobacterium*-mediated transformation. The amino acid contents in the pSB8-transgenic potato showed a 2.5- to 4-fold increase in lysene, methionine, and tyrosine, while the tissue-specifc pSB8G-transgenic potatoes showed a four- to eight-fold increase in these amino acids (Fig. 3.8).

Essential Phytochemicals

Besides being a major supplier of essential nutrients such as vitamins, amino acids, and minerals, plants are also an important source of phytochemicals that are known to be beneficial for health. Some examples of phytochemicals include indoles, isothiocyanates, and sulforaphane, found in vegetables such as broccoli; allylic sulfides, found in onions and garlic; and isoflavonoids, found mainly in soybeans. Since the

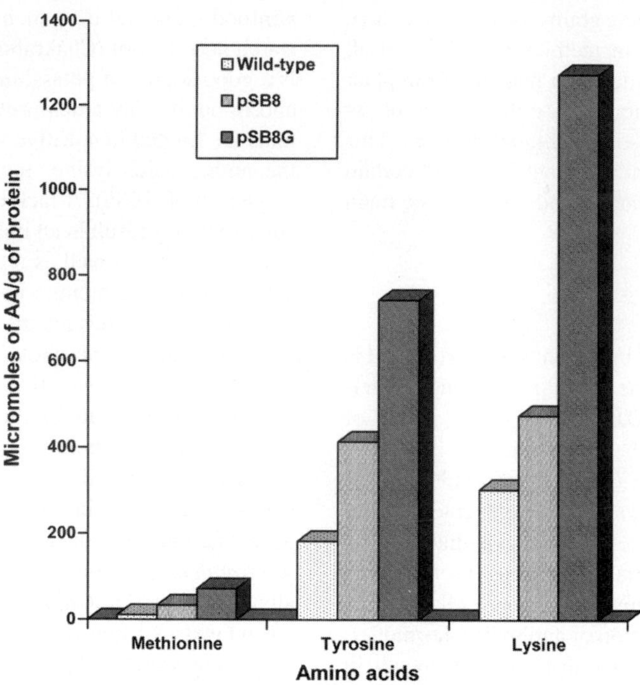

Figure 3.8. Comparison of amino acid composition from transgenic and wild-type potatoes. (From Chakraborty et al. 2000.)

intake of these phytochemicals is not always sufficient, scientists are trying to enhance the nutritional quality of plants through genetic engineering.

Isoflavonoids

Flavonoids, which include anthocyanins, condensed tannins, and isoflavonoids, are a class of phytochemicals that perform a range of important functions for plants, including pigmentation, feed deterrence, wood protection, fungi and insect defense, and induction of genes for root nodulation (Buchanan et al. 2001). Isoflavonoids (or isoflavones) are a type of phytoestrogen, or plant hormone, that has a chemical structure similar to human estrogen. The health benefits believed to be provided by isoflavonoids come from the weak estrogenic activity of these molecules in the human body (Jung et al. 2000). Isoflavonoids are found in soybeans, chickpeas, and many other legumes; however, soybeans are unique because they have the highest concentration of the two most beneficial isoflavonoids, genistein and daidzein (Eldridge and Kwolek 1983, Tsukamoto et al. 1995). In the studies conducted so far, isoflavonoids show great potential to fight many types of diseases. They help prevent the buildup of arterial plaque, which reduces the risk of coronary heart disease and stroke (FDA 1999); help reduce breast cancer (Peterson and Barnes 1991); help prevent prostate cancer by delaying cell growth (Messina and Barnes 1991); fight osteoporosis by stimulating bone formation (Civitelli 1997); and even relieve some menopausal symptoms (Nestel et al. 1999). The main source of isoflavonoids in human diet comes from the consumption of soybean and its products. Although present in high concentration in unprocessed soybean, isoflavonoid levels can decrease by 50% during seed processing for traditional soy foods (Wang and Murphy 1996). Increasing isoflavonoid concentrations in soybean could solve this problem. Another way to take advantage of isoflavonoids' health benefits is through the development of other crops that can produce these powerful compounds, thereby widening their consumption.

Isoflavonoids are synthesized by a branch in the degradation pathway of the amino acid phenylala-

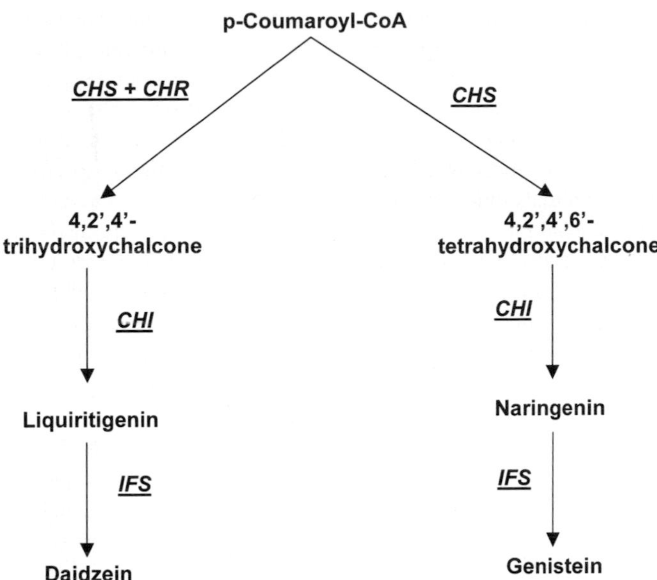

Figure 3.9. Biosynthetic pathway for the soybean isoflavonoids: daidzein and genistein. (Adapted from Jung et al. 2000.)

nine, and its first committed step is catalyzed by the enzyme isoflavone synthase (EC 1.14.13.53) (Fig. 3.9). Jung et al. (2000) identified two soybean genes encoding isoflavone synthase, *IFS1/IFS2*, and expressed these genes in *A. thaliana*, triggering the synthesis of the isoflavonoid genistein. Although *A. thaliana* does not synthesize isoflavonoids, it does have the substrate naringenin, which is an intermediate of the anthocyanin biosynthetic pathway. Naringenin can then be converted to the isoflavonoid genistein by a foreign isoflavone synthase. The soy isoflavone synthase gene *IFS1* was cloned in the plasmid pOY204 under the control of the 35S CaMV promoter and transferred into *A. thaliana* by *Agrobacterium*-mediated transformation. The introduced *ISF1* gene expressed and produced active isoflavone synthase in the transformed plant. The amount of genistein produced was approximately 2ng/μg of fresh plant weight (Jung et al. 2000).

BIOENGINEERED ANIMALS

With the development of transgenic technology, improvements in commercially important livestock species have become possible by transferring genes from related or unrelated species. Genetic improvement through biotechnology can be achieved in one generation, instead of the several generations required for traditional animal breeding methods. Although several methods of gene transfer have been developed, four methods are used today in the production of most transgenic animals: nuclear transfer, microinjection, viral vector infection, and embryonic stem cell transfer. The nuclear transfer method entails inserting the entire genetic material from the nucleus of a donor cell into a mature unfertilized egg whose nucleus has been removed. After that, the embryo is transferred into a foster mother, where it will develop into an animal that is genetically identical to the donor cell (Wolf et al. 2001). In microinjection, a segment of foreign DNA carrying one or more genes is injected into the male pronucleus of a fertilized egg. The egg needs to be in a single-cell stage to ensure that all somatic cells in the animal contain the transgene. The embryo is then transferred to the uterus of a surrogate mother (Wall 2002). In the retroviral infection technique, the gene is transferred with the help of a viral vector. Retroviruses are frequently used in the process of DNA transfer due to their natural ability to infect cells (Cabot et al. 2001). In the stem cell transfer technique, embryonic stem cells are collected from blastocysts and grown in culture. The cultured cells are then injected into the inner cell mass of an embryo in the blastocyst stage, which is then implanted into the foster mother, resulting in the production of a chimeric animal (Hochedlinger and Jaenisch 2003). It is important to remember that these methods do not create new species; they only offer tools for producing new strains of animals that carry novel genetic information. Some examples of genetically engineered animals include transgenic cows that produce milk with improved composition and transgenic swine that produce meat with lower fat content. The main goal of livestock genetic engineering programs is to increase production efficiency while delivering healthier animal food products.

Modified Milk in Transgenic Dairy Cattle

Bovine milk has been described as an almost perfect food because it is a rich source of vitamins, calcium, and essential amino acids (Karatzas and Turner 1997). Some of the vitamins found in milk include vitamins A, B, C, and D. Milk has greater calcium content than any other food source, and daily consumption of two servings of milk or other dairy products supplies all the calcium requirements of an adult person (Rinzler et al. 1999). Caseins represent about 80% of the total milk protein and have high nutritional value and functional properties (Brophy et al. 2003). The caseins have a strong affinity for cations such as calcium, magnesium, iron, and zinc. There are four types of naturally occurring caseins in milk: αS1, αS2, β, and κ (Brophy et al. 2003). They are clumped in large micelles, which determine the physicochemical properties of milk. Even small variations in the ratio of the different caseins influence micelle structure, which in turn can change the milk's functional properties. The amount of caseins in milk is an important factor for cheese manufacturing, since greater casein content results in greater cheese yield and improved nutritional quality (McMahon and Brown 1984). It has been estimated that enhancing the casein content in milk by 20% would result in an increase in cheese production, generating an additional $190 million/year for the dairy industry (Wall et al. 1997). Dairy cattle have only one copy of the genes that encode α (s1/s2), β, and κ-casein proteins, and out of the four caseins, κ and β are the most important (Bawden et al. 1994). Increased milk κ-casein content reduces the size of the micelle, resulting in improved heat stability. β-caseins are highly phosphorylated and bind to calcium phosphate, thus influencing milk calcium levels (Dalgleish et al. 1989, Jimenez Flores and Richardson 1988).

Research on modification of milk composition to improve nutritional or functional properties has been mostly done in transgenic mice. Mice are good models for the study of protein expression in mammary glands, but they do not always reflect the same protein expression levels as ruminants (Colman 1996). Brophy et al. (2003), using nuclear transfer technology, produced transgenic cows carrying extra copies of the genes *CSN2* and *CSN3*, which encode bovine β- and κ-caseins, respectively. Genomic clones containing *CSN2* and *CSN3* were isolated from a bovine genomic library. Previous studies conducted with mice revealed that *CSN3* had very low expression levels (Persuy et al. 1995). In order to enhance expression of *CSN3*, the researchers created a *CSN2/3* fusion construct, in which the *CSN3* gene was fused with the *CSN2* promoter. The *CSN2* genomic clone and the *CSN2/3* fusion construct were co-transfected into bovine fetal fibroblast (BFF) cells, where the two genes showed coordinated expression. The transgenic cells became the donor cells in the process of nuclear transfer, generating nine fully healthy and functional cows. Overexpression of *CSN2* and *CSN2/3* in the transgenic cows resulted in an 8–20% increase in β-casein and a 100% increase in κ-casein levels (Brophy et al. 2003).

Increased Muscle Growth in Cattle

Myostatin, also known as growth and differentiation factor 8 (GDF-8), is a member of the transforming growth factor β (TGF-β) family, which is responsible for negative regulation of skeletal muscle mass in mice, cattle, and possibly other vertebrates. Myostatin is expressed in embryo myoblasts and developing adult skeletal muscle; it is produced as a 375-amino-acid precursor molecule that is further processed by enzymatic cleavage of the N-terminus prodomain segment. The remaining C-terminus 109-amino-acid segment is the myostatin protein (Gleizes et al. 1997). The processed protein forms dimers that are biologically active. McPherron and Lee (1997), through alignment of myostatin amino acid sequences from baboon, bovine, chicken, human, murine, ovine, porcine, rat, turkey, and zebrafish, determined that the myostatin gene is highly conserved among vertebrates, which suggests that its function may be conserved as well. The researchers also demonstrated that the increase in muscle mass (double-muscling) phenotype observed in some breeds of cattle, such as Belgian Blue and Piedmontese, is due to mutations in the myostatin gene. In myostatin-null mice that had the myostatin gene knocked out by gene targeting, individual muscles weighed on average two to three times more than those in wild-type mice, and body weight was 30% higher. This difference was not due to an increase in fat amount, but to an increase in the cross-sectional area of the muscle fibers (hypertrophy) and an increase in the number of muscle fibers (hyperplasia) (McPherron and Lee 1997).

It is believed that myostatin prodomain may bind noncovalently with the mature myostatin, resulting in inhibition of the biological activity of myostatin (Thies et al. 2001). Based on the general molecular model of TGF proteins, Yang et al. (2001) hypothesized that overexpression of the prodomain segment would interfere with mature myostatin, resulting in the promotion of muscle development. The myostatin prodomain DNA was cloned into a pMEX-NMCS2 vector, which contained a rat myosin light chain 1 (MLC1) promoter, a SV40 poly adenylation sequence, and a MLC enhancer. This construct was then inserted into the mouse genome using the pronuclear microinjection technique. In comparison with wild-type mice, overexpression of the myostatin prodomain in the transgenic mice resulted in a 17–30% increase in body weight; a 22–44% increase in total carcass weight at 9 weeks of age (Fig. 3.10); and a significant decrease in epididymal fat pad weight, which is an indicator of body fat mass (Yang et al. 2001). No undesirable phenotypes or health or reproductive problems were observed in the transgenic animals. These results indicate that this same approach can be used to develop farm animals with enhanced growth performance, increased muscle mass, and decreased fat content, which would equate to a healthier meat product for consumers (Yang et al. 2001).

Figure 3.10. Comparison between wild-type and transgenic mice overexpressing myostatin prodomain. (Courtesy of Dr. J. Yang, Univ. of Hawaii.)

Reduced Fat Content in Transgenic Swine

In the human diet, ingested exogenous fats serve as the raw material for the synthesis of fat, cholesterol, and many phospholipids. Since fat energy content is two times greater than the energy obtained from carbohydrates and proteins, most of the energy that is stored in the body is in the form of fat. Fats are a group of chemical compounds that contain fatty acids. The most common fatty acids found in animal fats are palmitic acid, stearic acid, and oleic acid. The human body is able to synthesize these fats, but there is one more class of fatty acids called the essential fatty acids (linolenic acid, linoleic acid, and arachidonic acid), which the body cannot produce; they therefore must be obtained from diet (Campbell and Reece 2002). There are two main types of naturally occurring fatty acids: saturated and unsaturated. Saturated fatty acids (SFA) are mainly animal fats. They are called saturated because all the carbon chains are completely filled with hydrogen and there are no double bonds formed between the carbon atoms. Saturated fatty acids are believed to be "bad" fats since they raise both high-density lipoprotein (HDL) and low-density lipoprotein (LDL) cholesterol (Keys et al. 1965, NRC 1988). Unsaturated fats are found mainly in products derived from plant sources and are divided into two categories: monounsaturated fatty acids (MUFA), which have one double bond; and polyunsaturated fatty acids (PUFA), which have two or more double bonds in the carbon chain. It has been observed that the increased consumption of these "good" fats actually reduces LDL levels and enhances HDL levels (Grundy 1986, NRC 1988). It is now well established that a high fat diet (specially of SFA) not only increases the risk of heart disease but also the risk of breast, colon, and prostate cancer. Many health agencies, including the American Dietetic Association, the American Diabetes Association, and the American Heart Association, recommend that fat intake should be no more than 30% of the total daily calories.

In research conducted by the United State Department of Agriculture (USDA), scientists introduced a recombinant bovine growth hormone (rBGH) gene into pigs, with the purpose of understanding the relationship between rBGH expression and the amount of fatty acids in the animal (Solomon et al. 1994). Bovine growth hormone (BGH), also known as bovine somatotropin, which is produced in the pituitary gland, stimulates growth in immature cattle and enhances milk production in lactating cows (Leury et al. 2003). BGH is a protein hormone, and as such, it is broken down during digestion in the gastrointestinal tract, making it biologically inactive in humans (Etherton 1991). In 1993, based on rigorous scientific investigations, the U.S. Food and Drug Administration (FDA) concluded that products from transgenic-BGH and supplemented-BGH animals are safe for human consumption.

Bovine growth hormone has been shown to decrease fat content of transgenic pigs expressing an rBGH gene (Pursel et al. 1989). The transgenic pigs used in this study were created by pronuclear microinjection technique. The gene encoding rBGH was introduced into the pig genome under the control of the mouse metallothionein-I (MT) promoter. After rBGH transgenic lines of pigs were established, successive generations were produced by artificial insemination of nontransgenic females with sperm collected from rBGH transgenic males. To determine the effect of rBGH in the pigs' carcass composition, transgenic and nontransgenic (control) pigs were raised under the same conditions and fed the same type of diet. The animals were processed at five different live weights: 14, 28, 48, 68, and 92 kg. The entire left side of each carcass was ground, and random samples of tissue were collected and analyzed for fatty acid and cholesterol content. The researchers observed that as live body weight increased, carcasses from transgenic pigs showed a constant decline in the amount of total fat compared to control pigs (Table 3.1). Although the results did not demonstrate a difference in the cholesterol content of transgenic and control pigs, it was shown that transgenic pigs expressing BGH had a significant decrease in the levels of specific fatty acids compared with nontransgenic pigs in the control group (Fig. 3.11). These results indicate that consumers might greatly benefit from a pork product with a low fat content if regulation of BGH secretion levels can be precisely controlled during the fast growth stage of young pigs (Solomon et al. 1994).

Transgenic Poultry: Egg as Bioreactor

Mammals and birds have been the focus of intense research for their possible use as bioreactors. The use of mammals as bioreactors became possible with the creation of transgenic mice and the isolation of tissue-specific promoters (Gordon et al. 1980, Swift et al. 1984). Clark et al. (1987) were the first to propose the use of transgenic livestock mammary glands for the production of biopharmaceutical proteins in milk. Although expression of foreign protein in milk is high and milk production is large, there are some problems associated with the use of mammary glands as bioreactors, including the long time required to establish a stable line of transgenic founder animals and the high cost to purify foreign protein from milk (Ivarie 2003). Researchers have also long envisioned using chicken eggs for the expression of exogenous proteins. There are many advantages associated with the use of eggs as bioreactors, including the fact that a single ovalbumin gene controls most of the proteins in egg white (Gilbert 1984). Also, egg white has a relatively high protein content, is naturally sterile, and has a long shelf life (Tranter and Board 1982, Harvey et al.

Table 3.1. Comparison of Total Carcass Fat (g/100g) between rBGH Transgenic and Control Pigs, Measured at Different Live Weights

	Total Fat, g/100g	
Weight Group, kg	*Transgenic*	*Control*
14	6.19	10.04
28	7.62	12.32
48	8.16	16.58
68	5.97	26.78
92	4.49	29.07

Source: Adapted from Solomon et al. 1994.

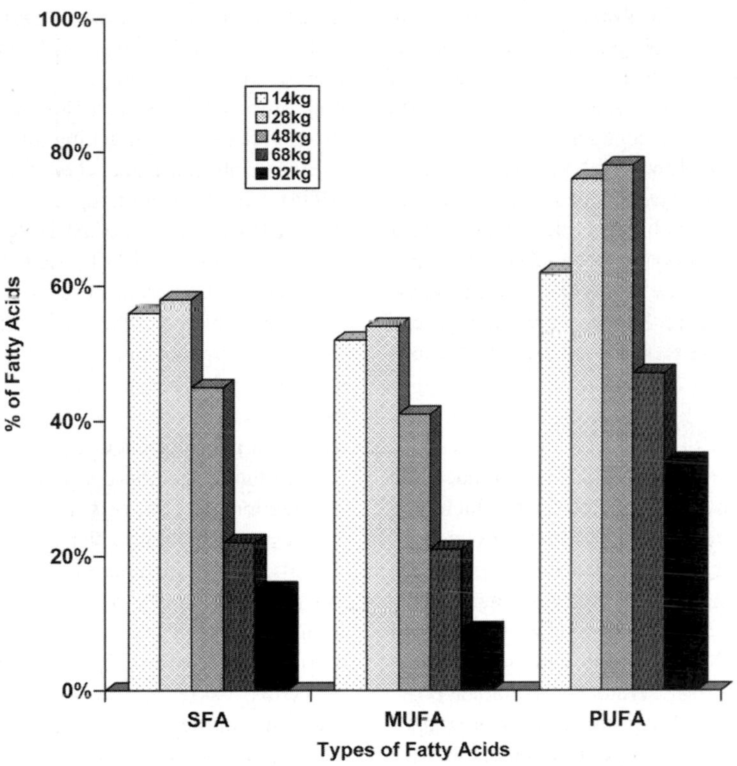

Figure 3.11. Fatty acids differences in the carcass composition of rBGH transgenic and wild-type pigs. (From Solomon et al. 1994.)

2002). There is an already established infrastructure for the production, harvesting, and processing of chicken eggs (Ivarie 2003).

Recently, a group of researchers from the biotech company AviGenics, Athens, Georgia, successfully introduced, expressed, and secreted a bacterial gene in the egg white of transgenic chicken (Harvey et al. 2002). The transgene chosen was the *E. coli* β-lactamase (EC 3.5.2.6) reporter gene because it is easily secreted and assayed from eukaryotic cells. A replication-deficient retroviral vector, named NLB, from the avian leucosis virus (ALV) was used to express the transgene. The β-lactamase coding sequence was inserted into the pNLB-CMV-BL viral vector and placed under the control of the ubiquitous cytomegalovirus (CMV) promoter. The protein β-lactamase was found to be biologically active and was secreted in the blood and egg white, and its expression levels remained constant across four generations of transgenic hens. These results demonstrate that it is technically possible to express and secret foreign proteins in the chicken egg, making it an attractive candidate for a bioreactor. The main work that needs to be done with the chicken model is to develop more efficient nonviral-based methods for creation of transgenic chicken and to identify, isolate, and characterize gene enhancers and promoters that have high activity and drive tissue-specific expression of proteins in adult oviducts (Harvey et al. 2002).

BIOENGINEERED FISH

Out of all the transgenic, domesticated animals that have been produced so far, fish are considered safest for human consumption and are expected to be the first transgenic animal to be approved as a food item (Niiler 2000). The company AquaBounty has an application under review with the FDA for the commercialization of Atlantic salmon carrying a growth hormone (GH) gene from Chinook salmon (Zbikowska 2003). The main obstacle to be overcome for the achievement of this goal is to better understand the potential risks involved with the release of transgenic fish in the wild, and at this point, not enough research has been conducted to answer these concerns (Muir and Howard 1999). One option to avoid proliferation of transgenic fish in the wild is to sterilize all transgenic fish, but a reliable method for 100% sterilization has not yet been achieved (Razak et al. 1999). Some of the transgenic strategies that are being developed to improve growth rate and increase the antifreeze property are described below.

Improving Fish Growth Rate

The fish growth hormone gene has been cloned and characterized from a number of fishes, including many salmon species (Du et al. 1992, Devlin et al. 1994). Researchers from the University of Southampton in the United Kingdom (Rahman et al. 1998) developed transgenic tilapia fish *(Oreochromis niloticus)* transformed with growth hormone genes of several salmonids. Rahman et al. used different types of constructs in their experiment, but the one that gave the best results was the construct with a Chinook salmon GH gene under the control of the ocean pout antifreeze promoter. The method used for the insertion of the transgene construct was the cytoplasmic microinjection of fertilized fish egg. The researchers reported the successful genomic integration of the construct in the founder (G0) tilapia and subsequent transfer of the transgene to the G1 and G2 generations. Transgenic tilapia expressing the transgene showed a growth rate three times greater than the wild-type tilapia and had a 33% higher food conversion ratio, which would reduce farmers' production cost. This transgenic tilapia also showed infertility at mature age, which is a desirable trait for commercial transgenic fish (Rahman et al. 1998).

Increasing Antifreeze Property in Fish

Many species of fish, such as ocean pout *(Macrozoarces americanus)* and winter flounder *(Pleuronectes americanus),* that inhabit below-freezing water in the northern regions produce and secrete specific proteins in their plasma to protect their bodies from freezing (Davies and Hew 1990). To this date, two types of these proteins that have been characterized are the antifreeze proteins (AFPs) and antifreeze glycoproteins (AFGs) (Davies and Sykes 1997). AFPs and AFGs lower the freezing temperature of the fish's serum and therefore protect the fish from freezing by attaching themselves to the ice surface, inhibiting ice crystal formation (DeVries 1984). There are four types of AFPs (I, II, III, IV) and at least one type of AFG identified at this time (Davies and Hew 1990, Deng et al. 1997). Most aquaculture-important species of fish, such as the Atlantic salmon and the tilapia, do not naturally produce any type of antifreeze protein and therefore cannot survive and be raised in areas of the world where water reaches sub-zero temperatures, which creates a major problem for sea cage farming along the northern Atlantic coast (Hew et al. 1995). The production of commercially important transgenic fish, especially salmon, that are freeze tolerant would greatly expand the area for fish farms, increase productivity, and reduce prices for consumers.

Flounder AFPs are small polypeptides that are part of the Type I AFPs, which have two different isoforms, "skin-type" and "liver-type." Skin-type AFPs are intracellular, mature proteins expressed in several peripheral tissues; liver-type AFPs are immature proteins that need to be further processed before being secreted into circulation and are found mainly in the liver tissue (Hew et al. 1986, Gong et al. 1996). Hew et al. (1999) used the liver-type AFP gene from winter flounder to generate a transgenic stable line of Atlantic salmon *(Salmo salar)* that demonstrated freeze tolerance capacity. By injecting the genes into the fertilized eggs, a single copy of the AFP gene was inserted and integrated into the salmon chromosome, generating transgenic founder fish with stable AFP expression and biologically active protein. The same levels of expression and protein activity were observed in up to three subsequent generations of transgenic salmon. Expression of AFP was liver specific and demonstrated seasonal variations similar to those in winter flounder, but the levels of AFP in the blood of these fish were low (250 μg/mL) compared with natural AFP concentrations in winter flounder (10–20 mg/mL) and therefore insufficient to provide freeze resistance to the salmon (Hew et al. 1999). The focus of current research has been to design gene constructs that will increase the copy numbers of the transgene and therefore enhance expression levels of AFP in appropriate tissues, conferring better antifreeze properties in farm fish.

BIOENGINEERED MICROORGANISMS

For over 5000 years, mankind has, knowingly and unknowingly, made use of spontaneous fermentation of a variety of food items, which include bread, alcoholic beverages, dairy products, vegetable products, and meat products. But it was more recently, just in the last century, that scientists realized that the process of fermentation was effected by the action of microorganisms and that each microorganism responsible for a specific food fermentation could be isolated and identified. Now, with advanced bioengineering techniques, it is possible to characterize with high precision important food strains, isolate and improve genes involved in the process of fermentation, and transfer desirable traits between strains or even between different organisms.

ELIMINATION OF CARCINOGENIC COMPOUNDS

Brewer's yeast *(Saccharomyces cerevisiae)* is one of the most important and widely used microorganisms in the food industry. This microorganism is cultured not only for the end products it synthesizes during fermentation, but also for the cells and the cell components (Aldhous 1990). Today, yeast is mainly used in the fermentation of bread and of alcoholic beverages. Recombinant DNA technologies have made it possible to introduce new properties into yeast, as well as eliminate undesirable by-products. One of the undesirable by-products formed during yeast fermentation of foods and beverages is ethylcarbamate, or urethane, which is a potential carcinogenic substance (Ough 1976). For this reason, the alcoholic beverage industry has dedicated a large amount of its resources to funding research oriented to the reduction of ethylcarbamate in its products (Dequin 2001). Ethylcarbamate is synthesized by the spontaneous reaction between ethanol and urea, which is produced from the degradation of arginine, found in large amounts in grapes. Yeasts, used in wine fermentation, possess the enzyme arginase that catalyzes degradation of arginine. If this enzyme can be blocked, arginine will no longer be degraded into urea, which in turn will not react with ethanol to form ethylcarbamate. In industrial yeast, the gene *CAR1* encodes the enzyme arginase (EC 3.5.3.1) (Dequin 2001). To reduce the formation of urea in sake, Kitamoto et al. (1991) developed a transgenic yeast strain in which the *CAR1* gene is inactivated. The researchers constructed the mutant yeast strain by introducing an ineffective *CAR1* gene, flanked by a DNA sequence homologous to regions of the arginase gene. Through homologous recombination, the ineffective gene was integrated into the active *CAR1* gene in the yeast chromosome, interrupting its function (Fig. 3.12). As a result, urea was eliminated and ethylcarbamate was no long formed during sake fermentation. This same procedure can be used to eliminate ethylcarbamate from other alcoholic beverages, including wine (Kitamoto et al. 1991).

INHIBITION OF PATHOGENIC BACTERIA

To increase safety, hygiene, and efficiency in the production of fermented foods, the use of starter and protective bacterial cultures is a common practice in the food industry today (Gardner et al. 2001). Starter culture is a liquid consisting of a blend of selected microorganisms, used to start a commercial fermentation. The difference between starter and protective cultures is that starter cultures give the food a desired aroma or texture, while protective cultures inhibit the growth of undesirable pathogenic microorganisms, but do not change the food property (Geisen and Holzapfel 1996). For the purpose of practicality during food processing, the same microorganism should be used for both starter and protective cultures, but unfortunately this is not always possible. Genetic engineering methods help improve available strains of microorganisms used in starter and protective cultures, so that new characteristics can be added and undesirable properties eliminated (Hansen 2002).

Genetic engineering research aimed at optimizing starter cultures is focused on three main goals: (1) to enhance process stability, (2) to increase efficiency, and (3) to improve product safety (Geisen and Holzapfel 1996). During the production of some fermented food, such as mold-ripened cheese, pH level rises in the culture due to lactic acid degradation by fungal activity. This alkaline media offers an ideal condition for the proliferation of foodborne pathogenic microorganisms such as *Listeria monocytogenes* (Lewus et al. 1991). The safety of food products could be greatly improved by the use of starter

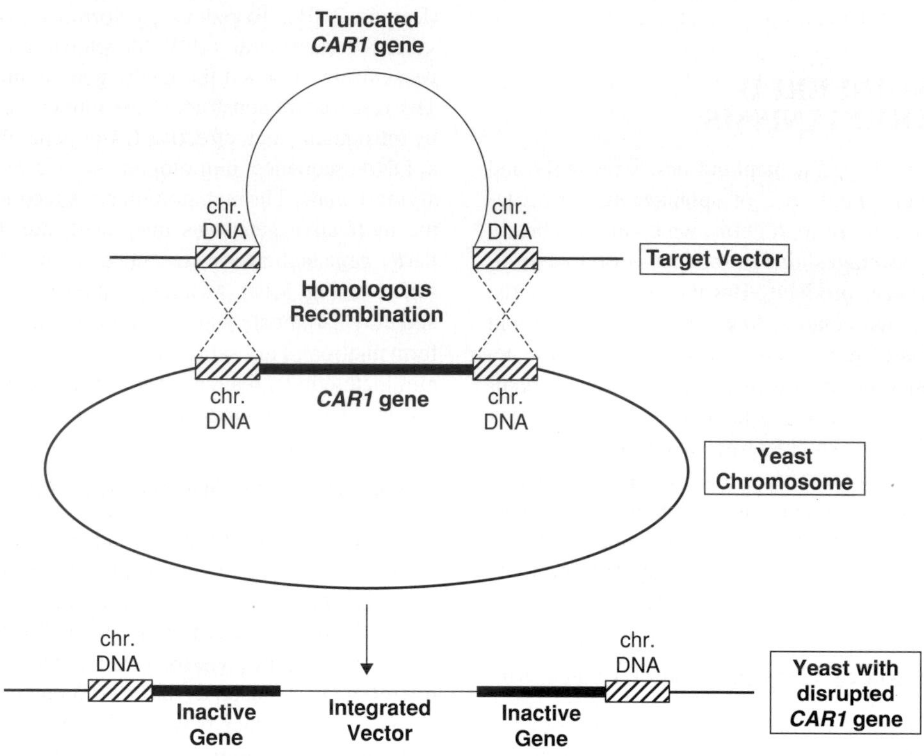

Figure 3.12. Gene disruption by homologous recombination.

cultures that can also serve as protective cultures and inhibit the growth of such harmful microorganisms. The enzyme lysozyme (EC 3.2.1.17) can be an effective agent for the inhibition of *Listeria* in food. Van de Guchte et al. (1992) integrated the gene responsible for lysozyme formation in a strain of the bacterium *Lactococcus lactis*. After genetic transformation, this bacterial strain was able to express and secrete lysozyme at high levels. The researchers cloned lysozyme-encoding genes from *E. coli* bacteriophages T4 and lambda in wide-host-range vectors and expressed in *L. lactis*. Biologically active lysozyme was produced and secreted by the transgenic *L. lactis* strains, suggesting that these bacteria can be used as both a starter and a protective culture (Van de Guchte et al. 1992).

Natural Sweetener Produced by Microorganisms

Techniques to enhance flavor in food have been known for a long time, but only recently has it been recognized that microorganisms can also be used in flavor production and enhancement. Today, many of the techniques for flavoring food and beverages make use of synthetic chemicals (Vanderhaegen et al. 2003). With increased public concern about the danger of using synthetic chemicals, flavors produced by biological methods, also called bioflavors, are becoming more popular with consumers (Armstrong and Yamazaki 1986, Cheetham 1993). The flavor and fragrance industry is estimated worldwide at $10 billion per year; and although thousands of natural volatile and synthetic fragrances are known, only a few hundred are regularly used and manufactured on an industrial scale (Somogyi 1996). There are several methods for the production of bioflavors including (1) product extraction from plant materials and (2) the use of specific bioengineered microorganisms for their biosynthesis. Biotechnological production of bioflavors using microorganisms has certain advantages such as large-scale production with low cost, nondependence on plant material, and preservation of natural resources (Krings and Berger 1998).

Xylitol, also called wood sugar, is made from xylose, which is found in the cell walls of most land plants (Nigam and Singh 1995). Pure xylitol is a white crystalline substance that looks and tastes like sugar, making it important for the food industry as a sweetener. One of the main advantages of xylitol over other sweeteners is that it can be used by diabetic patients, since its utilization is not dependent on insulin (Pepper and Olinger 1988). Xylitol is believed to reduce tooth decay rates by inhibiting *Streptococcus mutans*, the main bacteria responsible for cavities. Because xylitol is slowly absorbed and only partially utilized in the human body, it contains 40% fewer calories than regular sugar and other carbohydrates. In the United States, xylitol has been used since the 1960s, and it is approved as an additive for foods with special dietary purposes (Emodi 1978). Yeast *(S. cerevisiae)* is considered the ideal microorganism for commercial production of xylitol from xylose because of its well-established use in the fermentation industry. South African researchers (Govinden et al. 2001) isolated a xylose reductase (EC 1.1.1.21) gene (*XYL1*) from *Candida shehatae* and introduced it into *S. cerevisiae*. The *XYL1* gene from *Candida* was cloned into the yeast expression vector pJC1, behind the *PGK1* promoter, and the construct was transformed into yeast by electroporation. Xylitol production from xylose by the transformant was evaluated in the presence of different cosubstrates including glucose, galactose, and maltose. The highest xylitol yield (15 g/L from 50 g/L of xylose) was obtained with glucose as cosubstrate.

Production of Carotenoids in Microorganisms

Carotenoids are structurally diverse pigments found in microorganisms and plants. These pigments have a variety of biological functions, such as coloration, photo protection, light harvesting, and hormone production (Campbell and Reece 2002). Carotenoids are used as food colorants, animal feed supplements, and more recently, as nutraceuticals in the pharmaceutical industry. Recent studies have suggested many health benefits from the consumption of carotenoids. Carotenoids such as astaxanthin, β-carotene, and lycopene have high antioxidant properties, which may protect against many types of cancers, enhance the immune system, and help relieve the pain and inflammation of arthritis (Miki 1991, Jyocouchi et al. 1991, Giovannucci et al. 1995). There is an increased interest in extracting large amounts of carotenoids from natural sources. The 1999 world market for carotenoids was $800 million, with projections for $1 billion in 2005 (Business Communications Co. 2000). Although researchers have found certain microalgae such as *Haematococcus pluvialis* that produce high amounts of the carotenoid astaxanthin, extraction of carotenoids from these microalgae is difficult because of their thick cell wall. For this reason, genetic engineering methods have been applied to produce carotenoids in other microorganisms. The edible yeast *Candida utilis* is a good candidate for commercial carotenoid production. It is a "generally recognized as safe" (GRAS) organism, and large-scale production of peptides, such as glutathione, has already being successfully achieved in *C. utilis* (Boze et al. 1992).

In microorganisms and plants, carotenoids are synthesized from the precursor farnesyl pyrophosphate (FPP) (Fig. 3.13). Miura et al. (1998) developed a

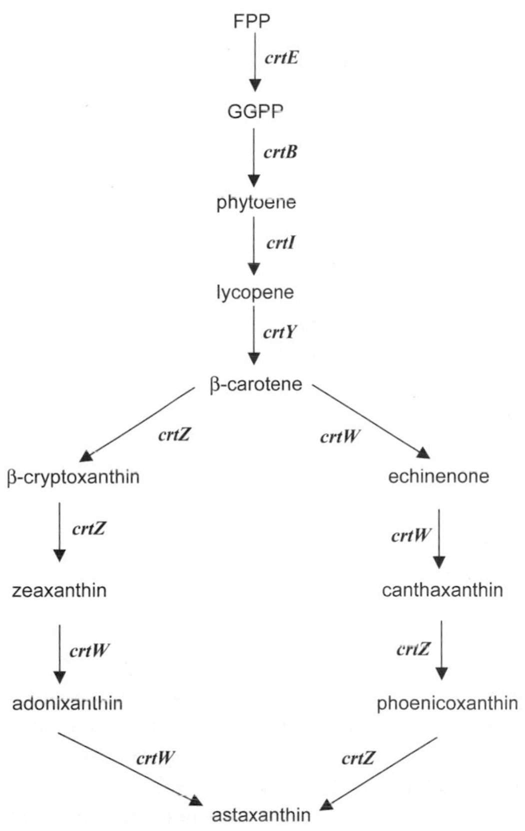

Figure 3.13. Biosynthetic pathway of the carotenoids lycopene, β-carotene, and astaxanthin.

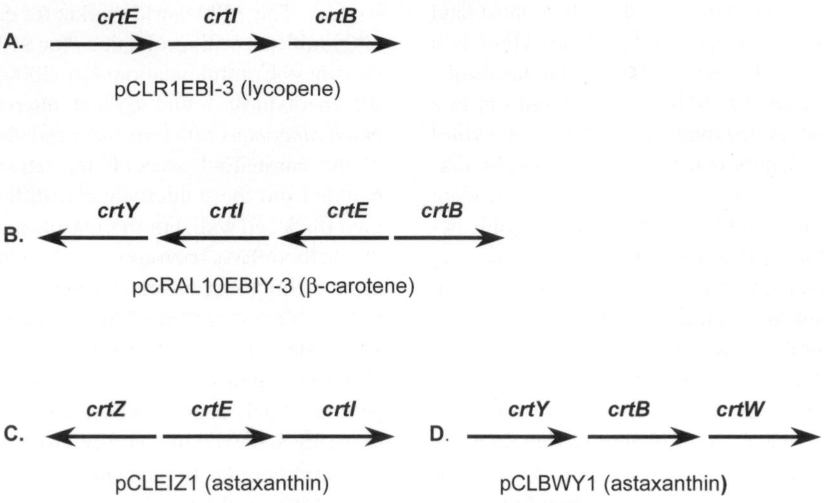

Figure 3.14. Lycopene, β-carotene, and astaxanthin plasmid constructs.

de novo biosynthesis of the carotenoids lycopene, β-carotene, and astaxanthin in *C. utilis* using cloned bacterial genes that encode enzymes of the biosynthetic pathway. They used four genes (*crtE, crtB, crtI, crtY*) from *Erwinia uredovora* and two genes (*crtZ, crtW*) from *Agrobacterium aurantiacum* for construction of four different plasmids. The plasmid pCLR1EBI-3 contained *crtE*, *crtI*, and *crtB*, required for the production of lycopene (Fig. 3.14a). For synthesis of β-carotene, the plasmid pCRAL-10EBIY-3 contained the genes *crtY*, *crtI*, *crtE*, and *crtB* (Fig. 3.14b). A dual plasmid system with pCLEIZ1 containing *crtE*, *crtI*, and *crtZ* and pCLBWY1 containing *crtW*, *crtY*, and *crtB* was used to produce astaxanthin (Fig. 3.14c,d). In order to integrate these genes into the yeast chromosome, the plasmids were linearized by restriction digest and transformed into *C. utilis* by electroporation. The resultant transgenic yeast produced significant amounts of lycopene (1.1 mg/g dry weight), β-carotene (0.4 mg/g dry weight), and astaxanthin (0.4 mg/g dry weight), in quantities similar to amounts found in microorganisms that naturally produce these carotenoids (Miura et al. 1998). These results indicate that *C. utilis* has a great potential for use in large-scale production of commercially important carotenoids.

DETECTION METHODS IN FOOD BIOTECHNOLOGY

Biotechnology plays an important role in maintaining the safety of the food supply. The development of reliable methods to ensure the traceability of genetic material in the food chain is of great value to food manufacturers and consumers. Consumer confidence in food biotechnology will increase if better traceability methods are in place. Modern biotechnology tools are also applied to develop sensitive, reliable, fast, and cheap methods for detection of harmful pathogenic organisms such as *E. coli* O157:H7 and the infectious agent for mad cow disease.

TRANSGENE DETECTION

Accomplishments in food biotechnology require continuous development of new products and their successful commercialization through consumer acceptance. One of the greatest demands made by consumer groups as a prerequisite for their support of transgenic plant use, is the development of reliable methods of detection of the transgene in human food products (James 2001). But as the number of genetically modified organisms (GMOs) approved for cultivation and commercialization grows, there is also an increased risk of transgenic material contamination in nontransgenic food products. One such well-publicized event took place in October 2000, when Safeway and Taco Bell recalled corn products because they were contaminated with small amounts of genetically engineered corn. For this and other reasons, the future success and acceptance of GMOs will depend on mechanisms for containment and proper detection of transgenic material. Among the different methods for detecting transgenic materials that are in use today, real-time quantitative PCR is the most powerful, accessible, and cost efficient (Higuchi et al. 1992). The main concern for the implementation of reliable detection methods is to determine what type of unique gene sequence should be amplified during the PCR screening. Signature sequences such as antibiotic resistance markers and promoters are the main elements used today for detection of GMOs, but they are not ideal since the same signature sequences can be found in more than one type of GMO. Also, there is an unproven concern that these signature sequences, especially antibiotic resistance markers, may cause health and environmental problems. To address this concern, the European Union, which has adopted stringent regulation on GMOs, banned the use of antibiotic gene as markers for transformation selection, by the year 2004. The European Union also established mandatory labeling of GMO foods with a 1% threshold level for the presence of transgenic material, which in turn encouraged more aggressive research on highly specific, precise, and sensitive methods for detection and quantification of GMOs in food products (European Commission 2000).

Researchers for the German company Icon Genetics developed a novel idea for universal identification of GMOs (Marillonet et al. 2003). They proposed the creation of a standardized procedure in which nontranscribed DNA-based technical information can be added to the transgene before it is inserted in the organism's genome. This artificial coding would be based on nucleotide triplets, just like amino acids codons, and each triplet would encode for one of the 26 Latin alphabetic letters, an Arabic numeral from 0 to 9, and one space character, giving a total of 37 characters (Table 3.2) (Marillonet et al. 2003). With these characters, the researchers could insert biologically neutral, nongenetic coding sequences that translate into unique information such as the name of the company, production date, place of production, product model, and serial number. The variable region where the information is encoded will be cloned between conserved sequences that contain primer-binding domains. To read the DNA-encoded information, one only needs to perform PCR and sequence the fragment.

Another PCR-based method for GMO detection involves the use of unique genomic sequences flanking the transgene. Hernandez et al. (2003), working with Monsanto's transgenic maize line MON810, which contains a gene encoding for the insecticide CryIA(b) endotoxin, identified a genomic sequence adjacent to the 3′-integration site of the transgenic plant by using a thermal asymmetric interlaced (TAIL)-PCR approach. PCR amplification of target DNA and real-time PCR product quantification are the two most used techniques for accurate DNA quantification. Real-time quantitative PCR can be used with different quantitative tools such as DNA-binding dyes (Morrison et al. 1998), fluorescent oligonucleotides (Whitcombe et al. 1999), molecular beacons (Tyagi and Kramer 1996), fluorescence resonance energy transfer (FRET) probes (Wittwer et al. 1997), and TaqMan probes (Heid et al. 1996). The main advantage of the TaqMan system is that it is highly specific because it uses three oligonucleotides in the PCR reaction. This detection system consists of two primers that are responsible for product amplification, and the TaqMan probe, a fluorogenic oligonucleotide that will anneal to the product. During amplification, Taq polymerase releases a 5′ fluorescent tag from the annealed TaqMan probe, which gives off a quantifiable fluorescence light. Higher light intensity translates into a greater amount of the target gene present in the food sample.

Food Pathogen Detection

Food poisoning may occur due to contamination of food by certain toxin-producing bacteria such as *Salmonella*, *Vibrio*, *Listeria*, and *E. coli*. The strain O157:H7 is the deadliest among all *E. coli* strains; it produces toxins called Shiga, which are encoded by two genes, *stx1* and *stx2*. Shiga toxins (Stx1 and Stx2) damage the lining of the large intestine, causing severe diarrhea and dehydration; and if absorbed into the bloodstream, the toxins can harm other

Table 3.2. Artificial Triplet Codons Encoding for Specific Alphabetical and Numeric Characters, Used in the Identification of GM Organisms

Characters	Codons			Characters	Codons	
1	TTA	CCT		*H*	CAC	AGA
2	TTG	CCC		*I*	CAA	AGG
3	CTT	ACC		*J*	GTC	
4	CTC	ACA		*K*	GTA	
5	CTA	ACG		*L*	CAG	AGT
6	CTG	ACT		*M*	GTG	
7	ATT	GCA		*N*	AAT	GGA
8	ATC	GCG		*O*	AAC	GGG
9	ATA	GCT		*P*	TCT	
0	TTC	CCG	GAG	*Q*	TCC	
space	TTT	CCA	AAG	*R*	AAA	GGT
				S	GAT	TGC
A	TAT	CGA		*T*	GAC	TGA
B	ATG			*U*	GAA	TGG
C	TAC	CGG		*V*	TCA	
D	TAA	CGT		*W*	TCG	
E	TAG	CGC		*X*	GCC	
F	GTT			*Y*	TGT	
G	CAT	AGC		*Z*	GGC	

Source: Adapted from Marillonnet et al. 2003.

organs, such as the kidney (Riley et al. 1983). In North America, the O157:H7 strain is found mostly in the intestines of healthy cattle. The Center for Disease Control (CDC) estimated that Shiga toxin–producing *E. coli* (STEC), such as O157:H7, causes 73,480 illness and 61 deaths per year in the United States alone and that 85% of these cases are attributed to foodborne transmission, especially from ground beef, unpasteurized milk, and roast beef (Mead et al. 1999). As reports of *E. coli* O157:H7 outbreaks have become more common, greater efforts have been made to develop fast and reliable methods for its detection. PCR has become the method of choice for pathogen detection because, contrary to culture isolation and serological tests, PCR methods provide fast, accurate, and highly sensitive results. Among the genes currently used as targets for PCR amplification of O157:H7 are Shiga toxin *(stx)* (Brian et al. 1992), intimin (Gannon et al. 1993), enterohemorrhagic *E. coli* hemolysin (Hall and Xu 1998), and β-glucuronidase (EC 3.2.1.31) (Feng 1993). In traditional PCR methods for detection of pathogens, which include gel electrophoresis, the number of samples that can be analyzed during one electrophoresis run is very small. For this reason, researchers from the Department of Nutrition and Food Science, University of Maryland, College Park, Maryland, and from the Center for Food Safety and Applied Nutrition, Food and Drug Administration, Washington, D.C., developed a simple, rapid, large-scale method for the analysis of PCR products with the use of enzyme-linked immunosorbent assay (ELISA) (Ge et al. 2002). This PCR-ELISA approach for the detection of *E. coli* O157:H7 and other STEC in food was based on the incorporation of digoxigenin-labeled deoxyuridine triphosphate (dUTP) and a biotin-labeled primer specific for *stx1* and *stx2* genes during PCR amplification. In this method, the biotin-labeled PCR products were bound to microtiter plate wells coated with streptavidin and then detected by ELISA using an anti-DIG-peroxidase conjugate (Fig. 3.15). To establish the specificity of the primers used in this PCR method, 39 different bacterial strains, including STEC and non-STEC strains such as *Salmonella*, were used. All of the STEC strains were positive and all non-STEC organisms were negative. The researchers observed that in comparison with the traditional gel electrophoresis with ethidium bromide staining method, the ELISA

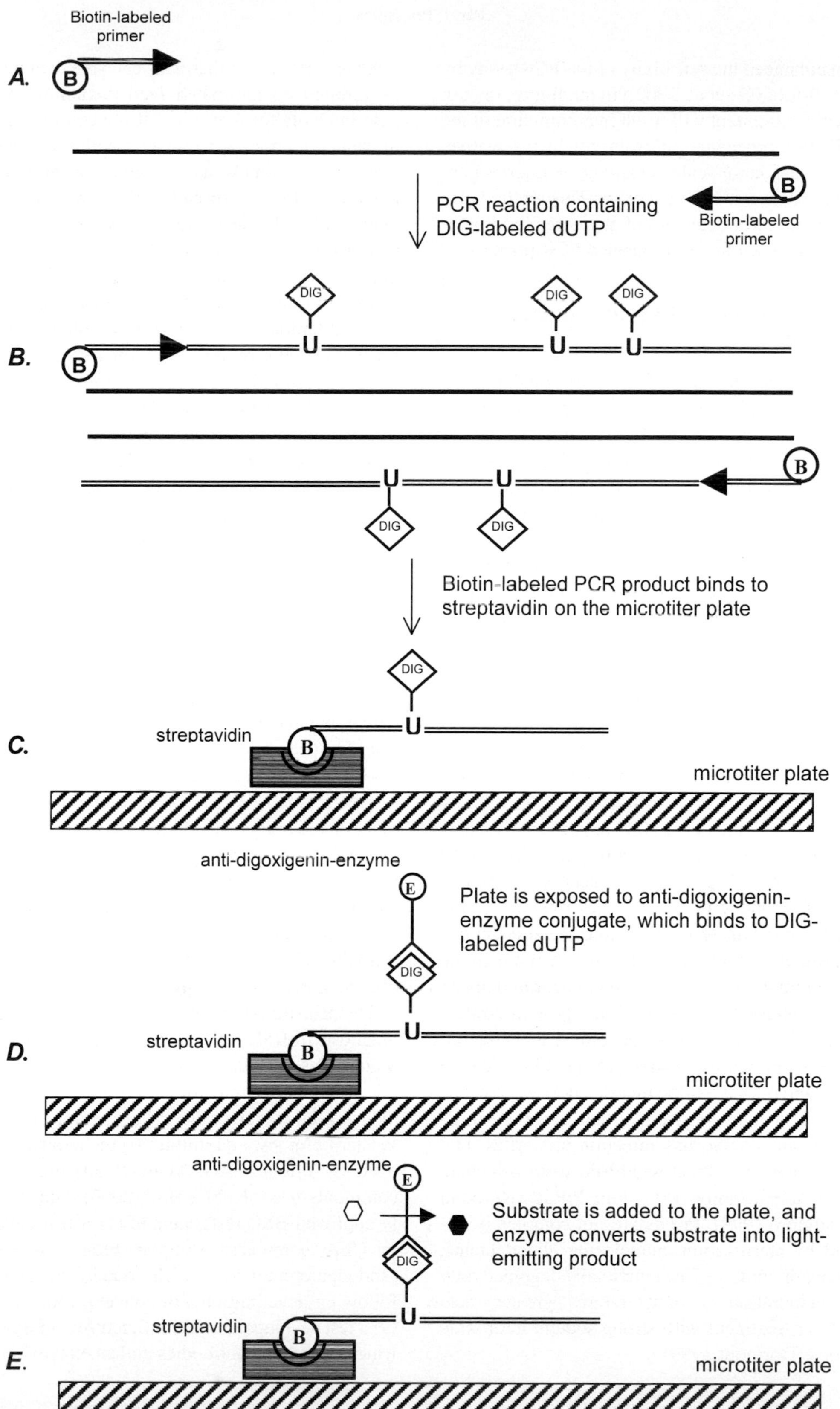

Figure 3.15. Diagram of digoxygenin-ELISA (DIG-ELISA) method.

system enhanced the sensitivity of the PCR assay by up to 100-fold (Ge et al. 2002). In the future, the use of robotic equipment will result in automation of the PCR-ELISA procedure, allowing for fast, sensitive, accurate, and large-scale screening of microorganisms that produce the Shiga toxins. This method also can be applied for detection of any other food pathogens, using specific biotin-labeled PCR primers.

Bovine Spongiform Encephalopathy Detection

The transmissible spongiform encephalopathies (TSEs), are a group of neurodegenerative diseases affecting many animals, including humans, and are characterized by the formation of microscopic "holes" in the brain tissue. Members of the TSE family include (1) the diseases that afflict humans: kuru, Gertsmann-Sträussler-Scheinker (GSS), fatal familial insomnia (FFI), and Creutzfeldt-Jakob disease (CJD); and (2) the diseases that afflict animals: scrapie (sheep and goats), wasting disease (elk and deer), mink encephalopathy (mink), feline spongiform encephalopathy (cats), and bovine spongiform encephalopathy (BSE) (Prusiner 1998). Bovine spongiform encephalopathy affects cattle and is commonly known as mad cow disease. The symptoms associated with BSE are weight loss, drooling, head waving, aggressive behavior, and eventually death. All TSEs, including BSE, are caused by a new infectious agent called a "prion protein" (PrP) (Prusiner 1982). Dr. Stanley Prusiner discovered prions in 1984 and was awarded the 1997 Medicine Nobel Prize for his research. Prions are endogenous glycoproteins found abundantly in the brain tissue of all mammals and may function as neuron helpers. They can manifest in two different protein conformations: (1) PrPc, the normal form, which is nonpathogenic, protease sensitive, and high in α-helical content, and (2) PrPSc, the misfolded form, which is disease-inducing, protease resistant, and high in β-sheet content. PrPSc has infection properties, and when it comes in contact with PrPc, it starts a chain reaction, transforming PrPc into PrPSc (Horiuchi and Caughey 1999). Prions are not completely destroyed by sterilization, autoclaving, disinfectants, radiation, or cooking. They are totally degraded only with incineration at temperatures greater than 1000°C or treatment with strong sodium hydroxide solutions (Dormont 1999).

Scientists believe that BSE can spread among cattle through contaminated feed containing the disease-inducing form of prion. It is a common practice in many countries to feed cattle with the remains of other farm animals as a source of protein. The hypothesis for the spread of BSE in cattle is that body parts of sheep infected with scrapie were included in cattle feed, and PrPSc jumped species to infect bovines (Bruce et al. 1997). The disease spread throughout the world when England sold BSE-contaminated cattle feed to other countries. Since it first appeared in 1986, the risk of BSE infection has resulted in the destruction of 3.7 million animals in the United Kingdom. In humans, CJD is an inherited disease caused by a mutation in the prion protein gene, *PRNP*, and affects one in a million people. It is believed that the human victims may have contracted a new variant of CJD (nv-CJD) from eating meat products contaminated with BSE. The pathogenic prion protein that causes BSE is nearly identical to the prion that causes nv-CJD (Johnson and Gibbs 1998). Between 1996 and 2003, 156 cases of nv-CJD have been suspected or confirmed in many countries, mainly in Great Britain.

One of the main problems in dealing with and containing the BSE epidemic is that there are no tests for detection of the disease in a live animal. Prions do not trigger any detectable specific immune response, and levels of abnormal prions in other parts of the body, such as the blood, are too low to detect. The only means of diagnosis in live animals is observation of BSE symptoms in the animal. Because of the lack of reliable live-detection methods, all animals in a herd that may have been in contact with BSE-contaminated feed must be destroyed, causing great losses for the cattle industry.

The primary laboratory method used to confirm a diagnosis of BSE is the microscopic examination of brain tissue after death of the animal. In addition to microscopic examination, there are several techniques used to detect the PrPSc, among which, the Western blot test and immunocytochemistry (developed by Prionics AG, Switzerland) are the most commonly used (Kübler et al. 2003). Other currently approved BSE tests include (1) a test developed by CEA, a research group in France, in which a sandwich immunoassay technique for PrPSc is done following denaturation and concentration steps and (2) a test developed by Enfer Scientific, in Ireland, in which polyclonal antibodies and an enzyme-coupled secondary antibody are used and detection is done by ELISA-chemiluminescence (Moynagh and Schimmel 2000). Although these tests are 100% reliable, they can only be done in postmortem brain tissue.

To improve the sensitivity of current detection methods and enable live-animal BSE detection, intense research has been done in developing assays that take advantage of the infectious capacity of the PrPSc to convert the normal prion proteins into pathogenic ones. A team of researchers from Serono Pharmaceutical Research Institute in Geneva cultured mutated prions in vitro for the first time and have devised a method to replicate them to high enough levels to detect the disease at an earlier stage (Saborio et al. 2001). The technique, called protein misfolding cyclic amplification (PMCA), is similar to a PCR reaction, in which amplification of small amounts of the pathogenic protein in a sample is achieved through multiple cycles of PrPSc incubation in the presence of excess PrPc, followed by disruption of the aggregates through sonication in the presence of detergents (Fig 3.16). This new technique has the potential to improve BSE and other TSE detection methods, to give a better understanding of prion diseases and to help in the search for drug targets for brain diseases. More recently, researchers from Chronix Biomedical in San Francisco and the Institute of Veterinary Medicine in Göttingen, Germany, claim to have developed the first BSE test that can be performed on live animals (Urnovitz 2003). This detection method, called the surrogate marker living test for BSE, is a real-time PCR test based on in vitro detection of specific RNA in the bovine serum. It uses blood samples from cows for the identification of a microvesicle RNA, and it is considered a surrogate marker test because it detects blood RNA and not prion proteins. RNA is found in the blood fraction that contains primarily microvesicles, and while microvesicles are found in both healthy and diseased animals, their RNA contents appear to be different. Cattle that show reactivity to this test should be subjected to a second more specific test for conclusive diagnosis of mad cow disease (Urnovitz 2003).

CONCLUSION

Research on transgenic organisms for the food industry has been intended mostly to benefit producers. The creation of pathogen-, insect-, and herbicide-resistant transgenic plants have led to increased productivity and decreased costs of producing many food crops. Similarly, the introduction of foreign genes in transgenic salmon enabled it to grow three times faster then wild-type salmon. The benefits of the next generation of biotechnology research will be directed toward consumers and will entail the creation of designer foods with enhanced characteristics such as better nutrition, taste, quality, and safety. A good example of such research is the use of biotechnology to decrease the amount of saturated fatty acids in vegetable oils.

Among the many possibilities available in the field of food biotechnology, future research in the area of plant genetic engineering is aimed at increasing the shelf life of fresh fruits and vegetables, creating plants that produce sweet proteins, developing caffeine-free coffee and tea, and improving the flavor of fruits and vegetables. Great progress has been achieved in using plants as bioreactors for manufacturing and as a delivery system for vaccines. A variety of crops such as tobacco, potato, tomato, and banana has been used to produce experimental vaccines against infectious diseases. Advances in genetic engineering will enable the creation of transgenic plants that produce proteins essential for the production of pharmaceuticals such as growth hormones, antigens, antibodies, enzymes, collagen, and blood proteins.

The applications of genetic engineering in the food industry are not limited to the manipulation of plant genomes. Animals and microorganisms also have been extensively researched to produce better food products. In the area of animal bioengineering, the main focus has been on the use of the mammary gland and the egg as bioreactors in the manufacturing of biopharmaceuticals. So far, the focus of animal bioengineering research has been directed toward the benefit of the producers. In the future, transgenic animal research will be shifted towards the production of healthier meat products, with decreased amounts of fat and cholesterol.

Microorganisms, especially yeast, have been vital to the food processing industry for centuries. Among their many qualities, yeast plays an important role in the production of fermented foods, enzymes, and proteins. Future research on genetic engineering of microorganisms may be focused on the large-scale production of biologically active

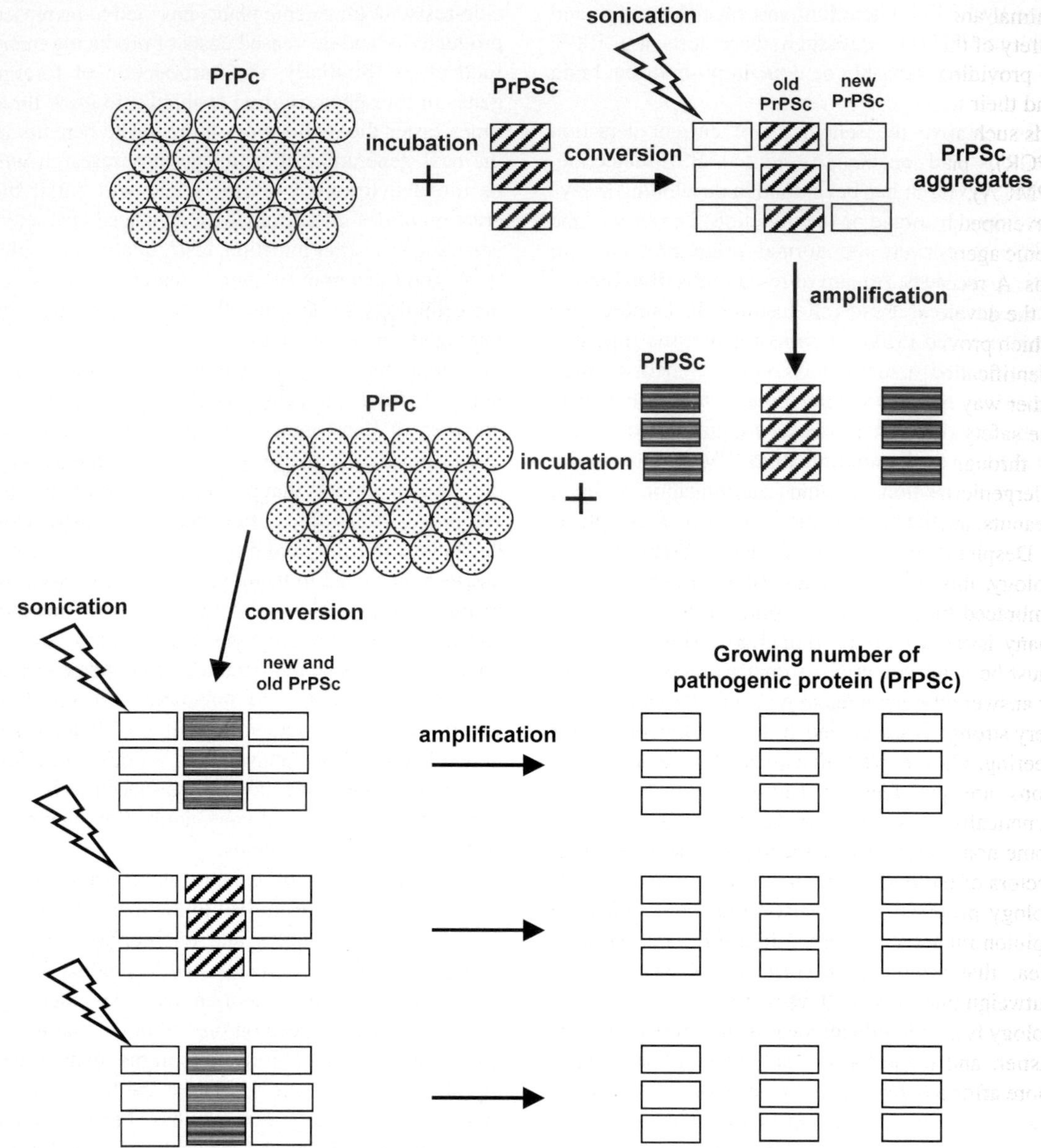

Figure 3.16. Diagram of protein-misfolding cyclic amplification (PMCA). (From Saborio et al. 2001.)

components that provide health benefits, but are found in low quantities in plants. Among such components that have anticancer properties are lycopenes found in tomato, glucosinolates found in broccoli, ellagic acids found in strawberry, and isoflavonoids found in soybean. Further biotechnology advances will enhance the value and scope of the use of microorganisms in the food and pharmaceutical industry.

Besides the production of novel compounds and the improvement of existing ones, biotechnology plays an important role in food safety. Microbial contamination is a major concern in the food industry. New biotechnology methods are being developed to help decrease the amount of microbes on animal and plant products and thereby improve the safety of the food supply. In addition, biotechnology is providing many tools to detect microorganisms and their toxins. Some new and old detection methods such as ELISA tests, polymerase chain reaction (PCR), protein misfolding cyclic amplification (PMCA), DNA probes, and biosensors have been developed to detect the presence of infectious pathogenic agents such as bacteria, viruses, fungi, and prions. A recent improvement in the detection method is the development of the real-time PCR technology, which provides a sensitive and more reliable tool for identification of pathogens in food products. Another way by which biotechnology helps to improve the safety of food products for human consumption is through rapid identification and extraction of allergenic proteins in food items such as shellfish, peanuts, and soybeans.

Despite the benefits provided by modern biotechnology, this relatively new science is not yet fully embraced by the general population. There are still many issues of safety, reliability, and efficacy that must be overcome before scientists can convincingly answer all public concerns. At this time, there is a very strong polarization on the issue of genetic engineering. On one side, some scientists and corporations are pushing for the commercialization of genetically engineered products; on the other side, some nongovernment agencies, representing certain sectors of the population, are trying to stop biotechnology products from being used. This division in opinion may subside in the future when it becomes clear that the benefits of biotechnology products outweigh their risks and when modern food biotechnology is able to deliver a variety of more nutritious, tastier, and safer food products to all people at a more affordable price.

REFERENCES

Agius F, Gonzales-Lamothe R, Caballero J, Munoz-Blanco J, Botella M, Valpuesta V. 2003. Engineering increased vitamin C levels in plants by over-expression of a D-galacturonic acid reductase. Nature Biotechnology 21:177–181.

Aldhous P. 1990. Genetic engineering. Modified yeast fine for food. Nature 344:186.

Andlauer W, Furst P. 1998. Antioxidative power of phytochemicals with special reference to cereals. Cereal Foods World 43:356–359.

Andrews SC, Arosio P, Bottke W, Briat JF, von Darl M, Harrison PM, Laulhere JP, Levi S, Lobreaux S, Yewdall SJ. 1992. Structure, function and evolution of ferritins. J Inorg Biochem 47:161–174.

Armstrong DW, Yamazaki H. 1986. Natural flavours production: A biotechnological approach. Trends Biotechnol 4:264–268.

Armstrong GA. 1994. *Eubacteria* show their true colors: Genetics of carotenoid pigment biosynthesis from microbes to plants. J Bacteriol 176:4795–4802.

Bartley GE, Scolnik PA, Giuliano G. 1994. Molecular biology of carotenoid biosynthesis in plants. Ann Rev Plant Physiol Plant Mol Biol 45:287–301.

Basta SS, Soekirman, Karyadi D, Scrimshaw NS. 1979. Iron deficiency anemia and the productivity of adult males in Indonesia. Am J Clin Nutr 32:916–925.

Bawden WS, Passey, RJ, MacKinlay, AG. 1994. The genes encoding the major milk-specific proteins and their use in transgenic studies and protein engineering. Biotechnol Genet Eng Rev 12:89.

Beard JL, Burton JW, Theil EC. 1996. Purified ferritin and soybean meal can be sources of iron for treating iron deficiency in rats. J Nutr 126:154–160.

Boze H, Moulin G, Glazy P. 1992. Production of food and fodder yeasts. Crit Rev Biotechnol 12:65–86.

Brian MJ, Frosolono M, Murray BE, Miranda A, Lopez EL, Gomez HF, Cleary TG. 1992. Polymerase chain reaction for diagnosis of enterohemorrhagic *Escherichia coli* infection and hemolytic-uremic syndrome. J Clin Microbiol 30:1801–1806.

Brinch-Pedersen H, Sorensen LD, Holm, PB. 2002. Engineering crop plants: Getting a handle on phosphate. Trends in Plant Science 7:118–125.

Brophy B, Smolenski G, Wheeler T, Wells D, L'Huillier P, Laible G. 2003. Cloned transgenic cattle produce milk with higher levels of β-casein and κ-casein. Nature Biotechnology 21:157–162.

Bruce M.E, Will RG, Ironside, JW, McConnell, I, Drummond D, Suttie A, McCardle L, Chree A, Hope, J, Birkett C, Cousens S, Fraser H, Bostock CJ. 1997. Transmissions to mice indicate that the new variant CJD is caused by the BSE agent. Nature 389:498–501.

Buchanan BB, Gruissem W, Jones RL. 2001. Biochemistry and Molecular Biology of Plants, 3rd edition. Rockville: American Society of Plant Physiologists. Pp.1288–1309.

Business Communications Company RGA-110. 2000. The Global Market for Carotenoids. Connecticut: Business Communications Company Norwalk.

Cabot RA, Kuhholzer B, Chan AW, Lai L, Park KW, Chong KY, Schatten G, Murphy CN, Abeydeera LR,

Day BN., Prather RS. 2001. Transgenic pigs produced using in vitro matured oocytes infected with a retroviral vector. Anim Biotechnol 12:205–214.

Cahoon EB, Hall SE, Ripp K.G, Ganzke TS, Hitz WD, Coughlan SJ. 2003. Metabolic redesign of vitamin E biosynthesis in plants for tocotrienol production and increased antioxidant content. Nature Biotechnology 21:1082–1087.

Campbell NA, Reece JB. 2002. Biology, 6th edition. San Francisco: Pearson Education. Pp. 69–70.

Chakraborty S, Chakraborty N, Datta A. 2000. Increased nutritive value of transgenic potato by expressing a nonallergenic seed albumin gene from *Amaranthus hypocondriacus*. PNAS 97:3724–3729.

Cheetham PSJ. 1993. The use of biotransformations for the production of flavours and fragrance*s*. Trends Biotechnol 11:478–488.

Chen Z, Young T, Ling J, Chang S, Gallie D. 2003. Increasing vitamin C content of plants through enhanced ascorbate recycling. PNAS 100:3525–3530.

Civitelli R. 1997. In vitro and in vivo effects of ipriflavone on bone formation and bone biomechanics. Calcif Tissue Int 61:S12–S14.

Clark AJ, Simons P, Wilmut I, Lathe R. 1987. Pharmaceuticals from transgenic livestock. Trends Biotechnol 5:20–24.

Collakova E, DellaPenna D. 2001. Isolation and functional analysis of homogentisatephytyltransferase from *Synechocystis* sp. PCC 6803 and *Arabidopsis*. Plant Physiol 127:1113–1124.

Colman A. 1996. Production of proteins in the milk of transgenic livestock: Problems, solutions, and successes. Am J Clin Nutr 63:S639–S645.

Cooper KH. 1996. Advanced Nutrition Therapies. Nashville: Thomas Nelson Publishers.

Dalgleish DG, Horne DS, Law AJ. 1989. Size-related differences in bovine casein micelles. Biochem Biophys Acta 991:383–387.

Daniell H, Dhingra A. 2002. Multigene engineering: Dawn of an exciting new era in biotechnology. Curr Opin Biotechnol 13:136–141.

Davey MW, Van Monatgu M, Sanmatin M, Kanellis A, Smirnoff N, Benzie IJJ, Strain JJ, Favell D, Fletcher J. 2000. Plant L-ascorbic acid: Chemistry, function, metabolism, bio-availability and effects of processing. J Sci Food Agric 80:825–860.

Davies PL, Hew CL. 1990. Biochemistry of fish antifreeze proteins. FASEB J 4:2460–2468.

Davies PL, Sykes BD. 1997. Antifreeze proteins. Current Opinion in Structural Biology 7:828–834.

Deng G, Andrews DW, Laursen RA. 1997. Amino acid sequence of a new type of antifreeze protein from the longhorn sculpin *Myoxocephalus octodecimspinosis*. FEBs Letters 402:17–20.

Dequin S. 2001. The potential of genetic engineering for improving brewing, wine-making and baking yeast. Appl Microbil Biotechnol 56:577–588.

Devlin RH, Yesaki TY, Biagi CA, Donaldson EM, Swanson P, Chan W. 1994. Extraordinary salmon growth. Nature 371:209–210.

DeVries AL. 1984. Role of glycopeptides and peptides in inhibition of crystallization of water in polar fishes. Phil Trans R Soc Lond B304:575–588.

Dormont D. 1999. Agents that cause transmissible subacute spongiform enchaphalopathies. Biomed Pharmacother 53:3–8.

Dove A. 2002. Uncorking the biomanufacturing bottleneck. Nature Biotechnology 20:777–779.

Du SJ, Gong ZY, Fletcher, GL, Shears MA, King MJ, Idler DR, Hew CL. 1992. Growth enhancement in transgenic Atlantic salmon by the use of an "all-fish" chimeric growth hormone gene constructs. Bio/ Technology 10:176–181.

Eldridge AC, Kwolek WF. 1983. Soybean isoflavones: Effect of environment and variety on composition. J Agric Food Chem 31:394–396.

Emodi A. 1978. Xylitol: Its properties and food applications. Food Technol 32:20–32.

Etherton TD. 1991. Clinical review 21: The efficacy and safety of growth hormone for animal agriculture. J Clin Endocrinol Metab 72:957A–957C.

European Commission. 2000. Commission regulation (EC) No. 49/2000 of the Commission amending Council Regulation (EC) 1139/98 of January 10, 2000, concerning the compulsory indication on the labeling of certain foodstuffs produced from genetically modified organisms of particulars other than those provided for in Directive 79/112/EEC Off J Eur Communities L6:13–14.

Falk J, Andersen G, Kernebeck B, Krupinska K. 2003. Constitutive overexpression of barley 4-hydroxyphenylpyruvate dioxygenase in tobacco results in elevation of the vitamin E content in seeds but not in leaves. FEBS Letters 540:35–40.

Feng P. 1993. Identification of *Escherichia coli* serotype O157:H7 by DNA probe specific for an allele of *uidA* gene. Mol Cell Probes 151–154.

Fickler J. 1995. The amino acid composition of feedstuffs. New York: Degussa Corporation.

Food and Drug Administration (FDA). 1999. Food labeling: Health claims; soy protein and coronary heart disease; final rule. Federal Register 64 FR 57699.

Gannon VP, Rashed M, King, RK, Thomas EJ. 1993. Detection and characterization of the *eae* gene of

Shiga-like toxin-producing *Escherichia coli* using polymerase chain reaction. J Clin Microbiol 1268–1274.

Gardner NJ, Savard T, Obermeier P, Caldwell G, Champagne CP. 2001. Selection and characterization of mixed starter cultures for lactic acid fermentation of carrot, cabbage, beet and onion vegetable mixtures. Int J Food Microbiol 64:261–75.

Ge B, Zhao S, Hall R, Meng J. 2002. A PCR-ELISA for detecting Shiga toxin-producing *Escherichia coli*. Microbes Infect 4:285–290.

Geisen R, Holzapfel WH. 1996. Genetically modified starter and protective cultures. Int J Food Micro 30:315–324.

Gibson DM, Ullah AB. 1990. Phytases and their action on phytic acid. In: Inositol Metabolism in Plants. New York: Wiley-Liss Inc. Pp. 77–92.

Giddings G, Allison G, Brooks D, Carter A. 2000. Transgenic plants as factories for biopharmaceuticals. Nature Biotechnology 18:1151–1155.

Gilbert AB. 1984. Egg albumen and its formation. In: Physiology and Biochemistry of the Domestic Fowl. Academic Press. Pp. 1291–1329.

Giovannucci E, Ascherio A, Rimm EB, Stampfer MJ, Colditz GA, Willett WC. 1995. Intake of carotenoids and retinal in relation to risk of prostate cancer. J Natl Cancer Inst 87:1767–1776.

Gleizes PE, Munger JS, Nunes I, Harpel JG, Mazzieri R, Noguera I, Rifkin DB. 1997. TGF-beta latency: Biological significance and mechanisms of activation. Stem Cells 15:190–197.

Gong Z, Ewart KV, Hu Z, Fletcher GL, Hew CL. 1996. Skin antifreeze protein genes of the winter flounder, *Pleuronectes americanus,* encode distinct and active polypeptides without the secretory signal and prosequences. J Biol Chem 271:4106–4112.

Gordon JW, Scangos GA, Plotkin DJ, Barbosa JA, Ruddle FH. 1980. Genetic transformation of mouse embryos by microinjection of purified DNA. Proc Natl Acd Sci 77:7380–7384.

Goto F, Yoshihara T, Shigemoto N, Toki S, Takaiwa F. 1999. Iron fortification of rice seed by the soybean *ferritin* gene. Nature Biotechnology 17:282–286.

Govinden R, Pillay B, van Zyl WH, Pillay D. 2001. Xylitol production by recombinant *Saccharomyces cerevisiae* expressing the *Pichia stipitis* and *Candida shehatae XYL1* genes. Appl Microbiol Biotechnol 55:76–80.

Grundy SM. 1986. Comparison of monounsaturated fatty acids and carbohydrates for lowering plasma cholesterol. N Engl J Med 314:745–748.

Grusack MA, DellaPenna D. 1999. Improving the nutrient composition of plants to enhance human nutrition and health. Annu Rev Plant Physiol Plant Mol Biol 50:133–161.

Hall RH, Xu JG, inventors. 1998. Rapid and sensitive detection of O157:H7 and other enterohemorrhagic *E. coli*. U.S. Patent 5,756,293.

Hallberg L, Brune M, Rossander L. 1989. Iron absorption in man: Ascorbic acid and dose-dependent inhibition by phytate. Am J Clin Nutr 49:140–144.

Hansen EB. 2002. Commercial bacterial starter cultures for fermented foods of the future. Int J Food Microbiol 78:119–131.

Harvey AJ, Speksnijder G, Baugh LR, Morris JA, Ivarie R. 2002. Expression of exogenous protein in the egg white of transgenic chickens. Nature Biotechnology 20:396–399.

Heid CA, Stevens J, Livak KJ, Williams PM. 1996. Real time quantitative PCR. Genome Res 6:986–994.

Hernandez M, Pla M, Esteve T, Prat S, Puigdomenech P, Ferrando A. 2003. A specific real-time quantitative PCR detection system for event MON810 in maize YieldGard based on the 3′-transgene integration sequence. Transgenic Research 12:179–189.

Hess J L. 1993. Antioxidants in Higher Plants. Boca Raton: CRC Press Inc. 112–134.

Hew CL, Fletcher GL, Davies PL. 1995. Transgenic salmon: Tailoring the genome for food production. J Fish Biol 47:1–19.

Hew CL, Poon R, Xiong F, Gauthier S, Shears M, King M, Davies PL, Fletcher GL. 1999. Liver-specific and seasonal expression of transgenic Atlantic salmon harboring the winter flounder antifreeze protein gene. Transgenic Research 8:405–414.

Hew CL, Scott GK, Davies PL. 1986. Molecular biology of antifreeze. In: Living in the Cold: Physiological and Biochemical Adaptation. New York: Elsevier Press. Pp. 117–123.

Higuchi R, Dollinger G, Walsh PS, Griffith R. 1992. Simultaneous amplification and detection of specific DNA sequences. Biotechnology 10:413–417.

Hochedlinger K., Jaenisch R. 2003. Nuclear transplantation, embryonic stem cells, and the potential for cell therapy. N Engl J Med 349:275–286.

Holm PB, Kritiansen KN. Pedersen HB. 2002. Transgenic approaches in commonly consumed cereals to improve iron and zinc content and bioavailability. Journal of Nutrition. 132:514S–516S.

Horiuchi M, Caughey B. 1999. Prion protein interconversions and the transmissible spongiform encephalopathies. Structure Fold Des 7:R231–R240.

Ivarie R. 2003. Avian transgenesis: Progress towards the promise. Trends in Biotechnology 21:14–19.

James C. 2001. Global Review of Commercialized Transgenic Crops. ISAAA Briefs 24.

———. 2003. Global Review of Commercialized Transgenic Crops: 2002 Feature: Bt Maize. ISAAA Briefs 29.

Jaynes J M, Yang MS, Espinoza N, Dodds JH. 1986. Plant protein improvement by genetic engineering: Use of synthetic genes. Trends Biotech 4:314–320.

Jimenez Flores R, Richardson T. 1988. Genetic engineering of the caseins to modify the behavior of milk during processing: A review. J Dairy Sci 71:2640–2654.

Johnson R, Gibbs CJ. 1998. Creutzfeldt-Jakob disease and related transmissible spongiform encephalopathies. New Engl J Med 339:1994–2004.

Jung W, Yu O, Lau SM, O'Keefe DP, Odell J, Fader G, McGonigle B. 2000. Identification and expression of isoflavone synthase, the key enzyme for biosynthesis of isoflavones in legumes. Nature Biotechnology 18:208–212.

Jyocouchi H, Hill J, Tomita Y, Good RA. 1991. Studies of immunomodulation actions of carotenoids. I. Effects of β-carotene and astaxanthin on murine lymphocyte functions and cell surface marker expression in vivo culture system. Nutr Cancer 6:93–105.

Kamal-Eldin A, Appelqvist LA. 1996. The chemistry and antioxidant properties of tocopherols and tocotrienols. Lipids 31:671–701.

Karatzas CN, Turner J.D. 1997. Toward altering milk composition by genetic manipulation: Current status and challenges. J Dairy Sci 80:2225–2232.

Keys A, Anderson T, Grande F. 1965. Serum cholesterol response to changes in diet. IV. Particular saturated fatty acids in the diet. Metabolism 14:776.

Kitamoto K, Oda K, Gomi K, Takashi K. 1991. Genetic engineering of sake yeast producing no urea by successive disruption of *arginase* gene. Appl Environ Microbiol 57:2568–2575.

Klein TM, Wolf ED, Wu R, Stanford JC. 1987. High velocity microprojectiles for delivering nucleic acids into living cells. Nature 327:70–73.

Kleter GA, van der Krieken WM, Kok EJ, Bosch D, Jordi W, Gilissen LJ. 2001. Regulation and exploitation of genetically modified crops. Nature Biotechnology 19:1105–1110.

Krings U, Berger RG. 1998. Biotechnological production of flavours and fragrances. Appl Microbiol Biotechnol 49:1–8.

Kübler, E, Oesch B, Raeber AL. 2003. Diagnosis of prion diseases. British Medical Bulletin 66:267–279.

Lantzsch HJ, Menke KH, Scheuermann SE. 1992. Comparative study of phosphorous utilization from wheat, barley and corn diets by young rats and pigs. J Anim Physiol Anim Nutr 67:123–132.

Lee TT, Wang MM, Hou RC, Chen LJ, Su RC, Wang CS, Tzen JT. 2003. Enhanced methionine and cysteine levels in transgenic rice seeds by the accumulation of sesame 2S albumin. Biosci Biotechnol Biochem 67:1699–1705.

Leury BJ, Baumgard LH, Block SS, Segoale N, Ehrhardt RA, Rhoads RP, Bauman DE, Bell AW, Boisclair YR. 2003. Effect of insulin and growth hormone on plasma leptin in periparturient dairy cows. Am J Physiol Regul Integr Comp Physiol 285: R1107–R1115.

Lewus CB, Kaiser A, Montville TJ. 1991. Inhibition of food-borne bacterial pathogens by bacteriocins from lactic acid bacteria isolated from meats. Appl Environ Microbiol 57:1683–1688.

Lott JNA. 1984. Accumulation of seed reserves of phosphorous and other minerals. In: Seed Physiology. Academic Press. Pp. 139–166.

Lucca P, Hurrell R, Potrykus I. 2002. Fighting iron deficiency anemia with iron-rich rice. J Am Coll. Nutr 21:184S–190S.

Maberly GF, Trowbridge FL, Yip R, Sullivan KM., West CE. 1994. Programs against micronutrient malnutrition: Ending hidden hunger. Annu Rev Public Health 15:277–301.

Marillonet S, Klimyuk V, Gleba Y. 2003. Encoding technical information in GM organisms. Nature Biotechnology 21:224–226.

McMahon DJ, Brown RJ. 1984. Composition, structure, and integrity of casein micelles: A review. J Dairy Sci 67:499.

McPherron AC, Lee S. 1997. Double muscling in cattle due to mutations in the *myostatin* gene. Proc Natl Acad Sci 94:12457–12461.

Mead PS, Slutsker L, Dietz V, McCaig LF, Bresee JS, Shapiro C, Griffin PM., Tauxe RV. 1999. Food-related illness and death in the United States. Emerg Infect Dis 5:607–625.

Messina M, Barnes S. 1991. The role of soy products in reducing cancer risks. J Natl Cancer Inst 83:541–546.

Miflin B, Napier J, Shewry P. 1999. Improving plant product quality. Nature Biotechnology 17:BV13–BV14.

Miki W. 1991. Biological functions and activity of animal carotenoids. Pure Appl Chem 63:141–146.

Miura Y, Kondo K, Saito T, Shimada H, Fraser PD, Misawa N. 1998. Production of the carotenoids lycopene, β-carotene, and astaxanthin in the food yeast *Candida utilis*. Appl Envir Micrbiol 64:1226–1229.

Morrison TM, Weiss JJ, Wittwer CT. 1998. Quantification of low-copy transcripts by continuous SYBR green I monitoring during amplification. Biotechniques 24:954–962.

Moynagh J, Schimmel H. 2000. Tests for BSE evaluated. Nature 400:105.

Muir WM, Howard RD. 1999. Possible ecological risks of transgenic organism release when transgenes affect mating success: Sexual selection and the Trojan gene hypothesis. Proc Natl Acad Sci USA 96: 13853–13856.

Munné-Bosch S, Alegre L. 2002. The function of tocopherols and tocotrienols in plants. Crit Rev Plant Sci 21:31–57.

Murakawa H, Bland CE, Willis WT, Dallman PR. 1987. Iron deficiency and neutrophil function: Different rates of correction of the depressions in oxidative burst and myeloperoxidase activity after iron treatment. Blood 695:1464–1468.

Murphy JF, Riordan J, Newcombe RG, Coles EC, Person JF. 1986. Relation of hemoglobin levels in first and second trimesters to outcome of pregnancy. Lancelet 1:992–995.

National Research Council (NRC). 1988. Designing foods: Animal Product Options in the Marketplace. Washington: National Academy Press.

Nestel PJ, Pomeroy S, Kay S, Komesaroff P, Behrsing J, Cameron J.D, West L. 1999. Isoflavones from red clover improve systemic arterial compliance but no plasma lipids in menopausal women. J Clin Endocrinol Metab 84:895–898.

Nigam P, Singh D. 1995. Processes for fermentative production of xylitol—A sugar substitute. Process Biochem 30:117–124.

Niiler E. 2000. FDA researchers consider first transgenic fish. Nature Biotechnology 18:143.

Ough CS. 1976. Ethylcarbamate in fermented beverages and foods. I. Naturally occurring ethylcarbamate. J Agric Food Chem 24:323–328.

Pepper T, Olinger PM. 1988. Xylitol in sugar-free confections. Food Technol 10:98–106.

Persuy MA, Legrain S, Printz C, Stinnakre MG, Lepourry L, Brignon G, Mercier JC. 1995. High-level, stage- and mammary-tissue-specific expression of a caprine κ-casein-encoding minigene driven by a β-casein promoter in transgenic mice. Gene 165: 291–296.

Peterson G, Barnes S. 1991. Genistein inhibition of the growth of human breast cancer cells: Independence from estrogen receptors and the multi-drug resistance gene. Biochem Biophys Res Commun 179: 661–667.

Potrykus I. 2001. Golden rice and beyond. Plant Physiol 125:1157–1161.

Prusiner SB. 1982. Novel proteinaceous infections particles cause scrapie. Science 216:136–144.

———. 1998. Prions. Proc Natl Acad Sci USA 95:13363–13383.

Pursel VG, Pinkert CA, Miller KF, Bolt DJ, Campbell RG, Palmiter RD, Brinster RL, Hammer RE. 1989. Genetic engineering of livestock. Science 244:1281–1288.

Rahman MA, Mak R, Ayad H, Smith A, Maclean N. 1998. Expression of a novel piscine growth hormone gene results in growth enhancement in transgenic tilapia *(Oreochromis niloticus)*. Transg Research 7:357–369.

Razak SA, Hwang GL, Rahman MA, Maclean N. 1999. Growth performance and gonadal development of growth enhanced transgenic tilapia *Oreochromis niloticus (L.)* following heat-shock-induced triploidy. Mar Biotechnol 1:533–544.

Ravindran V, Bryden WL, Kornegay ET. 1995. Phytates: Occurrence, bioavailability and implications in poultry nutrition. Poultry Avian Biol Rev 6:125–143.

Riley LW, Remis RS, Helgerson SD, McGee HB, Wells JG, Davis BR, Hebert RJ, Olcott ES, Johnson LM, Hargrett NT, Blake PA, Cohen M.L. 1983. Hemorrhagic colitis associated with a rare *Escherichia coli* serotype. New Engl J Med 308:681–685.

Rinzler CA, Jensen MD, Brody JE. 1999. The New Complete Book of Food: A Nutritional, Medical, and Culinary Guide. New York: Facts on File, Inc.

Saborio GP, Permanne B, Soto C. 2001. Sensitive detection of pathological prion protein by cyclic amplification of protein misfolding. Nature 411:810–813.

Schultz G, Soll J, Fiedler E, Schulze-Siebert D. 1985. Synthesis of prenylquinones in chloroplasts. Physiolo Plant 64:123–129.

Sharma HC, Crouch JH, Sharma KK, Seetharama N, Hash CT. 2002. Applications of biotechnology for crop improvement: Prospects and constraint. Plant Science 163:381–395.

Shintani D, DellaPenna D. 1998. Elevating the vitamin E content of plants through metabolic engineering. Science 282:2098–2100.

Smirnoff N, Conklin P, Loewus F. 2001. Biosynthesis of ascorbic acid: A renaissance. Annu Rev Plant Physiol Plant Molec Biol 52:437–467.

Soll J, Kemmerling M, Schultz G. 1980. Tocopherol and plastoquinone synthesis in spinach chloroplasts subfractions. Arch Biochem Biophy 204:544–550.

Solomon M.B, Pursel VG, Paroczay EW, Bolt DJ. 1994. Lipid composition of carcass tissue from transgenic pigs expressing a bovine growth hormone gene. J Anim Sci 72:1242–1246.

Somogyi LP. 1996. The flavour and fragrance industry: Serving a global market. Chem Ind 4:170–173.

Southgate EM, Davey MR, Power JB, Marchant R. 1995. Factors affecting the genetic engineering of plants by microprojectile bombardment. Biotechnol Adv 13:631–651.

Swift GH, Hammer RE, MacDonald RJ, Brinster RL. 1984. Tissue-specific expression of the rat pancreatic *elastase I* gene in transgenic mice. Cell 38:639–646.

Taylor PG, Martinez C, Romano EL, Layrisse M. 1986. The effect of cysteine-containing peptides released during meat digestion on iron absorption in humans. Am J Clin Nutr 43:68–71.

Taylor SL, Hefle SL. 2002. Genetically engineering foods: Implications for food allergy. Curr Opin Allergy Clin Immunol 2:249–252.

Theriault A, Chao, J-T, Wang Q, Gapor A, and Adeli K. 1999. Tocotrienol: A review of its therapeutical potential. Clin Biochem 32:309–319.

Thies RS, Chen T, Davies MV, Tomkinson KN, Pearson AA, Shakey QA, Wolfman NM. 2001. GDF-8 propeptide binds to GDF-8 and antagonizes biological activity by inhibiting GDF-8 receptor binding. Growth Factors 18:251–259.

Tranter HS, Board RB. 1982. The antimicrobial defense of avian eggs: Biological perspectives and chemical basis. J Appl Biochem 4:295–338.

Tsukamoto C, Shimada S, Igita K, Kudou S, Kokubun M, Okubo K, Kitamura K. 1995. Factors affecting isoflavone content in soybean seeds: Changes in isoflavones, saponins and composition of fatty acids at different temperatures during seed development. J Agric Food Chem 43:1184–1192.

Tyagi S, Kramer FR. 1996. Molecular beacons: Probes that fluoresce upon hybridization. Nature Biotechnology 14:303–308.

United Nations Population Fund (UNFPA) 1995. The State of World Population 1995. UNFPA 67:16–17.

Urnovitz HB. 2003. Mad cow disease: A case for studying living animals. Redflagdaily (serial online) www.redflagsweekly.com

Vain P, de Buyser J, Bui Trang, V, Haicour R, Henry Y. 1995. Foreign delivery into monocotyledonous species. Biotechnol Adv 13:653–671.

Van de Guchte M, van der Wal FJ, Kok J, Venema G. 1992. Lysozyme expression in *Lactococcus lactis*. Appl Microbiol Biotechnol 37:216–224.

Vanderhaegen B, Neven H, Coghe S, Verstrepen KJ, Derdelinckx, G, Verachtert H. 2003. Bioflavoring and beer refermentation. Appl Microbiol Biotechnol 62:140–150.

Vasil IK. 2003. The science and politics of plant biotechnology—A personal perspective. Nature Biotechnology 21:849–851.

Wall RJ. 2002. Pronuclear microinjection. Cloning Stem Cells 3:193–204.

Wall RJ, Kerr DE, Bondioli KR. 1997. Transgenic dairy cattle: Genetic engineering on a large scale. J Dairy Sci 80:2213–2224.

Walter T, De Anraca I, Chadud P, Perales CG. 1986. Iron deficiency anemia: Adverse effects on infant psychomotor development. Pediatrics 84:7–17.

Wang H-J, Murphy PA. 1996. Mass balance study of isoflavones during soybean processing. J Agric Food Chem 44:2377–2383.

Washko PW, Welch RW, Dhariwal KR, Wang Y, Levine M. 1992. Ascorbic acid and dehydroascorbic acid analyses in biological samples. Anal Biochem 204: 1–14.

Wheeler G, Jones M, Smirnoff N. 1998. The biosynthetic pathway of vitamin C in higher plants. Nature 393:365–369.

Whitcombe D, Theaker J, Guy SP, Brown T, Little S. 1999. Detection of PCR products using self-probing amplicons and fluorescence. Nature Biotechnology 17:804–807.

WHO/UNICEF/IVACG Task Force. 1997. Vitamin A Supplements: A Guide to Their Use in the Treatment and Prevention of Vitamin A Deficiency and Xerophthalmia, 2nd edition. Geneva, Switzerland.

Willmitzer L, Depicker A, Dhaese P, De Greve H, Hernalsteens JP, Holsters M, Leemans J, Otten L, Schroder J, Schroder G, Zambryski P, van Montagu M, Schell J. 1983. The use of Ti-plasmids as plant-directed gene vectors. Folia Biol 29:106–114.

Wittwer CT, Hermann MG, Moss AA, Rasmussen RP. 1997. Continuous fluorescence monitoring of rapid cycle DNA amplification. Biotechniques 22:130–138.

Wolf DP, Mitalipov S, Norgren Jr, RB. 2001. Nuclear transfer technology in mammalian cloning. Archives of Medical Research 32:609–613.

World Health Organization (WHO). 1992. National strategies for overcoming micronutrient malnutrition. Document A45/3 Geneva, Switzerland.

Wyss M, Pasamontes L, Remy R, Kohler J, Kusznir E, Gadient M, Muller F, van Loon APGM. 1998. Comparison of the thermostability properties of three acid phosphatases from molds: *Aspergillus fumigatus* phytase, *A. niger* phytase, and *A. niger* pH 2.4 acid phosphatase. Appl Environ Microbiol 64:4446–4451.

Yang J, Ratovitski T, Brady JP, Solomon MB, Wells KD, Wall RJ. 2001. Expression of myostatin pro domain results in muscular transgenic mice. Mol Repro Dev 60:351–361.

Ye X, Al-Babili S, Klöti A, Zhang J, Lucca P, Beyer P, Potrykus I. 2000. Engineering provitamin A (β-carotene) biosynthetic pathway into (carotenoid-free) rice endosperm. Science 287:303–305.

Zambryski P. 1988. Basic process underlying *Agrobacterium*-mediated DNA transfer to plant cells. Annu Rev Gent 22:1–30.

Zbikowska HM. 2003. Fish can be first—Advances in fish transgenesis for commercial applications. Transgenic Research 12:379–389.

Zheng Z, Sumi K, Tanaka K, Murai N. 1995. The bean seed storage protein β-phaseolin is synthesized, processed, and accumulated in the vacuolar type-II protein bodies of transgenic rice endosperm. Plant Physiol 109:777–786.

4
Browning Reactions

M. Villamiel, M. D. del Castillo, and N. Corzo

INTRODUCTION

Browning reactions are some of the most important phenomena occurring in food during processing and storage. They represent an interesting research for the implications in food stability and technology as well as in nutrition and health. The major groups of reactions leading to browning are enzymatic phenol oxidation and so-called nonenzymatic browning (Manzocco et al. 2001).

ENZYMATIC BROWNING

Enzymatic browning is one of the most important color reactions that affect fruits, vegetables, and seafood. It is catalyzed by the enzyme polyphenol oxidase (1,2 benzenediol; oxygen oxidoreductase, EC 1.10.3.1), which is also referred to as phenoloxidase, phenolase, monophenol oxidase, diphenol oxidase, and tyrosinase. Phenoloxidase enzymes catalyze the oxidation of phenolic constituents to quinones, which finally polymerize to colored melanoidins (Marshall et al. 2000).

Enzymatic browning reactions may affect fruits, vegetables, and seafood in either positive or negative ways. These reactions, for instance, may contribute to the overall acceptability of foods such as tea, coffee, cocoa, and dried fruits (raisins, prunes, dates, and figs). Products of enzymatic browning play key physiological roles. Melanoidins, produced as a consequence of polyphenol oxidase activity, may exhibit antibacterial, antifungal, anticancer, and antioxidant properties. Polyphenol oxidases impart remarkable physiological functions for developing of aquatic organisms, such as wound healing and hardening of the shell (sclerotization) after molting in insects and in crustaceans such as shrimp and lobster. The mechanism of wound healing in aquatic organisms is similar to that which occurs in plants in that the compounds produced as a result of the polymerization of quinones, melanins or melanoidins, exhibit both antibacterial and antifungal activities. In addition, enzymatic reactions are considered desirable during fermentation (Fennema 1976). Despite these positive effects, enzymatic browning is considered one of the most devastating reactions for many exotic fruits and vegetables, in particular tropical and subtropical varieties. Enzymatic browning is especially undesirable during processing of fruit slices and juices. Lettuce and other green leafy vegetables; potatoes and other starchy staples such as sweet potato, breadfruit, and yam; mushrooms; apples; avocados; bananas; grapes; olive (Sciancalepore 1985); peaches; pears (Vamosvigyazo and Nadudvari-markus 1982); and a variety of other tropical and subtropical fruits and vegetables are susceptible to browning. Crustaceans are also extremely vulnerable to enzymatic browning. Since enzymatic browning can affect the color, flavor, and nutritional value of these foods, it can cause tremendous economic losses (Marshall et al. 2000).

A better understanding of the mechanism of enzymatic browning in fruits, vegetables, and seafood; the properties of the enzymes involved; and their substrates and inhibitors may be helpful for controlling browning development, avoiding economic losses, and providing high quality foods. Based on this knowledge, new approaches for the control of enzymatic browning have been proposed. These subjects will be reviewed in this chapter.

Properties of Polyphenol Oxidase (PPO)

PPO (EC 1.10.3.1; *o*-diphenol oxidoreductase) is an oxidoreductase able to oxidize phenol compounds employing oxygen as a hydrogen acceptor. The abundance of phenolics in plants may be the reason for naming this enzyme polyphenol oxidase (Marshall et al. 2000). The molecular weight for polyphenol oxidase in plants ranges between 57 and 62 kDa (Hunt et al. 1993, Newman et al. 1993). Polyphenol oxidase catalyzes two basic reactions: (1) hydroxylation to the *o*-position adjacent to an existing hydroxyl group of the phenolic substrate (monophenol oxidase activity) and (2) oxidation of diphenol to *o*-benzoquinones (diphenol oxidase activity) (Fig. 4.1).

In plants, the ratio of monophenol to diphenol oxidase activity is usually in the range of 1:40 to 1:10 (Nicolas et al. 1994). Monophenol oxidase activity is considered more relevant for insect and crustacean systems due to its physiological significance in conjunction with diphenolase activity (Marshall et al. 2000).

Polyphenol oxidase is also referred to as tyrosinase to describe both monophenol and diphenol oxidase activities in either animals or plants. The monophenol oxidase acting in plants is also called cresolase due to its ability to employ cresol as a substrate (Marshall et al. 2000). This enzyme is able to metabolize aromatic amines and *o*-aminophenols (Toussaint and Lerch 1987).

The oxidation of diphenolic substrates to quinones in the presence of oxygen is catalyzed by diphenol

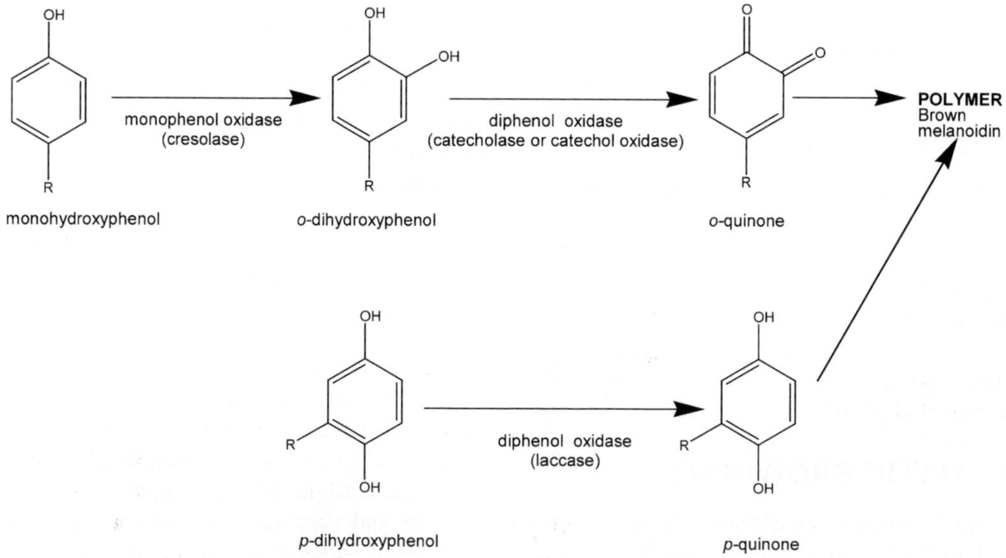

Figure 4.1. Polyphenol oxidase pathway.

oxidase activity (Fig. 4.1). The p-diphenol oxidase–catalyzed reaction follows a Bi Bi mechanism (Whitaker 1972). Diphenolase activity may be due to two different enzymes: catecholase (catechol oxidase) and laccase (Fig. 4.1). Laccase (p-diphenol oxidase, EC 1.10.3.2)(DPO) is a type of copper-containing polyphenol oxidase. It has the unique ability to oxidize p-diphenols. Phenolic substrates, including polyphenols, methoxy-substituted phenols, diamines, and a considerable range of other compounds, serve as substrates for laccase (Marshall et al. 2000). Laccases occur in many phytopathogenic fungi, higher plants (Mayer and Harel 1991), peaches (Harel et al. 1970), and apricots (Dijkstra and Walker 1991).

Catechol oxidase and laccase are distinguishable on the basis of both their phenolic substrates (Fig. 4.1) and their inhibitor specificities (Marshall et al. 2000).

Catecholase activity is more important than cresolase action in food because most of the phenolic substrates in food are dihydroxyphenols (Mathew and Parpia 1971).

This bifunctional enzyme, polyphenol oxidase (PPO), containing copper in its structure has been described as an oxygen and four-electron transferring phenol oxidase (Jolley et al. 1974). Figure 4.2 shows a simplified mechanism for the hydroxylation and oxidation of phenols by PPO. Both mechanisms involve the two copper moieties on the PPO.

Polyphenol oxidase, active between pH 5 and 7, does not have a very sharp pH optimum. At lower pH values of approximately 3, the enzyme is irreversibly inactivated. Reagents that complex or remove copper from the prosthetic group of the enzyme inactivate the enzyme (Fennema 1976).

Substrates

Although tyrosine is the major substrate for certain phenolases, other phenolic compounds such as caffeic acid and chlorogenic acid also serve as substrates (Fennema 1976). Structurally, they contain an aromatic ring bearing one or more hydroxyl groups, together with a number of other substituents

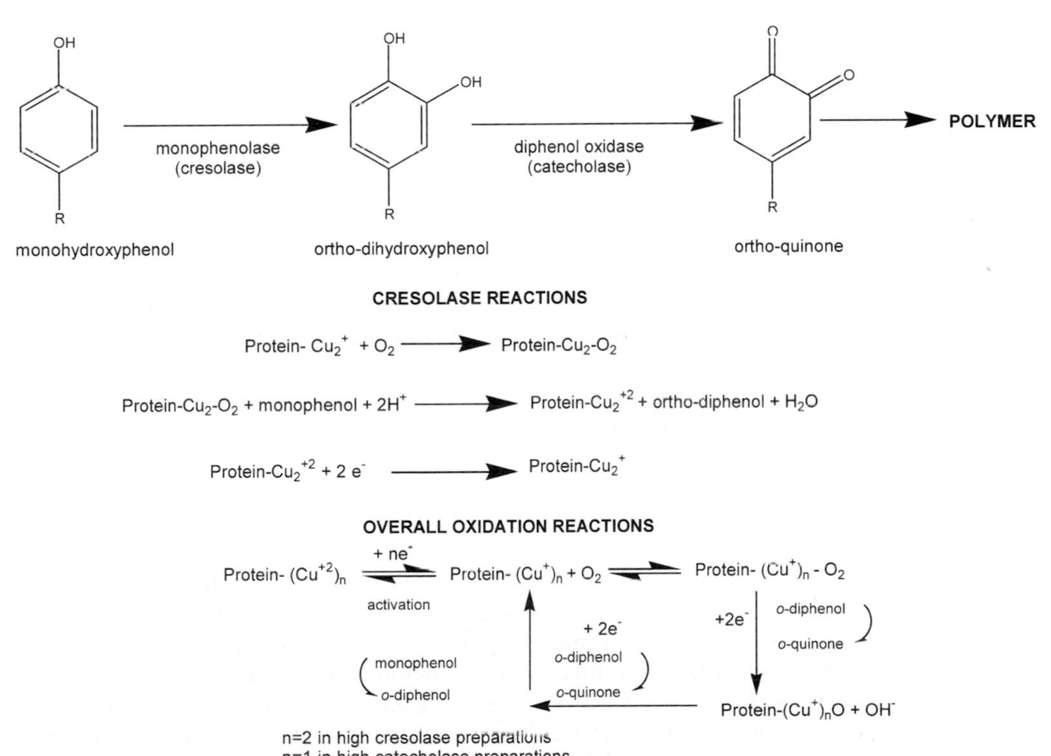

Figure 4.2. Simplified mechanism for the hydroxylation and oxidation of diphenol by phenoloxidase.

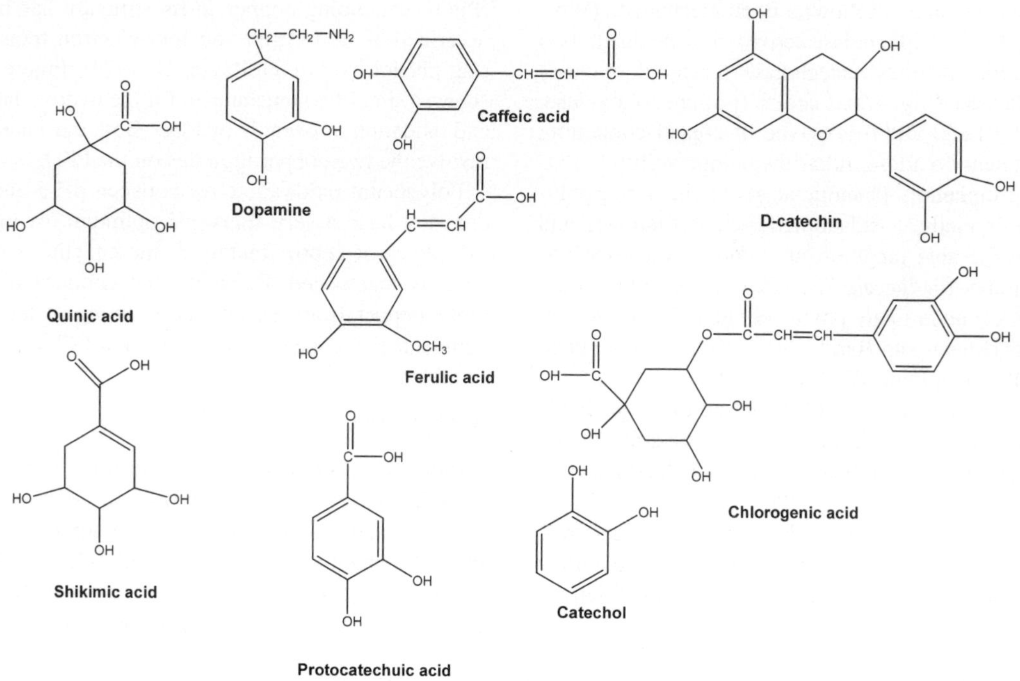

Figure 4.3. Structure of common phenols present in foods Maillard reaction.

(Marshall et al. 2000). Structures of common phenolic compounds present in foods are shown in Figure 4.3. Phenolic subtrates of PPO in fruits, vegetables, and seafood are listed in Table 4.1. The substrate specificity of PPO varies in accordance with the source of the enzyme. Phenolic compounds and polyphenol oxidase are, in general, directly responsible for enzymatic browning reactions in damaged fruits during postharvest handling and processing. The relationship of the rate of browning to phenolic content and PPO activity has been reported for various fruits. In addition to serving as PPO substrates, phenolic compounds act as inhibitors of PPOs (Marshall et al. 2000).

Control of Browning

Enzymatic browning may cause a decrease in the market value of food products originating from plants and crustaceans (Perez-Gilabert and García-Carmona 2000, Kubo et al. 2000, Subaric et al. 2001). Processing such as cutting, peeling, and bruising is enough to cause enzymatic browning. The rate of enzymatic browning is governed by the active polyphenol oxidase content of the tissues; the phenolic content of the tissue and the pH, temperature, and oxygen availability within the tissue. Consequently, the methods addressed to inhibiting the undesired browning are focused on either inhibiting or preventing PPO activity in foods. The methods are predicated on eliminating from the reaction one or more essential components (oxygen, enzyme, copper, or substrates). Numerous techniques have been applied to preventing enzymatic browning. Table 4.2 gives a list of procedures and inhibitors that may be employed for controlling enzymatic browning in foods. According to Marshall et al. (2000), there are six categories of PPO inhibitors applicable to control of enzymatic browning: reducing agents, acidulants, chelating agents, complexing agents, enzyme inhibitors, and enzyme treatments. The inhibition of enzymatic browning generally proceeds via direct inhibition of the PPO, nonenzymatic reduction of *o*-quinones, and chemical modification or removal of phenolic substrates of polyphenol oxidase.

Table 4.1. Phenolic Substrates of PPO in Foods

Source	Phenolic Substrates
Apple	Chlorogenic acid (flesh), catechol, catechin (peel), caffeic acid, 3,4-dihydroxyphenylalanine (DOPA), 3,4-dihydroxy benzoic acid, p-cresol, 4-methyl catechol, leucocyanidin, p-coumaric acid, flavonol glycosides
Apricot	Isochlorogenic acid, caffeic acid, 4-methyl catechol, chlorogenic acid, catechin, epicatechin, pyrogallol, catechol, flavonols, p-coumaric acid derivatives
Avocado	4-methyl catechol, dopamine, pyrogallol, catechol, chlorogenic acid, caffeic acid, DOPA
Banana	3,4-dihydroxyphenylethylamine (Dopamine), leucodelphinidin, leucocyanidin
Cacao	Catechins, leucoanthocyanidins, anthocyanins, complex tannins
Coffee beans	Chlorogenic acid, caffeic acid
Eggplant	Chlorogenic acid, caffeic acid, coumaric acid, cinnamic acid derivatives
Grape	Catechin, chlorogenic acid, catechol, caffeic acid, DOPA, tannins, flavonols, protocatechuic acid, resorcinol, hydroquinone, phenol
Lettuce	Tyrosine, caffeic acid, chlorogenic acid derivatives
Lobster	Tyrosine
Mango	Dopamine-HCl, 4-methyl catechol, caffeic acid, catechol, catechin, chlorogenic acid, tyrosine, DOPA, p-cresol
Mushroom	Tyrosine, catechol, DOPA, dopamine, adrenaline, noradrenaline
Peach	Chlorogenic acid, pyrogallol, 4-methyl catechol, catechol, caffeic acid, gallic acid, catechin, Dopamine
Pear	Chlorogenic acid, catechol, catechin, caffeic acid, DOPA, 3,4-dihydroxy benzoic acid, p-cresol
Plum	Chlorogenic acid, catechin, caffeic acid, catechol, DOPA
Potato	Chlorogenic acid, caffeic acid, catechol, DOPA, p-cresol, p-hydroxyphenyl propionic acid, p-hydroxyphenyl pyruvic acid, m-cresol
Shrimp	Tyrosine
Sweet potato	Chlorogenic acid, caffeic acid, caffeylamide
Tea	Flavanols, catechins, tannins, cinnamic acid derivatives

Source: Reproduced from Marshall et al. 2000.

Sulfites are the most efficient multifunctional agents in the control of enzymatic browning of foods. The use of sulfites has become increasingly restricted because they have been considered a cause of severe reactions in asthmatics. The best alternative to sulfite in control of browning is L-ascorbic acid or its stereoisomer erythorbic acid; 4-hexylresorcinol is also a good substitute for sulfite. Sulfhydryl amino acids such as cysteine and reduced glutathione, and inorganic salts like sodium chloride, kojic acid, and oxalic acid (Pilizota and Subaric 1998, Son et al. 2000, Burdock et al. 2001) cause an effective decrease in undesirable enzymatic browning in foods. A decrease of enzymatic browning is achieved by use of various chelating agents, which either directly form complexes with polyphenol oxidase or react with its substrates. P-cyclodextrin, or example, is an inhibitor that reacts with the copper-containing prosthetic group of PPO, forming an inclusion complex (Pilizota and Subaric 1998).

Thiol compounds like 2-mercaptoethanol may act as an inhibitor of the polymerization of *o*-quinone and as a reductant involved in the conversion of *o*-quinone to *o*-dihydroxyphenol (Negishi and Ozawa 2000).

Crown compounds have the potential to reduce the enzymatic browning caused by catechol oxidase due to their ability to complex with the copper present in its prosthetic group. The inhibition effect of crown compounds, macrocyclic esters, benz18-crown-6 with sorbic acid, and benz18-crown-6 with potassium sorbate has been proved for fresh-cut apples (Subaric et al. 2001).

Because of safety regulations and the consumer's demand for natural food additives, much research has been devoted to the search for natural and safe

Table 4.2. Inhibitors and Processes Employed in the Prevention of Enzymic Browning

Inhibition Targeted toward the Enzyme		
Processing		*Enzymes Inhibitors*
Heating 1) Steam and water blanching (70–105°C) 2) Pasteurization (60–85°C)		Chelating agents 1) Sodium azide 2) Cyanide 3) Carbon monoxide 4) Halide salts ($CaCl_2$, NaCl) 5) Tropolone 6) Ascorbic acid 7) Sorbic acid 8) Polycarboxylic acids (citric, malic, tartaric, oxalic and succinic acids) 9) Polyphosphates (ATP and pyrophosphate) 10) Macromolecules (porphyrins, proteins, polysaccharides) 11) EDTA 12) Kojic acid
Cooling 1) Refrigeration 2) Freezing (−18°C) Dehydration		Aromatic carboxylic acids 1) Benzoic acids 2) Cinnamic acids Aliphatic alcohol
Physical methods Freeze-drying Spray drying Radiative drying Solar drying Microwave drying	*Chemical methods* Sodium chloride and other salts Sucrose and other sugars Other sugar Glycerol Propylene glycol Modified corn syrup	Peptides and Amino acids
Irradiation 1) Gamma rays up to 1 kGy (Cobalt 60 or Cesium 137) 2) X-rays 3) Electron beams 4) Combined treatments using irradiation and heat		Substituted resorcinols
High pressure (600–900 Mpa) Supercritical carbon dioxide (58 atm, 43°C) Ultrafiltration		Honey (peptide ~ 600 Da and antioxidants) Proteases Acidulants Citric acid (0.5–2% w/v) Malic acid Phosphoric acid
Ultrasonication Employment of edible coating		Chitosan

Source: Adapted from Marshall et al. 2000.

Table 4.2. (Continued)

Inhibition Targeted toward the Substrate		Inhibition Targeted toward the Products
Removal of Oxygen	Removal of Phenols	
Processing 1) Vaccum treatment 2) Immersion in water, syrup, brine	Complexing agents 1) Cyclodextrins 2) Sulphate polysaccharides 3) Chitosan	Reducing agents 1) Sulphites ($SO_2, SO_3^{2-}, HSO_3^-, S_2O_5^{2-}$). 2) Acorbic acid and analogs 3) Cysteine and other thiol compounds
Reducing agents 1) Ascorbic acid 2) Erythorbic acid 3) Butylated hydroxyanisole (BHA) 4) Butylated hydroxytoluene (BTH) 5) Tertiarybutyl hydroxyquinone 6) Propyl gallate	Enzymatic modification 1) *O*-methyltransferase 2) Protocatechuate 3, 4-dioxigenase	Amino acids, peptides and proteins Chitosan Maltol

antibrowning agents. Honey (Chen et al. 2000); papaya latex extract (De Rigal et al. 2001); banana leaf extract, either alone or in combination with ascorbic acid and 4-hexylresorcinol (Kaur and Kapoor 2000); onion juice (Hosoda and Iwahashi 2002); onion oil (Hosoda et al. 2003); solutions containing citric acid, calcium chloride, and garlic extract (Ihl et al. 2003); Maillard reaction products obtained by heating of hexoses in presence of cysteine or glutathione (Billaud et al. 2003, Billaud et al. 2004); resveratrol, a natural ingredient of red wine possessing several biological activities, and other hydroxystilbene compounds including its analog oxyresveratrol (Kim et al. 2002); and hexanal (Corbo et al. 2000) are some examples of natural inhibitors of PPO. In most cases, the inhibiting activity of a plant extract is due to more than one component. Moreover, a good control of enzymatic browning may involve endogenous antioxidants (Mdluli and Owusu-Apenten 2003).

Regulation of the biosynthesis of polyphenols (Hisaminato et al. 2001) and the use of a commercial glucose oxidase–catalase enzyme system for oxygen removal (Parpinello et al. 2002) have been described as essential and effective ways of controlling enzymatic browning.

Commonly, an effective control of enzymatic browning can be achieved by a combination of antibrowning agents. A typical combination might consist of a chemical reducing agent such as ascorbic acid, an acidulant such as citric acid, and a chelating agent like EDTA (ethylenediaminetetraacetic acid) (Marshall et al. 2000).

A great emphasis is put on research to develop methods for preventing enzymatic browning, especially in fresh-cut (minimally processed) fruits and vegetables. Technological processing including microwave blanching alone or in combination with chemical antibrowning agents (Severini et al. 2001, Premakumar and Khurduya 2002), CO_2 treatments (Rocha and Morais 2001, Kaaber et al. 2002), pretreatments employing sodium or calcium chloride and lactic acid followed by conventional blanching (Severini et al. 2003), high-pressure treatments combined with thermal treatments, and chemical antibrowning agents such as ascorbic acid (Prestamo et al. 2000, Ballestra et al. 2002) have been employed to prevent enzymatic browning in foods.

The use of edible whey protein isolate–bees wax coating (Perez-Gago et al. 2003), a high-oxygen atmosphere (e.g., $> 70\%$ O_2) (Jacxsens et al. 2001), or an atmosphere of 90.5% N_2 + 7% CO_2 + 2.5%

O_2 (Soliva-Fortuny et al. 2001) seem to be adequate approaches to improving the shelf life of foods by inhibition of PPO.

Within the emergent tools for the control of PPO, the application of genetic techniques should be mentioned. Transgenic fruits carrying an antisense PPO gene show a reduction in the amount and activity of PPO, and the browning potential of transgenic lines is reduced compared with the nontransgenic ones (Murata et al. 2000, 2001). These procedures may be used to prevent enzymatic browning in a wide variety of food crops without the application of various food additives (Coetzer et al. 2001).

NONENZYMATIC BROWNING

The Maillard Reaction

Nonenzymatic browning is the most complex reaction in food chemistry due to the large number of food components able to participate in the reaction through different pathways, giving rise to a complex mixture of products (Olano and Martínez-Castro 1996). It is referred to as the Maillard reaction when it takes place between free amino groups from amino acids, peptides, or proteins and the carbonyl group of a reducing sugar.

The Maillard reaction is one of the main reactions causing deterioration of proteins during processing and storage of foods. This reaction can promote nutritional changes such as loss of nutritional quality (attributed to the destruction of essential amino acids) or reduction of protein digestibility and amino acid availability (Malec et al. 2002).

The Maillard reaction covers a whole range of complex transformations (Fig. 4.4) that lead to the formation of numerous volatile and nonvolatile compounds. It can be divided into three major phases, the early, intermediate, and advanced stages. The early stage (Fig. 4.5) consists of the condensation of primary amino groups of amino acids, peptides, or

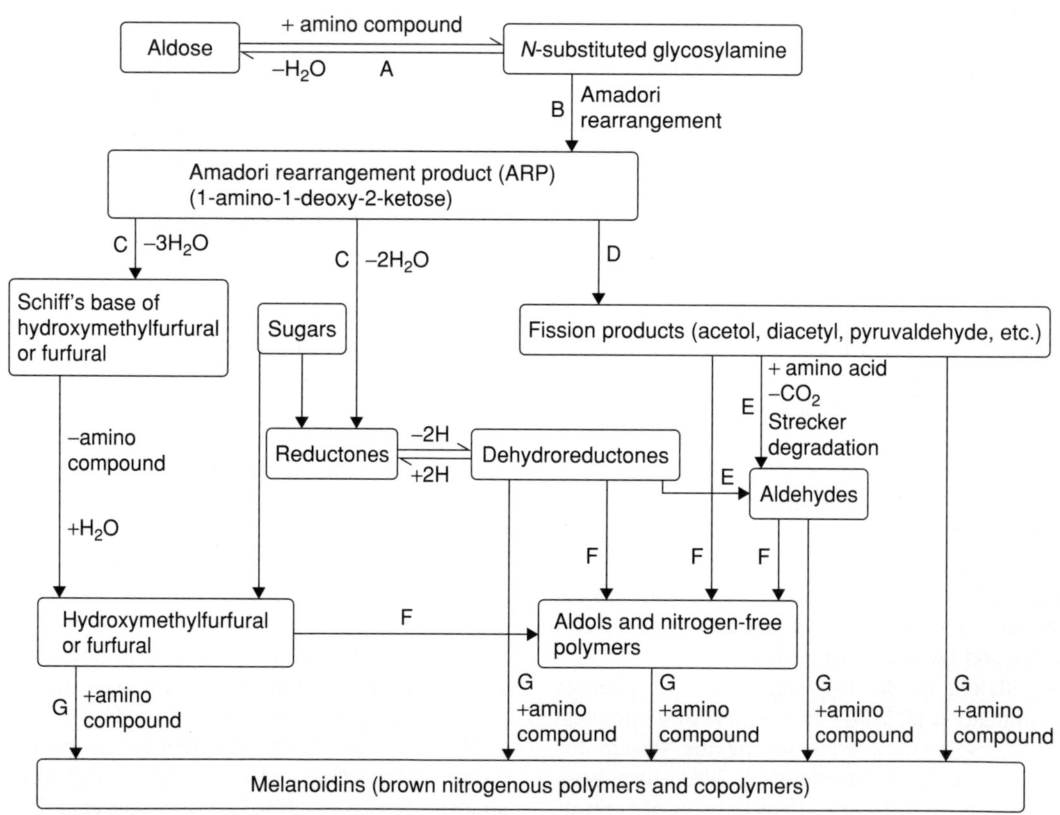

Figure 4.4. Scheme of different stages of Maillard reaction (Hodge 1953, Ames 1990).

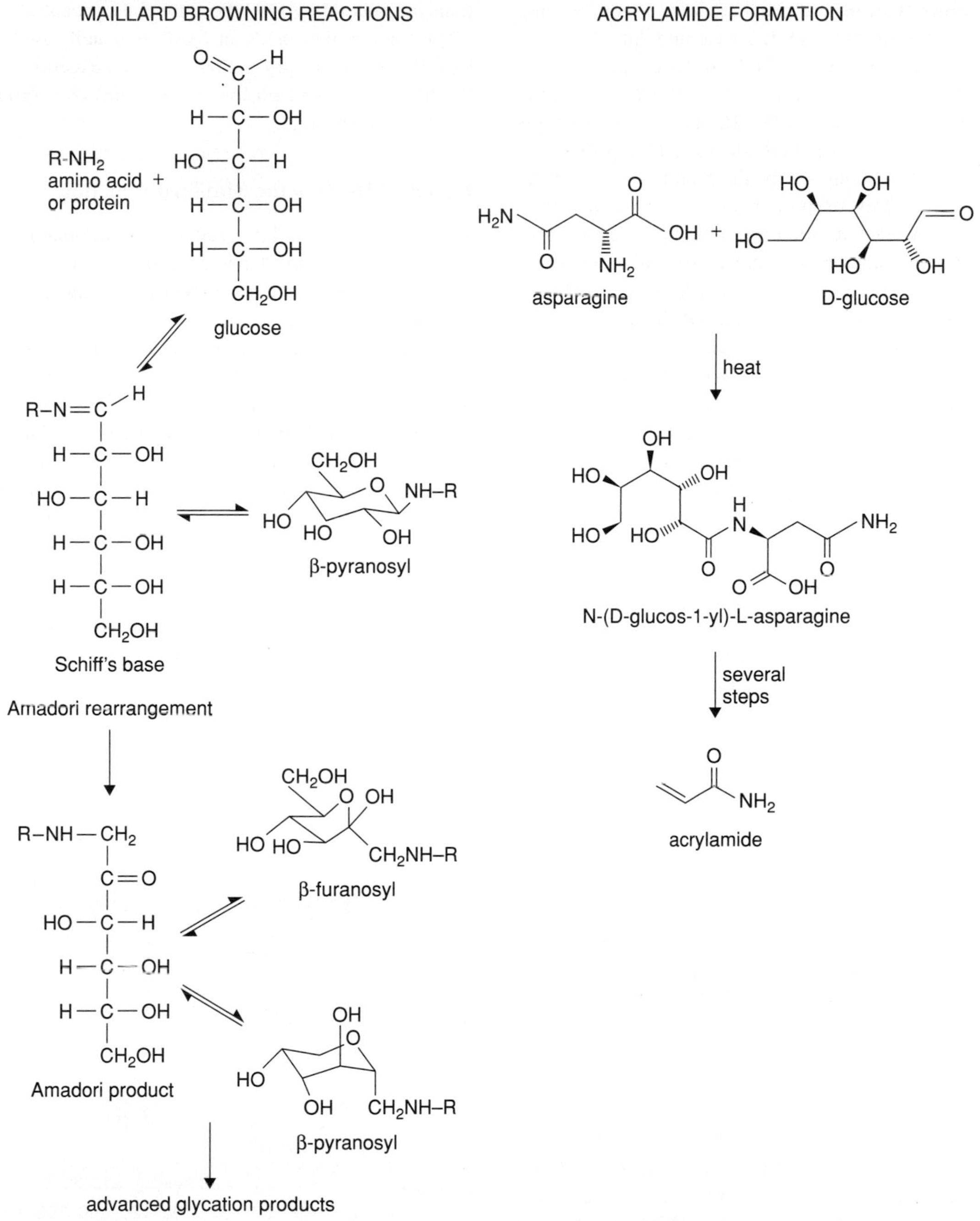

Figure 4.5. Initial stages of the Maillard reaction and acrylamide formation (Friedman 2003).

proteins with the carbonyl group of reducing sugars (aldose), with loss of a molecule of water, leading, via formation of a Schiff's base and Amadori rearrangement, to the scalled Amadori product (1-amin1-deoxy-2-ketose), a relatively stable intermediate (Feather et al. 1995). The Heyns compound is the analogous compound when a ketose is the starting sugar. In many foods, the ε-amino group of the lysine residues of proteins is the most important source of reactive amino groups, but due to blockage these lysine residues are not available for digestion, and consequently the nutritive value decreases (Brands and van Boekel 2001, Machiels and Istasse 2002). Amadori compounds are precursors of numerous compounds important in the formation of characteristic flavors, aromas, and brown polymers. They are formed before the occurrence of sensory changes; therefore, their determination provides a very sensitive indicator for early detection of quality changes caused by the Maillard reaction (Olano and Martínez-Castro 1996).

The intermediate stage leads to breakdown of Amadori compounds (or other products related to the Schiff's base) and the formation of degradation products, reactive intermediates (3-deoxyglucosone), and volatile compounds (formation of flavor). The 3-deoxyglucosone participates in cross-linking of proteins at much faster rates than glucose itself, and further degradation leads to two known advanced products: 5-hydroxymethyl-2-furaldehyde and pyrraline (Feather et al. 1995).

The final stage is characterized by the production of nitrogen-containing brown polymers and copolymers known as melanoidins (Badoud et al. 1995). The structure of melanoidins is largely unknown, and to date, there are several proposals about it. Melanoidins have been described as low molecular weight (LMW) colored substances that are able to cross-link proteins via ε-amino groups of lysine or arginine to produce high molecular weight (HMW) colored melanoidins. Also, it has been postulated that they are polymers consisting of repeating units of furans and/or pyrroles formed during the advanced stages of The Maillard reaction and linked by polycondensation reactions (Martins and van Boekel 2003).

In foods, predominantly glucose, fructose, maltose, lactose, and to some extent reducing pentoses are involved with amino acids and proteins in forming fructoselysine, lactuloselysine, or maltuloselysine. In general, primary amines are more important than the secondary ones because the concentration of primary amino acids in foods is usually higher than that of secondary amino acids (an exception is the high amount of proline in malt and corn products) (Ledl 1990).

Factors Affecting the Maillard Reaction

The rate of the Maillard reaction and the nature of the products formed depend on the chemical environment of food including the water activity (a_w), pH, and chemical composition of the food system, temperature being the most important factor (Carabasa-Giribert and Ibarz-Ribas 2000). In order to predict the extent of chemical reactions in processed foods, a knowledge of kinetic reactions is necessary to optimize the processing conditions. Since foods are complex matrices, these kinetic studies are often carried out using model systems in which sugars and amino acids react under simplified conditions. Model system studies may provide guidance regarding the directions in which to modify the food process and to find out which reactants may produce specific effects of the Maillard reaction (Lingnert 1990).

The reaction rate is significantly affected by the pH of the system; it generally increases with pH (Namiki et al. 1993, Ajandouz and Puigserver 1999).

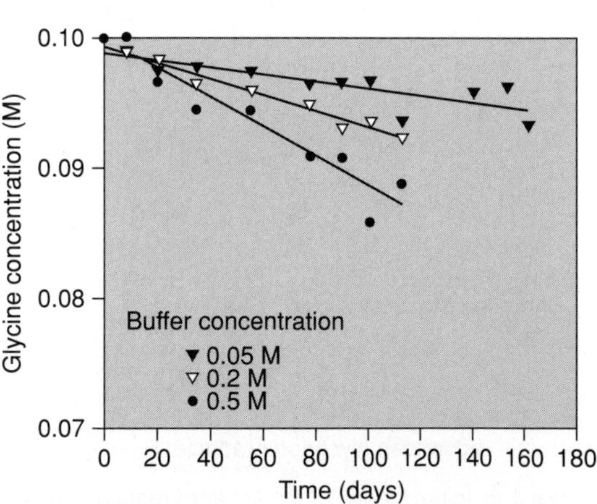

Figure 4.6. Effect of phosphate buffer concentration on the loss of glycine in 0.1M glucose/glycine solutions at pH 7 and 25°C (Bell 1997).

Bell (1997) studied the effect of buffer type and concentration on initial degradation of amino acids and formation of brown pigments in model systems of glycine and glucose stored for long periods at 25°C. The loss of glycine was faster at high phosphate buffer concentrations (Fig. 4.6), showing the catalytic effect of the phosphate buffer concentration on the Maillard reaction.

The type of reducing sugar has a great influence on Maillard reaction development. Pentoses (e.g., ribose) react more readily than hexoses (e.g., glucose), which, in turn, are more reactive than disaccharides (e.g., lactose) (Ames 1990). A study on brown development (absorbance 420 nm) in a heated model of fructose and lysine showed that browning was higher than in model systems with glucose (Ajandouz et al. 2001).

Participation of amino acids in the Maillard reaction is variable; lysine was the most reactive amino acid (Fig. 4.7) in the heated model system of glucose and lysine, threonine, and methionine in buffer phosphate at different pH values (4–12) (Ajandouz and Puigserver 1999). The influence of type of amino acid and sugar in the Maillard reaction development also has been studied more recently (Carabasa-Giribet and Ibarz-Ribas 2000, Mundt and Wedzicha 2003).

Studies on the effect of time and temperature of treatment on Maillard reaction development have been also conducted in different model systems, and it has been shown that an increase in temperature increases the rate of Maillard browning (Ryu et al. 2003, Martins and van Boekel 2003).

Concentration and ratio of reducing sugar to amino acid have a significant impact on the reaction. Browning reaction increased with increasing glycine:glucose ratios in the range 0.1:1 to 5:1 in a model orange juice system at 65°C (Wolfrom et al. 1974). In a model system of intermediate moisture (a_w, 0.52), Warmbier et al. (1976) observed an increase of browning reaction rate when the molar ratio of glucose to lysine increased from 0.5:1 to 3.0:1.

Water activity (a_w) is another important factor influencing Maillard reaction development; thus, this reaction occurs less readily in foods with high a_w values. At high a_w values, reactants are diluted, while at low a_w values the mobility of reactants is

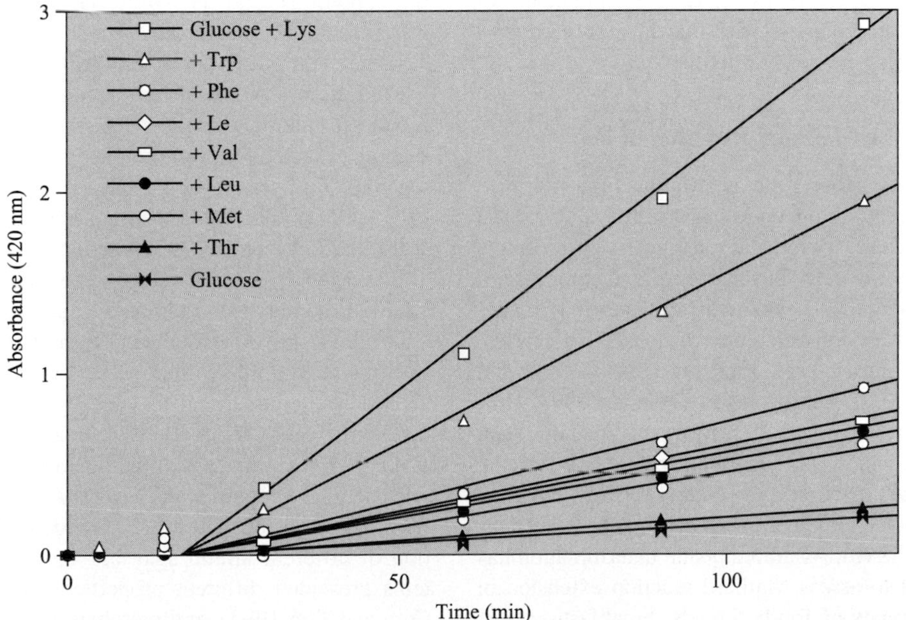

Figure 4.7. Brown color development in aqueous solutions containing glucose alone or in the presence of an essential amino acid when heated to 100°C at pH 7.5 as a function of time (Ajandouz and Puigserver 1999).

limited, despite their presence at increased concentrations (Ames 1990). Numerous studies have demonstrated a browning rate maximum at a_w values from 0.5 to 0.8 in dried and intermediate-moisture foods (Warmbier et al. 1976, Tsai et al. 1991, Buera and Karel 1995).

Due to the complex composition of foods, it is unlikely that the Maillard reaction involves only single compounds (mono- or disaccharides and amino acids). For this reason, several studies on factors (pH, T, a_w) that influence the Maillard reaction development have been carried out using more complex model systems: heated starch-glucose-lysine systems (Bates et al. 1998), milk-resembling model systems (lactose or glucose-caseinate systems) (Morales and van Boekel 1998), and a lactose-casein model system (Malec et al. 2002). Brands and van Boekel (2001) studied the Maillard reaction using heated monosaccharide (glucose, galactose, fructose, and tagatose)-casein model systems to quantify and identify the main reaction products and to establish the reaction pathways.

Studies on mechanisms of degradation, via the Maillard reaction, of oligosaccharides in a model system with glycine were performed by Hollnagel and Kroh (2000, 2002). The reactivity of di- and trisaccharides under quasi water-free conditions decreased in comparison with that of glucose due to the increasing degree of polymerization.

Study of the Maillard Reaction in Foods

During food processing, the Maillard reaction produces desirable and undesirable effects. Processes such as baking, frying, and roasting are based on the Maillard reaction for flavor, aroma, and color formation (Lignert 1990). Maillard browning may be desirable during manufacture of meat, coffee, tea, chocolate, nuts, potato chips, crackers, and beer and in toasting and baking bread (Weenen 1998, Burdulu and Karadeniz 2003). In other processes such as pasteurization, sterilization, drying, and storage, the Maillard reaction often causes detrimental nutritional (lysine damage) and organoleptic changes (Lingnert 1990). Available lysine determination has been used to assess Maillard reaction extension in different types of foods: breads, breakfast cereals, pasta, infant formula (dried and sterilized), and so on (Erbersdobler and Hupe 1991); dried milks (El and Kavas 1997); heated milks (Ferrer et al. 2003); and infant cereals (Ramírez-Jimenez et al. 2004).

Sensory changes in foods due to the Maillard reaction have been studied in a wide range of foods including honey (Gonzales et al. 1999), apple juice concentrate (Burdulu and Karadeniz 2003), and white chocolate (Vercet 2003).

Other types of undesirable effects produced in processed foods by Maillard reaction may include the formation of mutagenic and cancerogenic compounds (Lingnert 1990, Chevalier et al. 2001). Frying or grilling of meat and fish may generate low (ppb) levels of mutagenic/carcinogenic heterocyclic amines via Maillard reaction. The formation of these compounds depends on cooking temperature and time, cooking technique and equipment, heat, mass transport, and/or chemical parameters. Recently, Tareke et al. (2002) reported their findings on the carcinogen acrylamide in a range of cooked foods. Moderate levels of acrylamide (5–50 μg/kg) were measured in heated protein-rich foods, and higher levels (150–4000 μg/kg) were measured in carbohydrate-rich foods such a potato, beet root, certain heated commercial potato products, and crisp bread. Ahn et al. (2002) tested different types of commercial foods and some foods heated under home cooking conditions, and they observed that acrylamide was absent in raw or boiled foods, but it was present at significant levels in fried, grilled, baked, and toasted foods. Although the mechanism of acrylamide formation in heated foods is not yet clear, several authors have put forth the hypothesis that the reaction of asparagine (Fig. 4.5), a major amino acid of potatoes and cereals (Mottram et al. 2002, Weisshaar and Gutsche 2002), or methionine (Stadler et al. 2002) with reducing sugars (glucose, fructose) via Maillard reaction could be the pathway. In 2003, a lot of research was conducted to study the mechanism of acrylamide formation, to develop sensible analytical methods, and to quantify acrylamide in different types of foods (Becalski et al. 2003, Jung et al. 2003, Roach et al. 2003, Yasuhara et al. 2003).

Beneficial properties of Maillard products have been also described. Resultant products of the reaction of different amino acid and sugar model systems presented different properties: antimutagenic (Yen and Tsai 1993), antimicrobial (Chevalier et al. 2001), and antioxidative (Manzocco et al. 2001, Wagner et al. 2002). In foods, antioxidant effects of

Maillard reaction products have been found in honey (Antoni et al. 2000) and in tomato purees (Anese et al. 2002).

Control of the Maillard Reaction in Foods

For a food technologist, one of the most important objectives must be to limit nutritional damage of food during processing. In this sense, many studies have been performed find useful heat-induced markers derived from the Maillard reaction, and most of them have been proposed to control and check the heat treatments and/or storage of foods. There are many indicators of different stages of the Maillard reaction, but this review cites one of the most recent indicators proposed to control the early stages of the Maillard reaction during food processing: the 2-furoylmethyl amino acids as an indirect measure of Amadori compound formation.

Determination of the level of Amadori compounds provides a very sensitive indicator for early detection (before detrimental changes occur) of quality changes caused by the Maillard reaction as well as a retrospective assessment of the heat treatment or storage conditions to which a product has been subjected (Olano and Martínez-Castro 1996, del Castillo et al. 1999).

Evaluating for Amadori compounds can be carried out through furosine [ε-N-(2-furoylmethyl)-L-lysine] measurement. This amino acid is formed by acid hydrolysis of the Amadori compound ε-N-(1-deoxy-D-fructosyl)-L-lysine. It is considered a useful indicator of the damage in processed foods or foods stored for long periods: milks (Resmini et al. 1990, Villamiel et al. 1999), eggs (Hidalgo et al. 1995), cheese (Villamiel et al. 2000), honey (Villamiel et al. 2001), infant formulas (Guerra-Hernandez et al. 2002), jams and fruit-based infant foods (Rada-Mendoza et al. 2002), fresh filled pasta (Zardetto et al. 2003), prebaked breads (Ruiz et al., 2004), and cookies, crackers, and breakfast cereals (Rada-Mendoza et al. 2004).

In the case of foods containing free amino acids, free Amadori compounds can be present, and acid hydrolysis gives rise to the formation of the corresponding 2-furoylmethyl derivatives. For the first time, 2-furoylmethyl derivatives of different amino acids (arginine, asparagine, proline, alanine, glutamic acid, and γ-amino butyric acid) have been detected and have been used as indicators of the early stages of Maillard reaction in stored dehydrated orange juices (del Castillo et al. 1999). These compounds were proposed as indicators to evaluate quality changes either during processing or during subsequent storage. Later, most of these compounds were also detected in different foods: commercial orange juices (del Castillo et al. 2000), processed tomato products (Sanz et al. 2000), dehydrated fruits (Sanz et al. 2001), and commercial honey samples (Sanz et al. 2003).

Caramelization

During nonenzymatic browning of foods, various degradation products are formed via caramelization of carbohydrates, without amine participation (Ajandouz and Puigserver 1999, Ajandouz et al. 2001). Caramelization occurs when surfaces are heated strongly (e.g., during baking and roasting), during the processing of foods with high sugar content (e.g., jams and certain fruit juices) or in wine production (Kroh 1994). Caramelization is desirable to obtain caramel-like flavor and/or development of brown color in certain types of foods. Caramel flavoring and coloring, produced from sugar with different catalysts, is one of the most widely used additives in the food industry. However, caramelization is not always a desirable reaction due to the possible formation of mutagenic compounds (Tomasik et al. 1989) and the excessive changes in sensory attributes that could affect the quality of certain foods.

Caramelization is catalyzed under acidic or alkaline conditions (Namiki 1988), and many of the products formed are similar to those resulting from the Maillard reaction.

Caramelization of reducing carbohydrates starts with the opening of the hemiacetal ring followed by enolization, which proceeds via acid- and base-catalyzed mechanisms, leading to the formation of isomeric carbohydrates (Fig. 4.8). The interconversion of sugars through their enediols increases with increasing pH and is called the Lobry de Bruyn-Alberda van Ekenstein transformation (Kroh 1994).

In acid media, low amounts of isomeric carbohydrates are formed; however, dehydration is favored, leading to furaldehyde compounds: 5-(hydroxymethyl)-2-furaldehyde (HMF) from hexoses (Fig. 4.9) and 2-furaldehyde from pentoses. With unbuffered

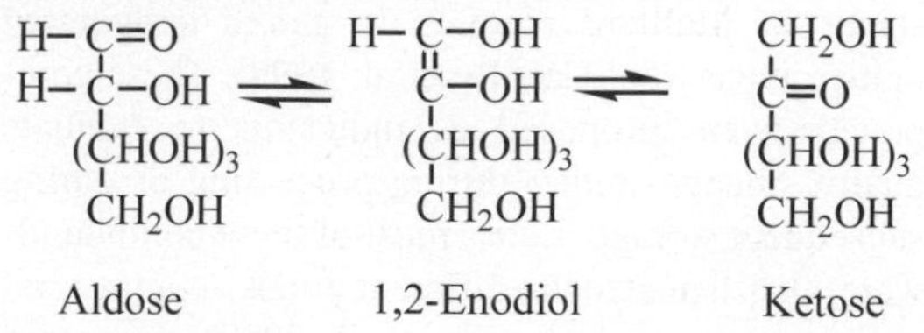

H–C=O | H–C–OH | CH_2OH
H–C–OH ⇌ C–OH ⇌ C=O
$(CHOH)_3$ | $(CHOH)_3$ | $(CHOH)_3$
CH_2OH | CH_2OH | CH_2OH

Aldose 1,2-Enodiol Ketose

Figure 4.8. The Lobry de Bruyn-Alberda van Ekenstein transformation.

acids as catalysts, higher yields of HMF are produced from fructose than from glucose. Also, only the fructose moiety of sucrose is largely converted to HMF under the unbuffered conditions that produce the highest yields. The enolization of glucose can be greatly increased in buffered acidic solutions. Thus, higher yields of HMF are produced from glucose and sucrose when a combination of phosphoric acid and pyridine is used as the catalyst than when phosphoric acid is used alone (Fenemma 1976).

In alkaline media, dehydration reactions are slower than in neutral or acid media, but fragmentation products such as acetol, acetoin, and diacetyl are detected. In the presence of oxygen, oxidative fission takes place, and formic, acetic, and other organic acids are formed.

All of these compounds react to produce brown polymers and flavor compounds (Olano and Martínez-Castro 1996).

In general, caramelization products consist of volatile and nonvolatile (90–95%) fractions of low and high molecular weights that vary depending on temperature, pH, duration of heating, and starting material (Defaye et al. 2000). Although it is known that caramelization is favored at temperatures higher than 120°C and at a pH greater than 9 and less than 3, depending on the composition of the system (pH and type of sugar), caramelization reactions may also play an important role in color formation in systems heated at lower temperatures. Thus, some studies have been conducted at the temperatures of accelerated storage conditions (45–65°C) and pH values from 4 to 6 (Buera et al. 1987a,b). These authors studied the changes of color due to caramelization of fructose, xylose, glucose, maltose, lactose, and sucrose in model systems of 0.9 a_w and found that fructose and xylose browned much more rapidly than the other sugars and that lowering the pH inhibited caramelization browning of sugar solutions.

In a study on the kinetics of caramelization of several monosaccharides and disaccharides, Diaz and Clotet (1995) found that at temperatures of 75–95°C, browning increased rapidly with time and to a higher final value with increasing temperature, this effect being more marked in the monosaccharides than in the disaccharides. In all sugars studied, increase of browning was greater at $a_w = 1$ than at $a_w = 0.75$.

The effect of sugars, temperature, and pH on caramelization was evaluated by Park et al. (1998). Reaction rate was highest with fructose, followed by sucrose. As reaction temperature increased from 80 to 110°C, reaction rate was greatly increased. With respect to pH, the optimum value for caramelization was 10.

Although most studies on caramelization have been conducted in model systems of mono- and disaccharides, a number of real food systems contain oligosaccharides or even polymeric saccharides; therefore, it is also of great interest to know the contribution of these carbohydrates to the flavor and color of foods. Kroh et al. (1996) reported the break-

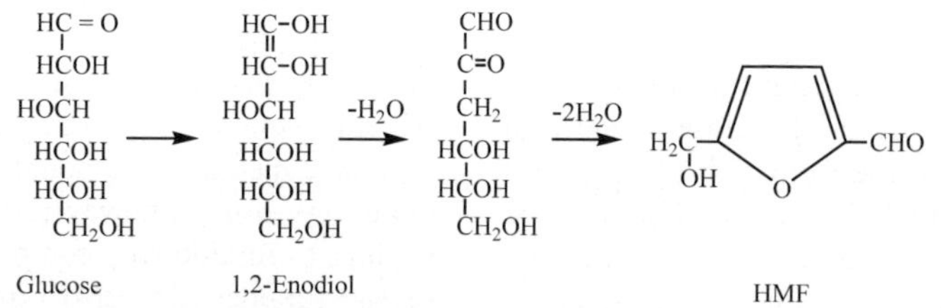

HC=O / HCOH / HOCH / HCOH / HCOH / CH_2OH → HC–OH / HC–OH / HOCH / HCOH / HCOH / CH_2OH $\xrightarrow{-H_2O}$ CHO / C=O / CH_2 / HCOH / HCOH / CH_2OH $\xrightarrow{-2H_2O}$ $H_2C(OH)$–[furan]–CHO

Glucose 1,2-Enodiol HMF

Figure 4.9. 1,2 Enolization and formation of hydroxymethyl furfural (HMF).

down of oligo- and polysaccharides to nonvolatile reaction products. Homoki-Farkas et al. (1997) studied, through an intermediate compound (methylglyoxal), the caramelization of glucose, dextrin 15, and starch in aqueous solutions at 170°C under different periods of time. The highest formation of methylglyoxal was in glucose and the lowest in starch systems. The authors attributed the differences to the number of reducing end groups. In the case of glucose, when all molecules were degraded, the concentration of methylglyoxal reached a maximum and began to transform, yielding low and high molecular weight color compounds. Hollhagel and Kroh (2000, 2002) investigated the degradation of maltoligosaccharides at 100°C through α-dicarbonyl compounds such as 1,4-dideoxyhexosulose, and they found that this compound is a reactive intermediate and precursor of various heterocyclic volatile compounds that contribute to caramel flavor and color.

Perhaps, as mentioned above, the most striking feature of caramelization is its contribution to the color and flavor of certain food products under controlled conditions. In addition, it is necessary to consider other positive characteristics of this reaction such as the antioxidant activity of the caramelization products. Kirigaya et al. (1968) suggested that high molecular weight and colored pigments might play an important role in the antioxidant activity of caramelization products. However, Rhee and Kim (1975) reported that caramelization products from glucose have antioxidant activity that consists mainly of colorless intermediates, such as reductones and dehydroreductones, produced in the earlier stages of caramelization.

In addition, the effect of caramelized sugars on enzymatic browning has been studied by several authors. Pitotti et al. (1995) reported that the antibrowning effect of some caramelization products is in part related to their reducing power. Lee and Lee (1997) obtained caramelization products by heating a sucrose solution at 200°C under various conditions to study the inhibitory activity of these products on enzymatic browning. The reducing power of caramelization products and their inhibitory effect on enzymatic browning increased with prolonged heating and with increased amounts of caramelization products. Caramelization was investigated in solutions of fructose, glucose, and sucrose heated at temperatures up to 200°C for 15–180 minutes. Browning intensity increased with heating time and temperature. The effect of the caramelized products on polyphenol oxidase (PPO) was evaluated, and the greatest PPO inhibitory effect was demonstrated by sucrose solution heated to 200°C for 60 minutes (Lee and Han 2001). More recently, Billaud et al. (2003) found caramelization products from hexoses with mild inhibitory effects on PPO, particularly after prolonged heating at 90°C.

Ascorbic Acid Browning

Ascorbic acid (vitamic C) plays an important role in human nutrition as well as in food processing (Chauhan et al. 1998). Its key effect as an inhibitor of enzymatic browning has been previously discussed in this chapter.

Browning of ascorbic acid can be briefly defined as the thermal decomposition of ascorbic acid under both aerobic and anaerobic conditions, by oxidative or nonoxidative mechanisms, in either the presence or absence of amino compounds (Wedzicha 1984).

Nonenzymatic browning is one of the main reasons for the loss of commercial value in citrus products (Manso et al. 2001). These damages, degradation of ascorbic acid followed by browning, also concern noncitrus foods such asparagus, broccoli, cauliflower, peas, potatoes, spinach, apples, green beans, apricots, melons, strawberries, corn, and dehydrated fruits (Belitz and Grosch 1997).

Pathway of Ascorbic Acid Browning

The exact route of ascorbic acid degradation is highly variable and dependent upon the particular system. Factors that can influence the nature of the degradation mechanism include temperature, salt and sugar concentration, pH, oxygen, enzymes, metal catalysts, amino acids, oxidants or reductants, initial concentration of ascorbic acid, and the ratio of ascorbic acid to dehydroascorbic acid (DHAA; Fennema 1976).

Figure 4.10 shows a simplified scheme of ascorbic acid degradation. When oxygen is present in the system, ascorbic acid is degraded primarily to DHAA. Dehydroascorbic acid is not stable and spontaneously converts to 2,3-diketo-L-gulonic acid (Lee and Nagy 1996). Under anaerobic conditions, DHAA is not formed and undergoes the generation of diketogulonic acid via its keto tautomer, followed by β

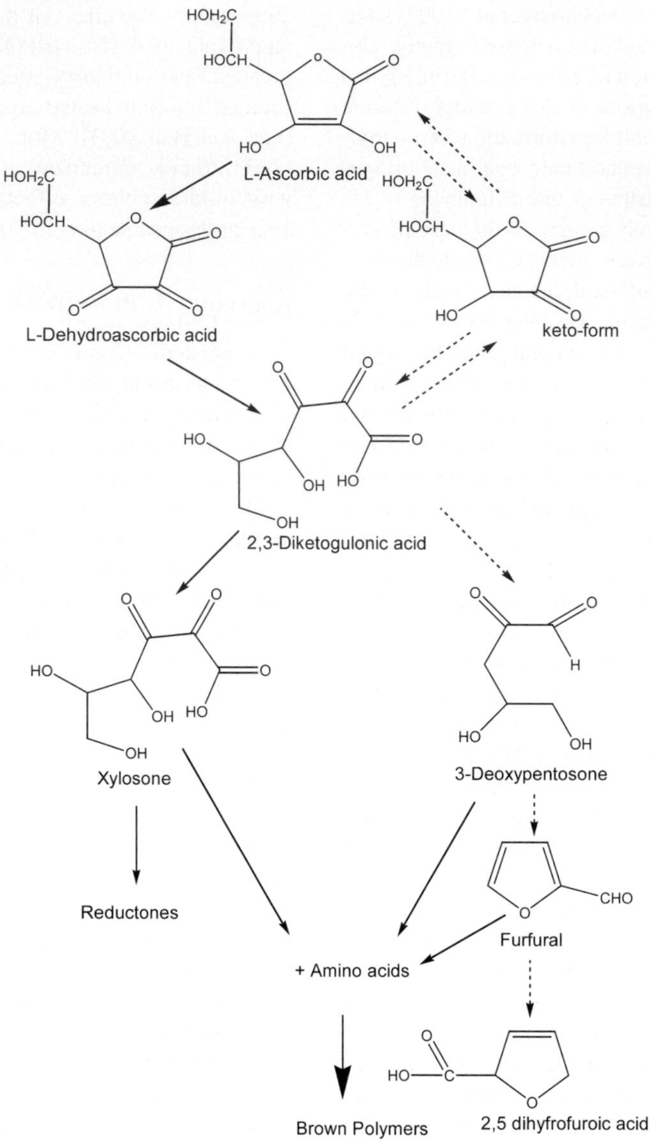

Figure 4.10. Pathways of ascorbic acid degradation (solid line, anaerobic route; dashed line, aerobic route).

elimination at C-4 from this compound and decarboxylation to give rise to 3-deoxypentosone, which is further degraded to furfural. Under aerobic conditions, xylosone is produced by simple decarboxylation of diketogulonic acid and is later converted to reductones. In the presence of amino acids, ascorbic acid, DHAA, and their oxidation products furfural, reductones, and 3-deoxypentosone may contribute to the browning of foods by means of a Maillard-type reaction (Fennema 1976, Belitz and Grosch 1997). Formation of Maillard-type products has been detected in both model systems and foods containing ascorbic acid (Kacem et al. 1987; Ziderman et al. 1989; Loschner et al. 1990, 1991; Mölnar-Perl and Friedman 1990; Yin and Brunk 1991; Davies and Wedzicha 1992, 1994; Pischetsrieder et al.

1995, 1997; Rogacheva et al. 1995; Koseki et al. 2001).

The presence of metals, especially Cu^{2+} and Fe^{3+}, causes great losses of Vitamin C. Catalyzed oxidation goes faster than spontaneous oxidation. Anaerobic degradation, which occurs more slowly than uncatalyzed oxidation, is maximum at pH 4 and minimum at pH 2 (Belitz and Grosch 1997).

Ascorbic acid oxidation is nonenzymatic in nature, but oxidation of ascorbic acid is sometimes catalyzed by enzymes. Ascorbic acid oxidase is a copper-containing enzyme that catalyzes oxidation of vitamin C. The reaction is catalyzed by copper ions. The enzymatic oxidation of ascorbic acid is important in the citrus industry. The reaction takes place mainly during extraction of juices. Therefore, it becomes important to inhibit ascorbic oxidase by holding juices for only short times and at low temperatures during the blending stage, by deaerating the juice to remove oxygen, and finally by pasteurizing the juice to inactivate the oxidizing enzymes.

Enzymatic oxidation also has been proposed as a mechanism for the destruction of ascorbic acid in orange peels during preparation of marmalade. Boiling the grated peel in water substantially reduces the loss of ascorbic acid (Fennema 1976).

Tyrosinase (PPO) may also possess ascorbic oxidase activity. A possible role of the ascorbic acid–PPO system in the browning of pears has been proposed (Espin et al. 2000).

In citrus juices, nonenzymatic browning is from reactions of sugars, amino acids, and ascorbic acid (Manso et al. 2001). In freshly produced commercial juice, filled into Tetra Brik cartons, it has been demonstrated that nonenzymatic browning was mainly due to carbonyl compounds formed from L-ascorbic acid degradation. Contribution from sugar-amine reactions is negligible, as is evident from the constant total sugar content of degraded samples. The presence of amino acids and possibly other amino compounds enhance browning (Roig et al. 1999).

Both oxidative and nonoxidative degradation pathways are operative during storage of citrus juices. Since large quantities of DHAA are present in citrus juices, it can be speculated that the oxidative pathway must be dominant (Lee and Nagy 1996, Rojas and Gerschenson 1997a). A significant relationship between DHAA and browning of citrus juice has been found (Kurata et al. 1973; Sawamura et al. 1991, 1994). The rate of nonoxidative loss of ascorbic acid is often one-tenth or up to one-thousandth the rate of loss under aerobic conditions (Lee and Nagy 1996). In aseptically packed orange juice, the aerobic reaction dominates first and is fairly rapid, while the anaerobic reaction dominates later and is quite slow (Nagy and Smoot 1977, Tannenbaum 1976). A good prediction of ascorbic acid degradation and the evolution of the browning index of orange juice stored under anaerobic conditions at 20–45°C may be performed employing the Weibull model (Manso et al. 2001).

Furfural, which is formed during anaerobic degradation of ascorbic acid, has a significant relationship to browning (Lee and Nagy 1988); its formation has been suggested as an adequate index for predicting storage temperature abuse in orange juice concentrates and as a quality deterioration indicator in single-strength juice (Lee and Nagy 1996). However, furfural is a very reactive aldehyde that forms and decomposes simultaneously; therefore, it would be difficult to use as an index for predicting quality changes in citrus products (Fennema 1976). In general, ascorbic acid would be a better early indicator of quality.

Control of Ascorbic Acid Browning

Sulfites (Wedzicha and Mcweeny 1974, Wedzicha and Imeson 1977), thiol compounds (Naim et al. 1997), maltilol (Koseki et al. 2001), sugars, and sorbitol (Rojas and Gerschenson 1997b) may be effective in suppressing ascorbic acid browning. Doses to apply these compounds greatly depend on factors such as concentration of inhibitors and temperatures. L-cysteine and sodium sulfite may suppress or accelerate ascorbic acid browning as a function of their concentration (Sawamura et al. 2000). Glucose, sucrose, and sorbitol protect L-ascorbic acid from destruction at low temperatures (23, 33, and 45°C), while at higher temperatures (70, 80, and 90°C) compounds with active carbonyls promoted ascorbic acid destruction. Sodium bisulfite was only significant in producing inhibition at lower temperature ranges (23, 33, and 45°C) (Rojas and Gerschenson 1997).

Although the stability of ascorbic acid generally increases as the temperature of the food is lowered, certain investigations have indicated that there may be an accelerated loss on freezing or frozen storage.

However, in general, the largest losses for noncitrus foods will occur during heating (Fennema 1976).

The rapid removal of oxygen from packages is an important factor in sustaining a higher concentration of ascorbic acid and lower browning of citrus juices over long-term storage. The extent of browning may be reduced by packing in oxygen-scavenging film (Zerdin et al. 2003).

Modified-atmosphere packages (Howard and Hernandez-Brenes 1998), microwave heating (Villamiel et al. 1998, Howard et al. 1999), ultrasound-assisted thermal processing (Zenker et al. 2003), pulsed electric field processing (Min et al. 2003), and carbon dioxide assisted high-pressure processing (Boff et al. 2003) are some examples of technological processes that allow ascorbic acid retention and, consequently, a prevention of undesirable browning.

Lipid Browning

Protein-Oxidized Fatty Acid Reactions

The organoleptic and nutritional characteristics of several foods are affected by lipids, which can participate in chemical modifications during processing and storage. Lipid oxidation occurs in oils and lard, but also in foods with low amounts of lipids, such as products from vegetable origin. This reaction occurs in both unprocessed and processed foods, and although in some cases it is desirable, for example, in the production of typical cheeses or of fried-food aromas (Nawar 1985), in general, it can lead to undesirable odors and flavors (Nawar 1996). Quality properties such as appearance, texture, consistency, taste, flavor, and aroma can be adversely affected (Eriksson 1987). Moreover, toxic compound formation and loss of nutritional quality can also be observed (Frankel 1980; Gardner 1989; Kubow 1990, 1992).

Although the lipids can be oxidized by both enzymatic and nonenzymatic reactions, the latter is the main involved reaction. This reaction proceeds via typical free radical mechanisms, with hydroperoxides as the initial products. As hydroperoxides are quite unstable, a network of dendritic reactions, with different reaction pathways and a diversity of products, can take place (Gardner 1989). The enzymatic oxidation of lipids occurs sequentially. Lipolytic enzymes can act on lipids to produce polyunsaturated fatty acids that are then oxidized by either lipoxygenase or cyclooxygenase to form hydroperoxides or endoperoxides, respectively. Later, these compounds suffer a series of reactions to produce, among other compounds, long-chain fatty acids that are responsible for important characteristics and functions (Gardner 1995).

Via polymerization, brown-colored oxypolymers can be produced subsequently from the lipid oxidation derivatives (Khayat and Schwall 1983). However, due to the electrophilic character of the carbonyl compounds produced during lipid oxidation, interaction with nucleophiles such as the free amino group of amino acids, peptides, or proteins can also take place, producing end products different from those formed during oxidation of pure lipids (Gillatt and Rossell 1992). When lipid oxidation occurs in the presence of amino acids, peptides, or proteins, not all the lipids have to be oxidized and degraded before oxidized lipid–amino acid reactions take place. In fact, both reactions occur simultaneously (Hidalgo and Zamora 2002).

The interaction between oxidized fatty acids and amino groups has been related to the browning detected during the progressive accumulation of lipofuscins (age-related yellow-brown pigments) in man and animals (Yin 1996). In foods, evidence of this reaction has been found during storage and processing of some fatty foods (Hidalgo et al. 1992, Nawar 1996), in salted sun-dried fish (Smith and Hole 1991), boiled and dried anchovy (Takiguchi 1992), smoked tuna (Zotos et al. 2001), meat and meat products (Mottram 1998), and rancid oils and fats with amino acids or proteins (Yamamoto and Kogure 1969, Okumura and Kawai 1970, Gillatt and Rossell 1992). For instance, Nielsen et al. (1985) found that peroxidized methyl linoleate can react with lysine, tryptophan, methionine, and cysteine in whey proteins.

Several studies have been carried out in model systems with the aim to investigate the role of lipids in nonenzymatic browning. The role of lipids in these reactions seems to be similar to the role of carbohydrates during the Maillard reaction (Hidalgo and Zamora 2000). Similarly to the Maillard reaction, oxidized lipid–protein interactions comprise a huge number of several related reactions. The isolation and characterization of the involved products is very difficult, mainly in the case of intermediate

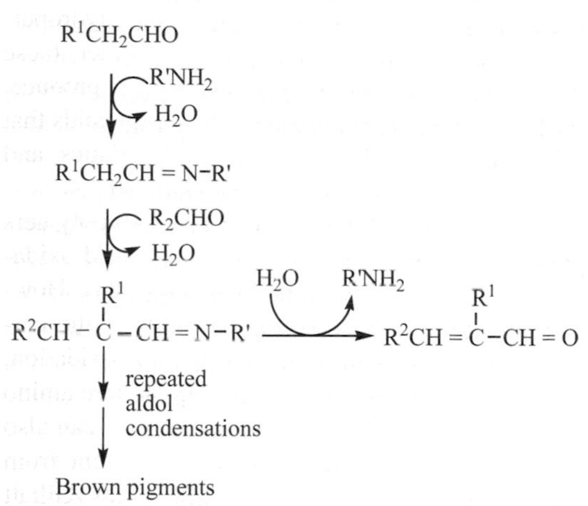

Figure 4.11. Formation of brown pigments by aldolic condensation (Hidalgo and Zamora 2000).

products, which are unstable and are present in very low concentrations.

According to the mechanism proposed for the protein browning caused by acetaldehyde, the carbonyl compounds derived from unsaturated lipids readily react with protein-free amino groups, following the scheme of Figure 4.11, to produce, by repeated aldol condensations, the formation of brown pigments (Montgomery and Day 1965, Gardner 1979, Belitz and Grosch 1997).

More recently, another mechanism based on the polymerization of the intermediate products 2-(1-hydroxyalkyl) pyrroles has been proposed (Zamora and Hidalgo 1994, Hidalgo and Zamora 1995). These authors, studying different model systems, tried to explain, at least partially, the nonenzymatic browning and fluorescence produced when proteins are present during the oxidation of lipids (Fig. 4.12). The 2-(1-hydroxyalkyl) pyrroles (I) have been found to originate from the reaction of 4,5-epoxy-2-alkenals (formed during lipid peroxidation) with the amino group of amino acids and/or proteins, and their formation is always accompanied by the production of N-substituted pyrroles (II). These last compounds are relatively stable and have been found in 22 fresh food products (cod, cuttlefish, salmon, sardine, trout, beef, chicken, pork, broad bean, broccoli, chickpea, garlic, green pea, lentil, mushroom, soybean, spinach, sunflower, almond, hazelnut, peanut, and walnut) (Zamora et al. 1999). However, the N-substituted 2-(1-hydroxyalkyl) pyrroles are unstable and polymerize rapidly and spontaneously to produce brown macromolecules with fluorescent melanoidin-like characteristics (Hidalgo and Zamora 1993). Zamora et al. (2000) observed that the formation of pyrroles is a step immediately prior to the formation of color and fluorescence. Pyrrole formation and perhaps some polymerization finish before maximum color and fluorescence are achieved.

Although melanoidins starting from either carbohydrates or oxidized lipids would have analogous chemical structures, carbohydrate-protein and oxidized lipid–protein reactions are produced under different conditions. Hidalgo et al. (1999) studied the effect of pH and temperature in two model systems:(1) ribose and bovine serum albumin and (2) methyl linoleate oxidation products and bovine serum albumin; they observed that from 25 to 50°C the latter system exhibited higher browning than the former. Conversely, the browning produced in the

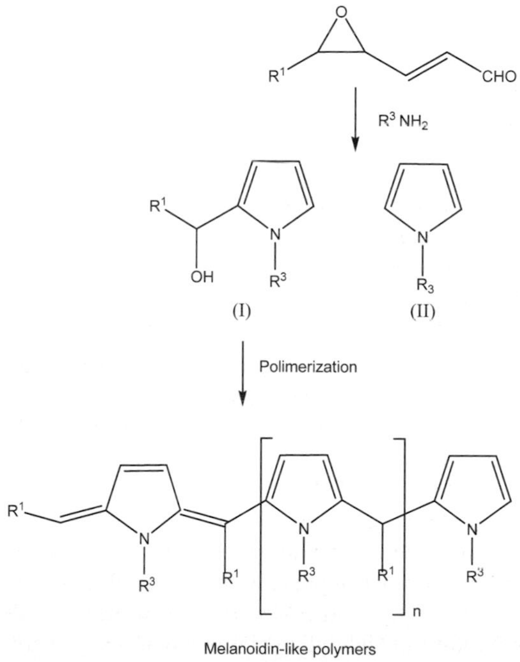

Figure 4.12. Mechanism for nonenzymatic browning produced as a consequence of 2-(1-hydroxyalkyl)pyrrole polymerisation (Hidalgo et al. 2003).

carbohydrate model system was increased at temperatures in the range 80–120°C. The effect of pH on browning was similar in both oxidized lipid-protein and carbohydrate-protein model systems.

Some oxidized lipid-amino acid reaction products have been shown to have antioxidant properties when they are added to vegetable oils (Zamora and Hidalgo 1993, Alaiz et al. 1995, Alaiz et al. 1996). All of the pyrrole derivatives, with different substituents in the pyrrole ring, play an important role in the antioxidant activity of foods, being the sum of the antioxidant activities of the different compounds present in the sample (Hidalgo et al. 2003). Alaiz et al. (1997), in a study on the comparative antioxidant activity of Maillard reaction compounds and oxidized lipid–amino acid reaction products, observed that both reactions seem to contribute analogously to increasing the stability of foods during processing and storage.

Zamora and Hidalgo (2003a) studied the role of the type of fatty acid (methyl linoleate and methyl linolenate) and the protein (bovine serum albumin)–lipid ratio on the relative progression of the process involved when lipid oxidation occurs in the presence of proteins. These authors found that methyl linoleate was only slightly more reactive than the methyl linolenate for bovine serum albumin, producing a higher increase of protein pyrroles in the protein and increased browning and fluorescence. In relation to the influence of the protein-lipid ratio on the advance of the reaction, the results observed in this study pointed out that a lower protein-lipid ratio increases sample oxidation and protein damage, as a consequence of the antioxidant activity of the proteins. These authors also concluded that the changes produced in the color of protein-lipid samples were mainly due to oxidized lipid-protein reactions and not a consequence of polymerization of lipid oxidation products.

Analogous to the Maillard reaction, oxidized lipid and protein interaction can cause a loss of nutritional quality due to the destruction of essential amino acids such as tryptophan, lysine, and methionine and essential fatty acids. Moreover, a decrease in digestibility and inhibition of proteolytic and glycolytic enzymes can also be observed. In a model system of 4,5(E)-epoxy-2(E)-heptenal and bovine serum albumin, Zamora and Hidalgo (2001) observed denaturation and polymerization of the protein, and the proteolysis of this protein was impaired as compared with the intact protein. These authors suggested that the inhibition of proteolysis observed in oxidized lipid–damaged proteins may be related to the formation and accumulation of pyrrolized amino acid residues.

To date, although most of the studies have been conducted using model systems, the results obtained point out the importance of lipids during browning, and in general, it is possible to suggest that the role of lipids during these reactions could be similar to the role of carbohydrates in the Maillard reaction or phenols in enzymatic browning. The complexity of the reaction is attributable to several fatty acids that can give rise to a number of lipid oxidation products that are able to interact with free amino groups. As a summary, Figure 4.13 shows an example of a general pathway for pyrrole formation during polyunsaturated fatty acid oxidation in the presence of amino compounds.

Nonenzymatic Browning of Aminophospholipids

In addition to the above-mentioned studies on the participation of lipids in the browning reactions, several reports have addressed amine-containing phospholipid interactions with carbohydrates. Due to the role of these membranous functional lipids in the maintenance of cellular integrity, most of the studies have been conducted in biological samples (Bucala et al. 1993, Requena et al. 1997, Lertsiri et al. 1998, Oak et al. 2000, Oak et al. 2002). The glycation of membrane lipids can cause inactivation of receptors and enzymes, cross-linking of membrane lipids and proteins, membrane lipid peroxidation, and consequently, cell death. Amadori compounds derived from the interaction between aminophospholipids and reducing carbohydrates are believed to be key compounds for generating oxidative stress, causing several diseases.

In foods, this reaction can be responsible for deterioration during processing. Although nonenzymatic browning of aminophospholipids was detected for the first time in dried egg by Lea (1957), defined structures from such reactions were reported later by Utzmann and Lederer (2000). These authors demonstrated the interaction of phosphatidylethanolamine (PE) with glucose in model systems; moreover, they

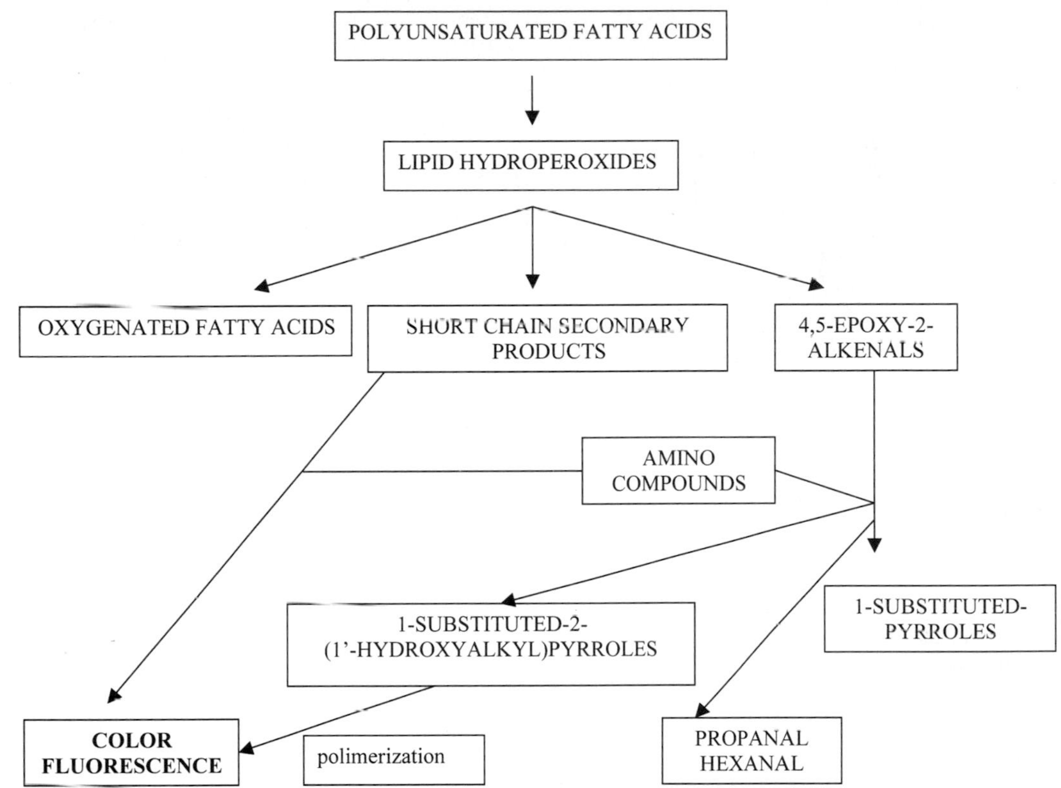

Figure 4.13. General pathways of pyrrole formation during polyunsaturated fatty acid oxidation (Zamora and Hidalgo 1995).

found the corresponding Amadori compound in spray-dried egg yolk powders and lecithin products derived therefrom and proposed the Amadori compound content of these products as a possible quality criterion. Oak et al. (2002) detected the Amadori compounds derived from glucose and lactose in several processed foods with high amounts of carbohydrates and lipids such as infant formula, chocolate, mayonnaise, milk, and soybean milk; of these, infant formula contained the highest levels of Amadori-PEs. However, these compounds were not detected in other foods due to differences in composition and the relatively low temperatures used during the processing.

As an example, Figure 4.14 shows a scheme for the formation of the Amadori compounds derived from glucose and lactose with PE. Carbohydrates react with the amino group of PE to form an unstable Schiff base, which undergoes an Amadori rearrangement to yield the stable PE-linked Amadori product (Amadori-PE).

On the other hand, similar to proteins, phospholipids such as PE and ethanolamine can react with secondary products of lipid peroxidation such as 4,5-epoxyalkenals. Zamora and Hidalgo (2003b) studied this reaction in model systems and characterized different polymers responsible for brown color and fluorescence development, confirming that lipid oxidation products are able to react with aminophospholipids in a manner analogous to their reactions with protein amino groups; therefore, both amino phospholipids and proteins might compete for lipid oxidation products.

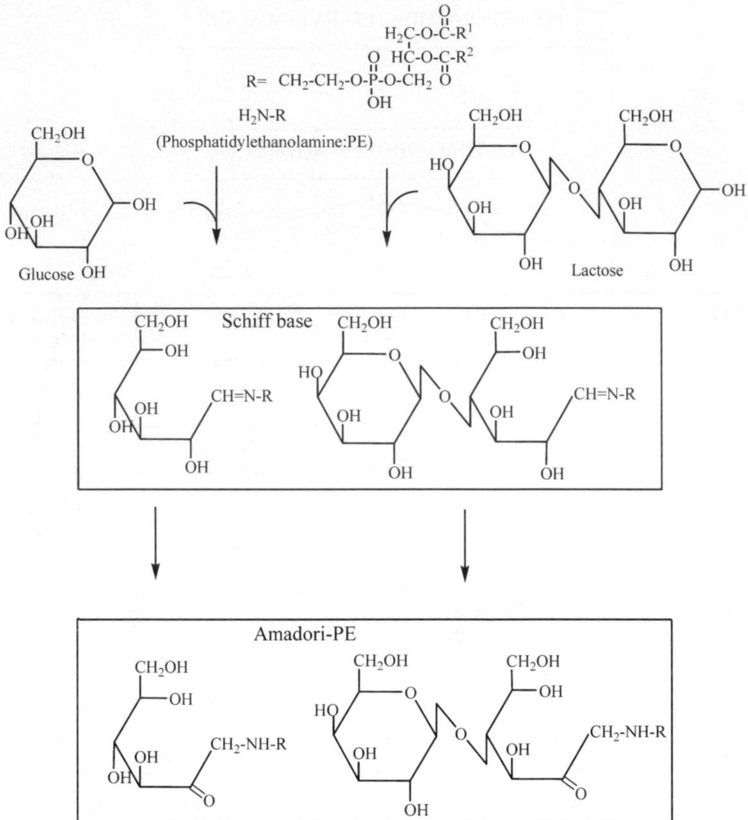

Figure 4.14. Scheme for the glycation with glucose and lactose of phosphatidylethanolamine (PE) (Oak et al. 2002).

REFERENCES

Ahn JS, Castle L, Clarke DB, Lloyd, AS, Philo M R, Speck DR. 2002. Verification of the findings of acrylamide in heated foods. Food Additives 19(12): 1116–1124.

Ajandouz EH, Puigserver A. 1999. Non-enzymatic browning reaction of essential amino acids: Effect of pH on caramelization and Maillard reaction kinetics. Journal of Agricultural and Food Chemistry 47(5): 1786–1793.

Ajandouz EH, Tchiakpe LS, Dalle Ore F, Benajiba A, Puigserver A. 2001. Effects of pH on caramelization and Maillard reaction kinetics in fructose-lysine model systems. Journal of Agricultural and Food Chemistry 66(7): 926–931.

Alaiz, M, Hidalgo, FJ, Zamora, R. 1997. Comparative antioxidant activity of Maillard- and oxidized lipid-damaged bovine serum albumin. Journal of Agricultural and Food Chemistry 45(8): 3250–3254.

Alaiz, M, Zamora, R, Hidalgo, FJ. 1996. Antioxidative activity of pyrrole, imidazole, dihydropyridine, and pyridinium salt derivatives produced in oxidized lipid/amino acid browning reactions. Journal of Agricultural and Food Chemistry 44(3): 686–691.

———. 1995. Antioxidative activity of ε-2-octenal/amino acids reaction products. Journal of Agricultural and Food Chemistry 43(3): 795–800.

Ames, JM. 1990. Control of the Maillard reaction in food systems. Trends in Food Science and Technology 1 (July): 150–154.

Anese, M, Falcone, P, Fogliano, V, Nicoli, MC, Massini, R. 2002. Effect of equivalent thermal treatments on the colour and the antioxidant activity of tomato purees. Journal of Food Science 67(9): 3442–3446.

Antoni, SM, Han, IY, Rieck, JR, Dawson, PL. 2000. Antioxidative effect of Maillard products formed from honey at different reaction times. Journal of Agriculture and Food Chemistry 48(9): 3985–3989.

Badoud, R, Fay, LB, Hunston, F, Pratz, G. 1995. Periodate oxidative degradation of Amadori compounds. Formation of N^{ε}-Carboxymethyllysine and N-Carboxymethylamino Acids as markers of the early Maillard reaction. In: TCh Lee, HJ Kim, editors, Chemical Markers for Processed and Stored Foods. ACS Symposium series 631, Chicago, Illinois. Pp. 208–220.

Ballestra P, Verret C, Largeteau A, Demazeau G, El Moueffak A. 2002. Effect of high-pressure treatment on polyphenoloxidase activity of Agaricus bisporus mushroom. High Pressure Research 22(3–4): 677–680 Sp. Iss.

Bates, L, Ames, JM, MacDougall DB, Taylor PC. 1998. Laboratory reaction cell to model Maillard colour developments in a starch-glucose-lysine system. Journal of Food Science 63(6): 991–996.

Becalski A, Lau BPY, Lewis D, Seaman SW. 2003 Acrylamide in foods: Occurrence, sources and modelling. Journal of Agricultural and Food Chemistry 51(3): 802–808.

Belitz H-D, Grosch W. 1997. Food Chemistry, 2nd edition. Springer: Berlin, Germany. Pp. 232–236.

Bell LN. 1997. Maillard reaction as influenced by buffer type and concentration. Food Chemistry 59(1): 143–147.

Billaud C, Brun-Merimee S, Louarme L, Nicolas J. 2004. Effect of glutathione and Maillard reaction products prepared from glucose or fructose with glutathione on polyphenoloxidase from apple-I: Enzymic browning and enzyme activity inhibition. Food Chemistry 84(29): 223–233.

Billaud C, Roux E, Brun-Merimee S, Maraschin C, Nicolas J. 2003. Inhibitory effect of unheated and heated D-glucose, D-fructose and L-cysteine solutions and Maillard reaction product model systems on polyphenoloxidase from apple. I. Enzymic browning and enzyme activity inhibition using spectrophotometric and polarographic methods. Food Chemistry 81(1): 35–50.

Boff JM, Truong TT, Min DB, Shellhammer TH. 2003. Effect of thermal processing and carbon dioxide-assisted high-pressure processing on pectin methylesterase and chemical changes in orange juice. Journal of Food Science 68(4): 1179–1184.

Brands CMJ, van Boekel MAJS. 2001. Reaction of monosaccharides during heating of sugar-casein systems: Building of a reaction network model. Journal of Agricultural and Food Chemistry 49(10): 4667–4675.

Bucala R, Makita Z, Koschinsky T, Cerami A, Vlassara H. 1993. Lipid advanced glycosylation: Pathway for lipid oxidation in vivo. Proceedings of the National Academy of Science (USA) 90(July): 6434–6438.

Buera MP, Karel M. 1995. Effect of physical changes on the rates of non-enzymic browning and related reactions. Food Chemistry 52(2): 167-173.

Buera MP, Chirife J, Resnik SL, Lozano RD. 1987a. Non-enzymic browning in liquid model systems of high water activity: Kinetics of colour changes due to caramelization of various single sugars. Journal of Food Science 52(4): 1059–1062,1073.

Buera MP, Chirife J, Resnik SL, Wetzler G. 1987b. Non-enzymic browning in liquid model systems of high water activity: Kinetics of colour changes due to Maillard's reaction between different single sugars and glycine and comparison with caramelization browning. Journal of Food Science 52(4): 1063–1067.

Burdock FA, Soni MG, Carabin IG. 2001. Evaluation of health aspects of kojic acid in food. Regulatory Toxicology and Pharmacology 33(1): 80–101.

Burdulu HS, Karadeniz F. 2003. Effect of storage on no enzymatic browning of apple juice concentrates. Food Chemistry 80(1): 91–97.

Carabasa-Giribert M, Ibarz-Ribas A. 2000. Kinetics of color development in aqueous glucose systems at high temperatures. Journal of Food Engineering 44(3): 181–189.

Chauhan AS, Ramteke RS, Eipeson WE. 1998. Properties of ascorbic acid and its application in food processing: A critical appraisal. Journal of Food Science and Technology- Mysore 35(5): 381–392.

Chen L, Mehta A, Berenbaum M, Zangerl AR, Engeseth NJ. 2000. Honeys from different floral sources as inhibitors of enzymic browning in fruit and vegetables homogenates. Journal of Agricultural and Food Chemistry 48(10): 4997–5000.

Chevalier, F, Hobart JM, Genot C, Haertle T. 2001. Scavenging of free radicals, antimicrobial, and cytotoxic activities of the Maillard reaction products of beta-lactoglobulin glycated with several sugars. Journal of Agricultural and Food Chemistry 49(10): 5031–5038.

Coetezer C, Corsini D, Love S, Pavek J, Tumer N. 2001. Control of enzymic browning in potato (Solanum tuberosum L.) by sense and antisense RNA from tomato polyphenol oxidase. Journal of Agricultural and Food Chemistry 49(2): 652–657.

Corbo, MR, Lanciotti R, Gardini F, Sinigaglia M, Guerzoni ME. 2000. Effects of hexanal, trans-2-hexanal, and storage temperature on shelf life of fresh sliced apples. Journal of Agricultural and Food Chemistry 48(6): 2401–2408.

Davies CGA, Wedzicha BL. 1992. Kinetics of the inhibition of ascorbic acid browning by sulfite. Food Additives and Contaminants 9(5): 471–477.

———. 1994. Ascorbic acid browning. The incorporation of C (1) from ascorbic acid into melanoidins. Food Chemistry 49(2): 165–167.

De Rigal D, Cerny M, Richard-Forget F, Varoquaux P. 2001. Inhibition of endive (Cichorium endivia L.) polyphenoloxidase by a Carica papaya latex preparation. International Journal of Food Science and Technology 36(69): 677–684.

Defaye J, Fernandez JMG, Ratsimba V. 2000. The molecules of caramelization: Structure and methodologies of detection and evaluation. Actualite Chimique 11(Nov):24–27.

del Castillo MD, Villamiel M, Olano A, Corzo N. 2000. Use of 2-furoylmethyl derivatives of GABA and arginine as indicators of the initial steps of Maillard reaction in orange juice. Journal of Agricultural and Food Chemistry 48 (9): 4217–4220.

del Castillo MD, Corzo N, Olano A. 1999. Early stages of Maillard reaction in dehydrated orange juice. Journal of Agricultural and Food Chemistry 47(10): 4388–4390.

Diaz N, Clotet R. 1995. Kinetics of the caramelization of simple sugar solutions. Alimentaria (259):35–38.

Dijkstra L, Walker JRL. 1991. Enzymic browning in apricots (Prunus armeniaca). Journal of Science of Food and Agriculture 54:229–234.

El SN, Kavas A. 1997. Available lysine in dried milk after processing. International Journal of Food Science and Nutrition 48(2): 109–111.

Erbersdobler HF, Hupe A. 1991 Determination of lysine damage and calculation of lysine bioavailability in several processed foods. Zeitschrift für Ernährungswissenschfat 30 (1): 46–49.

Eriksson CE. 1987. Oxidation of lipids in food systems. In: HWS Chan, editor, Autoxidation of Unsaturated Lipids. Academic Press: London. Pp. 207–231.

Espin JC, Veltman RH, Wichers HJ. 2000. The oxidation of L-ascorbic acid catalysed by pear tyrosinase. Physiologia Plantarum 109(1): 1–6.

Feather MS, Mossine V, Hirsch J. 1995. The use of aminoguanidine to trap and measure decarbonyl intermediates produced during the Maillard reaction. In: TCh Lee, HJ Kim, editors, Chemical Markers for Processed and Stored Foods. ACS Symposium series 631, Chicago, Illinois. Pp. 24–31.

Fennema OR. 1976. In: OR Fennema, editor, Principles of Food Science. New York, Marcel Dekker.

Ferrer E, Alegria A, Farre R, Abellan P, Romero F. 2003. Fluorometric determination of chemically available lysine: Adaptation, validation and application to different milk products. Nahrung-Food 47(6): 403–407.

Frankel EN. 1980. Lipid oxidation. Progress in Lipid Research 19(1/2): 1–22.

Friedman M. 2003. Chemistry, biochemistry, and safety of acrylamide. A review. Journal of Agricultural and Food Chemistry 51 (16): 4504–4526.

Gardner HW. 1979. Lipid hydro peroxide reactivity with proteins and amino acids: A review. Journal of Agricultural and Food Chemistry 27(2): 220–229.

———. 1989. Oxygen radical chemistry of polyunsaturated fatty acids. Free Radical Biology and Medicine 7(1): 65–86.

———. 1995. Biological roles and biochemistry of the lipoxygenase pathway. HortScience 30(2): 197–205.

Gillatt PN, Rossell JB. 1992. The interaction of oxidized lipids with proteins. In: FB Padley, editor, Advances in Applied Lipid Research. JAI Press, Greenwich, Connecticut. Pp. 65–118.

Gonzales AP, Burin L, Buera MD. 1999. Color changes during storage of honeys in relation to their composition and initial color. Food Research International 32(3): 185–191.

Guerra-Hernandez E, Leon-Gomez C, Garcia-Villanova B, Corzo N, Romera JMG. 2002. Effect of storage on non-enzymic browning of liquid infant milk formulae. Journal of the Science and Food Agriculture 82(5): 587–592.

Harel E, Mayer AM, Lerner HR. 1970. Changes in the levels of catechol oxidase and laccase activity in developing peaches. Journal of Science and Food Agricultural 21:542–544.

Hidalgo A, Rossi M, Pompei C. 1995. Furosine as a freshness parameter of shell egg. Journal of Agricultural and Food Chemistry 43(6): 1673–1677.

Hidalgo FJ, Zamora R, Alaiz M. 1992. Modificaciones producidas en las proteínas alimentarias por su interacción con lípidos peroxidados. II. Mecanismos conocidos de la interacción lípido (oxidado)-proteína. Grasas y Aceites 43(1): 31–38.

Hidalgo FJ, Alaiz M, Zamora R. 1999. Effect of pH and temperature on comparative non-enzymic browning of proteins produced by oxidized lipids and carbohydrates. Journal of Agricultural and Food Chemistry 47(2): 742–747.

Hidalgo FJ, Nogales F, Zamora R. 2003. Effect of the pyrrole polymerization mechanism on the antioxidative activity of no enzymatic browning reactions. Journal of Agricultural and Food Chemistry 51(19): 5703–5708.

Hidalgo FJ, Zamora R. 1993. Fluorescent pyrrole products from carbonyl-amine reactions. Journal of Biological Chemistry 268(22): 16190–16197.

———. 1995. Characterization of the products formed during microwave irradiation of the non-enzymic browning lysine (E)-4,5-epoxy-(E)-2-heptanal model system. Journal of Agricultural and Food Chemistry 43(4): 1023–1028.

———. 2000. The role of lipids in non-enzymic browning. Grasas y Accites 51(1–2): 35–49.

———. 2002. Methyl linoleate oxidation in the presence of bovine serum albumin. Journal of Agricultural and Food Chemistry 50(19): 5463–5467.

Hisaminato H, Murata M, Homma S. 2001. Relationship between the enzymatic browning and phenylalanine ammonia-lyase activity of cut lettuce, and the prevention of browning by inhibitors of polyphenol biosynthesis. Bioscience Biotechnology and Biochemistry 65(5): 1016–1021.

Hodge JE. 1953. Dehydrated foods. Chemistry of browning reactions in model systems. Journal of Agricultural and Food Chemistry 1(15): 928–943.

Hollnagel A, Kroh LW. 2002. 3-Deoxypentosulose: An alpha-dicarbonyl compound predominating in nonenzymic browning of oligosaccharides in aqueous solutions. Journal of Agricultural and Food Chemistry 50(6): 1659–1664.

———. 2000. Degradation of oligosaccharides in nonenzymic browning by formation of α-dicarbonyl compounds via a "peeling off" mechanism. Journal of Agricultural and Food Chemistry 48(12): 6219–6226.

Homoki-Farkas P, Örsi F, Kroh LW. 1997. Methylglyoxal determination from different carbohydrates during heat processing. Food Chemistry 59(1): 157–163.

Hosoda H, Iwahashi I. 2002. Inhibition of browning if apple slice and juice by onion juice. Journal of the Japanese Society for Horticultural Science 71(3): 452–454.

Hosoda H, Ohmi K, Sakaue K, Tanaka K. 2003. Inhibitory effect of onion oil on browning of shredded lettuce and its active components. Journal of the Japanese Society for Horticultural Science 75(5): 451–456.

Howard LA, Wong AD, Perry AK, Klein BP. 1999. Beta-carotene and ascorbic acid retention in fresh and processed vegetables. Journal of Food Science 64(5): 929–936.

Howard LR, Hernandez-Brenes C. 1998. Antioxidant content and market quality of jalapeno pepper rings as affected by minimal processing and modified atmosphere packaging. Journal of Food Quality 21 (4): 317–327.

Hunt MD, Eannetta NT, Yu H, Newman SM, Steffens JC. 1993. cDNA cloning and expression of potato polyphenol oxidase. Plant Molecular Biology 21(1): 59–68.

Ihl M, Aravena L, Scheuermann E, Uquiche E, Bifani V. 2003. Effect of immersion solutions on shelf life of minimally processed lettuce. Food Science and Technology 36 (6): 591–599.

Jacxsens L, Devlieghere F, Van der Steen C, Debevere J. 2001. Effect of high oxygen modified atmosphere packaging on microbiological growth and sensorial qualities of fresh cut produce. International Journal of Food Microbiology 71(2–3): 197–210.

Jolley RL, Evans LH, Makino N, Mason HS. 1974. Oxytyrosinase. Journal of Biological Chemistry 249(2): 335–345.

Jung MY, Choi DS, Jun JW. 2003. A novel technique for limitation of acrylamide formation in fried and baked corn chips and in French fries. Journal of Food Science 68(4): 1287–1290.

Kaaber L, Martinsen BK, Brathen E, Shomer A. 2002. Browning inhibition and textural changes of prepeeled potatoes caused by anaerobic conditions. Food Science and Technology 35(6): 526–531.

Kacem B, Cornell JA, Marshall MR, Shiremen RB, Mathews RF. 1987. Non-enzymatic browning in aseptically packaged orange drinks: Effect of ascorbic acid, amino acids and oxygen. Journal of Food Science 52(6): 1668–1672.

Kaur C, Kapoor HC. 2000. Inhibition of enzymic browning in apples, potatoes and mushrooms. Journal of Scientific and Industrial Research 59(5): 389–394.

Khayat A, Schwall D. 1983. Lipid oxidation in seafood. Food Technology 37(7): 130–140.

Kim YM, Yun J, Lee CK, Lee H, Min KR, Kim Y. 2002. Oxyresveratrol and hydroxystilbene compounds. Inhibitory effect on tyrosinase and mechanism of action. The Journal of Biological Chemistry 277(18): 16340–16344.

Kirigaya N, Kato H, Fujimaki M. 1968. Studies on the antioxidant of no enzymatic browning reaction products. Part 1: Relation of colour intensity and reductones with antioxidant activity of browning reaction products. Agricultural and Biological Chemistry. 32(3): 287.

Koseki H, Akima C, Ohasi K, Sakai T. 2001. Effect of sugars on decomposition and browning of vitamin C during heating storage. Journal of the Japanese Society for Food Science and Technology 48(4): 268–276.

Kroh LW. 1994. Caramelization in foods and beverages. Food Chemistry 51(4): 373–379.

Kroh LW, Jalyschko W, Häseler J. 1996. Non-volatile reaction products by heat-induced degradation of alpha-glucans. Part I: Analysis of oligomeric maltodextrins and anhydrosugars. Starch/Stärke 48(11/12): 426–433.

Kubo I, Kinst-Hori I, Chaudhuri SK, Kubo Y, Sanchez Y, Ogura T. 2000. Flavonols from Heterotheca inuloides: Tyrosinase inhibitory activity and structural criteria. Biorganic and Medicinal Chemistry 8(7): 1749–1755.

Kubow S. 1990. Toxicity of dietary lipid peroxidation products. Trends in Food Science and Technology 1(3): 67–70.

———. 1992. Routes of formation and toxic consequences of lipid oxidation products in foods. Free Radical Biology and Medicine 12(1): 63–81.

Kurata T, Fujimaki M, Sakurai ZF. 1973. Red pigment produced by reaction of dehydro-L-ascorbic acid with alpha-amino-acids. Agricultural and Biological Chemistry 37(6): 1471–1477.

Lea CH. 1957. Deteriorative reactions involving phospholipids and lipoproteins. Journal of Science of Food and Agriculture 8:1–13.

Ledl, F. 1990. Chemical pathways of the Maillard reaction. In: PA Finot, HU Aeschbacher, RF Hurrell, R Liardon, editors, The Maillard Reaction in Food Processing, Human Nutrition and Physiology. Birkhäuser Verlag, Basel, Switzerland. Pp. 19–42.

Lee GC, Han SC. 2001. Inhibition effects of caramelization products from sugar solutions subjected to different temperature on polyphenol oxidase. Journal of the Korean Society of Food Science and Nutrition 30(6): 1041–1046.

Lee GC, Lee CY. 1997. Inhibitory effect of caramelization products on enzymic browning. Food Chemistry 60(2): 231–235.

Lee HS, Nagy S. 1988. Quality changes and nonenzymic browning intermediates in grapefruit juice during storage. Journal of Food Science 53(1): 168–172, 180.

———. 1996. Chemical degradative indicators to monitor the quality of processed and stored citrus products. In: TCh Lee, HJ Kim, editors, Chemical Markers for Processed Foods. ACS Symposium Series 631, American Chemical Society, Washington, DC. Pp. 86–106.

Lertsiri S, Shiraishi M, Miyazawa T. 1998. Identification of deoxy-D-fructosyl phosphatidylethanolamine as a non-enzymatic glycation product of phosphatidylethanolamine and its occurrence in human blood plasma and red blood cells. Bioscience Biotechnology and Biochemistry 62(5): 893–901.

Lingnert H. 1990. Development of the Maillard reaction during food processing. In: PA Finot, HU Aeschbacher, RF Hurrell, R Liardon, editors, The Maillard Reaction in Food Processing, Human Nutrition and Physiology. Birkhäuser Verlag, Basel, Switzerland. Pp. 171–185.

Loschner J, Kroh L, Westphal G, Vogel J. 1991. L-Ascorbic-acid-A carbonyl component of non-enzymic browning reactions. 2. Amino-carbonyl reactions of L-ascorbic acid. Zeitschrift fur Lebensmittel Untersuchung und Forschung 192(4): 323–327.

Loschner J, Kroh L, Vogel J. 1990. L-Ascorbic acid- A carbonyl component on non-enzymic browning reactions. Browning of L-ascorbic acid in aqueous model system. Zeitschrift fur Lebensmittel Untersuchung und Forschung 191(4–5): 302–305.

Machiels D, Istasse L. 2002. Maillard reaction: Importance and applications in food chemistry. Annales de Medecine Veterinaire 146(6): 347–352.

Malec LS, Gonzales ASP, Naranjo GB, Vigo MS. 2002. Influence of water activity and storage temperature on lysine availability of a milk-like system. Food Research International 35(9): 849–853.

Manso MC, Oliveira FAR, Oliveira JC, Frías JM. 2001. Modelling ascorbic acid thermal degradation and browning in orange juice under aerobic conditions. International Journal of Food Science and Technology 36(3): 303–312.

Manzocco L, Calligaris S, Mastrocola D, Nicoli MC, Lerici CR. 2001. Review of non-enzymic browning and antioxidant capacity in processed foods. Trends in Food Science and Technology 11(9–10): 340–346.

Martins SIFS, van Boekel MAJS. 2003. Melanoidins extinction coefficient in the glucose/glycine Maillard reaction. Food Chemistry 83(1): 135–142.

Marshall MR, Kim J, Wei C-I. 2000. Enzymatic browning in Fruits, Vegetables and Seafoods. http://www.fao.org/waicent/faoinfo/agricult/ags/Agsi/ENZYMEFINAL/COPYRIGH.HTM. FAO, Rome, Italy.

Mathew AG, Parpia HA. 1971. Food browning as a polyphenol reaction. Advances in Food Research 19:75–145.

Mayer AM, Harel E. 1991. Phenoloxidase and their significance in fruit and vegetables. In: PF Fox, Food Enzymology. London, Elsevier. P. 373.

Mdluli KM, Owusu-Apenten R. 2003. Enzymic browning in marula fruit 1: Effect of endogenous antioxidants on marula fruit polyphenol oxidase. Journal of Food Biochemistry 27(1): 67–82.

Min S, Jin ZT, Min SK, Yeom H, Zhang QH. 2003. Commercial-scale pulsed electric field processing of orange juice. Journal of Food Science 68(4): 1265–1271.

Mölnar-Perl I, Friedman M. 1990. Inhibition of browning by sulfur amino acids. 3. Apples and Potatoes. Journal of Agricultural and Food Chemistry 38: 1652–1656.

Montgomery MW, Day EA. 1965. Aldehyde-amine condensation reaction: A possible fate of carbonyl in foods. Journal of Food Science 30(5): 828–832.

Morales FJ, van Boekel MAJS. 1998. A study on advanced Maillard reaction in heated casein/sugar solutions: Color formation. International Dairy Journal 8 (10–11): 907–915.

Mottram DS. 1998. Flavour formation in meat and meat products: A review. Food Chemistry 62(4): 415–424.

Mottram DS, Wedzicha BL, Dodson AT. 2002. Acrylamide is formed in the Maillard reaction. Nature 419(3 October): 448.

Mundt S, Wedzicha BL. 2003. A kinetic model for the glucose-fructose-glycine browning reaction. Journal of Agricultural and Food Chemistry 51(12): 3651–3655.

Murata M, Haruta M, Murai N, Tanikawa N, Nishimura M, Homma S, Itho A. 2000. Transgenic apple (Malus x domestica) shoot showing low browning potential. Journal of Agricultural and Food Chemistry 48(11): 5243–5248.

Murata M, Nishimura M, Murai N, Haruta M, Homma S, Itoh Y. 2001. A transgenic apple callus showing reduced polyphenol oxidase activity and lower browning potential. Bioscience Biotechnology and Biochemistry 65(2): 383–388.

Nagy S, Smoot J. 1977. Temperature and storage effects on percent retention and percent united-states recommended dietary allowance of vitamin C in canned single-strength orange juice. Journal of Agricultural and Food Chemistry 25(1): 135–138.

Naim M, Schutz O, Zehavi, U, Rouseff RL, Haleva Toledo, E. 1997. Effects of orange juice fortification with thiols on p-vinylguaiacol formation, ascorbic acid degradation, browning, and acceptance during pasteurisation and storage under moderate conditions. Journal of Agricultural and Food Chemistry 45(5): 1861–1867.

Namiki M. 1988. Chemistry of Maillard reactions: Recent studies on the browning reaction. Mechanism and the development of antioxidant and mutagens. Advances in Food Research 32:116–170.

Namiki M, Oka M, Otsuka M, Miyazawa T, Fujimoto K, Namiki K. 1993. Weak chemiluminiscence at an early-stage of the Maillard reaction. Journal of Agricultural and Food Chemistry 41(10): 1704–1709.

Nawar WW. 1985. Lipids. In OR Fennema, editor, Food Chemistry. Marcel Dekker: New York. Pp. 139–244.

———. 1996. Lipids. In: OR Fennema, editor, Food Chemistry, 3rd edition. Marcel Dekker: New York. Pp. 225–319.

Negishi O, Ozawa T. 2000. Inhibition of enzymic browning and protection of sulfhydryl enzymes by thiol compounds. Phytochemistry 54(5): 481–487.

Newman SM, Eannetta NT, Yu H, Prince JP, Carmen de Vicente M, Tanksley SD, Steffens JC. 1993. Organization of the tomato polyphenol oxidase gene family. Plant Mololecular Biology 21(6): 1035–1051.

Nicolas JJ, Richard-Forget FC, Goupy PM, Amoit MJ, Aubert SY. 1994. Enzymic browning reactions in apple and apple products. Critical Review of Food Science and Nutrition 32(2): 109–157.

Nielsen HK, Loliger J, Hurrel F. 1985. Reaction of proteins with oxidized lipids. 1. Analytical measurement of lipid oxidation and amino acid losses in a whey protein-methyl lineolate model system. British Journal of Nutrition 53(1): 61–73.

Oak J-H, Miyazawa T, Nakagawa K. 2002. UV analysis of Amadori-glycated phosphatidylethanolamine in foods and biological samples. Journal of Lipids Research 43(3): 523–529.

Oak J-H, Nakagawa K, Miyazawa T. 2000. Synthetically prepared Amadori-glycated phosphatidylethanolamine can trigger lipid peroxidation via free radical reactions. FEBS Letter 481(1): 26–30.

Okumura S, Kawai H. 1970. Amino acids for seasoning food. Japanese Patent 7039622.

Olano A, Martínez-Castro I. 1996. Nonenzymatic browning. In: ML Nollet, editor, Handbook of Food Analysis, vol. 2. Marcel Dekker, New York. Pp. 1683–1721.

Park CW, Kang KO, Kim WJ. 1998. Effects of reaction conditions for improvement of caramelization rate. Korean Journal of Food Science and Technology 30(4): 983–987.

Parpinello GP, Chinnici F, Versari A, Riponi C. 2002. Preliminary study on glucose oxidase-catalase enzyme system to control the browning of apple and pear purees. Food Science and Technology 35(3): 239–243.

Perez-Gago MB, Serra M, Alonso M, Mateos M, del Rio M. 2003. Effect of solid content and lipid content of whey protein isolate-beeswax edible coatings on color change of fresh cut apples. Journal of Food Science 68(7): 2186–2191.

Perez-Gilabert M, García-Carmona F. 2000. Characterization of catecholase and cresolase activities of eggplant polyphenol oxidase. Journal of Agricultural and Food Chemistry 48(3): 695–700.

Pilizota V, Subaric D. 1998. Control of enzymic browning of foods. Food Technology and Biotechnology 36(3): 219–227.

Pischetsrieder M, Larisch B, Muller U, Severin T. 1995. Reaction of ascorbic with aliphatic amines. Journal of Agricultural and Food Chemistry 43(12): 3004–3006.

Pischetsrieder M, Larisch B, Seidel W. 1997. Immunochemical detection of oxalic acid monoamides that are formed during the oxidative reaction of L-ascorbic and proteins. Journal of Agricultural and Food Chemistry 45(6): 2070–2075.

Pitotti A, Elizalde BE, Anese M. 1995. Effect of caramelization and Maillard reaction products on peroxide activity. Journal of Food Biochemistry 18(6): 445–457.

Premakumar K, Khurduya DS. 2002. Effect of microwave blanching on nutritional qualities of banana puree. Journal of Food Science and Technology-Mysore 39(3): 258–260.

Prestamo G, Sanz PD, Arroyo G. 2000. Fruit preservation under high hydrostatic pressure. High Pressure Research 19(1–6): 535–542.

Rada-Mendoza M, Olano A, Villamiel M. 2002. Furosine as indicator of Maillard reaction in jams and fruit-based infant foods. Journal of Agriculture and Food Chemistry 50(14): 4141–4145.

Rada-Mendoza M, García-Baños JL, Villamiel M, Olano A. 2004. Study on nonenzymatic browning in cookies, crackers and breakfast cereals by maltulose and furosine determination. Journal of Cereal Science 39(2): 167–173.

Ramírez-Jimenez A, Garcia-Villanova B, Guerra-Hernandez E. 2004. Effect to storage conditions and inclusión of milk on available lysine in infant cereals. Food Chemistry 85(2): 239–244.

Requena JR, Ahmed MU, Fountain CW, Degenhardt TP, Reddy S, Perez C, Lyons TJ, Jenkins AJ, Baynes JW, Thorpe SR. 1997. Carboxymethylethanolamine, a biomarker of phospholipid modification during the Maillard reaction in vivo. The Journal of Biological Chemistry 272(28): 14473–17479.

Resmini P, Pellegrino L, Batelli G. 1990. Accurate quantification of furosine in milk and dairy products by direct HPLC method. Italian Journal of Food Science 2(3): 173–183.

Rhee C, Kim, DH. 1975. Antioxidant activity of acetone extracts obtained from a caramelization-type browning reaction. Journal of Food Science 40(3): 460–462.

Roach JAG, Andrzejewski D, Gay ML, Nortrup D, Musser SM. 2003. Rugged LC-MS/MS survey analysis for acrylamide in foods Journal of Agricultural and Food Chemistry 51(26): 7547–7554.

Rocha AMCN, Morais AMMB. 2001. Influence of controlled atmosphere storage on polyphenoloxidase activity in relation to color changes of minimally processed 'Jonagored' apple. International Journal of Food Science and Technology 36(4): 425–432.

Rogacheva SM, Kuntcheva MJ, Panchev IN, Obretenov TD. 1995. L-ascorbic acid in no enzymatic reactions. 1. Reaction with glycine. Zeitschrift fur Lebensmittel Untersuchung und Forschung 200(1): 52–58.

Roig MG, Bello JF, Rivera ZS, Kennedy JF. 1999. Studies on the occurrence of non-enzymic browning during storage of citrus juice. Food Research International 32(9): 609–619.

Rojas AM, Gerschenson LN. 1997a. Ascorbic acid destruction in sweet aqueous model systems. Food Science and Technology 30(6): 567–572.

———. 1997b. Influence of system composition on ascorbic acid destruction at processing temperatures. Journal of the Science of Food and Agriculture 74(3): 369–378.

Ruiz JC, Guerra-Hernandez E, Garcia-Villanova B. 2004. Furosine is a useful indicator in pre-baked breads. Journal of the Science of Food and Agriculture 84(4): 336–370.

Ryu SY, Roh HJ, Noh BS, Kim SY, Oh DK, Lee WR, Kim SS. 2003. Effects of temperature and pH on the non-enzymic browning reaction of tagatose-glycine model system. Food Science and Biotechnology 12(6): 675–679.

Sanz ML, Corzo N, Olano A. 2003. 2-Furoylmethyl amino acids and hydroxymethylfurfural as indicators of honey quality Journal of Agricultural and Food Chemistry 51(15): 4278–4283.

Sanz ML, del Castillo MD, Corzo N, Olano A. 2000. Presence of 2 furoylmethyl derivatives in hydrolyzates of processed tomato products. Journal of Agricultural and Food Chemistry 48(2): 468–471.

———. 2001. Formation of Amadori compounds in dehydrated fruits. Journal of Agricultural and Food Chemistry 49(11): 5228-5231.

Sawamura M, Nakagawa T, Katsuno S, Hamaguchi H, Ukeda H. 2000. The effect of antioxidants on browning and on degradation products caused by dehydroascorbic acid. Journal of Food Science 65(1): 20–23.

Sawamura M, Takemoto K, Li ZF. 1991. C-14 Studies on browning of dehydroascorbic acid in an aqueous solution. Journal of Agricultural and Food Chemistry 39(10): 1735–1737.

Sawamura M, Takemoto K, Matsuzaki Y, Ukeda H, Kusunose H. 1994. Identification of 2 degradation products from aqueous dehydroascorbic acid. Journal of Agricultural and Food Chemistry 42(5): 1200–1203.

Sciancalepore V. 1985. Enzymic browning in 5 olive varieties. Journal of Food Science 54(4): 1194–1195.

Severini C, Baiano A, De Pilli T, Romaniello R, Derossi A. 2003. Prevention of enzymic browning in sliced potatoes by blanching in boiling saline solutions. Food Science and Technology 36(7): 657–665.

Severini C, de Pilli T, Baiano A, Mastrocola D, Massini R. 2001. Preventing enzymic browning of potato by microwave blanching. Sciences des Aliments 21(2): 149–160.

Smith G, Hole, M. 1991. Browning of salted sun-dried fish. Journal of the Science of Food and Agriculture 55(2): 291–301.

Soliva-Fortuny RC, Grigelmo-Miguel N, Odrizola-Serrano I, Gorinstein S, Martín-Belloso O. 2001. Browning evaluation of ready-to-eat apples as affected by modifies atmosphere packaging. Journal of Agricultural and Food Chemistry 49(8): 3685–3690.

Son SM, Moon KD, Lee CY. 2000. Kinetic study of oxalic inhibition on enzymic browning. Journal of Agricultural and Food Chemistry 48(6): 2071–2074.

Stadler RH, Blank I, Varga N, Robert F, Hau J, Guy PA, Robert MC, Riediker S. 2002. Acrylamide from Maillard reaction products. Nature 419(3 October): 449.

Subaric D, Pilizota V, Lovric T, Vukovic R, Erceg A. 2001. Effectiveness of some crown compounds on inhibition of polyphenoloxidase in model systems an in apple. Acta Alimentaria 30(1): 81–87.

Takiguchi A. 1992. Lipid oxidation and brown discoloration in niboshi during storage at ambient temperature and low temperatures. Bulletin of the Japanese Society of Scientific Fisheries 58(3): 489–494.

Tannenbaum, S.R. 1976. In: OR Fennema, editor, Principles of Food Chemistry, vol. 1. Marcel Dekker, Inc., New York and Basel. Pp. 357–360.

Tareke A, Rydberg P, Karlsson P, Eriksson S, Törnqvist M. 2002. Analysis of acrylamide, a carcinogen formed in heated foodstuffs. Journal of Agricultural and Food Chemistry 50(17): 4998–5006.

Tomasik P, Palasinski M, Wiejak S. 1989. The thermal decomposition of carbohydrates. Part I: The decomposition of mono-, di- and oligosaccharides. Advances in Carbohydrate Chemistry and Biochemistry 47:203–270.

Toussaint O, Lerch K. 1987. Catalytic oxidation of 2-aminophenols and ortho hydroxylation of aromatic amines by tyrosinase. Biochemistry 26(26): 8567–8571.

Tsai ChH, Kong MS, Pan BS. 1991. Water activity and temperature effects on non-enzymic browning of amino acids in dried squid and simulated model system. Journal of Food Science 56(3): 665–670, 677.

Utzmann CM, Lederer MO. 2000. Identification and quantification of aminophospholipid-linked Maillard compounds in model systems and egg yolk products. Journal of Agricultural and Food Chemistry 48(4): 1000–1008.

Vamosvigyazo L, Nadudvarimarkus V. 1982. Enzymic browning, polyphenol content, polyphenol oxidase and preoxidase-activities in pear cultivars. Acta Alimentaria 11(2): 157–168.

Vercet A. 2003. Browning of white chocolate during storage. Food Chemistry 81(3): 371–377.

Villamiel M, Arias M, Corzo N, Olano A. 1999. Use of different thermal indices to assess the quality of pasteurized milks. Zeitschrift fur Lebensmittel Untersuchung und Forschung 208(3): 169–171.

———. 2000. Survey of the furosine content in cheeses marketed in Spain. Journal of Food Protection 63(7): 974–975.

Villamiel M, del Castillo MD, Corzo N, Olano A. 2001. Presence of furosine in honeys. Journal of the Science of Food and Agriculture 81(8): 790–793.

Villamiel M, del Castillo MD, San Martín C, Corzo N. 1998. Assessment of the thermal treatment of orange juice during continuous microwave and conventional heating. Journal of the Science of Food and Agriculture 78(2): 196–200.

Wagner KH, Derkists S, Herr M, Schuh W, Elmadfa I. 2002. Antioxidative potential of melanoidins isolated from a roasted glucose-glycine model. Food Chemistry 78(3): 375–382.

Warmbier HC, Schnickel RA, Labuza TP. 1976. Nonenzymatic browning kinetics in an intermediate moisture model system. Effect of glucose to lysine ratio. Journal of Food Science 41(5): 981–983.

Wedzicha BL. 1984. Chemistry of sulfur dioxide in foods. London, Elsevier Applied Science Publishers, June.

Wedzicha BL, Imeson AP. 1977. Yield of 3-deoxy-4-sulfopentulose in sulfite inhibited non-enzymic browning of ascorbic acid. Journal of the Science of Food and Agriculture 28(8): 669–672.

Wedzicha BL, Mcweeny DJ. 1974. Non-enzymic browning reactions of ascorbic acid and their inhibition. Identification of 3-deoxy-4-sulphopentosulose in dehydrated, sulfited cabbage after storage. Journal of the Science of Food and Agriculture 25(5): 589–593.

Weenen H. 1998. Reactive intermediates and carbohydrate fragmentation in Maillard chemistry. Food Chemistry 62(4): 339–401.

Weisshaar R, Gutsche B. 2002. Formation of acrylamide in heated potato products. Model experiments pointing to asparagines as precursor. Deutsche Lebbensmittel 98 Jahrgang (11): 397–400.

Whitaker JR. 1972. Polyphenol oxidase. In: JR Whitaker, editor, Principles of Enzymology for the Food Sciences. New York, Marcel Dekker. Pp. 571–582.

Wolfrom ML, Kashimura N, Horton D. 1974. Factors affecting the Maillard browning reaction between sugars and amino acids studies on the no enzymatic browning of dehydrated orange juice. Journal of Agricultural and Food Chemistry 22 (5): 796–799.

Yamamoto A, Kogure M. 1969. Studies on new applications of amino acids in food industries. 1. Elimination of rancid odor in rice by L-Lysine. Journal of Food Science and Technology, Nihon Shokuhin Kogyo Gakkai-shi 16(9): 414–419.

Yasuhara A, Tanaka Y, Hengel M, Shibamoto T. 2003. Gas chromatography investigation of acrylamide formation in browning model systems. Journal of Agricultural and Food Chemistry 51(14): 3999–4003.

Yen, GC, Tsai LC. 1993. Antimutagenicity of a partially fractionated Maillard reaction-product. Food Chemistry 47(1): 11–15.

Yin D. 1996. Biochemical basis of lipofuscin, ceroid and age pigment-like fluorophores. Free Radical Biology and Medicine 21(6): 871–888.

Yin DZ, Brunk UT. 1991. Oxidized ascorbic acid and reaction-products between ascorbic acid and amino acids might constitute part of age pigments. Mechanism of Ageing and Development 61(1): 99–112.

Zamora R, Alaiz M, Hidalgo FJ. 1999. Determination of ε-N-pyrrolylnorleucine in fresh food products. Journal of Agricultural and Food Chemistry 47(5): 1942–1947.

———. 2000. Contribution of pyrrole formation and polymerization to the no enzymatic browning produced by amino-carbonyl reactions. Journal of Agricultural and Food Chemistry 48(8): 3152–3158.

Zamora R, Hidalgo FJ. 1993. Antioxidant activity of the compounds produced in the no enzymatic browning reaction of E-4,5-epoxy-E-2-heptenal and lysine. In: C Benedito de Barber, C Collar, MA Martínez-Anaya,J Morell, editors, Progress in Food Fermentation. IATA, CSIC: Valencia, Spain. Pp. 540–545.

———. 1994. Modification of lysine amino groups by the lipid peroxidation product 4,5(E)-epoxy-2(E)-heptenal. Lipids 29(4): 243–249.

———. 1995. Linoleic acid oxidation in the presence of amino compounds produces pyrroles by carbonyl amine reactions. Biochimica et Biophysica Acta 1258(3): 319–327.

———. 2001. Inhibition of proteolysis in oxidized lipid-damaged proteins. Journal of Agricultural and Food Chemistry 49(12): 6006–6011.

———. 2003a. Comparative methyl linoleate and methyl linolenate oxidation in the presence of bovine serum albumin at several lipid/protein ratios. Journal of Agricultural and Food Chemistry 51(16): 4661–4667.

———. 2003b. Phosphatidylethanolamine modification by oxidative stress product 4,5(E)-epoxy-2(E)-heptenal. Chemical Research in Toxicology 16(12): 1632–1641.

Zardetto S, Dalla Rosa M, Di Fresco S. 2003. Effects of different heat treatment on the furosine content in fresh filata pasta. Food Research International 36(9–10): 877–883.

Zenker M, Heinz V, Knorr D. 2003. Application of ultrasound-assisted thermal processing for preservation and quality retention of liquid foods. Journal of Food Protection 66(9): 1642–1649.

Zerdin K, Rooney ML, Vermue J. 2003. The vitamin C content of orange juice packed in an oxygen scavenger material. Food Chemistry 82(3): 387–395.

Ziderman II, Gregorskik S, Lopez SV, Friedman M. 1989. Thermal interaction of ascorbic acid and sodium ascorbate with proteins in relation to no enzymatic browning and Maillard reaction of foods. Journal of Agricultural and Food Chemistry 37(6): 1480–1486.

Zotos A, Petridis D, Siskos I, Gougoulias C. 2001. Production and quality assessment of a smoked tuna (Euthynnus affinis) product. Journal of Food Science 66(8): 1184–1190.

Part II
Water, Enzymology, Biotechnology, and Protein Cross-linking

Chapter 5
Water Chemistry and Biochemistry*

C. Chieh

INTRODUCTION

Water, the compound H_2O, is the most common food ingredient. Its rarely used chemical names are hydrogen oxide or dihydrogen monoxide. So much of this compound exists on the planet earth that it is often taken for granted. Water is present in solid, liquid, and gas forms in the small range of temperatures and pressures near the surface of the earth. Moreover, natural waters always have substances dissolved in them, and only elaborate processes produce pure water.

The chemistry and physics of water are organized studies of water: its chemical composition, formation, molecular structure, rotation, vibration, electronic energies, density, heat capacity, temperature dependency of vapor pressure, and its collective behavior in condensed phases (liquid and solid). In a broader sense, the study of water also includes interactions of water with atoms, ions, molecules, and biological matter. The knowledge of water forms the foundation for biochemistry and food chemistry. Nearly every aspect of biochemistry and food chemistry has something to do with water, because water is intimately linked to life, including the origin of life.

Science has developed many scientific concepts as powerful tools for the study of water. Although the study of water reveals a wealth of scientific concepts, only a selection of topics about water will be covered in the limited space here.

Biochemistry studies the chemistry of life at the atomic and molecular levels. Living organisms consist of many molecules. Even simple bacteria consist of many kinds of molecules. The interactions of the assembled molecules manifest life phenomena such as the capacity to extract energy or food, respond to stimuli, grow, and reproduce. The interactions follow chemical principles, and water chemistry is a key for the beginning of primitive life forms billions of years ago. The properties of water molecules give us clues regarding their interactions with other atoms, ions, and molecules. Furthermore, water vapor in the atmosphere increases the average temperature of the atmosphere by 30 K (Wayne 2000), making the earth habitable. Water remains important for human existence, for food production, preservation, processing, and digestion.

Water is usually treated before it is used by food industries. After usage, wastewaters must be treated before they are discharged into the ecological system. After we ingest foods, water helps us to digest, dissolve, carry, absorb, and transport nutrients to their proper sites. It further helps hydrolyze, oxidize, and utilize the nutrients to provide energy for various cells, and eventually, it carries the biological waste and heat out of our bodies. Oxidations of various foods also produce water. How and why water performs these functions depends very much on its molecular properties.

THE COMPOUND WATER

A **compound** is a substance that is made up of two or more basic components called **chemical elements** (e.g., hydrogen, carbon, nitrogen, oxygen, iron) commonly found in food. **Water** is one of the tens of millions of compounds in and on earth.

The chemical equation and thermal dynamic data for the formation of water from hydrogen and oxygen gas is

$$2H_2(g) + O_2(g) = 2H_2O(l),\ \Delta H^0 = -571.78\ \text{kJ}$$

The equation indicates that 2 mol of gaseous hydrogen, $H_2(g)$, react with 1 mol of gaseous oxygen, $O_2(g)$, to form 2 mol of liquid water. If all reactants and products are at their **standard states** of 298.15 K and 101.325 kPa (1.0 atm), formation of 2 mol of water releases 571.78 kJ of energy, as indicated by the negative sign for ΔH^0. Put in another way, the **heat of formation** of water, ΔH_f^0, is -285.89 (= $-571.78/2$) kJ/mol. Due to the large amount of energy released, the water vapor formed in the reaction is usually at a very high temperature compared with its standard state, liquid at 298.15 K. The heat of formation includes the heat that has to be removed when the vapor is condensed to liquid and then cooled to 298.15 K.

The reverse reaction, that is, the decomposition of water, is endothermic, and energy, a minimum of 285.89 kJ per mole of water, must be supplied. More energy is required by electrolysis to decompose water because some energy will be wasted as heat.

A hydrogen-containing compound, when fully oxidized, also produces water and energy. For example, the oxidation of solid (s) sucrose, $C_{12}H_{22}O_{11}$, can be written as

$$C_{12}H_{22}O_{11}(s) + 12O_2(g) = 12CO_2(g) + 11H_2O(l),\ \Delta H^0 = -5640\ \text{kJ}$$

The amount of energy released, -5640 kJ, is called the **standard enthalpy of combustion** of sucrose. The chemical energy derived this way can also be used to produce high-energy biomolecules. When oxidation is carried out in human or animal bodies, the oxidation takes place at almost constant and low temperatures. Of course, the oxidation of sucrose takes place in many steps to convert each carbon to CO_2.

THE POLAR WATER MOLECULES

During the 20th century, the study of live organisms evolved from physiology and anatomy to biochemistry and then down to the molecular level of intermolecular relations and functions. Atoms and molecules are the natural building blocks of matter, including that of living organisms. Molecular shapes, structures, and properties are valuable in genetics, biochemistry, food science, and molecular biology, all involving water. Thus, we have a strong desire to know the shape, size, construction, dimension, symmetry, and properties of water molecules, because they are the basis for the science of food and life.

The structure of water molecules has been indirectly studied using X-ray diffraction, spectroscopy,

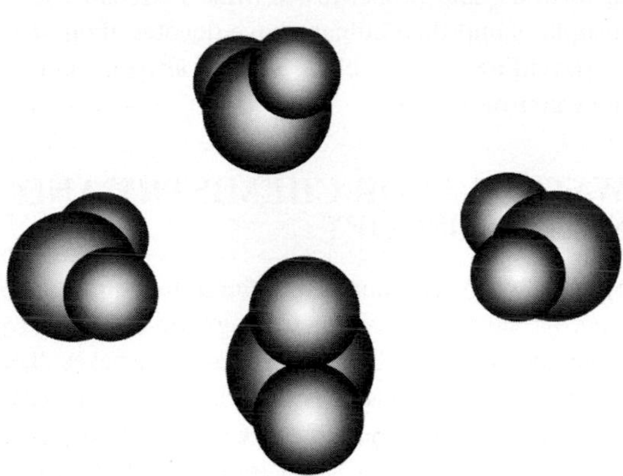

Figure 5.1. Some imaginative models of the water molecule, H_2O.

theoretical calculations, and other methods. Specific molecular dimensions from these methods differ slightly, because they measure different properties of water under different circumstances. However, the O–H bond length of 95.72 pm (1 pm = 10^{-12} m) and the H–O–H bond angle of 104.52° have been given after careful review of many recent studies (Petrenko and Whitworth 1999). The atomic radii of H and O are 120 and 150 pm, respectively. The bond length is considerably shorter than the sum of the atomic radii. Sketches of the water molecule are shown in Figure 5.1 (spherical atoms assumed).

Bonding among the elements H, C, N, and O is the key for biochemistry and life. Each carbon atom has the ability to form four chemical bonds. Carbon atoms can bond to other carbon atoms as well as to N and O atoms, forming chains, branched chains, rings, and complicated molecules. These carbon-containing compounds are called **organic compounds**, and they include foodstuff. The elements N and O have one and two more electrons, respectively, than carbon , and they have the capacity to form only three and two chemical bonds with other atoms. **Quantum mechanics** is a theory that explains the structure, energy, and properties of small systems such as atoms and molecules. It provides excellent explanations for the electrons in the atom as well as the bonding of molecules, making the observed hard facts appear trivial (Bockhoff 1969).

The well-known inert elements helium (He) and neon (Ne) form no chemical bonds. The elements C, N, and O have four, five, and six electrons more than He, and these are called **valence electrons** (VE). Quantum mechanical designation for the VE of C, N, and O are $2s^2 2p^2$, $2s^2 2p^3$, and $2s^2 2p^4$, respectively. The C, N, and O atoms share electrons with four, three, and two hydrogen atoms, respectively. The formation of methane (CH_4), ammonia (NH_3), and water (H_2O) gave the C, N, and O atoms in these molecules eight VE. These molecules are related to each other in terms of bonding (Fig. 5.2).

The compounds CH_4, NH_3, and H_2O have zero, one, and two **lone pairs** (electrons not shared with hydrogen), respectively. Shared electron pairs form single bonds. The shared and lone pairs dispose themselves in space around the central atom symmetrically or slightly distorted when they have both bonding and lone pairs.

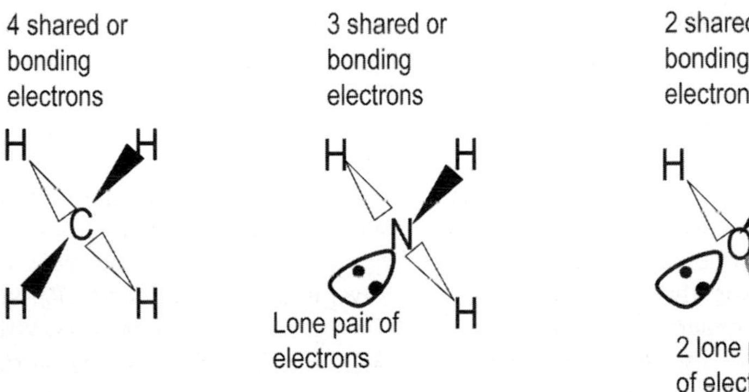

Figure 5.2. Molecules of CH_4, NH_3, and H_2O.

The lone pairs are also the negative sites of the molecule, whereas the bonded H atoms are the positive sites. The discovery of protons and electrons led to the idea of charge distribution in molecules. Physicists and chemists call NH_3 and H_2O **polar molecules** because their centers of positive and negative charge do not coincide. The polarizations in nitrogen and oxygen compounds contribute to their important roles in biochemistry and food chemistry. Elements C, N, and O play important and complementary roles in the formation of life.

Electronegativity is the ability of an atom to attract bonding electrons towards itself, and electronegativity increases in the order of H, C, N, and O (Pauling 1960). Chemical bonds between two atoms with different electronegativity are polar because electrons are drawn towards the more electronegative atoms. Thus, the polarity of the bonds increases in the order of H–C, H–N, and H–O, with the H atoms as the positive ends. The directions of the bonds must also be taken into account when the polarity of a whole molecule is considered. For example, the four slightly polar H–C bonds point toward the corner of a regular tetrahedron, and the polarities of the bonds cancel one another. The symmetric CH_4 molecules do not have a net dipole moment, and CH_4 is nonpolar. However, the asymmetric NH_3 and H_2O molecules are polar. Furthermore, the lone electron pairs make the NH_3 and H_2O molecules even more polar. The lone pairs also make the H–N–H, and H–O–H angles smaller than the 109.5° of methane. The chemical bonds become progressively shorter from CH_4 to H_2O as well. These distortions cause the dipole moments of NH_3 and H_2O to be 4.903×10^{-30} Cm (= 1.470 D) and 6.187×10^{-30} Cm (= 1.855 D), respectively. The tendency for water molecules to attract the positive sites of other molecules is higher than that of the ammonia molecule, because water is the most polar of the two.

Bond lengths and angles are based on their equilibrium positions, and their values change as water molecules undergo vibration and rotation or when they interact with each other or with molecules of other compounds. Thus, the bond lengths, bond angles, and dipole moments change slightly from the values given above. Temperature, pressure, and the presence of electric and magnetic fields also affect these values.

Using atomic orbitals, valence bond theory, and molecular orbital theory, quantum mechanics has given beautiful explanations regarding the shapes, distortions, and properties of these molecules. Philosophers and theoreticians have devoted their lives to providing a comprehensive and artistic view of the water molecules.

WATER VAPOR CHEMISTRY AND SPECTROSCOPY

Spectroscopy is the study of the absorption, emission, or interaction of electromagnetic radiation by molecules in solid, liquid, and gaseous phases. The spectroscopic studies of vapor, in which the H_2O molecules are far apart from each other, reveal a wealth of information about individual H_2O molecules.

Electromagnetic radiation (light) is the transmission of energy through space via no medium by the oscillation of mutually perpendicular electric and magnetic fields. The oscillating **electromagnetic waves** move in a direction perpendicular to both fields at the speed of light ($c = 2.997925 \times 10^8$ m/s). Max Planck (1858–1947) thought the waves also have particle-like properties except that they have no mass. He further called the light particles **photons,** meaning bundles of light energy. He assumed the **photon's energy,** *E*, to be proportional to its frequency. The proportional constant h (= 6.62618×10^{-34} J/s), now called the **Planck constant** in his honor, is universal. The validity of this assumption was shown by Albert Einstein's photoelectric-effect experiment.

Max Planck theorized that a bundle of energy converts into a light wave. His theory implies that small systems can be only at certain energy states called **energy levels**. Due to quantization, they can gain or lose only specific amounts of energy. Spectroscopy is based on these theories. Water molecules have quantized energy levels for their rotation, vibration, and electronic transitions. Transitions between energy levels result in the emission or absorption of photons.

The electromagnetic spectrum has been divided into several regions. From low energy to high energy, these regions are long radio wave, short radio wave, microwave, infrared (IR), visible, ultraviolet (UV), X rays, and gamma rays. Visible light of various colors is actually a very narrow region within the spectrum. On the other hand, IR and UV regions are very large, and both are often further divided into near and far, or A and B, regions.

Microwaves in the electromagnetic spectrum range from 300 MHz (3×10^8 cycles/s) to 300 GHz (3×10^{11} cycles/s). The water molecules have many rotation modes. Their pure rotation energy levels are very close together, and the transitions between pure rotation levels correspond to microwave photons. **Microwave spectroscopy** studies led to, among other valuable information, precise bond lengths and angles.

Water molecules vibrate, and there are some fundamental vibration modes. The three fundamental vibration modes of water are symmetric stretching (for $^1H_2{}^{16}O$), ν_1, 3657 cm^{-1}, bending ν_2, 1595 cm^{-1}, and asymmetric stretching ν_3, 3756 cm^{-1}. These modes are illustrated in Figure 5.3. Vibration energy levels are represented by three integers, ν_1, ν_2, and ν_3, to represent the combination of the basic modes. The frequencies of fundamental vibration states differ in molecules of other isotopic species (Lemus 2004).

Water molecules absorb photons in the IR region, exciting them to the fundamental and combined overtones. As pointed out earlier, water molecules also rotate. The rotation modes combine with any and all vibration modes. Thus, transitions corresponding to the vibration-rotation energy levels are very complicated, and they occur in the **infrared** (frequency range 3×10^{11} to 4×10^{14} Hz) region of the electromagnetic spectrum. High-resolution **IR spectrometry** is powerful for the study of water in the atmosphere and for water analyses (Bernath 2002a).

Visible light spans a narrow range, with wavelengths between 700 nm (red) and 400 nm (violet) (frequency 4.3–7.5 $\times$ 10^{14} Hz, wave number 14,000–25,000 cm^{-1}, photon energy 2–4 eV). It is interesting to note that the sun surface has a temperature of about 6000 K, and the visible region has the highest intensity of all wavelengths. The solar emission spectrum peaks at 630 nm (16,000 cm^{-1}, 4.8×10^{14} Hz), which is orange (Bernath 2002b).

Water molecules that have energy levels corresponding to very high overtone vibrations absorb photons of visible light, but the absorptions are very weak. Thus, visible light passes through water vapor with little absorption, resulting in water being transparent. On the other hand, the absorption gets progressively weaker from red to blue (Carleer et al. 1999). Thus, large bodies of water appear slightly blue.

Because visible light is only very weakly absorbed by water vapor, more than 90% of light passes through the atmosphere and reaches the earth's surface. However, the water droplets in clouds (water aerosols) scatter, refract, and reflect visible light, giving rainbows and colorful sunrises and sunsets.

Like the IR region, the **ultraviolet** (UV, 7×10^{14} to 1×10^{18} Hz) region spans a very large range in the electromagnetic spectrum. The photon energies are rather high, > 4 eV, and they are able to excite the electronic energy states of water molecules in the gas phase.

There is no room to cover the molecular orbitals (Gray 1964) of water here, but by analogy to electrons in atomic orbitals one can easily imagine that molecules have molecular orbitals or energy states. Thus, electrons can also be promoted to higher empty

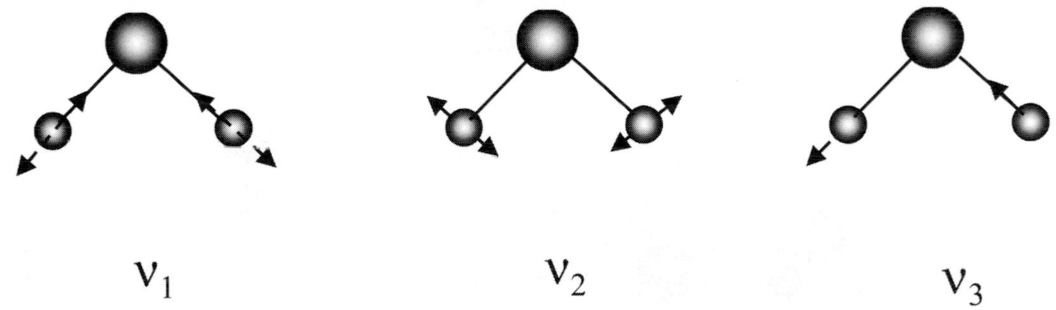

Figure 5.3. The three principle vibration modes of the water molecule, H_2O: ν_1, symmetric stretching; ν_2, bending; and ν_3, asymmetric stretching.

molecular orbitals after absorption of light energy. Ultraviolet photons have sufficiently high energies to excite electrons into higher molecular orbitals. Combined with vibrations and rotations, these transitions give rise to very broad bands in the UV spectrum. As a result, gaseous, liquid, and solid forms of water strongly absorb UV light (Berkowitz 1979). The absorption intensities and regions of water vapor are different from those of ozone, but both are responsible for UV absorption in the atmosphere. Incidentally, both triatomic water and ozone molecules are bent.

Hydrogen Bonding and Polymeric Water in Vapor

Attraction between the lone pairs and hydrogen among water molecules is much stronger than any dipole-dipole interactions. This type of attraction is known as the hydrogen bond (O–H—O), a very prominent feature of water. Hydrogen bonds are directional and are more like covalent bonds than strong dipole-dipole interactions. Each water molecule has the capacity to form four hydrogen bonds, two by donating its own H atoms and two by accepting H atoms from other molecules. In the structure of ice, to be described later, all water molecules, except those on the surface, have four hydrogen bonds.

Attractions and strong hydrogen bonds among molecules form **water dimers** and polymeric **water clusters** in water vapor. Microwave spectroscopy has revealed their existence in the atmosphere (Goldman et al. 2001, Huisken et al. 1996).

As water dimers collide with other water molecules, trimer and higher polymers form. The directional nature of the hydrogen bond led to the belief that water clusters are linear, ring, or cage-like rather than aggregates of molecules in clusters (see Fig. 5.4). Water dimers, chains, and rings have one and two hydrogen-bonded neighbors. There are three neighbors per molecule in cage-like polymers. Because molecules are free to move in the gas and liquid state, the number of nearest neighbors is between four and six. Thus, water dimers and clusters are entities between water vapor and condensed water (Bjorneholm et al. 1999).

By analogy, when a few water molecules are intimately associated with biomolecules and food mole-

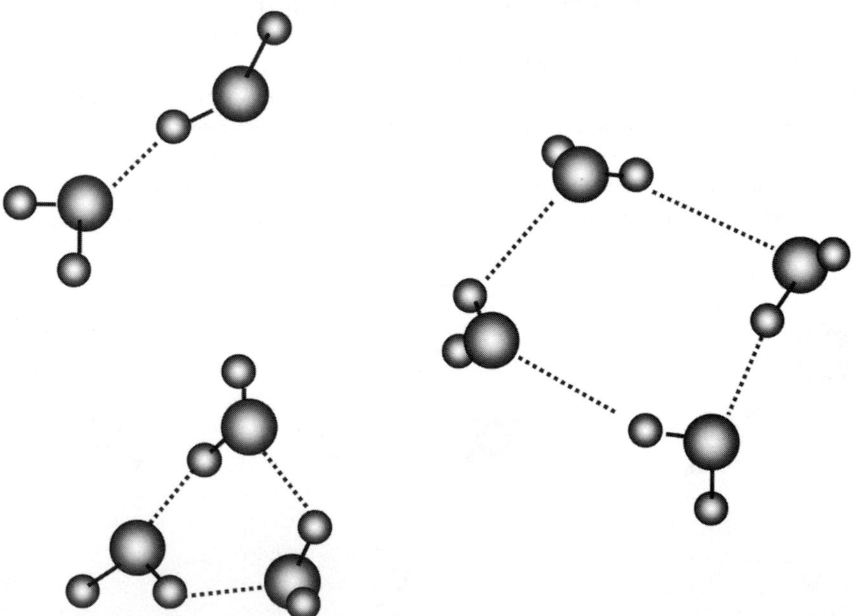

Figure 5.4. Hydrogen bonding in water dimers and cyclic forms of trimer and tetramer. Linear and transitional forms are also possible for trimers, tetramers, and polymers.

cules, their properties would be similar to those of clusters.

CONDENSED WATER PHASES

Below the critical temperature of 647 K (374°C) and under the proper pressure, water molecules condense to form a liquid or solid—**condensed water**. Properties of water, ice, and vapor must be considered in freezing, pressure-cooking, and microwave heating. In food processing, these phases transform among one another. The transitions and the properties of condensed phases are manifestations of microscopic properties of water molecules. However, condensation modifies microscopic properties such as bond lengths, bond angles, vibration, rotation, and electronic energy levels. The same is true when water molecules interact with biomolecules and food molecules. All phases of water play important parts in biochemistry and food science.

Water has many anomalous properties, which are related to polarity and hydrogen bonding. The melting point (mp), boiling point (bp), and critical temperature are abnormally high for water. As a rule, the melting and boiling points of a substance are related to its molecular mass; the higher the molar mass, the higher the melting and boiling points. Melting and boiling points of water (molar mass 18, mp 273 K, bp 373 K) are higher than those of hydrogen compounds of adjacent elements of the same period, NH_3 (molar mass 17, mp 195 K, bp 240 K) and HF (molar mass 20, mp 190 K, bp 293 K). If we compare the hydrogen compounds of elements from the same group (O, S, Se, and Te), the normal boiling point of H_2O (373 K) is by far the highest among H_2S, H_2Se, and H_2Te. Much energy (21 kJ mol^{-1}) is required to break the hydrogen bonds. The strong hydrogen bonds among water molecules in condensed phase result in anomalous properties, including the high enthalpies (energies) of fusion, sublimation, and evaporation given in Table 5.1. Internal energies and entropies are also high.

Densities of water and ice are also anomalous. Ice at 273 K is 9% less dense than water, but solids of most substances are denser than their liquids. Thus, ice floats on water, extending 9% of its volume above water. Water is the densest at 277 K (4°C). Being less dense at the freezing point, still water freezes from the top down, leaving a livable environment for aquatic organisms. The hydrogen bonding and polarity also lead to aberrant high surface tension, dielectric constant, and viscosity.

Table 5.1. Properties of Liquid Water at 298 K

Heat of formation ΔH_f	285.89 kJ mol^{-1}
Density at 3.98°C	1.000 g cm^3
Density at 25°C	0.9970480 g cm^3
Heat capacity	4.17856 J $g^{-1}K^{-1}$
$\Delta H_{vaporization}$	55.71 kJ mol^{-1}
Dielectric constant	80
Dipole moment	6.24×10^{-30} C m
Viscosity	0.8949 mPa s
Velocity of sound	1496.3 m s
Volumetric thermal expansion coefficient	0.0035 cm^3 $g^{-1}K^{-1}$

We illustrate the phase transitions between ice, liquid (water), and vapor in a phase diagram, which actually shows the equilibria among the common phases. Experiments under high pressure observed at least 13 different ices, a few types of amorphous solid water, and even the suggestion of two forms of liquid water (Klug 2002, Petrenko and Whitworth 1999). If these phases were included, the phase diagram for water would be very complicated.

SOLID H_2O

At 273.16 K, ice, liquid H_2O, and H_2O vapor at 611.15 Pa coexist and are at equilibrium; the temperature and pressure define the **triple-point** of water. At the normal pressure of 101.3 kPa (1 atm), ice melts at 273.15 K. The temperature for the equilibrium water vapor pressure of 101.3 kPa is the boiling point, 373.15 K.

Under ambient pressure, ice often does not begin to form until it is colder than 273.15 K, and this is known as **supercooling**, especially for ultrapure water. The degree of supercooling depends on volume, purity, disturbances, the presence of dust, the smoothness of the container surface, and similar factors. Crystallization starts by **nucleation**, that is, formation of ice-structure clusters sufficiently large that they begin to grow and become crystals. Once ice begins to form, the temperature will return to the freezing point. At 234 K (−39°C), tiny drops of ultrapure water would suddenly freeze, and this is known as **homogeneous nucleation** (Franks et al. 1987). Dust particles and roughness of the surface promote nucleation and help reduce supercooling for ice and frost formation.

At ambient conditions, **hexagonal ice (Ih)** is formed. Snowflakes exhibit the hexagonal symmetry. Their crystal structure is well known (Kamb 1972). Every oxygen atom has four hydrogen bonds around it, two formed by donating its two H atoms, and two by accepting the H atoms of neighboring molecules. The hydrogen bonds connecting O atoms are shown in Figure 5.5. In normal ice, Ih, the positions of the H atoms are random or disordered. The hydrogen bonds O–H—O may be slightly bent, leaving the H–O–H angle closer to 105° than to 109.5°, the ideal angle for a perfect tetrahedral arrangement. Bending the hydrogen bond requires less energy than opening the H–O–H angle. Bending of the hydrogen bond and the exchange of H atoms among molecules, forming H_3O^+ and OH^- in the solids, give rise to the disorder of the H atoms. These rapid exchanges are in a dynamic equilibrium.

In the structure of Ih, six O atoms form a ring; some of them have a chair form, and some have a boat form. Two configurations of the rings are marked by spheres representing the O atoms in Figure 5.5. Formation of the hydrogen bond in ice lengthens the O–H bond distance slightly from that in a single isolated water molecule. All O atoms in Ih are completely hydrogen bonded, except for the molecules at the surface. Maximizing the number of hydrogen bonds is fundamental to the formation of solid water phases. Pauling (1960) pointed out that formation of hydrogen bonds is partly an electrostatic attraction. Thus, the bending of O–H—O is expected. Neutron diffraction studies indicated bent hydrogen bonds.

Since only four hydrogen bonds are around each O atom, the structure of Ih has rather large channels at the atomic scale. Under pressure, many other types of structures are formed. In liquid water, the many tetrahedral hydrogen bonds are formed with immediate neighbors. Since water molecules constantly exchange hydrogen-bonding partners, the average number of nearest neighbors is usually more than four. Therefore, water is denser than Ih.

Other Phases of Ice

Under high pressures water forms many fascinating H_2O solids. They are designated by Roman numerals (e.g., ice XII; Klug 2002, Petrenko and Whitworth 1999). Some of these solids were known as early as 1900. Phase transitions were studied at certain temperatures and pressures, but metastable phases were also observed.

At 72 K, the disordered H atoms in Ih transform into an ordered solid called ice XI. The oxygen atoms of Ih and ice XI arrange in the same way, and both ices have a similar density, 0.917 Mg m^{-3}.

Under high pressure, various denser ices are formed. Ice II was prepared at a pressure about 1 GPa (1 GPa = 10^9 Pa) in 1900, and others with densities ranging from 1.17 to 2.79 Mg m^{-3} have been prepared during the 20th century. These denser ices consist of hydrogen bond frameworks different from

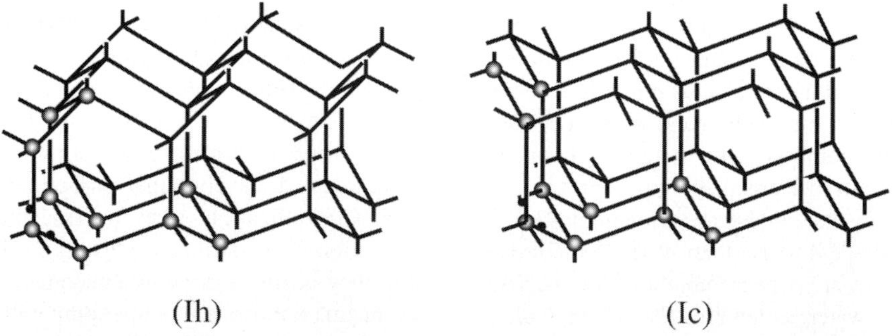

Figure 5.5. The crystal structures of ice Ih and Ic. Oxygen atoms are placed in two rings in each to point out their subtle difference. Each line represents a hydrogen bond O–H—O, and the H atoms are randomly distributed such that on average, every O atom has two O–H bonds of 100 pm. The O–H—O distance is 275 pm. The idealized tetrahedral bond angles around oxygen are 1095°.

Ih and XI, but each O atom is hydrogen-bonded to four other O atoms.

Cubic ice, Ic, has been produced by cooling vapor or droplets below 200 K (Mayer and Hallbrucker 1987, Kohl et al. 2000). More studies showed the formation of Ic between 130 and 150 K. Amorphous (glassy) water is formed below 130 K, but above 150 K Ih is formed. The hydrogen bonding and intermolecular relationships in Ih and Ic are the same, but the packing of layers and symmetry differ (see Fig. 5.5). The arrangement of O atoms in Ic is the same as that of the C atoms in the diamond structure. Properties of Ih and Ic are very similar. Crystals of Ic have cubic or octahedral shapes, resembling those of salt or diamond. The conditions for their formation suggest their existence in the upper atmosphere and in the Antarctic.

As in all phase transitions, energy drives the transformation between Ih and Ic. Several forms of amorphous ice having various densities have been observed under different temperatures and pressures. Unlike crystals, in which molecules are packed in an orderly manner, following the symmetry and periodic rules of the crystal system, the molecules in **amorphous ice** are immobilized from their positions in liquid. Thus, amorphous ice is often called frozen water or glassy water.

When small amounts of water freeze suddenly, it forms amorphous ice or glass. Under various temperatures and pressures, it can transform into high-density (1.17 Mg/m^3) amorphous water, and very high-density amorphous water. Amorphous water also transforms into various forms of ice (Johari and Anderson 2004). The transformations are accompanied by energies of transition. A complicated phase diagram for ice transitions can be found in *Physics of Ice* (Petrenko and Whitworth 1999).

High pressures and low temperatures are required for the existence of other forms of ice, and currently these conditions are seldom involved in food processing or biochemistry. However, their existence is significant for the nature of water. For example, their structures illustrate the deformation of the ideal tetrahedral arrangement of hydrogen bonding presented in Ih and Ic. This feature implies flexibility when water molecules interact with foodstuffs and with biomolecules.

Vapor Pressure of Ice Ih

The equilibrium vapor pressure is a measure of the ability or potential of the water molecules to escape from the condensed phases to form a gas. This potential increases as the temperature increases. Thus, vapor pressures of ice, water, and solutions are important quantities. The ratio of equilibrium vapor pressures of foods divided by those of pure water is called the **water activity**, which is an important parameter for food drying, preservation, and storage.

Ice sublimes at any temperature until the system reaches equilibrium. When the vapor pressure is high, molecules deposit on the ice to reach equilibrium. Solid ice and water vapor form an equilibrium in a closed system. The amount of ice in this equilibrium and the free volume enclosing the ice are irrelevant, but the water vapor pressure or partial pressure matters. The equilibrium pressure is a function of temperature, and detailed data can be found in handbooks, for example, the *CRC Handbook of Chemistry and Physics* (Lide 2003). This handbook has a new edition every year. More recent values between 193 and 273 K can also be found in *Physics of Ice* (Petrenko and Whitworth 1999).

The equilibrium vapor pressure of ice Ih plotted against temperature *(T)* is shown in Figure 5.6. The line indicates equilibrium conditions, and it separates the pressure-temperature *(P-T)* graph into two domains: vapor tends to deposit on ice in one, and ice sublimes in the other. This is the ice Ih—vapor portion of the phase diagram of water.

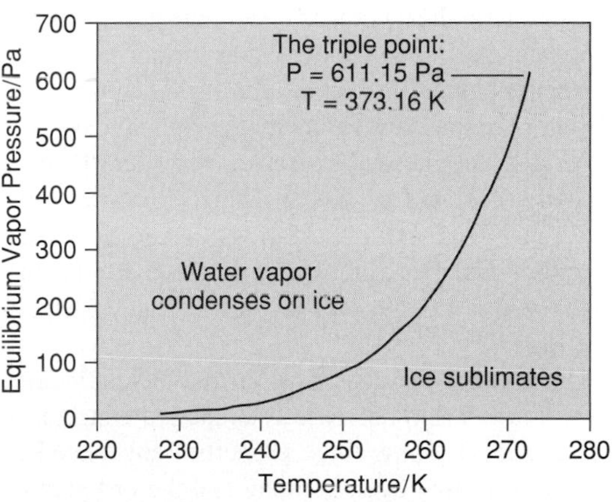

Figure 5.6. Equilibrium vapor pressure (Pa) of ice as a function of temperature (K).

Plot of ln P versus $1/T$ shows a straight line, and this agrees well with the Clausius-Clapeyron equation. The negative slope gives the enthalpy of the phase transition, ΔH, divided by the gas constant R (= 8.3145 J/K/mol).

$$\frac{(d(\ln P))}{(d(1/T))} = \frac{-\Delta H}{R}$$

This simplified equation gives an estimate of the vapor pressure (in Pa) of ice at temperatures in a narrow range around the triple point.

$$P = 611.15\exp\left(-6148\left(\frac{1}{T} - \frac{1}{273.16}\right)\right)$$

The value 6148 is the enthalpy of sublimation ($\Delta_{sub}H$ = 51.1 J/mol, varies slightly with temperature) divided by the gas constant R. Incidentally, the enthalpy of sublimation is approximately the sum of the enthalpy of fusion (6.0 J/mol) for ice and the heat of vaporization (45.1 J/mol, varies with temperature) of water at 273 K.

LIQUID H_2O—WATER

We started the chapter by calling the compound H_2O water, but most of us consider **water** the liquid H_2O. In terms of food processing, the liquid is the most important state. Water is contained in food, and it is used for washing, cooking, transporting, dispersing, dissolving, combining, and separating components of foods. Food drying involves water removal, and fermentation uses water as a medium to convert raw materials into commodities. Various forms of water ingested help digest, absorb, and transport nutrients to various part of the body. Water further facilitates biochemical reactions to sustain life. The properties of water are the basis for its many applications.

Among the physical properties of water, the heat capacity (4.2176 J $g^{-1}K^{-1}$ at 273.15 K) varies little between 273.15 and 373.15 K. However, this value decreases, reaches a minimum at about 308 K, and then rises to 4.2159–4.2176 J $g^{-1}K^{-1}$ at 373.15 K (see Fig. 5.7).

The viscosity, surface tension, and dielectric constant of liquid H_2O decrease as temperature increases (see Fig. 5.7). These three properties are related to the extent of hydrogen bonding and the ordering of the dipoles. As thermal disorder increases with rising temperature, these properties decrease. To show the variation, the properties at other temperatures are divided by the same property at 273 K. The ratios are then plotted as a function of tempertures. At 273.15 K (0°C), all the ratios are unity (1). The thermal conductivity, on the other hand, increases with temperature. Thus, the thermal conductivity at 373 K (679.1 W $K^{-1}m^{-1}$) is 1.21 times that at 273 K (561.0 W $K^{-1}m^{-1}$). Warm water better conducts heat. Faster moving molecules transport energy faster. The variations of these properties play important roles in food processing or preparation. For example, as we shall see later, the dielectric constant is a major factor for the microwave heating of food, and heat conductivity plays a role cooking food.

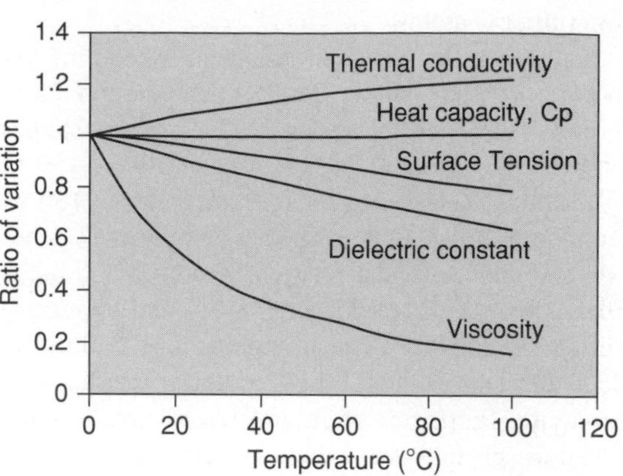

Figure 5.7. Variation of viscosity (1.793 mPa s), dielectric constant (87.90), surface tension (75.64 mN/m), heat capacity Cp (4.2176 J $g^{-1}K^{-1}$), and thermal conductivity (561.0 W $K^{-1}m^{-1}$) of water from their values at 273.16 K to 373.16 K (0 and 100°C). Values at 273.15°K are given.

Densities of other substances are often determined relative to that of water. Therefore, density of water is a primary reference. Variation of density with temperature is well known, and accurate values are carefully measured and evaluated especially between 273 and 313 K (0–40°C). Two factors affect water density. Thermal expansion reduces its density, but the reduced number of hydrogen bonds increases its density. The combined effects resulted in the highest density at approximately 277 K (4°C). Tanaka et al. (2001) has developed a formula to calculate the density within this temperature range, and the *CRC Handbook of Chemistry and Physics* (Lide 2003) has a table listing these values. The variation of

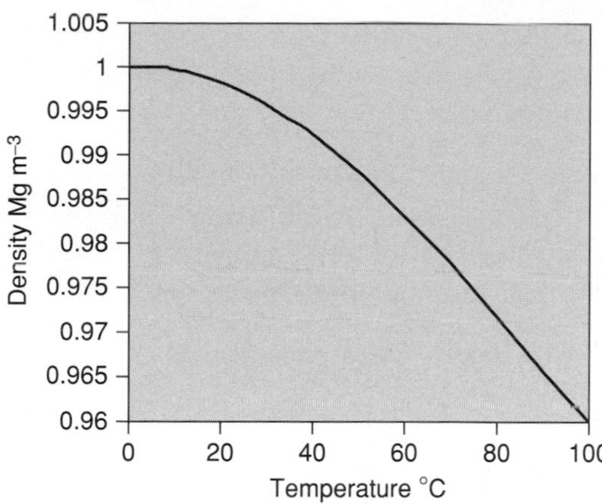

Figure 5.8. Density of water Mg m^{-3} as a function of temperature (°C).

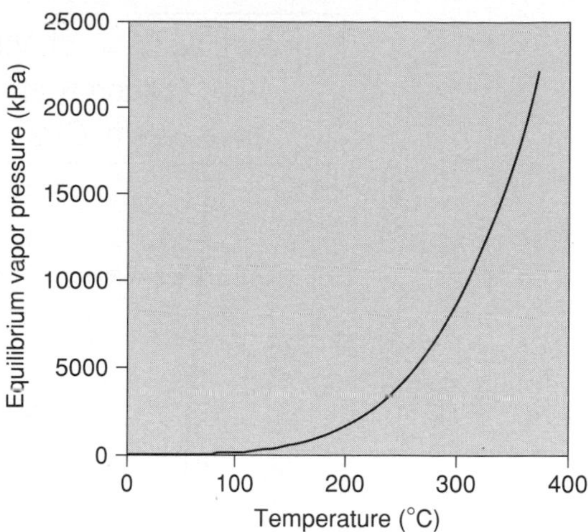

Figure 5.9. Equilibrium vapor pressure of water as a function of temperature.

water density between the freezing point and the boiling point is shown in Figure 5.8. The densities are 0.9998426, 0.9999750, and 0.9998509 Mg m^{-3} at 0, 4, and 8°C, respectively. The decrease in density is not linear, and at 100°C, the density is 0.95840 Mg m^{-3}, a decrease of 4% from its maximum.

Vapor Pressure of Liquid H_2O

Equilibrium vapor pressure of water, Figure 5.9, increases with temperature, similar to that of ice. At the triple point, the vapor pressures of ice Ih and water are the same, 0.611 kPa, and the **boiling point** (373.15 K, 100°C) is the temperature at which the vapor pressure is 101.325 kPa (1 atm). At slightly below 394 K (121°C), the vapor pressure is 202.65 kPa (2.00 atm). At 473 and 574 K, the vapor pressures are 1553.6 and 8583.8 kPa, respectively. The vapor pressure rises rapidly as temperature increases. The lowest pressure to liquefy vapor just below the **critical temperature**, 373.98°C, is 22,055 kPa (217.67 atm), and this is known as the **critical pressure**. Above 373.98°C, water cannot be liquefied, and the fluid is called **supercritical water**.

The partial pressure of H_2O in the air at any temperature is the **absolute humidity**. When the partial pressure of water vapor in the air is the equilibrium vapor pressure of water at the same temperature, the **relative humidity** is 100%, and the air is saturated with water vapor. The partial vapor pressure in the air divided by the equilibrium vapor pressure of water at the temperature of the air is the **relative humidity**, expressed as a percentage. The temperature at which the vapor pressure in the air becomes saturated is the **dew point**, at which dew begins to form. Of course when the dew point is below 273 K or 0°C, ice crystals (frost) begin to form. Thus, the relative humidity can be measured by finding the dew point and then dividing the equilibrium vapor pressure at the dew point by the equilibrium vapor pressure of water at the temperature of the air. The transformations between solid, liquid, and gaseous water play important roles in hydrology and in transforming the earth's surface. Solar energy causes phase transitions of water that make the weather.

Transformation of Solid, Liquid, and Vapor

Food processing and biochemistry involve transformations among solid, liquid, and vapor of water. Therefore, it is important to understand ice-water, ice-vapor, and water-vapor transformations and their equilibria. These transformations affect our daily lives as well. A map or diagram is helpful in order to comprehend these natural phenomena. Such a map, representing or explaining these transformations, is called a **phase diagram** (see Fig. 5.10). A sketch

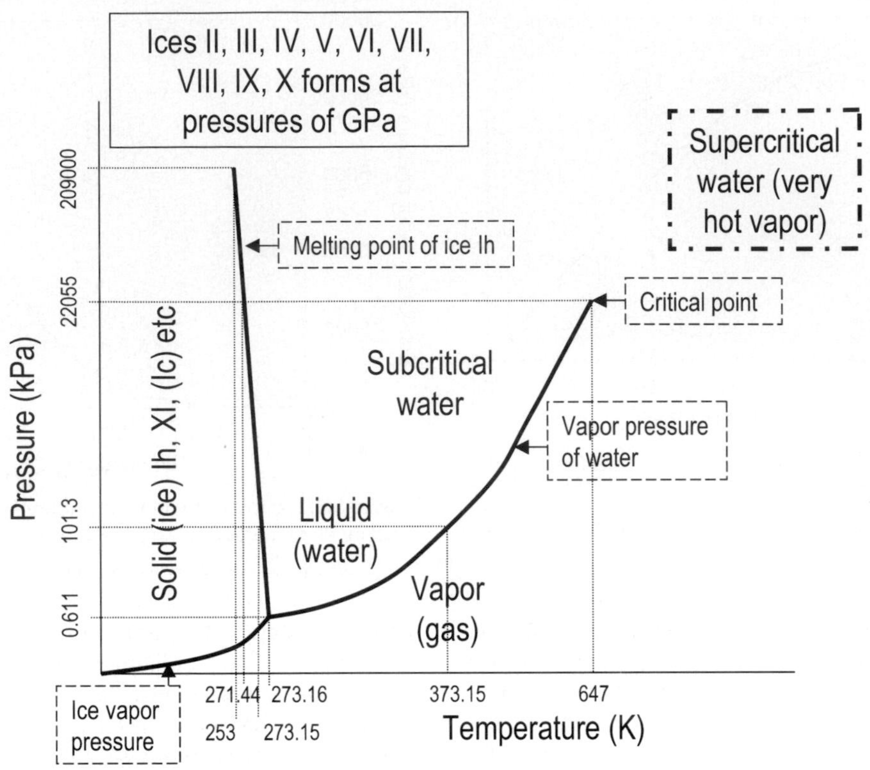

Figure 5.10. A sketch outlining the phase diagram of ice, water, and vapor.

must be used because the range of pressure involved is too large for the drawing to be on a linear scale.

The curves representing the equilibrium vapor pressures of ice and water as functions of temperature meet at the **triple point** (see Fig. 5.10). The other end of the vapor pressure curve is the critical point. The melting points of ice Ih are 271.44, 273.15, and 273.16 K at 22,055 kPa (the critical pressure), 101.325 kPa, and 0.611 kPa (the triple point), respectively. At a pressure of 200,000 kPa, Ih melts at 253 K. Thus, the line linking all these points represents the melting point of Ih at different pressures. This line divides the conditions (pressure and temperature) for the formation of solid and liquid. Thus, the phase diagram is roughly divided into regions of solid, liquid, and vapor.

Ice Ih transforms into the ordered ice XI at low temperature. In this region and under some circumstance, Ic is also formed. The transformation conditions are not represented in Figure 5.10, and neither are the transformation lines for other ices. These occur at much higher pressures in the order of gigapascals. A box at the top of the diagram indicates the existence of these phases, but the conditions for their transformation are not given. Ices II–X, formed under gigapascals pressure, were mentioned earlier. Ice VII forms at greater than 10 GPa and at a temperature higher than the boiling point of water.

Formation and existence of these phases illustrate the various hydrogen bonding patterns. They also show the many possibilities of H_2O-biomolecule interactions.

Subcritical and Supercritical Waters

Water at temperatures between the boiling and critical points (100–373.98°C) is called **subcritical water**, whereas the phase above the critical point is

supercritical water. In the 17th century, Denis Papin (a physicist) generated high-pressure steam using a closed boiler, and thereafter pressure canners have been used to preserve food. Pressure cookers were popular during the 20th century. Pressure cooking and canning use subcritical water. Plastic bags are gradually replacing cans, and food processing faces new challenges.

Properties of subcritical and supercritical water such as dielectric constant, polarity, surface tension, density, viscosity, and others, differ from those of normal water. These properties can be tuned by adjusting water temperature. At high temperature, water is an excellent solvent for nonpolar substances such as those for flavor and fragrance. Supercritical water has been used for wastewater treatment to remove organic matter, and this application will be interesting to the food industry if companies are required to treat their wastes before discharging them into the environment. The conditions for supercritical water cause polymers to depolymerize and nutrients to degrade. More research will tell whether supercritical water will convert polysaccharides and proteins into useful products.

Supercritical water is an oxidant, which is desirable for the destruction of substances. It destroys toxic material without the need of a smokestack. Water is a "green" solvent and reagent, because it causes minimum damage to the environment. Therefore, the potential for supercritical water is great.

AQUEOUS SOLUTIONS

Water dissolves a wide range of natural substances. Life began in natural waters, **aqueous solutions**, and continuation of life depends on them.

Water, a polar solvent, dissolves polar substances. Its polarity and its abilities to hydrogen bond make it a nearly universal solvent. Water-soluble polar substances are **hydrophilic**, whereas nonpolar substances insoluble in water are **hydrophobic** or **lipophilic.** Substances whose molecules have both polar and nonpolar parts are **amphiphilic**. These substances include detergents, proteins, aliphatic acids, alkaloids, and some amino acids.

The high dielectric constant of water makes it an ideal solvent for ionic substances, because it reduces the attraction between positive and negative ions in **electrolytes**: acids, bases, and salts. Electrolyte solutions are intimately related to food and biological sciences.

Colligative Properties of Aqueous Solutions

Vapor pressure of an aqueous solution depends only on the concentration of the solute, not on the type and charge of the solute. Vapor pressure affects the melting point, the boiling point, and the osmotic pressure, and these are **colligative properties**.

In a solution, the number of moles of a component divided by the total number of moles of all components is the **mole fraction** of that component. This is a useful parameter, and the mole fraction of any pure substance is unity. When 1 mol of sugar dissolves in 1.0 kg (55.56 mol) of water, the mole fraction of water is reduced to 0.9823, and the mole fraction of sugar is 0.0177. Raoult's law applies to aqueous solutions. According to it, the vapor pressure of a solution at temperature T is the product of mole fraction of the solvent and its vapor pressure at T. Thus, the vapor pressure of this solution at 373 K is 99.53 kPa (0.9823 atm). In general, the vapor pressure of a solution at the normal boiling point is lower than that of pure water, and a higher temperature is required for the vapor pressure to reach 101.3 kPa (1 atm). The net increase in the boiling temperature of a solution is known as the **boiling point elevation**.

Ice formed from a dilute solution does not contain the solute. Thus, the vapor pressure of ice at various temperatures does not change, but the vapor pressure of a solution is lower than that of ice at the freezing point. Further cooling is required for ice to form, and the net lowering of vapor pressure is the **freezing point depression**.

The freezing and boiling points are temperatures at ice-water and water-vapor (at 101.3 kPa) equilibria, not necessarily the temperatures at which ice begins to form or boiling begins. Often, the temperature at which ice crystals start to form is lower than the melting point, and this is known as **supercooling**. The degree of supercooling depends on many other parameters revealed by a systematic study of heterogeneous nucleation (Wilson et al. 2003). Similarly, the inability to form bubbles results in **superheating**. Overheated water boils explosively due to the sudden formation of many large bubbles.

Supercooling and superheating are nonequilibrium phenomena, and they are different from freezing point depression and boiling point elevation.

The temperature differences between the freezing and boiling points of solutions and those of pure water are proportional to the concentration of the solutions. The proportionality constants are **molar freezing point depression** ($K_f = -1.86$ K mol^{-1} kg) and the **molar boiling point elevation** ($K_b = 0.52$ K mol^{-1} kg), respectively. In other words, a solution containing one mole of nonvolatile molecules per kilogram of water freezes at 1.86° below 273.15 K, and boils 0.52° above 373.15 K. For a solution with concentration *C*, measured in moles of particles (molecules and positive and negative ions counted separately) per kilogram of water, the boiling point elevation or freezing point depression ΔT can be evaluated.

$$\Delta T = KC$$

where *K* represents K_f or K_b for freezing point depression or boiling point elevation, respectively. This formula applies to both cases. Depending on the solute, some aqueous solutions may deviate from Raoult's law, and the above formulas give only estimates.

In freezing or boiling a solution of a nonvolatile solute, the solid and vapor contain only H_2O, leaving the solute in the solution. A solution containing volatile solutes will have a total vapor pressure due to all volatile components in the solution. Its boiling point is no longer that of water alone. Freezing point depression and boiling point elevation are both related to the vapor pressure. In general, a solution of nonvolatile substance has a lower vapor pressure than that of pure water. The variations of these properties have many applications, some in food chemistry and biochemistry.

It is interesting to note that some fish and insects have antifreeze proteins that depress the freezing point of water to protect them from freezing in the arctic sea (Marshall et al. 2004).

Osmotic pressure is usually defined as the pressure that must be applied to the side of the solution to prevent the flow of the pure solvent passing a semipermeable membrane into the solution. Experimental results show that osmotic pressure is equal to the product of total concentration of molecules and ions *(C)*, the gas constant (R = 8.3145 J/mol/K), and the temperature (*T* in K):

$$\text{Osmotic pressure} = CRT.$$

The expression for osmotic pressure is the same as that for ideal gas. Chemists calculate the pressure using a concentration based on mass of the solvent. Since solutions are never ideal, the formula gives only estimates. The unit of pressure works out if the units for *C* are in mol/m^3, since 1 J = 1 Pa m^3. Concentration based on mass of solvent differs only slightly from that based on volume of the solution.

An **isotonic solution (isosmotic)** is one that has the same osmotic pressure as another. Solutions with higher and lower osmotic pressures are called **hypertonic (hyperosmotic)**, and **hypotonic (hypo-osmotic)**, respectively. Raw food animal and plant cells submerged in isotonic solutions with their cell fluids will not take up or lose water even if the cell membranes are semipermeable. However, in plant, soil, and food sciences, **water potential gradient** is the driving force or energy directing water movement. Water moves from high-potential sites to low-potential sites. Water moves from low osmotic pressure solutions to high osmotic pressure solutions. For consistency and to avoid confusion, negative osmotic pressure is defined as **osmotic potential**. This way, osmotic potential is a direct component of water potential for gradient consideration. For example, when red blood cells are placed in dilute solutions, their osmotic potential is negative. Water diffuses into the cells, resulting in the swelling or even bursting of the cells. On the other hand, when cells are placed in concentrated (hypertonic) saline solutions, the cells will shrink due to water loss.

Biological membranes are much more permeable than most man-made phospholipid membranes because they have specific membrane-bound proteins acting as water channels. Absorption of water and its transport throughout the body is more complicated than osmosis.

Solution of Electrolytes

Solutions of acids, bases, and salts contain ions. Charged ions move when driven by an electric potential, and electrolyte solutions conduct electricity. These ion-containing substances are called **electrolytes**. As mentioned earlier, the high dielectric constant of water reduces the attraction of ions within ionic solids and dissolves them. Furthermore, the polar water molecules surround ions, forming **hy-**

drated ions. The concentration of all ions and molecular substances in a solution contributes to the osmotic potential.

Water can also be an acid or a base, because H_2O molecules can receive or provide a proton (H^+). Such an exchange by water molecules in pure water, forming **hydrated protons** (H_3O^+ or $(H_2O)_4H^+$), is called **self-ionization**. However, the extent of ionization is small, and pure water is a very poor conductor.

Self-Ionization of Water

The self-ionization of water is a dynamic equilibrium,

$$H_2O(l) \leftrightarrow H^+(aq) + OH^-(aq),$$

$$K_w = [H^+][OH^-] = 10^{-14} \text{ at 298 K and 1 atm}$$

where $[H^+]$ and $[OH^-]$ represent the molar concentrations of H^+ (or H_3O^+) and OH^- ions, respectively, and K_w is called the **ion product of water**. Values of K_w under various conditions have been evaluated theoretically (Marshall and Franck 1981, Tawa and Pratt 1995). Solutions in which $[H^+] = [OH^-]$ are said to be **neutral**. Both **pH** and **pOH,** defined by the following equations, have a value of 7 at 298 K for a neutral solution.

$$\mathbf{pH} = -\log_{10}[H^+] = \mathbf{pOH} = -\log_{10}[OH^-] = 7 \text{ (at 298 K)}$$

The H^+ represents a hydrated proton (H_3O^+), which dynamically exchanges a proton with other water molecules. The self-ionization and equilibrium are present in water and all aqueous solutions.

Solutions of Acids and Bases

Strong acids $HClO_4$, $HClO_3$, HCl, HNO_3, and H_2SO_4 completely ionize in their solutions to give H^+ (H_3O^+) ions and anions ClO_4^-, ClO_3^-, Cl^-, NO_3^-, and HSO_4^-, respectively. Strong bases NaOH, KOH, and $Ca(OH)_2$ also completely ionize to give OH^- ions and Na^+, K^+, and Ca^{2+} ions, respectively. In an acidic solution, $[H^+]$ is greater than $[OH^-]$. For example, in a 1.00 mol/L HCl solution at 298 K, $[H^+] = 1.00$ mol/L, pH = 0.00, $[OH^-] = 10^{-14}$ mol/L.

Weak acids such as formic acid (HCOOH), acetic acid (HCH_3COO), ascorbic acid ($H_2C_6H_6O_6$), oxalic acid ($H_2C_2O_4$), carbonic acid (H_2CO_3), benzoic acid (HC_6H_5COO), malic acid ($H_2C_4H_4O_5$), lactic acid ($HCH_3CH(OH)COO$), and phosphoric acid (H_3PO_4) also ionize in their aqueous solutions, but not completely. The ionization of acetic acid is represented by the equilibrium

$$HCH_3COO(aq) \leftrightarrow H^+(aq) + CH_3COO^-(aq),$$

$$K_a = \frac{[H^+][CH_3COO^-]}{[HCH_3COO]} = 1.75 \times 10^{-5} \text{ at 298 K}$$

where K_a, as defined above, is the **acid dissociation constant**.

The solubility of CO_2 in water increases with partial pressure of CO_2, according to Henry's law. The chemical equilibrium for the dissolution is

$$H_2O(l) + CO_2(g) \leftrightarrow H_2CO_3(aq)$$

Of course, H_2CO_3 dynamically exchanges H^+ and H_2O with other water molecules, and this weak **diprotic acid** ionizes in two stages with acid dissociation constants K_{a1} and K_{a2}:

$$H_2O + CO_2(aq) \leftrightarrow H^+(aq) + HCO_3^-(aq),\ K_{a1} = 4.30 \times 10^{-7} \text{ at 298 K}$$

$$HCO_3^-(aq) \leftrightarrow H^+(aq) + CO_3^{2-}(aq),\ K_{a2} = 5.61 \times 10^{-11}.$$

Constants K_{a1} and K_{a2} increase as temperature rises, but the solubility of CO_2 decreases. At 298 K, the pH of a solution containing 0.1 mol/L H_2CO_3 is 3.7. At this pH, acidophilic organisms survive and grow, but most pathogenic organisms are neutrophiles, and they cease growing. Soft drinks contain other acids—citric, malic, phosphoric, ascorbic, and others. They lower the pH further.

All three hydrogen ions in phosphoric acid (H_3PO_4) are ionizable, and it is a **triprotic acid**. Acids having more than one dissociable H^+ are called **polyprotic acids**.

Ammonia and many nitrogen-containing compounds are weak bases. The ionization equilibrium of NH_3 in water and the **base dissociation constant** K_b are

$$NH_3 + H_2O \leftrightarrow NH_4O \leftrightarrow NH_4^+(aq) + OH^-(aq),$$

$$K_b = \frac{[H^+][OH^-]}{[NH_4OH]} = 1.70 \times 10^{-5} \text{ at 298 K.}$$

Other weak bases react with H_2O and ionize in a similar way.

The ionization or dissociation constants of inorganic and organic acids and bases are extensive, and they have been tabulated in various books (for example Perrin 1965, 1982; Kortüm et al. 1961).

Titration

Titration is a procedure for quantitative analysis of a solute in a solution by measuring the quantity of a reagent used to completely react with it. This method is particularly useful for the determination of acid or base concentrations. A solution with a known concentration of one reagent is added from a burette to a definite amount of the other. The **end point** is reached when the latter substance is completely consumed by the reagent from the burette, and this is detected by the color change of an indicator or by pH measurements. This method has many applications in food analysis.

The titration of strong acids or bases utilizes the rapid reaction between H^+ and OH^-. The unknown quantity of an acid or base may be calculated from the amount used to reach the end point of the titration.

The variation of pH during the titration of a weak acid using a strong base or a weak base using a strong acid is usually monitored to determine the end point. The plot of pH against the amount of reagent added is a **titration curve**. There are a number of interesting features on a titration curve. Titration of a weak acid HA using a strong base NaOH is based on two rapid equilibria:

$H^+ + OH^- = H_2O$, $K = \dfrac{[H_2O]}{K_w} = 5.56 \times 10^{15}$ at 298 K.

As the H^+ ions react with OH^-, more H^+ ions are produced due to the equilibrium

$HA = H^+ + A^-$, K_a (ionization constant of the acid).

Before any NaOH solution is added, HA is the dominant species; when half of HA is consumed, [HA] = $[A^-]$, which is called the **half equivalence point**. At this point, the pH varies the least when a little H^+ or OH^- is added, and the solution at this point is the **most effective buffer solution,** as we shall see later. The pH at this point is the same as the pK_a of the weak acid. When an equivalent amount of OH^- has been added, the A^- species dominates, and the solution is equivalent to a salt solution of NaA. Of course, the salt is completely ionized.

Polyprotic acids such as ascorbic acid H_2-$(H_6C_6O_6)$ (Vitamin C; $K_{a1} = 7.9 \times 10^{-5}$, $K_{a2} = 1.6 \times 10^{-12}$) and phosphoric acid H_3PO_4 ($K_{a1} = 6.94 \times 10^{-3}$, $K_{a2} = 6.2 \times 10^{-8}$, K_{a3} 2.1×10^{-12}) have more than one mole of H^+ per mole of acid. A titration curve of these acids will have two and three end points for ascorbic and phosphoric acids, respectively, partly due to the large differences in their dissociation constants (K_{a1}, K_{a2}, etc.). In practice, the third end point is difficult to observe in the titration of H_3PO_4. Vitamin C and phosphoric acids are often used as food additives.

Many food components (e.g., amino acids, proteins, alkaloids, organic and inorganic stuff, vitamins, fatty acids, oxidized carbohydrates, and compounds giving smell and flavor) are weak acids and bases. The pH affects their forms, stability, and reactions. When pH decreases by 1, the concentration of H^+, $[H^+]$, increases 10-fold, accompanied by a 10-fold decrease in $[OH^-]$. The H^+ and OH^- are very active reagents for the esterification and hydrolysis reactions of proteins, carbohydrates, and lipids, as we shall see later. Thus, the acidity, or pH, not only affects the taste of food, it is an important parameter in food processing.

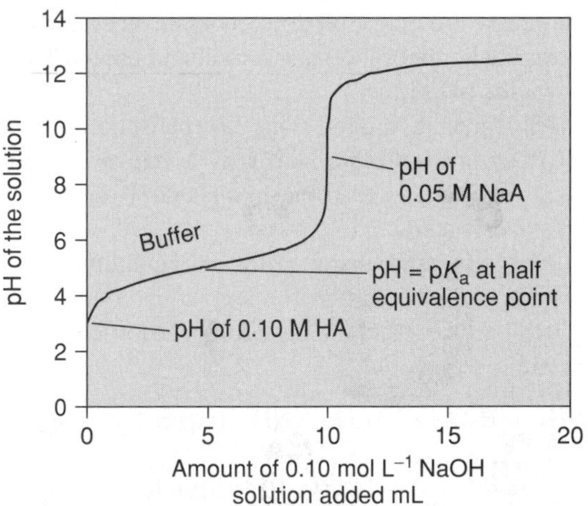

Figure 5.11. Titration curve of a 0.10 mol/L (or M) weak acid HA ($K_a = 1 \times 10^{-5}$) using a 0.10 mol/L strong base NaOH solution.

Solutions of Amino Acids

Amino acids have an amino group (NH_3^+), a carboxyl group (COO^-), a H, and a side chain (R)

attached to the asymmetric alpha carbon. They are the building blocks of proteins, polymers of amino acids. At a pH called the **isoelectric point**, which depends on the amino acid in question, the dominant species is a **zwitterion**, $RHC(NH_3^+)(COO^-)$, which has a positive and a negative site, but no net charge. For example, the isoelectric point for glycine is pH = 6.00, and its dominant species is $H_2C(NH_3^+)COO^-$. An amino acid exists in at least three forms due to the following ionization or equilibria:

$RHC(NH_3^+)(COOH) = RHC(NH_3^+)(COO^-) + H^+, K_{a1}$

$RHC(NH_3^+)(COO^-) = RHC(NH_2)(COO^-) + H^+, K_{a2}$.

Most amino acids behave like a diprotic acid with two dissociation constants, K_{a1} and K_{a2}. A few amino acids have a third ionizable group in their side chains.

Among the 20 common amino acids, the side chains of eight are nonpolar, and those of seven are polar, containing –OH, >C=O, or –SH groups. Aspartic and glutamic acid contain acidic –COOH groups in their side chains, whereas arginine, histidine, and lysine contain basic –NH or $-NH_2$ groups. These have four forms due to adding or losing protons at different pH values of the solution, and they behave as triprotic acids. For example, aspartic acid [Asp = $(COOH)CH_2C(NH_3^+)(COO^-)$] has these forms:

$AspH^+ = Asp + H^+$

$Asp = Asp^- + H^+$

$Asp^- = Asp^{2-} + H^+$

Proteins, amino acid polymers, can accept or provide several protons as the pH changes. At its **isoelectric point** (a specific pH), the protein has no net charge and is least soluble because electrostatic repulsion between its molecules is lowest, and the molecules coalesce or precipitate, forming a solid or gel.

Solutions of Salts

Salts consist of positive and negative ions, and these ions are hydrated in their solutions. Positive, hydrated ions such as $Na(H_2O)_6^+$, $Ca(H_2O)_8^{2+}$, and $Al(H_2O)_6^{3+}$ have six to eight water molecules around them. Figure 5.12 is a sketch of the interactions of water molecules with ions. The water molecules point the negative ends of their dipoles towards positive ions, and their positive ends towards negative ions. Molecules in the hydration sphere constantly and dynamically exchange with those around them. The number and lifetimes of hydrated water molecules have been studied by various methods. These studies reveal that the hydration sphere is one layer deep, and the lifetimes of these hydrated water molecules are in the order of picoseconds (10^{-12} seconds). The larger negative ions also interact with the polar water molecules, but not as strongly as do

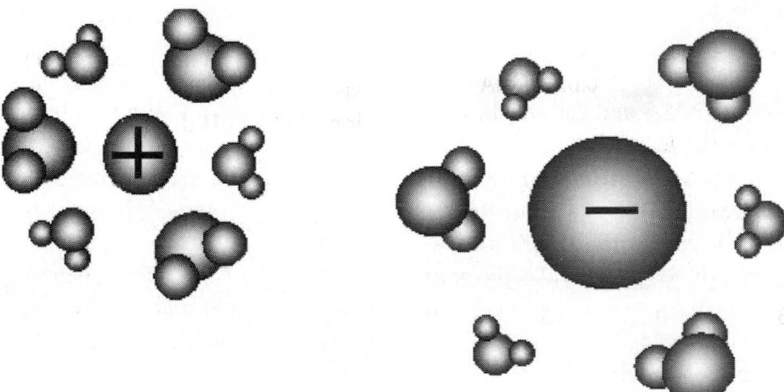

Figure 5.12. The first hydration sphere of most cations $M(H_2O)_6^+$, and anions $X(H_2O)_6^{-1}$. Small water molecules are below the plane containing the ions, and large water molecules are above the plane.

cations. The presence of ions in the solution changes the ordering of water molecules even if they are not in the first hydration sphere.

The hydration of ions releases energy, but breaking up ions from a solid requires energy. The amount of energy needed depends on the substance, and for this reason, some substances are more soluble than others. Natural waters in oceans, streams, rivers, and lakes are in contact with minerals and salts. The concentrations of various ions depend on the solubility of salts (Moeller and O'Connor 1972) and the contact time.

All salts dissolved in water are completely ionized, even those formed in the reaction between weak acids and weak bases. For example, the common food preservative sodium benzoate (NaC_6H_5COO) is a salt formed between a strong base NaOH and weak benzoic acid ($K_a = 6.5 \times 10^{-5}$). The benzoate ions, $C_6H_5COO^-$, in the solution react with water to produce OH^- ions giving a slightly basic solution:

$C_6H_5COO^- + H_2O \leftrightarrow C_6H_5COOH + OH^-$,
$K = 1.6 \times 10^{-10}$, at 298 K.

Ammonium bicarbonate, NH_4HCO_3, was a leavening agent before modern baking powder was popular. It is still called for in some recipes. This can be considered a salt formed between the weak base NH_4OH and the weak acid H_2CO_3. When NH_4HCO_3 dissolves in water, the ammonium and bicarbonate ions react with water:

$NH_4^+ + H_2O \leftrightarrow NH_3(aq) + H_3O^+$,
$K = 5.7 \times 10^{-10}$, at 298 K

$HCO_3^- + H_2O \leftrightarrow H_2CO_3 + OH^-$,
$K = 2.3 \times 10^{-8}$, at 298 K

$H_2CO_3 = H_2O + CO_2(g)$.

Upon heating, ammonia (NH_3) and carbon dioxide (CO_2) become gases for the leavening action. Thus, during the baking or frying process, ammonia is very pungent and unpleasant. When sodium bicarbonate is used, only CO_2 causes the dough to rise. Phosphoric acid, instead of NH_4^+, provides the acid in baking powder.

Buffer Solutions

A solution containing a weak acid and its salt or a weak base and its salt is a **buffer solution**, since its pH changes little when a small amount acid or base is added. For example, nicotinic acid ($HC_6H_4NO_2$, niacin, a food component) is a weak acid with $K_a = 1.7 \times 10^{-5}$ ($pK_a = -\log_{10}K_a = 4.76$):

$HC_6H_4NO_2 = H^+ + C_6H_4NO_2^-$,

$$K_a = \frac{[H^+][C_6H_4NO_2^-]}{[HC_6H_4NO_2]}.$$

$$pH = pK_a + \log_{10}\left\{\frac{[C_6H_4NO_2^-]}{[HC_6H_4NO_2]}\right\}$$ (Henderson-Hesselbalch equation).

In a solution containing niacin and its salt, $[C_6H_4NO_2^-]$ is the concentration of the salt, and $[HC_6H_4NO_2]$ is the concentration of niacin. The pair, $HC_6H_4NO_2$ and $C_6H_4NO_2^-$, are called **conjugate acid and base**, according to the Bronsted-Lowry definition for acids and bases. So, for a general acid and its conjugate base, the pH can be evaluated using the Henderson-Hesselbalch equation:

$$pH = pK_a = \log_{10}\left\{\frac{[base]}{[acid]}\right\}$$ (Henderson-Hesselbalch Equation)

Adding H^+ converts the base into its conjugate acid, and adding OH^- converts the acid into its conjugate base. Adding acid and base changes the ratio [base]/[acid], causing a small change in the pH if the initial ratio is close to 1.0. Following this equation, the most effective buffer solution for a desirable pH is to use an acid with a pK_a value similar to the desired pH value and to adjust the concentration of the salt and acid to obtain the ratio that gives the desired pH. For example, the pK_a for $H_2PO_4^-$ is 7.21, and mixing KH_2PO_4 and K_2HPO_4 in the appropriate ratio will give a buffer solution with pH 7. However, more is involved in the art and science of making and standardizing buffer solutions. For example, the ionic strength must be taken into account.

The pH of blood from healthy persons is 7.4. The phosphoric acid and bicarbonate ions in blood and many other soluble biomaterials in the intercellular fluid play a buffering role in keeping the pH constant. The body fluids are very complicated buffer solutions, because each conjugate pair in the solution has an equilibrium of its own. These equilibria plus the equilibrium due to the self-ionization of water stabilize the pH of the solution. Buffer solutions abound in nature: milk, juice, soft drinks, soup, fluid contained in food, and water in the ocean, for example.

Hydrophilic and Hydrophobic Effects

The **hydrophilic effect** refers to the hydrogen bonding, polar-ionic and polar-polar interactions with water molecules, which lower the energy of the system and make ionic and polar substances soluble. The lack of strong interactions between water molecules and lipophilic molecules or the nonpolar portions of amphiphilic molecules is called the **hydrophobic effect**, a term coined by Charles Tanford (1980).

When mixed with water, ionic and polar molecules dissolve and disperse in the solution, whereas the nonpolar or hydrophobic molecules huddle together, forming groups. At the proper temperature, groups of small and nonpolar molecules surrounded by water cages form stable phases called **hydrates** or **clathrates**. For example, the clathrate of methane forms stable crystals at temperatures below 300 K (Sloan 1998). The hydrophobic effect causes the formation of micelles and the folding of proteins in enzymes so that the hydrophobic parts of the long chain huddle together on the inside, exposing the hydrophilic parts to the outside to interact with water.

Hydrophilic and hydrophobic effects together stabilize three-dimensional structures of large molecules such as enzymes, proteins, and lipids. Hydrophobic portions of these molecules stay together, forming pockets in globular proteins. These biopolymers minimize their hydrophobic surface to reduce their interactions with water molecules. Biological membranes often have proteins bonded to them, and the hydrophilic portions extend to the intra- and intercellular aqueous solutions. These membrane-bound proteins often transport specific nutrients in and out of cells. For example, water, amino acid, and potassium-sodium ion transporting channels are membrane-bound proteins (Garrett and Grisham 2002).

Hydrophilic and hydrophobic effects, together with the ionic interaction, cause long-chain proteins called **enzymes** to fold in specific conformations (three-dimensional structures) that catalyze specific reactions. The pH of the medium affects the charges of the proteins. Therefore, the pH may alter enzyme conformations and affect their functions. At a specific pH, some enzymes consist of several subunits that aggregate into one complex structure in order to minimize the hydrophobic surface in contact with water. Thus, the chemistry of water is intimately mingled with the chemistry of life.

During food processing, proteins are denatured by heat, acid, base, and salt. These treatments alter the conformation of the proteins and enzymes. Denatured proteins lose their life-maintaining functionality. Molecules containing hydrophilic and hydrophobic parts are emulsifiers that are widely used in the food industry.

Hydrophilic and hydrophobic effects cause nonpolar portions of phospholipids, proteins, and cholesterol to assemble into micelles and bilayers, or biological membranes (Sloan 1998). The membrane conformations are stable due to their low energy, and they enclose compartments with components to perform biological functions. Proteins and enzymes attached to the membranes communicate and transport nutrients and wastes for cells, keeping them alive and growing.

Hard Waters and Their Treatments

Waters containing dissolved CO_2(same as H_2CO_3) are acidic due to the equilibria

$$H^+(aq) + HCO_3^-(aq) \leftrightarrow H_2CO_3(aq) \Delta H_2O + CO_2(g)$$

$$HCO_3^-(aq) \leftrightarrow H^+(aq) + CO_3^{2-}(aq).$$

Acidic waters dissolve $CaCO_3$ and $MgCO_3$, and waters containing Ca^{2+}, Mg^{2+}, HCO_3^-, and CO_3^{2-} are **temporary hard waters,** as the hardness is removable by boiling, which reduces the solubility of CO_2. When CO_2 is driven off, the solution becomes less acidic due to the above equilibria. Furthermore, reducing the acidity increases the concentration of CO_3^{2-}, and solids $CaCO_3$ and $MgCO_3$ precipitate:

$$Ca^{2+}(aq) + CO_3^{2-}(aq) \leftrightarrow CaCO_3(s)$$

$$Mg^{2+}(aq) + CO_3^{2-}(aq) \leftrightarrow MgCO_3(s)$$

Water containing less than 50 mg/L of these substances is considered soft; 50–150 mg/L moderately hard; 150–300 mg/L hard; and more than 300 mg/L very hard.

For water softening by the **lime treatment**, the amount of dissolved Ca^{2+} and Mg^{2+} is determined first; then an equal number of moles of lime, $Ca(OH)_2$, is added to remove them, by these reactions:

$$Mg^{2+} + Ca(OH)_2(s) \leftrightarrow Mg(OH)_2(s) + Ca^{2+}$$

$$Ca^{2+} + 2\ HCO_3^- + Ca(OH)_2(s) \leftrightarrow 2\ CaCO_3(s) + 2\ H_2O.$$

Permanent hard waters contain sulfate (SO_4^{2-}), Ca^{2+}, and Mg^{2+} ions. Calcium ions in the sulfate solution can be removed by adding sodium carbonate due to the reaction:

$$Ca^{2+} + Na_2CO_3 \rightarrow CaCO_3(s) + 2Na^+.$$

Hard waters cause scales or deposits to build up in boilers, pipes, and faucets—problems for food and other industries. Ion exchange using resins or zeolites is commonly used to soften hard waters. The calcium and magnesium ions in the waters are taken up by the resin or zeolite that releases sodium or hydrogen ions back to the water. Alternatively, when pressure is applied to a solution, water molecules, but not ions, diffuse through the semipermeable membranes. This method, called **reverse osmosis**, has been used to soften hard waters and desalinate seawater.

However, water softening replaces desirable calcium and other ions with sodium ions. Thus, soft waters are not suitable for drinking. Incidentally, calcium ions strengthen the gluten proteins in dough mixing. Some calcium salts are added to the dough by bakeries to enhance bread quality.

Ionic Strength and Solubility of Foodstuff

Ions are attracted to charged or polar sites of large biomolecules. Cations strongly interact with large molecules such as proteins. At low concentrations, they may neutralize charges on large organic molecules, stabilizing them. At high concentrations, ions compete with large molecules for water and destabilize them, resulting in decreased solubility. The concentration of electrolytes affects the solubility of foodstuffs.

One of the criteria for concentration of electrolytes is **ionic strength**, I, which is half of the sum (Σ) of all products of the concentration (C_i) of the ith ion and the square of its charge (Z_i^2):

$$I = \tfrac{1}{2} \Sigma\, C_i Z_i^2.$$

However, solubility is not only a function of ionic strength; it also depends very much on the anions involved.

The **salting-in phenomenon** refers to increases of protein solubility with increased concentrations of salt at low ionic strength. The enhancement of broth flavor by adding salt may be due to an increase of soluble proteins or amino acids in it. At high ionic strength, however, the solubilities of some proteins decrease; this is the **salting-out phenomenon**. Biochemists often use potassium sulfate, K_2SO_4, and ammonium sulfate, $(NH_4)_2SO_4$, for the separation of amino acids or proteins because the sulfate ion is an effective salting-out anion. The sulfate ion is a **stabilizer**, because the precipitated proteins are stable. Table salt is not an effective salting-out agent. Damodaran (1996) and Voet and Voet (1995) discuss these phenomena in much more detail.

WATER AS REAGENT AND PRODUCT

Water is the product from the oxidation of hydrogen, and the standard cell potential (ΔE^o) for the reaction is 1.229 V.

$$2H_2(g) + O_2(g) = 2H_2O(l),\ \Delta E^o = 1.229\ V$$

Actually, all hydrogen in any substance produces water during combustion and oxidation. On the other hand, water provides protons (H^+), hydroxide ions (OH^-), hydrogen atoms (H), oxygen atoms (O), and radicals (H·, ·OH) as reagents. The first two of these (H^+ and OH^-) also exhibit acid-base properties, as described earlier. Acids and bases promote hydrolysis and condensation reactions.

In **esterification** and **peptide synthesis**, two molecules are joined together, or condensed, releasing a water molecule. On the other hand, water breaks ester, peptide, and glycosidic bonds in a process called **hydrolysis**.

ESTERIFICATION, HYDROLYSIS, AND LIPIDS

Organic acids and various alcohols present in food react to yield esters in aqueous solutions. Esters, also present in food, hydrolyze to produce acids and alcohols. Water is a reagent and a product in these reversible equilibria. Figure 5.13 shows the Fisher esterification and hydrolysis reactions and the role of water in the series of intermediates in these equi-

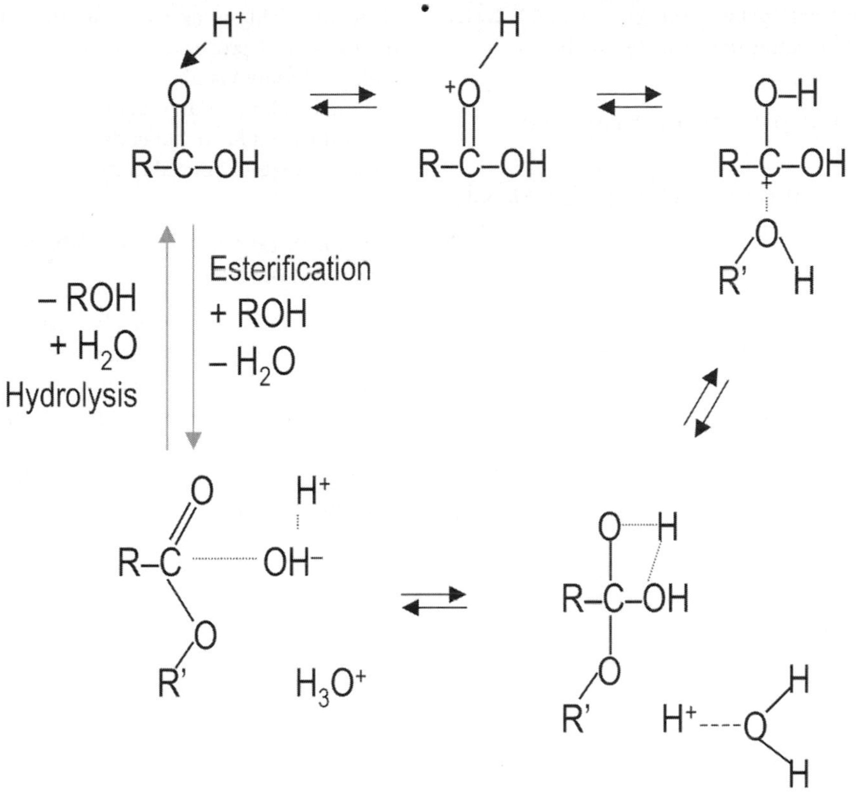

Figure 5.13. Esterification and hydrolysis in aqueous solutions.

libria. In general, esterification is favored in acidic solutions, and hydrolysis is favored in neutral and basic solutions.

The protonation of the slightly negative carbonyl oxygen (>C=O) of the carboxyl group (C(=O)OH) polarizes the C=O bond, making the carbon atom positive, to attract the alcohol group R′OH. Water molecules remove protons and rearrange the bonds in several intermediates for the simple overall reaction

$$RC(=O)OH + HOR' \leftrightarrow RC(=O)OR' + H_2O$$

In basic solutions, the OH^- ions are attracted to the slightly positive carbon of the carbonyl group. The hydrolysis is the reverse of esterification.

Hydrolysis of glycerol esters (glycerides: fat and oil) in basic solutions during soap making is a typical example of hydrolysis. Triglycerides are hydrophobic, but they can undergo partial hydrolysis to become amphiphilic diglycerides or monoglycerides. Esterification and hydrolysis are processes in metabolism.

Lipids, various water-insoluble esters of fatty acids, include glycerides, phospholipids, glycolipids, and cholesterol. Oils and fats are mostly triglycerides, which is a glycerol molecule (CH_2OH–CHOH–CH_2OH) esterified with three fatty acids [$CH_3(-CH_2)_nCOOH$, n = 8–16]. Some of the triglycerides are partially hydrolyzed in the gastrointestinal tract before absorption, but most are absorbed with the aid of bile salts, which emulsify the oil and facilitate its absorption. Many animals biosynthesize lipids when food is plentiful, as lipids provide the highest amount of energy per unit mass. Lipids, stored in fat cells, can be hydrolyzed, and upon further oxidation, they produce lots of energy and water. Some animals utilize fat for both energy and water to overcome the limitation of food and water supplies during certain periods of their lives.

Catalyzed by enzymes, esterification, and hydrolysis in biological systems proceed at much faster rates than when catalyzed by acids and bases.

WATER IN DIGESTION AND SYNTHESES OF PROTEINS

The digestion and the formation of many biopolymers (such as proteins and carbohydrates) as well as the formation and breakdown of lipids and esters involve reactions very similar to those of esterification and hydrolysis.

The digestion of proteins, polymers of amino acids, starts with chewing, followed by hydrolysis with the aid of protein-cleaving enzymes (proteases) throughout the gastrointestinal tract. Then the partially hydrolyzed small peptides and hydrolyzed individual amino acids are absorbed in the intestine. Water is a reagent in these hydrolyses (Fig. 5.14).

The **deoxyribonucleic acids** (DNAs) store the genetic information, and they direct the synthesis of **messenger ribonucleic acids** (mRNAs), which in turn direct the protein synthesis machinery to make various proteins for specific body functions and structures. This is an oversimplified description of the biological processes that carry out the polymerization of amino acids.

Proteins and amino acids also provide energy when fully oxidized, but carbohydrates are the major energy source in normal diets.

WATER IN DIGESTION AND SYNTHESIS OF CARBOHYDRATES

On earth, water is the most abundant inorganic compound, whereas carbohydrates are the most abundant class of organic compounds. Carbohydrates require water for their synthesis and provide most of the energy for all life on earth. They are also part of the glycoproteins and the genetic molecules of DNA. **Carbohydrates** are compounds with a deceivingly simple general formula $(CH_2O)_n$, $n \geq 3$, that appears to be made up of carbon and water, but although their chemistry fills volumes of thick books and more, there is still much for carbohydrate chemists to discover.

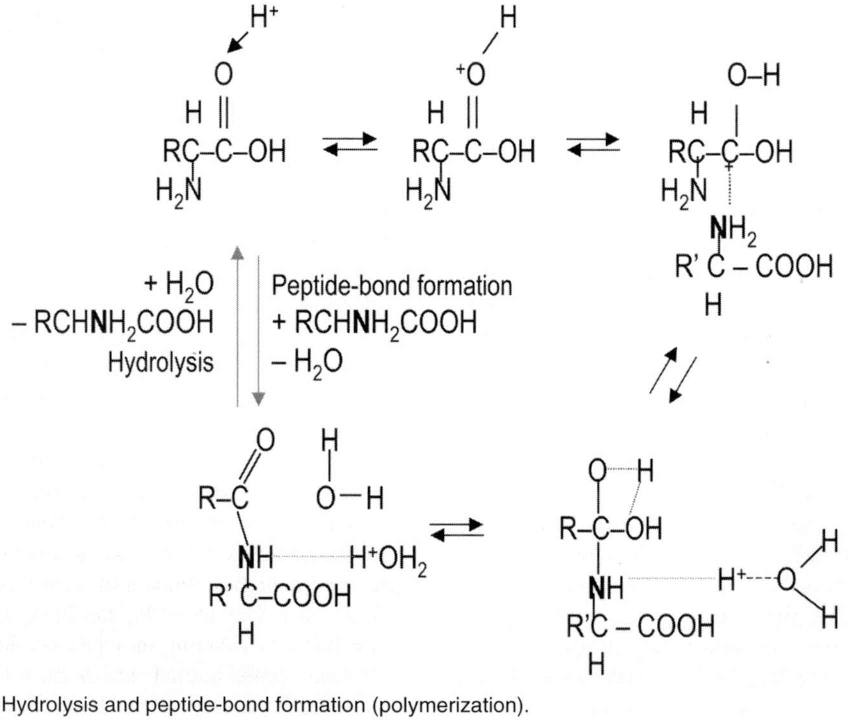

Figure 5.14. Hydrolysis and peptide-bond formation (polymerization).

Energy from the sun captured by plants and organisms converts carbon dioxide and water to high-energy carbohydrates,

$$6CO_2 + 12H_2O^* \rightarrow (CH_2O)_6 + 6H_2O + 6O^*_2$$

The stars (*) indicate the oxygen atoms from water released as oxygen gas (O^*_2). Still, this is an oversimplified equation for photosynthesis, but we do not have room to dig any deeper. The product $(CH_2O)_6$ is a **hexose**, a six-carbon simple sugar, or **monosaccharide,** that can be fructose, glucose, or another simple sugar. Glucose is the most familiar simple sugar, and its most common structure is a cyclic structure of the chair form. In glucose, all the OH groups around the ring are at the equatorial positions, whereas the small H atoms are at the axial locations (Fig. 5.15). With so many OH groups per molecule, glucose molecules are able to form several hydrogen bonds with water molecules, and thus most monosaccharides are soluble in water.

The **disaccharides** sucrose, maltose, and lactose have two simple sugars linked together, whereas starch and fiber are polymers of many glucose units. A disaccharide is formed when two OH groups of separate monosaccharides react to form an –O– link, called a **glycosidic bond**, after losing a water molecule.

$$C_6H_{12}O_6 + C_6H_{12}O_6 = C_6H_{11}O_5\text{–O–}C_6H_{11}O_5 + H_2O$$

Disaccharides are soluble in water due to their ability to form many hydrogen bonds.

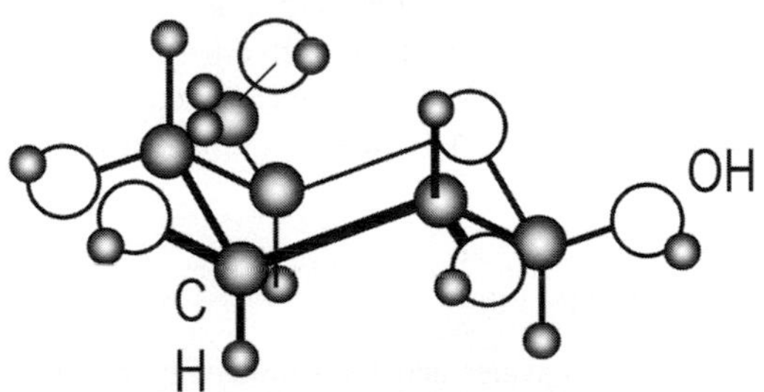

Figure 5.15. Chair form cyclic structure of glucose $C_6H_{12}O_6$. For glucose, all the OH groups are in the equatorial position, and these are possible H-donors for hydrogen bonding with water molecules. They are also possible sites to link to other hexoses.

Plants and animals store glucose as long-chain **polysaccharides** in **starch** and **glycogen**, respectively, for energy. Starch is divided into amylose and amylopectin. **Amylose** consists of linear chains, whereas **amylopectin** has branched chains. Due to the many interchain hydrogen bonds in starch, hydrogen bonding to water molecules develops slowly. Small starch molecules are soluble in water. Suspensions of large starch molecules thicken soup and gravy, and starch is added to food for desirable texture and appearance. Water increases the molecular mobility of starch, and starch slows the movement of water molecules. Food processors are interested in a quantitative relationship between water and starch and the viscosity of the suspension.

Glycogen, animal starch, is easily hydrolyzed to yield glucose, which provides energy when required. In the hydrolysis of polysaccharides, water molecules react at the glycosidic links. Certain enzymes catalyze this reaction, releasing glucose units one by one from the end of a chain or branch.

Polysaccharide chains in cellulose are very long, 7000 to 15,000 monosaccharides, and interchain hydrogen bonds bind them into fibers, which further stack up through interfiber hydrogen bonds. Many interchain hydrogen bonds make the penetration of water molecules between chains a time-consuming process. Heating speeds up the process.

Water, Minerals, and Vitamins

Most minerals are salts or electrolytes. These are usually ingested as aqueous solutions of electrolytes, discussed earlier. Ions (Ca^{2+}, Mg^{2+}, Na^+, K^+, Fe^{2+}, Zn^{2+}, Cu^{2+}, Mn^{2+}; Cl^-, I^-, S^{2-}, Se^{2-}, $H_2PO_4^-$, etc.) present in natural water are leached from the ground. Some of them are also present in food, because they are essential nutrients for plants and animals that we use as food. A balance of electrolytes in body fluid must be maintained. Otherwise, shock or fainting may develop. For example, the drinking water used by sweating athletes contains the proper amount of minerals. In food, mineral absorption by the body may be affected by the presence of other molecules. For example, vitamin D helps the absorption of calcium ions.

Small amounts of a group of organic compounds not synthesized by humans, but essential to life, are called vitamins; their biochemistry is very complicated and interesting; many interact with enzymes, and others perform vital functions by themselves. Regardless of their biological function and chemical composition, vitamins are divided into water-soluble and fat-soluble groups. This division, based on polarity, serves as a guide for food processing. For example, food will lose water-soluble vitamins when washed or boiled in water, particularly after cutting (Hawthorne and Kubatova 2002).

The water-soluble vitamins consist of a complex group of vitamin Bs, vitamin C, biotin, lipoic acid, and folic acid. These molecules are either polar or have the ability to form hydrogen bonds. Vitamins A (retinal), D2, D3, E, and K are fat soluble, because major portions of their molecules are nonpolar organic groups.

Vitamin C, L-ascorbic acid or 3-oxo-L-gulofuranolactone, has the simplest chemical formula ($C_6H_8O_6$) among vitamins. This diprotic acid is widely distributed in plants and animals, and only a few vertebrates, including humans, lack the ability to synthesize it.

Vitamin B complex is a group of compounds isolated together in an aqueous solution. It includes thiamine (B1), riboflavin (B2), niacin (or nicotinic acid, B3), pantothenic acid (B5), cyanocobalamin (B12), and vitamin B6 (any form of pyridoxal, pyridoxine, or pyridoxamine). Biotin, lipoic acid, and folic acid are also part of the water-soluble vitamins. These vitamins are part of enzymes or coenzymes that perform vital functions.

FOOD CHEMISTRY OF WATER

Water ingestion depends on the individual, composition of the diet, climate, humidity, and physical activity. A nonexercising adult loses the equivalent of 4% of his or her body weight in water per day (Brody 1999). Aside from ingested water, water is produced during the utilization of food. It is probably fair to suggest that food chemistry is the chemistry of water, since we need a constant supply of water as long as we live.

Technical terms have special meanings among fellow food scientists. Furthermore, food scientists deal with dynamic and nonequilibrium systems, unlike most natural scientists who deal with static and equilibrium systems. There are special concepts and parameters useful only to food scientists. Yet, the fundamental properties of water discussed above lay a foundation for the food chemistry of water. With respect to food, water is a component, solvent, acid, base, and dispersing agent. It is also a medium for biochemical reactions, for heat and mass transfer, and for heat storage.

Food chemists are very concerned with water content and its effects on food. They need reliable parameters for references, criteria, and working objectives. They require various indicators to correlate water with special properties such as perishability, shelf life, mobility, smell, appearance, color, texture, and taste.

WATER AS A COMMON COMPONENT OF FOOD

Water is a food as well as the most common component of food. Even dry foods contain some water, and the degree of water content affects almost every aspect of food: stability, taste, texture, and spoilage.

Most food molecules contain OH, C=O, NH, and polar groups. These sites strongly interact with water molecules by hydrogen bonding and dipole-dipole interactions. Furthermore, dipole-ion, hydrophilic, and hydrophobic interactions also occur between water and food molecules. The properties of hydrogen-bonded water molecules differ from those in bulk water, and they affect the water molecules next to them. There is no clear boundary for affected and unaffected water molecules. Yet it is convenient to divide them into **bound water** and **free water**. This is a vague division, and a consensus definition is hard to reach. Fennema and Tannenbaum (1996) give a summary of various criteria for them, indicating a diverse opinion. However, the concept is useful, because it helps us understand the changes that occur in food when it is heated, dried, cooled, or refrigerated. Moreover, when water is the major ingredient, interactions with other ingredients modify the properties of the water molecules. These aspects were discussed earlier in connection with aqueous solutions.

WATER ACTIVITY

Interactions of water and food molecules mutually change their properties. Water in food is not pure water. Water molecules in vapor, liquid, and solid phases or in solutions and food react and interchange in any equilibrium system. The tendency to react and interchange with each other is called the **chemical potential,** μ. At equilibrium, the potential of water in all phases and forms must be equal at a given temperature, T. The potential, μ, of the gas phase may be expressed as:

$$\mu = \mu_w + RT\ln(p/p_w),$$

where R is the gas constant (8.3145 J $mol^{-1}K^{-1}$), p is the partial water vapor pressure, and p_w is the vapor pressure of pure water at T. The ratio p/p_w is called the **water activity** a_w ($= p/p_w$), although this term is also called **relative vapor pressure** (Fennema and Tannenbaum 1996). The difference is small, and for simplicity, a_w as defined is widely used for correlating the stability of foods. For ideal solutions and for most moist foods, a_w is less than unity ($a_w < 1.0$; Troller 1978).

The Clausius-Clapeyron equation, mentioned earlier, correlates vapor pressure, P, heat of phase transition, ΔH, and temperature, T. This same relationship can be applied to water activity. Thus, the plot of $\ln(a_w)$ versus $1/T$ gives a straight line, at least within a reasonable temperature range. Depending on the initial value for a_w or moisture content, the slope differs slightly, indicating the difference in the heat of phase transition due to different water content.

Both water activity and relative humidity are fractions of the pure-water vapor pressure. Water activity can be measured in the same way as humidity. Water contents have a sigmoidal relationship (Fig. 5.16). As water content increases, a_w increases: $a_w = 1.0$ for infinitely dilute solutions, $a_w > 0.7$ for dilute solutions and moist foods, and $a_w < 0.6$ for dry foods. Of course, the precise relationship depends on the food. In general, if the water vapor in the atmosphere surrounding the food is greater than the water activity of the food, water is adsorbed; otherwise, desorption takes place. Water activity reflects the combined effects of water-solute, water-surface, capillary, hydrophilic, and hydrophobic interactions.

Water activity is a vital parameter for food monitoring. A plot of a_w versus water content is called an **isotherm**. However, desorption and adsorption isotherms are different (Fig. 5.16) because this is a nonequilibrium system. Note that isotherms in most other literature plot water content against a_w, the reverse of the axes of Figure 5.16, which is intended to show that a_w is a function of water content.

Water in food may be divided into tightly bound, loosely bound, and nonbound waters. Dry foods contain tightly bound (monolayer) water, and a_w rises slowly as water content increases; but as loosely bound water increases, a_w increases rapidly and approaches 1.0 when nonbound water is present. Crisp crackers get soggy after adsorbing water from moist air, and soggy ones can be dried by heating or exposure to dry air.

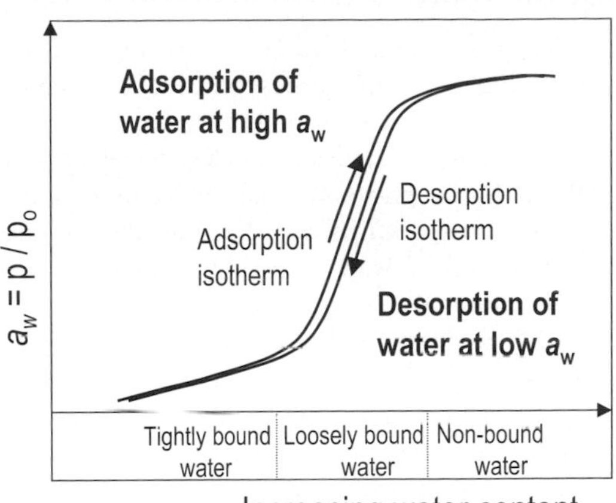

Figure 5.16. Nonequilibrium or hysteresis in desorption and adsorption of water by foodstuff. Arbitrary scales are used to illustrate the concept for a generic pattern.

Water activity affects the growth and multiplication of microorganisms. When $a_w < 0.9$, growth of most molds is inhibited. Growth of yeasts and bacteria also depends on a_w. Microorganisms cease growing if $a_w < 0.6$. **In a system, all components must be in equilibrium with one another, including all the microorganisms. Every type of organism is a component and a phase of the system, due to its cells or membranes. If the water activity of an organism is lower than that of the bulk food, water will be absorbed, and thus the species will multiply and grow. However, if the water activity of the organism is higher, the organism will dehydrate and become dormant or die.** Thus, it is not surprising that water activity affects the growth of various molds and bacteria. By this token, humidity will have the same effect on microorganisms in residences and buildings. Little packages of drying agent are placed in sealed dry food to reduce vapor pressure and prevent growth of bacteria that cause spoilage.

Aquatic Organisms and Drinking Water

Life originated in the water or oceans eons ago, and vast populations of the earliest unicellular living organisms still live in water today. Photosynthesis by algae in oceans consumes more CO_2 than the photosynthesis by all plants on land. Diversity of phyla (divisions) in the kingdoms of Fungi, Plantae, and Animalia live in water, ranging from single-cell algae to mammals.

All life requires food or energy. Some living organisms receive their energy from the sun, whereas others get their energy from chemical reactions. For example, the bacteria *Thiobacillus ferrooxidans* derive energy by catalyzing the oxidation of iron sulfide, FeS_2, using water as the oxidant (Barret et al. 1939). Chemical reactions provide energy for bacteria to sustain their lives and to reproduce. Many organisms feed on other organisms, forming a food chain. Factors affecting life in water include minerals, solubility of the mineral, acidity (pH), sunlight, dissolved oxygen level, presence of ions, chemical equilibria, availability of food, and electrochemical potentials of the material, among others.

Water used directly in food processing or as food is **drinking water**, and aquatic organisms invisible to the naked eye can be beneficial or harmful. The *Handbook of Drinking Water Quality* (De Zuane 1997) sets guidelines for water used in food services and technologies. Wastewater from the food industry needs treatment, and the technology is usually dealt with in industrial chemistry (Lacy 1992).

When food is plentiful, beneficial and pathogenic organisms thrive. Pathogenic organisms present in drinking water cause intestinal infections, dysentery, hepatitis, typhoid fever, cholera, and other diseases. Pathogens are usually present in waters that contain human and animal wastes that enter the water system via discharge, runoffs, flood, and accidents at sewage treatment facilities. Insects, rodents, and animals can also bring bacteria to the water system (Coler 1989, Percival et al. 2000). Testing for all pathogenic organisms is impossible, but some organisms have common living conditions. These are called **indicator bacteria**, because their absence signifies safety.

Water and State of Food

When a substance and water are mixed, they mutually dissolve, forming a homogeneous solution, or they partially dissolve in each other, forming solutions and other phases. At ambient pressure, various phases are in equilibrium with each other in isolated and closed systems. The equilibria depend on temperature. A plot of temperature versus composition showing the equilibria among various phases is a **phase diagram** for a two-component system. Phase diagrams for three-component systems are very complicated, and foods consist of many substances, including water. Thus, a strict phase diagram for food is almost impossible. Furthermore, food and biological systems are open, with a steady input and output of energy and substances. Due to time limits and slow kinetics, phases are not in equilibrium with each other. However, the changes follow a definite rate, and these are **steady states**. For these cases, plots of temperature against the composition, showing the existences of states (phases), are called **state diagrams.** They indicate the existence of various phases in multicomponent systems.

Sucrose (sugar, $C_{12}H_{22}O_{11}$) is a food additive and a sweetener. Solutions in equilibrium with excess solid sucrose are saturated, and their concentrations vary with temperature. The saturated solutions contain 64.4 and 65.4% at 0 and 10°C, respectively. The plot of saturated concentrations against temperature is **the equilibrium solubility curve,** ES, in Figure 5.17. The freezing curve, FE, shows the variation of freezing point as a function of temperature. Aqueous solution is in equilibrium with ice Ih along FE. At the **eutectic point,** E, the intersection of the solubility and freezing curves, solids Ih and sucrose coexist with a saturated solution. The eutectic point is the lowest melting point of water-sugar solutions. However, viscous aqueous sugar solutions or syrups may exist beyond the eutectic point. These conditions may be present in freezing and tempering (thawing) of food.

Dry sugar is stable, but it spoils easily if it contains more than 5% water. The changes that occur as a sugar solution is chilled exemplify the changes in some food components when foods freeze. Ice Ih forms when a 10% sugar solution is cooled below the freezing point. As water forms Ih, leaving sugar in the solution, the solution becomes more concentrated, decreasing the freezing point further along the freezing curve (FE) towards the eutectic point (E). However, when cooled, this solution may not reach equilibrium and yield sugar crystals at the eutectic point. Part of the reason for not having

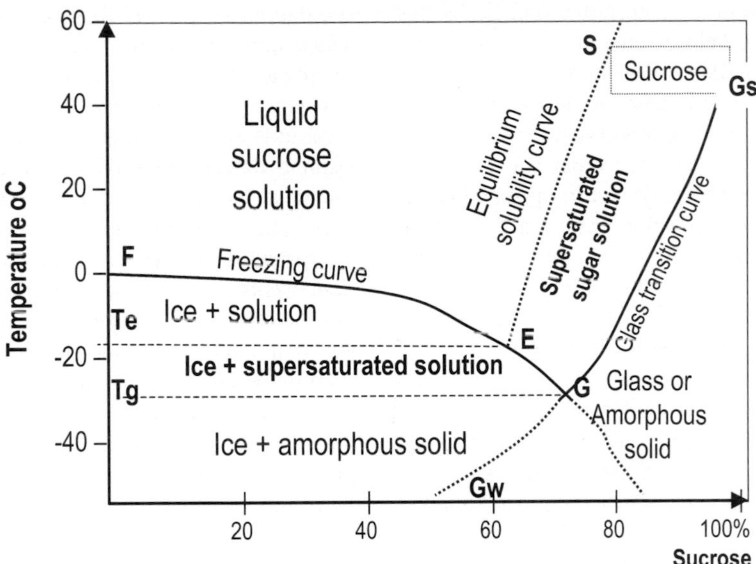

Figure 5.17. A sketch showing the phase and state diagram of water-sucrose binary system.

sucrose crystals is the high viscosity of the solution, which prevents molecules from moving and orienting properly to crystallize. The viscous solution reaches a glassy or amorphous state at the **glass transition temperature (Tg)**, point G. The glass state is a frozen liquid with extremely high viscosity. In this state, the molecules are immobile. The temperature, Tg, for glass transition depends on the rate of cooling (Angell 2002). The freezing of sugar solution may follow different paths, depending on the experimental conditions.

In lengthy experiments, Young and Jones (1949) warmed glassy states of water-sucrose and observed the warming curve over hours and days for every sample. They observed the eutectic mixture of 54% sucrose (Te = −13.95°C). They also observed the formation of phases $C_{12}H_{22}O_{11}\cdot 2.5H_2O$ and $C_{12}H_{22}O_{11}\cdot 3.5H_2O$ hydrated crystals formed at temperatures higher than the Te, which is for anhydrous sucrose. The water-sucrose binary system illustrates that the states of food components during freezing and thawing can be very complicated. Freshly made ice creams have wonderful texture, and the physics and chemistry of the process are even more interesting.

INTERACTION OF WATER AND MICROWAVE

Wavelengths of microwave range from meters down to a millimeter, their frequencies ranging from 0.3 to 300 GHz. A typical domestic oven generates 2.45 GHz microwaves, wavelength 0.123 m, and energy of photon 1.62×10^{-24} J (10 μeV). For industrial applications, the frequency may be optimized for the specific processes.

Percy L. Spencer (1894–1970), the story goes, noticed that his candy bar melted while he was inspecting magnetron testing at the Raytheon Corporation in 1945. As a further test, he microwaved popping corns, which popped. A team at Raytheon developed microwave ovens, but it took more than 25 years and much more effort to improve them and make them practical and popular. Years ago, boiling water in a paper cup in a microwave oven without harming the cup amazed those who were used to see water being heated in a fire-resistant container over a stove or fire. Microwaves simultaneously heat all the water in the bulk food.

After the invention of the microwave oven, many offered explanations on how microwaves heat food. Water's high dipole moment and high dielectric constant caused it to absorb microwave energy, leading to an increase in its temperature. Driven by the oscillating electric field of microwaves, water molecules rotate, oscillate, and move about faster, increasing water temperature to sometimes even above its boiling point. In regions where water has difficulty forming bubbles, the water is overheated. When bubbles suddenly do form in superheated water, an explosion takes place. Substances without dipole moment cannot be heated by microwaves. Therefore, plastics, paper, and ceramics won't get warm. Metallic conductors rapidly polarize, causing sparks due to arcing. The oscillating current and resistance of some metals cause a rapid heating. In contrast, water is a poor conductor, and the heating mechanism is very complicated. Nelson and Datta (2001) reviewed microwave heating in the *Handbook of Microwave Technology for Food Applications*.

Molecules absorb photons of certain frequencies. However, microwave heating is due not only to absorption of photons by the water molecules, but also to a combination of polarization and dielectric induction. As the electric field oscillates, the water molecules try to align their dipoles with the electric field. Crowded molecules restrict one another's movements. The resistance causes the orientation of water molecules to lag behind that of the electric field. Since the environment of the water molecules is related to their resistance, the heating rate of the water differs from food to food and region to region within the same container. Water molecules in ice, for example, are much less affected by the oscillating electric field in domestic microwave ovens, which are not ideal for thawing frozen food. The outer thawed layer heats up quickly, and it is cooked before the frozen part is thawed. Domestic microwave ovens turn on the microwave intermittently or at very low power to allow thermal conduction for thawing. However, microwaves of certain frequencies may heat ice more effectively for tempering frozen food. Some companies have developed systems for specific purposes, including blanching, tempering, drying, and freeze-drying.

The electromagnetic wave form in an oven or in an industrial chamber depends on the geometry of the oven. If the wave forms a standing wave in the oven, the electric field varies according to the wave pattern. Zones where the electric field varies with the largest amplitude cause water to heat up most rapidly, and the nodal zones where there are no oscillations of electric field will not heat up at all. Thus, uniform heating has been a problem with microwave heating, and various methods have been developed to partly overcome this problem. Also, foodstuffs attenuate microwaves, limiting their penetration depth into foodstuff. Uneven heating remains a challenge for food processors and microwave chefs, mostly due to the short duration of microwaving. On the other hand, food is also seldom evenly heated when conventionally cooked.

Challenges are opportunities for food industries and individuals. For example, new technologies in food preparation, packaging, and sensors for monitoring food temperature during microwaving are required. There is a demand for expertise in microwaving food. Industries microwave-blanche vegetables for drying or freezing to take advantage of its energy efficiency, time saving, decreased waste, and retention of water-soluble nutrients. The ability to quickly temper frozen food in retail stores reduces spoilage and permits selling fresh meat to customers.

Since water is the heating medium, the temperature of the food will not be much higher than the boiling point of the aqueous solutions in the food. Microwave heating does not burn food; thus, the food lacks the usual color, aroma, flavor, and texture found in conventional cooking. The outer layer of food is dry due to water evaporation. Retaining or controlling water content in microwaved food is a challenge.

When microwaved, water vapor is continually removed. Under reduced pressure, food dries or freeze-dries at low temperature due to its tendency to restore the water activity. Therefore, microwaving is an excellent means for drying food because of its savings in energy and time. Microwaves are useful for industrial applications such as drying, curing, and baking or parts thereof.

Microwave ovens have come a long way, and their popularity and improvement continue. Food industry and consumer attitudes about microwavable food have gone up and down, often due to misconceptions. Microwave cooking is still a challenge. The properties of water affect cooking in every way. Water converts microwave energy directly into heat, attenuates microwave radiation, transfers heat to various parts of the foodstuff, affects food texture, and interacts with various nutrients. All properties of water must be considered in order to take advantage of microwave cooking.

WATER RESOURCES AND THE HYDROLOGICAL CYCLE

Fresh waters are required to sustain life and maintain living standards. Therefore, fresh waters are called **water resources**. Environmentalists, scientists, and politicians have sounded alarms about limited water resources. Such alarms appear unwarranted because the earth has so much water that it can be called a water planet. Various estimates of global water distribution show that about 94% of earth's water lies in the oceans and seas. These salt waters sustain marine life and are ecosystems in their own right, but they are not fresh waters that satisfy human needs. Of the remaining 6%, most water is in solid form (at the poles and in high mountains before the greenhouse effect melts them) or underground. Less than 1% of earth's water is in lakes, rivers, and streams, and waters from these sources flow into the seas or oceans. A fraction of 1% remains in the atmosphere, mixed with air (Franks 2000).

A human may drink only a few liters of water in various forms each day, but 10 times more water is required for domestic usages such as washing and food preparation. A further equal amount is needed for various industries and social activities that support individuals. Furthermore, much more is required for food production, maintaining a healthy environment, and supporting lives in the ecosystems. Thus, one human may require more than 1000 L of water per day. In view of these requirements, a society has to develop policies for managing water resources both near and far as well as in the short and long terms. This chapter has no room to address the social and political issues, but facts are presented for readers to formulate solutions to these problems, or at least to ask questions regarding them. Based on these facts, is scarcity of world water resources a reality or not?

A major threat to water resources is climate change, because climate and weather are responsible for the **hydrologic cycle** of salt and fresh waters. Of course, human activities influence the climate in both short and long terms.

Based on the science of water, particularly its transformations among solid, liquid, and vapor phases under the influence of energy, we easily understand that heat from the sun vaporizes water from the ocean and land alike. Air movement carries the moisture (vapor) to different places than those from which it evaporated. As the vapor ascends, cooling temperature condenses the vapor into liquid drops. Cloud and rain eventually develop, and rain erodes, transports, shapes the landscape, creates streams and rivers, irrigates, and replenishes water resources. However, too much rain falling too quickly causes disaster in human life. On the other hand, natural water management for energy and irrigation has brought prosperity to society, easing the effects on humans of droughts and floods, when water does not arrive at the right time and place.

Water is a resource. Competition for this resource leads to "water war." Trade in food and food aid is equivalent to flow of water, because water is required for food production. Food and water management, including wastewater treatment, enable large populations to concentrate in small areas. Urban dwellers take these commodities for granted, but water enriches life both physically and mentally.

ACKNOWLEDGMENTS

The opportunity to put a wealth of knowledge in a proper perspective enticed me to a writing project, for which I asked more questions and found their answers from libraries and the Internet. I am grateful to all who have contributed to the understanding of water. I thank Professors L. J. Brubacher and Tai Ping Sun for their reading of the manuscript and for their suggestions. I am also grateful to other scholars and friends who willingly shared their expertise. The choice of topics and contents indicate my limitations, but fortunately, readers' curiosity and desire to know are limitless.

REFERENCES

Angell CA 2002. Liquid fragility and the glass transition in water and aqueous solutions. Chem. Rev. 102:2627–2650.

Barret J, Hughes MN, Karavaiko GI, Spencer PA. 1939. Metal Extraction by Bacterial Oxidation of Minerals. New York: Ellis Horwood.

Berkowitz J. 1979. Photoabsorption, Photoionization, and Photoelectron Spectroscopy. San Diego: Academic Press.

Bernath PF. 2002a. The spectroscopy of water vapour: Experiment, theory, and applications. Phys. Chem. Chem. Phys. 4:1501–1509.

———. 2002b. Water vapor gets excited. Science 297(Issue 5583): 943–945.

Bjorneholm O, Federmann F, Kakar S, Moller T. 1999. Between vapor and ice: Free water clusters studied by core level spectroscopy, J. Chem. Phys. 111(2): 546–550.

Bockhoff FJ. 1969. Elements of Quantum Theory. Reading, Masachusetts: Addison-Wesley, Inc.

Brody T. 1999. Nutritional Biochemistry, 2nd ed. San Diego: Academic Press.

Carleer M, Jenouvrier A, Vandaele A-C, Bernath PF, Merienne MF, Colin R, Zobov NF, Polyansky OL, Tennyson J, Savin VA. 1999. The near infrared, visible, and near ultraviolet overtone spectrum of water. J. Chem. Phys. 111:2444–2450.

Coler RA. 1989. Water pollution biology: A laboratory/field handbook. Lancaster, Pennsylvania: Technomic Publishing Co.

Damodaran S. 1996. Chapter 6, Amino acids, peptides and proteins. In: OR Fennema, editor, Food Chemistry, 3rd edition. New York: Marcel Dekker, Inc.

De Zuane J. 1997. Handbook of Drinking Water Quality, 2nd edition. Van Nostrand, Reinheld.

Fennema OR, Tannenbaum SR. 1996. Chapter 2, Water and ice. In: OR Fennema, editor, Food Chemistry, 3rd edition. New York: Marcel Dekker, Inc.

Franks F. 2000. Water—a Matrix of Life, 2nd edition. Cambridge: Royal Society of Chemistry.

Franks F, Darlington J, Schenz T, Mathias SF, Slade L, Levine H. 1987. Antifreeze activity of antarctic fish glycoprotein and a synthetic polymer. Nature 325: 146–147.

Garrett RH, Grisham CM. 2002. Principles of Biochemistry, with a Human Focus. Orlando, Florida: Harcourt College Publishers.

Goldman N, Fellers RS, Leforestier C, Saykally J. 2001. Water dimers in the atmosphere: Equilibrium constant for water dimerization from the VRT(ASP-W) potential surface. J. Phys. Chem. 105:515–519.

Gray HB. 1964. Chapter VII, Angular triatomic molecules. In: Electrons and Chemical Bonding. New York: Benjamin. Pp. 142–154.

Hawthorne SB, Kubatova A. 2002. Hot (subcritical) water extraction. In: J Pawliszyn, editor, A Comprehensive Analytical Chemistry XXXVII, Sampling and Sample Preparation for Field and Laboratory. New York: Elsevier. Pp. 587–608.

Huisken F, Kaloudis M, Kulcke A. 1996. Infrared spectroscopy of small size-selected water clusters. J. Chem. Phys. 104:17–25.

Johari GP, Anderson O. 2004. Water's polyamorphic transitions and amorphization of ice under pressure. J. Chem. Phys. 120:6207–6213.

Kamb B. 1972. Structure of the ice. In: Water and Aqueous Solutions—Structure, Thermodynamics and Transport Processes. New York: Wiley-Interscience. Pp. 9–25.

Klug DD. 2002. Condensed-matter physics: Dense ice in detail. Nature 420:749–751.

Kohl I, Mayer E, Hallbrucker A. 2000. The glassy water–cubic ice system: A comparative study by X-ray diffraction and differential scanning calorimetry. Phys. Chem. Chem. Phys. 2:1579–1586.

Kortüm G, Vogel W, Andrussow K.1961. Dissociation Constants of Organic Acids in Aqueous Solution. London: Butterworths.

Lacy WJ. 1992. Industrial wastewater and hazardous material treatment technology. In: JA Kent, editor, Riegel's Handbook of Industrial Chemistry, 9th edition. New York: Van Nostrand Reinhold. Pp. 31–82.

Lemus R. 2004. Vibrational excitations in H_2O in the framework of a local model. J. Mol. Spectrosc. 225: 73–92.

Lide DR, ed. 2003. CRC Handbook of Chemistry and Physics, 83rd edition. Cleveland, Ohio: CRC Press (There is an Internet version).

Marshall CB, Fletcher GL, Davis PL. 2004. Hyperactive antifreeze protein in a fish. Nature, 429:153–154.

Marshall WL, Franck EU. 1981. Ion Product of Water Substance, 0–1000°C, 1–10,000 bars, new international formulation and its background, J. Phys. Chem. Ref. Data 10:295–306.

Mayer E, Hallbrucker A. 1987. Cubic ice from liquid water. Nature 325:601–601.

Moeller T, O'Connor R. 1972. Ions in aqueous systems; an Introduction to Chemical Equilibrium and Solution Chemistry. New York: McGraw-Hill.

Nelson SO, Datta AK. 2001. Dielectric properties of food materials and electric field interactions. I: Handbook of Microwave Technology for Food Applications, ed. AK Datta, RC Anantheswaran. New York: Marcel Dekker, Inc.

Pauling L. 1960. The Nature of the Chemical Bond. Ithaca, New York: Cornell University Press.

Percival SL, Walker JT, Hunter PR. 2000. Microbiological Aspects of Biofilms and Drinking Water. Boca Raton: CRC Press.

Perrin DD. 1965. Dissociation Constants of Organic bases in Aqueous Solution. London: Butterworths.

———. 1982. Ionisation Constants of Inorganic Acids and Bases in Aqueous Solution. Toronto: Pergamon Press.

Petrenko VF, Whitworth RW. 1999. Physics of Ice. New York: Oxford University Press.

Sloan DE. 1998. Clathrate Hydrates of Natural Gases. New York: Marcel Dekker.

Tanaka M, Girard G, Davis R, Peuto A, Bignell N. 2001. Recommended table for the density of water between 0°C and 40°C based on recent experimental reports. Metrologia 38(4): 301–309.

Tanford C. 1980. The Hydrophobic Effect: Formation of Micelles and Biological Membranes, 2nd ed. New York: John Wiley and Sons.

Tawa GJ, Pratt LR. 1995. Theoretical calculation of the water ion product K_w, J. Am. Chem. Soc. 117:1625–1628.

Troller JA. 1978. Water Activity and Food. New York: Academic Press.

Voet D, Voet JG. 1995. Biochemistry, 2nd ed., Chapter 5. New York: John Wiley & Sons, Inc.

Wayne RP. 2000. Chemistry of Atmospheres, 3rd ed. New York: Oxford University Press.

Wilson PW, Heneghan AF, Haymetc ADJ. 2003. Ice nucleation in nature: Supercooling point (SCP) measurements and the role of heterogeneous nucleation. Cryobiology 46:88–98.

Young FE, Jones FT. 1949. Sucrose Hydrates. The sucrose-water phase diagram. J. of Physical and Colloid Chemistry 53:1334–1350.

6
Enzyme Classification and Nomenclature

*H. Ako and W. K. Nip**

INTRODUCTION

Before 1961, researchers reported on enzymes or enzymatic activities with names of their own preference. This situation caused confusion to others as various names could be given to the same enzyme. In 1956, the International Union of Biochemistry (IUB, later changed to International Union of Biochemistry and Molecular Biology, IUBMB) created the International Commission on Enzymes in consultation with the International Union of Pure and Applied Chemistry (IUPAC) to look into this situation. This Commission (now called the Nomenclature Committee of the IUBMB, NC-IUBMB) subsequently recommended classifying enzymes into six divisions (classes) with subclasses and sub-subclasses. General rules and guidelines were also established for classifying and naming enzymes. Each enzyme accepted to the Enzyme List was given a recommended name (trivial or working name; now called the common name), a systematic name, and an Enzyme Commission, or Enzyme Code (EC) number. The enzymatic reaction is also provided. A common name (formerly called recommended name) is assigned to each enzyme. This is normally the name most widely used for that enzyme, unless that name is ambiguous or misleading. A newly discovered enzyme can be given a common name and a systematic name, but not the EC number, by the researcher. EC numbers are assigned only by the authority of the NC-IUBMB.

The first book on enzyme classification and nomenclature was published in 1961. Some critical updates were announced as newsletters in 1984 (IUPAC-IUB and NC-IUB Newsletters 1984). The last (sixth) revision was published in 1992. Another update in electronic form was published in 2000 (Boyce and Tipton 2000). With the development of the Internet, most updated information on enzyme classification and nomenclature is now available through the website of the International Union of Biochemistry and Molecular Biology (http://www.chem.qmul.ac.uk/iubmb/enzyme.html). This chapter should be considered as an abbreviated version of enzyme classification and nomenclature, with examples of common enzymes related to food

*Corresponding contributor.

processing. Readers should visit the IUBMB enzyme nomenclature website for the most up-to-date details on enzyme classification and nomenclature.

CLASSIFICATION AND NOMENCLATURE OF ENZYMES

General Principles

- *First principle.* Names purporting to be names of enzymes, especially those ending in *-ase* should be used only for single enzymes, that is, single catalytic entities. They should not be applied to systems containing more than one enzyme.
- *Second principle.* Enzymes are classified and named according to the reaction they catalyze.
- *Third principle.* Enzymes are divided into groups on the basis of the type of reactions catalyzed, and this, together with the name(s) of the substrate(s), provides a basis for determining the systematic name and EC number for naming individual enzymes.

Common and Systematic Names

- The common name (recommended, trivial, or working name) follows immediately after the EC number.
- While the common name is normally that used in the literature, the systematic name, which is formed in accordance with definite rules, is more precise chemically. It should be possible to determine the reaction catalyzed from the systematic name alone.

Scheme of Classification and Numbering of Enzymes

The first Enzyme Commission, in its report in 1961, devised a system for the classification of enzymes that also serves as a basis for assigning EC numbers to them. These code numbers (prefixed by EC), which are now widely in use, contain four elements separated by periods (e.g., 1.1.1.1), with the following meaning:

- The first number shows to which of the six divisions (classes) the enzyme belongs,
- The second figure indicates the subclass,
- The third figure gives the sub-subclass, and
- The fourth figure is the serial number of the enzyme in its sub-subclass.

The main classes are

- *Class 1.* Oxidoreductases (dehydrogenases, reductases, or oxidases),
- *Class 2.* Transferases,
- *Class 3.* Hydrolases,
- *Class 4.* Lyases,
- *Class 5.* Isomerases (racemases, epimerases, cis-trans-isomerases, isomerases, tautomerases, mutases, cycloisomerases), and
- *Class 6.* Ligases (synthases).

Class 1. Oxidoreductases

Enzymes catalyzing oxidoreductions belong to this class. The reactions are of the form $AH_2 + B = A + BH_2$ or $AH_2 + B^+ = A + BH + H^+$. The substrate oxidized is regarded as the hydrogen or electron donor. All reactions within a particular sub-subclass are written in the same direction. The classification is based on the order "donor:acceptor oxidoreductase." The common name often takes the form "substrate dehydrogenase," wherever this is possible. If the reaction is known to occur in the opposite direction, this may be indicated by a common name of the form "acceptor reductase" (e.g., the common name of EC 1.1.1.9 is D-xylose reductase). "Oxidase" is used only in cases where O_2 is an acceptor. Classification is difficult in some cases because of the lack of specificity towards the acceptor.

Class 2. Transferases

Transferases are enzymes transferring a group (e.g., the methyl group or a glycosyl group), from one compound (generally regarded as donor) to another compound (generally regarded as acceptor). The classification is based on the scheme "donor:acceptor grouptransferase." The common names are normally formed as "acceptor grouptransferase." In many cases, the donor is a cofactor (coenzyme) carrying the group to be transferred. The aminotransferases constitute a special case (subclass 2.6): the reaction also involves an oxidoreduction.

Class 3. Hydrolases

These enzymes catalyze the hydrolysis of various bonds. Some of these enzymes pose problems because they have a very wide specificity, and it is not easy to decide if two preparations described by different authors are the same, or if they should be listed under different entries.

While the systematic name always includes "hydrolase," the common name is, in most cases, formed by the name of the substrate with the suffix *-ase*. It is understood that the name of this substrate with the suffix means a hydrolytic enzyme. The peptidases, subclass 3.4, are classified in a different manner from other enzymes in this class.

Class 4. Lyases

Lyases are enzymes cleaving C–C, C–O, C–N, and other bonds by means other than hydrolysis or oxidation. They differ from other enzymes in that two substrates are involved in one reaction direction, but only one in the other direction. When acting on the single substrate, a molecule is eliminated, leaving an unsaturated residue. The systematic name is formed according to "substrate group-lyase." In common names, expressions like decarboxylase, aldolase, and so on are used. "Dehydratase" is used for those enzymes eliminating water. In cases where the reverse reaction is the more important, or the only one to be demonstrated, "synthase" may be used in the name.

Class 5. Isomerase

These enzymes catalyze changes within one molecule.

Class 6. Ligases

Ligases are enzymes catalyzing the joining of two molecules with concomitant hydrolysis of the diphosphate bond in ATP or a similar triphosphate. The bonds formed are often high-energy bonds. "Ligase" is commonly used for the common name, but in a few cases, "synthase" or "carboxylase" is used. Use of the term "synthetase" is discouraged.

General Rules and Guidelines for Classification and Nomenclature of Enzymes

Table 6.1 shows the classification of enzymes by class, subclass, and sub-subclass, as suggested by the Nomenclature Committee of the International Union of Biochemistry and Molecular Biology (NC-IUBMB). The information is reformatted in table form instead of text form for easier reading and comparison.

Table 6.2 shows the rules for systematic names and guidelines for common names as suggested by NC-IUBMB. Table 6.3 shows the rules and guidelines for particular classes of enzymes as suggested by NC-IUBMB. The concept on reformatting used in Table 6.1 is also applied.

EXAMPLES OF COMMON FOOD ENZYMES

The food industry likes to use terms more easily understood by its people and is slow to adopt changes. For example, some commonly used terms related to enzymes are very general terms and do not follow the recommended guidelines established by NC-IUBMB. Table 6.4 is a list of enzyme groups commonly used by the food industry and researchers (Nagodawithana and Reed 1993).

Table 6.5 lists some common names, systematic names, and EC numbers for some common food enzymes.

Enzyme classification and nomenclature are now standardized procedures. Some journals already require that enzyme codes be used in citing or naming enzymes. Other journals are following the trend. It is expected that all enzymes will have enzyme codes in new research articles as well as (one hopes) in new reference books. However, for older literature, it is still difficult to identify the enzyme codes. It is hoped that Tables 6.4 and 6.5 will be useful as references.

ACKNOWLEDGMENT

The authors thank the International Union of Biology and Molecular Biology for permission to use of some of the copyrighted information on Enzyme Classification and Nomenclature. The authors also want to thank Prof. Keith Tipton, Department of Biochemistry, Trinity College, Dublin 2, Ireland, for his critical review of this manuscript.

Table 6.1. Classification of Enzymes by Class, Subclass, and Sub-subclass[a]

1. Oxidoreductases
   - *1.1 Acting on the CH–OH group of donors*
     - 1.1.1 With NAD^+ or $NADP^+$ as acceptor
     - 1.1.2 With a cytochrome as acceptor
     - 1.1.3 With oxygen as acceptor
     - 1.1.4 With a disulfide as acceptor
     - 1.1.5 With a quinone or similar compound as acceptor
     - 1.1.99 With other acceptors
   - *1.2 Acting on the aldehyde or oxo group of donors*
     - 1.2.1 With NAD^+ or $NADP^+$ as acceptor
     - 1.2.2 With a cytochrome as acceptor
     - 1.2.3 With oxygen as acceptor
     - 1.2.4 With a disulfide compound as acceptor
     - 1.2.7 With an iron-sulfur protein as acceptor
     - 1.2.99 With other acceptors
   - *1.3 Acting on the CH–CH group of donors*
     - 1.3.1 With NAD^+ or $NADP^+$ as acceptor
     - 1.3.2 With a cytochrome as acceptor
     - 1.3.3 With oxygen as acceptor
     - 1.3.5 With a quinone or related compound as acceptor
     - 1.3.6 With an iron-sulfur protein as acceptor
     - 1.3.99 With other acceptor
   - *1.4 Acting on the CH–NH_2 of donors*
     - 1.4.1 With NAD^+ or $NADP^+$ as acceptor
     - 1.4.2 With a cytochrome as acceptor
     - 1.4.3 With oxygen as acceptor
     - 1.4.4 With a disulfide as acceptor
     - 1.4.7 With an iron-sulfur protein as acceptor
     - 1.4.99 With other acceptors
   - *1.5 Acting on the CH–NH group of donors*
     - 1.5.1 With NAD^+ or $NADP^+$ as acceptor
     - 1.5.3 With oxygen as acceptor
     - 1.5.4. With disulfide as acceptor
     - 1.5.5. With a quinone or similar compound as acceptor
     - 1.5.8 With a flavin as acceptor
     - 1.5.99 With other acceptors
   - *1.6 Acting on NADH or NADPH*
     - 1.6.1 With NAD^+ or $NADP^+$ as acceptor
     - 1.6.2 With a heme protein as acceptor
     - 1.6.3 With oxygen as acceptor
     - 1.6.4 With a disulfide compound as acceptor
     - 1.6.5 With a quinone or similar compound as acceptor
     - 1.6.6 With a nitrogenous group as acceptor
     - 1.6.8 With a flavin as acceptor
     - 1.6.99 With other acceptors
   - *1.7 Acting on other nitrogenous compounds as donors*
     - 1.7.1 With NAD^+ or $NADP^+$ as acceptor
     - 1.7.2 With a cytochrome as acceptor
     - 1.7.3 With oxygen as acceptor
     - 1.7.7 With an iron-sulfur protein as acceptor

- 1.7.99 With other acceptors
- *1.8 Acting on a sulfur group of donors*
  - 1.8.1 With NAD^+ or $NADP^+$ as acceptor
  - 1.8.2 With a cytochrome as acceptor
  - 1.8.3 With oxygen as acceptor
  - 1.8.4 With a disulfide as acceptor
  - 1.8.5 With a quinone or related compound as acceptor
  - 1.8.7 With an iron-sulfur protein as acceptor
  - 1.8.98 With other, known, acceptors
  - 1.8.99 With other acceptors
  - 1.9.3 With oxygen as acceptor
  - 1.9.5 With a nitrogenous group as acceptor
  - 1.9.99 With other acceptors
- *1.10 Acting on diphenols and related substances as donors*
  - 1.10.1 With NAD^+ or $NADP^+$ as acceptor
  - 1.10.2 With a cytochrome as acceptor
  - 1.10.3 With oxygen as acceptor
  - 1.10.99 With other acceptors
- *1.11 Acting on hydrogen peroxide as acceptor*
  - 1.11.1 The peroxidases
- *1.12 Acting on hydrogen as donor*
  - 1.12.1 With NAD^+ or $NADP^+$ as acceptor
  - 1.12.2 With a cytochrome as acceptor
  - 1.12.5 With a quinone or similar compound as acceptor
  - 1.12.7 With an iron-sulfur protein as acceptor
  - 1.12.98 With other known acceptors
  - 1.12.99 With other acceptors
- *1.13 Acting on single donors with incorporation of molecular oxygen (oxygenases)*
  - 1.13.11 With incorporation of two atoms of oxygen
  - 1.13.12 With incorporation of one atom of oxygen (internal monoxygenases or internal mixed-function oxidases)
  - 1.13.99 Miscellaneous
- *1.14 Acting on paired donors with incorporation of molecular oxygen*
  - 1.14.11 With 2-oxoglutarate as one donor, and incorporation of one atom of oxygen into both donors
  - 1.14.12 With NADH or NADPH as one donor, and incorporation of two atoms of oxygen into one donor
  - 1.14.13 With NADH or NADPH as one donor, and incorporation of one atom of oxygen
  - 1.14.14 With reduced flavin or flavoprotein as one donor, and incorporation of one atom of oxygen
  - 1.14.15 With reduced iron-sulfur protein as one donor, and incorporation of one atom of oxygen
  - 1.14.16 With reduced pteridine as one donor, and incorporation of one atom of oxygen
  - 1.14.17 With reduced ascorbate as one donor, and incorporation of one atom of oxygen
  - 1.14.18 With another compound as one donor, and incorporation of one atom of oxygen
  - 1.14.19 With oxidation of a pair of donors resulting in the reduction of molecular oxygen to two molecules of water
  - 1.14.20 With 2-oxoglutarate as one donor, and the other dehydrogenated
  - 1.14.21 With NADH or NADPH as one donor, and the other dehydrogenated
  - 1.14.99 Miscellaneous (requires further characterization)
- *1.15 Acting on superoxide as acceptor*
- *1.16 Oxidizing metal ions*

(Continues)

Table 6.1. (Continued)

- 1.16.1 With NAD^+ or $NADP^+$ as acceptor
- 1.16.2 With oxygen as donor

1.17 Acting on -CH_2- groups
- 1.17.1 With NAD^+ or $NADP^+$ as acceptor
- 1.17.2 With oxygen as acceptor
- 1.17.4 With a disulfide compound as acceptor
- 1.17.99 With other acceptors

1.18 Acting on reduced ferredoxin as donor
- 1.18.1 With NAD^+ or $NADP^+$ as acceptor
- 1.18.3 With H^+ as acceptor (now EC 1.18.99)
- 1.18.6 With dinitrogen as acceptor
- 1.18.96 With other, known, acceptors
- 1.18.99 With H^+ as acceptor

1.19 Acting on reduced flavodoxin as donor
- 1.19.6 With dinitrogen as acceptor

1.20 Acting on phosphorus or arsenic in donors
- 1.20.1 With $NAD(P)^+$ as acceptor
- 1.20.4 With disulfide as acceptor
- 1.20.98 With other, known, acceptors
- 1.20.99 With other acceptors

1.21 Acting on X-H and Y-H to form an X-Y bond
- 1.21.3 With oxygen as acceptor
- 1.21.4 With disulfide as acceptor
- 1.21.99 With other acceptors

1.97 Other oxidoreductases

2. Transferases

2.1 Transferring one-carbon groups
- 2.1.1 Methyltransferases
- 2.1.2 Hydroxymethyl-, formyl-, and related transferases
- 2.1.3 Carboxyl- and carbamoyl transferases
- 2.1.4 Amidinotransferases

2.2 Transferring aldehyde or ketonic groups
- 2.2.1 Transketolases and transaldolase

2.3 Acyltransferases
- 2.3.1 Transferring groups other than amino-acyl groups
- 2.3.2 Aminoacyltransferases
- 2.3.3 Acyl groups converted into alkyl on transfer

2.4 Glycosyltransferases
- 2.4.1 Hexosyltransferases
- 2.4.2 Pentosyltransferases
- 2.4.99 Transferring other glycosyl groups

2.5 Transferring alkyl or aryl groups, other than methyl groups
(There is no subdivision in this section.)

2.6 Transferring nitrogenous groups
- 2.6.1 Transaminases
- 2.6.2 Amidinotransferases
- 2.6.3 Oximinotransferases
- 2.6.99 Transferring other nitrogenous groups

2.7 Transferring phosphorus-containing groups
- 2.7.1 Phosphotransferases with an alcohol group as acceptor

2.7.2 Phosphotransferases with a carboxy group as acceptor
2.7.3 Phosphotransferases with a nitrogenous group as acceptor
2.7.4 Phosphotransferases with a phosphate group as acceptor
2.7.6 Diphosphotransferases
2.7.7 Nucleotidyltransferases
2.7.8 Transferases for other substituted phosphate groups
2.7.9 Phosphotransferases with paired acceptors
2.8 Transferring sulfur-containing groups
2.8.1 Sulfurtransferases
2.8.2 Sulfotransferases
2.8.3 CoA-transferases
2.8.4 Transferring alkylthio groups
2.9 Transferring selenium-containing groups
2.9.1 Selenotransferases
3. Hydrolases
3.1 Acting on ester bonds
3.1.1 Carboxylic ester hydrolases
3.1.2 Phosphoric monoester hydrolases
3.1.3 Phosphoric diester hydrolases
3.1.4 Triphosphoric monoester hydrolases
3.1.5 Sulfuric ester hydrolases
3.1.6 Diphosphoric monoester hydrolases
3.1.7 Phosphoric triester hydrolases
3.1.11 Exodeoxyribonucleases producing 5′-phosphomonoesters
3.1.13 Exoribonucleases producing 5′-phosphomonoesters
3.1.14 Exoribonucleases producing 3′-phosphomonoesters
3.1.15 Exonucleases active with either ribo- or deoxyribonucleic acids and producing 5′-phosphomonoesters
3.1.16 Exonucleases active with either ribo- or deoxyribonucleic acids and producing 3′-phosphomonoesters
3.1.21 Endodeoxyribonucleases producing 5′-phosphomonoesters
3.1.22 Endodeoxyribonucleases producing 3′-phosphomonoesters
3.1.25 Site-specific endodeoxyribonucleases: specific for altered bases
3.1.26 Endoribonucleases producing 5′-phosphomonoesters
3.1.27 Endoribonucleases producing other than 5′-phosphomonoesters
3.1.30 Endonucleases active with either ribo- or deoxyribonucleic acids and producing 5′-phosphomonoesters
3.1.31 Endonulceases active with either ribo- or deoxyribonucleic acids and producing 3′-phophomonoesters
3.2 Glycosylases
3.2.1 Glycosidases, that is, enzymes hydrolyzing *O*- and *S*-glycosyl compounds
3.2.2 Hydrolyzing *N*-glycosy compounds
3.2.3 Hydrolyzing *S*-glycosyl compounds
3.3 Acting on ether bonds
3.3.1 Trialkylsulfonium hydrolases
3.3.2 Ether hydrolases
3.4 Acting on peptide bonds (peptidases)
3.4.11 Aminopeptidases
3.4.13 Dipeptidases
3.4.14 Dipeptidyl-peptidases and tripeptidyl-peptidases
3.4.15 Peptidyl-dipeptidases
3.4.16 Serine-type carboxypeptidases

(Continues)

Table 6.1. (Continued)

	3.4.17	Metallocarboxypeptidases
	3.4.18	Cysteine-type carboxypeptidases
	3.4.19	Omega peptidases
	3.4.21	Serine endopeptidases
	3.4.22	Cysteine endopeptidases
	3.4.23	Aspartic endopeptidases
	3.4.24	Metalloendopeptidases
	3.4.25	Threonine endopeptidases
	3.4.99	Endopeptidases of unknown catalytic mechanism
3.5	*Acting on carbon-nitrogen bonds, other than peptide bonds*	
	3.5.1	In linear amides
	3.5.2	In cyclic amides
	3.5.3	In linear amidines
	3.5.4	In cyclic amidines
	3.5.5	In nitriles
	3.5.99	In other compounds
3.6	*Acting on acid anhydrides*	
	3.6.1	In phosphorus-containing anhydrides
	3.6.2	In sulfonyl-containing anhydrides
	3.6.3	Acting on acid anhydrides; catalyzing transmembrane movement of substances
	3.6.4	Acting on acid anhydrides; involved in cellular and subcellular movements
	3.6.5	Acting on GTP; involved in cellular and subcellular movements
3.7	*Acting on carbon-carbon bonds*	
	3.7.1	In ketonic substances
3.8	*Acting on halide bonds*	
	3.8.3	In C-halide compounds
3.9	*Acting on phosphorus-nitrogen bonds*	
3.10	*Acting on sulfur-nitrogen bonds*	
3.11	*Acting on carbon-phosphorus bonds*	
3.12	*Acting on sulfur-sulfur bonds*	
3.13	*Acting on carbon-sulfur bonds*	
4. Lyases		
4.1	*Carbon-carbon lyases*	
	4.1.1	Carboxy-lyases
	4.1.2	Aldehyde-lyases
	4.1.3	Oxo-acid-lyases
	4.1.99	Other carbon-carbon lyases
4.2	*Carbon-oxygen lyases*	
	4.2.1	Hydro-lyases
	4.2.2	Acting on polysaccharides
	4.2.3	Acting on phosphates
	4.2.99	Other carbon-oxygen lyases
4.3	*Carbon-nitrogen lyases*	
	4.3.1	Ammonia-lyases
	4.3.2	Amidine-lyases
	4.3.3	Amine-lyases
	4.3.99	Other carbon-nitrogen lyases
4.4	*Carbon-sulfur lyases*	
4.5	*Carbon-halide lyases*	
4.6	*Phosphorus-oxygen lyases*	

4.99 Other lyases

5. Isomerases
   - *5.1 Racemases and epimerases*
     - 5.1.1 Acting on amino acids and derivatives
     - 5.1.2 Acting on hydroxy acids and derivatives
     - 5.1.3 Acting on carbohydrates and derivatives
     - 5.1.99 Acting on other compounds
   - *5.2 cis-trans-Isomerases*
   - *5.3 Intramolecular oxidoreductases*
     - 5.3.1 Interconverting aldoses and ketoses
     - 5.3.2 Interconverting keto and enol groups
     - 5.3.3 Transposing C=C bonds
     - 5.3.4 Transposing S-S bonds
     - 5.3.99 Other intramolecular oxidoreductases
   - *5.4 Intramolecular transferases*
     - 5.4.1 Transferring acyl groups
     - 5.4.2 Phosphotransferases (phosphomutases)
     - 5.4.3 Transferring amino groups
     - 5.4.4 Transferring hydroxy groups
     - 5.4.99 Transferring other groups
   - *5.5 Intramolecular lyases*
     - 5.5.1 Catalyze reactions in which a group can be regarded as eliminated from one part of a molecule, leaving a double bond, while remaining covalently attached to the molecule
   - *5.99 Other isomerases*
     - 5.99.1 Miscellaneous isomerases
6. Ligases
   - *6.1 Forming carbon-oxygen bonds*
     - 6.1.1 Ligases forming aminoacyl-tRNA and related compounds
   - *6.2 Forming carbon-sulfur bonds*
     - 6.2.1 Acid-thiol ligases
   - *6.3 Forming carbon-nitrogen bonds*
     - 6.3.1 Acid-ammonia (or amine) ligases (amide synthases)
     - 6.3.2 Acid-amino-acid ligases (peptide synthases)
     - 6.3.3 Cyclo-ligases
     - 6.3.4 Other carbon-nitrogen ligases
     - 6.3.5 Carbon-nitrogen ligases with glutamine as amido-*N*-donor
   - *6.4 Forming carbon-carbon bonds*
     - 6.5 Forming phosphoric ester bonds
     - 6.6 Forming nitrogen-metal bonds

Sources: Ref. NC-IUBMB. 1992. Enzyme Nomenclature, 6th ed. San Diego (CA): Academic Press, Inc. With Permission. NC-IUBMB Enzyme Nomenclature Website (www.iubmb.org).

[a]Readers should refer to the web: http://www.chem.qmul.ac.uk/iubmb/enzyme/rules.html for the most up-to-date changes.

Table 6.2. General Rules for Generating Systematic Names and Guidelines for Common Names[a]

Rules and Guidelines No.	Descriptions
1.	*Common names:* Generally accepted trivial names of substances may be used in enzyme names. The prefix D- should be omitted for all D-sugars and L- for individual amino acids, unless ambiguity would be caused. In general, it is not necessary to indicate positions of substitutes in common names, unless it is necessary to prevent two different enzymes having the same name. The prefix *keto-* is no longer used for derivatives of sugars in which -CHOH- has been replaced by –CO–; they are named throughout as dehydrosugars. *Systematic names:* To produce usable systematic names, accepted names of substrates forming part of the enzyme names should be used. Where no accepted and convenient trivial names exist, the official IUPAC rules of nomenclature should be applied to the substrate name. The 1, 2, 3 system of locating substitutes should be used instead of the α β γ system, although group names such as β-aspartyl-, γ-glutamyl- and also β-alanine-lactone are permissible; α and β should normally be used for indicating configuration, as in α-D-glucose. For nucleotide groups, *adenlyl* (not adenyl), etc. should be the form used. The name oxo acids (not keto acids) may be used as a class name, and for individual compounds in which –CH2– has been replaced by –CO–, oxo should be used.
2.	Where the substrate is normally in the form of an anion, its name should end in *-ate* rather than *-ic,* e.g., *lactate dehydrogenase*, not "lactic acid dehydrogenase."
3.	Commonly used abbreviations for substrates, e.g., ATP, may be used in names of enzymes, but the use of new abbreviations (not listed in recommendations of the IUPAC-IUB Commission on Biochemical Nomenclature) should be discouraged. Chemical formulae should not normally be used instead of names of substrates. Abbreviations for names of enzymes, e.g., GDH, should not be used.
4.	Names of substrates composed of two nouns, such as glucose phosphate, which are normally written with a space, should be hyphenated when they form part of the enzyme names, and thus become adjectives, e.g., *glucose-6-phosphate dehydrogenase* (EC 1.1.1.49). This follows standard practice in phrases where two nouns qualify a third; see for example, *Handbook of Chemical Society Authors*, 2nd ed., p. 14 (The Chemical Society, London, 1961).
5.	The use as enzyme names of descriptions such as *condensing enzyme, acetate-activating enzyme,* and *pH 5 enzyme* should be discontinued as soon as the catalyzed reaction is known. The word *activating* should not be used in the sense of converting the substrate into a substance that reacts further; all enzymes act by activating their substrates, and the use of the word in this sense may lead to confusion.
6.	*Common names:* If it can be avoided, a common name should not be based on a substance that is not a true substrate, e.g., enzyme EC 4.2.1.17 *(Enoyl-CoA hydratase)* should not be called "crotonase," since it does not act on crotonate.
7.	*Common names:* Where a name in common use gives some indication of the reaction and is not incorrect or ambiguous, its continued use is recommended. In other cases, a common name is based on the same principles as the systematic name (see below), but with a minimum of detail, to produce a name short enough for convenient use. A few names of proteolytic enzymes ending in *-in* are retained; all other enzyme names should end in *-ase*.

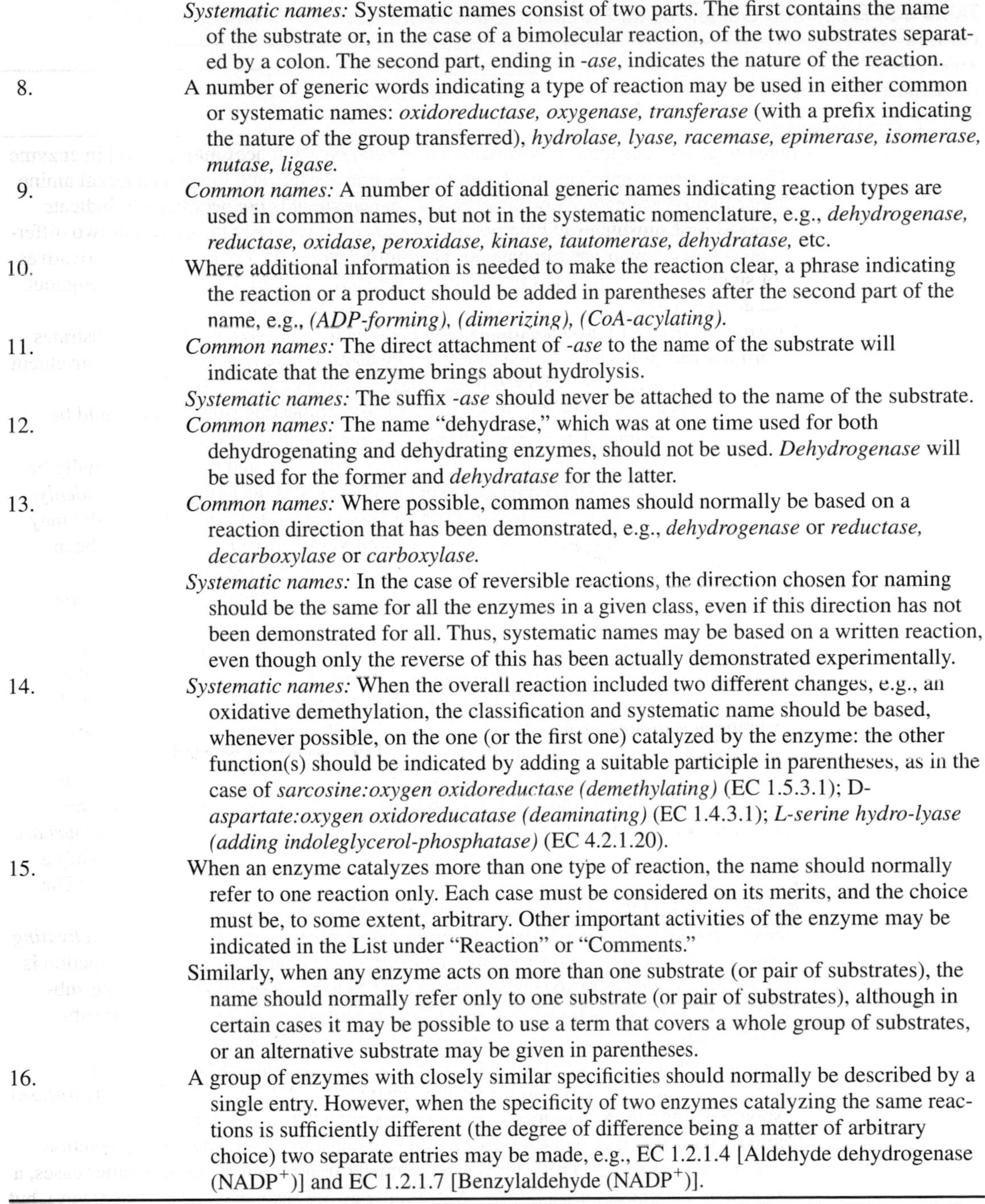

	Systematic names: Systematic names consist of two parts. The first contains the name of the substrate or, in the case of a bimolecular reaction, of the two substrates separated by a colon. The second part, ending in *-ase*, indicates the nature of the reaction.
8.	A number of generic words indicating a type of reaction may be used in either common or systematic names: *oxidoreductase, oxygenase, transferase* (with a prefix indicating the nature of the group transferred), *hydrolase, lyase, racemase, epimerase, isomerase, mutase, ligase.*
9.	*Common names:* A number of additional generic names indicating reaction types are used in common names, but not in the systematic nomenclature, e.g., *dehydrogenase, reductase, oxidase, peroxidase, kinase, tautomerase, dehydratase,* etc.
10.	Where additional information is needed to make the reaction clear, a phrase indicating the reaction or a product should be added in parentheses after the second part of the name, e.g., *(ADP-forming), (dimerizing), (CoA-acylating).*
11.	*Common names:* The direct attachment of *-ase* to the name of the substrate will indicate that the enzyme brings about hydrolysis.
	Systematic names: The suffix *-ase* should never be attached to the name of the substrate.
12.	*Common names:* The name "dehydrase," which was at one time used for both dehydrogenating and dehydrating enzymes, should not be used. *Dehydrogenase* will be used for the former and *dehydratase* for the latter.
13.	*Common names:* Where possible, common names should normally be based on a reaction direction that has been demonstrated, e.g., *dehydrogenase* or *reductase, decarboxylase* or *carboxylase.*
	Systematic names: In the case of reversible reactions, the direction chosen for naming should be the same for all the enzymes in a given class, even if this direction has not been demonstrated for all. Thus, systematic names may be based on a written reaction, even though only the reverse of this has been actually demonstrated experimentally.
14.	*Systematic names:* When the overall reaction included two different changes, e.g., an oxidative demethylation, the classification and systematic name should be based, whenever possible, on the one (or the first one) catalyzed by the enzyme: the other function(s) should be indicated by adding a suitable participle in parentheses, as in the case of *sarcosine:oxygen oxidoreductase (demethylating)* (EC 1.5.3.1); D-*aspartate:oxygen oxidoreducatase (deaminating)* (EC 1.4.3.1); *L-serine hydro-lyase (adding indoleglycerol-phosphatase)* (EC 4.2.1.20).
15.	When an enzyme catalyzes more than one type of reaction, the name should normally refer to one reaction only. Each case must be considered on its merits, and the choice must be, to some extent, arbitrary. Other important activities of the enzyme may be indicated in the List under "Reaction" or "Comments."
	Similarly, when any enzyme acts on more than one substrate (or pair of substrates), the name should normally refer only to one substrate (or pair of substrates), although in certain cases it may be possible to use a term that covers a whole group of substrates, or an alternative substrate may be given in parentheses.
16.	A group of enzymes with closely similar specificities should normally be described by a single entry. However, when the specificity of two enzymes catalyzing the same reactions is sufficiently different (the degree of difference being a matter of arbitrary choice) two separate entries may be made, e.g., EC 1.2.1.4 [Aldehyde dehydrogenase ($NADP^+$)] and EC 1.2.1.7 [Benzylaldehyde ($NADP^+$)].

Source: NC-IUBMB. 1992. Enzyme Nomenclature. 6th ed. San Diego, California: Academic Press, Inc. With permission.
[a]Readers should refer to the web: http://www.chem.qmul.ac.uk/iubmb/enzyme/rules.html for the most recent changes.

Table 6.3. Rules and Guidelines for Particular Classes of Enzymes[a]

Rules and Guidelines No.	Description
Class 1. Oxidoreductases	
1.	*Common names:* The terms *dehydrogenase* or *reductase* will be used much as hitherto. The latter term is appropriate when hydrogen transfer from the substance mentioned as donor in the systematic name is not readily demonstrated. *Transhydrogenase* may be retained for a few well-established cases. *Oxidase* is used only for cases where O_2 acts as an aceptor, and *oxygenase* only for those cases where the O2 molecule (or part of it) is directly incorporated into the substrate. *Peroxidase* is used for enzymes using H_2O_2 as acceptor. *Catalase* must be regarded as exceptional. Where no ambiguity is caused, the second reactant is not usually named; but where required to prevent ambiguity, it may be given in parentheses, e.g., EC 1.1.1.1, *alcohol dehydrogenase* and EC 1.1.1.2 *alcohol dehydrogenase ($NADP^+$).*
	Systematic names: All enzymes catalyzing oxidoreductions should be *oxidoreductases* in the systematic nomenclature, and the names formed on the pattern *donor:acceptor oxidoreductase.*
2.	*Systematic names:* For oxidoreductases using NAD^+ or $NADP^+$, the coenzyme should always be named as the acceptor except for the special case of Section 1.6 (enzymes whose normal physiological function is regarded as reoxidation of the reduced coenzyme). Where the enzyme can use either coenzyme, this should be indicated by writing $NAD(P)^+$.
3.	Where the true acceptor is unknown and the oxidoreductase has only been shown to react with artificial acceptors, the word *acceptor* should be written in parentheses, as in the case of EC 1.3.99.1, *succinate:(acceptor)oxidoreductase.*
4.	*Common names:* Oxidoreductases that bring about the incorporation of molecular oxygen into one\donor or into either or both of a pair of donors are named *oxygenase.* If only one atom of oxygen is incorporated, the term *monooxygenase* is used; if both atoms of O_2 are incorporated, the term *dioxygenase* is used.
	Systematic names: Oxidoreductases that bring about the incorporation of oxygen into one pair of donors should be named on the pattern *donor, donor:oxygen oxidoreductase (hydroxylating).*
Class 2. Transferases	
1.	*Common names:* Only one specific substrate or reaction product is generally indicated in the common names, together with the group donated or accepted. The forms *transaminase,* etc. may be used, if desired, instead of the corresponding forms *aminotransferase,* etc. A number of special words are used to indicate reaction types, e.g., *kinase* to indicate a phosphate transfer from ATP to the named substrate (not "phosphokinase"), *diphospho-kinase* for a similar transfer of diphosphate.
	Systematic names: Enzymes catalyzing group-transfer reactions should be named *transferase,* and the names formed on the pattern *donor:acceptor group-transferred-transferase,* e.g., *ATP:acetate phosphotransferase* (EC 2.7.2.1). A figure may be prefixed to show the position to which the group is transferred, e.g., *ATP:*D-*fructose 1-phospho-transferase* (EC 2.7.1.3). The spelling "transphorase" should not be used. In the case of the phosphotransferases, ATP should always be named as the donor. In the case of the transaminases involving 2-oxoglutarate, the latter should always be named as the acceptor.
2.	*Systematic names:* The prefix denoting the group transferred should, as far as possible, be noncommittal with respect to the mechanism of the transfer, e.g., *phospho-* rather than *phosphate-.*

Class 3. Hydrolases

1. *Common names:* The direct addition of *-ase* to the name of the substrate generally denotes a hydrolase. Where this is difficult, e.g., EC 3.1.2.1 (acetyl-CoA hydrolase), the word *hydrolase* may be used. Enzymes should not normally be given separate names merely on the basis of optimal conditions for activity. The *acid and alkaline phosphatases* (EC 3.1.3.1–2) should be regarded as special cases and not as examples to be followed. The common name *lysozyme* is also exceptional.

   Systematic names: Hydrolyzing enzymes should be systematically named on the pattern *substrate hydrolase*. Where the enzyme is specific for the removal of a particular group, the group may be named as a prefix, e.g., *adenosine aminohydrolase* (EC 3.5.4.4). In a number of cases, this group can also be transferred by the enzyme to other molecules, and the hydrolysis itself might be regarded as a transfer of the group to water.

Class 4. Lyases

1. *Common names:* The names *decarboxylase, aldolase,* etc. are retained; and *dehydratase* (not "dehydrase") is used for the hydro-lyases. "Synthetase" should not be used for any enzymes in this class. The term *synthase* may be used instead for any enzyme in this class (or any other class) when it is desired to emphasize the synthetic aspect of the reaction.

   Systematic names: Enzymes removing groups from substrates nonhydrolytically, leaving double bonds (or adding groups to double bonds) should be called *lyases* in the systematic nomenclature. Prefixes such as *hydro-, ammonia-* should be used to denote the type of reaction, e.g., (*S*)-*malate hydro-lyase* (EC 4.2.1.2). Decarboxylases should be regarded as *carboxy-lyases.* A hyphen should always be written before *lyase* to avoid confusion with hydrolases, carboxylases, etc.

2. *Common names:* Where the equilibrium warrants it, or where the enzyme has long been named after a particular substrate, the reverse reaction may be taken as the basis of the name, using *hydratase, carboxylase,* etc., e.g., *fumarate hydratase* for EC 4.2.1.2 (in preference to "fumarase,"which suggests an enzyme hydrolyzing fumarate).

   Systematic names: The complete molecule, not either of the parts into which it is separated, should be named as the substrate. The part indicated as a prefix to *-lyase* is the more characteristic and usually, but not always, the smaller of the two reaction products. This may either be the removed (saturated) fragment of the substrate molecule, as in *ammonia-, hydro-, thiol-lyase,* or the remaining unsaturated fragment, e.g., in the case of *carboxy-, aldehyde- or oxo-acid-lyases.*

3. Various subclasses of the lyases include a number of strictly specific or group-specific pyridoxal-5-phosphate enzymes that catalyze *elimination* reactions of β- or γ-substituted α-amino acids. Some closely related pyridoxal-5-phosphate-containing enzymes, e.g., *tryptophan synthase* (EC 4.2.1.20) and *cystathionine -synthase* (4.2.1.22) catalyse *replacement* reactions in which a β-, or γ-substituent is replaced by a second reactant without creating a double bond. Formally, these enzymes appeared to be transferases rather than lyases. However, there is evidence that in these cases the elimination of the β- or γ-substituent and the formation of an unsaturated intermediate is the first step in the reaction. Thus, applying rule 14 of the general rules for systematic names and guidelines for common names (Table 6.2), these enzymes are correctly classified lyases.

Class 5. Isomerases

In this class, the common names are, in general, similar to the systematic names that indicate the basis of classification.

1. *Isomerase* will be used as a general name for enzymes in this class. The types of isomerization will be indicated in systematic names by prefixes, e.g., *maleate cis-trans-isomerase* (EC 5.2.1.1), *phenylpyruvate keto-enol-isomerase* (EC 5.3.2.1), *3-oxosteroid*

(Continues)

Table 6.3. (Continued)

Rules and Guidelines No.	Description
	Δ^5-Δ^4-isomerase (EC 5.3.3.1). Enzymes catalyzing an aldose-ketose interconversion will be known as *ketol-isomerases,* e.g., L-*arabinose ketol-isomerase* (EC 5.3.1.4). When the isomerization consists of an intramolecular transfer of a group, the enzyme is named a *mutase,* e.g., EC 5.4.1.1 (lysolecithin acylmutase) and the *phosphomutases* in sub-subclass 5.4.2 (Phosphotransferases); when it consists of an intramolecular lyase-type reaction, e.g., EC 5.5.1.1 (muconate cycloisomerase), it is systematically named a *lyase (decyclizing).*
2.	Isomerases catalyzing inversions at asymmetric centers should be termed *racemases* or *epimerases,* according to whether the substrate contains one, or more than one, center of asymmetry: compare, e.g., EC 5.1.1.5 (lysine racemase) with EC 5.1.1.7 (diaminopimelate epimerase). A numerical prefix to the word *epimerase* should be used to show the position of the inversion.
Class 6. Ligases	
1.	*Common names:* Common names for enzymes of this class were previously of the type *XP synthetase.* However, as this use has not always been understood, and synthetase has been confused with synthase (see Class 4, item 1, above), it is now recommended that as far as possible the common names should be similar in form to the systematic name.
	Systematic names: The class of enzymes catalyzing the linking together of two molecules, coupled with the breaking of a diphosphate link in ATP, etc. should be known as *ligases.* These enzymes were often previously known as "synthetase"; however, this terminology differs from all other systematic enzyme names in that it is based on the product and not on the substrate. For these reasons, a new systematic class name was necessary.
2.	*Common name:* The common names should be formed on the pattern *X-Y ligase,* where X-Y is the substance formed by linking X and Y. In certain cases, where a trivial name is commonly used for XY, a name of the type *XY synthase* may be recommended (e.g., EC 6.3.2.11, *carnosine synthase*).
	Systematic names: The systematic names should be formed on the pattern *X:Y ligase (ADP-forming),* where X and Y are the two molecules to be joined together. The phrase shown in parentheses indicates both that ATP is the triphosphate involved and that the terminal diphosphate link is broken. Thus, the reaction is $X + Y + ATP = X–Y + ADP + P_i$.
3.	*Common name:* In the special case where glutamine acts an ammonia donor, this is indicated by adding in parentheses (*glutamine-hydrolyzing)* to a ligase name.
	Systematic names: In this case, the name *amido-ligase* should be used in the systematic nomenclature.

Source: NC-IUBMB. 1992. Enzyme Nomenclature. San Diego, California: Academic Press, Inc. With permission.
[a]Readers should refer to the web version of the rules, which are currently under revision to reflect the recent changes. http://www.chem.qmul.ac.uk/iubmb/enzyme/rules.html

Table 6.4. Common Enzyme Group Terms Used in the Food Industry

Group Term	Selected Enzymes in This Group
Carbohydrases	Amylases (EC 3.2.1.1, 3.2.1.2, 3.2.1.3) Pectic enzymes (see below) Lactases (EC 3.2.1.108, 3.2.1.23) Invertase (EC 3.2.1.26) α-Galactosidases (EC 3.2.1.22, 3.2.1.23) Cellulase (EC 3.2.1.4) Hemicellulases (EC not known) Dextranase (EC 3.2.1.11)
Proteases (Endopeptidases)	Serine proteases (EC 3.4.21) Cysteine proteases (EC 3.4.22) Aspartic proteases (EC 3.4.23) Mettaloproteases (EC 3.4.24)
Proteases (Exopeptidase)	Aminopeptidases (EC 3.4.11)
Proteases (Carboxypeptidases)	Carboxypeptidases (EC 3.4.16) Metallocarboxypeptidases (EC 3.4.17) Cysteine carboxypeptidases (EC 3.4.18) Dipeptide hydrolases (EC 3.4.13)
Oxidoreductases	Polyphenol oxidase (EC 1.10.3.1) Peroxidase (EC 1.11.1.7) Lactoperoxidase (EC 1.11.1.7) Catalase (EC 1.11.1.6) Sulfhydryl oxidase (EC 1.8.3.2) Glucose oxidase (EC 1.1.3.4) Pyranose oxidase (EC 1.1.3.10) Xanthine oxidase (EC 1.1.3.22) Lipooxygenase (EC 1.13.11.12) Dehydrogenases (see below) Alcohol oxidase (EC 1.1.3.13)
Pectolytic enzymes (Pectic enzymes)	Pectin (methyl) esterase (PE, EC 3.1.11.1) Polygalacturoniases (PG, EC 3.2.2.15 or 3.2.1.67) Pectate lyases or pectic acid lyases (PAL) (Endo-type, EC 3.2.1.15 or 4.2.2.2) (Exo-type, EC 3.2.1.67 or 4.2.2.9) Pectic lyase (PL, EC 4.2.2.10)
Heme enzymes	Catalase (EC 1.11.1.6) Peroxidase (EC 1.11.1.7) Phenol oxidase (Monophenol monooxygenase, EC 1.14.18.1 ?) Lipoperoxidase (EC 1.13.11.12)
Hemicellulase	Glucanases (EC 3.2.1.6, 3.2.1.39, 3.2.1.58, 3.2.1.75, 3.2.1.59, 3.2.1.71, 3.2.1.74, 3.2.1.84) Xylanase (EC 3.2.1.32)

(Continues)

Table 6.4. (Continued)

Group Term	Selected Enzymes in This Group
	Galactanases (EC 3.2.1.89, 3.2.1.23, 3.2.1.145) Mannanase (EC 3.2.1.25) Galactomannanases (EC not known) Pentosanases (EC not known)
Nucleolytic enzymes	Nucleases (Endonucleases, EC 3.1.30 or 3.1.27) (Exonucleases, EC 3.1.15 or 3.1.14) Phosphatases (Nonspecific phosphatases, EC 3.1.3) (Nucleotidases, EC 3.1.3) Nucleosidases (EC 3) (Inosinate nucleosidase, EC 3.2.2.12) Nucleodeaminases (Adenine deaminase, EC 3.5.4.2)
Dehydrogenase	Alcohol dehydrogenase (EC 1.1.1.1) Shikimate dehydrogenase (EC 1.1.1.25) Maltose dehydrogenase (EC 3.1 to 3.3) Isocitrate dehydrogenase (EC 1.1.1.41, 1.1.1.42) Lactate dehydrogenase (EC 1.1.1.3, 1.1.1.4, 1.1.1.27)

Source: Nagodawithana and Reed. 1993, NC–IUBMB 1992, NC–IUBMB website (www.iubmb.org)

Table 6.5. Recommended Names, Systematic Names, and Enzyme Codes (EC) for Some Common Food Enzymes

Recommended Name	Systematic Name	Enzyme Code (EC)
5′-nulceotidase	5′-ribonucleotide phosphohydrolase	3.1.3.5
α-amylase (Glycogenase)	1,4-α-D-glucan glucanohydrolase	3.2.1.1
α-galactosidase (Melibiase)	α-D-galactoside galactohydrolase	3.2.1.22
α-glucosidase (Maltase, glucoinvertase, glucosidosucrase, maltase-glucoamylase)	α-D-glucoside glucohydrolase	3.2.1.20
D-amino acid oxidase	D-amino-acid:oxygen oxidoreductase (deaminating)	1.4.3.3
D-lactate dehydrogenase (Lactic acid dehydrogenase)	(R)-lactate:NAD(+) oxidoreductase	1.1.1.28
L-amino acid oxidase	L-amino-acid:oxygen oxidoreductase (deaminating)	1.4.3.2
L-lactate dehydrogenase (Lactic acid dehydrogenase)	(S)-lactate:NAD(+) oxidoreductase	1.1.1.27

β-amylase (Saccharogen amylase, glycogenase)	1,4-α-D-glucan maltohydrolase	3.2.1.2
β-galactosidase (Lactase)	β-D-galactoside galactohydrolase	3.2.1.23
β-glucosidase (Gentiobiase, cellobiase, amyygdalase)	β-D-glucoside glucohydrolase	3.2.1.21
β-glucanase, see Hemicellulase		
Acid phosphatase (Acid phosphomonoesterase, phosphomonoesterase, glycerophosphatase)	Orthophosphoric-monoester phosphohydrolase (acid optimum)	3.1.3.2
Adenine deaminase (Adenase, adenine aminase)	Adenine aminohydrolase	3.5.4.2
Adenosine phosphate deaminase	Adenosine-phosphate aminohydrolase	3.5.4.17
AMP deaminase (Adenylic acid deaminase, AMP aminase)	AMP aminohydrolase	3.5.4.6
Alcalase, see Subtilisin Carlsberg		
Alcohol dehydrogenase (Aldehyde reductase)	Alcohol:NAD(+) oxidoreductase	1.1.1.1
Alcohol oxidase	Alcohol:oxygen oxidoreductase	1.1.3.13
Alkaline phosphatase (Alkaline phophomonoesterase, phosphomonoesterase, glycerolphosphatase)	Orthophosphoric-monoester phosphohydrolase(alkaline optimum)	3.1.3.1
Alliin lyase (Alliinase)	Alliin alkyl-sulfenate-lyase	4.4.1.4
Aminoacylase (Dehydropeptidase II, histozyme, acylase I, hippuricase, benzamidase)	*N-acyl-L-amino-acid amidohydrolase*	3.5.1.14
Asparate amino-lyase (Aspartase, fumeric aminase)	L-amino-acid:oxygen oxidoreductase (deaminating)	4.3.1.1
Aspergillus nuclease S (1) [Endonuclease S (1) (aspergillus), single-stranded-nucleate endo-nuclease, deoxyribonuclease S (1)]	None, see comments in original reference	3.1.30.1
Bacillus thermoproteolyticus neutral proteinase (Thermolysin)	None, see comments in original reference	3.4.24.4
Bromelain	None, see comments in original reference	3.4.22.4
Carboxy peptidase α (Carboxypolypeptidase)	Peptidyl-L-amino-acid hydrolase	3.4.17.1
Catalase	Hydrogen-peroxide:hydrogen-peroxide oxidoreductase	1.11.1.6
Catechol oxidase (Diphenol oxidase, o-diphenolase, phenolase, polyphenol oxidase, tyrosinase)	1,2-benzenediol:oxygen oxidoreductase	1.10.3.1
Cellulase (Endo-1,4-β-glucanase)	1,4-(1,3:1,4)-β-D-glucan 4-glucanohydrolase	3.2.1.4

(Continues)

Table 6.5. (Continued)

Recommended Name	Systematic Name	Enzyme Code (EC)
Chymosin (Rennin)	None, see comments in original reference	3.4.23.4
Chymotrypsin (Chymotrypsin α and β)	None, see comments in original reference	3.4.21.1
Debranching enzyme, see α-dextrin endo-1,6-α-glucosidase		
Dextranase	1,6-α-D-glucan 6-glucanohydrolase	3.2.1.11
α-dextrin endo-1,6-α-glucosidase (Limit dextrinase, debranching enzyme, amylopectin 6-gluco-hydrolase, pullulanase)	α-dextrin 6-glucanohydrolase	3.2.1.41
Diastase (obsolete term)		3.2.1.1 (obsolete)
Dipeptidase	Dipeptide hydrolase	3.4.13.11
Dipeptidyl peptidase I (Cathepsin C, dipeptidyl-amino-peptidase I, dipeptidyl transferase)	Dipeptidyl-peptide hydrolase	
Endopectate lyase, see Pectate lyase		
Exoribonuclease II (Ribonuclease II)	None, see comments in original reference	3.1.13.1
Exoribonuclease H	None, see comments in original reference	3.1.13.2
Ficin	None, see comments in original reference	3.4.22.3
Galactose oxidase	D-galactose-1,4-lactone:oxygen 2-oxidoreductase	1.1.3.9
Glucan-1,4-α-glucosidase (Glucoamylase, amyloglucosidase, γ-amylase, acid maltase, lysosomal α-glucosidase, exo-1,4-α-glucosidase)	1,4-α-D-glucan glucohydrolase	3.2.1.3
Glucose-6-phosphate isomerase [Phosphohexas isomerase, phosphohexomutase, oxoisomerase, hexosephosphate isomerase phosphosaccharomutase, phosphoglucoisomerase, phosphohexoisomerase, glucose isomerase (deleted, obsolete)]	D-glucose-6-phosphate keto-isomerase	5.3.1.9
Glucokinase	ATP:D-gluconate 6-phosphotransferase	2.7.1.12
Glucose oxidase (Glucose oxyhydrase)	β-D-glucose:oxygen 1-oxidoreductase	1.1.3.4
Gycerol kinase	ATP:glycerol 3-phosphotransferase	2.7.1.30
Glycogen (starch) synthase (UDPglucose-glycogen glucosyl-transferase)	UDPglucoase:glycogen glucosyltransferase	2.4.1.11
Inosinate nucleosidase (Inosinase)	5′-inosinate phosphoribohydrolase	3.2.2.2

β-fructofuranosidase (invertase, saccharase)	β-D-fructofuranoside fructohydrolase	3.2.1.26
Lactase	Lactose galactohydrolase	3.2.1.108
D-lactate dehydrogenase (Lactic acid dehydrogenase)	(R)-lactate:NAD(+) oxidoreductase	1.1.1.28
L-lactate dehydrogenase (Lactic acid dehydrogenase)	(S)-lactate:NAD(+) oxidoreductase	1.1.1.27
Lipoxygenase (Lipoxidase, carotene oxidase)	Linoleate:oxygen oxidoreductase	1.13.11.12
Lysozyme (Muramidase)	Peptidoglycan N-acetylmuramoylhydrolase	3.2.1.17
Malate dehydrogenase (Malic dehydrogenase)	(S)-malate:NAD(+) oxidoreductase	1.1.1.37
Microbial aspartic proteinase (Trypsinogen kinase)	None, see comments in original reference	3.4.23.6
Monophenol monoxygenase (Tyrosinase, phenolase, monophenol oxidase, cresolase)	Monophenol, L-dopa:oxygen oxidoreductase	1.14.18.1
Palmitoyl-CoA hydrolase (Long-chain fatty-acyl-CoA hydrolase)	Palmitoyl-CoA hydrolase	3.1.2.2
Papain (Papaya peptidase I)	None, see comments in original reference	3.4.22.2
Pectate lyase (Pectate transliminase)	Poly (1,4-α-D-galacturonide) lyase	4.2.2.2
Pectic esterase (Pectin demethoxylase, pectin methoxylase, pectin methylesterase)	Pectin pectylhydrolase	3.1.1.11
Pepsin A (Pepsin)	None, see comments in original reference	3.4.23.1
Peroxidase	Donor:hydrogen-peroxide oxidoreductase	1.11.1.7
Phosphodiesterase I (5′-exonuclease)	Oligonucleate 5’-nucleotidohydrolase	3.1.4.1
Phosphoglucomutase (Glucose phosphomutase)	α-D-glucose 1,6-phosphomutase	5.4.2.2
Phosphorylase (Muscle phosphorylase α and β, amylophosphorylase, polyphosphorylase)	1,4-α-D-glucan:orthophosphate a-D-glucosyltransferase	2.4.1.1
Plasmin (Fibrinase, fibrinolysin)	None, see comments in original reference	3.4.21.7
Polygalacturonase (Pectin depolymerase, pectinase)	Poly (1,4-α-D-galacturonide) glycanohydrolase	3.2.1.15
Pyranose oxidase (Glucoase-2-oxidase)	Pyranose:oxygen 2-oxidoreductase	1.1.3.10
Rennet, see Chymosin		
Pancreatic ribonuclease (RNase, RNase I, RNase A, pancreatic RNase, ribonuclease I)	None, see comments in original reference	3.1.2.75
Serine carboxypeptidase	Peptidyl-L-amino-acid hydrolase	3.4.16.1

(Continues)

Table 6.5. (Continued)

Recommended Name	Systematic Name	Enzyme Code (EC)
Shikimate dehydrogenase	Shikimate;NADP(+) 3-oxidoreductase	1.1.1.25
Staphylococcal cysteine protease (Staphylococcal proteinase II)	None, see comments in original reference	3.4.22.13
Starch (bacterial glycogen) synthase (ADPglucose-starch glucosyltransferase)	ADPglucose:1,4-α-D-glucan 4-α-D-glucosyltransferase	2.4.1.21
Streptococcal cysteine protease (Streptococcal proteinase, streptococcus peptidase a)	None, see comments in original reference	3.4.22.10
Subtilisin	None, see comments in original reference	3.4.21.14
Sucrose-phosphate synthase (UDPglucaose-fructose phosphate glucosyltransferase, sucrose-phosphate-udp glucosyl-transferase)	UDPglucose:D-fructose-6-phosphate 2-α-D-glucosyl-transferase	2.4.1.14
Sucrose synthase (UDPglucoase-fructose glucosyltransferase, sucrose-UDP glucosyltransferase)	UDPglucose:D-fructose 2-α-D-glucosyltransferase	2.4.1.13
Thiol oxidase	Thiol:oxigen oxidoreductase	1.8.3.2
Triacylglycerol lipase (Lipase, tributyrase, triglyceride lipase)		3.1.1.3
Trimethylamine-N-oxide aldolase	Trimethylamine-N-oxide formaldehydelase	4.1.2.32
Trypsin (α- and β-trypsin)	None, see comments in original reference	3.4.21.4
Tyrosinase, see Catechol oxidase		1.10.3.1
Xanthine oxidase (Hypoxanthin oxidase)	Xanthine:oxygen oxidoreductase	1.1.3.22

Sources: Nagadodawithana and Reed 1993, NC-IUBMB Enzyme Nomenclature website (www.iubmb.org).

REFERENCES

Nagodawithana T, Reed G. 1993. Enzymes in Food Processing. San Diego, California: Academic Press.

IUPAC-IUB Joint Commission on Biochemical Nomenclature (JCBN), and Nomenclature Commission of IUB (NC-IUB) Newsletters. 1984. Arch Biochem Biophys 229:237–245; Biochem J 217:I–IV; Biosci Rep 4:177–180; Chem. Internat (3): 24–25; Eur J Biochem 138:5–7; Hoppe-Seyler's Z Physiol Chem 365:I–IV.

NC-IUBMB. 1992. Enzyme Nomenclature 1992: Recommendations of the Nomenclature Committee of the International Union of Biochemistry and Molecular Biology, 6th edition. San Diego, California: Academic Press.

NC-IUBMB Enzyme Nomenclature website: http://www.chem.qmul.ac.uk/iubmb/enzyme/ or at www.iubmb.org.

Boyce S, Tipton KF. 2000. Enzyme classification and nomenclature. In: Nature Encyclopedia of Life Sciences. London: Nature Publishing Group. (http://www.els.net/[doi:10.1038/npg.els.0000710]).

7
Enzyme Activities†

*D. J. H. Shyu, J. T. C. Tzen, and C. L. Jeang**

INTRODUCTION

Long before human history, our ancestors, chimpanzees, might have already experienced the mild drunk feeling of drinking wine when they ate the fermented fruits that contained small amounts of alcohol. Archaeologists have also found some sculptured signs on 8000-year-old plates describing the beer-making processes. Chinese historians wrote about the lavish life of Tsou, a tyrant of Shang dynasty (around 1100 BC), who lived in a castle with wine-storing pools. All these indicated that people grasped the wine fermentation technique thousands of years ago. In the Chin dynasty (around 220 BC) in China, a spicy paste or sauce made from fermentation of soybean and/or wheat was mentioned. It is now called soybean sauce.

Although people applied fermentation techniques and observed the changes from raw materials to special products, they did not realize the mechanisms

*Corresponding author.

†This manuscript was reviewed by Dr. Hsien-Yi Sung, Department of Biochemical Science and Technology, National Taiwan University, Taipei, Taiwan, Republic of China, and Dr. Wai-Kit Nip, Department of Molecular Biosciences and Bioengineering, University of Hawaii at Manoa, Honolulu, Hawaii, USA.

that caused the changes. This mystery was uncovered by the development of biological sciences. Louis Pasteur claimed the existence and function of organisms that were responsible for the changes. In the same era, Justus Liebig observed the digestion of meat with pepsin, a substance found in stomach fluid, and proposed that a whole organism is not necessary for the process of fermentation.

In 1878, Kuhne was the first person to solve the conflict by introducing the word "enzyme," which means "in yeast" in Greek. This word emphasizes the materials inside or secreted by the organisms to enhance the fermentation (changes of raw material). Buchner (1897) performed fermentation by using a cell-free extract from yeast of significantly highlight, and these were molecules rather than lifeforms that did the work. In 1905, Harden and Young found that fermentation was accelerated by the addition of some small dialyzable molecules to the cell-free extract. The results indicated that both macromolecular and micromolecular compounds are needed. However, the nature of enzymes was not known at that time. A famous biochemist, Willstátter, was studying peroxidase for its high catalytic efficiency; since he never got enough samples, even though the reaction obviously happened, he hesitated (denied) to conclude that the enzyme was a protein. Finally, in 1926, Summer prepared crystalline urease from jack beans, analyzed the properties of the pure compound, and drew the conclusion that enzymes are proteins. In the following years, crystalline forms of some proteases were also obtained by Northrop and others. The results agreed with Summer's conclusion.

The studies on enzyme behavior were progressing in parallel. In 1894, Emil Fischer proposed a "lock and key" theory to describe the specificity and stereo relationship between an enzyme and its substrate. In 1902, Herri and Brown independently reported a saturation-type curve for enzyme reactions. They revealed an important concept in which the enzyme-substrate complex was an obligate intermediate formed during the enzyme-catalyzed reaction. In 1913, Michaelis and Menten derived an equation describing quantitatively the saturation-like behavior. At the same time, Monod and others studied the kinetics of regulatory enzymes and suggested a concerted model for the enzyme reaction. In 1959, Fischer's hypothesis was slightly modified by Koshland. He proposed an induced-fit theory to describe the moment the enzyme and substrate are attached, and suggested a sequential model for the action of allosteric enzymes.

For the studies on enzyme structure, Sanger and his colleagues were the first to announce the unveiling of the amino acid sequence of a protein, insulin. After that, the primary sequences of some hydrolases with comparatively small molecular weights, such as ribonuclease, chymotrypsin, lysozyme, and others, were defined. Twenty years later, Sanger won his second Nobel Prize for the establishment of the chain-termination reaction method for nucleotide sequencing of DNA. Based on this method, the deduction of the primary sequences of enzymes blossomed; to date, the primary structures of 55,410 enzymes have been deduced. Combining genetic engineering technology and modern computerized X-ray crystallography and/or NMR, about 15,000 proteins, including 9268 proteins and 2324 enzymes, have been analyzed for their three-dimensional structures [Protein Data Bank (PDB): http://www.rcsb.org/pdb]. Computer software was created, and protein engineering on enzymes with demanded properties was carried on successfully [Fang and Ford 1998, Igarashi et al. 1999, Pechkova et al. 2003, Shiau et al. 2003, Swiss-PDBViewer(spdbv): http://www.expasy.ch/spdbv/mainpage.htm, SWISS-MODELserver: http://www.expasy.org/swissmod/SWISS-MODEL.htm].

FEATURES OF ENZYMES

Most of the Enzymes Are Proteins

Proteins are susceptible to heat, strong acids and bases, heavy metals, and detergents. They are hydrolyzed to amino acids by heating in acidic solution and by the proteolytic action of enzymes on peptide bonds. Enzymes give positive results on typical protein tests, such as the Biuret, Millions, Hopkins-Cole, and Sakaguchi reactions. X-ray crystallographic studies revealed that there are peptide bonds between adjacent amino acid residues in proteins. The majority of the enzymes fulfill the above criterion; therefore, they are proteins in nature. However, the catalytic element of some well-known ribozymes is just RNAs in nature (Steitz and Moore 2003, Raj and Liu 2003).

Chemical Composition of Enzymes

For many enzymes, protein is not the only component required for its full activity. On the basis of the chemical composition of enzymes, they are categorized into several groups, as follows:

1. Polypeptide, the only component, for example, lysozyme, trypsin, chymotrypsin, or papain;
2. Polypeptide plus one to two kinds of metal ions, for example, α-amylase containing Ca^{2+}, kinase containing Mg^{2+}, and superoxide dismutase having Cu^{2+} and/or Zn^{2+};
3. Polypeptide plus a prosthetic group, for example, peroxidase containing a heme group;
4. Polypeptide plus a prosthetic group and a metal ion, for example, cytochome oxidase ($a+a_3$) containing a heme group and Cu^{2+};
5. Polypeptide plus a coenzyme, for example, many dehydrogenases containing NAD^+ or $NADP^+$;
6. Combination of polypeptide, coenzyme, and a metal ion, for example, succinate dehydrogenase containing both the FAD and nonheme iron.

Enzymes Are Specific

In the life cycle of a unicellular organism, thousands of reactions are carried out. For the multicellular higher organisms with tissues and organs, even more kinds of reactions are progressing in every moment. Less than 1% of errors that occur in these reactions will cause accumulation of waste materials (Drake 1999), and sooner or later the organism will not be able to tolerate these accumulated waste materials. These phenomena can be explained by examining genetic diseases; for example, Phenylketonuria (PKU) in humans, where the malfunction of phenylalanine hydroxylase leads to an accumulation of metabolites such as phenylalanine and phenylacetate, and others, finally causing death (Scriver 1995, Waters 2003). Therefore, enzymes catalyzing the reactions bear the responsibility for producing desired metabolites and keeping the metabolism going smoothly.

Enzymes have different types of specificities. They can be grouped into the following common types:

1. *Absolute specificity.* For example, urease (Sirko and Brodzik 2000) and carbonate anhydrase (Khalifah 2003) catalyze only the hydrolysis and cleavage of urea and carbonic acid, respectively. Those enzymes having small molecules as substrates or that work on biosynthesis pathways belong to this category.
2. *Group specificity.* For example, hexokinase prefers D-glucose, but it also acts on other hexoses (Cardenas et al. 1998).
3. *Stereo specificity.* For example, D-amino acid oxidase reacts only with D-amino acid, but not with L-amino acid; and succinate dehydrogenase reacts only with the fumarate (*trans* form), but not with maleate (*cis* form).
4. *Bond specificity.* For example, many digestive enzymes. They catalyze the hydrolysis of large molecules of food components, such as proteases, amylases, and lipases. They seem to have broad specificity in their substrates; for example, trypsin acts on all kinds of denatured proteins in the intestine, but prefers the basic amino acid residues at the C-terminal of the peptide bond. This broad specificity brings up an economic effect in organisms: they are not required to produce many digestive enzymes for all kinds of food components.

The following two less common types of enzyme specificity are impressive: (1) An unchiral compound, citric acid is formed by citrate synthase with the condensation of oxaloacetate and acetyl-CoA. An aconitase acts only on the moiety that comes from oxaloacetate. This phenomenon shows that, for aconitase, the citric acid acts as a chiral compound (Barry 1997). (2) If people notice the rare mutation that happens naturally, they will be impressed by the fidelity of certain enzymes, for example, amino acyl-tRNA synthethase (Burbaum and Schimmel 1991, Cusack 1997), RNA polymerase (Kornberg 1996), and DNA polymerase (Goodman 1997); they even correct the accidents that occur during catalyzing reactions.

Enzymes Are Regulated

The catalytic activity of enzymes is changeable under different conditions. These enzymes catalyze the set steps of a metabolic pathway, and their activities are responsible for the states of cells. Intermediate metabolites serve as modulators on enzyme activity; for example, at high concentrations of ADP, the catalytic activity of the phosphofructokinase, pyruvate kinase, and pyruvate dehydrogenase complex are enhanced; as [ATP] becomes high, they will be inhibited. These modulators bind at allosteric sites on enzymes (Hammes 2002). The structure and mechanism of allosteric regulatory enzymes such as aspartate transcarbamylase (Cherfils et al. 1990) and ribonucleoside diphosphate reductase (Scott et al. 2001) have been well studied. However, allosteric regulation is not the only way that the enzymatic activity is influenced. Covalent modification by protein kinases (Langfort et al. 1998) and phosphatases (Luan 2003) on enzymes will cause large fluctuations in total enzymatic activity in the metabolism pool. In addition, sophisticated tuning phenomena on glycogen phosphorylase and glycogen synthase through phosphorylation and dephosphorylation have been observed after hormone signaling (Nuttall et al. 1988, Preiss and Romeo 1994).

Enzymes Are Powerful Catalysts

The compound glucose will remain in a bottle for years without any detectable changes. However, when glucose is applied in a minimal medium as the only carbon source for the growth of *Salmonella typhimurium* (or *Escherichia coli*), phosphorylation of this molecule to glucose-6-phosphate is the first chemical reaction that occurs as it enters the cells. From then on, the activated glucose not only serves as a fuel compound to be oxidized to produce chemical energy, but also goes through numerous reactions to become the carbon skeleton of various micro- and macrobiomolecules. To obtain each end product, multiple steps have to be carried out. It may take less than 30 minutes for a generation to go through all the reactions. Only the existence of enzymes guarantees this quick utilization and disappearance of glucose.

ENZYMES AND ACTIVATION ENERGY

Enzymes Lower the Activation Energy

Enzymes are mostly protein catalysts; except for the presence of a group of ribonucleic acid–mediated reactions, they are responsible for the chemical reactions of metabolism in cells. For the catalysis of a reaction, the reactants involved in this reaction all require sufficient energy to cross the potential energy barrier, the activation energy (E_A), for the breakage of the chemical bonds and the start of the reaction. Few have enough energy to cross the reaction energy barrier until the reaction catalyst (the enzyme) forms a transition state with the reactants to lower the activation energy (Fig. 7.1). Thus, the enzyme lowers the barrier that usually prevents the chemical reaction from occurring and facilitates the reaction proceeding at a faster rate to approach equilibrium. The substrates (the reactants), which are specific for the enzyme involved in the reaction, combine with enzyme to lower the activation energy of the reaction. One enzyme type will combine with one specific type of substrate, that is, the active site of one particular enzyme will only fit one specific type of substrate, and this leads to the formation of and enzyme-substrate (ES) complex. Once the energy barrier is overcome, the substrate is then changed to another chemical, the product. It should be noted that the enzyme itself is not consumed or altered by the reaction, and that the overall free-energy change (ΔG) or the related equilibrium also remain constant. Only the activation energy (E_{Au}) of the uncatalyzed reaction decreases to that (E_{Ac}) of the enzyme-catalyzed reaction.

One of the most important mechanisms of the enzyme function that decreases the activation energy involves the initial binding of the enzyme to the substrate, reacting in the correct direction, and closing of the catalytic groups of the ES complex. The binding energy, part of the activation energy, is required for the enzyme to bind to the substrate and is determined primarily by the structure complementarities between the enzyme and the substrate. The binding energy is used to reduce the free energy of the transition-state ES complex but not to form a stable, not easily separated, complex. For a reaction to occur, the enzyme acts as a catalyst by efficiently binding to its substrate, lowering the energy barrier, and allowing the formation of product. The enzyme stabilizes the transition state of the catalyzed reaction, and the transition state is the rate-limiting state in a single-step reaction. In a two-step reaction, the step with the highest transition-state free energy is said to be the rate-limiting state. Though the reaction rate is the speed at which the reaction proceeds toward equilibrium, the speed of the reaction does not affect the equilibrium point.

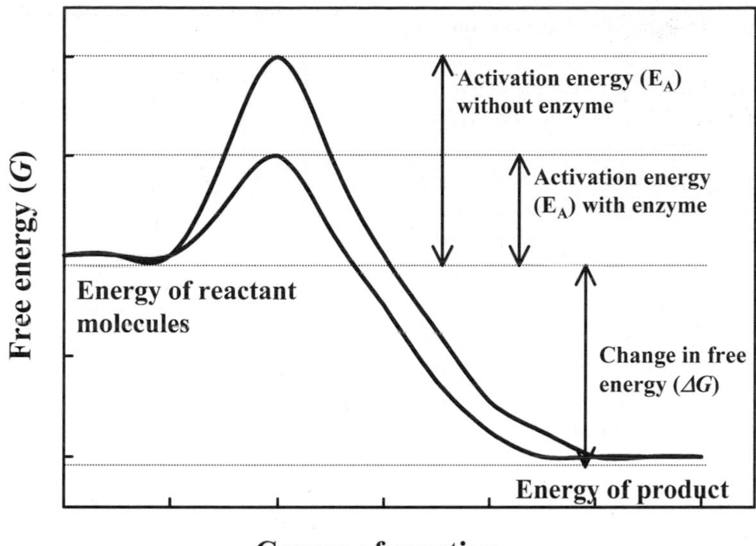

Figure 7.1. A transition state diagram, also called reaction coordinate diagram, shows the activation energy profile of the course of an enzyme-catalyzed reaction. One curve depicts the course of the reaction in the absence of the enzyme while the other depicts the course of reaction in the presence of the enzyme that facilitates the reaction. The difference in the level of free energy between the beginning state (ground state) and the peak of the curve (transition state) is the activation energy of the reaction. The presence of enzyme lowers the activation energy (E_A) but does not change the overall free energy (ΔG).

How Does an Enzyme Work?

An enzyme carries out the reaction by temporarily combining with its specific kind of substrate, resulting in a slight change of their structures to produce a very precise fit between the two molecules. The introduction of strains into the enzyme and substrate shapes allows more binding energy to be available for the transition state. Two models of this minor structure modification are the induced fit, in which the binding energy is used to distort the shape of the enzyme, and the induced strain, in which binding energy is used to distort the shape of the substrate. These two models, in addition to a third model, the nonproductive binding model, have been proposed to describe conformational flexibility during the transition state. The chemical bonds of the substrate break and form new ones, resulting in the formation of new product. The newly formed product is then released from the enzyme, and the enzyme combines with another substrate for the next reaction.

ENZYME KINETICS AND MECHANISM

Regulatory Enzymes

For the efficient and precise control of metabolic pathways in cells, the enzyme, which is responsible for a specific step in a serial reaction, must be regulated.

Feedback Inhibition

In living cells, a series of chemical reactions in a metabolic pathway occurs, and the resulting product of one reaction becomes the substrate of the next reaction. Different enzymes are usually responsible for each step of catalysis. The final product of each metabolic pathway will inhibit the earlier steps in the serial reactions. This is called feedback inhibition.

Noncompetitive Inhibition

An allosteric site is a specific location on the enzyme where an allosteric regulator, a regulatory molecule that does not directly block the active site of the enzyme, can bind. When the allosteric regulator attaches to the specific site, it causes a change in the conformation of enzyme active site, and the substrate therefore will not fit into the active site of enzyme; this results in the inhibition phenomenon. The enzyme cannot catalyze the reaction, and not only the amount of final product but also the concentration of the allosteric regulator decrease. Since the allosteric regulator (the inhibitor) does not compete with the substrate for binding to the active site of enzyme, the inhibition mechanism is called noncompetitive inhibition.

Competitive Inhibition

A regulatory molecule is not the substrate of the specific enzyme but shows affinity for attaching to the active site. When it occupies the active site, the substrate cannot bind to the enzyme for the catalytic reaction, and the metabolic pathway is inhibited. Since the regulatory molecule will compete with the substrate for binding to the active site on enzyme, this inhibitory mechanism of the regulatory molecule is called competitive inhibition.

Enzyme Kinetics

During the course of an enzyme-catalyzed reaction, the plot of product formation over time (product formation profile) reveals an initial rapid increase, approximately linear, of product, and then the rate of increase decreases to zero as time passes (Fig. 7.2A). The slope of initial rate (v_i or v), also called the steady state rate, appears to be initially linear in a plot where the product concentration versus reaction time follows the establishment of the steady state of a reaction. When the reaction is approaching equilibrium or the substrate is depleted, the rate of product formation decreases, and the slope strays away from linearity. As is known by varying the conditions of the reaction, the kinetics of a reaction will appear linear during a time period in the early phase of the reaction, and the reaction rate is measured during this phase. When the substrate concentrations are varied, the product formation profiles will display linear substrate dependency at this phase (Fig. 7.2B). The rates for each substrate concentration are measured as the slope of a plot of product formation over time. A plot of initial rate as

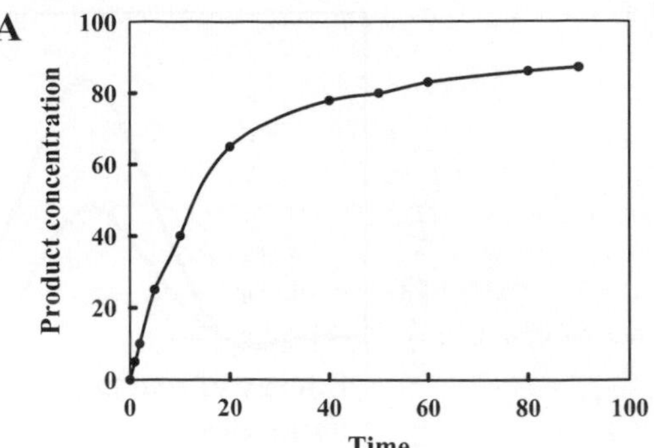

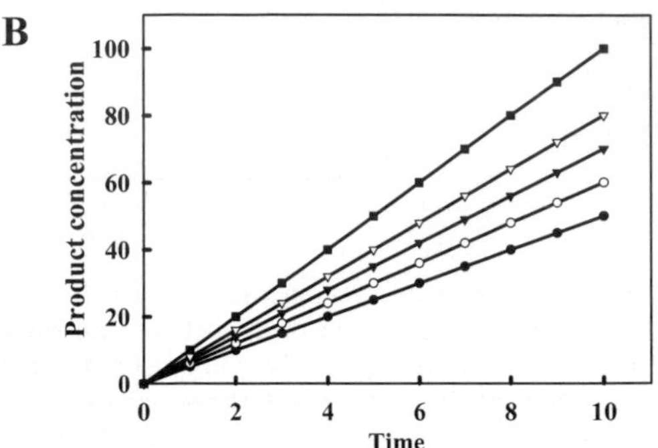

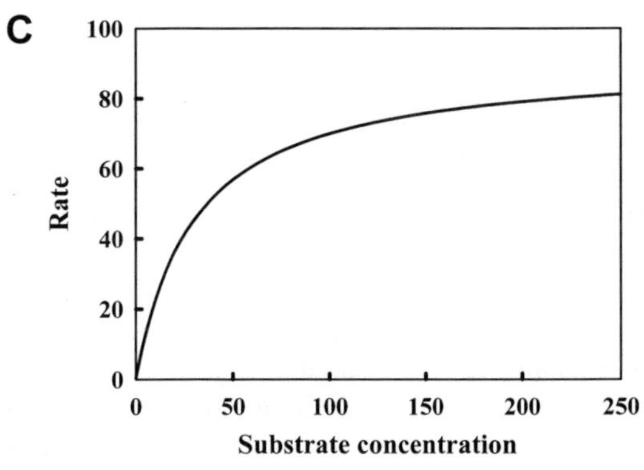

Figure 7.2. **(A)** Plot of progress curve during an enzyme-catalyzed reaction for product formation. **(B)** Plot of progress curves during an enzyme-catalyzed reaction for product formation with a different starting concentration of substrate. (**C**) Plot of reaction rate as a function of substrate concentration measured from slopes of lines from (B).

a function of substrate concentration is demonstrated as shown in Figure 7.2C. Three distinct portions of the plot will be noticed: an initial linear relationship showing first-order kinetic at low substrate concentrations; an intermediate portion showing curved linearity dependent on substrate concentration, and a final portion revealing no substrate concentration dependency, i.e., zero-order kinetic, at high substrate concentrations. The interpretation of the phenomenon can be described by the following scheme:

> The initial rate will be proportional to the ES complex concentration if the total enzyme concentration is constant and the substrate concentration varies. The concentration of the ES complex will be proportional to the substrate concentration at low substrate concentrations, and the initial rate will show a linear relationship and substrate concentration dependency. All of the enzyme will be in the ES complex form when substrate concentration is high, and the rate will depend on the rate of ES transformation into enzyme-product and the subsequent release of product. Adding more substrate will not influence the rate, so the relationship of rate versus substrate concentration will approach zero (Brown 1902).

However, a lag phase in the progress of the reaction will be noticed in coupled assays due to slow or delayed response of the detection machinery (see below). Though v can be determined at any constant concentration of substrate, it is recommended that the value of substrate concentration approaching saturation (high substrate concentration) be used so that it can approach its limiting value V_{max} with greater sensitivity and prevent errors occurring at lower substrate concentrations.

Measurement of the rate of enzyme reaction as a function of substrate concentration can provide information about the kinetic parameters that characterize the reaction. Two parameters are important for most enzymes—K_m, an approximate measurement of the tightness of binding of the substrate to the enzyme; and V_{max}, the theoretical maximum velocity of the enzyme reaction. To calculate these parameters, it is necessary to measure the enzyme reaction rates at different concentrations of substrate, using a variety of methods, and analyze the resulting data. For the study of single-substrate kinetics, one saturating concentration of substrate in combination with varying substrate concentrations can be used to investigate the initial rate (v), the catalytic constant (k_{cat}), and the specific constant (k_{cat}/K_m). For investigation of multisubstrate kinetics, however, the dependence of v on the concentration of each substrate has to be determined, one after another; that is, by measuring the initial rate at varying concentrations of one substrate with fixed concentrations of other substrates.

A large number of enzyme-catalyzed reactions can be explained by the Michaelis-Menten equation:

$$v = \frac{V_{max}[S]}{(K_m + [S])} = \frac{k_{cat}\,[E]\,[S]}{(K_m + [S])}$$

where [E] and [S] are the concentrations of enzyme and substrate, respectively (Michaelis and Menten 1913). The equation describes the rapid equilibrium that is established between the reactant molecules (E and S) and the ES complex, followed by slow formation of product and release of free enzyme. It assumes that $k_2 \ll k_{-1}$ in the following equation, and K_m is the value of [S] when $v = V_{max}/2$.

$$E + S \underset{k_{-1}}{\overset{k_1}{\rightleftarrows}} ES \xrightarrow{k_2} E + P$$

Briggs and Haldane (1924) proposed another enzyme kinetic model, called the steady state model; steady state refers to a period of time when the rate of formation of ES complex is the same as rate of formation of product and release of enzyme. The equation is commonly the same as that of Michaelis and Menten, but it does not require $k_2 \ll k_{-1}$, and $K_m = (k_{-1}+k_2)/k_1$ because [ES] now is dependent on the rate of formation of the ES complex (k_1) and the rate of dissociation of the ES complex (k_{-1} and k_2). Only under the condition that $k_2 \ll k_{-1}$ will $K_s = k_{-1}/k_1$ and be equivalent to K_m, the dissociation constant of the ES complex $\{K_s = k_{-1}/k_1 = ([E] - [ES])[S]/[ES]\}$, otherwise $K_m > K_s$.

The K_m is the substrate concentration at half of the maximum rate of enzymatic reaction, and it is equal to the substrate concentration at which half of the enzyme active sites are saturated by the substrate in the steady state. So, K_m is not a useful measure of ES binding affinity in certain conditions when K_m is not equivalent to K_s. Instead, the specific constant (k_{cat}/K_m) can be substituted as a measure of substrate binding; it represents the catalytic efficiency of the enzyme. The k_{cat}, the catalytic constant, represents conversion of the maximum number of reactant molecules to product per enzyme active site per unit time, or the number of turnover events occurring per unit time; it also represents the turnover rate whose unit is the reciprocal of time. In the Michaelis-Menten approach, the dissociation of the EP complex is slow, and the k_{cat} contribution to this rate constant will be equal to the dissociation constant. However, in the Briggs-Haldane approach, the dissociation rate of ES complex is fast, and the k_{cat} is equal to k_2. In addition, the values of k_{cat}/K_m are used not only to compare the efficiencies of different enzymes but also to compare the efficiencies of different substrates for a specified enzyme.

Deviations of expected hyperbolic reaction are occasionally found due to such factors as experimental artifacts, substrate inhibition, existence of two or more enzymes competing for the same substrate, and the cooperativity. They show nonhyperbolic behavior that cannot fit well into the Michaelis-Menten approach. For instance, a second molecule binds to the ES complex and forms an inactive ternary complex that usually occurs at high substrate concentrations and is noticed when rate values are lower than expected. It leads to breakdown of the Michaelis-Menten equation. It is not only the Michaelis-Menten plot that is changed to show a nonhyperbolic behavior; it is also the Lineweaver-Burk plot that is altered to reveal the nonlinearity of the curve (see below).

A second example of occasionally occurring nonhyperbolic reactions is the existence of more than one enzyme in a reaction that competes for the same substrate; the conditions usually are realized when crude or partially purified samples are used. Moreover, the conformational change in the enzymes may be induced by ligand binding for the regulation of their activities, as discussed earlier in this chapter; many of these enzymes are composed of multimeric subunits and multiple ligand-binding sites in cases that do not obey the Michaelis-Menten equation. The sigmoid, instead of the hyperbolic, curve is observed; this condition, in which the binding of a ligand molecule at one site of an enzyme may influence the affinities of other sites, is known as cooperativity. The increase and decrease of affinities of enzyme binding sites are the effects of positive and negative cooperativity, respectively. The number of potential substrate binding sites on the enzymes can be quantified by the Hill coefficient, h, and the degree of cooperativity can be measured by the Hill equation (see page 164). More detailed illustrations and explanations of deviations from Michaelis-Menten behavior can be found in the literatures (Bell and Bell 1988, Cornish-Bowden 1995).

Data Presentation

Untransformed Graphics

The values of K_m and V_{max} can be determined graphically using nonlinear plots by measuring the initial rate at various substrate concentrations and then transforming them into different kinetic plots. First, the substrate stock concentration can be chosen at a reasonably high level; then a two-fold (stringently) or five- or 10-fold (roughly) serial dilution can be made from this stock. After the data obtained has been evaluated, a Michaelis-Menten plot of rate v_i as a function of substrate concentration [S] is subsequently drawn, and the values of both K_m and V_{max} can be estimated. Through the plot, one can check if a hyperbolic curve is observed, and if the values estimated are meaningful. Both values will appear to be infinite if the range of substrate concentrations is low; on the other hand, although the value of V_{max} can be approximately estimated, that of K_m cannot be determined if the range of substrate concentrations is too high. Generally, substrate concentrations covering 20–80% of V_{max}, which corresponds to a substrate concentration of 0.25–5.0 K_m, is appreciated.

Lineweaver-Burk Plots

Though nonlinear plots are useful in determining the values of K_m and V_{max}, transformed, linearized plots are valuable in determining the kinetics of multisubstrate enzymes and the interaction between enzymes and inhibitors. One of the best known plots is the double-reciprocal or Lineweaver-Burk plot (Lineweaver and Burk 1934). Inverting both sides of the Michaelis-Menten equation gives the Lineweaver-Burk plot: $1/v = (K_m/V_{max})\ (1/[S]) + 1/V_{max}$ (Fig. 7.3). The equation is for a linear curve when one plots $1/v$ against $1/[S]$ with a slope of K_m/V_{max} and a y intercept of $1/V_{max}$. Thus, the kinetic values can also be determined from the slope and intercept values of the plot. However, caution should be taken due to small experimental errors that may be amplified by taking the reciprocal (the distorting effect), especially in the measurement of v values at low substrate concentrations. The problem can be solved

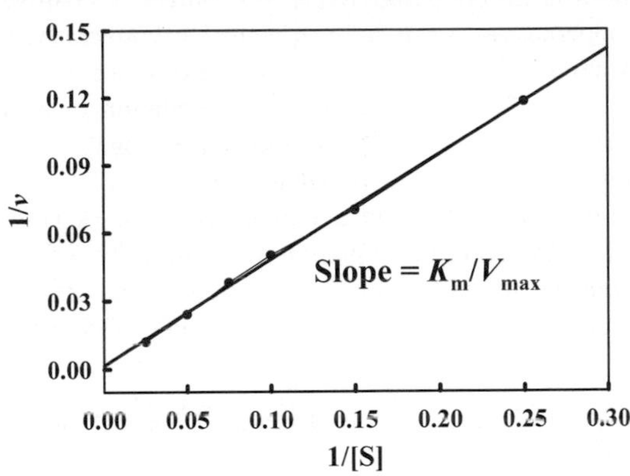

Figure 7.3. The Lineweaver-Burk double-reciprocal plot.

by preparing a series of substrate concentrations evenly spaced along the 1/[S] axis, that is, diluting the stock solution of substrate by two-, three-, four-, five-,... fold (Copeland 2000).

Eadie-Hofstee Plots

The Michaelis-Menten equation can be rearranged to give $v = -K_m\,(v/[S]) + V_{max}$ by first multiplying both sides by $K_m + [S]$, and then dividing both sides by [S]. The plot will give a linear curve with a slope of $-K_m$ and a y intercept of v when v is plotted against v/[S] (Fig. 7.4). The Eadie-Hofstee plot has the advantage of decompressing data at high substrate concentrations and being more accurate than the Lineweaver-Burk plot, but the values of v against [S] are more difficult to determine (Eadie 1942, Hofstee 1959). It also has the advantage of observing the range of v from zero to V_{max} and is the most effective plot for revealing deviations from the hyperbolic reaction when two or more components, each of which follows the Michaelis-Menten equation, although errors in v affect both coordinates (Cornish-Bowden 1996).

Hanes-Wolff Plots

The Lineweaver-Burk equation can be rearranged by multiplying both sides by [S]. This gives a linear plot $[S]/v = (1/V_{max})\,[S] + K_m/V_{max}$, with a slope of $1/V_{max}$, the y intercept of K_m/V_{max}, and the x intercept of $-K_m$ when $[S]/v$ is plotted as a function of [S] (Fig. 7.5). The plot is useful in obtaining kinetic values without distorting the data (Hanes 1932).

Eisenthal-Cornish-Bowden Plots (the Direct Linear Plots)

As in the Michaelis-Menten plot, pairs of rate values of v and the negative substrate concentrations [S] are applied along the y and x axis, respectively. Each pair of values is a linear curve that connects two points of values and extrapolates to pass the intersection. The location (x, y) of the point of intersection on the two-dimensional plot represents K_m and V_{max}, respectively (Fig. 7.6). The direct linear plots

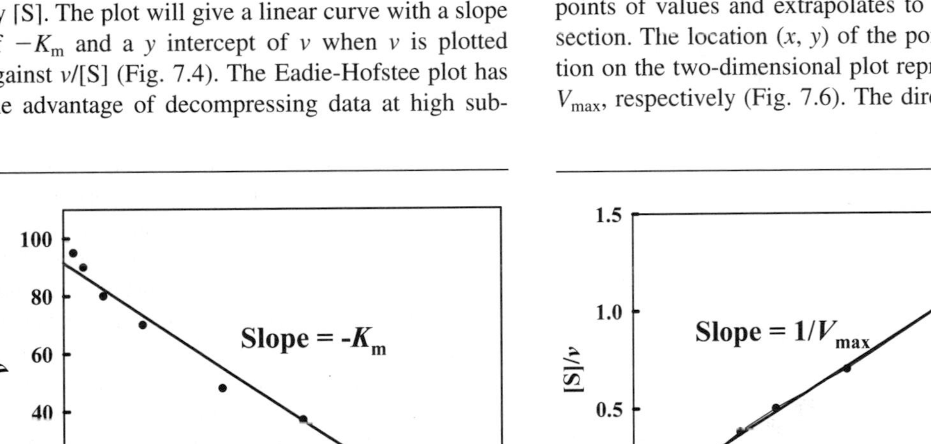

Figure 7.4. The Eadie-Hofstee plot.

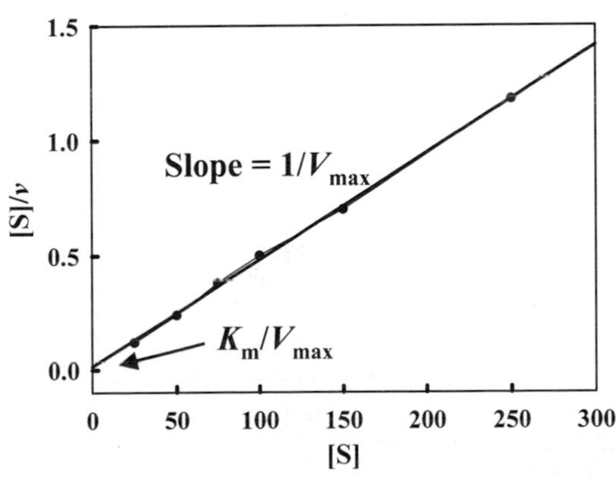

Figure 7.5 The Hanes-Wolff plot.

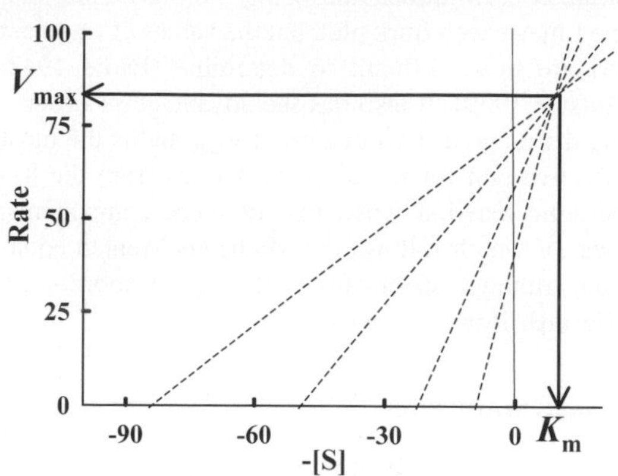

Figure 7.6. The Eisenthal-Cornish-Bowden direct plot.

(Eisenthal and Cornish-Bowden 1974) are useful in estimating the Michaelis-Menten kinetic parameters and are the better plots for revealing deviations from the reaction behavior.

Hill Plot

The Hill equation is a quantitative analysis measuring non–Michaelis-Menten behavior of cooperativity (Hill 1910). In a case of completely cooperative binding, an enzyme contains h binding sites that are all occupied simultaneously with a dissociation constant $K = [E][S]^h/[ES_h]$, where h is the Hill coefficient, which measures the degree of cooperativity. If there is no cooperativity, $h = 1$; there is positive cooperativity when $h > 1$, negative cooperativity when $h < 1$. The rate value can be expressed as $v = V_{max}[S]^h/(K + [S]^h)$, and the Hill equation gives a linear $\log(v_i/V_{max} - v) = h\log[S] - \log K$, when $\log(v/V_{max} - v)$ is plotted as a function of log[S] with a slope of h and a y intercept of -log$K$ (Fig. 7.7). However, the equation will show deviations from linearity when outside a limited range of substrate concentrations, that is, in the range of [S] = K.

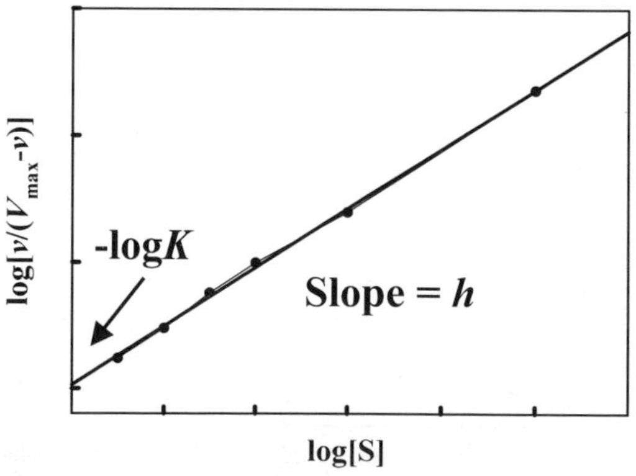

Figure 7.7. The Hill plot.

FACTORS AFFECTING ENZYME ACTIVITY

As discussed earlier in this chapter, the rate of an enzymatic reaction is very sensitive to reaction conditions such as temperature, pH, ionic strength, buffer constitution, and substrate concentration. To investigate the catalytic mechanism and the efficiency of an enzyme with changes in the parameters, and to evaluate a suitable circumstance for assaying the enzyme activity, the conditions should be kept constant to allow reproducibility of data and to prevent a misleading interpretation.

Enzyme, Substrate, and Cofactor Concentrations

In general, the substrate concentration should be kept much higher than that of enzyme in order to prevent substrate concentration dependency of the reaction rate at low substrate concentrations. The rate of an enzyme-catalyzed reaction will also show linearity to the enzyme concentration when substrate concentration is constant. The reaction rate will not reveal the linearity until the substrate is depleted, and the measurement in initial rate will be invalid. Preliminary experiments must be performed to determine the appropriate range of substrate concentrations over a number of enzyme concentrations to prevent the phenomenon of substrate depletion. In addition, the presence of inhibitor, activator, and cofactor in the reaction mixture also influence enzyme activity. The enzymatic activity will be low or undetectable in the presence of inhibitors or in the absence of activators or cofactors. In the case of cofactors, the activated enzymes will be proportional to the cofactor concentration added in the reaction mixture if enzymes are in excess. The rate of reaction will not represent the total amount of enzyme when the cofactor supplement is not sufficient, and the situation can be avoided by adding excess cofactors (Tipton 1992). Besides, loss of enzyme activity may be seen before substrate depletion, as when the enzyme concentration is too low and leads to the dissociation of the dimeric or multimeric enzyme. To test this possibility, addition of the same amount of enzyme can be applied to the reaction mixture to measure if there is the appearance of another reaction progression curve.

Effects of pH

Enzymes will exhibit maximal activity within a narrow range of pH and vary over a relatively broad range. For an assay of enzymatic catalysis, the pH of the solution of the reaction mixture must be maintained in an optimal condition to avoid pH-induced protein conformational changes, which lead to diminishment or loss of enzyme activity. A buffered solution, in which the pH is adjusted with a component with pK_a at or near the desired pH of the reaction mixture, is a stable environment that provides the enzyme with maximal catalytic efficiency. Thus, the appropriate pH range of an enzyme must be determined in advance when optimizing assay conditions.

To determine if the enzymatic reaction is pH dependent and to study the effects of group ionizations on enzyme kinetics, the rate as a function of substrate concentration at different pH conditions can be measured to simultaneously obtain the effects of pH on the kinetic parameters. It is known that the ionizations of groups are of importance, either for the active sites of the enzyme involved in the catalysis or for maintaining the active conformation of the enzyme. Plots of k_{cat}, K_m, and k_{cat}/K_m values at varying pH ranges will reflect important information about the roles of groups of the enzyme. The effect of pH dependence on the value of k_{cat} reveals the steps of ionization of groups involved in the ES complex and provides a pK_a value for the ES complex state; the value of K_m shows the ionizing groups that are essential to the substrate-binding step before the reaction. Moreover, the k_{cat}/K_m value reveals the ionizing groups that are essential to both the substrate-binding and the ES complex–forming steps, and provides the pK_a value of the free reactant

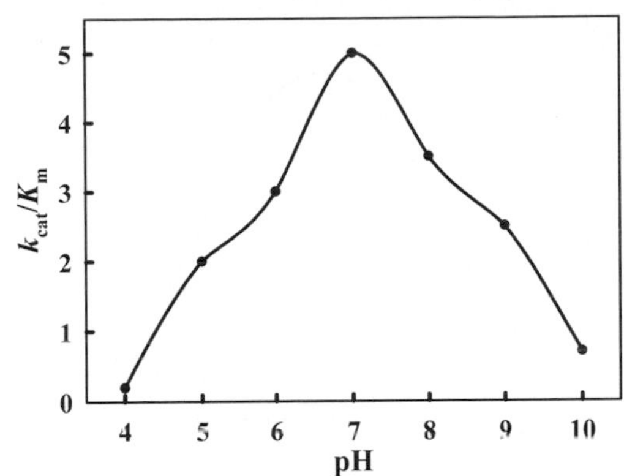

Figure 7.8. The effect of pH on the k_{cat}/K_m value of an enzymatic reaction that is not associated with the Henderson-Hasselbalch equation.

molecules state (Brocklehurst and Dixon 1977, Copeland 2000). From a plot of the k_{cat}/K_m value on the logarithmic scale as a function of pH, the pK_a value can be determined from the point of intersection of lines on the plot, especially for an enzymatic reaction that is not associated with the Henderson-Hasselbalch equation $pH = pK_a + \log([A^-]/[HA])$ (Fig. 7.8). Thus, the number of ionizing groups involved in the reaction can be evaluated (Dixon and Webb 1979, Tipton and Webb 1979).

Effects of Temperature

Temperature is one of the important factors affecting enzyme activity. For a reaction to occur at room temperature without the presence of an enzyme, small proportions of reactant molecules must have sufficient energy levels to participate in the reaction (Fig. 7.9A). When the temperature is raised above room temperature, more reactant molecules gain enough energy to be involved in the reaction (Fig. 7.9B). The activation energy is not changed, but the distribution of energy-sufficient reactants is shifted to a higher average energy level. When an enzyme is participating in the reaction, the activation energy is lowered significantly, and the proportion of reactant molecules at an energy level above the activation energy is also greatly increased (Fig. 7.9C). That means the reaction will proceed at a much higher rate.

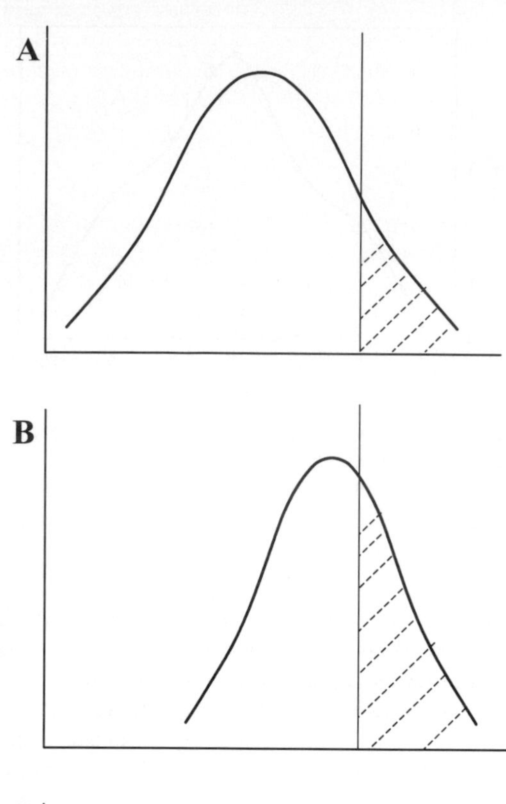

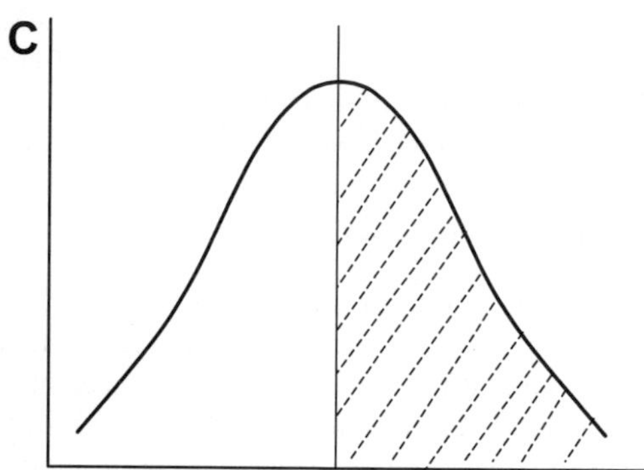

Figure 7.9. Plots of temperature effect on the energy levels of reactant molecules involved in a reaction. **(A)** The first plot depicts the distribution of energy levels of the reactant molecules at room temperature without the presence of an enzyme. **(B)** The second plot depicts that at the temperature higher than room temperature but in the absence of an enzyme. **(C)** The third plot depicts that at room temperature in the presence of an enzyme. The vertical line in each plot indicates the required activation energy level for a reaction to occur. The shaded portion of distribution in each plot indicates the proportion of reactant molecules that have enough energy levels to be involved in the reaction. The *x*-axis represents the energy level of reactant molecules, while the *y*-axis represents the frequency of reactant molecules at an energy level.

Most enzyme-catalyzed reactions are characterized by an increase in the rate of reaction and increased thermal instability of the enzyme with increasing temperature. Above the critical temperature, the activity of the enzyme will be reduced significantly; while within this critical temperature range, the enzyme activity will remain at a relatively high level, and inactivation of the enzyme will not occur. Since the rate of reaction increases due to the increased temperature by lowering the activation energy E_A, the relationship can be expressed by the Arrhenius equation: $k = A \exp(-E_A/RT)$, where A is a constant related to collision probability of reactant molecules, R is the ideal gas constant (1.987 cal/mol—deg), T is the temperature in degrees Kelvin (K = °C + 273.15), and k represents the specific rate constant for any rate, that is, k_{cat} or V_{max}. The equation can be transformed into: $\ln K = \ln A - E_A/RT$, where the plot of $\ln K$ against 1/T usually shows a linear relationship with a slope of $-E_A/R$, where the unit of E_A is cal/mol. The calculated E_A of a reaction at a particular temperature is useful in predicting the activation energy of the reaction at another temperature. And the plot is useful in judging if there is a change in the rate-limiting step of a sequential reaction when the line of the plot reveals bending of different slopes at certain temperatures (Stauffer 1989).

Taken together, the reaction should be performed under a constantly stable circumstance, with both temperature and pH precisely controlled from the start to the finish of the assay, for the enzyme to exhibit highly specific activity at appropriate acid-base conditions and buffer constitutions. Whenever possible, the reactant molecules should be equilibrated at the required assay condition following addition of the required components and efficient mixing to provide a homogenous reaction mixture. Because the enzyme to be added is usually stored at low temperature, the reaction temperature will not be significantly influenced when the enzyme volume added is at as low as 1–5% of the total volume of the reaction mixture. A lag phase will be noticed in the rate measurement as a function of time when the temperature of the added enzyme stock eventually influences the reaction. Significant temperature changes in the components stored in different environments and atmospheres should be avoided when the reaction mixture is mixed and the assay has started.

METHODS USED IN ENZYME ASSAYS

GENERAL CONSIDERATIONS

A suitable assay method is not only a prerequisite for detecting the presence of enzyme activity in the extract or the purified material: it is also an essential vehicle for kinetic study and inhibitor evaluation. Selection of an assay method that is appropriate to the type of investigation and the purity of the assaying material is of particular importance. It is known that the concentration of the substrate will decrease and that of product will increase when the enzyme is incubated with its substrate. Therefore, an enzyme assay is intended to measure either the decrease in substrate concentration or the increase in product concentration as a function of time. Usually the latter is preferred because a significant increase of signal is much easier to monitor. It is recommended that a preliminary test be performed to determine the optimal conditions for a reaction including substrate concentration, reaction temperature, cofactor requirements, and buffer constitutions, such as pH and ionic strength, to ensure the consistency of the reaction. Also an enzyme blank should be performed to determine if the nonenzymatic reaction is negligible or could be corrected. When a crude extract, rather than purified enzyme, is used for determining enzyme activity, one control experiment should be performed without adding substrate to determine if there are any endogenous substrates present and to prevent overcounting of the enzyme activity in the extract.

Usually the initial reaction rate is measured, and it is related to the substrate concentration. Highly sufficient substrate concentration is required to prevent a decrease in concentration during the assay period of enzyme activity, thus allowing the concentration of the substrate to be regarded as constant.

TYPES OF ASSAY METHODS

On the basis of measuring the decrease of the substrate or the increase of the product, one common characteristic property that is useful for distinguishing the substrate and the product is their absorbance spectra, which can be determined using one of the **spectrophotometric methods**. Observation of the change in absorbance in the visible or near UV spectrum is the method most commonly used in assaying enzyme activity. $NAD(P)^+$ is quite often a cofactor of dehydrogenases, and NAD(P)H is the resulting product. The absorbance spectrum is 340 nm for the product, NAD(P)H, but not for the cofactor, $NAD(P)^+$. A continuous measurement at absorbance 340 nm is required to continuously monitor the disappearance of the substrate or the appearance of the product. This is called a **continuous assay,** and it is also a **direct assay** because the catalyzed reaction itself produces a measurable signal. The advantage of continuous assays is that the progress curve is immediately available, as is confirmation that the initial rate is linear for the measuring period. They are generally preferred because they give more information during the measurement. However, methods based on fluorescence, the **spectrofluorometric methods**, are more sensitive than those based on the changes in the absorbance spectrum. The reaction is accomplished by the release or uptake of protons and is the basis of performing the assay. Detailed discussions on the assaying techniques are available below.

Sometimes when a serial enzymatic reaction of a metabolic pathway is being assayed, no significant products, the substrates of the next reaction, can be measured in the absorbance spectra due to rapid changes in the reaction. Therefore, there may be no suitable detection machinery available for this single enzymatic study. Nevertheless, there is still a measurable signal when one or several enzymatic reactions are coupled to the desired assay reaction. These are **coupled assays**, one of the particular **indirect assays** whose coupled reaction is not enzymatic. This means that the coupled enzymes and the substrates have to be present in excess to make sure that the rate-limiting step is always the reaction of the particular enzyme being assayed. Though a lag in the appearance of the preferred product will be noticed at the beginning of the assay, before reaching a steady state, the phenomenon will only appear for a short period of time if the concentration of the coupling enzymes and substrates are kept in excess. So the V_{max} of the coupling reactions will be greater than that of the preferred reaction. Besides, not only the substrate concentration but also other parameters may affect the preferred reaction (see above). To prevent this, control experiments can be performed to check if the V_{max} of the coupling reactions is actually much higher than that of the preferred reaction. For other indirect assays in which the desired catalyzed reaction itself does not produce a measurable signal, a suitable reagent that has no effect on enzyme activity can be added to the reaction mixture to react with one of the products to form a measurable signal.

Nevertheless, sometimes no easily readable differences between the spectrum of absorbance of the substrate and that of the product can be measured. In such cases, it may be possible to measure the appearance of the colored product, by the **chromogenic method** (one of the spectrofluorometric methods). The reaction is incubated for a fixed period of time. Then, because the development of color requires the inhibition of enzyme activity, the reaction is stopped and the concentration of colored product of the substrate is measured. This is the **discontinuous** or **end point assay**. Assays involving product separation that cannot be designed to continuously measure the signal change, such as electrophoresis, high performance liquid chromatography (HPLC), and sample quenching at time intervals, are also discontinuous assays. The advantage of discontinuous assays is that they are less time consuming for monitoring a large number of samples for enzyme activity. However, additional control experiments are needed to assure that the initiation rate is linear for the measuring period of time. Otherwise, the radiolabeled substrate of an enzyme assay is a highly sensitive method that allows the detection of radioactive product. Separation of the substrate and the product by a variety of extraction methods may be required to accurately assay the enzyme activity.

Detection Methods

Spectrophotometric Methods

Both the spectrophotometric method and the spectrofluorometric method discussed below use measurements at specific wavelengths of light energy [in the wavelength regions of 200–400 nm (UV, ultraviolet) and 400–800 nm (visible)] to determine how much light has been absorbed by a target molecule (resulting in changes in electronic configuration). The measured value from the spectrophotometric method at a specific wavelength can be related to the molecule concentration in the solution using a cell with a fixed path length (l cm) that obeys the Beer-Lambert law:

$$A = -\log T = \varepsilon c l,$$

where A is the absorbance at certain wavelength, T is the transmittance, representing the intensity of transmitted light, ε is the extinction coefficient, and c is the molar concentration of the sample. Using this law, the measured change in absorbance, ΔA, can be converted to the change in molecule concentration, Δc, and the change in rate, v_i, can be calculated as a function of time, Δt, that is, $v_i = \Delta A/\varepsilon l \Delta t$.

Choices of appropriate materials for spectroscopic cells (cuvettes) depend on the wavelength used; quartz cuvettes must be used at wavelengths less than 350 nm because glass and disposable plastic cuvettes absorb too much light in the UV light range. However, the latter glass and disposible plastic cuvettes can be used in the wavelength range of 350–800 nm. Selection of a wavelength for the measurement depends on finding the wavelength that produces the greatest difference in absorbance between the reactant and product molecules in the reaction. Though the wavelength of measurement usually refers to the maximal wavelength of the reactant or product molecule, the most meaningful analytical wavelength may not be the same as the maximum wavelength because significant overlap of spectra may be found between the reactant and product molecules. Thus, a different spectrum between two molecules can be calculated to determine the most sensitive analytical wavelengths for monitoring the increase in product and the decrease in substrate.

Spectrofluorometric Methods

When a molecule absorbs light at an appropriate wavelength, an electronic transition occurs from the ground state to the excited state; this short-lived transition decays through various high-energy, vibrational substrates at the excited electronic state by heat dissipation, and then relaxes to the ground state with a photon emission, the fluorescence. The emitted fluorescence is less energetic (longer wavelength) than the initial energy that is required to excite the molecules; this is referred to as the Stokes shift. Taking advantage of this, the fluorescence instrument is designed to excite the sample and detect the emitted light at different wavelengths. The ratio of quanta fluoresced over quanta absorbed offers a value of quantum yield *(Q)*, which measures the efficiency of the reaction leading to light emission. The light emission signals vary with concentration of fluorescent material by the Beer-Lambert law, where the extinction coefficient is replaced by the quantum yield, and gives: $I_f = 2.31\ I_0 \varepsilon c l Q$, where I_f and I_0 are the intensities of light emission and of incident light that excites the sample, respectively. However conversion of light emission values into concentration units requires the preparation of a standard curve of light emission signals as a function of the fluorescent material concentration, which has to be determined independently, due to the not strictly linear relationship between I_f and c (Lakowicz 1983).

Spectrofluromettric methods provide highly sensitive capacities for detection of low concentration changes in a reaction. However, quenching (diminishing) or resonance energy transfer (RET) of the measured intensity of the donor molecule will be observed if the acceptor molecule absorbs light and the donor molecule emits light at the same wavelength. The acceptor molecule will either decay to its ground state if quenching occurs or fluoresce at a characteristic wavelength if energy is transferred (Lakowicz 1983). Both situations can be overcome by using an appropriate peptide sequence, up to 10 or more amino acid residues, to separate the fluorescence donor molecule from the acceptor molecule if peptide substrates are used. Thus both effects can be relieved by cleaving the intermolecular bonding when protease activity is being assayed.

Care should be taken in performing the spectrofluorometric method. For instance, the construction of a calibration curve as the experiment is assayed is essential, and the storage conditions for the fluorescent molecules are important because of photodecomposition. In addition, the quantum yield is related to the temperature, and the fluorescence signal will increase with decreasing temperature, so a temperature-constant condition is required (Bashford and Harris 1987, Gul et al. 1998).

Radiometric Methods

The principle of the radiometric methods in enzymatic catalysis studies is the quantification of radioisotopes incorporated into the substrate and retained in the product. Hence, successful radiometric methods rely on the efficient separation of the radiolabled product from the residual radiolabeled substrate and on the sensitivity and specificity of the radioactivity detection method. Most commonly used radioisotopes, for example, ^{14}C, ^{32}P, ^{35}S, and ^{3}H, decay through emission of β particles; however, ^{125}I decays through emission of γ particles, whose loss is related to loss of radioactivity and the rate of decay (the half-life) of the isotope. Radioactivity expressed in Curies (Ci) decays at a rate of 2.22×10^{12} disintegrations per minute (dpm); expressed in Becquerels (Bq), it decays at a rate of 1 disintegrations per second (dps). The experimental units of radioactivity are counts per minute (cpm) measured by the instrument; quantification of the specific activity of the sample is given in units of radioactivity per mass or per molarity of the sample (e.g., μCi/mg or dpm/μmol).

Methods of separation of radiolabeled product and residual radiolabeled substrate include chromatography, electrophoresis, centrifugation, and solvent extraction (Gul et al. 1998). For the detection of radioactivity, one commonly used instrument is a scintillation counter that measures light emitted when solutions of *p*-terphenyl or stilbene in xylene or toluene are mixed with radioactive material designed around a photomultiplier tube. Another method is the autoradiography that allows detecting radioactivity on surfaces in close contact either with X-ray film or with plates of computerized phosphor imaging devices. Whenever operating the isotopes-containing experiments, care should always be taken for assuring safety.

Chromatographic Methods

Chromatography is applied in the separation of the reactant molecules and products in enzymatic reactions and is usually used in conjunction with other detection methods. The most commonly used chromatographic methods include paper chromatography, column chromatography, thin-layer chromatography (TLC), and high performance liquid chromatography (HPLC). Paper chromatography is a simple and economic method for the readily separation of large numbers of samples. By contrast, column chromatography is more expensive and has poor reproducibility. The TLC method has the advantage of faster separation of mixed samples, and like paper chromatography, it is disposable and can be quantified and scanned; it is not easily replaced by HPLC, especially for measuring small, radiolabeled molecules (Oldham 1992).

The HPLC method featured with low compressibility resins is a versatile method for the separation of either low molecular weight molecules or small peptides. Under a range of high pressures up to 5000 psi (which approximates to 3.45×10^7 Pa), the resolution is greatly enhanced with a faster flow rate and a shorter run time. The solvent used for elution, referred to as the mobile phase, should be an HPLC grade that contains low contaminants; the insoluble media is usually referred to as the stationary phase. Two types of mobile phase are used during elution; one is an isocratic elution whose composition is not changed, and the other is a gradient elution whose concentration is gradually increased for better resolution. The three HPLC methods most commonly used in the separation steps of enzymatic assays are reverse phase, ion-exchange, and size-exclusion chromatography.

The basis of reverse phase HPLC is the use of a nonpolar stationary phase composed of silica covalently bonded to alkyl silane, and a polar mobile phase used to maximize hydrophobic interactions with the stationary phase. Molecules are eluted in a solvent of low polarity (e.g., methanol, acetonitrile, or acetone mixed with water at different ratios) that is able to efficiently compete with molecules for the hydrophobic stationary phase.

The ion-exchange HPLC contains a stationary phase covalently bonded to a charged functional group; it binds the molecules through electrostatic interactions, which can be disrupted by the increasing ionic strength of the mobile phase. By modifying the composition of the mobile phase, differential elution, separating multiple molecules, is achieved.

In the size-exclusion HPLC, also known as gel filtration, the stationary phase is composed of porous beads with a particular molecular weight range of fractionation. However, this method is not recommended where molecular weight differences between substrates and products are minor, because of overlapping of the elution profiles (Oliver 1989).

Selection of the HPLC detector depends on the types of signals measured, and most commonly the UV/visible light detectors are extensively used.

Electrophoretic Methods

Agarose gel electrophoresis and polyacrylamide gel electrophoresis (PAGE) are widely used methods for separation of macromolecules; they depend, respectively, on the percentage of agarose and acrylamide in the gel matrix. The most commonly used method is the sodium dodecyl sulfate–polyacrylamide gel electrophoresis (SDS-PAGE) method; under denaturing conditions, the anionic detergent SDS is coated on peptides or proteins giving them equivalently the same anionic charge densities. Resolving of the samples will thus be based on molecular weight under an electric field over a period of time. After electrophoresis, peptides or proteins bands can be visualized by staining the gel with Coomassie Brilliant Blue or other staining reagents, and radiolabeled materials can be detected by autoradiography. Applications of electrophoresis assays are not only for detection of molecular weight and radioactivity differences, but also for detection of charge differences. For instance, the enzyme-catalyzed phosphorylation reactions result in phosphoryl transfer from substrates to products, and net charge differences between two molecules form the basis for separation by electrophoresis. If radioisotope ^{32}P-labeled phosphate is incorporated into the molecules, the reactions can be detected by autoradiography, by monitoring the radiolabel transfer after gel electrophoresis, or by immunological blotting with antibodies that specifically recognize peptides or proteins containing phosphate-modified amino acid residues.

Native gel electrophoresis is also useful in the above applications, where not only the molecular weight but also the charge density and overall molecule shape affect the migration of molecules in gels. Though SDS-PAGE causes denaturing to peptides and proteins, renaturation in gels is possible and can be applied to several types of in situ enzymatic activity studies such as activity staining and zymography (Hames and Rickwood 1990). Both methods assay enzyme activity after electrophoresis, but zymography is especially intended for proteolytic enzyme activity staining in which gels are cast with high concentrations of proteolytic enzyme substrates, for example, casein, gelatin, bovine serum albumin, collagen, and others. Samples containing proteolytic enzymes can be subjected to gel electrophoresis, but the renaturation step has to be performed if a denaturing condition is used; then the reaction is performed under conditions suitable for assaying proteolytic enzymes. The gel is then subjected to staining and destaining, but the entire gel background will not be destained because the gel is polymerized with protein substrates, except in clear zones where the significant proteolysis has occurred; the amount of staining observed will be greatly diminished due to the loss of protein. This process is also known as reverse staining. Otherwise, reverse zymography is a method used to assay the proteolytic enzyme inhibitor activity in gel. Similar to zymography, samples containing proteolytic enzyme inhibitor can be subjected to gel electrophoresis. After the gel renaturation step is performed, the reaction is assayed under appropriate conditions in the presence of a specific type of proteolytic enzyme. Only a specific type of proteolytic enzyme inhibitor will be resistant to the proteolysis, and after staining, the active proteolytic enzyme inhibitor will appear as protein band (Oliver et al. 1999).

Other Methods

The most commonly used assay methods are the spectrophotometric, spectrofluorometric, radiometric, chromatographic, and electrophoretic methods described above, but a variety of other methods are utilized as well. Immunological methods make use of the antibodies raised against the proteins (Harlow and Lane 1988). Polarographic methods make use of the change in current related to the change in concentration of an electroactive compound that undergoes oxidation or reduction (Vassos and Ewing 1983). Oxygen-sensing methods make use of the change in oxygen concentration monitored by an oxygen-specific electrode (Clark 1992), and pH-stat methods use measurements of the quantity of base or acid required to be added to maintain a constant pH (Jacobsen et al. 1957).

Selection of an Appropriate Substrate

Generally a low molecular mass, chromogenic substrate containing one susceptible bond is preferred for use in enzymatic reactions. A substrate containing many susceptible bonds or different functional groups adjacent to the susceptible bond may affect the cleavage efficiency of the enzyme, and this can result in the appearance of several intermediate sized products. Thus, they may interfere with the result and make interpretation of kinetic data difficult. Otherwise, the chromophore-containing substrate will readily and easily support assaying methods with absorbance measurement. Moreover, substrate specificity can also be precisely determined when an enzyme has more than one recognition site on both the preferred bond and the functional groups adjacent to it. Different sized chromogenic substrates can then be used to determine an enzyme's specificity and to quantify its substrate preference.

Unit of Enzyme Activity

The unit of enzyme activity is usually expressed as either micromoles of substrate converted or product formed per unit time, or unit of activity per milliliter under a standardized set of conditions. Though any unit of enzyme activity can be used, the Commission on Enzymes of the International Union of Biochemistry and Molecular Biology (IUBMB) has recommended that a unit of enzyme, Enzyme Unit or International Unit (U), be used. An Enzyme Unit is defined as that amount which will catalyze the conversion of one micromole of substrate per minute, $1\ U = 1\ \mu mol/min$, under defined conditions. The conditions include substrate concentration, pH, temperature, buffer composition, and other parameters that may affect the sensitivity and specificity of the reaction, and usually a continuous spectrophotometric method or a pH stat method is preferred. Another enzyme unit that now is not widely used is the International System of Units (SI unit) in which 1 katal (kat) $=$ 1 mol/sec, so 1 kat $=$ 60 mol/min $= 6 \times 10^7$ U.

In ascertaining successful purification of a specified enzyme from an extract, it is necessary to compare the specific activity of each step to that of the original extract; a ratio of the two gives the fold purification. The specific activity of an enzyme is usually expressed as units per milligram of protein when the unit of enzyme per milliliter is divided by milligrams of protein per milliliter, the protein concentration. The fold purification is an index reflecting only the increase in the specific activity with respect to the extract, not the purity of the specified enzyme.

REFERENCES

Barry MJ. 1997. Emzymes and symmetrical molecules. Trends Biochem Sci 22:228–230.

Bashford CL, Harris DA. 1987. Spectrophotometry and Spectrofluorimetry: A Practical Approach. Washington, DC: IRL Press.

Bell JE, Bell ET. 1988. Proteins and Enzymes. New Jersey: Prentice-Hall.

Briggs GE, Haldane JBS. 1925. A note on the kinetics of enzyme action. Biochem J 19:338–339.

Brocklehurst K, Dixon HBF. 1977. The pH-dependence of second-order rate constants of enzyme modification may provide free-reactant pKa values. Biochem J 167:859–862.

Brown AJ. 1902. Enzyme action. J Chem Soc 81:373–386.

Burbaum JJ, Schimmel P. 1991. Structural relationships and the classification of aminoacyl-tRNA synthetases. J Biol Chem 266:16965–16968.

Cardenas ML, Cornish-Bowden A, Ureta T. 1998. Evolution and regulatory role of the hexokinases. Biochim Biophys Acta 1401:242–264.

Cherfils J, Vachette P, Janin J. 1990. Modelling allosteric processes in *E. coli* aspartate transcarbamylase. Biochimie 72(8): 617–624.

Clark JB. 1992. In: Eisenthal R, Danson MJ, editors, Enzyme Assays, A Practical Approach. New York: Oxford University Press.

Copeland RA. 2000. Enzymes, 2nd edition. New York: Wiley-VCH Inc.

Cornish-Bowden A. 1995. Fundamentals of Enzyme Kinetics, 2nd edition. London: Portland Press.

Cornish-Bowden A. 1996. In: Engel PC, editor, Enzymology Labfax. San Diego: BIOS Scientific Publishers, Oxford and Academic Press. Pp. 96–113.

Cusack S. 1997. Aminoacyl-tRNA synthetases. Curr Opin Struct Biol 7(6): 881–889.

Dixon M, Webb EC. 1979. Enzymes. 3rd edition. New York: Academic Press.

Drake JW. 1999. The distribution of rates of spontaneous mutation over viruses, prokaryotes, and eukaryotes. Ann N Y Acad Sci. 870:100–107.

Eadie GS. 1942. The inhibition of cholinesterase by physostigmine and prostigmine. J Biol Chem 146: 85–93.

Eisenthal R, Cornish-Bowden A. 1974. The direct linear plot. A new graphical procedure for estimating enzyme parameters. Biochem J 139:715–720.

Fang TY, Ford C. 1998. Protein engineering of *Aspergillus awamori* glucoamylase to increase its pH optimum. Protein Eng 11:383–388.

Goodman MF. 1997. Hydrogen bonding revisited: Geometric selection as a principal determinant of DNA replication fidelity. Proc Natl Acad Sci USA 94:10493–10495.

Gul S, Sreedharan SK, Brocklehurst K. 1998. Enzyme Assays. Oxford: John Wiley and Sons Ltd.

Hames BD, Rickwood D. 1990. Gel Electrophoresis of Proteins: A Practical Approach, 2nd edition. Oxford: IRL Press.

Hammes GG. 2002. Multiple conformational changes in enzyme catalysis, Biochemistry 41(26): 8221–8228.

Hanes CS. 1932. Studies on plant amylases. I. The effect of starch concentration upon the velocity of hydrolysis by the amylase of germinated barley. Biochem J 26:1406–1421.

Harlow E, Lane D. 1988. Antibodies: A Laboratory Manual. New York. Cold Spring Harbor: CSHL Press.

Hill AV. 1910. The possible effects of the aggregation of molecules of hemoglobin on its dissociation curves. J Physiol 40, 4–7.

Hofstee BHJ. 1959. Non-inverted versus inverted plots in enzyme kinetics. Nature 184:1296–1298.

Igarashi K, Ozawa T, Ikawa-Kitatama K, Hayashi Y, Araki H, Endo K, Hagihara H, Ozaki K, Kawai S, Ito S. 1999. Thermostabilization by proline substitution in an alkaline, liquefying α-amylase from *Bacillus* sp. strain KSM-1378. Biosci Biotechnol Biochem 63:1535-1540.

Jacobsen CF et al. 1957. In: Glick D, editor, Methods of Biochemical Analysis, vol IV. New York: Interscience Publishers. Pp. 171–210.

Khalifah RG. 2003. Reflections on Edsall's carbonic anhydrase: Paradoxes of an ultra fast enzyme. Biophys Chem. 100:159–170.

Kornberg RD. 1996. RNA polymerase II transcription control. Trends Biochem Sci 21:325–327.

Lakowicz JR. 1983. Principles of Fluorescence Spectroscopy. New York: Plenum Publishing.

Langfort J, Ploug T, Ihlemann J, Enevoldsen LH, Stallknecht B, Saldo M, Kjaer M, Holm C, Galbo H. 1998. Hormone-sensitive lipase expression and regulation in skeletal muscle. Adv Exp Med Biol 441:219–228.

Luan S. 2003. Protein phosphatases in plants. Annu Rev Plant Biol 54:63–92.

Lineweaver H, Burk D. 1934. The determination of enzyme dissociation constants. J Am Chem Soc 56: 658–666.

Michaelis L, Menten ML. 1913. Die Kinetik der Invertinwirkun. Biochem Z 49:333–369.

Nuttall FQ, Gilboe DP, Gannon MC, Niewoehner CB, Tan AWH. 1988. Regulation of glycogen synthesis in the liver. Am J Med 85(Supp. 5A): 77–85.

Oldham KG. 1992. In: Eisenthal R, Danson MJ, editors. Enzyme Assays, A Practical Approach. New York: Oxford University Press. Pp. 93–122.

Oliver GW, Stetler-Stevenson WG, Kleiner DE. 1999. In: Sterchi EE, Stöcker W, editors, Proteolytic Enzymes: Tools and Targets. Berlin Heidelberg: Springer-Verlag Berlin. Pp. 63–76.

Oliver RWA. 1989. HPLC of Macromolecules: A Practical Approach. Oxford: IRL press.

Pechkova E, Zanotti G, Nicolini C. 2003. Three-dimensional atomic structure of a catalytic subunit mutant of human protein kinase CK2. Acta Crystallogr D Biol Crystallogr 59(Pt 12): 2133–2139.

Preiss J, Romeo T. 1994. Molecular biology and regulation aspects of glycogen biosynthesis in bacteria. Prog Nucl Acid Res Mol Biol 47:299–329.

Raj SML, Liu F. 2003. Engineering of RNase P ribozyme for gene-targeting application. Gene 313:59–69.

Scott CP, Kashlan OB, Lear JD, Cooperman BS. 2001. A quantitative model for allosteric control of purine reduction by murine ribonucleotide reductase. Biochemistry 40(6): 1651–1661.

Scriver CR. 1995. Whatever happened to PKU? Clin Biochem 28(2): 137–144.

Shiau RJ, Hung HC, Jeang CL. 2003. Improving the thermostability of raw-starch-digesting amylase from a *Cytophaga* sp. by site-directed mutagenesis. Appl Environ Microbiol 69(4): 2383–2385.

Sirko A, Brodzik R. 2000. Plant ureases: Roles and regulation. Acta Biochim Pol 47(4): 1189–1195.

Stauffer CE. 1989. Enzyme Assays for Food Scientists. New York: Van Nostrand Reinhold.

Steitz TA, Moore PB. 2003. RNA, the first macromolecular catalyst: The ribosome is a ribozyme. Trends Biochem Sci 28(8): 411–418.

Tipton KF. 1992. Principles of enzyme assay and kinetic studies. In: Eisenthal R, Danson MJ, editors, Enzyme Assays, A Practical Approach. New York: Oxford University Press. Pp. 1–58.

Tipton KF, Webb HBF. 1979. Effects of pH on enzymes. Methods Enzymol 63:183–233.

Vassos BH, Ewing GW. 1983. Electroanalytical Chemistry. New York: John Wiley and Sons.

Waters PJ. 2003. How PAH gene mutation cause hyperphenylalaninemia and why mechanism matters: Insights from in vitro expression. Hum Mutat 21(4): 357–369.

8
Enzyme Engineering and Technology

*D. Platis, G. A. Kotzia, I. A. Axarli, and N. E. Labrou**

*Corresponding author

PREFACE

Enzymes are proteins with powerful catalytic functions. They increase reaction rates sometimes by as much as a million-fold, but more typically by about a thousand-fold. Catalytic activity can also be exhibited, to a limited extent, by biological molecules other than the "classical" enzymes. For example, antibodies raised to stable analogs of the transition states of number of enzyme-catalyzed reactions can act as effective catalysts for those reactions (Hsieh-Wilson et al. 1996). In addition, RNA molecules can also act as catalysts for a number of different types of reactions (Lewin 1982). These antibodies and RNA catalysts are known as abzymes and ribozymes, respectively.

Enzymes have a number of distinct advantages over conventional chemical catalysts. Among these are their high productivity, catalytic efficiency, specificity, and ability to discriminate between similar parts of molecules (regiospecificity) or optical isomers (stereospecificity). Enzymes, in general, work under mild conditions of temperature, pressure, and pH. This advantage decreases the energy requirements and therefore reduces the capital costs.

However, there are some disadvantages in the use of enzymes, such as high cost and low stability. These shortcomings are currently being addressed mainly by employing protein-engineering approaches using recombinant DNA technology (Stemmer 1994, Ke and Madison 1997). These approaches aim at improving various properties, such as thermostability, specificity, and catalytic efficiency. The advent of designer biocatalysts enables production of not only process-compatible enzymes, but also novel enzymes able to catalyze new or unexploited reactions (Schmidt-Dannert et al. 2000, Umeno and Arnold 2004). This is just the start of the enzyme technology era.

ENZYME STRUCTURE AND MECHANISM

NOMENCLATURE AND CLASSIFICATION OF ENZYMES

Enzymes are classified according to the nature of the reaction they catalyze (e.g., oxidation/reduction, hydrolysis, synthesis, etc.) and subclassified according to the exact identity of their substrates and products. This nomenclature system was established by the Enzyme Commission (a committee of the International Union of Biochemistry). According to this system all enzymes are classified into six major classes:

1. *Oxidoreductases* catalyze oxidation-reduction reactions.
2. *Transferases* catalyze group transfer from one molecule to another.
3. *Hydrolases* catalyze hydrolytic cleavage of C—C, C—N, C—O, C—S, or O—P bonds. These are group transfer reactions, but the acceptor is always water.
4. *Lyases* catalyze elimination reactions, resulting in the cleavage of C—C, C—O, C—N, or C—S bonds or the formation of a double bond, or conversely, adding groups to double bonds.
5. *Isomerases* catalyze isomerization reactions, for example, racemization, epimerization, *cis-trans*-isomerization, tautomerization.
6. *Ligases* catalyze bond formation, coupled with the hydrolysis of a high-energy phosphate bond in ATP or a similar triphosphate.

The Enzyme Commission system consists of a numerical classification hierarchy of the form **EC a.b.c.d,** in which "a" represents the class of reaction catalyzed and can take values from 1 to 6, according to the classification of reaction types given above. "b" denotes the subclass, which usually specifies more precisely the type of the substrate or the bond cleaved, for example, by naming the electron donor of an oxidation-reduction reaction or by naming the functional group cleaved by a hydrolytic enzyme. "c" denotes the sub-subclass, which allows an even more precise definition of the reaction catalyzed. For example, sub-subclasses of oxidoreductases are denoted by naming the acceptor of the electron from its respective donor. "d" is the serial number of the enzyme within its sub-subclass. An example will be analyzed. The enzyme that oxidizes D-glucose using molecular oxygen catalyzes the following reaction:

β-D-glucose + O_2 —glucose oxidase→ D-glucono-1,5-lactone + H_2O_2

Hence, its systematic name is D-glucose:oxygen oxidoreductase, and its systematic number is EC 1.1.3.4.

The systematic names are often quite long, and therefore short trivial names and systematic numbers are often more convenient for enzyme designation. These shorter names are known as **recommended names**. The recommended names consist of the suffix *-ase* added to the substrate acted on. For example, for the enzyme mentioned above, the recommended name is glucose oxidase.

It should be noted that the system of nomenclature and classification of enzymes is based only on the reaction catalyzed and takes no account of the origin of the enzyme, that is, of the species or tissue from which it derives.

For additional information on enzyme nomenclature and classification, see Chapter 6 of this volume.

BASIC ELEMENTS OF ENZYME STRUCTURE

The Primary Structure of Enzymes

Enzymes are composed of **L-α-amino acids** joined together by a **peptide bond** between the carboxylic acid group of one amino acid and the amino group of the next:

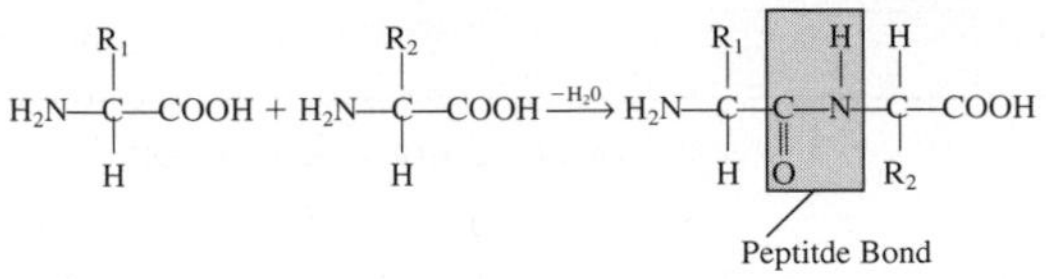

$$H_2N-\underset{H}{\overset{R_1}{C}}-COOH + H_2N-\underset{H}{\overset{R_2}{C}}-COOH \xrightarrow{-H_2O} H_2N-\underset{H}{\overset{R_1}{C}}-\underset{O}{\overset{}{C}}-\overset{H}{N}-\underset{R_2}{\overset{H}{C}}-COOH$$

Peptitde Bond

There are 20 common amino acids in proteins that are specified by the genetic code; in rare cases, others occur as the products of enzymatic modifications after translation. A common feature of all 20 amino acids is a central carbon atom (C_α) to which a hydrogen atom, an amino group ($—NH_2$) and a carboxyl group ($—COOH$) are attached. Free amino acids are usually zwitterionic at neutral pH, with the carboxyl group deprotonated and the amino group protonated. The structures of the most common amino acids found in proteins are shown in Table 8.1. Amino acids can be divided into four different classes depending on the structure of their side chains, which are called R groups: **nonpolar, polar uncharged, negatively charged** (at neutral pH), and **positively charged** (at neutral pH) (Richardson 1981). The properties of the amino acid side chains determine the properties of the proteins they constitute. The formation of a succession of peptide bonds generates the **main chain** or **backbone**.

The primary structure of a protein places several constraints on how it can fold to produce its three-dimensional structure (Cantor 1980, Fersht 1999). The backbone of a protein consists of a carbon atom C_α to which the side chain is attached, a NH group bound to C_α, and a carbonyl group C=O, where the carbon atom C is attached to C_α (Fig. 8.1). The peptide bond is planar, since it has a partial (~40%) double bond character with π electrons shared between the C—O and C—N bonds (Fig. 8.1). The peptide bond has a *trans*-conformation, that is, the

Table 8.1. Names, Symbol (One Letter and Three Letter Codes), and Chemical Structures of the Twenty Amino Acids Found in Proteins

Nonpolar Amino Acids		
Alanine	Ala (A)	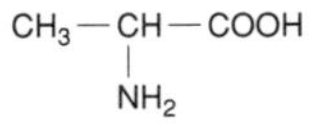 $CH_3-CH(NH_2)-COOH$
Valine	Val (V)	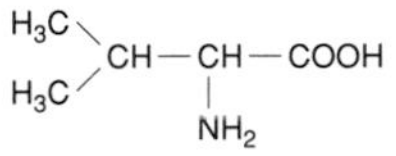 $(H_3C)_2CH-CH(NH_2)-COOH$
Leucine	Leu (L)	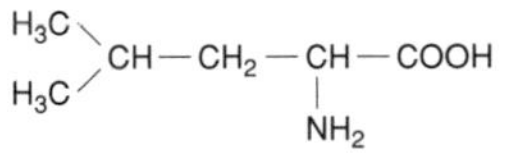 $(H_3C)_2CH-CH_2-CH(NH_2)-COOH$
Isoleucine	Ile (I)	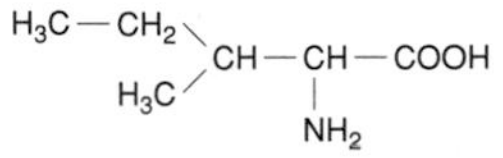 $H_3C-CH_2(H_3C)CH-CH(NH_2)-COOH$
Methionine	Met (M)	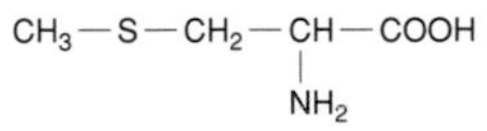 $CH_3-S-CH_2-CH(NH_2)-COOH$
Tryptophan	Trp (W)	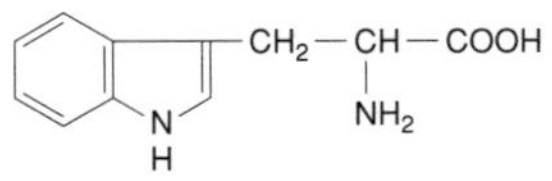 indole(N–H)$-CH_2-CH(NH_2)-COOH$
Phenylalanine	Phe (F)	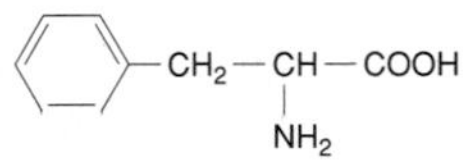 phenyl$-CH_2-CH(NH_2)-COOH$
Proline	Pro (P)	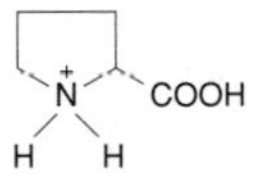 pyrrolidine ring, $\overset{+}{N}$ (H, H), COOH
Polar Uncharged Amino Acids		
Glycine	Gly (G)	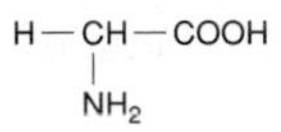 $H-CH(NH_2)-COOH$

(Continues)

Table 8.1. (Continued)

Asparagine	Asn (N)	$H_2N{-}C(=O){-}CH_2{-}CH(NH_2){-}COOH$
Glutamine	Gln (Q)	$H_2N{-}C(=O){-}CH_2{-}CH_2{-}CH(NH_2){-}COOH$
Serine	Ser (S)	$HO{-}CH_2{-}CH(NH_2){-}COOH$
Threonine	Thr (T)	$HO{-}CH_2(CH_3){-}CH(NH_2){-}COOH$
Cysteine	Cys (C)	$HS{-}CH_2{-}CH(NH_2){-}COOH$
Tyrosine	Tyr (Y)	$HO{-}C_6H_4{-}CH_2{-}CH(NH_2){-}COOH$
Negatively Charged Amino Acids		
Glutamate	Glu (E)	$HOOC{-}CH_2{-}CH_2{-}CH(NH_2){-}COOH$
Aspartate	Asp (D)	$HOOC{-}CH_2{-}CH(NH_2){-}COOH$
Positively Charged Amino Acids		
Arginine	Arg (R)	$H_2N{-}C(=NH){-}HN{-}CH_2{-}CH_2{-}CH_2{-}CH(NH_2){-}COOH$
Lysine	Lys (K)	$H_2N{-}(CH_2)_4{-}CH(NH_2){-}COOH$
Histidine	His (H)	imidazole (HN, N:)$-CH_2{-}CH(NH_2){-}COOH$

oxygen of the carbonyl group and the hydrogen of the NH group are in the *trans* position; the *cis* conformation occurs only in exceptional cases (Richardson 1981).

Enzymes have several "levels" of structure. The protein's sequence, that is, the order of amino acids, is termed as its **primary structure**. This is determined by the sequence of nucleotide bases in the gene that codes for the protein. Translation of the mRNA transcript produces a linear chain of amino acids linked together by a peptide bond between the carboxyl carbon of the first amino acid and the free amino group of the second amino acid. The first amino acid in any polypeptide sequence has a free amino group, and the terminal amino acid has a free carboxyl group. The primary structure is responsible for the higher levels of the enzyme's structure and therefore for the enzymatic activity (Richardson 1981, Price and Stevens 1999).

The Three-Dimensional Structure of Enzymes

Enzymes are generally very closely packed globular structures with only a small number of internal cavities, which are normally filled by water molecules. The polypeptide chains of enzymes are organized into ordered, hydrogen-bonded regions, known as **secondary structure** (Andersen and Rost 2003). In these ordered structures any buried carbonyl oxygen forms a hydrogen bond with an amino NH group.

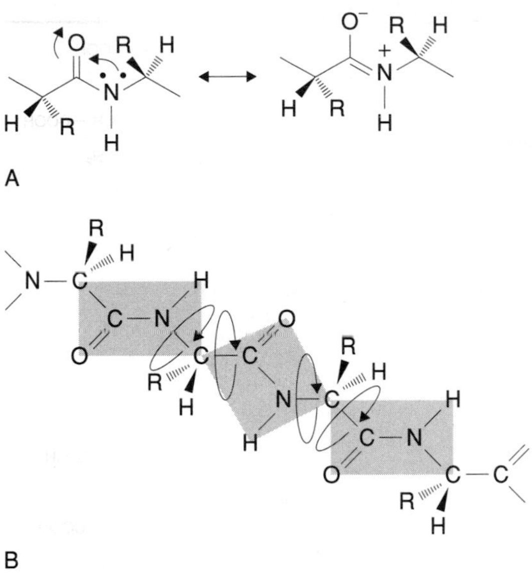

Figure 8.1. (**A**) The amide bond showing delocalization of electrons. (**B**) A tripeptide unit of a polypeptide chain showing the planar amide units and the relevant angles of rotation about the bonds to the central α-carbon atom.

This is done by forming α-helices and β-pleated sheets, as shown in Figure 8.2. The α-helix can be thought of as having a structure similar to a coil or spring (Surewicz and Mantsch 1988). The β-sheet can be visualized as a series of parallel strings laid on top of an accordion-folded piece of paper. These structures are determined by the protein's primary structure. Relatively small, uncharged, polar amino acids readily organize themselves into α-helices, while relatively small, nonpolar amino acids form β-sheets. Proline is a special amino acid because of its unique structure (Table 8.1). Introduction of proline into the sequence creates a permanent bend at that position (Garnier et al. 1990). Therefore, the presence of proline in an α-helix or a β-sheet disrupts the secondary structure at that point. The presence of a glycine residue confers greater than normal flexibility on a polypeptide chain. This is due to the absence of a bulky side chain, which reduces steric hindrance.

Another frequently observed structural unit is the β-turn (Fang and Shortle 2003). This occurs when the main chain sharply changes direction using a bend composed of four successive residues, often including proline and glycine. In these units the C=O group of residue i is hydrogen bonded to the NH of residue i + 3 instead of i + 4 as in the α-helix. Many different types of β-turn, which differ in terms of the number of amino acids and in conformation (e.g., Type I, Type II, Type III), have been identified (Sibanda et al. 1989).

The three-dimensional structure of a protein composed of a single polypeptide chain is known as its **tertiary structure**. Tertiary structure is determined largely by the interaction of R groups on the surface of the protein with water and with other R groups on the protein's surface. The intermolecular noncovalent attractive forces that are involved in stabilizing the enzyme's structure are usually classified into three types: ionic bonds, hydrogen bonds, and van der Waals attractions (Matthews 1993). Hydrogen bonding results from the formation of hydrogen bridges between appropriate atoms; electrostatic bonds are due to the attraction of oppositely charged groups located on two amino acid side chains. Van der Waals bonds are generated by the interaction between electron clouds. Another important weak force is created by the three-dimensional structure of water, which tends to force hydrophobic groups together in order to minimize their disruptive effect on the hydrogen-bonded network of water molecules. Apart from the peptide bond, the only other type of covalent bond involved in linking amino acids in enzymes is the disulphide (—S—S—) bond, which can be formed between two cysteine side chains under oxidizing conditions. The disulphide bond contributes significantly to the structural stability of an enzyme and, more precisely, to tertiary structural stabilization (see below for details) (Matthews 1993, Estape et al. 1998).

Detailed studies have shown that certain combinations of α-helices or β-sheets with turns occur in many proteins. These often-occurring structures have been termed **motifs**, or **supersecondary structure** (Kabsch and Sander 1983, Adamson et al. 1993). Examples of motifs found in several enzymes are shown in Figure 8.3. These protein folds represent highly stable structural units, and it is believed that they may form nucleating centers in protein folding (Richardson 1981).

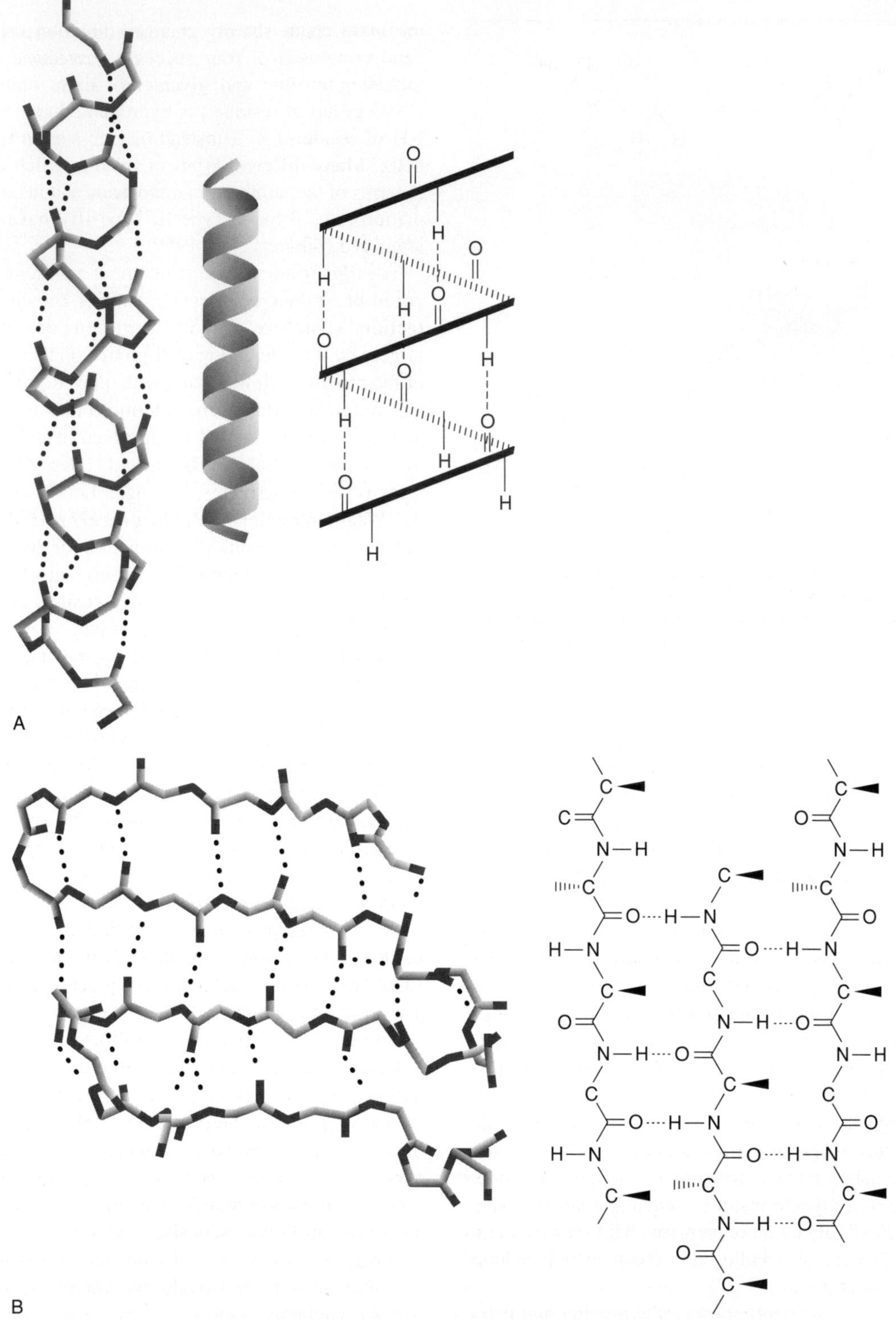

Figure 8.2. Representation of α-helix (**A**) and β-sheet (**B**). Hydrogen bonds are depicted by dotted lines.

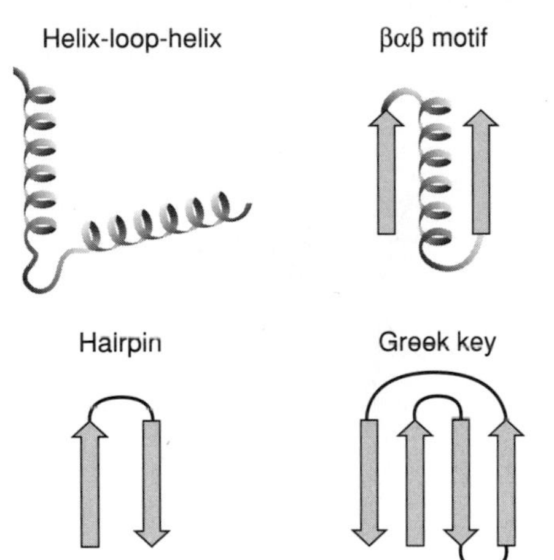

Figure 8.3. Common examples of motifs found in proteins.

Certain combinations of α-helices and β-sheets pack together to form compactly folded globular units, each of which is called protein domain, with a molecular mass of 15 to 20 kDa (Orengo et al. 1997). Domains may be composed of secondary structure and identifiable motifs and therefore represent a higher level of structure than motifs. The most widely used classification scheme of domains has been the four-class system (Fig. 8.4) (Murzin et al. 1995). The four classes of protein structure are as follows:

1. *All*-α proteins, which have only α-helix structure.
2. *All*-β proteins, which have β-sheet structure.
3. α/β proteins, which have mixed or alternating segments of α-helix and β-sheet structure.
4. α+β proteins, which have α-helix and β-sheet structural segments that do not mix but are separated along the polypeptide chain.

While small proteins may contain only a single domain, larger enzymes contain a number of domains. In fact most enzymes are dimers, tetramers, or polymers of several polypeptide chains. Each polypeptide chain is a **subunit**, and it may be identical to or different from the others. The side chains on each polypeptide chain may interact with each other as well as with water molecules to give a final enzyme structure. The overall organization of the subunits is known as the **quaternary structure,** and therefore the quaternary structure is a characteristic of multisubunit enzymes. The four levels of enzyme structure are illustrated in Figure 8.5.

Theory of Enzyme Catalysis and Mechanism

In order for a reaction to occur, the reactant molecules must possess a sufficient energy to cross a potential energy barrier, which is known as the **activation energy** (Fig. 8.6) (Hackney 1990). All reactant molecules have different amounts of energy, but only a small proportion of them have sufficient energy to cross the activation energy barrier of the reaction. The lower the activation energy, the more substrate molecules are able to cross the activation energy barrier. The result is that the reaction rate is increased.

Enzyme catalysis requires the formation of a specific reversible complex between the substrate and the enzyme. This complex is known as the **enzyme-substrate complex (ES)** and provides all the conditions that favor the catalytic event (Hackney 1990, Marti et al. 2004). Enzymes accelerate reactions by lowering the energy required for the formation of a complex of reactants that is competent to produce reaction products. This complex is known as the **transition state complex** of the reaction and is characterized by lower free energy than would be found in the uncatalyzed reaction:

$$E + S \rightleftharpoons ES \rightleftharpoons ES^* \rightleftharpoons EP \rightleftharpoons E + P$$

The enzyme-substrate complex (ES) must pass to the transition state (ES*). The transition state complex must advance to an enzyme-product complex (EP), which dissociates to free enzyme and product (P). This reaction's pathway goes through the transition states TS_1, TS_2, and TS_3. The amount of energy required to achieve the transition state is lowered; hence, a greater proportion of the molecules in the population can achieve the transition state and cross the activation energy barrier (Benkovic and Hammes-Schiffer 2003, Wolfenden 2003). Enzymes speed up the forward and reverse reactions proportionately, so that they have no effect on the equilibrium constant of the reactions they catalyze (Hackney 1990).

Substrate is bound to the enzyme by relatively weak noncovalent forces. The free energy of interaction of the ES complex ranges between −12 and −36 kJ/mol. The intermolecular attractive forces

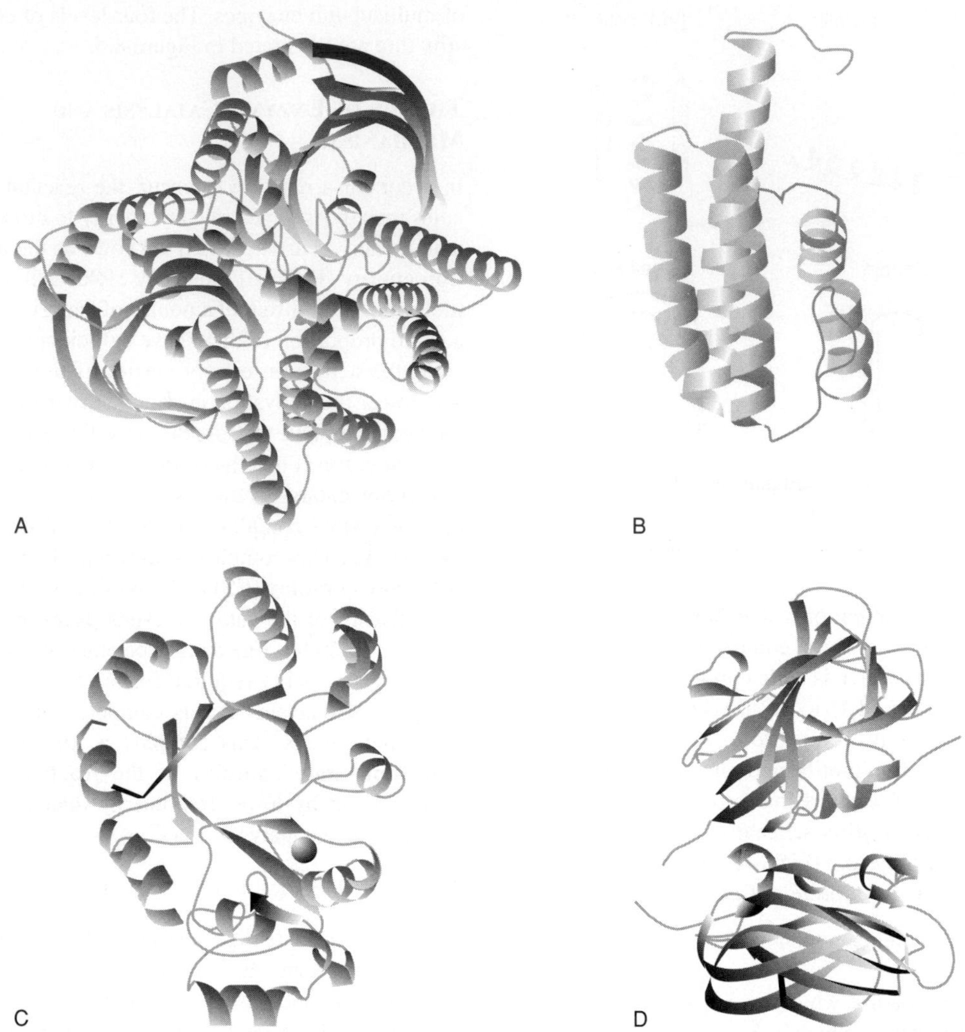

Figure 8.4. The four-class classification system of domains. (**A**) The $\alpha + \beta$ class, (structure of glycyl-tRNA synthetase α chain). (**B**) The all-α class (structure of the hypothetical protein (Tm0613) from *Thermotoga maritima.* (**C**) The α/β class (structure of glycerophosphodiester phosphodiesterase). (**D**) The all-β class (structure of allantoicase from *Saccharomyces cerevisiae.*

between enzyme and substrate, in general, are of three types: ionic bonds, hydrogen bonds, and van der Waals attractions.

The specific part of the protein structure that interacts with the substrate is known as the **substrate binding site** (Fig. 8.7). The substrate binding site is a three-dimensional entity suitably designed as a pocket or a cleft to accept the structure of the substrate, in three-dimensional terms. The binding residues are defined as any residue with any atom within 4 Å of a bound substrate. These binding residues that participate in the catalytic event are known as the **catalytic residues** and form the **active site**. According to Bartlett et al. (2002), a residue is

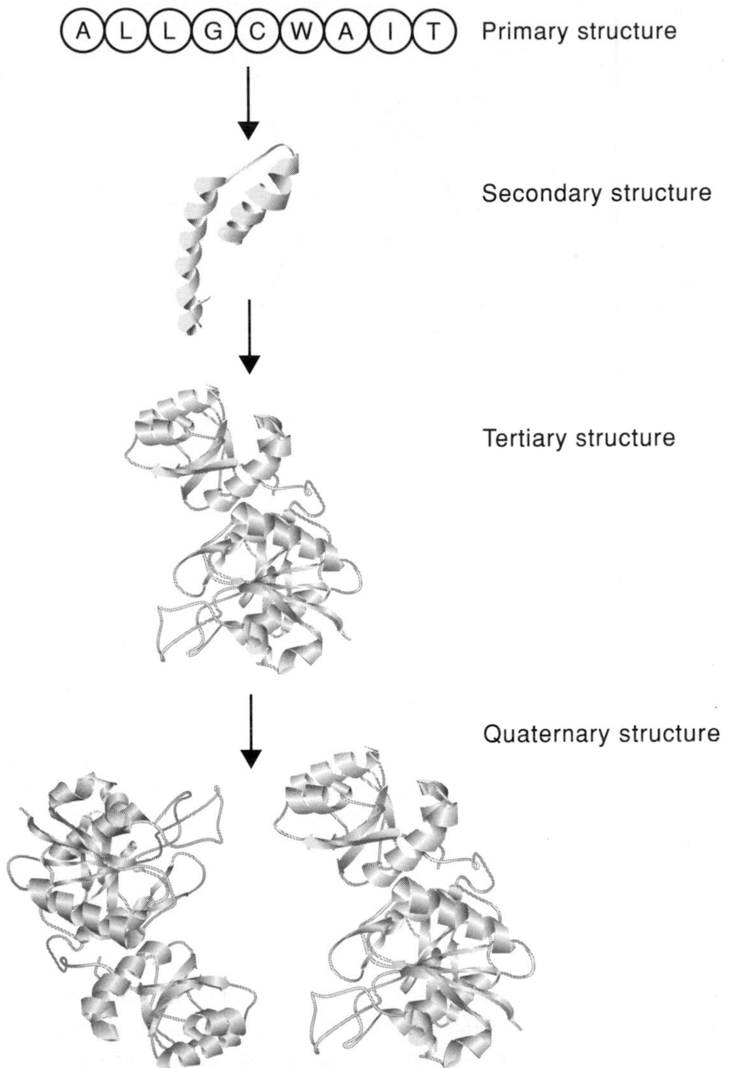

Figure 8.5. Schematic representation of the four levels of protein structure.

defined as catalytic if any of the following take place:

1. Direct involvement in the catalytic mechanism, for example, as a nucleophile.
2. Exertion of an effect, which aids catalysis, on another residue or water molecule that is directly involved in the catalytic mechanism.
3. Stabilization of a proposed transition-state intermediate.
4. Exertion of an effect on a substrate or cofactor that aids catalysis, for example, by polarizing a bond that is to be broken.

Despite the impression that the enzyme's structure is static and locked into a single conformation,

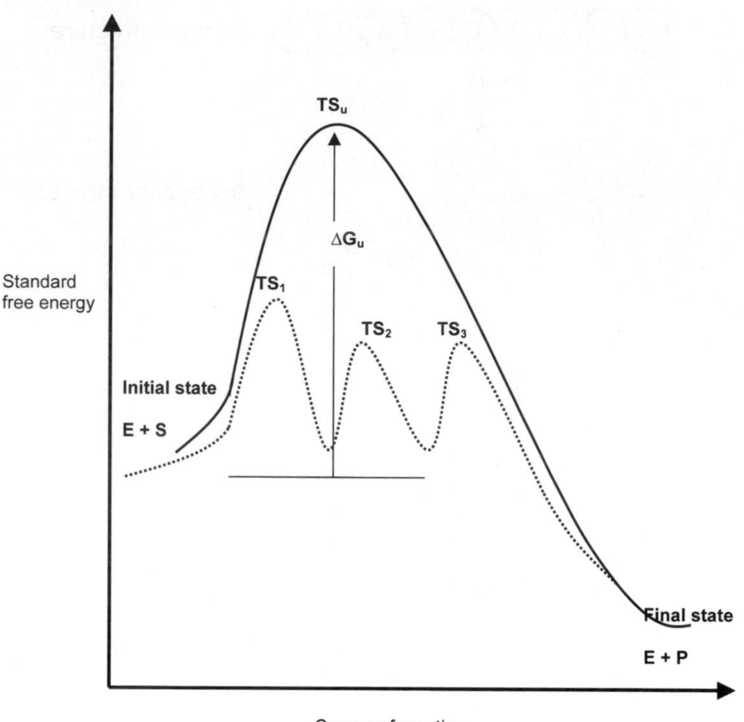

Figure 8.6. A schematic diagram showing the free energy profile of the course of an enzyme-catalyzed reaction involving the formation of enzyme-substrate (ES) and enzyme-product (EP) complexes. The catalyzed reaction pathway goes through the transition states TS_1, TS_2, and TS_3, with standard free energy of activation ΔG_c, whereas the uncatalyzed reaction goes through the transition state TS_u with standard free energy of activation ΔG_u.

several motions and conformational changes of the various regions always occur (Hammes 2002). The extent of these motions depends on many factors including temperature, the properties of the solvating medium, and the presence or absence of substrate and product (Hammes 2002). The conformational changes undergone by the enzyme play an important role in controlling the catalytic cycle. In some enzymes there are significant movements of the binding residues, usually on surface loops, and in other cases there are larger conformational changes. Catalysis takes place in the closed form, and the enzyme opens again to release the product. This favored model, which explains enzyme catalysis and substrate interaction, is the so-called **induced-fit hypothesis** (Anderson et al. 1979, Joseph et al. 1990). In this hypothesis the initial interaction between enzyme and substrate rapidly induces conformational changes in the shape of the active site, which results in a new shape of the active site that brings catalytic residues close to substrate bonds to be altered (Fig. 8.8). When binding of the enzyme to the substrate takes place, the shape adjustment triggers catalysis by generating transition-state complexes. This hypothesis helps to explain why enzymes only catalyze specific reactions (Anderson et al. 1979, Joseph et al. 1990). This basic cycle has been seen in many different enzymes including triosephosphate isomerase (TIM), which uses a small hinged loop to close the active site (Joseph et al. 1990), and kinases, which use two large lobes moving towards each other when the substrate binds (Anderson et al. 1979).

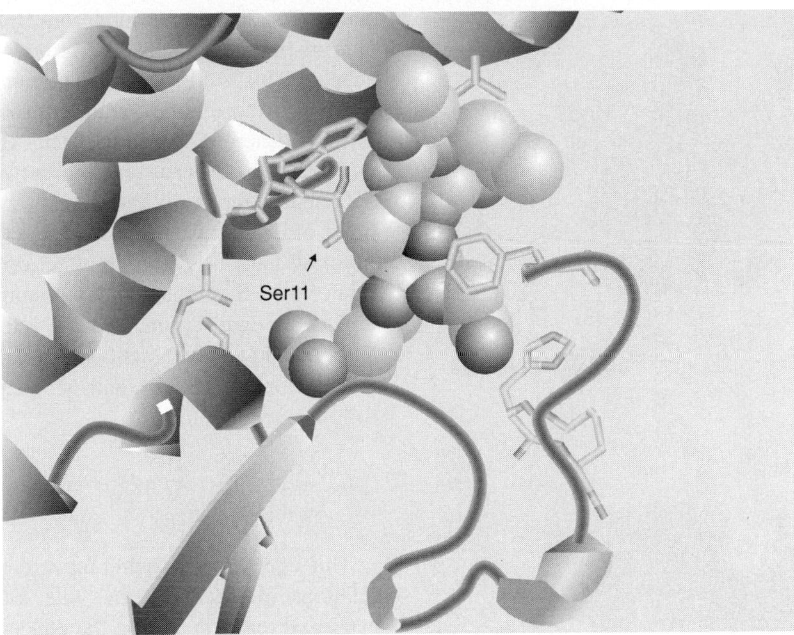

Figure 8.7. The substrate binding site of maize glutathione S-transferase. The binding residues are depicted as sticks, whereas the substrate is depicted in a space fill model. Only Ser 11 is involved directly in catalysis and is considered as catalytic residue.

Coenzymes, Prosthetic Groups, and Metal Ion Cofactors

Nonprotein groups can also be used by enzymes to affect catalysis. These groups, called **cofactors**, can be organic or inorganic and are divided into three classes: coenzymes, prosthetic groups, and metal ion cofactors (McCormick 1975). Prosthetic groups are tightly bound to an enzyme through covalent bonding. Coenzymes bind to an enzyme reversibly and associate and dissociate from the enzyme during each catalytic cycle; therefore, they may be considered as cosubstrates. An enzyme containing a cofactor or prosthetic group is termed as **holoenzyme**. Coenzymes can be broadly classified into three main groups: coenzymes that transfer groups onto substrate, coenzymes that accept and donate electrons, and compounds that activate substrates (Table 8.2). Metal ions such as Ca^{+2}, Mg^{+2}, Zn^{+2}, Mn^{+2}, Fe^{+2}, and Cu^{+2} may in some cases act as cofactors. These may be bound to the enzyme by simple coordination with electron-donating atoms of side chains (imidazole of His, -SH group of Cys, carboxylate O^- of Asp and Glu). In some cases, metals such as Mg^{+2} are associated with the substrate rather than the enzyme. For example Mg-ATP is the true substrate for kinases (Anderson et al. 1979). In other cases, metals may form part of a prosthetic group in which they are bound by coordinated bonds (e.g., heme, Table 8.2) in addition to side-chain groups. Usually in this case, metal ions participate in electron transfer reactions.

Kinetics of Enzyme-Catalyzed Reactions

The term enzyme kinetics implies a study of the velocity of an enzyme-catalyzed reaction and of the various factors that may affect this (Moss 1988). An extensive discussion of enzyme kinetics would stray too far from the central theme of this chapter, but some general aspects will be briefly considered.

The concepts underlying the analysis of enzyme kinetics continue to provide significant information

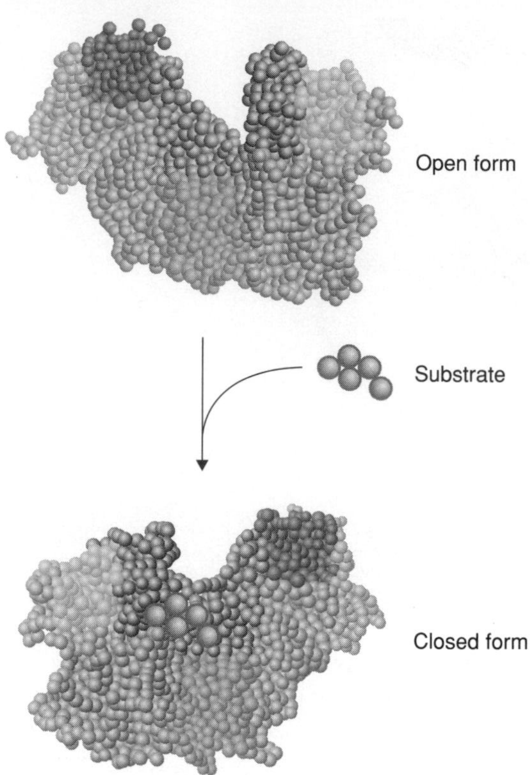

Figure 8.8. A schematic representation of the induced-fit hypothesis.

for understanding in vivo function and metabolism and for the development and clinical use of drugs aimed at selectively altering rate constants and interfering with the progress of disease states (Bauer et al. 2001). The central scope of any study of enzyme kinetics is knowledge of the way in which reaction velocity is altered by changes in the concentration of the enzyme's substrate and of the simple mathematics underlying this (Wharton 1983, Moss 1988, Watson and Dive 1994). As already discussed above, the enzymatic reactions proceed through an intermediate enzyme-substrate complex (ES) in which each molecule of enzyme is combined, at any given instant during the reaction, with one substrate molecule. The reaction between enzyme and substrate to form the enzyme-substrate complex is reversible. Therefore, the overall enzymatic reaction can be shown as:

$$E + S \underset{k_{-1}}{\overset{k_{+1}}{\rightleftharpoons}} ES \xrightarrow{k_{+2}} E + P$$

where k_{+1}, k_{-1} and k_{+2} are the respective rate constants. The reverse reaction concerning the conversion of product to substrate is not included in this scheme. This is allowed at the beginning of the reaction when there is no, or little, product present. In 1913, biochemists **Michaelis** and **Menten** suggested that, if the reverse reaction between E and S is sufficiently rapid, in comparison with the breakdown of ES complex to form product, the latter reaction will have a negligible effect on the concentration of the ES complex. Consequently, E, S, and ES will be in equilibrium, and the rates of formation and breakdown of ES will be equal. Based on these assumptions Michaelis and Menten produced the following equation:

$$v = \frac{V_{\max} \cdot [S]}{K_m + [S]}$$

This equation is a quantitative description of the relationship between the rate of an enzyme-catalyzed reaction (v) and the concentration of substrate [S]. The parameters $\boldsymbol{V_{max}}$ and $\boldsymbol{K_m}$ are constants at a given temperature and a given enzyme concentration. The K_m or Michaelis constant is the substrate concentration at which $v = V_{max}/2$, and its usual unit is M. The K_m provides information about the substrate **binding affinity** of the enzyme. A high K_m indicates a low affinity, and vice versa (Moss 1988, Price and Stevens 1999).

The V_{max} is the maximum rate of the enzyme-catalyzed reaction, and it is observed at very high substrate concentrations where all the enzyme molecules are saturated with substrate, in the form of ES complex. Therefore:

$$V_{max} = k_{cat}[E_t]$$

where $[E_t]$ is the total enzyme concentration and k_{cat} is the rate of breakdown of the ES complex (k_{+2}) in scheme 8.4, which is known as the **turnover number**. The k_{cat} represents the maximum number of substrate molecules that the enzyme can convert to product in a set time. The K_m depends on the particular enzyme and substrate being used and on the temperature, pH, ionic strength, and so on. Note, however, that K_m is independent of the enzyme concentration, whereas V_{max} is proportional to enzyme concentration. A plot of the initial rate (v) against initial substrate concentration ([S]) for a reaction obeying Michaelis-Menten kinetics has the form of

Table 8.2. The Structure of Some Common Coenzymes: Adenosine Triphosphate (ATP), Coenzyme A (CoA), Flavin Adenine Dinucleotide (FAD), Nicotinamide Adenine Dinucleotide (NAD^+), and Heme c.

Cofactor	Type of Reactions Catalyzed	Structure
Adenosine triphosphate (ATP)	Phosphate transfer reactions (e.g., kinases)	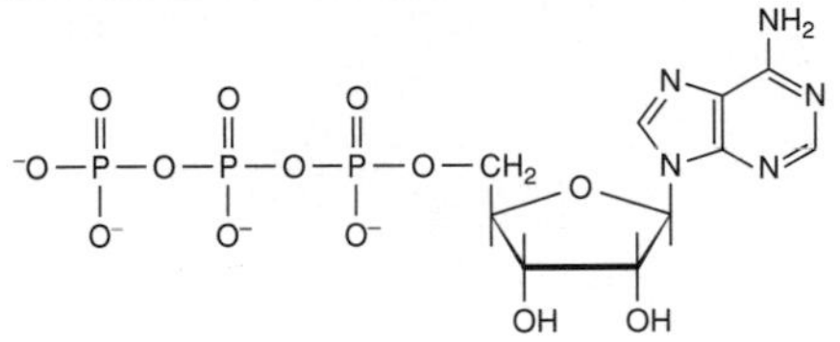
Coenzyme A (CoA)	Acyl transfer reactions (transferases)	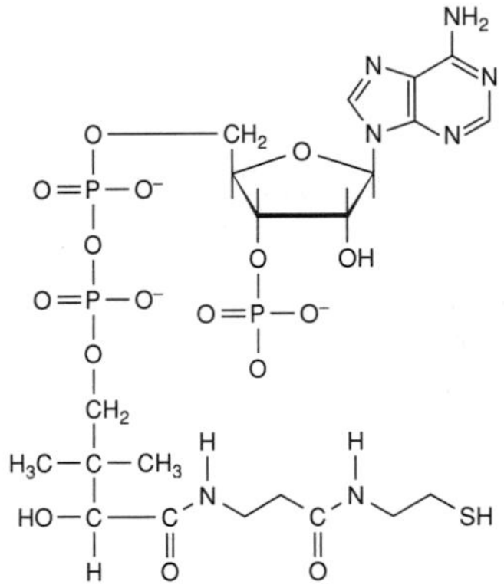
Flavin adenine dinucleotide (FAD)	Redox reactions (reductases)	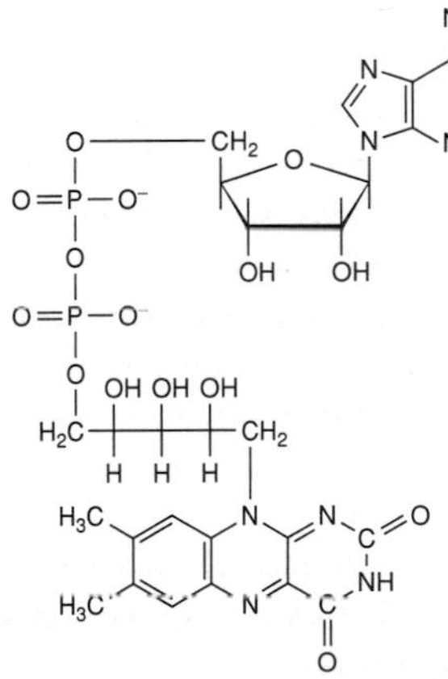

(Continues)

Table 8.2. (Continued)

Cofactor	Type of Reactions Catalyzed	Structure
Nicotinamide adenine dinucleotide (NAD^+)	Redox reactions (e.g., dehydrogenases)	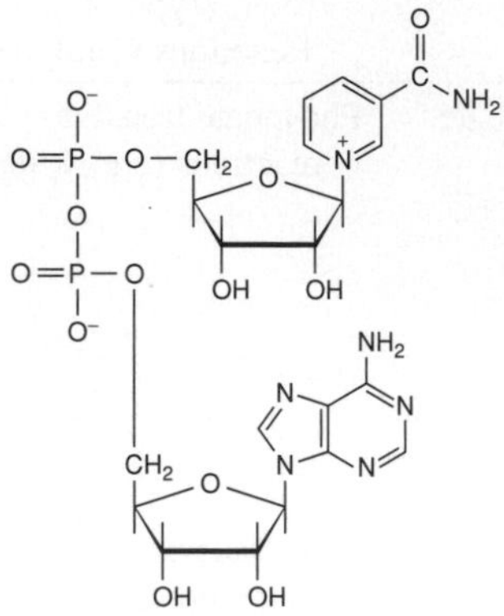
Heme c	Activate substrates	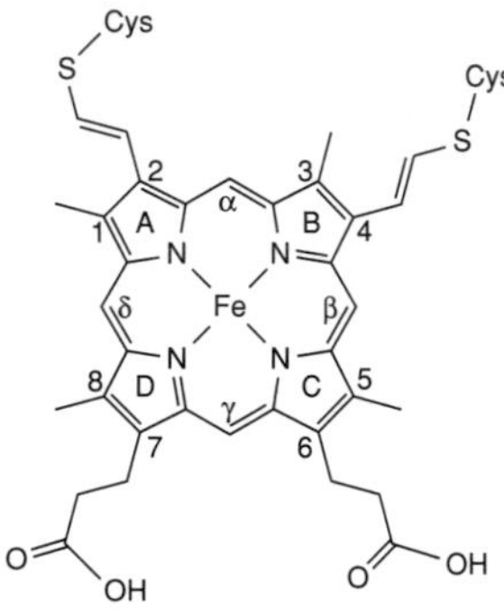

a rectangular hyperbola through the origin with asymptotes $v = V_{max}$ and $[S] = -K_m$ (Fig. 8.9A). The term *hyperbolic kinetics* is also sometimes used to characterize such kinetics.

There are several available methods for determining the parameters from the Michaelis-Menten equation. A better method for determining the values of V_{max} and K_m was formulated by Hans Lineweaver and Dean Burk and is termed the Lineweaver-Burk (LB) or **double reciprocal plot** (Fig. 8.9B). Specifically, it is a plot of $1/v$ versus $1/[S]$, according to the equation:

$$\frac{1}{v} = \frac{K_m}{V_{max}} \cdot \frac{1}{[S]} + \frac{1}{V_{max}}$$

Such a plot yields a straight line with a slope of K_m/V_{max}. The intercept on the $1/v$ axis is $1/V_{max}$ and the intercept on the $1/[S]$ axis is $-1/K_m$.

The rate of an enzymatic reaction is also affected by changes in pH and temperature (Fig. 8.10). When pH is varied, the velocity of reaction in the presence of a constant amount of enzyme is typically greatest over a relatively narrow range of pH. Since enzymes are proteins, they possess a large number of ionic groups, which are capable of existing in different ionic forms. The existence of a fairly narrow pH optimum for most enzymes suggests that one particular ionic form of the enzyme molecule, out of the many in which it can potentially exist, is the catalytically active one. The effect of pH changes on v is reversible, except after exposure to extremes of pH at which denaturation of the enzyme may occur.

The rate of an enzymatic reaction increases with increasing temperature. Although there are significant variations from one enzyme to another, on average, for each 10°C rise in temperature, the enzymatic activity is increased by an order of two. After

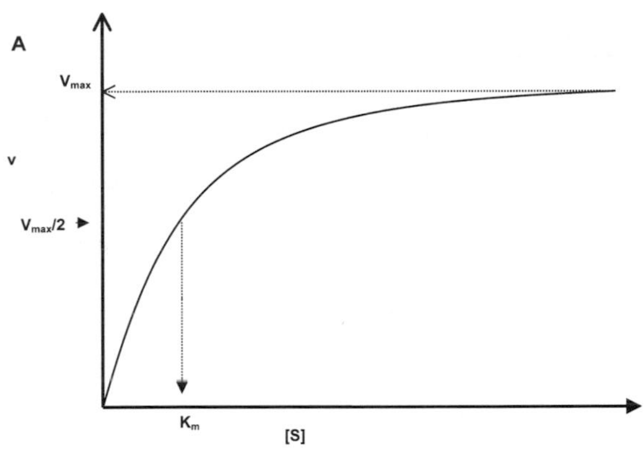

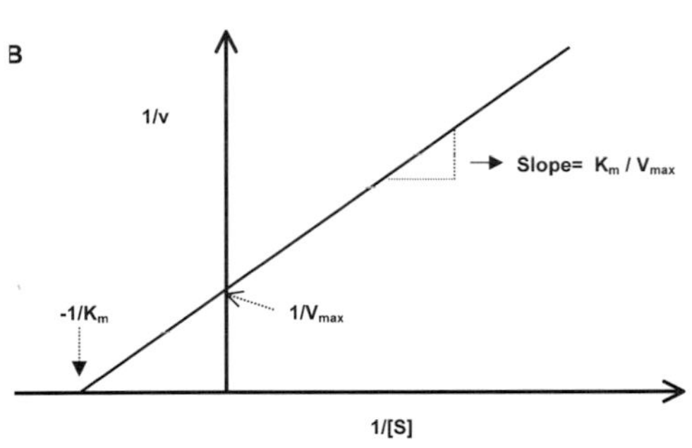

Figure 8.9. (**A**) A plot of the initial rate *(v)* against initial substrate concentration ([S]) for a reaction obeying the Michaelis-Menten kinetics. The substrate concentration, which gives a rate of half the maximum reaction velocity, is equal to the K_m. (**B**) The Lineweaver-Burk plot. The intercept on the $1/v$ axis is $1/V_{max}$, the intercept on the 1/[S] axis is $-1/K_m$, and the slope is K_m/V_{max}.

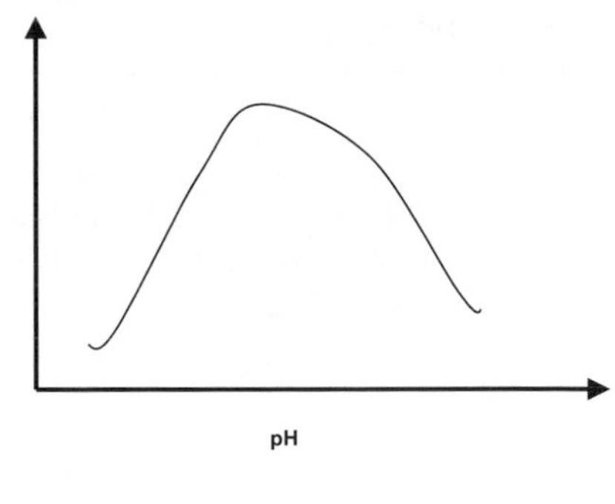

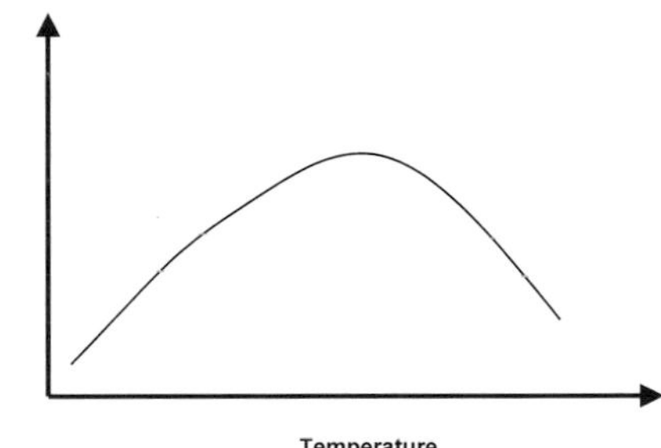

Figure 8.10. Relationship of the activity pH and activity temperature for a putative enzyme.

exposure of the enzyme to high temperatures (normally greater that 65°C), denaturation of the enzyme may occur and the enzyme activity decreases. The Arrhenius equation provides a quantitative description of the relationship between the rate of an enzyme-catalyzed reaction (V_{max}) and the temperature (T):

$$Log V_{\max} = \frac{-E_a}{2,303RT} + A$$

where E_a is the activation energy of the reaction, R is the gas constant, and A is a constant relevant to the nature of the reactant molecules.

The rates of enzymatic reactions are affected by changes in the concentrations of compounds other than the substrate. These modifiers may be activators (i.e., they may increase the rate of reaction) or inhibitors (i.e., their presence may inhibit the enzyme's activity). Activators and inhibitors are usually small molecules or even ions. Enzyme inhibitors fall into two broad classes: (1) those causing irreversible inactivation of enzymes and (2) those whose inhibitory effects can be reversed. Inhibitors of the first class, bind covalently to the enzyme so that physical methods of separating the two are ineffective. Reversible inhibition is characterized by the existence of equilibrium between enzyme and inhibitor (I):

$$E + I \rightleftharpoons EI$$

The equilibrium constant of the reaction, K_i, is given by the equation:

$$K_i = \frac{[ES]}{[E][I]}$$

Table 8.3. The Characteristic of Each Type of Inhibition and Their Effect on the Kinetic Parameters K_m and V_{max}

Inhibitor Type	Binding Site on Enzyme	Kinetic Effect
Competitive inhibitor	The inhibitor specifically binds at the enzyme's catalytic site, where it competes with substrate for binding: $E \underset{-S}{\overset{+S}{\rightleftarrows}} ES \rightarrow E + P$ $^{+I}\downarrow\uparrow_{-I}$ EI	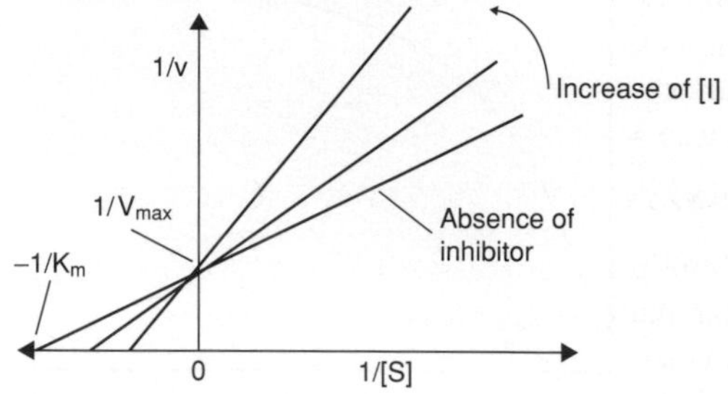 K_m is increased, V_{max} is unchanged. LB equation: $\frac{1}{v} = \frac{K_m}{V_{\max}} \cdot \left(1 + \frac{[I\}}{K_i}\right) \cdot \frac{1}{[S]} + \frac{1}{V_{\max}}$
Noncompetitive inhibitor	The inhibitor binds to E or to the ES complex (may form an EIS complex) at a site other than the catalytic. Substrate binding is unchanged, whereas EIS complex cannot form products: $E \underset{-S}{\overset{+S}{\rightleftarrows}} ES \rightarrow E + P$ $^{+I}\downarrow\uparrow_{-I} \quad ^{+I}\downarrow\uparrow_{-I}$ $EI \underset{-S}{\overset{+S}{\rightleftarrows}} EIS$	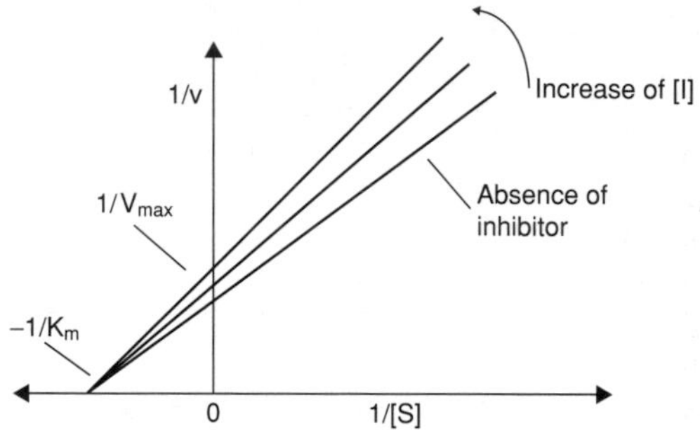 K_m is unchanged, V_{max} is decreased proportionately to inhibitor concentration. LB equation: $\frac{1}{v} = \frac{K_m}{V_{\max}} \cdot \left(1 + \frac{[I]}{K_i}\right) \cdot \frac{1}{[S]} + \frac{1}{V_{\max}}\left(1 + \frac{[I]}{K_i}\right)$
Uncompetitive inhibitor	Binds only to ES complexes at a site other than the catalytic site. Substrate binding alters enzyme structure, making inhibitor-binding site available: $E \underset{-S}{\overset{+S}{\rightleftarrows}} ES \rightarrow E + P$ $^{+I}\downarrow\uparrow_{-I}$ EIS	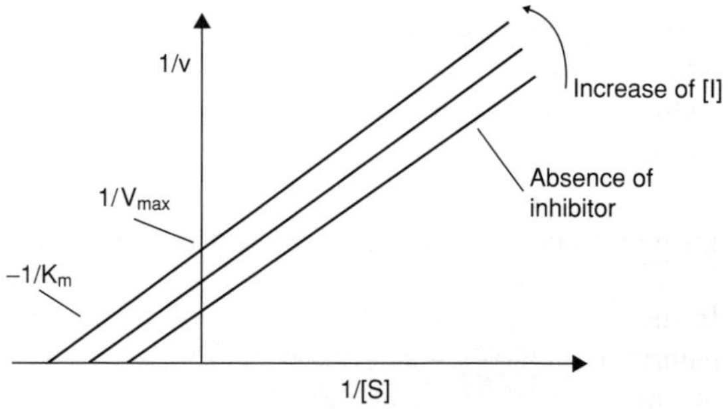 K_m and V_{max} are decreased. LB equation: $\frac{1}{v} = \frac{K_m}{V_{\max}} \cdot \frac{1}{[S]} + \frac{1}{V_{\max}}\left(1 + \frac{[I]}{K_i}\right)$

K_i is a measure of the affinity of the inhibitor for the enzyme. Reversible inhibitors can be divided into three main categories: (1) competitive inhibitors, (2) noncompetitive inhibitors, and (3) uncompetitive inhibitors. The characteristic of each type of inhibition and its effect on the kinetic parameters K_m and V_{max} are shown in Table 8.3.

ENZYME DYNAMICS DURING CATALYSIS

Multiple conformational changes and intramolecular motions appear to be a general feature of enzymes (Agarwal et al. 2002). The structures of proteins and other biomolecules are largely maintained by noncovalent forces and are therefore subject to thermal fluctuations ranging from local atomic displacements to complete unfolding. These changes are intimately connected to enzymatic catalysis and are believed to fulfill a number of roles in catalysis: enhanced binding of substrate, correct orientation of catalytic groups, removal of water from the active site, and trapping of intermediates. Enzyme conformational changes may be classified into four types (Gutteridge and Thornton 2004): (1) domain motion, where two rigid domains, joined by a flexible hinge, move relative to each other; (2) loop motion, where flexible surface loops (2–10 residues) adopt different conformations; (3) side chain rotation: rotation of side chains, which alters the position of the functional atoms of the side chain; and (4) secondary structure changes.

Intramolecular motions in biomolecules are usually very fast (picosecond to nanosecond) local fluctuations. The flexibility associated with such motions provides entropic stabilization of conformational states (Agarwal et al. 2002). In addition, there are also slower (microsecond to millisecond) and larger scale, thermally activated, transitions. Large-scale conformational changes are usually key events in enzyme regulation.

ENZYME PRODUCTION

In the past, enzymes were isolated primarily from natural sources, and thus a relatively limited number of enzymes were available to the industry (Eisenmesser et al. 2002). For example, of the hundred or so enzymes being used industrially, over one-half are from fungi and yeast, and over a third are from bacteria, with the remainder divided between animal (8%) and plant (4%) sources (Panke and Wubbolts 2002, van Beilen and Li 2002). Today, with the recent advances of molecular biology and genetic engineering, several expression systems have been developed, exploited, and used for the commercial production of several therapeutic (Walsh 2003), analytical, or industrial enzymes (Kirk et al. 2002, van Beilen and Li 2002). These systems have improved not only the availability of enzymes and the efficiency and cost with which they can be produced, but also their quality (Labrou et al. 2001, Labrou and Rigden 2001).

ENZYME HETEROLOGOUS EXPRESSION

There are two basic steps involved in the assembly of every heterologous expression system:

1. The introduction of the DNA encoding the gene of interest into the host cells, which requires: (1) identification and isolation of the gene of the protein to be expressed, (2) insertion of the gene into a suitable expression vector, and (3) introduction of the expression vector into the selected cell system that will accommodate the heterologous protein.
2. The optimization of protein expression by taking into account the effect of various factors such as growing medium, temperature, and induction period.

A variety of vectors able to carry the DNA into the host cells are available, ranging from plasmids, cosmids, phagemids, and viruses to artificial chromosomes of bacterial, yeast, or human origin (BAC, YAC, or HAC, respectively) (Ikeno et al. 1998, Sgaramella and Eridani 2004). The vectors are either integrated into the host chromosomal DNA or remain in an episomal form. In general, expression vectors have the following characteristics (Fig. 8.11):

- *Polylinker:* contains multiple restriction sites that facilitate the insertion of the desired gene into the vector.
- *Selection marker:* encodes for a selectable marker, allowing the vector to be maintained within the host cell under conditions of selective pressure (i.e., antibiotic).
- *Ori:* a sequence that allows for the autonomous replication of the vector within the cells.
- *Promoter:* inducible or constitutive; regulates RNA transcription of the gene of interest.

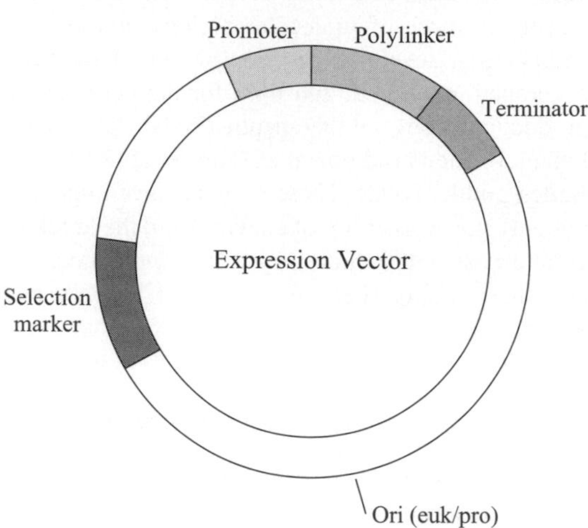

Figure 8.11. Generalized heterologous gene expression vector. The main characteristics are shown: polylinker sequence, promoter, terminator, selection marker, and origin of replication.

- *Terminator:* a strong terminator sequence that ensures that the RNA polymerase disengages and does not continue to transcribe other genes located downstream.

Vectors are usually designed with mixed characteristics for expression in both prokaryotic and eukaryotic host cells. Artificial chromosomes are designed for cloning of very large segments of DNA (100Kb), usually for mapping purposes, and contain host-specific telomeric and centromeric sequences. These sequences permit the proper distribution of the vectors to the daughter cells during cell division and increase chromosome stability (Fig. 8.12).

The Choice of Expression System

There are two main categories of expression systems, eukaryotic and prokaryotic. The choice of a suitable expression system involves the consideration of several important factors, such as protein yield, proper folding, posttranslational modifications (e.g., phosphorylation, glycosylation), and industrial applications of the expressed protein, as well as economic factors. For these reasons there is no universally applied expression system. A comparison of the most commonly used expression systems is shown in Table 8.4.

Bacterial Cells

Expression of heterologous proteins in bacteria remains the most extensively used approach for the production of heterologous proteins such as cytokines (Dracheva et al. 1995, Platis and Foster 2003), enzymes (Wardell et al. 1999, Labrou and Rigden 2001), antibodies (Humphreys 2003), and viral antigens (Ozturk and Erickson-Viitanen 1998, Piefer and Jonsson 2002) at both laboratory and industrial scales (Swartz 2001, Lesley et al. 2002, Choi and Lee 2004). Bacteria can be grown inexpensively and genetically manipulated very easily. They can reach very high densities rapidly and express high levels of recombinant proteins, reaching up to 50% of the total protein. However, in many cases, high-level expression correlates with poor quality. Often, the expressed protein is accumulated in the form of insoluble inclusion bodies (misfolded protein aggregates), and additional, sometimes labor intensive, genetic manipulation or resolubilization/ refolding steps are required (Balbas 2001, Panda 2003). Bacterial cells do not possess the eukaryotes' extensive posttranslational modification system (such as N- or O-glycosylation); this is a serious disadvantage when posttranslational modifications are essential to the protein's function (Zhang et al. 2004). However, they are capable of a surprisingly broad range of covalent modifications such as acetylation, amidation and deamidation, methylation, myristylation, biotinylation, and phosphorylation.

Mammalian Cells

Mammalian cells are ideal candidates for expression hosts when posttranslational modifications (N- and O-glycosylation, disulphide bond formation) are a critical factor for the efficacy of the expressed protein (Bendig 1988). Despite substantial limitations such as high cost, low yield, instability of expression, and lengthy production times, a significant number of proteins (e.g., cytokines) (Fox et al. 2004), antibodies (Schatz et al. 2003), enzymes (Kakkis et al. 1994), viral antigens (Holzer et al. 2003), and blood factors (Kaszubska et al. 2000) are produced in this system because it offers very high product fidelity. However, oligosaccharide processing is species and cell type dependent among mammalian cells, and differences between the glycosylation pattern in rodent cell lines and that in human

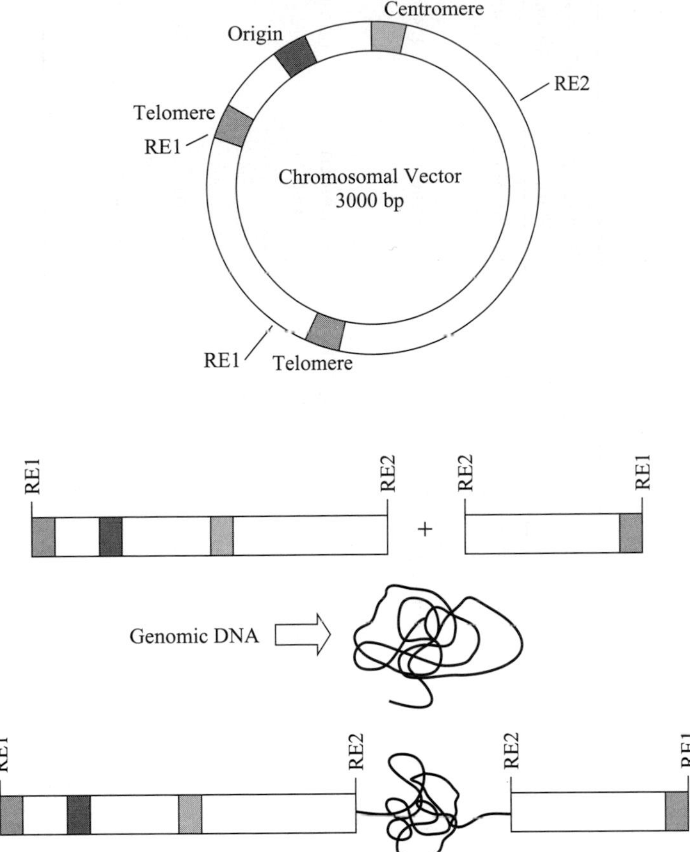

Figure 8.12. Artificial chromosome cloning system. Initially, the circular vector is digested with restriction endonucleases (RE) for linearization and then ligated with size-fractionated DNA (≈100 kb). The vector contains centromeric and telomeric sequences, which assure chromosome-like propagation within the cell, as well as selection marker sequences for stable maintenance.

tissues have been reported. The expressed proteins are usually recovered in a bioactive, properly folded form and secreted into the cell culture fluids.

Yeast

Yeast is a widely used expression system with many commercial, medical, and basic research applications. The fact that yeast is the most intensively studied eukaryote at the genetic/molecular level makes it an extremely advantageous expression system (Trueman 1995). Being a unicellular organism, it retains the advantages of bacteria (low cultivation cost, high doubling rate, ease of genetic manipulation, ability to produce heterologous proteins in large-scale quantities) combined with the advantages of higher eukaryotic systems (posttranslational modifications). The vast majority of yeast expression work has focused on the well-characterized baker's yeast *Saccharomyces cerevisiae* (Holz et al. 2003), but a growing number of non-*Saccharomyces* yeasts are becoming available as hosts for recombinant polypeptide production; these include *Hansenula polymorpha*, *Candida boidinii*, *Kluyveromyces lactis*, *Pichia pastoris* (Fischer et al. 1999b, Cregg et al. 2000, Cereghino et al. 2002), *Schizosaccharomyces pombe* (Giga-Hama and Kumagai 1999), *Schwanniomyces occidentalis*, and *Yarrowia*

Table 8.4. Comparison of the Main Expression Systems

	Bacteria	Yeast	Fungi	Insect Cells	Mammalian Cells	Transgenic Plants	Transgenic Animals
Developing time	Short	Short	Short	Intermediate	Intermediate	Intermediate	Long
Costs for downstream processing	+++	++	++	++	++	++	++
Levels of expression	High	Intermediate	Intermediate	Intermediate	Low	Low	Low
Recombinant protein stability	±	±	±	±	±	+++	±
Production volume	Limited	Limited	Limited	Limited	Limited	Unlimited	Unlimited
Posttranslational modifications (disulphide bond formation, glycosylation, etc.)	No	Yes	Yes	Yes	Yes	Yes	Yes
"Human-type" glycosylation	No	No	No	No	Yes	No	Yes
Folding capabilities	—	++	+++	+++	+++	+++	+++
Contamination level (pathogens, EPL, etc.)	++	—	—	—	++	—	++

lipolytica (Madzak et al. 2004). As in bacteria, expression in yeast relies on episomal or integrated multicopy plasmids with tightly regulated gene expression. Despite these advantages, expressed proteins are not always recovered in soluble form and may have to be purified from inclusion bodies. Posttranslational modifications in yeast differ greatly from those in mammalian cells (Kukuruzinska et al. 1987, Hamilton et al. 2003). This has sometimes proven to be a hindrance when high fidelity of complex carbohydrate modifications found in eukaryotic proteins appears to be important, as in many medical applications. Yeast cells do not add complex oligosaccharides and are limited to the high mannose–type carbohydrates. These higher order oligosaccharides are possibly immunogenic and could potentially interfere with the biological activity of the protein.

Filamentous Fungi

Filamentous fungi have been extensively used for studies of eukaryotic gene organization, regulation, and cellular differentiation. Additionally, fungi belonging to the genus *Aspergillus* and *Penicillium* are of significant industrial importance because of their applications in food fermentation and their ability to secrete a broad range of biopolymer degrading enzymes and to produce primary (organic acids) and secondary metabolites (antibiotics, vitamins). Extensive genetic knowledge as well as an already well developed fermentation technology has allowed for the development of heterologous protein expression systems (Berka and Barnett 1989, Archer and Peberdy 1997) expressing fungal (e.g., glucoamylase; Verdoes et al. 1993) or mammalian proteins of industrial and clinical interest (e.g., human interleukin-6: Contreras et al. 1991; lactoferrin: Ward et al. 1995; bovine chymosin: Ward et al. 1990) using filamentous fungi as hosts. However, the expression levels of mammalian proteins expressed in *Aspergillus* and *Trichoderma* species are low compared to those of homologous proteins. Significant advances in heterologous protein expression have dramatically improved expression efficiency by fusion of the heterologous gene to the 3′-end of a highly expressed homologous gene (mainly glucoamylase). Even so, limitations in protein folding, posttranslational modifications, translocation, and secretion, as well as secretion of extracellular proteases, could pose a significant hindrance for the production of bioactive proteins (Gouka et al. 1997, van den Hombergh et al. 1997).

Insect Cells

Recombinant baculoviruses are widely used as a vector for the expression of recombinant proteins in insect cells (Altmann et al. 1999, Kost and Condreay 1999, Kost and Condreay 2002), such as immunoglobulins (Hasemann and Capra 1990), viral antigens (Roy et al. 1994, Baumert et al. 1998), and transcription factors (Fabian et al. 1998). The recombinant genes are usually expressed under the control of the polyhedrin or *p10* promoter of the *Autographa californica* nuclear polyhedrosis virus (AcNPV) in cultured insect cells of *Spodoptera frugiperda* (Sf9 cells) or in insect larvae of *Lepidopteran* species infected with the recombinant baculovirus containing the gene of interest. The polyhedrin and *p10* genes possess very strong promoters and are highly transcribed during the late stages of the viral cycle. Usually, the recombinant proteins are recovered from the infected insect cells in soluble form and targeted in the proper cellular environment (membrane, nucleus, or cytoplasm). Insect cells have many posttranslational modification, transport, and processing mechanisms found in higher eukaryotic cells (Matsuura et al. 1987), although their glycosylation efficiency is limited, and they are not able to process complex oligosaccharides containing fucose, galactose, and sialic acid.

Dictyostelium discoideum

Recently, the cellular slime mold *Dictyostelium discoideum,* a well-studied single-celled organism, has emerged as a promising eukaryotic alternative system for the expression of recombinant proteins and enzymes (e.g., human antithrombin III; Tiltscher and Storr 1993) (Dittrich et al. 1994). Its advantage over other expression systems lies in its extensive posttranslational modification system (glycosylation, phosphorylation, acylation), which resembles that of higher eukaryotes (Jung and Williams 1997, Slade et al. 1997). It is a simple organism with a haploid genome of 5×10^7 bp and a life cycle that alternates between single cell and multicellular stages. Recombinant proteins are expressed from extrachromosomal plasmids (*Dictyostelium discoideum* is one of a few eukaryotes that have circular nuclear plasmids) rather than plasmids that are integrated in the genome (Ahern et al. 1988). The nuclear plasmids can be easily genetically manipulated and isolated in a one-step procedure, as in bacteria. This system is ideally suited for the expression of complex glycoproteins (Jung and Williams 1997), and although it retains many of the advantages of the bacterial (low cultivation cost) and mammalian systems (establishment of stable cell lines, glycosylation), the development of this system at an industrial scale is hampered by its relatively low productivity compared with bacterial systems.

Trypanosomatid Protozoa

A newly developed eukaryotic expression system is based on the protozoan lizard parasites of the *Leishmania* and *Trypanosoma* species. Their regulation and editing mechanisms are remarkably similar to those of higher eukaryotes and include the capability of "mammalian-like" glycosylation. It has a very rapid doubling time and can be grown to high densities in a relatively inexpensive medium. The recombinant gene is integrated into the small ribosomal subunit rRNA gene and can be expressed to high levels. Increased expression levels and additional promoter control can be achieved in T7 polymerase-expressing strains. Being a lizard parasite, these protozoa are not pathogenic to humans, which makes this system invaluable and highly versatile. Proteins and enzymes of significant interest, such as erythropoietin (EPO) (Breitling et al. 2002), interferon gamma (IFNγ) (Tobin et al. 1993), and interleukin 2 (IL-2)(La Flamme et al. 1995), have been successfully expressed in this system.

Transgenic Plants

The current protein therapeutics market is clearly an area of enormous interest from a medical and economic point of view. Recent advances in human genomics and biotechnology have made it possible to identify a plethora of potentially important drugs or drug targets. Transgenic technology has provided an alternative, more cost effective, bioproduction system than those previously used (*E. coli*, yeast, mammalian cells) (Larrick and Thomas 2001). The accumulated knowledge on plant genetic manipulation has been recently applied to the development of plant bioproduction systems (Fischer et al. 1999a, Fischer et al. 1999b, Fischer et al. 1999c, Russell 1999, Fischer et al. 2000, Daniell et al. 2001, Twyman et al. 2003). Expression in plants can be either constitutive or transient and can be directed to a specific tissue of the plant (depending on the type of promoter used). Expression of heterologous proteins in plants offers significant advantages, such as low production cost, high biomass production, unlimited supply, and ease of expandability. Plants also have high-fidelity expression, folding, and posttranslational modification mechanisms, which could produce human proteins of substantial structural and functional equivalency compared with proteins from mammalian expression systems (Gomord and Faye 2004). Additionally, plant-made human proteins of clinical interest (Fischer and Emans 2000), such as antibodies (Stoger et al. 2002, Schillberg et al. 2003a, Schillberg et al. 2003b), vaccines (Mason and Arntzen, 1995), and enzymes (Cramer et al., 1996; Fischer et al. 1999b, Sala et al. 2003) are free of potentially hazardous human diseases, viruses, or bacterial toxins. However, there is considerable concern regarding the potential hazards of contamination of the natural gene pool by the transgenes, and possible additional safety precautions will raise the production cost.

Transgenic Animals

Besides plants, transgenic technology has also been applied to many different species of animals (mice, cows, rabbits, sheep, goats, and pigs) (Janne et al. 1998, Rudolph 1999). The DNA containing the gene of interest is microinjected into the pronucleus of a single fertilized cell (zygote) and integrated into the genome of the recipient; therefore, it can be faithfully passed on from generation to generation. The gene of interest is coupled with a signal targeting protein expression towards specific tissues, mainly the mammary gland, and the protein can therefore be harvested and purified from milk. The proteins produced by transgenic animals are almost identical to human proteins, greatly expanding the applications of transgenic animals in medicine and biotechnology. Several human proteins of pharmaceutical value such as hemoglobin (Swanson et al. 1992, Logan and Martin 1994), lactoferrin (van Berkel et al. 2002), antithrombin III (Edmunds et al. 1998, Yeung 2000), protein C (Velander et al. 1992), and fibrinogen (Prunkard et al. 1996) have been produced in transgenic animals, and there is enormous interest for the generation of transgenic tissues suitable for transplantation in humans (only recently overshadowed by primary blastocyte technology). Despite the initial technological expertise required to produce a transgenic animal, the subsequent operational costs are low, and subsequent inbreeding ensures that the ability to produce the transgenic protein will be passed on to the transgenic animal's offspring. However, certain safety issues have arisen concerning the potential contamination of transgenically produced proteins by animal viruses or prions, which could possibly be passed on to the human population. Extensive testing required by the FDA substantially raises downstream costs.

Enzyme Purification

Once a suitable enzyme source has been identified, it becomes necessary to design an appropriate purification procedure to isolate the desired protein. The extent of purification required for an enzyme depends on several factors, the most important of which are the degree of enzyme purity required as well as the starting material, that is, the quantity of the desired enzyme present in the initial preparation (Lesley 2001, Labrou and Clonis 2002). For example, industrial enzymes are usually produced as relatively crude preparations. On the other hand, enzymes used for therapeutic or diagnostic purposes are generally subjected to the most stringent purification procedures, as the presence of molecular species other than the intended product may have an adverse clinical impact (Berthold and Walter 1994).

Purification of an enzyme usually occurs by a series of independent steps in which the various physicochemical properties of the enzyme of interest are utilized to separate it progressively from other unwanted constituents (Labrou and Clonis 2002, Labrou et al. 2004). The characteristics of proteins that are utilized in purification include solubility, ionic charge, molecular size, adsorption properties, and binding affinity to other biological molecules. Several methods that exploit differences in these properties are listed in Table 8.5.

Table 8.5. Protein Properties Used during Purification

Protein Property	Technique
Solubility	Precipitation
Size	Gel filtration
Charge	Ion exchange
Hydrophobicity	Hydrophobic interaction chromatography
Biorecognition	Affinity chromatography

Precipitation methods (usually employing $(NH_4)_2SO_4$, polyethyleneglycol, or organic solvents) are not very efficient methods of purification (Labrou and Clonis 2002). They typically give only a few-fold purification. However, with these methods the protein may be removed from the growth medium or from cell debris, where there are harmful proteases and other detrimental compounds that may affect protein stability. On the other hand **chromatography** is a highly selective separation technique (Regnier 1987, Fausnaugh 1990). A wide range of chromatographic techniques has been used for enzyme purification: size exclusion chromatography, ion-exchange chromatography, hydroxyapatite chromatography, hydrophobic interaction chromatography, reverse phase chromatography, and affinity chromatography (Labrou 2003). Of these, **ion-exchange** and **affinity chromatography** are the most common and probably the most important (Labrou and Clonis 1994).

Ion-Exchange Chromatography

Ion-exchange resins selectively bind proteins of opposite charge; that is, a negatively charged resin will bind proteins with a net positive charge and vice versa (Fig. 8.13). Charged groups are classified according to type (cationic and anionic) and strength (strong or weak); the charge characteristics of strong ion exchange media do not change with pH, whereas with weak ion exchange media they do. The most commonly used charged groups include diethylaminoethyl, a weakly anionic exchanger; carboxymethyl, a weakly cationic exchanger; quaternary ammonium, a strongly anionic exchanger; and methyl sulfonate, a strongly cationic exchanger (Table 8.6) (Levison 2003). The matrix material for the column is usually formed from beads of carbohydrate polymers such as agarose, cellulose, or dextrans (Levison 2003).

The technique takes place in five steps (Labrou 2000) (Fig. 8.13): (1) **equilibration** of the column to

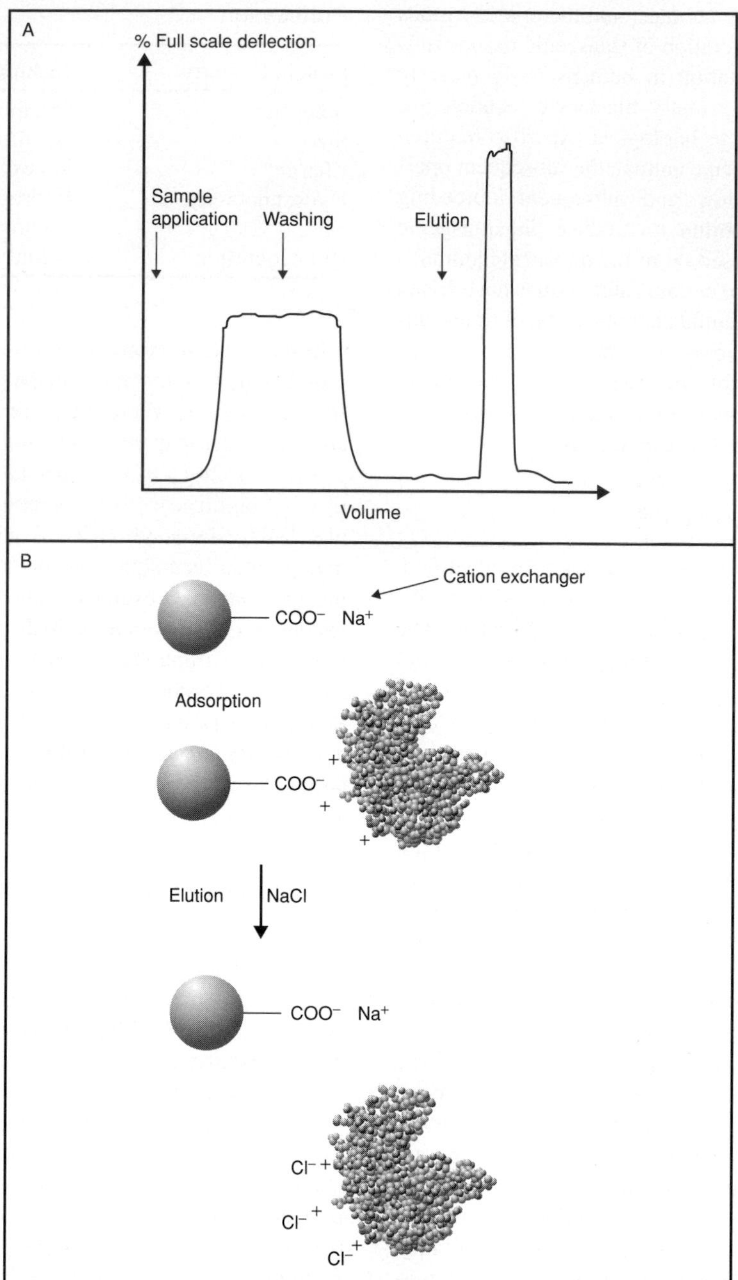

Figure 8.13. (**A**) Schematic diagram of a chromatogram showing the steps for a putative purification. (**B**) Schematic diagram depicting the principle of ion-exchange chromatography.

Table 8.6. Functional Groups Used on Ion Exchangers

Exchangers	Functional Group
Anion exchangers	
Diethylaminoethyl (DEAE)	—O—CH_2—CH_2—$N^+H(CH_2CH_3)_2$
Quaternary aminoethyl (QAE)	—O—CH_2—CH_2—$N^+(C_2H_5)_2$—CH_2—CHOH—CH_3
Quaternary ammonium (Q)	—O—CH_2—CHOH—CH_2O—CH_2—CHOH—$CH_2N^+(CH_3)_2$
Cation exchangers	Functional group
Carboxymethyl (CM)	—O—CH_2—COO^-
Sulphopropyl (SP)	—O—CH_2—CHOH—CH_2—O—CH_2—CH_2—$CH_2SO_3^-$

pH and ionic strength conditions suitable for target protein binding; (2) protein **sample application** to the column and reversible adsorption through counter-ion displacement; (3) **washing** of the unbound contaminating proteins, enzymes, nucleic acids, and other compounds; (4) introduction of **elution** conditions in order to displace bound proteins; and (5) **regeneration** and **reequilibration** of the adsorbent for subsequent purifications. Elution may be achieved either by increasing the salt concentration or by changing the pH of the irrigating buffer. Both methods are used in industry, but raising the salt concentration is by far the most common because it is easier to control (Levison 2003). Most protein purifications are done on anion exchange columns because most proteins are negatively charged at physiological pH values (pH 6–8).

Affinity Chromatography

Affinity chromatography is potentially the most powerful and selective method for protein purification (Fig. 8.14) (Labrou and Clonis 1994, Labrou 2003). According to the International Union of Pure and Applied Chemistry, affinity chromatography is defined as a liquid chromatographic technique that makes use of a "biological interaction" for the separation and analysis of specific analytes within a sample. Examples of these interactions include the binding of an enzyme with a substrate/inhibitor or of an antibody with an antigen or, in general, the interaction of a protein with a binding agent, known as the "**affinity ligand**" (Fig. 8.14) (Labrou 2002, 2003; Labrou et al. 2004). The development of an affinity chromatography–based purification step involves the consideration of the following factors: (1) selection of an appropriate ligand and (2) immobilization of the ligand onto a suitable support matrix to make an **affinity adsorbent**. The selection of the immobilized ligand for affinity chromatography is the most challenging aspect of preparing an affinity adsorbent. Certain factors need to be considered when selecting a ligand (Labrou and Clonis 1995, Labrou and Clonis 1996); these include (1) the specificity of the ligand for the protein of interest, (2) the reversibility of the interaction with the protein, (3) its stability against the biological and chemical operation conditions, and (4) the affinity of the ligand for the protein of interest. The binding site of a protein is often located deep within the molecule, and adsorbents prepared by coupling the ligands directly to the support exhibit low binding capacities. This is due to steric interference between the support matrix and the protein's binding site. In these circumstances a "spacer arm" is inserted between the matrix and the ligand to facilitate effective binding (Fig. 8.14). A hexyl spacer is usually inserted between ligand and support by substitution of 1,6-diaminohexane (Lowe 2001).

The ideal matrix should be hydrophilic, be chemically and biologically stable, and have sufficient modifiable groups to permit an appropriate degree of substitution with the enzyme. Sepharose is the most commonly used matrix for affinity chromatography on the research scale. Sepharose is a commercially available beaded polymer that is highly hydrophilic and generally inert to microbiological attack (Labrou and Clonis 2002). Chemically it is an agarose (poly-{β-1,3-D-galactose-α-1,4-(3,6-anhydro)-L-galactose}) derivative.

The selection of conditions for an optimum affinity chromatographic purification involves study of the following factors: (1) choice of adsorption conditions (e.g., buffer composition, pH, ionic strength)

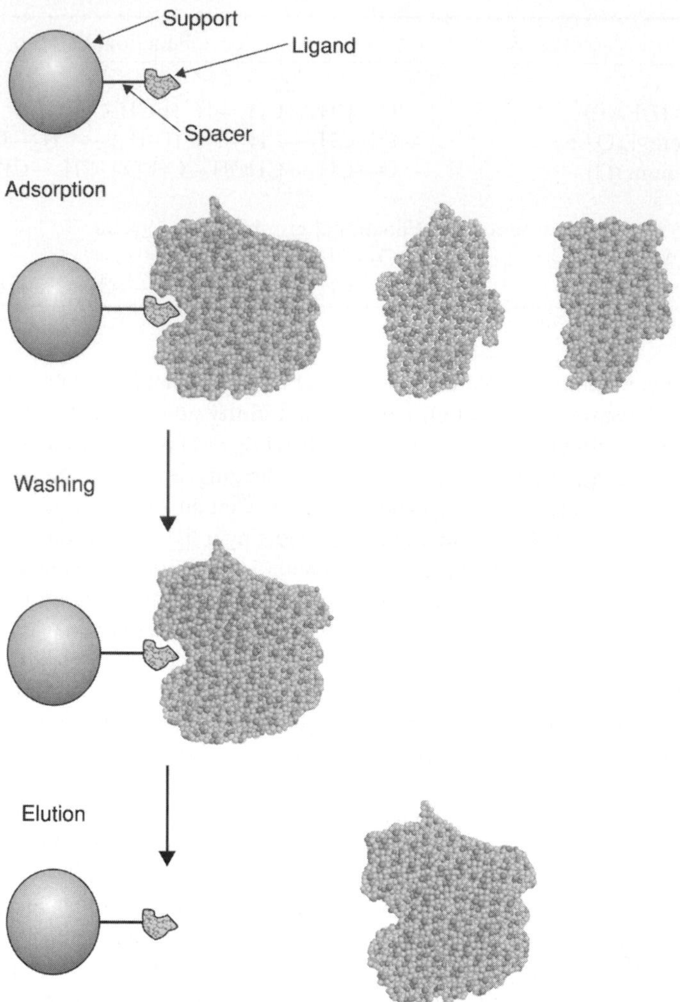

Figure 8.14. Schematic diagram depicting the principle of affinity chromatography.

to maximize the conditions required for the formation of a strong complex between the ligand and the protein to be purified, (2) choice of washing conditions to desorb nonspecifically bound proteins, and (3) choice of elution conditions to maximize purification (Labrou and Clonis 1995). The elution conditions of the bound macromolecule should be both tolerated by the affinity adsorbent and effective in desorbing the biomolecule in good yield and in the native state. Elution of bound proteins is performed in a nonspecific or biospecific manner. Nonspecific elution usually involves (1) changing the ionic strength (usually by increasing the buffer's molarity or including salt, e.g., KCl or NaCl) and pH (adsorption generally weakens with increasing pH), (2) altering the polarity of the irrigating buffer by employing, for example, ethylene glycol or other organic solvents, if the hydrophobic contribution in the protein-ligand complex is large. Biospecific elution is achieved by inclusion in the equilibration buffer of a suitable ligand, which usually competes with the immobilized ligand for the same binding site on the enzyme/protein (Labrou 2000). Any competing ligand may be used. For example, substrates, products, cofactors, inhibitors, or allosteric effectors are all potential candidates as long as they have higher affinity for the macromolecule than for the immobilized ligand.

Dye-ligand affinity chromatography represents a powerful affinity-based technique for enzyme and protein purification (Clonis et al. 2000, Labrou 2002, Labrou et al. 2004). The technique has gained broad popularity due to its simplicity and wide applicability to purifying a variety of proteins. The employed dyes used as affinity ligands are commercial textile chlorotriazine polysulfonated aromatic molecules, which are usually termed **triazine dyes** (Fig. 8.15). Such dye ligands have found wide application in the research market over the past 20 years as low specificity general affinity ligands to purify enzymes such as oxidoreductases, decarboxylases, glycolytic enzymes, nucleases, hydrolases, lyases, synthetases, and transferases (Scopes 1987). Anthraquinone triazine dyes are probably the most widely used dye ligands in enzyme and protein purification. The triazine dye Cibacron Blue 3GA (Fig. 8.15), especially, has been widely exploited as an affinity chromatographic tool to separate and purify a variety of proteins (Scopes 1987). With the aim of increasing the specificity of dye ligands, the biomimetic dye-ligand concept was introduced. According to this concept, new dyes that mimic natural ligands of the targeted proteins are designed by substituting the terminal 2-aminobenzene sulfonate moiety of the dye Cibacron Blue 3GA (CB3GA) with a substrate-mimetic moiety (Clonis et al. 2000; Labrou 2002,

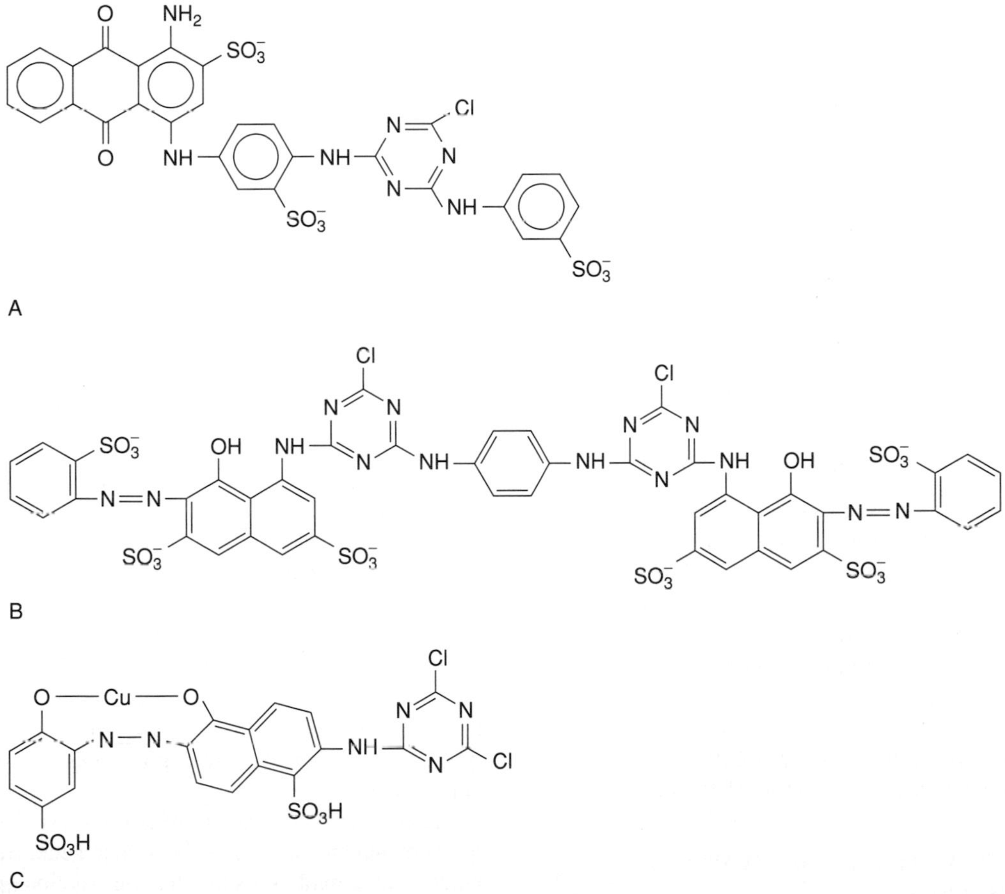

Figure 8.15. Structure of several representative triazine dyes: (**A**) Cibacron Blue 3GA, (**B**) Procion Red HE-3B, (**C**) Procion Rubine MX-B.

Table 8.7. Adsorbents and Elution Conditions of Affinity Tags

Affinity Tag	Matrix	Elution Condition
Poly-His	Ni^{2+}-NTA	Imidazole 20–250 mM or low pH
FLAG	Anti-FLAG monoclonal antibody	pH 3.0 or 2–5 mM EDTA
Strep-tag II	Strep-Tactin (modified streptavidin)	2.5 mM desthiobiotin
c-myc	Monoclonal antibody	Low pH
S	S-fragment of RNaseA	3 M guanidine thiocyanate, 0.2 M citrate pH 2, 3 M magnesium chloride
Calmodulin-binding peptide	Calmodulin	EGTA or EGTA with 1 M NaCl
Cellulose-binding domain	Cellulose	Guanidine HCl or urea > 4M
Glutathione S-transferase	Glutathione	5–10 mM reduced glutathione

2003; Labrou et al. 2004). These biomimetic dyes exhibit increased purification ability and specificity and provide useful tools for designing simple and effective purification protocols.

The rapid development of recombinant DNA technology since the early 1980s has changed the emphasis of classical enzyme purification work. For example, epitope tagging is a recombinant DNA method for inserting a specific protein sequence (affinity tag) into the protein of interest (Terpe 2003). This allows the expressed tagged protein to be purified by affinity interactions with a ligand that selectively binds to the affinity tag. Examples of affinity tags and their respective ligands used for protein and enzyme purification are shown in Table 8.7.

ENZYME ENGINEERING

Another extremely promising area of enzyme technology is enzyme engineering. New enzyme structures may be designed and produced in order to improve existing ones or create new activities. Over the past two decades, with the advent of protein engineering, molecular biotechnology has permitted not only the improvement of the properties of these isolated proteins, but also the construction of altered versions of these naturally occurring proteins with novel or tailor-made properties (Ryu and Nam 2000).

TAILOR-MADE ENZYMES BY PROTEIN ENGINEERING

There are two main intervention approaches for the construction of tailor-made enzymes: rational design and directed evolution (Chen 2001) (Fig. 8.16).

Rational design takes advantage of knowledge of the three-dimensional structure of the enzyme as well as structure/function and sequence information to predict, in a rational/logical way, sites on the enzyme that, when altered, would endow the enzyme with the desired properties. Once the crucial amino acids are identified, site-directed mutagenesis is applied, and the expressed mutants are screened for the desired properties. It is clear that protein engineering by rational design requires prior knowledge of the "hot spots" on the enzyme. Directed evolution (or molecular evolution) does not require such prior sequence or three-dimensional structural knowledge, as it usually employs random mutagenesis protocols to engineer enzymes that are subsequently screened for the desired properties. However, both approaches require efficient expression as well as sensitive detection systems for the protein of interest. During the selection process the mutations that have a positive effect are selected and identified. Usually, repeated rounds of mutagenesis are applied until enzymes with the desired properties are constructed.

Usually, a combination of both methods is employed by the construction of combinatorial libraries of variants, using random mutagenesis on selected (by rational design) areas of the parental "wild-type" protein (typically, binding surfaces or specific amino acids).

The industrial applications of enzymes as biocatalysts are numerous. Recent advances in genetic engineering have made possible the construction of enzymes with enhanced or altered properties (change of enzyme/cofactor specificity and enantioselectivity, altered thermostability, increased activity) to satisfy the ever-increasing needs of the industry for

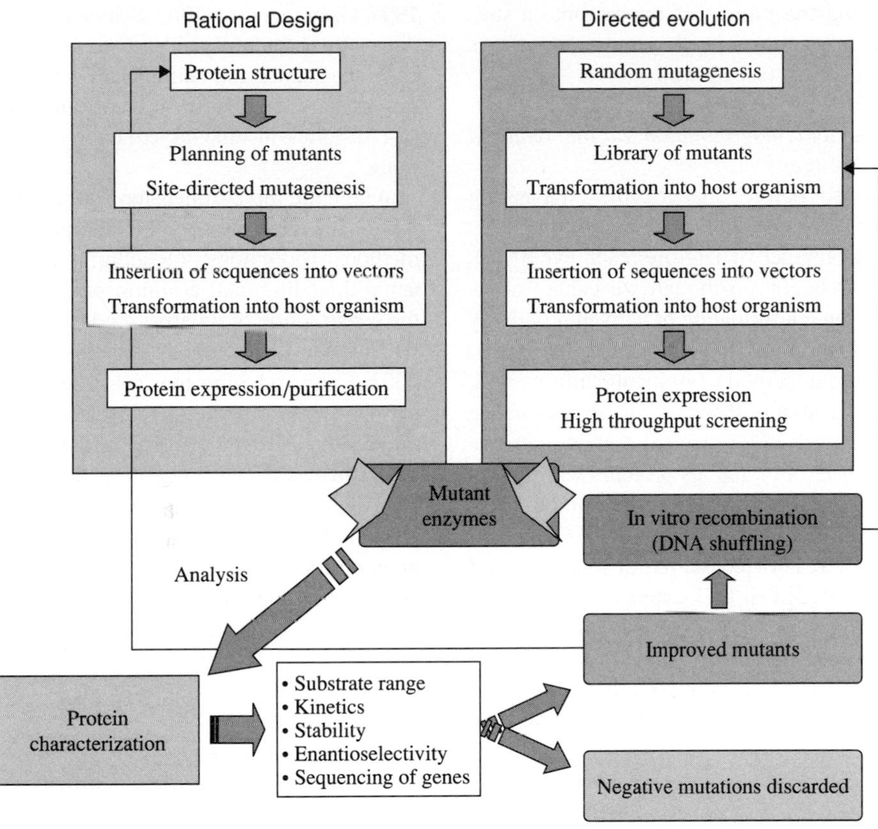

Figure 8.16. Comparison of rational design and directed evolution.

more efficient catalysts (Chen et al. 1999, Bornscheuer and Pohl 2001, Zaks 2001, Panke and Wubbolts 2002).

Rational Enzyme Design

The rational protein design approach is mainly used for the identification and evaluation of functionally important residues or sites in proteins. Although the protein sequence contains all the information required for protein folding and functions, today's state of technology does not allow for efficient protein design by simple knowledge of the amino acid sequence alone. For example, there are 10^{325} ways of rearranging amino acids in a 250-amino-acid-long protein, and prediction of the number of changes required to achieve a desired effect is an obstacle that initially appears impossible. For this reason, a successful rational design cycle requires substantial planning and could be repeated several times before the desired result is achieved. A rational protein design cycle requires the following:

1. *Knowledge of the amino acid sequence of the enzyme of interest and availability of an expression system that allows for the production of active enzyme.*

Isolation and characterization (annotation) of cDNAs encoding proteins with novel or preobserved properties has been significantly facilitated by advances in genomics (Schena et al. 1995, Zweiger and Scott 1997, Schena et al. 1998) and proteomics (Anderson and Anderson 1998, Anderson et al. 2000, Steiner and Anderson 2000) and is increasing rapidly. These cDNA sequences are stored in gene (NCBI) and protein databanks (Swiss-Prot; Release 44.3 of 16-Aug-04 contains 156,998 protein sequence entries; Apweiler 2001, Gasteiger et al. 2001, Apweiler et al. 2004). However, before the protein design cycle begins, a protein expression system has to be established. Introduction of the cDNA encoding the protein of interest into a suitable expression vector/host cell system is nowadays a standard procedure (see above).

2. *Structure/function analysis of the initial protein sequence and determination of the required amino acids changes.*

As mentioned before, the enzyme engineering process could be repeated several times until the desired result is obtained. Therefore, each cycle ends where the next begins. Although we cannot accurately predict the conformation of a given protein by knowledge of its amino acid sequence, the amino acid sequence can provide significant information. Initial screening should therefore involve sequence comparison analysis of the original protein sequence to other sequence homologous proteins with potentially similar functions by utilizing current bioinformatics tools (Andrade and Sander 1997, Fenyo and Beavis 2002). Areas of conserved or nonconserved amino acid residues can be located within the protein and could possibly provide valuable information concerning the identification of binding and catalytic residues. Additionally, such methods could also reveal information pertinent to the three-dimensional structure of the protein.

3. *Availability of functional assays for identification of changes in the properties of the protein.*

This is probably the most basic requirement for efficient rational protein design. The expressed protein has to be produced in a bioactive form and characterized for size, function, and stability in order to build a baseline comparison platform for the ensuing protein mutants. The functional assays should have the required sensitivity and accuracy to detect the desired changes in the protein's properties.

4. *Availability of the three-dimensional structure of the protein or capability of producing a reasonably accurate three-dimensional model by computer modeling techniques.*

The structures of thousands of proteins have been solved by various crystallographic techniques (X-ray diffraction, NMR spectroscopy) and are available in protein structure databanks (PDB). Current bioinformatics tools and elaborate molecular modeling software (Wilkins et al. 1999, Gasteiger et al. 2003) permit the accurate depiction of these structures, allow the manipulation of the amino acid sequence, and even predict with significant accuracy the result that a single amino acid substitution would have on the conformation and electrostatic or hydrophobic potential of the protein (Guex and Peitsch 1997, Gasteiger et al. 2003, Schwede et al. 2003). Additionally, protein-ligand interactions can, in some cases, be successfully simulated, which is especially important in the identification of functionally important residues in enzyme-cofactor/substrate interactions.

Where the three-dimensional structure of the protein of interest is not available, computer-modeling methods (homology modeling, fold recognition using threading, and ab initio prediction) allow for the construction of putative models based on known structures of homologous proteins (Schwede et al. 2003, Kopp and Schwede 2004). Additionally, comparison with proteins having homologous three-dimensional structure or structural motifs could provide clues as to the function of the protein and the location of functionally important sites. Even if the protein of interest shows no homology to any other known protein, current amino acid sequence analysis software could provide putative tertiary structural models. A generalized approach to predicting protein structure is shown in Figure 8.17.

5. *Genetic manipulation of the wild-type nucleotide sequence.*

A combination of previously published experimental literature and sequence/structure analysis information is usually necessary for the identification of functionally important sites in the protein. Once an adequate three-dimensional structural model of the protein of interest has been constructed, manipulation of the gene of interest is necessary for the construction of mutants. Polymerase chain reaction (PCR) mutagenesis is the basic tool for the genetic manipulation of the nucleotide sequences. The genetically redesigned proteins are engineered by the following:

a. Site-directed mutagenesis: alteration of specific aminoacid residues.

There are a number of experimental approaches designed for this purpose. The basic principle involves the use of synthetic oligonucleotides (oligonucleotide-directed mutagenesis) that are complementary to the cloned gene of interest but contain a single (or sometimes multiple) mismatched base(s) (Balland et al. 1985, Garvey and Matthews 1990, Wagner and Benkovic 1990). The cloned gene is either carried by a single-stranded vector (M13 oligonucleotide-directed mutagenesis) or a plasmid that is later denatured by alkali (plasmid DNA oligonucleotide-directed mutagenesis) or heat (PCR-amplified oligonu-

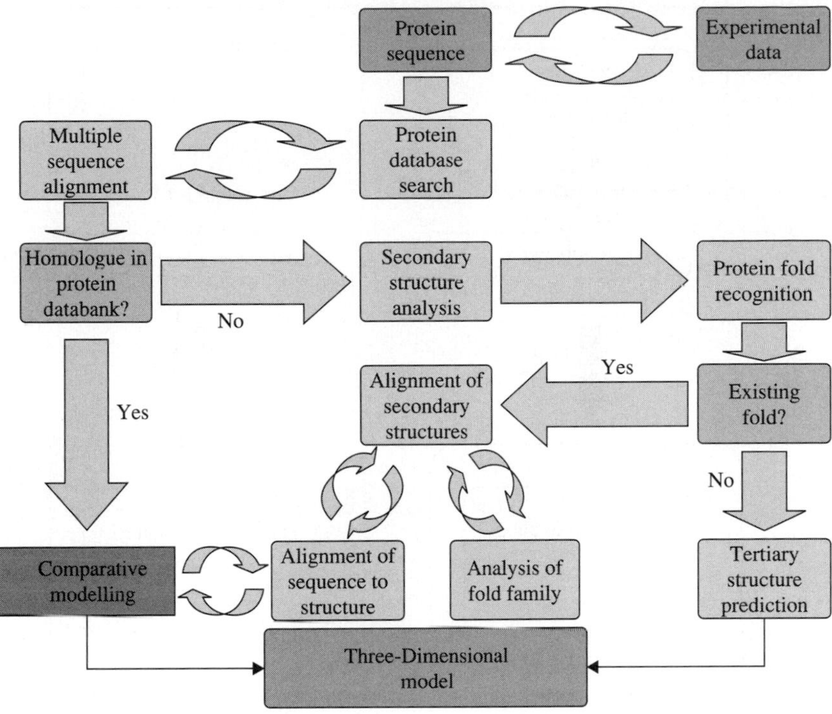

Figure 8.17. A generalized schematic for the prediction of three-dimensional protein structure.

cleotide-directed mutagenesis) in order for the mismatched oligonucleotide to anneal. The latter then serves as a primer for DNA synthesis catalyzed by externally added DNA polymerase for the creation of a copy of the entire vector, carrying however, a mutated base. PCR mutagenesis is the most frequently used mutagenesis method (Fig. 8.18). For example, substitution of specific aminoacid positions by site-directed mutagenesis altered the preference for coenzyme from $NADP^+$ to NAD^+ for *Chromatium vinosum* glutathione reductase (Scrutton et al., 1990).

So far substitution of a specific amino acid by another has been limited by the availability of only 20 naturally occurring amino acids. However, it is chemically possible to construct hundreds of designer-made amino acids. Incorporation of these novel protein building blocks could help shed new light onto the cellular and protein functions (Wang and Schultz 2002, Chin et al. 2003, Deiters et al. 2003).

b. Construction of deletion mutants: deletion of specified areas within or at the 5′/3′ ends (truncation mutants) of the gene.

c. Construction of insertion/fusion mutants: insertion of a functionally/structurally important epitope or fusion to another protein fragment. There are numerous examples of fusion proteins designed to facilitate protein expression and purification, display of proteins on surfaces of cells or phages, cellular localization, and metabolic engineering as well as protein-protein interaction studies (Nixon et al. 1998).

d. Domain swapping: exchanging of protein domains between homologous or heterologous proteins. For example, exchange of a homologous region between *Agrobacterium tumefaciens* β-glucosidase (optimum at pH 7.2–7.4 and 60°C) and *Cellvibrio gilvus* β-glucosidase (optimum at pH 6.2–6.4 and 35°C) resulted in a hybrid enzyme with optimal activity at pH 6.6–7.0 and 45–50°C (Singh et al. 1995).

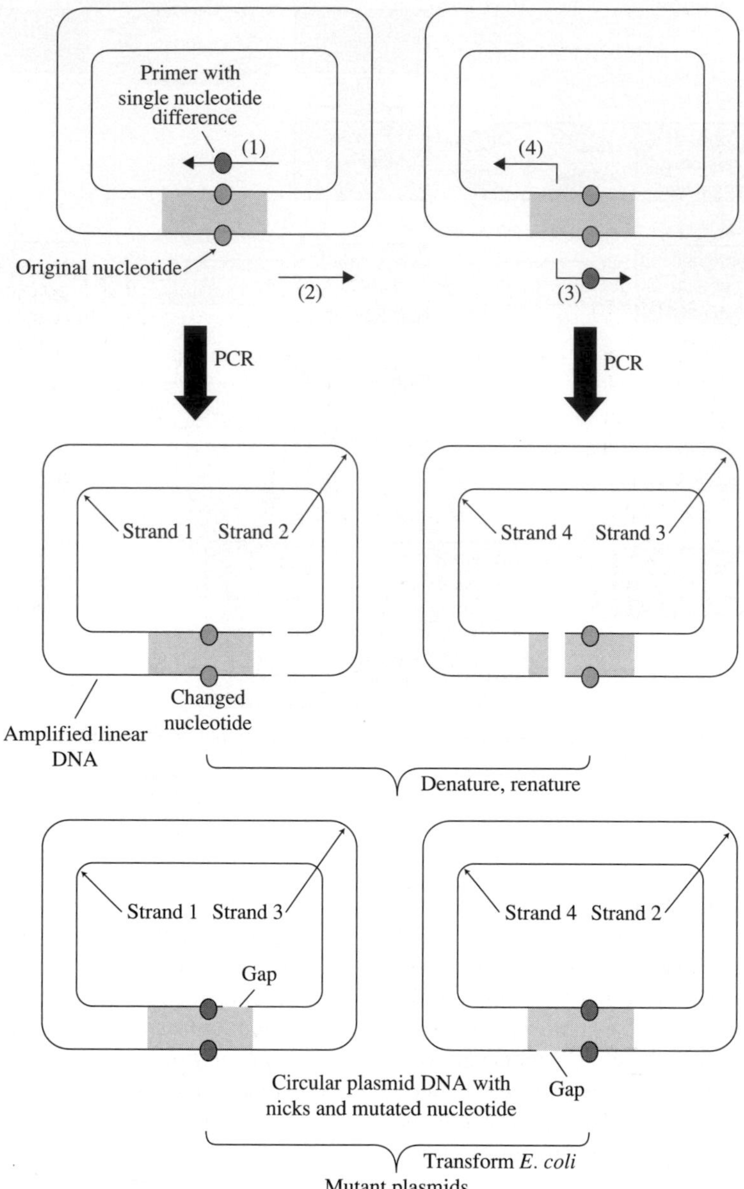

Figure 8.18. PCR oligonucleotide-directed mutagenesis. Two sets of primers are used for the amplification of the double-stranded plasmid DNA. The primers are positioned as shown and only one contains the desired base change. After the initial PCR step, the amplified PCR products are mixed together, denatured, and renatured to form, along with the original amplified linear DNA, nicked circular plasmids containing the mutations. Upon transformation into *E.coli,* the nicked plasmids are repaired by host cell enzymes, and the circular plasmids can be maintained.

Although site-directed mutagenesis is widely used, it is not always feasible due to the limited knowledge of the protein structure-function relationship and the approximate nature of computer-graphic modeling. In addition, rational design approaches can fail due to unexpected influences exerted by the substitution of one or more amino acid residues. Irrational approaches can therefore be preferable alternatives for engineering enzymes with highly specialized traits.

Directed Enzyme Evolution

Directed evolution by DNA recombination can be described as a mature technology for accelerating protein evolution. Evolution is a powerful algorithm with proven ability to alter enzyme function and especially to "tune" enzyme properties. The methods of directed evolution use the process of natural selection, but in a directed way (Altreuter and Clark 1999). The major step in a typical directed enzyme evolution experiment is first to make a set of mutants and then to find the best variants through a high-throughput selection or screening procedure. The process can be iterative, so that a "generation" of molecules can be created in a few weeks or even in a few days, with large numbers of progeny subjected to selective pressures not encountered in nature (Arnold 2001).

There are many methods to create combinatorial libraries using directed evolution. Some of these are random mutagenesis using mainly error-prone PCR (Ke and Madison 1997), DNA shuffling (Stemmer 1994, Crameri et al. 1998), StEP (staggered extension process; Zhao et al. 1998), RPR (random-priming in vitro recombination; Shao et al. 1998), and incremental truncation for the creation of hybrid enzymes (ITCHY; Lutz et al. 2001). The most frequently used methods for DNA shuffling are shown in Figure 8.19.

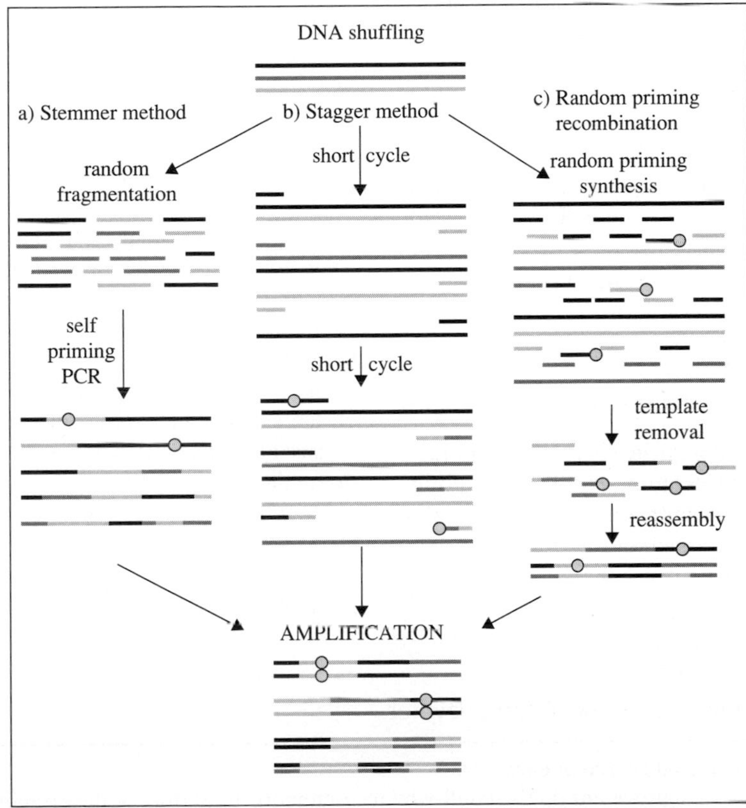

Figure 8.19. A schematic representation of the most frequently used methods for DNA shuffling.

Currently, directed evolution has gained considerable attention as a commercially important strategy for rapid design of molecules with properties tailored for the biotechnological and pharmaceutical market. Over the past four years, DNA family shuffling has been successfully used to improve enzymes of industrial and therapeutic interest (Kurtzman et al. 2001).

IMMOBILIZED ENZYMES

The term "**immobilized enzymes**" describes enzymes physically confined, localized in a certain region of space or attached on a support matrix (Abdul 1993). The main advantages of enzyme immobilization are listed in Table 8.8.

There are at least four main areas in which immobilized enzymes may find applications: industrial, environmental, analytical, and chemotherapeutic (Powell 1984, Liang et al. 2000). Environmental applications include wastewater treatment and the degradation of chemical pollutants of industrial and agricultural origin (Dravis et al. 2001). Analytical applications include biosensors. Biosensors are analytical devices that have a biological recognition mechanism (most commonly enzyme) that transduces a reaction into a signal, usually electrical, that can be detected by using a suitable detector (Phadke 1992). Immobilized enzymes, usually encapsulated, are also being used for their possible chemotherapeutic applications in replacing enzymes that are absent from individuals with certain genetic disorders (DeYoung 1989).

METHODS FOR IMMOBILIZATION

There are a number of ways in which an enzyme may be immobilized: **adsorption, covalent coupling, cross-linking, matrix entrapment,** or **encapsulation** (Podgornik and Tennikova 2002) (Fig. 8.20). These methods will be discussed in the following sections.

Adsorption

Adsorption is the simplest method and involves reversible interactions between the enzyme and the support material (Fig. 8.20A). The driving force causing adsorption is usually the formation of several noncovalent bonds such as salt links, van der Waals, hydrophobic, and hydrogen bonding (Calleri et al. 2004). The methodology is easy to carry out and can be applied to a wide range of support matrices such as alumina, bentonite, cellulose, anion and cation exchange resins, glass, hydroxyapatite, kaolinite, and others. The procedure consists of mixing together the enzyme and a support under suitable conditions of pH, ionic strength, temperature, and so on. The most significant advantages of this method are (1) absence of chemicals, resulting in little damage to the enzyme, and (2) reversibility, which allows regeneration with fresh enzyme. The main disadvantage of the method is the leakage of the enzyme from the support under many conditions of changes in the pH, temperature, and ionic strength. Another disadvantage is the nonspecific adsorption of other proteins or other substances to the support. This may modify the properties of the support or of the immobilized enzyme.

Covalent Coupling

The covalent coupling method is achieved by the formation of a covalent bond between the enzyme and the support (Fig. 8.20B). The binding is very strong, and therefore little leakage of enzyme from the support occurs (Calleri et al. 2004). The bond is formed between reactive electrophile groups present on the support and nucleophile side chains on the surface of the enzyme. These side chains are usually the amino group ($-NH_2$) of lysine, the imidazole group of histidine, the hydroxyl group (-OH) of serine and threonine, and the sulfydryl group (-SH) of cysteine. Lysine residues are found to be the most generally useful groups for covalent bonding of

Table 8.8. Advantages of Immobilized Enzymes

1. Repetitive use of a single batch of enzymes.
2. Immobilization can improve enzyme's stability by restricting the unfolding of the protein.
3. Product is not contaminated with the enzyme. This is very important in the food and pharmaceutical industries.
4. The reaction is controlled rapidly by removing the enzyme from the reaction solution (or vice versa).

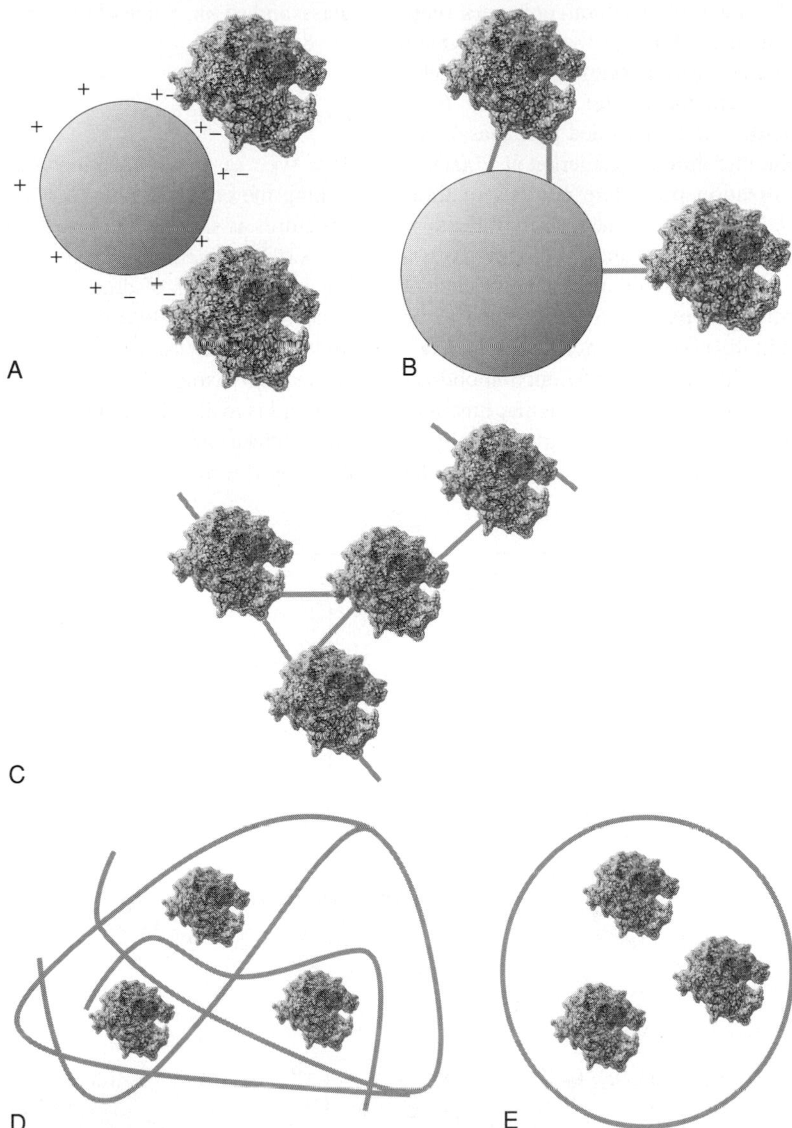

Figure 8.20. Representation of the methods by which an enzyme may be immobilized: (**A**) adsorption, (**B**) covalent coupling, (**C**) cross-linking, (**D**) matrix entrapment, and (**E**) encapsulation.

enzymes to insoluble supports, due to their widespread surface exposure and high reactivity, especially in slightly alkaline solutions.

It is important that the amino acids essential to the catalytic activity of the enzyme are not involved in the covalent linkage to the support (Dravis et al. 2001). This may be difficult to achieve, and enzymes immobilized in this fashion generally lose activity upon immobilization. This problem may be prevented if the enzyme is immobilized in the presence of saturating concentrations of substrate, product, or a competitive inhibitor to protect active site residues. This ensures that the active site remains "unreacted" during the covalent coupling and reduces the occurrence of binding in unproductive conformations.

Various types of beaded supports have been used successfully: for example, natural polymers (e.g., agarose, dextran, and cellulose), synthetic polymers (e.g., polyacrylamide, polyacryloyl trihydroxymethylacrylamide, polymethacrylate), inorganic (e.g., silica, metal oxides, and controlled pore glass), and microporous flat membranes (Calleri et al. 2004).

The immobilization procedure consists of three steps (Calleri et al. 2004): (1) activation of the support, (2) coupling of ligand, and (3) blocking of residual functional groups in the matrix. The choice of coupling chemistry depends on the enzyme to be immobilized and its stability. A number of methods are available in the literature for efficient immobilization of enzyme through a chosen particular functional side chain's group by employing glutaraldehyde, oxirane, cyanogen bromide, 1,1-carbonyldiimidazole, cyanuric chloride, trialkoxysilane to derivatize glass, and so on. Some of them are illustrated in Figure 8.21.

Cross-linking

This type of immobilization is achieved by cross-linking the enzymes to each other to form complex structures as shown in Figure 8.20C. It is therefore a support-free method and less costly than covalent linkage. Methods of cross-linking involve covalent bond formation between the enzymes using bi- or multifunctional reagent. Cross-linking is frequently carried out using glutaraldehyde, which is of low cost and is available in industrial quantities. To minimize close proximity problems associated with the cross-linking of a single enzyme, albumin and

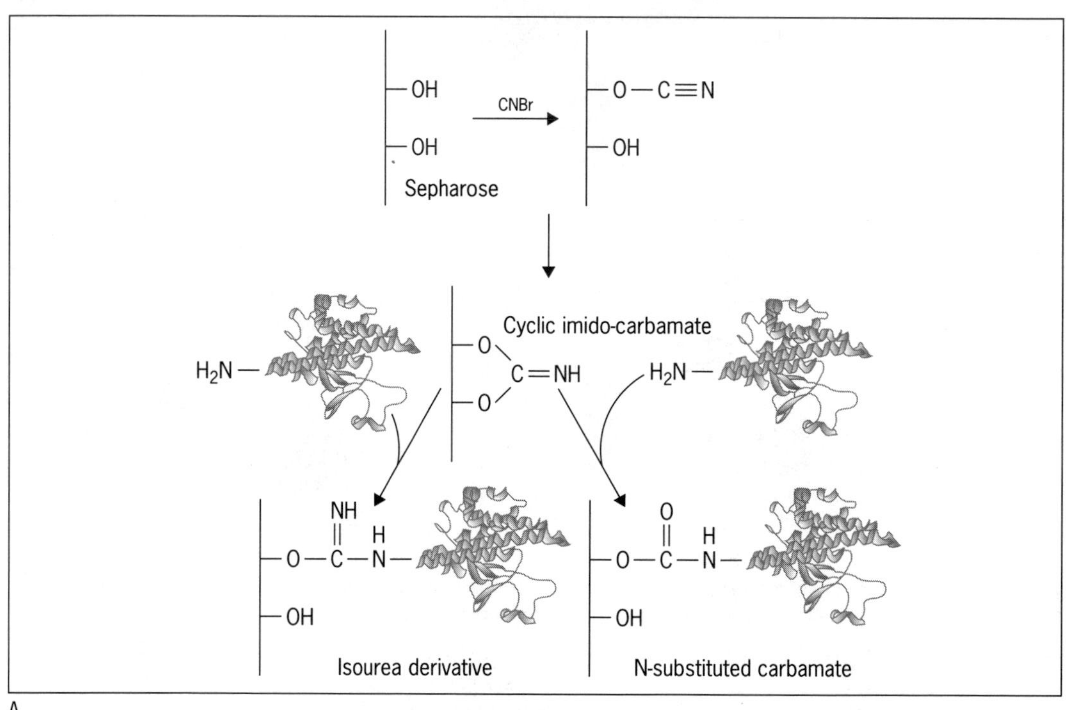

Figure 8.21. Commonly used methods for the covalent immobilization of enzymes. (**A**) Activation of hydroxyl support by cyanogen bromide. (**B**) Carbodiimides may be used to attach amino groups on the enzyme to carboxylate groups on the support or carboxylate groups on the enzyme to amino groups on the support. (**C**) Glutaraldehyde is used to cross-link enzymes or link them to supports. The product of the condensation of enzyme and glutaraldehyde may be stabilized against dissociation by reduction with sodium borohydride.

gelatin are usually used to provide additional protein molecules as spacers (Podgornik and Tennikova 2002).

Entrapment and Encapsulation

In the immobilization by entrapment, the enzyme molecules are free in solution, but restricted in movement by the lattice structure of the gel (Figure 8.20D) (Balabushevich et al. 2004). The entrapment method of immobilization is based on the localization of an enzyme within the lattice of a polymer matrix or membrane (Podgornik and Tennikova 2002). It is done in such a way as to retain protein while allowing penetration of substrate. Entrapment can be achieved by mixing an enzyme with chemical monomers that are then polymerized to form a cross-linked polymeric network, trapping the enzyme in

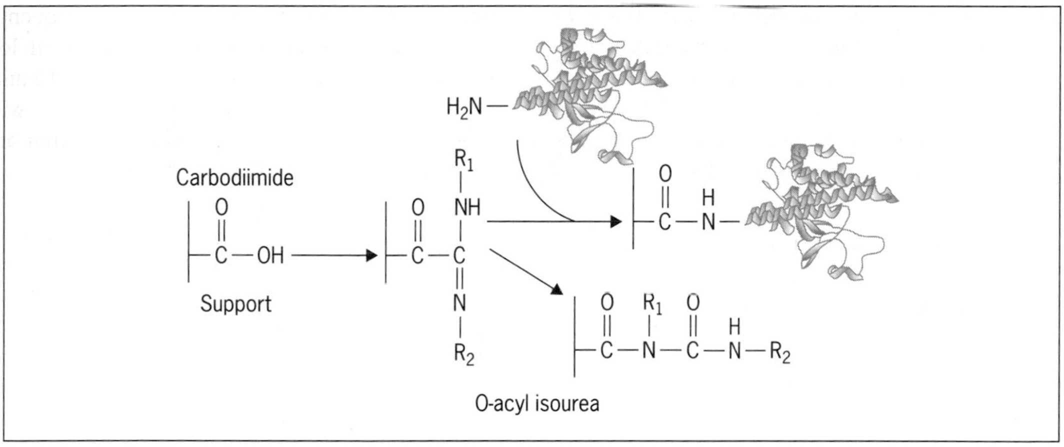

B

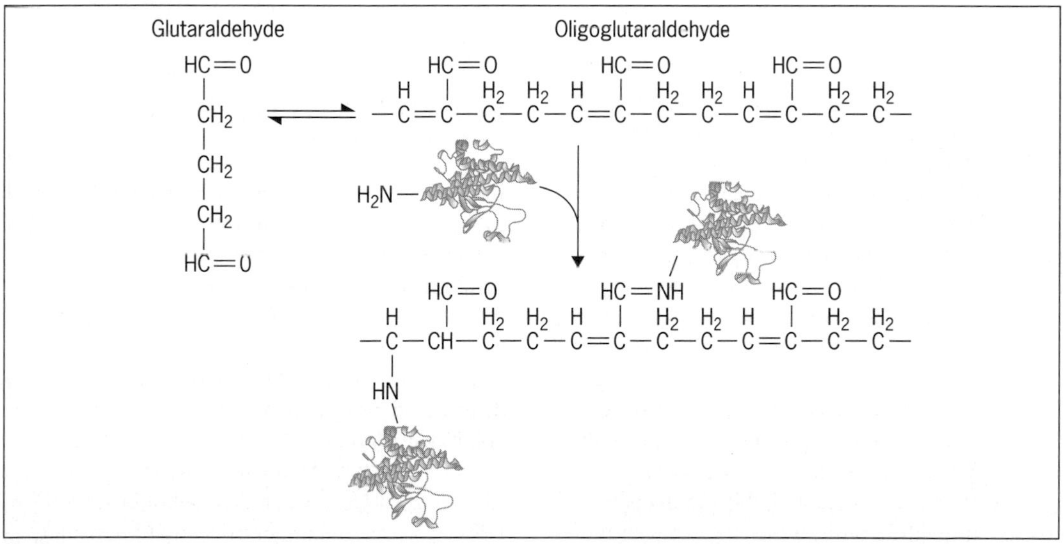

C

Figure 8.21. Continued

the interstitial spaces of lattice. Many materials, for example, alginate, agarose, gelatin, polystyrene, and polyacrylamide, have been used. As an example of this latter method, the enzyme's surface lysine residues may be derivatized by reaction with acryloyl chloride ($CH_2{=}CH{-}CO{-}Cl$) to give the acryloyl amides. This product may then be copolymerized and cross-linked with acrylamide ($CH_2{=}CH{-}CO{-}NH_2$) and bisacrylamide ($H_2N{-}CO{-}CH{=}CH{-}CH{=}CH{-}CO{-}NH_2$) to form a gel.

Encapsulation of enzymes can be achieved by enveloping the biological components within various forms of semipermeable membranes as shown in Figure 8.20E. Encapsulation is most frequently carried out using nylon and cellulose nitrate to construct microcapsules varying from 10 to 100 μm. In general, entrapment methods have found more application in the immobilization of cells.

New Approaches for Oriented Enzyme Immobilization: The Development of Enzyme Arrays

With the completion of several genome projects, attention has turned to the elucidation of the functional activities of the encoded proteins. Due to the enormous number of newly discovered open reading frames (ORF), progress in the analysis of the corresponding proteins depends on the ability to perform characterization in a parallel and high throughput (HTS) format (Cahill and Nordhoff 2003). This typically involves construction of protein arrays based on recombinant proteins. Such arrays are then analyzed for their enzymatic activities, for the ability to interact with other proteins or small molecules, and so forth. The development of enzyme array technology is hindered by the complexity of protein molecules. The tremendous variability in the nature of enzymes, and consequently in the requirements for their detection and identification, makes the development of protein chips a particularly challenging task. Additionally, enzyme molecules must be immobilized on a matrix in a way that preserves their native structures and makes them accessible to their targets (Cutler 2003). The immobilization chemistry must be compatible with preserving enzyme molecules in native states. This requires good control of local molecular environments for the immobilized enzyme molecule (Yeo et al. 2004). There is one major barrier in enzyme microarray development: the immobilization chemistry has to be such that it preserves the enzyme in a native state and allows optimal orientation for substrate interaction. This problem may be solved by the recently developed in vitro protein ligation (IPL) methodology. Central to this method is the ability of certain protein domains (inteins) to excise themselves from a precursor protein (Lue et al. 2004). In a simplified intein expression system, a thiol reagent induces cleavage of the intein-extein bond, leaving a reactive thioester group on the C-terminus of the protein of interest. This group can then be used to couple essentially any polypeptide with an N-terminal cysteine to the thioester-tagged protein by restoring the peptide bond. In another methodology, optimal orientation is based on the unique ability of protein prenyl transferases to recognize short but highly specific C-terminal protein sequences (Cys-A-A-X-, where A is any aliphatic amino acid), as shown in Figure 8.22. The enzyme accepts a spectrum of phosphoisoprenoid analogs while displaying a very strict specificity for the protein substrate. This feature is explored for protein derivatization. Several types of pyrophosphates (biotin analogs, photoreactive analogs, and benzophenone analogs; Figure 8.22) can be covalently attached to the protein tagged with the Cys-A-A-X motif. After modification the protein can be immobilized directly, either reversibly through biotin-avidin interaction on avidin-modified support or covalently through the photoreactive group on several supports.

ENZYME UTILIZATION IN INDUSTRY

Enzymes offer the potential for many exciting applications in industry. Some important industrial enzymes and their sources are listed in Table 8.9. In addition to the industrial enzymes listed below, a number of enzyme products have been approved for therapeutic use. Examples include tissue plasminogen activator and streptokinase for cardiovascular disease; adenosine deaminase for the rare severe combined immunodeficiency disease; β-glucocerebrosidase for Type 1 Gaucher disease; L-asparaginase for the treatment of acute lymphoblastic leukemia; DNAse for the treatment of cystic fibrosis; and neuraminidase, which is being targeted for the treatment of influenza (Cutler 2003).

There are also thousands of enzyme products used in small amounts for research and development

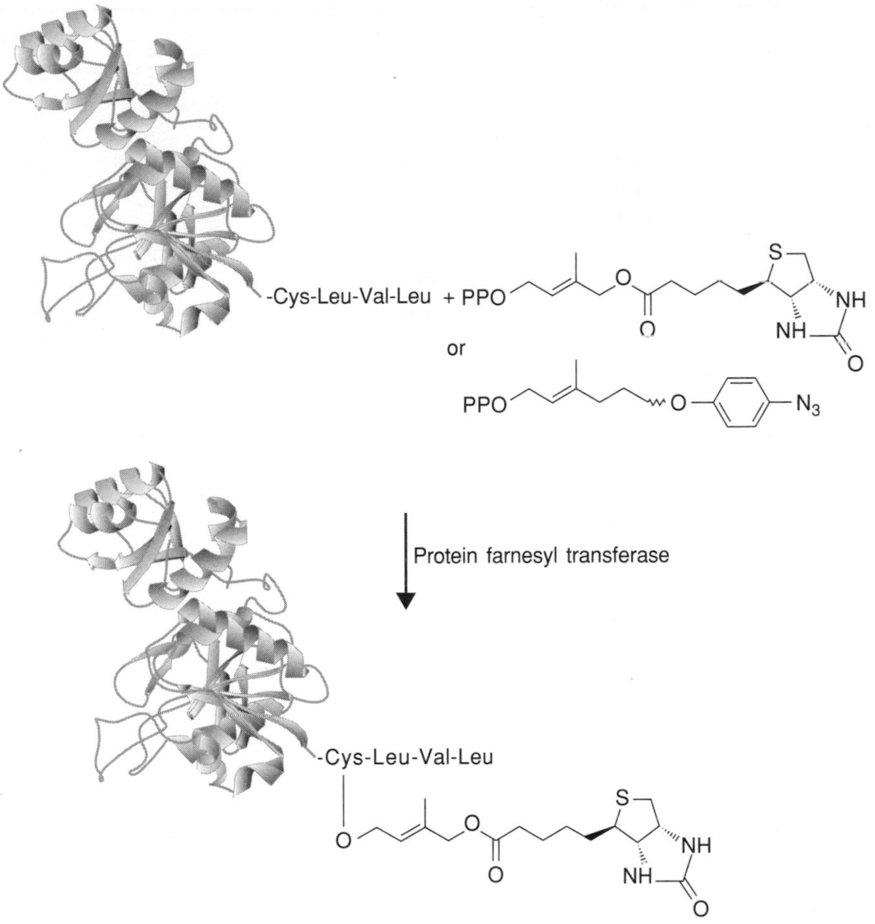

Figure 8.22. Principal scheme of using CAAX-tagged proteins for covalent modification with prenyl transferases.

in routine laboratory practice, and others that are used in clinical laboratory assays. This group also includes a number of DNA and RNA modifying enzymes (DNA and RNA polymerase, DNA ligase, restriction endonucleases, reverse transcriptase, etc.), which led to the development of molecular biology methods and were a foundation for the biotechnology industry (Yeo et al. 2004). The clever application of one thermostable DNA polymerase led to the polymerase chain reaction (PCR), and this has since blossomed into numerous clinical, forensic, and academic embodiments. Along with the commercial success of these enzyme products, other enzyme products are currently in commercial development.

Another important field of application of enzymes is in metabolic engineering. Metabolic engineering is a new approach involving the targeted and purposeful manipulation of the metabolic pathways of an organism, aiming at improving the quality and yields of commercially important compounds. It typically involves alteration of cellular activities by manipulation of the enzymatic functions of the cell using recombinant DNA and other genetic techniques. For example, the combination of rational pathway engineering and directed evolution has been successfully applied to optimize the pathways for the production of isoprenoids such as carotenoids (Schmidt-Dannert et al. 2000, Umeno and Arnold 2004).

Table 8.9. Some Important Industrial Enzymes and Their Sources

Enzyme	EC number	Source	Industrial Use
Chymosin	3.4.23.4	Abomasum	Cheese
α-amylase	3.2.1.1	Malted barley, *Bacillus, Aspergillus*	Brewing, baking
β-amylase	3.2.1.2	Malted barley, *Bacillus*	Brewing
Bromelain	3.4.22.4	Pineapple latex	Brewing
Catalase	1.11.1.6	Liver, *Aspergillus*	Food
Penicillin amidase	3.5.1.11	*Bacillus*	Pharmaceutical
Lipoxygenase	1.13.11.12	Soybeans	Food
Ficin	3.4.22.3	Fig latex	Food
Pectinase	3.2.1.15	*Aspergillus*	Drinks
Invertase	3.2.1.26	*Saccharomyces*	Confectionery
Pectin lyase	4.2.2.10	*Aspergillus*	Drinks
Cellulase	3.2.1.4	*Trichoderma*	Waste
Chymotrypsin	3.4.21.1	Pancreas	Leather
Lipase	3.1.1.3	Pancreas, *Rhizopus, Candida*	Food
Trypsin	3.4.21.4	Pancreas	Leather
β-glucanase	3.2.1.6	Malted barley	Brewing
Papain	3.4.22.2	Pawpaw latex	Meat
Asparaginase	3.5.1.1	*E. chrisanthemy, E. carotovora, Escherichia coli*	Human health
Xylose isomerase	5.3.1.5	*Bacillus*	Fructose syrup
Protease	3.4.21.14	*Bacillus*	Detergent
Aminoacylase	3.5.1.14	*Aspergillus*	Pharmaceutical
Raffinase	3.2.1.22	*Saccharomyces*	Food
Glucose oxidase	1.1.3.4	*Aspergillus*	Food
Dextranase	3.2.1.11	*Penicillium*	Food
Lactase	3.2.1.23	*Aspergillus*	Dairy
Glucoamylase	3.2.1.3	*Aspergillus*	Starch
Pullulanase	3.2.1.41	*Klebsiella*	Starch
Raffinase	3.2.1.22	*Mortierella*	Food
Lactase	3.2.1.23	*Kluyveromyces*	Dairy

The new era of the enzyme technology industry is growing at a constant rate. The potential economic, social, and health benefits that may be derived from this industry are unforeseen, and therefore future development of enzyme products will be unlimited.

REFERENCES

Abdul MM. 1993. Biocatalysis and immobilized enzyme/cell bioreactors. Promising techniques in bioreactor technology. Biotechnology (NY) 11:690–695.

Adamson JG, Zhou NE, Hodges RS. 1993. Structure, function and application of the coiled-coil protein folding motif. Curr Opin Biotechnol 4:428–437.

Agarwal PK, Billeter SR, Rajagopalan PT, Benkovic SJ, Hammes-Schiffer S. 2002. Network of coupled promoting motions in enzyme catalysis. Proc Natl Acad Sci USA 99:2794–2799.

Ahern KG, Howard PK, Firtel RA. 1988. Identification of regions essential for extrachromosomal replication and maintenance of an endogenous plasmid in *Dictyostelium.* Nucleic Acids Res 16:6825–6837.

Altmann, F, Staudacher E, Wilson IB, Marz L. 1999. Insect cells as hosts for the expression of recombinant glycoproteins. Glycoconj J 16:109–123.

Altreuter DH, Clark DS. 1999. Combinatorial biocatalysis: Taking the lead from nature. Curr Opin Biotechnol 10:130–136.

Andersen CA, Rost B. 2003. Secondary structure assignment. Methods Biochem Anal 44:341–363.

Anderson CM, Zucker FH, Steitz TA. 1979. Space-filling models of kinase clefts and conformation changes. Science 204:375–380.

Anderson NL, Anderson NG. 1998. Proteome and proteomics: New technologies, new concepts, and new words. Electrophoresis 19:1853–1861.

Anderson NL, Matheson AD, Steiner S. 2000. Proteomics: Applications in basic and applied biology. Curr Opin Biotechnol 11:408–412.

Andrade MA, Sander C. 1997. Bioinformatics: From genome data to biological knowledge. Curr Opin Biotechnol 8:675–683.

Apweiler R. 2001. Functional information in SWISS-PROT: The basis for large-scale characterisation of protein sequences. Brief Bioinform 2:9–18.

Apweiler R, Bairoch A, Wu CH. 2004. Protein sequence databases. Curr Opin Chem Biol 8:76–80.

Archer DB. 1994. Enzyme production by recombinant *Aspergillus*. Bioprocess Technol 19:373–393.

Archer DB, Peberdy JF. 1997. The molecular biology of secreted enzyme production by fungi. Crit Rev Biotechnol 17:273–306.

Arnold FH. 2001. Combinatorial and computational challenges for biocatalyst design. Nature 409:253–257.

Balabushevich NG, Zimina EP, Larionova NI. 2004. Encapsulation of catalase in polyelectrolyte microspheres composed of melamine formaldehyde, dextran sulfate, and protamine. Biochemistry (Moscow) 69:763–769.

Balbas P. 2001. Understanding the art of producing protein and nonprotein molecules in Escherichia coli. Mol Biotechnol 19:251–267.

Balland A, Courtney M, Jallat S, Tessier LH, Sondermeyer P, de la Salle, Harvey R, Degryse E, Tolstoshev P. 1985. Use of synthetic oligonucleotides in gene isolation and manipulation. Biochimie 67:725–736.

Bartlett GJ, Porter CT, Borkakoti N, Thornton JM. 2002. Analysis of catalytic residues in enzyme active sites. J Mol Biol 324:105–121.

Bauer C, Osman AM, Cercignani G, Gialluca N, Paolini M. 2001. A unified theory of enzyme kinetics based upon the systematic analysis of the variations of k(cat), K(M), and k(cat)/K(M) and the relevant DeltaG (0 not equal) values—possible implications in chemotherapy and biotechnology. Biochem Pharmacol 61:1049–1055.

Baumert TF, Ito S, Wong DT, Liang TJ. 1998. Hepatitis C virus structural proteins assemble into viruslike particles in insect cells. J Virol 72:3827–3836.

Bendig MM. 1988. The production of foreign proteins in mammalian cells. Genet Eng 91–127.

Benkovic SJ, Hammes-Schiffer S. 2003. A perspective on enzyme catalysis. Science 301:1196–1202.

Berka R.M, Barnett CC. 1989. The development of gene expression systems for filamentous fungi. Biotechnol Adv 7:127–154.

Berthold W, Walter J. 1994. Protein purification: Aspects of processes for pharmaceutical products. Biologicals 22:135–150.

Bornscheuer UT, Pohl M. 2001. Improved biocatalysts by directed evolution and rational protein design. Curr Opin Chem Biol 5:137–143.

Breitling R, Klingner S, Callewaert N, Pietrucha R, Geyer A, Ehrlich G, Hartung R, Muller A, Contreras R, Beverley SM, Alexandrov K. 2002. Nonpathogenic trypanosomatid protozoa as a platform for protein research and production. Protein Expr Purif 25:209–218.

Cahill DJ, Nordhoff E. 2003. Protein arrays and their role in proteomics. Adv Biochem Eng Biotechnol 83:177–187.

Calleri E, Temporini C, Massolini G, Caccialanza G. 2004. Penicillin G acylase-based stationary phases: Analytical applications. J Pharm Biomed Anal 35: 243–258.

Cantor CR. 1980. The conformation of biological macromolecules. W. H. Freeman, San Francisco.

Cereghino GP, Cereghino JL, Ilgen C, Cregg JM. 2002. Production of recombinant proteins in fermenter cultures of the yeast *Pichia pastoris*. Curr Opin Biotechnol 13:329-332.

Chen R. 2001. Enzyme engineering: Rational redesign versus directed evolution. Trends Biotechnol 19:13–14.

Chen W, Bruhlmann F, Richins RS, Mulchandani A. 1999. Engineering of improved microbes and enzymes for bioremediation. Curr Opin Biotechnol 10: 137–141.

Chin JW, Cropp TA, Anderson JC, Mukherji M, Zhang Z, Schultz PG. 2003. An expanded eukaryotic genetic code. Science 301:964–967.

Choi JH, Lee SY. 2004. Secretory and extracellular production of recombinant proteins using *Escherichia coli*. Appl Microbiol Biotechnol 64:625–635.

Clonis,YD, Labrou NE, Kotsira VP, Mazitsos C, Melissis S, Gogolas G. 2000. Biomimetic dyes as affinity chromatography tools in enzyme purification. J Chromatogr A 891:33–44.

Contreras R, Carrez D, Kinghorn JR, van den Hondel CA, Fiers W. 1991. Efficient KEX2-like processing of a glucoamylase-interleukin-6 fusion protein by *Aspergillus nidulans* and secretion of mature interleukin-6. Biotechnology (NY).

Cramer CL, Weissenborn DL, Oishi KK, Grabau EA, Bennett S, Ponce E, Grabowski GA, Radin DN. 1996. Bioproduction of human enzymes in transgenic tobacco. Ann NY Acad Sci 792:62–71.

Crameri A, Raillard SA, Bermudez E, Stemmer WP. 1998. DNA shuffling of a family of genes from diverse species accelerates directed evolution. Nature 391:288–291.

Cregg JM, Cereghino JL, Shi J, Higgins DR. 2000. Recombinant protein expression in *Pichia pastoris*. Mol Biotechnol 16:23–52.

Cutler P. 2003. Protein arrays: The current state-of-the-art. Proteomics. 3:3–18.

Daniell H, Streatfield SJ, Wycoff K. 2001. Medical molecular farming: Production of antibodies, biopharmaceuticals and edible vaccines in plants. Trends Plant Sci 6:219–226.

Deiters A, Cropp TA, Mukherji M, Chin JW, Anderson JC, Schultz PG. 2003. Adding amino acids with novel reactivity to the genetic code of *Saccharomyces cerevisiae*. J Am Chem Soc 125:11782–11783.

DeYoung JL. 1989. Development of pancreatic enzyme microsphere technology and US findings with Pancrease in the treatment of chronic pancreatitis. Int J Pancreatol 5(Suppl): 31–36.

Dittrich W, Williams KL, Slade MB. 1994. Production and secretion of recombinant proteins in *Dictyostelium discoideum*. Biotechnology (NY) 12:614–618.

Dracheva S, Palermo RE, Powers GD, Waugh DS. 1995. Expression of soluble human interleukin-2 receptor alpha-chain in *Escherichia coli*. Protein Expr Purif 6:737–747.

Dravis BC, Swanson PE, Russell AJ. 2001. Haloalkane hydrolysis with an immobilized haloalkane dehalogenase. Biotechnol Bioeng 75:416–423.

Edmunds T, Van Patten SM, Pollock J, Hanson E, Bernasconi R, Higgins E, Manavalan P, Ziomek C, Meade H, McPherson JM, Cole ES. 1998. Transgenically produced human antithrombin: Structural and functional comparison to human plasma-derived antithrombin. Blood 91:4561–4571.

Eisenmesser EZ, Bosco DA, Akke M, Kern D. 2002. Enzyme dynamics during catalysis. Science 295: 1520–1523.

Estape D, van den HJ, Rinas U. 1998. Susceptibility towards intramolecular disulphide-bond formation affects conformational stability and folding of human basic fibroblast growth factor. Biochem J 335 (Pt 2): 343–349.

Fabian JR, Kimball SR, Jefferson LS. 1998. Reconstitution and purification of eukaryotic initiation factor 2B (eIF2B) expressed in Sf21 insect cells. Protein Expr Purif 13:16–22.

Fang Q, Shortle D. 2003. Prediction of protein structure by emphasizing local side-chain/backbone interactions in ensembles of turn fragments. Proteins 53(Suppl 6): 486–490.

Fausnaugh JL. 1990. Protein purification and analysis by liquid chromatography and electrophoresis. Bioprocess Technol 7:57–84.

Fenyo D, Beavis RC. 2002. Informatics and data management in proteomics. Trends Biotechnol 20:S35–S38.

Fersht A. 1999. Structure and mechanism in protein science: A guide to enzyme catalysis and protein folding. W.H. Freeman, New York.

Fischer R, Drossard J, Commandeur U, Schillberg S, Emans N. 1999a. Towards molecular farming in the future: Moving from diagnostic protein and antibody production in microbes to plants. Biotechnol Appl Biochem 30(Pt 2): 101–108.

Fischer R, Drossard J, Emans N, Commandeur U, Hellwig S. 1999b. Towards molecular farming in the future: *Pichia pastoris*-based production of single-chain antibody fragments. Biotechnol Appl Biochem 30(Pt 2): 117–120.

Fischer R, Emans N. 2000. Molecular farming of pharmaceutical proteins. Transgenic Res 9:279–299.

Fischer R, Hoffmann K, Schillberg S, and Emans N. 2000. Antibody production by molecular farming in plants. J Biol Regul Homeost Agents 14:83–92.

Fischer R, Liao YC, Hoffmann K, Schillberg S, Emans N. 1999c. Molecular farming of recombinant antibodies in plants. Biol Chem 380:825–839.

Fischer R, Vaquero-Martin C, Sack M, Drossard J, Emans N, Commandeur U. 1999d. Towards molecular farming in the future: Transient protein expression in plants. Biotechnol Appl Biochem 30(Pt 2): 113–116.

Fox SR, Patel UA, Yap MG, Wang DI. 2004. Maximizing interferon-gamma production by Chinese hamster ovary cells through temperature shift optimization: Experimental and modeling. Biotechnol Bioeng 85:177–184.

Garnier J, Levin JM, Gibrat JF, Biou V. 1990. Secondary structure prediction and protein design. Biochem Soc Symp 57:11–24.

Garvey EP, Matthews CR. 1990. Site-directed mutagenesis and its application to protein folding. Biotechnology 14:37–63.

Gasteiger E, Gattiker A, Hoogland C, Ivanyi I, Appel RD, Bairoch A. 2003. ExPASy: The proteomics server for in-depth protein knowledge and analysis. Nucleic Acids Res 31:3784–3788.

Gasteiger E, Jung E, Bairoch A. 2001. SWISS-PROT: Connecting biomolecular knowledge via a protein database. Curr Issues Mol Biol 3:47–55.

Giga-Hama, Y, Kumagai H. 1999. Expression system for foreign genes using the fission yeast *Schizosaccharomyces pombe*. Biotechnol Appl Biochem 30(Pt 3): 235–244.

Gomord V, Faye L. 2004. Posttranslational modification of therapeutic proteins in plants. Curr Opin Plant Biol 7:171–181.

Gouka RJ, Punt PJ, van den Hondel CA. 1997. Efficient production of secreted proteins by *Aspergillus:* Progress, limitations and prospects. Appl Microbiol Biotechnol 47:1–11.

Guex N, Peitsch MC. 1997. SWISS-MODEL and the Swiss-PdbViewer: An environment for comparative protein modeling. Electrophoresis 18:2714–2723.

Gutteridge A, Thornton J. 2004. Conformational change in substrate binding, catalysis and product release: An open and shut case? FEBS Lett 567:67–73.

Hackney D. 1990. Binding energy and catalysis. In: The Enzymes, vol. XIX. Academic Press. P. 136.

Hamilton SR, Bobrowicz P, Bobrowicz B, Davidson RC, Li H, Mitchell T, Nett JH, Rausch S, Stadheim TA, Wischnewski H, Wildt S, Gerngross TU. 2003. Production of complex human glycoproteins in yeast. Science 301:1244–1246.

Hammes GG. 2002. Multiple conformational changes in enzyme catalysis. Biochemistry 41:8221–8228.

Hasemann CA, Capra JD. 1990. High-level production of a functional immunoglobulin heterodimer in a baculovirus expression system. Proc Natl Acad Sci USA 87:3942–3946.

Holz C, Prinz B, Bolotina N, Sievert V, Bussow K, Simon B, Stahl U, Lang C. 2003. Establishing the yeast *Saccharomyces cerevisiae* as a system for expression of human proteins on a proteome-scale. J Struct Funct Genomics 4:97–108.

Holzer GW, Mayrhofer J, Leitner J, Blum M, Webersinke G, Heuritsch S, Falkner FG. 2003. Overexpression of hepatitis B virus surface antigens including the preS1 region in a serum-free Chinese hamster ovary cell line. Protein Expr Purif 29:58–69.

Hsieh-Wilson LC, Schultz PG, Stevens RC. 1996. Insights into antibody catalysis: Structure of an oxygenation catalyst at 1.9-angstrom resolution. Proc Natl Acad Sci USA 93:5363–5367.

Humphreys DP. 2003. Production of antibodies and antibody fragments in *Escherichia coli* and a comparison of their functions, uses and modification. Curr Opin Drug Discov Devel 6:188–196.

Ikeno M, Grimes B, Okazaki T, Nakano M, Saitoh K, Hoshino H, McGill NI, Cooke H, Masumoto H. 1998. Construction of YAC-based mammalian artificial chromosomes. Nat Biotechnol 16:431–439.

Janne J, Alhonen L, Hyttinen JM, Peura T, Tolvanen M, Korhonen VP. 1998. Transgenic bioreactors. Biotechnol Annu Rev 4:55–74.

Joseph D, Petsko GA, Karplus M. 1990. Anatomy of a conformational change: Hinged "lid" motion of the triosephosphate isomerase loop. Science 249:1425–1428.

Jung E, Williams KL. 1997. The production of recombinant glycoproteins with special reference to simple eukaryotes including *Dictyostelium discoideum*. Biotechnol Appl Biochem 25(Pt 1): 3–8.

Kabsch W, Sander C. 1983. Dictionary of protein secondary structure: Pattern recognition of hydrogen-bonded and geometrical features. Biopolymers 22: 2577–2637.

Kakkis ED, Matynia A, Jonas AJ, Neufeld EF. 1994. Overexpression of the human lysosomal enzyme alpha-L-iduronidase in Chinese hamster ovary cells. Protein Expr Purif 5:225–232.

Kaszubska W, Zhang H, Patterson RL, Suhar TS, Uchic ME, Dickinson RW, Schaefer VG, Haasch, Janis RS, DeVries PJ, Okasinski GF, Meuth JL. 2000. Expression, purification, and characterization of human recombinant thrombopoietin in Chinese hamster ovary cells. Protein Expr Purif 18:213–220.

Ke SH, Madison EL. 1997. Rapid and efficient site-directed mutagenesis by single-tube "megaprimer" PCR method. Nucleic Acids Res 25:3371–3372.

Kirk O, Borchert TV, Fuglsang CC. 2002. Industrial enzyme applications. Curr Opin Biotechnol 13:345–351.

Kopp J, Schwede T. 2004. Automated protein structure homology modeling: A progress report. Pharmacogenomics. 5:405–416.

Kost TA, Condreay JP. 1999. Recombinant baculoviruses as expression vectors for insect and mammalian cells. Curr Opin Biotechnol 10:428–433.

———. 2002. Recombinant baculoviruses as mammalian cell gene-delivery vectors. Trends Biotechnol 20:173–180.

Kukuruzinska MA, Bergh ML, Jackson BJ. 1987. Protein glycosylation in yeast. Annu Rev Biochem 56:915–944.

Kurtzman AL, Govindarajan S, Vahle K, Jones JT, Heinrichs V, Patten PA. 2001. Advances in directed protein evolution by recursive genetic recombination: Applications to therapeutic proteins. Curr Opin Biotechnol 12:361–370.

La Flamme AC, Buckner FS, Swindle J, Ajioka J, Van Voorhis WC. 1995. Expression of mammalian cytokines by *Trypanosoma cruzi* indicates unique signal sequence requirements and processing. Mol Biochem Parasitol 75:25–31.

Labrou NE. 2000. Dye-ligand affinity chromatography for protein separation and purification. Methods Mol Biol 147:129–139.

———. 2002. Affinity Chromatography. In: MN Gupta, editor, Methods for Affinity-Based Separations of Enzymes and Proteins. Birkhiuser Verlag AG, Switzerland. Pp. 16–18.

———. 2003. Design and selection of ligands for affinity chromatography. J Chromatogr B Analyt Technol Biomed Life Sci 790:67–78.

Labrou NE, Clonis YD. 1994. The affinity technology in downstream processing. J Biotechnol 36:95–119.

———. 1995. Biomimetic dye affinity chromatography for the purification of bovine heart lactate dehydrogenase. J Chromatogr A 718:35–44.

———. 1996. Biomimetic-dye affinity chromatography for the purification of mitochondrial L-malate dehydrogenase from bovine heart. J Biotechnol 45: 185–194.

———. 2002. Chapter 8. Immobilised synthetic dyes in affinity chromatography. In: MA Vijayalakshmi, editor, Biochromatography–Theory and Practice. Taylor and Francis Publishers, London. Pp. 235–251.

Labrou, NE, Mazitsos K, Clonis Y. 2004. Dye-ligand and Biomimetic Affinity Chromatography. In: DS Hage, editor, Handbook of Affinity Chromatography. Marcel Dekker, Inc., New York (in press).

Labrou NE, Mello LV, Clonis YD. 2001. The conserved Asn49 of maize glutathione S-transferase I modulates substrate binding, catalysis and intersubunit communication. Eur J Biochem 268:3950–3957.

Labrou NE, Rigden DJ. 2001. Active-site characterization of *Candida boidinii* formate dehydrogenase. Biochem J 354:455–463.

Larrick JW, Thomas DW. 2001. Producing proteins in transgenic plants and animals. Curr Opin Biotechnol 12:411–418.

Lesley SA. 2001. High-throughput proteomics: Protein expression and purification in the postgenomic world. Protein Expr Purif 22:159–164.

Lesley SA, Graziano J, Cho CY, Knuth MW, Klock HE. 2002. Gene expression response to misfolded protein as a screen for soluble recombinant protein. Protein Eng 15:153–160.

Levison PR. 2003. Large-scale ion-exchange column chromatography of proteins. Comparison of different formats. J Chromatogr B Analyt Technol Biomed Life Sci 790:17–33.

Lewin R. 1982. RNA can be a catalyst. Science 218: 872–874.

Liang JF, Li YT, Yang VC. 2000. Biomedical application of immobilized enzymes. J Pharm Sci 89:979–990.

Logan JS, Martin MJ. 1994. Transgenic swine as a recombinant production system for human hemoglobin. Methods Enzymol 231:435–445.

Lowe CR. 2001. Combinatorial approaches to affinity chromatography. Curr Opin Chem Biol 5:248–256.

Lue RY, Chen GY, Hu Y, Zhu Q, Yao SQ. 2004. Versatile protein biotinylation strategies for potential high-throughput proteomics. J Am Chem Soc 126: 1055–1062.

Lutz S, Ostermeier M, Benkovic SJ. 2001. Rapid generation of incremental truncation libraries for protein engineering using alpha-phosphothioate nucleotides. Nucleic Acids Res 29:E16.

Madzak C, Gaillardin C, Beckerich JM. 2004. Heterologous protein expression and secretion in the non-conventional yeast *Yarrowia lipolytica:* A review. J Biotechnol 109:63–81.

Marti S, M Roca M, Andres J, Moliner V, Silla E, Tunon I, Bertran J. 2004. Theoretical insights in enzyme catalysis. Chem Soc Rev 33:98–107.

Mason HS, Arntzen CJ. 1995. Transgenic plants as vaccine production systems. Trends Biotechnol 13:388–392.

Matsuura Y, Possee RD, Overton HA, Bishop DH. 1987. Baculovirus expression vectors: The requirements for high level expression of proteins, including glycoproteins. J Gen Virol 68(Pt 5): 1233–1250.

Matthews BW. 1993. Structural and genetic analysis of protein stability. Annu Rev Biochem 62:139–160.

McCormick DB. 1975. Vitamins and Coenzymes, Part L. Methods in Enzymology 282.

Moss DW. 1988. Theoretical enzymology: Enzyme kinetics and enzyme inhibition. In: DL Williams, V Marks, editors, Principles of Clinical Biochemistry, 2nd edition. Bath Press. 423–440.

Murzin AG, Brenner SE, Hubbard T, Chothia C. 1995. SCOP: A structural classification of proteins database for the investigation of sequences and structures. J Mol Biol 247:536–540.

Nixon AE, Ostermeier M, Benkovic SJ. 1998. Hybrid enzymes: Manipulating enzyme design. Trends Biotechnol. 16:258–264.

Orengo CA, Michie AD, Jones S, Jones DT, Swindells MB, Thornton JM. 1997. CATH—A hierarchic classification of protein domain structures. Structure. 5: 1093–1108.

Ozturk DH, Erickson-Viitanen S. 1998. Expression and purification of HIV-I p15NC protein in *Escherichia coli*. Protein Expr Purif 14:54–64.

Panda AK. 2003. Bioprocessing of therapeutic proteins from the inclusion bodies of *Escherichia coli*. Adv Biochem Eng Biotechnol 85:43–93.

Panke S, Wubbolts MG. 2002. Enzyme technology and bioprocess engineering. Curr Opin Biotechnol 13: 111–116.

Phadke RS. 1992. Biosensors and enzyme immobilized electrodes. Biosystems 27:203–206.

Piefer AJ, Jonsson CB. 2002. A comparative study of the human T-cell leukemia virus type 2 integrase expressed in and purified from *Escherichia coli* and *Pichia pastoris*. Protein Expr Purif 25:291–299.

Platis D, Foster GR. 2003. High yield expression, refolding, and characterization of recombinant interferon alpha2/alpha8 hybrids in *Escherichia coli*. Protein Expr Purif 31:222–230.

Podgornik A, Tennikova TB. 2002. Chromatographic reactors based on biological activity. Adv. Biochem Eng Biotechnol 76:165–210.

Powell LW. 1984. Developments in immobilized-enzyme technology. Biotechnol Genet Eng Rev 2: 409–438.

Price N, Stevens L. 1999. Fundamentals of Enzymology, 3rd edition. Oxford University Press. Pp. 79–92.

Prunkard D., Cottingham I, Garner I, Bruce S, Dalrymple M, Lasser G, Bishop P, Foster D. 1996. High-level expression of recombinant human fibrinogen in the milk of transgenic mice. Nat Biotechnol 14:867–871.

Regnier FE. 1987. Chromatography of complex protein mixtures. J Chromatogr 418:115–143.

Richardson JS. 1981. The anatomy and taxonomy of protein structure. Adv Protein Chem 34:167–339.

Roy P, Bishop DH, LeBlois H, Erasmus BJ. 1994. Long-lasting protection of sheep against bluetongue challenge after vaccination with virus-like particles: Evidence for homologous and partial heterologous protection. Vaccine 12:805–811.

Rudolph NS. 1999. Biopharmaceutical production in transgenic livestock. Trends Biotechnol 17:367–374.

Russell DA. 1999. Feasibility of antibody production in plants for human therapeutic use. Curr Top Microbiol Immunol 240:119–138.

Ryu DD, Nam DH. 2000. Recent progress in biomolecular engineering. Biotechnol Prog 16:2–16.

Sala F, Manuela RM, Barbante A, Basso B, Walmsley AM, Castiglione S. 2003. Vaccine antigen production in transgenic plants: strategies, gene constructs and perspectives. Vaccine 21:803–808.

Schatz SM, Kerschbaumer RJ, Gerstenbauer G, Kral M, Dorner F, Scheiflinger F. 2003. Higher expression of Fab antibody fragments in a CHO cell line at reduced temperature. Biotechnol Bioeng 84:433–438.

Schena M, Heller RA, Theriault TP, Konrad K, Lachenmeier E, Davis RW. 1998. Microarrays: Biotechnology's discovery platform for functional genomics. Trends Biotechnol 16:301–306.

Schena M, Shalon D, Davis RW, Brown PO. 1995. Quantitative monitoring of gene expression patterns with a complementary DNA microarray. Science 270:467–470.

Schillberg S, Fischer R, Emans N. 2003a. 'Molecular farming' of antibodies in plants. Naturwissenschaften 90:145–155.

———. 2003b. Molecular farming of recombinant antibodies in plants. Cell Mol Life Sci 60:433–445.

Schmidt-Dannert C, Umeno D, Arnold FH. 2000. Molecular breeding of carotenoid biosynthetic pathways. Nat Biotechnol 18:750–753.

Schwede T, Kopp J, Guex N, Peitsch MC. 2003. SWISS-MODEL: An automated protein homology-modeling server. Nucleic Acids Res 31:3381–3385.

Scopes RK. 1987. Dye-ligands and multifunctional adsorbents: An empirical approach to affinity chromatography. Anal Biochem 165:235–246.

Scrutton NS, Berry A, Perham RN. 1990. Redesign of the coenzyme specificity of a dehydrogenase by protein engineering. Nature 343:38–43.

Sgaramella V, Eridani S. 2004. From natural to artificial chromosomes: An overview. Methods Mol Biol 240:1–12.

Shao Z, Zhao H, Giver L, Arnold FH. 1998. Random-priming in vitro recombination: An effective tool for directed evolution. Nucleic Acids Res 26:681–683.

Sibanda BL, Blundel TL, Thornton JM. 1989. Conformation of beta-hairpins in protein structures. A systematic classification with applications to modelling by homology, electron density fitting and protein engineering. J Mol Biol 206:759–777.

Singh A, Hayashi K, Hoa TT, Kashiwagi Y, Tokuyasu K. 1995. Construction and characterization of a chimeric beta-glucosidase Biochem J 305 (Pt 3):715–719.

Slade MB, Emslie KR, Williams KL. 1997. Expression of recombinant glycoproteins in the simple eukaryote *Dictyostelium discoideum*. Biotechnol. Genet Eng Rev 14:1–35.

Steine S, Anderson NL. 2000. Pharmaceutical proteomics. Ann NY Acad Sci 919:48–51.

Stemmer WP. 1994. Rapid evolution of a protein in vitro by DNA shuffling. Nature 370:389–391.

Stoge, E, Sack M, Fischer R, Christou P. 2002. Plantibodies: Applications, advantages and bottlenecks. Curr Opin Biotechnol 13:161–166.

Surewicz WK., Mantsch HH. 1988. New insight into protein secondary structure from resolution-enhanced infrared spectra. Biochim Biophys Acta 952:115–130.

Swanson ME, Martin MJ, O'Donnell JK, Hoover K, Lago W, Huntress V, Parsons CT, Pinkert CA, Pilder S, Logan JS. 1992. Production of functional human hemoglobin in transgenic swine. Biotechnology (NY) 10:557–559.

Swartz JR. 2001. Advances in *Escherichia coli* production of therapeutic proteins. Curr Opin Biotechnol 12:195–201.

Terpe K. 2003. Overview of tag protein fusions: From molecular and biochemical fundamentals to commercial systems. Appl Microbiol Biotechnol 60: 523–533.

Tiltscher H, Storr M. 1993. Immobilization of the slime mould *Dictyostelium discoideum* for the continuous production of recombinant human antithrombin III. Appl Microbiol Biotechnol 40:246–250.

Tobin JF, Reiner SL, Hatam F, Zheng S, Leptak CL, Wirth DF, Locksley RM. 1993. Transfected *Leishmania* expressing biologically active IFN-gamma. J Immunol 150:5059–5069.

Trueman LJ. 1995. Heterologous expression in yeast. Methods Mol Biol 49:341–354.

Twyman RM, Stoger E, Schillberg S, Christou P, Fischer R. 2003. Molecular farming in plants: Host systems and expression technology. Trends Biotechnol 21:570–578.

Umeno D, Arnold FH. 2004. Evolution of a pathway to novel long-chain carotenoids. J Bacteriol 186:1531–1536.

van Beilen JB, Li Z. 2002. Enzyme technology: An overview. Curr. Opin. Biotechnol. 13:338–344.

van Berke PH, Welling MM, Geerts M, van Veen HA, Ravensbergen B, Salaheddine M, Pauwels EK, Pieper F, Nuijens JH, Nibbering PH. 2002. Large scale production of recombinant human lactoferrin in the milk of transgenic cows. Nat Biotechnol 20:484–487.

van den Hombergh JP, van de Vondervoort PJ, Fraissinet-Tachet L, Visser J. 1997. *Aspergillus* as a host for heterologous protein production: The problem of proteases. Trends Biotechnol 15:256–263.

Velander WH, Page RL, Morcol T, Russell CG, Canseco R, J.M. Young JM, Drohan WN, Gwazdauskas FC, Wilkins TD, Johnson JL. 1992. Production of biologically active human protein C in the milk of transgenic mice. Ann NY Acad Sci 665:391–403.

Verdoes JC, Punt PJ, Schrickx JM, van Verseveld HW, Stouthamer AH, van den Hondel CA. 1993. Glucoamylase overexpression in *Aspergillus niger*: Molecular genetic analysis of strains containing multiple copies of the glaA gene. Transgenic Res 2:84–92.

Wagner CR, Benkovic SJ. 1990. Site directed mutagenesis: A tool for enzyme mechanism dissection. Trends Biotechnol 8:263–270.

Walsh G. 2003. Pharmaceutical biotechnology products approved within the European Union. Eur J Pharm Biopharm 55:3–10.

Wang L, Schultz PG. 2002. Expanding the genetic code. Chem Commun (Cambridge) 1–11.

Ward M, Wilson LJ, Kodama KH, Rey MW, Berka RM. 1990. Improved production of chymosin in *Aspergillus* by expression as a glucoamylase-chymosin fusion. Biotechnology (NY) 8:435–440.

Ward PP, Piddington CS, Cunningham GA, Zhou X, Wyatt RD, Conneely OM. 1995. A system for production of commercial quantities of human lactoferrin: A broad spectrum natural antibiotic. Biotechnology (NY) 13:498–503.

Wardell AD, Errington W, Ciaramella G, Merson J, McGarvey MJ. 1999. Characterization and mutational analysis of the helicase and NTPase activities of hepatitis C virus full-length NS3 protein. J Gen Virol 80(Pt 3): 701–709.

Watson JV, Dive C. 1994. Enzyme kinetics. Methods Cell Biol 41:469–507.

Wharton CW. 1983. Some recent advances in enzyme kinetics. Biochem Soc Trans 11:817–825.

Wilkins MR, Gasteiger E, Bairoch A, Sanchez JC, Williams KL, Appel RD, Hochstrasser DF. 1999. Protein identification and analysis tools in the ExPASy server. Methods Mol Biol 112:531–552.

Wolfenden R. 2003. Thermodynamic and extrathermodynamic requirements of enzyme catalysis. Biophys Chem 105:559–572.

Yeo DS, Panicker RC, Tan LP, Yao SQ. 2004. Strategies for immobilization of biomolecules in a microarray. Comb Chem High Throughput Screen 7:213–221.

Yeung PK. 2000. Technology evaluation: Transgenic antithrombin III (rhAT-III). Genzyme Transgenics Curr Opin Mol Ther 2:336–339.

Zaks A. 2001. Industrial biocatalysis. Curr Opin Chem Biol 5:130–136.

Zhang Z, Gildersleeve J, Yang YY, Xu R, Loo JA, Uryu S, Wong CH, Schultz PG. 2004. A new strategy for the synthesis of glycoproteins. Science 303:371–373.

Zhao H, Giver L, Shao Z, Affholter JA, Arnold FH. 1998. Molecular evolution by staggered extension process (StEP) in vitro recombination. Nat Biotechnol 16:258–261.

Zweiger G, Scott RW. 1997. From expressed sequence tags to 'epigenomics': An understanding of disease processes. Curr Opin Biotechnol 8:684–687.

9
Protein Cross-linking in Food

J. A. Gerrard

INTRODUCTION

Protein cross-links play an important role in determining the functional properties of food proteins. Manipulation of the number and nature of protein cross-links during food processing offers a means by which the food industry can manipulate the functional properties of food, often without damaging the nutritional quality. This chapter discusses advances in our understanding of protein cross-linking over the last decade and examines current and future applications of this chemistry in food processing. It builds on, and updates, two recent reviews in this area (Gerrard 2002, Miller and Gerrard 2004) in addition to earlier reviews on this subject (Matheis and Whitaker 1987, Feeney and Whitaker 1988, Singh 1991).

The elusive relationship between the structure and the function of proteins presents a particular challenge for the food technologist. Food proteins are often denatured during processing, so there is a need to understand the protein both as a biological entity with a predetermined function and as a randomly coiled biopolymer. To understand and manipulate food proteins thus requires a knowledge of both protein biochemistry and polymer science. If the protein undergoes chemical reaction during processing, both the "natural" function of the molecule, and the properties of the denatured polymeric state may be influenced. One type of chemical reaction that has major consequences for protein function in either their native or denatured states is protein cross-linking. It is, therefore, no surprise that protein cross-linking can have profound effects on the functional properties of food proteins.

This chapter sets out to define the different types of protein cross-links that can occur in food, before and after processing, and the consequences of these cross-links for the functional and nutritional properties of the foodstuff. Methods that have been employed to introduce cross-links into food deliberately are then reviewed, and future prospects for the use of this chemistry for the manipulation of food during processing are surveyed.

PROTEIN CROSS-LINKS IN FOOD

Protein cross-linking refers to the formation of covalent bonds between polypeptide chains within a protein (intramolecular cross-links) or between proteins (intermolecular cross-links) (Feeney and Whitaker 1988). In biology, cross-links are vital for maintaining the correct conformation of certain proteins, and may control the degree of flexibility of the polypeptide chains. As biological tissues age, further protein cross-links may form that often have deleterious consequences throughout the body, and play an important role in the many conditions of ageing (Zarina et al. 2000). Similar chemistry to that which occurs during ageing may take place if biological tissues are removed from their natural environment—for example, when harvested as food for processing.

Food processing often involves high temperatures, extremes in pH, particularly alkaline, and exposure to oxidizing conditions and uncontrolled enzyme chemistry. Such conditions can result in the introduction of protein cross-links, producing substantial changes in the structure of proteins, and therefore the functional (Singh 1991) and nutritional (Friedman 1999a,b,c) properties of the final product. A summary of protein cross-linking in foods is given in Figure 9.1, in which the information is organized according to the amino acids that react to form the cross-link. Not all amino acids participate in protein cross-linking, no matter how extreme the

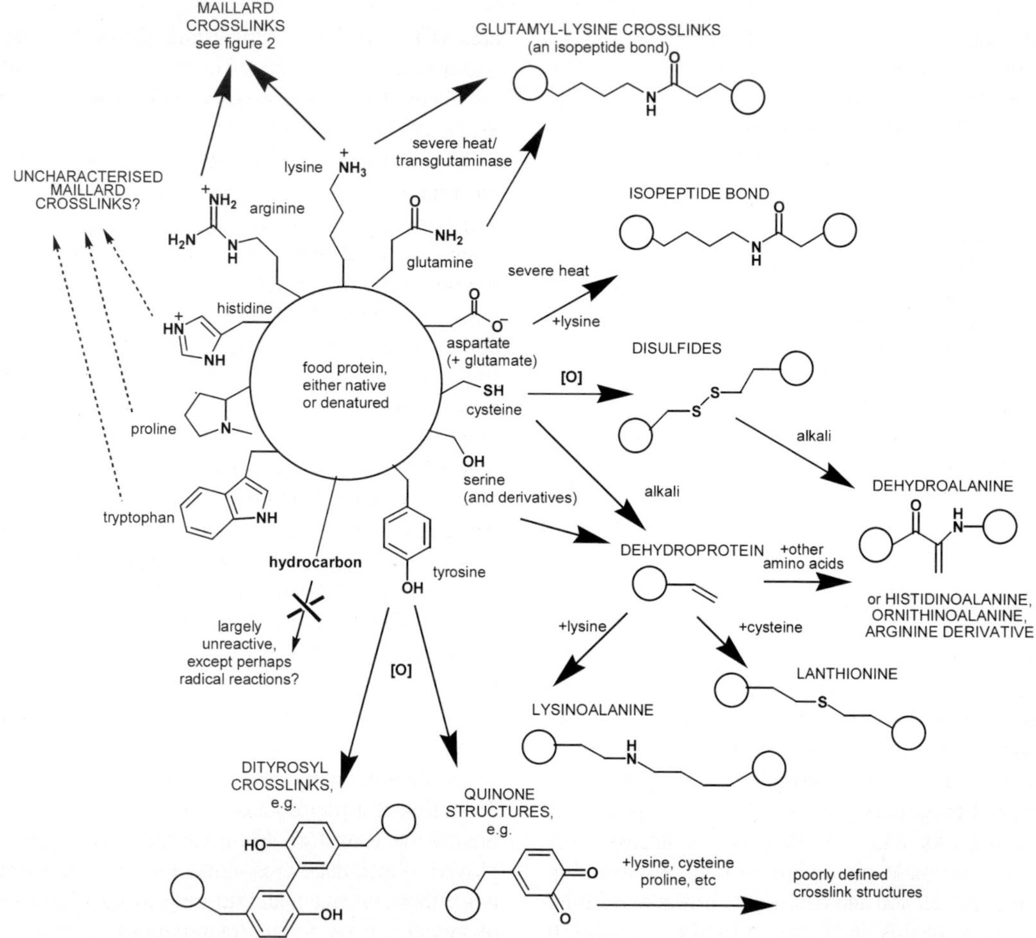

Figure 9.1. A summary of the cross-linking reactions that can occur during food processing, from Gerrard (2002). Further details are given in the text.

processing regime. Those that react do so with differing degrees of reactivity under various conditions.

Disulfide Cross-links

Disulfide bonds are the most common and well-characterized type of covalent cross-link in proteins in biology. They are formed by the oxidative coupling of two cysteine residues that are close in space within a protein. A suitable oxidant accepts the hydrogen atoms from the thiol groups of the cysteine residues, producing disulfide cross-links. The ability of proteins to form intermolecular disulfide bonds during heat treatment is considered to be vital for the gelling of some food proteins, including milk proteins, surimi, soybeans, eggs, meat, and some vegetable proteins (Zayas 1997). Gels are formed through the cross-linking of protein molecules, generating a three-dimensional solid-like network, which provides food with desirable texture (Dickinson 1997).

Disulfide bonds are thought to confer an element of thermal stability to proteins and are invoked, for example, to explain the heat stability of hen egg white lysozyme, which has four intramolecular disulfide cross-links in its native conformation (Masaki et al. 2001). This heat stability influences many of the properties of egg white observed during cooking. Similarly, the heat treatment of milk promotes the controlled interaction of denatured β-lactoglobulin with κ-casein, through the formation of a disulfide bond. This increases the heat stability of milk and milk products, preventing precipitation of β-lactoglobulin (Singh 1991). Disulfide bonds are also important in the formation of dough. Disulfide interchange reactions during the mixing of flour and water result in the production of a protein network with the viscoelastic properties required for bread making (Lindsay and Skerritt 1999). The textural changes that occur in meat during cooking have also been attributed to the formation of intermolecular disulfide bonds (Singh 1991).

Cross-links Derived from Dehydroprotein

Alkali treatment is used in food processing for a number of reasons, such as the removal of toxic constituents, and the solubilization of proteins for the preparation of texturized products. However, alkali treatment can also cause reactions that are undesirable in foods, and its safety has come into question (Friedman 1999a,c; Shih 1992; Savoie 1991). Exposure to alkaline conditions, particularly when coupled to thermal processing, induces racemization of amino acid residues and the formation of covalent cross-links, such as dehydroalanine, lysinoalanine, and lanthionine (Friedman 1999a,b,c). Dehydroalanine is formed from the base-catalyzed elimination of persulfide from an existing disulfide cross-link. The formation of lysinoalanine and lanthionone cross-links occurs through β-elimination of cysteine and phosphoserine protein residues, thereby yielding dehydroprotein residues. Dehydroprotein is very reactive with various nucleophilic groups including the ε-amino group of lysine residues and the sulfhydryl group of cysteine. In severely heat- or alkali-treated proteins, imidazole, indole, and guanidino groups of other amino acid residues may also react (Singh 1991). The resulting intra- and intermolecular cross-links are stable, but food proteins that have been extensively treated with alkali are not readily digested, reducing their nutritional value. Mutagenic products may also be formed (Friedman 1999a,c).

Cross-links Derived from Tyrosine

Various cross-links formed between two or three tyrosine residues have been found in native proteins and glycoproteins, for example in plant cell walls (Singh 1991). Dityrosine cross-links have recently been identified in wheat, and are proposed to play a role in formation of the cross-linked protein network in gluten (Tilley 2001). They have also been formed indirectly, by treating proteins with hydrogen peroxide or peroxidase (Singh 1991), and are implicated in the formation of caseinate films by gamma irradiation (Mezgheni et al. 1998). Polyphenol oxidase can also lead indirectly to protein cross-linking, due to reaction of cysteine, tyrosine, or lysine with reactive benzoquinone intermediates generated from the oxidation of phenolic substrates (Matheis and Whitaker 1987, Feeney and Whitaker 1988). Such plant phenolics have been used to prepare cross-linked gelatin gels to develop novel food ingredients (Strauss and Gibson 2004).

Cross-links Derived from the Maillard Reaction

The Maillard reaction is a complex cascade of chemical reactions, initiated by the deceptively simple condensation of an amine with a carbonyl group, often within a reducing sugar or fat breakdown product (Fayle and Gerrard 2002). During the course of the Maillard reaction, reactive intermediates such as α-dicarbonyl compounds and deoxysones are generated, leading to the production of a wide range of compounds, including polymerized brown pigments called melanoidins, furan derivatives, and nitrogenous and heterocyclic compounds (e.g., pyrazines) (Fayle and Gerrard 2002). Protein cross-links form a subset of the many reaction products, and the cross-linking of food proteins by the Maillard reaction during food processing is well established (Miller and Gerrard 2004). The precise chemical structures of these cross-links in food, however, are less well understood. Thus, surprisingly little is known about the extent of Maillard cross-linking in processed foods, the impact of this process on food quality, and how the reaction might be controlled to maximize food quality.

In biology and medicine, where the Maillard reaction is important during the ageing process, several cross-link structures have been identified, including those shown in Figure 9.2 (Ames 1992; Easa et al. 1996a,b; Fayle et al. 2000, 2001; Gerrard et al. 2002a,b,c; Gerrard et al. 1999; Gerrard et al. 1998a; Hill and Easa 1998; Hill et al. 1993; Mohammed et al. 2000). One of the first protein-derived Maillard reaction products isolated and characterized was the cross-link pentosidine (Sell and Monnier 1989, Dyer et al. 1991), a fluorescent moiety that is believed to form through the condensation of a lysine residue with an arginine residue and a reducing sugar. The exact mechanism of formation of pentosidine remains the subject of considerable debate (Chellan and Nagaraj 2001, Biemel et al. 2001).

Some of the earliest studies that assessed the effect of protein cross-linking on food quality examined the digestibility of a model protein following glycation. Kato et al. observed that following incubation of lysozyme with a selection of dicarbonyl compounds for 10 days, the digestibility of lysozyme by a pepsin-pancreatin solution was reduced to up to 30% relative to the nonglycated sample (Kato et al. 1986a). This trend of decreasing digestibility was concomitant with an observed increase in cross-linking of lysozyme. The increased resistance of cross-linked proteins to enzymes commonly involved in the digestion of proteins in the body is unfavorable from a nutritional standpoint. It has also been reported that the digestion process can be inhibited by the Maillard reaction (Friedman 1996b).

AGE Protein Cross-links Isolated to Date in Food

Information regarding the presence of specific Maillard protein cross-links in food is, to date, limited, with only a handful of studies in this area (Biemel et al. 2001, Henle et al. 1997, Schwarzenbolz et al. 2000). For example, compared to the extensive literature on the Maillard chemistry in vivo, relatively little has been reported on the existence of pentosidine in food. In a study by Henle et al., the pentosidine content of a range of foods was examined, with the highest values observed in roasted coffee (Henle et al. 1997). Overall concentrations, however, were considered low, among most of the commercial food products tested. It was, therefore, concluded that pentosidine does not have a major role in the polymerization of food proteins. Schwarzenbolz et al. showed pentosidine formation in a casein-ribose reaction carried out under high hydrostatic pressure, which could be relevant to some areas of food processing (Schwarzenbolz et al. 2000). Iqbal et al. have investigated the role of pentosidine in meat tenderness in broiler hens (Iqbal et al. 2000, 1999a,b, 1997).

Other Maillard cross-links have recently been detected in food, and attempts have been made to quantify the levels found. Biemel et al. examined the content of the lysine-arginine cross-links GODIC, MODIC, DODIC, and glucosepane in food. These cross-links were found to be present in proteins extracted from biscuits, pretzels, alt stick, and egg white, in the range of 7–151 mg/kg. The lysine-lysine imidazolium cross-links MOLD and GOLD were also isolated but were present at a lower concentrations compared with MODIC and GODIC (Biemel et al. 2001).

Melanoidins

Very advanced glycation end products that form as a result of food processing are dubbed melanoidins. This is a structurally diverse class of compounds, which until recently were very poorly characterized. However, a subset of food melanoidins undoubtedly includes those that cross-link proteins.

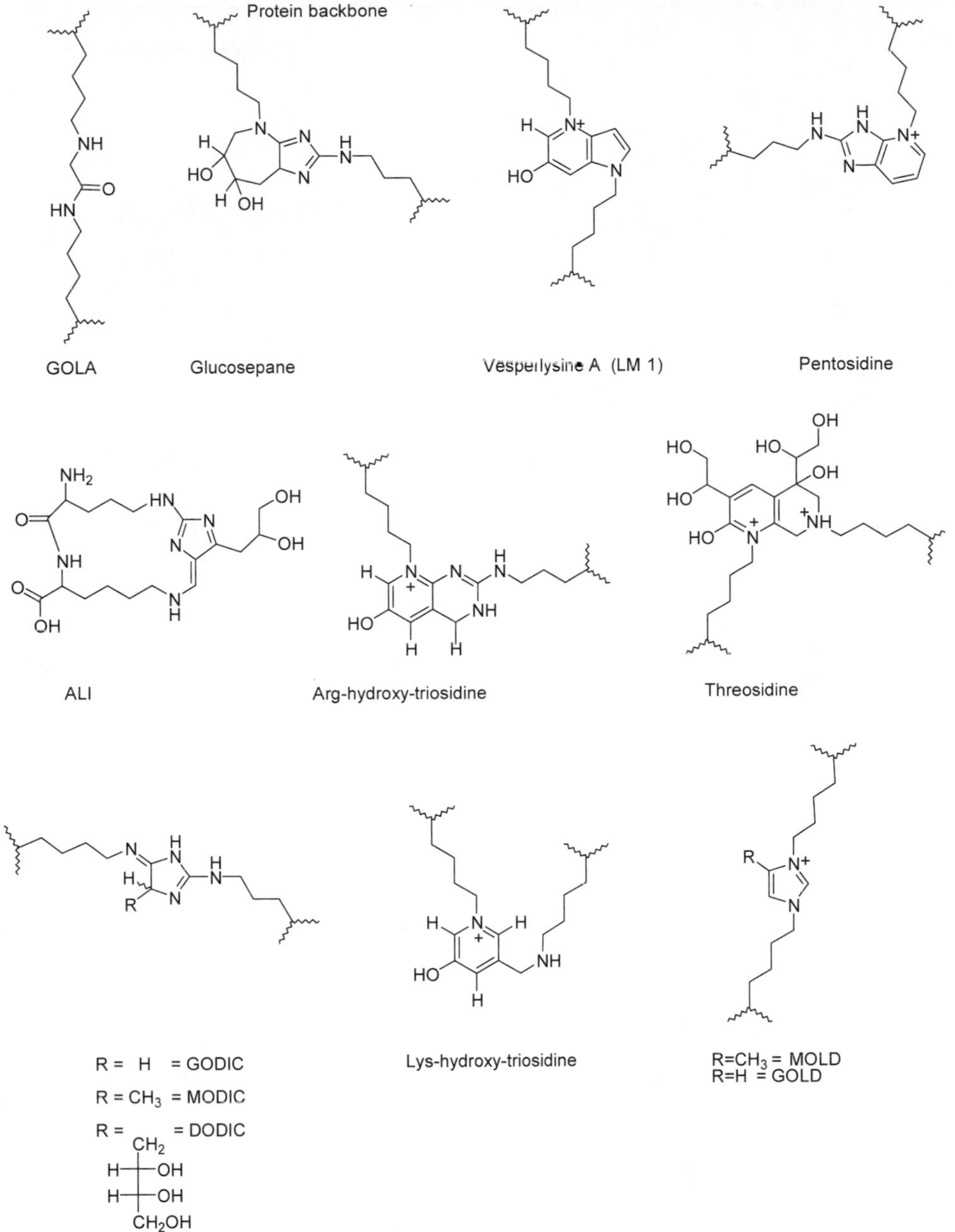

Figure 9.2. A selection of known protein cross-links derived from the Maillard reaction in the body, from Miller and Gerrard (2004). (Al-Abed and Bucala 2000, Brinkmann-Frye et al. 1998, Glomb and Pfahler 2001, Lederer and Buhler 1999, Lederer and Klaiber 1999, Nakamura et al. 1997, Prabhakaram et al. 1997, Sell and Monnier 1989, Tessier et al. 2003). The extent to which these cross-links are present in food is largely undetermined. *Abbreviations/acronyms:* GOLA, N^6-{2-[(5-amino-5-carboxypentyl)amino]-2-oxoethyl}lysine; ALI, arginine-lysine imidazole; GODIC, N^6-(2-{[(4S)-4-ammonio-5-oxido-5-oxopentyl]amino}-3,5-dihydro-4H-imidazol-4-ylidene)-L-lysinate; MODIC, N^6-(2-{[(4S)-4-ammonio-5-oxido-5-oxopentyl]amino}-5 methyl-3,5-dihydro-4H-imidazol-4-ylidene)-L-lysinate; DODIC, N^6-{2-{[(4S)-4-ammonio-5-oxido-5-oxopentyl]amino}-5[(2S,3R)-2,3,4-trihydroxybutyl]-3, 5-dihydro-4H-imidazol-4-ylidene)-L-lysinate; MOLD, methylglyoxal lysine dimer; GOLD, glyoxal lysine dimer.

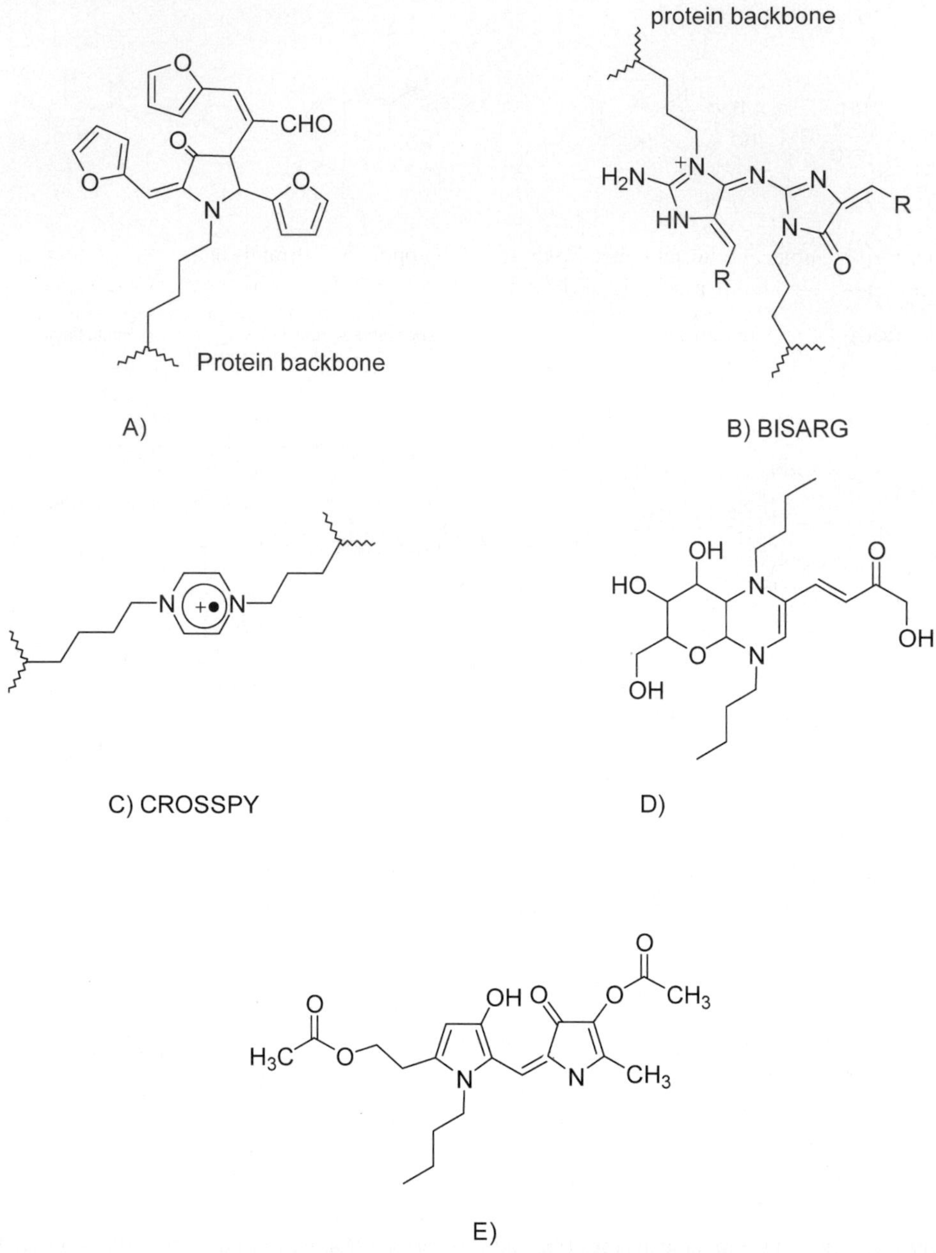

A) B) BISARG

C) CROSSPY D)

E)

Figure 9.3. Formation of melanoidin-type colorants on reaction of **(A)** Protein-bound lysine with furan-2-aldehyde (Hofmann 1998b). **(B)** Two protein-bound arginine residues with glyoxal in the presence of carbonyl compound (Hofmann 1998a). **(C)** Two protein-bound lysine residues with glycolaldehyde or glyoxal (in the presence of ascorbate) (Hofmann et al. 1999). Product of reaction of **(D)** Propylamine with glucose (Knerr et al. 2001). **(E)** Butylaminammonium acetate with glucose (Lerche et al. 2003). (R = remainder of the carbonyl compound).

Melanoidins can have molecular weights of up to 100 kDa (Hofmann 1998a). Due to their sheer chemical complexity, it has been difficult to isolate and characterize these molecules (Fayle and Gerrard 2002). However, some recent work in this area has led to some new hypotheses. Hofmann proposed that proteins may play an important role in the formation of these complex, high molecular weight melanoidins (Hofmann 1998b). Thus, it was proposed that a low molecular weight carbohydrate-derived colorant reacts with protein-bound lysine and/or arginine, forming a protein cross-link. Formation of color following polymerization of protein has indeed been observed following incubation with carbohydrates (Cho et al. 1984, 1986a,b; Okitani et al. 1984). Hofmann tested this hypothesis by reaction of casein with the pentose-derived intermediate furan-2-carboxaldehyde and subsequent isolation of a melanoidin type colorant compound from this reaction mixture (Fig. 9.3). Although non-cross-linking in nature, the results encouraged further studies to isolate protein cross-links that are melanoidins, and possibly deconvolute the chemistry of melanoidin from these data. Further studies by Hofmann in the same year proposed a cross-link structure BISARG, which was formed on reaction of *N*-protected arginine with glyoxal and furan-2-carboxaldehyde (Hofmann 1998a) (Fig. 9.3). This cross-link was proposed as plausible in food systems due to the large amount of protein furan-2-carboxaldehyde and glyoxal that may be present in food (Hofmann 1998a).

A lysine-lysine radical cation cross-link, CROSSPY (Fig. 9.3C), has been isolated from a model protein cross-link system involving bovine serum albumin and glycoladehyde, followed by thermal treatment (Hofmann et al. 1999). CROSSPY was also shown to form from glyoxal, but only in the presence of ascorbic acid, supporting the suggestion that reductones are able to initiate radical cation mechanisms resulting in cross-links (Hofmann et al. 1999). Electron paramagnetic resonance (EPR) spectroscopy of dark-colored bread crust revealed results that suggested CROSSPY was most likely associated with the browning bread crust (Hofmann et al. 1999). Encouragingly, when browning was inhibited, radical formation was completely blocked.

A model system containing propylamine and glucose was found to yield a yellow cross-link product under food processing conditions (Fig. 9.3D) (Knerr et al. 2001); however, this product is still to be isolated from foodstuffs. In another model study undertaken by Lerche et al., reacting butylaminammonium acetate (a protein-bound lysine mimic) with glucose resulted in the formation of a yellow product (Fig. 9.3E); this molecule is also still to be isolated from foodstuffs (Lerche et al. 2003).

The hypothesis that formation of melanoidins involves reaction of protein with carbohydrate is supported by Brands et al., who noted that following dialysis of a casein-sugar (glucose or fructose) system with 12 kDa cutoff tubing, around 70% of the brown colored products were present in the retentate (Brands et al. 2002).

Maillard-Related Cross-links

Although not strictly classified under the heading of Maillard chemistry, animal tissues such as collagen and elastin contain complex heterocyclic cross-links, formed from the apparently spontaneous reaction of lysine and derivatives with allysine, an aldehyde formed from the oxidative deamination of lysine catalyzed by the enzyme lysine oxidase (Feeney and Whitaker 1988). A selection of cross-links that form from this reaction is outlined in Figure 9.4. The extent to which these cross-links occur in food has not been well studied, although their presence in gelatin has been discussed, along with the presence of pentosidine in these systems (Cole and Roberts 1996, Cole and Roberts 1997).

Cross-links Formed via Transglutaminase Catalysis

An enzyme that has received extensive recent attention for its ability to cross-link proteins is transglutaminase. Transglutaminase catalyzes the acyl-transfer reaction between the γ-carboxyamide group of peptide-bound glutamine residues and various primary amines. As represented in Figure 9.1, the ε-amino groups of lysine residues in proteins can act as the primary amine, yielding inter- and intramolecular ε-*N*-(γ-glutamyl)lysine cross-links (Motoki and Seguro 1998). The formation of this cross-link does not reduce the nutritional quality of the food, as the lysine residue remains available for digestion (Seguro et al. 1996).

Transglutaminase is widely distributed in most animal tissues and body fluids and is involved in

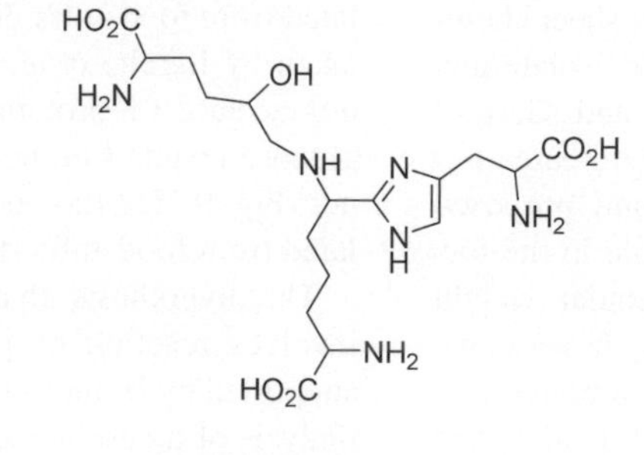

Histidinohydroxylysinonorleucine

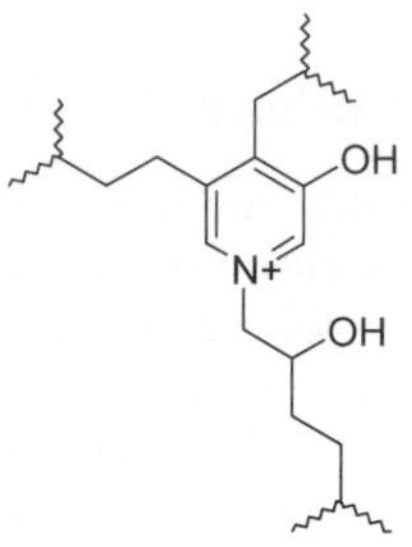

Hydroxypyridinium (pyridinoline)

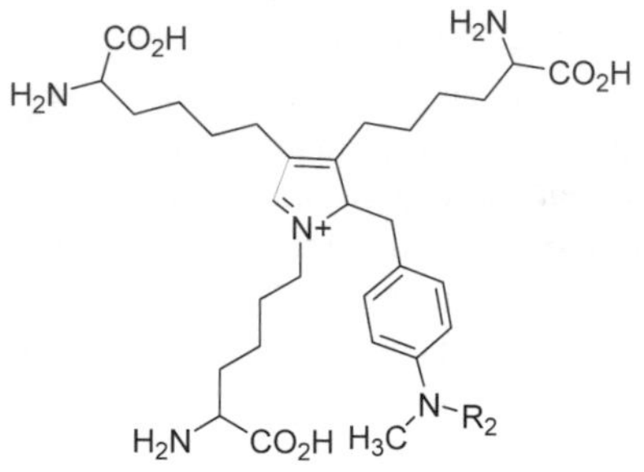

Figure 9.4. Trifunctional cross-links reported to result from the spontaneous reaction of allysine with lysine (Brady and Robins 2001, Eyre et al. 1984, Yamauchi et al. 1987).

biological processes such as blood clotting and wound healing. ε-*N*-(γ-glutamyl)lysine cross-links can also be produced by severe heating (Motoki and Seguro 1998), but are most widely found where a food is processed from material that contains naturally high levels of the enzyme. The classic example here is the gelation of fish muscle in the formation of surumi products, a natural part of traditional food processing of fish by the Japanese suwari process, although the precise role of endogenous transglutaminase in this process is still under debate (An et al. 1996, Motoki and Seguro 1998) and is an area of active research (Benjakul et al. 2004a,b). ε-*N*-(γ-glutamyl)lysine bonds have been found in various raw foods including meat, fish, and shellfish. Transglutaminase-cross-linked proteins have thus long been ingested by man (Seguro et al. 1996). The increasing applications of artificially adding this enzyme to a wide range of processed foods are discussed in detail below.

Other Isopeptide Bonds

In foods of low carbohydrate content, where Maillard chemistry is inaccessible, severe heat treatment can result in the formation of isopeptide cross-links during food processing, via condensation of the ε-amino group of lysine with the amide group of an asparagine or glutamine residue (Singh 1991). This chemistry has not been widely studied in the context of food.

MANIPULATING PROTEIN CROSS-LINKING DURING FOOD PROCESSING

A major task of modern food technology is to generate new food structures with characteristics that please the consumer, using only a limited range of ingredients. Proteins are one of the main classes of molecule available to confer textural attributes, and the cross-linking and aggregation of protein molecules is an important mechanism for engineering food structures with desirable mechanical properties (Dickinson 1997). The cross-linking of food proteins can influence many properties of food, including texture, viscosity, solubility, emulsification, and gelling properties (Motoki and Kumazawa 2000, Kuraishi et al. 2000). Many traditional food textures are derived from a protein gel, including those of yogurt, cheese, sausage, tofu, and surimi. Cross-linking provides an opportunity to create gel structures from protein solutions, dispersions, colloidal systems, protein-coated emulsion droplets, or protein-coated gas bubbles, and create new types of food or improve the properties of traditional ones (Dickinson 1997). In addition, the judicious choice of starting proteins for cross-linking can produce food proteins of higher nutritional quality through cross-linking of different proteins containing complementary amino acids (Kuraishi et al. 2000).

Chemical Methods

An increasing understanding of the chemistry of protein cross-linking opens up opportunities to control these processes during food processing. Many commercial cross-linking agents are available, for example, from Pierce (2001). These are usually double-headed reagents developed from molecules that derivatize the side chains of proteins (Matheis and Whitaker 1987, Feeney and Whitaker 1988) and generally exploit the lysine and/or cysteine residues of proteins in a specific manner. Doubt has recently been cast on the accuracy with which reactivity of these reagents can be predicted (Green 2001), but they remain widely used for biochemical and biotechnological applications.

Unfortunately, these reagents are expensive and not often approved for food use, so their use has not been widely explored (Singh 1991). They do, however, prove useful for "proof of principle" studies to measure the possible effects of introducing specific new cross-links into food. If an improvement in functional properties is seen after treatment with a commercial cross-linking agent, then further research effort is merited to find a food-approved, cost effective means by which to introduce such cross-links on a commercial scale. Such proof of principle studies include the use of glutaraldehyde to demonstrate the potential effects of controlled Maillard cross-linking on the texture of wheat-based foods (Gerrard et al. 2002a,b,c). The cross-linking of hen egg white lysozyme with a double-headed reagent has also been used to show that the protein is rendered more stable to heat and enzyme digestion, with the foaming and emulsifying capacity reduced (Matheis and Whitaker 1987, Feeney and Whitaker 1988). Similarly, milk proteins cross-linked with formaldehyde showed greater heat stability (Singh 1991).

Food preparation for consumption often involves heating, which can result in a deterioration of the functional properties of these proteins (Bouhallab et al. 1999, Morgan et al. 1999a, Shepherd et al. 2000). In an effort to protect them from denaturation, particularly in the milk processing industry, some have harnessed the Maillard reaction to produce more stable proteins following incubation with monosaccharide (Aoki et al. 1999, 2001; Bouhallab et al. 1999; Chevalier et al. 2002; Handa and Kuroda 1999; Matsudomi et al. 2002; Morgan et al. 1999a; Shepherd et al. 2000). Although increases in protein stability at high temperatures and improved emulsifying activity have been observed (Aoki et al. 1999, Bouhallab et al. 1999, Shepherd et al. 2000), dimerization and oligomerization of these proteins has also been noted (French et al. 2002, Aoki et al. 1999, Bouhallab et al. 1999, Chevalier et al. 2002, French et al. 2002, Morgan et al. 1999a, Pellegrino et al. 1999). In β-lactoglobulin, it has been suggested that oligomerization is initiated by glycation of a β-lactoglobulin monomer, resulting in a conformational change in this protein that engenders a propensity to form stable covalent homodimers (Morgan et al. 1999b). From this point, it is thought that polymerization occurs via hydrophobic interactions between unfolded homodimers and modified monomers (Morgan et al. 1999b). Experiments undertaken by Bouhallab et al. showed that this final polymerization process was not due to intermolecular cross-linking via the Maillard reaction, but did confirm that increased solubility of the protein at high temperatures (65–90°C) was associated with polymerization (Bouhallab et al. 1999), presumably invoking disulfide linkages. Some recent data, however, suggest that Maillard cross-links may after all be important in the polymerization process (Chevalier et al. 2002). Interestingly, Pellegrino et al. have observed that an increase in pentosidine formation coincided with heat-induced covalent aggregation of β-casein, which lacks cysteine, but could not account for the extent of aggregation observed, suggesting the presence of other cross-links that remain undefined (Pellegrino et al. 1999).

The glycation approach has also been employed as a tool to improve gelling properties of dried egg white (Handa and Kuroda 1999, Matsudomi et al. 2002), a protein that is used in the preparation of surimi and meat products (Chen 2000, Weerasinghe et al. 1996). Some have suggested that the polymerization may occur via protein cross-links formed as a result of the Maillard reaction, but disulfide bonds may still play some role (Handa and Kuroda 1999); the exact mechanism remains undefined (Matsudomi et al. 2002).

The effect of the Maillard reaction and particularly protein cross-linking on food texture has received some attention (Gerrard et al. 2002a). Introduction of protein cross-links into baked products has been shown to improve a number of properties that are valued by the consumer (Gerrard et al. 1998b, 2000). In situ studies revealed that following addition of glutaraldehyde to dough, the albumin and globulin fraction of the extracted wheat proteins were cross-linked (Gerrard et al. 2002b). Inclusion of glutaraldehyde during bread preparation resulted in the formation of a dough with an increased dough relaxation time, relative to the commonly used flour improver ascorbic acid (Gerrard et al. 2002c). These results confirm that chemical cross-links are important in the process of dough development, and suggest that they can be introduced via Maillard-type chemistry.

The Maillard reaction has also been used to modify properties in tofu. Kaye et al. reported that following incubation with glucose, a Maillard network formed within the internal structure of tofu, resulting in a loss in tofu solubility and a reduction in tofu weight loss (Kaye et al. 2001). Changes in tofu structure have also been observed in the author's laboratory when including glutaraldehyde, glyceraldehydes, or formaldehyde in tofu preparation (Yasir, unpublished data).

The covalent polymerization of milk during food processing has been reported (Singh and Latham 1993). This phenomenon was shown to be sugar dependent, as determined in model studies with β-casein. Further, pentosidine formation paralleled protein aggregation over time at 70°C (Pellegrino et al. 1999).

The Maillard reaction has been studied under dry conditions in order to gain an understanding of the details of the chemistry (French et al. 2002, Kato et al. 1988). Kato et al. reported the formation of protein polymers following incubation of galactose or talose with ovalbumin under desiccating conditions: an increase in polymerization, relative to the protein only control, was observed (Kato et al. 1986b). This study was extended using the milk sugar lactose, as milk can often be freeze dried for shipping and storage purposes. In this study, it was shown, in a dry reaction mixture, that ovalbumin was polymerized following incubation with lactose and glucose (Kato et al. 1988). In model studies with the milk protein β-lactoglobulin, the formation of protein dimers has been observed on incubation with lactose (French et al. 2002).

Proof of principle has thus been obtained in several systems, demonstrating that reactive cross-linking molecules are able to cross-link food proteins within the food matrix, leading to a noticeable change in the functional properties of the food. However, much work remains to be done in order to generate sufficient quantities of cross-linking intermediates during food processing to achieve a controlled change in functionality.

Enzymatic Methods

The use of enzymes to modify the functional properties of foods is an area that has attracted considerable interest, since consumers perceive enzymes to be more "natural" than chemicals. Enzymes are also favored as they require milder conditions, have high specificity, are only required in catalytic quantities, and are less likely to produce toxic products (Singh 1991). Thus, enzymes are becoming commonplace in many industries for improving the functional properties of food proteins (Chobert et al. 1996, Poutanen 1997).

Due to the predominance of disulfide cross-linkages in food systems, enzymes that regulate disulfide interchange reactions are of interest to food researchers. One such enzyme is protein disulfide isomerase (PDI). PDI catalyzes thiol/disulfide exchange, rearranging "incorrect" disulfide cross-links in a number of proteins of biological interest (Hillson et al. 1984, Singh 1991). The reaction involves the rearrangement of low molecular weight sulfhydryl compounds (e.g., glutathione, cysteine, and dithiothreitol) and protein sulfhydryls. It is thought to proceed by transient breakage of the protein disulfide bonds by the enzyme, and reaction of the exposed active cysteine sulfhydryl groups with other appropriate residues to reform native linkages (Singh 1991).

PDI has been found in most vertebrate tissues and in peas, cabbage, yeast, wheat, and meat (Singh 1991). It has been shown to catalyze the formation of disulfide bonds in gluten proteins synthesized in vitro. Early reports suggested that a high level of activity corresponded to a low bread-making quality (Grynberg et al. 1977). The use of oxidoreduction enzymes, such as PDI, to improve product quality is an area of interest to the baking industry (van Oort 2000, Watanabe et al. 1998) and the food industry in general (Hjort 2000). The potential of enzymes such as protein disulfide isomerase to catalyze their interchange has been extensively reviewed by Shewry and Tatham (1997).

Sulfhydryl (or thiol) oxidase catalyzes the oxidative formation of disulfide bonds from sulfhydryl groups and oxygen and occurs in milk (Matheis and Whitaker 1988, Singh 1991). Immobilized sulfhydryl oxidase has been used to eliminate the "cooked" flavor of ultra high temperature treated milk (Swaisgood 1980). It is not a well-studied enzyme, although the enzyme from chicken egg white has received recent attention in biology (Hoober 1999). Protein disulfide reductase catalyzes a further sulfhydryl-disulfide interchange reaction and has been found in liver, pea, and yeast (Singh 1991).

Peroxidase, lipoxygenase, and catechol oxidase occur in various plant foods and are implicated in the deterioration of foods during processing and storage. They have been shown to cross-link several food proteins, including bovine serum albumin, casein, β-lactoglobulin, and soy, although the uncontrolled nature of these reactions casts doubt on their potential for food improvement (Singh 1991). Lipoxygenase, in soy flour, is used in the baking industry to improve dough properties and baking performance. It acts on unsaturated fatty acids, yielding peroxy free radicals and starting a chain reaction. The cross-linking action of lipoxygenase has been attributed to both the free radical oxidation of free thiol groups to form disulfide bonds and to the generation of reactive cross-linking molecules such as malondialdehyde (Matheis and Whitaker 1987).

All enzymes discussed so far have been overshadowed in recent years by the explosion in research on the enzyme transglutaminase. Due largely to its ability to induce the gelation of protein solutions, transglutaminase has been investigated for uses in a diverse range of foods and food-related products. The use of this enzyme has been the subject of a series of recent reviews, covering both the scientific and patent literature (Nielsen 1995, Zhu et al. 1995, Motoki and Seguro 1998, Kuraishi et al. 2001). These are briefly highlighted below.

Transglutaminase

The potential of transglutaminase in food processing was hailed for many years before a practical source of the enzyme became widely available. The production of a microbially derived enzyme by Ajinomoto, Inc., proved pivotal in paving the way for industrial applications (Motoki and Seguro 1998). In addition, the transglutaminases that were discovered in the early years were calcium ion dependent, which imposed a barrier for their use in foods that did not contain a sufficient level of calcium. The commercial preparation is not calcium dependent, and thus finds much wider applicability (Motoki and Seguro 1998). The production of microbial transglutaminase, derived from *Streptoverticillium mobaraense,* is described by Zhu (1995), and methods with which to purify and assay the enzyme are reviewed by Wilhelm (1996). The commercial enzyme operates effectively over the pH range 4–9, from 0–50°C (Motoki and Seguro 1998).

There is a seemingly endless list of foods in which the use of transglutaminase has been successfully used (De Jong and Koppelman 2002): seafood, surimi, meat, dairy, baked goods, sausages (as a potential replacement for phosphates and other salts), gelatin (Kuraishi et al. 2001), and noodles and pasta (Larre et al. 1998, 2000). It is finding increasing use in restructured products, such as those derived from scallops and pork (Kuraishi et al. 1997). In all cases, transglutaminase is reported to improve firmness, elasticity, water-holding capacity, and heat stability (Kuraishi et al. 2001). It also has potential to alleviate the allergenicity of some proteins (Watanabe et al. 1994). Dickinson (1997) reviewed the application of transglutaminase to cross-link different kinds of colloidal structures in food and enhance their solid-like character in gelled and emulsified systems, controlling rheology and stability. These applications are particularly appealing in view of the fact that cross-linking by transglutaminase is thought to protect nutritionally valuable lysine residues in food from various deteriorative reactions (Seguro et al. 1996). Furthermore, the use of transglutaminase potentially allows production of food proteins of higher nutritional quality, through cross-linking of different proteins containing complementary amino acids (Zhu et al. 1995).

The use of transglutaminase in the dairy industry has been explored extensively. The enzyme has been trialed in many cheeses, from Gouda to Quark, and the use of transglutaminase in ice cream is reported to yield a product that is less icy and more easily scooped (Kuraishi et al. 2001). Milk proteins form emulsion gels, which are stabilized by cross-linking, opening new opportunities for protein-based spreads, desserts, and dressings (Dickinson and Yamamoto 1996). The use of transglutaminase has been explored in an effort to improve the functionality of whey proteins (Truong et al. 2004); for example, recent developments include the use of transglutaminase to incorporate whey protein into cheese (Cozzolino et al. 2003).

Soy products have also benefited from the introduction of transglutaminase, with the enzyme providing manufacturers with a greater degree of texture control. The enzyme is reported to enhance the quality of tofu made from old crops, giving a product with increased water-holding capacity, a good consistency, a silky and firmer texture, and a texture that is more robust in the face of temperature change (Kuraishi et al. 2001, Soeda 2003). Transglutaminase has also been used to incorporate soy protein into new products, such as chicken sausages (Muguruma et al. 2003).

New foods are being created using transglutaminase; for example, imitation shark fin for the Southeast Asian market has been generated by cross-linking gelatin and collagen (Zhu et al. 1995). Cross-linked proteins have also been tested as fat substitutes in products such as salami and yogurt (Nielson 1995), and the use of transglutaminase-cross-linked protein films as edible films has been patented (Nielson 1995).

Not surprisingly, the rate of cross-linking by transglutaminase depends on the particular structure of the protein acting as substrate. Most efficient cross-linking occurs in proteins that contain a glutamine residue in a flexible region of the protein, or within a reverse turn (Dickinson 1997). Casein is a very good substrate, but globular proteins such as ovalbumin and β-lactoglobulin are poor substrates (Dickinson 1997). Denaturation of proteins increases their reactivity, as does chemical modification by disruption of disulfide bonds, or by adsorption at an oil-water interface (Dickinson 1997). Many of the reported substrates of transglutaminase have actually been acetylated and/or denatured with reagents such as dithiothreitol under regimes that are not food approved (Nielson 1995). More work is needed to find ways to modify certain proteins in a food-allowed manner, to render them amenable to cross-linking by transglutaminase in a commercial setting.

While the applications of transglutaminase have been extensively reported in the scientific and patent literature, the precise mode of action of the enzyme in any one food-processing situation remains relatively unexplored. The specificity of the enzyme suggests that in mixtures of food proteins, certain proteins will react more efficiently than others, and there is value in understanding precisely which protein modifications exert the most desirable effects. Additionally, transglutaminase has more than one activity: as well as cross-linking, the enzyme may catalyze the incorporation of free amines into proteins by attachment to a glutamine residue. Furthermore, in the absence of free amine, water becomes the acyl acceptor, and the γ-carboxamide groups are deamidated to glutamic acid residues (Ando et al. 1989). The extent of these side reactions in foods and the consequences of any deamidation to the functionality of food proteins have yet to be fully explored.

Perhaps the most advanced understanding of the specific molecular effects of transglutaminase in a food product is seen in yogurt, where the treated product has been analyzed by gel electrophoresis, and specific functional effects have been correlated to the loss of β-casein, with the α-casein remaining. The specificity of the reaction was found to alter according to the exact transglutaminase source (Kuraishi et al. 2001). The specific effects of transglutaminase in baked goods (Gerrard et al. 1998, 2000) have also been analyzed at a molecular level. In particular, the enzyme produced a dramatic increase in the volume of croissants and puff pastries, with desirable flakiness and crumb texture (Gerrard et al. 2001). These effects were later correlated with cross-linking of the albumins and globulins and high molecular weight glutenin fractions by transglutaminase (Gerrard et al. 2001). Subsequent research suggested that the dominant effect was attributable to cross-linking of the high molecular weight glutenins (Gerrard et al. 2002).

FUTURE APPLICATIONS OF PROTEIN CROSS-LINKING

Although protein cross-linking is often considered to be detrimental to the quality of food, it is increasingly clear that it can also be used as a tool to improve food properties. The more we understand of the chemistry and biochemistry that take place during processing, the better placed we are to exploit it—minimizing deleterious reactions and maximizing beneficial ones.

Food chemists are faced with the task of understanding the vast array of reactions that occur during the preparation of food. Food often has complex structures and textures, and through careful manipulation of specific processes that occur during food preparation, these properties can be enhanced to generate a highly marketable product (Dickinson 1997). The formation of a structural network within food is critical for properties such as food texture. From a biopolymer point of view, this cross-linking process has been successfully applied in the formation of protein films, ultimately for use as packaging. Enhancing textural properties, emulsifying and foaming properties, by protein cross-linking has been a subject of interest of many working in this area, as exemplified by the number of studies detailed above.

Cross-linking using chemical reagents remains challenging, in terms of both controlling the chemistry and gaining consumer acceptance. However, as the extensive recent use of transglutaminase dramatically illustrates, protein cross-linking using enzymes has huge potential for the improvement of traditional products and the creation of new ones. Transglutaminase itself will no doubt find yet more application as its precise mode of action becomes better understood, especially if variants of the enzyme with a broader substrate specificity are found.

Whether other enzymes, which cross-link by different mechanisms, can find equal applicability remains open to debate. The thiol exchange enzymes, such as protein disulfide isomerase, may offer advantages to food processors if their mechanisms can be unraveled, and then controlled within a foodstuff. Other enzymes, such as lysyl oxidase, have not yet been used, but have potential to improve foods (Dickinson 1997), especially in the light of recent work characterizing this class of enzymes (Buffoni and Ignesti 2000), which may allow their currently unpredictable effects to be better understood.

ACKNOWLEDGEMENTS

I thank all the postgraduates and postdoctoral fellows who have worked with me on the Maillard reaction and protein cross-linking of food, in particular: Dr Siân Fayle, Paula Brown, Dr. Indira Rasiah, Dr. Susie Meade, Dr. Antonia Miller, and Suhaimi Yasir.

REFERENCES

Al-Abed Y, Bucala R. 2000 Structure of a synthetic glucose derived advanced glycation end product that is immunologically cross-reactive with its naturally occurring counterparts. Bioconjug. Chem. 11:39–45.

Amadori M 1931. Atti R Acad Naz Lincei Mem Cl Sci Fis Mat Nat 13:72.

Ames JM. 1992. The Maillard reaction in foods. In: BJF Hudson, editor, Biochemistry of Food Proteins, pp. 99–153. Elsevier Applied Science.

An HJ, et al. 1996. Roles of endogeneous enzymes in surimi gelation. Trends Food Sci Technol 7:321–327.

Ando H, et al. 1989. Purification and characteristics of a novel transglutaminase derived from microorganisms. Agric. Biol. Chem. 53:2613–2617.

Aoki T, Hiidome Y, Kitahata K, Sugimoto Y, Ibrahim HR, Kato Y. 1999. Improvement of heat stability and emulsifying activity of ovalbumin by conjugation with glucuronic acid through the Maillard reaction. Food Res Intern 32:129–133.

Aoki T, Hiidome Y, Sugimoto Y, Ibrahim HR, Kato Y. 2001. Modification of ovalbumin with oligogalacturonic acids through the Maillard reaction. Food Res Intern 34:127–132.

Benjakul S, Visessanguan W, Chantarasuwan C. 2004a. Cross-linking activity of sarcoplasmic fraction from bigeye snapper *(Priacanthus tayenus)* muscle. Food Sci Technol37:79–85.

———. 2004b. Effect of porcine plasma protein and setting on gel properties of surimi produced from fish caught in Thailand. Food Sci Technol 37:177–185.

Biemel KM, Buhler, HP, Reihl O, Lederer MO. 2001. Identification and quantitative evaluation of the lysine-arginine crosslinks GODIC, MODIC, DODIC, and glucosepan in foods. Nahrung 45:210–214.

Biemel KM, Friedl AD, Lederer MO. 2002. Identification and quantification of major Maillard crosslinks in human serum albumin and lens protein: Evidence for glucosepane as the dominant compound. J Biol Chem 277:24907–24915.

Bouhallab S, Morgan F, Henry G, Molle D, Leonil J. 1999. Formation of stable covalent dimer explains the high solubility at pH 4.6 of lactose-β-lactoglobulin conjugates heated near neutral pH. J Agric Food Chem 47:1489–1494.

Brady JD, Robins SP. 2001. Structural characterization of pyrrolic cross-links in collagen using a biotinylated Ehrlich's reagent. J Biol Chem 276:18812–18818.

Brands CMJ, Wedzicha BL, van Boekel M. 2002. Quantification of melanoidin concentration in sugar-casein systems. J Agric Food Chem 50:1178-1183.

Brinkmann-Frye E, Degenhardt TP, Thorpe SR, Baynes JW. 1998. Role of the Maillard reaction in aging of tissue proteins. J Biol Chem 273:18714–18719.

Buffoni F, Ignesti G. 2000. The copper-containing amine oxidases: Biochemical aspects and functional role. Mol Genet Metabolism 71:559–564.

Chellan P, Nagaraj RH. 2001. Early glycation products produce pentosidine cross-links on native proteins: Novel mechanism of pentosidine formation and propagation of glycation. J Biol Chem 276:3895–3903.

Chen HH. 2000. Effect of non muscle protein on the thermogelation of horse mackerel surimi and the resultant cooking tolerance of kamaboko. Fish Sci 66: 783–788.

Chevalier F, Chobert, JM, Popineau Y, Nicolas, MG, Haertle T. 2001. Improvement of functional properties of β-lactoglobulin glycated through the Maillard reaction is related to the nature of the sugar. Int Dairy J 11:145–152.

Chevalier F, Chobert JM, Dalgalarrondo M, Choiset Y, Haertle T. 2002. Maillard glycation of β-lactoglobulin induces conformation changes. Nahrung 46:58–63.

Cho RK, Okitani A, Kato H. 1984. Chemical properties and polymerizing ability of the lysozyme monomer isolated after storage with glucose. Agric Biol Chem 48:3081–3089.

———. 1986a. Polymerization of acetylated lysozyme and impairment of their amino acid residues due to α-dicarbonyl and α-hydroxycarbonyl compounds. Agric Biol Chem 50:1373–1380.

———. 1986b. Polymerization of proteins and impairment of their arginine residues due to intermediate compounds in the Maillard reaction. Dev Food Sci 13:439–448.

Chobert JM et al. 1996. Recent advances in enzymatic modifications of food proteins for improving their functional properties. Nahrung 40:177–182.

Cole CGB, Roberts JJ. 1996. Gelatine fluorescence and its relationship to animal age and gelatine colour. SAJ Food Sci Nutr 8:139–143.

———. 1997. The fluorescence of gelatin and its implications. Imaging Sci J 45:145–149.

Cozzolino A, Di Pierro P, Mariniello L, Sorrentino A, Masi P, Porta R. 2003. Incorporation of whey proteins into cheese curd by using transglutaminase. Biotechnol Appl Biochem 38:289–295.

De Jong GAH, Koppelman SJ. 2002. Transglutaminase catalyzed reactions: Impact on food applications. J Food Sci 67:2798–2806.

Dickinson E. 1997. Enzymic crosslinking as a tool for food colloid rheology control and interfacial stabilization. Trends Food Sci Technol 8:334–339.

Dickinson E, Yamamoto Y. 1996. Rheology of milk protein gels and protein-stabilized emulsion gels cross-linked with transglutaminase. J Agric Food Chem 44: 1371–1377.

Dyer DG, Blackledge JA, Thorpe SR, Baynes JW. 1991. Formation of pentosidine during nonenzymatic browning of proteins by glucose. Identification of glucose and other carbohydrates as possible precursors of pentosidine in vivo. J Biol Chem 266:11654–11660.

Easa AM, Armstrong HJ, Mitchell JR, Hill SE, Harding SE, Taylor AJ. 1996a. Maillard induced complexes of bovine serum albumin—a dilute solution study. Int J Biol Macromol 18:297–301.

Easa AM, Hill SE, Mitchell JR, Taylor AJ. 1996b. Bovine serum albumin gelation as a result of the Maillard reaction. Food Hydrocolloids 10:199–202.

Eyre D, Paz M, Gallop PM. 1984. Cross-linking in collagen and elastin. Ann Rev Biochem 53:717–748.

Fayle SE, Gerrard JA. 2002. The Maillard Reaction. Cambridge: Royal Society of Chemistry.

Fayle SE, Gerrard JA, Simmons L, Meade SJ, Reid EA, Johnston AC. 2000. Crosslinkage of proteins by dehydroascorbic acid and its degradation products. Food Chem 70:193–198.

Fayle SE, Healy JP, Brown PA, Reid EA, Gerrard JA, Ames JM. 2001. Novel approaches to the analysis of the Maillard reaction of proteins. Electrophoresis 22:1518–1525.

Feeney RE, Whitaker JR. 1988. Importance of cross-linking reactions in proteins. Adv Cereal Sci Tech IX:21–43.

French SJ, Harper WJ, Kleinholz NM, Jones RB, Green-Church KB. 2002. Maillard reaction induced lactose attachment to bovine β-lactoglobulin: Electrospray ionization and matrix-assisted laser desorption/ionization examination. J Agric Food Chem 50: 820–823.

Friedman M. 1996a. Food browning and its prevention: An overview. J Agric Food Chem 44:631–653.

———. 1996b. The impact of the Maillard reaction on the nutritional value of food proteins. In: R Ikan, editor, The Maillard Reaction: Consequences for the Chemical and Life Sciences, p. 119. John Wiley and Sons.

———. 1999a. Chemistry, biochemistry, nutrition, and microbiology of lysinoalanine, lanthionine, and histidinoalanine in food and other proteins. J Agric Food Chem 47:1295–1319.

———. 1999b. Lysinoalanine in food and in antimicrobial proteins. Adv Exp Med Biol 459:145–159.

———. 1999c. Chemistry, nutrition, and microbiology of D-amino acids. J Agric Food Chem 47:3457–3479.

Gerrard JA. 2002. Protein-protein crosslinking in food: Methods, consequences, applications. Trends in Food Science and Technology 13:391–399.

Gerrard JA, Brown PK, Fayle SE. 2002a. Maillard crosslinking of food proteins I: The reaction of glutaraldehyde, formaldehyde and glyceraldehyde with ribonuclease. Food Chem 79:343–349.

Gerrard JA, Brown PK, Fayle SE. 2002b. Maillard crosslinking of food proteins II: The reactions of glutaraldehyde, formaldehyde and glyceraldehyde with wheat proteins in vitro and in situ. Food Chem 80:35–43.

Gerrard JA, Brown PK, Fayle SE. 2002c. Maillard crosslinking of food proteins III: The effects of glutaraldehyde, formaldehyde and glyceraldehyde upon bread and croissants. Food Chem 80:45–50.

Gerrard JA, Fayle SE, Sutton KH. 1999. Covalent protein adduct formation and protein crosslinking resulting from the Maillard reaction between cyclotene and a model food protein. J Agric Food Chem 47: 1183–1188.

Gerrard JA, Fayle SE, Sutton KH, Pratt AJ. 1998a. Dehydroascorbic acid mediated crosslinkage of proteins using Maillard chemistry—Relevance to food processing. In: J O'Brien, HE Nursten, MJC Crabbe, JM Ames, editors, The Maillard Reaction in Foods and Medicine, pp. 127–132. Cambridge: The Royal Society of Chemistry Press.

Gerrard JA, Fayle SE, Wilson AJ, Newberry MP, Ross M, Kavale S. 1998b. Dough properties and crumb strength of white pan bread as affected by microbial transglutaminase. J Food Sci 63:472–475.

Gerrard JA, Newberry MP, Ross M, Wilson AJ, Fayle SE, Kavale S. 2000. Pastry lift and croissant volume as affected by microbial transglutaminase. J Food Sci 65:312–314.

Gerrard JA et al. 2001. Effects of microbial transglutaminase on the wheat proteins of bread and croissant dough. J Food Sci 66:782–786.

Glomb MA, Pfahler C. 2001. Amides are novel protein modifications formed by physiological sugars. J Biol Chem 276:41638–41647.

Green NS et al. 2001. Quantitative evaluation of the lengths of homobifunctional protein cross-linking reagents used as molecular rulers. *Protein Sci.* 10, 1293-1304.

Grynberg A et al. 1977. De la presence d'une proteine disulfide isomerase dans l'embryon de ble. CR Hebd Seances Acad Sci Paris 284:235–238.

Handa A, Kuroda N. 1999. Functional improvements in dried egg white through the Maillard reaction. J Agric Food Chem 47:1845–1850.

Henle T, Schwarzenbolz U, Klostermeyer H. 1997. Detection and quantification of pentosidine in foods. Z. Lebensm-Unters Forsch A 204:95–98.

Hill S, Easa AM. 1998. Linking proteins using the Maillard reaction and the implications for food processors. In: J O'Brien, HE Nursten, MJC Crabbe, JM Ames, editors, The Maillard Reaction in Foods and Medicine, pp. 133–138. Cambridge: Royal Society of Chemistry Press.

Hill SE, Armstrong HJ, Mitchell JR. 1993. The use and control of chemical reactions to enhance gelation of macromolecules in heat processed foods. Food Hydrocolloids 7:289–294.

Hillson DA, et al. 1984. Formation and isomerisation in proteins: protein disulfide isomerase. Meth Enzymol 107:281–304.

Hjort CM. 2000. Protein disulfide isomerase variants that preferably reduce disulfide bonds and their manufacture,p. 82. In: PCT Intl. Appl.

Hofmann T. 1998a. 4-alkylidene-2-imino-5-[4-alkylidene-5-oxo-1,3-imidazol-2- inyl]azamethylidene-1, 3-imidazolidine—A novel colored substructure in melanoidins formed by Maillard reactions of bound arginine with glyoxal and furan-2-carboxaldehyde. J Agric Food Chem 46:3896–3901.

———. 1998b. Studies on melanoidin-type colorants generated from the Maillard reaction of protein-bound lysine and furan-2-carboxaldehyde: Chemical characterization of a red colored domain. Z Lebens-Unters Forsch A 206:251–258.

Hofmann T, Bors W, Stettmaier K. 1999. Radical-assisted melanoidin formation during thermal processing of foods as well as under physiological conditions. J Agric Food Chem 47:391–396.

Hoober KL, et al. 1999. Sulfhydryl oxidase from egg white—A facile catalyst for disulfide bond formation in proteins and peptides. J Biol Chem 274: 22147–22150

Iqbal M, Kenney PB, Al-Humadi NH, Klandorf H. 2000. Relationship between mechanical properties and pentosidine in tendon: Effects of age, diet restriction, and aminoguanidine in broiler breeder hens. Poult Sci 79:1338–1344.

Iqbal M, Kenney PB, Klandorf H. 1999a. Age-related changes in meat tenderness and tissue pentosidine: Effect of diet restriction and aminoguanidine in broiler breeder hens. Poult Sci 78:1328–1333.

Iqbal M, Probert LL, Alhumadi NH, Klandorf H. 1999b. Protein glycosylation and advanced glycosylated endproducts (AGEs) accumulation: An avian solution? J. Gerontol. A Biol Sci Med Sci 54:B171–B176.

Iqbal M, Probert LL, Klandorf H. 1997. Effect of dietary aminoguanidine on tissue pentosidine and reproductive performance in broiler breeder hens. Poult Sci 76:1574–1579.

Kato H, Chuyen NV, Utsunomiya N, Okitani A. 1986a. Changes of amino acids composition and relative digestibility of lysozyme in the reaction with α-dicarbonyl compounds in aqueous system. J Nutr Sci Vitaminol 32:55–65.

Kato Y, Matsuda T, Kato N, Nakamura R. 1988. Browning and protein polymerization induced by amino carbonyl reaction of ovalbumin with glucose and lactose. J Agric Food Chem 36:806–809.

Kato Y, Matsuda T, Kato N, Watanabe K, Nakamura R. 1986b. Browning and insolubilization of ovalbumin by the Maillard reaction with some aldohexoses. J Agric Food Chem 34:351–355.

Kaye YK, Easa AM, Ismail N. 2001. Reducing weight loss of retorted soy protein tofu by using glucose- and microwave-pre-heating treatment. Int J Food Sci Technol 36:387–392.

Knerr T, Lerche H, Pischetsrieder M, Severin T. 2001. Formation of a novel colored product during the Maillard reaction of D-glucose. J Agric Food Chem 49:1966–1970.

Kuraishi C, et al. 1997. Production of restructured meat using microbial transglutaminase without salt or cooking. J Food Sci 62:488–490.

———. 2000. Application of transglutaminase for food processing. Hydrocolloids 2:281–285.

———. 2001. Transglutaminase: Its utilization in the food industry. Food Rev Int 17:221–246.

Larre C, et al. 1998. Hydrated gluten modified by a transglutaminase. Nahrung Food 42:155–157.

———. 2000. Biochemical analysis and rheological properties of gluten modified by transglutaminase. Cereal Chem 77:32–38.

Lederer MO, Buhler HP. 1999. Cross-linking of proteins by Maillard processes—Characterization and detection of a lysine-arginine cross-link derived from D-glucose. Bioorg Med Chem 7:1081–1088.

Lederer MO, Klaiber RG. 1999. Cross-linking of proteins by Maillard processes: Characterization and detection of lysine-arginine cross-links derived from glyoxal and methylglyoxal. Bioorg Med Chem 7: 2499–2507.

Lerche H, Pischetsrieder M, Severin T. 2003. Maillard reaction of D-glucose: Identification of a colored product with hydroxypyrrole and hydroxypyrrolinone rings connected by a methine group. J Agric Food Chem 51:4424–4426.

Lindsay MP, Skerritt JH. 1999. The glutenin macropolymer of wheat flour doughs: Structure-function perspectives. Trends Food Sci Technol 10:247–253.

Maillard L-C. 1912. Action des acides amines sur les sucres; formation des melanoidines par voie methodique CR Acad Sci Ser 2 154:66.

Masaki K, et al. 2001. Thermal stability and enzymatic activity of a smaller lysozyme from silk moth *(Bombyx mori).* J Protein Chem 20:107–113.

Matheis G, Whitaker JR. 1987. Enzymic cross-linking of proteins applicable to food. J Food Biochem 11:309–327.

Matsudomi N, Nakano K, Soma A, Ochi A. 2002. Improvement of gel properties of dried egg white by modification with galactomannan through the Maillard reaction. J Agric Food Chem 50:4113–4118.

Mezgheni E, et al. 1998. Formation of sterilized edible films based on caseinates—Effects of calcium and plasticizers. J Agric Food Chem 46:318–324.

Miller AG, Gerrard JA. 2004. The Maillard reaction and food protein crosslinking. Progress in Food Biopolymer Research 1 (in press).

Mohammed ZH, Hill SE, Mitchell JR. 2000. Covalent crosslinking in heated protein systems. J Food Sci 65:221–226.

Morgan F, Leonil J, Molle D, Bouhallab S. 1999a. Modification of bovine β-lactoglobulin by glycation in a powdered state or in an aqueous solution: Effect on association behaviour and protein conformation. J Agric Food Chem 47:83–91.

Morgan F, Molle D, Henry G, Venien A, Leonil J, Peltre G, Levieux D, Maubois JL, Bouhallab S. 1999b. Glycation of bovine β-lactoglobulin: Effect on the protein structure. Int J Food Sci Technol 34:429–435.

Motoki M, Kumazawa Y. 2000. Recent trends in transglutaminase technology for food processing. Food Sci Technol Res 6(3): 151–160.

Motoki M, Seguro K. 1998. Transglutaminase and its use for food processing. Trends Food Sci Technol 9: 204–210.

Muguruma M, Tsuruoka K, et al. 2003. Soybean and milk proteins modified by transglutaminase improves chicken sausage texture even at reduced levels of phosphate. Meat Science 63:191–197.

Nakamura K, Nakazawa Y, Ienaga K. 1997. Acid-stable fluorescent advanced glycation end products: Vesperlysines a, b and c are formed as crosslinked products in the Maillard reaction between lysine or protein with glucose. Biochem Biophys Res Commun 232:227–230.

Nielsen PM. 1995. Reactions and potential industrial applications of transglutaminase—Review of literature and patents. Food Biotechnol 9:119–156.

Okitani A, Cho RK, Kato H. 1984. Polymerization of lysozyme and impairment of its amino acid residues caused by reaction with glucose. Agric Biol Chem 48:1801–1808.

O'Sullivan MM, Kelly AL, et al. 2002a. Effect of transglutaminase on the heat stability of milk: A possible mechanism. Journal of Dairy Science 85:1–7.

———. 2002b. Influence of transglutaminase treatment on some physico-chemical properties of milk. Journal of Dairy Research 69:433–442.

Pellegrino L, van Boekel M, Gruppen H, Resmini P, Pagani MA. 1999. Heat-induced aggregation and covalent linkages in β-casein model systems. Int. Dairy J 9:255–260.

Pierce. 2001. Crosslinker selection guide. Rockford Illinois Pierce Chemical Company, available on line at http://www.piercenet.com.

Poutanen K. 1997. Enzymes: An important tool in the improvement of the quality of cereal foods. Trends Food Sci Technol 8:300–306.

Prabhakaram M, Cheng Q, Feather MS, Ortwerth B. 1997. Structural elucidation of a novel lysine-lysine crosslink generated in a glycation reaction with L-threose. Amino Acids 12:225–236.

Rich LM, Foegeding EA. 2000. Effects of sugars on whey protein isolate gelation. J Agric Food Chem 48:5046–5052.

Roos N, Lorenzen PC, et al. 2003. Cross-linking by transglutaminase changes neither the in vitro proteolysis nor the in vivo digestibility of caseinate. Kieler Milchwirtschaftliche Forschungsberichte 55: 261–276.

Savoie L, et al. 1991. Effects of alkali treatment on in vitro digestibility of proteins and release of amino acids. J Sci Food Agric 56:363–372.

Schwarzenbolz U, Klostermeyer H, Henle T. 2000. Maillard-type reactions under high hydrostatic pressure: Formation of pentosidine. Eur Food Res Technol 211:208–210.

Seguro K, et al. 1996. The epsilon-(gamma-glutamyl) lysine moiety in crosslinked casein is an available source of lysine for rats. J Nutr 126:2557–2562.

Sell DR, Monnier VM. 1989. Structure elucidation of a senescence crosslink from human extracellular matrix. J Biol Chem 264:21597–21602.

Shepherd R, Robertson A, Ofman D. 2000. Dairy glycoconjugate emulsifiers: casein-maltodextrins. Food Hydrocolloids 14:281–286.

Shewry PR, Tatham AS. 1997. Disulfide bonds in wheat gluten proteins. J Cer Sci 25:207–222.

Shih FF 1992. Modification of food proteins by nonenzymatic methods. In: BJF Hudson, editor, Biochemistry of Food Proteins, pp. 235–248. Elsvier Applied Science.

Singh H. 1991. Modification of food proteins by covalent crosslinking. Trends Food Sci Technol 2:196–200.

Singh H, Latham J. 1993. Heat stability of milk: Aggregation and dissociation of protein at ultra-high temperatures. Int Dairy J 3:225-235.

Soeda T. 2003. Effects of microbial transglutaminase for gelation of soy protein isolate during cold storage. Food Sci Technol Res 9:165–169.

Strauss G, Gibson SA. 2004. Plant phenolics as cross-linkers of gelatin gels and gelatin-based coacervates for use as food ingredients. Food Hydrocolloids 18:81–89.

Swaisgood HE. 1980. Sulfhydryl oxidase: Properties and applications. Enzyme Micro Technol 2:265–272.

Tessier FJ, Monnier VM, Sayre LM, Kornfield JA. 2003. Triosidines: Novel Maillard reaction products and cross-links from the reaction of triose sugars with lysine and arginine residues. Biochem J 369: 705–719.

Tilley KA, et al. 2001. Tyrosine cross-links: Molecular basis of gluten structure and function. J Agric Food Chem 49:2627–2632.

Truong VD, Clare DA, et al. 2004. Cross-linking and rheological changes of whey proteins treated with microbial transglutaminase. Journal of Agricultural and Food Chemistry 52:1170–1176.

van Oort M. 2000. Peroxidases in breadmaking. In: JD Schofield, editor, Wheat Structure, Biochemistry and Functionality. Cambridge: Royal Society of Chemistry.

Watanabe E, et al. 1998. The effect of protein disulfide isomerase on dough rheology assessed by fundamental and empirical testing. Food Chem 61:481–486.

Watanabe M, et al. 1994. Controlled enzymatic treatment of wheat proteins for production of hypoallergenic flour. Biosci Biotechnol Biochem 58:388–390.

Weerasinghe VC, Morrissey MT, An HJ. 1996. Characterization of active components in food-grade proteinase inhibitors for surimi manufacture. J Agric Food Chem 44:2584–2590.

Wilhelm B, et al. 1996. Transglutaminases—Purification and activity assays. J Chromatogr B: Biomedical Applications 684:163–177.

Yamauchi M, London RE, Guenat C, Hashimoto F, Mechanic GL. 1987. Structure and formation of a stable histidine-based trifunctional cross-link in skin collagen. J Biol Chem 262:11428–11343.

Zarina S, Zhao HR, Abraham EC. 2000. Advanced glycation end products in human senile and diabetic cataractous lenses. Mol Cell Biochem 210:29–34.

Zayas JF. 1997. Functionality of Proteins in Food. Springer-Verlag, Berlin.

Zhu Y, et al. 1995. Microbial transglutaminase. A review of its production and application in food processing. Appl Microbiol Biotechnol 44:277–282.

10
Chymosin in Cheese Making

V. V. Mistry

INTRODUCTION

Cheeses are classified into those that are aged for the development of flavor and body over time and those that are ready for consumption shortly after manufacture. Many cheese varieties are also expected to develop certain functional characteristics either immediately after manufacture or by the end of the aging period. It is fascinating that what starts in the cheese vat, as a white liquid that is composed of many nutrients, emerges as a completely transformed compact mass, cheese, sometimes colored yellow, sometimes white, and sometimes with various types of mold on the surface or within the cheese mass (Kosikowski and Mistry 1997). Some cheeses may also exhibit a unique textural character that turns stringy when heated, and others may have shiny eyes. Each cheese also has unique flavor qualities. Thus, cheese is a complex biological material whose characteristics are specifically tailored by the cheese maker through judicious blends of enzymes, starter bacteria, acid, and temperature.

Milk is a homogeneous liquid in which components exist in soluble, emulsion, or colloidal form. The manufacture of cheese involves a phase change from liquid (milk) to solid (cheese). This phase change occurs under carefully selected conditions that alter the physicochemical environment of milk such that the milk system is destabilized and is no longer homogeneous. In many fresh cheeses, this conversion takes place with the help of acid through isoelectric precipitation of casein and subsequent temperature treatments and draining for the conversion of liquid to solid. In the case of most ripened cheeses, this conversion occurs enzymatically at higher pH and involves the transformation of the calcium paracaseinate complex through controlled lactic acid fermentation. Later on, the products of this reaction and starter bacteria interact for the development

Reviewers of this article were Dr. Ashraf Hassan, Department of Dairy Science, South Dakota State University, Brookings, and Dr. Lloyd Metzger, Department of Food Science and Nutrition, University of Minnesota, St. Paul.

of flavor and texture. All this is the result of the action of the enzyme chymosin (Andren 2003).

CHYMOSIN

Chymosin, rennet, and rennin are often used interchangeably to refer to this enzyme. The latter, rennin, should not be confused with renin, which is an enzyme associated with kidneys and does not clot milk. Chymosin is the biochemical name of the enzyme that was formerly known as rennin. It belongs to the group of aspartic acid proteinases, EC 3.4.23, that have a high content of dicarboxylic and hydroxyamino acids and a low content of basic amino acids. Its molecular mass is 40 kDa (Andren 2003, Foltmann 1993).

Rennet is the stomach extract that contains the enzyme chymosin in a stabilized form that is usable for cheese making (Green 1977). While the amount of chymosin that is required for cheese making is very small, this enzyme industry has undergone an interesting transformation over time. It is believed that in the early days of cheese making, milk coagulation occurred either by filling dried stomachs of calves or lambs with milk or by immersing pieces of such stomachs in milk (Kosikowski and Mistry 1997). The chymosin enzyme imbedded within the stomach lining diffused into the milk and coagulated it. This crude process of extracting coagulating enzymes was eventually finessed into an industry that employed specific methods to extract and purify the enzyme and develop an extract from the fourth stomach of the calf or lamb. Extracts of known enzyme activity and predictable milk clotting properties then became available as liquids, concentrates, powders, and blends of various enzymes. Live calves were required for the manufacture of these products. Because of religious and economic reasons, another industry also had emerged for manufacturing alternative milk clotting enzymes from plants, fungi, and bacteria. These products remain popular and meet the needs of various kosher and other religious needs. Applications of genetic technology in rennet manufacture were then realized, and in 1990 a new process utilizing this technology was approved in the United States for manufacturing rennet.

RENNET MANUFACTURE

For the manufacture of traditional rennet, calves, lambs, or kids that are no more than about 2 weeks old and fed only milk are used (Kosikowski and Mistry 1997). As calves become older and begin eating other feeds, the proportion of bovine pepsin in relation to chymosin increases. Extracts from milk-fed calves that are 3 weeks old contain over 90% chymosin, and the balance is pepsin. As the calves age and are fed other feeds such as concentrates, the ratio of chymosin to pepsin drops to 30:70 by 6 months of age. In a full-grown cow there are only traces of chymosin.

Milk-fed calves are slaughtered, and the unwashed stomachs (vells) are preserved for enzyme extraction by emptying their contents, blowing them into small balloons and drying them. The vells may also be slit opened and dry salted or frozen for preservation. Air-dried stomachs give lower yields of chymosin than frozen stomachs; 12–13 air-dried stomachs make 1 L of rennet standardized to 1:10,000 strength, but only 7–8 frozen stomachs would be required for the same yield.

Extraction of chymosin and production of rennet begin by extracting, for several days, chopped or macerated stomachs with a 10% sodium chloride solution containing about 5% boric acid or glycerol. Additional salt up to a total of 16–18% is introduced followed by filtration and clarification. Mucine and grass particles in suspension are removed by introducing 1% of potash alum, followed by an equal amount of potassium phosphate. The suspension is adjusted to pH 5.0 to activate prochymosin (zymogen) to chymosin, and the enzyme strength is standardized, so that one part coagulates 15,000 or 10,000 parts of milk. Sodium benzoate, propylene glycol, and salt are added as preservatives for the final rennet. Caramel color is also usually added. The finished rennet solutions must be kept cold and protected from light.

Powdered rennet is manufactured by saturating a rennet suspension with sodium chloride or acidifying it with a food-grade acid. Chymosin precipitates and secondary enzymes such as pepsin remain in the original suspension. The chymosin-containing precipitate is dried to rennet powder (Kosikowski and Mistry 1997).

A method has been developed for manufacturing rennet without slaughtering calves. A hole (fistula) is surgically bored in the side of live calves, and at milk feeding time, excreted rennet is removed. After the calf matures, the hole is plugged, and the animal is returned to the herd. This method has not been commercialized but may be of value where religious practices do not allow calf slaughter.

Rennet paste, a rich source of lipolytic enzymes, has been a major factor in the flavor development of ripened Italian cheeses such as Provolone and Romano. Conceivably, 2100 years ago, Romans applied rennet paste to cheese milks to develop a typical "picante" flavor in Romano and related cheeses, for even then calf rennet was used to set their cheese milks. In more modern times many countries, including the United States, applied calf rennet paste to induce flavor, largely for Italian ripened cheeses.

Farntiam et al. (1964) successfully produced, from goats, dried preparations of a pregastric-oral nature, also rich in lipolytic enzymes. They applied them to milk for ripened Italian cheese with excellent results. Since then, lipase powders, of various character and strength, have largely replaced rennet paste.

Chymosin Production by Genetic Technology

Genetic technology has been used for the commercial production of a 100% pure chymosin product from microbes. This type of chymosin is often called fermentation-produced chymosin. Uchiyama et al. (1980) and Nishimori et al. (1981) published early on the subject, and in 1990 Pfizer, Inc., received U.S. Food and Drug Administration approval to market a genetically transformed product, Chy-Max, with GRAS stature (Duxbury 1990). Other brands using other microorganisms have also been approved. Table 10.1 lists some examples of such products.

The microbes used for this type of rennet include nonpathogenic microorganisms *Escherichia coli* K-12, *Kluyveromyces marxianus* var. *lactis* and *Aspergillus niger* var. *awamori*. Prochymosin genes obtained from young calves are transferred through DNA plasmid intervention into microbial cells. Fermentation follows to produce prochymosin, cell destruction, activation of the prochymosin to chymosin, and harvesting/producing large yields of pure, 100% chymosin. This product, transferred from an animal, is considered a plant product, as microbes are of the plant kingdom. Thus, they are acceptable to various religious groups.

These products have been widely studied to evaluate their impact on the quality and yield of cheese. An initial concern was that because of the absence of pepsin from these types of rennets, proper cheese flavor might not develop. Most studies have generally concluded that there are no significant differences in flavor, texture, composition, and yield compared with calf rennet controls (Banks 1992, Barbano and Rasmussen 1992, Biner et al. 1989, Green et al. 1985, Hicks et al. 1988, IDF 1992, Ortiz de Apodaca et al. 1994).

When these products were first introduced, it was expected that a majority of cheese makers would convert to these types of rennets because they are virtually identical to traditional calf rennet and much cheaper. Except for some areas, this has not generally happened because of some concerns towards genetically modified products. Further, some protected cheeses, such as those under French AOC regulations or Italian DOP regulations, require the

Table 10.1. Examples of Some New Chymosin Products

Product Name	Company and Type
AmericanPure	Sanofi Bio-Industries Calf rennet purified via ion exchange
Chy-max	Pfizer (Now Chr. Hansen) Fermentation—Using *Escherichia coli* K-12
Chymogen	Chr. Hansen Fermentation—Using *Aspergillus niger* var. *awamori*
ChymoStar	Rhône-Poulenc (Now Danisco) Fermentation—Using *Aspergillus niger* var. *awamori*
Maxiren	Gist-Brocades Fermentation—Using *Kluyveromyces marxianus* var. *lactis*
Novoren[a] (Marzyme GM)	Novo Nordisk Fermentation—using *Aspergillus oryzae*

[a]Novoren is not chymosin: it is the coagulating enzyme of *Mucor miehei* cloned into *As. oryzae*. The other products listed above are all 100% chymosin.

use of only calf rennet. A method has been published by the International Dairy Federation to detect fermentation-produced chymosin (Collin et al. 2003). This method uses immunochemical techniques (ELISA) to detect such chymosin in rennet solutions. It cannot be used for cheese.

Thus, traditional calf rennet and rennet substitutes derived from *Mucor Miehei, Cryphonectria, Parasitica,* and others are still used, especially for high cooking temperature cheeses such as Swiss and Mozzarella. Since the introduction of fermentation chymosin, the cost of traditional calf rennet has dropped considerably, and they both now cost approximately the same. Fungal rennets are still available at approximately 65% of the cost of the recombinant chymosin.

A coagulating enzyme of *Mucor miehei* cloned into *Aspergillus oryzae* was developed by Novo Nordisk of Denmark. This product hydrolyzes kappa-casein only at the 105-106 bond, and reduces protein losses to whey.

Rennet Production by Separation of Bovine Pepsin

Traditional calf rennet contains about 5% pepsin and almost 95% chymosin. Industrial purification of standard rennet to 100% chymosin rennet by the removal of pepsin was developed by Sanofi-Bioingredients Co. The process involves separation of the pepsin by ion exchange. Chymosin has no charge at pH 4.5, but bovine pepsin is negatively charged. Traditional commercial calf rennet is passed through an ion-exchange column containing positively charged ions. Bovine pepsin, due to its negative charge, is retained, and chymosin passes through and is collected, resulting in a 100% chymosin product (Pszczola 1989).

RENNET SUBSTITUTES

Improved farming practices, including improved genetics, have led over the years to significant increases in the milk production capacity of dairy cattle. As a result, while total milk production in the world has increased, the total cow population has declined. Hence, there has been a reduction in calf populations and the availability of traditional calf rennet. Furthermore, the increased practice of raising calves to an older age for meat production further reduced the numbers available for rennet production. It is for these reasons that a shortage of calf rennet occurred and substitutes were sought. Various proteolytic enzymes from plant, microbial, and animal sources were identified and developed for commercial applications. These enzymes should possess certain key characteristics to be successful rennet substitutes: (1) the clotting-to-proteolytic ratio should be similar to that of chymosin (i.e., the enzyme should have the capacity to clot milk without being excessively proteolytic); (2) the proteolytic specificity for beta-casein should be low because otherwise bitterness will occur in cheese; (3) the substitute product should be free of contaminating enzymes such as lipases; and (4) cost should be comparable to or lower than that of traditional rennet.

Animal

Pepsin derived from swine shows proteolytic activity between pH 2 and 6.5, but by itself it has difficulty in satisfactorily coagulating milk at pH 6.6. For this reason, it is used in cheese making as a 1:1 blend with rennet. Pepsin, used alone as a milk coagulator, shows high sensitivity to heat and is inclined to create bitter cheese if the concentrations added are not calculated and measured exactly (Kosikowski and Mistry 1997).

Plant

Rennet substitutes from plant sources are the least widely used because of their tendency to be excessively proteolytic and to cause formation of bitter flavors. Most such enzymes are also heat stable and require higher setting temperatures in milk. Plant sources include ficin from the fig tree, papain from the papaya tree, and bromelin from pineapple. A notable exception to these problems is the proteolytic enzymes from the flower of thistle *(Cynara cardunaculus),* as reported by Vieira de Sa and Barbosa (1970) and *Cynara humilis* (Esteves and Lucey 2002). These enzymes have been used successfully for many years in Portugal to make native ripened Serra cheese with excellent flavor and without bitterness.

In India, enzymes derived from *Withania coagulans* have been used successfully for cheese making, but commercialization has been minimal, especially with the development of genetically derived chymosin.

Microbial

Substitutes from microbial sources have been very successful and continue to be used. Many act like trypsin and have an optimum pH activity between 7 and 8 (Green 1977, Kosikowski and Mistry 1997). Microorganisms, including *Bacillus subtilis*, *B. cereus*, *B. polymyxa*, *Cryphonectria parasitica* (formerly *Endothia parasitica*), and *Mucor pusillus* Lindt (also known as *Rhizomucor pusillus)* and *Rh. miehei* have been extracted for their protease enzymes. The bacilli enzyme preparations were not suited for cheese making because of excessive proteolytic activity while the fungal-derived enzymes gave good results, but not without off flavors such as bitter. Commercial enzyme preparations isolated from *Cr. parasitica* led to good quality Emmental cheese without bitter flavor, but some cheeses, such as Cheddar, that use lower cooking temperatures reportedly showed bitterness. Enzyme preparations of *Rh. Miehei,* at recommended levels, produced Cheddar and other hard cheeses of satisfactory quality without bitter flavor. Fungal enzyme preparations are in commercial use, particularly in North America. In cottage cheese utilizing very small amounts of coagulating enzymes, shattering of curd was minimum in starter-rennet set milk and maximum in starter–microbial milk coagulating enzyme set milk (Brown 1971).

Enzymes derived from *Rh. miehei* and *Rh. pusillus* are not inactivated by pasteurization. Any residual activity in whey led to hydrolysis of whey proteins during storage of whey powder. This problem has been overcome by treating these enzymes with peroxides to reduce heat stability. Commercial preparations of *Rh. miehei* and *Rh. pusillus* are inactivated by pasteurization. The heat stability of *Cr. parasitica*–derived enzyme is similar to that of chymosin.

These fungal enzymes are also milk-clotting asparatic enzymes, and all except *Cr. parasitica* clot milk at the same peptide bond as chymosin. *Cryphonectria* clots kappa-casein at the 104-105 bond.

CHYMOSIN ACTION ON MILK

Chymosin produces a smooth curd in milk, and it is relatively insensitive to small shifts in pH that may be found in milk due to natural variations, does not cause bitterness over a wide range of addition, and is not proteolytically active if the cheese supplements other foods. Chymosin coagulates milk optimally at pH 6.0–6.4 and at 20–30°C in a two-step reaction, although the optimum pH of the enzyme is approximately 4 (Kosikowski and Mistry 1997, Lucey 2003). Optimum temperature for coagulation is approximately 40°C, but milk for cheese making is coagulated with rennet at 31–32°C because at this temperature the curd is rheologically most suitable for cheese making. Above pH 7.0, activity is lost. Thus, mastitic milk is only weakly coagulable, or does not coagulate at all.

Chymosin is highly sensitive to shaking, heat, light, alkali, dilutions, and chemicals. Stability is highest when stored at 7°C and pH 5.4–6.0 under dark conditions. Liquid rennet activity is destroyed at 55°C, but rennet powders lose little or no activity when exposed to 140°C. Standard single-strength rennet activity deteriorates at about 1% monthly when held cold in dark or plastic containers.

Single-strength rennet is usually added at 100–200 mL per 1000 kg of milk. It serves to coagulate milk and to hydrolyze casein during cheese ripening for texture and flavor development.

It should be noted that bovine chymosin has greater specificity for cow's milk than chymosin derived from kid or lamb. Similarly, kid chymosin is better suited for goat milk.

Milk contains fat, protein, sugar, salts, and many minor components in true solution, suspension, or emulsion. When milk is converted into cheese, some of these components are selectively concentrated as much as 10-fold, but some are lost to whey. It is the fat, casein, and insoluble salts that are concentrated. The other components are entrapped in the cheese serum or whey, but only at about the levels at which they existed in the milk. These soluble components are retained, depending on the degree that the serum or whey is retained in the cheese. For example, in a fresh Cheddar cheese, the serum portion is lower in volume than in milk. Thus, the soluble component percentage of the cheese is smaller.

In washed curd cheeses such as Edam or Brick the above relationship does not hold, and lactose, soluble salts, and vitamins in the final cheese are reduced considerably. Approximately 90% of living bacteria in the cheese-milk, including starter bacteria, are trapped in the cheese curd. Natural milk enzymes and others are, in part, preferentially absorbed on fat and protein, and thus a higher concentration remains with the cheese. In rennet coagulation of milk and the subsequent removal of much of the whey or serum, a selective separation of the milk components occurs, and in the resulting concentrated curd mass, many biological agents become active, marking the beginning of the final product, cheese.

Milk for ripened cheese is coagulated at a pH above the isoelectric point of casein by special proteolytic enzymes, which are activated by small amounts of lactic acid produced by added starter bacteria. The curds are sweeter and more shrinkable and pliable than those of fresh, unripened cheeses, which are produced by isoelectric precipitation. These enzymes are typified by chymosin that is found in the fourth stomach, or abomasum, of a young calf.

The isoelectric condition of a protein is that at which the net electric charge on a protein surface is zero. In their natural state in milk, caseins are negatively charged, and this helps maintain the protein in suspension. Lactic acid neutralizes the charge on the casein. The casein then precipitates as a curd at pH 4.6, the isoelectric point, as in cottage or cream cheese and yogurt. For some types of cheeses this type of curd is not desirable because it would be too acid, and the texture would be too firm and grainy. For ripened cheeses such as Cheddar and Swiss, a sweeter curd is desired, so casein is precipitated as a curd near neutrality, pH 6.2, by chymosin in 30–45 minutes at 30–32°C. This type of curd is significantly different than isoelectric curd and much more suitable for ripened cheeses. Milk proteins other than casein are not precipitated by chymosin. The uniform gel formed is made up of modified casein with fat entrapped within the gel.

Milk Coagulation and Protein Hydrolysis by Chymosin

The process of curd formation by chymosin is a complex process and involves various interactions of the enzyme with specific sites on casein, temperature and acid conditions, and calcium. Milk is coagulated by chymosin into a smooth gel capable of extruding whey at a uniformly rapid rate. Other common proteolytic enzymes such as pepsin, trypsin, and papain coagulate milk too, but may cause bitterness and loss of yield and are more sensitive to pH and temperature changes.

An understanding of the structure of casein in milk has contributed to the explanation of the mechanism of chymosin action in milk. Waugh and von Hippel (1956) began to unravel the heterogeneous nature of casein with their observations on kappa-casein. Casein is comprised of 45–50% alpha-s-casein, 25–35% beta-casein, 8–15% kappa-casein, and 3–7% gamma-casein. Various subfractions of alpha-casein have also been identified, as have genetic variants. Each casein fraction differs in sensitivity to calcium, solubility, amino acid makeup and electrophoretic mobility.

An understanding of the dispersion of casein in milk has provoked much interest. Several models of casein micelles have been proposed including the coat-core, internal structure, and subunit models. With the availability of newer analytical techniques such as three-dimensional X-ray crystallography, work on casein micelle modeling is continuing, and additional understanding is being gained. The subunit model (Rollema 1992) is widely accepted. It suggests that the casein micelle is made up of smaller submicelles that are attached by calcium phosphate bonds. Hence, the micelle consists of not only pure casein (approximately 93%), but also minerals (approximately 7%), primarily calcium and phosphate. Most of the kappa-casein is located on the surface of the micelle. This casein has few phosphoserine residues and hence is not affected by ionic calcium. It is located on the surface of the micelle and provides for stability of the micelle and protects the calcium-sensitive and hydrophobic interior, including alpha- and beta- caseins from ionic calcium. These two caseins have phosphoserine residues and high calcium-binding affinity. The carbohydrate, *N*-acetly neuramic acid, which is attached to kappa-casein, makes this casein hydrophilic. Thus, the casein micelle is suspended in milk as long as the properties of kappa-casein remain unchanged. More recently the dual-bonding model has been proposed (Horne 1998).

Introducing chymosin to the cheese-milk, normally at about 32°C, destabilizes the casein micelle in a two-step reaction, the first of which is enzymatic (or primary) and the second, nonenzymatic (or secondary) (Kosikowski and Mistry 1997, Lucey 2003). These two steps are separate but cannot be visually distinguished: only the appearance of a curd signifies the completion of both steps. The primary phase was probably first observed by Hammersten in the late 1800s (Kosikowski and Mistry 1997) and must occur before the secondary phase begins.

In the primary phase, chymosin cleaves the phenylalanine-methionine bond (105-106) of kappa-casein, thus eliminating its stabilizing action on calcium-sensitive alpha-s- and beta-caseins. In the secondary phase, the micelles without intact kappa-casein aggregate in the presence of ionic calcium in milk and form a gel (curd). This mechanism may be summarized as follows (Fig. 10.1):

$$\text{kappa-Casein} \xrightarrow{\text{chymosin}} \text{para-kappa-Casein} + \text{glycomacropeptide (Soluble in whey)}$$

$$\text{Para-kappa-Casein} \xrightarrow[\text{pH 6.4-6.0}]{Ca^{++}} \text{Dicalcium para-kappa-Casein}$$

Figure 10.1. Destabilization of the casein micelle by introduction of chymosin.

The macropeptide, also known as caseinmacropeptide, contains approximately 30% amino sugar, hence the name glycomacropeptide (Lucey 2003). In the ultrafiltration processes of cheese making, this macropeptide is retained with whey proteins to increase cheese yields significantly. In the conventional cheese-making process, the macropeptide is found in whey.

The two-step coagulation does not fully explain the presence of the smooth curd or coagulum. Beau, many years ago, established that a strong milk gel arose because the fibrous filaments of paracasein cross-linked to make a lattice. Bonding at critical points between phosphorus, calcium, free amino groups, and free carboxyl groups strengthened the lattice, with its entrapped lactose and soluble salts. A parallel was drawn between the gel formed when only lactic acid was involved as in fresh, unripened cheese and when considerable chymosin was involved as in hard, ripened cheese. Both gels were considered originally as starting with a fibrous protein cross bonding, but the strictly lactic acid curd was considered weaker, because not having been hydrolyzed by a proteolytic enzyme, it possessed less bonding material. The essential conditions for a smooth gel developing in either case were sufficient casein, a quiescent environment, moderate temperature, and sufficient time for reaction.

FACTORS AFFECTING CHYMOSIN ACTION IN MILK

Rennet coagulation in milk is influenced by many factors that ultimately have an impact on cheese characteristics (Kosikowski and Mistry 1997). The cheese maker is usually able to properly control these factors, some of which have a direct affect on the primary phase and others on the secondary.

Chymosin Concentration

Chymosin concentration affects gel firmness (Lomholt and Qvist 1999), and enzyme kinetics have been used to study these effects. An increase in the chymosin concentration (larger amounts of rennet added to milk) reduces the total time required for rennet clotting, measured by the appearance of the first curd particle. As a result, the secondary phase of rennet action will also proceed much earlier, with the net result of an increase in the rate of increase in gel firmness. This property of chymosin is sometimes used to control the rennet coagulation in concentrated milks, which have a tendency to form firmer curd because of the higher protein content. Lowering the amount of rennet reduces the rate of curd firming. It should also be noted that with an alteration in the amount of rennet added to milk, there also will be a change in the amount of residual chymosin in the cheese.

Temperature

The optimal condition for curd formation in milk with chymosin is 40–45°C, but this temperature is not suitable for cheese making. Rennet coagulation for cheese making generally occurs at 30—35°C for proper firmness. At lower temperatures rennet clotting rate is significantly reduced, and at refrigeration temperatures virtually no curd is obtained.

pH

Chymosin action in milk is optimal at approximately pH 6. This is obtained with starter bacteria. If the pH is lowered further rennet clotting occurs at a faster rate, and curd firmness is reduced. Lowering pH also changes the water-holding capacity, which will have an impact on curd firmness.

Calcium

Calcium has a significant impact on rennet curd, though it does not have a direct effect on the primary phase of rennet action. Addition of ionic calcium, as in the form of calcium chloride, for example, reduces the rate of rennet clot formation time and also increases rennet curd firmness. Similarly if the calcium content of milk is lowered by approximately 30%, coagulation does not occur. Milks that have a tendency to form weak curd may be fortified with calcium chloride prior to the addition of rennet.

Milk Processing

Heating milk to temperatures higher than approximately 70°C leads to delayed curd formation and weak rennet curd (Vasbinder et al. 2003). In extreme cases, there may be no curd at all. These effects are a result of the formation of a complex involving disulphide linkages between kappa-casein and beta-lactoglobulin under high heat treatment. Under these conditions the 105-106 bond in kappa-casein is inaccessible for chymosin action. This effect is generally not reversible. Under mild overheating conditions, the addition of 0.02% calcium chloride may help obtain firm rennet curd. Maubois et al. have developed a process for reversing this effect by the use of ultrafiltration. Increasing the protein content by ultrafiltration before or after UHT treatment restores curd-forming ability (Maubois et al. 1972). According to Ferron-Baumy et al. (1991), such a phenomenon results from lowering the zeta potential of casein micelles on ultrafiltration.

Homogenization has a distinct impact on milk rennet coagulation properties. Homogenized milk produces softer curd, but when only the cream portion of milk is homogenized, rennet curd becomes firmer (Nair et al. 2000). This is possibly because of the reduction of fat globule size due to homogenization and coating of the fat globule surface with casein. These particles then act as casein micelles.

Concentrating milk prior to cheese making is now a common practice. Techniques include evaporative concentration, ultrafiltration, and microfiltration. Each of these procedures increases the casein concentration in milk; thus the rate of casein aggregation during the secondary phase increases. Rennet coagulation usually occurs at a lower degree of kappa-casein hydrolysis, and rennet curd is generally firmer. For example, in unconcentrated milk approximately 90% kappa-casein must be hydrolyzed by chymosin before curd formation, but in ultrafiltered milk only 50% must be hydrolyzed. (Dalgleish 1992). Under high concentration conditions, rennet curd is extremely firm, and difficulties are encountered in cutting curd using traditional equipment. In cheeses manufactured from highly concentrated milks, rennet should be properly mixed in the milk mixture to prevent localization of rennet action.

Genetic Variants

Protein polymorphism (genetic variants) of kappa-casein has been demonstrated to have an effect on rennet coagulation (Marzialli and Ng-Kwai-Hang 1986). Protein polymorphism refers to a small variation in the makeup of proteins due to minor differences in the amino acid sequence. Examples include kappa-casein AA, AB, or BB, beta-lactoglobulin AA, AB, and so on. Milk with kappa-casein BB variants forms a firmer rennet curd because of increased casein content associated with this variant of kappa-casein. Some breeds of cows, Jerseys in particular, have larger proportions of this variant. The BB variant of beta-lactoglobulin is also associated with higher casein content in milk.

EFFECT OF CHYMOSIN ON PROTEOLYSIS IN CHEESE

Once rennet curd has been formed and a fresh block of cheese has been obtained, the ripening process begins. During ripening, numerous biochemical reactions occur and lead to unique flavor and texture development. During this process residual chymosin that is retained in the cheese makes important contributions to the ripening process (Kosikowski and Mistry 1997, Sousa et al. 2001). As the cheese takes form, chymosin hydrolyzes the paracaseins into peptides optimally at pH 5.6 and creates a peptide pool for developing flavor complexes (Kosikowski and Mistry 1997).

The amount of rennet required to coagulate milk within 30 minutes can be considerably below that required to properly break down the paracaseins during cheese ripening. Consequently, using too little rennet in cheese making retards ripening, as discerned by the appearance of the cheese and its end products. A lively, almost translucent looking cheese, after ripening several months, indicates adequate rennet; an opaque, dull looking cheese indicates inadequate amounts. Apparently, conversion to peptides and later hydrolysis to smaller molecules by bacterial enzymes give the translucent quality. Residual rennet in cheese hydrolyzes alpha-s-1-casein to alpha-s-1-I-casein, which leads to a desirable soft texture in the aged cheese (Lawrence et al. 1987).

Electrophoretic techniques (Ledford et al. 1966) have demonstrated that in most ripening Cheddar cheese from milk coagulated by rennet, para-beta-casein largely remains intact while para-alpha-casein is highly degraded. According to Fox et al. (1993), if there is too much moisture in cheese or too little salt, the residual chymosin will produce bitter peptides due to excessive proteolysis. In cheeses with high levels of beta-lactoglobulin, as in cheeses made by ultrafiltration, proteolysis by residual chymosin is retarded because of partial inhibition of chymosin by the whey protein (Kosikowski and Mistry 1997).

Edwards (1969) found that milk-coagulating enzymes from the mucors hydrolyze the paracaseins somewhat similarly to rennet, while enzymes from *Cr. parasitica* and the papaya plant completely hydrolyze para-alpha- and para-beta-caseins during ripening.

Proteolytic activity of rennet substitutes, especially microbial and fungal substitutes, is greater than that of chymosin. Cheese yield losses, though small, do occur with these rennets. Barbano and Rasmussen (1992) reported that fat and protein losses to whey were higher with *Rh. miehei* and *Rh. pusillus* compared with those for fermentation-derived chymosin. As a result, the cheese yield efficiency was higher with the latter than with the former two microbial rennets.

EFFECT OF CHYMOSIN ON CHEESE TEXTURE

The formation of a rennet curd is the beginning of the formation of a cheese mass. Thus, a ripened cheese assumes its initial biological identity in the vat or press. This is where practically all of the critical components are assembled and the young cheese becomes ready for further development.

Undisturbed fresh rennet curd is not cheese because it still holds considerable amounts of water and soluble constituents that must be removed before any resemblance to a cheese occurs. This block of curd is cut into thousands of small cubes. Traditionally, this is done with wire knives (Fig. 10.2) but in large automated vats, built-in blades rotate in one direction at a selected speed to systematically cut the curd. Then the whey, carrying with it

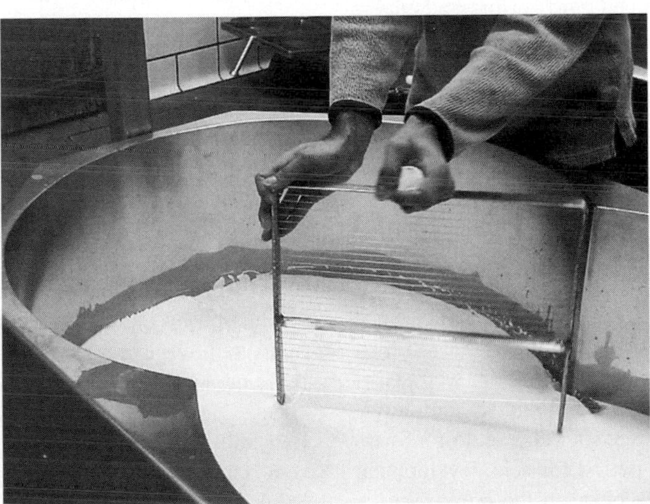

Figure 10.2. Cutting of rennet curd.

lactose, whey proteins, and soluble salts, streams from the cut pores, accelerated by gentle agitation, slowly increasing temperature, and a rising rate of lactic acid production. These factors are manipulated to the degree desired by the cheese maker. Eventually the free whey is separated from curd.

Optimum acid development is essential for forming rennet curd and creating the desired cheese mass. This requires viable, active microbial starters to be added to milk before the addition of rennet. These bacteria should survive the cheese making process.

Beta-D-galactosidase (lactase) and phosphorylating enzymes of the starter bacteria hydrolyze lactose and initiate the glucose-phosphate energy cycles for the ultimate production of lactic acid through various intermediary compounds. Lactose is hydrolyzed to glucose and galactose. Simultaneously, the galactose is converted in the milk to glucose, from which point lactic acid is produced by a rather involved glycolytic pathway. In most cheese milks, the conversion of glucose to lactic acid is conducted homofermentatively by lactic acid bacteria, and the conversions provide energy for the bacteria. For each major cheese type, lactic acid must develop at the correct time, usually not too rapidly, nor too slowly, and in a specific concentration.

The cheese mass, which evolves as the whey drains and the curds coalesce, contains the insoluble salts, $CaHPO_4$ and $MgHPO_4$, which serve as buffers in the pressed cheese mass, and should the acid increase too rapidly during cooking, they will dissolve in the whey. In the drained curds or pressed cheese mass under these circumstances, since no significant pool of insoluble divalent salt remains for conversion into buffering salts, the pH will lower to 4.7—4.8 from an optimum 5.2, leading to a sour acid–ripened cheese. Thus, conservation of a reserve pool of insoluble salts is necessary until the cheese mass is pressed. Thereafter, the insoluble salts are changed into captive soluble salts by the lactic acid and provide the necessary buffering power to maintain optimum pH at a level that keeps the cheese sweet. Such retention also is aided or controlled by the skills of the cheese maker. For Cheddar, excess wetness of the curd before pressing or too rapid an acid development in the whey causes sour cheese. For a Swiss or Emmental, the time period for acid development differs; most of it takes place in the press, but the principles remain the same (Kosikowski and Mistry 1997).

The development of proper acid in a curd mass controls the microbial flora. Sufficient lactic acid produced at optimum rates favors lactic acid bacteria and discriminates against spoilage or food poisoning bacteria such as coliforms, clostridia, and coagulase-positive staphylococci. But its presence does more than that, for it transforms the chemistry of the curd to provide the strong bonding that is necessary for a smooth, integrated cheese mass.

As discussed earlier, chymosin action on milk results in curd mass involving dicalcium paracasein. Dicalcium paracasein is not readily soluble, stretchable, or possessive of a distinguished appearance. However, if sufficient lactic acid is generated, this compound changes. The developing lactic acid solubilizes considerable calcium, creating a new compound, monocalcium paracasein. The change occurs relatively quickly, but there is still a time requirement. For example, a Cheddar cheese curd mass, salted prior to pressing, may show an 8:2 ratio of dicalcium paracasein to monocalcium paracasein; after 24 hours in the press, it is reversed, 2:8. Monocalcium paracasein has interesting properties. It is soluble in warm 5% salt solution, it can be stretched and pulled when warmed, and it has a live, glistening appearance. The buildup of monocalcium paracasein makes a ripened cheese pliable and elastic. Then, as more lactic acid continues to strip off calcium, some of the monocalcium paracasein changes to free paracasein as follows:

Dicalcium paracasein + lactic acid → monocalcium paracasein + calcium lactate

Monocalcium paracasein + lactic acid → free paracasein + calcium lactate

Free paracasein is readily attacked by many enzymes, contributing to a well-ripened cheese. The curd mass becomes fully integrated upon the uniform addition of sodium chloride, the amount of which varies widely for different cheese types. Salt directly influences flavor and arrests sharply the acid production by lactic acid starter bacteria. Also, salt helps remove excess water from the curd during pressing and lessens the chances for a weak-bodied cheese. Besides controlling the lactic acid fermentation, salt partially solubilizes monocalcium paracasein. Thus, to a natural ripened cheese or to one that is heat processed, salt helps give a smoothness and plasticity of body that is not fully attainable in its absence.

Texture of cheese is affected significantly during ripening, especially during the first 1–2 weeks according to Lawrence et al. (1987). During this period, the alpha-s-1-casein is hydrolyzed to alpha-s-1-I by residual chymosin, making the body softer and smoother. Further proteolysis during ripening continues to influence texture.

REFERENCES

Andren A. 2003. Rennets and coagulants. In: H Roginski, JW Fuquay, PF Fox, editors. Encyclopedia of Dairy Sciences. London: Academic Press, London. Pp. 281–286.

Banks JM. 1992.Yield and quality of Cheddar cheese produced using a fermentation-derived calf chymosin. Milchwissenschaft 47:153–201.

Barbano DM, Rasmussen RR. 1992. Cheese yield performance of fermentation-produced chymosin and other milk coagulation. J Dairy Sci 75:1–12.

Biner VEP, Young D, Law BA. 1989. Comparison of Cheddar cheese made with a recombinant calf chymosin and with standard calf rennet. J Dairy Res 56:657–64.

Brown GD. 1971. Microbial enzymes in the production of Cottage cheese. MS Thesis. Ithaca NY: Cornell University.

Collin J-C, Repelius C, Harboe HK. 2003. Detection of fermentation produced chymosin. In: Bulletin 380. Brussels Belgium: Int Dairy Fed. Pp. 21–24.

Dalgleish DG. 1992. Chapter 16. The enzymatic coagulation of milk. In: PF Fox, editor. Advanced Dairy Chemistry. Vol 1, Proteins. Essex England: Elsevier Sci Publ.Ltd.

Duxbury DD. 1990. Cheese enzyme is first rDNA technology derived food ingredients granted GRAS approved by FDA. Food Proc (June): 46.

Edwards JL. 1969. Bitterness and proteolysis in Cheddar cheese made with animal, microbial, or vegetable rennet enzymes. PhD Thesis. Ithaca NY: Cornell University.

Esteves CLC, Lucey JA. 2002. Rheological properties of milk gels made with coagulants of plant origin and chymosin. Int Dairy J 12:427–434.

Ferron-Baumy C, Maubois JL, Garric G, Quiblier J P. 1991. Coagulation présure du lait et des rétentats d'ultrafiltration. Effets de divers traitements thermiques. Lait 71:423–434.

Farntiam MG, inventor. 1964, September 27. Cheese modifying enzyme product. U.S. patent 3,531,329.

Foltmann B. 1993. General and molecular aspects of rennets. In: PF Fox, editor, Cheese: Chemistry, Physics and Microbiology Vol. 1, 2nd edition,, pp. 37–68. New York: Chapman and Hall.

Fox PF, Law J, McSweeney PH, Wallace J. 1993. Biochemistry of cheese ripening. In: PF Fox, editor. Cheese: Chemistry, Physics and Microbiology, 2nd edition, vol. 1. New York: Chapman and Hall.

Green ML. 1977. Review of progress of dairy science: Milk coagulants. J Dairy Res 44:159–88.

Green ML, Angel S, Lowe PA, Marston AO. 1985. Cheddar cheesemaking with recombinant calf chymosin synthesized in Escherichia. J Dairy Res 52:281.

Hicks CL, O'Leary J, Bucy J. 1988. Use of recombinant chymosin in the manufacture of Cheddar and Colby cheeses. J Dairy Sci 71:1127–1131.

Horne DS. 1998. Casein interactions: Casting light on the black boxes, the structure in dairy products. Int Dairy J 8:171–177.

IDF. 1992. Fermentation produced enzymes and accelerated ripening in cheesemaking. Bulletin No. 269. Brussels Belgium: Int Dairy Fed.

Kosikowski FV, Mistry VV. 1997. Cheese and Fermented Milk Foods. Vol 1, Origins and Principles. Great Falls, Virginia: FV Kosikowski LLC.

Lawrence RC, Creamer LK, Gilles J. 1987. Texture development during cheese ripening. J Dairy Sci 70:1748–1760.

Ledford RA, O'Sullivan AC, Nath KR. 1966. Residual casein fractions in ripened cheese determined by polyacrylamide-gel electrophoresis. J Dairy Sci 49: 1098–1101.

Lomholt SB, Qvist KB. 1999. Gel firming rate of rennet curd as a function of rennet concentration. Int Dairy J 9:417–418.

Lucey JA. 2003. Rennet coagulation in milk. In: H Roginski, JW Fuquay, PF Fox, editors, Encyclopedia of Dairy Sciences, pp. 286–293. London: Academic Press.

Marzialli AS, Ng-Kwai-Hang KF. 1986. Relationship between milk protein polymorphisms and cheese yielding capacity. J Dairy Sci 69:1193–1201.

Maubois JL, Mocquot G, Vassal L, inventors. 1972. Procédé traitement du lait et de sous-produits laitiers. *French Patent no.* 2 166 315.

Nair MG, Mistry VV, Oommen BS. 2000. Yield and functionality of Cheddar cheese as influenced by homogenization of cream. Int Dairy J 10:647–657.

Nishimori K, Kawaguchi Y, Hidaka M, Vozumi T, Beppu T. 1981. Cloning in *Escherichia coli* of the structured gene of pro-rennin the precursor of calf milk—clotting rennin. J Biochem 90:901–904.

Ortiz de Apodaca MJ, Amigo L, Ramos M. 1994. Study of milk-clotting and proteolytic activity of calf rennet, fermentation-produced chymosin, vegetable and microbial coagulants. Milchwissenschaft 49:13–16.

Pszczola DF. 1989. Rennet containing 100% chymosin increases cheese quality and yield. Food Technol 43(6): 84–89.

Rollema HS. 1992. Chapter 3. Casein association and micelle formation. In: PF Fox, editor. Advanced Dairy Chemistry. Vol. 1, Proteins. Essex England: Elsevier Sci Publ Ltd.

Sousa M J, Ardo Y, McSweeney PLH. 2001. Advances in the study of proteolysis during cheese ripening. Int Dairy J 11:327–345.

Uchiyama H, Vozumi T, Beppu T, Arimad K. 1980. Purification of pro rennin MRNA and its translation in vitro. Agric Biol Chem 44:1373–1381.

Vasbinder AJ, Rollema HS, de Kruif CG. 2003. Impaired rennetability of heated milk; study of enzymatic hydrolysis and gelation kinetics. J Dairy Sci 86:1548–1555.

Vieira de sa F, Barbosa M. 1970. Cheesemaking experiments using a clotting enzyme from cardon *(Cynara cardunculus)* Proceed 18th Int Dairy Cong 1E:288.

Waugh DG, Von Hippel PH. 1956. Kappa-casein and the stabilization of casein micelles. J Amer Chem Soc 78:4576–4582.

11
Starch Synthesis in the Potato Tuber

P. Geigenberger and A. R. Fernie*

INTRODUCTION

Starch is the most important carbohydrate used for food and feed purposes and represents the major resource of our diet. The total yield of starch in rice, corn, wheat, and potato exceeds 10^9 tons per year (Kossmann and Lloyd 2000, Slattery et al. 2000). In addition to its use in a nonprocessed form, due to the low cost incurred, extracted starch is processed in many different ways. Processed starch is subsequently used in multiple forms, for example in high-fructose syrup, as a food additive, or for various technical processes based on the fact that as a soluble macromolecule it exhibits high viscosity and adhesive properties (Table 11.1). The considerable importance of starch has made increasing the content and engineering the structural properties of plant starches major goals of both classical breeding and biotechnology over the last few decades (Smith et al. 1997, Sonnewald et al. 1997, Regierer et al. 2002). Indeed, since the advent and widespread adoption of transgenic approaches some 15 years ago gave rise to the discipline of molecular plant physiology, much information has been obtained concerning the potential to manipulate plant metabolism. For this chapter we intend to review genetic manipulation of starch metabolism in potato *(Solanum tuberosum).* Potato is one of the most important crops worldwide, ranking fourth in annual production behind the cereal species rice *(Oryza sativa),* wheat *(Triticum aestivum),* and maize *(Zea mais).* Although in Europe and North America the consumption of potatoes is mainly in the form of processed foodstuffs such as fried potatoes and chips, in less developed countries it represents an important staple food and is grown by many subsistence farmers. The main reasons for the increasing popularity of the potato in developing countries are the high nutritional value of the tubers combined with the simplicity of its propagation by vegetative amplification (Fernie and Willmitzer 2001). Since all potato varieties are true tetraploids and display a high degree of heterozygosity, genetics have played only a minor role in metabolic studies in this species. However, because the potato is a member of the *Solanaceae* family, it was amongst the first crop plants to be accessible to transgenic approaches. Furthermore, due to its relatively large size and metabolic homogeneity, the potato tuber represents a convenient experimental system for biochemical studies (Geigenberger 2003a).

In this chapter we will describe transgenic attempts to modify starch content and structure in potato tubers that have been carried out in the last two decades. In addition to describing biotechnologically significant results we will also detail fundamental research in this area that should enable future

*Corresponding author.

Table 11.1. Industrial Uses of Starch

Starch and Its Derivates	Processing Industry	Application
Amylose and amylopectin (polymeric starch)	Food	Thickener, texturants, extenders, low calorie snacks
	Paper	Beater sizing, surface sizing, coating
	Textile	Wrap sizing, finishing, printing
	Polymer	Absorbents, adhesives, biodegradable plastics
Products of starch hydrolysis, such as glucose, maltose, or dextrins	Food	Sweeteners or stabilizing agents
	Fermentation	Feedstock to produce ethanol, liquors, spirits, beer, etc.
	Pharmaceutical	Feedstock to produce drugs and medicine
	Chemical	Feedstock to produce organic solvents or acids

Source: Adapted from Jansson et al. 1997, with modification.

biotechnology strategies. However, we will begin by briefly describing starch, its structure, and its synthesis.

WHAT IS STARCH?

For many years it has been recognized that the majority of starches consist of two different macromolecules, amylose and amylopectin (Fig. 11.1), which are both polymers of glucose and are organized into grains that range in size from 1 μm to more than 100 μm. Amylose is classically regarded as an essentially linear polymer wherein the glucose units are linked through α-1-4-glucosidic bonds. In contrast, although amylopectin contains α-1-4-glucosidic bonds, it also consists of a high proportion of α-1-6-glucosidic bonds. This feature of amylopectin makes it a more branched, larger molecule than amy-

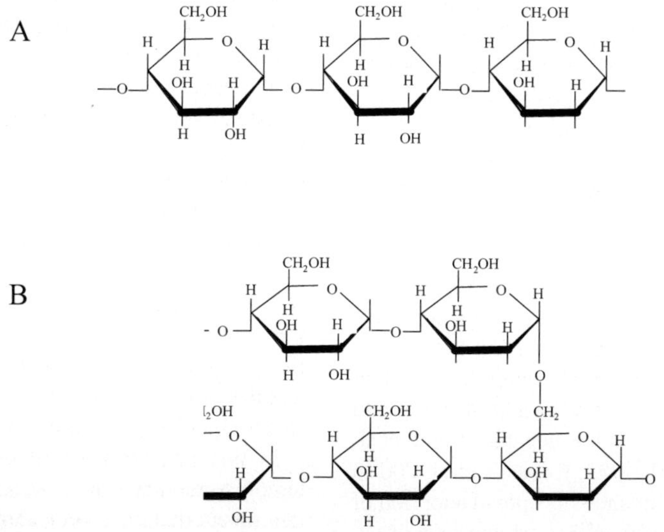

A

B

Figure 11.1. Starch structure: Amylose **(A)** is classically regarded as an essentially linear polymer wherein the glucose units are linked through α-1-4-glucosidic bonds. In contrast, although amylopectin contains α-1-4-glucosidic bonds, it also consists of a high proportion of α-1-6-glucosidic bonds **(B)**.

lose, having a molecular weight of 10^7–10^8 as opposed to 5×10^5–10^6. These molecules can thus be fractionated by utilizing differences in their molecular size as well as in their binding behavior. Moreover, amylose is able to complex lipids, while amylopectin can contain covalently bound phosphate, adding further complexity to their structure. In nature, amylose normally accounts between 20 and 30% of the total starch. However, the percentage of amylose depends on the species and the organ used for starch storage. The proportion of amylose to amylopectin and the size and structure of the starch grain give distinct properties to different extracted starches (properties important in food and industrial purposes; Dennis and Blakeley 2000). Starch grain size is also dependent on species and organ type. It is well established that starch grains grow by adding layers, and growth rings within the grain may represent areas of fast and slower growth (Pilling and Smith 2003); however, very little is known about how these highly ordered structures are formed in vivo. The interested reader is referred to articles by Buleon et al. (1998) and Kossmann and Lloyd (2000).

ROUTES OF STARCH SYNTHESIS AND DEGRADATION AND THEIR REGULATION

With the possible exception of sucrose, starch is the most important metabolite of plant carbohydrate metabolism. It is by far the most dominant storage polysaccharide and is present in all major organs of most plants, in some instances at very high levels. Due to the high likelihood of starch turnover, its metabolism is best considered as the balance between the antagonistic operation of pathways of synthesis and degradation.

To investigate the regulation of starch synthesis in more detail, growing potato tubers have been used as a model system. Unlike many other tissues the entry of sucrose into metabolism is relatively simple, in that it is unloaded symplasmically from the phloem, degraded via sucrose synthase to fructose and UDP-glucose, which are converted to hexose monophosphates by fructokinase and UDPglucose pyrophosphorylase, respectively (Geigenberger 2003a). In contrast to sucrose degradation, which is localized in the cytosol, starch is synthesized predominantly, if not exclusively, in the plastid. The precise pathway of starch synthesis depends on the form in which carbon crosses the amyloplast membrane (Fig. 11.2). This varies between species and has been the subject of considerable debate (Keeling et al. 1998, Hatzfeld and Stitt 1990, Tauberger et al. 2000). Categorical evidence that carbon enters potato tuber, *Chenopodium rubrum* suspension cell, maize endosperm, and wheat endosperm amyloplasts in the form of hexose monophosphates rather than triose phosphates was provided by determination of the degree of randomization of radiolabel in glucose units isolated from starch following incubation of the various tissues with glucose labeled at the C1 or C6 positions (Keeling et al. 1988, Hatzfeld and Stitt 1990). The cloning of a hexose monophosphate transporter from potato and the finding that the cauliflower homolog is highly specific for glucose-6-phosphate provides strong support for this theory (Kammerer et al. 1998). Further evidence in support of glucose-6-phosphate import was provided by studies of transgenic potato lines in which the activity of the plastidial isoform of phosphoglucomutase was reduced by antisense inhibition, leading to a large reduction in starch content of the tubers (Tauberger et al. 2000). These data are in agreement with the observations that heterotrophic tissues lack plastidial fructose-1,6-bisphosphatase expression and activity (Entwistle and Rees 1990, Kossmann et al. 1992). The results of recent transgenic and immunolocalization experiments have indicated that the substrate for uptake is most probably species specific, with clear evidence for the predominant route of uptake in the developing potato tuber being in the form of glucose-6-phosphate. By contrast, in barley, wheat, oat, and possibly maize the predominant form of uptake, at least during early stages of seed endosperm development, is as ADP-glucose (Neuhaus and Emes 2000).

Irrespective of the route of carbon import, ADP-glucose pyrophosphorylase (AGPase, EC 2.7.7.27) plays an important role in starch synthesis, catalyzing the conversion of glucose-1-phosphate and ATP to ADP-glucose and inorganic pyrophosphate. Inorganic pyrophosphate is subsequently metabolized to inorganic phosphate by a highly active inorganic pyrophosphatase within the plastid. AGPase is generally considered as the first committed step of starch biosynthesis since it produces ADP-glucose, the direct precursor for the starch polymerizing reactions catalyzed by starch synthase (EC 2.4.1.21) and branching enzyme (EC 2.4.1.24; Fig. 11.3). These three enzymes appear to be involved in starch

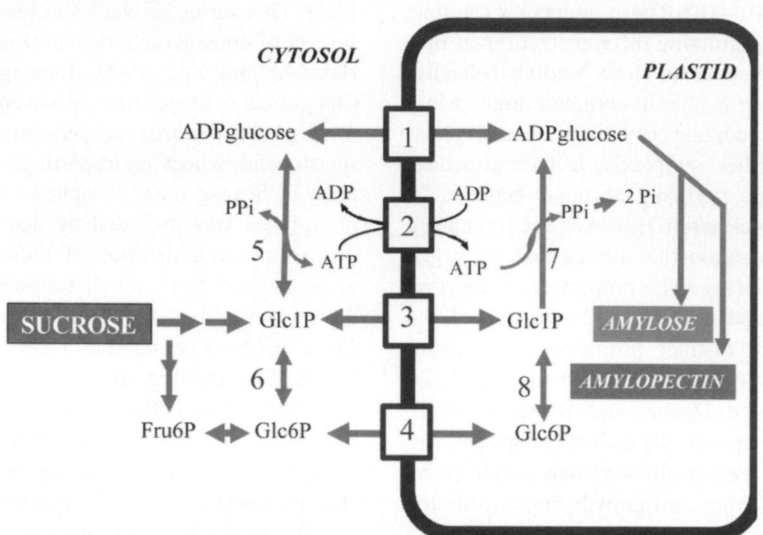

Figure 11.2. Pathways of sucrose to starch conversion in plants: **1**, ADP-glucose transporter, **2**, ATP/ADP translocator, **3**, glucose-1-phoshate (Glc1P) translocator, **4**, glucose-6-phosphate (Glc6P) translocator, **5**, cytosolic ADP-glucose pyrophosphorylase (AGPase), **6**, cytosolic phosphoglucomutase, **7**, plastidial AGPase, **8**, plastidial phosphoglucomutase. In growing potato tubers, incoming sucrose is degraded by sucrose synthase to fructose and UDPglucose and subsequently converted to fructose-6-phosphate (Fru6P) and Glc1P by fruktokinase and UDPglucose pyrophosphorylase, respectively (not shown in detail). The conversion of Fru6P to Glc6P in the cytosol is catalyzed by phosphoglucoisomerase, and the cleavage of PP_i to 2 P_i in the plastid is catalyzed by inorganic pyrophosphatase. In potato tubers, there is now convincing evidence that carbon enters the plastid almost exclusively via the Glc6P translocator, whereas in cereal endosperm the predominant form of uptake is as ADPglucose.

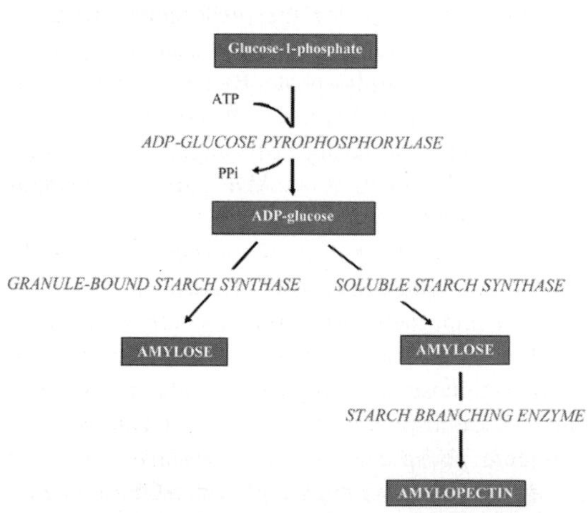

Figure 11.3. Routes of amylose and amylopectin synthesis within potato tuber amyloplasts. For simplicity, the possible roles of starch-degrading enzymes in trimming amylopectin have been neglected in this scheme.

synthesis in all species. With the exceptions of the cereal species described above, which also have a cytosolic isoform of AGPase, these reactions are confined to the plastid. The activities of AGPase and starch synthase are sufficient to account for the rates of starch synthesis in a wide variety of photosynthetic and heterotrophic tissues (Smith et al. 1997). Furthermore, changes in the activities of these enzymes correlate with changes in the accumulation of starch during development of storage organs (Smith et al. 1995). AGPase from a range of photosynthetic and heterotrophic tissues is well established to be inhibited by phosphate, which induces sigmoidal kinetics, and to be allosterically activated by 3-phosphoglycerate (3PGA), which relieves phosphate inhibition (Preiss 1988; Fig. 11.4). There is clear evidence that allosteric regulation of AGPase is important in vivo to adjust the rate of starch synthesis to changes in the rate of respiration that go along with changes in the levels of 3PGA, and an impres-

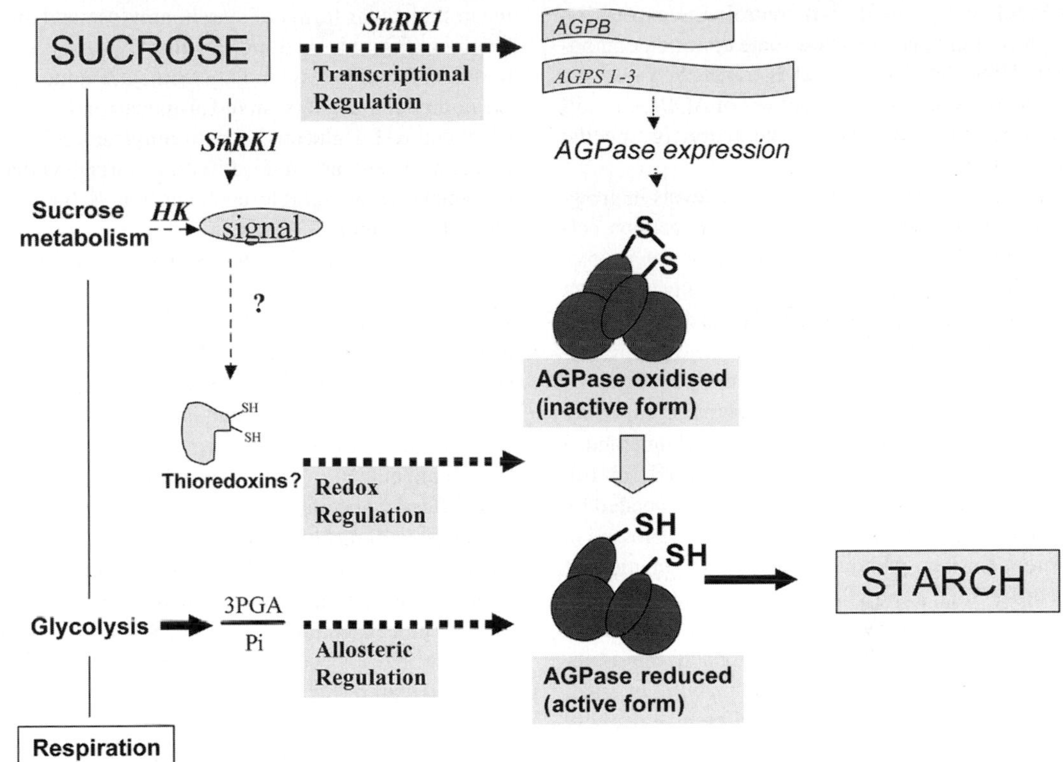

Figure 11.4. Regulation of starch synthesis in potato tubers: ADPGlc-pyrophosphorylase (AGPase) is a key regulatory enzyme of starch biosynthesis. It is regulated at different levels of control, involving allosteric regulation, regulation by posttranslational redox modification and transcriptional regulation. Redox regulation of AGPase represents a novel mechanism regulating starch synthesis in response to changes in sucrose supply (Tiessen et al. 2002). There are at least two separate sugar-sensing pathways leading to posttranslational redox activation of AGPase, one involving an SNF1-like protein kinase (SnRK1), the other involving hexokinase (HK) (Tiessen et al. 2003).

sive body of evidence has been provided that there is a strong correlation between the 3PGA and ADP-glucose levels and the rate of starch synthesis under a wide variety of environmental conditions (Geigenberger et al. 1998, Geigenberger 2003a).

More recently, an important physiological role for posttranslational redox regulation of AGPase has been established (Tiessen et al. 2002). In this case, reduction of an intermolecular cysteine bridge between the two small subunits of the heterotetrameric enzyme leads to a dramatic increase of activity, due to increased substrate affinities and sensitivity to allosteric activation by 3PGA (Fig. 11.4). Redox activation of AGPase *in planta* correlated closely with the potato tuber sucrose content across a range of physiological and genetic manipulations (Tiessen et al. 2002), indicating that redox modulation is part of a novel regulatory loop that directs incoming sucrose towards storage starch synthesis (Tiessen et al. 2003). Crucially, it allows the rate of starch synthesis to be increased in response to sucrose supply and independently of any increase in metabolite levels (Fig. 11.4), and it is therefore an interesting target for approaches to improving starch yield (see below). There are at least two separate sugar signaling pathways leading to posttranslational redox activation of AGPase, one involving an SNF1-like protein kinase (SnRK1), the other involving hexokinase (Tiessen et al. 2003). Both hexokinase and SnRK1 have previously been shown to be involved in the transcriptional regulation of many plant genes. Obviously, the transduction pathway that regulates the reductive activation of AGPase in plastids and the regulatory network that controls the expression of genes in the cytosol share some common components. How the sugar signal is transferred into the plastid and leads to redox changes of AGPase is still unknown, but may involve interaction of specific thioredoxins with AGPase.

Subcellular analyses of metabolite levels in growing potato tubers have shown that the reaction catalyzed by AGPase is far from equilibrium in vivo (Tiessen et al. 2002), and consequently, the flux through this enzyme is particularly sensitive to regulation by the above-mentioned factors. It is interesting to note that plants contain multiple forms of AGPase, which is a tetrameric enzyme comprising two different polypeptides, a small subunit and a large subunit, both of which are required for full activity (Preiss 1988). Both subunits are encoded by multiple genes, which are differentially expressed in different tissues. Although the precise function of this differential expression is currently unknown, it seems likely that these isoforms will differ in their capacity to bind allosteric regulators. If this is indeed the case, then different combinations of small and large subunits should show different sensitivity to allosteric regulation—such as those observed in tissues from cereals.

While there is a wealth of information on the regulation of the plastidial isoforms of AGPase—the only isoform present in the potato tuber—very little is known about cytosolic isoforms in other species. Several important studies provide evidence that the ADP-glucose produced in the cytosol can be taken up by the plastid (Sullivan et al. 1991, Shannon et al. 1998). From these studies and from characterization of mutants unable to transport ADP-glucose, it would appear that this is a predominant route for starch synthesis within maize. Despite these findings, the physiological significance of cytosolic ADP-glucose production remains unclear for a range of species. However, it has been calculated that the AGPase activity of the plastid is insufficient to account for the measured rates of starch synthesis in barley endosperm, suggesting that at least some of the ADP-glucose required for this process is provided by cytosolic production (Thorbjornsen et al. 1996).

It is interesting to note that the involvement of the various isoforms of AGPase in starch biosynthesis is strictly species dependent, whereas the various starch-polymerizing activities are ever present and responsible for the formation of the two different macromolecular forms of starch, amylose and amylopectin (Fig. 11.3). Starch synthases catalyze the transfer of the glucosyl moiety from ADP-glucose to the nonreducing end of an α-1,4-glucan and are able to extend α-1,4-glucans in both amylose and amylopectin. There are four different starch synthase isozymes—three soluble and one that is bound to the starch granule.

Starch branching enzymes (SBE), meanwhile, are responsible for the formation of α-1,6 branch points within amylopectin (Fig. 11.3). The precise mechanism by which this is achieved is unknown; however, it is thought to involve cleavage of a linear α-1,4-linked glucose chain and reattachment of the chain to form an α-1,6 linkage. Two isozymes of starch branching enzyme, SBEI and SBEII, are present and differ in specificity. The former preferentially branches unbranched starch (amylose), while the latter preferentially branches amylopectin. Furthermore, in vitro studies indicate that SBEII transfers smaller glucan chains than does SBEI and would therefore be expected to create a more highly branched starch (Schwall et al. 2000). The fact that the developmental expression of these isoforms correlates with the structural properties of starch during pea embryo development is in keeping with this suggestion (Smith et al. 1997). Apart from that, isoforms of isoamylase (E.C. 3.2.1.68) might be involved in debranching starch during its synthesis (Smith et al. 2003).

While the pathways governing starch synthesis are relatively clear, those associated with starch degradation remain somewhat controversial (Smith et al. 2003). The degradation of plastidial starch can proceed via phosphorolytic or hydrolytic cleavage mechanisms involving α-1-4-glucan phosphorylases or amylases, respectively. The relative importance of these different routes of starch degradation has been a matter of debate for many years. The question is whether they are, in fact, independent pathways, since oligosaccharides released by hydrolysis can be further degraded by amylases or, alternatively, by phosphorylases. In addition to this, the mechanisms responsible for the initiation of starch grain degradation in the plastid remain to be resolved, since starch grains have been found to be very stable and relatively resistant against enzymatic action in vitro.

More recently, molecular and genetic approaches have allowed rapid progress in clarifying the route of starch degradation in leaves of the model species *Arabidopsis thaliana* (Fig. 11.5). The phosphorylation of starch granules by glucan water dikinase (GWD, R1) has been found to be essential for the initiation of starch degradation in leaves and tubers (Ritte et al. 2002, Ritte et al. 2004). Transgenic potato plants (Lorberth et al. 1998) and *Arabidopsis* mutants (Yu et al. 2001) with decreased GWD activity showed a decrease in starch-bound phosphate and were severely restricted in their ability to degrade starch, leading to increased starch accumulation in leaves. The underlying mechanisms are still unknown. Phosphorylation of starch may change the structure of the granule surface to make it more susceptible to enzymatic attack or may regulate the extent to which degradative enzymes can attack the granule.

It is generally assumed that the initial attack on the starch granule is catalyzed by α-amylase (endoamylase). This enzyme catalyzes the internal cleavage of glucan chains from amylose or amylopectin, yielding branched and unbranched α-1,4-glucans, which are then subject to further digestion (Fig. 11.5). Debranching enzymes (isoamylase and pullulanase) are needed to convert branched glucans into

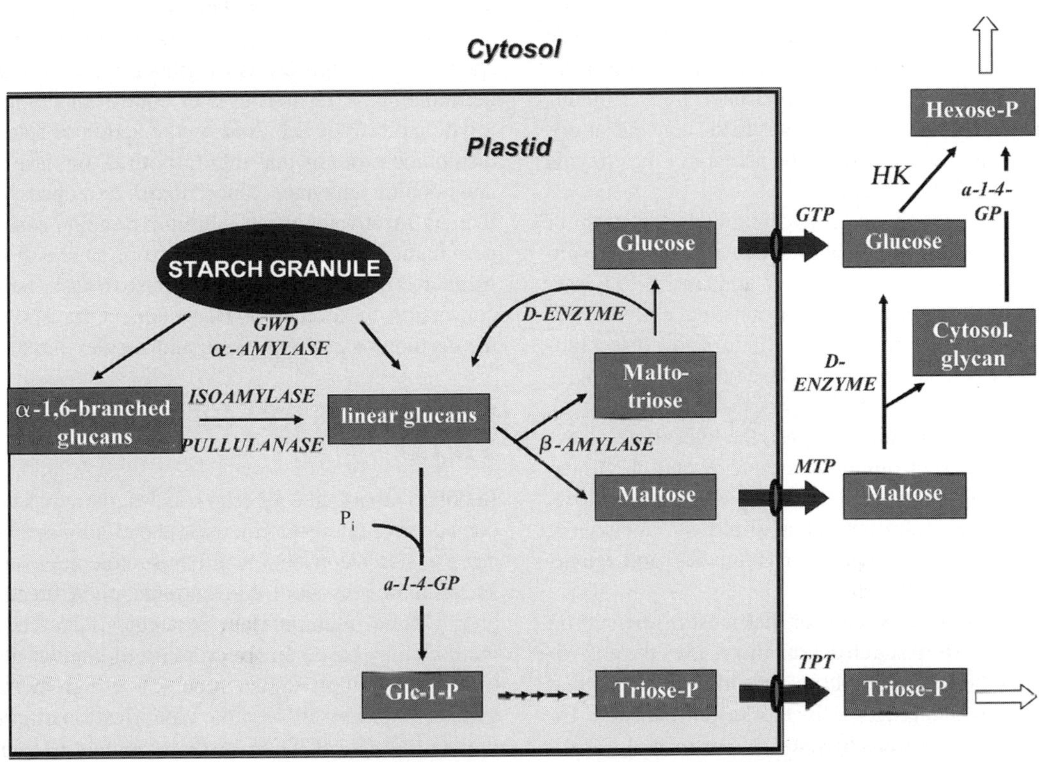

Figure 11.5. Outline of the pathway of plastidial starch degradation. The scheme is mainly based on recent molecular studies in *Arabidopsis* leaves (Ritte et al. 2002, Smith et al. 2003, Chia et al. 2004). After phosphorylation of glucans on the surface of the starch granule via glucan water dikinase (GWD; R1-protein), starch granules are attacked most probably by α-amylase and the resulting branched glucans subsequently converted to unbranched α-1,4-glucans via debranching enzymes (isoamylase and pullulanase). Linear glucans are metabolized by the concerted action of β-amylase and disproportionating enzyme (D-enzyme, glucan transferase) to maltose and glucose. The phosphorolytic degradation of linear glucans to Glc-1-P (glucose-1-phosphate) by α-1-4-glucan phosphorylase (α-1-4-GP) is also possible, but seems of minor importance under normal conditions. Most of the carbon resulting from starch degradation leaves the chloroplast via a maltose transporter in the inner membrane. Subsequent cytosolic metabolism of maltose involves the combined action of glucan transferase (D-enzyme), α-1-4-glucan phosphorylase (α-1-4-GP), and hexokinase (HK). Glucose transporter (GTP) and triose-P/P_i-translocator (TPT) are also shown.

linear glucans by cleaving the α-1,6 branch points. Further metabolism of linear glucans could involve phosphorolytic or hydrolytic routes. In the first case, α-glucan phosphorylase leads to the phosphorolytic release of Glc1-P, which can be further metabolized to triose-P within the chloroplast and subsequently exported to the cytosol via the triose-P/P_i translocator. Recent results show that the contribution of α-glucan phosphorylase to plastidial starch degradation is relatively small. Removal of the plastidial form of phosphorylase in *Arabdopsis* did not affect starch degradation in leaves of *Arabidopsis* (Zeeman et al. 2004) and potato (Sonnewald et al. 1995). It has been suggested that the phosphorolytic pathway could be more important to degrading starch under certain stress conditions (i.e., water stress; Zeeman et al. 2004). However, no regulatory properties have been described for glucan phosphorylase in plants, other than the effect of changes in the concentrations of inorganic phosphate on the activity of the enzyme (Stitt and Steup 1985).

In the second case, hydrolytic degradation of linear glucans in the plastid can involve the combined four action enzymes: α-amylase, β-amylase, α-glucosidase and disproportionating enzyme (D-enzyme, glucan transferase). There is now direct molecular evidence that β-amylase (exoamylase) plays a significant role in this process (Scheidig et al. 2002). This enzyme catalyzes the hydrolytic cleavage of maltose from the nonreducing end of a linear glucan polymer that is larger than maltotriose. Maltotriose is further metabolized by D-enzyme, producing new substrate for β-amylase and releasing glucose (Fig. 11.5).

Recent studies document that most of the carbon that results from starch degradation leaves the chloroplast in the form of maltose, providing evidence that hydrolytic degradation is the major pathway for mobilization of transitory starch (Weise et al. 2004). Elegant studies with *Arabidopsis* mutants confirmed this interpretation and identified a maltose transporter in the chloroplast envelope that is essential for starch degradation in leaves (Niittylä et al. 2004). The further metabolism of maltose to hexose-phosphates is then performed in the cytosol and is proposed to involve cytosolic forms of glycosyltransferase (D-enzyme; Lu and Sharkey 2004, Chia et al. 2004), α-glucan phosphorylase (Duwenig et al. 1997), and hexokinase, similar to maltose metabolism in the cytoplasm of *E. coli* (Boos and Shuman 1998). It will be interesting to find the potato homolog of the maltose transporter and to investigate its role during starch degradation in tubers.

Despite recent progress in clarifying the route of starch degradation in *Arabidopsis* leaves and potato tubers, the regulation of this pathway still remains an open question. More information is available concerning cereal seeds, where the enzymes involved in starch hydrolysis have been found to be especially active during seed germination, when starch is mobilized within the endosperm, which at this stage of development represents a nonliving tissue. The most studied enzyme in this specialized system is α-amylase, which is synthesized in the surrounding aleurone layer and secreted into the endosperm. This activity and that of α-glucosidase increase in response to the high levels of gibberellins present at germination. A further level of control of the amylolytic pathway is achieved by the action of specific disulphide proteins that inhibit both α-amylase and debranching enzyme. Thioredoxin *h* reduces and thereby inactivates these inhibitor proteins early in germination. Glucose liberated from starch in this manner is phosphorylated by a hexokinase, before conversion to sucrose and subsequent transport to the developing embryo (Beck and Ziegler 1989).

MANIPULATION OF STARCH YIELD

In potato tubers, like all crop species, there has been considerable interest to increase the efficiency of sucrose to starch conversion and thus to increase starch accumulation by both conventional plant breeding and genetic manipulation strategies. Traditional methodology based on the crossing of haploid potato lines and the establishment of a high density genetic map have allowed the identification of quantitative trait loci (QTL) for starch content (Schäfer-Pregl et al. 1998); however, this is outside the scope of this chapter, and the interested reader is referred to Fernie and Willmitzer (2001). Transgenic approaches in potato have focused primarily on the modulation of sucrose import (Leggewie et al. 2003) and sucrose mobilization (Trethewey et al. 1998) or the plastidial starch biosynthetic pathway (see Table 11.2); however, recently more indirect targets have been tested, which are mostly linked to the supply of energy for starch synthesis (see Tjaden et al. 1998, Jenner et al. 2001, Regierer et al. 2002). To date, the

Table 11.2. Summary of Transgenic Approaches to Manipulate Starch Structure and Yield in Potato Tubers

Target	Transgenic Approach	Starch Content/ Yield	Starch Structure	Grain/Tuber Morphology	Application	Reference
Plastidial ATP/ADP translocator	Antisense inhibition	Starch content decreased to 20%	Less amylose	Grains size decreased to 50%, strange tuber shape	—	Tjaden et al. 1998
Plastidial ATP/ADP translocator	Overexpression of AATP	Starch content increased by up to 30%	More amylose	Grains with angular shape	High starch with more amylose	Tjaden et al. 1998, Geigenberger et al. 2001
Plastidial adenylate kinase	Antisense inhibition	Starch content and yield increased by up to 60%	Not determined	Not determined	High starch	Regierer et al. 2003
ADPGlc pyrophosphorylase	Antisense inhibition of AGPB	Starch content decreased down to 10% of WT-level	Amylose content decreased by 40%	Smaller grains, smaller tubers, increased tuber number	—	Müller-Röber et al. 1992, Lloyd et al. 1999b
ADPGlc pyrophosphorylase	Overexpression of *glgC*	Starch content increased by up to 30%	Not determined	Not determined	High starch	Stark et al. 1991
Granule bound starch synthase	Antisense inhibition of GBSSI	No effect	Starch is free of amylose	No effect	Amylose free starch	Kuipers et al. 1994
Granule-bound starch synthase	Overexpression of GBSS I	No effect	Amylose increased up to 25.5%	No effect	—	Kuipers et al. 1994

(Continues)

Table 11.2. (Continued)

Target	Transgenic Approach	Starch Content/ Yield	Starch Structure	Grain/Tuber Morphology	Application	Reference
Soluble starch synthase	Antisense inhibition of SS III	No effect	Amylopectin chain length altered, more P in starch	Granules cracked with deep fissures	—	Abel et al. 1996, Marshall et al. 1996; Fulton et al. 2002
Soluble and granule-bound starch synthases	Antisense inhibition of GBSS I, SS II & SS III	No effect	Less amylose, shorter amylopectin chains	Concentric granules	Freeze/thaw-stable starch	Jobling et al. 2002, Fulton et al. 2002
Starch branching enzymes	Antisense inhibition of SBE I and SBE II	Decreased yield	Amylose increased to > 70%, low amylopectin, five-fold more P in starch	Granule morphology altered	High amylose and high P starch	Schwall et al. 2000
Starch water dikinase (R1)	Antisense inhibition of R1	No effect	Less P in starch	Not determined	Low P starch	Lorberth et al. 1998

most successful transgenic approaches have resulted from the overexpression of a bacterial AGPase (Stark et al. 1991) and the *Arabidopsis* plastidial ATP/ADP translocator (Tjaden et al. 1998), and the antisense inhibition of a plastidial adenylate kinase (Regierer et al. 2002) in potato tubers.

The majority of previous attempts to improve the starch yield of potato tubers concentrated on the expression of a more efficient pathway of sucrose degradation, consisting of a yeast invertase, a bacterial glucokinase, and a sucrose phosphorylase (Trethewey et al. 1998, 2001). However, although the transgenics exhibited decreased levels of sucrose and elevated hexose phosphates and 3-PGA with respect to wild type, these attempts failed. Tubers of these plants even contained less starch than the wild type, but showed higher respiration rates. Recent studies have shown that as a consequence of the high rates of oxygen consumption, oxygen tensions fall to almost zero within growing tubers of these transformants, possibly as a consequence of the high energy demand of the introduced pathway, and this results in a dramatic decrease in the cellular energy state (Bologa et al. 2003, Geigenberger 2003b). This decrease is probably the major reason for the unexpected observation that starch synthesis decreases in these lines. In general, oxygen can fall to very low concentrations in developing sink organs like potato tubers and seeds, even under normal environmental conditions (Geigenberger 2003b, Vigeolas et al. 2003, van Dongen et al. 2004). The consequences of these low internal oxygen concentrations for metabolic events during storage product formation have been ignored in metabolic engineering strategies. Molecular approaches to increase internal oxygen concentrations could provide a novel and exiting route for crop improvement.

Another failed attempt at increasing tuber starch accumulation was the overexpression of a heterologous sucrose transporter from spinach under the control of the CaMV 35S promoter (Leggewie et al. 2003). The rationale behind this attempt was that it would increase carbon partitioning toward the tuber; however, in the absence of improved photosynthetic efficiency this was not the case.

With respect to the plastidial pathway for starch synthesis, much attention has been focused on AGPase. Analysis of potato lines exhibiting different levels of reduction of AGPase due to antisense inhibition have been used to estimate flux control coefficients for starch synthesis of between 0.3 and 0.55 for this enzyme (Geigenberger et al. 1999a, Sweetlove et al. 1999), showing that AGPase is colimiting for starch accumulation in potato tubers. In addition to having significantly reduced starch content in the tubers, these lines also exhibit very high tuber sucrose content, produce more but smaller tubers per plant, and produce smaller starch grains than the wild type (Müller-Röber et al. 1992; see also later discussion and Table 11.2). To date one of the most successful approaches for elevating starch accumulation in tubers was that of Stark et al. (1991), who overexpressed an unregulated bacterial AGPase. This manipulation resulted in up to a 30% increase in tuber starch content. However, it should be noted that the expression of exactly the same enzyme within a second potato cultivar did not significantly affect starch levels (Sweetlove et al. 1996), indicating that these results might be highly context dependent. Another more promising route to increase starch yield would be to manipulate the regulatory network leading to posttranslational redox activation of AGPase (see Fig. 11.4). Based on future progress in this field, direct strategies can be taken to modify the regulatory components leading to redox regulation of starch synthesis in potato tubers.

More recent studies have shown that the adenylate supply to the plastid is of fundamental importance to starch biosynthesis in potato tubers (Loef et al. 2001, Tjaden et al. 1998). Overexpression of the plastidial ATP/ADP translocator resulted in increased tuber starch content, whereas antisense inhibition of the same protein resulted in reduced starch yield, modified tuber morphology, and altered starch structure (Tjaden et al. 1998). Furthermore, incubation of tuber discs in adenine resulted in a considerable increase in cellular adenylate pool sizes and a consequent increase in the rate of starch synthesis (Loef et al. 2001). The enzyme adenylate kinase (EC 2.7.4.3) interconverts ATP and AMP into 2 ADP. Because adenylate kinase is involved in maintaining the levels of the various adenylates at equilibrium, it represents an interesting target for modulating the adenylate pools in plants. For this reason, a molecular approach was taken to downregulate the plastidial isoform of this enzyme by the antisense technique (Regierer et al. 2002). This manipulation led to a substantial increase in the levels of all adenylate pools (including ATP) and, most importantly, to a record increase in tuber starch content up to 60%

above wild type. These results are particularly striking because this genetic manipulation also resulted in a dramatic increase in tuber yield during several field trials of approximately 40% higher than that of the wild type. When taken in tandem, these results suggest a doubling of starch yield per plant. In addition to the changes described above, more moderate increases in starch yield were previously obtained by targeting enzymes esoteric to the pathway of starch synthesis, for example, plants impaired in their expression of the sucrose synthetic enzyme, sucrose phosphate synthase (Geigenberger et al. 1999b). This enzyme exerts negative control on starch synthesis since it is involved in a futile cycle of sucrose synthesis and degradation and leads to a decrease in the net rate of sucrose degradation in potato tubers (Geigenberger et al. 1997).

Although these results are exciting from a biotechnological perspective and they give clear hints as to how starch synthesis is coordinated in vivo, they do not currently allow us to establish the mechanisms by which they operate. It is also clear that these results, while promising, are unlikely to be the only way to achieve increases in starch yield. Recent advances in transgenic technologies now allow the manipulation of multiple targets in tandem (Fernie et al. 2001), and given that several of the successful manipulations described above were somewhat unexpected, the possibility that further such examples will be uncovered in the future cannot be excluded. One obvious future target would be to reduce the expression levels of the starch degradative pathway since, as described above, starch content is clearly a function of the relative activities of the synthetic and degradative pathways. Despite the fact that a large number of *Arabidopsis* mutants has now been generated that are deficient in the pathway of starch degradation, the consequence of such deficiencies has not been investigated in a crop such as potato tubers. Furthermore, there are no reports to date of increases in starch yield in heterotrophic tissues displaying mutations in the starch degradative pathway.

A further phenomenon in potatoes that relates to starch metabolism is that of cold-induced sweetening, where the rate of degradation of starch to reducing sugars is accelerated. As raw potatoes are sliced and cooked in oil at high temperature, the accumulated reducing sugars react with free amino acids in the potato cell, forming unacceptably brown- to black-pigmented chips or fries via a nonenzymatic, Maillard-type reaction. Potatoes yielding these unacceptably colored products are generally rejected for purchase by the processing plant. If a "cold-processing potato" (i.e., one which has low sugar content even in the cold) were available, energy savings would be realized in potato-growing regions where outside storage temperatures are cool. In regions where outside temperatures are moderately high, increased refrigeration costs may occur. This expense would be offset, however, by removal of the need to purchase dormancy-prolonging chemicals, by a decreased need for disease control, and by improvement of long-term tuber quality. Although such a cold-processing potato is not yet on the market, several manipulations potentially fulfill this criterion, perhaps most impressively the antisense inhibition of GWD, which is involved in the initiation of starch degradation (Lorberth et al. 1998). Further examples on this subject are excellently reviewed in a recent paper by Sowokinos (2001).

While only a limited number of successful manipulations of starch yield have been reported to date, far more successful manipulations have been reported with respect to engineering starch structure. These will be reviewed in the following section.

MANIPULATION OF STARCH STRUCTURE

In addition to attempting to increase starch yield there have been many, arguably more, successful attempts to manipulate its structural properties. Considerable natural variation exists between the starch structures of crop species, with potato starch having larger granules, less amylose, a higher proportion of covalently bound phosphate, and less protein and lipid content than cereal starches. The level of phosphorylation strongly influences the physical properties of starch, and granule size is another important factor for many applications; for example, determining starch noodle processing and quality (Jobling et al. 2004). In addition, the ratio between the different polymer types can affect the functionality of different starches (Slattery et al. 2000). High amylose starches are used in fried snack products to create crisp and evenly brown snacks, as gelling agents, and in photographic films, whereas high amylopectin starches are useful in the food industry (to improve uniformity, stability, and texture) and in the paper and adhesive industries. Both, conventional breeding and transgenic approaches have been utilized for modification of starch properties (specifically amylose, amylopectin, and phosphate content; Sene et al. 2000, Kossmann and Lloyd 2000). However, in contrast to the situation described above for yield, the use of natural mutants has been far more prevalent in this instance.

Quantitative trait loci (QTL) analyses have recently been adopted for identifying the genetic factors underlying starch structure. Recombinant inbred lines were produced from parental maize lines of differing starch structure, and the amylose, amylopectin, and water-soluble fractions were analyzed. The loci linked to these traits were located on a genetic map consisting of RFLP (restriction fragment length polymorphism) markers (Sene et al. 2000). Using this strategy, candidate genes were identified that influenced starch structure. It is clear that application of these techniques may facilitate future breeding approaches to generate modified starch properties.

In addition to plant breeding approaches the use of transgenesis has been of fundamental importance for understanding and influencing starch structure (see Table 11.2). Potato tubers with decreased expression of AGPase (Müller-Röber et al. 1992) had decreased amylose contents, down to about 60% of that found in wild type, and smaller starch granules (Lloyd et al. 1999b). The reduction in amylose content is most likely due to the decreased levels of ADP-glucose found in these plants, since this leads to selective restriction of granule-bound versus soluble starch synthases, the latter having a higher affinity for ADPGlc (Frydman and Cardini 1967). Similarly, when the plastidial adenylate supply was altered by changes in the expression of the amyloplastidial ATP/ADP translocator, not only the starch content, but also its structural properties were altered (Tjaden et al. 1998, Geigenberger et al. 2001). Tubers of ATP/ADP-translocator overexpressing lines had higher levels of ADP-glucose and starch with higher amylose content, whereas the opposite was true for antisense lines. Furthermore, microscopic examination of starch grains revealed that their size in antisense tubers was considerably decreased (by 50%) in comparison with the wild type.

Antisensing granule-bound starch synthase I (GBSSI) in potato tubers resulted in a starch that was almost free of amylose (Kuipers et al. 1994). This type of starch has improved paste clarity and stability with potential applications in the food and paper industries (Jobling 2004). In contrast to this, the amylose content of potato starch could not be increased above a value of 25.5% on overexpression of GBSS, hinting that this enzyme is not limiting in amylose production (Flipse et al. 1994, 1996). It has additionally been demonstrated that amylose can be completely replaced by a branched material that exhibits properties in between those of amylose and amylopectin following plastid-targeted expression of bacterial glycogen synthases (Kortstee et al. 1998).

The roles of the other isoforms of starch synthase (SS) are, if anything, less clear. A dramatic reduction in the expression of SSI in potato had absolutely no consequences on starch structure (Kossmann et al. 1999). The effects of modifying SSII were observed in pea seeds mutated at the *rug*5 locus, which encodes this protein. These mutants display a wrinkled phenotype that is often caused by decreased embryo starch content. Starch granules in these plants exhibit a striking alteration in morphology in that they have a far more irregular shape and exhibit a reduction in medium chain length glucans coupled to an increase in short- and long-chain glucans. Despite the dramatic changes observed in the pea mutant, the down-regulation of SSII in the potato tuber had only minor effects, including a 50% reduction in the phosphate content of the starch and a slight increase in short chain length glucans (Kossmann et al. 1999, Lloyd et al. 1999a). However, it is worth noting that SSII only contributes 15% of the total soluble SS activity within the tuber. Finally, a third class of SS has also been identified and has been reported to be the major form in potato tubers (Marshall et al. 1996). When this isoform (SSIII) was downregulated by the creation of transgenic potato plants, deep fissures were observed in the starch granule under the electron microscope, most probably due to GBSS making longer glucan chains in this background (Fulton et al. 2002). However, the starch was not altered in its amylose content, nor did it display major differences in chain length distribution of short chains within the amylopectin (Marshall et al. 1996, Lloyd et al. 1999a). However, dramatic changes were observed when the structure of the side chains was studied, with an accumulation of shorter chains observed in the transgenic lines (Lloyd et al. 1999a). In addition to this, the phosphate content of the starch was doubled in this transformant (Abel et al. 1996).

Given that some of the changes observed on alteration of a single isoform of starch synthase were so dramatic, several groups have looked into the effects of simultaneously modifying the activities of more than one isoform by the generation of chimeric antisense constructs. When potato tubers were produced in which the activities of SSII and SSIII were simultaneously repressed (Edwards et al. 1999, Lloyd et al. 1999b), the effects were not additive. The amylopectin from these lines was somewhat different from that resulting from inhibition of the single isoforms, with the double antisense plants exhibiting grossly modified amylopectin consisting of more short and extra long chains but fewer medium length chains, leading to a gross alteration in the structure of the starch granules. In another study, the parallel reduction of GBSSI, SSII, and SSIII resulted in starch with less amylose and shorter amylopectin chains, which conferred additional freeze-thaw stability of starch with respect to the wild type (Jobling et al. 2002). This is beneficial since it replaces the necessity for expensive chemical substitution reactions, and in addition, its production may require less energy, as this starch cooks at much lower temperatures than normal potato starches.

Simultaneously inhibiting two isoforms of starch-branching enzyme to below 1% of the wild-type activities resulted in highly modified starch in potato tubers. In this starch, normal, high molecular weight amylopectin was absent, whereas the amylose content was increased to levels above 70%, comparable to that in the highest commercially available maize starches (Schwall et al. 2000; Table 11.2). There was also a major effect on starch granule morphology. In addition, the phosphorus content of the starch was increased more than five-fold. This unique starch, with its high amylose, low amylopectin, and high phosphorus levels, offers novel properties for food and industrial applications. A further example of altered phosphate starch is provided by research into potato plants deficient in GWD (previously known as R1, Lorberth et al. 1998, Ritte et al. 2002). The importance of starch phosphorylation becomes clear when it is considered that starch phosphate monoesters increase the clarity and viscosity of starch pastes and decrease the gelatinization and retrogradation rate. Interestingly, crops that produce high-amylose starches are characterized by considerably lower yield (Jobling 2004). Additional transgenic or breeding approaches, as discussed in the previous section, will be required to minimize the yield penalty that goes along with the manipulation of crops for altered starch structure and functionality.

As discussed above for starch yield, the importance of starch degradative processes in determination of starch structural properties is currently poorly understood. However, it represents an interesting avenue for further research.

CONCLUSIONS AND FUTURE PERSPECTIVES

In this chapter we have reviewed the pathway organization of starch synthesis within the potato tuber and detailed how it can be modulated through transgenesis to result in higher starch yield or the production of starches of modified structure. While several successful examples exist for both types of manipulation, these are yet to reach the field. It is likely that such crops, once commercially produced, will yield both industrial and nutritional benefits to society. In addition to this, other genetic and biochemical factors may well also influence starch yield and structure, and further research is required in this area in order to fully optimize these parameters in crop species. In this context, transgenic approaches complementary to conventional breeding will allow more "fine-tuning" of starch properties.

REFERENCES

Abel GJW, Springer F, Willmitzer L, Kossmann J. 1996. Cloning and functional analysis of a cDNA encoding a novel 139 kDa starch synthase from potato (*Solanum tuberosum* L). Plant J 10:981–991.

Beck E, Ziegler P. 1989. Biosynthesis and degradation of starch in higher plants. Annu Rev Plant Physiol Plant Mol Biol 40:95–117.

Boos W, Shuman H. 1998. Maltose/maltodextrin system of *Escherichia coli*: Transport, metabolism, and regulation. Microbiol Mol Biol Rev 62:204–229.

Bologa KL, Fernie AR, Leisse A, Loureiro ME, Geigenberger P. 2003. A bypass of sucrose synthase leads to low internal oxygen and impaired metabolic performance in growing potato tubers. Plant Physiol 132:2058–2072.

Buleon A, Colonna P, Planchot V, Ball S. 1998. Starch granule structure and biosynthesis. *Int J Biol Macromol* 23:85–112.

Chia T, Thorneycroft D, Chapple A, Messerli G, Chen J, Zeeman SC, Smith SM, Smith AM. 2004. A cytosolic glycosyltransferase is required for conversion of starch to sucrose in *Arabidopsis* leaves at night. Plant J (in press).

Dennis DT, Blakeley SD. 2000. Carbohydrate metabolism. In: BB Buchanan, W Gruissem, RL Jones, editors, Biochemistry and Molecular Biology of Plants, pp. 630–675.Rockville: American Society of Plant Physiologists.

Duwenig E, Steup M, Willmitzer L, Kossmann J. 1997. Antisense inhibition of cytosolic phosphorylase in potato plants (*Solanum tuberosum* L.) affects tuber sprouting and flower formation with only little impact on carbohydrate metabolism. Plant J 12:323–333.

Edwards A, Fulton DC, Hylton CM, Jobling SA, Gidley M, Roessner U, Martin C, Smith AM. 1999. A combined reduction in the activity of starch synthases II and III of potato has novel effects on the starch of tubers. Plant J 17:251–261.

Entwistle G, Rees T. 1990. Lack of fructose-1,6-bisphosphatase in a range of higher plants that store starch. Biochem J 271:467–472.

Fernie AR, Roessner U, Leisse A, Lubeck J, Trethewey RN, Willmitzer L. 2001. Simultaneous antagonistic modulation of enzyme activities in transgenic plants through the expression of a chimeric transcript. Plant Physiol Biochem 39:825–830.

Fernie AR, Willmitzer L. 2001. Molecular and biochemical triggers of potato tuber development. Plant Physiol 127:1459–1465.

Flipse E, Huisman JG, de Vries BJ, Bergervoet JEM, Jacobsen E, Visser RGF. 1994. Expression of a wild type GBSS introduced into an amylose free potato mutant by *Agrobacterium tumefaciens* and the inheritance of the inserts at the microscopic level. Theo Appl Gen 88:369–375.

Flipse E, Keetels CJAM, Jacobsen E, Visser RGF. 1996. The dosage effect of the wild type GBSS allele is linear for GBSS activity but not for amylose content: Absence of amylose has a distinct influence on the physio-chemical properties of starch. Theo Appl Gen 92:121–127.

Frydman RB, Cardini CE. 1967. Studies on the biosynthesis of starch. J Biol Chem 242:312–317.

Fulton DC, Edwards A, Pilling E, Robinson HL, Fahy B, Seale R, Kato L, Donald AM, Geigenberger P, Martin C, Smith AM. 2002. Role of granule-bound starch synthase in determination of amylopectin structure and starch granule morphology in potato. J Biol Chem 277:10834–10841.

Geigenberger P. 2003a. Regulation of sucrose to starch conversion in growing potato tubers. J Exp Bot 54: 457–465.

———. 2003b. Response of plant metabolism to too little oxygen. Curr Opin Plant Biol 6:247–256.

Geigenberger P, Geiger M, Stitt M. 1998. High-temperature perturbation of starch synthesis by decreased levels of glycerate-3-phosphate in growing potato tubers. Plant Physiol 117:1307–1316.

Geigenberger P, Muller-Rober B, Stitt M. 1999a. Contribution of adenosine 5'-diphosphoglucose pyrophosphorylase to the control of starch synthesis is decreased by water stress in growing potato tubers. Planta 209:338–345.

Geigenberger P, Reimholz R, Deiting U, Sonnewald U, Stitt M. 1999b. Decreased expression of sucrose phosphate synthase strongly inhibits the water stress-induced synthesis of sucrose in growing potato tubers. Plant J 19:119–129.

Geigenberger P, Reimholz R, Geiger M, Merlo L, Canale V, Stitt M. 1997. Regulation of sucrose and starch metabolism in potato tubers in response to short-term water deficit. Planta 201:502–518.

Geigenberger P, Stamme C, Tjaden J, Schulz A, Quick PW, Betsche T, Kersting HJ, Neuhaus HE. 2001. Tuber physiology and properties of starch from tubers of transgenic potato plants with altered plastidic adenylate transporter activity. Plant Physiol 12:1667–1678.

Hatzfeld WD, Stitt M. 1990. A study of the rate of recycling of triose phosphates in heterotrophic *Chenopodium rubrum* cells, potato tubers and maize endosperm. Planta 180:198–204.

Jenner HL, Winning BM, Millar AH, Tomlinson KL, Leaver CJ, Hill SA. 2001. NAD-malic enzyme and the control of carbohydrate metabolism in potato tubers. Plant Physiol 126:1139–1149.

Jansson C, Chuanxin S, Puthigae S, Deiber A, Ahlandsberg S. 1997. Cloning, characterisation and modification of genes encoding starch branching enzymes in barley. In: PJ Frazier, P Richmond, AM Donald, editors, Starch Structure and Functionality, pp. 196–203. Cambridge: The Royal Society of Chemistry.

Jobling SA. 2004. Improving starch for food and industrial applications. Curr Opin Plant Biol 7 (in press).

Jobling SA, Westcott RJ, Tayal A, Jeffcoat R, Schwall GP. 2002. Production of a freeze-thaw-stable potato starch by antisense inhibition of three starch synthase genes. Nature Biotech 20:295–299.

Kammerer B, Fischer K, Hilpert B, Schubert S, Gutensohn M, Weber A, Flugge UI. 1998. Molecular characterization of a carbon transporter in plastids from heterotrophic tissues: The glucose 6-phosphate phosphate antiporter. Plant Cell 10:105–117.

Keeling PL, Wood JR, Tyson RH, Bridges IG. 1988. Starch biosynthesis in developing wheat grains. Evidence against the direct involvement of triose phosphates in the metabolic pathway. Plant Physiol 87: 311–319.

Kortstee AJ, Suurs LCJM, Vermeesch AMS, Keetels CJAM, Jacobsen E, Visser RGF. 1998. The influence of an increased degree of branching on the physiochemical properties of starch from genetically modified potato. Carb Poly 37:173–184.

Kossmann J, Müller-Röber B, Dyer TA, Raines CA, Sonnewald U, Willmitzer L. 1992. Cloning and expression analysis of the plastidic fructose-1,6-bisphosphatase coding sequence: Circumstantial evidence for the transport of hexoses into chloroplasts. Planta 188:7–12.

Kossmann J, Abel GJW, Springer F, Lloyd JR, Willmitzer L. 1999. Cloning and functional analysis of a cDNA encoding a starch synthase from potato (*Solanum tuberosum* L.) that is predominantly expressed in leaf tissue. Planta 208:503–511.

Kossmann J, Lloyd J. 2000. Understanding and influencing starch biochemistry. Crit Rev Plant Sci 19: 171–226.

Kuipers AGJ, Jacobsen E, Visser RGF. 1994. Formation and deposition of amylose in the potato tuber starch granule are affected by the reduction of granule bound starch synthase gene expression. Plant Cell 6:43–52.

Leggewie G, Kolbe A, Lemoine R, Roessner U, Lytovchenko A, Zuther E, Kehr J, Frommer WB, Riesmeier JW, Willmitzer L, Fernie AR. 2003. Overexpression of the sucrose transporter SoSUT1 in potato results in alterations in leaf partitioning and in tuber metabolism but has little impact on tuber morphology. Planta 217:158–167.

Lloyd JR, Landschütze V, Kossmann J. 1999a. Simultaneous antisense inhibition of two starch synthase isoforms in potato tubers leads to accumulation of grossly modified amylopectin. Biochem J 338:515–521.

Lloyd JR, Springer F, Buleon A, Müller-Röber B, Willmitzer L, Kossmann J. 1999b. The influence of alterations in ADPglucose pyrophosphorylase activities on starch structure and composition in potato tubers. Planta 209:230–238.

Loef I, Stitt M, Geigenberger P. 2001. Increased levels of adenine nucleotides modify the interaction between starch synthesis and respiration when adenine is supplied to discs from growing potato tubers. Planta 212:782–791.

Lorberth R, Ritte G, Willmitzer L, Kossmann J. 1998. Inhibition of a starch-granule-bound protein leads to modified starch and the repression of cold sweetening. Nature Biotech 16:473–477.

Lu Y, Sharkey TD. 2004. The role of amylomaltase in maltose metabolism in the cytosol of photosynthetic cells. Planta 218:466–473.

Marshall J, Sidebottom C, Debet M, Martin C, Smith AM, Edwards A. 1996. Identification of the major isoform of starch synthase in the soluble fraction of potato tubers. Plant Cell 8:1121–1135.

Müller-Röber B, Sonnewald U, Willmitzer L. 1992. Inhibition of ADPglucose pyrophosphorylase in transgenic potatoes leads to sugar storing tubers and influences tuber formation and the expression of tuber storage protein genes. EMBO J 11:1229–1238.

Neuhaus HE, Emes MJ. 2000. Nonphotosynthetic metabolism in plastids. Annu Rev Plant Physiol Plant Mol Biol 51:111–140.

Niittylä T, Messerli G, Trevisan M, Chen J, Smith AM, Zeeman SC. 2004. A previously unknown maltose transporter essential for starch degradation in leaves. Science 303:87–89.

Pilling E, Smith AM. 2003. Growth ring formation in the starch granules of potato tubers. Plant Physiol 132:365–371.

Preiss J. 1988. Biosynthesis of starch and its regulation. In: J Preiss, editor, The Biochemistry of Plants, pp. 181–254.San Diego: Academic Press.

Regierer B, Fernie AR, Springer F, Perez-Melis A, Leisse A, Koehl K, Willmitzer L, Geigenberger P, Kossmann J. 2002. Starch content and yield increase as a result of altering adenylate pools in transgenic plants. Nature Biotech 20:1256–1260.

Ritte G, Lloyd JR, Eckermann N, Rottmann A, Kossmann J, Steup M. 2002. The starch-related R1 protein is an alpha-glucan, water dikinase. Proc Natl Acad Sci USA 99:7166–7171.

Ritte G, Scharf A, Eckermann N, Haebel S, Steup M. 2004. Phosphorylation of transitory starch is increased during degradation. Plant Physiology (in press).

Schäfer-Pregl R, Ritter E, Concilio L, Hesselbach J, Lovatti L, Walkemeier B, Thelen H, Salamini F, Gebhardt C. 1998. Analysis of quantitative trait loci (QTLs) and quantitative trait alleles (QTAs) for potato tuber yield and starch content. Theo Appl Gen 97:834–846.

Scheidig A, Fröhlich A, Schulze S, Lloyd JR, Kossmann J. 2002. Down regulation of a chloroplast-targeted beta-amylase leads to a starch-excess phenotype in leaves. Plant J 30:581–591.

Schwall GP, Safford R, Westcott RJ, Tayal A, Shi Y-C, Gidley MJ, Jobling SA. 2000. Production of a very-high amylose potato starch by inhibition of SBE A and B. Nature Biotech 18:551–554.

Sene M, Causse M, Damerval C, Thevenot C, Prioul JL. 2000. Quantitative trait loci affecting amylose, amylopectin and starch content in maize recombinant inbred lines. Plant Physiol Biochem 38:459–472.

Shannon JC, Pien FM, Cao H, Liu KC. 1998. Brittle-1, an adenylate translocator, facilitates transfer of extraplastidial synthesized ADPglucose into amyloplasts of maize endosperms. Plant Physiol 117:1235–1252.

Slattery CJ, Kavakli IH, Okita TW. 2000. Engineering starch for increased quantity and quality. Trends Plant Sci 5:291–298.

Smith AM, Denyer K, Martin CR. 1995. What controls the amount and structure of starch in storage organs. Plant Physiol 107:673–677.

———. 1997. The synthesis of the starch granule. Annu Rev Plant Physiol Plant Mol Biol 48:65–87.

Smith AM, Zeeman SC, Thorneycroft D, Smith SM. 2003. Starch mobilisation in leaves. J Exp Bot 54: 577–583.

Sonnewald U, Basner A, Greve B, Steup M. 1995. A second L-type isozyme of potato glucan phophorylase: Cloning, antisense inhibition and expression analysis. Plant Molecular Biology 21:567–576.

Sonnewald U, Hajiraezaei MR, Kossmann J, Heyer A, Trethewey RN, Willmitzer L. 1997. Expression of a yeast invertase in the apoplast of potato tubers increases tuber size. Nature Biotech 15:794–797.

Stark DM, Timmermann KP, Barry GF, Preiss J, Kishore GM. 1991. Regulation of the amount of starch in plant tissues. Science 258:287–292.

Sowokinos JR. 2001. Biochemical and molecular control of cold-induced sweetening in potatoes. Am J Pot Res 78:221–236.

Stitt M, Steup M. 1985. Starch and sucrose degradation. In: R Douce, DA Day, editors, Encyclopedia of Plant Physiology, vol. 18, pp. 347–390. Berlin: Springer-Verlag.

Sullivan TD, Stretlow LI, Illingworth CA, Phillips RL, Nelson OE. 1991. Analysis of maize *Brittle*-1 alleles and a defective suppressor mutator inducible allelle. Plant Cell 3:1337–1348.

Sweetlove LJ, Burrell MM, ap Rees T. 1996. Characterisation of transgenic potato *(Solanum tuberosum)* tubers with increased ADPglucose pyrophosphorylase. Biochem J 320:487–492.

Sweetlove LJ, Müller-Röber B, Willmitzer L, Hill SA. 1999. The contribution of adenosine 5'-diphosphoglucose pyrophosphorylase to the control of starch synthesis in potato tubers. Planta 209:330–337.

Tauberger E, Fernie AR, Emmermann M, Renz A, Kossmann J, Willmitzer L, Trethewey RN. 2000. Antisense inhibition of plastidial phosphoglucomutase provides compelling evidence that potato tuber amyloplasts import carbon from the cytosol in the form of glucose-6-phosphate. Plant J 23:43–53.

Thorbjornsen T, Villard P, Denyer K, Olsen OA, Smith AM. 1996. Distinct isoforms of ADPglucose pyrophosphorylase occur inside and outside the amyloplast in barley endosperm. Plant J 16:531–540.

Tiessen A, Hendriks JHM, Stitt M, Branscheid A, Gibon Y, Farre EM, Geigenberger P. 2002. Starch synthesis in potato tubers is regulated by post-translational redox modification of ADP-glucose pyrophosphorylase: A novel regulatory mechanism linking starch synthesis to the sucrose supply. Plant Cell 14:2191–2213.

Tiessen A, Prescha K, Branscheid A, Palacios N, McKibbin R, Halford NG, Geigenberger P. 2003. Evidence that SNF1-related kinase and hexokinase are involved in separate sugar-signalling pathways modulating post-translational redox activation of ADP-glucose pyrophosphorylase in potato tubers. Plant J 35:490–500.

Tjaden J, Möhlmann T, Kampfenkel K, Henrichs G, Neuhaus HE. 1998. Altered plastidic ATP/ADP-transporter activity influences potato (*Solanum tuberosum* L.) tuber morphology, yield and composition of tuber starch. Plant J 16:531–540.

Trethewey RN, Fernie AR, Bachmann A, Fleischer-Notter H, Geigenberger P, Willmitzer L. 2001. Expression of a bacterial sucrose phosphorylase in potato tubers results in a glucose-independent induction of glycolysis. Plant Cell Environ 24:357–365.

Trethewey RN, Geigenberger P, Riedel K, Hajirezaei MR, Sonnewald U, Stitt M, Riesmeier JW, Willmitzer L. 1998. Combined expression of glucokinase and invertase in potato tubers leads to a dramatic reduction in starch accumulation and a stimulation of glycolysis. Plant J 15:109–118.

Van Dongen JT, Roeb GW, Dautzenberg M, Froehlich A, Vigeolas H, Minchin PEH, Geigenberger P. 2004. Phloem import and storage metabolism are highly coordinated by the low oxygen concentrations within developing wheat seeds. Plant Physiol (in press).

Vigeolas, H, van Dongen JT, Waldeck P, Hühn D, Geigenberger P. 2003. Lipid metabolism is limited by the prevailing low oxygen concentrations within developing seeds of oilseed rape. Plant Physiol 133: 2048–2060.

Weise SE, Weber APM, Sharkey TD. 2004. Maltose is the major form of carbon exported from the chloroplast at night. Planta 218:474–482.

Yu T-S, Kofler H, Häusler RE, Hille D, Flügge U-I, Zeeman SC, Smith AM, Kossmann J, Lloyd J, Ritte G, Steup M, Lue W-L, Chen J, Weber A. 2001. The *Arabidopsis sex1* mutant is defective in the R1 protein, a general regulator of starch degradation in plants, and not in the chloroplast hexose transporter. Plant Cell 13:1907–1918.

Zeeman SC, Thorneycroft D, Schupp N, Chapple A, Weck M, Dunstan H, Haldimann P, Bechtold N, Smith AM, Smith SM. 2004. The role of plastidial α-glucan phosphorylase in starch degradation and tolerance of abiotic stress in *Arabidopsis* leaves. Plant Physiol (in press).

12
Pectic Enzymes in Tomatoes

*M. S. Kalamaki, N. G. Stoforos, and P. S. Taoukis**

INTRODUCTION

In this chapter, the enzymes that act on the pectin fraction of tomato cell walls are presented. In the introduction, background information on the structure of the pectin network in cell walls of higher plants as well as the method used for fractionation analysis of pectic substances are given. Next, pectic enzymes in tomato in terms of the genes that code pectin modifying enzymes, their expression profiles during fruit ripening, and the activities of the protein are discussed. Furthermore, attempts to alter the expression of these enzymes using genetic engineering in order to retard fruit deterioration at the later stages of ripening, increase fruit firmness and shelf life, and improve processing characteristics of tomato products are described. Finally, the inactivation kinetics (by heat and high hydrostatic pressure) of two pectic enzymes, polygalacturonase (PG) and pectin methylesterase (PME), are discussed.

STRUCTURE OF THE PECTIN NETWORK IN CELL WALLS OF HIGHER PLANTS

The primary cell wall of higher plants constitutes a complex network, mainly constructed of polysaccharides and structural proteins. Two coextensive structural networks can be identified (Fig. 12.1), the cellulose-matrix glycan (hemicellulose) network, consisting of cellulose microfibrils held together by matrix glycans, and the pectin network (Carpita and Gibeaut 1993). The cell wall provides shape and structural integrity to the cell, functions as a protective barrier to pathogen invasion, and provides signal molecules important in cell-to-cell signaling, and in plant-symbiote and plant-pathogen interactions.

Pectin is an elaborate network of highly hydrated polysaccharides, which fills in the spaces between the microfibrils in the cellulose-matrix glycan network. Pectins can be classified in three major groups:

*Corresponding author.

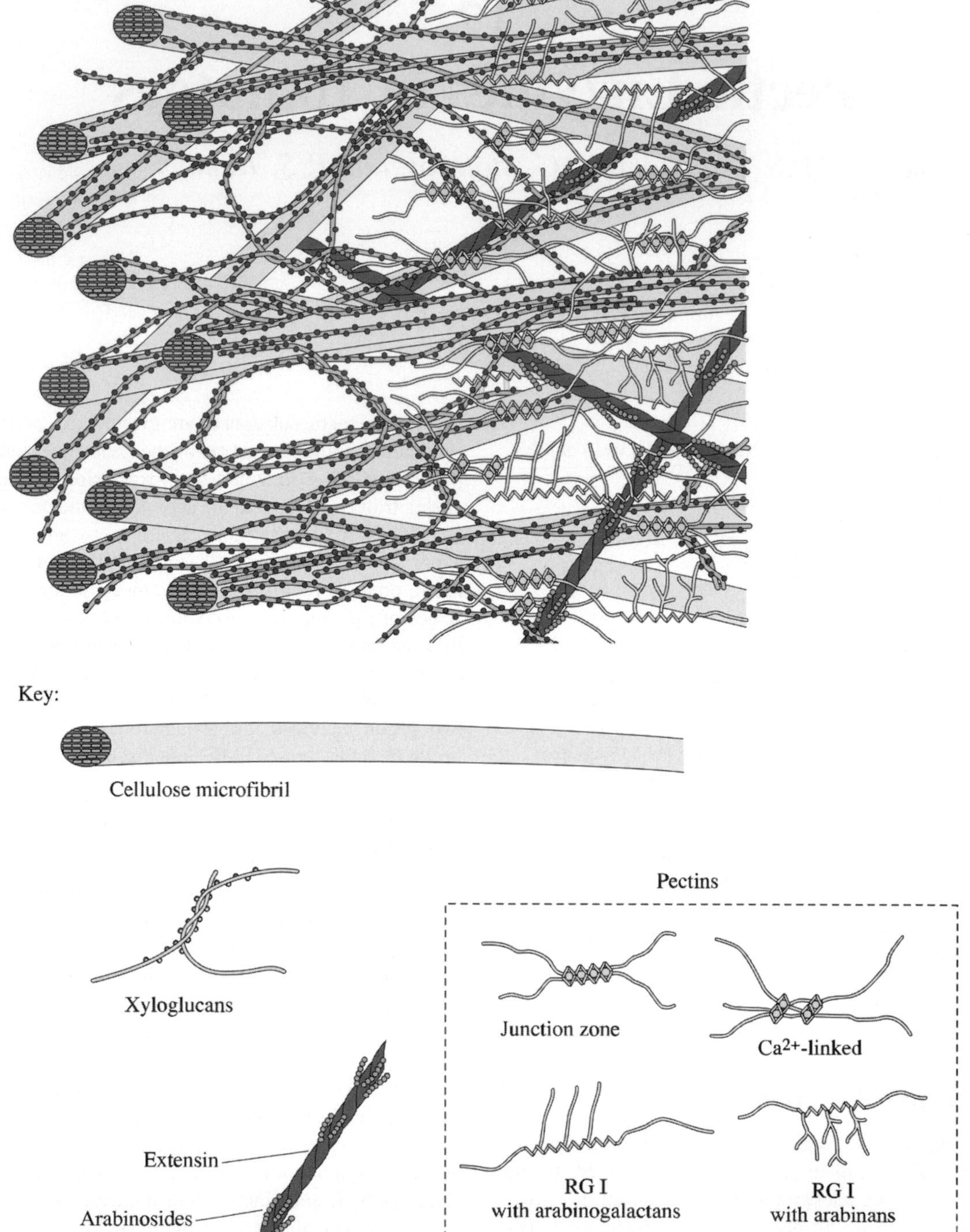

Figure 12.1. Current structural model of cell walls of flowering plants. (Reprinted from Buchanan et al. 2000.)

homogalacturonan (HGA), rhamnogalacturonan I (RGI), and rhamnogalacturonan II (RGII). Homogalacturonan is a homopolymer of (1 → 4) α-D-galacturonic acid (Fig. 12.2A). Galacturonosyl residues are esterified to various extents at the carboxyl group with methanol. Nonesterified regions of HGA chains can be associated with each other by ionic interactions via calcium ion bridges, forming egg-box like structures (Fig. 12.2A). Rhamnogalacturonan I is composed of repeating disaccharide units of (1 → 2)-α-D-rhamnose -(1 → 4)-α-D-galacturonic acid. Rhamnosyl residues on the backbone of this polymer can be further substituted with β-D-galactose- and α-L-arabinose-rich side chains (Fig. 12.2B). Rhamnogalacturonan II (RGII) is a minor constituent of the cell wall consisting of an HGA backbone that exhibits a complex substitution pattern with many diverse sugars, including the unique sugar aceric acid. RGII can form dimers via borate diesters at the apiose residues. Pectins are covalently interlinked, in part via diferulic acid, and further cross-linked to cellulose, matrix glycans, and other cell-wall polymers (Thompson and Fry 2000).

Pectin often consists more than 50% of the fruit cell wall. During ripening, pectin is extensively modified by de novo synthesized enzymes that bring about fruit softening. The pectin backbone is de-esterified, allowing calcium associations and gelling to occur (Blumer et al. 2000). Pectin is solubilized from the wall complex, followed by a decrease in the average polymer size (Brummell and Labavitch 1997, Chun and Huber 1997). These changes are accompanied by loss of galactosyl and arabinosyl moieties from pectin side chains (Gross 1984).

Fractionation of Pectic Substances

Methods for analyzing cell-wall polymeric substances are based on deconstruction of the cell wall by sequentially extracting its components. Tomato

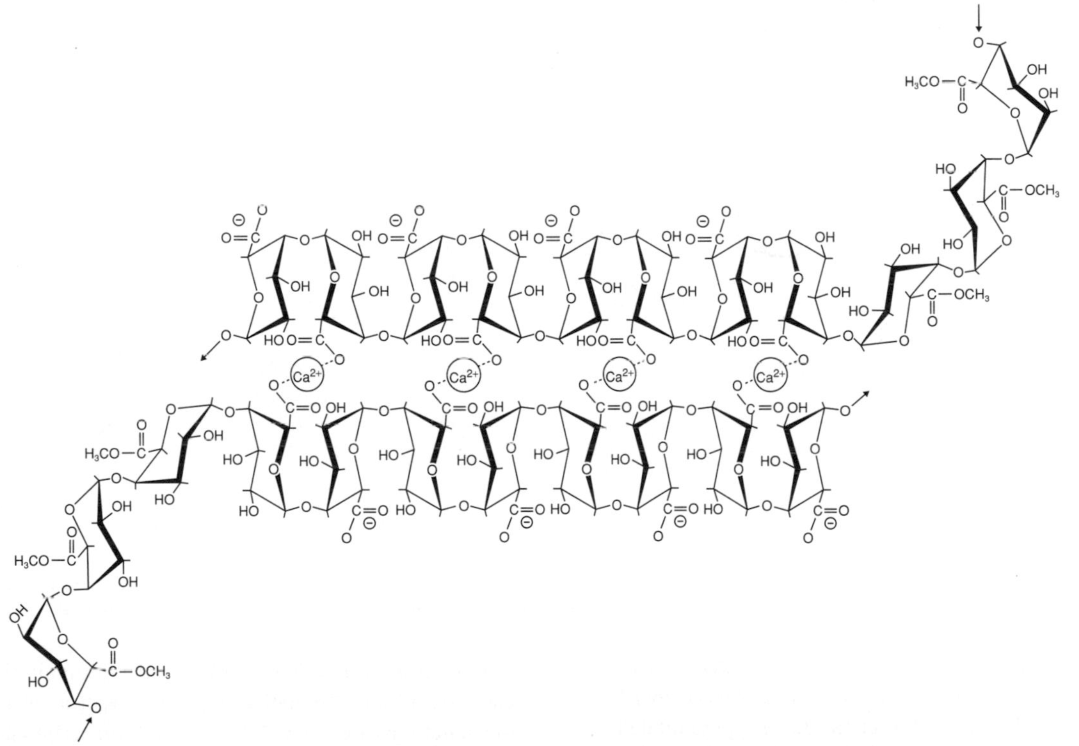

Figure 12.2. A. Structure of homogalacturonan showing calcium ion associations. **B.** Structure of rhamnogalacturonan I, with 5-arabinan *(A)*, 4-galactan *(B)*, and type I arabinogalactan *(C)* side chains. (Reprinted from Buchanan et al. 2000.)

(Continues)

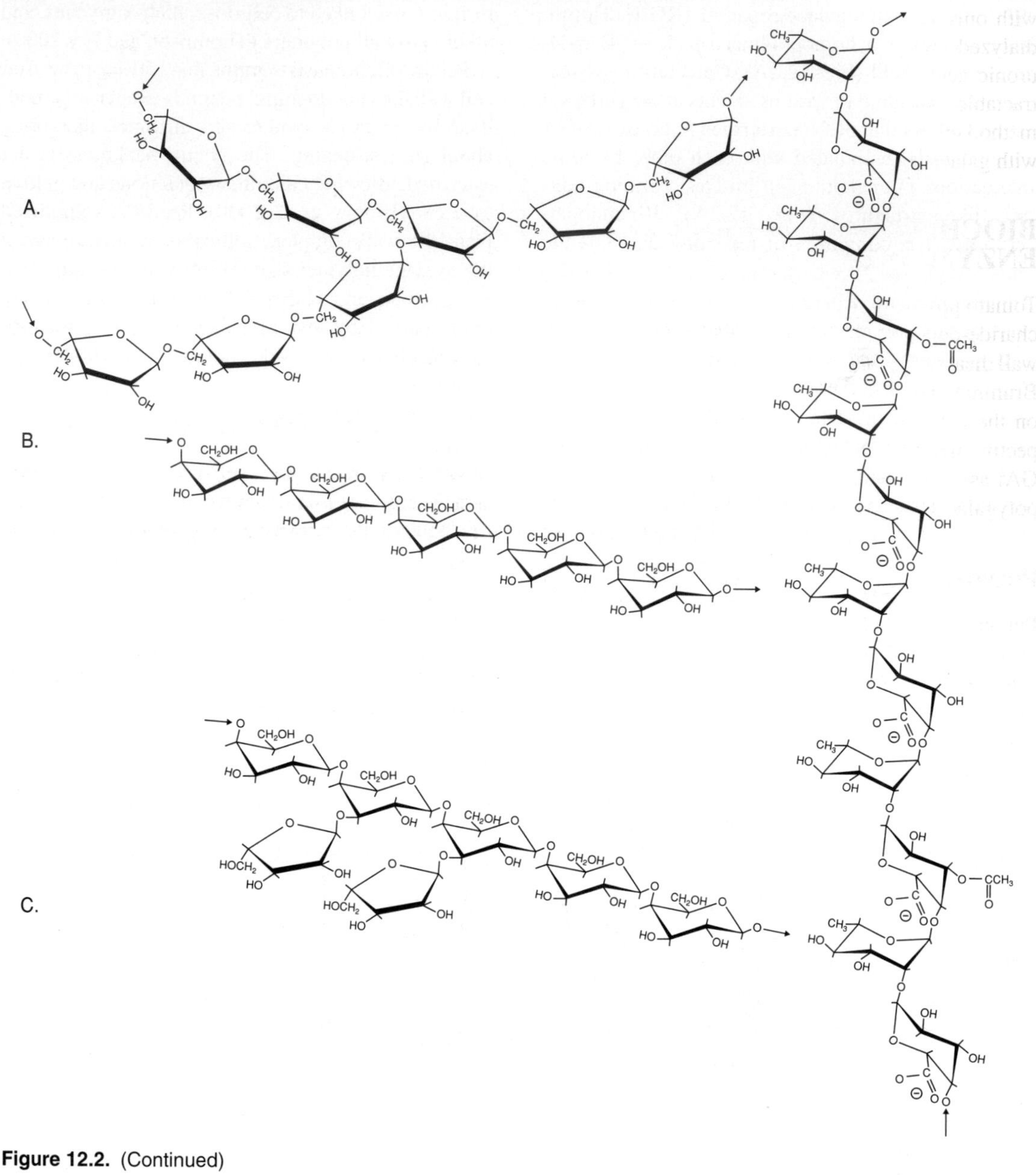

A.

B.

C.

Figure 12.2. (Continued)

pericarp is boiled in ethanol or extracted with tris-buffered phenol to inactivate enzymes (Huber 1992), and cell-wall polysaccharides are precipitated with ethanol. Alcohol-insoluble solids (AIS) are sequentially extracted with 50 mM *trans*-1,2-diaminocyclohexane-*N,N,N',N'*-tetraacetic acid (CDTA), 50 mM sodium acetate buffer (pH = 6.5) to obtain the chelator-soluble (or ionically bound) pectin. Next, the pellet is extracted with 50 mM sodium carbonate, 20 mM sodium borohydride, which results in breaking of ester bonds, to obtain the carbonate-soluble (covalently bound) pectin. The remaining residue consists primarily of cellulose and matrix glycans with only about 5% remaining pectin. Extracts are dialyzed exhaustively against water at 4°C, and uronic acid (UA) content of dialyzed extracts (extractable polyuronide) is determined using the method of Blumenkrantz and Asboe-Hansen (1973), with galacturonic acid as a standard.

BIOCHEMISTRY OF PECTIC ENZYMES IN TOMATO

Tomato possesses an elaborate ensemble of polysaccharide-degrading enzymes that is involved in cell-wall disassembly during ripening (Giovannoni 2001, Brummell and Harpster 2001). The enzymes that act on the pectin network are polygalacturonase (PG), pectin methylesterase (PME), β-galactosidase (β-GALase), and putatively, pectate lyase (PL) and exo-polygalacturonate lyase (PGL).

POLYGALACTURONASE

Polygalacturonases are enzymes that act on the pectin fraction of cell walls and can be divided into endo-PGs and exo-PGs. Exo-PG (EC3.2.1.67) removes a single galacturonic acid residue from the nonreducing end of HGA. Endo-PG [poly(1,4-α-D-galacturonide)glycanohydrolase, EC 3.2.1.15] is responsible for pectin depolymerization by hydrolyzing glycosidic bonds in demethylesterified regions of homogalacturonan (Fig. 12.3).

Endo-PGs are encoded by large multigene families with distinct temporal and spatial expression profiles (Hadfield and Bennett 1998). In ripening tomato fruit, the mRNA of one gene is accumulated at high levels, constituting 2% of the total polyadenylated RNA (DellaPenna et al. 1986), in a fashion paralleling PG protein accumulation. PG gene expression during ripening is ethylene responsive, with very low levels of ethylene (0.15 μl/L) being sufficient to induce PG mRNA accumulation and PG activity (Sitrit and Bennett 1998). PG mRNA levels continue to increase as ripening progresses, and persist at the overripe stage. Immunodetectable PG protein accumulation follows the same pattern as mRNA accumulation during ripening and fruit senescence. This mRNA encodes a predicted polypeptide 457 amino acids long that contains a signal sequence of 24 amino acids targeting it to the cell's endomembrane system for further processing and secretion. The mature protein is produced by cleaving a 47 amino acid amino-terminal prosequence and a sequence of 13 amino acids from the carboxyl terminus followed by glycosylation. Two isoforms, differing only with respect to the degree of glycosylation, are identified: PG2A, with a molecular mass of 43 kDa, and PG2B, with a molecular mass of 45 kDa (Sheehy et al. 1987, DellaPenna and Bennett 1988, Pogson et al. 1991). Another isoform (PG1) accumulates during the early stages of tomato fruit ripening, when PG2 levels are low. PG1 has a molecular mass of 100 kDa and is composed of one or

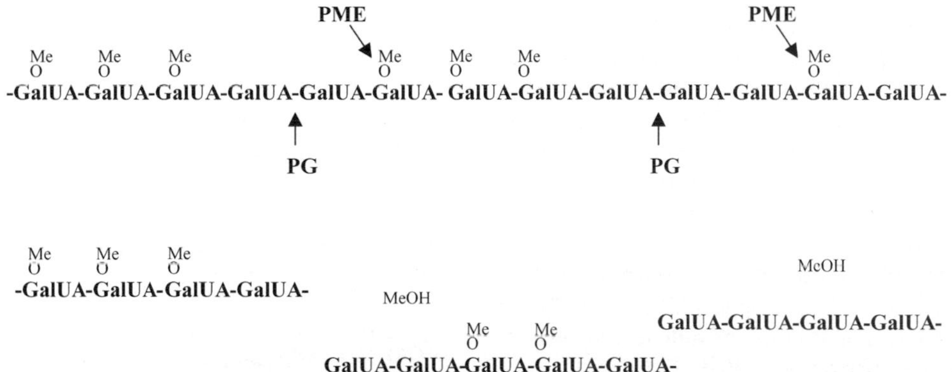

Figure 12.3. Action pattern of PG and PME on pectin backbone. Endo-PG hydrolyzes glycosidic bonds in demethylesterified regions of homogalacturonan. PME catalyzes the removal of methyl ester groups (OMe) from galacturonic acid (GalUA) residues of pectin.

possibly two PG2A or PG2B subunits and the PG β-subunit (also called a converter).

The PG β-subunit is an acidic, heavily glycosylated protein. The precursor protein contains a 30 amino acid signal peptide, an N-terminus propeptide of 78 amino acids, the mature protein domain, and a large C-terminus propeptide. The mature protein presents a repeating motif of 14 amino acids and has a size of approximately 37 to 39 kDa. This difference in size results from different glycosylation patterns or different posttranslational processing at the carboxyl terminus of the protein (Zheng et al. 1992). The PG β-subunit protein is encoded by a single gene. β-subunit mRNA levels increase during fruit development and reach a maximum at 30 days after pollination (i.e., just before the onset of ripening), then decrease to undetectable levels during ripening, whereas immunodetectable β-subunit protein persists throughout fruit development and ripening (Zheng et al. 1992, 1994). The presence of the PG β-subunit alters the physicochemical properties of PG2. Although the PG β-subunit does not possess any glycolytic activity, binding to PG2 leads to modification of PG2 activity with respect to pH optima, heat stability, and Ca^{++} requirements (Knegt et al. 1991, Zheng et al. 1992). Since the β-subunit is localized at the cell wall long before PG2 starts accumulating, it may serve as an anchor to localize PG2 to certain areas of the cell wall (Moore and Bennett 1994, Watson et al. 1994). Another proposed action of the β-subunit is that of limiting access of PG to its substrate or restricting PG activity by binding to the PG protein (Hadfield and Bennett 1998).

Generally, PG-mediated pectin disassembly contributes to fruit softening at the later stages of ripening and during fruit deterioration. Overall, PG activity is neither sufficient nor necessary for fruit softening, as is evident from data using transgenic tomato lines with suppressed levels of PG mRNA accumulation (discussed later in this chapter) and studies on ripening-impaired tomato mutants. It is evident from data in other fruit, especially melon (Hadfield and Bennett 1998), persimmon (Cutillas-Iturralde et al. 1993), and apple (Wu et al. 1993) that very low levels of PG may be sufficient to catalyze pectin depolymerization in vivo.

Pectin Methylesterase

Pectin methylesterase is a deesterifying enzyme (EC 3.1.1.11) catalyzing the removal of methyl ester groups from galacturonic acid residues of pectin, thus leaving negatively charged carboxylic residues on the pectin backbone (Fig. 12.3). Demethylesterification of galacturonan residues leads to a change in the pH and charge density on the HGA backbone. Free carboxyl groups from adjacent polygalacturonan chains can then associate with calcium or other divalent ions to form gels (Fig. 12.2A). PME protein is encoded by at least four genes in tomato (Turner et al. 1996, Gaffe et al. 1997). The PME polypeptide is 540–580 amino acids long and contains a signal sequence targeting it to the apoplast. The mature protein has a molecular mass of 34–37 kDa and is produced by cleaving an amino-terminal prosequence of approximately 22 kDa (Gaffe et al. 1997). PME is found in multiple isoforms in fruit and other plant tissues (Gaffe et al. 1994). PME isoforms have pIs in the range of 8–8.5 in fruit and around 9.0 in vegetative tissues (Gaffe et al. 1994). Tomato fruit PME is active throughout fruit development and influences accessibility of PG to its substrate. PME transcript accumulates early in tomato fruit development and peaks at the mature green fruit stage, followed by a decline in transcript levels. In contrast, PME protein levels increase in developing fruit at the early stages of ripening and then decline (Harriman et al. 1991, Tieman et al. 1992). Pectin is synthesized in a highly methylated form, which is then demethylated by the action of PME. In ripening tomato fruit, the methylester content of pectin is reduced from an initial 90% at the mature green stage to about 35% in red ripe fruit (Koch and Nevins 1989). Evidence supports the hypothesis that demethylesterification of pectin, which allows Ca^{++} cross-linking to occur, may restrict cell expansion. Constitutive expression of a petunia PME gene in potato resulted in diminished PME activity in some plants. A decrease in PME activity in young stems of transgenic potato plants correlated with an increased growth rate (Pilling et al. 2000). The action of PME is required for PG action on the pectin backbone (Wakabayashi et al. 2003).

β-Galactosidase

β-galactosidase (EC 3.2.1.23) is an exo-acting enzyme that catalyzes cleavage of terminal galactose residues from pectin β-(1,4)-D-galactan side chains (Fig. 12.2B). Loss of galactose from wall polysaccharides occurs throughout fruit development, and it accelerates with the onset of ripening (Gross 1984, Seymour et al. 1990). In tomato fruit, β-galactosidase is encoded by a small multigene family of at least seven members, *TBG1–TBG7* (Smith et al. 1998, Smith and Gross 2000). Based on sequence analysis, it was found that β-galactosidase genes encode putative polypeptides with a predicted molecular mass between 89.8 and 97 kDa, except for TBG4, which is predicted to be shorter by 100 amino acids at its carboxyl terminus (Smith and Gross 2000). A signal sequence targeting these proteins to the apoplast was only identified in *TBG4, TBG5,* and *TBG6,* whereas *TBG7* is predicted to be targeted to the chloroplast. These seven genes showed distinct expression patterns during fruit ripening, and transcripts of all clones except *TBG2* were detected in other tomato plant tissues (Smith and Gross 2000). β-galactosidase transcripts were also detected in the ripening-impaired *ripening-inhibitor* (*rin*), *nonripening* (*nor*), and *never-ripe* (*nr*) mutant tomato lines, which do not soften during ripening; however, the expression profile differed from that in wild-type fruit. *TGB4* mRNA accumulation was impaired in comparison with wild-type levels, whereas accumulation of *TBG6* transcript persisted up to 50 days after pollination in fruit from the three mutant lines (Smith and Gross 2000). Presence of β-galactosidase transcripts in these lines suggests that β-galactosidase activity alone cannot lead to fruit softening. Expression of the *TBG4* clone, which encodes β-galactosidase isoform II, in yeast resulted in production of active protein. This recombinant protein was able to hydrolyze synthetic substrates (p-nitrophenyl-β-D-galactopyranoside) and lactose as well as galactose-containing wall polymers (Smith and Gross 2000). Heterologous expression of *TBG1* in yeast also resulted in the production of a protein with exo-galactanase/β-galactosidase activity, also active in cell-wall substrates (Carey et al. 2001).

Pectate Lyase

Pectate lyases (pectate transeliminase, EC 4.2.2.2) are a family of enzymes that catalyze the random cleavage of demethylesterified polygalacturonate by β-elimination, generating oligomers with 4,5-unsaturated reducing ends (Yoder et al. 1993). Pectate lyases (PLs) were thought to be of microbial origin and were commonly isolated from macerated plant tissue infected with fungal or bacterial pathogens (Yoder et al. 1993, Barras et al. 1994, Mayans et al. 1997, and others). In plants, presence of PL-like sequences was first identified in mature tomato flowers, anthers, and pollen (Wing et al. 1989). Presence of PL in pollen of other species (Albani et al. 1991, Taniguchi et al. 1995, Wu et al. 1996, Kulikauskas and McCormick 1997) as well as in other tissues (Domingo et al. 1998, Milioni et al. 2001, Pilatzke-Wunderlich and Nessler 2001) has been well documented. Recently, PL-like sequences were identified in ripening banana (Pua et al. 2001, Marín-Rodríguez et al. 2003) and strawberry fruit (Medina-Escobar et al. 1997, Benítez-Burraco et al. 2003). Transgenic strawberry plants that expressed a pectin lyase antisense construct driven by the cauliflower mosaic virus 35S (CaMV35S) promoter were produced and evaluated (Jiménez-Bermúdez et al. 2002). Fruit from PL-suppressed lines were firmer than controls, with the level of PL suppression being correlated to internal fruit firmness (Jiménez-Bermúdez et al. 2002). Although pectate lyase activity in ripening tomato fruit has not been detected, several putative pectin lyase sequences exist in the tomato expressed sequence tag (EST) databases (Marín-Rodríguez et al. 2002), suggesting that these enzymes may contribute to cell-wall disassembly during tomato fruit ripening.

GENETIC ENGINEERING OF PECTIC ENZYMES IN TOMATO

Cell-wall disassembly in ripening fruit is an important contributor to the texture of fresh fruit. Modification of cell-wall enzymatic activity during ripening, using genetic engineering, can impact cell-wall polysaccharide metabolism, which in turn can influence texture. Texture of fresh fruit, in turn, influences processing characteristics and final viscosity of processed tomato products. The development of transgenic tomato lines in which the expression of single or multiple genes is altered has allowed the role of specific enzymes to be evaluated both in fresh fruit and processed products.

Genetic Engineering of PG in Tomato Fruit

Polygalactronase (PG) was the first target of genetic modification aiming to retard softening in tomatoes (Sheehy et al. 1988, Smith et al. 1988). Transgenic plants were constructed in which expression of PG was suppressed to 0.5–1% of wild-type levels by the expression of an antisense PG transgene under the control of the cauliflower mosaic virus 35S (CaMV35S) promoter. Fruit with suppressed levels of PG ripened normally as wild-type plants did. An increase in the storage life of the fruit was observed in the PG-suppressed fruit (Schuch et al. 1991, Kramer et al. 1992, Langley et al. 1994). The influence of PG suppression on individual classes of cell-wall polymers was investigated (Carrington et al. 1993, Brummell and Labavitch 1997). Sequential extractions of cell-wall constituents revealed a decrease in the extent of depolymerization of CDTA-soluble polyuronides during fruit ripening. Naturally occurring mutant tomato lines exist that carry mutations affecting the normal process of ripening. One of these lines, the *ripening-inhibitor (rin),* possesses a single locus mutation that arrests the response to ethylene. *rin* fruit fails to synthesize ethylene and remains green and firm throughout ripening (DellaPenna et al. 1989). Transgenic *rin* plants that expressed a functional PG gene driven by the ethylene-inducible E8 promoter were obtained. After exposure to propylene, PG mRNA accumulation as well as active PG protein was attained, but the fruit did not soften. In these fruits, depolymerization and solubilization of cell-wall pectin was at near wild-type levels (Giovannoni et al. 1989, DellaPenna et al. 1990). Altogether, results obtained from analyzing transgenic tomato fruit with altered levels of PG indicate that PG-mediated depolymerization of pectin does not influence softening at the early stages of ripening, and it is not enough to cause fruit softening. However, its outcome is manifested at the later stages of ripening and in fruit senescence.

Suppression of β-subunit expression by the introduction of a β-subunit cDNA in the antisense orientation resulted in reduction of immunodetectable β-subunit protein levels and PG1 protein levels to $< 1\%$ of PG1 in control fruit. In transgenic lines, PG2 expression remained unaltered (Watson et al. 1994). Suppression of β-subunit expression resulted in increased CDTA polyuronide levels in transgenic fruit during ripening. The size distribution profiles of CDTA-extractable polyuronides in green fruit of transgenic and control plants were nearly identical at the mature green stage. However, at subsequent ripening stages, an increased number of smaller size polyuronides were observed in suppressed β-subunit fruit compared with control fruit (Watson et al. 1994).

Genetic Engineering of PG and Food Processing

Apart from the influence of transgenic modification of PG gene expression on the ripening characteristics and shelf life of fresh tomato fruit, the processing characteristics of the fruit were also modified by the introduction of the PG transgene. A shift in the partitioning of polyuronides in water and CDTA extracts of cell walls from pastes prepared from control and suppressed PG lines was observed. The majority of polyuronides were solubilized in the water extract in control pastes, whereas in pastes from PG-suppressed fruit, a greater proportion was solubilized in CDTA. Water and CDTA extracts from control paste contained a larger amount of smaller sized polyuronides compared with the corresponding extracts from PG-suppressed fruit (Fig. 12.4). However, the molecular size profiles from carbonate-soluble pectin were indistinguishable in the two lines. Tomato paste prepared from PG-suppressed fruit exhibited enhanced gross and serum viscosity (Brummell and Labavitch 1997, Kalamaki 2003). The enhancement of viscosity characteristics of pastes from transgenic PG fruit was attributed to the presence of pectin polymers of higher molecular size and wider size distribution in the serum. Although PG seems to play only a small role in fruit softening, the viscosity of juice and reconstituted paste prepared from antisense PG lines is substantially increased, largely due to reduced breakdown of soluble pectin during processing (Schuch et al. 1991, Kramer et al. 1992, Brummell and Labavitch 1997, Kalamaki et al. 2003a).

Genetic Engineering of PME in Tomato Fruit

Pectin methylesterase (PME) protein is accumulated during tomato fruit development, and protein levels increase with the onset of ripening. Production of transgenic tomato lines in which the expression of PME has been suppressed via the introduction of an antisense PME gene driven by the CaMV35S promoter has been reported (Tieman et al. 1992, Hall et al. 1993). PME activity was reduced to less than 10% of wild-type levels, and yet fruit ripened normally.

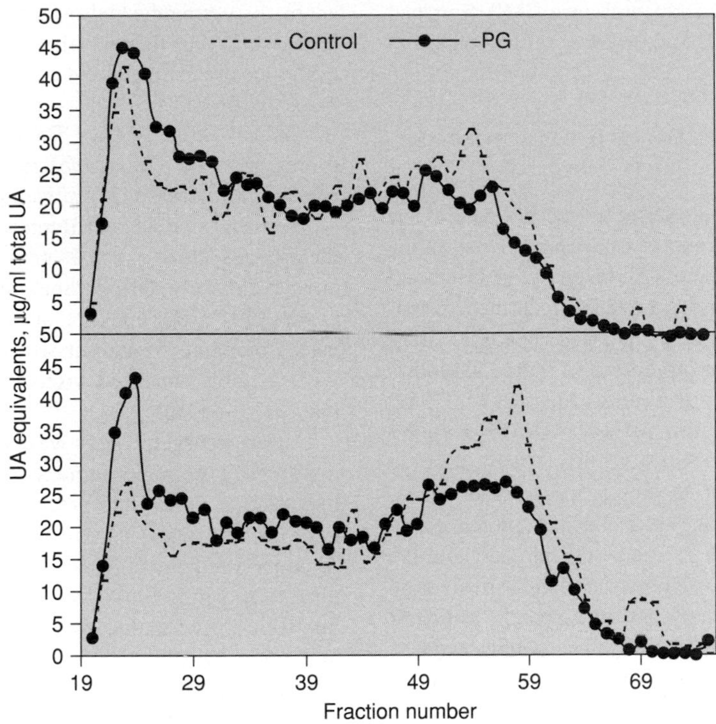

Figure 12.4. Size exclusion chromatography of CDTA-soluble pectin fractions from juice *(upper)* and paste *(lower)* prepared from control and suppressed PG fruit. Larger polysaccharide molecules elute close to the void volume of the column (fraction 20), whereas small molecules elute at the total column volume (fraction 75). (Redrawn from Kalamaki 2003.)

The degree of methylesterification increased significantly in antisense PME fruit, by 20–40% compared with wild-type fruit. Additionally, the amount of chelator-soluble polyuronides decreased in transgenic lines, a result that was expected since demethylesterification is necessary for calcium-mediated cross-linking of pectin. Furthermore, chelator-soluble polyuronides from red ripe tomato pericarp exhibited a larger amount of intermediate size polymers in transgenic lines than in controls, indicating a decrease in pectin depolymerization in the chelator-soluble fraction. As described above, PME action precedes that of PG. An increase of about 15% in soluble solids was also observed in transgenic fruit pericarp cell walls (Tieman et al. 1992). Expression of the antisense PME construct also influenced the ability of the fruit tissue to bind divalent cations. Calcium accumulation in the wall was not influenced; however, two-thirds of the total calcium in transgenic fruit was present as soluble calcium, compared with one-third in wild-type fruit (Tieman and Handa 1994). Pectin with a higher degree of methylesterification was found in PME-suppressed fruit, thus presented fewer sites for calcium binding in the cell wall.

Genetic Engineering of PME and Food Processing

From a food-processing standpoint, the effect of these modifications in PME activity in tomato cell walls during ripening was further investigated. Juice prepared from transgenic antisense PME fruit had about a 35–50% higher amount of total uronic acids than control fruit. Pectin extracted from juice processed by a hot break, a microwave break, or a cold break method was methylesterified at a higher degree in PME-suppressed juices. In addition, pectin from transgenic fruit was of larger molecular mass compared with controls (Thakur et al. 1996a). Juice and paste prepared from antisense PME fruit had higher viscosity and lower serum separation (Thakur et al. 1996b).

GENETIC ENGINEERING OF β-GALACTOSIDASE IN TOMATO FRUIT

In tomato fruit, a considerable loss of galactose is observed during the course of ripening. This is attributed to the action of enzymes with exo-galactanase/β-galactosidase activity. In order to further investigate their role in ripening, β-galactosidase *TBG1* gene expression was suppressed in transgenic tomato by the expression of a sense construct of 376 bp of the *TBG1* gene driven by the CaMV35S promoter (Carey et al. 2001). Although plants with different expression levels of the transgene were obtained, there was no effect on β-galactosidase protein activity and fruit softening. The results suggest that the *TBG1* gene product may not be readily involved in fruit softening but may act on a specific cell-wall substrate (Carey et al. 2001). The β-galactosidase isoform II, encoded by *TBG4,* has been found to accumulate in ripening tomato fruit (Smith et al. 1998). The activity of TBG4 has been suppressed by the expression of a 1.5 kb antisense construct of TBG4 (Smith et al. 2002).

DOUBLE TRANSGENIC TOMATO LINES IN FOOD PROCESSING

As reported above, modifying the expression of individual cell-wall hydrolases does not alter fruit softening substantially, suggesting that cell-wall modifying enzymes may act as a consortium to bring about fruit softening during ripening. Generation of double transgenic lines further aids in understanding the influence of red ripe fruit characteristics on the physicochemical properties of processed products.

SUPPRESSED PG AND SUPPRESSED EXP1

The influence of the suppression of the expression of two enzymes in the same transgenic line has been reported (Powell et al. 2003, Kalamaki et al. 2003a). Tomato lines with simultaneous suppression of the expression of both PG and ripening-associated expansin have been produced and their chemical and physical properties characterized (Kalamaki et al. 2003a, Powell et al. 2003). Expansins are proteins that lack wall hydrolytic activity and are proposed to act by disrupting the hydrogen bonding between cellulose microfibrils and the cross-linking glycan (xyloglucan) matrix (see Fig. 12.1). Expansins are involved in cell-wall disassembly during ripening (Brummell et al. 1999a). Their role in ripening is proposed to be that of loosening the wall and increasing the accessibility of wall polymers to hydrolytic enzymes (Rose and Bennett 1999). In tomato, the product of one expansin gene *(Exp1)* is accumulated exclusively during tomato fruit ripening (Rose et al. 1997, Brummell et al. 1999b). Juices and pastes prepared from fruit with modified levels of Exp1 exhibit enhanced viscosity attributes (Kalamaki et al. 2003b).

Double transgenic lines were generated by crossing single transgenic homozygous lines. Fruit was harvested at the mature green stage and allowed to ripen at 20°C, and texture was determined as the force required to compress the blossom end of the tomato by 2 mm. Fruits from the double suppressed line were firmer than controls and single transgenic lines at all ripening stages. The same results were observed in fruit ripened on vine. At the red ripe stage, fruits from the double suppressed line were 20% firmer than controls. The flow properties of juice prepared from control, single transgenic, and double transgenic lines at the mature green/breaker, pink, and red ripe stages were evaluated using a Bostwick consistometer. Average juice viscosity decreased as ripening progressed in all genotypes. Juice prepared from fruit of the suppressed PG and the double suppressed lines at the red ripe stage exhibited higher viscosity than the control (Powell et al. 2003).

In another experiment, an elite processing variety was used as the wild-type background to suppress the expression of PG, Exp1, and both PG and Exp1 in the same line (Kalamaki et al. 2003a). Juice and paste were prepared from control, single suppressed, and double suppressed fruit, and their flow properties were characterized. In paste diluted to 5° Brix, an increase in Bostwick consistency of about 18% was observed for the PG-suppressed and Exp1-suppressed genotypes. Diluted paste from the double suppressed genotype exhibited the highest viscosity; however, this increase in consistency of the double transgenic line was only an additional 4% compared with the single transgenic lines. Analysis of particle size distributions at 5° Brix in juices and pastes of the different genotypes indicates that suppression of PG or Exp1 results in small increases in the number of particles below a diameter of 250 μm (Fig. 12.5). Suppression of Exp1 also results in the appearance of some larger particles, above 1400 μm in diameter. The presence of these larger fragments shows that suppression of Exp1 affects the way the cells and cell walls rupture during processing. Biochemical characterization of cell-wall polymers did not show large differences in the amounts of extractable polyuronide or neutral sugar in sequential cell-wall extracts of juices and diluted concentrates of the various genotypes. Only minor size differences appear in individual polymer sizes in sequential extracts. In suppressed PG genotypes, larger pectin sizes were observed in the CDTA extract. The most dramatic difference was in the carbonate extracts, where the double suppressed line contained larger pectin polymers at all concentration levels relative to other genotypes (Kalamaki 2003). Concurrent

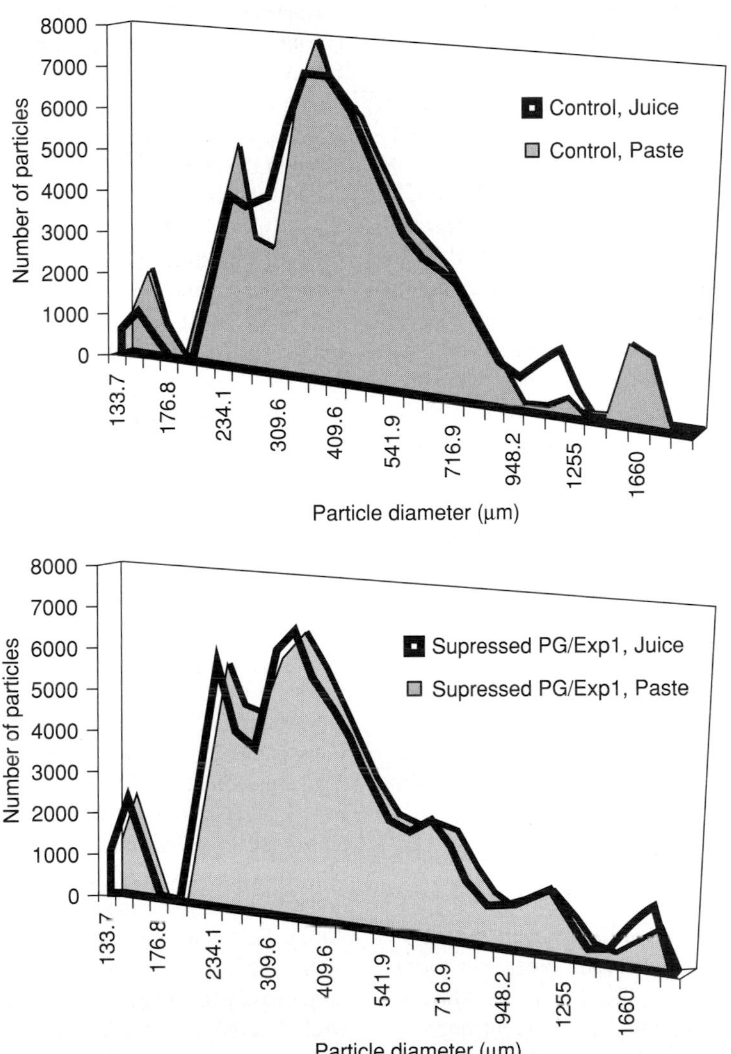

Figure 12.5. Particle size determination in juice and paste from control and suppressed PG/Exp1 fruit. (Redrawn from Kalamaki et al. 2003a.)

suppression of PG and Exp1 in the same transgenic line showed an overall increase in viscosity compared with the control, but only a 4% additional increase compared with the single transgenic lines. Particle size distribution in juices and pastes from the double transgenic line are more polydisperse (Fig. 12.5), and particles appear to be more rigid. In this line, an increase in the size of carbonate-soluble pectin polymers was observed. This size increase could suggest differences in the degree of polymer cross-linking in the particles, resulting in altered particle properties. By modifying cell-wall metabolism during ripening, the processing qualities of the resulting juices and pastes were influenced. Hence, reducing the activity of these cell-wall enzymes may be a route to the selection of improved processing tomato varieties.

Suppressed PG and Suppressed PME

The activities of both PG and PME were suppressed in the same line, and juices were evaluated and compared to single transgenics and controls (Errington et al. 1998). Small differences in Bostwick consistency were observed in hot-break juices, with the suppressed PG juices having the highest viscosity. In cold-break juices from PG-suppressed fruit, a time dependent increase in viscosity was observed. Since a similar increase was not observed in the double transgenic line, it was concluded that in order for the increase in viscosity of cold-break juice to occur, both absence of PG and continued action of PME is required. Absence of PG activity will lead to a larger size of pectin, whereas continuing demethylesterification of pectin chains by PME will increase calcium associations between pectin chains, leading to gel formation and thus improved viscosity. In conclusion, simultaneous transgenic suppression of PG and PME expression did not result in an additive effect.

TOMATO PROCESSING

The majority of tomatoes are consumed in a processed form such as juice, paste, pizza and pasta sauce, and various diced or sliced products. Most of the products are concentrated to different degrees and stored in a concentrated form until ready to use. Industrial concentrates are then diluted to reach the desired final product consistency.

Textural properties of tomato fruit are important contributors to the overall quality in both fresh market and processing tomatoes (Barrett et al. 1998). In some processed products, the most important quality attribute is viscosity (Alviar and Reid 1990). It was recognized early that the structure most closely associated with viscosity is the cell wall (Whittenberger and Nutting 1957). Both the concentration and type of cell-wall polymers in the serum fraction and the pulp (particle fraction) are important contributors to viscosity. In serum, the amount and size of the soluble cell wall polymers influence serum viscosity (Beresovsky et al. 1995), whereas in pulp, the size distribution, the shape, and the degree of deformability of cell-wall fragments influence viscosity (Den Ouden and Van Vliet 1997). Viscosity is influenced partly by factors that dictate the chemical composition and physical structure of the juice such as fruit variety, cultivation conditions, and the ripening stage of the fruit at harvest. However, it is also influenced by processing factors such as break temperature (Xu et al. 1986), finisher screen size (Den Ouden and Van Vliet 1997), mechanical shearing during manufacture, and degree of concentration (Marsh et al. 1978). These processes result in changes in the microstructure of the fruit cell wall that are manifested as changes in viscosity (Xu et al. 1986). Since the integrity of cell-wall polymers in the juice is imperative, plant pectic enzymes usually have to be inactivated during processing in order to diminish their activity and prevent pectin degradation. Pectin degradation could lead to increased softening in, for example, pickled vegetable production or peach canning, and loss of viscosity in processed tomato products (Crelier et al. 2001). Although several enzymes are reported to act on cell-wall polymers, the main depolymerizing enzyme is PG. Therefore, inactivation of PG activity during tomato processing is essential for viscosity retention. However, there are cases where residual activity of other pectic enzymes is desirable, as for example, with PME. Retention of PME activity with concurrent and complete inactivation of PG could lead to products with higher viscosity (Errington et al. 1998, Crelier et al. 2001). Selective inactivation of PG is not possible with conventional heat treatment since PME is rather easily inactivated by heat at ambient pressure, while PG requires much more severe heat treatment for complete inactivation. PME is inactivated by heating at 82.2°C for 15 seconds at ambient pressure, while for PG inactivation a temperature of 104.4°C for 15 seconds is required in canned tomato pulp (Luh and Daoud 1971). In order to achieve this selective inactivation a combination of heating and high hydrostatic pressure can be used during processing or alternatively, the expression of a particular enzyme can be suppressed using genetic engineering in fruit.

Thermal Inactivation

Thermal processing (i.e., exposure of the food to elevated temperatures for relatively short times) has been used for almost 200 years to produce shelf-stable products by inactivating microbial cells, spores, and enzymes in a precisely defined and controlled procedure. Kinetic description of the destructive effects of heat on both desirable and undesirable attributes is essential for proper thermal process design.

At constant temperature and pressure conditions PME thermal inactivation follows first-order kinetics (Crelier et al. 2001, Fachin et al. 2002, Stoforos et al. 2002):

$$-\frac{dA}{dt} = kA, \quad k = f(T, P, \ldots) \qquad (12.1)$$

where A is the enzyme activity at time t, k the reaction rate constant, and P and T the pressure and temperature process conditions.

Ignoring the pressure dependence, Equation 12.1 leads to

$$A = A_o e^{-k_T t} \qquad (12.2)$$

where A_o is the initial enzyme activity and the reaction rate constant k_T, a function of temperature, is adequately described by Arrhenius kinetics through Equation 12.3.

$$k_T = k_{T_{ref}} \exp\left[-\frac{E_a}{R}\left(\frac{1}{T} - \frac{1}{T_{ref}}\right)\right] \qquad (12.3)$$

where k_{Tref} is the reaction rate constant at a constant reference temperature T_{ref}, E_a is the activation energy, and R is the universal gas constant (8.314 J/(mol·K)).

PG thermal inactivation follows first-order kinetics (Crelier et al. 2001, Fachin et al. 2003), as suggested by Equation 12.2 above, or a fractional conversion model (Eq. 12.4) that suggests a residual enzyme activity at the end of the treatment.

$$A = A_\infty + (A_o - A_\infty)e^{-k_T t} \qquad (12.4)$$

where A_∞ is the residual enzyme activity after prolonged heating.

From the data presented by Crelier et al. (2001) the effect of processing temperature on crude tomato juice PG and PME thermal inactivation rates at ambient pressure is illustrated in Figure 12.6. The higher resistance of PG, compared with PME, to thermal inactivation is evident (Fig. 12.6). Furthermore, PG

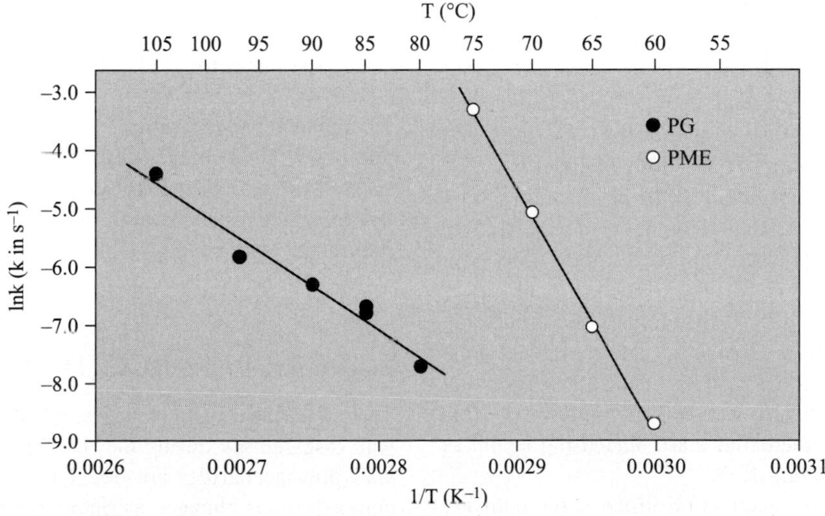

Figure 12.6. Effect of processing temperature on PG and PME thermal inactivation rates at ambient pressure. (Redrawn from Crelier and others 2001.)

inactivation is less sensitive to temperature changes, compared with PME inactivation (as can be seen by comparing the slopes of the corresponding curves in Fig. 12.6). Values for the activation energies (E_a, see Eq. 12.3) equal to 134.5 ± 15.7 kJ/mol for the case of PG and 350.1 ± 6.0 kJ/mol for the case of PME thermal inactivation have been reported (Crelier et al. 2001). It must be noted that the literature values presented here are restricted to the system used in the particular study and are mainly reported here for illustrative purposes. Thus, for example, the origin and the environment (e.g., pH) of the enzyme can influence the heat resistance (k or D—the decimal reduction time—values) of the enzyme as well as the temperature sensitivity (E_a or z—the temperature difference required for 90% change in D-values) of the enzyme thermal inactivation rates.

High-Pressure Inactivation

High hydrostatic pressure processing of foods [i.e., processing at elevated pressures (up to 1000 MPa) and low to moderate temperatures (usually less than 100°C)] has been introduced as an alternative nonthermal technology that causes inactivation of microorganisms and denaturation of several enzymes with minimal destructive effects on the quality and the organoleptic characteristics of the product. The improved product quality attained during high-pressure processing of foods, and the potential for production of a variety of novel foods with particular, desirable characteristics, have made the high-pressure technology attractive (Farr 1990, Knorr 1993).

As far as high-pressure enzyme inactivation goes, PG is easily inactivated at moderate pressures and temperatures (Crelier et al. 2001, Shook et al. 2001, Fachin et al. 2003), while PME inactivation at elevated pressures reveals an antagonistic (protective) effect between pressure and temperature (Crelier et al. 2001, Shook et al. 2001, Fachin et al. 2002, Stoforos et al. 2002). Depending on the processing temperature, PME inactivation rate at ambient pressure (0.1 MPa) is high, rapidly decreases as pressure increases, practically vanishes at pressures of 100–500 MPa, and thereafter starts increasing again, as illustrated in Figure 12.7.

High-pressure inactivation kinetics for both PG and PME follow first-order kinetics (Crelier et al. 2001; Stoforos et al. 2002; Fachin et al. 2002, 2003).

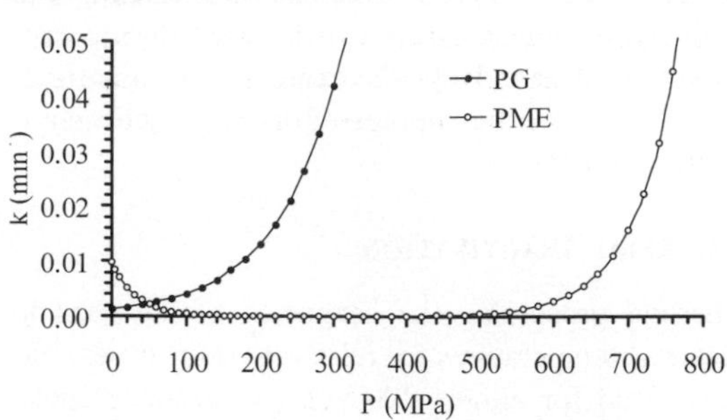

Figure 12.7. Schematic representation of the effect of processing pressure on PG and PME inactivation rates during high-pressure treatment (at 60°C).

Values for the reaction rate constants, k, for high-pressure inactivation of PME and PG as a function of processing temperature, at selected conditions, are given in Table 12.1 (Crelier et al. 2001).

Through the activation volume concept (Johnson and Eyring 1970), the pressure effects on the reaction rate constants can be expressed as

$$k_P = k_{P_{ref}} \exp\left[-\frac{V_a}{R}\frac{(P - P_{ref})}{T}\right] \tag{12.5}$$

where $k_{P_{ref}}$ is the reaction rate constant at a constant reference pressure, P_{ref}, and V_a is the activation volume.

Models to describe the combined effect of pressure and temperature on tomato PME or PG inactivation have been presented in the literature (Crelier et al. 2001, Stoforos et al. 2002, Fachin et al. 2003). Based on literature data (Crelier et al. 2001), a schematic representation of high-pressure inactivation of tomato PG and PME is presented in Figure 12.7. From data like these, one can see the possibilities of selective inactivation of the one or the other enzyme by appropriately optimizing the processing conditions (pressure, temperature, and time).

FUTURE PERSPECTIVES

Texture in ripe tomato fruit is largely dictated by cell-wall disassembly during the ripening process. Cell-wall polysaccharides are depolymerized, and their composition is changed as ripening progresses. The coordinated and synergistic activities of many proteins are responsible for fruit softening, and although

Table 12.1. First-Order Reaction Rate Constants, k (sec^{-1}), for High-Pressure Inactivation of PME and PG as a Function of Processing Temperature

P (MPa) \ T	First-Order Reaction Rate Constants, k (sec^{-1})							
	PG				PME			
	60°C	65°C	70°C	75°C	30°C	40°C	50°C	60°C
0.1	161×10^{-6}	872×10^{-6}	6250×10^{-6}	3680×10^{-6}				
100	7.58×10^{-6}	163×10^{-6}	883×10^{-6}	574×10^{-6}				52.6×10^{-6}
300	9.70×10^{-6}	112×10^{-6}	51.4×10^{-6}	18.3×10^{-6}				309×10^{-6}
400				38.6×10^{-6}	193×10^{-6}	351×10^{-6}	701×10^{-6}	2510×10^{-6}
500				103×10^{-6}	3890×10^{-6}	4090×10^{-6}	3620×10^{-6}	7020×10^{-6}
600	43.8×10^{-6}	44.3×10^{-6}	70.2×10^{-6}	52.5×10^{-6}		1720×10^{-6}	4810×10^{-6}	7530×10^{-6}
800	1500×10^{-6}	2330×10^{-6}	2920×10^{-6}	2710×10^{-6}				

Source: Crelier et al. 2001.

expansins and polygalacturonases are among the more abundantly expressed proteins in ripening tomato fruit, the modification of their expression has not been sufficient to account for all of the cell-wall changes associated with softening or for the overall extent of fruit softening. The texture of fresh fruit directly influences the rheological characteristics of processed tomato products. Viscosity is influenced by modifications of the cell-wall architecture but also by changes that may affect polysaccharide mobility and interactions of different components in juices and pastes. Since cell-wall metabolism in ripening fruit is a complex processes and many enzymes have been identified that contribute to this process, further insight to the effect of these enzymes on fruit texture and processing attributes can be gained by the simultaneous suppression or overexpression of combinations of enzymes. Researchers are now preparing double transgenic lines with both PG and β-galactosidase gene expression suppressed (John M. Labavitch, pers. comm., 2004). Moreover, our knowledge of cell-wall enzymes increases, and other genes are continuously identified that contribute to pectin depolymerization and fruit softening (Ramakrishna et al. 2003). Manipulation of the expression of multiple ripening-associated genes in transgenic tomato lines will further shed light on the ripening physiology, postharvest shelf life, and processing qualities of tomato fruit. Finally, the interest exists and efforts are made to design optimal processes for selective enzyme inactivation and thus for production of products with desirable characteristics, by introducing pressure as an additional (to time and temperature) processing variable.

ACKNOWLEDGEMENTS

The authors would like to thank Asst. Professor P. Christakopoulos, National Technical University of Athens, School of Chemical Engineering, Biosystems Technology Laboratory, and Dr. C. Mallidis, Institute of Technology of Agricultural Products, National Agricultural Research Foundation of Greece for reviewing the manuscript and for their constructive comments.

REFERENCES

Albani D, Altosaar I, Arnison PG, Fabijanski SF. 1991. A gene showing sequence similarity to pectin esterase is specifically expressed in developing pollen of *Brassica napus* sequences in its 5’ flanking region are conserved in other pollen-specific promoters. Plant Mol Biol 16:501–513.

Alviar MSB, Reid DS. 1990. Determination of rheological behavior of tomato concentrates using back extrusion. J Food Sci 55:554–555.

Barras F, van Gijsegem F, Chatterjee AK. 1994. Extracellular enzymes and soft-rot Erwinia. Annu Rev Phytopathol 32:201–234.

Barrett DM, Garcia E, Wayne JE. 1998. Textural modification of processing tomatoes. Crit Rev Food Sci Nutr 38:173–258.

Benítez-Burraco A, Blanco-Portales R, Redondo-Nevado J, Bellido ML, Moyano E, Caballero JL, Muñoz-Blanco J. 2003. Cloning and characterization of two ripening-related strawberry (*Fragaria x ananassa* cv Chandler) pectate lyase genes. J Exp Bot 54:633–645.

Beresovsky N, Kopelman IJ, Mizrahi S. 1995. The role of pulp interparticle interaction in determining tomato juice viscosity. J Food Process Pres 19:133–146.

Blumenkrantz N, Asboe-Hansen G. 1973. New method for quantitative determination of uronic acids. Anal Biochem 54:484–489.

Blumer JM, Clay RP, Bergmann CW, Albersheim P, Darvill A. 2000. Characterization of changes in pectin methylesterase expression and pectin esterification during tomato fruit ripening. Can J Bot 78: 607–618.

Brummell DA, Harpster MH, Civello PM, Palys JM, Bennett AB, Dunsmuir P. 1999a. Modification of expansin protein abundance in tomato fruit alters softening and cell wall polymer metabolism during ripening. Plant Cell 11:2203–2216.

Brummell DA, Harpster MH, Dunsmuir P. 1999b. Differential expression of expansin gene family members during growth and ripening of tomato fruit. Plant Mol Biol 39:161–169.

Brummell DA, Harpster MH. 2001. Cell wall metabolism in fruit softening and quality and its manipulation in transgenic plants. Plant Mol Biol 47:311–340.

Buchanan BB, Gruissem W, Jones RL. 2000. Biochemistry and Molecular Biology of Plants. Rockville, Maryland: American Society of Plant Physiologists. 1367 pp.

Brummell DA, Labavitch JM. 1997. Effect of antisense suppression of endopolygalacturonase activity on polyuronide molecular weight in ripening tomato fruit and in fruit homogenates. Plant Physiol 115: 717–725.

Carey AT, Smith DL, Harrison E, Bird CR, Gross KC, Seymour GB, Tucker GA. 2001. Down-regulation of a ripening-related β-galactosidase gene *(TBG1)* in trasgenic tomato fruits. J Exp Bot 52:663–668.

Carpita NC, Gibeaut DM. 1993. Structural models of primary cell walls in flowering plants consistency of molecular structure with the physical properties of the walls during growth. Plant J 3:1–30.

Carrington CMS, Greve LC, Labavitch JM. 1993. Cell wall metabolism in ripening fruit. 6. Effect of the antisense polygalacturonase gene on cell wall changes accompanying ripening in transgenic tomatoes. Plant Physiol 103:429–434.

Chun JP, Huber DJ. 1997. Polygalacturonase isoenzyme 2 binding and catalysis in cell walls from tomato fruit: pH and β-subunit effects. Physiol Plantarum 101:283–290.

Crelier S, Robert M-C, Claude J, Juillerat M-A. 2001. Tomato *(Lycopersicon esculentum)* pectin methylesterase and polygalacturonase behaviors regarding heat- and pressure-induced inactivation. J Agr Food Chem 49:5566–5575.

Cutillas-Iturralde A, Zarrra I, Lorences EP. 1993. Metabolism of cell wall polysaccharides from persimmon fruit: Pectin solubilization during fruit ripening occurs in apparent absence of polygalacturonase activity. Physiol Plantarum 89:369–375.

DellaPenna D, Alexander DC, Bennett AB. 1986. Molecular cloning of tomato fruit polygalacturonase: Analysis of polygalacturonase mRNA levels during ripening. Proc Natl Acad Sci USA 83:6420–6424.

DellaPenna D, Bennett AB. 1988. *In vitro* synthesis and processing of tomato fruit polygalacturonase. Plant Physiol 86:1057–1063.

DellaPenna D, Lashbrook CC, Toenjes K, Giovannoni JJ, Fischer RL, Bennett AB. 1990. Polygalacturonase isoenzymes and pectin depolymerization in transgenic *rin* tomato fruit. Plant Physiol 94:1882–1886.

DellaPenna D, Lincoln JE, Fischer RL, Bennett AB. 1989. Transcriptional analysis of polygalacturonase and other ripening associated genes in Rutgers, *rin*, *nor*, and *nr* tomato fruit. Plant Physiol 90:1372–1377.

Den Ouden FWC, Van Vliet T. 1997. Particle size distribution in tomato concentrate and effects on rheological properties. J Food Sci 62:565–567.

Domingo C, Roberts K, Stacey NJ, Connerton I, Ruíz-Terán F, McCann MC. 1998. A pectate lyase from *Zinnia elegans* is auxin inducible. Plant J 13:17–28.

Errington N, Tucker GA, Mitchell JR. 1998. Effect of genetic down-regulation of polygalacturonase and pectin esterase activity on rheology and composition of tomato juice. J Sci Food Agr 76:515–519.

Fachin D, Van Loey AM, Nguyen BL, Verlent I, Indrawati, Hendrickx ME. 2002. Comparative study of the inactivation kinetics of pectinmethylesterase in tomato juice and purified form. Biotechnol Progr 18:739–744.

———. 2003. Inactivation kinetics of polygalacturonase in tomato juice. Innovative Food Science and Emerging Technologies 4:135–142.

Farr D. 1990. High-pressure technology in the food industry. Trends Food Sci Tech 1:14–16.

Gaffe J, Tieman DM, Handa AK. 1994. Pectin methylesterase isoforms in tomato *Lycopersicon esculentum* tissue. Effects of expression of a pectin methylesterase antisense gene. Plant Physiol 105:199–203.

Gaffe J, Tiznado ME, Handa AK. 1997. Characterization and functional expression of a ubiquitously expressed tomato pectin methylesterase. Plant Physiol 114:1547–1556.

Giovannoni J. 2001. Molecular biology of fruit maturation and ripening. Annu Rev Plant Physiol Plant Mol Biol 52:725–749.

Giovannoni JJ, DellaPenna D, Bennett AB, Fischer RL. 1989. Expression of a chimeric polygalacturonase gene in transgenic *rin* (ripening inhibitor) tomato fruit results in polyuronide degradation but not fruit softening. Plant Cell 1:53–63.

Gross KC. 1984. Fractionation and partial characterization of cell walls from normal and non-ripening mutant tomato fruit. Physiol Plantarum 62:25–32.

Hadfield KA, Bennett AB. 1998. Polygalacturonases—Many genes in search of a function. Plant Physiol 117:337–343.

Hall LN, Tucker GA, Smith CJS, Watson CF, Seymour GB, Bundick Y, Boniwell JM, Fletcher JD, Ray JA, Schuch W, Bird CR, Grierson D. 1993. Antisense inhibition of pectin esterase gene expression in transgenic tomatoes. Plant J 3:121–129.

Harriman RW, Tieman DM, Handa AK. 1991. Molecular cloning of tomato pectin methylesterase gene and its expression in Rutgers, ripening inhibitor, nonripening, and never ripe tomato fruits. Plant Physiol 97:80–87.

Huber DJ. 1992. Inactivation of pectin depolymerase associated with isolated tomato fruit cell wall: Implications for the analysis of pectin solubility and molecular weight. Physiol Plantarum 86:25–32.

Jiménez-Bermúdez S, Redondo-Nevado J, Muñoz-Blanco J, Caballero JL, López-Aranda JM, Valpuesta V, Pliego-Alfaro F, Quesada MA, Mercado JA.

2002. Manipulation of strawberry fruit softening by antisense expression of a pectate lyase gene. Plant Physiol 128:751–759.

Johnson FH, Eyring H. 1970. The kinetic basis of pressure effects in biology and chemistry. In: AM Zimmerman, editor. High Pressure Effects on Cellular Processes. New York: Academic Press. Pp. 1–44.

Kalamaki MS. 2003. The Influence of Transgenic Modification of Gene Expression during Ripening on Physicochemical Characteristics of Processed Tomato Products. [DPhil. Dissertation]. Davis, CA: Univ. of California. UMI Digital Dissertations, Ann Arbor, MI. 155 pp.

Kalamaki MS, Harpster MH, Palys MJ, Labavitch JM, Reid DS, Brummell DA. 2003a. Simultaneous transgenic suppression of LePG and LeExp1 influences rheological properties of juice and concentrates from a processing tomato variety. J Agr Food Chem 51: 7456–7464.

Kalamaki MS, Powell ALT, Struijs K, Labavitch JM, Reid DS, Bennett AB. 2003b. Transgenic overexpression of expansin influences particle size distribution and improves viscosity of tomato juice and paste. J Agr Food Chem 51:7465–7471.

Koch JL, Nevins DJ. 1989. Tomato fruit cell wall. I. Use of purified tomato polygalacturonase and pectinmethylesterase to identify developmental changes in pectins. Plant Physiol 91:816–822.

Knegt E, Vermeer E, Pak C, Bruinsma J. 1991. Function of the polygalacturonase convertor in ripening tomato fruit. Physiol Plantarum 82:237–242.

Knorr D. 1993. Effects of high-hydrostatic-pressure processes on food safety and quality. Food Technol 47(6): 156–161.

Kramer M, Sanders RA, Bolkan H, Waters CM, Sheehy RE, Hiatt WR. 1992. Postharvest evaluation of transgenic tomatoes with reduced levels of polygalacturonase: Processing, firmness, and disease resistance. Postharvest Biol Tec 1:241–255.

Kulikauskas R, McCormick S. 1997. Identification of the tobacco and Arabidopsis homologues of the pollen-expressed LAT59 gene of tomato. Plant Mol Biol 34:809–814.

Langley KR, Martin A, Stenning R, Murray AJ, Hobson GE, Schuch WW, Bird CR. 1994. Mechanical and optical assessment of the ripening of tomato fruit with reduced polygalacturonase activity. J Sci Food Agr 66:547–554.

Luh BS, Daoud HN. 1971. Effect of break temperature and holding time on pectin and pectic enzymes in tomato pulp. J Food Sci 36:1039–1043.

Marín-Rodríguez MC, Orchard J, Seymour GB. 2002. Pectate lyases, cell wall degradation and fruit softening. J Exp Bot 53:2115–2119.

Marín-Rodríguez MC, Smith DL, Manning K, Orchard J, Seymour GB. 2003. Pectate lyase gene expression and enzyme activity in ripening banana fruit. Plant Mol Biol 51:851–857.

Marsh GL, Buhlert J, Leonard S. 1978. Effect of degree of concentration and of heat treatment on consistency of tomato pastes after dilution. J Food Process Pres 1:340–346.

Mayans O, Scott M, Connerton I, Gravesen T, Benen J, Visser J, Pickersgill R, Jenkins J. 1997. Two crystal structures of pectin lyase A from *Aspergillus niger* reveal a pH driven conformational change and striking divergence in the substrate-binding clefts of pectin and pectate lyases. Structure 5:677–689.

Medina-Escobar N, Cárdenas J, Moyano E, Caballero JL, Muñoz-Blanco J. 1997. Cloning molecular characterisation and expression pattern of a strawberry ripening-specific cDNA with sequence homology to pectate lyase from higher plants. Plant Mol Biol 34:867–877.

Milioni D, Sado P-E, Stacey NJ, Domingo C, Roberts K, McCann MC. 2001. Differential expression of cell-wall-related genes during the formation of tracheary elements in the *Zinnia* mesophyll cell system. Plant Mol Biol 47:221–238.

Moore T, Bennett AB. 1994. Tomato fruit polygalacturonase isozyme 1. Plant Physiol 106:1461–1469.

Pilatzke-Wunderlich I, Nessler CL. 2001. Expression and activity of cell-wall-degrading enzymes in the latex of opium poppy, *Papaver somniferum* L. Plant Mol Biol 45:567–576.

Pilling J, Willmitzer L, Fisahn J. 2000. Expression of a *Petunia inflata* pectin methyl esterase in *Solanum tuberosum* L. enhances stem elongation and modifies cation distribution. Planta 210:391–399.

Pogson BJ, Brady CJ, Orr GR. 1991. On the occurrence and structure of subunits of endopolygalacturonase isoforms in mature-green and ripening tomato fruits. Aust J Plant Physiol l8:65–79.

Powell ALT, Kalamaki MS, Kurien P, Gurrieri S, Bennett AB. 2003. Simultaneous transgenic suppression of LePG and LeExp1 influences fruit texture and juice viscosity in a fresh-market tomato variety. J Agr Food Chem 51:7450–7455.

Pua E-C, Ong C-K, Liu P, Liu J-Z. 2001. Isolation and expression of two pectate lyase genes during fruit ripening of banana *(Musa acuminata)*. Physiol Plantarum 113:92–99.

Ramakrishna W, Deng Z, Ding CK, Handa AK, Ozminkowski RH,Jr. 2003. A novel small heat shock protein gene, *vis1*, contributes to pectin depolymerization and juice viscosity in tomato fruit. Plant Physiol 131:725–735.

Rose JKC, Bennett AB. 1999. Cooperative disassembly of the cellulose-xyloglucan network of plant cell walls: Parallels between cell expansion and fruit ripening. Trends Plant Sci 4:176–183.

Rose JKC, Lee HH, Bennett AB. 1997. Expression of a divergent expansin gene is fruit-specific and ripening-regulated. Proc Natl Acad Sci USA 94:5955–5960.

Schuch W, Kanczler J, Robertson D, Hobson G, Tucker G, Grierson D, Bright S, Bird C. 1991. Fruit quality characteristics of transgenic tomato fruit with altered polygalacturonase activity. Hortscience 26:1517–1520.

Seymour GB, Colquhoun IJ, DuPont MS, Parsley KR, Selvendran RR. 1990. Composition and structural features of cell wall polysaccharides from tomato fruits. Phytochemistry 29:725–731.

Sheehy RE, Kramer M, Hiatt WR. 1988. Reduction of polygalacturonase activity in tomato fruit by antisense RNA. Proc Natl Acad Sci USA 85:8805–8809.

Sheehy RE, Pearson J, Brady CJ, Hiatt WR. 1987. Molecular characterization of tomato fruit polygalacturonase. Mol Gen Genet 208:30–36.

Shook CM, Shellhammer TH, Schwartz SJ. 2001. Polygalacturonase, pectinmethylesterase, and lipoxygenase activities in high-pressure-processed diced tomatoes. J Agr Food Chem 49:664–668.

Sitrit Y, Bennett AB. 1998. Regulation of tomato fruit polygalacturonase mRNA accumulation by ethylene: A re-examination. Plant Physiol 116:1145–1150.

Smith CJS, Watson CF, Ray J, Bird CR, Morris PC, Schuch W, Grierson D. 1988. Antisense RNA inhibition of polygalacturonase gene expression in transgenic tomatoes. Nature 334:724–726.

Smith DL, Abbott JA, Gross KC. 2002. Down-regulation of tomato beta-galactosidase 4 results in decreased fruit softening. Plant Physiol 129:1755–1762.

Smith DL, Gross KC. 2000. A family of at least seven β-galactosidase genes is expressed during tomato fruit development. Plant Physiol 123:1173–1183.

Smith DL, Starrett DA, Gross KC. 1998. A gene coding for tomato fruit β-galactosidase II is expressed during fruit ripening. Plant Physiol 117:417–423.

Stoforos NG, Crelier S, Robert M-C, Taoukis PS. 2002. Kinetics of tomato pectin methylesterase inactivation by temperature and high pressure. J Food Sci 67:1026–1031.

Taniguchi Y, Ono A, Sawatani M, Nanba M, Kohno K, Usui M, Kurimoto M, Matuhasi T. 1995. *Cry j* I, a major allergen of Japanese cedar pollen, has a pectate lyase enzyme activity. Allergy 50:90–93.

Thakur BR, Singh RK, Handa AK. 1996a. Effect of an antisense pectin methylesterase gene on the chemistry of pectin in tomato *(Lycopersicon esculentum)* juice. J Agr Food Chem 44:628–630.

Thakur BR, Singh RK, Tieman DM, Handa AK. 1996b. Tomato product quality from transgenic fruits with reduced pectin methylesterase. J Food Sci 61:85–108.

Thompson JE, Fry SC. 2000. Evidence for covalent linkage between xyloglucan and acidic pectins in suspension-cultured rose cells. Planta 211:275–286.

Tieman DM, Handa AK. 1994. Reduction in pectin methylesterase activity modifies tissue integrity and cation levels in ripening tomato (*Lycopersicon esculentum* Mill.) fruits. Plant Physiol 106:429–436.

Tieman DM, Harriman RW, Ramomohan G, Handa AK. 1992. An antisence pectin methylesterase gene alters pectin chemistry and soluble solids in tomato fruit. Plant Cell 4:667–679.

Turner LA, Harriman RW, Handa AK. 1996. Isolation and nucleotide sequence of three tandemly arranged pectin methylesterase genes (Accession Nos. U70675, U70676 and U70677) from tomato. Plant Physiol 112:1398.

Wakabayashi K, Hoson T, Huber DJ. 2003. Methyl de-esterification as a major factor regulating the extent of pectin depolymerization during fruit ripening: A comparison of the action of avocado *(Persea americana)* and tomato *(Lycopersicon esculentum)* polygalacturonases. J Plant Physiol 160:667–673.

Watson CF, Zheng LS, DellaPenna D. 1994. Reduction of tomato polygalacturonase β-subunit expression affects pectin solubilization and degradation during fruit ripening. Plant Cell 6:1623–1634.

Whittenberger RT, Nutting GC. 1957. Effect of tomato cell structures on consistency of tomato juice. Food Technol 11:19–22.

Wing RA, Yamaguchi J, Larabell SK, Ursin VM, McCormick S. 1989. Molecular and genetic characterization of two pollen-expressed genes that have sequence similarity to pectate lyases of the plant pathogen *Erwinia*. Plant Mol Biol 14:17–28.

Wu Q, Szakacs-Dobozi M, Hemmat M, Hrazdina G. 1993. En-dopolygalacturonase in apples *(Malus domestica)* and its expression during fruit ripening. Plant Physiol 102:219–225.

Wu Y, Qiu X, Du S, Erickson L. 1996. PO149 a new member of pollen pectate lyase-like gene family from alfalfa. Plant Mol Biol 32:1037–1042.

Xu S-Y, Shoemaker CF, Luh BS. 1986. Effect of break temperature on rheological properties and microstructure of tomato juices and pastes. J Food Sci 51:399–407.

Yoder MD, Keen NT, Jurnak F. 1993. New domain motif: The structure of pectate lyase C, a secreted plant virulence factor. Science 260:1503–1507.

Zheng L, Heupel RC, DellaPenna D. 1992. The β-subunit of tomato fruit polygalacturonase 1: Isolation characterization, and identification of unique structural features. Plant Cell 4:1147–1156.

Zheng L, Watson CF, DellaPenna D. 1994. Differential expression of the two subunits of tomato polygalacturonase isoenzyme 1 in wild-type and *rin* tomato fruit. Plant Physiol 105:1189–1195.

Part III
Muscle Foods

13
Biochemistry of Raw Meat and Poultry

F. Toldrá and M. Reig

BACKGROUND INFORMATION

In recent years there has been a decline in consumption of beef, accompanied by a slight increase in pork and an increased demand for chicken. There are many reasons behind these changes including the consumer concern for safety and health, changes in demographic characteristics, changes in consumer life-styles, availability and convenience, price, and so on (Resurrección 2003). The quality perception of beef changes depending on the country. So, Americans are concerned with cholesterol, calorie content, artificial ingredients, convenience characteristics, and price (Resurrección 2003). In the United States, this changing demand has influenced the other meat markets. Beef has been gradually losing market share to pork and, especially, chicken (Grunert 1997).

But a significant percentage of meat is used as raw material for further processing into different products such as cooked, fermented, and dry-cured meats. Some of the most well-known meat products are bacon, cooked ham, fermented and dry-fermented sausages, and dry-cured ham (Flores and Toldrá 1993, Toldrá, 2002).

The processing meat industry faces various problems, but one of the most important is the variability in the quality of meat as a raw material. Genetics, age and sex, intensive or extensive production systems, type of feeding and slaughter procedures including preslaughter handling, stunning methods, and postmortem treatment have an influence on important biochemical traits with a direct effect on the meat quality and its aptitude as raw material.

STRUCTURE OF MUSCLE

A good knowledge of the structure of muscle is essential for the use of muscle as meat. The structure is important for the properties of the muscle, and its changes during postmortem events influence the quality properties of the meat.

Muscle has different colors within the range of white to red, depending on the proportion of fibers. There are different classifications for fibers. Based on color, they can be classed as red, white, or intermediate (Moody and Cassens 1968): (1) red fibers are characterized by a higher content of myoglobin and higher numbers of capillaries and mitochondria, and they exhibit oxidative metabolism; (2) white fibers contain low amounts of myoglobin and exhibit glycolytic metabolism; and (3) intermediate fibers exhibit intermediate properties. Red muscles contain a high proportion of red fibers and are mostly related to locomotion, while white muscles contain a higher proportion of white fibers and are engaged in support tasks (Urich 1994). Other classifications are based on the speed of contraction (Pearson and Young 1989): Type I for slow-twitch oxidative fibers, type IIA for fast-twitch oxidative fibers, and IIB for fast-twitch glycolytic fibers.

The skeletal muscle contains a great number of fibers. Each fiber, which is surrounded by connective tissue, contains around 1000 myofibrils, all of them arranged in a parallel way and responsible for contraction and relaxation. They are embedded in a liquid known as sarcoplasm, which contains the sarcoplasmic (water-soluble) proteins. Each myofibril contains clear dark lines, known as Z-lines, regularly located along the myofibril (see Fig. 13.1). The distance between two consecutive Z-lines is known as a sarcomere. In addition, myofibrils contain thick and thin filaments, partly overlapped and giving rise to alternating dark (A band) and light (I band) areas. The thin filaments extend into the Z-line that serves as linkage between consecutive sarcomeres. All these filaments are composed of proteins known as myofibrillar proteins.

Muscle, fat, bones, and skin constitute the main components of carcasses; muscle is the major compound and is, furthermore, associated to the term of meat. The average percentage of muscle in relation to live weight varies depending on the species, de-

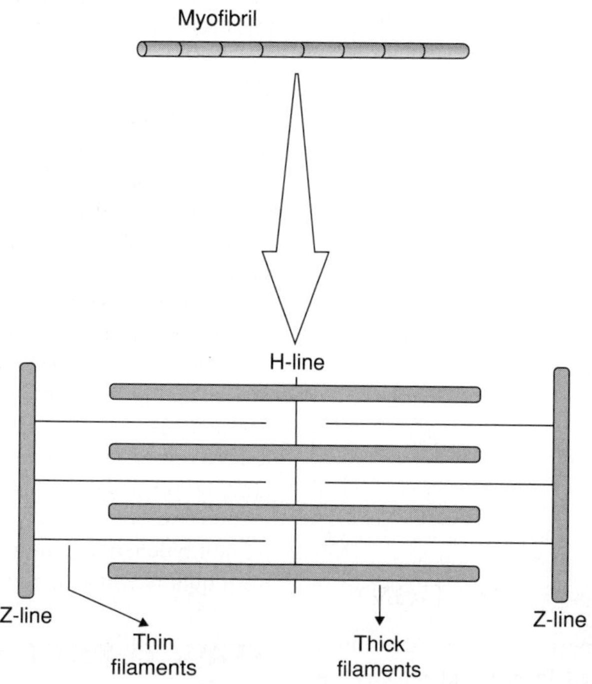

Figure 13.1. Structure of a myofibril, with details of the main filaments in the sarcomere.

gree of fatness, and dressing method: 35% for beef, 32% for veal, 36% for pork, 25% for lamb, 50% for turkey, and 39% for broiler chicken. The muscle to bone ratio is also an important parameter representative of muscling: 3.5 for beef, 2.1 for veal, 4.0 for pork, 2.5 for lamb, 2.9 for turkey and 1.8 for poultry (Kauffman 2001).

MUSCLE COMPOSITION

Essentially, meat is basically composed of water, proteins, lipids, minerals, and carbohydrates. Lean muscle tissue contains approximately 72–74% moisture, 20–22 % protein, 3–5% fat, 1% ash, and 0.5% of carbohydrate. An example of typical pork meat composition is shown in Table 13.1. These proportions are largely variable, especially in the lipid content, which depends on species, amount of fattening, inclusion of the adipose tissue, and so on. There is an inverse relationship between the percentages of protein and moisture and the percentage of fat so that meats with high content of fat have lower content of moisture and proteins. A brief description of proteins and lipids as major components of meat is given below.

MUSCLE PROTEINS

Proteins constitute the major compounds in the muscle, approximately 15–22%, and have important roles for the structure, normal function, and integrity of the muscle. Proteins experience important changes during the conversion of muscle to meat that mainly affect tenderness; and additional changes occur during further processing, through the generation of peptides and free amino acids as a result of the proteolytic enzymatic chain. There are three main groups of proteins in the muscle (see Table 13.2): myofibrillar proteins, sarcoplasmic proteins, and connective tissue proteins.

Myofibrillar Proteins

These proteins are responsible for the basic myofibrillar structure, and thus they contribute to the

Table 13.1. Example of the Approximate Composition of Pork Muscle *Longissimus dorsi*, Expressed as Means and Standard Deviation (SD)

	Units	Mean	SD
Gross composition			
Moisture	g/100g	74.5	0.5
Protein	g/100g	21.4	2.1
Lipid	g/100g	2.7	0.3
Carbohydrate	g/100g	0.5	0.1
Ash	g/100g	0.9	0.1
Proteins			
Myofibrillar	g/100g	9.5	0.8
Sarcoplasmic	g/100g	9.1	0.6
Connective	g/100g	3.0	0.4
Lipids			
Phospholipids	g/100g	0.586	0.040
Triglycerides	g/100g	2.12	0.22
Free fatty acids	g/100g	0.025	0.005
Some minor compounds			
Cholesterol	mg/100g	46.1	6.1
Haem content	mg/100g	400	30
Dipeptides	mg/100g	347.6	35.6
Free amino acids	mg/100g	90.2	5.8

Sources: From Aristoy and Toldrá 1998; Hernández et al. 1998; Toldrá 1999, unpublished).

continuity and strength of the muscle fibers. They are soluble in high ionic strength buffers, and the most important of them are listed in Table 13.2. Myosin and actin are by far the most abundant and form part of the structural backbone of the myofibril. Tropomyosin and troponins C, T, and I are considered as regulatory proteins because they play an important role in muscle contraction and relaxation (Pearson 1987). There are many proteins in the Z-line region (although in a low percentage) that serve as bridges between the thin filaments of adjacent sarcomeres. Titin and nebulin are two very large

Table 13.2. Type, Localization and Main Role of Major Muscle Proteins

Cellular Localization	Main Proteins	Main Role in Muscle
Myofibril	Myosin and actin	Cytoskeletal proteins providing support to the myofibril and responsible for contraction-relaxation of the muscle
	Tropomyosin	Regulatory protein associated to troponins than cause its movement towards the F actin helix to permit contraction
	Troponins T, C, I	Regulatory proteins. Tn-T binds to tropomyosin, Tn-C binds Ca^{2+} and initiates the contractile process and Tn-I inhibits the actin-myosin interaction in conjunction with tropomyosin and Tn-T and Tn-C
	Titin and nebulin	Large proteins located between Z-lines and thin filaments. They provide myofibrils with resistance and elasticity
	α, β, γ and eu-actinin	Proteins regulating the physical state of actin. α-actinin acts as a cementing substance in the Z-line. γ-actinin inhibits the polymerization of actin at the nucleation step. β-actinin is located at the free end of actin filaments, preventing them from binding each other. Eu-actinin interacts with both actin and α-actinin
	Filamin,synemin, vinculin, zeugmatin, Z nin, C, H, X, F, I proteins	Proteins located in the Z-line that contribute to its high density
	Desmin	Protein that links adjacent myofibrils through Z-lines
	Myomesin, creatin kinase and M protein	Proteins located in the center of the sarcomere forming part of the M-line
Sarcoplasm	Mitochondrial enzymes	Enzymes involved in the respiratory chain
	Lysosomal enzymes	Digestive hydrolases very active at acid pH (cathepsins, lipase, phospholipase, peptidases, glucohydrolases, etc.)
	Other cytosolic enzymes	Neutral proteases, lipases, glucohydrolases, ATP-ases, etc.
	Myoglobin	Natural pigment of meat
	Hemoglobin	Protein present from remaining blood within the muscle
Connective tissue	Collagen	Protein giving support, strength and shape to the fibers
	Elastin	Protein that gives elasticity to tissues like capillaries, nerves, tendons, etc.

Sources: From Bandman 1987, Pearson and Young 1989.

proteins, present in a significant proportion, that are located in the void space between the filaments and the Z-line and contribute to the integrity of the muscle cells (Robson et al. 1997). Desmin is located on the external area of the Z-line and connects adjacent myofibrils at the level of the Z-line.

Sarcoplasmic Proteins

These are water-soluble proteins, comprising about 30–35% of the total protein in muscle. Sarcoplasmic proteins contain a high diversity of proteins (summarized in Table 13.2.), mainly metabolic enzymes (mitochondrial, lysosomal, microsomal, nuclear, or free in the cytosol) and myoglobin. Some of these enzymes play a very important role in postmortem meat and during further processing, as described in Chapter 14. Minor amounts of hemoglobin may be found in the muscle if blood has not been drained properly. Myoglobin is the main sarcoplasmic protein, responsible for the red meat color of meat as well as the typical pink color of drippings. The amount of myoglobin depends on many factors. Red fibers contain higher amounts of myoglobin than white fibers. The species is very important, and thus beef and lamb contain more myoglobin than pork and poultry. For a given species, the myoglobin content in the muscle increases with the age of the animal.

Connective Tissue Proteins

Collagen, reticulin, and elastin constitute the main stromal proteins in connective tissue. There are several types (I to V) of collagen containing different polypeptide chains (up to 10 α chains). Type I collagen is the major component of the epimysium and perimysium that surround the muscles. Types III, IV, and V collagen are found in the endomysium, which provides support to the muscle fiber (Eskin 1990). There are a high number of cross-linkages in the collagen fibers that increase with age, and this is why meat is tougher in older animals. Elastin is found in lower amounts, usually in capillaries, tendons, nerves, and ligaments.

NONPROTEIN COMPOUNDS

Dipeptides

Muscle contains three natural dipeptides: carnosine (β-alanyl-L-histidine), anserine (β—alanyl-L-1-methylhistidine), and balenine (β—alanyl-L-3-methylhistidine). These dipeptides perform somephysiological functions in muscle, for example, as buffers, antioxidants, neurotransmitters, and modulators of enzyme action (Chan and Decker 1994, Gianelli et al. 2000). Dipeptide content is especially higher in muscles with glycolytic metabolism (see Table 13.3), but it varies with the animal species, age, and diet (Aristoy and Toldrá 1998, Aristoy et al. 2003). Beef and pork have a higher content of carnosine and are lower in anserine, lamb has similar amounts of carnosine and anserine, and poultry is very rich in anserine (see Table 13.3). Balenine is present in minor amounts in pork muscle but at very low concentrations in other animal muscle, except in

Table 13.3. Example of the Composition in Dipeptides of the Porcine Glycolytic Muscle *Longissimus dorsi* and Oxidative Muscle *Trapezius*

	Carnosine (mg/100 g muscle)	Anserine (mg/100 g muscle)
Effect of muscle metabolism		
Glycolytic *(M. Longissimus dorsi)*	313	14.6
Oxidative *(M. Trapezius)*	181.0	10.7
Animal species		
Pork (loin)	313.0	14.5
Beef (top loin)	372.5	59.7
Lamb (neck)	94.2	119.5
Chicken (pectoral)	180.0	772.2

Sources: From Aristoy and Toldrá 1998, Aristoy et al. 2003.

marine mammalians such as dolphins and whales (Aristoy et al. 2003).

Free Amino Acids

The action of muscle aminopeptidases contributes to the generation of free amino acids in living muscle. An example of the typical content of free amino acids, at less than 45 minutes postmortem, in glycolytic and oxidative porcine muscles is shown in Table 13.4. It can be observed that most of the amino acids are present in significantly higher amounts in the oxidative muscle (Aristoy and Toldrá 1998). The free amino acid content is relatively low just postmortem, but it is substantially increased during postmortem storage due to the action of the proteolytic chain, which is very active and stable during meat aging.

Muscle and Adipose Tissue Lipids

Skeletal muscle contains a variable amount of lipids, between 1 and 13%. Lipid content mainly depends on the degree of fattening and the amount of adipose tissue. Lipids can be found within the muscle (intramuscular), between muscles (intermuscular), and in adipose tissue. Intramuscular lipids are mainly composed of triacylglycerols, which are stored in fat cells, and phospholipids, which are located in cell membranes. The amount of cholesterol in lean meat is around 50–70 mg/100 g. Intermuscular and adipose tissue lipids are mainly composed of triacylglycerols and small amounts of cholesterol, around 40–60 mg/100g (Toldrá 2004).

Triacylglycerols

Tri-acylglycerols are the major constituents of fat, as shown in Table 13.1. The fatty acid content mainly depends on age, production system, type of feed, and environment (Toldrá et al. 1996b). Monogastric animals such as swine and poultry tend to reflect the fatty acid composition of the feed in their fat. In the case of ruminants, the nutrient and fatty acid composition are somehow standardized due to biohydro-

Table 13.4. Example of the Composition in Free Amino Acids of the Glycolytic Muscle *Longissimus dorsi* and Oxidative Muscle *Trapezius*

Amino acids	*M. Longissimus dorsi* (mg/100 g muscle)	*M. Trapezius* (mg/100 g muscle)
Essential		
Histidine	2.90	4.12
Threonine	2.86	4.30
Valine	2.78	2.09
Methionine	0.90	1.01
Isoleucine	1.52	1.11
Leucine	2.43	1.82
Phenylalanine	1.51	1.25
Lysine	2.57	0.22
Non-essential		
Aspartic acid	0.39	0.74
Glutamic acid	2.03	5.97
Serine	2.02	4.43
Asparragine	0.91	1.63
Glycine	6.01	12.48
Glutamine	38.88	161.81
Alanine	11.29	26.17
Arginine	5.19	5.51
Proline	2.83	4.45
Tyrosine	2.11	1.63
Ornithine	0.83	0.83

Source: From Aristoy and Toldrá 1998.

genation by the microbial population of the rumen (Jakobsen 1999). The properties of the fat will depend on its fatty acid composition. A great percentage of the triacylglycerols are esterified to saturated and monounsaturated fatty acids (see neutral muscle fraction and adipose tissue data in Table 13.5). When triacylglycerols are rich in polyunsaturated fatty acids such as linoleic and linolenic acids, fats tend to be softer and prone to oxidation. These fats may even have an oily appearance when kept at room temperature.

Phospholipids

These compounds are present in cell membranes, and although present in minor amounts (see Table 13.1), they have a strong relevance to flavor development due to their relatively high proportion of polyunsaturated fatty acids (see polar fraction in Table 13.5). Major constituents are phosphatidylcholine (lecithin) and phosphatidylethanolamine. The phospholipid content may vary depending on the genetic type of the animal and the anatomical location of the muscle (Armero et al 2002, Hernández et al. 1998). For instance, red oxidative muscles have a higher amount of phospholipids than white glycolytic muscles.

CONVERSION OF MUSCLE TO MEAT

A great number of chemical and biochemical reactions take place in living muscle. Some of these reactions continue, while others are altered due to changes in pH, the presence of inhibitory compounds, the release of ions into the sarcoplasm, and so on during the early postmortem time. In a few hours, these reactions are responsible for the conversion of muscle to meat; this process is basically schematized in Figure 13.2 and consists of the following steps: Once the animal is slaughtered, the blood circulation is stopped, and the importation of nutrients and the removal of metabolites to the muscle cease. This fact has very important and drastic consequences. The first consequence is the reduction of the oxygen concentration within the muscle cell because the oxygen supply has stopped. An immediate consequence is a reduction in mitochondrial

Table 13.5. Example of Fatty Acid Composition (Expressed as Percent of Total Fatty Acids) of Muscle *Longissimus dorsi* and Adipose Tissue in Pigs Fed with a Highly Unsaturated Feed (Neutral and Polar Fractions of Muscle Lipids Also Included)

	Muscle			
Fatty Acid	Total	Neutral	Polar	Adipose Tissue
Myristic acid (C 14:0)	1.55	1.97	0.32	1.40
Palmitic acid (C 16:0)	25.10	26.19	22.10	23.78
Estearic acid (C 18:0)	12.62	11.91	14.49	11.67
Palmitoleic acid (C 16:1)	2.79	3.49	0.69	1.71
Oleic acid (C 18:1)	36.47	42.35	11.45	31.64
C 20:1	0.47	0.52	0.15	0.45
Linoleic acid (C 18:2)	16.49	11.38	37.37	25.39
C 20:2	0.49	0.43	0.66	0.78
Linolenic acid (C 18:3)	1.14	1.17	0.97	2.64
C 20:3	0.30	0.10	1.04	0.10
Arachidonic acid (C 20:4)	2.18	0.25	9.83	0.19
C 22:4	0.25	0.08	0.84	0.07
Total SFA	39.42	40.23	37.03	37.02
Total MUFA	39.74	46.36	12.26	33.81
Total PUFA	20.84	13.41	50.70	29.17
Ratio M/S	1.01	1.15	0.33	0.91
Ratio P/S	0.53	0.33	1.37	0.79

activity and cell respiration (Pearson 1987). Under normal aerobic values (see an example of resting muscle in Fig. 13.3), the muscle is able to produce 12 moles of ATP per mole of glucose, and thus the ATP content is kept around 5–8 μmol/g of muscle (Greaser 1986). ATP constitutes the main source of energy for the contraction and relaxation of the muscle structures as well as other biochemical reactions in postmortem muscle. As the redox potential is reduced towards anaerobic values, ATP generation is more costly. So only 2 moles of ATP are produced per mole of glucose under anaerobic conditions (an example of a stressed muscle is shown in Fig. 13.3). The extent of anaerobic glycolysis depends on the reserves of glycogen in the muscle (Greaser 1986). Glycogen is converted to dextrines, maltose, and finally, glucose through a phosphorolytic pathway; glucose is then converted into lactic acid with the synthesis of 2 moles of ATP (Eskin 1990). Additionally, the enzyme creatin kinase may generate some additional ATP from ADP and creatine phosphate at very early postmortem times, but only while creatin phosphate remains. The main steps in glycolysis are schematized in Figure 13.4.

The generation of ATP is strictly necessary in the muscle to supply the required energy for muscle contraction and relaxation and to drive the Na/K pump of the membranes and the calcium pump in the sarcoplasmic reticulum. The initial situation in postmortem muscle is rather similar to that in the stressed muscle, but with an important change: the absence of blood circulation. Thus, there is a lack of nutrient supply and waste removal (see Fig. 13.5). Initially, the ATP content in postmortem muscle does not drop substantially because some ATP may be formed from creatin phosphate through the action

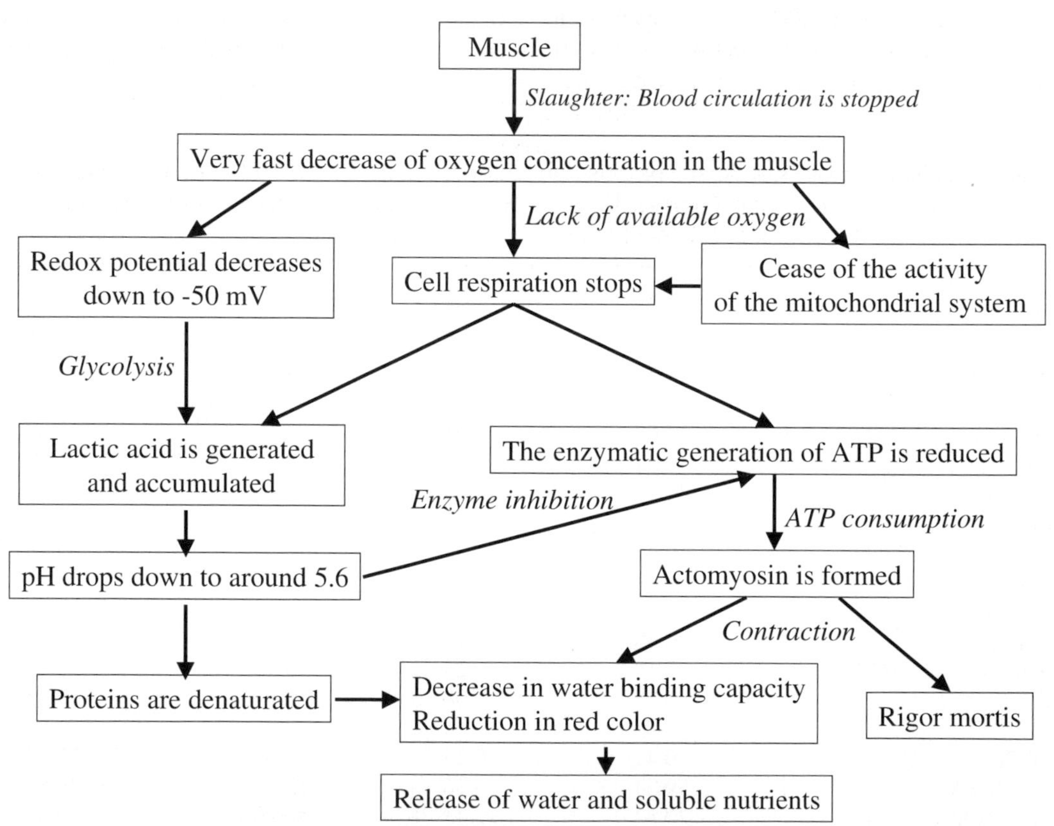

Figure 13.2. Summary of main changes during conversion of muscle to meat.

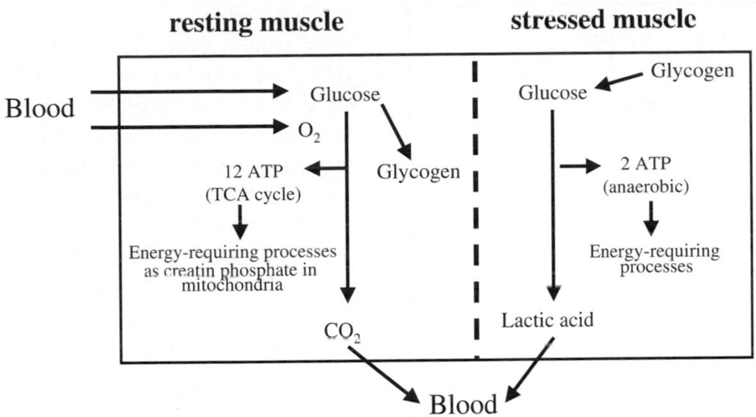

Figure 13.3. Comparison of energy generation between resting and stressed muscles.

Enzymes	Reactions	Other	Comments
Hexokinase	Glucose → glucose-6-P	ATP→ ADP	Requires Mg^{2+}
Phosphoglucoisomerase	Glucose-6-P → Fructose-6-P		Requires Mg^{2+}
Phosphofructokinase	Fructose-6-P → Fructose-1,6-biP	ATP → ADP	Inhibited by excess of ATP Requires Mg^{2+}
Aldolase	Fructose-1,6-biP → Dihydroxyacetone-3-P ↓↑ Fructose-1,6-biP → Glyceraldehyde-3-P		
Triose phosphate dehydrogenase	Glyceraldehyde-3-P → 1,3diphosphoglycerol	$2NAD^+$ → 2NADH	
Phosphoglycerokinase	1,3 diphosphoglycerol → 3-phosphoglycerol	2ADP → 2ATP	Requires Mg^{2+}
Phosphoglyceromutase	3-phosphoglycerol → 2-phosphoglycerol		Requires Mg^{2+}
Enolase	2-phosphoglycerol → Phosphoenolpyruvate	+ H_2O	
Pyruvate kinase	Phosphoenolpyruvate → Pyruvic acid	2ADP → 2ATP	If O_2 available, produces CO_2 via TCA cycle Requires K^+, Mg^{2+}
Lactate dehydrogenase	Pyruvic acid → Lactic acid	NADH → NAD^+	Only under anaerobic conditions

Figure 13.4. Main steps in glycolysis during early postmortem. (Adapted from Greaser 1986.)

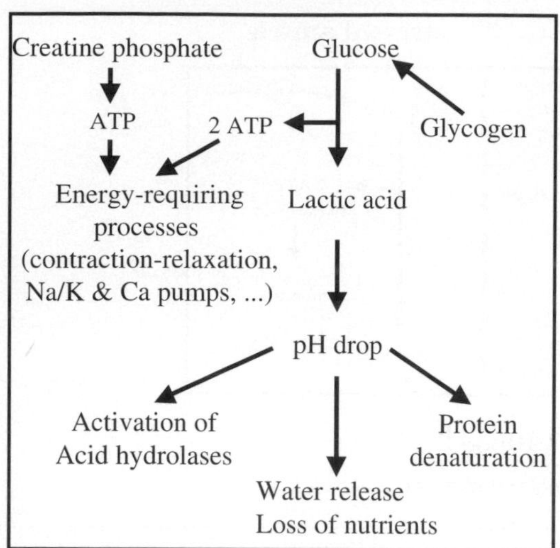

Figure 13.5. Scheme of energy generation in postmortem muscle.

of the enzyme creatine kinase and through anaerobic glycolysis. As mentioned above, once creatine phosphate and glycogen are exhausted, ATP drops within a few hours to negligible values by conversion into ADP, AMP, and other derived compounds such as 5′-inosine-monophosphate (IMP), 5′-guanosine-monophosphate (GMP), and inosine (see Fig. 13.6). An example of the typical content of ATP breakdown products in pork at 2 and 24 hours postmortem is shown in Table 13.6. The reaction rates depend on the metabolic status of the animal prior to slaughter. For instance, reactions proceed very quickly in pale, soft, exudative (PSE) muscle, where ATP can be almost fully depleted within a few minutes. The rate is also affected by the pH and temperature of the meat (Batlle et al. 2000, 2001). For instance, the ATP content in beef *Sternomandibularis* kept at 10–15°C is around 5 μmol/g at 1.5 hours postmortem and decreases to 3.5 μmol/g at 8–9 hours postmortem. However, when that muscle is kept at 38°C, ATP content is below 0.5 μmol/g at 6–7 hours postmortem.

Once the ATP concentration is exhausted, the muscle remains contracted, as no more energy is available for relaxation. The muscle develops a rigid condition known as rigor mortis, in which the cross-bridge of myosin and actin remains locked, forming actomyosin (Greaser 1986). The postmortem time necessary for the development of rigor mortis is variable, depending on the animal species, size of carcass, amount of fat cover, and environmental conditions such as the temperature of the chilling tunnel and the air velocity (see pork pieces after cutting in a slaughterhouse in Fig. 13.7). The rates of enzymatic reactions are strongly affected by temperature. In this sense, the carcass cooling rate will affect glycolysis rate, pH drop rate, and the time course of rigor onset (Faustman 1994). The animal species and size of carcass have a great influence on the cooling rate of the carcass. Furthermore, the location in the carcass is also important because surface muscles cool more rapidly than deep muscles (Greaser 2001). So, when carcasses are kept at 15°C, the time required for rigor mortis development may be about 2–4 hours in poultry, 4–18 hours in pork, and 10–24 hours in beef.

Muscle glycolytic enzymes hydrolyze the glucose to lactic acid, which is accumulated in the muscle because muscle waste substances cannot be eliminated due to the absence of blood circulation. This

Table 13.6. Example of Nucleotide and Nucleoside Content in Pork Postmortem Muscle at 2 and 24 Hours

Compound	2 h Postmortem (μmol/g muscle)	24 h Postmortem (μmol/g muscle)
ATP	4.39	—
ADP	1.08	0.25
AMP	0.14	0.20
ITP+GTP	0.18	—
IMP	0.62	6.80
Inosine	0.15	1.30
Hypoxanthine	0.05	0.32

Source: From Batlle et al. 2001.

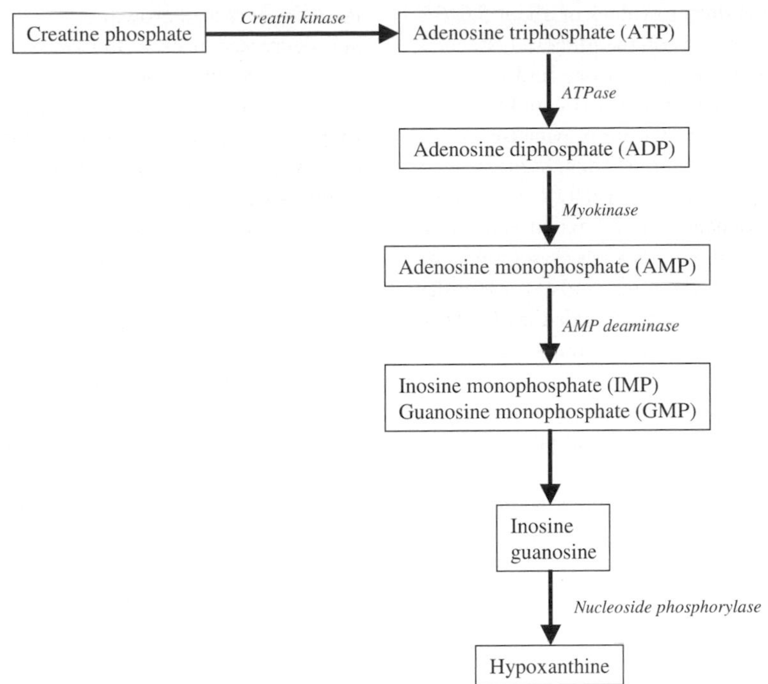

Figure 13.6. Main ATP breakdown reactions in early postmortem muscle.

Figure 13.7. Pork hams after cutting in a slaughterhouse, ready for submission to a processing plant. (By courtesy of Industrias Cárnicas Vaquero SA, Madrid, Spain.)

lactic acid accumulation produces a relatively rapid (in a few hours) pH drop to values of about 5.6–5.8. The pH drop rate depends on the glucose concentration, the temperature of the muscle, and the metabolic status of the animal previous to slaughter. Water binding decreases with pH drop because of the change in the protein's charge. Then, some water is released out of the muscle as a drip loss. The amount of released water depends on the extent and rate of pH drop. Soluble compounds such as sarcoplasmic proteins, peptides, free amino acids, nucleotides, nucleosides, B vitamins, and minerals may be partly lost in the drippings, affecting nutritional quality (Toldrá 2004).

The pH drop during early postmortem has a great influence on the quality of pork and poultry meats. The pH decrease is very fast, below 5.8 after 2 hours postmortem, in muscles from animals with accelerated metabolism. This is the case of the pale, soft, exudative pork meats (PSE) and red, soft, exudative pork meats (RSE). ATP breakdown also proceeds very quickly in these types of meats, with almost full ATP disappearance in less than 2 hours (Batlle et al 2001). Red, firm, normal meat (RFN) experiences a progressive pH drop down to values around 5.8–6.0 at 2 hours postmortem. In this meat, full ATP breakdown may take up to 8 hours. Finally, the dark, firm, dry pork meat (DFD) and dark cutting beef meat are produced when the carbohydrates in the animal are exhausted from before slaughter, and thus almost no lactic acid can be generated during early postmortem due to the lack of a substrate. Very low or almost negligible glycolysis is produced, and the pH remains high in these meats, which constitutes a risk from the microbiological point of view. These meats constitute a risk because they are prone to contamination by foodborne pathogens and must be carefully processed, with extreme attention to good hygienic practices.

FACTORS AFFECTING BIOCHEMICAL CHARACTERISTICS

EFFECT OF GENETICS

Genetic Type

The genetic type has an important relevance for quality, not only due to differences among breeds, but also to differences among animals within the same breed. Breeding strategies have been focused towards increased growth rate and lean meat content and decreased backfat thickness. Although grading traits are really improved, poorer meat quality is sometimes obtained. Usually, large ranges are found for genetic correlations between production and meat quality traits, probably due to the reduced number of samples when analyzing the full quality of meat, or to a large number of samples but with few determinations of quality parameters. This variability makes it necessary to combine the results from different research groups to obtain a full scope (Hovenier et al. 1992).

Current pig breeding schemes are usually based on a backcross or on a three- or four-way cross. For instance, a common cross in the European Union is a three-way cross, where the sow is a Landrace × Large White (LR × LW) crossbreed. The terminal sire is chosen depending on the desired profitability per animal, and there is a wide range of possibilities. For instance, the Duroc terminal sire grows faster and shows a better food conversion ratio but accumulates an excess of fat; Belgian Landrace and Pietrain are heavily muscled but have high susceptibility to stress and thus usually present a high percentage of exudative meats; or a combination of Belgian Landrace × Landrace gives good conformation and meat quality (Toldrá 2002).

Differences in tenderness between cattle breeds have also been observed. For instance, after 10 days of ageing, the steaks from an Angus breed were more tender than steaks from a ½ Angus–½ Brahman breed, and more tender than steaks from a ¼ Angus–¾ Brahman breed (Johnson et al. 1990). In other studies, it was found that meat from Hereford cattle was more tender than that from Brahman cattle (Wheeler et al. 1990). Differences in the activity of proteolytic enzymes, especially calpains, that are deeply involved in the degradation of Z-line proteins appear to have a major role in the tenderness differences between breeds.

The enzyme fingerprints, which include the assay of many different enzymes such as endo- and exoproteases as well as lipases and esterases, are useful for predicting the expected proteolysis and lipolysis during further meat processing (Armero et al. 1999a, 1999b). These enzymatic reactions, extensively described in Chapter 14, are very important for the development of sensory characteristics such as tenderness and flavor in meat and meat products.

Genes

Some genes have been found to have a strong correlation to certain positive and negative characteristics of meat. The dominant RN^- allele, also known as the Napole gene, is common in the Hampshire breed of pigs and causes high glycogen content and an extended pH decline. The carcasses are leaner, and the eating quality is better in terms of tenderness and juiciness, but the more rapid pH fall increases drip loss by about 1%, while the technological yield is reduced by 5–6% (Rosenvold and Andersen 2003, Josell et al. 2003). The processing industry is not interested in pigs with this gene because most pork meat is used for further processing, and the meat from carriers of the RN^- allele gives such a low technological yield (Monin and Sellier 1985).

Pigs containing the halothane gene are stress susceptible, a condition also known as porcine stress syndrome (PSS). These pigs are very excitable in response to transportation and environmental situations, have a very high incidence of PSE, and are susceptible to death due to malignant hyperthermia. These stress-susceptible pigs may be detected through the application of the halothane test, observing their reaction to inhalation anesthesia with halothane (Cassens 2000). These pigs give a higher carcass yield and leaner carcasses, which constitutes a direct benefit for farmers. However, the higher percentage of PSE, with high drip loss, poor color, and deficient technological properties, makes it unacceptable to the meat processing industry. These negative effects recently convinced major breeding companies to remove the halothane gene from their lines (Rosenvold and Andersen 2003).

Incidence of Exudative Meats

The detection of exudative meats at early postmortem time is of primary importance for meat processors to avoid further losses during processing. It is evident that PSE pork meat is not appealing to the eye of the consumer because it has a pale color, abundant dripping in the package, and a loose texture (Cassens 2000). Exudative pork meat also generates a loss of the nutrients that are solubilized in the sarcoplasm and lost in the drip and an economic loss due to the loss of weight as a consequence of its poor binding properties if the meat is further processed.

PSE meat is the result of protein denaturation at acid pH and relatively high postmortem temperatures. There are several classification methodologies, for example, the measurement of pH or conductivity at 45 minutes postmortem. Other methodologies involve the use of more data and thus give a more accurate profile (Warner et al. 1993, 1997; Toldrá and Flores 2000). So exudative meats are considered when pH measured at 2 hours postmortem (pH_{2h}) is lower than 5.8 and drip loss (DL) is higher than 6%. Drip loss, which is usually expressed as a percent, gives an indication of water loss (difference in weight between 0 and 72 hours): a weighed muscle portion is hung within a sealed plastic bag for 72 hours under refrigeration, then reweighed (Honikel 1997). The color parameter, L, is higher than 50 (pale color) for PSE meats and between 44 and 50 for red exudative (RSE) meats. Meats are considered normal when pH_{2h} is higher than 5.8, L is between 44 and 50, and drip loss is below 6%. Meats are classified as dark, firm, and dry (DFD) when L is lower than 44 (dark red color), drip loss is below 3%, and pH measured at 24 hours postmortem (pH_{24h}) remains high. Typical pH drops are shown in Figure 13.8.

There are some measures such as appropriate transport and handling, adequate stunning, and chilling rate of carcasses that can be applied to prevent, or at least reduce, the incidence of negative effects in exudative meats. Even though the problem is well known and there are some available corrective measures, exudative meats still constitute a problem. A survey carried out in the United States in 1992 revealed that 16% of pork carcasses were PSE, 10% DFD, and about 58% of questionable quality, mainly RSE, indicating little progress in the reduction of the problem (Cassens 2000). A similar finding was obtained in a survey carried out in Spain in 1999, where 37% of carcasses were PSE, 12% RSE, and 10% DFD (Toldrá and Flores 2000).

Effect of the Age and Sex

The content in intramuscular fat content increases with the age of the animal. In addition, the meat tends to be more flavorful and colorful, due to an increased concentration of volatiles and myoglobin, respectively (Armero et al. 1999b). Some of the muscle proteolytic and lipolytic enzymes are affected by age. Muscles from heavy pigs (11 months old)

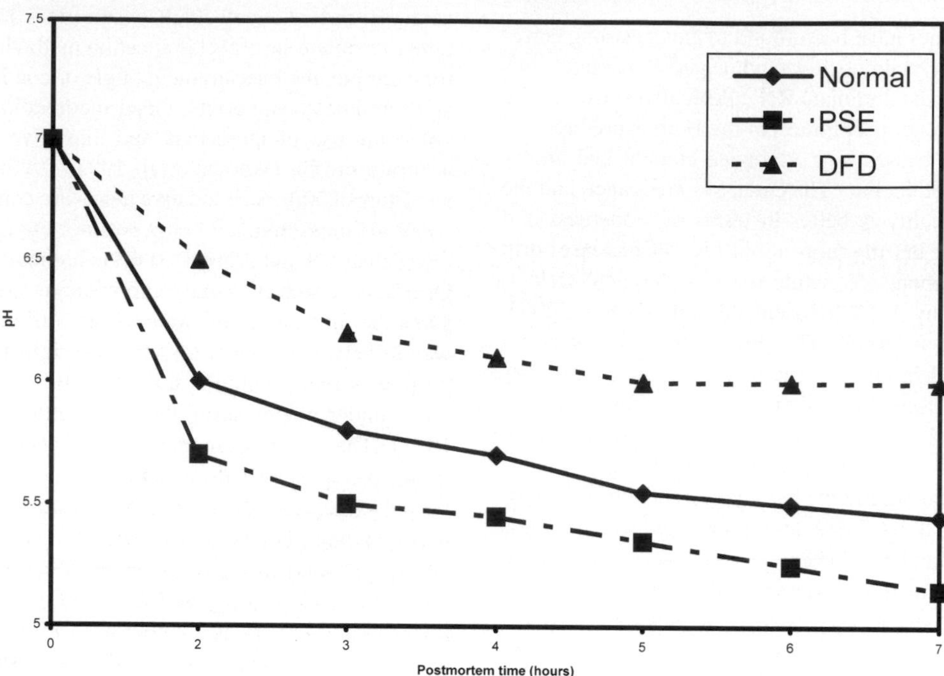

Figure 13.8. Typical postmortem pH drop of normal, PSE, and DFD pork meats (Toldrá, unpublished).

are characterized by a greater peptidase to proteinase ratio and a higher lipase, dipeptidylpeptidase IV, and pyroglutamyl aminopeptidase activity. On the other hand, the enzyme activity in light pigs (7–8 months old) shows two groups. The larger one is higher in moisture content and cathepsins B and B + L and low in peptidase activity, while the minor one is intermediate in cathepsin B activity and high in peptidase activity (Toldrá et al. 1996a). In general, there is a correlation between the moisture content and the activity of cathepsin B and B+L (Parolari et al. 1994). So, muscles with higher moisture content show higher levels of cathepsin B and B+L activity. This higher cathepsin activity may produce an excess of proteolysis in processed meat products with long processing times (Toldrá 2002).

A minor effect of sex is observed. Meats from barrows contain more fat than those from gilts. They present higher marbling, and the subcutaneous fat layer is thicker (Armero et al. 1999a). In the case of muscle enzymes, only very minor differences have been found. Sometimes, meats from entire males may give some sexual odor problems due to high contents of androstenone or escatol.

Effect of the Type of Feed

A great research effort has been exerted since the 1980s for the manipulation of the fatty acid composition of meat, to achieve nutritional recommendations, especially an increase in the ratio between polyunsaturated and saturated fatty acids (PUFA: SFA ratio). More recently, nutritionists recommend that PUFA composition should be manipulated towards a lower n-6:n-3 ratio.

Fats with a higher content of PUFA have lower melting points that affect the fat firmness. Softer fats may raise important problems during processing if the integrity of the muscle is disrupted by any mechanical treatment (chopping, mincing, stuffing, etc.). The major troubles are related to oxidation and generation of off-flavors (rancid aromas) and color deterioration (trend towards yellowness in the fat) (Toldrá and Flores 2004).

Pigs and poultry are monogastric animals that incorporate part of the dietary fatty acids practically unchanged into the adipose tissue and cellular membranes, where desaturation and chain elongation processes may occur (Toldrá et al. 1996, Jakobsen 1999). The extent of incorporation may vary depending on the specific fatty acid and the type of feed. Dietary oils and their effects on the proportions in fatty acid composition have been studied. The use of canola or linseed oils produce a substantial increase in the content of linolenic acid (C 18:3), which is a n-3 fatty acid. In this way, the n-6:n-3 ratio can be reduced from 9 to 5 (Enser et al. 2000). Other dietary oils such as soya, peanut, corn, and sunflower increase the content of linoleic acid (C 18:2), an n-6 fatty acid. Although it increases the total PUFA content, this fatty acid does not contribute to decrease the n-6:n-3 ratio, just the reverse. A similar trend is observed in the case of poultry, where the feeds with a high content of linoleic acid such as grain, corn, plant seeds, or oils also increase the n-6:n-3 ratio (Jakobsen 1999). As in the case of pork, the use of feeds containing fish oils or algae, enriched in n-3 fatty acids such as eicosapentaenoic (C 22:5 n-3) and docosahexanoic (C 22:6 n-3) acids, can enrich the poultry meat in n-3 fatty acids and reduce the n-6:n-3 ratio from around 8.4 to 1.7 (Jakobsen 1999).

The main problem arises from oxidation during heating, because some volatile compounds such as hexanal are typically generated, producing rancid aromas (Larick et al. 1992). The rate and extent of oxidation of muscle foods mainly depends on the level of PUFA, but they are also influenced by early postmortem events such as pH drop, carcass temperature, ageing, and other factors. It must be pointed out that the increased linoleic acid content is replacing the oleic acid to a large extent (Monahan et al. 1992). Feeds rich in saturated fats such as tallow yield the highest levels of palmitic, palmitoleic, stearic, and oleic acids in pork loin (Morgan et al. 1992). Linoleic and linolenic acid content may vary as much as 40% between the leanest and the fattest animals (Enser et al. 1988). An example of the effect of feed type on the fatty acid composition of subcutaneous adipose tissue of pigs is shown in Table 13.7. The PUFA content is especially high in phospholipids, located in subcellular membranes such as mitochondria, microsomes, and so on, making them vulnerable to peroxidation because of the proximity of a range of prooxidants such as myoglobin, cytochromes, nonheme iron, and trace elements (Buckley et al. 1995). Muscle contains several antioxidant systems, for example, those of superoxide dismutase and glutathione peroxidase, and ceruplasmin and transferrin, although they are weakened during postmortem storage.

An alternative for effective protection against oxidation consists in the addition of natural antioxidants like vitamin E (α-tocopheryl acetate); this has constituted a common practice in the last decade. This compound is added in the feed as an antioxidant and is accumulated by animals in tissues and subcellular structures, including membranes, substantially increasing its effect. The concentration and time of supplementation are important. Usual levels are around 100–200 mg/kg in the feed for several weeks prior to slaughter. The distribution of vitamin E in the organism is variable, being higher in the muscles of the thoracic limb, neck, and thorax and lower in the muscles of the pelvic limb and back (O'Sullivan et al. 1997). Dietary supplementation with this lipid-soluble antioxidant improves the oxidative stability of the meat. Color stability in beef, pork, and poultry is improved by protection of myoglobin against oxidation (Houben et al. 1998, Mercier et al. 1998). The water-holding capacity in pork is improved by protecting the membrane phospholipids against oxidation (Cheah et al. 1995, Dirinck et al. 1996). The reduction in drip loss by vitamin E is observed even in frozen pork meat, upon thawing. Oxidation of membrane phospholipids causes a loss in membrane integrity and affects its function as a semipermeable barrier. As a consequence, there is an increased passage of sarcoplasmic fluid through the membrane, known as drip loss (Monahan et al. 1992).

The fatty acid profile in ruminants is more saturated than in pigs, and thus the fat is firmer (Wood et al. 2003). The manipulation of fatty acids in beef is more difficult due to the rumen biohydrogenation. More than 90% of the polyunsaturated fatty acids are hydrogenated, leaving a low margin for action to increase the PUFA:SFA ratio above 0.1. However, meats from ruminants are rich in conjugated linoleic acid (CLA), mainly 9-*cis*,11-*trans*-octadecadienoic acid, which exerts important health-promoting biological activity (Belury 2002).

In general, a good level of nutrition increases the amount of intramuscular fat. On the other hand,

Table 13.7. Effect of Type of Feed on Total Fatty Acid Composition (Expressed as Percent of Total Fatty Acids) of Pork Muscle Lipids

Fatty Acid\ Feed Enriched in	Barley + Soya Bean Meal[a]	Safflower Oil[b]	Tallow Diet[c]	High Oleic Sunflower Oil[d]	Canola Oil[e]
Myristic acid (C 14:0)	—	—	1.37	0.05	1.6
Palmitic acid (C 16:0)	23.86	27.82	24.15	6.35	20.6
Estearic acid (C 18:0)	10.16	12.53	11.73	4.53	9.8
Palmitoleic acid (C 16:1)	3.00	3.56	3.63	0.45	3.6
Oleic acid (C 18:1)	39.06	37.81	46.22	71.70	45.9
C 20:1	—	0.01	0.29	0.26	
Linoleic acid (C 18:2)	17.15	14.60	8.95	15.96	12.3
C 20:2	—	0.01	0.44	—	0.4
Linolenic acid (C 18:3)	0.91	0.01	0.26	0.71	3.0
C 20:3	0.21	0.01	0.25	—	0.1
Arachidonic acid (C 20:4)	4.26	2.14	2.13	—	0.74
Eicosapentanoic acid (C 22:5)	0.64	0.01	—	—	—
Hexadecanoic acid (C 22:6)	0.75	0.01	—	—	—
Total SFA	34.02	40.35	37.83	10.93	33.6
Total MUFA	42.06	42.38	50.26	72.41	49.5
Total PUFA	23.92	16.79	11.91	16.67	16.6
Ratio M/S	1.24	1.05	1.33	6.62	1.47
Ratio P/S	0.70	0.42	0.32	1.53	0.49

Sources: [a]From Morgan et al. 1992, [b]Larick et al. 1992, [c]Leszczynski et al. 1992, [d]Rhee et al. 1988, [e]Miller et al. 1990.

food deprivation may result in an induced lipolysis that can be rapidly detected (in just 72 hours) through a higher content of free fatty acids and monoacylglycerols, especially in glycolytic muscles (Fernández et al. 1995). Fasting within 12–15 hours preslaughter is usual to reduce the risk of microbial cross-contamination during slaughter.

CARCASS CLASSIFICATION

CURRENT GRADING SYSTEMS

Meat grading constitutes a valuable tool for the classification of a large number of carcasses into classes and grades with similar characteristics such as quality and yield. The final purpose is to evaluate specific characteristics to determine carcass retail value. In addition, the weight and category of the carcass are useful for establishing the final price to be paid to the farmer. Carcasses are usually evaluated for conformation, carcass length, and backfat thickness. The carcass yields vary depending on the degree of fatness and the degree of muscling. The grade is determined based on both degrees. The grading system is thus giving information on quality traits of the carcass that help producers, processors, retailers, and consumers.

Official grading systems are based on conformation, quality, and yield. Yield grades indicate the quantity of edible meat in a carcass. In the United States, beef carcasses receive a grade for quality (prime, choice, good, standard, commercial, utility, and cutter) and a grade for predicted yield of edible meat (1 to 5). There are four grades for pork carcasses (U.S. No. 1 to U.S. No. 4) based on backfat thickness and expected lean yield. The lean yield is predicted by a combination of backfat thickness measured at the last rib and the subjective estimation of the muscling degree. In the case of poultry, there are three grades (A to C) based on the bilateral symmetry of the sternum, the lateral convexity and distal extension of the pectoral muscles, and the fat cover on the pectoral muscles (Swatland 1994).

In Europe, beef, pork, and lamb carcasses are classified according to the EUROP scheme (Council re-

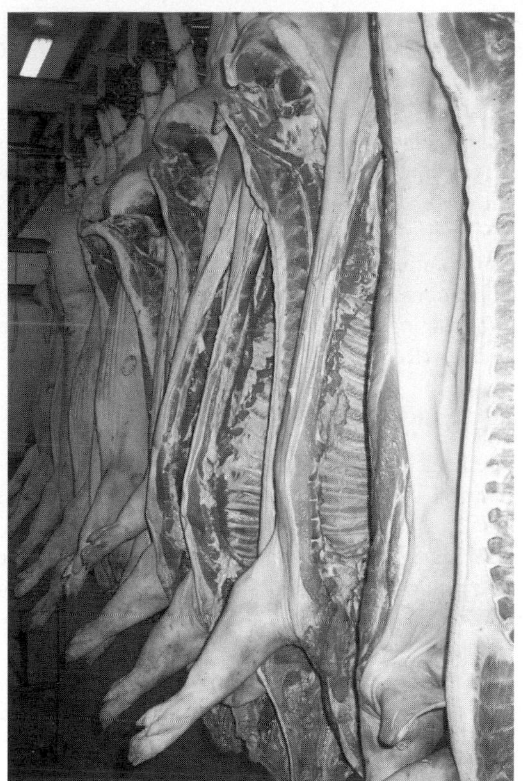

Figure 13.9. Classification of pork carcasses in the slaughterhouse. (By courtesy of Industrias Cárnicas Vaquero SA, Madrid, Spain.)

gulation 1208/81, Commission Directives 2930/81, 2137/92, and 461/93). These European Union Directives are compulsory for all the member states. Carcass classification is based on the conformation according to profiles, muscle development, and fat level. Each carcass is classified by visual inspection, based on photos corresponding to each grade (see Fig. 13.9). The six conformation classification ratings are S (superior), E (excellent), U (very good), R (good), O (fair) and P (poor). S represents the highest quality level and must not present any defect in the main pieces. The five classification ratings for fat level are 1 (low), 2 (slight), 3 (average), 4 (high), and 5 (very high). The grading system for each carcass consists in a letter for conformation, which is given first, and a number for the fat level. In some countries, additional grading systems may be added. For instance, in France the color is measured and rated 1 to 4 from white to red color.

New Grading Systems

New grading systems are being developed that take advantage of the rapid developments in video image analysis and other new physical techniques as well as those in biochemical assay tests. New methodologies based on physical methods for the on-line evaluation of meat yield and meat quality, applied as schematized in Figure 13.10, include near infrared reflectance, video image analysis, ultrasound, texture analysis, nuclear magnetic resonance, and magnetic resonance spectroscopy techniques.

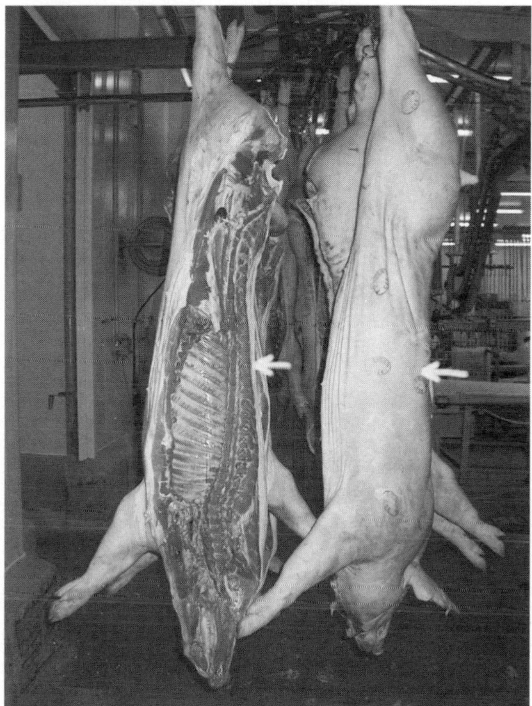

Figure 13.10. Example of the application of physical-based methods for on-line evaluation of meat yield and quality. At the slaughterhouse the hand-held devices are applied to the carcass between the last three and four ribs (place marked with an arrow). The signal is then received in the unit and computer processed. The carcass quality is estimated and, depending on the technique, is also classified by yield. (By courtesy of Industrias Cárnicas Vaquero SA, Madrid, Spain.)

Physical Techniques

Near infrared reflectance (NIR) is applied for the rapid and nondestructive analysis of meat composition in fat, protein, and water (Byrne et al. 1998, Rodbotten et al. 2000). This technique has also been applied to aged meat, giving a good correlation with texture, and thus constituting a good predictor for meat tenderness. Video image analysis is very useful for the measurement of carcass shape, marbling, and meat color. Conductivity has been in use for several years to predict meat composition and quality. Ultrasounds are based on the measurement of different parameters such as velocity, attenuation, and backscattering intensity and may constitute a valuable tool for the measurement of meat composition (Got et al. 1999, Abouelkaram et al. 2000). Texture analysis, as the image processing of the organization of grey pixels of digitized images, can be used for the classification of photographic images of meat slices (Basset et al. 2000). This technique appears to give good correlation with fat and collagen, which are especially visible under UV light, and would allow classification according to three factors: muscle type, age, and breed (Basset et al. 1999).

Nuclear magnetic resonance has good potential as a noninvasive technique for better characterization and understanding of meat features. Thus, magnetic resonance imaging can give a spatial resolution that characterizes body composition. This technique is well correlated to important meat properties such as pH, cooking yield, and water-holding capacity (Laurent et al. 2000). Magnetic resonance spectroscopy may be useful to determine the fatty acid composition of animal fat. This technique may have further applications; for example, in the possible use of ^{23}Na imaging to follow brine diffusion in cured meat products (Renou et al. 1994).

Biochemical Assay Techniques

Biochemical assay methodologies are based on biochemical compounds that can be used as markers of meat quality. The mode of operation is essentially schematized in Figure 13.11. Some of the most promising techniques include assay of proteolytic muscle enzymes and use of peptides as biochemical markers.

The assay of certain proteolytic muscle enzymes such as calpain I, alanyl aminopeptidase, or dipep-

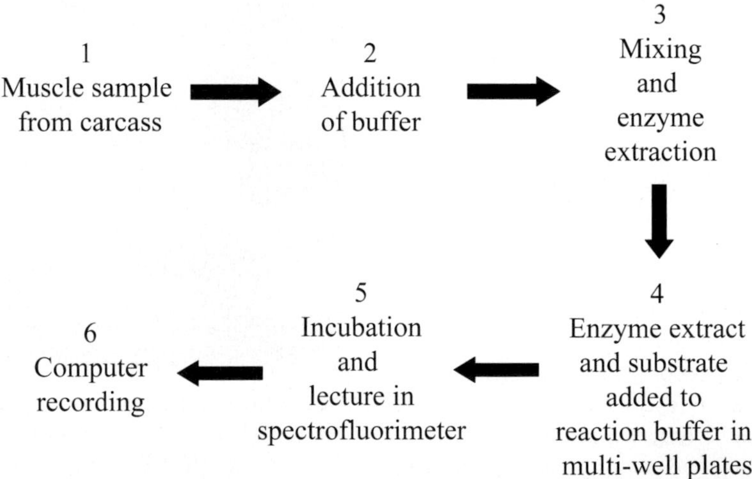

Figure 13.11. Example of application of biochemical-based methods for evaluation of meat quality. A sample of the carcass **(1)** is mixed with buffer **(2)** for enzyme extraction and homogenization **(3)**. Enzyme extracts are placed in the wells of a multiwell plate and synthetic substrates, previously dissolved in reaction buffer, are added **(4)**. The released fluorescence, which is proportional to the enzyme activity, is read by a multiwell plate spectrofluorimeter **(5)** and computer recorded **(6)**.

tidylpeptidase IV at just 2 hours postmortem, has shown good ability to predict the water-holding capacity of the meat (Toldrá and Flores 2000). Modified assay procedures, based on the use of synthetic fluorescent substrates, have been developed to allow relatively fast and simple measurements of enzyme activity with enough sensitivity.

Some peptides have been proposed as biochemical markers for meat tenderness, which is particularly important in beef. These are peptides with molecular masses ranging from 1282 to 5712 kDa, generated from sarcoplasmic and myofibrillar proteins (Stoeva et al. 2000). The isolation and identification of these peptides is tedious and time consuming, but once the full sequence is known, ELISA test kits can be developed, and this would allow rapid assay and online detection at the slaughterhouse.

In summary, there are many biochemical and chemical reactions of interest in raw meats that contribute to important changes in meat and affect its quality. Most of these changes, as have been described in this chapter, have an important role in defining the aptitude of the meat as raw material for further processing.

GLOSSARY

Ageing—Holding meat at refrigeration temperatures (0–5°C) for brief period of time (several days or few weeks) to improve meat tenderness and palatability.

ATP—Adenosine triphosphate, a high energy compound in metabolism.

DFD—Pork meat with dark, firm, and dry characteristics due to a lack of carbohydrates in muscle and thus poor glycolysis and reduced latic acid generation. These meats have pH values above 6.0 after 24 hours postmortem and are typical of exhausted stressed pigs before slaughtering. It is known as dark-cutting in the case of beef.

Drip loss—This indicates the amount of water lost as drippings during postmortem treatments. It can be easily measured by hanging a weighed muscle portion within a sealed plastic bag for 72 hours under refrigeration. The difference in weights between 0 and 72 hours, expressed as percent, gives the drip loss.

Glycolysis—Enzymatic breakdown of carbohydrates with the formation of pyruvic acid and lactic acid and the release of energy in the form of ATP.

Krebs cycle—Also known as citric acid cycle. It consists in a sequence of enzymatic reactions to provide energy for storage in the form of ATP.

Lactate dehydrogenase—Enzyme that catalyzes the oxidation of pyruvic acid to lactic acid.

Lipolysis—Enzymatic breakdown of lipids with the formation of free fatty acids

Lysosome—Cell organelle surrounded by a single membrane that contains different types of digestion enzymes with hydrolytic activity.

Mitochondria—Cytoplasmic organelles, surrounded by a membrane system. Its main function is to recover energy through the Krebs cycle and the respiratory chain and convert it by phosphorylation into ATP.

Peroxide value—Term used to measure rancidity and expressed as millimoles of peroxide taken up by 1000 g of fat.

Proteolysis—Enzymatic breakdown of proteins with the formation of peptides and free amino acids.

PSE—Meat with pale, soft, and exudative characteristics resulting from an accelerated glycolysis and thus rapid lactic acid generation. The pH drop is very fast, reaching values as low as 5.6 in just 1 hour postmortem.

Rigor mortis—Stiffening and rigidity of the muscle after death. It takes a few hours to develop, depending on the species and temperature.

Skeletal muscle fiber—Elongate, thick-walled with a characteristic striated or banded pattern; multinucleate.

Water holding capacity—This expresses the capacity of the muscle to retain water.

REFERENCES

Abouelkaram S, Suchorski K, Buquet B, Berge P, Culioli J, Delachartre P, Basset O. 2000. Effects of muscle texture on ultrasonic measurements. Food Chem. 69:447–455.

Aristoy MC, Toldrá F. 1998. Concentration of free amino acids and dipeptides in porcine skeletal muscles with different oxidative patterns. Meat Sci 50:327–332.

Aristoy MC, Soler C, Toldrá F. 2003. A simple, fast and reliable methodology for the analysis of histidine dipeptides as markers of the presence of animal origin proteins in feeds for ruminants. Food Chem. (In press.)

Armero E, Barbosa JA, Toldrá F, Baselga M, Pla M. 1999a. Effect of the terminal sire and sex on pork muscle cathepsin (B, B+L and H), cysteine proteinase inhibitors and lipolytic enzyme activities. Meat Sci 51:185–189.

Armero E, Flores M, Toldrá F, Barbosa JA, J Olivet J, Pla M, Baselga M. 1999b. Effects of pig sire types and sex on carcass traits, meat quality and sensory quality of dry-cured ham. J Sci Food Agric 79:1147–1154.

Armero E, Navarro JL, Nadal MI, Baselga M, Toldrá F. 2002. Lipid composition of pork muscle as affected by sire genetic type. J Food Biochem 26:91–102.

Bandman E. 1987. Chemistry of animal tissues. Part 1—Proteins. In: JF Price and BS Schweigert, editors, The Science of Meat and Meat Products. Trumbull, Connecticut: Food and Nutrition Press, Inc. Pp. 61–102.

Basset O, Buquet B, Abouelkaram S, Delachartre P, Culioli J. 2000. Application of texture image analysis for the classification of bovine meat. Food Chem. 69:437–445.

Basset O, Dupont F, Hernandez A, Odet C, Abouelkaram S, Culioli J. 1999. Texture image analysis: Application to the classification of bovine muscles from meat slice images. Optical Eng 38:1956–1959.

Batlle N, Aristoy MC, Toldrá F. 2000. Early postmortem detection of exudative pork meat based on nucleotide content. J Food Sci 65:413–416.

———. 2001. ATP metabolites during ageing of exudative and non exudative pork meats. J Food Sci 66:68–71.

Belury, MA. 2002. Dietary conjugated linoleic acid in health: Physiological effects and mechanisms of action. Ann Rev Nutr 22:505–531.

Buckley DJ, Morrisey PA, Gray JI. 1995. Influence of dietary vitamin E on the oxidative stability and quality of pig meat. J Anim Sci 73:3122–3130.

Byrne CE, Downey G, Troy DJ, Buckley DJ. 1998. Non-destructive prediction of selected quality attributes of beef by near infrared reflectance spectroscopy between 750 and 1098 nm. Meat Sci. 49:399–409.

Cassens RG. 2000. Historical perspectives and current aspects of pork meat quality in the USA. Food Chem 69:357–363.

Chan KM, Decker EA. 1994. Endogenous skeletal muscle antioxidants. Crit Rev Food Sci Nutr 34: 403–426.

Cheah KS, Cheah AM, Krausgill DI. 1995. Effect of dietary supplementation of vitamin E on pig meat quality. Meat Sci 39:255–264.

Demeyer DI, Toldrá F. 2003. Fermentation. In: W Jensen, C Devine, M Dikemann, editors, Encyclopedia of Meat Sciences. London: Elsevier Science, Ltd. (In press.)

Dirinck P, De Winne A, Casteels M, Frigg M. 1996. Studies on vitamin E and meat quality: 1. Effect of feeding high vitamin E levels on time-related pork quality. J Agric Food Chem 44:65–68.

Enser M, Hallet KG, Hewett B, Fursey GA, Wood JD, Harrington G. 1998. Fatty acid content and composition of UK beef and lamb muscle in relation to production system and implications for human nutrition. Meat Sci 49:329–341.

Enser M, Richardson RI, Wood JD, Gill BP, Sheard PR. 2000. Feeding linseed to increase the n-3 PUFA of pork: Fatty acid composition of muscle, adipose tissue, liver and sausages. Meat Sci 55:201–212.

Eskin NAM. 1990. Biochemical changes in raw foods: Meat and fish. In: Biochemistry of Foods, 2nd edition, pp. 3–68. San Diego: Academic Press.

Faustman LC. 1994. Postmortem changes in muscle foods. In: DM Kinsman, AW Kotula and BC Breidenstein, editors, Muscle Foods, pp. 63–78. New York: Chapman and Hall.

Fernández X, Mourot J, Mounier A, Ecolan P. 1995. Effect of muscle type and food deprivation for 24 hours on the compoisition of the lipid fraction in muscles of Large White pigs. Meat Sci 41:335–343.

Flores J, Toldrá F. 1993. Curing: Processes and applications. In: R MacCrae, R Robinson, M Sadle, G Fullerlove, editors, Encyclopedia of Food Science, Food Technology and Nutrition, pp. 1277–1282. London: Academic Press.

Gianelli MP, Flores M, Moya VJ, Aristoy MC, Toldrá F. 2000. Effect of carnosine, anserine and other endogenous skeletal peptides on the activity of porcine muscle alanyl and arginyl aminopeptidases. J Food Biochem 24:69–78.

Got F, Culioli J, Berge P, Vignon X, Astruc T, Quideau JM, Lethiecq M. 1999. Effects of high intensity high frequency ultrasound on ageing rate, ultrastructure and some physico-chemical properties of beef. Meat Sci 51:35–42.

Greaser ML. 1986. Conversion of muscle to meat. In: PJ Bechtel, editor, Muscle as Food, pp. 37–102. Orlando: Academic Press.

———. 2001. Postmortem muscle chemistry. In: YH Hui, WK Nip, RW Rogers, OA Young, editors, Meat Science and Applications, pp. 21–37. New York: Marcel Dekker, Inc.

Grunert KG. 1997. What's in a steak? A cross-cultural study on the quality perception of beef. Food Qual Pref 8:157–174.

Hernández P, Navarro JL, Toldrá F. 1998. Lipid composition and lipolytic enzyme activities in porcine skeletal muscles with different oxidative pattern. Meat Sci 49:1–10.

Honikel KO. 1997. Reference methods supported by OECD and their use in Mediterranean meat products. Food Chem 59:573–582.

Houben JH, Eikelenboom G, Hoving-Bolink AH. 1998. Effect of the dietary supplementation with vitamin E on colour stability and lipid oxidation in packaged, minced pork. Meat Sci 48:265–273.

Hovenier R, Kanis E, Asseldonk Th, Westerink NG. 1992. Genetic parameters of pig meat quality traits in a halothane negative population. Meat Sci 32: 309–321.

Jakobsen K. 1999. Dietary modifications of animal fats: Status and future perspectives. Fett/Lipid 101: 475–483.

Johnson MH, Calkins CR, Huffman RD, Johnson DD, Hargrove DD. 1990. Differences in cathepsin B+L and calcium-dependent protease activities among breed type and their relationship to beef tenderness. J Anim Sci 68:2371–2379.

Josell A, Seth G, Tornberg E. 2003. Sensory and meat quality traits of pork in relation to post-slaughter treatment and RN genotype. Meat Sci 66:113–124.

Kauffman RG. 2001. Meat composition. In: YH Hui, WK Nip, RW Rogers and OA Young, editors, Meat Science and Applications, pp. 1–19. New York: Marcel Dekker, Inc.

Larick DK, Turner BE, Schoenherr WD, Coffey MT, Pilkington DH. 1992. Volatile compound contents and fatty acid composition of pork as influenced by linoleic acid content of the diet. J Anim Sci 70:1397–1403.

Laurent W, Bonny JM, Renou JP. 2000. Muscle characterization by NMR imaging and spectroscopic techniques. Food Chem. 69:419–426.

Leszczynski DE, Pikul J, Easter RA, McKeith FK, McLaren DG, Novakofski J, Bechtel PJ, Jewell DE. 1992. Characterization of lipid in loin and bacon from finishing pigs fed full-fat soybeans or tallow. J Anim Sci 70:2175–2181.

Mercier Y, Gatellier P, Viau M, Remignon H, Renerre M. 1998. Effect of dietary fat and vitamin E on colour stability and on lipid and protein oxidation in turkey meat during storage. Meat Sci 48:301–318.

Miller MF, Shackelford SD, Hayden KD, Reagan JO. 1990. Determination of the alteration in fatty acid profiles, sensory characteristics and carcass traits of swine fed elevated levels of monounsaturated fats in the diet. J Anim Sci 68:1624–1631.

Monahan FJ, Asghar A, Gray DJ, Buckley DJ, Morrisey PA. 1992. Influence of dietary vitamin E (alpha tocopherol) on the color stability of pork chops. Proc 38th Int Congress of Meat Science and Technology, pp. 543–544, Clermont-Ferrand, France, August 1992.

Monin G, Sellier P. 1985. Pork of low technological quality with a normal rate of muscle pH fall in the inmediate post-mortem period: the case of the Hampshire breed. Meat Sci 13:49–63.

Moody WG, Cassens RG. 1968. Histochemical differentiation of red and white muscle fibers. J Anim Sci 27:961–966.

Morgan CA, Noble RC, Cocchi M, McCartney R. 1992. Manipulation of the fatty acid composition of pig meat lipids by dietary means. J Sci Food Agric 58:357–368.

O'Sullivan MG, Kerry JP, Buckley DJ, Lynch PB, Morrisey PA. 1997. The distribution of dietary vitamin E in the muscles of the porcine carcass. Meat Sci 45:297–305.

Parolari G, Virgili R, Schivazzappa C. 1994. Relationship between cathepsin B activity and compositional parameters in dry-cured ham of normal and defective texture. Meat Sci 38:117–122.

Pearson AM. 1987. Muscle function and postmortem changes. In: JF Price, BS Schweigert, editors, The Science of Meat and Meat Products, pp. 155–191. Westport, Connecticut: Food and Nutrition Press.

Pearson AM, Young RB. 1989. Muscle and Meat Biochemistry, pp. 1–261, San Diego: Academic Press.

Renou JP, Benderbous S, Bielicki G, Foucat L, Donnat JP. 1994. ^{23}Na magnetic resonance imaging: Distribution of brine in muscle. Magnetic Resonance Imaging 12:131–137.

Resurrección AVA. 2003. Sensory aspects of consumer choices for meat and meat products. Meat Sci 66:11–20.

Rhee KS, Davidson TL, Knabe DA, Cross HR, Ziprin YA, Rhee KC. 1988. Effect of dietary high-oleic sunflower oil on pork carcass traits and fatty acid profiles of raw tissues. Meat Sci 24:249–260.

Robson RM, Huff-Lonergan E, Parrish Jr FC, Ho CY, Stromer MH, Huiatt TW, Bellin RM, Sernett SW. 1997. Postmortem changes in the myofibrillar and other cytoskeletal proteins in muscle. Proceedings of the 50th Annual Reciprocal Conference, Ames, Iowa. 50:43–52.

Rodbotten R, Nilsen BN, Hildrum KI. 2000. Prediction of beef quality attributes from early post mortem near infrared reflectance spectra. Food Chem 69:427–436.

Rosenvold A, Andersen HJ. 2003. Factors of significance for pork quality—A review. Meat Sci 64:219–237.

Stoeva S, Byrne CE, Mullen AM, Troy DJ, Voelter W. 2000. Isolation and identification of proteolytic fragments from TCA soluble extracts of bovine M. *Longissimus dorsi*. Food Chem 69:365–370.

Swatland HJ. 1994. Structure and development of meat animals and poultry, pp. 143–199. Lancaster: Technomic Publishing Co.

Toldrá F. 1992. The enzymology of dry-curing of meat products. In: FJM Smulders, F Toldrá, J Flores, M Prieto, editors, New Technologies for Meat and Meat Products, pp. 209–231. Nijmegen (The Netherlands): Audet.

———. 1998. Proteolysis and lipolysis in flavour development of dry-cured meat products. Meat Sci 49:s101–s110.

———. 2002. Dry-Cured Meat Products, pp. 1–238. Trumbull, Connecticut: Food and Nutrition Press.

———. 2003. Muscle foods: Water, structure and functionality. Food Sci Technol Int 9:173–177.

———. 2005. Meat chemistry and biochemistry. In: YH Hui et al., editors, Handbook of Food Science, Technology and Engineering. New York: Marcel-Dekker, Inc. (In press.)

Toldrá F, Flores M. 1998. The role of muscle proteases and lipases in flavor development during the processing of dry-cured ham. CRC Crit Rev Food Sci Nutr 38:331–352.

———. 2000. The use of muscle enzymes as predictors of pork meat quality. Food Chem 69:387–395.

———. 2004. Meat quality factors. In: L Nollet, editor, Handbook of Food Analysis, 1961–1978. New York: Marcel Dekker, Inc.

Toldrá F, Flores M, Aristoy MC, Virgili R, Parolari G. 1996a. Pattern of muscle proteolytic and lipolytic enzymes from light and heavy pigs. J Sci Food Agric 71:124–128.

Toldrá F, Reig M, Hernández P, Navarro JL. 1996b. Lipids from pork meat as related to a healthy diet. Recent Res Devel Nutr 1:79-86.

Toldrá F, Verplaetse A. 1995. Endogenous enzyme activity and quality for raw product processing. In K Lündstrom, I Hansson and E Winklund, editors, Composition of Meat in Relation to Processing, Nutritional and Sensory Quality, pp. 41–55. Uppsala (Sweden): Ecceamst.

Urich K. 1994. Comparative Animal Biochemistry, pp. 526–623. Berlin: Springer Verlag.

Warner RD, Kauffman RG, Russell RL. 1997. Muscle protein changes post mortem in relation to pork quality traits. Meat Sci 45:339–372.

Warner RD, Russell RL, Kauffman RG. 1993. Quality attributes of major porcine muscles: A comparison with the *Longissimus lumborum*. Meat Sci 33:359–372.

Wheeler TL, Savell JW, Cross HR, Lunt DK, Smith SB. 1990. Mechanisms associated with the variation in tenderness of meat from Brahman and Hereford cattle. J Anim Sci 68:4206–4220.

Wood JD, Richardson RI, Nute GR, Fisher AV, Campo MM, Kasapidou E, Sheard PR, Enser M. 2003. Effects of fatty acids on meat quality. Meat Sci 66: 21–23

14
Biochemistry of Processing Meat and Poultry

F. Toldrá

BACKGROUND INFORMATION

There are a wide variety of meat products that are attractive to consumers because of their characteristic color, flavor, and texture. This perception varies depending on local traditions and heritage. Most of these products have been produced for many years or even centuries based on traditional practices. For instance, cured meat products reached America with settlers. Pork was cured in New England for consumption in the summer. Curers expanded these products by trying different recipes based on the use of additives such as salt, sugar, pepper, spices, and so forth, and smoking (Toldrá 2002).

Although scientific literature on biochemical changes during meat conditioning (ageing) and in some meat products were abundantly reported during the 1970s and 1980s, little information was available on the origin of the biochemical changes in other products such as cooked, dry-fermented, and dry-cured meats. The need to improve the processing and quality of these meat products prompted research in the last decades on endogenous enzyme systems that play important roles in these processes, as has been later demonstrated (Flores and Toldrá 1993). It is important to remember that the potential role of a certain enzyme in a specific observed or reported biochemical change can only be established if all the following requirements are met (Toldrá 1992): (1) the enzyme is present in the skeletal muscle or adipose tissue, (2) the enzyme is able to degrade in vitro the natural substance (i.e., a protein in the case of a protease, a tri-acylglycerol in the case of a lipase, etc.), (3) the enzyme and substrate are located closely enough in the real meat product for an effective interaction, and (4) the enzyme exhibits enough stability during processing for the changes to be developed.

DESCRIPTION OF THE MUSCLE ENZYMES

There are a wide variety of enzymes in the muscle. Most of them have an important role in the in vivo muscle functions, but they also serve an important role in biochemical changes such as the proteolysis and lipolysis that occur in postmortem meat, and during further processing of meat. Some enzymes are located in the lysosomes, while others are free in the cytosol or linked to membranes. The muscle enzymes with most important roles during meat processing are grouped by families and described below.

MUSCLE PROTEASES

Proteases are characterized by their ability to degrade proteins, and they receive different names depending on respective mode of action (see Fig. 14.1). They are endoproteases or proteinases when they are able to hydrolyze internal peptide bonds, but they are exopeptidases when they hydrolyze external peptide bonds, either at the amino termini or the carboxy termini.

Neutral Proteinases: Calpains

Calpains are cysteine endopeptidases consisting of heterodimers of 110 kDa, composed of an 80 kDa catalytic subunit and a 30 kDa subunit of unknown function. They are located in the cytosol, around the Z-line area. Calpains have received different names in the scientific literature, such as calcium-activated neutral proteinase, calcium-dependent protease, and calcium-activated factor. Calpain I is also called μ-calpain because it needs micromolar amounts (50–70 μM) of Ca^{2+} for activation. Similarly, calpain II is called m-calpain because it requires millimolar amounts (1–5 mM) of Ca^{2+}. Both calpains show maximal activity around pH 7.5. Calpain activity decreases very quickly when pH decreases to 6.0, or even reaches ineffective activity at pH 5.5 (Etherington 1984). Calpains have shown good ability to degrade important myofibrillar proteins, such as titin, nebulin, troponins T and I, tropomyosin, C-protein, filamin, desmin, and vinculin, which are responsible for the fiber structure. On the other hand, they are not active against myosin, actin, α-actinin and troponin C (Goll et al. 1983, Koohmaraie 1994).

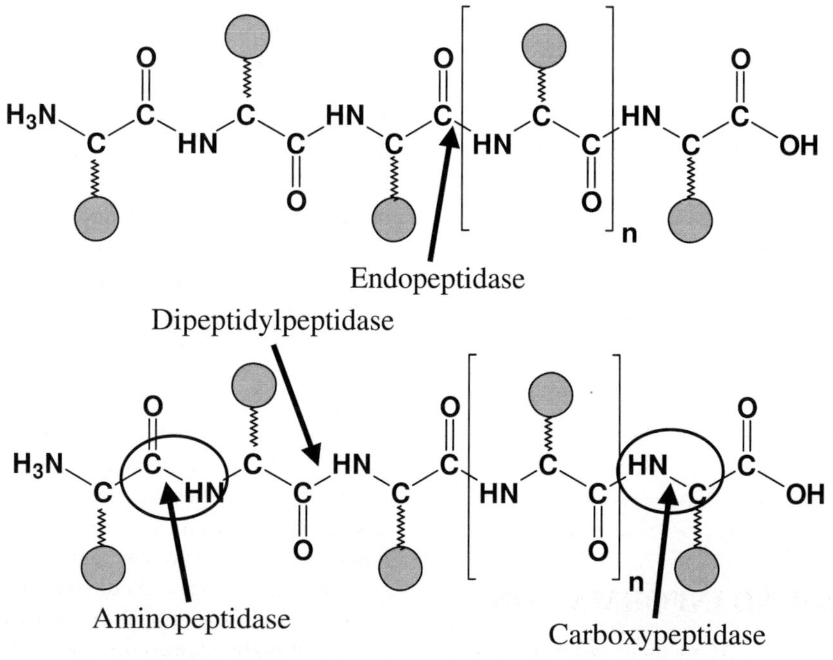

Figure 14.1. Mode of action of the different types of muscle proteases.

The stability of calpain I in postmortem muscle is very poor because it is readily autolyzed, especially at high temperatures, in the presence of the released Ca^{2+} (Koohmaraie 1994). Calpain II appears as more stable, just 2–3 weeks before losing its activity (Koohmaraie et al. 1987). In view of this rather poor stability, the importance of calpains should be restricted to short-term processes. A minor contribution, just at the beginning, has been observed in long processes such as dry curing of hams (Rosell and Toldrá 1996) or in fermented meats where the acid pH values makes calpain activity rather unlikely (Toldrá et al. 2001).

Calpastatin is a polypeptide (between 50 and 172 kDa) acting as an endogenous reversible and competitive inhibitor of calpain in the living muscle. In postmortem muscle, calpastatin regulates the activity of calpains, through a calcium dependent interaction, although only for a few days, because it is destroyed by autolysis (Koohmaraie et al. 1987). The levels of calpastatin vary with animal species, and pork muscle has the lowest level (Valin and Ouali 1992).

Lysosomal Proteinases: Cathepsins

There are several acid proteinases in the lysosomes that degrade proteins in a nonselective way. The most important are cathepsins B, H, and L, which are cysteine proteinases, and cathepsin D, which is an aspartate proteinase. The optimal pH for activity is slightly acid (pH around 6.0) for cathepsins B and L, acid (pH around 4.5) for cathepsin D, and neutral (pH 6.8) for cathepsin H (Toldrá et al. 1992). Cathepsins require a reducing environment such as that found in postmortem muscle to express their optimal activity (Etherington 1987). All of them are of small size, within the range 20–40 kDa, and are thus able to penetrate into the myofibrillar structure. Cathepsins have shown a good ability to degrade different myofibrillar proteins. Cathepsins D and L are very active against myosin heavy chain, titin, M and C proteins, tropomyosin, and troponins T and I (Matsakura et al. 1981, Zeece and Katoh 1989). Cathepsin L is extremely active in degrading both titin and nebulin. Cathepsin B is able to degrade myosin heavy chain and actin (Schwartz and Bird 1977). Cathepsin H exhibits both endo- and aminopeptidase activity, and this is the reason for its classification as an aminoendopeptidase (Okitani et al. 1981). In the muscle, there are endogenous inhibitors against cysteine peptidases. These inhibitory compounds, known as cystatins, are able to inhibit cathepsins B, H, and L. Cystatin C and chicken cystatin are the most well known cystatins.

Proteasome Complex

The proteasome is a multicatalytic complex with different functions in living muscle, even though its role in postmortem muscle is still not well understood. The 20S proteasome has a large molecular mass, 700 kDa, and a cylinder-shaped structure with several subunits. Its activity is optimal at pH above 7.0, but it rapidly decreases when pH decreases, especially below 5.5. It exhibits three major activities: chymotrypsin-like activity, trypsin-like activity, and peptidyl-glutamyl hydrolyzing activity (Coux et al. 1996). This multiple activity behavior is the reason why there was originally some confusion among laboratories over its identification. The 20S proteasome concentration is higher in oxidative muscles than in glycolytic ones (Dahlman et al. 2001). This enzyme has shown degradation of some myofibrillar proteins such as troponin C and myosin light chain and could be involved in postmortem changes in slow twitch oxidative muscles or in high pH meat, where an enlargement of the Z-line with more or less density loss is observed (Sentandreu et al. 2003).

Exoproteases: Peptidases

There are several peptidases in the muscle with the ability to release small peptides of importance for taste. Tripeptidylpeptidases (TPPs) are enzymes capable of hydrolyzing different tripeptides from the amino termini of peptides, while Dipeptidylpeptidases (DPPs) are able to hydrolyze different dipeptide sequences. There are two TPPs and four DPPs, and their molecular masses are relatively high, between 100 and 200 kDa, or even as high as 1000 kDa for TPP II, and have different substrate specificities. TPP I is located in the lysosomes, has an optimal acid pH (4.0), and is able to hydrolyze tripeptides Gly-Pro-X, where X is an amino acid, preferentially of hydrophobic nature. TPP II has optimal neutral pH (6.5–7.5) and wide substrate specificity, except when Pro is present on one of both sides of the hydrolyzed bond. DPPs I and II are located in the lysosomes and have optimal acid pH (5.5). DPP I has a special preference for hydrolyzing the dipeptides Ala-Arg and Gly-Arg, while DPP II prefers a terminal Gly-Pro sequence. DPP III is located in the cytosol and has special preference for terminal Arg-Arg and Ala-Arg sequences. DPP IV is linked to the plasma membrane and prefers a terminal Gly-Pro sequence. Both DPP III and IV have an optimal pH around 7.5–8.0. All these peptidases have been purified and fully characterized in porcine skeletal muscle (Toldrá 2002).

Exoproteases: Aminopeptidases

There are five aminopeptidases, known as leucyl, arginyl, alanyl, pyroglutamyl, and methionyl aminopeptidases, based on their respective preference or requirement for a specific N-terminal amino acid. They are able, however, to hydrolyze other amino acids, although at a lower rate (Toldrá 1998). Aminopeptidases are metalloproteases with a very high molecular mass and complex structures. All of them are active at neutral or basic pH. Alanyl aminopeptidase, also known as the major aminopeptidase because it exhibits very high activity, is characterized by its preferential hydrolysis of alanine, but it is also able to act against a wide spectrum of amino acids such as aromatic, aliphatic, and basic aminoacyl bonds. Methionyl aminopeptidase has preference for methionine, alanine, lysine, and leucine, but also has a wide spectrum of activity. This enzyme is activated by calcium ions. Arginyl aminopeptidase, also known as aminopeptidase B, hydrolyzes basic amino acids such as arginine and lysine (Toldrá and Flores 1998).

Carboxypeptidases are located in the lysosomes and have optimal acid pH. They are able to release free amino acids from the carboxy termini of peptides and proteins. Carboxypeptidase A has preference for hydrophobic amino acids, while carboxypeptidase B has a wide spectrum of activity (McDonald and Barrett 1986).

Lipolytic Enzymes

Lipolytic enzymes are characterized by their ability to degrade lipids, and they receive different names depending on their mode of action (see Fig. 14.2). They are known as lipases when they are able to release long-chain fatty acids from tri-acylglycerols, while they are know as esterases when they act on short-chain fatty acids . Lipases and esterases are located either in the skeletal muscle or in the adipose tissue. Phospholipases, mainly found in the skeletal muscle, hydrolize fatty acids at positions 1 or 2 in phospholipids.

Muscle Lipases

Lysosomal acid lipase and acid phospholipase are located in the lysosomes. Both have an optimal acid pH (4.5–5.5) and are responsible for the generation of long-chain free fatty acids. Lysosomal acid lipase has preference for the hydrolysis of tri-acylglycerols at positions 1 or 3 (Fowler and Brown 1984). This enzyme also hydrolyzes di- and monoacylglycerols but at a lower rate (Imanaka et al. 1985, Negre et al. 1985). Acid phospholipase hydrolyzes phospholipids at position 1 at the water-lipid interface.

Phospholipase A and lysophospholipase have optimal pH in the basic region and regulate the hydrolysis of phospholipids, at positions 1 and 2, respectively. The activities of these enzymes are higher in oxidative muscles than in glycolytic muscles, and this fact would explain the high content of free fatty acids in oxidative muscles. The increase in activity is about 10- to 25-fold for phospholipase A and four- to five-fold for lysophospholipase (Alasnier and Gandemer 2000).

Acid and neutral esterases are located in the lysosomes and cytosol, respectively, and are quite stable (Motilva et al. 1992). Esterases are able to hydrolyze short-chain fatty acids from tri-, di-,and monoacylglycerols, but they exert poor action due to the lack of adequate substrate.

Adipose Tissue Lipases

Hormone-sensitive lipase is the most important enzyme present in adipose tissue. This enzyme is responsible for the hydrolysis of stored adipocyte lipids. It has a high specificity and preference for the hydrolysis of long-chain tri- and diacylglycerols (Belfrage et al. 1984). This enzyme has positional specificity since it hydrolyzes fatty acids at positions 1 or 3 in tri-acylglycerols four times faster than it hydrolyzes fatty acids in position 2 (Belfrage et al. 1984). The hormone-sensitive lipase has a molecular mass of 84 kDa and neutral optimal pH, around

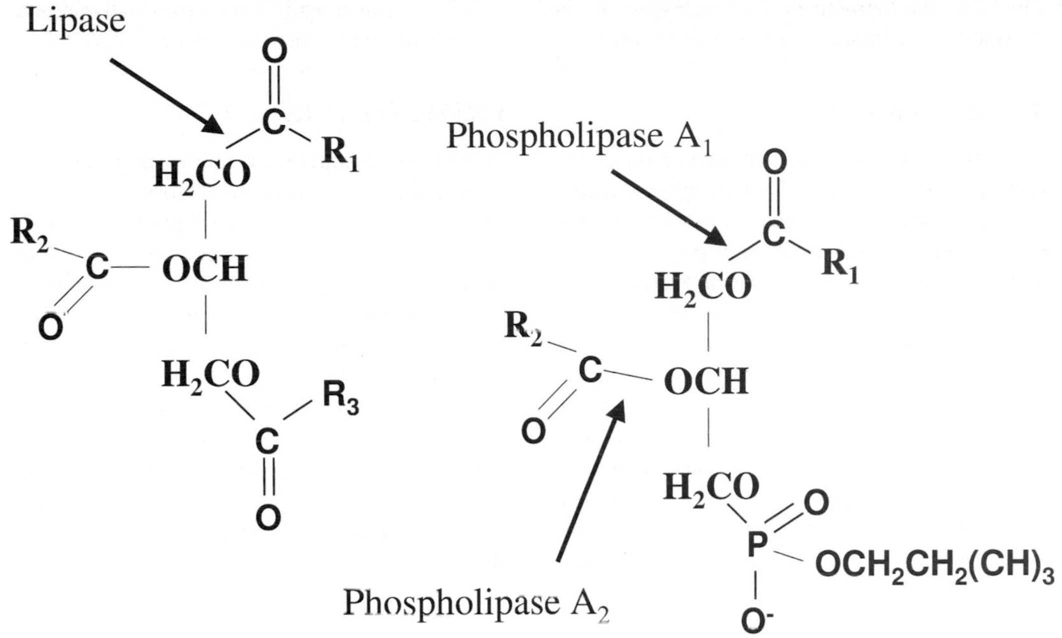

Figure 14.2. Mode of action of muscle lipases and phospholipases.

7.0. The mono-acylglycerol lipase is mainly present in the adipocytes, and very little is present in stromal and vascular cells. It has a molecular mass of 160 kDa and hydrolyzes medium- and long-chain mono-acylglycerols resulting from previous hydrolysis by the hormone-sensitive lipase (Tornqvist et al. 1978). Lipoprotein lipase is located in the capillary endothelium and is able to hydrolyze the acylglycerol components at the luminal surface of the endothelium (Smith and Pownall 1984), with preference for fatty acids at position 1 over those at position 3 (Fielding and Fielding 1980). Lipoprotein lipase is an acylglycerol lipase responsible for the degradation of lipoprotein triacylglycerol. Its molecular mass is around 60 kDa and it has an optimal basic pH. Unsaturated mono-acylglycerols are more quickly hydrolyzed than saturated compounds (Miller et al. 1981).

The lipolysis phenomenon in adipose tissue is not so complex as in muscle. The hormone-sensitive lipase hydrolyzes tri- and diacylglycerols, as a rate-limiting step. The resulting mono-acylglycerols from this reaction or from lipoprotein lipase (Belfrage et al. 1984) are then further hydrolyzed by the mono-acylglycerol lipase. The end products are glycerol and free fatty acids.

Acid and neutral esterases are also present in adipose tissue (Motilva et al. 1992). During mobilization of depot lipids, esterases can participate by mobilizing stored cholesteryl esters. Esterases can also degrade lipoprotein cholesteryl esters taken up from the plasma (Belfrage et al. 1984).

Muscle Oxidative and Antioxidative Enzymes

Oxidative Enzymes

Lipoxygenase contains iron and catalyzes the incorporation of molecular oxygen into polyunsaturated fatty acids, especially arachidonic acid, and esters containing a Z,Z-1,4-pentadien (Marczy et al. 1995). They receive different names, 5-, 12-, or 15-lipoxygenase, depending on the position where oxygen is introduced. The final product is a conjugated hydroperoxide. They usually require millimolar concentrations of Ca^{+2}, and their activity is stimulated by ATP (Yamamoto 1992). Lipoxygenase has been found to be stable during frozen storage and is responsible for rancidity development in chicken, especially in the muscle *Gastrocnemius* (Grosman et al. 1988).

Antioxidative Enzymes

Antioxidative enzymes and their regulation in the muscle constitute a defense system against oxidative susceptibility (i.e., an increased concentration of polyunsaturated fatty acids) and physical stress (Young et al. 2003). Glutathione peroxidase contains a covalently bound selenium atom that is essential for its activity. This enzyme catalyzes the dismutation of alkyl hydroperoxides by reducing agents like phenols. Its activity has been reported to be lower in oxidative muscles than in glycolytic muscles (Daun et al. 2001). Superoxide dismutase is a copper metalloenzyme, and catalase is an iron metalloenyzme. Both enzymes catalyze the dismutation of hydrogen peroxide to less harmful hydroxides. These enzymes influence the shelf life of the meat and protect against the prooxidative effects of chloride during further processing.

PROTEOLYSIS

Proteolysis constitutes an important group of reactions during the processing of meat and meat products. In fact, proteolysis has a high impact on texture, and thus meat tenderness, because it contributes to the breakdown of the myofibrillar proteins responsible for muscle network, but proteolysis also generates peptides and free amino acids that have a direct influence on taste and also act as substrates for further reactions contributing to aroma (Toldrá 1998, 2002). In general, proteolysis has several consecutive stages (see Fig. 14.3) as follows: (1) action of calpains and cathepsins on major myofibrillar proteins, generating protein fragments and intermediate size polypeptides; (2) these generated fragments and

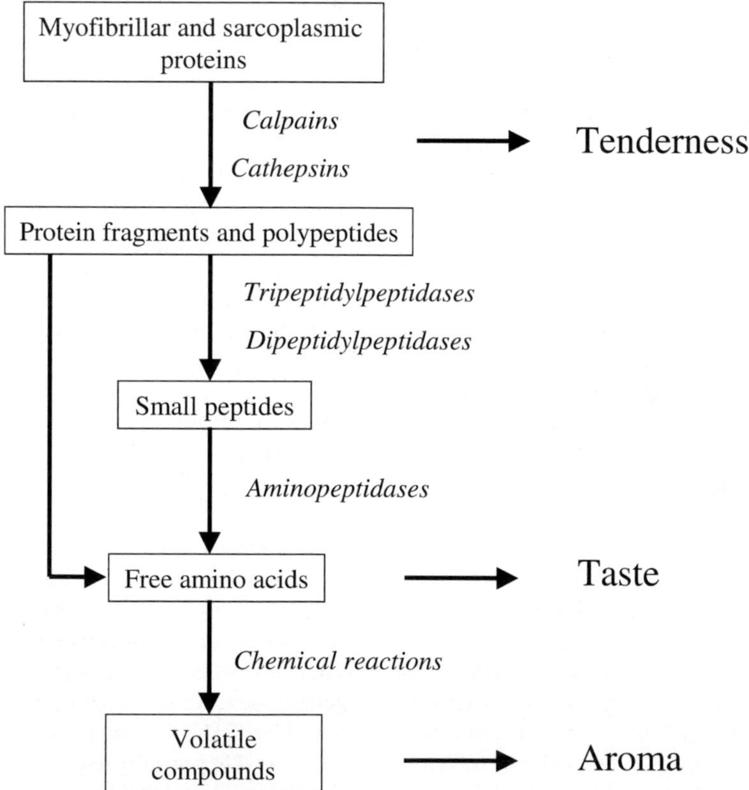

Figure 14.3. General scheme of proteolysis during the processing of meat and meat products.

polypeptides are further hydrolzyed to small peptides by di- and tripeptidylpeptidases; and (3) dipeptidases, aminopeptidases, and carboxypeptidases are the last proteolytic enzymes that act on previous polypeptides and peptides to generate free amino acids. The progress of proteolysis varies depending on the processing conditions, the type of muscle, and the amount of endogenous proteolytic enzymes, as described below.

Proteolysis in Aged Meat and Cooked Meat Products

During meat ageing, there is proteolysis of important myofibrillar proteins such as troponin T by calpains, with the associated release of a characteristic 30 kDa fragment, which is associated with meat tenderness; nebulin; desmin; titin; troponin I; myosin heavy chain; and proteins at the Z-line level (Yates et al. 1983). A 95 kDa fragment is also characteristically generated (Koohmaraie 1994). Examples of cathepsin and aminopeptidase activity during beef ageing are shown in Figures 14.4 and 14.5, respectively.

The fastest ageing rate is observed in chicken, followed by pork, lamb, and beef. For instance, it has been reported that chicken myofibrils are easily damaged by cathepsin L, while beef myofibrils are much more resistant (Mikami et al. 1987). The reasons for this difference are the differences between species in enzymatic activity, inhibitor content, and susceptibility to proteolysis of the myofibrillar structure. These factors are also strongly linked to the type of muscle metabolism, which has a strong influence on the ageing rate. In fact, proteolysis is faster in fast-twitch white fibers (which contract rapidly) than in slow-twitch red fibers (which contract slowly), that is, ageing rate increases with increasing speed of contraction but decreases with the increasing level of heme iron (Ouali 1991). Even though the effect of muscle type is estimated to be 10-fold lower than the effect of temperature, it is three-fold higher than animal effects (Dransfield 1980–81). There are also some physical and chemical conditions in postmortem muscle, listed in Tables 14.1 and 14.2, respectively, that can affect enzyme activity. Of these conditions, the most significant are pH, which decreases once the animal is slaughtered, and osmolality, which increases from 300 to around 550 mOsm within 2 days, due to the release of ions to the cytosol. The osmolality has been observed to be higher in fast-twitch muscle, which also experiences a faster ageing rate (Valin and Ouali 1992). Age also decreases the ageing rate, because collagen content

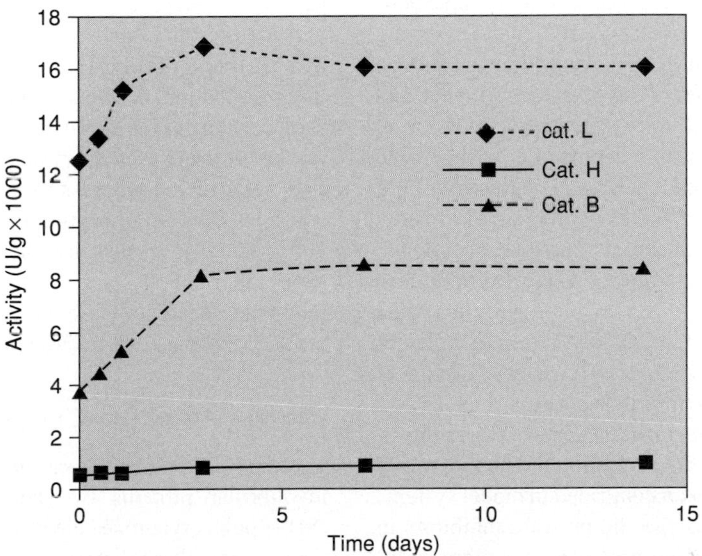

Figure 14.4. Evolution of muscle cathepsins during the ageing of beef. (Toldrá, unpublished data.)

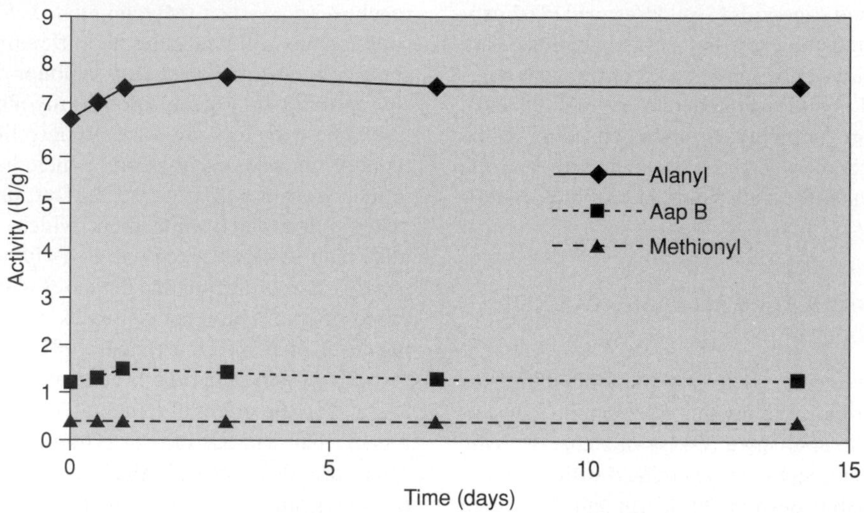

Figure 14.5. Evolution of muscle aminopeptidases during the ageing of beef. (Adapted from Goransson et al. 2002.)

as well as the cross-links that make the muscle more heat stable and mechanically resistant, increase with age.

Proteolysis in Fermented Meats

The progressive pH decrease by lactic acid (generated during the fermentation by lactic acid bacteria), the added salt (2–3%), and the heating/drying conditions during the fermentation and ripening/drying affect protein solubility. The reduction may reach 50–60% in the case of myofibrillar proteins and 20–47% for sarcoplasmic proteins (Klement et al. 1974). There is a variable degree of contribution to proteolysis by both endogenous proteases and those of microbial origin. This contribution mainly depends on the raw materials, the type of product, and the processing conditions. The pH drop during the fermentation stage is very important. So, when pH drops below 5.0, the proteolytic activity of endogenous cathepsin D becomes very intense (Toldrá et al. 2001). Several myofibrillar proteins, such as myosin and actin, are degraded, and some fragments of 135, 38, 29, and 13 kDa are formed. The major role of cathepsin D has been confirmed in model systems using antibiotics and specific protease inhibitors in order to inhibit bacterial proteinases or other endogenous muscle proteinases (Molly et al. 1997). A minor role is played by other muscle cathepsins (B, H, and L) and bacterial proteinases. Peptides and small protein fragments are produced during fermentation, heating (smoking), and ripening. The generation of free amino acids, as final products of proteolysis, depends on the pH reached in the product (as aminopeptidases are affected by low pH values), concentration of salt (these enzymes are inhibited or activated by salt), and processing conditions (time, temperature, and water activity that affect the enzyme activity) (Toldrá 1998). All these factors affect the contribution of the different aminopeptidases, as described in Tables 14.1 and 14.2; thus, the pH reached at the fermentation stage ($pH < 5.0$) is decisive because it reduces substantially both muscle and microbial aminopeptidase activity (Sanz et al. 2002). Finally, it must be taken into account that some microorganisms, grown during fermentation, might have decarboxylase (an enzyme able to generate amines from amino acids) activity.

Proteolysis in Dry-Cured Ham

The analysis of muscle sarcoplasmic proteins and myofibrillar proteins by sodium dodecyl sulfate (SDS)-polyacrylamide electrophoresis reveals an intense proteolysis during the process. This proteolysis appears to be more intense in myofibrillar than

Table 14.1. Physical Factors Affecting Proteolytic Activity during Meat and Meat Product Processing

Factor	Typical Trend	Effect on Proteases
pH	Near neutral pH in dry-cured ham and cooked meat products	Favors the activity of calpain, cathepsins B and L, DPP III and IV, TPP II, and aminopeptidases
	Slightly acid in aged meat	Favors the activity of cathepsins B, H, and L
	Acid pH in dry-fermented sausages	Favors the activity of cathepsin D, DPP I, DPP II and TPP I
Time	Short in aged meat, cooked meat products, and fermented products	Short action for enough enzyme action except for calpains that contribute to tenderness
	Medium in dry-fermented sausages	Time allows significant biochemical changes
	Long in dry-cured ham	Very long time for important biochemical changes
Temperature	High increase in cooking and smoking	Enzymes are strongly activated by cooking temperatures, although stability decreases rapidly and time of cooking is short
	Mild temperatures in fermentation and dry curing	Enzymes have time enough for their activity even with the use of mild temperatures
Water activity	Slightly reduced in cooked meats	Enzymes have good conditions for activity
	Substantially reduced in dry meats	Restricted enzyme activity as a_w drops
Redox potential	Anaerobic values in postmortem meat	Most of the muscle enzymes need reducing conditions for activity
Osmolality	High in fast-twitch muscle; may increase from 300 to 550 mOsm within 2 days	Enhances proteases activity, and these muscles are aged at faster rate

in sarcoplasmic proteins. The patterns for myosin heavy chain, myosin light chains 1 and 2, and troponins C and I show a progressive disappearance during the processing (Toldrá et al. 1993). Several fragments of 150, 95, and 16 kDa and in the ranges of 50–100 kDa and 20–45 kDa are formed (Toldrá 2002). The analysis of ultrastructural changes by both scanning and transmission electron microscopy shows weakening of the Z-line as well as important damage to the fibers, especially at the end of salting (Monin et al. 1997). An excess of proteolysis may create unpleasant textures because of intense structural damage. The result is a poor firmness that is poorly rated by sensory panelists and consumers. This excess of proteolysis is frequently due to the breed type and/or age, which have a marked influence on some enzymes, or just a higher level of cathepsin B activity (Toldrá 2004a). A high residual cathepsin B activity and/or low salt content, a strong inhibitor of cathepsin activity, are correlated with the increased softness. The action of calpains is restricted to the initial days of processing due to their poor stability. Cathepsin D would contribute during the initial 6 months, and cathepsins B, L, and H,

Table 14.2. Chemical Factors Affecting Proteolytic Activity during Meat and Meat Product Processing

Factor	Typical Trend	Effect on Proteases
NaCl	Low in aged meat	No effect on enzyme activity
	Medium concentration in cooked meat products	Partial inhibition of most proteases, except calpain and aminopeptidase B, that are chloride-activated at low NaCl concentration
	High concentration in dry-cured hams	Strong inhibition of almost all proteases
Nitrate and nitrite	Concentration around 125 ppm in cured meat products	Slight inhibitory effect on the enzyme activity, except cathepsin B, that is activated
Ascorbic acid	Concentration around 500 ppm in cured meat products	Slight inhibition of m-calpain, cathepsin H, leucyl aminopeptidase, and aminopeptidase B
Glucose	Poor concentration in aged meat and dry-cured ham	No effect
	High concentration (up to 2 g/L) in fermented meats	Slight activation of leucyl aminopeptidase and cathepsins B, H, and D

which are very stable and have an optimal pH closer to that in ham, would act during the full process (Toldrá 1992). An example of the evolution of these enzymes is shown in Figure 14.6.

Numerous peptides are generated during processing: mainly in the range 2700–4500 Da during post-salting and early ripening, and below 2700 Da during ripening and drying (Aristoy and Toldrá 1995). Some of the smaller tri- and dipeptides recently have been sequenced. Dipeptidylpeptidase I and tripeptidylpeptidase I appear to be the major enzymes involved in the release of di- and tripeptides, respec-

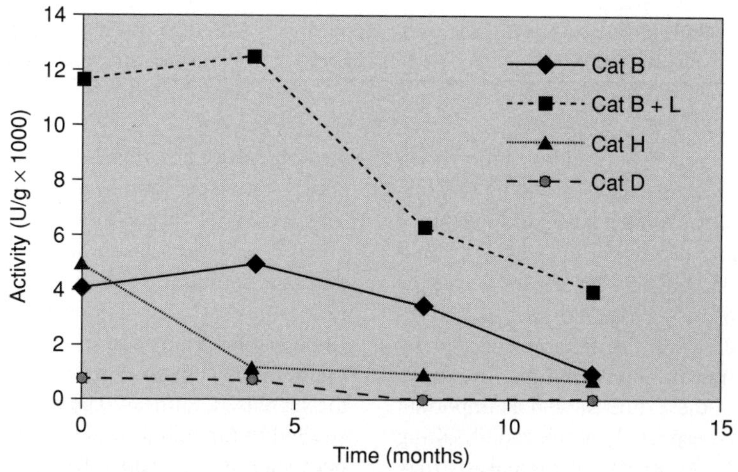

Figure 14.6. Evolution of cathepsins during the processing of dry-cured ham. (Toldrá, unpublished data.)

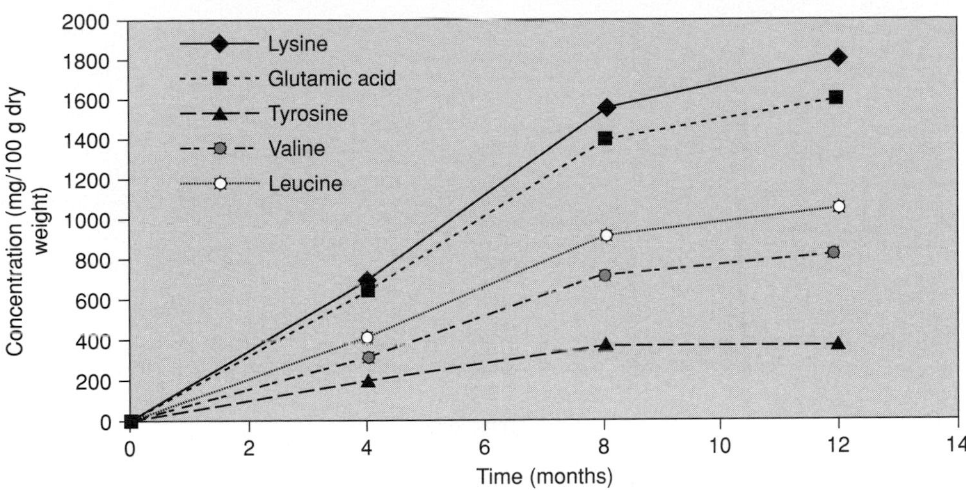

Figure 14.7. Example of the generation of some free amino acids during the processing of dry-cured ham. (Adapted from Toldrá et al. 2000.)

tively, due to their good activity, stability, and an optimal pH near that in ham. The other peptidases would play a minor role (Sentandreu and Toldrá 2002). The generation of free amino acids during the processing of dry-cured ham is very high (Toldrá 2004b). Alanine, leucine, valine, arginine, lysine, and glutamic and aspartic acids are some of the amino acids generated in higher amounts. An example of generation is shown in Figure 14.7. The final concentrations depend on the length of the process and the type of ham (Toldrá et al. 2000). Based on the specific enzyme characteristics and the process conditions, alanyl and methionyl aminopeptidases appear to be the most important enzymes involved in the generation of free amino acids, while arginyl aminopeptidase would mostly generate arginine and lysine (Toldrá 2002).

NUCLEOTIDE BREAKDOWN

The disappearance of ATP is very fast; in fact, it only takes a few hours to reach negligible levels. Many enzymes are involved in the degradation of nucleotides and nucleosides, as described in Chapter 13. The main changes in the nucleotide breakdown products occur during a few days postmortem, as shown in Figure 14.8. So, adenosine triphosphate (ATP) and adenosine monophosphate (AMP), which are intermediate degradation compounds, also disappear within 24 h postmortem. Inosine monophosphate (IMP) reaches a maximum by 1 day postmortem, but some substantial amount is still recovered after 7 days postmortem. On the other hand, inosine and hypoxanthine, as final products of these reactions, increase up to 7 days postmortem (Batlle et al. 2001).

GLYCOLYSIS

Glycolysis consists in the hydrolysis of carbohydrates, mainly glucose, either that remaining in the muscle or that formed from glycogen, to give lactic acid as the end product. As lactic acid accumulates in the muscle, pH falls from neutral values to acid values around 5.3–5.8. The glycolytic rate, or speed of pH fall, depends on the animal species and the metabolic status. On the other hand, the glycolytic potential, which depends on the amount of stored carbohydrates, gives an indication of ultimate pH. The pH drop due to the lactic acid accumulation is perhaps the main consequence of glycolysis, and it has very important effects on meat processing because pH affects numerous chemical and biochemical (all the enzymes) reactions. There are many enzymes involved in the glycolytic chain; some of the most important are phosphorylase, phosphofructokinase, pyruvic kinase, and lactate dehydrogenase (Demeyer and Toldrá 2004). Lactate dehydrogenase

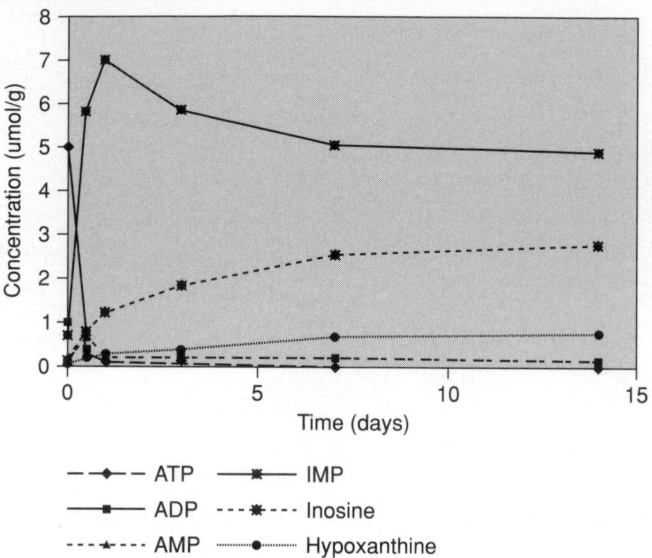

Figure 14.8. Evolution of nucleotides and nucleosides during the ageing of pork meat. (Adapted from Batlle et al. 2001.)

is involved in the last step, which consists in the conversion of pyruvic acid into lactic acid. The contribution of glycolysis to pH drop is restricted to a few hours postmortem, although it is also important in fermented meats where sugar is added for microorganism growth (see Chapter 15). The evolution of some glycolytic enzymes is shown in Figure 14.9.

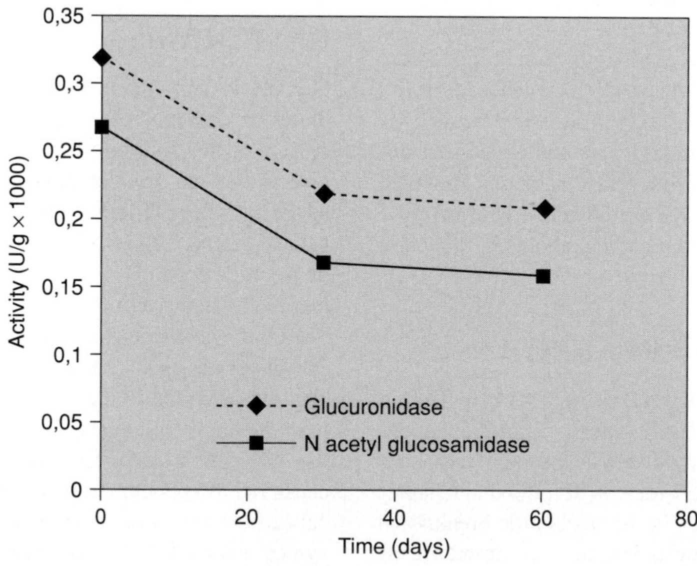

Figure 14.9. Evolution of β-glucuronidase and N-acetyl-β-glucosaminidase during the processing of a dry-fermented sausage. (Toldrá, unpublished data.)

LIPOLYSIS

Lipolysis makes an important contribution to the quality of meat products by the generation of free fatty acids, some of which have a direct influence on flavor, and others that, with polyunsaturations, may be oxidized to volatile aromatic compounds, acting as flavor precursors. The general scheme for lipolysis in muscle and adipose tissue is shown in Figure 14.10. In addition, the breakdown of triacylglycerols affects the texture of the adipose tissue, and an excess of lipolysis/oxidation may contribute to the development of rancid aromas or yellowish colors in fat (Toldrá 1998).

LIPOLYSIS IN AGED MEAT AND COOKED MEAT PRODUCTS

The relative amounts of released free fatty acids depends not only on the enzyme preference, but also on many other factors such as raw materials (especially affected by feed composition), type of process, and extent of cooking (Toldrá 2002). The generation rate observed during in vitro incubations of muscle lipases with pure phospholipids is the following (in order of importance): linoleic > oleic > linolenic > palmitic > stearic > arachidonic acid (Toldrá, unpublished results). Lipolysis may also vary depending on the type of muscle, as oxidative and glucolytic muscles exhibit different lipase activity (Flores et al. 1996). The generation of free short-chain fatty acids is very low due to the lack of adequate substrate (Motilva et al. 1993b). The physical and chemical conditions in the muscle and adipose tissue, especially during cooking, may affect the enzyme activity (see Tables 14.3 and 14.4). The evolution of muscle lipases during ageing of pork meat is shown in Figure 14.11.

LIPOLYSIS IN FERMENTED MEATS

Depending on the raw materials, type of product, and processing conditions, the degree of contribution of endogenous and microbial origin lipases will vary. The relative importance of both enzyme systems has been checked with mixtures of antibiotics and antimycotics used in sterile meat model systems (Molly et al. 1997). The results showed a minimal effect of antibiotics, and it was concluded that lipolysis is mainly brought about by muscle and adipose tissue lipases (60–80% of total free fatty acids generated), even though some variability might be

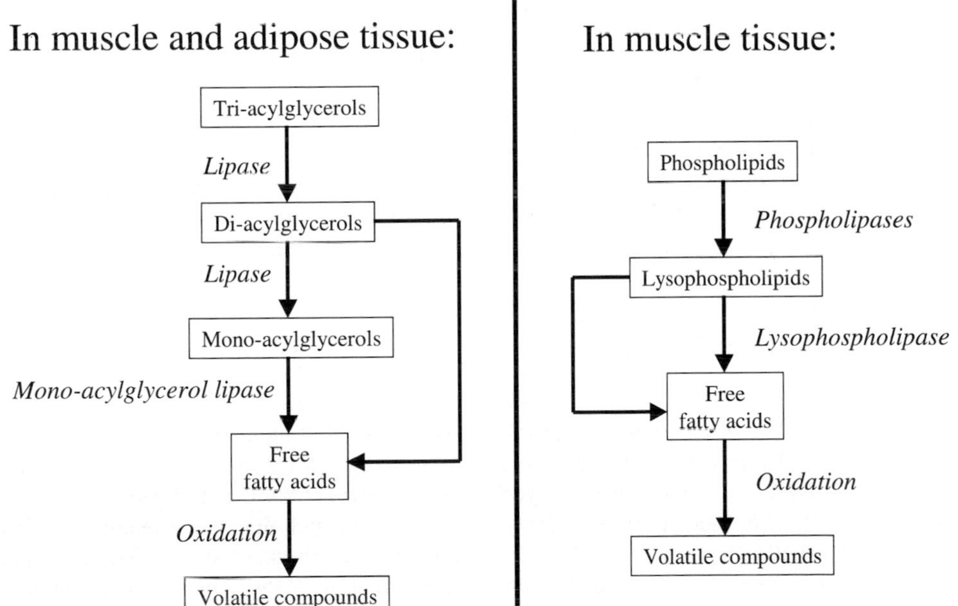

Figure 14.10. General scheme of lipolysis during the processing of meat and meat products.

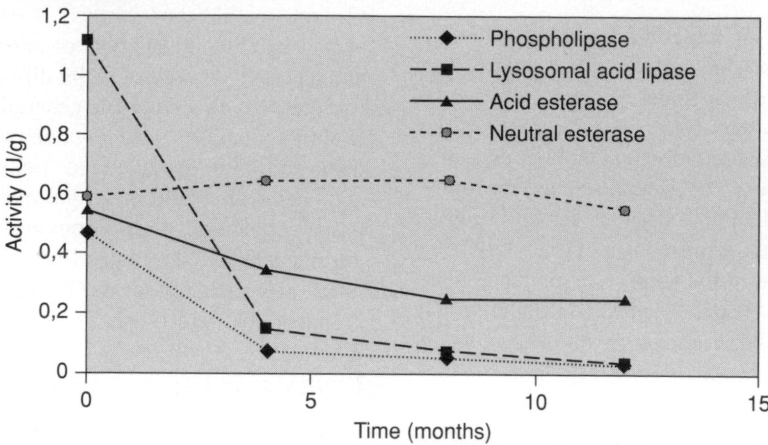

Figure 14.11. Evolution of muscle lipases during the processing of dry-cured ham.(Toldrá, unpublished data.)

found, depending on the batch and the presence of specific strains (Molly et al. 1997). Other authors have also observed that fatty acids are released in higher amounts when starters are added, but that there is significant lipolysis in the absence of microbial starters (Montel et al. 1993, Hierro et al. 1997). When pH drops during fermentation, the action of muscle lysosomal acid lipase and acid phospholipase becomes very important. Some strains are selected as starters based on their contribution to lipolysis. So, *Micrococaceae* present a highly variable amount of extra- and intracellular lipolytic enzymes, dependant on the strain and type of substrate. The action of the extracellular enzymes on the hydrolysis of tri-acylglycerols becomes more important after 15–20 days of ripening (Ordoñez et al. 1999). Other microorganisms used as starters are *Staphylococcus warneri,* which gives the highest lipolytic activity, and *S. Saprophyticus*; *S. carnosus* and *S. xylosus* present poor and variable lipolytic activity (Montel et al. 1993). Lactic acid bacteria have poor lipolytic activity, mostly intracellular. Many molds and yeasts such as *Candida*, *Debaryomices*, *Cryptococcus,* and *Trichosporum* have been isolated from fermented sausages, and all of them exhibit lipolytic acivity (Ordoñez et al. 1999, Ludemann et al. 2004, Sunesena and Stahnke 2004).

The increases in the levels of free fatty acids show a great variability depending on the raw materials, type of sausage, and processing conditions (Toldrá et al. 2002). The increase may reach 2.5–5% of the total fatty acids, and the rate of release decreases in the following order: oleic > palmitic > stearic > linoleic (Demeyer et al. 2000). The release of polyunsaturated fatty acids from phospholipids is more pronounced during ripening (Navarro et al. 1997). Some short-chain fatty acids such as acetic acid may increase, especially during early stages of ripening. The generation of volatile compounds with impact on the aroma of fermented sausages depends on the type of processing and the starter culture added (Talon et al. 2002, Tjener et al. 2004). Some of the most important are aliphatic saturated and unsaturated aldehydes, ketones, methyl-branched aldehydes and acids, free short-chain fatty acids, sulfur compounds, some alcohols, terpenes (from spices), and some nitrogen-derived volatile compounds (Stahnke 2002).

Lipolysis in Dry-Cured Ham

The generation of free fatty acids in the muscle is correlated with the period of maximal phospholipid degradation (Motilva et al. 1993b, Buscailhon et al. 1994). Furthermore, a decrease in linoleic, arachidonic, oleic, palmitic, and estearic acids from phospholipids is observed at early stages of processing (Martin et al. 1999). This fact corroborates muscle phospholipases as the most important enzymes involved in muscle lipolysis. The amount of generated

Table 14.3. Physical Factors Affecting Lipolytic Activity during Meat and Meat Products Processing

Factor	Typical Trend	Effect on Lipolytic Enzymes
pH	Near neutral pH in dry-cured ham and cooked meat products	Favors the activity of neutral lipase, neutral esterase, and hormone-sensitive lipase
	Slightly acid in aged meat	Favors the activity of lysosomal acid lipase, acid esterase, and acid phospholipase
	Acid pH in dry-fermented sausages	Favors the activity of lysosomal acid lipase, acid esterase, and acid phospholipase
Time	Short in aged meat, cooked meat products, and fermented products	Short action for enough enzyme action but significant oxidation of the released fatty acids
	Medium in dry fermented sausages and long in dry-cured ham	Time allows significant biochemical changes in dry sausages and very important in dry-cured ham
Temperature	High increase in cooking and smoking	Enzymes are strongly activated by cooking temperatures, although for a short time
	Mild temperatures in fermentation and dry curing	Enzymes have time enough for their activity even with the use of mild temperatures; lipases are active even during freezing storage of the raw meats
Water activity	Slightly reduced in cooked meats	Enzymes have very good conditions for activity
	Substantially reduced in dry meats	Neutral muscle lipase and esterases are affected; rest of lipolytic enzymes remain unaffected by a_w drop
Redox potential	Anaerobic values in postmortem meat	Slight effect on lysosomal acid lipase and phospholipase; more intense for rest of lipases

free fatty acids increases with ageing time, up to 6 months of processing (see Fig. 14.12) and is higher in the external muscle *(Semimembranosus),* which contains more salt and is more dehydrated, than in the internal muscle *(Biceps femoris).* As lipase activity is mainly influenced by pH, salt concentration, and water activity (Motilva and Toldrá 1993), it appears that the observed lipid hydrolysis is favored by the same variables (salt increase and a_w reduction), as shown in Tables 14.3 and 14.4, that enhance enzyme activity in vitro (Vestergaard et al. 2000, Toldrá et al. 2004).

In the case of the adipose tissue, triacylglycerols form the major part of this tissue (around 90%) and are mostly hydrolyzed by neutral lipases to di- and monoacylglycerols and free fatty acids (see Fig. 14.12), especially up to 6 months of processing (Motilva et al. 1993a). The amount of triacylglycerols decreases from about 90% to 76% (Coutron-Gambotti and Gandemer 1999). There is a preferential hydrolysis of polyunsaturated fatty acids, although some of them may not accumulate due to further oxidation during processing. Triacylglycerols that are rich in oleic and linoleic acids and are liquid

Table 14.4. Chemical Factors Affecting Lipolytic Activity during Meat and Meat Product Processing

Factor	Typical Trend	Effect on Lipolytic Enzymes
NaCl	Low in aged meat	No effect on enzyme activity
	Medium concentration in cooked meat products	Partial inhibiton of neutral muscle lipase and esterases; adipose tissue lipases not affected; activation of lysosomal acid lipase and acid phospholipase
	High concentration in dry-cured hams	Slight activation of lysosomal acid lipase and acid phospholipase
Nitrate and nitrite	Concentration around 125 ppm in cured meat products	No significant effect
Ascorbic acid	Concentration around 500 ppm in cured meat products	Slight inhibition of all lipolytic enzymes
Glucose	Poor concentration in aged meat and dry-cured ham	No effect
	High concentration (up to 2 g/L) in fermented meats	No effect

at 14–18°C are more hydrolyzed than triacylglycerols that are rich in saturated fatty acids such as palmitic acid and are solid at those temperatures (Coutron-Gambotti and Gandemer 1999). This means that the physical state of the triacylglycerols would increase the lipolysis rate by favoring the action of lipases at the water-oil interface. The rate of release of individual fatty acids is as follows: linoleic > oleic > palmitic > stearic > arachidonic (Toldrá 1992). The generation rate remains high up to 10 months of processing, when the accumulation of free fatty acids remains asymptotic or even decreases as a

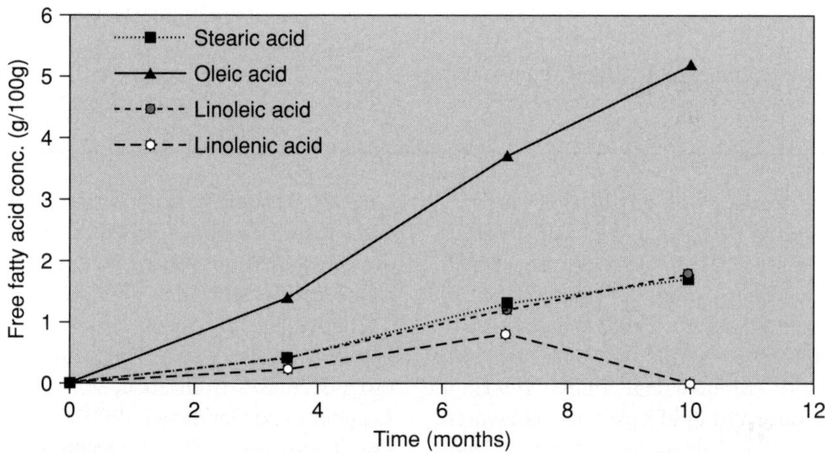

Figure 14.12. Example of the generation of some free fatty acids in the adipose tissue during the processing of dry-cured ham. (Adapted from Motilva et al. 1993b.)

consequence of further oxidative reactions. Oleic, linoleic, estearic, and palmitic acids are those accumulated in higher amounts because they are present in great amounts in the triacylglycerols and have a better stability against oxidation.

OXIDATIVE REACTIONS

The lipolysis and generation of free polyunsaturated fatty acids, susceptible to oxidation, constitute a key stage in flavor generation. The susceptibility of fatty acids to oxidation and the rate of oxidation depends on their unsaturation (Shahidi 1998a). So, linolenic acid (C 18:3) is more susceptible than linoleic acid (C 18:2), which is more susceptible than oleic acid (C 18:1). The animal species have different susceptibility to autoxidation in the following order: poultry > pork > beef > lamb (Tichivangana and Morrisey 1985). Oxidation has three consecutive stages (Shahidi 1998b). The first stage, initiation, consists in the formation of a free radical. This reaction can be enzymatically catalyzed by muscle lypoxygenase or chemically catalyzed by light, moisture, heat, and/or metallic cations. The second stage, propagation, consists in the formation of peroxide radicals by reaction of the free radicals with oxygen. When peroxide radicals react with double bonds, they form primary oxidation products, or hydroxyperoxides, that are very unstable. Their breakdown produces many types of secondary oxidation products by a free radical mechanism. Some of these secondary oxidation products are potent flavor-active compounds that can impart off-flavor to meat products during cooking or storage. The oxidative reactions finish by inactivation of free radicals when they react with each other (last stage). Thus, the result of these oxidative reactions consists in the generation of volatile compounds responsible for final product aroma. It is important to have a good control of these reactions because sometimes, oxidation may give undesirable volatile compounds with unpleasant off-flavors.

OXIDATION TO VOLATILE COMPOUNDS

As mentioned above, some oxidation is needed to generate volatile compounds with desirable flavor properties. For instance, a characteristic aroma of dry-cured meat products is correlated with the initiation of lipid oxidation (Buscailhon et al. 1994, Toldrá and Flores 1998). However, an excess of oxidation may lead to off-flavors, rancidity, and yellow colors in fat.

The primary oxidation products, or hydroperoxides, are flavorless, but the secondary oxidation products have a clear contribution to flavor. There are a wide variety of volatile compounds formed by oxidation of the unsaturated fatty acids. The most important are (1) aliphatic hydrocarbons that result from autoxidation of the lipids; (2) alcohols, mainly originated by oxidative decomposition of certain lipids; (3) aldehydes, which can react with other components to produce flavor compounds; and (4) ketones produced through either β-keto acid decarboxylation or fatty acid β-oxidation. Other compounds, like esters, may contribute to characteristic aromas (Shahidi et al. 1998a).

Oxidation rates may vary depending on the type of product or the processing conditions. For instance, TBA (thiobarbituric acid), a chemical index used as an indication of oxidation, increases more markedly in products such as Spanish chorizo than in French saucisson or Italian salami (Chasco et al. 1993). On the other hand, processing conditions such as curing or smoking also give a characteristic flavor to the product.

ANTIOXIDANTS

The use of spices such as paprika and garlic, which are rich in natural antioxidants, protects the product from certain oxidations. The same applies to antioxidants such as vitamin E that are added in the feed to prevent undesirable oxidative reactions in polyunsaturated fatty acids. Nitrite constitutes a typical curing agent that generally retards the formation of off-flavor volatiles that can mask the flavor of the product, and allows extended storage of the product (Shahidi 1998). Nitrite acts against lipid oxidation through different mechanisms: (1) binding of heme and prevention of the release of the catalytic iron, (2) binding of heme and nonheme iron and inhibition of catalysis, and (3) stabilization of lipids against oxidation. Smoking also contains some antioxidant compounds such as phenols that protect the external part of the product against undesirable oxidations. The muscle antioxidative enzymes also exert some contribution to the lipid stability against oxidation. In the case of fermented meats, the microbial enzyme catalase degrades the peroxides formed during the processing of fermented sausages (Toldrá et al. 2001). Thus, this enzyme contributes to stabilizing the color and flavor of the final sausage. Catalase increases its activity with cell growth to a maximum at the onset or during the stationary phase, but it is mainly formed during the ripening stage. Large amounts of salt exert an inhibitory effect on catalase activity, especially at low pH values. The catalase activity is different depending on the strain. For instance, *S. carnosus* has a high catalase activity in anaerobic conditions, while *S. warneri* has a low catalase activity (Talon et al. 1999).

GLOSSARY

Aminopeptidases—Exopeptidases that catalyze the release of an amino acid from the amino terminus of a peptide.

Catalase—Enzyme able to catalyze the decomposition of hydrogen peroxide into molecular oxygen and water.

Cathepsins—Enzymes located in lysosomes and able to hydrolyze myofibrillar proteins to polypeptides.

Decarboxylases—Enzymes able to transform an amino acid into an amine.

Glycolysis—Enzymatic breakdown of carbohydrates with the formation of pyruvic acid and lactic acid and the release of energy in the form of ATP.

Lactate dehydrogenase—Enzyme that catalyzes the oxidation of pyruvic acid to lactic acid.

Lipolysis—Enzymatic breakdown of lipids with the formation of free fatty acids

Lysosomal acid lipase—Enzyme that catalyzes the release of fatty acids by hydrolysis of triacylglycerols at positions 1 and 3.

Peroxide value—Term used to measure rancidity and expressed as millimoles of peroxide taken up by 1000 g of fat.

Proteolysis—Enzymatic breakdown of proteins with the formation of peptides and free amino acids.

Water activity *(a_w)*—The ratio of the equilibrium water vapor pressure over the system and the vapor pressure of pure water at the same temperature; indicates the availability of water in a food.

REFERENCES

Alasnier C, Gandemer G. 2000. Activities of phospholipase A and lysophospholipases in glycolytic and oxidative skeletal muscles in the rabbit. J Sci Food Agric 80:698–704.

Aristoy MC, Toldrá F. 1995. Isolation of flavor peptides from raw pork meat and dry-cured ham. In: G Charalambous, editor, Food Flavors: Generation, Analysis and Process Influence, pp. 1323–1344. Amsterdam (The Netherlands): Elsevier Science Pub.

Batlle N, Aristoy MC, Toldrá F. 2001. ATP metabolites during ageing of exudative and non exudative pork meats. J Food Sci 66:68–71.

Belfrage P, Fredrikson G, Stralfors P, Tornquist H. 1984. Adipose tissue lipases. In: B Borgström, HL Brockman, editors, Lipases, pp. 365–416. London: Elsevier Science Pub.

Buscailhon S, Gandemer G, Monin G. 1994. Time-related changes in intramuscular lipids of French dry-cured ham. Meat Sci 37:245–255.

Chasco J, Berianin MJ, Bello J. 1993. A study of changes in the fat content of some varieties of dry sausage during the curing process. Meat Sci 34:191–204.

Coutron-Gambotti C, Gandemer G. 1999. Lipolysis and oxidation in subcutaneous adipose tissue during dry-cured ham processing. Food Chem 64:95–101.

Coux O, Tanaka K, Goldberg A. 1996. Structure and function of the 20S and 26S proteasomes. Ann Rev Biochem 65:801–847.

Dahlmann, Ruppert T, Kloetzel PM, Kuehn L. 2001. Subtypes of 20S proteasomes from skeletal muscle. Biochimie 83:295–299.

Daun C, Johansson M, Onning G, Akesson B. 2001. Glutathione peroxidase activity, tissue and soluble selenium content in beef and pork in relation to meat ageing and pig RN phenotype. Food Chem 73:313–319.

Demeyer DI, Raemakers M, Rizzo A, Holck A, De Smedt A, Ten Brink B, Hagen B, Montel C, Zanardi E, Murbrek E, Leroy F, Vanderdriessche F, Lorentsen K, Venema K, Sunesen L, Stahnke L, De Vuyst L, Talon R, Chizzolini R, Eerola S. 2000. Control of bioflavor and safety in fermented sausages: First results of a European project. Food Research Int 33: 171–180.

Demeyer DI, Toldrá F. 2004. Fermentation. In: W Jensen, C Devine, M Dikemann, editors, Encyclopedia of Meat Sciences, pp. 467–474. London: Elsevier Science, Ltd.

Dransfield, E, Jones RCD, MacFie HJH. 1980–81. Quantifying changes in tenderness during storage of beef meat. Meat Sci 5:131–137.

Etherington DJ. 1984. The contribution of proteolytic enzymes to postmortem changes in muscle. J Anim Sci 59:1644–1650.

———. 1987. Conditioning of meat factors influencing protease activity. In: A Romita, C Valin, AA

Taylor, editors Accelerated Processing of Meat, pp. 21–28. London: Elsevier Applied Sci.

Fielding CJ, Fielding PE. 1980. Characteristics of triacylglycerol and partial acylglycerol hydrolysis by human plasma lipoprotein lipase. Biochim Biophys Acta 620:440–446.

Flores J, Toldrá F. 1993. Curing: Processes and applications. In: R MacCrae, R Robinson, M Sadle, G Fullerlove, editors, Encyclopedia of Food Science, Food Technology and Nutrition, pp.1277–1282. London: Academic Press.

Flores M, Alasnier C, Aristoy MC, Navarro JL, Gandemer G, Toldrá F. 1996. Activity of aminopeptidase and lipolytic enzymes in five skeletal muscles with various oxidative patterns. J Sci Food Agric 70:127–130.

Flores M, Spanier AM, Toldrá F. 1998. Flavour analysis of dry-cured ham. In: F Shahidi, editor, Flavor of Meat, Meat Products and Seafoods, pp. 320–341. London: Blackie Academic and Professional.

Fowler SD, Brown WJ. 1984. Lysosomal acid lipase. In: B Borgström, HL Brockman, editors, Lipases, pp. 329–364. London: Elsevier Science Pub.

Goll DE, Otsuka Y, Nagainis PA, Shannon JD, Sathe SK, Mugurama M. 1983. Role of muscle proteinases in maintenance of muscle integrity and mass. J Food Biochem 7:137–177.

Goransson A, Flores M, Josell A, Ferrer JM, Trelis MA, Toldrá F. 2002. Effect of electrical stimulation on the activity of muscle exoproteases during beef ageing. Food Sci Tech Int 8:285–289.

Grossman S, Bergman M, Sklan D. 1988. Lipoxygenase in chicken muscle. J Agric Food Chem 36:1268–1270.

Hierro E, de la Hoz L, Ordoñez JA. 1997. Contribution of microbial and meat endogenous enzymes to the lipolysis of dry fermented sausages. J Agric Food Chem 45:2989–2995.

Imanaka T, Yamaguchi M, Ahkuma S, Takano 1985. Positional specifity of lysosomal acide lipase opurified from rabbit liver. J Biochem 98:927–931

Klement JT, Cassens RG, Fennema OW. 1974. The effect of bacterial fermentation on protein solubility in a sausage model system. J Food Sci 39:833–835.

Koohmaraie M. 1994. Muscle proteinases and meat ageing. Meat Sci 36:93–104.

Koohmaraie M., Seideman SC, Schollmeyer JE, Dutson TR, Grouse JD. 1987. Effect of postmortem storage on Ca^{2+} Dependent proteases, their inhibitor and myofibril fragmentation. Meat Sci 19:187–196.

Ludemann V, Pose G, Pollio ML, Segura J. 2004. Determination of growth characteristics and lipolytic and proteolytic activities of *Penicillium* strains isolated from Argentinian salami. Int J Food Microbiol 96:13–18.

Marczy JS, Simon ML, Mozsik L, Szajani B. 1995. Comparative study on the lipoxygenase activities of some soybean cultivars. J Agric Food Chem 43:313–315.

Martin L, Córdoba JJ, Ventanas J, Antequera T. 1999. Changes in intramuscular lipids during ripening of Iberian dry-cured ham. Meat Sci 51:129–134.

Matsukura U, Okitani A, Nishimura T, Katoh H. 1981. Mode of degradation of myofibrillar proteins by an endogenous protease, cathepsin L. Biochim Biophys Acta 662:41–47.

McDonald JK, Barrett AJ. (eds.) 1986. Mammalian proteases. A glossary and bibliography. Vol 2. Exopeptidases. London: Academic Press.

Mikami M, Whiting AH, Taylor MAJ, Maciewicz RA, Etherington DJ. 1987. Degradation of myofibrils from rabbit, chicken and beef by cathepsin L and lysosomal lysates. Meat Sci 21:81–87.

Miller CH, Parce JW, Sisson P, Waite M. 1981. Specificity of lipoprotein lipase and hepatic lipase towrds monoacylglycerols varying in the acyl composition. Biochim Biophys Acta 665:385–392.

Molly K, Demeyer DI, Johansson G, Raemaekers M, Ghistelinck M, Geenen I. 1997. The importance of meat enzymes in ripening and flavor generation in dry fermented sausages. First results of a European project. Food Chem 54:539–545.

Monin G, Marinova P, Talmant A, Martin JF, Cornet M, Lanore D, Grasso F. 1997. Chemical and structural changes in dry-cured hams (Bayonne hams) during processing and effects of the dehairing technique. Meat Sci 47:29–47.

Montel MC, Talon R, Berdague JL, Cantonnet M. 1993. Effects of starter cultures on the biochemical characteristics of French drysausages. Meat Sci 35:229–240.

Motilva MJ, Toldrá F. 1993. Effect of curing agents and water activity on pork muscle and adipose subcutaneous tissue lipolytic activity. Z Lebensm Unters Forsch 196:228–231.

Motilva MJ, Toldrá F, Aristoy MC, Flores J. 1993a. Subcutaneous adipose tissue lipolysis in the processing of dry-cured ham. J Food Biochem 16:323–335.

Motilva MJ, Toldrá F, Flores J. 1992. Assay of lipase and esterase activities in fresh pork meat and dry-cured ham. Z Lebensm Unters Forsch 195:446–450.

Motilva MJ, Toldrá F, Nieto P, Flores J. 1993b. Muscle lipolysis phenomena in the processing of dry-cured ham. Food Chem 48:121–125.

Navarro JL, Nadal MI, Izquierdo L, Flores J. 1997. Lipolysis in dry cured sausages as affected by processing conditions. Meat Sci 45:161–168.

Negre AE, Salvayre RS, Dagan A, Gatt S. 1985. New fluorimetric assay of lysosomal acid lipase and its application to the diagnosis of Wolman and cholesteryl ester storage diseases. Clin Chim Acta 149:81–88.

Okitani A, Nishimura T, Kato H. 1981. Characterization of hydrolase H, a new muscle protease possesing aminoendopeptidase activity. Eur J Biochem 115:269–274.

Ordoñez JA, Hierro EM, Bruna JM, de la Hoz L. 1999. Changes in the components of dry-fermented sausages during ripening. Crit Rev Food Sci Nutr 39: 329–367.

Ouali A. 1991. Sensory quality of meat as affected by muscle biochemistry and modern technologies. In: LO Fiems, BG Cottyn, DI Demeyer, editors, Animal Biotechnology and the Quality of Meat Production, pp. 85–105. Amsterdam, The Netherlands: Elsevier Science Pub, BV.

Rosell CM, Toldrá F. 1996. Effect of curing agents on m-calpain activity throughout the curing process. Z Lebensm Unters Forchs 203:320–325.

Sanz Y, Sentandreu MA, Toldrá F. 2002. Role of muscle and bacterial exopeptidases in meat fermentation. In: F Toldrá, editor, Research Advances in the Quality of Meat and Meat Products, pp. 143–155. Trivandrum (India): Research Signpost.

Schwartz WN, Bird JWC. 1977. Degradation of myofibrillar proteins by cathepsins B and D. Biochem J 167:811–820.

Sentandreu MA, Coulis G, Ouali A. 2003. Role of muscle endopeptidases and their inhibitors in meat tenderness. Trends Food Sci Technol 13:398–419.

Sentandreu MA, Toldrá F. 2002. Dipeptidylpeptidase activities along the processing of Serrano dry-cured ham. Eur Food Res Technol 213:83–87

Shahidi F. 1998a. Assesment of lipid oxidation and off-flavour development in meat, meat products and seafoods. In: F Shahidi, editor, Flavor of Meat, Meat Products and Seafoods, pp. 373–394. London: Blackie Academic and Professional.

———. 1998b. Flavour of muscle foods—An overview. In: F. Shahidi, editor, Flavor of Meat, Meat Products and Seafoods, pp. 1–4. London: Blackie Academic and Professional.

Smith LC, Pownall HJ. 1984. Lipoprotein lipase. In: B Borgström, HL Brockman, Lipases, pp. 263–305. London: Elsevier, Science Pub.

Stahnke LH. 2002. Flavour formation in fermented sausage. In: F Toldrá, editor, Research Advances in the Quality of Meat and Meat Products, pp. 193–223. Trivandrum (India): Research Signpost.

Sunesena LO, Stahnke LH. 2004. Mould starter cultures for dry sausages—Selection, application and effects. Meat Sci 65:935–948.

Talon R, Leroy-Sétrin S, Fadda S. 2002. Bacterial starters involved in the quality of fermented meat products. In: F Toldrá, editor, Research Advances in the Quality of Meat and Meat Products, pp. 175–191. Trivandrum (India): Research Signpost.

Talon R, Walter D, Chartier S, Barriere C, Montel MC. 1999. Effect of nitrate and incubation conditions on the production of catalase and nitrate reductase by staphylococci. Int J Food Microbiol. 52:47–56.

Tichivagana JZ, Morrisey PA. 1985. Metmyoglobin and inorganic metals as prooxidants in raw and cooked muscle systems. Meat Sci 15:107–116.

Tjener K, Stahnke LH, Andersen L, Martinussen J. 2004. Growth and production of volatiles by *Staphylococcus carnosus* in dry sausages: Influence of inoculation level and ripening time. Meat Sci 67:447–452.

Toldrá F. 1992. The enzymology of dry-curing of meat products. In: FJM Smulders, F Toldrá, J Flores, M Prieto, editors, New Technologies for Meat and Meat Products, pp. 209–231. Nijmegen (The Netherlands): Audet.

———. 1998. Proteolysis and lipolysis in flavour development of dry-cured meat products. Meat Sci 49:s101–s110.

———. 2002. Dry-Cured Meat Products, pp. 1–238. Trumbull, Connecticut: Food and Nutrition Press.

———. 2004a. Dry-cured ham. In: YH Hui, LM Goddik, J Josephsen, PS Stanfield, AS Hansen, WK Nip, F Toldrá, editors, Handbook of Food and Beverage Fermentation Technology, pp. 369–384. New York: Marcel Dekker, Inc.

———. 2004b. Curing: Dry. In: W Jensen, C Devine, M Dikemann, editors, Encyclopedia of Meat Sciences, pp. 360–365. London: Elsevier Science, Ltd.

Toldrá F, Flores M. 1998. The role of muscle proteases and lipases in flavor development during the processing of dry-cured ham. CRC Crit Rev Food Sci Nutr 38:331–352.

Toldrá F, Gavara G, Lagarón JM. 2004. Packaging and quality control. In: YH Hui, LM Goddik, J Josephsen, PS Stanfield, AS Hansen, WK Nip, F Toldrá, editors, Handbook of Food and Beverage Fermentation Technology, pp. 445–458. New York: Marcel Dekker, Inc.

Toldrá F, Rico E, Flores J. 1992. Activities of pork muscle proteases in cured meats. Biochimie 74:291–296.

———. 1993. Cathepsin B, D, H and L activity in the processing of dry-cured-ham. J Sci Food Agric 62: 157–161.

Toldrá F, Sanz Y, Flores M. 2001. Meat fermentation technology. In: YH Hui, WK Nip, RW Rogers, OA Young, editors, Meat Science and Applications pp. 537–561. New York: Marcel Dekker, Inc.

Tornqvist H, Nilsson-Ehle P, Belfrage P. 1978. Enzymes catalyzing the hydrolysis of long-chain monoacylglycerols in rat adipose tissue. Biochim Biophys Acta 530:474–486.

Valin C, Ouali A. 1992. Proteolytic muscle enzymes and postmortem meat tenderisation. In: FJM Smulders, F Toldrá, J Flores, M Prieto, editors, New Technologies for Meat and Meat Products, pp. 163–179. Nijmegen (The Netherlands): Audet.

Vestergaard CS, Schivazzappa C, Virgili R. 2000. Lipolysis in dry-cured ham maturation. Meat Sci 55:1–5.

Yamamoto S. 1992. Mammalian lipoxygenases: Molecular structures and functions. Biochim Biophys Acta 1128:117–131.

Yates LD, Dutson TR, Caldwell J, Carpenter ZL. 1983. Effect of temperature and pH on the post-mortem degradation of myofibrillar proteins. Meat Sci 9: 157–179.

Young JF, Rosenvold K, Stagsted J, Steffensen CL, Nielsen JH, Andersen HJ. 2003. Significance of preslaughter stress and different tissue PUFA levels on the oxidative status and stability of porcine muscle and meat. J Agric Food Chem 51:6877–6881.

Zeece MG, Katoh K. 1989. Cathepsin D and its effects on myofibrillar proteins: A review. J Food Biochem. 13: 157–161.

15

Chemistry and Biochemistry of Color in Muscle Foods

J. A. Pérez-Alvarez and J. Fernández-López

GENERAL ASPECTS OF MUSCLE-BASED FOOD COLOR

The first impression that a consumer receives concerning a food product is established visually, and among the properties observed are color, form, and surface characteristics.

Color is the main aspect that defines a food's quality, and a product may be rejected simply because of its color, even before other properties, such as aroma, texture, and taste, can be evaluated. This is why the appearance (optical properties, physical form, and presentation) of meat products at the sales point is of such importance for the meat industry (Lanari et al. 2002). As regards the specific characteristics that contribute to the physical appearance of meat, color is the quality that most influences consumer choice (Krammer 1994).

The relation between meat color and quality has been the subject of study since the 1950s, indeed, since Urbain (1952) described how consumers had learned through experience that the color of fresh meat is bright red; and any deviation from this color (nonuniform or anomalous coloring) is unacceptable (Diestre 1992). The color of fresh meat and associated adipose tissue is, then, of great importance for its commercial acceptability, especially in the cases of beef and lamb (Cornforth 1994) and in certain countries, for example, the United States and Canada, and there have been many studies to identify the factors controlling its stability. Adams and Huffman (1972) affirmed that consumers relate the color of meat to its freshness. In poultry, the consumers of many countries also associate meat color with the way in which the animal was raised (intensive or extensive) and fed (cereals, animal feed, etc.).

Color as quality factor for meat can be appreciated in different ways in different countries; for example, in Denmark, pork meat color holds fifth place among qualities that affect consumers' purchase decision (Bryhni et al. 2002). Sensorial quality, especially color and appearance (Brewer and Mckeith 1999), of meat can be affected by both internal and external factors.

Food technologists, especially those concerned with the meat industry, have a special interest in the color of food for several reasons. First, because of the need to maintain a uniform color throughout processing; second, to prevent any external or internal agent from acting on the product during its processing, storage, and display; third, to improve or optimize a product's color and appearance; and, lastly, to attempt to bring the product's color into line with what the consumer expects.

Put simply, the color of meat is determined by the pigments present in it. These can be classified into four types: (1) biological pigments (carotenes and hemopigments), which are accumulated or synthesized in the organism antemortem (Lanari et al. 2002); (2) pigments produced as a result of damage during manipulation or inadequate processing conditions; (3) pigments produced postmortem (through enzymatic or nonenzymatic reactions) (Montero et al. 2001); and (4) pigments resulting from the addition of natural or artificial colorants (Fernández-López et al. 2002).

As a quality parameter, color has been widely studied in fresh meat (MacDougall 1982, Cassens et al. 1995, Faustman et al. 1996) and cooked products (Anderson et al. 1990, Fernández-Ginés et al. 2003, Fernández-López et al. 2003a). However, dry-cured meat products have received less attention (Pérez-Alvarez 1996, Pagán-Moreno et al. 1998, Aleson et al. 2003) because in this type of product color formation takes place during the different processing stages (Pérez-Alvarez et al. 1997, Fernández-López et al. 2000); recently, a new heme pigment has been identified in this type of product (Parolari et al. 2003; Wakamatsu et al. 2004a,b).

From a practical point of view, color plays a fundamental role in the animal production sector, especially in meat production (beef and poultry, basically) (Zhou et al. 1993, Esteve 1994, Verdoes et al. 1999, Irie 2001), since in many countries of the European Union (e.g., Spain and Holland) paleness receives a wholesale premium.

CHEMICAL AND BIOCHEMICAL ASPECTS OF MUSCLE-BASED FOOD COLOR

Of the major components of meat, proteins are the most important since they are only provided by essential amino acids, which are very important for the organism's correct functioning; proteins also make a technological contribution during processing, and some are responsible for such important attributes as color. These are the so-called chromoproteins, and they are mainly composed of a porphyrinic group conjugated with a transition metal, principally iron (metalloporphyrin), which forms conjugation complexes (heme groups) (Whitaker 1972) that are responsible for color. However, carotenes and carotenoproteins (organic compounds with isoprenoid-type conjugated systems) exist alongside chromoproteins and also play an important part in meat color.

There are also some enzymatic systems whose coenzymes or prostetict groups possess chromophoric properties (peroxidases, cytochromes, and flavins) (Faustman et al. 1996). However, their contribution to meat color is slight. Below, are described the principal characteristics of the major compounds that impart color to meat.

CAROTENES

Carotenes are responsible for the color of beef, poultry meat and skin, fish, and shellfish; in the latter case, this is of great economic importance. The color of the fat is also important, in carcass grading. Furthermore, carotenoids can be used as food coloring agents (Verdoes et al. 1999).

An important factor to be taken into account with these compounds is that they not be synthesized by the live animal; they are obtained by assimilation (Pérez-Álvarez et al. 2000b). In fats, fatty acid composition can affect their color. When the ratio of *cis*-monounsaturated to saturated fatty acids is high, the fat exhibits a greater yellow color (Zhou et al. 1993). In the case of the carotenes present in fish tissues, these come from the ingestion of zooplankton and algae, and the levels are sometimes very high. The shells of many crustaceans, for example, lobster *(Panilurus argus),* also contain these compounds.

The pigments responsible for color in fish, particularly salmonids (trout and salmon, among others) are astaxantine and cantaxantine, although they are also present in tunids and are one of the most important natural pigments of marine origin.

In the case of shellfish, their color depends on the so-called carotenoproteins, which are proteins with a prostetic group that may contain various types of carotene (Minguez-Mosquera 1997), which are themselves water soluble (Shahidi and Matusalach-Brown 1998).

In fish-derived products, the carotene content has previously been used as a quality parameter on its own; however, it has been demonstrated that this is not appropriate, and that other characteristics generally influence color (Little et al. 1979).

The carotene content and its influence on color is perhaps one of the characteristics that has received most attention (Swatland 1995). In the case of meat, especially beef, an excess of carotenes may actually lower the quality (Irie 2001), as occurs sometimes when classifying carcasses. The Japanese system for beef carcass classification classifies as acceptable fats with a white, slightly off-white, or slightly reddish white color, while pink-yellowish and dark yellow are unacceptable (Irie 2001). It is precisely the carotenes that are responsible for these last two colorations.

However, in other animal species, such as chicken, the opposite effect is observed, since a high carotene (xantophile) concentration is much appreciated by consumers (Esteve 1994), yellow being associated with traditional or "home-reared" feeding (Pérez-Álvarez et al. 2000b).

The use of the carotenoid canthaxanthin as a coloring agent in poultry feeds is designed to result in the desired coloration of poultry meat skins. The carotenoids used include citranaxanthin, capsanthin, and capsorubin, but canthaxanthin shows superior pigmenting properties and stability during processing and storage (Blanch 1999).

Farmed fish, especially colored fish (salmon and rainbow trout, for example), are now a major industry; for example, Norway exports a great part of its salmon. To improve its color and brilliance, 0.004–0.04 weight percent (wt%) proanthocyanidin is added to fish feed containing carotenoids (Sakiura 2001). For rainbow trout carotenoid concentrations could be 10.7 or 73 ppm canthaxanthin, or 47 or 53 ppm astaxanthin.

Hemoproteins

Of the hemoproteins present in the muscle postmortem, myoglobin (Mb) is the one mainly responsible for color, since hemoglobin (Hb) arises from the red cells that are not eliminated during the bleeding process and are retained in the vascular system, basically in the capillaries (incomplete exsanguination; the average amount of blood remaining in meat joints is 0.3%) (Warris and Rhodes 1977). However, the contribution of red cells to color does not usually exceed 5% (Swatland 1995). There is wide variation in amounts of hemoglobin from muscle tissue of bled and unbled fish. Myoglobin content was minimal as compared with hemoglobin content in fish light muscle and white fish whole muscle. Hemoglobin made up 65 and 56% by weight of the total heme protein in dark muscle from unbled and bled fish, respectively (Richards and Hultin 2002).

Myoglobin, on average, represents 1.5% by weight of the proteins of the skeletal muscle, while Hb represents about 0.5%, the same as the cytochromes and flavoproteins combined. Myoglobin is an intracellular (sarcoplasmic) pigment apparently distributed uniformly within muscles (Ledward 1992, Kanner 1994). It is red in color and water soluble, and it is found in the red fibers of both vertebrates and invertebrates (Knipe 1993, Park and Morrisey 1994), where it fulfills the physiological role of intervening in the oxidative phosphorylization chain in the muscle (Moss 1992).

Structure of Myoglobin

Structurally, Mb can be described as a monomeric globular protein with a very compact, well-ordered structure that is specifically, almost triangularly, folded and bound to a heme group (Whitaker 1972). It is structurally composed of two groups: a proteinaceous group and a heme group.

The protein group has only one polypeptidic chain composed of 140–160 amino acid residues, measuring 3.6 nm and weighing 16,900 Da in vertebrates (Lehninger 1981). It is composed of eight relatively straight segments (where 70% of the amino acids are found), separated by curvatures caused by the incorporation into the chain of proline and other amino acids that do not form alpha-helices (such as serine and isoleukin). Each segment is composed of a portion of alpha-helix, the largest of 23 amino acids and the shortest of seven amino acids, all dextrogyrotating.

Myoglobin's high helicoidal content (forming an ellipsoid of $44 \times 44 \times 25$ Å) and lack of disulphide bonds (there is no cysteine) make it an atypical globular protein. The absence of these groups makes the molecule highly stable (Whitaker 1972). Although the three-dimensional structure seems irregular and asymmetric, it is not totally anarchic, and all the molecules of Mb have the same conformation.

One very important aspect of the protein part of Mb is its lack of color. However, the variations presented by its primary structure and the amino acid composition of the different animal and fish species destined for human consumption are the cause of the different colorations of meat and their stability when the meats are displayed in the same retail illumination conditions (Lorient 1982, Lee et al. 2003a).

The heme group of Mb (as in Hb and other proteins) is, as mentioned above, a metalloporphyrin. These molecules are characterized by their high degree of coloration as a result of their conjugated cyclic tetrapyrrolic structure (Kalyanasundaram 1992). The heme group is composed of a complex, organic annular structure, protoporphyrin, to which an iron atom in ferrous state is united (Fe II). This atom has six coordination bonds, four with the flat protoporphyrin molecule (forming a flat square complex) and two perpendicular to it. The sixth bond is open and acts as a binding site for the oxygen molecule.

Protoporphyrin is a system with a voluminous flat ring composed of four pyrrolic units connected by methyl bridges (=C–). The Fe atom, with a coordination number of 6, lies at the center of the tetrapyrrol ring and is complexed to four pyrrolic nitrogens. The heme group is complexed to the polypeptidic chain (globin) through a specific hystidine residue (imadazolic ring) occupying the fifth position of the Fe atom (Davidsson and Henry 1978).

The heme group is bound to the molecule by hydrogen bridges, which are formed between the propyonic acid side chains and other side chains. Other aromatic rings exist near, and almost parallel to, the heme group, which may also form pi (Π) bonds (Stauton-West et al. 1969).

The Hb contains a porphyrinic heme group identical to that of Mb and equally capable of undergoing reversible oxygenation and deoxygenation. Indeed, it is functionally and structurally paired with Mb, and its molecular weight is four times greater since it contains four peptidic chains and four heme groups. The Hb, like Mb, has its fifth ligand occupied by the imidazol group of a histidine residue, while the sixth ligand may or may not be occupied. It should be mentioned that positions 5 and 6 of other hemoproteins (cytochromes) are occupied by R groups of specific amino acid residues of the proteins and therefore cannot bind to oxygen (O_2), carbon monoxide (CO), or cyanide (CN^-), except a_3, which in its biological role, usually binds to oxygen.

One of the main differences between fish and mammalian Mb is that fish Mb had two distinct endothermic peaks, indicating multiple states of structural unfolding, whereas mammalian Mb followed a two-state unfolding process. Changes in alpha-helix content and tryptophan fluorescence intensity with temperature were greater for fish Mb than for mammalian Mb. Fish Mb shows labile structural folding, suggesting greater susceptibility to heat denaturation than that of mammalian Mb (Saksit et al. 1996).

Helical contents of frozen-thawed Mb were practically the same as those of unfrozen Mb, regardless of pH. Frozen-thawed Mb showed a higher autoxidation rate than unfrozen Mb. During freezing and thawing, Mb suffered some conformational changes in the nonhelical region, resulting in a higher susceptibility to both unfolding and autoxidation (Chow et al. 1989). In tuna fish, Mb stability was in the order bluefin tuna *(Thunnus thynnus)* > yellowfin tuna *(Thunnus albacares)* > bigeye tuna *(Thunnus obesus);* autoxidation rates were in the reverse order. The pH dependency of Mb from skipjack tuna *(Katsuwonus pelamis)* and mackerel *(Scomber scombrus)* were similar. Lower Mb stability was associated with higher autoxidation rates (Chow 1991).

Chemical Properties of Myoglobin

The chemical properties of Mb center on its ability to form ionic and covalent groups with other molecules. Its interaction with several gases and water depends on the oxidation state of the Fe of the heme group (Fox 1966) since this may be in either its ferrous (Fe II) or its ferric (Fe III) state. Upon oxidation, the Fe of the heme group takes on a positive charge (Kanner 1994) and, typically, binds with negatively charged ligands, such as nitrites, the agents responsible for the nitrosation reactions in cured meat products.

When the sixth coordination ligand is free Mb is usually denominated **deoxymyoglobin** (DMb), which is purple in color. However, when this site is occupied by oxygen, the oxygen and the Mb form a noncovalent complex, denominated **oxymyoglobin** (OMb), which is cherry or bright red (Lanari and Cassens 1991). When the oxidation state of the iron atom is modified to the ferric state and the sixth position is occupied by a molecule of water, the Mb is denominated **metmyoglobin** (MMb), which is brown.

There are several possible causes for generation of MMb, and these may include the ways in which tunids, meat, and meat products are obtained, transformed, or stored (MacDougall 1982, Lee et al. 2003b, Mancini et al. 2003). Among the most important factors are low pH, the presence of ions, and high temperatures during processing (Osborn et al. 2003); the growth and/or formation of metabolites from the microbiota (Renerre 1990); the activity of endogenous reducing enzymes (Arihara et al. 1995, Osborn et al. 2003); and the levels of endogenous (Lanari et al. 2002) or exogenous antioxidants, such as ascorbic acid or its salts, tocopherols (Irie et al. 1999), or plant extracts (Fernández-López et al. 2003b, Sánchez-Escalante et al. 2003). This change in the oxidation state of the heme group will result in the group being unable to bind with the oxygen molecule (Arihara et al. 1995).

DMb is able to react with other molecules to form colored complexes, many of which are of great economic relevance for the meat industry. The most characteristic example is the reaction of DMb with nitrite, since its incorporation generates a series of compounds with distinctive colors: red in dry-cured meat products or pink in heat-treated products. The products resulting from the incorporation of nitrite are denominated cured, and such products are of enormous economic importance worldwide (Pérez-Alvarez 1996). The reaction mechanism is based on the propensity of nitric oxide (NO, generated in the reaction of nitrite in acid medium, readily gives up electrons) to form strong coordinated covalent bonds; it forms an iron complex with the DMb heme group independent of the oxidation state of the heme structure. The compound formed after the nitrification reaction is denominated nitrosomyoglobin (NOMb).

As mentioned above, the presence of reducing agents such as hydrosulfhydric acid (H_2S) and ascorbates lead to the formation of undesirable pigments in both meat and meat products. These green pigments are called sulphomyoglobin (SMb) and colemyoglobin (ColeMb), respectively, and are formed as a result of bacterial activity and an excess of reducing agents in the medium. The formation of SMb is reversible, but that of ColeMb is an irreversible mechanism, since it is rapidly oxidized between pH 5 and 7, releasing the different parts of the Mb (globin, iron, and the tetrapyrrolic ring).

From a chemical point of view, it should be borne in mind that the color of Mb, and therefore of the meat or meat products, not only depends on the molecule that occupies the sixth coordination site, but also on the oxidation state of the iron atom (ferrous or ferric), the type of bond formed between the ligand and the heme group (coordinated covalent, ionic, or none), and the state of the protein (native or denatured form), not to mention the state of the porphyrin of the heme group (intact, substituted, or degraded) (Pérez-Alvarez 1996).

During heat treatment of fish flesh, aggregation of denatured fish proteins is generally accompanied by changes in light scattering intensity. Results demonstrate the use of changes in relative light scattering intensity for studying structural unfolding and aggregation of proteins under thermal denaturation (Saksit et al. 1998). When fatty fish meat like *Trachurus japonicus* was heat treated, MMb content increased linearly, and the percentages of denatured myoglobin and apomyoglobin increased rapidly when mince was exposed to heat, but when temperature reached 60°C the linearity was broken. Results indicated that color stability of Mmb was higher than that of Mb and that the thermal stability of heme was higher than that of apomyoglobin (Hui et al. 1998).

Both Mb and ferrous iron accelerated lipid oxidation of cooked, water-extracted fish meat. EDTA (ethylenediaminetetraacetic acid) inhibited the lipid oxidation accelerated by ferrous iron, but not that accelerated by Mb. Also, with cooked, nonextracted mackerel meat, EDTA noticeably inhibited lipid oxidation. Nonheme iron catalysis seemed to be related in part to lipid oxidation in cooked mackerel meat. Addition of nitrite in combination with ascorbate resulted in marked inhibition of lipid oxidation in the cooked mackerel meat. From these results, it was postulated that nitric oxide ferrohemochromogen, formed from added nitrite and Mb (present in the mackerel meat) in the presence of a reducing agent, possesses an antioxidant activity, which is attributable in part to its function as a metal chelator (Ohshima et al. 1988).

Tuna fish meat color can be improved when the flesh is treated or packaged with a modified atmosphere in which CO is included. Normally, the rate of penetration of CO or CO_2 in fish meat such as tuna, cod, or salmon, under different packaging conditions, is measured by monitoring pressure changes in a closed constant volume chamber with constant volume and temperature. Alternatively, however, the specific absorption spectrum of carboxymyoglobin

(MbCO), within the visible range, can be obtained and used as an indicator of MbCO formation. Mb extracts from tuna muscle treated with CO exhibited higher absorbance at 570 than at 580 nm (Chau et al. 1997). Therefore, the relation between absorbance at 570 nm and absorbance at 580 nm could be used to determine extent of CO penetration of tuna steaks placed in a modified atmosphere in which CO was included. Penetration of CO into tuna muscle was very slow. After approximately 1–4 hours, CO had penetrated 2–4 mm under the surface, and after 8 hours, CO had penetrated 4–6 mm (Chau et al. 1997).

In dry-cured meat products such as Parma ham (produced without nitrite or nitrate), the characteristic bright red color (Wakamatsu et al. 2004a) is caused by Zn-protoporphyrin IX (ZPP) complex, a heme derivative. This type of pigment can be formed by endogenous enzymes as well as microorganisms (Wakamatsu et al. 2004b). Spectroscopic studies of Parma ham during processing revealed a gradual transformation of muscle myoglobin, initiated by salting and continuing during ageing. Pigments became increasingly lipophilic during processing, suggesting that a combination of drying and maturing yields a stable red color (Parolari et al. 2003). Electron spin resonance spectra showed that the pigment in dry-cured Parma ham is at no stage a nitrosyl complex of ferrous myoglobin such as that found in brine-cured ham and Spanish Serrano hams (Moller et al. 2003). These authors also establish that the heme moiety is present in the acetone-water extract and that Parma ham pigment is gradually transformed from a myoglobin derivative into a nonprotein heme complex, that is thermally stable in acetone-water solution. Adamsen et al. (2003) also demonstrated that the heme moieties of Parma ham pigments have antioxidative properties.

Cytochromes

Cytochromes are metalloproteins with a prostetic heme group, whose putative role in meat coloration is undergoing revision (Boyle et al. 1994, Faustman et al. 1996): initially, they were not thought to play a very important role (Ledward 1984). These compounds are found in low concentrations in the skeletal muscle, and in poultry they do not represent more than 4.23% of the total hemeoproteins present (Pikul et al. 1986). It has now been shown that the role of cytochrome (especially its concentration) in poultry meat color is fundamental, when the animal has been previously exposed to stress (Ngoka and Froning 1982, Pikul et al. 1986). Cytochromes are most concentrated in cardiac muscle, so that when this organ is included in meat products, heart contribution to color, not to mention the reactions that take place during elaboration processes, must be taken into consideration (Pérez-Álvarez et al. 2000b).

COLOR CHARACTERISTICS OF BLOOD

Animal blood is little used in the food industry because of the dark color it imparts to the products to which it is added. For solving the negative aspects of blood incorporation, specifically food color related problems, several different processes and means have been employed, but they are not always completely satisfactory. The addition of 12% blood plasma to meat sausages leads to pale-colored products. Addition of discolored whole blood or globin (from which the hemoglobin's heme group has been eliminated) has also been used to address color problems.

From blood, natural red pigments can be obtained without using coloring agents such as nitrous acid salts; these pigments have zinc protoporphyrin as the metalloporphyrin moiety and can be used to produce a favorable color beef products, whale meat products, and fish products (including fish pastes) (Numata and Wakamatsu 2003).

There was wide variation in amounts of hemoglobin extracted from muscle tissue of bled and unbled fish, and the residual level in the muscle of bled fish was substantial. Myoglobin content was minimal as compared with hemoglobin content in mackerel light muscle and trout whole muscle. Hemoglobin made up 65 and 56% by weight of the total heme protein in dark muscle from unbled and bled mackerel, respectively. The blood-mediated lipid oxidation in fish muscle depends on various factors, including hemoglobin concentration, hemoglobin type, plasma volume, and erythrocyte integrity (Richards and Hultin 2002).

The presence of blood, Hb, Mb, Fe^{2+}, Fe^{3+}, or Cu^{2+} can stimulate lipid oxidation in the fillets of icefish (Rehbein and Orlick 1990).

FAT COLOR

From a technological point of view, fat fulfills several functions, although as regards color its principal role is in the brightness of meat products. Processes such as "afinado" during the elaboration of dry-cured ham involve temperatures at which fat melts, so that it infiltrates the muscle mass and increases its brilliance (Sayas 1997). When the fat is finely chopped, it "dilutes" the red components of the color, thus decreasing the color intensity of the finished product (Pérez-Alvarez et al. 2000). However, fats do not play such an important role in fine pastes since, after emulsification, the fat is masked by the matrix effect of the emulsion, so that it contributes very little to the final color.

The color of fat basically depends on the feed that the live animal received (Esteve 1994, Irie 2001). In the case of chicken and ostrich, the fat has a "white" appearance (common in Europe) when the animal has been fed with "white" cereals or other ingredients not containing xanthophylls, since these are accumulated in subcutaneous fat and other fatty deposits. However, when the same species are fed on maize (rich in xanthophylls), the fatty deposits take on a yellow color.

Beef or veal fat, that is dark, hard (or soft), excessively bright, or shiny lowers the carcass and cut price. Fat with a yellowish color in healthy animals reflects a diet containing beta-carotene (Swatland 1988). While fat color evaluation has traditionally been a subjective process, modern methods include such techniques as optical fiber spectrophotometry (Irie 2001).

Another factor influencing fat color is the concentration of the Hb retained in the capillaries of the adipose tissues (Swatland 1995). As in meat, the different states of Hb may influence the color of the meat cut. OMb is responsible for the yellowish appearance of fat, since it affects different color components (yellow-blue and red-green).

The different states of hemoglobin present in adipose tissue may react in a similar way to those in meat, so that fat color should be measured as soon as possible to avoid possible color alterations.

When the Hb in the adipose tissue reacts with nitrite incorporated in the form of salt, nitrosohemoglobin (NOHb), a pigment that imparts a pink color to fat, is generated. This phenomenon occurs principally in dry-cured meat products with a degree of anatomical integrity, such as dry-cured ham or shoulder (Sayas 1997).

When fat color is measured, its composition should be kept in mind since its relation with fatty acids modifies its characteristics, making it more brilliant or duller in appearance. The fat content of the conjunctive tissue must also be borne in mind: collagen may present a glassy appearance because, at acidic pH, it is "swollen," imparting a transparent aspect to the product.

ALTERATIONS IN MUSCLE-BASED FOOD COLOR

The color of meat and meat products may be altered by several factors, including exposure to light (source and intensity), microbial growth, rancidity, and exposure to oxygen. Despite the different alterations in color that may take place, only a few have been studied; these include the pink color of boiled uncured products, premature browning, and melanosis in crustaceans.

PINK COLOR OF UNCURED MEAT PRODUCTS

The normal color of a meat product that has been heat treated but not cured is "brown," although it has recently been observed that these products show an anomalous coloration (red or pink) (Hunt and Kropf 1987). This problem is of great economic importance in "grilled" products since this type of color is not considered desirable.

This defect may occur both in meats with a high hemoprotein content such as beef and lamb (red) and in those with a low hemoprotein concentration, including chicken and turkey (pink) (Conforth et al. 1986). One of the principal causes of this defect is the use of water rich in nitrates, which are reduced to nitrites by nitrate-reducing bacteria, which react with the Mb in meat to form NOMb (Nash et al. 1985). The same defect may occur in meat products containing paprika, which according to Fernández-López (1998), contains nitrates that, once incorporated in the product, may be similarly reduced by microorganisms. Conforth et al. (1991) mention that several nitrogen oxides may be generated in gas and electric ovens used for cooking ham, and that these nitrogen oxides will react with the Mb to generate nitrosohemopigments. Also produced in ovens is CO, which reacts with Mb during thermal treatment to form a pink-colored pigment, carboxyhemochrome. It has also been described how the use of adhesives formed from starchy substances produces the same undesirable pink color in cooked products (Scriven et al. 1987).

The same anomalous pink color may be generated when the pH of the meat is high (because of the addition of egg albumin to the ingredients) (Froning et al. 1968) and when the cooking temperature during processing is too low. These conditions favor the development of a reducing environment that maintains the iron of the Mb in its ferrous form, imparting a reddish/pink color (as a function of the concentration of hemopigments) instead of the typical grayish brown color of heat-treated, uncured meat products.

Cooking uncured meat products, such as roast beef, at low temperatures (less than 60°C) may produce a reddish color inside the product, which some consumers may like. This internal coloring is not related to the formation of nitrosopigments, but results from the formation of OMb, a phenomenon that occurs because there exist in the muscle MMb-reducing enzymatic systems that are activated at temperatures below 60°C (Osborn et al. 2003).

Microbial growth may also cause the formation of a pink color in cooked meats since these reduce the oxidoreduction potential of the product during their growth. This is important when the microorganisms that develop in the medium are anaerobes, since they may generate reducing substances that reduce the heme iron. When extracts of *Pseudomonas* cultures are applied, the MMb may be reduced to Mb (Faustman et al. 1990).

Melanosis

Melanosis or blackspot, involving the appearance of a dark, even black, color, may develop postmortem in certain shellfish during chilled and frozen storage (Slattery et al. 1995). Melanosis is of huge economic importance since the coloration may suggest a priori in the eyes of the consumer that the product is in bad condition, despite the fact that the formation of the pigments responsible involves no health risk.

Melanosis is an objectionable surface discoloration of such high value shellfish as lobsters that is caused by enzymic formation of precursors of phenolic pigments. Blackspot is a process regulated by a complex biochemical mechanism, whereby the phenols present in a food are oxidized to quinones in a series of enzymatic reactions caused by polyphenol oxidase (PPO)(Ogawa et al. 1984). This is followed by a polymerization reaction, which produces pigments of a high molecular weight and dark color. Melanosis is produced in the exoskeleton of crustaceans, first in the head and gradually spreading towards the tail. Melanosis of shell and hyperdermal tissue in some shellfish, such as lobsters, was related to stage of molt. The molting fluid is considered to be the source of the natural activator(s) of pro-PPO. Polyphenol oxidase (catechol oxidase) can be isolated from shellfish cuticle (Ali et al. 1994) and is still active during iced or refrigerated storage. Sulphites can control the process (Ferrer et al. 1989), although their use is prohibited in many countries. Ficin (Taoukis et al. 1990), 4-hexylresorcinol, also functioned as a blackspot inhibitor, alone and in combination with L-lactic acid (Benner et al. 1994).

Premature Browning

Hard-to-cook patties show persistent internal red color and are associated with high pH (> 6) raw meat. Pigment concentration affects red color intensity after cooking (residual undenatured myoglobin), so this phenomenon is often linked to high pH dark cutting meat from older animals. Premature browning is a condition in which ground beef (mince) looks well done at a lower than expected temperature (Warren et al. 1996).

Premature browning (PMB) of ground beef is a condition in which myoglobin denaturation appears to occur on cooking at a temperature lower than expected; it may indicate falsely that an appropriate internal core temperature of 71°C has been achieved (Suman et al. 2004). The relationship between cooked color and internal temperature of beef muscle is inconsistent and depends on pH and animal maturity. Increasing the pH may be of benefit in preventing premature browning, but may increase incidence of red color in well-cooked meat (cooked over internal temperature of 71.1°C) (Berry 1997). When pale, soft, exudative (PSE) meat was used in patty processing, patties containing OMb easily exhibited premature browning. One reason for this behavior is that the percentage of Mb denaturation increased as cooking temperature rose (Lien et al. 2002).

COLOR AND SHELF LIFE OF MUSCLE-BASED FOODS

Meat and meat products are susceptible to degradation during storage and throughout the retail process. In this respect, color is one of the most important quality attributes for indicating the state of preservation in meat.

Any energy received by food can initiate its degradation, but the rate of any reaction depends on the exact composition of the product (Jensen et al. 1998), environmental factors (light, temperature, presence of oxygen), and the presence of additives.

Transition metals such as copper and iron are very important in the oxidative/antioxidative balance of meat. When the free ions of these two metals interact, they reduce the action of certain agents, such as cysteine, ascorbate, and alpha-tocopherol, oxidizing them and significantly reducing the antioxidant capacity in muscle (Zanardi et al. 1998).

Traditionally, researchers have determined the discoloration of meat using as criterion the brown color of the product, calculated as percent MMb (Mancini et al. 2003). These authors demonstrated that in the estimation of the shelf life of beef or veal (considered as discoloration of the product), the diminution in the percent of OMb is a better tool than the increase in percentage of MMb.

Occasionally, when the meat cut contains bone (especially in pork and beef), the hemopigments (mainly Hb) present in the medulla lose color because the erythrocytes are broken during cutting and accumulate on the surface of the bone hemoglobin. When exposed to light and air, the color of the Hb changes from the bright red (oxyhemoglobin: OHb) characteristic of blood to brown (metahemoglobin MHb) or even black (Gill 1996). This discoloration basically takes place during long periods of storage, especially during shelf life display (Mancini et al. 2004). This characteristic is aggravated if the product is kept in a modified atmosphere rich in oxygen (Lanari et al. 1995). These authors also point out that the effect of bone marrow discoloration is minimized by the effect of bacterial growth in modified atmosphere packaging.

As in the case of fresh meat, the shelf life of meat products is limited by discoloration (Mancini et al. 2004). This phenomenon is important in this type of product because they are normally displayed in illuminated cabinets. Consequently, the possibility of photooxidation of nitrosated pigments (NOMb) needs to be taken into account. During this process, the molecule is activated because it absorbs light; this may subsequently deactivate the NOMb and give the free electrons to the oxygen to generate MMb and free nitrite.

In model systems of NOMb photooxidation, addition of solutions of dextrose, an important component of the salts used for curing cooked products and in meat emulsions, can diminish the effect of NOMb photooxidation.

When a meat product is exposed to light or is stored in darkness, the use of ascorbic acid or its salts may help stabilize the product's color. Such behavior has been described both in model systems of NOMb (Walsh and Rose 1956) and in dry-cured meat products (e.g., *longanizas,* Spanish dry-fermented sausage). However, when sodium isoascorbate or erythorbate is used in *longaniza* production, color stability is much reduced during the retail process (Ruíz-Peluffo et al. 1994).

The discoloration of white meats such as turkey is characterized by color changes that go from pink-yellow to yellow-brown, while in veal and beef the changes go from purple to grayish brown. In turkey, it has been demonstrated that the presence or absence of lipid oxidation depends on, among other things, the concentration of vitamin E in the tissues. The color and lipid oxidation are interrelated since it has been seen that lipid oxidation in red and white muscle depends on the predominant form of catalyzing iron, Mb, or free iron (Mercier et al. 1998).

Compared with red meat, tuna flesh tends to undergo more rapid discoloration during refrigerated storage. Discoloration due to oxidation of Mb in red fish presented a problem, even at low temperatures. This low color stability might be related to the lower activity or poorer stability of MMb reductase in tuna flesh (Ching et al. 2000). Another reason for the low color stability is that aldehydes produced during lipid oxidation can accelerate tuna OMb oxidation in vitro (Lee et al. 2003b). Tuna flesh could be immersed in MMb reductase solution to extend the color stability of tuna fish. Also, the use of this enzyme can reduce MMb formation during refrigerated storage of tuna (Tze et al. 2001).

Yellowtail *(Seriola quinqueradiata)* fillets stored in gas barrier film packs filled with N_2 and placed in cold storage at 0–5°C, stayed fresh for 4–7 days. N_2 or CO_2 packaging did not prevent discoloration in frozen tuna fillets; better results were achieved by thawing the frozen tuna meat in an O_2 atmosphere (Oka 1989). Packaging in atmospheres containing 4 or 9% O_2 was inferior to packaging in air, as these atmospheres promoted MMb formation. Packaging in 70% O_2 maintained the fresh red color of tuna dorsal muscle for storage periods less than 3 days (Tanaka et al. 1996). In order to change the dark brown color into a bright red color, processors sometimes treat tuna with 100% carbon monoxide (CO) during modified atmosphere packaging. Since Mb can react with CO rapidly even at low CO concentrations (Chi et al. 2001), modified atmosphere packaging with 100% CO may result in high CO residues in the flesh and that may cause health problems.

MICROORGANISMS AND MUSCLE-BASED FOOD COLOR

Although the real limiting factor in the shelf life of fresh meat is the microbial load, consumers choose fresh meat according to its color. The bacterial load is usually the most important cause of discoloration in fresh meat and meat products (sausages and other cooked products), and slaughter, cutting, and packaging must be strictly controlled.

Bacterial contamination decisively affects the biochemical mechanisms responsible for the deterioration of meat (Renerre 1990). Is it important to take into account that, just as with the bacterial load, the effect of discoloration on meat is more pronounced in meats that are more strongly pigmented (beef) than in less pigmented meats such as pork and chicken (Gobantes and Oliver 2000).

Another variable affecting color stability in meat is the quantity of microorganisms present (Houben et al. 1998); concentrations in excess of 10^6/g have a strong effect. Although antioxidants, such as ascorbic acid, slow lipid oxidation and consequently improve color stability, when bacterial growth affects color these substances have little effect (Zerby et al. 1999).

REFERENCES

Adams DC, Huffman RT. 1972. Effect of controlled gas atmospheres and temperature on quality of packaged pork. J Food Sci 37:869–872.

Adamsen CE, Hansen ML, Moller JKS, Skibsted LH. 2003. Studies on the antioxidative activity of red pigments in Italian-type dry-cured ham. Eur Food Res Technol 217(3): 201–206.

Aleson L, Fernández-López J, Sayas-Barberá E, Sendra E, Pérez-Alvarez. 2003. Utilization of lemon albedo in dry-cured sausages. J Food Sci 68:1826–1830.

Ali MT, Gleeson RA, Wei CI, Marshall MR. 1994. Activation mechanisms of pro-phenoloxidase on melanosis development in Florida spiny lobster *(Panulirus argus)* cuticle. J Food Sci 59(5): 1024–1030.

Anderson HJ, Bertelsen G, Skibsled LH. 1990. Colour and colour stability of hot processed frozen minced beef. Result from chemical model experiments tested under storage conditions. Meat Sci 28(2): 87–97.

Arihara K, Itoh M, Kondo Y. 1995. Significance of metmyoglobin reducing enzyme system in myocytes. Proc. 41th International Congress of Meat Sci and Technology, San Antonio,Texas. C70:378–379.

Benner RA, Miget R, Finne G, Acuff GR. 1994. Lactic acid/melanosis inhibitors to improve shelf life of brown shrimp *(Penaeus aztecus).* J Food Sci 59(2): 242–245.

Berry BW. 1997. Color of cooked beef patties as influenced by formulation and final internal temperature. Food Res Int 30(7): 473–478

Blanch A. 1999. Getting the colour of yolk and skin right. World Poultry 15(9): 32–33.

Boyle RC, Tappel AL, Tappel AA, Chen H, Andersen HJ. 1994. Quantitation of haeme proteins from spectra of mixtures. J Agric Food Chem 42:100–104.

Brewer MS, Mckeith FK. 1999. Consumer–rated quality characteristics as related to purchase intent of fresh pork. J Food Sci 64:171–174.

Bryhni EA, Byrne DV, Rødbotten M, Claudi-Magnussen C, Agerhem H, Johansson M, Lea P, Martens M. 2002. Consumer perceptions of pork in Denmark, Norway and Sweden. Food Qual Pref 13:257–266.

Cassens RG, Demeyer D, Eikelenboom G, Honikel KO, Johansson G, Nielsen T, Renerre M, Richardson Y, Sakata R. 1995. Recommendation of reference method for assessment of meat color. Proc of 41th International Congress of Meat Sci and Technology, San Antonio, Texas. C86:410–411.

Chau JC, Sue ML, Mei LT. 1997. Characteristics of reaction between carbon monoxide gas and myoglobin in tuna flesh. J Food Drug Anal 5(3): 199–206.

Chi CY, Kuang HL, Chau JC. 2001. Effect of carbon monoxide treatment applied to tilapia fillet. Taiwanese J Agric Chem Food Sci 39(2): 117–121.

Ching YP, Tze KC, Ming LH, Shan TJ. 2000. Effect of polyethylene package on the metmyoglobin reductase activity and color of tuna muscle during low temperature storage. Fish Sci 66(2): 84–89.

Chow CJ. 1991. Relationship between the stability and autoxidation of myoglobin. J Agric Food Chem 39 (1):22–26.

Chow CJ, Ochiai Y, Watabe S, Hashimoto K. 1989. Reduced stability and accelerated autoxidation of tuna myoglobin in association with freezing and thawing. J Agric Food Chem 37(5): 1391–1395.

Conforth D. 1994. Colour-its basis and importance. In: AM Pearson, TR Dutson, editors, Advances in Meat Research. London: Chapman Hall. 9:34–78.

Conforth DP, Calkins CR, Faustman CT. 1991. Methods for identification and prevention of pink color in cooked meats. Proc of Reciprocal Meat Conference. 44:53–58.

Conforth DP, Vahabzadhe CE, Bartholomew DT. 1986. Role of reduced haemochromes in pink colour defect of cooked turkey. J Food Sci 51:1132–1135

Davidsson I, Henry JB. 1978. Diagnóstico Clínico por el Laboratorio, pp. 310–316. Barcelona: Salvat.

Diestre A. 1992. Principales problemas de la calidad de la carne en el porcino. Alim Eq Tecnol 98:73–78.

Esteve E. 1994. Alimentación animal y calidad de la carne. Eurocarne 31:71–77

Faustman C, Chan WKM, Lynch MP, Joo ST. 1996. Strategies for increasing oxidative stability of (fresh) meat color. Proc 49th Annual Reciprocal Meat Conference, Provo, Utah, pp. 73–78.

Faustman C, Johnson JL, Cassens RG, Doyle MP. 1990. Color reversion in beef: Influence of psychotrophic bacteria. Fleischwirt 70:676.

Fernández-Ginés JM, Sayas-Barberá E, Navarro C, Sendra E, Pérez-Alvarez JA 2003. Effect of storage conditions on quality characteristics of bologna sausages made with citrus fiber. J Food Sci 68:710–715.

Férnández-López J. 1998. Estudio del color por métodos objetivos en sistemas modelo de pastas de embutidos crudo-curados. [PhD Thesis]. Murcia, Spain: Universidad de Murcia. 310 pp.

Fernández-López J, Pérez-Alvarez JA, Aranda-Catalá V. 2000. Effect on mincing degree on color properties in pork meat. Color Res Appl 25:376–380.

Fernández-López J, Pérez-Alvarez JA, Sayas-Barberá ME, López-Santoveña F. 2002. Effect of paprika *(Capsicum annum)* on color of Spanish-type sausages during the resting stage. J Food Sci 67:2410–2414

Fernández-López JM, Sayas-Barberá E, Sendra E, Pérez-Alvarez JA. 2003a. Physical, chemical and sensory properties of bologna sausage made with ostrich meat. J Food Sci 68:1511–1515.

Fernández-López J, Sevilla L, Sayas-Barberá E, Navarro C, Marín F, Pérez-Alvarez. 2003b. Evaluation of antioxidant potential of hyssop (*Hyssopus officinalis* L.) and rosemary (*Rosmarinus officinalis* L.) extract in cooked pork meat. J Food Sci68:660–664.

Ferrer OJ, Otwell WS, Marshall MR. 1989. Effect of bisulfite on lobster shell phenoloxidase. J Food Sci 54(2): 478–480.

Fox JB. 1966. The chemistry of meat pigments. J Agric Food Chem 14:207.

Froning GW, Hargus G, Hartung TE. 1968. Color and texture of ground turkey meat products as affected by dry egg white solids. Poult Sci 47:1187.

Fuentes MM, Palacios R, Garcés MM, Caballero ML, Moneo I. 2004. Isolation and characterization of a heat-resistant beef allergen: myoglobin. Allergy 59(3): 327–331.

Gill CO. 1996. Extending the storage life of raw chilled meats. Meat Sci 43(Suppl): S99–S109.

Gobantes IY, Oliver MA. 2000. Problemática de la estabilidad del color en carne de vacuno: La adición de vitamina E y el envasado bajo atmósfera protectora. Eurocarne 85:57–62.

Harada K. 1991. Fish colour. Aust Fish 50(4): 18–19.

Houben JH, Eikelenboom G, Hoving-Bolink AH. 1998. Effect of the dietary supplementation with vitamin E on colour stability and lipid oxidation in packaged, minced pork. Meat Sci 48 (3/4): 265–273.

Hui HC, Wen HH, Tzu HH. 1998. Thermal stability of myoglobin and color in horse mackerel mince. Food Science Taiwan 25(4): 419–427.

Hunt MC, Kropf DH. 1987. Colour and appearance. In: AM Pearson, TR Dutson, editors, Restructured Meat and Poultry Products, Advances in Meat Research Series. New York: Van Nostran Reinhold. 3:125–159.

Irie M. 2001. Optical evaluation of factors affecting appearance of bovine fat. Meat Sci 57:19–22.

Irie M, Fujita K, Suduo K. 1999. Changes in meta color and alpha-tocopherol concentrations in plasma and tissues from Japanese beef cattle fed by two methods of vitamin E supplementation. Asian Aust J Animal Sci 11:810–814.

Jensen C, Laurisen C, Bertelsen G. 1998. Dietary vitamin E: Quality and storage stability of pork and poultry. Trends Food Sci Technol 9:62–72.

Kalyanasundaram K. 1992. Photochemistry of polypyridine and porphyrin complexes, pp. 25–56. New York: Academic Press.

Kanner J. 1994. Oxidative processes in meat products: Quality implications. Meat Sci 36:169–189.

Knipe L. 1993. Basic science of meat processing. Cured meat short course. Ames: Meat Laboratory, Iowa State University.

Krammer A. 1994. Use of color measurements in quality control of food. Food Technol 48(10): 62–71.

Lanari MC, Brewster M, Yang A, Tume RK. 2002. Pasture and grain finishing affect the color stability of beef. J Food Sci 67:2467–2473.

Lanari MC, Cassens RG. 1991. Mitochondrial activity and beef muscle color stability. J Food Sci 56:1476–1479.

Lanari MC, Schaefer DM, Scheller KK. 1995. Dietary vitamin E supplementation and discoloration of pork bone and muscle following modified atmosphere packaging. Meat Sci 41(3): 237–250.

Ledward DA. 1984. Haemoproteins in meat and meat products. In: DA Ledward, editor, Developments in Food Proteins, pp. 33–68. London: Applied Science.

———. 1992. Colour of raw and cooked meat. In: DA Ledward, DE Johnston, MK Knight, editors, The Chemistry of Muscle Based Foods, pp. 128–144. Cambridge: The Royal Society of Chemistry.

Lee EJ, Love J, Ahn DU. 2003a. Effects of antioxidants on consumer acceptance of irradiated turkey meat. J Food Sci 68:1659–1663.

Lee S, Joo ST, Alderton AL, Hill DW, Faustman C. 2003b. Oxymyoglobin and lipid oxidation in yellowfin tuna *(Thunnus albacares)* loins. J Food Sci 68(5): 1164–1168.

Lehninger AL. 1981. Bioquímica. Las bases moleculares de la estructura y función celular, pp. 230–233. Barcelona: Omega.

Lien R, Hunt MC, Anderson S, Kropf DH, Loughin TM, Dikeman ME, Velazco J. 2002. Effects of endpoint temperature on the internal color of pork patties of different myoglobin form, initial cooking state, and quality. J Food Sci 67(3): 1011–1015.

Little AC, Martinse C, Sceurman L. 1979. Color assessment of experimentally pigmented rainbow trout. Col Res Applic 4(2): 92–95.

Lorient D. 1982. Propiedades funcionales de las proteínas de origen animal. In: CM Bourgeois, P Le Roux, editors, Proteínas Animales, pp. 215–225. México: El Manual Moderno.

MacDougall DB. 1982. Changes in the colour and opacity of meat. Food Chem 9:(1/2): 75–88.

Mancini RA, Hunt MC, Hachmeister KA, Kropf DH, Jhonson DE. 2004 Ascorbic acid minimizes lumbar vertebrae discoloration. Meat Sci 68(3): 339–345.

Mancini RA, Hunt MC, Kropf DH. 2003. Reflectance at 610 nm estimates oxymyoglobin content on the surface of ground beef. Meat Sci 64:157–162.

McClements DJ, Chantrapoornchai W, Claydesdale F. 1998. Prediction of food emulsion color using light scattering theory. J Food Sci 63(6): 935–939.

Mercier Y, Gatellier P, Viau M, Remington H, Renerre M. 1998. Effect of dietary fat and vitamin E on colour stability and on lipid and protein oxidation in turkey meat during storage. Meat Sci 48(3/4): 301–318.

Minugéz-Mosquera MI. 1997. Clorofilas y carotenoides en tecnología de alimentos, pp. 15–17. Sevilla: Ed. Universidad de Sevilla.

Moller JKS, Adamsen CE, Skibsted LH. 2003. Spectral characterisation of red pigment in Italian-type dry-cured ham. Increasing lipophilicity during processing and maturation. Eur Food Res Technol 216(4): 290–296.

Montero P, Ávalos A, Pérez-Mateos M. 2001. Characterization of polyphenoloxidase of prawns *(Penaeus japonicus).* Alternatives to inhibition: Additives and high pressure treatment. Food Chem 75:317–324.

Moss BW. 1992. Lean meat, animal welfare and meat quality. In: DA Ledward, DE Johnston, MK Knight, editors, The Chemistry of Muscle Based Foods, pp. 62–76. Cambridge: The Royal Society of Chemistry.

Nash DM, Proudfoot FG, Hulan HW. 1985. Pink discoloration in broiler chicken. Poult Sci 64:917–919.

Ngoka DA, Froning GW. 1982. Effect of free struggle and preslaughter excitement on color of turkey breast muscles. Poult Sci 61:2291–2293.

Numata M, Wakamatsu J. 2003. Natural red pigments and foods and food materials containing the pigments. Patent WO 03/063615 A1.

Ogawa M, Perdiago NB, Santiago ME, Kozima TT. 1984. On physiological aspects of black spot appearance in shrimp. Bull Jap Soc Scien Fish 50(10): 1763–1769.

Ohshima T, Wada S, Koizumi C. 1988. Influences of haeme pigment, non-haeme iron, and nitrite on lipid oxidation in cooked mackerel meat. Bull Jap Soc Scien Fish 54(12): 2165–2171.

Oka H. 1989. Packaging for freshness and the prevention of discoloration of fish fillets. Pack Technol Sci 2(4): 201–213.

Osborn HM, Brown H, Adams JB, Ledward DA. 2003. High temperature reduction of metmyoglobin in aqueous muscle extracts. Meat Sci 65:631–637.

Pagán-Moreno MJ, Gago-Gago MA, Pérez-Alvarez JA, Sayas-Barberá ME, Rosmini MR, Perlo F, Aranda-Catalá V. 1998. The evolution of colour parameters during "chorizo" processing. Fleischwt 78(9): 987–989.

Park JW, Morrisey MT. 1994. The need for developing the surimi standar. In: G Sylvia, A Shriver, MT

Morrisey, editors, Quality Control and Quality Assurance Seafood, p. 265. Corvallis: Oregon Sea Grant.

Parolari G, Gabba L, Saccani G. 2003. Extraction properties and absorption spectra of dry cured hams made with and without nitrate. Meat Sci 64(4): 483–490.

Pérez-Alvarez JA. 1996. Contribución al estudio objetivo del color en productos cárnicos crudo-curados. [PhD Thesis]. Valencia, Spain: Universidad Politécnica de Valencia. 210 pp.

Pérez-Alvarez JA, Fernández-López J, Sayas Barberá ME. 2000. Fundamentos físicos, químicos, ultraestructurales y tecnológicos en el color de la carne. In: M Rosmini, JA Pérez-Alvarez, J Fernández-López, editors, Nuevas tendencias en la tecnología e higiene de la industria cárnica, pp. 51–71. Elche: Miguel Hernández de Elche.

Pérez-Alvarez JA, Fernández-López J, Sayas-Barberá ME, Cartagena R. 1998a. Utilización de vísceras como materias primas en la elaboración de productos cárnicos: Güeña. Alimentaria 291:63–70.

———. 1998b. Caracterización de los parámetros de color de diferentes materias primas usadas en la industria cárnica. Eurocarne 63:115–122.

Pérez-Alvarez JA, Sánchez-Rodriguez ME, Fernández-López J, Gago-Gago MA, Ruíz-Peluffo MC, Rosmini MR, Pagán-Moreno MJ, Lopez-Santoveña F, Aranda-Catalá V. 1997. Chemical and color characteristics of "Lomo embuchado" during salting seasoning. J Muscle Food 8(4): 395–411.

Pikul J, Niewiarowicz A, Kupijaj H. 1986. The cytochrome c content of various poultry meats. J Sci Food Agric 37:1236–1240.

Rehbein H, Orlick B. 1990. Comparison of the contribution of formaldehyde and lipid oxidation products to protein denaturation and texture deterioration during frozen storage of minced ice-fish fillet *(Champsocephalus gunnari* and *Pseudochaenichthys georgianus).* Int J Ref 13(5): 336–341.

Renerre M. 1990. Review: Factors involved in discoloration of beef meat. Int J Food Sci and Technol 25:613–630.

Richards MP, Hultin HO. 2002. Contributions of blood and blood components to lipid oxidation in fish muscle. J Agric Food Chem 50(3): 555–564.

Ruíz-Peluffo MC, Pérez-Alvarez JA, Aranda-Catalá V. 1994. Longaniza de pascua: Influencia del ascorbato e isoascorbato de sodio sobre las propiedades fisicoquímicas y de color. In: P Fito, J Serra, E Hernández, D Vidal, editors, Anales de Investigación del Master en Ciencia e Ingeniería de Alimentos. Valencia: Reproval. 6:675–688.

Sakiura T. 2001. Cultured fish carotenoid and polyphenol added feed for improving fish body color tone and fish meat brilliance. U.S. Patent 2001/0043982 A1.

Saksit C, Nieda H, Ogawa M, Tamiya T, Tsuchiya T. 1998. Changes in light scattering intensity of fish holo-, and apo- and reconstituted myoglobins under thermal denaturation. Fish Sci 64(5): 846–87.

Saksit C, Ogawa M, Tamiya T, Tsuchiya T. 1996. Studies on thermal denaturation of fish myoglobins using differential scanning calorimetry, circular dichroism, and tryptophan fluorescence. Fish Sci 62(6): 927–932.

Sánchez-Escalante A, Djenane D, Torrescano G, Beltran JA, Roncales P. 2003. Antioxidant action of borage, rosemary, oregano and ascorbic acid in beef patties packaged in modified atmospheres. J Food Sci 68:339–344.

Sayas ME. 1997. Contribuciones al proceso tecnológico de elaboración del jamón curado: aspectos físicos, fisicoquímicos y ultraestructurales en los procesos de curado tradicional y rápido. [PhD Thesis]. Valencia, Spain: Universidad Politécnica de Valencia. 175 pp.

Scriven F, Sporns P, Wolfe F. 1987. Investigations of nitrite and nitrate levels in paper materials used to package fresh meats. J Agric Food Chem 35:188–192.

Shahidi F, Matusalach-Brown JA. 1998. Carotenoid pigments in seafoods and aquaculture. Crit Rev Food Sci 38(1): 1–67.

Slattery SL, Williams DJ, Cusack A. 1995. A sulphite-free treatment inhibits blackspot formation in prawns. Food-Australia. 47(11): 509–514.

Stauton-West E, Todd WR, Mason HS, Van Bruggen JT. 1969. Bioquímica Médica, pp.616–623. México: Interamericana.

Suman SP, Faustman C, Lee S, Tang J, Sepe HA, Vasudevan P, Annamalai T, Manojkumar M, Marek P, DeCesare M, Venkitanarayanan KS. 2004. Effect of muscle source on premature browning in ground beef. Meat Sci 68(3): 457–461.

Swatland HJ. 1988. Carotene reflectance and the yellows of bovine adipose tissue measured with a portable fiber-optic spectrophotometer. J Sci Food Agric 46:195–200.

———. 1995. Reversible pH effect on pork paleness in a model system. J Food Sci 60(5): 988–991.

Tanaka M, Nishino H, Satomi K, Yokoyama M, Ishida Y. 1996. Gas-exchanged packaging of tuna fillets. Nip Sui Gak 62(5): 800–805.

Taoukis PS, Labuza TP, Lillemo JH, Lin SW. 1990. Inhibition of shrimp melanosis (black spot) by ficin. Lebens Wis Technol 23(1): 52–54.

Tze KC, Ching YP, Fang PN, Shann TJ. 2001. Effect of met-myoglobin reductase on the color stability of blue fin tuna during refrigerated storage. Fish Sci 67(4): 694–702.

Urbain MW. 1952. Oxygen is key to the color of meat. Nat Prov 127:140–141.

Verdoes JC, Krubasik P, Sandmann G, van Ooyen AJJ. 1999. Isolation and functional characterisation of a novel type of carotenoid biosynthetic gene from Xanthophyllomyces dendrorhous. Mol Gen Genet 262(3): 453–461.

Wakamatsu J, Nishimura T, Hattori A. 2004a. A Zn-porphyrin complex contributes to bright red colour in Parma ham. Meat Sci 67(1): 95–100.

Wakamatsu J, Okui J, Ikeda Y, Nishimura T, Hattori A. 2004b. Establishment of a model experiment system to elucidate the mechanism by which Zn-protoporphyrin IX is formed in nitrite-free dry-cured ham. Meat Sci 68(2): 313–317.

Walsh KA, Rose D. 1956. Factors affecting the oxidation of nitric oxide myoglobin. J Agric Food Chem 4(4): 352–355.

Warren KE, Hunt MC, Kropf DH, Hague MA. Waldner CL, Stroda SL, Kastner, CL. 1996. Chemical properties of ground beef patties exhibiting normal and premature brown internal cooked color. J Muscle Foods 7(3): 303–314.

Warris PD, Rhodes DN. 1977. Haemoglobin concentrations in beef. J Sci Food Agric 28:931–934.

Wen LC, Chau JC, Ochiai Y. 1996. Effects of washing media and storage condition on the color of milkfish meat paste. Fish Sci 62(6): 938–944.

Whitaker JR. 1972. Principles of enzymology for the food sciences, pp. 225–250. New York: Marcel Decker.

Williams HG, Davidson GW, Mamo JC. 2003. Heat-induced activation of polyphenoloxidase in wester rock lobster *(Panulinus cygnus)* hemolymph: Implications for heat processing. J Food Sci 68:1928–1932.

Xin F, Shun W. 1993. Enhancing the antioxidant effect of alpha-tocopherol with rosemary in inhibiting catalyzed oxidation caused by Fe^{2+} and hemoprotein. Food Res Int 26(6): 405–411.

Zanardi E, Novelli E, Nanni N, Ghiretti GP, Delbono G, Campani G, Dazzi G, Madarena G, Chizzolini R. 1998. Oxidative stability and dietary treatment with vitamin E, oleic acid and copper of fresh and cooked pork chops. Meat Sci 49(3): 309–320.

Zerby HN, Belk KE, Sofos JN, Mcdowell LR, Williams SN, Smith GC. 1999. Display life of fresh beef containing different levels of vitamin E and initial microbial contamination. J Muscle Foods 10: 345–355.

Zhou GH, Yang A, Tume RK. 1993. A relationship between bovine fat colour and fatty acid composition. Meat Sci 35(2): 205–12.

16
Biochemistry of Seafood Processing*

Y. H. Hui, N. Cross, H. G. Kristinsson, M. H. Lim, W. K. Nip, L. F. Siow, and P. S. Stanfield

*The information in this chapter has been derived from *Food Processing Manual,* copyrighted and published by Science Technology System, West Sacramento, California. 2004(c). Used with permission.

INTRODUCTION

Most of us like to eat seafood, especially when it is fresh, although processed products are also favorites of many consumers. In addition to sensory attributes, the preference for seafood now includes attention to its health benefits. Thus, when processing seafood, it is important to understand the scientific and technical reasons that are responsible for the sensory and health attributes of seafood, in addition to how the manufacturing process can affect the basic quality of seafood. The same knowledge may lead to a reduction in the perishability of seafood, a problem that has existed since the beginning of time, though it has been partially solved by many modern techniques including canning, freezing, and dehydration.

The nutritive value of seafood is well known for its content of protein and some minerals and, recently, its fat quality. For the past 20 years considerable attention has been paid to the content of polyunsaturated omega-3 fatty acids in fish, especially their prevalence in fatty fish. Of course, one must not forget the relatively high levels of cholesterol in some fish and shellfish, especially crustaceans and squid. The biochemical changes in macronutrients of seafood are of interest to nutritionists, food processors, and consumers. Their importanc in public health cannot be overestimated if one considers how seafood contributes to the nutrition and health of the human body. A short discussion of the nutritive values of fish serves as an introduction.

Perishability of seafood is important economically as well as for its safety. The price of seafood can drop drastically if it is not handled properly and is handled without refrigeration or other forms of preservation. The biochemical aspects of seafood will be studied using the following approaches:

- *Glycogen in seafood.* Biochemical changes in its glycogen, which exists in small quantities, initiates the biochemical process, in the same way that it occurs in other animal products.
- *Nitrogenous compounds in seafood.* The degradation of protein will be discussed, with a special reference to sarcoplasmic, myofibrillar, and stromal protein. This discussion is accompanied by a brief mention of nonprotein nitrogenous substances.
- *Lipids in seafood.* Fat can undergo many biochemical changes, both qualitatively and quantitatively. Some changes can cause rancidity, especially during the storage of fatty fish. Such biochemical changes in seafood's protein and nonprotein components usually reduce its economic value and may also create safety problems.
- *Pigments in seafood.* Biochemical changes will be discussed with reference to epithelial discoloration, hemoglobin, hemocyanin, myoglobin, carotenoids, and melaninosis.
- *Quality indices in seafood.* The monitoring of the quality of seafood has been the subject of intense scientific research. Currently, the techniques available are lactic acid formation with lowering of pH, nucleotide catabolism, degradation of myofibrillar proteins, collagen degradation, dimethylamine formation, free fatty acid accumulation, and tyrosine accumulation.
- *Processing methods for seafood.* In the last 25 years, great advances have taken place in the processing of seafood. The basic observation is simple. No matter what method is used or invented to process the products, their biochemistry is affected. Three types of processing techniques will be discussed in this chapter: freezing, drying, and heating.

When studying this chapter, the reader should refer to Chapter 1 of this book for more details on food biochemistry.

NUTRITIVE COMPOSITION OF THE MAJOR GROUPS OF SEAFOOD AND THEIR HEALTH ATTRIBUTES

As mentioned earlier, seafood is nutritious. It contains nutrients, some of which are essential to human health. The nutrients are divided into two groups: macronutrients and micronutrients. This first section covers the macronutrients (water, protein, lipids) and their changes during processing and preservation. This is followed by a discussion of the micronutrients (vitamins and minerals). The implications of undesirable metals in seafood are then discussed. Issues related to health attributes in seafood are presented last.

COMPOSITION

Macronutrients

The macronutrients found in seafood include protein, fats and oils, and water. All other nutrients found in seafood are considered micronutrients and are of minor significance. Fish contains 63–84% water, 14–24% protein, and 0.5–17% lipid by weight. Fish, like other muscle sources of protein, are devoid of carbohydrates. The amount of protein is similar in pelagic and demersel fish. However, pelagic fish, commonly called fatty fish, are higher in lipids (9–17 g/100 g) than demersal fish that live on the bottom of the ocean (0.3–1.6 g/100 g).

The primary source of fish protein is muscle, and the protein quality is comparable to other animal protein from milk, eggs, and beef. Muscle from fish and shellfish has very little connective tissue and is readily hydrolyzed upon heating, resulting in a product that is tender and easy to chew.

The forms of lipid in fish are triglycerides or triacylglycerols. Triglycerides in pelagic fish contain the long-chain polyunsaturated fatty acids 20:5ω3 eicosapentoic acid (EPA) and 22:6ω3, docosahexanoic acid (DHA), which have many health benefits including normal development of the brain and retina in infants and prevention of heart disease in adults.

Pelagic fish store lipids in the head and muscle, whereas lipids in demersal fish are stored primarily in the liver and peritoneal lining and under the skin, except in halibut, which has both liver and muscle stores. The lipid content of pelagic fish varies with the season, feeding ground, water salinity, and spawning. Fish do not feed during spawning but depend on lipid stores for energy. For example, the lipid content of salmon may be 13% when the salmon begins traveling upstream to spawning grounds in the spring and only 5% at the end of the spawning season in the fall. Herring show more seasonal variation in lipid content, with a peak level of 20% in the spring, a drop to 10–15% during the fall spawning season, and lows of 3–4% in the winter, because of colder sea temperatures and a scarcity of food. Water replaces the lipid depleted from muscle tissue and results in fish of poor eating quality.

Livers of species of the Gadidae (cod) family contain 50% lipids and are a concentrated source of vitamins A and D and omega-3 fatty acids. Oil is commercially extracted from the liver and purified for medicinal use and animal feed. Livers from other species of fish (haddock, coley, ling, shark, huss, halibut, and tuna) are also commercially processed to extract the fish oil.

In general, processing does not change the nutrient content of fish. Drying and smoking fish reduces the water content. The process can also reduce the lipid content of the fish if extremely high temperatures are used. Drying also increases the concentration of all nutrients on a weight basis. Sodium nitrite (usually with sodium nitrate) that is added during the process of drying or smoking increases the content of sodium considerably. The salmon carcass used for smoking is lower in lipids than whole salmon.

Breading fish adds carbohydrate and yields a final product with lower concentrations of protein, vitamins, and minerals than unbreaded fish.

Freezing and canning do not change the nutrient composition of fish except for canned tuna. Most canned tuna is low in lipids since the lighter colored tuna is lower in fat and is preferred by consumers over the deep brown-red colored tuna. Tuna is also cooked prior to canning, which removes some of the lipid.

Tables 16.1 and 16.2 compare the water, protein, and lipid content of different forms of fish.

The macronutrients in crustaceans and mollusks differ from those in bony fish since all mollusks contain carbohydrate, and lobster and scallops contain small amounts of carbohydrate. Carbohydrate in shellfish is in the form of glycogen. The protein content is more variable than in bony fish, and the lipid content is comparable to that in pelagic fish. See Table 16.3. Refer to Tables 16.4 and 16.5 for comparison of micronutrients. Ranges are used in Tables 16.1–16.5 because data from many sources are so variable.

Micronutrients

Micronutrients cover vitamins and minerals. Seafood is an important dietary source of vitamins A and D and the B vitamins B_6 and B_{12}. Although seafood is not high in thiamin or riboflavin, the amounts are comparable to those found in beef. Fish and shellfish have little or no vitamin C or folate.

Cod liver oil has been used to supplement diets with vitamins A and D; one teaspoon (5 g) of cod liver oil provides 12.5 μg of vitamin D and 1500 μg of vitamin A.

Mineral contributions of fish and shellfish include iodine and selenium; a 100 g portion contains more than 50% of the Dietary Reference Intake (DRI) for iodine (DRI= 150 μg/day) and selenium (DRI = 55 μg/day) respectively. Generally, meat, eggs, and dairy products are poor sources of iodine and selenium. Iron and zinc are found in moderate amounts in most seafood. Clams and oysters are concentrated sources of zinc, with 24 and 91 mg/100 g serving, respectively, compared with a DRI of 15 mg/day. Seafood, similar to other dietary protein sources is also an excellent source of potassium. Fresh seafood contains varying amounts of sodium, from a low of 31 mg/100 g of trout to a high of 856 mg/100 g of crab. Sodium is also added during canning and smoking fish and other seafood. In general, seafood is a poor source of calcium, except for whole canned salmon and sardines: when their bones are included, they will provide 200–400 mg calclium/100 g portion.

Table 16.1. Distribution of Water, Protein, and Lipids in 100 g of Raw Edible Fish including Bones and Cartilage

Fish	Water (g)	Protein (g)	Lipid (g)
Cod	79–81	17–19	0.6–0.8
Herring	78–80	18–20	0.5–0.7
Ling	79–80	17–19	0.6–0.8
Mackerel	63–65	17–19	15–17
Monkfish	82–84	14–16	0.3–0.5
Pilchard/sardine	66–68	19–21	8–10
Plaice	78–80	15–17	1.3–1.5
Saithe/coley	79–81	17–19	0.9–1.1
Salmon	66–68	19–21	10–12
Skate	79–81	14–16	0.3–0.5
Sole	80–82	16–18	1.4–1.6
Sprat	68–70	16–18	9–10
Trout (rainbow)	75–77	18–20	5–6
Tuna	69–71	22–24	4–5
Whiting	79–81	17–19	0.5–0.8

Table 16.2. Distribution of Water, Protein, and Lipid in 100 g of Cooked, Edible fish

Fish	Water (g)	Protein (g)	Lipid (g)
Haddock (smoked)	71–72	23–24	0.8–1.0
Haddock (steamed)	78–79	20–21	0.5–0.7
Herring (grilled)	63–64	20–21	11–12
Mackerel (smoked)	47–48	18–20	30–31
Mackerel (grilled)	58–59	20–21	17–18
Salmon (smoked)	64–65	25–26	4–5
Salmon (steamed)	64–65	21–22	11–12
Salmon (breaded nuggets)	60–61	12–13	11–12
Tuna (canned in water)	75–76	25–26	0.9–0.9
Tuna (canned in oil)	64–65	26–27	0.8–0.9
Tuna (raw, fresh, or frozen)	70–71	23–24	4–5

Table 16.3. Distribution of Water, Protein, and Lipid in 100 g of Raw or Cooked Nonfish Seafood

Edible Nonfish Seafood	Water (g)	Protein (g)	Lipid (g)	Carbohydrate(g)
Crustacean varieties				
Crab (steamed)	77–78	19–20	1.5–1.6	0
Crayfish (steamed)	79–80	16–17	1–2	0
Lobster (steamed)	66–67	26–27	1–2	3–4
Prawn (raw)	79–80	17–18	0.5–0.7	0
Scallops (steamed)	73–74	23–24	1–2	3–4
Shrimp (boiled)	62–63	23–24	2–3	trace
Mollusk varieties				
Clam (steamed)	63–64	25–26	1–2	5–6
Cuttlefish (steamed)	61–62	32–33	1–2	1–2
Mussel (steamed)	61–62	23–24	4–5	7–8
Octopus (steamed)	60–61	29–30	2–3	4–5
Oyster (steamed)	64–65	18–19	4–5	9–10
Whelk (steamed)	32–33	47–48	0.8–0.9	15–16

Table 16.4. Vitamin Content of Fish and Shellfish per 100 grams of Edible Fish or Shellfish (Raw Unless Otherwise Stated)

Seafood	Vitamins					
	A	D	E	B_1	B_2	B_6
	μg	μg	mg	mg	mg	mg
Fish						
Cod	1–3	0.5–1	0.2–0.5	0.02–0.05	0.03–0.06	0.15–0.20
Haddock	0.02–0.04	0.02–0.04	0.3–0.5	0.03–0.06	0.05–0.08	0.3–0.5
Herring	39–50	17–21	0.5–0.9	0.0–0.02	0.18–0.30	0.30–0.50
Mackerel	40–50	3–6	0.3–0.6	0.08–0.16	0.22–0.30	0.3–0.5
Salmon, Atlantic	11–15	6–10	1.7–2.0	0.21–0.30	0.09–0.15	0.6–0.9
Trout, Rainbow	45–55	8–11	0.5–0.9	0.1–0.3	0.05–0.15	0.3–0.5
Tuna	20–30	6–9	trace	0.05 0.15	0.1–0.2	0.2–0.5
Crustacean						
Crab	trace	trace	trace	0.05–0.10	0.7–0.9	0.05–0.20
Lobster (boiled)	trace	trace	1.0–1.6	0.05–0.10	0.5–1.0	0.01–0.10
Prawns	trace	trace	2–3	0.0–0.7	0.05–0.20	0.02–0.07
Shrimp (boiled)	trace	trace	trace	trace	trace	0.05–1.0
Mollusk						
Mussel	trace	trace	0.6–0.9	trace	0.2–0.4	0.00–1.0
Oyster	60–80	0.5–1.5	0.5–1.0	0.1–0.3	0.1–0.3	0.1–0.2

Table 16.5. Mineral Content of Fish and Shellfish per 100 grams of Edible Fish or Shellfish (Raw Unless Otherwise Stated)

Seafood	Minerals						
	Sodium	Potassium	Calcium	Iron	Zinc	Iodine	Selenium
	mg	mg	mg	mg	mg	μg	μg
Fish							
Cod	50–70	300–400	5–10	0.0–0.2	0.2–0.5	100–110	20–32
Haddock	50–70	350–400	10–18	0.0–0.2	0.2–0.5	200–250	25–30
Herring	119–130	300–350	50–70	0.9–1.5	0.5–1.3	25–35	30–40
Mackerel	60–70	270–320	9–13	0.6–1.0	0.4–0.8	120–160	25–35
Salmon, Atlantic	40–50	350–370	19–23	0.5–1.0	0.5–1.5	70–82	20–30
Trout (Rainbow)	40–50	400–450	15–20	0.1–0.5	0.3–0.8	10–16	15–20
Tuna	45–50	350–450	12–20	1.0–1.6	0.5–1.0	25–35	55–60
Crustacean							
Crab	400–450	230–280	trace	1.3–1.8	5,0–6,0	trace	15–20
Lobster (boiled)	300–350	250–280	60–70	0.5–1.0	2.0–3.0	95–105	120–140
Shrimp (boiled)	3500–3900	350–450	300–350	1.5–2.0	2.0–2.5	90–110	40–50
Prawn	150–220	300–400	75–80	1–2	1–2	1–3	10–20
Mollusk							
Mussel	250–310	300–350	30–42	5–7	2–3	120–160	45–55
Oyster	500–550	250–300	120–160	5–7	55–65	55–65	20–30

Other Components

Seafood, like other animal products, contains cholesterol. The cholesterol content in fish and mollusks ranges from 40 to 100 mg/100 g portion and is lower than in other meat such as beef, pork, and chicken. The shellfish, shrimp, and prawns, however, are quite high in cholesterol—195 mg /100 g portion of shrimp. The cholesterol content of shrimp is of interest to individuals with cardiovascular disease who are advised to limit dietary cholesterol to 300–400 mg/day. However, diets high in total fat and saturated fat also increase the risk for cardiovascular disease, and both shrimp and prawns contain negligible amounts of fat.

Health Attributes

In the history of mankind, fish has been considered to have special health attributes, especially in various ethnic populations. Recently, scientific research has confirmed certain claims and raised controversy in relation to others. Seafood is a rich source of omega-3 fatty acids and may be good for the heart. Fish is being recommended for people of all ages to prevent a variety of disorders. Fish is also being promoted for women during pregnancy, for children at risk for asthma, and in the elderly to reduce the risk of Alzheimer's disease. However, pregnant women and children have been advised to limit fish to prevent the risk of mercury toxicity. Natural habitats are becoming polluted with mercury that is taken up and stored in some varieties of fish. Fish farming is a rapidly growing enterprise and has the benefit of controlling the food and water supply to avoid possible contamination of fish with toxic materials.

Fish and shellfish are an important part of a healthy diet. Fish and shellfish contain high quality protein and other essential nutrients, are low in saturated fat, and contain two omega-3 fatty acids, eicosapentoic acid (EPA) and docosahexanoic acid (DHA). The American Heart Association recommends that healthy adults eat at least two servings of fish a week (12 ounces) especially varieties that are high in omega-3 fatty acids, such as mackerel, lake trout, herring, sardines, albacore tuna, and salmon.

Fish oil can help reduce deaths from heart disease, according to evidence in reports from the Agency for Healthcare Research and Quality (AHRQ, www.ahrq.gov/clinic/epcindex.htm#dietsup). A systematic review of the available literature found evidence that long-chain omega-3 fatty acids, the beneficial component ingested by eating fish or taking a fish oil supplement, reduce not only the risk for a heart attack and other problems related to heart and blood vessel disease in persons who already have these conditions, but also their overall risk of death. Although omega-3 fatty acids do not alter total cholesterol, HDL cholesterol, or LDL cholesterol, evidence suggests that they can reduce levels of triglycerides—a fat in the blood that may contribute to heart disease.

The AHRQ review reported evidence that fish oil can help lower high blood pressure slightly, may reduce the risk of coronary artery reblockage after angioplasty, may increase exercise capability among patients with clogged arteries, and may possibly reduce the risk of irregular heart beats—particularly in individuals with a recent heart attack. Alpha-linolenic acid (ALA), a type of omega-3 fatty acid from plants such as flaxseed, soybeans, and walnuts, may help reduce deaths from heart disease, but to a much lesser extent than fish oil.

Based on the evidence to date, it is not possible to conclude whether omega-3 fatty acids help improve respiratory outcomes in children and adults who have asthma. Omega-3 fatty acids appear to have mixed effects on people with inflammatory bowel disease, kidney disease, and osteoporosis, and no discernible effect on rheumatoid arthritis.

Depending on the stage of life, consumers need to be aware of both the benefits and risks of eating fish. Fish may contain mercury, which can harm an unborn baby or young child's developing nervous system. For most people, the risk from mercury by eating fish and shellfish is not a health concern. However, the Food and Drug Administration (FDA) and the Environmental Protection Agency (EPA) are advising women who may become pregnant, pregnant women, nursing mothers, and young children to avoid some types of fish known to be high in mercury content.

The risks from mercury in fish and shellfish depend on the amount of fish and shellfish eaten and the levels of mercury in the fish and shellfish. Nearly all fish and shellfish contain traces of mercury, but two varieties of fish high in omega-3 fatty acids (mackerel and albacore tuna) may contain high amounts of mercury. Larger fish that have lived longer have the highest levels of mercury because they have had more time to accumulate it. These large fish (swordfish, shark, king mackerel, and tilefish) pose the greatest risk. Other types of fish and shellfish may be eaten in the amounts recommended by the FDA and EPA. (www.FDA.gov)

By following these recommendations for selecting and eating fish or shellfish, women and young children will receive the benefits of eating fish and shellfish and be confident that they have reduced their exposure to the harmful effects of mercury.

- Avoid eating fish that contain high levels of mercury: shark, swordfish, king mackerel, or tilefish.
- Eat up to 12 ounces (two average meals) a week of a variety of fish lower in mercury: shrimp, canned light tuna, salmon, pollock, and catfish. However, limit intakes of albacore ("white") tuna to 6 ounces (one average meal) per week because albacore has more mercury than canned light tuna.
- Check local advisories about the safety of fish caught by family and friends in your local lakes, rivers, and coastal areas. Some kinds of fish and shellfish caught in local waters may have higher or much lower than average levels of mercury. This depends on the levels of mercury in the water in which the fish are caught. Those fish with much lower levels may be eaten more frequently and in larger amounts. If no advice is available, eat up to 6 ounces (one average meal) per week of fish you catch from local waters, but don't consume any other fish during that week.

For young children, these same recommendations can be followed, except that a serving size for children is smaller than for adults, 2–3 ounces for children instead of 6 ounces.

BIOCHEMISTRY OF GLYCOGEN DEGRADATION

When the fish or crustacean animals are being caught, they struggle vigorously in the fishing gear and on board, causing antemortem exhaustion of energy reserves, mainly glycogen, and high-energy phosphates. Asphyxia phenomena set in, with gradual formation of anoxial conditions in the muscle. Tissue enzymes continue to metabolize energy reserves. Degradation of high-energy phosphates eventually produces hypoxanthine, followed by formation of formaldehyde, ammonia, inorganic phosphate, and ribose phosphates. The degradation of glycogen in fish follows the Embden-Meyerhof-Parnas pathway via the amylolytic route, catalyzed by endogenous enzymes. This results in an accumulation of lactic acid and a reduction of pH (~7.2 to ~5.5), contracting the tissues and inducing rigor mortis (Hobbs 1982, Hultin 1992a, Sikorski et al. 1990a). The rate of glycolysis is temperature dependent and is slowed by a lower storage temperature.

BIOCHEMISTRY OF PROTEIN DEGRADATION

Approximately 11–27% of seafood (fish, crustaceans, and mollusks) consists of crude proteins. The types of seafood proteins, similar to other muscle foods, may be classified as sarcoplasmic, myofibrillar, and stromal. The sarcoplasmic proteins, mainly albumins, account for approximately 30% of the total muscle proteins. A large proportion of sarcoplasmic proteins are composed of hemoproteins. The myofibrillar proteins are myosin, actin, actomyosin, and troponin; these account for 40–60% of the total crude protein in fish. The rest of the muscle proteins, classified as stromal, consist mainly of collagenous material (Shahidi 1994).

Sarcoplasmic Proteins

Sarcoplasmic proteins are soluble proteins in the muscle sarcoplasm. They include a large number of proteins such as myoglobin, enzymes, and other albumins. The enzymatic degradation of myoglobin is discussed in the section on pigment degradation. Sarcoplasmic enzymes are responsible for quality deterioration of fish after death and before bacterial spoilage. The significant enzyme groups are hydrolases, oxidoreductases, and transferases. Other sarcoplasmic proteins are the heme pigments, parvalbumins, and antifreeze proteins (Haard et al. 1994). Glycolytic deterioration in seafood was discussed in Chapter 1 in this book and will not be repeated here. Changes in the heme pigments will be covered below in the section on Biochemical Changes in Pigments during Handling, Storage, and Processing in this chapter. Hydrolytic deterioration of seafood myofibrillar and collagenous proteins are discussed below.

Myofibrillar Protein Deterioration

The most common myofibrillar proteins in the muscles of aquatic animals are myosin, actin, tropomyosin, and troponins C, I, and T (Suzuki 1981). Myofibrillar proteins undergo changes during rigor mortis, resolution of rigor mortis, and long-term frozen storage. The integrity of the myofibrillar protein molecules and the texture of fish products are affected by these changes. These changes have been demonstrated in various research reports and reviews (Haard 1992a,b, 1994; Jiang 2000; Kye et al. 1988; Martinez 1992; Sikorski 1994 a,b; Sikorski and Pan 1994; Sikorski et al. 1990a). The degradation of myofibrillar proteins in seafood causes these proteins to lose their integrity and gelation power in ice-stored seafood. The cooked seafood will no longer possess the characteristic firm texture of very fresh seafood; it will show a mushy or soft (sometimes mislabeled as tender) mouthfeel. For frozen seafoods, these degradations are accompanied by a loss in the functional characteristics of muscular proteins, mainly solubility, water retention, gelling ability, and lipid emulsifying properties. This situation gets even worse when the proteins are cross-linked due to the presence of formaldehyde formed from trimethylamine degradation. The cooked products become tough, chewy, and stringy or fibrous. Repeated freezing and thawing make the situation even worse. Readers should refer to the review by Sikorski and Kolakowska (1994) for a detailed discussion of the topic.

Stromal Protein Deterioration

The residue remaining after extraction of sarcoplasmic and myofibrillar proteins is known as stromal protein. It is composed of collagen and elastin from connective tissues (Sikorski and Borderias 1994). Degradation causes textural changes in these seafoods (honeycombing in skipjack tuna and mackerel and mushiness in freshwater prawn) (Frank et al. 1984, Nip et al. 1985, Pan et al. 1986). Bremner (1992) reviewed the role of collagen in fish flesh structure, postmortem aspects, and the implications for fish processing, using electron microscopic illustrations. Jiang (2000) reviewed the proteinases involved in the textural changes of postmortem muscle and surimi-based products. Microstructural changes in ice-stored freshwater prawns have been revealed (Nip and Moy 1988). These textural changes are due to the degradation of collagenous matter and definitely influence the quality of seafood. For example, in freshwater prawn, the development of mushy texture downgrades its quality. In tuna, the development of honeycombing is an undesirable defect. In the postcooking examination of tuna before filling into the cans, the appearance of honeycombing is a sign of mishandling of the raw material. In more extensive cases, it may even cause the rejection of the precooked fish for further processing into canned tuna.

BIOCHEMICAL CHANGES IN NONPROTEIN NITROGENOUS COMPOUNDS

Reviews on the postmortem degradation of nonprotein nitrogenous (NPN) compounds are available (Haard et al. 1994, Sikorski et al. 1990a, Sikorski and Pan 1994). Such compounds in the meat of marine animals vary among species, with the habitat, and with life cycle; more importantly, they play a role in the postmortem handling processes (Sikorski 1994a, Sikorski et al. 1990a). Bykowski and Kolodziejski (1983) reported that the white meat generally contains less NPN compounds than the dark meat. For example, in the meat of white fish, the NPN generally made up 9–15% of the total nitrogen, in clupeids 16–18%, in muscles of mollusks and crustaceans 20–50%, and in some shark up to 55%. Ikeda (1979) showed that about 95% of the total amount of NPN in the muscle of marine fish and shellfish is composed of free amino acids, imidazole dipeptides, trimethylamine oxide (TAMO) and its degradation products, urea, guanidine compounds, nucleotides and the products of their postmortem changes, and betaines.

The content of free amino acids in the body of oysters is higher in the winter than it is in the summer (Sakaguchi and Murata 1989).

The endogenous enzymatic breakdown of TMAO to dimethylamine (DMA) and then formaldehyde, and the bacterial reduction of TMAO to TMA have been most extensively studied. It should be noted that the production of DMA and formaldehyde takes place mainly in anaerobic conditions (Lundstrom et al. 1982).

Shark muscles contain fairly high amounts of urea, and ammonia may accumulate due to the activity of endogenous urease after death. This problem renders shark meat with ammonia odor unacceptable. Fortunately, this problem can be overcome by complete bleeding of the shark at the tail, gutting, filleting, and thorough washing right after catch to reduce the amount of substrate (urea). This procedure has been successfully practiced to provide acceptable shark meat for consumers.

After death, the fish muscle also produces large amounts of ammonia due to degradation of adenosine triphosphate to adenosine monophosphate (AMP), followed by deamination of AMP.

It should be noted that both ammonia production and degradation of TMAO can be either endogenous or contributed by bacteria. Ammonia, TMA, small amounts of DMA, and methyamine constitute the "total volatile base," an indicator of freshness commonly used for seafood.

Postmortem enzymatic breakdown of nucleotides in fish may have a positive or negative impact on the flavor of seafood. The production of inosine monophosphate (IMP) at a certain concentrations in dried fish product can enhance the flavor (Murata and Sakaguchi 1989). Hypoxanthine contributes to bitterness and may add undesirable notes to the product (Lindsay 1991).

LIPIDS IN SEAFOODS

Lipid Composition

Lipids play an important nutritional role in seafoods, but at the same time they contribute greatly to quality changes in many species. Just as aquatic species vary widely biologically, the lipid content and fatty acid types are greatly variable between species and sometimes within the same species. This can create a challenge for the seafood harvester and processor since varying amounts of lipids and the presence (or absence) of certain fatty acids can lead to significantly different effects on quality and shelf life. Seafood can be roughly classified into four categories based on fat content (O'Keefe 2000). Lean species (e.g., cod, halibut, Pollock, snapper, shrimp, and scallops) have fat contents below 2%; low fat species (e.g., yellowfin tuna, Atlantic sturgeon, and smelt) have fat contents between 2 and 4%; medium fat species (e.g., catfish, mullet, trout, and salmon) are classified as having 4–8% total fat; and fatty species (e.g., herring, mackerel, eel, and sablefish) are classified as having more than 8% total fat. This classification is, however, not valid for many species, since they can fluctuate widely in composition between seasons. For example, Atlantic mackerel can have less than 4% fat

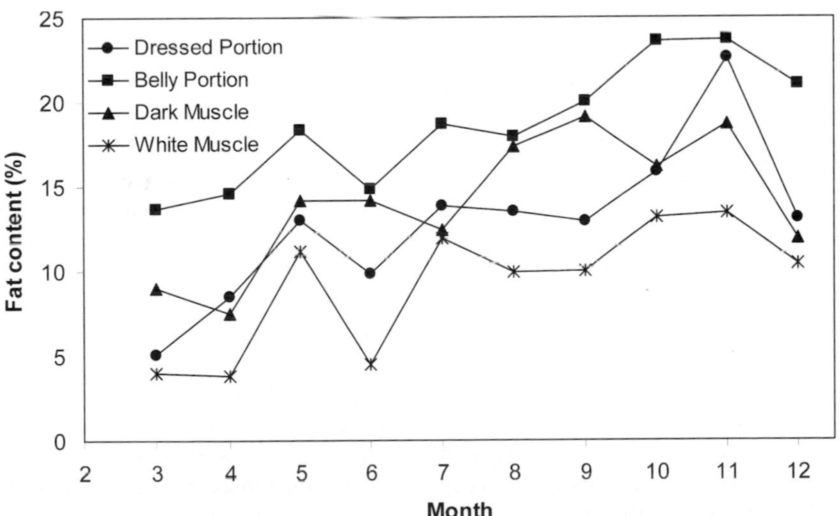

Figure 16.1. Seasonal variation in the fat content in different parts of Atlantic mackerel. (Adapted from Leu et al. 1981.)

during late spring shortly after spawning, while it can have about 24% fat during late November due to active feeding (Leu et al. 1981; Fig. 16.1).

Different parts of the aquatic animal will also have different fat contents. Ohshima et al. (1993a) reported the following descending order for the fat content of mackerel and sardines:

- Skin (including subcutaneous lipids),
- Viscera,
- Dark muscle, and
- White muscle.

Some species, most notably salmonids, have larger fat deposits in the belly flaps than in other parts of the fish. Distribution of fat in many fish seems to be in descending fat levels from the head to the tail (Icekson et al. 1998, Kolstad et al. 2004). Diet will have a significant impact on the fat content and fatty acid profiles of aquatic animals. Aquacultured fish on a controlled diet, for example, do not show the same seasonal variation in fat content as wild species. It is also possible to modify the fat content and fat type via diet. For example, adding fish oil to fish feed will increase the fat content of fish (Lovell and Mohammed 1988), usually proportionately to the level of fat/oil in the diet (Solberg 2004). It has also been shown with many species (salmon being the most investigated) that the fatty acid composition is a reflection of the fatty acid composition in the fish diet (Jobling 2004). It is therefore possible to selectively increase the nutritional value of aquacultured fish by increasing the level and improving the ratio of nutritionally beneficial fatty acids (Lovell and Mohammed 1988). Conversely, one can decrease the level of unstable fatty acids to increase the shelf life of the product. A good example is catfish (Table 16.6), which in the wild has more unstable (but more nutritionally beneficial) fatty acids than grain-fed aquacultured catfish (St. Angelo 1996).

Although varying levels of total fat content in fish can influence quality, fatty acid type rather than quantity appears to be more important. The major classes of lipids in fish are triglycerides and phospholipids. Triglycerides, the majority of lipids in fish (except very lean fish), are neutral lipids. They exist as either large droplets within the adipose tissue or smaller droplets between or within muscle cells (Undeland 1997). These lipids are deposited and stored as an energy source for fish (Huss 1994). The phospholipids, which are polar lipids, are one of the main building blocks of muscle cell membranes and are usually present only at low levels, 0.5–1.1% (O'Keefe 2000). Some muscles of very lean species such as cod, pollock, and whiting can contain less than 1% lipids, most of which are in the form of membrane phospholipids.

Table 16.6. Fatty Acid Composition of Extracted Lipids from Wild and Pond—Reared Channel Catfish

Fatty Acids	Wild (%)	Pond–Reared (%)
16:0	20.63	19.62
16:1ω7	7.48	4.15
18:0	8.04	6.67
18:1ω	23.72	42.45
18:2ω6	2.9	12.35
18:3ω3	2.78	1.66
18:4ω3	0.6	0.96
20:1ω9	0.83	0.97
20:3ω6	0.36	0.98
20:4ω6	6.82	1.88
20:4ω3	0.62	0.16
20:5ω3	7.02	1.49
24:1ω9	0.34	0.16
22:4ω6	0.68	0.14
22:5ω6	1.34	0.51
22:5ω3	3.05	0.98
22:6ω6	13.56	4.97
Saturated	28.67	26.28
Monounsaturated	31.64	47.64
Polyunsaturated	39.77	26.07
ω-6 PUFA	12.13	15.85
ω-3 PUFA	27.64	10.22
ω3/ω6	2.54	0.62

Source: Adapted from Chanmugam et al. 1986.

Compared to other animals, aquatic animals are especially rich in long-chain polyunsaturated fatty acids (PUFAs), which are found in higher numbers in the membrane phospholipids than in storage triglycerides (Hultin and Kelleher 2000). Aquatic animals adapted to cold water also have higher levels of PUFAs than those adapted to warmer waters, since more unsaturation is essential to maintain membrane fluidity and function at low temperatures. It has been reported that about half of the membrane phospholipids in cold-water fish are the omega-3 fatty acids eicosapentoic acid (20:5ω3) and docosahexanoic acid (22:6ω3) (Shewfelt 1981). These fatty acids have been found to be highly active physiologically, for example, leading to decreased plasma triacylglycreols, cholesterol, and blood pressure (Wijendran and Hayes 2004) and modulating immunological activities (e.g., inflammation) (Klurfeld 2002), to name a few important functions.

Lipids and Quality Problems

Even though the phospholipids are at very low levels compared to neutral fats, they are believed to lead to more quality problems such as lipid oxidation, which is a major cause of quality deterioration in many aquatic foods. This is in part because of their higher level of unsaturation, but also is due to their significantly larger surface area compared with neutral lipids (Hultin and Kelleher 2000). Undeland et al. (2002) tested this hypothesis in a model system with cod membrane lipids and varying levels of neutral menhaden lipids and hemoglobin as a prooxidant. The oxidation was found to be independent of the amount of neutral lipids in the system. Their larger surface area also makes the unstable phospholipids more exposed to a variety of prooxidants (and antioxidants) such as heme proteins, radicals, iron, and copper. These prooxidants are especially rich in the metabolically more active dark muscle, which is in turn more susceptible to lipid oxidation than white muscle. These prooxidants, especially the heme proteins hemoglobin and myoglobin, can lead to high levels of lipid oxidation products, especially as pH of the muscle is reduced (which occurs postmortem) or when muscle is minced and these components are mixed in with lipids and oxygen (Hultin 1994, Undeland et al. 2004). The secondary products formed as a result of lipid oxidation have highly unpleasant off-odors and flavors, and they can also adversely affect the texture and color of the muscle. Several studies have suggested a link between lipid oxidation and protein oxidation (e.g., Srinivasan et al. 1996, Tironi et al. 2002, King and Li 1999). Oxidation of proteins can lead to protein cross-linking and thus textural defects, including muscle toughening and loss of water-holding capacity. King and Li (1999) proposed that lipid oxidation products may also induce protein denaturation, which in turn could cause textural problems.

Although oxidation is often delayed during frozen storage of fish muscle, thawed fish can oxidize more rapidly than fresh fish since muscle cells are disrupted, leading to an increase in free iron and copper, among other changes (Hultin 1994, Benjakul and Bauer 2001). Much frozen seafood has also been found to accumulate high levels of free fatty acids over time as a result of lipase and phospholipase activity (Undeland 1997, Reddy et al. 1992). These free fatty acids may or may not be more susceptible to lipid oxidation (Shewfelt 1981), but they can lead to protein denaturation and cross-linking and thus adverse effects on texture and water-holding capacity (Reddy et al. 1992).

Conflicting results have been reported on the effect of thermal processing on lipids. Canning has been reported to increase oxidation (Auborg 2001). Short-term heating to 80°C, in contrast to long-term heating, has been reported to reduce oxidation due to inactivation of lipoxygenases, while long-term heating may accelerate nonenzymatic oxidation reactions (Wang et al. 1991). Shahidi and Spurvey (1996) reported that cooked dark muscle of mackerel oxidized more than raw dark muscle, while interestingly, the opposite was found for white muscle. In a study where herring was cooked by various methods (microwave, boiling, grilling, and frying), there was essentially no effect on lipid oxidation (Regulska-Ilow and Ilow 2002). That same study also demonstrated that fish lipids, including the nutritionally important omega-3 fatty acids, were highly stable under the different cooking methods (Fig. 16.2). Very high temperatures (e.g., 550°C) can, however, lead to decomposition of fish fatty acids and thus to decreased nutritional value (Sathivel et al. 2003). Changes in seafood fatty acid profiles can occur with frying. It has been reported that there may be an exchange of fatty acids from the frying medium and the muscle (Sebredio et al. 1993). This can either lower or increase the nutritional quality and stability of the seafood, depending on the type of frying oil and the type of seafood.

Novel processing methods such as irradiation and high-pressure processing have been reported to lead to increased levels of lipid oxidation. High-pressure treatment has been reported to have relatively small effects on purified lipids from fish, while high-pressure treatment of fish muscle leads to high levels of lipid oxidation products (Ohshima et al. 1993a). This increase in oxidation is hypothesized, in part, to be via pressure-induced denaturation of heme proteins, which are potent prooxidants (Yagiz, Kristinsson,

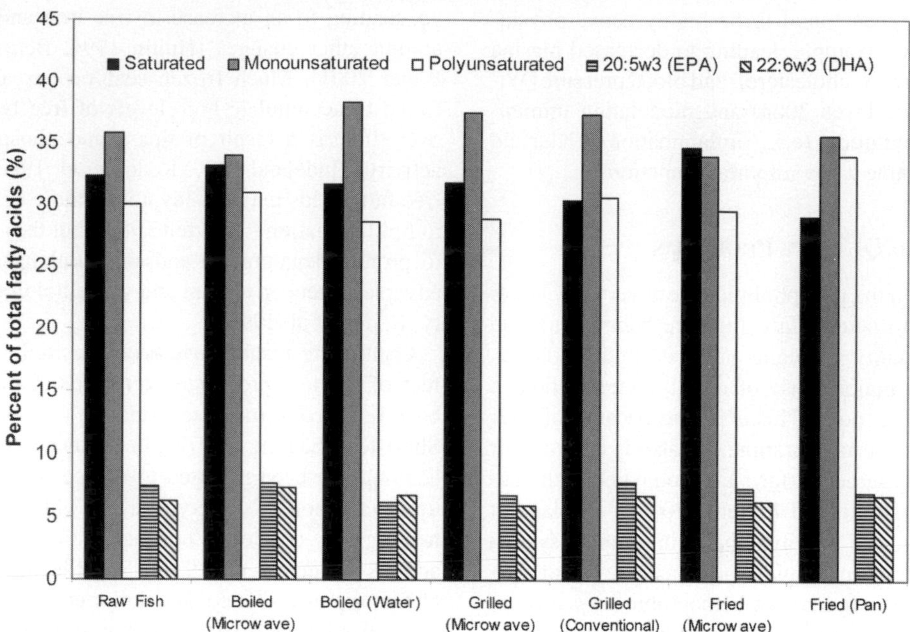

Figure 16.2. Effect of different thermal treatments on fatty acids in herring. (Adapted from Regulska-Ilow and Ilow 2002.)

Marshall, and Balaban, unpublished data). Increased pressure and time also appears to lead to increased oxidation. Lipid hydrolysis has been found to be increased by medium levels of pressure, while it is reduced at high pressures, likely due to lipase inactivation (Ohshima et al. 1993b). Several studies have been conducted on the effect irradiation has on seafood quality, although this process is not allowed in many countries. It has been observed that low doses of gamma irradiation (<5 kGy) may have an accelerating effect on the oxidative stability of some fish species and not others during post-irradiation refrigerated storage. Al-Kahtani et al. (1996) reported that irradiation (1.5–10 kGy) led to more oxidation in tilapia and Spanish mackerel during refrigerated storage than in untreated fish. Fatty acid changes were also observed in that study. The likely cause for increased lipid oxidation on irradiation is the formation of radicals, more at higher levels of irradiation.

Minimization of Lipid-Derived Quality Problems

To maintain or extend the sensory and nutritional qualities of seafood, it is important to minimize or delay the undesirable reactions of lipids, such as oxidation and hydrolysis. Time is of the essence with these reactions, and thus interventions should be done as early as possible to be able to extend quality as much as possible. The simplest means of controlling oxidation is maintaining low temperatures, since enzymatic and nonenzymatic oxidation reactions are greatly influenced by temperature (Hultin 1994). As mentioned before, very low temperatures such as freezing will accelerate lipid hydrolysis. Hydrolysis as well as oxidation will, however, be reduced if products are kept at extremely low frozen storage temperatures (e.g., below −40°C) compared with conventional frozen storage (about −20°C or higher).

At the harvest level, bleeding fish can lead to significantly lower levels of oxidation since heme proteins are reduced (Richards et al. 1998). Special care should be taken to prevent tissue disruption in storage since both oxidation and hydrolysis will be increased. Washing fish fillets and fish mince will also remove significant amounts of heme proteins (Richards et al. 1998, Kelleher et al. 1992). Another effective way to reduce oxidation of unstable species is the removal of dark muscle, or deep skinning. This removes not only a large fraction of the heme proteins, but also a large amount of oxidatively unstable lipids present in the dark muscle itself and below the skin. Protection from oxygen is another effective means for reducing oxidation. This can be achieved through modified atmosphere packaging, where oxygen is either reduced (for lean species) or completely removed (for fatty species). Vacuum packaging of seafood is also highly effective for reducing oxidation (Flick et al. 1992). Filleting fish under water (where O_2 is low) has also been reported to lead to less oxidation on storage than filleting in air (Richards et al. 1998). Special gases such as carbon monoxide have been found to significantly reduce oxidation of several species, even after fillets are removed from the gas, most likely since the gas reduces the prooxidative activities of heme proteins (Kristinsson et al. 2003).

Various antioxidative components can be highly effective in delaying lipid oxidation. All aquatic animals have a number of different indigenous antioxidants, and it is important to avoid conditions that can negatively effect or destroy these (Hultin 1994). It has been well researched and is well known that antioxidants can be added to seafood during processing to increase their oxidative stability (Fig. 16.3). Addition of antioxidants early during processing is also important. Since many of the prooxidants in seafood are in the aqueous phase and the lipids constitute a nonpolar phase, a combination of polar and nonpolar antioxidants has been found to be very effective. Tocopherol (vitamin E), a nonpolar antioxidant, and ascorbate (vitamin C), a polar antioxidant, are among the most popular added antioxidants. Sometimes metal chelators such as EDTA are added. It is worth mentioning, however, that under certain conditions antioxidants can act as prooxidants. For example, this is true for ascorbic acid and EDTA, both of which can stimulate iron-mediated lipid oxidation. For example, when the ratio of EDTA: iron is < 1, it is prooxidative, but when the ratio is > 1, it is antioxidative (Halliwell and Gutteridge 1998). Since ascorbate is a reducing agent, it can also reduce iron under certain conditions and thus make it more prooxidative (Halliwell and Gutteridge 1988).

BIOCHEMICAL CHANGES IN PIGMENTS DURING HANDLING, STORAGE, AND PROCESSING

The main pigments in seafood can be classified as heme proteins (hemoglobin and myoglobin) in redmeat (warm-blooded) fish, hemocyanin in coldblooded shellfish such as crustaceans and mollusks,

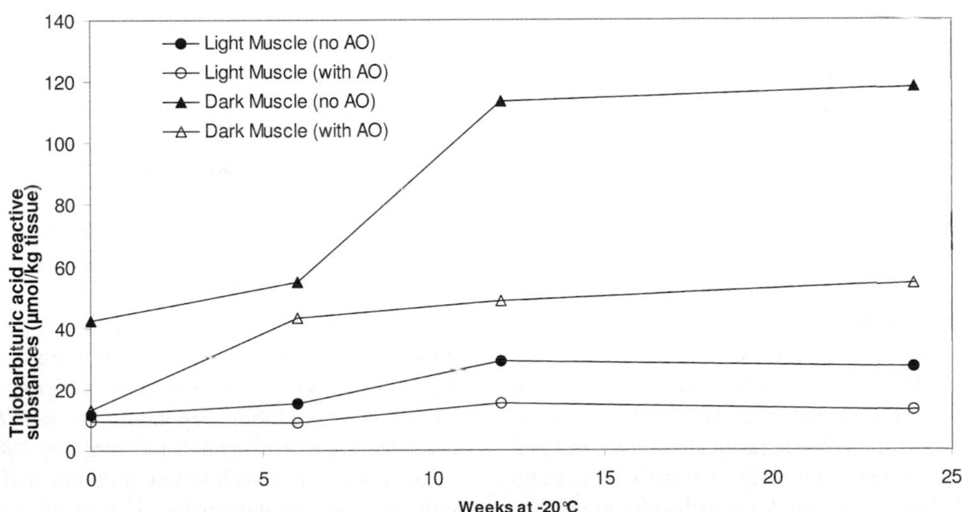

Figure 16.3. The effect of absence or presence of antioxidants on the oxidative stability of washed dark and white mackerel muscle during frozen storage (−20°C). The antioxidant TBHQ was added during grinding while ascorbate and EDTA were added during grinding and washing. (Adapted from Kelleher et al. 1992.)

and carotenoids in some important fish and shellfish products. The changes during handling, storage, and processing greatly affect the quality of these seafoods. Epithelial pigments and carotenoids also undergo changes during postmortem ice-chilled storage.

Epithelial Discoloration

The market value of squid is related to the contraction state of its epidermal chromatophores, called ommochromes. In recently harvested, prerigor squid, the pigment is dispersed throughout the chromatophores: hence the dark red-brown appearance of the epithelial tissue. Following rigor mortis, the pigment cells contract, giving the skin a pale, light coloration dotted with dark flecks. The continued storage of squid results in a structural deterioration of the ommochrome membrane, leading to bleeding of pigment and downgrading of quality. The dark brown coloration characteristic of very fresh squid can be retained during processing, for example, by freezing or dehydration. However, improper storage of frozen or dried squid will result in chromatophore disruption and red discoloration of the meat (Hink and Stanley 1985).

The frozen storage of some fish may result in subcutaneous yellowing of flesh below the pigmented skin (Thompson and Thompson 1972). Apparently, freezing or other processes that disrupt chromatophores can lead to the release of carotenoids and their migration to the subcutaneous fat layer. Subcutaneous yellowing that occurs during the prolonged storage of frozen fish can originate from other causes, that is, yellowing associated with lipid oxidation and carbonyl-amine reactions.

Hemoglobin

Normally, hemoglobin contributes less to the appearance of seafood than myoglobin because it is lost easily during handling and storage, while myoglobin is retained in the intracellular structure. The amount of hemoglobin present also affects to a greater extent the color appearance of light red and dark red muscles of red-meat fish flesh (Wang and Amiro 1979). For example, in yellowfin tuna *Neothunnus macropterus*, hemoglobin concentrations ranged from 12 to 50 mg% in light-red muscle and 50 to 380 mg% in dark-red muscle (Livingston and Brown 1981). The amount of residual hemoglobin in fish muscle is obviously influenced by the bleeding efficacy at the time of catch. Method of catch also affects the residual hemoglobin in the fish muscle. For example, the percent of total heme as hemoglobin in the meat of yellowfin caught by bait boat and purse seine was 24 and 32%, respectively (Barrett et al. 1965).

The green meat of raw or frozen broadbill swordfish *(Xiphias gladus)* is believed to be due to the combination of hemoglobin with hydrogen sulfide generated from the fairly extensive decomposition of the meat (Amano and Tomiya 1953).

Hemocyanin

Hemocyanin, not hemoglobin, is present in the blood of shellfish, that is, crustaceans and mollusks. Hemocyanins are copper-containing proteins, as compared with iron-containing proteins in hemoglobins, and they combine reversely with oxygen. The contribution of hemocyanins to seafood quality is not very well understood. It is suspected that the blue discoloration of canned crabmeat is associated with a high content of hemocyanin. The average copper content of blue meat (e.g., 2.8mg%) is higher than meat of normal color (e.g., 0.5mg%) (Ghiretti 1956).

Myoglobin

Myoglobin in fish muscle is retained in the intracellular structure. In fish muscle, the red, white, and intermediate fibers tend to be more distinctively segregated than they are in muscle from land animals. The myoglobin content in muscle of yellowfin tuna was found to range from 37 to 128 mg% in the light-colored muscle and from 530 to 22,400 mg% in the dark-colored muscle (Wolfe et al. 1978). In cod, the deep-seated, dark-colored muscle is richer in myoglobin than superficial dark-colored muscle (Brown 1962, Love et al. 1977).

Myoglobin in fish is easily oxidized to a brown-colored metmyoglobin. The discoloration of tuna during frozen storage is associated with the formation of metmyoglobin, depending on temperature and location (Tichivangana and Morrissey 1985).

Greening is a discoloration problem associated with cooking various tunas. The problem arises from the formation of a sulfhydryl adduct of myoglobin in the presence of an oxidizing agent (e.g., trimethylamine oxide, TMAO). This greening problem can be effectively prevented by the use of a reducing agent (e.g., sodium hydrosulfite or another legally permitted additive) (Tomlinson 1966, Koizumi and Matsura 1967, Grosjean et al. 1969).

Carotenoids

The carotenoids contribute to the attractive yellow, orange, and red color of several important fish and shellfish products. The more expensive seafoods such as lobster, shrimp, salmon, and red snapper have orange-red integument and/or flesh from carotenoid pigments. For example, the red color of salmon is directly related to the price of the product. Carotenoids in seafoods may be easily oxidized by a lipogenase-like enzyme (Schwimmer 1981).

Melaninosis (Melanin Formation)

Melaninosis (melanin formation) or "blackspot" in shrimp during postmortem storage is caused by phenol oxidase and has economic implications (Simpson et al. 1987). Consumers consider shrimp with blackspot to be defective. This problem can be overcome by the use of appropriate reducing or inhibiting agent(s).

A comprehensive review of the biochemistry of color and color change in seafood was presented by Haard (1992a). Readers should refer to this reference for detailed information on this subject.

BIOCHEMICAL INDICES

Postharvest biochemical events in fish can be classified into two phases: metabolic (enzymatic) and microbial (Eskin et al. 1971). A discussion of microbial events is not the objective of this chapter. If interested, please consult references in this chapter or other sources. Abundant literature is available. Physical and instrumental methods for assessing seafood quality were reviewed by Sorenson (1992).

The metabolic (enzymatic) changes result from the activity of enzymes remaining in the fish flesh after death. Metabolites from these enzymatic changes can be used as indices of freshness and can be monitored by biochemical or chemical methods. The major metabolites coming from the actions of inherent enzymes in the fish themselves are lactic acid, nucleotide catabolites, collagen and myofibrillar protein degradation products, dimethylamine formation, free fatty acid accumulation, and tyrosine. Methodology for the analysis of these metabolites can be found in government publications such as Official Methods of Analysis of the American Association of Official Chemists in the United States.

Lactic Acid Formation with Lowering of pH

It is well known that, in animals, lactic acid accumulates during postmortem changes because of glycolytic conversion of storage glycogen in the muscle after cessation of respiration, and finfish, crustaceans, and shellfish are no exception. The metabolic pathway of glycogen degradation in animals was presented in Chapter 1, Food Biochemistry—An Introduction. The accumulation of lactic acid can cause a drop in pH. Both lactic acid and pH can be measured without difficulty using modern instrumentation (Jacober and Rand 1982).

Nucleotide Catabolism

Postmortem dephosphorylation of nucleotides by autolysis in fish has been studied for many years as an index of quality. Nucleotide degradation commences with death and proceeds at a temperature-dependent rate (Spinelli et al. 1964, Eskin et al. 1971). Adenosine nucleotides are rapidly deaminated to inosine monophosphate (IMP) and further degraded from inosine to hypoxanthine during storage (Jones et al. 1964). There is no hypoxanthine in freshly caught fish and marine invertebrates. Accumulation of the metabolite hypoxanthine has attracted considerable attention for many years as an index of fish freshness. Hypoxanthine content can be determined by applying the colorimetric xanthine oxidase (EC 1.2.3.2) test. The rate of postmortem degradation of adenosine nucleotides in marine fish and invertebrates differs with the species and muscle type. For this reason, hypoxanthine analysis is of limited use for the evaluation of quality in certain species. It is most useful for the analysis of pelagics, redfish, salmon, and squid but is of little value for the estimation of lean fish quality (Gill 1992).

Because hypoxanthine is a metabolite fairly close to the end of the nucleotide degradation, Saito et al. (1959) proposed the use of K-value, defined as the ratio of inosine (HxR) plus hypoxanthine (Hx) to the total amount of adenosine triphosphate (ATP) and related compounds (ADP, AMP, IMP, HxR, and Hx) in a fish muscle extract. Many researchers have demonstrated that K-value is related to fish freshness.

However, its limitation is that it is difficult to analyze all these nucleotides and their metabolites by simple procedures. The use of high performance liquid chromatography (HPLC) offers accurate and reliable results but is not suitable for routine analysis. Since adenosine nucleotides rapidly break down and disappear within 24 hours postmortem, the K-value can be simplified as (HxR + Hx)/(IMP + HxR + Hx). Karube et al. (1984) called it the K_1 value and reported that this value is strongly correlated to the K-value proposed by Saito et al. (1959). Like the K-value, the K_1 value is species dependent (Huynh et al. 1990).

Degradation of Myofibrillar Proteins

During chilled storage of fish and marine invertebrates, it is common to notice textural changes before bacterial spoilage. It is believed that these textural changes are caused by the autolytic degradation of myofibrillar proteins. This was demonstrated in ice-stored freshwater prawn (Kye et al. 1988) and in other finfish (Shewfelt 1980). Analysis of myofibrillar proteins is tedious and time-consuming using electrophoresis, but it will give a definite picture of protein degradation.

Collagen Degradation

In the tuna canning industry, it was long recognized that the appearance of honeycombing in precooked tuna was an index of deterioration or mishandling. It was demonstrated that the appearance of honeycombing was related to collagen degradation (Frank et al. 1984). Mackerel was also shown to develop honeycombing when poorly handled (Pan et al. 1986). Freshwater prawn developed mushiness rapidly due to poor handling after catch. It was also demonstrated that this mushiness problem was related to collagen degradation (Nip et al. 1985). It should be noted that all these problems are observable only after the fish or prawn have been cooked and are only detected visibly. It is believed that these problems developed before bacterial spoilage of the fish. Analysis of collagen degradation is also tedious, involving hydrolysis of the extracted collagen and analysis of hydroxyproline.

Dimethylamine Formation

Trimethylamine oxide (TMAO) is commonly found in large quantities in marine species of fish, especially the elasmobranch (Jiang and Lee 2004) and gadoid (Bonnell 1994) species. After death, TMAO is readily degraded to dimethylamine (DMA) through a series of reactions during iced and frozen storage. DMA is typically observed in frozen gadoid species such as cod, hake, haddock, whiting, red hake, and polluck (Castell et al. 1973). TMAO degradation with DMA formation was enhanced by the presence of an endogenous enzyme (TMAOase) in the fish tissues, as observed in cod muscle by Amano and Yamada (1965). It should be noted that TMAO degradation can also be bacterial. Readers should consult the reviews by Regenstein et al. (1982), Hebard et al. (1982), Hultin (1992b), or other literature elsewhere.

Free Fatty Acid Accumulation

Postmortem lipid degradation in seafood, especially fatty fish, proceeds mainly due to enzymatic hydrolysis, with the accumulation of free fatty acids. About 20% of lipids are hydrolyzed during the shelf life of iced fish. The amount of free fatty acids is more or less doubled during that period, mostly from phospholipids, followed by triglycerides, cholesterol esters, and wax esters (Sikorski et al. 1990b, Haard 1990).

Tyrosine Accumulation

The accumulation of lactic acid accompanied with a drop in pH causes the liberation and activation of inherent acid cell proteases, cathepsins (Eskin et al. 1971). Tyrosine has been reported to accumulate in stored fish due to autolysis. Its use as an index of freshness has been proposed by Shenouda et al. (1979) because of its simplicity in analysis. However, it was shown that the pattern of tyrosine accumulation was similar to that of the total volatile base (TVB), an indicator of bacterial spoilage. Therefore, its use as a biochemical index (before bacterial spoilage) is not sufficiently sensitive and specific enough to assess total fish quality (Simpson and Haard 1984).

BIOCHEMICAL AND PHYSICOCHEMICAL CHANGES IN SEAFOOD DURING FREEZING AND FROZEN STORAGE

Freezing is widely used to preserve and maintain the quality of food products for an extended period of time because microbial growth and enzymatic and biochemical reaction kinetics are reduced at low temperatures. However, ice crystals formed as a result of freezing may damage cells and disrupt the texture of food products, and the concentrated unfrozen matrix may result in changes in pH, osmotic pressure, and ionic strength. These changes can affect biochemical and physicochemical reactions such as protein denaturation, lipid oxidation, and enzymatic degradation of trimethylamine oxide (TMAO) in frozen seafood. It is therefore essential to understand these reactions in order to extend the shelf life and improve the quality attributes of seafood.

PROTEIN DENATURATION

During freezing or frozen storage, changes in the physical state of water and the presence of lipids create an environment that induces protein denaturation. This denaturation can be caused by one or more of the following factors: (1) ice crystal formation, (2) dehydration effect, (3) increase in solute concentration, (4) interaction of protein with intact lipids and free fatty acids (FFA), and (5) interaction of protein with oxidized lipids (Shenouda 1980). Denaturation of protein changes the texture and functional properties of protein. The texture of fish may become more fibrous and tough as a result of the loss of protein solubility and water-holding capacity. These textural changes give rise to undesirable sensory attributes, which are often described as sponginess, dryness, rubbery texture, and loss of juiciness (Haard 1992a). Tseng et al. (2003) suggested that to retain good eating qualities, that is, to maintain tenderness and cooking yield, and reduce lipid oxidation, red claw crayfish *(Cherax quadricarinatus)* should not be subjected to more than three freeze-thaw cycles.

Ice Crystal Effect

Ice may form inter- and intracellularly during freezing, which ruptures membranes and changes the structure of the muscle cells (Mazur 1970, 1984; Friedler et al. 1988). At a slow freezing rate, fluids in the extracellular spaces freeze first, thus increasing the concentration of extracellular solutes and drawing water osmotically from the unfrozen cell through the semipermeable cellular membrane (Mazur 1970, 1984). The diffusion of water from the internal cellular spaces to the extracellular spaces results in drip, collected from frozen muscle tissues when thawed (Jiang and Lee 2004). Drip contains proteins, peptides, amino acids, lactic acid, purines, vitamin B complex, and various salts (Sulzbacher and Gaddis 1968), and their concentration in drip increases with storage time (Einen et al. 2002). On the other hand, when the tissue is frozen rapidly, the cellular fluids do not have enough time to migrate out to the extracellular spaces and will freeze as small crystals uniformly distributed throughout the tissue (Mazur 1984). Thus, rapid freezing has a less detrimental effect on the cell or tissue (Coggins and Chamul 2004). However, if the storage temperature fluctuates, the intracellular ice recrystallizes, forming large ice crystals and causing cell disruption, which leads to drip loss as well. Mechanical damage from ice crystals on the texture of food during freezing and frozen storage has been studied in kiwi fruit (Fuster et al. 1994), lamb meat (Payne and Young 1995), squid mantle muscle (Ueng and Chow 1998), pork (Ngapo et al. 1999), and beef (Farouk et al. 2003).

During storage, if the product is not wrapped or is improperly packaged, ice may sublime into the headspace and produce a product defect called freezer burn (Blond and Le Meste 2004, Coggins and Chamul 2004). The sublimation of ice crystals leaves behind small cavities and causes the surface of fish to appear greyish. It is more pronounced when the storage temperature is high (Blond and Le Meste 2004). Freezer burn increases the rate of rancidity and discoloration because of the greater exposed surface, resulting in a "woody" (Blond and Le Meste 2004), tough, and dry texture in fish, making and leaving the fish less acceptable (Coggins and Chamul 2004).

Dehydration Effect

Native protein is stabilized by the folding of hydrophobic chains into the protein molecules and is held together by many other forces, including hydrogen bonding, dipole-dipole interactions, electrostatic interactions, and disulphite linkages (Benjakul and Bauer 2000). The formation of ice during freezing removes water from the protein molecules, thus disrupting the hydrogen bonding network between the native proteins and water molecules and exposing the hydrophobic or hydrophilic sites of the protein molecules (Shenouda 1980). The exposed hydrophobic or hydrophilic sites of the protein molecules interact with each other to form hydrophobic-hydrophobic and hydrophilic-hydrophilic bonds, either within the protein molecules, resulting in deconformation of the native three-dimensional structure of protein, or between adjacent protein molecules, causing protein-protein interactions and resulting in aggregation (Shenouda 1980). Xiong (1997) proposed that protein aggregates to maintain its lowest free energy as water forms ice, thus resulting in protein denaturation. Matsumoto (1979) suggested that redistribution of water during freezing allows protein molecules to move closer together and aggregate through intermolecular interactions. Lim and Haard (1984) found that the loss of protein solubility as a result of protein denaturation in Greenland halibut during frozen storage was mostly due to the noncovalent, hydrophobic interactions in protein molecules. Buttkus (1970) proposed that the formation of intermolecular S-S bonds is the major cause of protein denaturation.

Solute Concentration Effect

As ice forms, the concentration of mineral salts and soluble organic substances in the unfrozen matrix increases. As a result, salts and other compounds that are only slightly soluble (such as phosphate) may precipitate out, which will change the pH (Einen et al. 2002) and ionic strength of the unfrozen matrix and cause conformational changes in proteins. Ions in the concentrated matrix will compete with the existing electrostatic bonds and cause the breakdown of some of the electrostatic bonds (Dyer and Dingle 1961, Shenouda 1980). Takahashi et al. (1993) found, in their freeze denaturation study of carp myofibrils with KCl or NaCl, that freeze denaturation above −13°C is caused by the concentrated salt solution.

Reaction of Protein with Intact Lipids

There are different views in the literature on the effect of intact lipids (i.e., lipids that have not been subjected to partial or total hydrolysis or oxidation) on fish proteins. On one hand, they seem to protect proteins; on the other, they form lipoprotein complexes, which affect protein properties (Shenouda 1980, Mackie 1993). Dyer and Dingle (1961) found that lean fish (fat content less than 1%) showed a rapid decrease in protein (actomyosin) extractability when compared with fatty fish species (3–10 % lipids). They therefore hypothesized that moderate levels of lipids may reduce protein denaturation during frozen storage. In contrast, Shenouda and Piggot (1974) observed a detrimental effect of intact lipids on protein denaturation in their study of a model system, which involved incubating lipid and protein extracted from the same fish at 4°C overnight. They showed that when fish actin (G-form) was incubated with fish polar or neutral lipids, high molecular weight protein aggregates formed. They suggested that during freezing, lipid and protein components form lipoprotein complexes, which change the textural quality of muscle tissue.

Reaction of Proteins with Oxidized Lipids

During frozen storage, lipid oxidation products cause proteins to become insoluble and harder (Takama 1974). When proteins are exposed to peroxidized lipids, peroxidized lipid–protein complexes will form through hydrophobic interactions or hydrogen bonds (Narayan et al. 1964), thus causing conformational changes in the protein. The unstable free radical intermediates of lipid peroxidation remove hydrogen from protein, forming a protein radical, which could initiate various reactions such as cross-linking with other proteins or lipids and formation of protein-protein and protein-lipid aggregates (Karel et al. 1975, Schaich and Karel 1975, Gardner 1979). Roubal and Tappel (1966) found that peroxidized protein cross-links into a range of oligomers, which are associated with protein insolubility. Careche and Tejada (1994) found that oleic and myristic acid had a detrimental effect on the ATPase activity, protein solubility, and viscosity of Hake muscle during frozen storage.

Secondary products from lipid oxidation such as aldehydes react chemically with the amino groups of proteins through the formation of Shiff base adducts, which fluoresce (Leake and Karel 1985, Kikugawa et al. 1989). Ang and Hultin (1989) suggested that formaldehyde might interact with protein side chains and form aggregates without causing cross-linking.

Lipid Oxidation and Hydrolysis

Lipids degrade by two mechanisms: hydrolysis (lipolysis) and oxidation (Shenouda 1980, Shewfelt

1981). In frozen foods, enzymes generally catalyze lipid hydrolysis. Phospholipase A and lipases from muscle tissues are responsible for the hydrolysis of phospholipids and lipids in frozen fish (Shewfelt 1981, De Koning et al. 1987). As the storage time increases and the frozen storage temperature elevates, free fatty acids from the hydrolysis of lipids start to accumulate (Dyer and Dingle 1961). Free fatty acids have a detrimental effect on both the textural properties and flavor of fish.

Lipid oxidation is considered one of the major factors that limit the shelf life of frozen seafood. Fish lipids are known for their susceptibility to oxidation, particularly during frozen storage. Oxidation of polyunsaturated fatty acids yields various oxidative products including a mixture of aldehydes, epoxides, and ketones, which give fish a rancid flavor (Gardner 1979). The low flavor threshold of most aldehydes formed during lipid oxidation means they are easily perceived by the consumer and therefore reduce the acceptability of the products. Rancid flavor in salmon is caused by the formation of volatile products such as (E,Z)-2,6-nonadienal (cucumber odor), (Z)-3-hexanal (green odor), and (Z,Z)-3,6-nonadienal (fatty odor) (Milo and Grosch 1996). The characteristic "seaweed" odor of fresh fish tissue results from the volatile compounds formed during rapid degradation of site-specific hydroperoxides (Josephson and Lindsey 1986). The content of lipid hydroperoxides and free fatty acids in salmon increases during storage, and these changes are fastest when stored at −10°C (Refsgaard et al. 1998). Cod samples stored for 18 months at −15°C had hepta-trans-2-enal and hepta-trans-2,*cis*-dienal. These compounds were described as cold storage flavor (cardboard, musty) with a very low flavor threshold (Coggins and Chamul 2004). Brake and Fennema (1999) found that the rate decrease for thiobarbituric acid reactive substances (TBARS) was abrupt below glass transition temperature (Tg'), whereas the rates of decrease for lipid hydrolysis and peroxide values were moderate to small, respectively, in frozen minced Mackerel. They suggested that the TBARS reduction rate is more diffusion limited than those for lipid hydrolysis and peroxide.

Color changes are an indicator of food quality deterioration. Nonenzymic browning occurs as a consequence of chemical reactions between peroxidizing lipids in the presence of protein. Fluorescence Schiff base adducts formed as a result of chemical reactions between tetrameric dialdehyde and amino groups of protein (Haard 1992a,b). Aubourg (1998) observed a higher fluorescence ratio in a formaldehyde and fatty fish model system compared with the model for formaldehyde and lean fish. In addition to flavor and color changes, lipid oxidation products, including free fatty acids and aldehydes, decrease protein solubility and cause undesirable changes in the functional properties of proteins (Sikorski et al. 1976), as described in the previous section.

Degradation of Trimethylamine Oxide

Trimethylamine oxide (TMAO), a source of formaldehyde, is present naturally in many marine animals as an osmoregulator and as a means of excreting nitrogen (Hebard et al. 1982). After death, TMAO is readily degraded to dimethylamine (DMA) and formaldehyde in the presence of the endogenous enzyme (TMAOase) in the fish tissues (Bremmer 1977, Hebard et al. 1982). It has been postulated that formaldehyde binds covalently to various functional groups in proteins and hence results in a deconformation of the protein, followed by cross-linking between the protein peptide chains via methylene bridges (Sikorski et al. 1976). The interaction of formaldehyde with muscle protein accelerates muscle protein denaturation (Crawford et al. 1979, Ciarlo et al. 1985, Sotelo et al. 1995). Research using both mechanical and sensory tests has shown that formaldehyde formed from the degradation of TMAO increases the firmness and decreases juiciness of mince prepared from white muscle or fillets of gadoid and nongadoid fish species (Rehbein 1988). The presence of formaldehyde also causes a noticeable decrease in the extractability of total proteins, particularly the myofibrillar group (Lim and Haard 1984, Benjakul and Bauer 2000). Ishikawa et al. (1978) suggested that depletion of TMAO accelerates the autoxidation reaction of lipids. Therefore, it is clear that the degradation of TMAO both increases the toughness of muscle protein and accelerates the oxidation and hydrolysis of lipids.

Summary

During freezing, the formation of ice changes the order of water molecules in the food environment, causes dehydration and solute concentration, and thus disturbs the conformation of protein, leading to protein denaturation. Lipid degradation and enzymatic degradation of TMAO during freezing affect the textural and sensory properties of frozen seafood. Therefore, it is important to understand the mechanisms of various biochemical and physicochemical reactions occurring during freezing and frozen storage so that the quality of frozen food can be maintained.

BIOCHEMISTRY OF DRIED, FERMENTED, PICKLED, AND SMOKED SEAFOOD

Salting, fermenting, marinating (pickling), drying, and smoking of fish and marine invertebrates increase the shelf life and develop in the products' desirable sensory properties. Extension of storage life is achieved mainly through the combined effects of

- A reduction in water activity from the addition of salt,
- A decrease in microbial load by the application of heat,
- The presence of inherent preservatives such as acetic acid
- The use of chemical preservatives such as ascorbic acid, BHA, and BHT, and
- The antibacterial and antioxidant activities of various smoke components.

The enzymatic and spoilage processes are controlled not only by these chemical components, but also by the temperature, pH, and availability of oxygen. Desirable changes in sensory properties are thus developed as a result of these carefully controlled chemical and enzymatic processes (Doe and Olley 1990, Fuke 1994, Haard 1994, Miler and Sikorski 1990, Shenderyuk and Bykowski 1990, Sikorski and Ruiter 1994, Shewan 1944, Perez-Villarreal 1992).

There are some changes in proteins. In the salting of fish, salt will penetrate slowly into the tissues, affecting the stability of the native proteins and reducing their extractability. Heavily salted fish, when compared with less salted fish, has disadvantages:

- More water loss due to osmosis,
- Tougher texture, and
- Less developed flavor.

It should be remembered that enzymes are also proteins, and their activities are affected by salt concentration.

Another effect of salting is the changes in texture of the final product. It is believed that the calcium and magnesium ions present as impurities in salt may penetrate the fish, giving rise to a soft, "mushy" texture in the fillet. This may be undesirable for most salted fish, but this effect is considered highly desirable in some Chinese salted products (tender salted threadfin, tender salted mackerel, and others), and Scandinavian products (such as kryddersild, tidbit, and gaffelbiter) (Shewan 1944). Salting of whole fish should be controlled precisely, as overripening will result in excessively soft products with sensory properties that are undesirable to most consumers. However, changes in salted fish fillets depend mostly on endogenous muscle proteases.

Fermented fish paste and sauce are popular products prepared and consumed in southern China and Southeast Asian countries as a source of nutrients and as condiments. Generally, whole fish or shrimp are used as the raw materials for the preparation of these products. It is believed that extensive proteolysis occurs under carefully controlled conditions. It is difficult to differentiate the endogenous and bacterial actions of proteolysis because of the use of whole fish or shrimp and the way these products are produced. Sikorski and Ruiter (1994) summarized some of the work on proteolysis in fermented fish products. Cathepsins A and C as well as trypsin-like enzyme are endogenous proteases that appear to contribute to the fish sauce production both in yield and quality (Orejana and Liston 1982, Raksakulkthai et al. 1986, Rosario and Maldo 1984).

Chiou et al. (1989) reported that cathepsin D–like and aminopeptidase activities release a large amount of free amino acids during roe processing and contribute to the flavor. In the drying of squid, endogenous cathepsin C appears to contribute to desirable qualities (texture and flavor) of traditional products (Haard 1983). Simpson and Haard (1984) reported that added trypsin appears to be a key enzyme contributing to the texture and flavor of matjes herring.

Marinating fish (mainly herring) by means of salt and acetic acid is one of the oldest ways of preserving food in European countries. The acid condition of the marinades, with pH 4–4.5, makes the tissue cathepsins much more active. This results in the degradation of muscle proteins into peptides and amino acids. This permits the marinade to create the proper flavor and texture in the product (Meyers 1965).

Smoke curing means the smoking of presalted fish. The action of the smoke constituents produces a unique smoky odor, taste, and color. The process also contributes to the tenderizing action on the fish tissue. In cold-smoked fish, this tenderization is caused predominately by the action of endogenous proteolytic enzymes because of their activity at such ambient temperatures.

BIOCHEMISTRY OF THERMAL-PROCESSED PRODUCTS

Exposure of the fish to elevated temperatures is detrimental to tissue structure, and results in very undesirable effects (Haard 1994). Extreme texture softening in fish was believed to be the result of cysteine protease acting on the fish muscle at elevated temperatures (Konegaya 1984). Heat-stable alkaline proteases and neutral proteases (modori or gel-degradation) are active when the temperature reaches 60–70°C (Lin and Lanier 1980, Kinoshita et al. 1990). Degradation of connective tissue of mackerel appears to be associated with muscle proteases and pyloric ceca collagenase (Pan et al. 1986).

The rate of oxidation of desirable myoglobin and oxymyoglobin of the red muscles of tuna to brown metmyoglobin depends on the species of the fish and on the storage temperature (Mattews 1983). Color deterioration in iced and frozen stored bonito, yellowfin, and skipjack tuna caught in Seychelles waters was demonstrated in this study.

Food biochemistry plays a role in the production of thermally processed seafood. During the canning of seafood, it is a common practice to precook the raw materials, for example, fish and crustaceans. The accompanying thermal treatment coagulates the proteins in the muscles for easy handling and removal of nonedibles as well as for quality inspection in the later steps. It also inactivates the enzymes that can cause biochemical deterioration of the raw materials, when exposed to elevated temperatures and extended storage times. For example, in manufacturing canned tuna, the fish is precooked and permitted to cool completely, sometimes overnight, before the following steps: (1) removal of skin, bones, dark meats, and viscera and (2) inspection for the presence of defects, e.g, honeycombed and/or burnt tissues.

It should be noted that honeycombed tissues are only detectable in the cooked fish. In the production of canned crabmeat, the precooked crab is cooled before the extraction of crabmeat. Without precooking to coagulate the crabmeat, it is almost impossible to separate the crabmeat from its shell efficiently. The precooking also inactivates those enzymes that can degrade the quality of the product. This is especially important in products like crab and shrimp, as the deteriorating enzymes can act very quickly on these tissues at elevated temperatures. However, there are exceptions where the raw fish is not precooked in order to preserve the premium quality. This includes the production of canned salmon, where sections of raw salmon with skin and bones are stuffed into the can before sealing and processing.

Thermal processing such as mild heating also is applied in the production of some dried seafood products such as fish, shrimp and squid. The heating process inactivates the deteriorating enzymes in the raw materials. This stops the enzymatic reactions from occurring during drying or dehydrating processes that expose the intermediate products to ambient or elevated temperatures for extended periods. For example, in the production of dried, shaved bonito, the raw bonito is first precooked in brine before the processes of recovery of loin tissues, drying to coagulate the tissues, and shaving of the dried product. Without this precooking process, the fish tissue will deteriorate during the long drying process. For dried shrimp, the majority of the product is produced with the precooking process to inactivate the deteriorating enzymes before drying or dehydrating. However, a small amount of dried shrimp is produced without precooking to produce specialty products. For dried squid, it is usually produced without precooking to develop the unique flavor from enzymatic reactions during drying or dehydration. However, a small amount is produced by precooking the product prior to the drying process.

Thermal processing of seafood such as canning and mild heat treatment attempts to produce a final product with long shelf life and favorable consumer acceptance. For more details, refer to the review by Aubourgh (2001).

REFERENCES

Alasalvar C, Taylor T. 2002. Seafoods—Quality, Technology and Nutraceutical Applications. Berlin/New York: Springer. 224 pp.

Al-Kahtani HA, Abu-Tarbouch HM, Bajabe AS, Atia M, Abou-Arab AA, El-Mojaddidi MA. 1996. Chemical changes after irradiation and post-irradiation storage in tilapia and Spanish mackerel. Journal of Food Science 61(4): 729–733.

Amano K, Tomiya F. 1953. Studies on the green discoloration of frozen swordfish (*Xiphia gladis* L.) II. Some experiments on the cause of discoloration. Bull Japanese Soc Scientific Fisheries 19(5): 671–678.

Amano K, Yamada K. 1965. The biological formation of formaldehyde in cod fish. In: R Kreuzer, editor, The Technology of Fish Utilization, pp. 73–87. London: Fishing News Books, Ltd.

Ang JF, Hultin HO. 1989. Denaturation of cod myosin during freezing after modification with formaldehyde. Journal of Food Science 54:814–818.

Angelo, Allen J. St. 1996. Lipid oxidation in foods. Critical Reviews in Food Science and Human Nutrition 36(3): 175–224.

Aubourg SP. 1998. Influence of formaldehyde in the formation of fluorescence related to fish deterioration. Zeitschrift fur Lebensmittel-Untersuchung und-Forschung A—Food Research and Technology 206: 29–32.

———. 2001. Review: Loss of quality during the manufacture of canned fish products. Food Science and Technology International 7(3): 199–215.

Barrett I, Brinner L, Brwon WD, Doiev A, Kwon TW, Little A, Olcott HS, Schaefer MB, Schrader P. 1965. Changes in tuna quality and associated biochemical changes during handling and storage aboard vessels. Food Technol 19(2): 108–117.

Benjakul S, Bauer F. 2000. Physicochemical and enzymatic changes of cod muscle proteins subjected to different freeze-thaw cycles. Journal of the Science of Food and Agriculture. 80:1143–1150.

———. 2001. Biochemical and physiochemical changes in catfish (*Silurus glanis* Linne) muscle as influenced by different freeze-thaw cycles. Food Chemistry 72(2): 207–217.

Blond G, Le Meste M. 2004. Principles of Frozen Storage. In: YH Hui, P Cornillon, IG Legaretta, MH Lim, KD Murrell, WK Nip, editors, Handbook of Frozen Foods, pp. 25–53. New York: Marcel Dekker, Inc.

Bonnell AD. 1994. Quality Assurance in Seafood Processing: A Practical Guide, pp. 74–75.London: Chapman and Hall.

Brake NC, Fennema OR. 1999. Lipolysis and lipid oxidation in frozen minced mackerel as related to Tg', molecular diffusion, and presence of gelatin. Journal of Food Science, 64: 25–32.

Bremmer HA. 1977. Storage trials on the mechanically separated flesh of three Australian mid-water fish species. 1. Analytical tests. Food Technology in Australia. 29:89–93.

———. 1992. Fish flesh structure and the role of collagen–Its post-mortem aspects and implications for fish processing. In: HH Huss, M Jakobsen, J Liston, editors, Quality Assurance in the Fish Industry, pp. 39–62. Amsterdam: Elsevier.

Brown WD. 1962. The concentration of myoglobin hemoglobin in tuna flesh. Jour Food Sci 27(1): 26—28.

Buttkus H. 1970. Accerelated denaturation of myosin in frozen solution. Journal of Food Science, 35:558–562.

Bykowski P, Kolodziejski W. 1983. Wlasciwosci Miesa z Kryla odskorupionego Metoda Rolkowa. Bull Sea Fisheries Inst, Gdynia. 14(5–6): 53–57.

Careche M, Tejada M. 1994. Hake natural actomyosin interaction with free fatty acids during frozen storage. Journal of the Science of Food and Agriculture 64:501–507.

Castell C, Neal W, Date J. 1973. Comparison of changes in TMA, DMA, and extarctable protein in iced and frozen stored gadoid fillets. Jour Fish Res Board Canada 30:1246–1250.

Chanmugam O, Boudreau M, Hwang DH. 1986. Differences in the ω3 fatty acid contents in pond-reared and wild fish and shellfish. Journal of Food Science 51(6): 1556–1557.

Chiou TK, Matsui T, Konosu S. 1989. Proteolytic activities of mullet and Alaska pollack roes and their changes during processing. Nippon Suisan Gakkaishi 55:805–809.

Ciarlo AS, Boeri RL, Giannini DH. 1985. Storage life of frozen blocks of Patagonian Hake *(Merluccius hubbsi)* filleted and minced. Journal of Food Science 50:723–726.

Coggins PC, Chamul RS. 2004. Food Sensory Attributes. In: YH Hui, P Cornillon, IG Legaretta, MH Lim, KD Murrell, WK Nip, editors, Handbook of Frozen Foods, pp. 93–147. New York: Marcel Dekker, Inc.

Crawford D, Law D, Babbitt J, McGill L. 1979. Comparative stability and desirability of frozen Pacific hake fillet and minced flesh blocks. Journal of Food Science, 44:363–367.

De Koning AJ, Milkovitch A, Mol TH. 1987. The origin of free fatty acids formed in frozen cape hake mince *(Merluccius capensis, castelnau)* during cold storage at −18°C. Journal of the Science of Food and Agriculture 39:79–84.

Doe P, Olley J. 1990. Drying and dried fish products. In: ZE Sikorski, editor. Seafood: Resources, Nutritional Composition, and Preservation, pp. 125–145. Boca Raton: CRC Press, Inc.

Dyer WJ, Dingle JR. 1961. Fish proteins with special reference to freezing. In: G Borgstrom, editor, Fish as Food, pp. 275–327. New York: Academic Press.

Einen O, Guerin T, Fjaera SO, Skjervold PO. 2002. Freezing of pre-rigor fillets of Atlantic salmon. Aquaculture 212:129–140.

Eskin NA, Henderson HM, Townsend RJ. 1971. Biochemistry of Foods. New York: Academic Press.

Farouk MM, Wieliczko KJ, Merts I. 2003. Ultra-fast freezing and low storage temperatures are not necessary to maintain the functional properties of manufacturing beef. Meat Science 66:171–179.

Flick GJ, Hong GP, Knobl GM. 1992. Lipid oxidation of seafood during storage. In: AJ St Angelo, editor, Lipid Oxidation in Food, pp. 183–207. Washington: American Chemical Society.

Frank HA, Rosenfield ME, Yoshinaga DH, Nip WK. 1984. Relationship between honeycombing and collagen breakdown in skipjack tuna. Marine Fish Rev 46(2): 40–42.

Friedler S, Giudice LC, Lamb EJ. 1988. Cryopreservation of embryos and ova. Fertility and Sterility 49:743–763.

Fuke S. 1994. Taste-active compounds of seafoods with special references to umami substances. In: F Shahidi, JR Botta, editor. Seafoods: Chemistry, Processing Technology and Quality, pp. 115–139. London: Blackie Academic and Professional.

Fuster C, Prestamo G, Cano MP. 1994. Drip loss, peroxidase and sensory changes in kiwi fruit slices during frozen storage. Journal of the Science of Food and Agriculture 64:23–29.

Gardner HW. 1979. Lipid hydroperoxide reactivity with proteins and amino acids: A review. Journal of Agricultural and Food Chemistry 27:220–229.

Ghiretti F. 1956. The decomposition of hydrogen peroxide by hemocyanin and by its dissociation products. Archives of Biochemistry and Physiology. 63:165–176.

Gill TA. 1992. Biochemical and chemical indices of seafood quality. In: HH Huss, M Jakobsen, J Liston, editors, Quality Assurance in the Fish Industry, pp. 377–388. Amsterdam: Elsevier.

Grosjean O, Cobb BF, Mebine B, Brown WD. 1969. Formation of a green pigment from tuna myoglobin. Jour Food Sci 34:404–407.

Haard NF. 1983. Dehydration of Atlantic short finned squid. In: BS Pan, editor, Properties and Processing of Marine Foods, pp. 36–60. Keelung (Taiwan, Republic of China): National Taiwan Ocean University.

———. 1990. Biochemical reactions in fish muscle during frozen storage. In: EG Bligh, editor, Seafood Science and Technology, pp. 46–57. Oxford: Fishing News Books (Blackwell Scientific Publications, Ltd.).

———. 1992a. Biochemistry and chemistry of color and color change in seafoods. In: GJ Flick, RE Martin, editors, Advances in Seafood Biochemistry, pp. 305–360. Lancester, Pennsylvania: Technomic Publishing Co., Inc.

———. 1992b. Biochemical reactions in fish muscle during frozen storage. In: EG Bligh, editor, Seafood Science and Technology, p. 176–209. Oxford: Fishing New Books.

———. 1994. Protein hydrolysis in seafoods. In: F Shahidi, JR Botta, editors, Seafoods: Chemistry, Processing Technology and Quality, pp. 10–33. London: Blackie Academic and Professional.

Haard NF, Simpson BK, Pan BS. 1994. Sarcoplasmic proteins and other nitrogenous compounds. In: ZE Skorski, BS Pan, F Shahidi, editors, Seafood Proteins, pp. 13–39. New York: Chapman and Hall.

Halliwell B, Gutteridge JMC. 1988. Free Radicals in Biology and Medicine. Oxford: Claredon Press.

Hebard CE, Flick GF, Martin RE. 1982. Occurrence and significance of trimethylamine oxide and its derivatives in fish and shellfish. In: RE Martin, GJ Flick, CE Hebard, DR Ward, editors, Chemistry and Biochemistry of Marine Food Products, pp.149–304. Westport, Connecticut: AVI Publishing Company.

———. 1982. Occurrence and significance of trimethylamine oxide and its derivatives in fish and shellfish. In: RE Martin, GJ Flick, CE Hebard, editors, Chemistry and Biochemistry in Marine Food Products, pp. 149–304. Westport, Connecticut: AVI Publishing.

Hink MJ, Stanley DW. 1985. Colour measurement of the squid *(Illex illecebrosus)* and its relationship to quality and chromatophore structure. Canadian Institute of Food Sci Technol Jour 18(3): 233–241.

Hobbs G. 1982. Changes in fish after catching. In: S Aitkin, IM Mackie, JH Merritt, ML Windsor, editors, Fish Handling and Processing, pp. 20–27. Edinburgh: Her Majesty's Stationery Office.

Hultin HO. 1992a. Biochemical deterioration of fish muscle. In: HH Huss, M Jakobsen, J Liston, editors. Quality assurance in the fish industry, pp. 125–138. Amsterdam: Elsevier.

———. 1992b. Trimethylamine-N-oxide (TMAO) demethylation and protein denaturation in fish muscle. In: GJ Flick, RE Martin, editors, Advances in Seafood Biochemistry, pp. 25–42. Lancaster, Connecticut: Technomic Publishing Co., Inc.

———. 1994. Oxidation of lipids in seafoods. In: S Fereidoon Shahidi, J Richard Botta, editors. Seafoods: Chemistry, Processing Technology and Quality, pp. 49–74. Glasgow: Blackie Academic and Professional.

Hultin HO, Kelleher SD. 2000. Surimi processing from dark muscle fish. In: JW Park, editor, Surimi and Surimi Seafood, pp. 59–77. New York: Marcel Dekker.

Huss HH. 1994. Quality and Quality Changes in Fresh Fish. Rome: Food and Agricultural Organizaton of the United Nations.

Huynh MD, Mackey J, Gawley R. 1990. Freshness assessment of Pacific fish species using K-value. In: EG Bligh, editor, Seafood Science and Technology, pp. 258–268. Oxford: Fishing News Books (Blackwell Scientific Publications, Ltd.).

Icekson I, Drabkin V, Aizendorf S, Gelman A. 1998. Lipid oxidation levels in different parts of the mackerel, *Scomber scombrus*. Journal of Aquatic Food Product Technology 7(2): 17–29.

Ikeda S. 1979. Other organic components and inorganic components. In: JJ Connell and the staff of Torry Research Station, editors. Advances of Fish Science and Technology, pp. 111–112. Farnham, Surrey: Fish News Books.

Ishikawa Y, Yuki E, Kato H, Fukimaki M. 1978. Synergistic effect of trimethylamine oxide on the inhibition of the autoxidation of methyl linoleate by γ-tocopherol. Agricultural and Biological Chemistry 42:703–709.

Jacober LF, Rand AG. 1982. Biochemical evaluation of seafood. In: RE Martin, GJ Flick, CE Hebard, DR Ward, editors, Chemistry and Biochemistry of Marine Food Products, pp. 347–365. Westport, Connecticut: AVI Publsihing Company.

Jiang ST. 2000. Enzymes and their effects on seafood texture, pp. 411–450. In: NF Haard, BK Simpson, editors. Seafood enzymes. New York: Marcel Dekker, Inc.

Jiang ST, Lee TC. 2004. Frozen seafood and seafood products: Principles and applications. In: YH Hui, P Cornillon, IG Legareta, M Lim, KD Murrell, WK Nip, editors. Handbook of Frozen Foods, pp. 245–294. New York: Marcel Dekker, Inc.

Jobling M. 2004. Are modifications in tissue fatty acid profiles following a change in diet the result of dilution? Test of a simple dilution model. Aquaculture 232(1–4): 551–562.

Jones NR, Murray J, Livingston EI, Murray CK. 1964. Rapid estimation of hypoxanthine concentrations as indices of freshness of chilled-stored fish. Jour Sci Food Agric 15(11): 763–773.

Josephson DB, Lindsey RC. 1986. Enzymic generation of fresh fish volatile aroma compounds. In: TH Parliment, R Croteau, editors, Biogenesis of Aromas, p. 201. ACS Symposium No. 317.

Karel M, Schaich K, Roy RB. 1975. Interaction of peroxidizing methyl linoleate with some proteins and amino acids. Journal of Agricultural and Food Chemistry 23:159–163.

Karube I, Matsuoka H, Susuki S, Watanabe E, Toyama K. 1984. Determination of fish freshness with an enzyme sensor system. J Agric Food Chem 32:314–319.

Kelleher SD, Silva LA, Hultin HO, Wilhelm KA. 1992. Inhibition of lipid oxidation during processing of washed, minced Atlantic mackerel. Journal of Food Science 57(5): 1103–1108,1119.

Kikugawa K, Kato T, Iwata A. 1989. A tetrameric dialdehyde formed in the reaction of butyraldehyde and benzylamine: A possible intermediary component for protein cross-linking induced by lipid oxidation. Lipids 24:962–969.

King AJ, Li SJ. 1999. Association of malonaldehyde with rabbit myosin subfragment 1. In: YL Xiong, C-T Ho, F Shahidi, editors. Quality Attributes of Muscle Foods, pp. 277–286. New York: Kluwer Academic/Plenum Publishers.

Kinoshita M, Toyohara H, Shimizu Y. 1990. Diverse distribution of four distinct types of modori (gel degradation)-inducing proteinases among fish species. Nippon Suisan Gakkaishi 56:1485–1492.

Klurfeld DM. 2002. Dietary fats, ecosanoids, and the immune system. In: CC Akoh, DB Min, editors, Food Lipids: Chemistry, Nutrition and Biochemistry, 2nd edition, pp. 589–601. New York: Marcel Dekker.

Koizumi C, Matsura F. 1967. Studies on 'green' tuna. IV. Effect of cysteine on greening of myoglobin in the presence of trimenthylamine oxide. Bull Japanese Soc Scientific Fisheries 33(9): 839–842.

Kolstad K, Vegusdal A, Baeverfjord G, Einen O. 2004. Quantification of fat deposits and fat distribution in Atlantic halibut (*Hippoglossus hippoglossus* L.) using computerized X-ray tomography (CT). Aquaculture 229(1–4): 255–264.

Konegaya S. 1984. Studies on the jellied meat of fish, with special reference to that of yellowfin tuna. Bull Tokai Reg Fish Lab 116:39–47.

Kristinsson HG, Hultin HO. 2004. The effect of acid and alkali unfolding and subsequent refolding on the prooxidative activity of trout hemoglobin. Journal of Agricultural and Food Chemistry 52(17): 5482–5490.

Kristinsson HG, Mony S, Demir N, Balaban MO, Otwell WS. The effect of carbon monoxide and filtered smoke on the properties of aquatic muscle and selected muscle components, 2003 June 10–14, pp. 27–29. Reykjavik, Iceland: Icelandic Fisheries Laboratory.

Kye HW, Nip WK, My JH. 1988. Changes in myofibrillar proteins and texture in freshwater prawn during ice-storage. Marine Fish Rev 50(1): 53–56.

Leake L, Karel M. 1985. Nature of fluorescent compounds generated by exposure of protein to oxidized lipids. Journal of Food Biochemistry 9:117–136.

Leu S-S, Jhaveri SN, Karakoltsidis PA, Constantinides SM. 1981. Atlantic mackerel (*Scomber scombrus* L.): Seasonal variation in proximate composition and distribution of chemical nutritients. Journal of Food Science 46:1635–1638.

Lim HK, Haard NF. 1984. Protein insolubilization in frozen Greenland halibut *(Reinhardtius Hippoglossoides).* Journal of Food Biochemistry 8:163–187.

Lin TS, Lanier TC. 1980. Properties of an alkaline protease from the skeletal muscle of Atlantic croaker. Jour Food Biochem 4(1): 17–28.

Lindsay RC. 1991. Chemical basis of the quality of seafood flavors and aromas. Marine Technol Soc Japan 25:16–22.

Livingston DJ, Brown WD. 1981. The chemistry of myoglobin and its reactions. Food Technol 35:244–252.

Love RM, Munro LJ, Robertson I. 1977. Adaptation of the dark muscle of cod to swimming activity. Jour Fish Biol 11:431–436.

Lovell RT, Mohammed T. 1988. Content of omega-3 fatty acids can be increased in farm-raised catfish. Highlights of Agricultural Research 35(3): 16.

Lundstrom RC, Correia FF, Wilhelm KA. 1982. Dimethylamine production of fresh red hake *(Urophycis chuss):* The effect of packaging material, oxygen permeability and cellular damage. Jour Food Biochem 6:229–241.

Mackie IM. 1993. The effects of freezing on flesh proteins. Food Reviews International 9:575–610.

Martinez I. 1992. Fish myosin degradation upon storage. In: HH Huss, M Jakobsen, J Liston, editors. Quality Assurance in the Fish Industry, pp. 389–397. Amsterdam: Elsevier.

Matsumoto JJ. 1979. Denaturation of fish muscle proteins during frozen storage. In: O Fennema, editor, Proteins at Low Temperatures, pp. 205–224. Washington, DC: American Chemical Society.

Mattews AD. 1983. Muscle color deterioration in iced and frozen stored bonito, yellowfin and skipjack tuna caught in Seychelles waters. Jour Food Technol 18:387–392.

Mazur P. 1970. The freezing of biological systems. Science 168:939–949.

———. 1984. Freezing of living cells: Mechanisms and implications. American Journal of Physiology 247:C125–C142.

Meyers V. 1965. Chapter 5. Marinades. In: G Borgstrom, editor, Fish as Food, vol. III, pp. 165–193. New York: Academic Press.

Miler KBM, Sikorski ZE. 1990. Smoking. In: ZE Sikorski, editor. Seafood: Resources, Nutritional Composition, and Preservation, pp. 163–180. Boca Raton: CRC Press, Inc.

Milo C, Grosch W. 1996. Changes in the odorants of boiled salmon and cod as affected by the storage of the raw material. Journal of Agricultural and Food Chemistry 44:2366–2371.

Murata M, Sakaguchi M. 1989. The effects of phosphatase treatment of yellowtail muscle extracts and subsequent addition of IMP on flavor intensity. Nippon Suisan Gakkaishi 55:1599–1603.

Narayan KA, Sugai M, Kummerow FA. 1964. Complex formation between oxidized lipids + egg albumin. Journal of the American Oil Chemists' Society 41:254–259.

Ngapo TM Babare IH, Reynolds J, Mawson RF. 1999. Freezing and thawing rate effects on drip loss from samples of pork. Meat Science 53:149–158.

Nip WK, Lan CY, Moy JH. 1985. Partial characterization of a collagenolytic enzyme fraction from the hepatopancreas of the freshwater prawn, *Macrobrachium rosenbergii*. Jour Food Sci 50:1187–1188.

Nip WK, Moy JH. 1988. Microstructural changes of iced-chilled and cooked freshwater prawn, *Macrobrachium rosenbergii*. Jour Food Sci 53:319–322.

Ohshima T, Fujita Y, Koizumi C. 1993a. Oxidative stability of sardine and mackerel lipids with reference to synergism between phospholipid and α-tocopherol. Journal of the American Oil Chemist Society 7(3): 269–276.

Ohshima T, Ushio H, Koizumi C. 1993b. High pressure processing of fish and fish products. Trends in Food Science and Technology 4:370–375.

O'Keefe SF. 2000. Fish and shellfish. In: GL Christen, JS Smith, editors. Food Chemistry: Principles and Applications, pp. 399–420. West Sacramento: Science Technology Systems.

Orejana FM, Liston J. 1982. Agents of proteolysis and its inhibition of patis (fish sauce) fermentation. Jour Food Sci 47(1): 198–203, 209.

Pan BS, Kuo JM, Luo LJ, Yang HM. 1986. Effect of endogenous proteinases on histamine and honeycombing in mackerel. Jour Food Biochem 10:305–319.

Payne SR, Young OA. 1995. Effects of pre-slaughter administration of antifreeze proteins on frozen meat quality. Meat Science 41:147–155.

Perez-Villarreal PR. 1992. Ripening of the salted anchovy *(Engraulis encrasicholus):* Study of the sensory, biochemical and microbiological aspects. In: HH Huss, M Jakobsen, J Listin, editors, pp. 157–167. Quality Assurance in the Fish Industry. Amsterdam: Elsevier.

Raksakulkthai N, Lee YZ, Haard NF. 1986. Influence of mincing and fermentation aids on fish sauce prepared from male, inshore capelin, *Mallotus villosus*. Canadian Inst Food Sci Technol Jour 19:28–33.

Reddy GVS, Srikar LN, Sudhakara NS. 1992. Deteriorative changes in pink perch mince during frozen storage. International Journal of Food Science and Technology 27(3): 271–276.

Refsgaard HH, Brockhoff PM, Jensen B. 1998. Sensory and chemical changes in farmed Atlantic salmon *(Salmo salar)* during frozen storage. Journal of Agricultural and Food Chemistry 46:3473–3479.

Regenstein JM, Schlosser MA, Samson A, Fey M. 1982. Chemical changes of trimethylamine oxide during fresh and frozen storage of fish. In: RE Martin, GJ Flick, CE Hebard, DR Ward, editors, Chemistry and Biochemistry of Marine Food Products, pp.137–148. Westport, Connecticut: AVI Publsihing Company.

Regulska-Ilow B, Ilow R. 2002. Comparison of the effects of microwave cooking and conventional cooking methods on the composition of fatty acids and fat quality indicators in herring. Nahrung/Food 46(6): 383–388.

Rehbein H. 1988. Relevance of trimethylamine oxide demethylase activity and haemoglobin content of formaldehyde production and texture deterioration in frozen stored minced fish muscle. Journal of the Science of Food and Agriculture 43:261–276.

Richards MP, Kelleher SD, Hultin HO. 1998. Effect of washing with or without antioxidants on quality retention of mackerel fillets during refrigerated and frozen storage. Journal of Agricultural and Food Chemistry 46(10): 4363–4371.

Rosario RR, Maldo SM. 1984. Biochemistry of patis formation. I. Activitiy of cathepsins in patis hydrolysis. Philippine Agricuturist 67(2): 167–175.

Roubal WT, Tappel AL. 1966. Damage to proteins enzymes and amino acids by peroxidizing lipids. Archives of Biochemistry and Biophysics 113: 5–8.

Saito T, Arai K, Matuyoshi M. 1959. A new method for estimating the freshness of fish. Jap Soc Sci Fish 24:749–750.

Sakaguchi M, Murata M. 1989. Seasonal variations of free amino acids in oyster whole body and adductor muscle. Nippon Suisan Gakkaishi 55:2037–2041.

Sathiyela S, Prinyawiwatkulb W, Negulescuc II, Kingb JM, Basnayaked BFA. 2003. Thermal degradation of FA and catfish and menhaden oils at different refining steps. Journal of the American Oil Chemist Society 80(11): 1131–1134.

Schaich KM, Karel M. 1975. Free radicals in lysozyme reacted with peroxidizing methyl linoleate. Journal of Food Science 40:456–459.

Schwimmer S. 1981. Source Book of Food Enzymology, p. 967. Westport, Connecticut: AVI Publishing Co., Inc.

Sebedio JL, Ratnayake WMN, Ackman RG, Prevost J. 1993. Stability of polyunsaturated omega-3 fatty acids during deep fat frying of Atlantic mackerel (*Scomber scombrus* L.). Food Research International 26:163–172.

Shahidi F. 1994. Seafood proteins and preparation of protein concentrates. In: F Shahidi, JR Botta, editors, Seafoods: Chemistry, Processing Technology and Quality, pp. 3–9. London: Blackie Academic and Professional.

Shahidi F, Ghadi SV. 1991. Influence of gamma-irradiation and ice storage on fat oxidation in 3 Indian fish. International Journal of Food Science and Technology 26(4): 397–401.

Shahidi F, Spurvey SA. 1996. Oxidative stability of fresh and heat-processed dark and light muscles of mackerel *(Scomber scombrus).* Journal of Food Lipids 3(1): 13–25.

Shenderyuk VI, Bykowski PJ. 1990. Sating and marinating of fish. In: ZE Sikorski, editor. Seafood: Resources, Nutritional Composition, and Preservation, pp. 147–162. Boca Raton: CRC Press, Inc.

Shenouda S, Monteclavo J, Jhaveri S, Costantinides SM. 1979 Technical studies on ocean pout, an unexploited fish species, for direct human consumption. Jour Food Sci 44:164–168.

Shenouda SYK. 1980. Theories of protein denaturation during frozen storage of fish flesh. Advances in Food Research 26:275–311.

Shenouda SYK, Piggot GM. 1974. Lipid-protein interaction during aqueous extraction of fish protein: Myosin-lipid interaction. Journal of Food Science 39:726–734.

Shewan JM. 1944. The effect of smoke curing and salt curing on the composition, keeping quality and culinary properties of fish. Proc Nutr Soc (1–2): 105–112.

Shewfelt R. 1980. Fish muscle hydrolysis—A review. Jour Food Biochem 5:79–94.

Shewfelt RL. 1981. Fish muscle lipolysis—A review. Journal of Food Biochemistry 5:79–100.

Sikorski ZE. 1994a. The contents of proteins and other nitrogenous compounds in marine animals. In: ZE Skorski, BS Pan, F Shahidi, editors, Seafood Proteins, pp. 6–12. New York: Chapman and Hall.

Sikorski ZE. 1994b. The myofibrillar proteins in seafoods. In: ZE Skorski, BS Pan, F Shahidi, editors, Seafood Proteins, pp. 40–57. New York: Chapman and Hall.

Sikorski ZE, Borderias JA. 1994. Collagen in the muscles and skin of marine animals. In: ZE Skorski, BS Pan, F Shahidi, editors, Seafood Proteins, pp. 58–70. New York: Chapman and Hall.

Sikorski ZE, Kolakowska A. 1994. Changes in proteins in frozen stored fish. In: ZE Skorski, BS Pan, F Shahidi, editors, Seafood Proteins, pp. 99–112. New York: Chapman and Hall.

Sikorski ZE, Kolakowska A, Burt JR. 1990a. Postharvest biochemical and microbial changes. In: ZE Sikorski, editor. Seafood Resources: Nutritional Composition, and Preservation, pp. 55–75. Boca Raton: CRC Press.

Sikorski ZE, Kolakowska A, Pan BS. 1990b. The nutritive composition of the major groups of marine food organisms. In: ZE Sikorski, editor, Seafood resources: Nutritional, Compositional and Preservation, pp. 29–54. Boca Raton: CRC Press.

Sikorski Z, Olley J, Kostuch S. 1976. Protein changes in frozen fish. CRC Critical Reviews in Food Science and Nutrition 8:97–129.

Sikorski ZE, Pan BS. 1994. The involvement of proteins and nonprotein nitrogen in postmortem changes in seafoods. In: ZE Skorski, BS Pan, F Shahidi, editors, Seafood Proteins, pp. 71-83. New York: Chapman and Hall.

Sikorski ZE, Ruiter A. 1994. Changes in proteins and nonprotein nitrogen compounds in cured, fermented, and dried seafoods. In: ZE Sikorski, BS Pan, F Shahidi, editors. Seafood Proteins, pp. 113–126. New York: Chapman and Hall.

Simpson BK, Haard NF. 1984. Trypsin from Greenland cod *(Gadus ogac)* as a food processing aid. Jour Applied Biochem 6:135–143.

Simpson BK, Marshall MR, Otwell WS. 1987. Phenoloxidase from shrimp *(Paneus setiferus):* Purification and some properties. Jour Agric Food Chem 35:918–921.

Solberg C. 2004. The influence of dietary oil content on the growth and chemical composition of Atlantic salmon *(Salmo salar)*. Aquaculture Nutrition 10(1): 31–37.

Sorenson NK. 1992. Physical and instrumental methods for assessing seafood quality. In: M Jackobsen, J Listin, editors, Quality Assurance in the Fish Industry, pp. 321–332.Amsterdam: Elsevier.

Sotelo CG, Pineiro C, Perezmartin RI. 1995. Denaturation of fish proteins during frozen storage-role of formaldehyde. Zeitschrift fur Lebensmittel-Untersu chung Und-Forschung A—Food Research and Technology 200:14–23.

Spinelli J, Eklund M, Miyauchi D. 1964. Measurement of hypoxanthin in fish as a method of assessing freshness. Jour Food Sci 79:710–714.

Srinivasan S, Xiong YL, Decker EA. 1996. Inhibition of protein and lipid oxidation in beef heart surimi-like materials by antioxidants and combinations of pH, NaCl, and buffer type in the washing media. Journal of Agricultural and Food Chemistry 44(1): 119–125.

Sulzbacher WL, Gaddis AM. 1968. Meats: Preservation of quality by freezer storage. In: DK Tressler, WB Van Arsdel, MJ Copley, editors, The Freezing Preservation on Foods, vol. II, pp. 159–178. Westport, Connecticut: AVI Publishing Company, Inc.

Suzuki T. 1981. Fish and Krill Processing Technology, pp. 10–13.London: Applied Science Publishers.

Takahashi K, Inoue N, Shinano H. 1993. Effect of storage temperature on freeze denaturation of Carp myofibrils with KCl or NaCl. Nippon Suisan Gakkaishi 59:519–527.

Takama K. 1974. Insolubilization of rainbow trout actomyosin during storage at −20°C. 1. Properties of insolubilized proteins formed by reaction of propanal or caproic acid with actomyosin. Bulletin of the Japanese Society of Scientific Fisheries 40:585–588.

Thompson HC, Thompson MH. 1972. Inhibition of flesh browning and skin color fading in frozen fillets of yelloweye snapper *(Lutjanus vivanus)* NOAA Technical Report NMFS SSRF-544. 6 pp.

Tichivangana JZ, Morrissey PA. 1985. Metmyoglobin and inorganic metals as pro-oxidants in raw and cooked muscle systems. Meat Sci 15(2): 107–116.

Tironi VA, Tomas MC, Anon MC. 2002. Structural and functional changes in myofibrillar proteins of sea salmon *(Pseudopercis semifasciata)* by interaction with malonaldehyde. Journal of Food Science 67: 930–935.

Tomlinson N. 1966. Bulletin 150: Greening in tuna and related species. Ottawa: Fisheries Research Board of Canada. 21 pp.

Tseng YC, Xiong YLL, Feng JL, Ramirez-Suarez JC, Webster CD, Thompson KR, Muzinic LA. 2003. Quality changes in Australian red claw crayfish

(Cherax quadricarinatus) subjected to multiple freezing-thawing cycles. Journal of Food Quality, 26:285–298.

Ueng YE, Chow CJ. 1998. Textural and histological changes of different squid mantle muscle during frozen storage. Journal of Agricultural and Food Chemistry 46:4728–4733.

Undeland I. 1997. Lipid oxidation in fish—Causes, changes and measurements. In: G Olafsdóttir, J Luten, P Dalgaard, M Careche, V Verrez-Bagnis, E Martinsdóttir, K Heia, editors, Methods to Determine the Freshness of Fish in Research and Industry, pp. 241–256. International Institute of Refrigeration, Paris, France.

Undeland I, Hultin HO, Richards MP. 2002. Added triacylglycerols do not hasten hemoglobin-mediated lipid oxidation in washed minced cod muscle. Journal of Agricultural and Food Chemistry 50(23): 6847–6853.

Undeland I, Kristinsson HG, Hultin HO. 2004. Hemoglobin-mediated oxidation of washed minced cod muscle phospholipids: Effect of pH and hemoglobin source. Journal of Agricultural and Food Chemistry 52(14): 4444–4451.

Wang JCC, Amiro ER. 1979. A fluorometric method for the micro-quantitative determination of hemoglobin and myoglobin in fish muscle. Jour Sci Food Agric 30(11): 1089–1096.

Wang Y-J, Miller L, Addis P. 1991. Effect of heat inactivation of lipoxygenase on lipid oxidation in lake herring *(Coregonus artedii)*. Journal of the American Oil Chemist Society 68(10): 752–757.

Wijendran V, Hayes KC. 2004. Dietary n-6 and n-3 fatty acid balance and cardiovascular health. Annual Review of Nutrition 24:597–615.

Wolfe SK, Watt DA, Brown WD. 1978. Analysis of myoglobin derivatives in meat or fish samples using absorption spectrophotometry. Jour Agric Feed Chem 26(1): 217–219.

Xiong YL. 1997. Protein denaturation and functionality losses. In: MC Erickson, YC Hung, editors, Quality in Frozen Food, pp. 111–140. New York: Chapman and Hall.

17
Seafood Enzymes

M. K. Nielsen and H. H. Nielsen*

* Corresponding author.

INTRODUCTION

"Freshness" serves as a basic parameter of quality to a greater extent in seafood than in muscle foods of other types. The attention directed at the freshness of seafood and seafood products is appropriate since certain changes in seafood occur soon after capture and slaughter, often leading to an initial loss in prime quality. Such changes are not necessarily related to microbial spoilage, but may be due to the activity of endogenous enzymes. However, not all postmortem endogenous enzyme activities in seafood lead to a loss in quality: many autolytic changes are important for obtaining the desired texture and taste.

Despite the biological differences between aquatic and terrestrial animals, the biochemical reactions of most animals are identical and are catalyzed by homologous enzymes; only few reactions are unique for certain species. Few reactions are catalyzed by analog enzymes that differ in structure and evolutionary origin. The aquatic habitat can be extreme in various respects including temperature, hydrostatic pressure, and osmotic pressure, compared with the environment of terrestrial animals. These conditions all have thermodynamic implications and can affect the kinetics and stability of marine enzymes. Also, because of the large biological diversity of marine animals and such seasonally determined factors as spawning and the availability of food, seafood species can differ markedly in enzyme activity.

Seafood is a gastronomic term that in principle includes all edible aquatic organisms. In this chapter, it is mainly the "classical" seafood groups—fish, shellfish, and molluscs—that are dealt with. Muscle enzymes are mostly considered, since muscle is the organ of which most seafood products primarily consist. Various other types of seafood enzymes, such as those used in digestion, are only dealt with to a lesser extent. Further information on seafood enzymes can be found in the book *Seafood Enzymes,*edited by Haard and Simpson (2000), and the references therein.

Cold Adaptation of Enzymes

Most seafood species are poikilotherme and are thus forced to adapt to the temperature of their habitat, which may range from the freezing point of oceanic seawater, of approximately −1.9°C, to temperatures similar to those of the warm-blooded species. Since most of the aquatic environment has a temperature of below 5°C, the enzyme systems of seafood species are often adapted to low temperatures. Although all enzyme reaction rates are positively correlated with temperature, the enzyme systems of most cold-adapted species exhibit, at low temperatures, catalytic activities similar to those that homologous mesophilic enzymes of warm-blooded animals exhibit at higher in vivo temperatures. This is primarily due to the higher turnover number (k_{cat}) of the cold-adapted enzymes: substrate affinity, estimated by the Michaelis-Menten constant (K_M), and other regulatory mechanisms are generally conserved (Somero 2003).

Several studies reviewed by Simpson and Haard (1987) have found the level of activity of cold-adapted marine enzymes to be higher than that of homologous bovine enzymes; they have also shown marine enzymes to be much lower in thermal stability. Although the attempt has been made to explain the temperature adaptation of the enzymes of marine species in terms of molecular flexibility in a general sense, studies of the molecular structure of such enzymes have not revealed any common structural features that can provide a universal explanation of temperature adaptation. Rather, there appear to be various minor structural adjustments of varying character that provide the necessary flexibility of the protein structure for achieving cold adaption (Smalås et al. 2000, Fields 2001, de Backer et al. 2002).

Since the habitat temperature of most marine animals used as seafood tends to be near to the typical cold storage temperature of approximately 0–5°C, the enzymatic reaction rates are not thermally depressed at storage temperatures as they usually would be in the meat of warm-blooded animals. With a habitat temperature of seafood of about 5°C and Q_{10} values in the normal range of 1.5–2.0, the enzyme reactions during ice storage can thus be expected to proceed at rates four to nine times as high as in warm-blooded animals under similar conditions.

Effects of Pressure on Enzyme Reactions

Although the oceans, with their extreme depth, represent the largest habitat on earth, it is often forgotten that hydrostatic pressure is an important biological variable, affecting the equilibrium constants of all reactions involving changes in volume. Regarding enzyme reactions, pressure affects the molecular stability of enzymes and the intermolecular interactions between enzymes and substrates, cofactors, and macromolecules. Pressure alterations may therefore change both the concentration of catalytically active enzymes and their kinetic parameters, such as the Michaelis-Menten constant, K_M.

Siebenaller and Somero (1978) showed that the apparent K_M values of a lactate dehydrogenase (LDH) enzyme from a deep-sea fish species are less sensitive to pressure changes than its homologous enzyme from a closely related fish species living in shallow water. Similar relations between the pressure sensitivity of LDH and habitat pressure have been found for the LDH of other marine species (Siebenaller and Somero 1979, Dahlhoff and Somero 1991). In this case, only the LDH substrate interaction was found to be affected by pressure, whereas the turnover number appeared to be unaffected by pressure.

Very high pressures, of 10^8 Pa (ca. 1000 bar), corresponding to the pressure in the deepest trenches on earth can lead to denaturization of proteins. Some deep-sea fish have been found to accumulate particularly high levels of trimethylamine-*N*-oxide (TMAO). TMAO stabilizes most protein structures and may counteract the effect of pressure on proteins. Adaptation to high pressure thus seems to be achieved both by evolution of pressure-resistant proteins and by adjustments of the intracellular environment.

The metabolic rates of deep-sea fish tend to be lower than those of related species living in shallow waters. Several lines of evidence suggest that the decrease in metabolic rate with increasing depth is largely due to the food consumption levels being lower and to a reduced need for locomotor capability in the darkness (Childress 1995, Gibbs 1997). The latter makes intuitive sense if one considers the often far from streamlined, and sometimes bizarre, anatomy of deep-sea fish species.

Osmotic Adaptation of Enzymes

Marine teleosts (bonefish) are hypoosmotic, having intracellular solute concentrations that are similar to those of terrestrial vertebrates. By contrast, elasmobranches (sharks and rays) are isoosmotic, employing the organic compounds urea and TMAO as osmolytes. Although the concentration of osmolytes can be high, the stability and functionality of enzymes remain mostly unaffected due to the compatability and balanced counteracting effects of the osmolytes (Yancey et al. 1982).

ENZYMATIC REACTIONS IN THE ENERGY METABOLISM OF SEAFOOD

Most key metabolic pathways such as the glycolysis are found in all animal species, and the majority of animal enzymes are thus homologous. Nevertheless, the energy metabolism of seafood species shows a number of differences from that of terrestrial animals, and these differences contribute to the characteristics of seafood products.

The red and white fish muscle cells are anatomically separated into minor portions of red muscle and major portions of white muscle tissues (see Fig. 17.2 in the section about postmortem proteolysis in fresh fish). The energy metabolism and the histology of the two muscle tissues differ, the anaerobic white muscle having only few mitochondria and blood vessels. When blood circulation stops after death, the primary effect on the white muscle metabolism is therefore not the seizure of oxygen supply but the accumulation of waste products that are no longer removed.

Adenosine-5'-triphosphate (ATP) is the predominant nucleotide in the muscle cells of rested fish and represents a readily available source of metabolic energy. It regulates numerous biochemical processes and continues to do so during postmortem processes. Both the formation and degradation of ATP continue postmortem, the net ATP level being the result of the difference between the rates of the two processes. In the muscle of rested fish, the ATP content averages 7–10 μmol/g tissue (Cappeln et al. 1999, Gill 2000). This concentration is low in comparison with the turnover of ATP, and the ATP pool would thus be quickly depleted if not continously replenished.

Postmortem Generation of ATP

In anaerobic tissues such as postmortem muscle tissue, ATP can be generated from glycogen by glycolysis. Smaller amounts of ATP may also be generated by other substances, including creatine or arginine phosphate, glucose, and adenosine diphosphate (ADP).

Creatine phosphate provides an energy store that can be converted into ATP by a reversible transfer of a high-energy phosphoryl group to ADP catalyzed by sarcoplasmic creatine kinase (EC 2.7.3.2). The creatine phosphate in fish is usually depleted within hours after death (Chiba et al. 1991). In many invertebrates, including shellfish, creatine phosphate is often substituted by arginine phosphate, and ATP is thus generated by arginine kinase (EC 2.7.3.3) (Ellington 2001, Takeuchi 2004).

Glucose can enter glycolysis after being phosphorylated by hexokinase (EC 2.7.1.1). Although glucose represents an important energy resource in living animals, it is in scarce supply in postmortem muscle tissue due to the lack of blood circulation, and it contributes only little to the postmortem formation of ATP.

ATP is also formed by the enzyme adenylate kinase (EC 2.7.4.3), which catalyzes the transfer of a phosphoryl group between two molecules of ADP, producing ATP and adenosine monophosphate (AMP). As AMP is formed, adenylate kinase participates also in the degradation of adenine nucleotides, as will be described later.

The glycogen deposits in fish muscle are generally smaller than the glycogen stores in the muscle of rested mammals. Nevertheless, muscle glycogen is normally responsible for most of the postmortem ATP formation in fish. The degradation of glycogen and the glycolytic formation of ATP and lactate proceed by pathways that are similar to those in mammals. Glycogen is degraded by glycogen phosphorylase

(EC 2.4.1.1), producing glucose-1-phosphate. The highly regulated glycogen phosphorylase is inhibited by glucose-1-phosphate and glucose and is activated by AMP. Glucose-1-phosphate is subsequently transformed into glucose-6-phosphate by phosphoglucomutase (EC 5.4.2.2). Glucose-6-phosphate is further degraded by the glycolysis, leading to the formation of ATP, pyruvate, and NADH. NADH originates from the reduction of NAD^+ during glycolysis. In the absence of oxygen, NAD^+ is regenerated by lactate dehydrogenase (EC 1.1.1.27), as pyruvate is reduced to lactate. The formation of lactate forms a metabolic blind end and causes pH to fall, the final pH level being reached when the formation of lactate has stopped. The postmortem concentration of lactate reflects, by and large, the total degradation of ATP. Since in living fish, however, removal of lactate from the large white muscle often proceeds only slowly, most lactate formed just before death may remain in the muscle postmortem. Thus, the final pH level in the fish is affected only a little by the stress to which the fish was exposed and the struggle it went through during capture (Foegeding et al. 1996).

The accumulation of lactate in fish is limited by the amount of glycogen in the muscle. Since glycogen is present in fish muscle at only low concentrations, the final pH level of the fish meat is high in comparison to that of other types of meat, typically 6.5–6.7. In species that have more muscular glycogen, this substance rarely limits the formation of lactate, which may continue until the catalysis stops for other reasons, at a pH of approximately 5.5–5.6 in beef (Hultin 1984). The final pH depends not only on the lactate concentration, but also on the pH buffer capacity of the tissue. In the neutral and slightly acidic pH range, the buffer capacity of the muscle tissue in most animals, including fish, originates primarily from histidyl amino acid residues and phosphate groups (Somero 1981, Okuma and Abe 1992). Some fish species, in order to decrease muscular acidosis, have evolved particularly high pH buffer capacities. For example, pelagic fish belonging to the tuna family (genus *Thunnus*) possess a very high glycolytic capacity (Storey 1992, and references therein) and also accumulate high concentrations of histidine and anserine (*N*-(β-alanyl)-*N*-methylhistidine) (van Waarde 1988, Dickson 1995). Since tuna and a few other fish species can effectively retain metabolic heat, their core temperature at the time of capture may significantly exceed the water temperature. The high muscle temperature of such species expectedly accelerates autolytic processes, including proteolysis, and may, in combination with the lowering of pH, promote protein denaturation (Haard 2002).

Some species of carp, including goldfish, are able to survive for several months without oxygen. The metabolic adaptation this requires includes substituting ethanol for lactate as the major end product of glycolysis. Ethanol has the advantage over lactate that it can be easily excreted by the gills and does not lead to acidosis (Shoubridge and Hochachka 1980). Under anoxic conditions, NAD^+ is thus regenerated by muscular alcohol dehydrogenase (EC 1.1.1.1).

Postmortem Degradation of Nucleotides

ATP can be metabolized by various ATPases. In muscle tissues, most of the ATPase activity is represented by Ca^{2+}-dependent myosin ATPase (EC 3.6.1.32) that is directly involved in contraction of the myofibrils. Thus, release of Ca^{2+} from its containment in the sarcoplasmic reticulum dramatically increases the rate of ATP breakdown.

The glycogen consumption during and after slaughter is highly dependent on the fishing technique and the slaughtering process employed. In relaxed muscle cells the sarcoplasmic Ca^{2+} levels are low, but upon activation by the neuromuscular junctions, Ca^{2+} is released from the sarcoplasmic reticulum, activating myosin ATPase. Since the muscular work performed during capture is often intensive, the muscular glycogen stores may be almost depleted even before slaughtering takes place. The activation of myosin ATPase by spinally mediated reflexes continues after slaughter, but the extent to which it occurs can be reduced by employing rested harvest techniques such as anesthetization (Jerrett and Holland 1998, Robb et al. 2000); by destroying the nervous system by use of appropriate methods such as *iki jime,* where the brain is destroyed by a spike (Lowe 1993); or by use of still other techniques (Chiba et al. 1991, Berg et al. 1997).

ATPases hydrolyze the terminal phosphate ester bond, forming ADP. ADP is subsequently degraded by adenylate kinase, forming AMP. The degradation of ADP is favored by the removal of AMP by AMP deaminase (EC 3.5.4.6), producing inosine monophosphate (IMP) and ammonia. AMP deaminase is a key regulator of the intracellular adenine nucleotide pool and as such is a highly regulated enzyme inhibited by inorganic phosphate, IMP and NH_3 and activated by ATP.

It is widely accepted that IMP contributes to the desirable taste of fresh seafood (Fluke 1994, Gill 2000, Haard 2002). IMP is present in small amounts in freshly caught relaxed fish but builds up during the depletion of ATP (Berg et al. 1997). IMP can be hydrolyzed to form inosine, yet the rate of hydrolysis in fish is generally low, following zero-order kinetics, indicating that the IMP-degrading enzymes are fully substrate saturated (Tomioka et al. 1987, Gill 2000, Itoh and Kimura 2002). However, a large species variation in the degradation of IMP has been shown, and IMP may thus persist during storage for several days or even weeks (Dingle and Hines 1971, Gill 2000, Haard 2002). The formation of inosine is often rate limiting for the overall breakdown of nucleotides and is considered to constitute the last purely autolytic step in the nucleotide catabolism of chilled fish. The formation of inosine usually indicates a decline in the prime quality of seafood (Surette et al. 1988, Gill 2000, Haard 2002). IMP can be hydrolyzed by various enzymes, 5'-nucleotidase (EC.3.1.3.5) being regarded as the most important of these in the case of chilled fish (Gill 2000, Haard 2002). Several forms of 5'-nucleotidase exist, of which a soluble sarcoplasmic form has been documented in fish (Itoh and Kimura 2002).

The degradation of inosine continues via hypoxanthine and its oxidized products xanthine and uric acid. These reactions, all of them related to spoilage, are catalyzed by both endogenous and microbial enzymes, the activity of the latter depending upon the spoilage flora that are present and thus differing much between species and products. Inosine can be degraded by either nucleoside phosphorylase (EC 2.4.2.1) or inosine nucleosidase (EC 3.2.2.2), leading to the formation of different by-products (see Fig. 17.1). The final two-step oxidation of

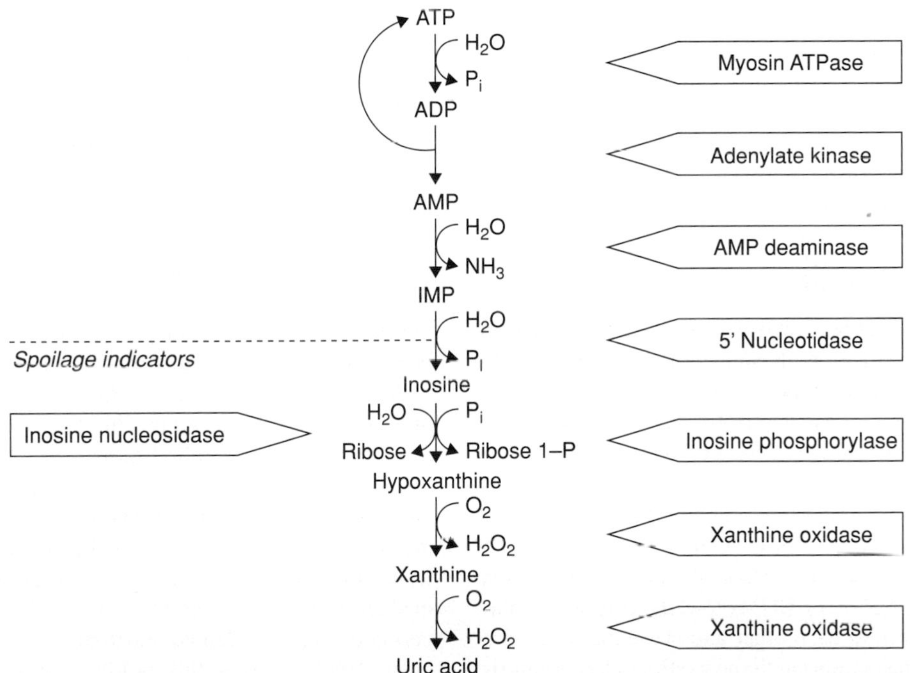

Figure 17.1. The successive reactions of the main pathway for nucleotide degradation in fish. Substances listed below the dotted line are generally considered to be indicators of spoilage. P_i is inorganic phosphate.

hypoxanthine is catalyzed by one and the same enzymes, xanthine oxidase (EC 1.1.3.22). The oxidation requires a reintroduction of oxygen in the otherwise anaerobic pathway, which often limits the extent of the xanthine oxidase reaction and causes inosine and hypoxanthine to accumulate. Hypoxanthine is an indicator of spoilage and may contribute to bitter taste (Hughes and Jones 1966). The hydrogen peroxide produced by the oxidation of hypoxanthine can be expected to cause oxidation, but the technological implications of this are minor, since it is only after the quality of the fish has declined beyond the level of acceptability that these reactions take place on a large scale.

The relation between the accumulation of breakdown products of ATP and the postmortem storage period of seafood products has resulted in various freshness indicators being defined. Saito et al. (1959) defined a freshness indicator *K* as a simplified way of expressing the state of nucleotide degradation by a single number:

$$K = 100 \times \frac{[\text{Inosine}] + [\text{Hypoxanthine}]}{\left(\begin{array}{c} [\text{ATP}] + [\text{ADP}] + [\text{AMP}] + [\text{IMP}] \\ + [\text{Inosine}] + [\text{Hypoxanthine}] \end{array} \right)}$$

Higher *K* values correspond to a lower quality of fish. The *K* value often increases linearly during the first period of storage on ice, but the *K* value at sensory rejection differs between species. An alternative and simpler K_i value correlates well with the *K* value of wild stock fish landed by traditional methods (Karube et al. 1984).

$$K_i = 100 \times \frac{[\text{Inosine}] + [\text{Hypoxanthine}]}{[\text{IMP}] + [\text{Inosine}] + [\text{Hypoxanthine}]}$$

Both *K* values reflect fish quality well, provided the numerical values involved are not compared directly with the values from other species. Still *K* values necessarily remain less descriptive than the concentrations of the degradation products themselves.

ATP is an important regulator of biochemical processes in all animals and continues to be so during the postmortem processes. One of the most striking postmortem changes is that of rigor mortis. When the concentration of ATP is low, the contractile filaments lock into each other and cause the otherwise soft and elastic muscle tissue to stiffen. Rigor mortis is thus a direct consequence of ATP depletion. The biochemical basis and physical aspects of rigor mortis are reviewed by Hultin (1984) and Foegeding et al. (1996).

ATP depletion and the onset of rigor mortis in fish are highly correlated with the residual glycogen content at the time of death and with the rate of ATP depletion. The onset of rigor mortis may occur only a few minutes after death or may be postponed for up to several days when rested harvest techniques, gentle handling, and optimal storage conditions are employed (Azam et al. 1990, Lowe et al. 1993, Skjervold et al. 2001). The strength of rigor mortis is greater when it starts soon after death (Berg et al. 1997). Although the recommended temperature for chilled storage is generally close to 0°C, in some fish from tropical and subtropical waters, the degradation of nucleotides has been found to be slower, and the onset of rigor mortis to be further delayed when the storage temperature is elevated to 5, 10, and even 20°C (Saito et al. 1959; Iwamoto et al. 1987, 1991).

Rigor mortis impedes filleting and processing of fish. In the traditional fishing industry, therefore, filleting is often postponed until rigor mortis has been resolved. In aquaculture, however, it has been shown that rested harvesting techniques can postpone rigor mortis sufficiently to allow prerigor filleting to be carried out (Skjervold et al. 2001). Fillets from prerigor filleted fish shorten, in accordance with the rigor contraction of the muscle fibers, by as much as 8%. Prerigor filleting has few problems with gaping, since the rigor contraction of the fillet is not hindered by the backbone (Skjervold et al. 2001). Gaping, which represents the formation of fractures between segments of the fillets, is described further in the section Postmortem Proteolysis in Fresh Fish.

All enzyme reactions are virtually stopped during freezing at low temperatures. The ATP content of frozen prerigor fish can thus be stabilized. During the subsequent thawing of prerigor fish muscle, the leakage of Ca^{2+} from the organelles results in a very high level of myosin Ca^{2+}–ATPase activity and a rapid consumption of ATP. This leads to a strong form of rigor mortis called thaw rigor. Thaw rigor results in an increase in drip loss, in flavor changes, and in a dry and tough texture (Jones 1965) and gaping (Jones 1969). Thaw rigor can be avoided by controlled thawing, holding the frozen products at intermediate freezing temperatures above −20°C for a period of time (Mcdonald and Jones 1976, Cappeln et al. 1999). During this holding period, ATP is degraded at moderate rates, allowing a slow onset of rigor mortis in the partially frozen state.

Rigor mortis is a temporary condition, even though in the absence of ATP, the actin-myosin complex is locked. The resolution of rigor mortis is due to structural decay elsewhere in the muscular structure, as will be discussed later.

ENZYMATIC DEGRADATION OF TRIMETHYLAMINE-*N*-OXIDE

The substance trimethylamine-*N*-oxide (TMAO) is found in all marine seafood species but occurs in some freshwater fish as well (Anthoni et al. 1990, Parab and Rao 1984, Niizeki et al. 2002). It contributes to cellular osmotic pressure, as previously described, but several other physiological functions of TMAO have also been suggested. TMAO itself is a harmless and nontoxic constituent, yet it forms a precursor of undesirable breakdown products. In seafood products, TMAO can be degraded enzymatically by two alternative pathways as described below.

THE TRIMETHYLAMINE-*N*-OXIDE REDUCTASE REACTION

Many common spoilage bacteria reduce TMAO to trimethylamine (TMA) by means of the enzyme trimethylamine-*N*-oxide reductase (EC 1.6.6.9):

$$O{=}N(CH_3)_3 + NADH \rightarrow N(CH_3)_3 + NAD^+ + H_2O$$

TMA has a strong fishy odor, and TMAO reductase activity is responsible for the typical off-odor of spoiled fish. Since TMAO reductase is of microbial origin, the formation of TMA occurs primarily under conditions such as cold storage that allow microbial growth to take place. TMA is thus an important spoilage indicator of fresh seafood products. The formation of TMA is further discussed in the food microbiology literature (e.g., Barrett and Kwan 1985, Dalgaard 2000).

THE TRIMETHYLAMINE-*N*-OXIDE ALDOLASE REACTION

TMAO is also the precursor of the formation of dimethylamine (DMA) and formaldehyde:

$$O{=}N(CH_3)_3 \rightarrow H{-}N(CH_3)_2 + H{-}C({=}O){-}H$$

This reaction is catalyzed by trimethylamine-*N*-oxide aldolase (TMAOase or TMAO demethylase, EC 4.1.2.32) but may also to some extent be nonenzymatic, catalyzed by iron and various reductants (Vaisey 1956, Spinelli and Koury 1981, Nitisewojo and Hultin 1986, Kimura et al. 2002). In most cases in which significant amounts of formaldehyde and DMA are accumulated, however, species possessing TMAOase enzyme activity are involved. The reaction leads to cleavage of a C-N bond and the elimination of an aldehyde, resulting in the classification of TMAOase as a lyase (EC 4.1.2.32) and the IUBMB name aldolase.

DMA is a reactive secondary amine with a milder odor than TMA. Formaldehyde is highly reactive and strongly affects the texture of fish meat by making it tougher, harder, more fibrous, and less juicy, as well as increasing the drip loss in the thawed products. The quality changes are associated with a loss in protein solubility and in particular the solubility of the myofibrillar proteins. For reviews, see Sikorski and Kostuch (1982), Hultin (1992), Mackie (1993), Sikorski and Kolakowska (1994) and Sotelo et al. (1995).

Formaldehyde can react with a number of chemical groups, including some protein amino acid residues and terminal amino groups, resulting in denaturation and possibly in the cross-linking of proteins, both of which are believed to be the cause of the observed effects on seafood products. The formaldehyde concentration in severely damaged seafood products may reach 240 μg/g (Nielsen and Jørgensen 2004). These concentrations are generally considered nontoxic, but may still exceed various national trade barrier limits.

Whereas TMAO is widespread, TMAOase is only found in a limited number of animals, many of which belong to the order of gadiform fish (pollock, cod, etc.). In gadiform species the highest levels of enzyme activity are found in the inner organs (kidney, spleen, and intestine), while the enzyme activity in the large white muscle is low. The formation of DMA and formaldehyde in various tissues of certain nongadiform fish, as well as of crustaceans and mollusks, has also been reported, as reviewed by Sotelo and Rehbein (2000), although the taxonomic distribution of the enzyme has not been investigated systematically. The TMAOase content of gadiform fish exhibits large individual variation, probably due to the influence of biological factors that have not yet been adequately studied (Nielsen and Jørgensen 2004).

The enzyme is stable and tolerates both high salt concentrations and freezing. The accumulation of DMA and formaldehyde progresses only slowly, and most of the accumulation is produced during prolonged frozen storage or cold storage of the salted fish (e.g., salted cod and baccalao). During freezing, the enzymatic reaction proceeds as long as liquid water and substrate are available. In practice, the formation of DMA and formaldehyde is found at temperatures down to approximately −30°C (Sotelo and Rehbein 2000). At higher temperatures the rate of formation of formaldehyde is also higher; thus, freezing of gadiform fish at insufficiently low temperatures may result in dramatic changes in the solubility of myofibrillar proteins within a few weeks (Nielsen and Jørgensen 2004).

Despite the TMAOase concentration of the white muscle being low, it is the enzyme activity of the white muscle that is responsible for the accumulation of formaldehyde in whole fish and in fillets (Nielsen and Jørgensen 2004). In the case of minced products, even minor contamination by TMAOase-rich tissues leads to a marked rise in the rate of accumulation (Dingle et al. 1977, Lundstrøm et al. 1982, Rehbein 1988, Rehbein et al. 1997). Despite its dependency upon other factors, such as cofactors, the rate of accumulation of formaldehyde can to a large extent be predicted from the TMAOase enzyme activity of fish meat alone (Nielsen and Jørgensen 2004).

The physiological function of TMAOase remains unknown. The formation of formaldehyde could have a digestive function, yet this would not explain the high TMAOase content of the kidney and spleen. Accordingly, it has been speculated that TMAO may not even be its primary natural substrate (Sotelo and Rehbein 2000).

The in vitro reaction rate is low without the presence of a number of redox-active cofactors. Although the overall TMAOase reaction is not a redox reaction, its dependency on redox agents shows that the reaction mechanism must include redox steps. Little is known about the in vivo regulation of TMAOase, but studies suggest the importance of nucleotide coenzymes and iron (Hultin 1992). TMAOase appears to be membrane bound and is only partly soluble in aqueous solutions. This has complications for its purification and is one of the main reasons why the complete characterization of the enzyme remains yet to be done.

POSTMORTEM PROTEOLYSIS IN FRESH FISH

Postmortem proteolysis is an important factor in many changes in seafood quality. During cold storage, the postmortem proteolysis of myofibrillar and connective tissue proteins contributes to deterioration in texture. Despite the natural tenderness of seafood, texture is an important quality parameter, in fish and shellfish alike. Deterioration in seafood quality through proteolysis involves a softening of the muscle tissue. This reduces the cohesiveness of the muscle segments in fillets, which promotes gaping, a formation of gaps and slits between muscle segments. The negative character of this effect contrasts with the effect of postmortem proteolysis on the meat of cattle and pigs, in which the degradation of myofibrillar proteins produces a highly desired tenderization in the conversion of muscle to meat.

The proteolytic enzymes that cause a softening of seafood are of basically the same classes as those found in terrestial animals. The special effects of proteolysis on seafood result from the combined action of the homologous proteolytic enzymes on the characteristic muscle structures of seafood. The muscle of fish differs on a macrostructural level from that of mammals, fish muscle being segmented into muscle blocks called myotomes. A myotome consists of a single layer of muscle fibers arranged side by side and separated from the muscle fibers of the adjacent myotomes by collagenous sheets termed the myocommata (Bremner and Hallett 1985). See Figure 17.2.

A myotome resembles a mammalian skeletal muscle in terms of its intracellular structure (see Fig. 17.3) and the composition of the extracellular matrix. Each muscle fiber is surrounded by fine collagen fibers, the endomysium, joined with a larger network of collagen fibers, the perimysium, which is contiguous with the myocommata (Bremner and Hallet 1985). Although the endomysium, perimysium, and myocommata are considered to be discrete areas of the extracellular matrix, they join to form a single weave.

Such phenomena in seafood as softening, gaping, and the resolution of rigor are believed to be caused by hydrolysis of the myofibrillar and the extracellular matrix proteins. Initially the disintegration of the attachment between the myocommata and muscle fibers leads to the resolution of rigor (Taylor et al.

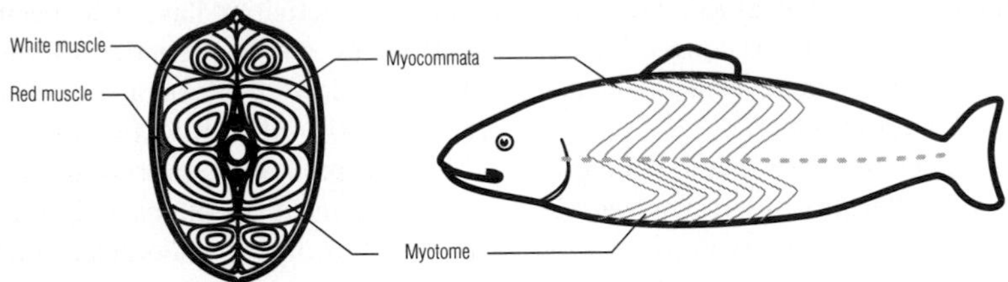

Figure 17.2. Gross anatomy of fish muscle. (Drawing courtesy of A. S. Matforsk, Norwegian Food Research Institute.)

2002). Endogenous proteolytic enzymes that cleave muscle proteins under physiological conditions and at neutral pH can be a factor in the resolution of rigor mortis. The continuation of this process is regarded as one of the main causes of gaping (Taylor et al. 2002, Fletcher et al. 1997).

The softening of fish muscle appears to be the result of multiple changes in muscle structure. Histological studies have shown that the attachment between muscle fibers in fish muscle is broken during ice storage and has been associated with a loss of "hardness" as measured by instrumental texture analysis (Taylor et al. 2002). During cold storage, the junction between the myofibrils and connective tissue of the myocommata is hydrolyzed (Bremner 1999, Fletcher et al. 1997, Taylor et al. 2002), and the collagen fibers of the perimysium surrounding the bundles of muscle fibers are degraded (Ando et al. 1995, Sato et al. 2002).

It appears that cleavage of the costameric proteins that link the myofibrils (the sarcomere) with the sarcolemma (the cell membrane), and of the basement membrane, which is attached to the fine collagen fibers of the endomysium (see Fig. 17.3), leads to muscle fibers being detached, which results in softening. Studies have shown that the costameric proteins found in fish muscle are already degraded 24 hours postmortem (Papa et al. 1997). This emphasizes the importance of early postmortem changes.

It has been shown that changes in the collagen fraction of the extracellular matrix and the softening of fish are related. Collagen V, a minor constituent of the pericellular connective tissue of fish muscle, becomes soluble when rapid softening takes place, whereas no solubilization is observed in fish that do not soften (Sato et al. 2002). Recent studies also indicate a degradation of the most abundant collagen in fish muscle, collagen I, during 24 hours of cold

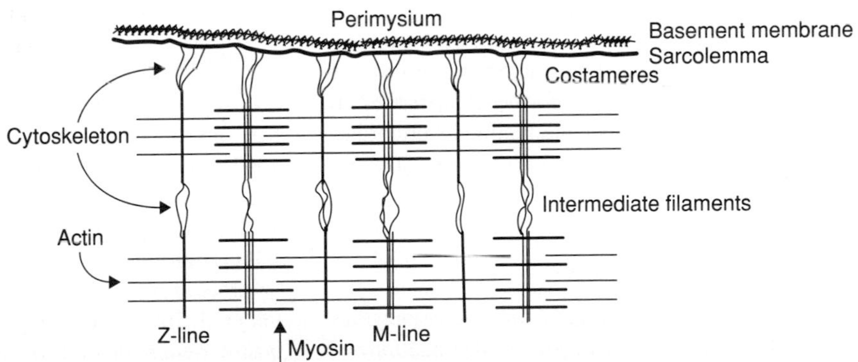

Figure 17.3. A longitudinal view of the intracellular structure of a skeletal muscle fiber showing the contractile elements (actin and myosin), the cytoskeleton, and the attachment to the extracellular matrix (endomysium and perimysium). (Adapted from Lødemel 2004.)

storage (Shigemura et al. 2004). This indicates that the degradation of collagen and the resulting weakening of the intramuscular pericellular connective tissue also play a role in the textural changes in fish muscle that occur early postmortem.

A number of studies have revealed structural changes in the myofibrillar proteins in fish muscle during cold storage. Analysis of myofibrillar proteins from the muscle of salmon stored at 0°C for as long as 23 days has shown that several new protein fragments form (Lund and Nielsen 2001). Other studies suggest that predominantly proteins of the cytoskeletal network, such as the high molecular weight proteins titin and nebulin, are degraded (Busconi et al. 1989, Astier et al. 1991). The extent of degradation of the muscle myofibrillar proteins varies among species. It has been found, for example, that the intermediate filament protein desmin is clearly degraded during the cold storage of turbot and sardines but that no degradation occurs during the cold storage of sea bass and brown trout (Verrez-Bagnis et al. 1999).

As research shows, both myofibrillar and extracellular matrix proteins in the muscle of many fish are degraded during storage, and textural changes can be expected to result from degradation of proteins from both structures.

Proteases in Fish Muscle

The structural and biochemical changes just described can be considered to largely represent the concerted action of different endogenous proteolytic enzymes. Proteolytic enzymes of all major classes have been documented in the muscle of various fish species. An overview of the different proteases believed to play a role in the postmortem proteolysis of seafood is presented below. Further information on the proteases found in fish and marine invertebrates can be found in reviews by Kolodziejska and Sikorski (1995, 1996).

Matrix Metalloproteinases

Matrix metalloproteinases are extracellular enzymes involved in the in vivo catabolism (degradation) of the extracellular matrix (degradation of the helical regions of the collagens). They have been isolated from rainbow trout (Saito et al. 2000), Japanese flounder (Kinoshita et al. 2002) and Pacific rockfish (Bracho and Haard 1995). Collagenolytic and gelatinolytic activities have also been detected in the muscle of winter flounder (Teruel and Simpson 1995), yellowtail (Kubota et al. 1998), ayu (Kubota et al. 2000), salmon, and cod (Lødemel and Olsen 2003). Type I collagen is solubilized and degraded by matrix metalloproteinases in rainbow trout (Saito et al. 2000), Japanese flounder (Kubota et al. 2003), and Pacific rockfish (Bracho and Haard 1995), suggesting that these proteases can participate in postmortem textural changes.

Cathepsins

Lysosomal proteinases such as cathepsins B, D, and L have been isolated from a number of fish species, including herring (Nielsen and Nielsen 2001), mackerel (Aoki and Ueno 1997), and tilapia (Jiang et al. 1991). In vitro studies show that certain cathepsins are capable of cleaving myofibrillar proteins (Ogata et al. 1998, Nielsen and Nielsen 2001) and may also participate in degradation of the extracellular matrix since they can cleave both nonhelical regions of the collagen (Yamashita and Konagaya 1991) and collagen that has already been partly degraded by matrix metalloproteinases. Since cathepsins in the muscle of living fish are located in the lysosomes, they are not originally in direct contact with either the myofibrils or the extracellular matrix, although it has been shown that the enzymes leak from the lysosomes in fish muscle postmortem, and from lysosomes in bovine muscles postmortem as well (Geromel and Montgomery 1980, Ertbjerg et al. 1999).

Calpains

Calpains have been reported in the muscle of various fish species (Wang and Jiang 1991, Watson et al. 1992). They are only active in the neutral pH range, although research shows that they also remain active at the slightly acidic postmortem pH (Wang and Jiang 1991) and are capable of degrading myofibrillar proteins in vitro (Verrez-Bagnis et al. 2002, Geesink et al. 2000). This suggests a participation in the postmortem hydrolysis of fish muscle. Watson et al. (1992) found evidence for calpains being involved in the degradation of myofibrillar proteins in tuna, leading to the muscle becoming pale and grainy, which is referred to as "burnt tuna."

20S Proteasome

The 20S proteasome enzyme is a 700 kDa multicatalytic proteinase with three major catalytic sites. The term 20S refers to its sedimentation coefficient. In the eukaryotic cells, the enzyme exists in the cytoplasm, either in a free state or associated with large regulatory complexes. In vivo, it is involved in nonlysosomal proteolysis and apoptosis. Research strongly indicates that this proteasome is involved in meat tenderization in cattle (Sentandreu et al. 2002). Although proteasomes have been detected in the muscle of such fish species as carp (Kinoshita et al. 1990a), white croaker (Busconi et al. 1992), and salmon (Stoknes and Rustad 1995), their postmortem activity in fish muscle has not been clarified.

Heat-Stable Alkaline Proteinases

Kinoshita et al. (1990b) reported the existence of up to four distinct heat-stable alkaline proteinase (HAP) in fish muscle: two sarcoplasmic proteinases activated at 50 and 60°C, and two myofibril-associated proteinases activated at 50 and 60°C, respectively. The distribution of the four proteinases was found to be quite diverse among the 12 fish species that were studied. The mechanisms activating these proteinases in vivo and their precise physiological functions are not clear.

The participation of one or the other protease in the many different degradation scenarios that occur has still been only partially elucidated. However, various studies have found a close relationship between protein degradation in fish muscle and the activity of specific proteases. A high level of activity of cathepsin L has been found in chum salmon during spawning, a period during which the fish exhibit an extensive softening in texture (Yamashita and Konagaya 1990). Similarly, Kubota et al. (2000) found an increase in gelatinolytic activity in the muscle of ayu during spawning, a period involving a concurrent marked decrease in muscle firmness. Also, in the muscle of hake, a considerably higher level of proteolytic activity during the prespawning period than in the postspawning period was found (Perez-Borla et al. 2002).

The possible existence of a direct link between protease activity and texture has been explored in situ by perfusing protease inhibitors into fish muscle and later measuring changes in texture during cold storage. The activity of metalloproteinase and of trypsin-like serine protease was found in this way to play a role in the softening of flounder (Kubota et al. 2001) and of tilapia (Ishida et al. 2003).

POSTMORTEM HYDROLYSIS OF LIPIDS IN SEAFOOD DURING FROZEN AND COLD STORAGE

Changes in the lipid fraction of fish muscle during storage can lead to changes in quality. Both the content and the composition of the lipids in fish muscle can vary considerably between species and from one time of year to another, and also differ greatly depending upon whether white or red muscle fibers are involved. As already mentioned, these two types of muscle fibers are separated from each other, the white fibers generally constituting most of the muscle as a whole, although fish species vary considerably in the amounts of dark meat, which has a higher myoglobin and lipid content. In species like tuna and in small and fatty pelagic fish, the dark muscle can constitute up to 48% of the muscle as a whole, whereas in lean fish such as cod and flounder, the dark muscle constitutes only a small percentage of the muscle (Love 1970). Since triglycerides are deposited primarily in the dark muscle, providing fatty acids as substrate to aerobic metabolism, whereas the phospholipids represent most of the lipid fraction of the white muscle, phospholipids constitute a major part of the lipid fraction in lean fish (Lopez-Amaya and Marangoni 2000b).

Not much research on lipid hydrolysis in fresh fish during ice storage has been carried out, research having concentrated more on changes in the lipid fraction during frozen storage. This could be due to freezing being the most common way of storing and processing seafood and to lipid hydrolysis playing no appreciable role in ice-stored fish before microbial spoilage becomes extreme. Knudsen (1989), however, detected an increase in free fatty acids in cod muscle during 11 days of ice storage, indicative of the occurrence of lipolytic enzyme activity, an increase that was most pronounced during days 5 to 11. Ohshima et al. (1984) reported a similar delay in the increase in free fatty acids in cod muscle stored in ice for 30 days. Both results are basically consistent with the observation of Geromel and Montgomery (1980) of no lysosomal lipase activity being evident in trout muscle after seven days on ice. In contrast the authors reported that slow freezing and fluctuations in temperature during frozen storage were found to result in the release of acid lipase from the lysosomes of the dark muscle of rainbow trout.

Several researchers have reported an increase in free fatty acids during frozen storage of muscle of different fish species such as trout (Ingemansson et al. 1995), salmon (Refsgaard et al. 1998, 2000), rayfish (Fernandez-Reiriz et al. 1995), tuna, cod, and prawn (Kaneniwa et al. 2004). The release of free fatty acids during frozen storage can induce changes in texture by stimulation of protein denaturation and through off-flavors being produced by lipid oxidation (Lopez-Amaya and Marangoni 2000b). Refsgaard et al. (1998, 2000) observed a marked increase in free fatty acids in salmon stored at −10 and −20°C. This increase in free fatty acid content was connected with changes in sensory attributes, suggesting that lipolysis plays a significant role in deterioration of the quality of salmon during frozen storage.

Kaneniwa et al. (2004) detected a large variation in the formation of free fatty acids among nine species of fish and shellfish stored at −10°C for 30 days. Once again, this demonstrates the large variation in enzyme activity in seafood species. Findings of Ben-gigirey et al. (1999) indicate that the temperature at which fish are stored has a clear influence on the lipase activity occurring in the muscle. They noted, for example, that the formation of free fatty acids in the muscle of albacore tuna during storage for the period of a year was considerably higher at −18°C than at −25°C.

Two classes of lipases, the lysosomal lipases and the phospholipases, are apparently involved in the hydrolysis of lipids in fish muscle during storage. Nayak et al. (2003) found significant differences among four fish species (rohu, oil sardine, mullet, and Indian mackerel) in the degree of red muscle lipase activity. This is quite in line with differences in lipid hydrolysis among species that Kaneniwa et al. (2004) reported.

Although both the lipase activity and the formation of free fatty acids in fish muscle are well documented, only a few studies have actually isolated and characterized the muscle lipases and phospholipases involved. Aaen et al. (1995) have isolated and characterized an acidic phospholipase from cod muscle, and Hirano et al. (2000) a phospholipase from the white muscle of bonito. Similarly, triacylglycerol lipase from salmon and from rainbow trout has been isolated and characterized by Sheridan and Allen (1984) and by Michelsen et al. (1994), respectively. Knowledge of the properties of the lipolytic enzymes in the muscle in seafood and of the responses the enzymes show to various processing parameters, however, is sparse. Extensive reviews of work done on the lipases and phospholipases in seafood has been presented by Lopez-Amaya and Marangoni (2000a,b).

ENDOGENOUS ENZYMATIC REACTIONS DURING THE PROCESSING OF SEAFOOD

During seafood processing such as salting, heating, fermentation, and freezing, endogenous enzymes can be active and contribute to the sensory characteristics of the final product. Such endogenous enzyme activities are sometimes necessary in order to obtain the desired taste and texture.

Salting of Fish

Salting has been used in many countries for centuries as a means of preserving fish. Today, the primary purpose of salting fish is no longer only to preserve them. Instead salting enables fish products with sensory attributes that are sought after, such as salted herring and salted cod, as produced on the northern European continent and in Scandinavia. Enzymatic degradation of muscle proteins during salting is a factor that contributes to the development of the right texture and taste of the products.

Ripening of Salt-Cured Fish

Spiced sugar–salted herring in its traditional form is made by mixing approximately 100 kg of headed, ungutted herring with 15 kg of salt and 7 kg of sugar in barrels, usually adding spices as well. After a day or two, after a blood brine has been formed, saturated brine is added, after which the barrels are stored at 0–5°C for up to a year. During this period a ripening of the herring takes place, and it achieves its characteristic taste and texture (Stefánsson et al. 1995).

During the ripening period, both intestinal and muscle proteases participate in the degradation of muscle protein, contributing to the characteristic softening of the fillet and liberating free amino acids and small peptides that help create the characteristic flavor of the product (Nielsen 1995, Olsen and Skåra 1997). Studies have shown that intestinal trypsin- and chymotrypsin-like enzymes migrate into the fillet, where they play an active role in the degradation of muscle proteins during storage (Engvang and Nielsen 2000, Stefánsson et al. 2000). Furthermore, Nielsen (1995) shows that muscle amino peptidases are also active in the salted herring during storage.

In southern Europe, a similar product based on the use of whole sardines or anchovies is produced. In contrast to the salted herring from Scandinavia, these salted sardines and anchovies are stored at ambient temperature, which can vary between 18°C and 30°C (Nunes et al. 1997). Nunes et al. (1997) found that proteases from both the intestines and the muscles participated in the ripening of sardines *(Sardina pilchardus)*. Hernadez-Herrero et al. (1999) reported an increase in proteinase activity as well as in protein hydrolysis during the storage of salted anchovies *(Engraulis encrasicholus)* and found a close relationship between proteolysis and the development of the sensory characteristics of the product.

Production of Fish Sauce and Fish Paste

Fish sauce and fish paste are fermented fish products produced mainly in Southeast Asia, where they are highly appreciated food flavorings. Fish sauce is the liquefied protein fraction, and fish paste is the "solid" protein fraction obtained from the prolonged hydrolysis of heavily salted small pelagic fish. Production takes place in closed tanks at ambient tropical temperatures during a period of several months (Gildberg 2001, Saishiti 1994). The hydrolysis represents the combined action of the fishes' own digestive proteases and of enzymes from halotolerant lactic acid bacteria (Saishiti 1994). Orejani and Liston (1981) concluded, on the basis of inhibitor studies, that a trypsin-like protease is one of the enzymes responsible for the hydrolysis. Vo et al. (1984) detected a high and stable level of activity of intestinal amino peptidase during the production of fish sauce. Del Rosario and Maldo (1984) measured the activity of four different proteases in fish sauce produced from horse mackerel. During a four-month period of measurement, they found the activity of cathepsins A, C, and B to be stable and that of cathepsin D to decrease. Raksakulthai and Haard (1992) found that cathepsin C obtained from capelin was active in the presence of 20–25% salt, suggesting that this enzyme is involved in the hydrolysis of capelin fish sauce. These studies indicate that several different proteases need to be active and that their concerted action is necessary to achieve the pronounced hydrolysis required for production of such fish sauces.

It is notable that the similarly high storage temperatures present in southern Europe do not lead to a solubilization of salted sardines and anchovies. Ishida et al. (1994) reported that at 35°C salted Japanese anchovies *(E. japonica)* degraded to a marked degree, whereas at this temperature salted anchovies from southern Europe *(E. encrasicholus)* were structurally stable. Also, they detected a thermostable trypsin-like proteinase in the muscle of both salted Japanese anchovies and European anchovies, but its activity was much higher in the Japanese anchovies. An explanation of the difference between the European and the Asian products might be a large difference between the hydrolytic enzyme activity of the respective raw materials.

Production of Surimi

Surimi is basically a myofibrillar protein concentrate that forms a gel due to cross-linking of its actomyosin molecules (An et al. 1996). It is made from minced fish flesh obtained mainly from pelagic white fish of low fat content, such as Alaskan pollack and Pacific whiting. The mince is washed several times with water to remove undesired elements such as connective tissue and lipids. The particulate is then stabilized by cryoprotectants before being frozen (Park and Morrisey 2000).

Surimi is used as a raw material for the manufacture of various products, such as imitation crabmeat and shellfish substitutes. The manufacture of surimi products involves the use of a slow temperature-setting process, which can be in the temperature range of 4–40°C, followed by a heating process involving temperatures of 50–70°C, which results in the gel strength being enhanced (Park 2000). Surimi-based products are very important fish products in the Asian and Southeast Asian countries, where Japan has the largest production and marketing of surimi-based products. The quality and the price are closely dependent on the gel strength (Park 2000).

The enzyme transglutaminase plays an important role in the gelation process through catalyzing the cross-linking of the actomyosin (An et al. 1996). Some of the differences in gelling capability among different species are due to the properties and levels of activity of muscle transglutaminase (An et al. 1996, Lanier 2000). Studies have shown that the addition of microbial transglutaminase can increase the gel strength obtained in fish species having low transglutaminase activity (Perez-Mateos et al. 2002, Sakamoto et al. 1995). The effect of transglutaminase on the processing of surimi has been reviewed extensively by An et al. (1996) and by Ashie and Lanier (2000).

A softening of the gel during the temperature-setting and heating processes can occur due to autolysis of the myosin and actomyosin through the action of endogenous heat-stable proteinases. Whether or not this occurs is partly a function of the species used for the surimi production. Two groups of proteinases have been identified as being responsible for the softening: cysteine cathepsin and heat-stable alkaline proteinases (HAP).

The presence of HAP in the muscle of different fish species used for surimi production and the effects it has on degradation of the fish gel during the heating process have been taken up in a number of studies (Makinodan et al. 1985, Toyahara et al. 1990, Cao et al. 1999). The finding of Kinoshita et al. (1990b) that the fish species differ in the amount of HAP in muscle, could partly explain why softening of the gel is more pronounced in some fish species than in others.

A study by An et al. (1994) indicated that the cysteine proteinase cathepsin L contributes to degradation of the myofibrils in surimi at a temperature of about 55°C during the heating process, a result substantiated by Ho et al. (2000), who also measured the softening of mackerel surimi upon the addition of mackerel cathepsin L.

TECHNOLOGICAL APPLICATIONS OF ENZYMES FROM SEAFOOD

Utilization of enzymes from seafood as technical aids in both seafood processing and other areas of food and feed processing has been an area of active research for many years. There are two factors that have provided the primary motivation for such research: (1) the cold-adaptation properties of seafood enzymes and (2) the increasing production of marine by-products used as potential sources of enzymes. Although the results have been promising, many of the potential technologies are still in their initial stages of development and are not yet fully established industrially.

IMPROVED PROCESSING OF ROE

Roe is considered by many to be a seafood delicacy. Russian caviar, produced from the roe of sturgeon, is the form best known, although roe produced from a variety of other species, such as salmon, trout, herring, lumpfish, and cod, has also gained wide acceptance. The roe is originally covered by a two-layer membrane (chorion) termed the roe sack. In some species, mechanical or manual separation of the roe from the sack results in damage to the eggs and in yields as low as 50% (Gildberg 1993). Pepsin-like proteases isolated from the intestines of seafood species, as well as collagenases from the hepatopancreas of crabs, have been shown to cleave the linkages between the sac and the eggs without damaging the eggs. Such enzyme treatment has been reported to increase the yield from 70% to 90% (Gildberg et al. 2000, and references therein).

PRODUCTION OF FISH SILAGE

Fish silage is a liquid nitrogenous product made from small pelagic fish or fish by-products mixed with acid. It is used as a source of protein in animal feed (Aranson 1994, Gildberg 1993). In its manufacture, the fish material is mixed with 1–3.5% formic acid solution, reducing pH to 3–4. This is optimal for the intestinal proteases and aspartic muscle proteases contained in the fish material, allowing the solubilization of the fish material to proceed as an autolytic process driven by both types of protease (Gildberg et al. 2000). Gildberg and Almas (1986) have reported the existence of two very active pepsins (I and II) in silage manufactured from cod viscera. They were able to show that fish by-products having low protease activity could be hydrolyzed and used for silage by adding protease-rich cod viscera.

DESKINNING AND DESCALING OF FISH

Deskinning fish enzymatically can increase the edible yield as compared with that achieved by mechanical deskinning (Gildberg et al. 2000). It also provides the possibility of utilizing alternative species such as skate, the skin of which is very difficult to remove mechanically without ruining the flesh (Stefánsson and Steingrimsdottir 1990). It has been shown that herring can be deskinned enzymatically by use of acid proteases obtained from cod viscera (Joakimsson 1984). Enzymatic removal of the skin of other species has been reported as well. Kim et al. (1993) have described removal of the skin of filefish by use of collagenase extracted from the intestinal organs of the fish. Crude protease extract obtained from minced arrowtooth flounder has been found to be effective in solubilizing the skin of pollock (Tschersich and Choudhury 1998).

Removing squid skin can be a difficult task. Skinning machines only remove the outer skin of the squid tubes, leaving the tough rubbery inner membrane. Strom and Raa (1991) reported a gentler and more efficient enzymatic method of deskinning the squid, using digestive enzymes from the squid itself. Also, a method for deskinning the squid by making use of squid liver extract has been developed by Leuba et al. (1987).

For certain markets, such as Japanese sashimi restaurants and fresh fish markets, skin-on fillets without the scales are demanded. Also, fish skin is used in the leather industry, where descaling is likewise necessary. Obtaining a prime quality product requires gentle descaling. Mechanical descaling can be difficult to accomplish without the fish flesh being damaged, especially in the case of certain soft-fleshed species, where enzymatic descaling results in a gentle descaling (Svenning et al. 1993). Digestive enzymes of fish have proved to be useful for removing scales gently (Gildberg et al. 2000).

Improved Production of Fish Sauce

As already indicated, the original process for manufacturing fish sauce is carried out at high ambient temperature, involving the use of an autolytic process catalyzed by endogenous proteases. The rate of the hydrolysis depends on the content of digestive enzymes in the fish. There has been an obvious interest, however, in shortening the time required for producing the fish sauce, and use has also been made of other fish species, such as Arctic capelin and Pacific whiting. Arctic capelin is usually caught during the winter, when the fish has a low feed intake and its digestive enzyme content thus is low. Research has shown, however, that supplementing Arctic capelin with cod intestines or squid pancreas, both of which are rich in digestive enzymes, allows an acceptable fish sauce to be produced during the winter (Gildberg 2001, Raksakulthai et al. 1986). Tungkawachara et al. (2003) showed that fish sauce produced from a mixture of Pacific whiting and surimi by-products (head, bone, guts, and skin from Pacific whiting) has the same sensory quality as a commercial anchovy fish sauce.

Seafood Enzymes Used in Biotechnology

The poor temperature stability of seafood enzymes is a useful property that has lead to the production of enzymes useful in gene technology, where only very small amounts of enzymes are needed. Alkaline phosphatase from cold-water shrimp *(Pandalus borealis)* is more heat labile than alkaline phosphatases from mammals and can be denaturated at 65°C for 15 minutes (Olsen et al. 1991). The heat-labile enzyme is therefore more suitable as a DNA-modifying enzyme in gene-cloning technology, where higher temperatures can denaturate the DNA. The enzyme is recovered for commercial use from shrimp-processing wastewater in Norway. Other seafood enzymes with heat-labile properties, such as Uracil-DNA *N*-glycosylase from cod (Lanes et al. 2000), and shrimp nuclease, are likewise produced commercially as recombinant enzymes for gene-cloning technology.

Potential Applications of Seafood Enzymes in the Dairy Industry

Research has shown that digestive proteases from fish, due to their specificity, can be useful as rennet substitutes for calf chymosin in cheese making (Brewer et al. 1984, Tavares et al. 1997, Shamsuzzaman and Haard 1985). Due to their heat lability, they can be useful for preventing oxidized flavor from developing in milk (Simpson and Haard 1984). Simpson and Haard (1984) found that cod and bovine trypsin are equally effective in preventing copper-induced off-flavors from developing in milk. But the cod enzyme has the advantage of being completely inactivated after pasteurization at 70°C for 45 minutes, whereas 47% of the bovine trypsin is still active. These studies show that cold-adaption properties of marine enzymes can be an advantage in the processing of different foods.

REFERENCES

Aaen B, Jessen F, Jensen B. 1995. Partial purification and characterization of a cellular acidic phospholipase A_2 from cod *(Gadus morhua)* muscle. Comp Biochem Physiol 110B:547–554.

An H, Peters MY, Seymour TA. 1996. Roles of endogenous enzymes in surimi gelation. Trends in Food Sci Technol 7:321–327.

An H, Weerasinghe V, Seymour TA, Morrissey MT. 1994. Cathepsin degradation of Pacific whiting surimi proteins. J Food Sci 59(5):1013–1033.

Ando M, Yoshimoto MY, Inabu K, Nakagawa T, Makinodan Y. 1995. Post-mortem change of the three-dimensional structure of collagen fibrillar network in fish muscle pericellular connective tissues corresponding to post-mortem tenderization. Fish Sci 61:327–330.

Anthoni U, Børresen T, Christophersen C; Gram L; Nielsen PH. 1990. Is trimethylamine oxide a reliable indicator for the marine origin of fish. Comp Biochem Phys 97B:569–571.

Aoki T, Ueno R. 1997. Involvement of cathepsins B and L in the *post-mortem* autolysis of mackerel muscle. Food Res Intern 30(8): 585–591.

Arason S. 1994. Production of fish silage. In: AM Martin, editor. Fisheries processing–Biotechnological Applications, pp. 244–269. London: Chapman and Hall.

Ashie INA, Lanier TC. 2000. Transglutaminases in seafood processing. In: NF Haard, BK Simpson, editors. Seafood Enzymes, pp. 147–166. New York: Marcel Dekker, Inc.

Astier C, Labbe JP, Roustan C, Benyamin Y. 1991. Sarcomeric disorganization in post-mortem fish muscles. Comp Biochem Physiol 100B:459–465.

Azam K, Strachan NJC, Mackie IM, Smith J, Nesvadba P. 1990. Effect of slaughter method on the progress of rigor of rainbow trout Salmo gairdneri as measured by an image processing system. Int J Food Sci Technol 25:477–482.

Barrett EL, Kwan HS. 1985: Bacterial reduction of trimethylamine oxide. Ann Rev Microbiol 39:131–149.

Ben-gigirey B, Vieites Baptista de Sousa JM, Villa TG, Barroz-velazquez J. 1999. Chemical changes and visual appearance of albacore tuna as related to frozen storage. J Food Sci 64(1): 20–24.

Berg T, Erikson U, Nordtvedt TS. 1997. Rigor mortis assessment of Atlantic salmon *(Salmo salar)* and effects of stress . J Food Sci 62(3): 439–446.

Bracho GE, Haard NF. 1995. Identification of two matrix metalloproteinase in the skeletal muscle of Pacific rockfish (*Sebastes* sp.). J Food Biochem 19:299–319.

Bremner HA. 1999. Gaping in fish flesh. In: K Sato, M Sakaguchi, HA Bremner, editors. Extra cellular matrix of Fish and Shellfish, pp. 81–94. Trivandrum India: India Research Signpost.

Bremner HA, Hallett IC. 1985. Muscle fiber–connective tissue junctions in the fish blue grenadier *(Macruronus novaezelandia)*. A scanning electron microscope study. J Food Sci 50:975–80.

Brewer P, Helbig N, Haard NF. 1984. Atlantic cod pepsin—Characterization and use as a rennet substitute. Can Inst Food Sci Technol J 17(1): 38–43.

Busconi L, Folco EJ, Martone CB, Sanchez JJ. 1989. Postmortem changes in cytoskeletal elements of fish muscle. J of Food Biochem 13:443–451.

Busconi L, Folco EJ, Studdert C, Sanchez JJ. 1992. Purification and characterization of a latent form of multicatalytical proteinase from fish muscle. Comp Biochem Physiol 102B:303–309.

Cappeln G, Nielsen J, Jessen F. 1999. Synthesis and degradation of adenosine triphosphate in cod *(Gadus morhua)* at subzero temperatures. J Sci Food Agric 79(8): 1099–1104.

Cao MJ, Hara K, Osatomi K, Tachibana K, Izumi T, Ishihara T. 1999. Myofibril-bound serine proteinase (MBP) and its degradation of myofibrillar proteins. J Food Sci 64(4): 644–647.

Chiba A, Hamaguchi M, Kosaka M, Tokuno T, Asai T, Chichibu S. 1991. Quality evaluation of fish meat by ^{31}phosphorus-nuclear magnetic-resonance. J Food Sci 56(3): 660–664.

Childress JJ. 1995. Are there physiological and biochemical adaptations of metabolism in deep-sea animals?. Trends Ecol Evol 10(1): 30–36.

Dahlhoff, Somero GN. 1991. Pressure and temperature adaptation of cytosolic malate dehydrogenase of shallow and deep-living marine invertebrates. Evidence for high body temperatures in hydrothermal vent animals. J Exp Biol 159:473–487.

Dalgaard P. 2000. Fresh and lightly preserved seafood. In: CMD Man, AA Jones, AN Aspen, editors. Shelf-life Evaluation of Foods, 2nd edition, pp. 110–139. Gaithersburg, Maryland: Aspen Publishers, Inc.

de Backer M, McSweeney S, Rasmussen HB, Riise BW, Lindley P, Hough E. 2002. The 1.9 Å crystal structure of heat-labile shrimp alkaline phosphatase. J Mol Biol 318:1265–1274.

Del Rosario RR, Maldo SM. 1984. Biochemistry of patis formation I. Activity of cathepsins in patis hydrolysates. Phil. Agric 67:167–75.

Dickson KA. 1995. Unique adaptations of the metabolic biochemistry of tunas and billfishes for the life in the pelagic environment. Env Biol Fish 42:65-97.

Dingle JR, Hines JA. 1971. Degradation of inosine 5′monophosphate in the skeletal muscle of several North Atalantic fishes. J Fish Res Bd Can 28(8): 1125–1131.

Dingle JR, Keith RA, Lall B. 1977. Protein instability in frozen storage induced in minced muscle of flatfishes by mixture with muscle of red hake. Can Inst Food Sci Technol J 10:143–46.

Ellington WR. 2001. Evolution and physiological roles of phosphagen systems. Ann Rev Phys 63:289–325.

Engvang K, Nielsen HH. 2000. *In situ* activity of chymotrypsin in sugar-salted herring during cold storage. J Sci Food Agric 80:1277–1283.

Ertbjerg P, Larsen LM, Møller AJ. 1999. Effect of prerigor lactic acid treatment on lysosomal enzyme release in bovine muscle. J Sci Food Agric 79:98–100.

Fernandez-Reiriz MJ, Pastoriza L, Sampedro G, Herrera JJ. 1995. Changes in lipids of whole and minced rayfish *(Raja clavata)* muscle during frozen storage. Z Lebensm Unters Forsch. 200(6): 420–424.

Fields PA. 2001. Protein function at thermal extremes: balancing stability and flexibility. Comp Biochem Physiol 129A:417–431.

Fletcher GC, Hallett IC, Jerrett AR, Holland AJ. 1997. Changes in the fine structure of the myocommata-muscle fibre junction related to gaping in rested and exercised muscle from king salmon *(Oncorrhynchus tshawytscha)* Lebensm-Wiss Technol 30(3): 246–252.

Fluke S. 1994. Taste active compounds of seafoods with a special reference to umani substances. In: F Shahidi, JR Botta, editors. Seafoods: Chemistry, Processing Technology and Quality, pp. 115–139. Glasgow, Great Britain: Blackie Academic and Professional.

Foegeding EA, Lanier TC, Hultin HO. 1996. Characteristics of edible muscle tissues. In: OR Fenema, editor. Food Chemistry, 3rd edition, pp. 879–942. New York: Marcel Dekker, Inc.

Geromel EJ, Montgomery MW. 1980. Lipase release from lysosomes of rainbow trout (*Salmo gairdneri*) muscle subjected to low temperatures. J Food Sci 45:412–419.

Geesink GH, Morton JD, Taylor RG, Christiansen JA. 2000. Partial purification and characterization of Chinook salmon *(Oncorhynchus tshawytscha)* calpains and an evaluation of their role in postmortem proteolysis. J Food Sci 77:2685–2692.

Gibbs, AG. 1997. Biochemistry at depth. In: DJ Randall, AP Farrell, editors. Deep-Sea Fishes. Fish Physiology, vol. 16, pp. 239–271. San Diego, California: Academic Press.

Gildberg A. 1993. Enzymic processing of marine raw materials. Process Biochem 28:1–15.

———. 2001. Utilisation of male Arctic capelin and Atlantic cod intestines for fish sauce production—Evaluation of fermentation conditions. Biores. Technol. 76:119–123.

Gildberg A, Almas KA. 1986. Utilization of fish viscera. In: ML Maguer, P Jelen, editors. Food Engineering and Process Applications 2, Unit Operations, pp. 425–435. London: Elsevier Applied Science.

Gildberg A, Simpson BK, Haard NF. 2000 Uses of enzymes from marine organisms. In: NF Haard, BK Simpson, editors. Seafood Enzymes, pp. 619–639. New York: Marcel Dekker, Inc.

Gill T. 2000. Nucleotide-degrading enzymes. In: NF Haard, BK Simpson, editors. Seafood Enzymes, pp. 37–68. New York: Marcel Dekker, Inc.

Haard N. 2002. The role of enzymes in determining seafood color, flavor and texture. In: HA Bremner, editor. Safety and Quality Issues in Fish Processing, pp. 220–253. Cambridge: Woodhead Publishing Limited and CRC Press LLC.

Haard NF, Simpson BK (eds.). 2000. Seafood Enzymes. New York: Marcel Dekker, Inc. 681 pp.

Hernandez-Herrero MM, Roig-Sagues AX, Lopez-Sabater EI, Rodriguez-Jerez JJ, Mora-Ventura MT. 1999. Protein hydrolysis and proteinase activity during the ripening of salted anchovy (*Engraulis encrasicholus* L.): A microassay method for determing the protein hydrolysis. J Agric Food Chem 47:3319–3324.

Hirano K, Okada E, Tanaka T, Satouchi K. 2000. Purification and regiospecificity of multiple enzyme activities of phospholipase A_1 from bonito muscle. Biochim Biophys Acta 1483:325–333.

Ho ML, Chen GH, Jiang ST. 2000. Effects of mackerel cathepsins L and L-like, and calpain on the degradation of mackerel surimi. Fish Sci 66:558–568.

Hughes RB, Jones NR. 1966. Measurement of hypoxanthine concentration in canned herring as an index of freshness of raw material with a comment on flavour relations. J Sci Food Agric 17(9): 434–436.

Hultin HU. 1984. Post mortem biochemistry of meat and fish. J Chem Ed 61(4): 289–298.

Hultin HO. 1992. Trimethylamine-*N*-oxide (TMAO) demethylation and protein denaturation in fish muscle. In: GJ Flick, Jr, RE Martin, editors. Advances in Seafood Biochemistry. Composition and Quality, pp. 25–42.Lancaster, Pennsylvania: Technomic Publishing Company, Inc.

Ingemansson T, Kaufmann P, Ekstrand B. 1995. Multivariate evaluation of lipid hydrolysis and oxidation data from light and dark muscle of frozen stored rainbow trout *(Oncorhynchus mykiss).* J Agric Food Chem 43:2046–2052.

Ishida M, Niizeki S, Nagayama F. 1994. Thermostable proteinase in salted anchovy muscle. J Food Sci 59(4): 781–785.

Ishida N, Yamashita M, Koizumi N, Terayama M, Ineno T, Minami T. 2003. Inhibition of post-mortem muscle softening following in situ perfusion of protease inhibitors in tilapia. Fish Sci 69:632–638.

Itoh R, Kimura K. 2002. Occurrence of IMP-GMP 5'-nucleotidase in three fish species: A comparative study on *Trachurus japonicus, Oncorhynchus masou* and *Triakis scyllium.* Comp Biochem Physiol 132B (2): 401–408.

Iwamoto M, Yamanaka H, Watabe S, Hashimoto K. 1987. Effect of storage temperature on rigor-mortis and ATP degradation in plaice *Paralichthys olivaceus* muscle. J Food Sci 52(6): 1514–1517.

———. 1991. Changes in ATP and related breakdown compounds in the adductor muscle of itayagai scallop *Pecten albicans* during storage at various temperatures. Nipp Suis Gakkai 57(1): 153–156.

Jiang ST, Wang YT, Chen CS. 1991. Purification and characterization of a proteinase identified as cathepsin D from tilapia muscle (*Tilapia nilotica* × *Tilapia aurea*). J Agric Food Chem 39: 1597–1601.

Jerrett AR, Holland AJ, Cleaver SE. 1998. Rigor contractions in "rested" and "partially exercised" chinook salmon white muscle as affected by temperature. J Food Sci 63(1): 53–56.

Joakimsson K. 1984. Enzymatic deskinning of herring. [DPhil thesis]. Tromsø, Norway: Institute of Fisheries, University of Tromsø.

Jones NR. 1965. Freezing Fillets at Sea. In: Fish Quality at Sea, pp. 81–95. London: Grampian Press Ltd.

———. 1969. Fish as a raw material for freezing. Factors influencing the quality of products frozen at sea. In: R Kreuzer, editor. Freezing and Irradiation of Fish, pp. 31–39. London: Fishing News (Books) Limited.

Kaneniwa M, Yokoyama M, Murata Y, Kuwahara R. 2004. Enzymatic hydrolysis of lipids in muscle of fish and shellfish during cold storage. In: F Shahidi, AM Spanier, CT Ho, T Braggins, editors. Quality of Fresh and Processed Foods, pp. 113–119. New York: Kluwer Academic Publishers.

Karube I, Matsuoka H, Suzuki S, Watanabe E, Toyama K. 1984. Determination of fish freshness with an enzyme sensor system. J Agric Food Chem 32(2): 314–319.

Kim SK, Byun HG, Choi KD, Roh HS, Lee WH, Lee EH. 1993. Removal of skin from filefish using enzymes. Bull Korean Fish Soc 26:159–172.

Kimura M, Seki N, Kimura I. 2002. Enzymic and nonenzymic cleavage of trimethylamine-*N*-oxide *in vitro* at subzero temperatures. Nipp Suis Gakkai 68:85–91.

Kinoshita M, Toyahara H, Shimizu Y. 1990a. Induction of carp muscle multicatalytic proteinase activities by sodium dodecyl sulfate and heating. Comp Biochem Physiol 96B:565–569.

———. 1990b. Diverse distribution of four distinct types of modori (gel degradation)-inducing proteinases among fish species. Bull Japan Soc Sci Fish 56(9): 1485–1492.

Kinoshita M, Yabe T, Kubota M, Takeuchi K, Kubota S, Toyohara H, Sakaguchi M. 2002. cDNA cloning and characterization of two gelatinases from Japanese flounder. Fish Sci 68:618–626.

Knudsen LB, 1989. Protein-lipid interaction in fish. [DPhil thesis]. Lyngby: Danish Institute for Fisheries Research, Technical University of Denmark. 188 pp

Kolodziejska I, Sikorski ZE. 1995. Muscle cathepsins of marine fish and invertebrates. Pol J Food Nutr Sci 4/45(3): 3–10.

———. 1996. Neutral and alkaline muscle proteases of marine fish and invertebrates A review. J Food Biochem 20:349–363.

Kubota S, Toyahara H, & Sakaguchi M. 1998. Occurrence of gelatinolytic activities in yellowtail tissues. Fish Sci 64:439–442.

Kubota S, Kinoshito M, Yokoyama Y, Toyohara H, Sakaguchi M. 2000. Induction of gelatino activities in ayu muscle at the spawning stage. Fish Sci 66: 574–578.

Kubota M, Kinoshita M, Kubota S, Yamashita M, Toyahara H, Sagaguchi M. 2001. Possible implication of metalloproteinases in post-mortem tenderization of fish muscle. Fish Sci 67:965–968.

Kubota M, Kinoshita M, Takeuchi K, Kubota S, Toyohara H, Sakaguchi M. 2003. Solubilization of type I collagen from fish muscle connective tissue by matrix metalloproteinase-9 at chilled temperature. Fish Sci 69:1053–1059.

Lanes O, Guddal PH, Gjellesvik DR, Willassen NP. 2000. Purification and characterization of a cold-adapted uracil-DNA N glycosylase from Atlantic cod *(Gadus morhua).* Comp Biochem Physiol 127B: 399–410.

Lanier TC. 2000. Surimi gelation chemistry. In: JW Park, editor. Surimi and Surimi Seafood, pp. 237–266. New York: Marcel Dekker, Inc.

Leuba JL, Meyer I, Andersen EM, inventors; Nestec S A, assignee. 1987 (Oct 20). Method for dissolving squid membranes. U.S. patent 4,701,339.

Lødemel JB. 2004. Gelatinolytic activities, matrix metalloproteinases and tissue inhibitors of metalloproteinases in fish [Dphil thesis]. Tromsø, Norway: University of Tromsø, Dept. of Marine Biotechnology. 46 pp.

Lødemel JB, Olsen RL. 2003. Gelatinolytic activities in muscle of Atlantic cod *(Gadus morhua)*, spotted wolffish (*Anarhichas minor*) and Atlantic salmon *(Salmo salar)*. J Sci Food Agric 83:1031–1036.

Lopez-Amaya C, Marangoni AG. 2000a. Phospholipases. In seafood processing. In: NF Haard, BK Simpson, editors. Seafood Enzymes, pp. 91–119. New York: Marcel Dekker Inc.

———. 2000b. Phospholipases in seafood processing. In: NF Haard, BK Simpson, editors. Seafood Enzymes, pp. 121–146. New York: Marcel Dekker, Inc.

Love RM. 1970. The Chemical Biology of Fishes, p. 547. New York: Academic Press.

Lowe TE, Ryder JM, Carragher JF, Wells RMG. 1993. Flesh quality in snapper, *Pagrus auratus*,affected by capture stress. J Food Sci 58:770–773.

Lund KE, Nielsen HH. 2001. Proteolysis in salmon *(Salmo salar)* during cold storage: Effects of storage time and smoking process. J Food Biochem 25:379–395.

Lundstrøm RC, Correia FF, Wilhelm KA. 1982. Enzymatic dimethylamine and formaldehyde production in minced American plaice and blackback flounder mixed with a red hake TMAOase active fraction. J Food Sci 47:1305–1310.

Mackie IM. 1993. The effects of freezing on flesh proteins. Food Rev Int 9:575–610.

Mcdonald I, Jones NR. 1976. Control of thaw rigor by manipulation of temperature in cold store. J Food Technol 11:69–76.

Makinodan Y, Toyohara H, Niwa E. 1985. Implication of muscle alkaline proteinase in the textural degradation of fish meat gel. J Food Sci 50:1351–1355.

Michelsen KG, Harmon JS, Sheridan MA. 1994. Adipose tissue lipolysis in rainbow trout, *Oncorhynchus mykiss*, is modulated by phosphorylation of triacylglycerol lipase. Comp Biochem Physiol 107B:509–513.

Nayak J, Viswanathan Nair PG, Ammu K, Mathew S. 2003. Lipase activity in different tissues of four species of fish: Rohu *(Labeo rohita Hamilton)*, oil sardine *(Sardinella longceps linnaeus)*, mullet *(Liza subviridis Valenciennes)* and India mackerel *(Rastrelliger kanagurta Cuvier)*. J Sci Food Agric 83: 1139–1142.

Nielsen HH. 1995. Proteolytic enzyme activites in salted herring during cold storage. [DPhil thesis]. Lyngby: Dept. of Biotechnology, Technical University of Denmark. 131 pp.

Nielsen LB, Nielsen HH. 2001. Purification and characterization of cathepsin D from herring muscle *(Clupea harengus)*. Comp Biochem Physiol 128B: 351–363.

Nielsen MK, Jorgensen BM. 2004. Quantitative relationship between trimethylamine oxide aldolase activity and formaldehyde accumulation in white muscle from gadiform fish during frozen storage. J Agric Food Chem 52(12): 3814–3822.

Niizeki N, Daikoku T, Hirata T, El-Shourbagy I, Song X, Sakaguchi M. 2002. Mechanism of biosynthesis of trimethylamine oxide from choline in the teleost tilapia, *Oreochromis niloticus*, under freshwater conditions. Comp Biochem Physiol 131B:371–386.

Nitisewojo P, Hultin HO. 1986. Characteristics of TMAO degrading systems in Atlantic short finned squid *(Illex illecebrosus)*. J Food Biochem 10:93–106.

Nunes ML, Campos RM, Batista I. 1997. Sardine ripening: Evolution of enzymatic, sensorial and biochemical aspects. In: JB Luten, T Børresen, J Oehlenschläger, editors. Seafood from Producer to Consumer, Integrated Approach to Quality, pp. 319–330. Amsterdam: Elsevier Science B.V.

Ogata H, Aranishi F, Hara K, Osatomi K, Ishihara T. 1998. Proteolytic degradation of myofibrillar components by carp cathepsin L. J Sci Food Agric 76: 499–504.

Ohshima T, Wada S, Koizuma C. 1984. Enzymatic hydrolysis of phospholipids in cod flesh during storage in ice. Bull Japan Soc Sci Fish 50(1): 107–114.

Okuma E, Abe H. 1992. Major buffering constituents in animal muscle. Comp Biochem Physiol 102A(1): 37–41.

Olsen RL, Øverbø K, Myrnes B. 1991. Alkaline phosphatase from the hepatopancreas of shrimp *(Pandalus borealis)*: A dimeric enzyme with catalytically active subunits. Comp Biochem Physiol 99B:755–761.

Olsen SO, Skåra T. 1997. Chemical changes during ripening of North Sea herring. In: JB Luten, T Børresen, J Oehlenschläger, editors. Seafood from Producer to Consumer, Integrated Approach to Quality, pp. 305–318. Amsterdam: Elsevier Science B.V.

Orejani FL, Liston J. 1981. Agents of proteolysis and its inhibition in patis (fish sauce) fermentation. J Food Sci 47:198–203.

Papa I, Tayler RG, Ventre F, Lebart MC, Roustan C, Ouali A, Benyamin Y. 1997. Dystrophin cleavage and sarcolemna detachment are early postmortem changes of bass *(Dicentrachus labrax)* white muscle. J Food Sci 62:917–921.

Parab SN, Rao SB. 1984. Distribution of trimethylamine oxide in some marine and freshwater fish. Curr Sci 53:307–309.

Park JW. 2000. Surimi Seafood—Products, Market and Manufacturing. In: JW Park, editor. Surimi and Surimi Seafood, pp. 201–235.New York: Marcel Dekker, Inc.

Park JW, Morrissey MT. 2000. Manufacturing of surimi from light muscle fish. In: JW Park, editor. Surimi and Surimi Seafood, pp. 23–58.New York: Marcel Dekker Inc.

Perez-Mateos M, Montero P, Gomez-Guillen MC. 2002. Addition of microbial transglutaminase and protease inhibitors to improve gel properties of frozen squid muscle. Eur Food Res Technol 214:377–381.

Perez-Borla O, Roura SI, Monteechia CL, Roldan H, Crupkin M. 2002. Proteolytic activity of muscle in pre- and post-spawning hake *(Merluccius hubbsi marini)* after frozen storage. Lebensm-Wiss Technol 35:325–330.

Raksakulthai N, Haard NF. 1992. Fish sauce from capelin *(Mallotus villosus):* Contribution of cathepsin C to the fermentation. ASEAN Food J 7:147–151.

Raksakulthai N, Lee YZ, Haard NF. 1986. Effect of enzyme supplements on the production on the production of fish sauce prepared from male capelin *(Mallotus villosus).* Can Inst Food Sci Technol J 19:28–33.

Refsgaard HHF, Brockhoff PMB, Jensen B. 1998. Sensory and chemical changes in farmed Atlantic salmon *(Salmo salar)* during frozen storage. J Agric Food Chem 46:3473–3479.

———. 2000. Free polyunsaturated fatty acids cause taste deterioration of salmon during frozen storage. J Agric Food Chem 48:3280–3285.

Rehbein H.1988. Relevance of trimethylamine oxide demethylase activity and haemoglobin content to formaldehyde production and texture deterioration in frozen stored minced fish muscle. J Sci Food Agric 43:261–276.

Rehbein H, Schubring R, Havemeister W, Sotelo CG, Nielsen MK, Jørgensen BM, Jessen F. 1997. Relation between TMAOase activity and content of formaldehyde in fillet minces and belly flap minces from gadoid fishes. Inf Fischwirtsch 44:114–118.

Robb DHF, Kestin SC, Warriss PD. 2000. Muscle activity at slaughter: I. Changes in flesh colour and gaping in rainbow trout. Aquaculture 182(3–4): 261–269.

Saishiti P. 1994. Traditional fermented fish: Fish sauce production. In: AM Martin, editor. Fisheries Processing—Biotechnological Applications, pp. 111–129. London: Chapman and Hall.

Saito T, Arai M, Matsuyoshi M. 1959. A new method for estimating the freshness of fish. Bull Jpn Soc Sci Fish 24(9): 749–750.

Saito M, Sato K, Kunisaki N, Kimura S. 2000. Characterization of a rainbow trout matrix metalloproteinase capable of degrading type I collagen. Eur J Biochem 267:6943–6950.

Sakamoto H, Kumazawa Y, Toiguchi S, Seguro K, Soeda T, Motoki M. 1995. Gel strength enhancement by addition of microbial transglutaminase during onshore surimi manfacture. J Food Sci 60(2): 300–304.

Sato K, Uratsuji S, Sato M, Mochizuki S, Shigemura Y, Ando M, Nakamura Y, Ohtsuki K. 2002. Effect of slaughter method on degradation of intramuscular type V collagen during short-term chilled storage of chub mackerel *Scomber japonicus*. J Food Biochem 26:415–429.

Sentandreu MA, Coulis G, Ouali A. 2002. Role of muscle endopeptidases and their inhibitors in meat tenderness. Trends Food Sci Technol 13:398–419.

Shamsuzzaman K, Haard NF. 1985. Milk clotting and cheese making properties of a chymosin-like enzyme from harp seal mucosa. J Food Biochem 9: 173-92.

Sheridan MA, Allen WV. 1984. Partial purification of a triacylglycerol lipase isolated from steelhead trout *(Salmo gairdneri)* adipose tissue. Lipids 19(5): 347–352.

Shigemura Y, Ando M, Harada K, Tsukamasa Y. 2004. Possible degradation of type I collagen in relation to yellowtail muscle softening during chilled storage. Fish Sci 70:703–709.

Shoubridge EA, Hochachka PW. 1980. Ethanol—Novel end product of vertebrate anaerobic metabolism. Science 209(4453): 308–309.

Siebenaller JF, Somero GN. 1978. Pressure-adaptive differences in lactate dehydrogenases of congeneric marine fishes living at different depths. Science 201: 255–257.

———. 1979. Pressure-adaptive differences in the binding and catalytic properties of muscle-type (M_4)

lactate dehydrogenase of shallow- and deep-living marine fishes. J Comp Physiol 129:295–300.

Sikorski Z, Kostuch S. 1982. Trimethylamine *N*-oxide demethylase—its occurrence, properties, and role in technological changes in frozen fish. Food Chem 9:213–222.

Sikorski ZE, Kolakowska A. 1994. Changes in proteins in frozen stored fish. In: ZE Sikorski, BS Pan, F Shahidi, editors. Seafood Proteins, pp. 99–231. New York: Chapmann and Hall.

Simpson BK, Haard NF. 1984. Trypsin from Greenland cod as a food-processing aid. J Appl Biochem 6:135–143.

———. 1987. Cold-adapted enzymes from fish. In: D Knorr, editor. Food Biotechnology, pp. 495–527. New York: Marcel Dekker, Inc.

Skjervold PO, Fjaera SO, Ostby PB, Einen O. 2001. Live-chilling and crowding stress before slaughter of Atlantic salmon (*Salmo salar*). Aquaculture 192(2-4): 265–280.

Smalås AO, Leiros HKS, Os V, Willassen NP. 2000. Cold adapted enzymes. Biotechnol Ann Rev 6:1–57.

Somero GN. 1981. pH-temperature interactions on proteins—Principles of optimal pH and buffer system-design. Mar Biol Lett 2(3): 16–78.

———. 2003. Protein adaptations to temperature and pressure: complementary roles of adaptive changes in amino acid sequence and internal milieu. Comp Biochem Physiol 136B:577–591.

Sotelo CG, Pineiro C, Pérez-Martáin RI. 1995. Denaturation of fish proteins during frozen storage: role of formaldehyde. Z Lebensm Unters Forsch 200:14–23.

Sotelo CG, Rehbein H. 2000. TMAO-degrading enzymes. Utilization and influence on post harvest seafood quality. In: NF Haard, BK Simpson, editors. Seafood Enzymes, pp. 167–190. New York. Marcel Dekker, Inc.

Spinelli J, Koury BJ. 1981. Some new observations on the pathways of formation of dimethylamine in fish muscle and liver. J Agric Food Chem 29:327–331.

Stefánsson G, Nielsen HH, Gudmundsdóttir G. 1995. Ripening of spice-salted herring. TemaNord 613. Copenhagen, Denmark: Nordic Council of Ministers.

Stefánsson G, Nielsen HH, Skåra T, Oehlenschläger J, Schubring R, Luten J, Derrick S, Gudmunsdóttir G. 2000. Frozen herring as raw material for spice-salting. J Sci Food Agric 80:1319–1324.

Stefánsson G, Steingrimsdottir U. 1990. Application of enzymes for fish processing in Iceland—Present and future aspects. In: MN Voight, JR Botta, editors. Advances in Fisheries Technology and Biotechnology for Increased Profitability, pp. 237–250. Lancaster: Technomic publication.

Stoknes I, Rustad T. 1995. Purification and characterization of a multicatalytic proteinase from Atlantic salmon *(Salmo salar)* muscle. Comp Biochem Physiol 111B(4): 587–596.

Storey KB. 1992: The basis of enzymatic adaptation. In: EE Bittar, editor. Chemistry of the Living Cell. Fundamentals of Medical Cell Biology, vol. 3A, pp. 137–156. Amsterdam: Elsevier.

Strom T, Raa J. 1991. From basic research to new industries within marine biotechnology: Successes and failures in Norway. In: HK Kuanf, K Miwa, MB Salim, editors. Proceeding of a Seminar on Advances in Fishery Post-Harvest Technology in Southeast Asia, pp. 63–71.Singapore: Changi Point.

Svenning R, Stenberg E, Gildberg A, Nilsen K. 1993. Biotechnological descaling of fish. INFOFISH Int 6/93:30–31.

Surette M, Gill TA, Leblanc PJ. 1988. Biochemical basis of postmortem nucleotide catabolism in cod *(Gadus morhua)* and its relationship to spoilage. J Agric Food Chem 36:19–22.

Takeuchi M, Mizuta C, Uda K, Fujimoto N, Okamoto M and Suzuki T. 2004. Unique evolution of Bivalvia arginine kinases. Cellular and Molecular Life Sciences 61(1): 110–117.

Tavares JFP, Baptista JAB, Marcone MF. 1997. Milk-coagulating enzymes of tuna fish waste as a rennet substitute. Int J Food Sci Nutr 48(3): 169–176.

Taylor RG, Fjaera SO, Skjervold PO. 2002. Salmon fillet texture is determined by myofiber-myofiber and myofiber myocommata attachment. J Food Sci 67: 2067–2071.

Teruel SRL, Simpson BK. 1995. Characterization of the collagenolytic enzyme fraction from winter flounder *(Pseudopleuronectes americanus)*. Comp Biochem Physiol 112B:131–136.

Tomioka K, Kuragano T, Yamamoto H, Endo K. 1987. Effect of storage temperature on the dephosphorylation of mucleotides in fish muscle. Nipp Suis Gakkai 53(3): 503–507.

Toyohara H, Sakata T, Yamashita K, Kinoshita M, Shimizu Y. 1990. Degradation of oval-filefish meat gel caused by myofibrillar proteinase(s). J Food Sci 52:364–368.

Tschersich P, Choudhury GS. 1998. Arrowtooth flounder *(Atheresthes stomias)* protease as a processing aid. J Aqua Food Prod Technol 7(1): 77–89.

Tungkawachara S, Park JW, Choi YJ. 2003. Biochemical properties and consumer acceptance of Pacific whiting fish sauce. J. Food Sci 68(3): 855–860.

Vaisey EB. 1956. The non-enzymatic reduction of trimethylamine oxide to trimethylamine, dimethylamine, and formaldehyde. Can J Biochem Physiol 34:1085–1090.

van Waarde A. 1988. Biochemistry of non-protein nitrogenous compounds in fish including the use of amino acids for anaerobic energy production. Comp Biochem Physiol 91B:207–228.

Verrez-Bagnis V, Ladrat C, Noel J, Fleurance J. 2002. In vitro proteolysis of myofibrillar and sarcoplasmic proteins of European sea bass (*Dicentrarchus labrax* L.) by an endogenous m-calpain. J Sci Food Agric 82:1256–1262.

Verrez-Bagnis V, Noel J, Sauterreau C, Fleurance J. 1999. Desmin degration in postmortem fish muscle. J Food Sci 64:240–242.

Vo VT, Kusakabe I, Murakami K. 1984. The aminopeptidase activity in fish sauce. Agric Biol Chem 48(2): 525–527.

Wang JH, Jiang ST. 1991. Properties of calpain II from tilapia muscle *(Tilapia nilotica* × *Tilapia aurea)* Agric Biol Chem 55:339–345.

Watson CL, Morrow HA, Brill RW. 1992. Proteolysis of skeletal muscle in yellowfin tuna *(Thunnus albacares):* Evidence of calpain activation. Comp Biochem Physiol 103B:881–887.

Yamashita M, Konagaya S. 1990. Participation of cathepsin L into extensive softening of the muscle of chum salmon caught during spawning migration. Bull Japan Soc Sci Fish 56:1271–1277.

———. 1991. Hydrolytic action of salmon cathepsins B and L to muscle structural proteins in respect of muscle softening. Bull Japan Soc Sci Fish 57:1917–1922.

Yancey PH, Clark ME, Hand SC, Bowlus RD, Somero GN. 1982. Living with water stress: Evolution of osmolyte systems. Science 217:1214–1222.

18
Proteomics: Methodology and Application in Fish Processing

O. T. Vilhelmsson, S. A. M. Martin, B. M. Poli, and D. F. Houlihan*

INTRODUCTION

Proteomics is most succinctly defined as "the study of the entire proteome or a subset thereof," the proteome being the expressed protein complement of the genome. Unlike the genome, the proteome varies among tissues, as well as with time, in reflection of the organism's environment and its adaptation thereto. Proteomics can therefore give a snapshot of the organism's state of being and, in principle at least, map the entirety of its adaptive potential and mechanisms. As with all living matter, foodstuffs are in large part made up of proteins. This is especially true of fish and meat, where the bulk of the food matrix is constructed from proteins. Furthermore, the construction of the food matrix, both on the cellular and tissuewide levels, is regulated and brought about by proteins. It stands to reason, then, that proteomics is a tool that can be of great value to the food scientist, giving valuable insight into the composition of the raw materials; quality involution within the product before, during, and after processing or storage; and the interactions of the proteins with one another, with other food components, or with the human immune system after consumption. In this chapter, a brief overview of "classical" proteomics methodology is presented, and present and future applications in relation to fish and seafood processing and quality are discussed.

*Corresponding author.

PROTEOMICS METHODOLOGY

Unlike nucleic acids, proteins are an extremely variegated group of compounds in terms of their chemical and physical properties. It is not surprising, then, that a field that concerns itself with "the systematic identification and characterization of proteins for their structure, function, activity and molecular interactions" (Peng et al. 2003) should possess a toolkit containing a wide spectrum of methods that continue to be developed at a brisk pace. While high-throughput, gel-free methods, for example, those based on liquid chromatography tandem mass spectrometry (LC-MS/MS) (Peng et al. 2003), surface-enhanced laser desorption/ionization (Hogstrand et al. 2002), or protein arrays (Lee and

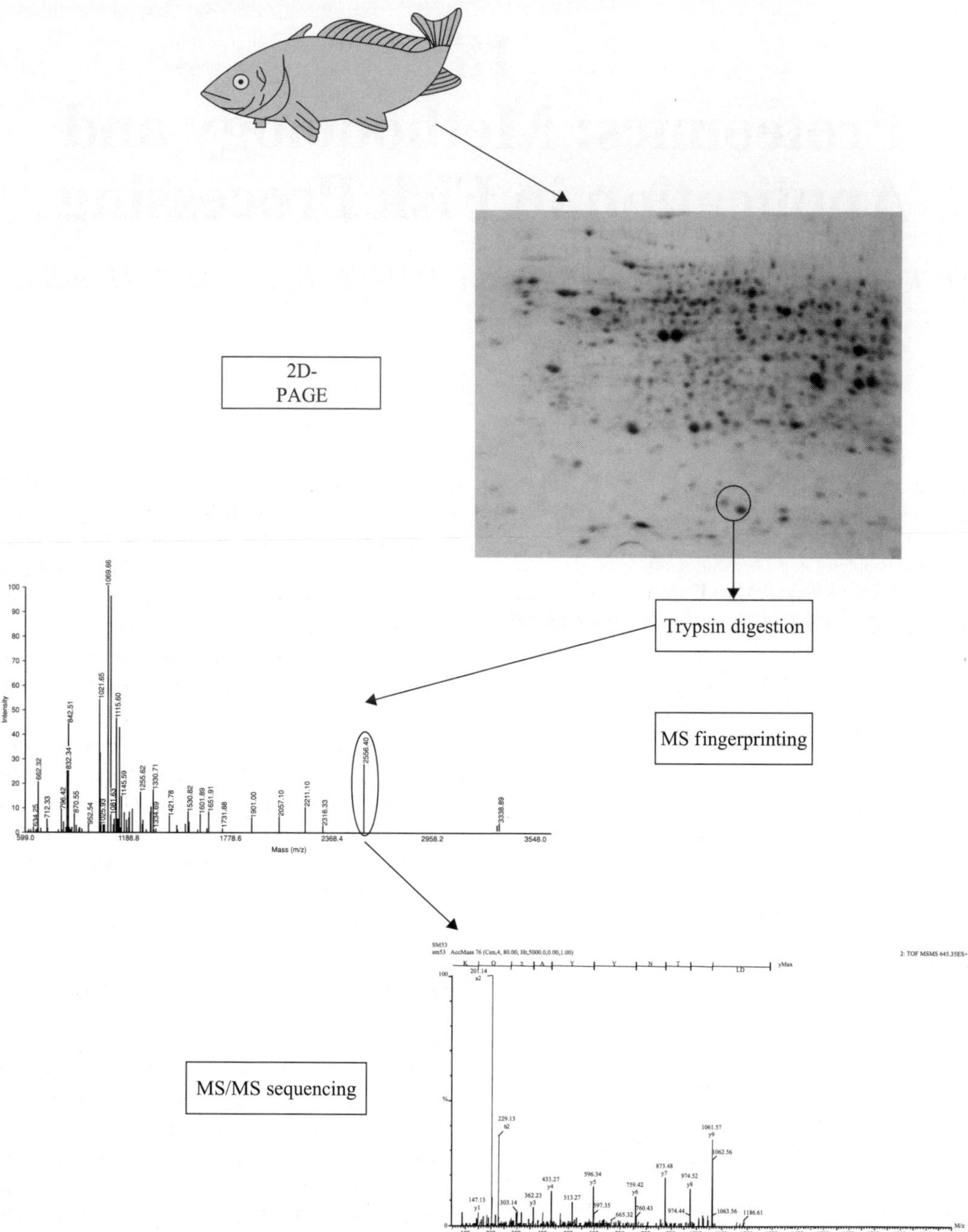

Figure 18.1. An overview over the 'classic approach' in proteomics. First, a protein extract (crude or fractionated) from the tissue of choice is subjected to two-dimensional polyacrylamide gel electrophoresis. Once a protein of interest has been identified, it is excised from the gel and subjected to digestion by trypsin (or other suitable protease). The resulting peptides are analyzed by mass spectrometry, yielding a peptide mass fingerprint. In many cases this is sufficient for identification purposes, but if needed, peptides can be dissociated into smaller fragments, and small partial sequences can be obtained by tandem mass spectrometry. See text for further details.

Nagamune 2004), hold great promise and are deserving of discussion in their own right, the classic process of two-dimensional electrophoresis (2DE) followed by protein identification via peptide mass fingerprinting of trypsin digests (Fig. 18.1) remains the workhorse of most proteomics work, largely because of its high resolution, simplicity, and mass accuracy. This "classic approach" will therefore be the main focus of this chapter. A number of reviews on the advances and prospects of proteomics within various fields of study are available. Some recent ones include Aebersold and Mann (2003), Cash (2002), Cash and Kroll (2003), Graves and Haystead (2003), Huber et al. (2003), Kvasnicka (2003), Phizicky et al. (2003), Piñeiro et al. (2003), Pusch et al. (2003), Takahashi et al. (2003), and Tyers and Mann (2003).

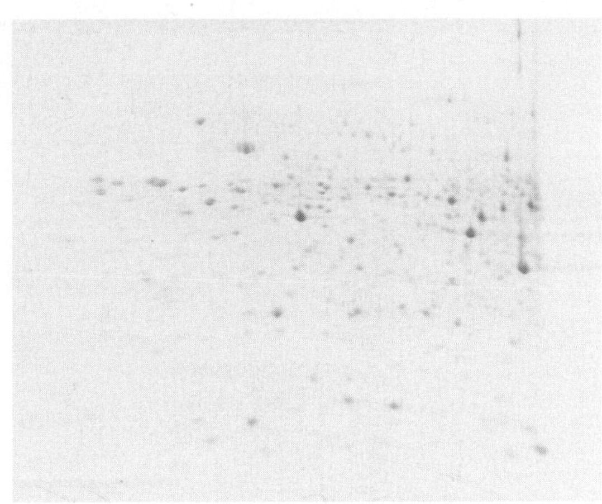

Figure 18.2. A 2DE protein map of rainbow trout *(Oncorhynchus mykiss)* liver proteins with pH between 4 and 7 and molecular mass about 10–100 (S. Martin, unpublished). The proteins are separated according to their pH in the horizontal dimension and according to their mass in the vertical dimension. Isoelectrofocusing was by pH 4–7 immobilized pH gradient (IPG) strip, and the second dimension was in a 10–15% gradient polyacrylamide slab gel.

Two-Dimensional Electrophoresis

Two-dimensional electrophoresis (2DE), the cornerstone of most proteomics research, is the simultaneous separation of hundreds, or even thousands, of proteins on a 2D polyacrylamide slab gel. The potential of a 2D protein separation technique was realized early on, and considerable development efforts took place in the 1960s (Kaltschmidt and Wittmann 1970, Margolis and Kenrick 1969). The method most commonly used today was developed by Patrick O'Farrell and is described in his seminal and thorough 1975 paper (O'Farrell 1975). O'Farrell's method first applies a process called isoelectric focusing, where an electric field is applied to a tube gel on which the protein sample and carrier ampholytes have been deposited. This separates the proteins according to their molecular charge. The tube gel is then transferred onto a polyacrylamide slab gel, and the isoelectrically focused proteins are further separated according to their molecular mass by conventional sodium dodecyl sulfate–polyacrylamide gel electrophoresis (SDS-PAGE), yielding a two-dimensional map (Fig. 18.2) rather than the familiar banding pattern observed in one-dimensional SDS-PAGE. The map can be visualized and individual proteins quantified by radiolabeling or by using any of a host of protein dyes and stains, such as Coomassie blue, silver stains, or fluorescent dyes. By comparing the abundance of individual proteins on a number of gels (Fig. 18.3), up- or downregulation of these proteins can be inferred. It is worth emphasizing that great care must be taken that the proteome under investigation is reproducibly represented on the 2DE gels, and that individual variation in specific protein abundance is taken into consideration by running gels from a sufficient number of samples and performing the appropriate statistics. Pooling samples may also be an option, depending on the type of experiment. Although a number of refinements have been made to 2DE since O'Farrell's paper was published, most notably the introduction of immobilized pH gradients (IPGs) for isoelectrofocusing (Görg et al. 1988), the procedure remains essentially as outlined above. For more detailed, up-to-date descriptions of methods, the reader is referred to any of a number of excellent books and laboratory manuals, such as Berkelman and Stenstedt (1998), Link (1999), Walker (2002), and Westermeier and Naven (2002).

Some Problems and Their Solutions

The high resolution and good sensitivity of 2DE are what make it the method of choice for most proteomics work, but the method nevertheless has

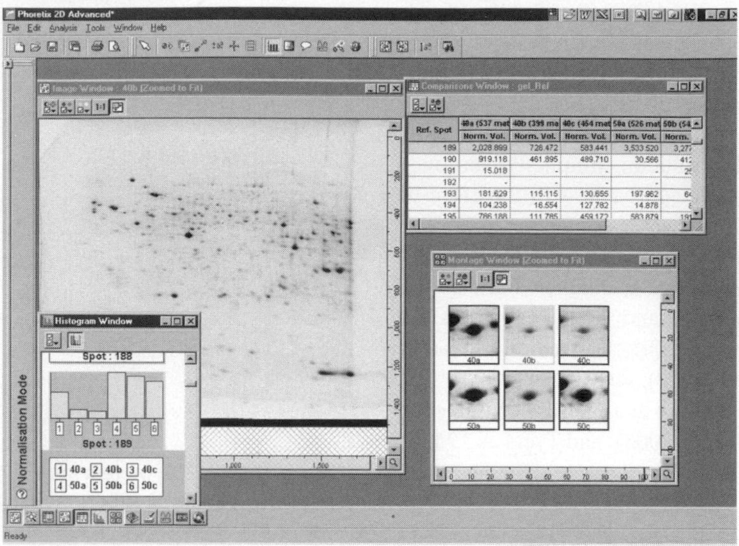

Figure 18.3. A screenshot from the 2DE analysis program Phoretix 2-D (NonLinear Dynamics, Gateshead, Tyne and Wear, United Kingdom) showing some steps in the analysis of a two-dimensional protein map. Variations in abundance of individual proteins, as compared with a reference gel, can be observed and quantified.

several drawbacks. The most significant of these have to do with the diversity of proteins and their expression levels. For example, hydrophobic proteins do not readily dissolve in the buffers used for isoelectrofocusing. This problem can be overcome, though, using nonionic or zwitterionic detergents, allowing for 2DE of membrane and membrane-associated proteins (Babu et al. 2004, Chevallet et al. 1998, Henningsen et al. 2002, Herbert 1999). Vilhelmsson and Miller (2002), for example, were able to use "membrane protein proteomics" to demonstrate the involvement of membrane-associated metabolic enzymes in the osmoadaptive response of the foodborne pathogen *Staphylococcus aureus*. A 2DE gel image of *S. aureus* membrane-associated gels is shown in Figure 18.4.

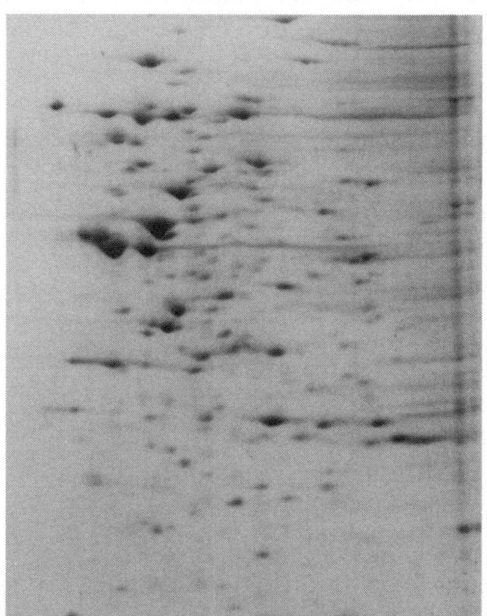

Figure 18.4. A 2DE membrane proteome map from *Staphylococcus aureus,* showing proteins with pH between 3 and 10 and molecular mass about 15–100 (O. Vilhelmsson and K. Miller, unpublished). Isoelectrofocusing was in the presence of a mixture of pH 5–7 and pH 3–10 carrier ampholytes, and the second dimension was in a 10% polyacrylamide slab gel with a 4% polyacrylamide stacker.

Similarly, resolving alkaline proteins, particularly those with pH above 10, on 2D gels has been problematic in the past. Although the development of highly alkaline, narrow-range IPGs (Bossi et al. 1994) allowed reproducible two-dimensional resolution of alkaline proteins (Görg et al. 1997), their representation on wide-range 2DE of complex mixtures such as cell extracts remained poor. Improvements in resolution and representation of alkaline proteins on wide-range gels have been made (Görg et al. 1999); nevertheless an approach that involves several gels, each of a different pH range, from the same sample is advocated for representative inclusion of alkaline proteins when studying entire proteomes (Cordwell et al. 2000). Indeed, Cordwell et al. were able to significantly improve the representation of alkaline proteins in their study on the relatively highly alkaline *Helicobacter pylori* proteome by using both pH 6–11 and pH 9–12 IPGs (Bae et al. 2003).

A second drawback of 2DE has to do with the extreme difference in expression levels of the cell's various proteins, which can be as much as 10,000-fold. This leads to swamping of low abundance proteins by high abundance ones on the two-dimensional map, rendering analysis of low abundance proteins difficult or impossible. For applications such as species identification or the study of major biochemical pathways, where the proteins of interest are present in relatively high abundance, this does not present a problem. However, when investigating, for example, regulatory cascades, the proteins of interest are likely to be present in very low abundance and may at times be undetectable because of the dominance of high abundance proteins. Simply increasing the amount of sample is usually not an option, as it will give rise to overloading artifacts in the gels (O'Farrell 1975). In transcriptomic studies, where a similar disparity can be seen in the abundance of RNA transcripts present, this problem can be overcome by amplifying the low abundance transcripts using the polymerase chain reaction (PCR), but no such technique is available for proteins. The remaining option, then, is fractionation of the protein sample in order to weed out the high abundance proteins, allowing a larger sample of the remaining proteins to be analyzed. A large number of fractionation protocols, both specific and general, are available. Thus, Østergaard et al. used acetone precipitation to reduce the abundance of hordeins present in barley *(Hordeum vulgare)* extracts (Østergaard et al. 2002), whereas Locke et al. used preparative isoelectrofocusing to fractionate Chinese snow pea *(Pisum sativum macrocarpon)* lysates into fractions covering three pH regions (Locke et al. 2002). The fractionation method of choice will depend on the specific requirements of the study and on the tissue being studied. Discussion of some fractionation methods can be found in Butt et al. (2001), Corthals et al. (1997), Dreger (2003), Issaq et al. (2002), Lopez et al. (2000), Millea and Krull (2003), Pieper et al. (2003), and Rothemund et al. (2003).

IDENTIFICATION BY PEPTIDE MASS FINGERPRINTING

Identification of proteins on 2DE gels is most commonly achieved via mass spectrometry of trypsin digests. Briefly, the spot of interest is excised from the gel and digested with trypsin (or another protease), and the resulting peptide mixture is analyzed by mass spectrometry. The most popular mass spectrometry method is matrix-assisted laser desorption/ionization–time-of-flight (MALDI-TOF) mass spectrometry (Courchesne and Patterson 1999), where peptides are suspended in a matrix of small, organic, UV-absorbing molecules (such as 2,5-dihydroxybenzoic acid), followed by ionization by a laser at the excitation wavelength of the matrix molecules and acceleration of the ionized peptides in an electrostatic field into a flight tube where the time of flight of each peptide is measured, giving its expected mass.

The resulting spectrum of peptide masses (Fig. 18.5) is then used for protein identification by searching against expected peptide masses calculated from data in protein sequence databases, such as Swiss Prot or the National Center for Biotechnology Information (NCBI) nonredundant protein sequences data base, using the appropriate software. Several programs are available, many with a web-based open-access interface. The ExPASy Tools website (http://www.expasy.org/tools) contains links to most of the available software for protein identification and several other tools. Attaining a high identification rate is problematic in fish and seafood proteomics due to the relative paucity of available protein sequence data for these animals. As can be seen in Table 18.1, this problem is surprisingly acute for species of commercial importance. To circumvent this problem, it is possible to take advantage of the available nucleotide sequences, which in many cases are more extensive than the protein sequences available, to obtain a tentative identity. How useful this method is will depend on the length and quality of the available nucleotide sequences. It is important to realize, however, that an identity obtained in this manner is less reliable than that obtained through protein sequences and should be regarded only as tentative in the absence of corroborating evidence (such as two-dimensional immunoblots, correlated activity measurements, or transcript abundance). In their work on the rainbow trout *(Oncorhynchus mykiss)* liver proteome, Martin et al. (2003b) and

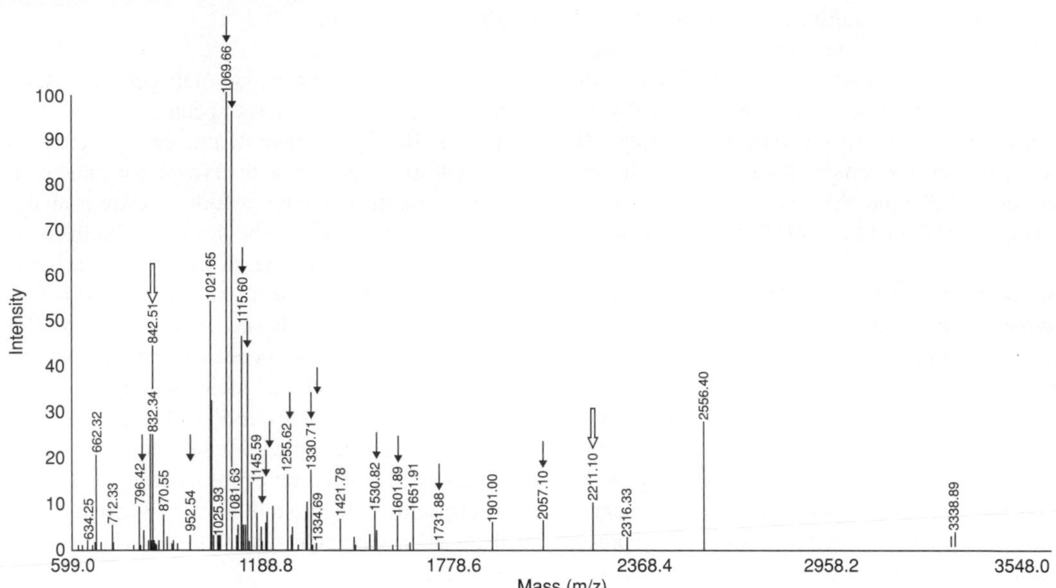

Figure 18.5. A trypsin digest mass spectrometry fingerprint of a rainbow trout liver protein spot, identified as apolipoprotein A I-1 (S. Martin, unpublished). The open arrows indicate mass peaks corresponding to trypsin self-digestion products and were, therefore, excluded from the analysis. The solid arrows indicate the peaks that were found to correspond to expected apolipoprotein A I-1 peptides.

Vilhelmsson et al. (2004) were able to attain an identification rate of about 80% using a combination of search algorithms that included the open-access Mascot program (Perkins et al. 1999) and a licensed version of Protein Prospector MS-Fit (Clauser et al. 1999), searching against both protein databases and a database containing all salmonid nucleotide sequences. In those cases where both the protein and nucleotide databases yielded results, a 100% agreement was observed between the two methods.

A more direct, if rather more time consuming, way of obtaining protein identities is by direct sequence comparison. Until recently, this was accomplished by *N*-terminal or internal (after proteolysis) sequencing by Edman degradation of eluted or electroblotted protein spots (Erdjument-Bromage et al. 1999, Kamo and Tsugita 1999). Today, the method of choice is tandem mass spectrometry (MS/MS). In the peptide mass fingerprinting discussed above, each peptide mass can potentially represent any of a large number of possible amino acid sequence combinations. The larger the mass (and longer the sequence), the higher is the number of possible combinations. In MS/MS one or several peptides are separated from the mixture and dissociated into fragments that then are subjected to a second round of mass spectrometry, yielding a second layer of information. Correlating this spectrum with the candidate peptides identified in the first round narrows down the number of candidates. Furthermore, several short stretches of amino acid sequence will be obtained for each peptide, which, when combined with the peptide and fragment masses obtained, enhances the specificity of the method even further (Chelius et al. 2003, Wilm et al. 1996, Yu et al. 2003b). Mass spectrometry methods in proteomics are reviewed in Yates (1998).

SEAFOOD PROTEOMICS AND THEIR RELEVANCE TO PROCESSING AND QUALITY

Two-dimensional electrophoresis–based proteomics have found a number of applications within food science. Among early examples are such applications as characterization of bovine caseins (Zeece et al.

Table 18.1. Some Commercially or Scientifically Important Fish and Seafood Species and the Availability of Protein and Nucleotide Sequence Data as of June 7, 2004

	Protein Sequences	Nucleotide Sequences
***Actinopterygii* (ray-finned fishes)**	77,396	1,586,862
Elopomorpha	1,215	1,473
Anguilliformes (eels and morays)	966	1,354
European eel (*Anguilla anguilla*)	114	199
Clupeomorpha	180	337
Clupeiformes (herrings)	180	337
Atlantic herring (*Clupea harengus*)	29	35
European pilchard (*Sardina pilchardus*)	17	44
Ostariophysii	21,562	771,661
Cypriniformes (carps)	18,890	722,727
Zebrafish (*Danio rerio*)	13,659	704,204
Siluriformes (catfishes)	1,674	47,635
Channel catfish (*Ictalurus punctatus*)	532	35,240
Protacanthopterygii	4,392	257,953
Salmoniformes (salmons)	4,230	257,923
Atlantic salmon (*Salmo salar*)	686	90,577
Rainbow trout (*Oncorhynchus mykiss*)	1,480	159,907
Arctic charr (*Salvelinus alpinus*)	90	251
Paracanthopterygii	1,880	2,335
Gadiformes (cods)	1,445	1,528
Atlantic cod (*Gadus morhua*)	905	936
Alaska pollock (*Theragra chalcogramma*)	124	136
Saithe (*Pollachius virens*)	16	26
Tetraodontiformes (puffers and filefishes)	29,387	305,449
Pufferfish (*Takifugu rubripes*)	948	89,901
Green pufferfish (*Tetraodon nigroviridis*)	28,149	215,158
Zeiformes (dories)	171	57
John Dory (*Zeus faber*)	34	29
Scorpaeniformes (scorpionfishes/flatheads)	634	1,388
Redfish (*Sebastes marines*)	3	7
Lumpsucker (*Cyclopterus lumpus*)	3	14
***Chondrichthyes* (cartilagenous fishes)**	2,389	2,224
Carcharhiniformes (ground sharks)	480	399
Lesser spotted dogfish (*Scyliorhinus canicula*)	208	104
Blue shark (*Prionace glauca*)	8	4
Lamniformes (mackerel sharks)	178	239
Basking shark (*Cetorhinus maximus*)	16	16
Rajiformes (skates)	275	304
Thorny skate (*Raja radiata*)	39	6
Blue skate (*Raja batis*)	1	0
Little skate (*Raja erinacea*)	162	152
***Mollusca* (mollusks)**	11,229	35,187
Bivalvia	3,072	15,926
Blue mussel (*Mytilus edulis*)	535	591

(Continues)

Table 18.1. (*Continued*)

	Protein Sequences	Nucleotide Sequences		Protein Sequences	Nucleotide Sequences
Haddock (*Melanogrammus aeglefinus*)	56	61	Bay scallop (*Argopecten irradians*)	99	2,106
Lophiiformes (anglerfishes)	197	82	*Gastropoda*	7,036	17,484
Monkfish (*Lophius piscatorius*)	6	9	Common whelk (*Buccinum undatum*)	4	15
Acanthopterygii	45,732	550,100	Abalone (*Haliotis tuberculata*)	11	158
Perciformes (perch-likes)	9,532	60,715	*Cephalopoda*	931	1,490
Gilthead sea bream (*Sparus aurata*)	139	325	Northern European squid (*Loligo forbesi*)	30	39
European sea bass (*Dicentrachus labrax*)	150	264	Common cuttlefish (*Sepia officinalis*)	52	44
Atlantic mackerel (Scomber scombrus)	8	23	Common octopus (*Octopus vulgaris*)	58	79
Albacore (*Thunnus alalunga*)	40	124			
Bluefin tuna (*Thunnus thynnus*)	85	178	***Crustacea* (crustaceans)**	6,295	24,638
Spotted wolffish (*Anarhichas minor*)	30	16	*Caridea*	689	916
Beryciformes (*sawbellies*)	345	181	Northern shrimp (*Pandalus borealis*)	11	8
Orange roughy (*Hoplostethus atlanticus*)	0	12	*Astacidea* (lobsters and crayfishes)	646	3,507
Pleuronectiformes (flatfishes)	957	7,392	American lobster (*Homerus americanus*)	160	2,140
Atlantic halibut (*Hippoglossus hippoglossus*)	38	699	European crayfish (*Astacus astacus*)	26	11
Witch (*Glyptocephalus cynoglossus*)	5	22	Langoustine (*Nephrops norvegicus*)	18	18
Plaice (*Pleuronectes platessa*)	50	216	*Brachyura* (short-tailed crabs)	556	1,213
Winter flounder (*Pseudopleuronectes americanus*)	131	1,347	Edible crab (*Cancer pagurus*)	34	7
Turbot (*Scophthalmus maximus*)	49	112	Blue crab (*Callinectes sapidus*)	45	30

Source: NCBI TaxBrowser (http://www.ncbi.nlm.nih.gov/Taxonomy/taxonomyhome.html/).

1989), wheat flour baking quality factors (Dougherty et al. 1990), and soybean protein bodies (Lei and Reeck 1987). In recent years, proteomic investigations on fish and seafood products, as well as in fish physiology, have gained considerable momentum, as can be seen in recent reviews (Parrington and Coward 2002, Piñeiro et al. 2003). Herein, recent and future developments in fish and seafood proteomics as relates to issues of concern in fish processing or other quality considerations are discussed, paying particular attention to the as yet little exploited potential for investigating the antemortem proteome for the benefit of postmortem quality involution.

Tracking Quality Changes Using Proteomics

A persistent problem in the seafood industry is postmortem degradation of fish muscle during chilled storage, which has deleterious effects on the fish flesh texture, yielding a tenderized muscle. This phenomenon is thought to be primarily due to autolysis of muscle proteins, but the details of this protein degradation are still somewhat in the dark. However, degradation of myofibrillar proteins by calpains and cathepsins (Ladrat et al. 2000, Ogata et al. 1998) and degradation of the extracellular matrix by the matrix metalloproteases and matrix serine proteases, which are capable of degrading collagens, proteoglycans, and other matrix components (Lødemel and Olsen 2003, Woessner 1991), are thought to be among the main culprits. Whatever the mechanism, it is clear that these quality changes are species dependent (Papa et al. 1996, Verrez-Bagnis et al. 1999) and, furthermore, appear to display seasonal variations (Ingólfsdóttir et al. 1998, Ladrat et al. 2000). For example, whereas desmin is degraded postmortem in sardine and turbot, no desmin degradation was observed in sea bass and brown trout (Verrez-Bagnis et al. 1999). Of further concern is the fact that several commercially important fish muscle processing techniques, such as curing, fermentation, and production of surimi and conserves, occur under conditions conducive to endogenous proteolysis (Pérez-Borla et al. 2002). As with postmortem protein degradation during storage, autolysis during processing seems to be somewhat specific. Indeed, the myosin heavy chain of the Atlantic cod (*Gadus morhua*) was shown to be significantly degraded during processing of "salt fish" *(bachalhau),* whereas actin was less affected (Thorarinsdottir et al. 2002). Problems of this kind, where differences are expected to occur in the number, molecular mass, and pH of the proteins present in a tissue, are well suited to investigation using 2DE-based proteomics. It is also worth noting that protein isoforms other than proteolytic ones, whether they be encoded in structural genes or brought about by posttranslational modification, usually have a different molecular weight or pH and can, therefore, be distinguished on 2DE gels. Thus, specific isoforms of myofibrillar proteins, many of which are correlated with specific textural properties in seafood products, can be observed using 2DE or other proteomic methods (Martinez et al. 1990, Piñeiro et al. 2003).

Several 2DE studies have been performed on postmortem changes in seafood flesh (Kjærsgård and Jessen 2003, Martinez and Jakobsen Friis 2004, Martinez et al. 2001a, Martinez et al. 1990, Morzel et al. 2000, Verrez-Bagnis et al. 1999) and have demonstrated the importance and complexity of proteolysis in seafood during storage and processing. For example, Martinez et al. (1992) used a 2DE approach to demonstrate different protein compositions of surimi made from prerigor versus postrigor cod, and they found that 2DE could distinguish between the two. Kjærsgård and Jessen, who used 2DE to study changes in the abundance of several muscle proteins during storage of the Atlantic cod *(Gadus morhua),* proposed a general model for postmortem protein degradation in fish flesh in which initially calpains are activated due to the increase in calcium levels in the muscle tissue. Later, as pH decreases and ATP is depleted, with the consequent onset of rigor mortis, cathepsins and the proteasome are activated sequentially (Kjærsgård and Jessen 2003).

Antemortem Effects on Quality and Processability

Malcolm Love started his 1980 review paper on biological factors affecting fish processing (Love 1980) with a lament for the easy life of poultry processors who, he said, had the good fortune to work on a product reared from hatching under strictly controlled environmental and dietary conditions "so that plastic bundles of almost identical foodstuff for man can be lined up on the shelf of a shop." Since the time of Love's review, the advent of aquaculture has made attainable, in theory at least, just such a utopian vision. As every food processor knows, the quality of the raw material is among the most crucial variables that affect the quality of the final product. In fish processing, therefore, the animal's own individual physiological status will to a large extent dictate where quality characteristics will fall within the constraints set by the species' physical and biochemical makeup. It is well known that an organism's phenotype, including quality characteristics, is determined by environmental as well as genetic factors. Indeed, Huss noted in his review (Huss 1995) that product quality differences within the same fish species can depend on feeding and rearing conditions, differences wherein can affect postmortem biochemical processes in the product, which in turn, affect the involution of quality characteristics in the fish product. The practice of rearing fish in aquaculture, as opposed to catching wild fish, therefore raises the tantalizing prospect of managing the quality characteristics of the fish flesh antemortem, where individual physiological characteristics, such as those governing gaping tendency, flesh softening during storage, and so on, are optimized. To achieve that goal, the interplay between these physiological parameters and environmental and dietary variables needs to be understood in detail. With the ever-increasing resolving power of molecular techniques, such as proteomics, this is fast becoming feasible.

Antemortem Metabolism and Postmortem Quality in Trout

In mammals, antemortem protease activities have been shown to affect meat quality and texture (Kristensen et al. 2002, Vaneenaeme et al. 1994). For example, an antemortem upregulation of calpain activity in swine *(Sus scrofa)* will affect postmortem proteolysis and, hence, meat tenderization (Kristensen et al. 2002). In beef *(Bos taurus),* a correlation was found between ante- and postmortem activities of some proteases, but not others (Vaneenaeme et al. 1994). As discussed in the above section, postmortem proteolysis is a matter of considerable importance in the fish and seafood industry, and any antemortem effects thereon are surely worth investigating.

In a recent study on the feasibility of substituting fishmeal in rainbow trout diets with protein from plant sources, 2DE-based proteomics were among techniques used (Martin et al. 2003a,b; Vilhelmsson et al. 2004). Concomitantly, various quality characteristics of fillet and body were also measured (De Francesco et al. 2004, Parisi et al. 2004). Among the findings was that, according to a triangular sensory test using a trained panel, cooked trout that had been fed the plant protein diet had higher hardness, lower juiciness, and lower odor intensity than those fed the fishmeal-containing diet, indicating an effect of antemortem metabolism on product texture. Furthermore, the amount and composition of free amino acids in the fish flesh was significantly affected by the diet, as was the postmortem development of the free amino acid pool. For example, while abundance of arginine was found to decrease during storage of flesh from fishmeal-fed fish, it increased during storage of plant protein–fed fish (Table 18.2). The diets had been formulated to have a nearly identical amino acid composition, and therefore these results may be taken to indicate altered postmortem proteolytic activity in the plant protein–fed fish as compared with the fishmeal-fed ones.

In the proteomics part of the study, the liver proteome was chosen for investigation, since the liver is the primary seat of many of the fish's key metabolic pathways. This makes a direct comparison of the proteomic and quality characteristics results difficult; nevertheless, some interesting observations can be made. The study identified a number of metabolic pathways sensitive to plant protein substitution in rainbow trout feed, for example, pathways involved in cellular protein degradation, fatty acid breakdown, and NADPH metabolism (Table 18.3). In the context of this chapter, the effects on the proteasome are particularly noteworthy. The proteasome is a multisubunit enzyme complex that catalyzes proteolysis via the ATP-dependent ubiquitin-proteasome pathway, which in mammals, is thought to be responsible for a large fraction of cellular proteolysis (Craiu et al. 1997, Rock et al. 1994). In rainbow trout, the ubiquitin-proteasome pathway has been shown to be downregulated in response to starvation (Martin et al. 2002) and to have a role in regulating protein deposition efficiency (Dobly et al. 2004).

Correlating the findings of these two parts of the study, it seems likely that the difference in texture and postmortem free amino acid pool development are affected by antemortem proteasome activity,

Table 18.2. Free Amino Acids and NH_3 Levels in Muscle of Trout Fed a Fishmeal–Containing Diet (FM) or a Plant Protein–Containing Diet (PP) at 0 and 9 Days after Death

FAA + NH_3 (mg/100 g fresh muscle)	0 d FM (n = 3) (Mean ± SD)	9 d FM (n = 3) (Mean ± SD)	0 d PP (n = 3) (Mean ± SD)	9 d PP (n = 3) (Mean ± SD)
D,O-phosphoserine	0.52±0.19	0.47±0.22	0.49±0.05	1.10±0.25
O-phosphoethanolamine	ND[a]	0.22±0.44	ND	ND
Taurine	91.05±48.32	96.68±26.21	66.01±22.48	95.00±27.40
Aspartic acid	0.52±0.30	0.32±0.18	0.21±0.22	0.74±0.30
Threonine	6.36±1.24	6.81±1.37	6.94±2.65	11.94±1.91
Serine	4.57±1.66	6.41±2.03	2.92±1.39	4.00±0.40
Asparagine	7.33±2.43	ND	5.03±0.83	ND
Glutamic acid	8.61±2.87	13.03±2.61	6.41±1.33	13.92±5.51
Glutamine	22.74±2.51	3.24±3.78	14.79±3.27	ND
Proline	2.03±0.90	5.12±2.27	17.10±2.14	14.59±10.11
Glycine	61.18±7.43	93.99±21.81	76.50±42.51	81.85±30.06
Alanine	11.48±6.80	16.88±2.18	13.33±1.73	24.33±5.14
Citrulline	0.22±0.30	0.22±0.26	1.76±0.94	0.19±0.24
2-amino-n-butyric acid	0.21±0.08	0.19±0.06	0.41±0.12	0.38±0.06
Valine	3.55±0.85	4.35±1.02	3.84±0.62	5.64±1.08
Cystine	ND	0.03±0.06	ND	ND
Methionine	1.99±0.63	2.60±0.67	1.82±0.21	2.16±0.26
Cystathionine	ND	0.17±0.06	0.13±0.05	0.17±0.02
Leucine	1.79±0.54	2.35±0.61	1.67±0.37	2.88±0.65
Isoleucine	2.88±0.55	3.72±0.77	2.81±0.43	4.49±0.68
Phenylalanine	1.06±0.45	1.89±0.72	1.26±0.15	2.02±0.28
Tyrosine	1.65±0.74	2.11±0.60	1.46±0.75	2.03±0.23
β-alanine	3.03±1.72	4.09±0.88	5.64±2.29	8.63±3.78
D-2-amino-isobutyric acid	0.09±0.16	ND	0.06±0.11	0.12±0.21
D-homocystine	0.02±0.03	0.03±0.06	0.15±0.15	0.09±0.08
D-4-amino-butyric acid	1.65±0.12	1.36±0.38	4.10±3.92	5.22±1.30
Tryptophane	ND	ND	ND	ND
Ethanolamine	ND	ND	ND	ND
Allo-hyd	ND	ND	ND	ND
D-hydroxylysine	ND	ND	ND	ND
NH_3	43.48±1.34	47.68±5.12	47.16±5.17	52.33±1.37
Ornithine	2.74±1.29	1.55±0.66	1.85±0.53	5.08±5.00
Lysine	31.54±24.53	18.17±6.74	15.43±5.37	22.46±3.09
Histidine	32.03±15.42	39.29±3.91	147.49±15.71	127.40±21.45
3-methyhistidyne	0.06±0.02	0.06±0.05	0.14±0.25	0.08±0.14
1-methyhistidyne	0.57±0.35	3.88±1.25	0.39±0.20	5.58±2.34
Anserine+L-carnosine	466.46±50.44	398.46±23.85	515.03±132.25	451.93±21.90
Arginine	9.50±4.64	5.60±2.68	4.77±1.20	7.86±1.35
Σ	873.62	834.55	1022.85	1013.70

[a]ND − identity not determined.

Table 18.3. Protein Spots Affected by Dietary Plant Protein Substitution in Rainbow Trout as Judged by 2DE and Their Identities as Determined by Trypsin Digest Mass Fingerprinting

Spot Reference No.	pH	MW (kDa)	Normalized Volume Diet FM[a]	Normalized Volume Diet PP100[a]		Fold Difference	P
Downregulated							
128	6.3	66	303 ± 57	60 ± 19	Vacuolar ATPase β-subunit	5	0.026
291	6.4	42	521 ± 37	273 ± 30	β-ureidopropionase	2	0.004
356	6.3	38	161 ± 37	44 ± 19	Transaldolase	4	0.031
747	5.6	43	101 ± 19	19 ± 11	β-actin	2	0.040
760	6.3	39	41 ± 6	21± 5	ND[b]	2	0.040
766	4.8	27	12 ± 1	6 ± 1	ND	2	0.004
Upregulated							
80	4.4	82	9 ± 4	47 ± 8	"Unknown protein"	5	0.007
87	5.7	75	58 ± 14	262 ± 21	Transferrin	5	< 0.001
138	5.5	67	99 ± 16	267 ± 39	Hemopexin-like	3	0.009
144	5.4	63	26 ± 6	265 ± 66	L-Plastin	10	0.018
190	5.9	54	6 ± 2	50 ± 9	Malic enzyme	9	0.018
199	5.9	53	60 ± 16	156 ± 13	Thyroid hormone receptor	3	0.020
275	6.1	45	1 ± 0.6	11 ± 0.6	NSH[b]	9	< 0.001
387	5.6	35	97 ± 3	251 ± 49	Electron transferring flavoprotein	3	0.035
389	5.8	35	192 ± 45	414 ± 54	Electron transferring flavoprotein	2	0.027
399	6.8	33	59 ± 12	130 ± 10	Aldolase B	2	0.028
457	4.7	29	26 ± 7	57 ± 5	14-3-3 B2 protein	2	0.021
461	4.7	27	75 ± 9	190 ± 12	Proteasome alpha 2	3	0.004
517	4.4	22	15 ± 6	135 ± 29	Cytochrome *c* oxidase	9	0.013
539	4.9	19	7 ± 3	18 ± 3	ND	3	0.033
551	4.1	17	40 ± 11	143 ± 28	ND	4	0.018
563	5.2	15	814 ± 198	3762 ± 984	Fatty acid binding protein	5	0.039

639	6.4	84	10 ± 6	28 ± 5	NSH	3	0.047
648	6.1	55	17 ± 5	154 ± 46	Hydroxymethylglutaryl-CoA synthase	9	0.040
678	5.3	48	26 ± 7	69 ± 15	Proteasome 26S ATPase subunit 4	3	0.044
746	4.4	46	45 ± 13	107 ± 15	"similar to catenin"	2	0.012
754	4.1	15	6 ± 2	36 ± 4	ND	7	<0.001
761	6.1	36	44 ± 21	204 ± 34	Transaldolase	5	0.006
764	6.2	65	0	102 ± 17	NSH	>10	N/A
770	5.0	21	4 ± 1	18 ± 4	ND	4	0.026

[a]Values are mean normalized protein abundance (± SE). Data were analyzed by the Student's *t* test (n = 5). In diet FM, protein was provided in the form of fishmeal; in diet PP100, protein was provided by a cocktail of plant product with an equivalent amino acid composition to fishmeal.

[b]NSH = no significant homology detected; ND = identity not determined.

although further studies are needed to verify that statement.

Potential for Further Antemortem Protein Degradation Studies

We are not aware of any proteomic studies, other than that discussed above, into the link between antemortem protein metabolism and postmortem quality in fish and seafood. However, given the substantial importance of protein degradation to the quality and processability of fish and seafood, it may be worthwhile to consider the potential for application of proteomics within this field of study. In addition to having a hand in controlling autolysis determinants, protein turnover is a major regulatory engine of cellular structure, function, and biochemistry. Cellular protein turnover involves at least two major systems: the lysosomal system and the ubiquitin-proteasome system (Hershko and Ciechanover 1986, Mortimore et al. 1989). The 20S proteasome has been found to have a role in regulating the efficiency with which rainbow trout deposit protein (Dobly et al. 2004). It seems likely that the manner in which protein deposition is regulated, particularly in muscle tissue, has profound implications for the quality and processability of the fish flesh.

Protein turnover systems, such as the ubiquitin-proteasome or the lysosome systems, are suitable for rigorous investigation using proteomic methods. For example, lysosomes can be isolated and the lysosome subproteome queried to answer the question of whether and to what extent lysosome composition varies among fish expected to yield flesh of different quality characteristics. Proteomic analysis on lysosomes has been successfully performed in mammalian (human) systems (Journet et al. 2000, Journet et al. 2002).

An exploitable property of proteasome-mediated protein degradation is the phenomenon of polyubiquitination, whereby proteins are targeted for destruction by the proteasome by covalent binding to multiple copies of ubiquitin (Ciechanover 1994, Hershko and Ciechanover 1986). By targeting these ubiquitin-labeled proteins, it is possible to observe the ubiquitin-proteasome "degradome," that is, to observe which proteins are being degraded by the proteasome at a given time or under given conditions. Peng et al. have developed methods to study the ubiquitin-proteasome degradome in the yeast *Saccharomyces cerevisiae* using multidimensional LC-MS/MS (Peng et al. 2003).

Some proteolysis systems, such as that of the matrix metalloproteases, may be less directly amenable to proteomic study. Activity of matrix metalloproteases is regulated via a complex network of specific proteases (Brown et al. 1993, Okumura et al. 1997, Wang and Lakatta 2002). Monitoring of the expression levels of these regulatory enzymes and how they vary with environmental or dietary variables may be more conveniently carried out using transcriptomic methods.

Can Antemortem Proteomics Shed Light on Gaping Tendency?

A well-known quality issue when farmed fish are compared with wild catch is that of gaping, a phenomenon caused by cleavage by matrix proteases of myocommatal collagen cross-links, which results in weakening and rupturing of connective tissue (Børresen 1992, Foegeding et al. 1996). Gaping can be a serious quality issue in the fish processing industry as, apart from the obvious visual defect, it causes difficulties in mechanical skinning and slicing (Love 1992) of the fish. Weakening of collagen, and hence, gaping, is facilitated by low pH. Well-fed fish, such as those reared in aquaculture, tend to yield flesh of comparatively low pH, which thus tends to gape (Einen et al. 1999, Foegeding et al. 1996). Gaping is therefore a cause for concern with aquaculture-reared fish, particularly of species with high natural gaping tendency, such as the Atlantic cod. Gaping tendency varies considerably among wild fish caught in different areas (Love et al. 1974), and thus, it is conceivable that gaping tendency can be controlled with dietary or other environmental manipulations. Proteomics and transcriptomics, with their capacity to monitor multiple biochemical processes simultaneously, are methodologies eminently suitable to finding biochemical or metabolic markers that can be used for predicting features such as gaping tendendency of different stocks reared under different dietary or environmental conditions.

SPECIES AUTHENTICATION

Food authentication is an area of increasing importance, both economically and from a public health standpoint. Taking into account the large differences in the market value of different fish species and the increased prevalence of processed product on the market, it is perhaps not surprising that species authentication is fast becoming an issue of supreme commercial importance. Along with other molecular techniques, such as DNA-based species identification (Mackie et al. 1999, Martinez et al. 2001b, Sotelo et al. 1993) and isotope distribution techniques for determining geographical origin (Campana and Thorrold 2001), proteomics are proving to be a powerful tool in this area, particularly for addressing questions on the health status of the organism, stresses or contamination levels at the place of breeding, and postmortem treatment (Martinez and Jakobsen Friis 2004). Martinez et al. (2003) recently reviewed proteomic and other methods for species authentication in foodstuffs. Since, unlike the genome, the proteome is not a static entity, but changes between tissues and with environmental conditions, proteomics can potentially yield more information than genomic methods, possibly indicating freshness and tissue information in addition to species. Therefore, although it is likely that DNA-based methods will remain the methods of choice for species authentication in the near term, proteomic methods are likely to develop rapidly and find commercial uses within this field. In many cases, the proteomes of even closely related fish species can be easily distinguishable by eye from one another on 2D gels (Fig. 18.6), indicating that diagnostic

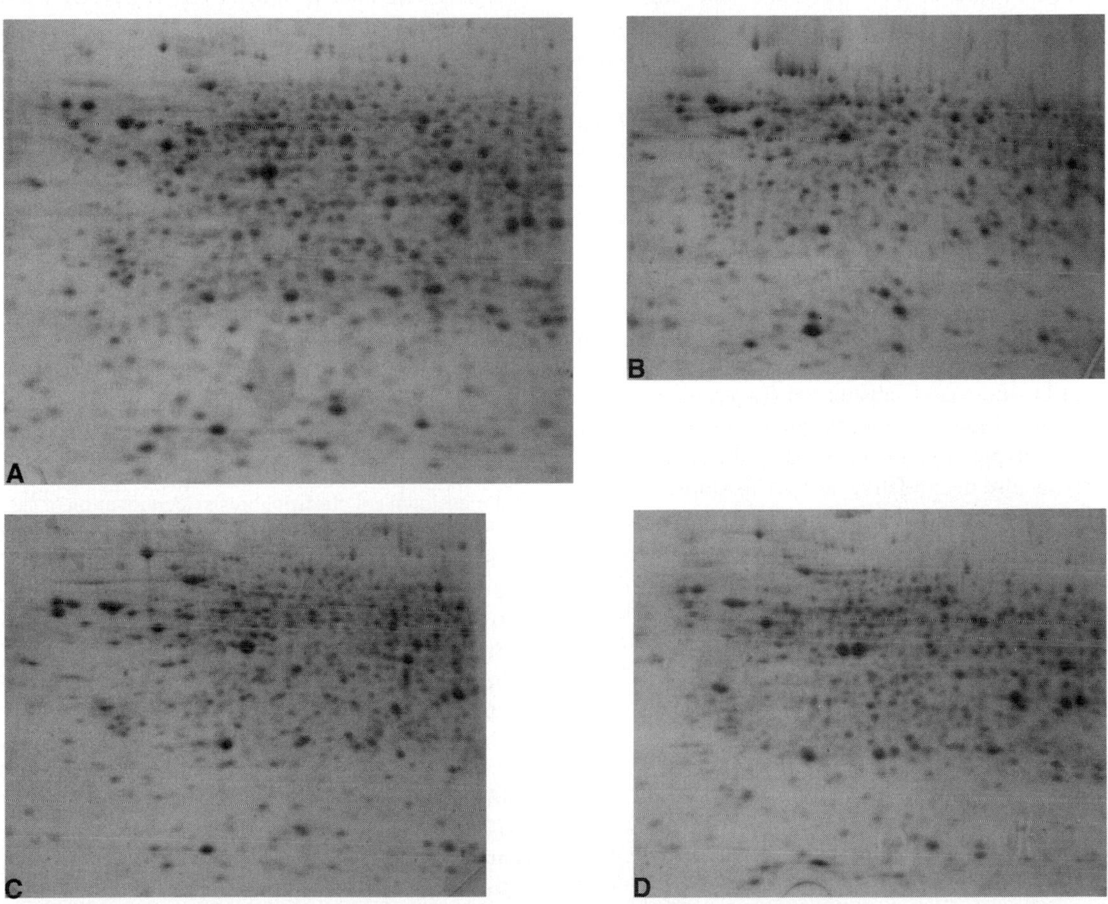

Figure 18.6. 2DE liver proteome maps of four salmonid fish (S. Martin and O. Vilhelmsson, unpublished). Running conditions are as in Figure 18.2. **A**. Brown trout *(Salmo trutta)*, **B**. Arctic charr *(Salvelinus alpinus)*, **C**. rainbow trout *(Oncorhynchus mykiss)*, **D**. Atlantic salmon *(Salmo salar)*.

protein spots may be used to distinguish closely related species.

From early on, proteomic methods have been recognized as a potential method of fish species identification. During the 1960s one-dimensional electrophoretic techniques were developed to identify the raw flesh of various species (Cowie 1968, Mackie 1969, Tsuyuki et al. 1966); this was soon followed by methods to identify species in processed or cooked products (Mackie 1972, Mackie and Taylor 1972). These early efforts were reviewed in 1980 (Hume and Mackie 1980, Mackie 1980).

More recently, 2DE-based methods have been developed to distinguish various closely related species, such as the gadoids or several flat fishes (Piñeiro et al. 1999, Piñeiro et al. 1998, Piñeiro et al. 2001). Piñeiro et al. have found that Cape hake *(Merluccius capensis)* and European hake *(Merluccius merluccius)* can be distinguished on 2D gels from other closely related species by the presence of a particular protein spot that they identified, using nanoelectrospray ionization mass spectrometry, as nucleoside diphosphate kinase (Piñeiro et al. 2001). Lopez et al., studying three species of European mussels, *Mytilus edulis, Mytilus galloprovincialis* and *Mytilus trossulus*, found that *M. trossulus* could be distinguished from the other two species on foot extract 2D gels by a difference in a tropomyosin spot. They found the difference to be due to a single T to D amino acid substitution (Lopez et al. 2002). Recently, Martinez and Jakobsen Friis went further and attempted to identify not only the species present, but also their relative ratios in mixtures of several fish species and muscle types (Martinez and Jakobsen Friis 2004). They concluded that such a strategy would become viable once a suitable number of markers have been identified, although detection of species present in very different ratios is problematic.

IDENTIFICATION AND CHARACTERIZATION OF ALLERGENS

Food safety is a matter of increasing concern to food producers and should be included in any consideration of product quality. Among issues within this field that are of particular concern to the seafood producer is that of allergenic potential. Allergic reactions to seafood affect a significant part of the population: about 0.5% of young adults are allergic to shrimp (Woods et al. 2002). Seafood allergies are caused by an immunoglobulin E–mediated response to particular proteins, including structural proteins such as tropomyosin (Lehrer et al. 2003). Proteomics provide a highly versatile toolkit to identify and characterize allergens. As yet, these have seen little use in the study of seafood allergies, although an interesting and elegant approach has been reported by Yu et al. (Yu et al. 2003a) at National Taiwan University. These authors, studying the cause of shrimp allergy in humans, performed 2DE on crude protein extracts from the tiger prawn, *Penaeus monodon*, blotted the 2D gel onto a polyvinyl difluoride (PVDF) membrane, and probed the membranes with serum from confirmed shrimp allergic patients. The allergens were then identified by MALDI-TOF mass spectrometry of tryptic digests. The allergen was identified as a protein with close similarity to arginine kinase. The identity was further corroborated by cloning and sequencing the relevant cDNA. A final proof was obtained by purifying the protein, demonstrating that it had arginine kinase activity and reacted to serum IgE from shrimp allergic patients and, furthermore, induced skin reactions in sensitized shrimp allergic patients.

REFERENCES

Aebersold R, Mann M. 2003. Mass spectrometry-based proteomics. Nature 422:198–207.

Babu GJ, Wheeler D, Alzate O, Periasamy M. 2004. Solubilization of membrane proteins for two-dimensional gel electrophoresis: identification of sarcoplasmic reticulum membrane proteins. Anal Biochem 325:121–125.

Bae SH, Harris AG, Hains PG, Chen H, Garfin DE, Hazell SL, Paik YK, Walsh BJ, Cordwell SJ. 2003. Strategies for the enrichment and identification of basic proteins in proteome projects. Proteomics 3: 569–579.

Berkelman T, Stenstedt T 1998. 2-D Electrophoresis Using Immobilized pH Gradients: Principles and Methods. Uppsala, Sweden: Amersham Biosciences. 104 pp.

Børresen T. 1992. Quality aspects of wild and reared fish. In: HH Huss, M Jakobsen, J Liston, editors. Quality Assurance in the Fish Industry, pp. 1–17. Amsterdam: Elsevier Science Publishers.

Bossi A, Righetti PG, Vecchio G and Severinsen S. 1994. Focusing of alkaline proteases (subtilisins) in pH 10–12 immobilized pH gradients. Electrophoresis 15:1535–1540.

Brown PD, Kleiner DE, Unsworth EJ, Stetler-Stevenson WG. 1993. Cellular activation of the 72 kDa type IV procollagenase/TIMP-2 complex. Kidney Int 43: 163–170.

Butt A, Davison MD, Smith GJ, Young JA, Gaskell SJ, Oliver SG, Beynon RJ. 2001. Chromatographic separations as a prelude to two-dimensional electrophoresis in proteomics analysis. Proteomics 1:42–53.

Campana SE, Thorrold SR. 2001. Otoliths, increments, an elements: Keys to a comprehensive understanding of fish populations? Can J Fish Aquat Sci 58:30–38.

Cash P. 2002. Proteomics: The protein revolution. Biologist 49:58–62.

Cash P, Kroll JS. 2003. Protein characterization by two-dimensional gel electrophoresis. Meth Mol Med 71:101–118.

Chelius D, Zhang T, Wang GH, Shen RF. 2003. Global protein identification and quantification technology using two-dimensional liquid chromatography nanospray mass spectrometry. Anal Chem 75:6658–6665.

Chevallet M, Santoni V, Poinas A, Rouquie D, Fuchs A, Kieffer S, Rossignol M, Lunardi J, Garin J, Rabilloud T. 1998. New zwitterionic detergents improve the analysis of membrane proteins by two-dimensional electrophoresis. Electrophoresis 19:1901–1909.

Ciechanover A. 1994. The ubiquitin-proteasome proteolytic pathway. Cell 79:13–21.

Clauser KR, Baker PR and Burlingame AL. 1999. Role of accurate measurement (± 10 ppm) in protein identification strategies employing MS or MS/MS and database searching. Anal Chem 71:2871–2882.

Cordwell SJ, Nouwens AS, Verrills NM, Basseal DJ, Walsh BJ. 2000. Subproteomics based upon protein cellular location and relative solubilities in conjunction with composite two-dimensional electrophoresis gels. Electrophoresis 21:1094–1103.

Corthals GL, Molloy MP, Herbert BR, Williams KL, Gooley AA. 1997. Prefractionation of protein samples prior to two-dimensional electrophoresis. Electrophoresis 18:317–323.

Courchesne PL, Patterson SD. 1999. Identification of proteins by matrix-assisted laser desorption/ionization mass spectrometry using peptide and fragment ion masses. In: AJ Link, editor. 2-D Proteome Analysis Protocols, pp. 487–511. Totowa, New Jersey: Humana Press.

Cowie WP. 1968. Identification of fish species by thin slab polyacrylamide gel electrophoresis. J Sci Food Agric 19:226–229.

Craiu A, Akopian T, Goldberg A, Rock KL. 1997. Two distinct proteolytic processes in the generation of a major histocompatibility complex class I-presented peptide. Proc Natl Acad Sci USA 94:10850–10855.

De Francesco M, Parisi G, Médale F, Lupi P, Kaushik SJ, Poli BM. 2004. Effect of long-term feeding with a plant protein mixture based diet on growth and body/fillet quality traits of large rainbow trout *(Oncorhynchus mykiss)*. Aquaculture. (In press)

Dobly A, Martin SAM, Blaney S and Houlihan DF. 2004. Efficiency of conversion of ingested proteins into growth; protein degradation assessed by 20S proteasome activity in rainbow trout, *Oncorhynchus mykiss*. Comp Biochem Physiol A 137:75–85.

Dougherty DA, Wehling RL, Zeece MG, Partridge JE. 1990. Evaluation of selected baking quality factors of hard red winter wheat flours by two-dimensional electrophoresis. Cereal Chem 67:564–569.

Dreger M. 2003. Subcellular proteomics. Mass Spectrometry Reviews 22:27–56.

Einen O, Mørkøre T, Rørå AMB and Thomassen MS. 1999. Feed ration prior to slaughter—A potential tool for managing product quality of Atlantic salmon *(Salmo salar)*. Aquaculture 178:149–169.

Erdjument-Bromage H, Lui M, Lacomis L, Tempst P. 1999. Characterizing proteins from 2-DE gels by internal sequence analysis of peptide fragments. In: AJ Link, editor. 2-D Proteome Analysis Protocols, pp. 467–472. Totowa, New Jersey: Humana Press.

Foegeding EA, Lanier TC, Hultin HO. 1996. Characteristics of edible muscle tissues. In: OR Fennema, editor. Food Chemistry, pp. 879–942. New York: Marcel Dekker, Inc.

Görg A, Obermaier C, Boguth G, Csordas A, Diaz J-J, Madjar J-J. 1997. Very alkaline immobilized pH gradients for two-dimensional electrophoresis of ribosomal and nuclear proteins. Electrophoresis 18:328–337.

Görg A, Obermaier C, Boguth G, Weiss W. 1999. Recent developments in two-dimensional gel electrophoresis with immobilized pH gradients: Wide pH gradients up to pH 12, longer separation distances and simplified procedures. Electrophoresis 20:712–717.

Görg A, Postel W, Gunther S. 1988. The current state of two-dimensional electrophoresis with immobilized pH gradients. Electrophoresis 9:531–546.

Graves PR, Haystead TA. 2003. A functional proteomics approach to signal transduction. Recent Prog Horm Res 58:1–24.

Henningsen R, Gale BL, Straub KM, DeNagel DC. 2002. Application of zwitterionic detergents to the solubilization of integral membrane proteins for two-dimensional gel electrophoresis and mass spectrometry. Proteomics 2:1479–1488.

Herbert B. 1999. Advances in protein solubilization for two-dimensional electrophoresis. Electrophoresis 20: 660–663.

Hershko A, Ciechanover A. 1986. The ubiquitin pathway for the degradation of intracellular proteins. Prog Nucl Acid Res Mol Biol 33:19–56.

Hogstrand C, Balesaria S, Glover CN. 2002. Application of genomics and proteomics for study of the integrated response to zinc exposure in a non-model fish species, the rainbow trout. Comp Biochem Physiol B 133:523–535.

Huber LA, Pfaller K, Vietor I. 2003. Organelle proteomics: Implications for subcellular fractionation in proteomics. Circulation Res 16:962–968.

Hume A, Mackie I. 1980. The use of electrophoresis of the water-soluble muscle proteins in the quantitative analysis of the species components of a fish mince mixture. In: JJ Connell, editor. Advances in Fish Science and Technology. Aberdeen, United Kingdom: Fishing News Books, Ltd.

Huss HH. 1995. Quality and Quality Changes in Fresh Fish. Rome: FAO. 195 pp.

Ingólfsdóttir S, Stefánsson G, Kristbergsson K. 1998. Seasonal variations in physicochemical and textural properties of North Atlantic cod *(Gadus morhua)* mince. J. Aquat Food Prod Technol 7:39–61.

Issaq HJ, Conrads TP, Janini GM, Veenstra TD. 2002. Methods for fractionation, separation and profiling of proteins and peptides. Electrophoresis 23:3048–3061.

Journet A, Chapel A, Kieffer S, Louwagie M, Luche S, Garin J. 2000. Towards a human repertoire of monocytic lysosomal proteins. Electrophoresis 21:3411–3419.

Journet A, Chapel A, Kieffer S, Roux F, Garin J. 2002. Proteomic analysis of human lysosomes: Application to monocytic and breast cancer cells. Proteomics 2:1026–1040.

Kaltschmidt E, Wittmann H-G. 1970. Ribosomal proteins. VII. Two-dimensional polyacrylamide gel electrophoresis for fingerprinting of ribosomal proteins. Anal Biochem 36:401–412.

Kamo M, Tsugita A. 1999. N-terminal amino acid sequencing of 2-DE spots. In: AJ Link, editor. 2-D Proteome Analysis Protocols, pp. 461–466. Totowa, New Jersey: Humana Press.

Kjærsgård IVH and Jessen F. 2003. Proteome analysis elucidating post-mortem changes in cod *(Gadus morhua)* muscle proteins. J. Agric Food Chem 51: 3985–3991.

Kristensen L, Therkildsen M, Riis B, Sørensen MT, Oksbjerg N, Purslow PP, Ertbjerg P. 2002. Dietary-induced changes of muscle growth rate in pigs: Effects on in vivo and postmortem muscle proteolysis and meat quality. J Anim Sci 80:2862–2871.

Kvasnicka F. 2003. Proteomics: General strategies and application to nutritionally relevant proteins. J Chromatogr B 787:77–89.

Ladrat C, Chaplet M, Verrez-Bagnis V, Noël J, Fleurence J. 2000. Neutral calcium-activated proteases from European sea bass (*Dicentrachus labrax* L.) muscle: Polymorphism and biochemical studies. Comp Biochem Physiol B 125:83–95.

Lee BH, Nagamune T. 2004. Protein microarrays and their applications. Biotechnol Bioproc Eng 9:69–75.

Lehrer SB, Ayuso R, Reese G. 2003. Seafood allergy and allergens: a review. Marine Biotechnology 5: 339–348.

Lei MG, Reeck GR. 1987. Two dimensional electrophoretic analysis of isolated soybean protein bodies and of the glycosylation of soybean proteins. J Agric Food Chem 35:296–300.

Link AJ (ed.). 1999. 2-D proteome analysis protocols. Totowa: New Jersey Humana Press. 601 pp.

Locke VL, Gibson TS, Thomas TM, Corthals GL, Rylatt DB. 2002. Gradiflow as a prefractionation tool for two-dimensional electrophoresis. Proteomics 2:1254–1260.

Lødemel JB, Olsen RL. 2003. Gelatinolytic activities in muscle of Atlantic cod *(Gadus morhua),* spotted wolffish *(Anarhichas minor)* and Atlantic salmon *(Salmo salar).* J Sci Food Agric 83:1031–1036.

Lopez JL, Marina A, Alvarez G, Vazquez J. 2002. Application of proteomics for fast identification of species-specific peptides from marine species. Proteomics 2:1658–1665.

Lopez MF, Kristal BS, Chernokalskaya E, Lazarev A, Shestopalov AI, Bogdanova A, Robinson M. 2000. High-throughput profiling of the mitochondrial proteome using affinity fractionation and automation. Electrophoresis 21:3427–3440.

Love RM. 1980. Biological factors affecting processing and utilization. In: JJ Connell, editor. Advances in Fish Science and Technology, pp. 130–138. Aberdeen: Fishing News Books, Ltd.

———. 1992. Biochemical dynamics and the quality of fresh and frozen fish. In: GM Hall, editor. Fish Processing Technology. London: Blackie Academic and Professional.

Love RM, Robertson G, Smith GL, Whittle KJ. 1974. The texture of cod muscle. J Texture Stud 5:201–212.

Mackie I. 1980. A review of some recent applications of electrophoresis and isoelectric focusing in the identification of species of fish in fish and fish products. In: JJ Connell, editor. Advances in Fish Science and Technology. Aberdeen, United Kingdom: Fishing News Books, Ltd.

Mackie IM. 1969. Identification of fish species by a modified polyacrylamide disc electrophoresis technique. J Assoc Public Anal 5:83–87.

———. 1972. Some improvements in the polyacrylamide disc electrophoretic method of identifying species of cooked fish. J Assoc Public Anal 8:18–20.

Mackie IM, Pryde SE, Gonzales-Sotelo C, Medina I, Pérez-Martín R, Quinteiro J, Rey-Mendez M and Rehbein H. 1999. Challenges in the identification of species of canned fish. Trends Food Sci Technol 10:9-14

Mackie IM and Taylor T. 1972. Identification of species of heat-sterilized canned fish by polyacrylamide disc electrophoresis. Analyst 97:609–611.

Margolis J, Kenrick KG. 1969. Two-dimensional resolution of plasma proteins by combination of polyacrylamide disc and gradient gel electrophoresis. Nature 221:1056–1057.

Martin SA, Blaney S, Bowman AS, Houlihan DF. 2002. Ubiquitin-proteasome-dependent proteolysis in rainbow trout *(Oncorhynchus mykiss):* Effect of food deprivation. Pflugers Arch 445:257–266.

Martin SAM, Vilhelmsson O, Houlihan DF. 2003a. Rainbow trout liver proteome—Dietary manipulation and protein metabolism. In: WB Souffrant, CC Metges, editors. Progress in Research on Energy and Protein Metabolism, pp. 57–60. Wageningen, The Netherlands: Wageningen Academic Publishers.

Martin SAM, Vilhelmsson O, Médale F, Watt P, Kaushik S, Houlihan DF. 2003b. Proteomic sensitivity to dietary manipulations in rainbow trout. Biochim Biophys Acta 1651:17–29.

Martinez I, Aursand M, Erikson U, Singstad TE, Veliyulin E, van den Zwaag C. 2003. Destructive and non-destructive analytical techniques for authenticationand composition analyses of foodstuffs. Trends Food Sci Technol 14:489–498.

Martinez I, Jakobsen Friis T. 2004. Application of proteome analysis to seafood authentication. Proteomics 4:347–354.

Martinez I, Jakobsen Friis T, Careche M. 2001a. Post mortem muscle protein degradation during ice-storage of Arctic *(Pandalus borealis)* and tropical *(Penaeus japonicus* and *Penaeus monodon)* shrimps: A comparative electrophoretic and immunological study. J Sci Food Agric 81:1199–1208.

Martinez I, Jakobsen Friis T, Seppola M. 2001b. Requirements for the application of protein sodium dedecyl sulfate-polyacrylamide gel electrophoresis and randomly amplified polymorphic DNA analyses to product speciation. Electrophoresis 22:1526–1533.

Martinez I, Ofstad R, Olsen RI. 1990. Electrophoretic study of myosin isoforms in white muscles of some teleost fishes. Comp Biochem Physiol B 96:221–227.

Martinez I, Solberg C, Lauritzen C, Ofstad R. 1992. Two-dimensional electrophoretic analyses of cod (*Gadus morhua*, L.) whole muscle proteins, water soluble fraction and surimi. Effect of the addition of $CaCl_2$ and $MgCl_2$ during the washing procedure. Appl Theor Electrophoresis 2:201–206.

Millea KM, Krull IS. 2003. Subproteomics in analytical chemistry: Chromatographic fractionation techniques in the characterization of proteins and peptides. J Liq Chromatogr R T 26:2195–2224.

Mortimore GE, Pösö AR, Lardeaux BR. 1989. Mechanism and regulation of protein degradation in liver. Diabetes Metab Rev 5:49–70.

Morzel M, Verrez-Bagnis V, Arendt EK, Fleurence J. 2000. Use of two-dimensional electrophoresis to evaluate proteolysis in salmon *(Salmo salar)* muscle as affected by a lactic fermentation. J Agric Food Chem 48:239–244.

O'Farrell PH. 1975. High resolution two-dimensional electrophoresis of proteins. J Biol Chem 250:4007–4021.

Ogata H, Aranishi F, Hara K, Osatomi K and Ishihara T. 1998. Proteolytic degradation of myofibrillar components by carp cathepsin L. J Sci Food Agric 76: 499–504.

Okumura Y, Sato H, Seiki M, Kido H. 1997. Proteolytic activation of the precursor of membrane type 1 matrix metalloproteinase by human plasmin. A possible cell surface activator. FEBS Lett 402:181–184.

Østergaard O, Melchior S, Roepstorff P, Svensson B. 2002. Initial proteome analysis of mature barley seeds and malt. Proteomics 2:733–739.

Papa I, Alvarez C, Verrez-Bagnis V, Fleurence J, Benyamin Y. 1996. *Post mortem* release of fish white muscle a-actinin as a marker of disorganisation. J Sci Food Agric 72:63–70.

Parisi G, De Francesco M, Médale F, Scappini F, Mecatti M, Kaushik SJ, Poli BM. 2004. Effect of total replacement of dietary fish meal by plant protein sources on early post mortem changes in the biochemical and physical parameters of rainbow trout. Vet Res Comm. (In press.)

Parrington J, Coward K. 2002. Use of emerging genomic and proteomic technologies in fish physiology. Aquat Living Resour 15:193–196.

Peng J, Elias JE, Schwartz D, Thoreen CC, Cheng D, Marsischky G, Roelofs J, Finley D, Gygi SP. 2003. A proteomics approach to understanding protein ubiquitination. Nature Biotechnol 21:921–926.

Pérez-Borla O, Roura SI, Montecchia CL, Roldán H, Crupkin M. 2002. Proteolytic activity of muscle in pre- and post-spawning hake *(Merluccius hubbsi Marini)* after frozen storage. Lebensm-Wiss u-Technol 35:325–330.

Perkins DN, Pappin DJC, Creasy DM, Cottrell JS. 1999. Probability-based protein identification by searching sequence databases using mass spectrometry data. Electrophoresis 20:3551–3567.

Phizicky E, Bastiaens PIH, Zhu H, Snyder M, Fields S. 2003. Protein analysis on a proteomic scale. Nature 422:208–215.

Pieper R, Su Q, Gatlin CL, Huang ST, Anderson NL, Steiner S. 2003. Multi-component immunoaffinity subtraction chromatography: An innovative step towards a comprehensive survey of the human plasma proteome. Proteomics 3:422–432.

Piñeiro C, Barros-Velázquez J, Sotelo CG, Gallardo JM. 1999. The use of two-dimensional electrophoresis for the identification of commercial flat fish species. Z Lebensm Unters Forsch 208:342–348.

Piñeiro C, Barros-Velázquez J, Sotelo CG, Pérez-Martín R, Gallardo JM. 1998. Two-dimensional electrophoretic study of the water-soluble protein fraction in white muscle of gadoid fish species. J Agric Food Chem 46:3991–3997.

Piñeiro C, Barros-Velázquez J, Vázquez J, Figueras A, Gallardo JM. 2003. Proteomics as a tool for the investigation of seafood and other marine products. J Proteome Res 2:127–135.

Piñeiro C, Vázquez J, Marina AI, Barros-Velázquez J, Gallardo JM. 2001. Characterization and partial sequencing of species-specific sarcoplasmic polypeptides from commercial hake species by mass spectrometry following two-dimensional electrophoresis. Electrophoresis 22:1545–1552.

Pusch W, Flocco MT, Leung SM, Thiele H, Kostrzewa M. 2003. Mass spectrometry-based clinical proteomics. Pharmacogenomics 4:463–476.

Rock KL, Gramm C, Rothstein L, Clark K, Stein R, Dick L, Hwang D, Goldberg AL. 1994. Inhibitors of the proteasome block the degradation of most cell proteins and the generation of peptides presented on MHC class I molecules. Cell 78:761–771.

Rothemund DL, Locke VL, Liew A, Thomas TM, Wasinger V, Rylatt DB. 2003. Depletion of the highly abundant protein albumin from human plasma using the Gradiflow. Proteomics 3:279–287.

Sotelo CG, Piñeiro C, Gallardo JM, Pérez-Martín RI. 1993. Fish species identification in seafood products. Trends Food Sci Technol 4:395–401.

Takahashi N, Kaji H, Yanagida M, Hayano T and Isobe T. 2003. Proteomics: Advanced technology for the analysis of cellular function. J Nutr 133:2090S–2096S.

Thorarinsdottir KA, Arason S, Geirsdottir M, Bogason SG, Kristbergsson K. 2002. Changes in myofibrillar proteins during processing of salted cod *(Gadus morhua)* as determined by electrophoresis and differential scanning calorimetry. Food Chem 77:377–385.

Tsuyuki H, Uthe JF, Roberts E, Clarke LW. 1966. Comparative electropherograms of *Coregonis clupeoformis, Salvelinus namaycush, S. alpinus, S. malma* and *S. fontinalis* from the family Salmonidae. J Fish Res Board Can 23:1599–1606.

Tyers M, Mann M. 2003. From genomics to proteomics. Nature 422:193–197.

Vaneenaeme C, Clinquart A, Uytterhaegen L, Hornick JL, Demeyer D, Istasse L. 1994. Postmortem proteases activity in relation to muscle protein-turnover in Belgian Blue bulls with different growth-rates. Sci Alim 14:475–483.

Verrez-Bagnis V, Noël J, Sautereau C, Fleurence J. 1999. Desmin degradation in postmortem fish muscle. J Food Sci 64:240–242.

Vilhelmsson O, Miller KJ. 2002. Synthesis of pyruvate dehydrogenase in *Staphylococcus aureus* is stimulated by osmotic stress Appl Environ Microbiol 68: 2353–2358.

Vilhelmsson OT, Martin SAM, Médale F, Kaushik SJ, Houlihan DF. 2004. Dietary plant protein substitution affects hepatic metabolism, but does not invoke a stress response, in rainbow trout. Br J Nutr 92:71–80.

Walker JM (ed.). 2002. The Protein Protocols Handbook. Totowa: New Jersey Humana Press. 1176 pp.

Wang M, Lakatta EG. 2002. Altered regulation of matrix metalloproteinase-2 in aortic remodeling during aging. Hypertension 39:865–873.

Westermeier R, Naven T (ed.). 2002. Proteomics in Practice. Weinheim Wiley-VCH. 318 pp.

Wilm M, Shevchenko A, Houthaeve T, Breit S, Schweigerer L, Fotsis T, Mann M. 1996. Femtomole sequencing of proteins from polyacrylamide gels by nano-electrospray mass spectrometry Nature 379: 466–469.

Woessner JF. 1991. Matrix metalloproteases and their inhibitors in connective-tissue remodelling. FASEB J. 5:2145–2154.

Woods RK, Thien F, Raven J, Walters EH, Abramson M. 2002. Prevalence of food allergies in young adults and their relationship to asthma, nasal allergies, and eczema. Ann Allerg Asthma Immunol 88: 183–189.

Yates JR. 1998. Mass spectrometry and the age of the proteome. J Mass Spectrom 33:1–19.

Yu C-J, Lin Y-F, Chiang B-L, Chow L-P. 2003a. Proteomics and immunological analysis of a novel shrimp allergen, pen m 2. J Immunol 170:445–453.

Yu YL, Huang ZY, Yang PY, Rui YC, Yang PY. 2003b. Proteomic studies of macrophage-derived foam cell from human U937 cell line using two-dimensional gel electrophoresis and tandem mass spectrometry. J Cardiovasc Pharmacol 42:782–789.

Zeece MG, Holt DL, Wehling RL, Liewen MB, Bush LR. 1989. High-resolution two-dimensional electrophoresis of bovine caseins. J Agric Food Chem 37:378–383.

Part IV
Milk

19
Chemistry and Biochemistry of Milk Constituents

P. F. Fox and A. L. Kelly

INTRODUCTION

Milk is a fluid secreted by female mammals, of which there are approximately 4500 species, to meet the complete nutritional, and some of the physiological, requirements of the neonate of the species. Because nutritional requirements are species specific and change as the neonate matures, it is not surprising that the composition of milk shows very large interspecies differences; for example, the concentrations of fat, protein, and lactose range from 1 to 50%, 1 to 20%, and 0 to 10%, respectively. Interspecies differences in the concentrations of many of the minor constituents of milk are even greater than those of the macro constituents. The composition of milk changes markedly during lactation, reflecting the changing nutritional requirements of the neonate. Milk composition also changes markedly during mastitis and physiological stress. In this chapter, the typical characteristics of the principal, and some of the minor, constituents of bovine milk will be described.

Sheep and goats were domesticated about 8000 BC, and their milk has been used by humans ever since. However, cattle, especially breeds of *Bos taurus*, are now the dominant dairy animals. Total recorded world milk production is approximately 600×10^6 metric tons per year(mt/yr), of which about 85% is bovine, 11% is buffalo, and 2% each is from sheep and goats. Camels, mares, reindeer, and yaks are important dairy animals in limited geographical regions with specific cultural and/or climatic conditions. This article will be restricted to the constituents and properties of bovine milk. The constituents of the milk of the other main dairy species are generally similar to those of bovine milk, although minor differences exist, and the technological properties of the milk of these species differ significantly.

Milk is a very flexible raw material; several thousand dairy products are produced around the world in a great diversity of flavors and forms, including about 1400 varieties/variants of cheese. The principal dairy products and the percentage of milk used in their production are liquid (beverage) milk, about 40%; cheese, about 33%; butter, about 32%; whole milk powder, about 6%; skimmed milk powder, about 9%; concentrated milk products about 2%; fermented milk products, about 2%; casein, about 2%; infant formulas, about 0.3%. The flexibility of milk as a raw material is a result of the properties, many of them unique, of the constituents of milk, of which the principal ones are very easily isolated, permitting the production of valuable food ingredients. Milk is free of off-flavors, pigments, and toxins, which greatly facilitates its use as a food or as a raw material for food production.

The processibility and functionality of milk and milk products are determined by the chemical and physicochemical properties of its principal constituents, that is, lactose, lipids, proteins, and salts, which will be described in this chapter. The exploitation and significance of the chemical and physicochemical properties of milk constituents in the production and properties of the principal groups of dairy foods, that is, liquid milk products, cheese, butter, fermented milks, functional milk proteins, and lactose will be described in Chapter 20. Many of the principal problems encountered during the processing of milk are caused by variability in the concentrations and properties of the principal constituents, arising from several factors, including breed, individuality of the animal (i.e., genetic factors), stage of lactation, health of the animal (especially mastitis), and nutritional status. Synchronized calving, as practiced in New Zealand, Australia, and Ireland to take advantage of cheap grass as the principal component of the cow's diet, has a very marked effect on the composition and properties of milk (see O'Brien et al. 1999 a,b,c; Mehra et al. 1999). However, much of the variability can be offset by standardizing the composition of milk using various methods (e.g., centrifugation, ultrafiltration, or supplementation) or by modifying the process technology.

The chemical and physicochemical properties of the principal constituents of milk are well characterized and described. The very extensive literature includes textbooks by Walstra and Jenness (1984), Wong et al. (1988), Fox (1992, 1995, 1997, 2003), Fox and McSweeney (1998, 2003), and Walstra et al. (1999).

LACTOSE

INTRODUCTION

Lactose is a reducing disaccharide comprised of glucose and galactose, linked by a β-1-4-O-glycosidic bond (Fig. 19.1). It is unique to milk and is synthesized in the mammary gland from glucose transported from the blood; one molecule of glucose is epimerized to galactose, as UDP-galactose, via the Leloir pathway and is condensed with a second molecule of glucose by a two-component enzyme, lactose synthetase. Component A is a general UDP-galactosyl transferase (UDP-GT; EC 2.4.1.2.2) that transfers galactose from UDP-galactose to any of a range of sugars, peptides, or lipids. Component B is the whey protein, α-lactalbumin (α-la) , in the presence of which the K_M of UDP-GT for glucose is reduced 1000-fold and lactose is the principal product synthesized. There is a good correlation between the concentrations of lactose and α-la in milk. Lactose is responsible for approximately 50% of the osmotic pressure of milk, which is equal to that of blood and varies little; therefore, the concentration of lactose in milk is tightly controlled and is independent of breed, individuality, and nutritional factors, but decreases as lactation advances and especially during a mastitic infection, in both cases due to the influx of NaCl from the blood. The physiological function of α-la is probably to control the synthesis of lactose and thus maintain the osmotic pressure of milk at a relatively constant level.

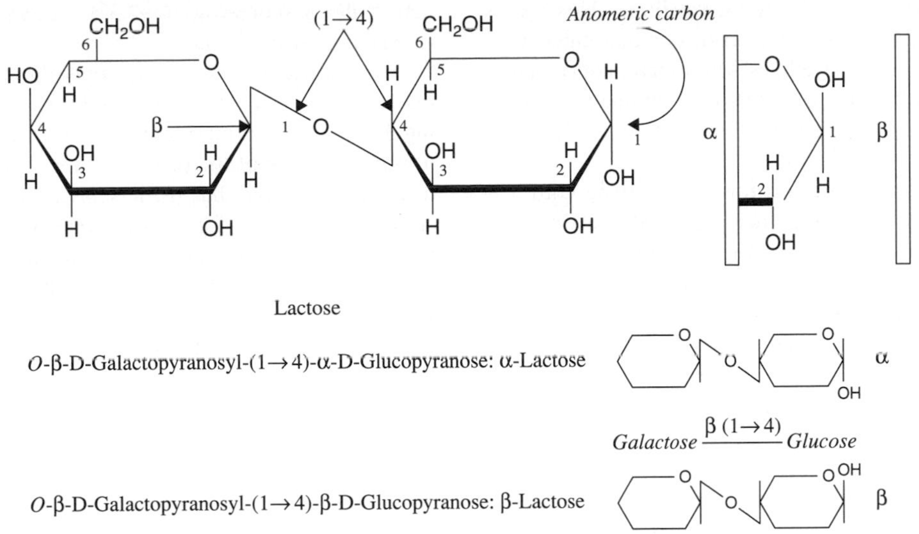

Lactose

O-β-D-Galactopyranosyl-(1→4)-α-D-Glucopyranose: α-Lactose

Galactose β (1→4) Glucose

O-β-D-Galactopyranosyl-(1→4)-β-D-Glucopyranose: β-Lactose

Figure 19.1. Structures of lactose.

The concentration of lactose in milk ranges from almost 0 (in marine mammals) to about 10% in the milk of some monkeys; midlactation bovine milk contains about 4.8% lactose and human milk about 7.0%.

Lactose is the principal sugar in milk, but the milk of most, if not all, species also contains oligosaccharides, up to hexasaccharides, derived from lactose (the nonreducing end of the oligosaccharides is lactose, and many contain fucose and N-acetylneuraminic acid). About 130 oligosaccharides have been identified in human milk; the milk of elephants and bears also contains high levels of oligosaccharides. The oligosaccharides are considered to be important sources of certain monosaccharides, especialy fucose and N-acetlyglucosamine, for neonatal development, especially of the brain (Urashima et al. 2001).

Chemical and Physicochemical Properties of Lactose

Among sugars, lactose has a number of distinctive characteristics, some of which cause problems in milk products during processing and storage; however, some of its characteristics are exploited to its advantage.

The functional aldehyde group at the C-1 position of the glucose moiety exists mainly in the hemiacetal form and, consequently, C-1 is a chiral, asymmetric carbon. Therefore, like all reducing sugars, lactose can exist as two anomers, α and β, which have markedly different properties. From a functional viewpoint, the most important of these properties are differences in solubility and crystallization characteristics between the isomers; α-lactose crystallizes as a monohydrate, while crystals of β-lactose are anhydrous. The solubility of α- and β-lactose in water at 20°C is about 7 and 50 g/100 mL, respectively. However, the solubility of α-lactose is much more temperature-dependent than that of β-lactose, and the solubility curves intersect at approximately 93.5°C (see Fox and McSweeney 1998).

At equilibrium in aqueous solution, lactose exists as a mixture of α- and β-anomers in the approximate ratio 37:63. When an excess of α-lactose is added to water, approximately 7 g/100 mL dissolve immediately, some of which mutarotates to give an α:β ratio of 37:63, leaving the solution unsaturated with respect to both α- and β-lactose. Further α-lactose then dissolves, some of which mutarotates to β-lactose. Solubilization and mutarotation continue until two conditions exist, that is, approximately 7 g/100 mL of dissolved α-lactose and an α:β ratio of 37:63, giving a final solubility of about 18.2 g/100 mL. When β-lactose is added to water, approximately 50 g/100 mL dissolve initially, but about 18.5 g of this mutarotate to α-lactose, which exceeds its solubility, and some α-lactose crystallizes. This upsets the α:β ratio, and more β-lactose mutarotates to α-lactose, which crystallizes. Mutarotation of β-lactose and crystallization of α-lactose continue until about 7 and 11.2 g of α- and β-lactose, respectively, are in solution.

Although lactose has low solubility in comparison with other sugars, once dissolved, it crystallizes with difficulty and forms supersaturated solutions. Highly supersaturated solutions crystallize spontaneously, but if the solution is only slightly supersaturated (one- to two-fold), lactose crystallizes slowly and forms large, sharp, tomahawk-shaped crystals of α-lactose. If the dimensions of the crystals exceed approximately 15 μm, they are detectable on the tongue and palate as a sandy texture. Crystals of β-lactose are smaller and are monoclinic in shape. In the metastable zone, crystallization of lactose is induced by seeding with finely powdered lactose (Fig. 19.2). Since the solubility of α-lactose is lower than that of the β-anomer at < 93.5°C, α-lactose is the normal commercial form.

When concentrated milk is spray-dried, the lactose does not have sufficient time to crystallize during drying, and an amorphous glass is formed. If the moisture content of the powder is kept low, the lactose glass is stable, but if the moisture content increases to about 6%, for example, on exposure of the powder to a high humidity atmosphere, the lactose will crystallize as α-lactose monohydrate. If extensive crystallization occurs, an interlocking mass of crystals is formed, resulting in "caking" of the powder, which is a particularly serious problem in whey powders owing to the high content of lactose (approximately 70%).

The problem is avoided by precrystallizing as much as possible of the lactose before drying, which is achieved by seeding the concentrated solution with finely powdered lactose. Spray-dried milk pow-

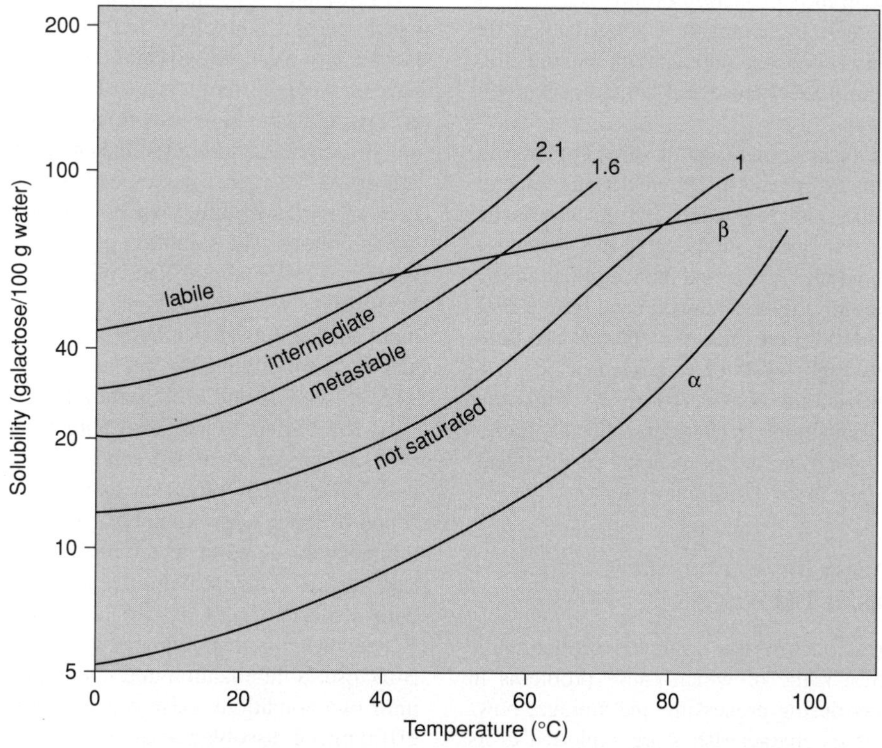

Figure 19.2. Initial solubility of α- and β-lactose, final solubility at equilibrium (line 1) and supersaturation by a factor of 1.6 and 2.1 (α-lactose excluding water of crystallization). (Fox and McSweeney 1998.)

der has poor wettability because the small particles swell on contact with water, thereby blocking the channels between the particles (see Kelly et al. 2003 for review). The wettability (often incorrectly referred to as "solubility") of spray-dried milk powder may be improved by controlling the drying process to produce milk powder with coarser, more easily wetted particles; such powders are said to be "instantized" and are produced by agglomerating the fine powder particles, in effect by controlling the caking process. In the case of whole milk powder, instantization processes must also overcome the intrinsic hydrophobic nature of milk fat; this is normally achieved by adding the amphiphilc agent lecithin.

The crystallization of lactose in frozen milk products results in destabilization of the casein, which aggregates when the product is thawed. In this case, the effect of lactose is indirect; when milk is frozen, pure water freezes, and the concentration of solutes in the unfrozen water is increased. Since milk is supersaturated with calcium phosphate ($\approx$ 66 and 57%, respectively, of the Ca and PO_4 are insoluble and occur in the casein micelles as colloidal calcium phosphate; see Milk Salts section). When the amount of water becomes limiting, soluble Ca $(H_2PO_4)_2$ and $CaHPO_4$ crystallize as $Ca_3(PO_4)_2$, with the concomitant release of H^+ and a decrease in pH to approximately 5.8. During frozen storage, lactose crystallizes as lactose monohydrate, thus reducing the amount of solvent water and aggravating the problems of calcium phosphate solubility and pH decline. Thorough crystallization of lactose before freezing alleviates, but does not eliminate, this problem. Preheating milk prior to freezing also alleviates the problem, but prehydrolysis of lactose to the more soluble sugars glucose and galactose, using β-galactosidase, appears to be the best solution.

Although lactose is hygroscopic when it crystallizes, properly crystallized lactose has very low hygroscopicity; consequently, it is a very effective component of icing sugar.

Lactose has low sweetness (16% as sweet as sucrose in a 1% solution). This limits its usefulness as a sweetener (the principal function of sugars in foods) but makes it is a very useful diluent, for example, for food colors, flavors, enzymes, and so on, when a high level of sweetness is undesirable.

Being a reducing sugar, lactose can participate in the Maillard reaction, with very undesirable consequences in all dairy products: for example, brown color, off-flavors, reduced solubility, and reduced nutritional value.

Food Applications of Lactose

Total milk production ($\approx 600 \times 10^6$ mt/yr) contains about 30×10^6 mt of lactose. Most of this lactose is consumed as a constituent of milk but whey, a by-product of the manufacture of cheese and, to a lesser extent, of casein, contains $8–9 \times 10^6$ mt of lactose. About 400,000 mt of lactose are isolated/prepared per year. A number of high-lactose food products are also produced, for example, approximately 2,000,000 mt of whey permeate powder, regular whey powder, and electrodialyzed whey powder; these serve as crude sources of lactose for several food products, including infant formulas. Thus, most available lactose is now utilized in some form or another, and little is wasted.

Although some of its properties, especially its low sweetness and low solubility, limit the usefulness of lactose as a sugar, other properties, that is, very low hygroscopicity if properly crystallized, low sweetness, and reducing properties, make it a valuable ingredient for the food and pharmaceutical industries. In the pharmaceutical industry, lactose is widely used as a diluent in pelleting operations. The principal application of lactose in the food industry is in the humanization of infant formulas; human milk contains approximately 7% lactose, compared with about 4.8% in bovine milk. For this application, demineralized whey is widely used—it is cheaper and more suitable than purified lactose because it also supplies whey proteins, which help bring the casein:whey protein ratio of the formula closer to the value, 40:60, found in human milk, compared with 80:20 in bovine milk. It is necessary to demineralize the whey because bovine milk contains about four times as much inorganic salts as human milk.

Lactose is also used as an agglomerating/free flowing agent in foods (e.g., butter powders) in the confectionery industry to improve the functionality of shortenings, as an anticaking agent in icing mixtures at high humidity, or as a reducing sugar if Maillard browning is desired. The low sweetness of lactose limits its widespread use as a sugar, but is advantageous in many applications. Lactose absorbs compounds and may be used as a diluent for food flavors and colors or to trap flavors.

Lactose Derivatives

A number of more useful and more valuable products may be produced from lactose. The most significant are

- *Lactulose (galactosyl-*β-1,4-fructose). A sugar not found in nature, which is produced from lactose by heating, especially under slightly alkaline conditions. The concentration of lactulose in milk is a useful index of the severity of the heat treatment to which milk is subjected, for example, in-container sterilization > indirect UHT > direct UHT > HTST pasteurization. Lactulose is not hydrolyzed by intestinal β-galactosidase and enters the large intestine, where it promotes the growth of *Bifidobacterium* spp. It also has a laxative effect; more than 20,000 mt are produced annually.
- *Glucose-galactose syrups.* Produced by acid or enzymatic (β-galacosidase) hydrolysis (see Chapter 20); the technology for the production of such hydrolyzates has been developed but the product is not cost competitive with other sugars (sucrose, glucose, glucose-fructose).
- *Galactooligsaccharides.* β-galactosidase has transferase as well as hydrolytic activity, and under certain conditions, the former predominates, leading to the formation of galactooligosaccharides containing up to 6 monosaccharides linked by glycosidic bonds, which are not hydrolyzed by the enzymes secreted by the human small intestine. The undigested oligosaccharides enter the large intestine, where they have bifidogenic properties and are considered to have promising food applications. These oligosaccharides are quite distinct from the naturally occurring oligosaccharides referred to earlier.
- *Ethanol.* Produced commercially by the fermentation of lactose by *Kluyveromyces lactis.*

Other derivatives that have limited but potentially important applications include lactitol, lactobionic acid, lactic acid, acetic acid, propionic acid, lactosyl urea, and single-cell proteins. Most of these derivatives can be produced by fermentation of sucrose, which is cheaper than lactose, or by chemical synthesis. However, lactitol and lactobionic acid are derived specifically from lactose and may have economic potential. Lactitol is a synthetic sugar alcohol produced by reduction of lactose; it is not metabolized by higher animals but is relatively sweet, and hence has potential for use as a noncalorific sweetener. It has also been reported to reduce blood cholesterol levels, to reduce sucrose absorption, and to be anticarcinogenic. Lactobionic acid has a sweet taste, which is unusual for an acid and therefore should have some interesting applications.

Nutritional Aspects of Lactose

Lactose is responsible for two enzyme-deficiency syndromes: lactose intolerance and galactosemia. The former is due to a deficiency of intestinal β-galactosidase, which is rare in infants but common in adults, except northwestern Europeans and a few African tribes. Since humans are unable to absorb disaccharides, including lactose, from the small intestine, unhydrolyzed lactose enters the large intestine, where it is fermented by bacteria, leading to flatulence and cramps, and to the absorption of water from the intestinal mucosa, which causes diarrhea. These conditions cause discomfort and perhaps death.

The problems caused by lactose intolerance can be avoided by

- Excluding lactose-containing products from the diet, which is inadvertently the normal practice in regions of the world where lactose intolerance is widespread;
- Removing lactose from milk, for example, by ultrafiltration;
- Hydrolysis of the lactose by adding β-galactosidase at the factory or in the home. The technology for the production of lactose-hydrolyzed milk and dairy products is well developed, but is of commercial interest mainly for lactose-intolerant individuals in Europe or North America. Because the consumption of milk is very limited in Southeast Asia, the use of β-galactosidase is of little interest, although lactose intolerance is widespread.

Galactosemia is caused by the inability to catabolize galactose, owing to a deficiency of either of two enzymes, galactokinase or galactose-1-phophate uridyl transferase. A deficiency of galactokinase leads to the accumulation of galactose, which is catabolized via alternative routes, one of which leads to the accumulation of galactitol in various tissues, including the eye, where it causes cataracts over a period of about 20 years. A deficiency of galactose-1-phosphate uridyl transferase leads to abnormalities in membranes of the brain and to mental retardation unless galactose is excluded from the diet within a few weeks of birth. Both forms of galactosemia occur at a frequency of 1 per 50,000 births.

Lactose in Fermented Dairy Products

The fermentation of lactose to lactic acid by lactic acid bacteria (LAB) is a critical step in the manufacture of all fermented dairy products (cheese, fermented milks, and lactic butter). The fermentation pathways are well established (see Cogan and Hill 1993). Lactose is not a limiting factor in the manufacture of fermented dairy products; only about 20% of the lactose is fermented in the production of these products. Individuals suffering from lactose intolerance may be able to consume fermented milks without ill effects, possibly because LAB produce β-galactosidase, and emptying of the stomach is slower than for fresh milk products, thus releasing lactose more slowly into the intestine.

In the manufacture of cheese, most (96–98%) of the lactose is removed in the whey. The concentration of lactose in fresh curd depends on its concentration in the milk and on the moisture content of the curd and varies from about 1.7% by weight in fresh Cheddar curd to about 2.4% by weight in fresh Camembert. The metabolism of residual lactose in the curd to lactic acid has a major effect on the quality of mature cheese (Fox et al. 1990, 2000). The resultant lactic acid may be catabolized to other compounds, for example, to CO_2 and H_2O by surface mold in Camembert, or to propionic acid, acetic acid, and CO_2 in Emmental-type cheeses. Excessive lactic acid in cheese curd may lead to a low pH and a number of defects, such as a strong, acid, harsh taste, an increase in brittleness, and a decrease in firmness. The pH of full-fat Cheddar is inversely related to the lactose/lactic acid content of the curd. Excess residual lactose may also be fermented by heterofermentative lactobacilli, with the production of CO_2, leading to an open texture.

In the manufacture of some cheese varieties, for example, Dutch cheese, the curds are washed to reduce lactose content and thereby regulate the pH of the pressed curd at about 5.3. For Emmental, the curd-whey mixture is diluted with water by about 20%, again to reduce the lactose content of the curd, maintain the pH at about 5.3 and keep the calcium concentration high, which is important for the textural properties of this cheese. For Cheddar, the level of lactose, and hence lactic acid, in the curd is not controlled. Hence, changes in the concentration of lactose in milk can result in marked changes in the quality of such cheeses. Marked changes occur in the concentration of lactose in milk throughout lactation. The lactose content of bulk herd milk from randomly calved cows varies little throughout the year, but differences can be quite large when calving of cows is synchronized; for example, in Ireland, the level of lactose in creamery milk varies from about 4.8% in May to about 4.2% in October. To overcome seasonal variations in the lactose content of milk, the level of wash water used for Dutch-type cheeses is related to the concentrations of lactose and casein in the milk. Ideally, the lactose-to-protein ratio in any particular variety should be standardized (e.g., by washing the curd) to minimize variations in the level of concentration of lactic acid, in the pH, and in the quality of the cheese.

MILK LIPIDS

Definition and Variability

The lipid fraction of milk is defined as those compounds that are soluble in apolar solvents (ethyl/petroleum ether or chloroform/methanol) and is comprised mainly of triglycerides (98%), with approximately 1% phospholipids and small amounts of diglycerides, monoglycerides, cholesterol, cholesterol esters, and traces of fat-soluble vitamins and other lipids. The lipids occur as globules, 0.1–20 μm in diameter, each surrounded by a membrane, the milk fat globule membrane (MFGM), which serves as an emulsifier. The concentration of total and individual lipids varies with breed, individual animal, stage of lactation, mastitic infection, plane of nutrition, interval between milkings, and point during milking when the sample is taken. Among the principal dairy breeds, Friesian/Holstein cows produce milk with the lowest fat content ($\approx$ 3.5%) and Jersey/Guernsey the highest ($\approx$ 6%). The fat content varies considerably throughout lactation; when synchronized calving is practiced, the fat content of bulk Friesian milk varies from about 3% in early lactation to $>$ 4.5% in late lactation. Such large variations in lipid content obviously affect the economics of milk production and the composition of milk products but can be modified readily by natural creaming, centrifugal separation, or addition of cream and hence need not affect product quality. Milk lipids exhibit variability in fatty acid composition and in the size and stability of the globules. These variations, especially fatty acid profile, are essentially impossible to standardize and hence are responsible for considerable variations in the rheological properties, color, chemical stability, and nutritional properties of fat-containing dairy products.

Fatty Acid Profile

Ruminant milk fat contains a wider range of fatty acids than any other lipid system—up to 400 fatty acids have been reported in bovine milk fat; the principal fatty acids are the homologous series of saturated fatty acids with an even number of C-atoms, $C_{4:0}$–$C_{18:0}$ and $C_{18:1}$. The outstanding features of the fatty acid profile of bovine milk fat are a high concentration of short- and medium-chain acids (ruminant milk fats are the only natural lipids that contain butanoic acid, $C_{4:0}$) and a low concentration of polyunsaturated fatty acids.

In ruminants, the fatty acids for the synthesis of milk lipids are obtained from triglycerides in chylomicrons in the blood or synthesized de novo in the mammary gland from acetate or β-hydroxybutyrate produced in the rumen. The triglycerides in chylomicrons are derived from the animal's feed or synthesized in the liver. Butanoic acid ($C_{4:0}$) is produced by the reduction of β-hydroxybutyrate, which is synthesized from dietary roughage by bacteria in the rumen and therefore varies substantially with the animal's diet. All $C_{6:0}$–$C_{14:0}$ and 50% of $C_{16:0}$ are synthesized in the mammary gland via the malonyl-CoA pathway from acetyl-CoA produced from acetate synthesized in the rumen. Essentially 100% of $C_{18:0}$, $C_{18:1}$, $C_{18:2}$, and $C_{18:3}$ and 50% of $C_{16:0}$ are derived from blood lipids (chylomicrons) and represent about 50% of total fatty acids in ruminant milk fat. Unsaturated fatty acids in the animal's diet are saturated by bacteria in the rumen unless they are protected, for example, by encapsulation.

When milk production is seasonal, for example, as in Australia, New Zealand, and Ireland, very significant changes occur in the fatty acid profile of milk fat throughout the production season (see Fox 1995, Fox and McSweeney 1998). These variations are reflected in the hardness of butter produced from such milk; the spreadability of butter produced in winter is much lower than that of summer butter. Owing to the lower degree of unsaturation, winter butter should be less susceptible to lipid oxidation than the more unsaturated summer product, but the reverse appears to be the case, probably owing to higher levels of prooxidants, for example, Cu and Fe, in winter milk.

Although a ruminant's diet, especially if grass based, is rich in polyunsaturated fatty acids (PUFA), these are hydrogenated by bacteria in the rumen, and consequently, ruminant milk fat contains very low levels of PUFA; for example, bovine milk fat contains approximately 2.4% $C_{18:2}$ compared with 13 or 12% in human and porcine milk fat, respectively. PUFAs are considered to be nutritionally desirable, and consequently, there has been interest in increasing the PUFA content of bovine milk fat. This can be done by feeding encapsulated PUFA-rich lipids or crushed PUFA-rich oil seed to the animal. Increasing the PUFA content also reduces the melting point (MP) of the fat and makes butter produced from it more spreadable. However, the lower MP fat may have undesirable effects on the rheological properties of cheese, and PUFA-rich dairy products are very susceptible to lipid oxidation. Although the technical feasibility of increasing the PUFA content of milk fat by feeding protected PUFA-rich lipids to the cow has been demonstrated, it is not economical to do so in most cases. Blending milk fat with PUFA-rich or $C_{18:1}$-rich vegetable oil appears to be much more viable and is now widely practiced commercially.

Conjugated Linoleic Acid

Linoleic acid (*cis*-9,*cis*-12-octadecadienoic acid) is the principal essential fatty acid and has been the focus of nutritional research for many years. However, conjugated isomers of linoleic acid (CLA) have attracted very considerable attention recently. CLA is a mixture of eight positional and geometric isomers of linoleic acid, which have a number of health-promoting properties, including anticarcinogenic and antiatherogenic activities, reduction of the catabolic effects of immune stimulation, and the ability to enhance growth and reduce body fat (see Parodi 1999, Yurawecz et al. 1999). Of the eight isomers of CLA, only the *cis*-9, *trans*-11 isomer is biologically active. This compound is effective at very low concentrations, 0.1 g/100 g diet.

Fat-containing foods of ruminant origin, especially milk and dairy products, are the principal sources of dietary CLA, which is produced as an intermediate during the biohydrogenation of linoleic acid by the rumen bacterium, *Butyrivibrio fibrisolvens.* Since CLA is formed from linoleic acid, it is not surprising that the CLA content of milk is affected by diet and season, being highest in summer when cows are on fresh pasture rich in PUFAs (Lock and Carnsworthy 2000, Lawless et al. 2000), and is higher in the fat of milk from cows on mountain than from those on lowland pasture (Collomb et al. 2002). The concentration of CLA in milk fat can be increased five- to seven-fold by increasing the level of dietary linoleic acid, for example, by duodenal infusion (Kraft et al. 2000) or by feeding a linoleic acid–rich oil (e.g., sunflower oil; Kelly et al. 1998)

A number of other lipids, for example, sphingomyelin, butanoic acid, and ether lipids, may have anticarcinogenic activity, but little information is available on these to date (Parodi 1997, 1999)

Structure of Milk Triglycerides

Glycerol for milk lipid synthesis is obtained in part from hydrolzed blood lipids (free glycerol and monoglycerides), partly from glucose, and a little from free blood glycerol. Synthesis of triglycerides within the cell is catalyzed by enzymes located on the endoplasmic reticulum. Esterification of fatty acids is not random (Table 19.1). The concentrations of $C_{4:0}$ and $C_{18:1}$ appear to be rate limiting because of the need to keep the lipid liquid at body temperature. Some notable features of the structure are as follows:

- Butanoic and hexanoic acids are esterified almost entirely, and octanoic and decanoic acids predominantly, at the *sn*-3 position.
- As the chain length increases up to $C_{16:0}$, an increasing proportion of the fatty acids is esterified at the *sn*-2 position; this is more marked for human than for bovine milk fat, especially in the case of palmitic acid ($C_{16:0}$).
- Stearic acid ($C_{18:0}$) is esterified mainly at *sn*-1.
- Unsaturated fatty acids are esterified mainly at the *sn*-1 and *sn*-3 positions, in roughly equal proportions.

The fatty acid distribution is significant from two viewpoints:

1. It affects the melting point and hardness of the fat, which can be reduced by randomizing the fatty acid distribution. Transesterification can be performed by treatment with $SnCl_2$ or enzymatically under certain conditions; increasing attention is being focused on enzymatic esterification as an acceptable means of modifying the hardness of butter.

Table 19.1. Composition of Fatty Acids (mol% of the total) Esterified to Each Position of The Triacyl-*sn*-glycerols in Bovine or Human Milk

Fatty Acid	Cow			Human		
	sn-1	*sn*-2	*sn*-3	*sn*-1	*sn*-2	*sn*-3
4:0	–	–	35.4	–	–	–
6:0	–	0.9	12.9	–	–	–
8:0	1.4	0.7	3.6	–	–	–
10:0	1.9	3.0	6.2	0.2	0.2	1.1
12:0	4.9	6.2	0.6	1.3	2.1	5.6
14:0	9.7	17.5	6.4	3.2	7.3	6.9
16:0	34.0	32.3	5.4	16.1	58.2	5.5
16:1	2.8	3.6	1.4	3.6	4.7	7.6
18:0	10.3	9.5	1.2	15.0	3.3	1.8
18:1	30.0	18.9	23.1	46.1	12.7	50.4
18:2	1.7	3.5	2.3	11.0	7.3	15.0
18:3	–	–	–	0.4	0.6	1.7

2. Pancreatic and many other lipases are specific for the fatty acids at the *sn*-1 and *sn*-3 positions. Therefore, $C_{4:0}$ to $C_{8:0}$ are released rapidly from milk fat; these are water soluble and are readily absorbed from the intestine. Medium- and long-chain acids are absorbed more effectively as 2-monoglycerides than as fatty acids; this appears to be quite important for the digestion of lipids by human infants, who have a limited ability to digest lipids due to the absence of bile salts. Infants metabolize human milk fat more efficiently than bovine milk fat, apparently due to the very high proportion of $C_{16:0}$ esterified at *sn*-2 in the former. The effect of transesterification on the digestibility of milk fat by infants merits investigation.

   Short-chain fatty acids ($C_{4:0}$–$C_{10:0}$) have a strong aroma and flavor, and their release by indigenous lipoprotein lipase and microbial lipases cause off-flavors in milk and many dairy products; this is referred to as hydrolytic rancidity.

Rheological Properties of Milk Fat

The melting characteristics of ruminant milk fat are such that at low temperatures (e.g., *ex*-refrigerator), it contains a high proportion of solid fat and has poor spreadability. The rheological properties of milk lipids may be modified by fractional crystallization; for example, one effective treatment involves removing the middle melting point fraction and blending high and low melting point fractions. Fractional crystallization is expensive and is practiced in industry to only a limited extent; in particular, securing profitable outlets for the middle melting point fraction is a major economic problem.

Alternatively, the rheological properties of milk fat may be modified by increasing the level of PUFAs, by feeding cows with protected PUFA-rich lipids, but this practice is also expensive. The melting characteristics of blends of milk fat and vegetable oils can be easily varied by changing the proportions of the different fats and oils in the blend. This procedure is economical and is widely practiced commercially; blending also increases the level of nutritionally desirable PUFAs. The rheological properties of milk fat–based spreads can also be improved by increasing the moisture content of the product; obviously, this is economical and nutritionally desirable in the sense that the caloric value is reduced, but the resultant product is less microbiologically stable than butter.

The melting characteristics and rheological properties of milk fat can also be modified by inter- and transesterification. Chemically catalyzed inter- and transesterification are not permitted in the food industry, but enzymatic catalysis may be acceptable. Lipases capable of such modifications on a commercial scale are available,but their use is rather limited. Enzymatic transesterification allows modification of the nutritional as well as the rheological properties of lipids. The nutritional and rheological properties of lipids can also be modified by the use of a desaturase, which converts $C_{18:0}$ to $C_{18:1}$ (these enzymes are a subject of ongoing research; see hppt://bioinfo.pbi.nrc.ca/covello/r-fattyacid.html, Meesapyodsuk et al. 2000). However, this type of enzyme does not seem to be available commercially yet.

Milk Fat as an Emulsion

An emulsion consists of two immiscible, mutually insoluble liquids, usually referred to as oil and water, in which one of the liquids is dispersed as small droplets (globules, the dispersed phase) in the other (the continuous phase). If the oil is the dispersed phase, the emulsion is referred to as an oil-in-water (o/w) emulsion; if water is the dispersed phase, the emulsion is referred to as a water-in-oil (w/o) emulsion. The dispersed phase is usually, but not necessarily, the phase present in the smaller amount. An emulsion is prepared by dispersing one phase into the other. Since the liquids are immiscible, they will remain discrete and separate if they differ in density, as is the case with lipids and water, the densities of which are 0.9 and 1.0, respectively; the lipid globules will float to the surface and coalesce. Coalescence is prevented by adding a compound that reduces the interfacial tension, γ, between the phases. Compounds capable of doing this have an amphiphatic structure, that is, they have both hydrophilic and hydrophobic regions (e.g., phospholipids, monoglycerides, diglycerides, proteins, soaps, and numerous synthetic compounds), and are known as emulsifiers or detergents. The emulsifier forms a layer on the surface of the globules, with its hydrophobic region penetrating the oil phase and its hydrophilic region in the aqueous phase. An emulsion thus stabilized will cream if left undisturbed, but the globules remain discrete and can be redispersed readily by gentle agitation.

In milk, the lipids exist as an o/w emulsion in which the globules range in size from about 0.1 to 20 μm, with a mean of 3–4 m. The mean size of the fat globules is higher in high fat than in low-fat milk, for example, Jersey compared with Friesian, and decreases with advancing lactation. Consequently, the separation of fat from milk is less efficient in winter than in summer, especially when milk production is seasonal, and it may not be possible to meet the upper limit for fat content in some products, for example, casein, during certain periods.

Stability of Milk Fat Globules

In milk, the emulsifier is the MFGM. On the inner side of the MFGM is a layer of unstructured lipoproteins, acquired within the secretory cells as the triglycerides move from the site of synthesis in the rough endoplasmic reticulum (RER) in the basal region of the cell towards the apical membrane. The fat globules are excreted from the cells by exocytosis, that is, they are pushed through and become surrounded by the apical cell membrane. Milk proteins and lactose are excreted from the cell by the reverse process: the proteins are synthesized in the RER and are transported to the Golgi region, where the synthesis of lactose occurs under the control of α-la. The milk proteins and lactose are encapsulated in the Golgi membrane; the vesicles move towards, and fuse with, the apical cell membrane, open, and discharge their contents into the alveolar lumen, leaving the vescicle (Golgi) membrane as part of the apical membrane, thereby replacing the membrane lost in the excretion of fat globules. Thus, the outer layer of the MFGM is composed of a trilaminar membrane, consisting of phospholipids and proteins, with a fluid mosaic structure. The MFGM contains many enzymes, which originate mainly from the Golgi apparatus: in fact, most of the indigenous enzymes in milk are concentrated in the MFGM, notable exceptions being plasmin and lipoprotein lipase (LPL), which are associated with the casein micelles. The trilaminar membrane is unstable and is shed during storage, and especially during agitation, into the aqueous phase, where it forms microsomes.

The stability of the MFGM is critical for many aspects of the milk fat system:

- The existence of milk as an emulsion depends on the effectiveness of the MFGM.
- Damage to the MFGM leads to the formation of nonglobular (free) fat, which may be evident as "oiling-off" on tea or coffee, cream plug, or age thickening. An elevated level of free fat in whole milk powder reduces its wettability. Problems related to, or arising from, free fat are more serious in winter than in summer, probably due to the reduced stability of the MFGM. Homogenization, which replaces the natural MFGM by a layer of proteins from the skim milk phase, principally caseins, eliminates problems caused by free fat.
- The MFGM protects the lipids in the core of the globule against lipolysis by LPL in the skim milk (adsorbed on the casein micelles). The MFGM may be damaged by agitation, foaming, freezing (e.g., on bulk tank walls), and especially by homogenization, allowing LPL access to the core lipids and leading to lipolysis and hydrolytic rancidity. This is potentially a major problem in the dairy industry unless milking machines, especially pipeline milking installations, are properly installed and serviced.
- The MFGM appears to be less stable in winter/late lactation than in summer/ midlactation; therefore, hydrolytic rancidity is more likely to be a problem in winter than in summer. An aggravating factor is that less milk is usually produced in winter than in summer, especially in seasonal milk production systems, which leads to greater agitation during milking and, consequently, a greater risk of damage to the MFGM.

Creaming

Since the specific gravity of lipids and skim milk is 0.9 and 1.036, respectively, the fat globules in milk held under quiescent conditions will rise to the surface under the influence of gravity, a process referred to as creaming. The rate of creaming, *V*, of fat globules is given by Stoke's equation:

$$V = \frac{2r^2(\rho^1 - \rho^2)\,g}{9\eta}$$

where r is the radius of the globule, ρ^1 is the specific gravity of skim milk, ρ^2 is the specific gravity of the fat globules, g is the acceleration due to gravity, and η is the viscosity of skim milk.

The values of r, ρ^1, ρ^2, and η suggest that a cream layer should form in milk after about 60 hours, but milk creams in about 30 minutes. The rapid rate of creaming is due to the strong tendency of the fat globules to agglutinate (stick together) due to the action of indigenous immunoglobulin M, which precipitates onto the fat globules when milk is cooled (hence the term, cryoglobulin). Considering the effect of globule size *(r)* on the rate of creaming, large globules rise faster than smaller ones and collide with, and adhere to, smaller globules, an effect promoted by cryoglobulins. Owing to the larger value of *r*, the clusters of globules rise faster than individual globules, and therefore the creaming process accelerates as the globules rise and clump. Ovine, caprine, and buffalo milk do not contain cryoglobulins and therefore cream much more slowly than bovine milk.

In the past, creaming was a very important physicochemical property of milk:

- The cream layer served as an index of fat content and hence of quality to the consumer.
- Creaming was the traditional method for preparing fat (cream) from milk for use in the manufacture of butter. Its significance in this respect declined with the development of the mechanical separator in 1878, but natural creaming is still used to adjust the fat content of milk for some cheese varieties, for example, Parmigiano-Reggiano. A high proportion (about 90%) of the bacteria in milk become occluded in the clusters of fat globules.

Homogenization of Milk

Today, creaming is of little general significance. In most cases, its effect is negative, and for most dairy products, milk is homogenized, that is, subjected to a high shearing pressure that reduces the size of the fat globules (average diameter < 1 μm), increases the fat surface area (four- to six-fold), replaces the natural MFGM with a layer of caseins, and denatures cryoglobulins, and hence prevents the agglutination of globules. Homogenization has several very significant effects on the properties of milk:

- If properly executed, creaming is delayed indefinitely due to the reduced size of the fat globules and the denaturation of cryoglobulins.
- Susceptiblity to hydrolytic rancidity is markedly increased because indigenous LPL has ready access to the triglycerides; consequently, milk must be heated under conditions sufficiently severe to inactivate LPL before (usually) or immediately after homogenization.
- Susceptibility to oxidative rancidity is reduced because prooxidants in the MFGM, for example, metals and xanthine oxidase, are distributed throughout the milk.
- The whiteness of milk is increased, due to the greater number of light-scattering particles.
- The strength and syneretic properties of rennet-coagulated milk gels for cheese manufacture are reduced; hence, cheese with a higher moisture content is obtained. Consequently, milk for cheese manufacture is not normally homogenized; an exception is reduced-fat cheese, in which a higher moisture content improves texture.
- The heat stability of whole milk and cream is reduced, the magnitude of the effect increasing directly with fat content and homogenization pressure; homogenization has no effect on the heat stability of skimmed milk.
- The viscosity of whole milk and cream is increased by single-stage homogenization due to the clustering of newly formed fat globules; the clumps of globules are dispersed by a second homogenization stage at a lower pressure, which may be omitted if an increased viscosity is desired.

Lipid Oxidation

The chemical oxidation of lipids is a major cause of instability in dairy products (and many other foods). Lipid oxidation is a free-radical, autocatalytic process principally involving the methylene group between a pair of double bonds in PUFAs. The process is initiated and/or catalyzed by polyvalent metals (especially Cu and Fe); UV light; ionizing radiation; or enzymes such as lipoxygenase in the case of plant oils, or xanthine oxidase (a major component of the MFGM) in milk. Oxygen is a primary reactant. The principal end products are unsaturated carbonyls, which cause major flavor defects; the reaction intermediates, that is, fatty acid free radicals, peroxy free radicals and hydroperoxides, have no flavor. Polymerization of free radicals and other species leads to the formation of pigmented products and to an increase in viscosity, but it is unlikely that polymerization-related problems occur to a significant extent in dairy products.

Lipid oxidation can be prevented or controlled by

- Avoiding metal contamination at all stages of processing through the use of stainless steel equipment.
- Avoiding exposure to UV light by using opaque packing (foil or paper).
- Packaging under an inert atmosphere, usually N_2.
- Use of O_2 or free radical scavengers, for example, glucose oxidase and superoxide dismutase (indigenous enzymes in milk), respectively.
- Use of antioxidants that break the free-radical chain reaction; synthetic antioxidants are not permitted in dairy products, but the level of natural antioxidants, for example, tocopherols (vitamin E), in milk may be increased by supplementing the animal's feed. Polyphenols are very effective antioxidants; their direct addition to dairy products is not permitted, but it may be possible to increase their concentration in milk by supplementing the animal's feed. Antioxidants are compounds that readily supply a $H^{\bullet}$ to fatty acid and peroxy radicals, leaving a stable oxidized radical. Many antioxidants are polyphenols, which give up a $H^{\bullet}$ and are converted to a quinone; examples are tocopherols (vitamin E) and catechins. At low concentrations, ascorbic acid is a good antioxidant, but at high concentations it functions as a prooxidant, apparently as a complex with Cu. Sulphydryl groups of proteins are also effective antioxidants; cream for butter making is usually heated to a high temperature to denature proteins and expose/activate sulphydryl groups.

Fat-Soluble Vitamins

Since the fat-soluble vitamins (A, D, E, and K) in milk are derived from the animal's diet, large seasonal variations in their concentration in milk can be expected. The breed of cow also has a significant effect on the concentration of fat-soluble vitamins in milk; high-fat milk (Jersey and Guernsey) has a higher content of these vitamins than Friesian or Holstein milk. Variations in the concentrations of fat-soluble vitamins in milk have a number of consequences:

- Nutritionally, milk contributes a substantial portion of the RDA for these vitamins to Western diets; it is common practice to fortify milk and butter with vitamins A and D.
- The yellow-orange color of high-fat dairy products depends on the concentrations of carotenoids and vitamin A present, and hence on the diet of the animal. New Zealand butter is much more highly colored than Irish butter, which in turn is much more yellow than American or German products. The differences are due in part to the greater dependence of milk production in New Zealand and Ireland on pasture and to the higher proportion of carotenoid-rich clover in New Zealand pasture and the higher proportion of Jersey cows in New Zealand herds.
- Goats, sheep, and buffalo do not transfer carotenoids to their milk, which is, consequently, whiter than bovine milk. Products produced from these milks are whiter than corresponding products made from bovine milk. The darker color of the bovine milk may be unattractive to consumers accustomed to caprine or ovine milk products. If necessary, the carotenoids in bovine milk may be bleached (by benzoyl peroxide) or masked (by chlorophyll or TiO_2).
- Vitamin E (tocopherols) is a potent antioxidant and contributes to the oxidative stability of dairy products. The tocopherol contents of milk and meat can be readily increased by supplementing the animal's diet with tocopherols, which is sometimes practiced.

MILK PROTEINS

Introduction

Technologically, the proteins of milk are its most important constituents. They play important, even essential, roles in all dairy products except butter and anhydrous milk fat. The roles played by milk proteins are nutritional, physiological, and physicochemical:

- *Nutritional:* All milk proteins.
- *Physiological:* Immunoglobulins, lactoferrin, lactoperoxidase, vitamin-binding proteins, protein-derived biologically active peptides.
- *Physicochemical:*
  - *Gelation.* Enzymatically, acid or thermally induced gelation in all cheeses, fermented milks, whey protein concentrates and isolates.
  - *Heat stability.* All thermally processed dairy products.

- *Surface activity.* Caseinates, whey protein concentrates, and isolates.
- *Rheology.* All protein-containing dairy products.
- *Water sorption:* Most dairy products, comminuted meat products.

Milk proteins have been studied extensively and are very well characterized at molecular and functional levels (for reviews see Fox and McSweeney 2003).

Heterogeneity of Milk Proteins

It has been known since 1830 that milk contains two types of protein, which can be separated by acidification to what we now know as pH 4.6. The proteins insoluble at pH 4.6 are called caseins and represent about 78% of the total nitrogen in bovine milk; the soluble proteins are called whey or serum proteins. As early as 1885, it was shown that there are two types of whey protein, globulins and albumins, which were thought to be transferred directly from the blood (the proteins of blood and whey have generally similar physicochemical properties and are classified as albumins and globulins). Initially, the term casein was not restricted to the acid-insoluble proteins in milk but was used to describe all acid-insoluble proteins; however, it was recognized at an early stage that the caseins are unique milk-specific proteins.

The casein fraction of milk protein was considered initially to be homogeneous, but from 1918 onwards, evidence began to accumulate that it is heterogeneous. Through the application of free boundary electrophoresis (FBE) in the 1930s and zone electrophoresis in the 1960s, in starch or polyacrylamide gels (SGE, PAGE) containing urea and a reducing agent, it has been shown that casein is in fact very heterogeneous. Bovine casein consists of four families of caseins: α_{s1}, α_{s2}, β, and κ, which represent about 38, 10, 36, and 12%, respectively, of whole casein. Urea-PAGE showed that each of the casein families exhibits microheterogeneity due to

- Genetic polymorphism, usually involving substitution of one or two amino acids;
- Variations in the degree of phosphorylation;
- Varations in the degree of glycosylation of κ-casein;
- Inter-molecular disulphide bond formation in α_{s2}- and κ-caseins; and
- Limited proteolysis, especially of β- and as α_{s2}-caseins, by plasmin; the resulting peptides include the γ- and λ-caseins and proteose peptones.

In the 1930s, FBE showed that both the globulin and albumin fractions of whey protein are heterogeneous, and in the 1950s, the principal constituents were isolated and characterized. It is now known that the whey protein fraction of bovine milk comprises four main proteins: β-lactoglobulin (β-lg), α-lactalbumin (α-la), immunoglobulins (Igs), and blood serum albumin (BSA), which represent about 40, 20, 10, and 10%, respectively, of the total whey protein in mature milk. The remaining 10% is mainly nonprotein nitrogen and trace amounts of several proteins, including about 60 indigenous enzymes. About 1% of total milk protein is part of the MFGM, including many enzymes. β-lg and α-la are synthesized in the mammary gland and are milk specific; they exhibit genetic polymorphism (in fact the genetic polymorphism of milk proteins was first demonstrated for β-lg in 1956). BSA, most of the Ig (IgG), and most of the minor proteins are transferred from the blood.

Methods for the isolation of the individual proteins were developed and gradually improved during the period 1950–1970, so that by about 1970 all the principal milk proteins had been purified to homogeneity.

The concentration of total protein in milk is affected by most of the factors that affect the concentration of fat, that is, breed, individuality, nutritional status, health, and stage of lactation, but with the exception of the last, the magnitude of most effects is less than in the case of milk fat. The concentration of protein in milk decreases very markedly during the first few days postpartum, mainly due to the decrease in Ig from approximately 10% in the first colostrum to 0.1% within about 1 week. The concentration of total protein continues to decline more slowly thereafter, to a minimum after about 4 weeks, and then it increases until the end of lactation. Data on variations in the groups of proteins throughout lactation have been published (see Mehra et al. 1999), but there are few data on variations in the concentrations of the individual principal proteins.

Molecular Properties of Milk Proteins

The six principal milk-specific proteins have been isolated and are very well characterized at the molecular level; their chemical composition is summarized in Table 19.2. The most notable features of the principal milk-specific proteins are discussed below.

The principal milk-specific proteins are quite small molecules (molecular weight of ≈ 15–25 kDa). All the caseins are phosphorylated, but to different and variable degrees, 1–13 mol P/mol protein; the phosphate groups are esterified as monoesters of serine residues.

The primary structures of the caseins have a rather uneven distribution of polar and apolar residues along their sequences. Clustering of the phosphoseryl residues is particularly marked, resulting in strong anionic patches in α_{s1}-, α_{s2}-, and β-caseins. κ-casein does not have a phosphoseryl cluster, but the N-terminal two-thirds of the molecule is quite hydrophobic, while its C-terminal is relatively hydrophilic—it contains no aromatic and no cationic residues. These features give the caseins an amphiphatic structure, making then very surface active, with good emulsifying and foaming properties. The amphiphatic structure of κ-casein is particularly significant and is largely responsible for its micelle-stabilizing properties. The distribution of amino acids in β-lg and α-la is quite random.

The two principal caseins, α_{s1}- and β-casein, are devoid of cysteine or cystine residues; the two minor caseins, α_{s2}- and κ-casein, contain two intermolecular disulphide bonds. β-lactoglobulin contains two intramolecular disulphide bonds and one sulphydryl group, which is buried and unreactive in the native protein but becomes exposed and reactive when the molecule is denatured; it reacts via sulphydryl-disulphide interactions with other proteins, especially κ-casein, with major consequences for many important properties of the milk protein system, especially heat stability and cheese-making properties. α-la has four intramolecular disulphide bonds.

All the caseins, especially β-casein, contain a high level of proline (in β-casein, 17 of the 209 residues are proline), which disrupts α- and β-structures; consequently, the caseins are rather unstructured molecules and are readily susceptible to proteolysis. However, theoretical calculations suggest that the caseins may have a considerable level of secondary and tertiary structure; to explain the differences between the experimental and theoretical indices of higher structures, it has been suggested that the caseins are very mobile, flexible, rheomorphic molecules (i.e., the casein molecules, in solution, are sufficiently flexible to adopt structures that are dictated by their environment; Holt and Sawyer 1993).

In contrast, the whey proteins are highly compact and structured, with high levels of α-helices, β-sheets, and β-turns. In β-lg, the β-sheets are in an antiparallel arrangement and form a β-barrel calyx. β-lactoglobulin is a member of the lipocalin family, which now includes 14 proteins (for review see Akerstrom et al. 2000)

The caseins are often regarded as very hydrophobic proteins, but they are not particularly so; however, they do have a high surface hydrophobicity, owing to their open structures. Also due to their open structure, the caseins are quite susceptible to proteolysis, which is as would be expected for proteins the putative function of which is to be a source of amino acids for the neonate. However, due to their hydrophobic patches, caseins have a high propensity to yield bitter hydrolyzates, even in cheese that undergoes relatively little proteolysis. In contrast, the highly structured whey proteins are very resistant to proteolysis in the native state and may traverse the intestinal tract of the neonate intact.

Some κ-casein molecules are glycosylated. The sugar moieties present are galactose, galactosamine, and N-acetylneuraminic acid (sialic acid), which occur as trisaccharides or tetrasaccharides. The oligosaccharides are attached to the polypeptide via threonine residues in the C terminal region of the molecule and vary from 0 to 4 mol/mol protein.

Probably because of their rather open structures, the caseins are extremely heat stable, for example, sodium caseinate can be heated at 140°C for 1 hour without obvious physical effects. The more highly structured whey proteins are comparatively heat labile, although in comparison with many other globular proteins, they are quite heat stable; they are completely denatured on heating at 90°C for 10 minutes.

Under the ionic conditions in milk, α-la exists as monomers with a molecular weight of 14.7 kDa. β-lactoglobulin exists as dimers (molecular weight ≈ 36 kDa) in the pH range 5.5–7.5; at pH values < 3.5 or > 7.5, it exists as monomers, while at pH 3.5–5.5, it exists as octamers. The caseins have a very strong tendency to associate. Even in sodium caseinate, the most soluble form of whole casein, the proteins exist mainly as decamers or larger aggregates at 20°C. At a low concentration, for example, $< 0.6\%$, at 4°C,

Table 19.2. Characteristics of the Principal Proteins in Cows' Milk

Protein	Molecular Weight	Amino Acid Residues			Phosphate Groups	Concentration (g/L)	Genetic Variants
		Total	Pro	Cys			
α_{s1}-casein	23,164	199	17	0	8	10.0	A, B, C, D, E, F, G, H
α_{s2}-casein	25,388	207	10	2	10–13	2.6	A, B, C, D
β-casein	23,983	209	35	0	5	9.3	A^1, A^2, A^3, B, C, D, E, F, G
κ-casein	19,038	169	20	2	1	3.3	A, B, C, D, E, F^S, F^I, G^S, H, I, J
β-lactoglobulin	18,277	162	8	5	0	3.2	A, B, C, D, E, F, G, H, I, J
α-lactalbumin	14,175	123	2	8	0	1.2	A, B, C

Source: Adapted from Fox 2003.

β-casein exists as monomers, but associates to form micelles if the concentration or temperature is increased. In the presence of Ca, for example, calcium caseinate, casein forms large aggregates of several million Daltons. In milk, the caseins exist as very complex structures, known as casein micelles, which are described in the next section, Casein Micelles.

The function of the caseins appears to be to supply amino acids to the neonate. They have no biological function stricto sensu, but their Ca-binding properties enable a high concentration of calcium phosphate to be carried in milk in a "soluble" form; the concentration of calcium phosphate far exceeds its solubility, and without the "solubilizing" influence of casein it would precipitate in the ducts of the mammary gland and cause atopic milk stones.

β-lactoglobulin has a hydrophobic pocket within which it can bind small hydrophobic molecules. It binds and protects retinol in vitro and perhaps functions as a retinol carrier in vivo. In the intestine, it exchanges retinol with a retinol-binding protein. It also binds fatty acids and thereby stimulates lipase; this is perhaps its principal biological function. All members of the lipocalin family have some form of binding function (see Akerstrom et al. 2000).

α-lactalbumin is a metalloprotein: it binds one Ca atom per molecule in a peptide loop containing four Asp residues. The apoprotein is quite heat labile, but the metalloprotein is rather heat stable; when studied by differential scanning calorimetry, α-la is, in fact, observed to be quite heat labile. However, the metalloprotein renatures on cooling, whereas the apoprotein does not. The difference in heat stability between the halo- and apoprotein may be exploited in the isolation of α-la on a large (i.e., potentially industrial) scale.

As discussed in the introduction to the Lactose section, α-la is an enzyme modifier protein in the lactose synthesis pathway; it makes UDP-galactosyl transferase highly specific for glucose as an acceptor for galactose, resulting in the synthesis of lactose.

Casein Micelles

α_{s1}-, α_{s2}- and β-caseins, which together represent about 85% of total casein, are precipitated by calcium at concentrations > 6 mM at temperatures > 20°C. Since milk contains approximlately 30 mmol/L Ca, it would be expected that most of the caseins would precipitate in milk. However, κ-casein is soluble at high concentrations of Ca, and it reacts with and stabilizes the Ca-sensitive caseins through the formation of casein micelles.

Since the stability, or perhaps instability, of the casein micelles is responsible for many of the unique properties and processibility of milk, their structure and properties have been studied intensively. It has been known for nearly 200 years that the casein in milk exists as large colloidal particles that are retained by porcelain-Chamberlain filters. During the early part of the 20th century there was interest in explaining the rennet coagulation of milk and hence in the structure and properties of the Ca-caseinate particles (for review see Fox and Kelly 2003). The term casein micelle appears to have been first used by Beau (1920). The idea that the rennet-induced coagulation of milk is due to the destruction of a protective colloid (Schutzcolloid) dates from the 1920s; initially, it was suggested that the whey proteins were the "protective colloid." The true nature of the protective colloid, the structure of the casein micelle, and the mechanism of rennet coagulation did not become apparent until the pioneering work on the identification, isolation, and characterization of κ-casein by Waugh and von Hippel (1956). Since then, the structure and properties of the casein micelle have been studied intensively. The evolution of views on the structure of the casein micelle has been described by Fox and Kelly (2003). Current knowledge on the composition, structure, and properties of the casein micelle and the key features thereof are summarized below.

The micelles are spherical colloidal particles, with a mean diameter of nearly 120 nm (range, 50–600 nm). They have a mean particle mass of about 10^5 kDa, that is, there are about 5000 casein molecules (20–25 kDa each) in an average micelle. On a dry weight basis, the micelles contain approximately 94% protein and 6% nonprotein species, mainly calcium and phosphate, with smaller amounts of Mg and citrate and traces of other metals; these are collectively called colloidal calcium phosphate (CCP). Under the conditions that exist in milk, the micelles are hydrated to the extent of about 2 g H_2O/g protein. There are approximately 10^{15} micelles/mL milk, with a total surface area of approximately 5×10^4 cm^2; the micelles are about 240 nm apart. Owing to their very large surface area, the surface properties of the micelles are of major significance, and because they are quite closely packed, even in unconcentrated milk, they collide frequently due to Brownian, thermal, and mechanical motion.

The micelles scatter light; the white appearance of milk is due mainly to light scattering by the casein micelles, with a contribution from the fat globules. The micelles are generally stable to most processes and conditions to which milk is normally subjected. The micelles may be reconstituted from spray-dried or freeze-dried milk without major changes in their properties. Freezing milk destabilizes the micelles, due to a decrease in pH and an increase in [Ca^{2+}]. On very severe heating, for example, at 140°C, the micelles shatter initially, then aggregate, and eventually, after about 20 minutes, the system coagulates (see Chapter 20).

The micelles can be sedimented by centrifugation: approximately 50% are sedimented by centrifugation at 20,000 × g for 30 minutes and approximately 95% by centrifugation at 100,000 × g for 60 minutes. The pelleted micelles can be redispersed by agitation, for example, by ultrasonication, in natural or synthetic milk ultrafiltrate; the properties of the redispersed micelles are not significantly different from those of native micelles.

The casein micelles are not affected by regular (≈ 20 MPa) or high-pressure (up to 250 MPa) homogenization (Hayes and Kelly 2003). The average size is increased by high-pressure treatment at 200 MPa, but micelles are disrupted (i.e., average size reduced by ≈ 50%) by treatment at a somewhat higher pressure, that is, ≥400 MPa (Huppertz et al. 2002, 2004).

The microstructure of the casein micelle has been the subject of considerable research during the past 40 years, that is, since the discovery and isolation of the micelle-stabilizing protein, κ-casein; however, there is still a lack of general consensus. Numerous models have been proposed, the most widely supported being the submicelle model, first proposed by Morr (1967) and refined several times since (Fox and Kelly 2003). Essentially, this model proposes that the micelle is built up from submicelles (molecular weight ≈ 500 kDa) held together by CCP and surrounded and stabilized by a surface layer, approximately 7 nm thick, rich in κ-casein but containing some of the other caseins also (Fig. 19.3a). It is proposed that the hydrophilic C-terminal region of κ-casein protrudes from the surface, creating a hairy layer around the micelle and stabilizing it through a zeta potential of about −20 mV and by steric stabilization. The principal direct experimental evidence for this model is provided by electron microscopy, which indicates a nonuniform electron density; this has been interpreted as indicating submicelles.

However, several authors have expressed reservations about the subunit model, and three alternative models have been proposed recently. Visser (1992) suggested that the micelles are spherical agglomerates of casein molecules, randomly aggregated and held together partly by salt bridges in the form of amorphous calcium phosphate and partly by other forces (e.g., hydrophobic interactions), with a surface layer of κ-casein. Holt (1992) proposed that the Ca-sensitive caseins are linked by microcrystals of CCP and surrounded by a layer of κ-casein, with its C-terminal region protruding from the surface (Fig. 19.3b). In the dual-binding model of Horne (2002), it is proposed that individual casein molecules interact via hydrophobic regions in their primary structures, leaving the hydrophilic regions free and with the hydrophilic C-terminal region of κ-casein protruding into the aqueous phase (Fig. 19.3c). Thus, the key structural features of the submicelle model are retained in the three alternatives, that is, the integrating role of CCP and a κ-casein-rich surface layer.

The micelles disintegrate

- When the CCP is removed, for example, by acidification to pH 4.6 and dialysis in the cold against bulk milk or by addition of trisodium citrate, also followed by dialysis; about 60% of the CCP can be removed without disintegration of the micelles.
- By raising the pH to about 9.0, which does not solubilize the CCP and presumably causes disintegration by increasing the net negative charge.
- By urea at > 5 M, which suggests that hydrogen and/or hydrophobic bonds are important for micelle integrity.
- When the micelles are precipitated by ethanol or other low molecular weight alcohols at ≥ about 35% at 20°C, but if the temperature is increased to ≥ 70°C, surprisingly, the precipitated casein dissolves and the solution becomes quite clear, indicating dissociation of the micelles (O'Connell et al. 2001a,b). Micelle-like particles reform on cooling and form a gel at about 4°C. It is not known if the subparticles formed by any of these treatments correspond to casein submicelles.

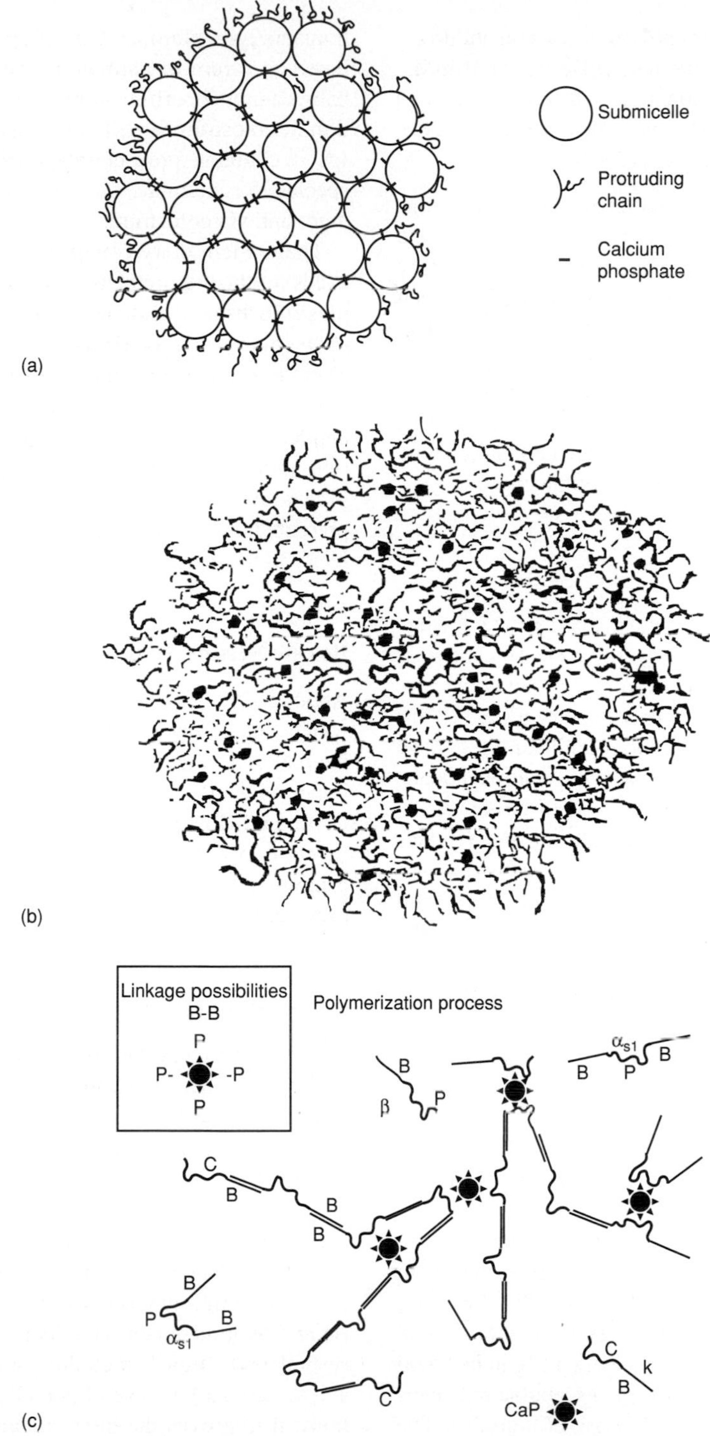

Figure 19.3. Models of casein micelle structure: **(a)** the submicelle model of Walstra (1999); **(b)** the Holt model (from Fox and McSweeney 1998); **(c)** the dual-binding model, showing interactions between α_{s1}-, β-, and κ-caseins. (Horne 2002.)

- The turbidity of skim milk increases on addition of sodium dodecyl sulphate (SDS) up to 21 mM, at which concentration a gel is formed; however, at ≥ 28 mM (~ 0.8%, w/v) the micelles disperse (see Lefebvre-Cases et al. 2001a,b).

There have been few studies on variations in micelle size throughout lactation and these have failed to show consistent trends. No studies on variability in the microstructure of the casein micelle have been published. We are not aware of studies on the effect of the nutritional or health status of the animal on the structure of the casein micelles, although their stability and behavior are strongly dependent on pH, milk salts, and whey proteins, which are affected by the health and nutritional status of the cow.

Minor Proteins

In addition to the caseins and the two principal whey proteins, milk contains several proteins at low or trace levels. Many of these minor proteins are biologically active (see Schrezenmeir et al. 2000); some are regarded as highly significant and have attracted considerable attention as neutraceuticals. When ways of increasing the value of milk proteins are discussed, the focus is usually on these minor proteins, but they are, in fact, of little economic value to the overall dairy industry. They are found mainly in the whey, but some are also located in the fat globule membrane. Reviews on the minor proteins include those by Fox and Flynn (1992) and Haggarty (2002).

Immunoglobulins

Colostrum contains approximately 10% (w/v) immunoglobulins (Igs), but this level declines rapidly to approximately 0.1% about 5 days postpartum. IgG1 is the principal Ig in bovine, caprine, or ovine milk, with lesser amounts of IgG2, IgA, and IgM; IgA is the principal Ig in human milk. The cow, sheep, and goat do not transfer Ig to the fetus in utero, and the neonate is born without Ig in its blood serum; consequently, it is very susceptible to bacterial infection, with a very high risk of mortality. The young of these species can absorb Ig from the intestine for several days after birth and thereby acquire passive immunity until they synthesize their own Ig, within a few weeks of birth. The human mother transfers Ig in utero, and the offspring is born with a broad spectrum of antibodies. Although the human baby cannot absorb Ig from the intestine, the ingestion of colostrum is still very important because the Igs it contains prevent intestinal infection. Some species, for example, the horse, transfer Ig both in utero and via colostrum.

The modern dairy cow produces colostrum far in excess of the requirements of its calf. Therefore, surplus colostrum is available for the recovery of Ig and other nutriceuticals (Paakanen and Aalto 1997). There is also considerable interest in hyperimmunizing cows against certain human pathogens, for example, rota virus, for the production of antibody-rich milk for human consumption, especially by infants; the Ig could be isolated from the milk and presented as a "pharmaceutical" or consumed directly in the milk.

Bovine Serum Albumin (BSA)

About 1–2% of the protein in bovine milk is BSA, which enters from the blood by leakage through intercellular junctions. As befits its physiological importance, BSA is very well characterized (see Carter and Ho 1994). BSA has no known biological function in milk, and considering its very low concentration, it probably has no technological significance either.

Metal-Binding Proteins

Milk contains several metal-binding proteins: the caseins (Ca, Mg, PO_4) are quantitatively the most important; others are α-la (Ca), xanthine oxidase (Fe, Mo), alkaline phosphatase (Zn, Mg), lactoperoxidase (Fe), catalase (Fe), glutathione peroxidase (Se), lactoferrin (Fe), and seroferrin (Fe).

Lactoferrin (Lf), a nonheme iron-binding glycoprotein (see Lonnerdal 2003), is a member of a family of iron-binding proteins, which includes seroferrin and ovotransferrin (conalbumin). It is present in several body fluids, including saliva, tears, sweat, and semen. Lf has several potential biological functions: it improves the bioavailability of Fe and has bacteriostatic (by sequestering Fe and making it unavailable to intestinal bacteria), antioxidant, antiviral, anti-inflammatory, immunomodulatory, and anticarcinogenic activity. Human milk contains a much higher level of Lf (≈ 20% of total N) than bovine milk, and therefore there is interest in fortifying bovine milk–based infant formulas with Lf. The pI of Lf is about 9.0, that is, it is cationic at the pH of milk, whereas most milk proteins are anionic, and can be isolated on an industrial scale by adsorption on a cation exchange resin. Hydrolysis of Lf by pepsin yields peptides called lactoferricins, which are more bacteriostatic than Lf, and their activity is independent of iron status. Bovine milk also contains a low level of serum transferrin.

Milk contains a copper-binding glycoprotein, ceruloplasmin, also known as ferroxidase (EC 1.16.3.1) (see Wooten et al. 1996). Ceruloplasmin is an α_2-globulin with a molecular weight of about 126 kDa; it binds six atoms of copper per molecule and may play a role in delivering essential copper to the neonate.

β_2-Microglobulin

β_2-microglobulin, initially called lactollin, was first isolated from bovine acid-precipitated casein by Groves et al. (1963). Lactollin, reported to have a molecular weight of 43 kDa, is a tetramer of β_2-microglobulin, which consists of 98 amino acids, with a calculated molecular weight of 11,636 Da. β_2-microglobulin, a component of the immune system (see Groves and Greenberg 1982) and is probably produced by proteolysis of larger proteins, mainly within the mammary gland; it has no known significance in milk.

Osteopontin

Osteopontin (OPN) is a highly phosphorylated acid ic glycoprotein, consisting of 261 amino acid residues with a calculated molecular weight of 29,283 Da(total molecular weight of the glycoprotein, ≈ 60 kDa). OPN has 50 potential calcium-binding sites, about half of which are saturated under normal physiological concentrations of calcium and magnesium.

OPN occurs in bone (it is one of the major non-collagenous proteins in bone), in many other normal and malignant tissues, and in milk and urine, and it can bind to many cell types. It is believed to have a diverse range of functions (Denhardt and Guo 1993, Bayless et al. 1997), but its role in milk is not clear.

Proteose Peptone 3

Bovine proteose peptone 3 (PP3) is a heat-stable phosphoglycoprotein that was first identified in the proteose-peptone (heat-stable, acid-soluble) fraction of milk. Unlike the other peptides in this fraction, which are proteolytic products of the caseins, PP3 is an indigenous milk protein, synthesized in the mammary gland. Bovine PP3 is a polypeptide of 135 amino acids, with five phosphorylation and three glycosylation sites. When isolated from milk, the PP3 fraction contains at least three components with molecular weights of approximately 28, 18, and 11 kDa; the largest of these is PP3, while the smaller components are fragments of PP3 generated by plasmin (see Girardet and Linden 1996). PP3 is present mainly in acid whey but some is present in the MFGM also. It has been proposed (Girardet and Linden 1996) to change the name to "lactophorin" or "lactoglycoporin." PP3 has also been referred to as "the hydophobic fraction of proteose peptone."

Owing to its strong surfactant properties (Campagna et al. 1998), PP3 can prevent contact between milk lipase and its substrates, thus preventing spontaneous lipolysis. Although its amino acid composition suggests that PP3 is not a hydrophobic protein, it behaves hydrophobically, possibly owing to the formation of an amphiphilic α-helix, one side of which contains hydrophilic residues, while the other side is hydrophobic. The biological role of PP3 is unknown.

Vitamin-Binding Proteins

Milk contains binding proteins for at least the following vitamins: retinol (vitamin A, i.e., β lg, see the Molecular Properties of Milk Proteins section, above), biotin, folic acid, and cobalamine (vitamin B_{12}). The precise role of these proteins is not clear, but they may improve the absorption of vitamins from the intestine or act as antibacterial agents by rendering vitamins unavailable to bacteria. The concentration of these proteins varies during lactation, but the influence of other factors such as individuality, breed, and nutritional status is not known. The activity of these proteins is reduced or destroyed on heating at temperatures somewhat higher than high temperature short time (HTST) pasteurization.

Angiogenins

Angiogenins induce the growth of new blood vessels, that is, angiogenesis. They have high sequence homology with members of the RNase A superfamily of proteins and have RNase activity. Angiogenesis is a complex biological process in which the ribonucleolytic activity of angiogenins is one of a number of essential biochemical steps that lead to the formation of new blood vessels (Strydom 1998).

Two angiogenins (ANG-1 and ANG-2) have been identified in bovine milk and blood serum. Both strongly promote the growth of new blood vessels in a chicken membrane assay. Bovine ANG-1 has 64% sequence identity with human angiogenin and 34% identity with bovine RNase A. The amino acid sequence of bovine ANG-2 has 57% identity with that of bovine ANG-1; ANG-2 has lower RNase activity than ANG-1.

The function(s) of the angiogenins in milk is unknown. They may be part of a repair system to protect either the mammary gland or the intestine of the neonate and/or part of the host defense system.

Kininogen

Two forms of kininogen have been identified in bovine milk, a high molecular weight form (> 68 kDa) and a low molecular weight form (16–17 kDa) (Wilson et al. 1989). Bradykinin, a biologically active peptide containing nine amino acids, which is released from the high molecular weight kininogen by the action of the enzyme, kallikrein, has been detected in the mammary gland and is secreted into milk, from which it has been isolated. Plasma kininogen is an inhibitor of thiol proteases and has an important role in blood coagulation. Bradykinin affects smooth muscle contraction, induces hypertension, and is involved in natriuresis and diuresis. The biological significance of bradykinin and kininogen in milk is unknown.

Glycoproteins

Many of the minor proteins discussed above are glycoproteins; in addition, several other minor glycoproteins have been found in milk and colostrums, but their identity and function have not been elucidated fully. Some of these glycoproteins belong to a family of closely related, highly acidic glycoproteins, called M-1 glycoproteins. Some glycoproteins stimulate the growth of bifidobacteria, presumably via their amino sugars. One of the M-1 glycoproteins found in colostrum, but which has not been detected in milk, is orosomucoid (α_1-acid glycoprotein), a member of the lipocalin family which is thought to modulate the immune system.

One of the high molecular weight glycoproteins in bovine milk is prosaposin, a neurotrophic factor that plays an important role in the development, repair, and maintenance of the nervous system (Patton et al. 1997). It is a precursor of saposins A, B, C, and D, which are sphingolipid activator proteins, but saposins have not been detected in milk. The physiological role of prosaposin in milk is not known, although the potent biological activity of saposin C, released by digestion, could be important for the growth and development of the young.

Proteins in the Milk Fat Globule Membrane

About 1% of the total protein in milk is in the milk fat globule membrane (MFGM). Most of the proteins are present at trace levels, including many of the indigenous enzymes in milk. The principal proteins in the MFGM include mucin, adipophilin, butyrophilin, and xanthine oxidase (Keenan and Mather 2002). Butyrophilin is a very hydrophobic protein and has similarities to the immunoglobulins (for review see Mather 2000).

Growth Factors

A great diversity of protein growth factors (hormones), including epidermal growth factor, insulin, insulin-like growth factors 1 and 2, three human milk growth factors (α_1, α_2 , and β), two mammary-derived growth factors (I and II), colony-stimulating factor, nerve growth factor, platelet-derived growth factor, and bombasin, are present in milk. It is not clear whether these factors play a role in the development of the neonate or in the development and functioning of the mammary gland, or both.

Indigenous Milk Enzymes

Milk contains about 60 indigenous enzymes, which represent a minor but very important part of the milk protein system (for review, see Fox and McSweeney 2003). The enzymes originate from the secretory cells or the blood. Many of the indigenous enzymes are concentrated in the MFGM and originate in the

Golgi membranes of the cell or the cell cytoplasm, some of which occasionally becomes entrapped as crescents inside the encircling membrane during exocytosis. Plasmin and lipoprotein lipase are associated with the casein micelles, and several enzymes are present in the milk serum; many of the enzymes in the milk serum are derived from the MFGM, which is shed as the milk ages.

There has been interest in indigenous milk enzymes since lactoperoxidase was discovered in 1881. Although present at relatively low levels, the indigenous enzymes are significant for several reasons:

Technological

- Plasmin causes proteolysis in milk and some dairy products; it may be responsible for age gelation in UHT milk and contributes to proteolysis in cheese during ripening, especially in varieties that are cooked at a high temperature and in which the coagulant is completely or extensively denatured, for example, Emmental, Parmesan, and Mozzarella.
- Lipoprotein lipase may cause hydrolytic rancidity in milk and butter but contributes positively to cheese ripening, especially in cheeses made from raw milk.
- Acid phosphatase can dephosphorylate casein and modify its functional properties; it may contribute to cheese ripening.
- Xanthine oxidase is a very potent prooxidant and may cause oxidative rancidity in milk; it reduces nitrate, used to control the growth of clostridia in several cheese varieties, to nitrite.
- Lactoperoxidase is a very effective bacteriocidal agent in the presence of a low level of H_2O_2 and SCN^- and is exploited for the cold sterilization of milk.

Indices of Milk Quality and History

- The standard assay to assess the adequacy of HTST pasteurization is the inactivation of alkaline phosphatase. Proposed assays for superpasteurization of milk are based on the inactivation of γ-glutamyltranspeptidase or lactoperoxidase.
- The concentration/activity of several enzymes in milk increases during mastitic infection, and some have been used as indices of this condition, e.g., catalase, acid phosphatase, and especially, N-acetylglucosaminidase.

Antibacterial

- Milk contains several bactericidal agents, two of which are lysozyme and lactoperoxidase.

MILK SALTS

When milk is heated in a muffle furnace at 500°C for approximately 5 hours, an ash derived mainly from the inorganic salts of milk and representing approximately 0.7% by weight of the milk, remains. However, the elements are changed from their original forms to oxides or carbonates, and the ash contains P and S derived from caseins, lipids, sugar phosphates, or high-energy phosphates. The organic salts, the most important of which is citrate, are oxidized and lost during ashing; some volatile metals, for example, sodium, are partially lost. Thus, ash does not accurately represent the salts of milk. However, the principal inorganic and organic ions in milk can be determined directly by potentiometric, spectrophotometric, or other methods. The typical concentrations of the principal elements, often referred to as macroelements, are shown in Table 19.3. Considerable variability occurs, due in part, to poor analytical methods and/or to samples from cows in very early or late lactation or suffering from mastitis. Milk also contains 20–25 elements at very low or trace levels. These microelements are very important from a nutritional viewpoint, and some, for example, Fe and Cu, are very potent lipid prooxidants.

Some of the salts in milk are fully soluble, but others, especially calcium phosphate, exceed their solubility under the conditions in milk and occur partly in the colloidal state, associated with the casein micelles; these salts are referred to as colloidal calcium phosphate (CCP), although some magnesium, citrate, and traces of other elements are also present in the micelles. The typical distribution of the principal organic and inorganic ions between the soluble and colloidal phases is summarized in Table 19.3. The actual form of the principal species can be determined or calculated after making certain assumptions; typical values are shown in Table 19.3.

The solubility and ionization status of many of the principal ionic species are interrelated, especially H^+, Ca^{2+}, PO_4^{3-}, and citrate^{3-}. These relationships have major effects on the stability of the caseinate system and, consequently, on the processing properties of milk. The status of various ionic species in milk can be modified by adding certain salts to milk,

Table 19.3. Distribution of Organic and Inorganic Ions between Soluble and Colloidal Phases of Bovine Milk

Species	Concentration (mg/L)	Soluble %	Soluble Form	Colloidal %
Sodium	500	92	Ionized	8
Potassium	1450	92	Ionized	8
Chloride	1200	100	Ionized	—
Sulphate	100	100	Ionized	—
Phosphate	750	43	10% bound to Ca and Mg 51% H_2PO4^- 39% HPO_4^{2-}	57
Citrate	1750	94	84% bound to Ca and Mg 14% $Citr^{3-}$ 1% $HCitr^{2-}$	6
Calcium	1200	34	35% Ca^{2+} 55% bound to citrate 10% bound to phosphate	66
Magnesium	130	67	Probably similar to calcium	33

Source: Modified from Fox and McSweeney 1998.

for example, $[Ca^{2+}]$ by PO_4^{3-} or citrate^{3-}; addition of $CaCl_2$ to milk affects the distribution and ionization status of calcium and phosphate and the pH of milk.

The precise nature and structure of CCP are uncertain. It is associated with the caseins, probably via the casein phosphate residues; it probably exists as microcrystals, which include PO_4 residues of casein. The simplest stoichiometry is $Ca_3(PO_4)_2$, but spectroscopic data suggest that $CaHPO_4$ is the most likely form.

The distribution of species between the soluble and colloidal phases is strongly affected by pH and temperature. As the pH is reduced, CCP dissolves and is completely soluble at less than about pH 4.9; the reverse occurs when the pH is increased. These pH-dependent shifts mean that acid-precipitated products, for example, acid casein and acid-coagulated cheeses, have a very low concentration of Ca.

The solubility of calcium phosphate decreases as the temperature is increased. Consequently, soluble calcium phosphate is transferred to the colloidal phase, with the release of H^+ and a decrease in pH:

$$CaHPO_4 \,/\, Ca\,(H_2PO_4)_2 \leftrightarrow Ca_3(PO_4)_2 + 3H^+$$

These changes are quite substantial but are at least partially reversible on cooling.

Since milk is supersaturated with calcium phosphate, concentration of milk by evaporation of water increases the degree of supersaturation and the transfer of soluble calcium phosphate to the colloidal state, with the concomitant release of H^+. Dilution has the opposite effect.

Milk salts equilibria are also shifted on freezing; as pure water freezes, the concentrations of solutes in unfrozen liquid are increased. Soluble calcium phosphate precipitates as $Ca_3(PO_4)_2$, releasing H^+ ions (the pH may decrease to 5.8). The crystallization of lactose as a monohydrate aggravates the situation by reducing the amount of solvent water.

There are substantial changes in the concentrations of the macroelements in milk during lactation, especially at the beginning and end of lactation and during mastitic infection (see White and Davies 1958, Keogh et al. 1982, O'Keeffe 1984, O'Brien et al. 1999a). Changes in the concentration of some of the salts in milk, especially calcium phosphate and citrate, have major effects on the physicochemical properties of the casein system and on the processibility of milk, especially rennet coagulability and related properties and heat stability.

VITAMINS

Milk contains all the vitamins in sufficient quantities to allow normal growth and maintenance of the neo-

Table 19.4. Typical Concentrations of Vitamins in Milk and Proportion of RDA Supplied by 1 L Milk

Vitamin	Average Level in 1 L Milk	% RDA in 1 L Milk
A	380 retional equivalents	38
B_1 (thiamine)	0.4 mg	33
B_2 (riboflavin)	1.8 mg	139
B_3 (niacin)	9.0 niacin equivalents	53
B_5 (pantothenic acid)	3.5 mg	70
B_6 (pyridoxine)	0.5 mg	39
B_7 (biotin)	30 μg	100
B_{11} (folic acid)	50 μg	13
B_{12} (cobalamin)	4 μg	167
C	15 mg	25
D	0.5 μg	10
E	1 mg	10
K	35 mg	44

Source: Adapted from Schaafma 2002.

nate. Cows' milk is a very significant source of vitamins, especially biotin (B_7), riboflavin (B_2), and cobalamine (B_{12}), in the human diet. Typical concentrations of all the vitamins and the percent of RDA supplied by 1 L of milk are shown in Table 19.4. For general information on the vitamins see Schaafma (2002); for specific aspects in relation to milk and dairy products, including stability during processing and storage, the reader is referred to a set of articles in Roginski et al. (2002).

In addition to their nutritional significance, four vitamins are of significance for other reasons, which have been discussed in the subsection Fat-Soluble Vitamins, above.

- Vitamin A (retinol) and especially carotenoids are responsible for the yellow-orange color of fat-containing products made from cows' milk.
- Vitamin E (tocopherols) is a potent antioxidant.
- Vitamin C (ascorbic acid) is an antioxidant or prooxidant, depending on its concentration.
- Vitamin B2 (riboflavin), which is greenish-yellow, is responsible for the color of whey or UF permeate. It cocrystallizes with lactose and is responsible for its yellowish color, which may be removed by recrystallization or bleached by oxidation. Riboflavin acts as a photocatalyst in the development of light-oxidized flavor in milk, which is due to the oxidation of methionine (not to the oxidation of lipids).

SUMMARY

Milk is a very complex fluid. It contains several hundred molecular species, mostly at trace levels. Most of the microconstituents are derived from blood or mammary tissue, but most of the macroconstituents are synthesized in the mammary gland and are milk specific. The constituents of milk may be present in true aqueous solution (e.g., lactose and most inorganic salts), in a colloidal solution (proteins, which may be present as individual molecules or as large aggregates of several thousand molecules, called micelles), or as an emulsion (lipids). The macroconstituents can be fractionated readily and are used as food ingredients. The natural function of milk is to supply the neonate with its complete nutritional requirements for a period (sometimes several months) after birth and with many physiologically important molecules, including carrier proteins, protective proteins, and hormones.

The properties of milk lipids and proteins may be readily modified by biological, biochemical, chemical, or physical means and thus converted into novel dairy products.

In this chapter, the chemical and physicochemical properties of milk sugar (lactose), lipids, proteins, and inorganic salts are discussed. The technology used to convert milk into a range of food products is described in Chapter 20.

REFERENCES

Akerstrom B, Flower DR, Salier JP. 2000. Lipocalins 2000. Biochim Biophys Acta 1482:1–356.

Bayless KJ, Davis GE, Meininger GA. 1997. Isolation and biological properties of osteopontin from bovine milk. Prot Expr Purif 9:309–314.

Beau M. 1920. Les mátiéres albuminoides du lait. Le Lait 1:16–26.

Campagna S, Vitoux B, Humbert G, Girardet JM, Linden G, Haertle T, Gaillard JL. 1998. Conformational studies of a synthetic peptide from the putative lipid-binding domain of bovine milk component PP3. J Dairy Sci 81:3139–3148.

Carter DC, Ho JX. 1994. Structure of serum albumin. Adv Protein Chem 45:153–203.

Cogan TM, Hill, C. 1993. Cheese starter cultures. In: PF Fox, editor. Cheese: Chemistry, Physics and Microbiology, vol. 1, 2nd edition. London: Chapman and Hall. Pp. 193–255.

Collomb M, Butikofer U, Sieber R, Jeangros B, Bosset JO. 2002. Composition of fatty acids in cow's milk fat produced in the lowlands, mountains and highlands of Switzerland using high-resolution gas chromatography. Int Dairy J 12:649–659.

Denhardt DT, Guo X. 1993. Osteopontin: A protein with diverse functions. FASEB J 7:1475–1482.

Fox PF (editor). 1992. Advanced Dairy Chemistry. Vol. 1. Proteins. 2nd edition. London: Elsevier Applied Science.

———. 1995. Advanced Dairy Chemistry. Vol. 2. Lipids. 2nd edition. London: Chapman and Hall.

———. 1997. Advanced Dairy Chemistry. Vol. 3. Lactose, Water, Salts and Vitamins. 2nd edition. London: Chapman and Hall.

Fox PF. 2003. The major constituents of milk. In: G Smit, editor. Dairy Processing: Improving Quality. Cambridge: Woodhead Publishing, Ltd.

Fox PF, Flynn A. 1992. Biological properties of milk proteins. In: Fox PF, editor. Advanced Dairy Chemistry. Vol. 1. Proteins. London: Elsevier Applied Sciences. Pp. 255–284.

Fox PF, Guinee TP, Cogan TM, McSweeney PLH. 2000. Fundamentals of Cheese Science. Gaithersburg, Maryland: Aspen Publishers, Ltd.

Fox PF, Kelly AL. 2004. The caseins. In: R Yada, editor. Proteins in Food Processing, Pp. 29–71. New York: Woodhead Publishing.

Fox PF, Lucey JA, Cogan TM. 1990. Glycolysis and related reactions during cheese manufacture and ripening. CRC Crit Rev Food Sci Nutr 29:237–253.

Fox PF, McSweeney PLH. 1998. Dairy Chemistry and Biochemistry. London: Chapman and Hall.

Fox PF, McSweeney PLH (editors). 2003. Advanced Dairy Chemistry. Vol. 1, Proteins. New York: Kluwer Academic-Plenum Publishers.

Girardet JM, Linden G. 1996. PP3 component of bovine milk: A phosphorylated whey glycoprotein. J Dairy Res 63:333–350.

Groves ML, Basch JJ, Gordon W. 1963. Isolation, characterisation and amino acid composition of a new crystalline protein, lactollin, from milk. Biochem 2: 814–817.

Groves ML, Greenberg R. 1982. β_2-microglobulin and its relationship to the immune system. J Dairy Sci 65:317–325.

Haggarty NW. 2002. Milk proteins; minor proteins, bovine serum albumin and vitamin-binding proteins. In: H Roginski, JW Fuquay, PF Fox, editors. Encyclopedia of Dairy Science. London: Academic Press. Pp. 1939–1946.

Hayes MG, Kelly AL. 2003. High pressure homogenisation of raw whole bovine milk. (a) Effects on fat globule size and other properties. J Dairy Res 70: 297–305.

Holt C. 1992. Structure and properties of bovine casein micelles. Adv Protein Chem 43:63–151.

Holt C, Sawyer L. 1993. Caseins as rheomorphic proteins. Interpretation of primary and secondary structures of α_{s1}, β- and κ-caseins. J Chem Soc Faraday Trans 89:2683–2692.

Horne DS. 2002. Caseins, micellar structure. In: H Roginski, JW Fuquay, PF Fox, editors. Encyclopedia of Dairy Science. London: Academic Press. Pp. 1902–1909.

Huppertz T, Fox PF, Kelly AL. 2004. High pressure treatment of bovine milk: Effects on casein micelles and whey proteins. J Dairy Res 71:98–106.

Huppertz T, Kelly AL, Fox PF. 2002. Effects of high pressure on constituents and properties of milk: A review. Int Dairy J 12:561–572.

Keenan TW, Mather IH. 2002. Milk fat globule membrane. In: H Roginski,JW Fuquay, PF Fox, editors. Encyclopedia of Dairy Science. London: Academic Press. Pp. 1568–1576.

Kelly AL, O'Connell JE, Fox PF. 2003. Manufacture and properties of milk powders. In: Advanced Dairy Chemistry. PF Fox, PLH McSweeney, editors. Vol. 1, Proteins. New York: Kluwer Academic-Plenum Publishers. Pp. 1027–1062.

Kelly ML, Berry JR, Dwyer DA, Griinari JM, Chournard PY, Van Amburg ME, Bauman DE. 1998. Dietary fatty acid content significantly affects conjugated linoleic acid concentrations in milk from lactating dairy cows. J Nutr 128:881–885.

Keogh MK, Kelly PM, O'Keeffe AM, Phelan JA. 1982. Studies of milk composition and its relationship to some processing criteria. I. Seasonal variation in the mineral levels in milk. Irish J Food Sci Technol 6:13–27.

Kraft J, Lebzien P, Flachowski G, Mockel P, Jahreis G. 2000. Duodenal infusion of conjugated linoleic acid mixture influences milk fat synthesis and milk conjugated linoleic acid content. In: Milk Composition. British Society of Animal Science, Occasional Publication No 25. Pp. 143–147.

Lawless F, Murphy, JJ, Fitzgerald S, O'Brien B, Devery R, Staunton C. 2000. Dietary effect on bovine milk fat conjugated linoleic acid content. In: Milk Composition. British Society of Animal Science, Occasional Publication No 25. Pp. 283–293.

Lefebvre-Cases E, Tarodo de la Fuente B, Cuq JL. 2001a. Effect of SDS on casein micelles: SDS-induced milk gel formation. J Food Sci 66:38–42.

———. Effect of SDS on acid milk coagulability. J Food Sci 66:555–560.

Lock AL, Garnsworthy PC. 2000. Independent effects of dietary linoleic and linolenic fatty acids on the conjugated linoleic acid content of cow's milk. Animal Sci 74:163–176.

Lonnerdal B. 2003. Lactoferrin. In: Advanced Dairy Chemistry. PF Fox, PLH McSweeney, editors. Vol. 1. Proteins. New York: Kluwer Academic-Plenum Publishers. Pp. 449–466.

Mather IH. 2000. A revised and proposed nomenclature for major proteins of the milk fat globule membrane. J Dairy Sci 83:203–247.

Meesapyodsuk D, Reed DW, Savile CK, Buist PH, Schäfer UA, Ambrose SJ, Covello PS. 2000. Substrate specificity, regioselectivity and cryptoregiochemistry of plant and animal omega-3-fatty acid desaturases. Biochem Soc Trans 28:632–635.

Mehra R, O'Brien B, Connolly JF, Harrington D. 1999. Seasonal variation in the composition of Irish manufacturing and retail milk. 2. Nitrogen fractions. Irish J Agric Food Res 38:65–74

Morr CV. 1967. Effect of oxalate and urea upon ultracentrifugation properties of raw and heated skim milk casein micelles. J Dairy Sci 50:1744–1751.

O'Brien B, Lennartsson T, Mehra R, Cogan TM, Connolly JF, Morrissey PA, Harrington D. 1999b. Seasonal variation in the composition of Irish manufacturing and retail milk. 3. Vitamins. Irish J Agric Food Res 38:75–85.

O'Brien B, Mehra R, Connolly JF, Harrington D. 1999a. Seasonal variation in the composition of Irish manufacturing and retail milk. 1. Chemical composition and renneting properties. Irish J Agric Food Res 38:53–64.

———. 1999c. Seasonal variation in the composition of Irish manufacturing and retail milk. 4. Minerals and trace elements. Irish J Agric Food Res 38:87–99.

O'Connell JE, Kelly AL, Auty MAE, Fox PF, de Kruif KG. 2001a. Ethanol-dependent heat-induced dissociation of casein micelles. J Agr Food Chem 49: 4420–4423.

O'Connell JE, Kelly AL, Fox PF, de Kruif KG. 2001b. Mechanism for the ethanol-dependent heat-induced dissociation of casein micelles. J Agr Food Chem 49:4424–4428.

O'Keeffe AM. 1984. Seasonal and lactational influences on moisture content of Cheddar cheese. Ir J Food Sci Technol 8:27–37.

Paakanen R, Aalto J. 1997. Growth factors and antimicrobial factors of bovine colostrums. Int Dairy J 7: 285–297.

Parodi PW. 1997. Cows' milk fat components as potential anticarcinogenic agents. J Nutr 127:1055–1060.

———. 1999. Conjugated linoleic acid and other anticarcinogenic agents in bovine milk fat. J Dairy Sci 82:1399–1349.

Patton S, Carson GS, Hiraiwa M, O'Brien JS, Sano A. 1997. Prosaposin, a neurotrophic factor: Presence and properties in milk. J Dairy Sci 80:264–272.

Roginski H, Fuquay, JW, Fox PF (editors). 2002. Encyclopedia of Dairy Science. London: Academic Press.

Schaafma G. 2002. Vitamins: General introduction. In: H Roginski, JW Fuquay, PF Fox, editors. Encyclopedia of Dairy Science. London: Academic Press. Pp. 2653–2657.

Schrezenmeir J, Korhonen H, Williams CM, Gill HS, Shah NP. 2000. Beneficial natural bioactive substances in milk and colostrums. Br J Nutr 84 (Suppl 1): S1–S166.

Strydom DJ. 1998. The angiogenins. Cell Mol Life Sci 54:811–824.

Urashima T, Saito T, Nakamura T, Messev M. 2001. Oligosaccharides of milk and colostrums in nonhuman mammals. Glycoconjugate J 18:357–371.

Visser H. 1992 A new casein micelle model and its consequences for pH and temperature effects on the properties of milk. In: H Visser, editor. Protein Interactions. Weinheim: VCH. Pp. 135–165.

Walstra P, Guerts T, Noomen A. 1999. Dairy Technology: Principles of Milk Properties and Processing. New York: Marcel Dekker.

Walstra P, Jenness R. 1984. Dairy Chemistry and Physics. New York: John Wiley and Sons.

Waugh DF, von Hippel PH. 1956. κ-casein and the stabilisation of casein micelles. J Am Chem Soc 78: 4576–4582.

White JCD, Davies DT. 1958. The relationship between the chemical composition of milk and the stability of the caseinate complex. I. General introduction, description of samples, methods and chemical composition of samples. J Dairy Res. 25:236–255.

Wilson WE, Lazarus LH, Tomer KB. 1989. Bradykinin and kininogens in bovine milk. J Biol Chem 264:17777–17783.

Wong NP, Jenness R, Kenney M, Marth EH. 1988. Fundamentals of Dairy Chemistry. New York: Van Nostrand Reinhold Co.

Wooten L, Shulze RA, Lancey RW, Lietzow M, Linder MC. 1996. Ceruloplasmin is found in milk and amniotic fluid and may have a nutritional role. Nutr Biochem 7:632–639.

Yurawecz MP, Kramer, JKG, Pariza MW. 1999. Advances in Conjugated Linoleic Acid Research. Minneapolis, Minnesota: American Oil Chemists Society.

20
Biochemistry of Milk Processing

A. L. Kelly and P. F. Fox

INTRODUCTION

As described in Chapter 19, milk is a very complex system; the continuous phase is an aqueous solution of a specific sugar, lactose; globular (whey/serum) proteins; inorganic salts; and hundreds of minor constituents (e.g., vitamins) at trace levels. Dispersed in the aqueous phase as an oil-in-water emulsion are lipids in the form of small globules and a second, unique, group of proteins, the caseins, which exist as large colloidal particles, known as casein micelles, with an average diameter of approximately 150 nm (range 50–600 nm), that contain, on average, about 5000 protein molecules.

Owing to its physical state, milk is an unstable system. The fat globules, which are less dense than the aqueous phase, rise to the surface, where they form a cream layer. The fat-rich cream can be skimmed off and used for the manufacture of butter, butter oil (ghee), or other fat-rich products. Alternatively, the fat-depleted lower layer (skimmed milk) may be run off, for example, through a valve at the bottom of a separating vat. The skimmed milk may be consumed directly as a beverage or used for the manufacture of a wide range of products. Gravity creaming and butter manufacture have been used since prehistoric times, but gravity creaming has been largely replaced by centrifugal separation of the fat since the development of the cream separator by Gustav de Laval in 1878. Gravity separation is still used to prepare reduced-fat milk for the manufacture of Parmegiano-Reggiano cheese. The physics of fat separation by gravity or centrifugal separation and the process of demulsifying milk fat to butter or butter oil will be described briefly later in this chapter.

Creaming (fat separation) is undesirable in many dairy products (e.g., liquid/beverage milk, concentrated milks) and is prevented by a process known as homogenization (the commonly used valve homogenizer was developed by Auguste Gaulin in 1905). Homogenization prevents creaming by reducing the size of the fat globules and preventing their agglomeration (clustering) by denaturing a particular minor protein, cryoglobulin (type M immunoglobulin).

The casein micelles are physicochemically stable but can be destabilized by a number of processes/treatments, which are exploited for the production of new dairy products. The most important of these are limited proteolysis and acidification, which are used in the production of cheese, fermented milks, and functional milk proteins. The mechanism and consequences of proteolysis and acidification will be described briefly. Thus, the three main phases of milk, that is, fat, casein, and the aqueous solution (whey) can be separated easily.

Milk is remarkably heat stable and can be sterilized by heat in standard or concentrated form, or dried to produce a wide range of dairy products. The principal processes and the resulting product families are summarized in Table 20.1. All these families of dairy products contain many products, for example, it is reported that about 1400 varieties of cheese are produced worldwide.

In addition to the above considerations, milk is a rich medium for the growth of a wide range of microorganisms. While this is exploited in the production of a range of fermented dairy products, it also means that milk can harbor microorganisms that may cause spoilage of products, or may present health risks to the consumer. Largely for the latter reason, very little milk is now consumed in the raw state, the vast majority being heat treated sufficiently severely to kill all pathogenic and food-poisoning bacteria.

In this chapter, several of the main processes used in the modern dairy industry will be discussed, and the relationships between the biochemical properties of milk and the processes applied explored.

THERMAL PROCESSING OF MILK

INTRODUCTION

The most common process applied to most food products is probably heat treatment, which is principally used to inactivate microorganisms (e.g., bacteria, yeasts, molds, and viruses) associated with foodborne disease, food poisoning, or spoilage. Secondary objectives of heat treating food include inactivation of enzymes in the tissue or fluid which would otherwise negatively influence product quality, and effecting changes in the structure of the food, e.g., through denaturation of proteins or gelatinization of starch.

A wide range of thermal processes is commonly applied to milk today, as summarized in Table 20.2.

The most widely used thermal process for milk is pasteurization. In 1864–1866, Louis Pasteur, the famous French scientist, discovered that the spoilage of wine and beer could be prevented by heating the product to around 60°C for several minutes; this process is now referred to as pasteurization. The historical development of the thermal processing of milk was reviewed by Westhoff (1978).

Although pasteurization was introduced to improve the stability and quality of food, it soon became apparent that it offered consumers protection against hazards associated with the consumption of raw milk (particularly the risk of transmission of tuberculosis from infected cows to humans); developments in the technology and widespread implementation occurred early in the 20th century. The first commercial pasteurizers, in which milk was heated at 74–77°C for an unspecified time period, were made by Albert Fesca in Germany in 1882 (Westhoff 1978). The first commercially operated milk pasteurizer (made in Germany) was installed in Bloomville, New York, in 1893 (Holsinger et al. 1997); the first law requiring pasteurization of liquid (beverage) milk was passed by the authorities in the city of Chicago in 1908.

Many early pasteurization processes used conditions not very different from those proposed by

Table 20.1. Diversity of Dairy Products Produced by Different Processes

Process	Primary (**bold**) and Secondary Products
Heat treatment (generally with homogenization)	**Market milk**
Centrifugal separation	**Standardized milk**
	Cream
	Butter, butter oil, ghee
	Creams of various fat content (pasteurized or UHT-treated); whipping, coffee, dessert creams
	Skim milk
	Skim milk powder, casein, cheese, protein concentrates
Concentration (e.g., evaporation or membrane separation)	In-container or UHT-sterilized concentrated milks, sweetened condensed milk
Concentration and drying	Whole milk powders, infant formulas
Enzymatic coagulation	**Cheese**
	Numerous varieties, processed cheese, cheese sauces and dips
	Rennet casein; cheese analogs
	Whey
	Whey powders; demineralized whey powders; whey protein concentrates; fractionated whey proteins; whey protein hydrolysates; lactose and lactose derivatives
Acid coagulation	**Cheese (fresh)**
	Acid casein; functional proteins
	Whey (as rennet whey)
Fermentation	Various fermented milk products, e.g., yogurt, buttermilk, acidophilus milk, bioyogurt
Freezing, aeration, and whipping	**Ice cream**

Source: Adapted from Fox and McSweeney 1998.

Table 20.2. Thermal Processes Commonly Applied to Liquid Milk Products

Process	Conditions	Heat Exchanger	Reason
Thermization	63°C for 15 s	Plate heat exchanger	Killing psychrotrophic bacteria before cold storage of milk
Pasteurization	72–74°C for 15–30 s	Plate heat exchanger	Killing of vegetative pathogenic bacteria and shelf-life extension
UHT (indirect)	135–140°C for 3–5 s	Plate heat exchanger	Production of safe long-shelf-life milk
		Scraped-surface heat exchanger	
		Tubular heat exchanger	
UHT (direct)	135–140°C for 3–5 s	Steam infusion system	Production of safe long-shelf-life milk
		Steam injection system	
Sterilization	118°C for 12 min	Batch or continuous retort	Production of sterile milk

Pasteur, generally heating milk to 62–65°C for at least 30 minutes, followed by rapid cooling to less than 10°C [now referred to as low temperature long time (LTLT) pasteurization]. However, high temperature short time (HTST) pasteurization, in which milk is treated at 72–74°C for 15 seconds in a continuous-flow plate heat exchanger, gradually became the standard industrial procedure for the heat treatment of liquid milk and cream; compared with the LTLT process, the HTST process has the advantages of reduced heat damage and flavor changes, and increased throughput (Kelly and O'Shea 2002).

Later developments in the thermal processing of milk led to the introduction in the 1940s of ultra-high-temperature (UHT) processes, in which milk is heated to a temperature in the range 135–140°C for 2–5 seconds (Lewis and Heppell 2000). UHT treatments can be applied using a range of heat exchanger technologies (e.g., indirect and direct processes), and essentially result in a sterile product that is typically shelf stable for at least 6 months at room temperature; eventual deterioration of product quality generally results from physicochemical rather than microbiological or enzymatic processes.

Sterilized milk products may also be produced using in-container retort systems; in fact, such processes were developed before the work of Pasteur. In 1809, Nicholas Appert developed an in-container sterilization process that he applied to the preservation of a range of food products, including milk. In-container sterilization is generally used for concentrated (condensed) milks; typical conditions involve heating at 115°C for 10–15 minutes. Although not sterile, a related class of product is preserved by adding a high level of sugar to concentrated milk; the sugar preserves the product through osmotic action (sweetened condensed milk was patented by Gail Borden in 1856).

The primary function of thermal processing of milk is to kill undesirable microorganisms; modern pasteurization is a very effective means for ensuring that liquid milk is free of the most heat resistant pathogenic bacteria likely to be present in raw milk. The original target species was *Mycobacterium tuberculosis,* but a little later, *Coxiella burnetti* became the target. In recent years, evidence for the survival of *Listeria monocytogenes* and, in particular, *Mycobacterium avium* ssp. *paratuberculosis* in pasteurized milk is of concern and is the subject of ongoing research (Grant et al. 2001, Ryser 2002). Most health risks linked to the consumption of pasteurized milk are probably due to postpasteurization contamination of the product.

Heat-Induced Changes in Milk Proteins

The caseins are very resistant to temperatures normally used for processing milk, and in general, unconcentrated milk is stable to the thermal processes to which it is exposed, for example, HTST pasteurization and UHT or in-container sterilization. However, very severe heating (e.g., at 140°C for a prolonged period) causes simultaneous dissociation and aggregation of the micelles and eventually coagulation.

However, the whey proteins are typical globular proteins and are relatively susceptible to changes on heating milk. β-lactoglobulin (β-lg) is susceptible to thermal denaturation, to an extent dependent on factors such as pH, protein concentration, and ionic strength (Sawyer 2003); denaturation exposes a highly reactive sulphydryl (-SH) group. Denatured β-lg can react, via sulphydryl-disulphide interchange reactions, with other proteins, including κ-casein, the micelle-stabilizing protein that is concentrated at the surface of the casein micelles. Thus, heating milk at a temperature higher than that used for minimal HTST pasteurization (which itself causes little denaturation of β-lg, although even slightly higher temperatures will cause progressively increased levels of denaturation) results in the formation of casein-whey protein complexes in milk, with major effects on many technologically important properties of milk.

Such complexes are found either at the surface of the micelles or in the serum phase, depending on the pH, which determines the likelihood of heat-induced dissociation of the κ-casein–β-lg complex from the casein micelle. Largely as a result of this, the heat stability of milk (expressed as the number of minutes at a particular temperature, for example, 140°C, before visible coagulation of the milk occurs—the heat coagulation time, HCT) is highly dependent on pH, and exhibits large differences over a narrow range of pH values (O'Connell and Fox 2003).

Most milk samples show a type A heat stability–pH (heat coagulation time, HCT-pH) profile, with a pronounced minimum and maximum, typically around pH 6.7 and 7.0, respectively, with decreasing stability on the acidic side of the maximum and increasing stability on the alkaline side of the minimum. The rare type B HCT-pH profile has no minimum, and heat stability increases progressively throughout the pH range 6.4–7.4.

In addition to pH, a number of factors affect the heat stability of milk (O'Connell and Fox 2002):

- Reduction in the level of Ca^{2+} or Mg^{2+} increases stability in pH range 6.5–7.5.
- Lactose hydrolysis increases heat stability throughout the pH range.
- Addition of κ casein eliminates the minimum in the type A profile.
- Addition of β-lactoglobulin to a type B milk converts it to a type A profile.
- Addition of phosphates increases heat stability.
- Reducing agents reduce heat stability and convert a type A to a type B profile.
- Alcohols and sulphydryl-blocking agents reduce the heat stability of milk.

Concentrated (e.g., evaporated) milk is much less thermally stable than unconcentrated milk, and its HCT-pH profile, normally assayed at 120°C, is quite different. It shows a maximum at approximately pH 6.4, with decreasing stability at higher and lower pH values (Fig. 20.1).

Severe heating has a number of effects on the casein molecules themselves, including dephosphorylation, deamindation of glutamine and asparagine residues, cleavage and formation of covalent cross-links; these changes may result in protein-protein interactions, which contribute to thermal instability.

Stability of UHT Milk on Storage

UHT milk is stable in long-term storage at ambient temperatures if microbiological sterility has been achieved by the thermal process and maintained by aseptic packaging in hermetically sealed containers. The shelf life of UHT milk is often limited by age gelation, which refers to a progressive increase in viscosity during storage, followed by the formation of a gel that cannot be redispersed (Lewis and Heppell 2000). The mechanism of age gelation is poorly understood, but it is probably due to physicochemical or biochemical factors, or both:

- *Physicochemical factors.* For example, dissociation of casein/whey protein complexes; changes in casein micelle structure and/or properties; cross-linking due to the Maillard reaction; removal or binding of calcium ions.
- *Biochemical factors.* For example, action of proteolytic enzymes, such as plasmin or heat-stable bacterial proteinases (i.e., from *Pseudomonas* species), which may hydrolyze κ-casein, inducing micelle coagulation.

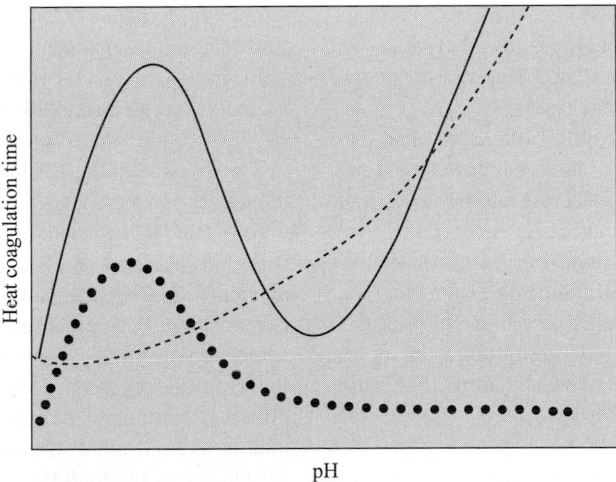

Figure 20.1. Heat coagulation time(HCT)-pH profiles of typical type A bovine milk (solid line), type B or serum protein–free milk (dashed line), as determined at 140°C, or concentrated milk (dotted line), as determined at 120°C.

Recent reviews on this subject include Nieuwenhuijse (1995), Datta and Deeth (2001) and Nieuwenhuijse and van Boekel (2003).

Heat-Induced Changes in Lactose in Milk and the Maillard Reaction

Heating lactose causes several chemical modifications, the nature and extent of which depend on environmental conditions and the severity of heating; changes include degradation to acids (with a concomitant decrease in pH), isomerization (e.g., to lactulose), production of compounds such as furfural, and interactions with amino groups of proteins (Maillard reaction).

In the Maillard reaction, lactose or lactulose reacts with an amino group, such as the -amino group of lysine residues, in a complex (and not yet fully understood) series of reactions with a variety of end products (O'Brien 1995, van Boekel 1998). The early stages of the Maillard reaction result in the formation of the protein-bound Amadori product, lactulosyllysine, which then degrades to a range of advanced Maillard products, including hydroxymethylfurfural, furfurals, and formic acid. The degradation of lactose to organic acids reduces the pH of milk. The most obvious result of the Maillard reaction is a change in the color of milk (browning), due to the formation of pigments called melanoidins, or advanced-stage Maillard products; extensive Maillard reactions also result in polymerization of proteins (van Boekel 1998). The Maillard reaction also changes the flavor and nutritive quality of dairy products, in the latter case through reduced digestibility of the caseins and loss of available lysine.

Moderately intense heating processes cause primarily the isomerization of lactose to lactulose. Lactulose, a disaccharide of galactose and fructose, is of interest as a bifidogenic factor and also as a laxative. More severe treatments (e.g., sterilization) will result preferentially in Maillard reactions.

There is particular interest in the use of products of heat-induced changes in lactose, such as lactulose and forosine, as indices of heat treatment of milk (Birlouez-Aragon et al. 2002).

Inactivation of Enzymes on Heating of Milk

Heat treatment of milk inactivates many enzymes, both indigenous and endogenous (i.e., of bovine or bacterial origin). The inactivation of enzymes is of interest both for the stability of heated milk products and as an index of heat treatment. Thermal inactivation characteristics of a number of indigenous milk enzymes are summarized in Table 20.3.

Because of their importance, the thermal inactivation kinetics of several milk enzymes have been studied in detail. A particular case of interest is that of alkaline phosphatase, which has for decades been used as an indicator of the adequacy of pasteurization of milk, because its thermal inactivation kinetics in milk closely approximates those of *Mycobacterium tuberculosis*. A negative result (residual activity below a set maximum value) in a phosphatase test (assay) is regarded as an indication that milk has been pasteurized correctly (Wilbey 1996). Detailed kinetic studies on the thermal inactivation of alkaline phosphatase under conditions similar to pasteurization have been published (McKellar et al. 1994, Lu et al. 2001).

To test for overpasteurization (excessive heating) of milk, a more heat stable enzyme, for example, lactoperoxidase, has been used as the indicator enzyme (Storch test); lactoperoxidase may also be used as an index of the efficacy of pasteurization of cream, which must be heated more severely than milk to ensure the killing of target bacteria, due to the protective effect of the fat therein.

Recently, other enzymes have been studied as indicators (sometimes called time temperature integrators, TTIs) of heat treatments (particularly at temperatures above 72°C) that may be applied to milk; these include catalase, lipoprotein lipase, acid phosphatase, N-acetyl-β-glucosaminidase and γ-glutamyl transferase (McKellar et al. 1996, Wilbey 1996).

The alkaline milk proteinase, plasmin, is resistant to pasteurization; indeed, its activity may increase during storage of pasteurized milk due to inactivation of inhibitors of plasminogen activators (Richardson 1983). Treatment under UHT conditions greatly reduces its activity (Enright et al. 1999); there is some evidence that the low level of plasmin activity in UHT milk contributes to the destabilization of the proteins, leading to defects such as age gelation, although this is not universally accepted (for review see Datta and Deeth 2001).

Since most indigenous milk enzymes are inactivated in UHT-sterilized milk products and in all in-container sterilized products, enzymes are not suitable indices of adequate processing, and chemical

Table 20.3. Thermal Inactivation Characteristics of Some Indigenous Milk Enzymes

Enzyme	Characteristics
Alkaline phosphatase	Inactivated by pasteurization; index of pasteurization; may reactivate on storage
Lipoprotein lipase	Almost completely inactivated by pasteurization
Xanthine oxidase	May be largely inactivated by pasteurization, depending on whether milk is homogenized or not
Lactoperoxidase	Affected little by pasteurization; inactivated rapidly around 80°C; used as index of flash pasteurization or pasteurization of cream
Sulphydryl oxidase	About 40% of activity survives pasteurization, completely inactivated by UHT
Superoxide dismutase	Largely unaffected by pasteurization
Catalase	Largely inactivated by pasteurization but may reactivate during subsequent cold storage
Acid phosphatase	Very thermostable; survives pasteurization
Cathepsin D	Largely inactivated by pasteurization
Amylases (α and β)	Thermal stability unclear; α more stable than β
Lysozyme	Largely survives pasteurization

Source: Adapted from Farkye and Imafidon 1995.

indices are more usually used (e.g., the concentration of lactulose or the extent of denaturation of β-lg). The use of a range of indicators for different heat treatments applied to milk was reviewed by Claeys et al. (2002).

The thermal inactivation of bacterial enzymes in milk is also of considerable significance. For example, psychotrophic bacteria of the genus *Pseudomonas* produce heat-stable lipases and proteases during growth in refrigerated milk; while pasteurization readily kills the bacterium, the enzymes survive such treatment and may contribute to the deterioration of dairy products made from such milk (Stepaniak and Sørhaug 1995). *Pseudomonas* enzymes partially survive in UHT milk and may be involved in its age gelation (Datta and Deeth 2001).

CHANGES ON EVAPORATION AND DRYING OF MILK

Dehydration, either partial (e.g., concentration to ~ 40–50% total solids, TS) or (almost) total (e.g., spray-drying to ~ 96–97% TS), is a common method for the preservation of milk. Reviews of the technology of the drying of milk and its effects on milk quality include Caríc and Kalab (1987), Písecky (1997), and Kelly et al. (2003).

CONCENTRATION OF MILK

The most common technology for concentrating milk is thermal evaporation in a multieffect (multistage) falling-film evaporator, which is used either as a prelude to spray-drying, to increase the efficiency of the latter, or to produce a range of concentrated dairy products (e.g., sterilized concentrated milk, sweetened condensed milk).

In many processes, a key stage in milk evaporation is preheating the feed. The minimum requirements of this step are to pasteurize the milk (e.g., skim milk, whole milk, filled milk) to ensure food safety and to bring its temperature to the boiling point in the first effect of the evaporator, for thermal efficiency. However, the functionality of the final product can be determined and modified by this processing step; preheating conditions range from those for conventional pasteurization to 90°C for 10 minutes or 120°C for 2 minutes. The most significant result of such heating is denaturation of β-lg, to a degree that is dependent on the severity of heating. During subsequent evaporation and, if applied, spray-drying, relatively little denaturation of the whey proteins occurs, as the temperature of the milk generally does not exceed 70°C (Singh and Creamer 1991). However, further association of whey proteins with the casein micelles can occur during evaporation, probably because the decrease in pH reduces protein charge, facilitating association reactions (Oldfield 1998)

Evaporation also increases the concentrations of lactose and salts in milk and induces a partially reversible transfer of soluble calcium phosphate to the colloidal form, with a concomitant decrease in pH (Le Graet and Brule 1982, Nieuwenhuijse et al.

1988, Oldfield 1998). The extent of transfer of phosphate to the colloidal phase, which is greater than that of calcium, depends on the temperature of preheating. The concentrations of soluble calcium and phosphorus in reconstituted milk powder are generally lower than those in the original milk, due to irreversible shifts induced during drying (Le Graet and Brule 1982).

Holding milk concentrate at > 60°C for an extended period before spray-drying can increase the viscosity of the concentrate, making it more difficult to atomize and thereby affecting the properties of the final powder.

Alternative technologies for concentrating milk are available, the most significant of which is probably membrane separation using reverse osmosis (RO). Reverse osmosis can achieve only a relatively low level of total solids (< 20%) and has a relatively low throughput, but is far less thermally severe than evaporation (Písecky 1997). Other membrane techniques used to concentrate or fractionate milk include nanofiltration (NF), which essentially removes only water, and ultrafiltration (UF), which allows standardization of the composition (e.g., protein and lactose contents) of the final powder (Mistry and Pulgar 1996, Horton 1997).

Spray-Drying of Milk

Today, milk powders are produced in large, highly efficient spray-dryers; the choice of dryer design (e.g., single versus multistage, with integrated or external fluidized bed drying steps, choice of atomizer, air inlet, and outlet temperatures) depends on the final product characteristics required. Due to a wide range of applications of milk powders, from simple reconstitution to use in cheese manufacture or incorporation as an ingredient into complex food products, the precise functionality required for a powder can vary widely.

Skim milk powders (SMP) are usually classified on the basis of heat treatment (largely meaning the intensity of preheating during manufacture), which influences their solubility and flavor (Caríc and Kalab 1987, Kyle 1993, Pellegrino et al. 1995). Instant powders (readily soluble in cold or warm water) are desirable for applications requiring reconstitution in the home, and are produced by careful production of agglomerated powders containing an extensive network of air spaces that can fill rapidly with water on contact (Písecky 1997). A number of techniques are used to produce agglomerated powders; these include (1) feeding fines (small powder particles) back to the atomization zone during drying (in straight-through agglomeration processes) and (2) wetting dry powder in chambers that promote sticking together of moistened particles, which when subsequently dried, produce porous agglomerated powder particles (rewet processes).

Milk powder contains occluded air (in vacuoles within individual powder particles) and interstitial air (entrapped between neighboring powder particles). The amount of occluded air depends on the heat treatment applied to the feed (whey protein denaturation determines foaming properties), the method of atomization, and outlet air temperature, while the interstitial air content of a powder depends on the size, shape, and surface geometry of individual powder particles. The drying process can be manipulated (e.g., by using multiple stages or by returning fines to the atomization zone) to increase the levels of interstitial and occluded air and instantize the powder (Kelly et al. 2003).

In the case of whole milk powders (WMP), a key characteristic is the state of the milk fat, whether fully emulsified or partially in free form (the latter is desirable for WMP used in chocolate manufacture). The production of instant WMP requires the use of an amphiphilic additive such as lecithin, as well as agglomeration, to overcome the intrinsic hydrophobicity of milk fat (Jensen 1975, Sanderson 1978). Lecithin may be added either during multistage drying of milk (between the main drying chamber and the fluidized bed dryer) or separately, in a rewet process.

During spray-drying of milk, there is relatively little change in milk proteins, apart from that caused by preheating and evaporation. The constituent most affected by the drying stage, per se, is probably lactose, which on the rapid removal of moisture from the matrix assumes an amorphous glassy state, which is hygroscopic and can readily absorb moisture if the powder is exposed to a high relative humidity. This can result in crystallization of lactose, with a concomitant uptake of water, which can cause caking and plasticization of the powder (Kelly et al. 2003).

Amorphous lactose is the principal constituent of both SMP and WMP and forms the continuous matrix in which proteins, fat globules, and air vacuoles are dispersed. In products containing a very high level of lactose (e.g., whey powders), the lactose may be precrystallized to avoid the formation of an amorphous glass. Usually, small lactose crystals are added to milk or whey concentrate prior to drying to promote crystallization under relatively mild conditions (i.e., in the liquid rather than the powder form); the added crystals act as nuclei for crystallization.

CHEESE AND FERMENTED MILKS

INTRODUCTION

About 35% of total world milk production is used to produce cheese ($\sim 15 \times 10^6$ metric tons per year [mt/yr]), mainly in Europe and North America. Cheese manufacture essentially involves coagulating the casein micelles to form a gel that entraps the fat globules, if present; when the gel is cut or broken, the casein network contracts (syneresis), expelling whey. The resulting curds may be consumed fresh as mild-flavored products, or ripened for a period ranging from 2 weeks (e.g., for Mozzarella) to > 2 years (e.g., for Parmagiano-Reggiano).

Cheese production is the most biochemically significant dairy process. However, since the action of rennet on milk and dairy fermentations are described in Chapters 10 and 26, respectively, only a brief summary is given here. For more detailed discussions of the science and technology of cheese making, see also Robinson and Wilbey (1998), Law (1999), Eck and Gillis (2000), and Fox et al. (2004).

The coagulation of milk for cheese is achieved by one of three methods:

1. Rennet-induced coagulation, which is used for most ripened cheeses and accounts for approximately 75% of total cheese production.
2. Acidification to pH of approximately 4.6 at 30–36°C by in situ production of acid by fermentation of lactose to lactic acid by lactic acid bacteria *(Lactococcus, Lactobacillus,* or *Streptococcus)* or direct acidification with acid or acidogen (e.g., gluconic acid-delta-lactone). Most acid-coagulated cheeses are consumed fresh and represent about 25% of total cheese production. Major examples of acid-coagulated cheeses are cottage, quark, and cream cheeses.
3. Acidification of milk, whey, or mixtures thereof to a pH of approximately 5.2 and heating to approximately 90°C. These cheeses are usually consumed fresh; common examples include ricotta and variants thereof, manouri, and some forms of queso blanco.

ACID-COAGULATED CHEESES

Acid-coagulated cheeses are at one end of the spectrum of fermented dairy products, the production of which is summarized in Figure 20.2a. Depending on the desired fat content of the final product, the start ing material may be low-fat cream, whole milk, semi-skimmed milk, or skimmed milk. The milk for cottage-type cheese is subjected to a mild heat treatment so that the synergetic properties of the coagulum are not impaired.

FERMENTED MILKS

In addition to acid-coagulated cheeses, a wide range of fermented milk products are produced worldwide today (Surono and Hasono 2002a), including

- Products of fermentation by mesophilic or thermophilic lactic acid bacteria (e.g., yogurt) and
- Products obtained by alcohol-lactic fermentation by lactic acid bacteria and yeast (e.g., kefir).

For yogurt, as for acid-coagulated cheese, the milk base (with a fat content depending on the final product) is usually supplemented with milk solids to enhance the viscosity and rigidity of the final coagulum, in some cases to simulate the viscosity of yogurt made from sheep's milk, which has a higher total solids content than cow's milk. The milk solids added may be whole or skim powder or whey powders, depending on the type of product required (González-Martínez et al. 2002, Remeuf et al. 2003).

Syneresis, which causes free whey in the product, is very undesirable in fermented milk products. Therefore, the milk or starting mix is subjected to a severe heat treatment (e.g., 90°C for 10 minutes) to denature the whey proteins, increase their water-binding capacity, and produce a firmer gel network, consisting of casein-whey protein complexes. The heat treatment also kills non-spore-forming pathogenic bacteria, rendering the product safe (Robinson 2002a). Homogenization of the mix also increases the firmness of the gel because the fat globules in

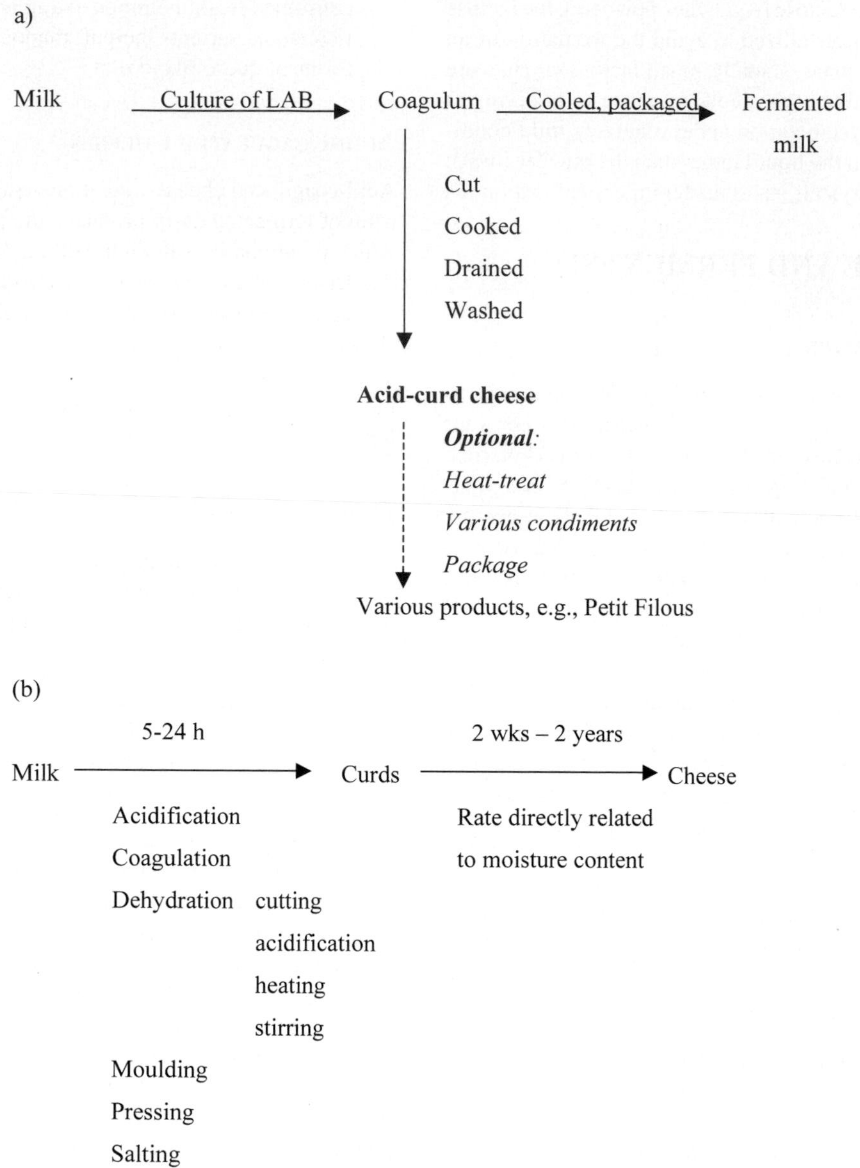

Figure 20.2. Principles of manufacture of **(a)** acid-coagulated and **(b)** rennet-coagulated cheese.

homogenized milk, stabilized mainly by a surface layer of adsorbed casein micelles, can become structural elements of the final acid gel.

The relative proportion of caseins and whey proteins in yogurt has a significant influence on the structure of the gel (Puvanenthiran et al. 2002). Hydrocolloid stabilizers (polysaccharides or proteins) may be added to the mix to modify the viscosity and/or prevent syneresis (Fizsman et al. 1999, Tamime and Robinson 1999). Sweetening agents (e.g., natural sugars or synthetic sweeteners such as aspartame) may be added to the mix, and fruit purees or fruit essence may be added either at the start or end of fermentation.

The starter culture for yogurt contains *Streptococcus thermophilus* and *Lactobacillus delbrueckii* subsp. *bulgaricus*. In recent years, many fermented milk products with added probiotic bacteria such as *Bifidobacteria* spp. or *Lactobacillus acidophilus* have been introduced, due to increasing consumer interest in the health benefits of such bacteria, and increasing clinical evidence of their efficacy (Surono and Hosono 2002b, Robinson 2002b). The texture and viscosity of yogurt can be modified by use of an exopolysaccharide(EPS)-producing starter culture (strains of both *S. thermophilus* and *L. delbrueckii* subsp. *bulgaricus* have been identified that secrete capsular heteropolysaccharides; De Vuyst et al. 2001, 2003; Hassan et al. 2003).

During fermentation, the production of lactic acid destabilizes the casein micelles, resulting in their coagulation (acid gelation) (Lucey and Singh 1998). The main point of difference between the production protocols for yogurt and acid-coagulated cheese is that when the pH has reached 4.6 and the milk has set into a coagulum, this coagulum is cut or broken for the latter but not for the former.

Depending on the desired characteristics of the finished product, the milk may be fermented in the final package (to produce a firm set yogurt) or in bulk tanks, followed by filling (to produce a more liquid stirred yogurt) (Tamime and Robinson 1999, Robinson 2002a). The rate of acidification for set or stirred yogurt differs due to differences in the level of inoculation and incubation temperature.

It is important to cool the yogurt rapidly when pH 4.6 is reached, to preserve the gel structure and prevent further starter activity (Robinson, 2002a). In some cases, the viscosity of the yogurt may be reduced by postfermentation homogenization to produce drinking yogurt.

Yogurt has a relatively short shelf life; common changes during storage involve syneresis, with whey expulsion, and further slow acidification by starter bacteria. To prevent the latter, and thereby extend the shelf life of yogurt, the final product may be pasteurized after acidification; to prevent damage to the gel structure at this point, the role of hydrocolloid or protein stabilizers is critical (Walstra et al. 1999).

Acid-Heat Coagulated Cheeses

These cheeses were produced initially in southern European countries from whey, as a means of recovering nutritionally valuable whey proteins; their production involves heating whey from rennet-coagulated cheese to about 90°C to denature and coagulate the whey proteins. Well-known examples are ricotta and manouri. Today, such cheeses are made from blends of milk and whey, and in this case, it is necessary to adjust the pH of the blend to about 5.2 using vinegar, citrus juice, or fermented milk.

Acid-heat coagulated cheese may also be produced from whole milk by acidifying to pH 5.2 and heating to 90°C. An example is US-style queso blanco, which does not melt on heating and hence has interesting functional properties for certain applications.

Rennet-Coagulated Cheeses

These cheeses, which represent about 75% of total production, are produced in a great diversity of shapes, flavors, and textures (approximately 1400 varieties worldwide). Their production can be divided into two phases: (1) conversion of milk to cheese curd and (2) ripening of the curd (Fig. 20.2b).

Coagulation

The coagulation of milk for the production of rennet-coagulated cheese exploits a unique characteristic of the casein system. As described in Chapter 19, the casein in milk exists as large (diameter 50–600 nm, mean 150 nm) colloidal particles, known as casein micelles. The micelles are stabilized by κ-casein, which is concentrated on the surface, with its hydrophobic N-terminal segment interacting with the α_{s1}-, α_{s2}- and β-caseins, and its hydrophilic C-terminal third protruding into the aqueous environment, forming a layer approximately 7 nm thick, which stabilizes the micelles by a zeta potential of about −20 mV and by steric stabilization. The stability of the micelles is lost when the surface κ-casein layer is destroyed by heat, alcohol, or proteinases (known as rennets).

Several proteinases can coagulate milk but the traditional, and most effective, rennets were NaCl extracts of the stomachs of young, milk-fed calves, kids, or lambs. The active enzyme in these rennets is chymosin; as the animal ages, the secretion of chymosin decreases and is replaced by pepsin. The supply of chymosin has been inadequate for about 50 years due to the increased production of cheese and the reduced availability of calf stomachs (due to the birth of fewer calves and the slaughter of calves at an older age). This shortage has led to a search for alternative coagulants (rennet substitutes).

Many, perhaps most, proteinases can coagulate milk under certain conditions, but almost all are unsuitable as rennets because they are too proteolytic, resulting in a reduced yield of cheese curd and off-flavored cheese. Only five successful rennet substitutes have been identified: bovine and porcine pepsins and acid proteinases from *Rhizomucor mehei, R. pusillus,* and *Cryphonectria parasitica*. Recently, the calf chymosin gene has been cloned in *Kluyveromyces lactis, E. coli,* and *Aspergillus niger,* and fermentation-produced chymosin is now used widely.

Chymosin and most of the other commercially successful rennets hydrolyze κ-casein specifically at the Phe_{105}-Met_{106} bond; *C. parasitica* proteinase cleaves κ-casein at Ser_{104}-Phe_{105}. The liberated hydrophilic C-terminal segment, known as the (glyco) caseinomacropeptide (CMP), diffuses into the surrounding aqueous phase, and the stability of the micelles is destroyed. When about 85% of the κ-casein has been hydrolyzed, the rennet-altered micelles aggregate to form a gel in the presence of a critical concentration of Ca^{2+} and at a temperature greater than about 18°C.

Acidification

The second characteristic step in cheese making is acidification of the milk and curd from a pH of approximately 6.7 to a value in the range 4.6–5.2, depending on the variety. Until relatively recently, and for some minor artisanal varieties still, acidification was due to the production of lactic acid from lactose by adventitious lactic acid bacteria (LAB). Today, cheese milk is inoculated with a culture (starter) of selected LAB for more controlled and reproducible acidification. If the cheese curds are cooked to < 40°C, a culture of *Lactococcus lactis* and/or *cremoris* is used, but for high-cooked cheese (whose curds are cooked to a high temperature; 50–55°C), a culture of *Lactobacillus delbrueckii* subsp. *bulgaricus,* possibly in combination with *Streptococcus thermophilus,* is used. Cheese starters have been refined progressively over the years, especially with respect to the rate of acidification, resistance to bacteriophages, and cheese-ripening characteristics. Today, mixtures of highly selected defined strains of LAB are used widely.

Acidification at the correct rate and time is essential for succesful cheese making; it affects at least the following aspects:

- Activity and stability of the coagulant during renneting;
- Strength of the rennet-induced gel;
- Rate and extent of syneresis of the gel when cut, and hence the composition of the cheese;
- Retention of rennet in the curd;
- Dissolution of colloidal calcium phosphate (the concentration of calcium in the cheese has a major effect on the texture and functionality of cheese);
- Inhibition of growth of undesirable microorganisms, especially pathogenic bacteria; and
- Activity of various enzymes in the cheese during ripening.

Postcoagulation Operations

If left undisturbed, a rennet-coagulated milk gel is quite stable, but if cut or broken, it contracts, producing curds and whey. By controlling the rate and extent of syneresis, the cheese maker controls the moisture content of cheese and thereby the rate and pattern of ripening and the quality and stability of the cheese. Syneresis is affected by

- Concentrations of fat, protein, and calcium;
- pH;
- Size of curd particles;
- Temperature of cooking;
- Stirring the curd-whey mixture and the curds after whey drainage;
- Pressing of the curd; and
- Salting (2 kg H_2O lost for each kilogram of NaCl taken up; NaCl should not be used to control moisture content).

When the desired degree of syneresis has occurred, as judged subjectively by the cheese maker, the curds and whey are separated, usually on some form of perforated metal screen. The curds are subjected to various treatments, which are more or less variety specific. These include cheddaring, kneading-stretching, molding, pressing, and salting.

Ripening

Rennet-coagulated cheese curd may be consumed at the end of the manufacturing process, and some is (e.g., junket or Burgos cheese), but most is ripened (matured) for a period ranging from about 2 weeks (mozzarella) to > 2 years (Parmegianno-Reggiano, extra-mature Cheddar), during which the characteristic flavor, texture, and functionality of the cheese develop. The principal changes in cheese during ripening are listed in Table 20.4.

Ripening is a very complex biochemical process, catalyzed by coagulants, indigenous milk enzymes, starter LAB and their enzymes, nonstarter LAB, secondary cultures, and exogenous enzymes:

- *Coagulant*: Depending on the coagulant used, the pH of the curd at whey drainage, the temperature to which the curds are cooked and the moisture content of the curds, 0–30% of the rennet added to the milk is retained in the curd.
- *Indigenous milk enzymes*: These are particularly important in raw-milk cheese, but many indigenous enzymes are sufficiently heat stable to withstand HTST pasteurization. Important indigenous enzymes include plasmin, xanthine oxidase, acid phosphatase, and in raw-milk cheese, lipoprotein lipase. Information on the significance of other indigenous enzymes is lacking.
- *Starter LAB and their enzymes*: These reach maximum numbers ($\sim 10^9$ cfu/g) at the end of curd manufacture (6–12 hours). They then die and lyse at strain-dependent rates, releasing their intracellular enzymes.
- *Non-starter LAB (NSLAB)*: These are adventitious LAB that contaminate the cheese milk at farm and/or factory. Traditionally, cheese makers relied on NSLAB for acid production during curd manufacture; this is still the case with some minor artisanal varieties, but starter LAB are now used in all factory and most artisanal cheese-making operations. Most NSLAB are killed by HTST pasteurization, and curd made from good quality pasteurized milk in modern enclosed equipment has only a few hundred NSLAB, mainly mesophilic lactobacilli, per gram at the start of ripening. However, they grow at a rate dependent mainly on the temperature to 10^7–10^8 cfu/g within about 3 months. In cheese made from raw milk, the NSLAB microflora is more diverse and reaches a higher number than in cheese made from pasteurized milk; the difference is probably mainly responsible for the more intense flavor of the former compared with that of the latter. This situation applies to all cheese varieties that have been investigated, and the NSLAB dominate the viable microflora of cheese ripened for > 2 months.

Table 20.4. Changes in Cheese during Ripening

Softening of the texture. Due to
- Proteolysis of the protein matrix, the continuous solid phase in cheese
- Increase in pH due to the catabolism of lactic acid and/or production of NH_3 from amino acids
- Migration of Ca to the surface in some cheeses, e.g., Camembert

Decrease in water activity. Due to
- Uptake of NaCl
- Loss of water through evaporation
- Production of low molecular weight compounds
- Binding of water to newly formed charged groups, e.g., NH_4^+ or COO^-

Changes in appearance. Examples:
- Mold growth
- Color, due to growth of *B. linens*
- Formation of eyes

Flavor development. Due to
- Formation of numerous sapid and aromatic compounds, as discussed above
- Release of flavorful compounds from the cheese matrix due to proteolysis

Changes in functionality, e.g., meltability, stretchability, water-binding properties, Maillard browning, and browning. Due to
- Proteolysis
- Catabolism of lactose

- *Secondary cultures*: Most cheese varieties develop a secondary microflora, which was originally adventitious but now develops mainly from added cultures. Examples are *Propionibacteria freudenrichii* subsp. *shermanii* (Swiss cheeses), *Penicillium roqueforti* (blue cheeses), *P. camamberti* and *Geotrichium candidum* (Camembert and Brie), *Brevibacterium linens, Arthrobacter* spp., *Corynebacterium* and yeasts (surface smear–ripened cheese), citrate-positive *Lc. lactis* and *Leuconostoc* spp. (Dutch-type cheeses). Most of the microorganisms used as secondary cultures are very active metabolically and many secrete very active proteolytic and lipolytic enzymes. Consequently, they dominate the ripening of varieties in which they are used. Traditionally, a secondary culture was not used for Cheddar-type cheese, but it is becoming increasingly common to add an adjunct culture, usually mesophilic lactobacilli, to accelerate ripening, intensify flavor, and perhaps create tailor-made flavor. Essentially, the objective is to reproduce the microflora, and consequently the flavor, of raw-milk cheese.
- *Exogenous enzymes*: For some varieties of Italian cheese, e.g., provolone and pecorino varieties, rennet paste is used as coagulant. The rennet paste contains a lipase, pregastric esterase (PGE), in addition to chymosin; PGE is responsible for extensive lipolysis and the characteristic piquant flavor of the cheese. Rennet extract used for most varieties lacks PGE.

During ripening, a very complex series of reactions, which fall into three groups, occur: (1) glycolysis of lactose and modification and/or catabolism of the resulting lactate, (2) lipolysis and the modification and catabolism of the resulting fatty acids, and (3) proteolysis and catabolism of the resulting amino acids.

Glycolysis and Related Events Fresh cheese curd contains about 1% lactose, which is converted to lactic acid (mainly the L-isomer) by the starter LAB, usually within 24 hours. Depending on the variety, the L-lactic acid is racemized to DL-lactic acid by NSLAB, or catabolized to CO_2 and H_2O in mold-ripened and in smear-ripened cheese. In Swiss-type cheese, lactic acid is converted to propionic and acetic acids, CO_2, and H_2O by *P. freudenrichii* subsp. *shermanii*; the CO_2 is responsible for the characteristic eyes in such cheeses.

Lipolysis Little lipolysis occurs in most cheese varieties, in which it is catalyzed by the weakly lipolytic LAB or NSLAB, and indigenous milk lipase if raw milk is used. Extensive lipolysis occurs in some hard Italian-type cheeses, for example, provolone and pecorino varieties, for which PGE is responsible, and in blue cheese, for which *P. roqueforti* is responsible. Fatty acids are major contributors to the flavor of provolone and pecorino cheeses, and some may be converted to lactones or esters, which have characteristic flavors. In blue-veined cheese, fatty acids are converted to methyl ketones by *P. roqueforti,* and these are mainly responsible for the characteristic peppery taste of such cheeses. Some of the methyl ketones may be reduced to secondary alcohols.

Proteolysis Proteolysis is the most complex and probably the most important of the three primary ripening reactions for the quality of cheese, especially internal bacterially ripened varieties. Initially, the caseins are hydrolyzed by chymosin and, to a lesser extent, by plasmin, in the case of β-casein; chymosin is almost completely inactivated in high-cook varieties, and in this case plasmin is the primary agent of primary proteolysis. The polypeptides produced by chymosin and plasmin are too large to affect flavor, but primary proteolysis has a major influence on the texture and functionality of cheese. The peptides produced by chymosin and plasmin are hydrolyzed to smaller peptides and amino acids by proteinases and peptidases of starter LAB and NSLAB. Small peptides contribute positively to the background brothy flavor of cheese, but some are bitter. Many amino acids also have a characteristic flavor, but more importantly, they serve as substrates for a great diversity of catabolic reactions catalyzed by enzymes of LAB, NSLAB, and the secondary culture. The flavor of internal bacterially ripened cheese is probably due mainly to compounds produced from amino acids.

WHEY PROCESSING

Range of Whey Products

For many centuries, whey was often viewed as an unwanted and valueless by-product of cheese manufacture; however, since around 1970, whey has been repositioned as a valuable source of a range of food ingredients, rather than as a disposal problem. While part of the impetus for this strategic reevaluation of the potential of whey was driven by environmental and other concerns about the disposal of a product of such a high biological oxygen demand, it has resulted in a previously unsuspected richness and diversity of products, and a significant new processing sector of the dairy industry (see Sienkiewicz and Riedel 1990, Jelen 2002).

The first stage in the production of any whey product is preliminary purification of the crude whey, recovered after cheese or casein manufacture. Fresh whey is classified as either sweet, with a pH > 5.6 and a low calcium content, or acid, with a pH of around 4.6 and a high calcium content; the difference in calcium content arises from the solubilization of colloidal calcium phosphate (CCP) as pH decreases. Common pretreatments include centrifugal clarification to remove curd particles (fines) and pasteurization (to inactivate starter lactic acid bacteria). For many whey protein–based products, removal of all lipids is desirable; this can be achieved by centrifugal separation to produce whey cream, which may be churned into whey butter. To obtain a very low level of lipids, calcium chloride may be added to the whey, with adjustment of the pH to more alkaline values, followed by heating and cold storage to precipitate and remove lipoprotein complexes (thermocalcic aggregation; Karleskind et al. 1995).

The next level of processing technology applied to the pretreated whey depends on the end product to be manufactured. The simplest whey products are probably whey beverages, consisting of clarified whey, typically blended with natural or concentrated fruit juices. Whey beverages have a nutritionally beneficial amino acid profile, plus an isotonic nature; although not widely commercialized, such drinks have been successful in some European countries (e.g., Rivella in Switzerland).

Whey may be concentrated by evaporation and spray-dried to whey powders. The key consideration in spray-drying whey is the constraints imposed by the fact that the major constituent of whey is lactose (typically > 75% of whey solids). Thus, processes for drying whey must include controlled crystallization of lactose to yield small crystals (30–50 μm). This is typically achieved by controlled cooling of concentrated (i.e., supersaturated with respect to lactose) whey under a programmed temperature regime, with careful stirring and addition of seed crystals, typically of α-lactose monohydrate, followed by holding under conditions sufficient to allow crystallization to proceed. In general, only about 70% of the lactose present will crystallize at this stage, and processes often include a postdrying crystallization stage, for example, on a belt attached to the spray-dryer, where the remainder crystallizes (Mulvihill and Ennis 2003).

The simplest dried whey product is whey powder, which is produced in a single-stage spray-dryer. Additional care must be taken in drying acid whey products, which may be sticky; the corrosive nature of the acids (e.g., HCl) used in their manufacture may also present processing difficulties, and such products may be neutralized prior to drying.

Whey powders may be unsuitable for use in certain food applications (e.g., infant formulas) due to their high mineral content; in such cases, whey is demineralized by ion exchange or electrodialysis before concentration and dialysis (for a discussion of the technologies involved, see Burling 2002). Demineralization of whey is also desirable for use in ice cream production, to reduce the salty taste of normal whey powder.

Whey Protein–Rich Products

Whey powder has a low protein content (~ 12–15%) and a high lactose content (~ 75%); many whey products have undergone at least some degree of purification to increase the level of a particular constituent, usually protein (Matthews, 1984). There is a family of protein-enriched whey-derived powders, which are differentiated based on the level of protein: whey protein concentrates (WPCs), with a protein content in the range 35–80%, and whey protein isolates (WPIs), with a protein content > 88%. Typically, WPCs are produced by ultrafiltration (UF), with protein being progressively concentrated in the retentate, while lactose, salts, and water are removed in the permeate (Ji and Haque 2003). Higher protein levels can be achieved by diafiltration, that is, dilution of the retentate followed by UF. The use of ion exchangers to adsorb the proteins from whey, followed by selective release into suitable buffer solutions, is required to achieve the high degree of purity required of WPIs.

Lactose Processing

Whey or whey permeate, produced as a by-product of UF processing, contains a high level of lactose.

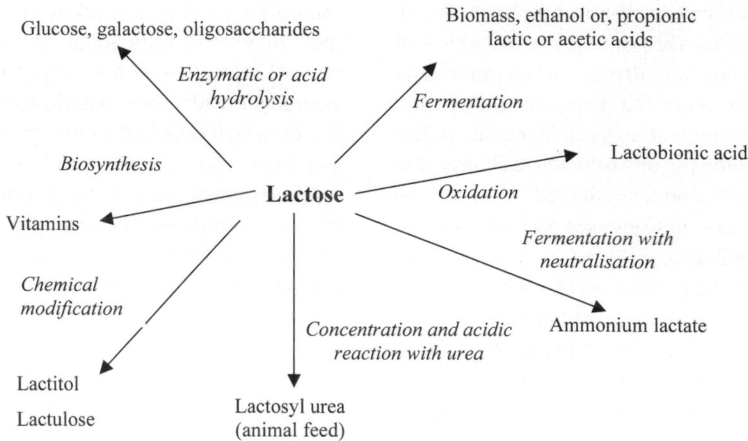

Figure 20.3. Food-grade derivatives of lactose.

This can be recovered by crystallization from a concentrated (supersaturated, 60–62% TS) preparation of either whey or whey permeate (for review see Muir 2002). The crystalline lactose is usually recovered using a decanter centrifuge, dried in a fluidized bed dryer, and ground to a fine powder, which is used in food applications, for example, confectionery products and infant formulas. Lactose is also used widely in pharmaceutical applications (e.g., as a diluent in drug tablets); for such applications, the lactose is generally further refined by redissolution in hot water and mixing with activated carbon, followed by filtration, crystallization,and drying.

Many derivatives of lactose can be produced; these derivatives are more valuable and useful than lactose (Fig. 20.3).

MILK PROTEINS: ISOLATION, FRACTIONATION, AND APPLICATIONS

In recent years, it has been increasingly recognized that milk proteins have functionalities that can be exploited in food systems other than conventional dairy products; these proteins have been increasingly recognized as desirable ingredients for a range of food products (Mulvihill and Ennis 2003). The very different properties of the two classes of milk proteins, the caseins and the whey proteins, present different technological challenges for recovery and are suitable for quite different applications. The whey proteins were discussed in Whey Processing section, and the caseins are discussed below.

Recovery and Application of Caseins

Technologies for the recovery of caseins from milk are based on the fact that relatively simple perturbations of the milk system can destabilize the caseins selectively, resulting in their precipitation and facilitating their recovery from milk (Mulvihill and Fox 1994, Mulvihill and Ennis 2003). As discussed in the Coagulation subsection (under Rennet-Coagulated Cheeses), the stability of the caseins in a colloidal micellar form is possible due to the amphiphilic nature of κ-casein. To overcome this stability and precipitate the caseins merely requires destruction of this stabilizing effect.

The two key principles used to destabilize the micelles are (1) acidification to the isoelectric point of the caseins (pH 4.6) or (2) the addition of enzymes, for example, chymosin, which hydrolyze κ-casein, removing the stabilizing glycomacropeptide. In both cases, the starting material for casein production is skim milk. For acid casein, acidification can be achieved either by addition of a mineral acid, usually HCl, or fermentation of lactose to lactic acid by a culture of lactic acid bacteria. When the isoelectric point is reached, a precipitate (rapid acidification) or a gel (slower acidification) is formed; the latter is cut/broken to initiate syneresis and expel whey. The mixture of casein and whey is stirred and cooked to enhance syneresis, and the casein is separated from the whey (either centrifugally or by sieving). To improve the purity of the casein, the curds are repeatedly washed with water to remove residual salts and lactose, and the final casein is dried, typically in specialized dryers such as attrition or ring dryers. For rennet casein, the milk gel is formed by adding a suitable coagulant to skim milk, but subsequent stages are similar to those for acid casein.

Both acid and rennet casein are insoluble products, which are used in the production of cheese analogs (rennet casein) and convenience food products, as well as in nonfood applications, such as the manufacture of glues, plastics, and paper glazing (acid casein).

To extend the range of applications of acid casein and improve its functionality, it may be converted to the alkali metal salt form (e.g., Na, K, Ca caseinates) by mixing a suspension of acid casein with the appropriate hydroxide and heating, followed by drying (typically with a low, i.e., ~ 20%, total solids concentration in the feed, due to high product viscosity, in a spray-dryer). Caseinates, most commonly sodium caseinate, have a range of useful functional properties, including emulsification and thickening; calcium caseinate has a micellar structure (Mulvihill and Ennis 2003).

Microfiltration (MF) technology can be used to produce powders enriched in micellar casein (Pouliot et al. 1996, Maubois 1997, Kelly et al. 2000, Garem et al. 2000). Whole casein may be resolved into β- and α_s/κ-casein-rich fractions by processes that exploit the dissociation of β-casein at a low temperature (reviewed by Mulvihill and Ennis 2003). Chromatographic or precipitation processes can also be used to purify the other caseins, although such processes are generally not easy to scale up for industrial-level production (Coolbear et al. 1996, Farise and Cayot 1998).

EXOGENOUS ENZYMES IN DAIRY PROCESSING

The dairy industry represents one of the largest markets for commercial enzyme preparations. The use of some enzymes in the manufacture of dairy products, for example, rennet for the coagulation of milk is, in fact, probably the oldest commercial application of enzyme biotechnology.

The use of rennets to coagulate milk and in the ripening of cheese, were discussed in Cheese and Fermented Milk, above. A further application of proteolytic enzymes in dairy products, that is, the production of dairy protein hydrolysates, is discussed in Protein Hydrolysates, below. The applications of enzymes not considered elsewhere, but of significance for dairy products, are discussed in this section.

β-Galactosidase

Hydrolysis of lactose (4-*O*-β-D-galactopyranosyl-D-glucopyranose), either at a low pH or enzymatically (by β-galactosidases), yields two monosaccharises, D-glucose and D-galactose (Mahoney 1997, 2002). A large proportion of the world's population suffers from lactose intolerance, leading to varying degrees of gastrointestinal distress if lactose-containing dairy products are consumed. However, glucose and galactose are readily metabolized; they are also sweeter than lactose. Glucose/galactose syrups are one of the simplest and most common products of the industrial lactose hydrolysis processes.

β-galactosidases are available from several sources, including *E. coli, Aspergillus niger, A. oryzae,* and *Kluyveromyces lactis* (Whitaker 1992, Mahoney 2002). β-galactosidase for use in dairy products must be isolated from a safe source and be acceptable to regulatory authorities; for this reason, genes for some microbial β-galactosidases have been cloned into safe hosts for expression and recovery. Of particular interest are enzymes that are active either at a low or a very high temperature, to avoid the microbiological problems associated with enzymatic treatment at temperatures that encourage the growth of mesophilic microorganisms (Vasiljevic and Jelen 2001). Microorganisms capable of producing thermophilic β-galactosidases include *Lactobacillus delbruckii* subsp. *bulgaricus* (Vasiljevic and Jelen 2001) and *Thermus thermophilus* (Maciunska et al. 1998), while cold-active β-galactosidase has been purified from psychrophilic bacteria (Nakagawa et al. 2003).

Enzymatic hydrolysis of lactose rarely leads to complete conversion to monosaccharides, due to feedback inhibition of the reaction by galactose, as well as to concurrent side reactions (due to transferase activity) that produce isomers of lactose and oligosaccharides (Chen et al. 2002). While initially considered to be undesirable by-products of lactose hydrolysis, galactooligosaccharides are now recognized to be bifidogenic factors, which enhance the growth of desirable probiotic bacteria in the intestine of consumers and suppress the growth of harmful anaerobic colonic bacteria (Shin et al. 2000).

Lactose has a number of other properties that cause difficulty in the processing of dairy products, such as its tendency to form large crystals on cooling of concentrated solutions of the sugar; lactose-hydrolyzed concentrates are not susceptible to such problems. For example, hydrolysis of lactose in whey concentrates can preserve these products through increasing osmotic pressure, while maintaining physical stability (Mahoney 1997, 2002).

For the hydrolysis of lactose in dairy products, β-galactosidase may be added in free solution, allowed sufficient time to react at a suitable temperature, and inactivated by heating the product. To control the reaction more precisely and avoid the uneconomical single use of the enzyme, immobilized enzyme technology (e.g., where the enzyme is immobilized on an inert support, such as glass beads) or systems where the enzyme is recovered by UF of the product after hydrolysis and reused, have been studied widely (Obon et al. 2000). A further technique with potential for application in lactose hydrolysis is the use of permeabilized bacterial or yeast cells (e.g., *Kluyveromyces lactis*) with β-galactosidase activity. In such processes, the cells are treated with agents (e.g., ethanol) that damage their cell membrane and allow diffusion of substrate and reaction products across the damaged membrane; the cell itself becomes the immobilization matrix, and the enzyme is active in its natural cytoplasmic environment (Fontes et al. 2001, Becerra et al. 2001). This provides a crude but convenient and inexpensive enzyme utilization strategy. Overall, however, few immobilized systems for lactose hydrolysis are used commercially, due to the technological limitations and high cost of such processes (Zadow 1993).

One of the more common applications of lactose hydrolysis is the production of low-lactose liquid milk, suitable for consumption by lactose-intolerant consumers; this may be achieved in a number of ways, including adding a low level of β-galactosidase to packaged UHT milk; alternatively, the consumer may add β-galactosidase to milk during domestic refrigerated storage (Modler et al. 1993). In the case of ice cream, lactose hydrolysis reduces the incidence of sandiness (due to lactose crystallization) during storage and, due to the enhanced sweetness, permits the reduction of sugar content. Lactose hydrolysis may also be applied in yogurt manufacture, to make reduced-calorie products.

Overall, despite considerable interest, industrial use of lactose hydrolysis by β-galactosidases has not been widely adopted, although such processes have found certain niche applications (Zadow 1993).

Transglutaminase

Transglutaminases (TGase; protein-glutamine:amine γ-glutamyl transferase) are enzymes that create inter- or intramolecular cross-links in proteins by catalyzing an acyl-group transfer reaction between the γ-carboxyamide group of peptide-bound glutamine residues and the primary amino group of lysine residues in proteins. TGase treatment can modify the properties of many proteins through the formation of new cross-links, incorporation of amines, or deamidation of glutamine residues (Motoki and Seguro 1998).

TGases are widespread in nature; for example, blood factor XIIIa, or fibrinoligase, is a TGase-type enzyme, and TGase-mediated cross-linking is involved in cellular and physiological phenomena such as cell growth and differentiation, as well as blood clotting and wound healing. TGases have been identified in animals, plants, and microbes (e.g., *Streptoverticillium mobaraense*, which is the source of much of the TGase used in food studies). TGase may be either calcium-independent (most microbial enzymes) or calcium-dependent (typical for mammalian enzymes).

TGase-catalyzed cross-linking can alter the solubility, hydration, gelation, rheological and emulsifying properties, rennetability, and heat stability of a variety of food proteins. The structure of individual proteins determines whether cross-linking by TGase is possible. Caseins are good substrates for TGase due to their open structure, but the whey proteins, due to their globular structure, require modification (e.g., heat-induced denaturation) to allow cross-linking (Ikura et al. 1980; Sharma et al. 2001; O'Sullivan et al. 2002a,b). When the whey proteins in milk are in the native state, the principal cross-linking reactions involve the caseins; heat-induced denaturation renders the whey proteins susceptible to cross-linking, both to each other and to casein molecules.

TGase treatment has several significant effects on the properties of milk and dairy products. For example, TGase treatment of fresh raw milk increases its heat stability; however, if milk is preheated under conditions that denature the whey proteins prior to TGase treatment, the increase in heat stability is even more marked (O'Sullivan et al. 2002a). This is probably due to the formation of cross-links between the caseins and the denatured whey proteins, brought into close proximity by the formation of disulphide bridges between β-lactoglobulin and micellar κ-casein. There has also been considerable interest in the effects of TGase treatment on the cheese-making properties of milk, in part due to the potential for increasing cheese yield. It has been suggested that TGase treatment of milk before renneting can achieve this effect. However, a number of recently published studies (Lorenzen 2000, O'Sullivan et al. 2002a) have indicated that the rennet coagulation properties of milk and the syneretic properties of TGase-cross-linked, renneted milk gels, as well as the proteolytic digestibility of casein, are impaired by cross-linking the proteins, which may alter cheese manufacture and ripening. Further studies are required to evaluate whether limited, targeted cross-linking may give desirable effects.

Other potential benefits of TGase treatment of dairy proteins include physical stablilization and structural modification of products such as yogurt, cream, and liquid milk products, particularly in formulations with a reduced fat content; and the production of protein products, such as caseinates or whey protein products, with modified or tailor-made functional properties (Dickinson and Yamamoto 1996, Lorenzen and Schlimme 1998, Færgamand and Qvist 1998, Færgamand et al. 1998).

Overall, it is likely that TGase will have commercial applications in modifying the functional characteristics of milk and dairy products. However, some important issues remain to be clarified. For example, knowledge of the heat inactivation kinetics of the enzyme is needed to facilitate control of the reaction through inactivation at the desired extent of cross-linking. There is also a dearth of information on the effects of processing variables (e.g., temperature, pH) on the nature and rate of cross-linking reactions. Finally, a clear commercial advantage for using TGase over other methods of manipulating the structure and texture of dairy products (such as addition of proteins or hydrocolloids) must be established.

Lipases

Lipases, that is, enzymes that produce free fatty acids (FFAs) and flavor precursors from triglycerides, diglycerides, and monoglycerides, are produced by a wide range of plants, animals, and microorganisms (Kilara 2002). For example, calves produce a salivary enzyme called pregastric esterase (PGE), which is present in high levels in the stomachs of calves, lambs, or kids slaughtered immediately after suckling (Kilcawley et al. 1998). Microbial lipases have been isolated from *Aspergillus niger*, *Aspergillus oryzae*, *Pseudomonas fluorescens*, and *Penicillium roqueforti*. Exogenous lipases may be used for a range of applications (Kilara 2002) including (1) modification of milk fat to improve physical properties and digestibility, reduce calorific value, and enhance flavor (Balcão and Malcata 1998); (2) production of enzyme-modified cheese flavors (e.g., "buttery," "blue cheese," or "yogurt"), and (3) production of lipolyzed creams for bakery applications.

The correct choice of lipase for particular applications is very important, as the FFA profile differs with the type of enzyme used, and different lipases vary in their pH and temperature optima, heat stability, and other characteristics. Calf PGE, for example, generates a buttery and slightly peppery flavor, while kid PGE generates a sharp peppery flavor (Birsbach 1992).

The intense flavor of blue cheese is derived from lipolysis and thus catabolism of free fatty acids; most other cheese varieties undergo very little lipolysis during ripening. A method for the production of blue cheese flavor concentrates was described by Tomasini et al. (1995). It has been reported that the use of exogenous lipases (e.g., PGE) can enhance the flavor of Cheddar cheese and accelerate the development of perceived maturity (for review see Kilcawley et al. 1998). Enzyme-modified cheese preparations (e.g., slurries or powders) can also be produced by incubation of cheese emulsified in water with commercial enzymes (such as those produced by Novo Nordisk); these have 5–20 times the flavor intensity of mild Cheddar cheese.

MILK LIPIDS

PRODUCTION OF FAT-BASED DAIRY PRODUCTS

Milk fat is a complex mixture of triglycerides, which in milk, is maintained as a stable oil-in-water (o/w) emulsion by the presence of the milk fat globule membrane (MFGM), which surrounds milk fat globules and protects the lipids therein from mechanical or enzymatic damage.

One of the oldest dairy products is butter, a fat-continuous (water-in-oil, w/o) emulsion containing 80–81% fat, not more than 16% H_2O, and usually, 1.5% added salt. Production of butter from cream requires destabilization of the emulsion and phase inversion, followed by consolidation of the fat and removal of a large part of the aqueous phase (Frede and Buchheim 1994, Frede 2002a); this is somewhat analogous to the manufacture of cheese, but the key component to be preferentially concentrated is fat rather than protein, and destruction of the factor resulting in its stability in milk is achieved by mechanical rather than enzymatic means. The manufacture of butter has been the subject of several recent reviews (e.g., Keogh 1995, Lane 1998, Ranjith and Rajah 2001, Frede 2002a).

In traditional butter manufacture, cream is churned (mixed) in a large partially filled rotating cylindrical, conical, or cubical churn, which damages the MFGM. Mechanically damaged fat globules become adsorbed on the surfaces of air bubbles (flotation churning) and gradually coalesce, being bound together by expressed free fat, to form butter grains. Eventually, the air entrapped within such grains is expelled, and churning generally progresses until the grains have grown to approximately the size of a pea (Frede and Buchheim 1994, Vanapilli and Coupland 2001). In a batch process, growth of grains is monitored visually and audibly (impact of masses of grains on the churn walls); at this point, the buttermilk (essentially skim milk but with a high content of sloughed milk fat membrane components such as phospholipids) is drained off and the butter worked by repeated falls and the resulting impaction within the rotating churn. After a set time, salt is added and worked throughout the butter.

With the exception of very small scale plants, most dairy factories today use continuous rather than batch butter makers. In the most common system, the Fritz system, each stage of the process occurs in a separate horizontal cylindrical chamber, with product passing vertically between stages. In the churning cylinder, rotating impellers churn the cream very rapidly, and in the second stage butter grains are consolidated and the buttermilk drained off. In the final stage, the butter mass is worked; this chamber is sloped upwards and, while augers transport the butter and squeeze it through a series of perforated plates, the buttermilk drains off in the opposite direction. Midway through the working stage, salt (generally as a concentrated brine) is added, and becomes distributed in small moisture droplets with a high local salt concentration, which acts to flavor and microbiologically stabilize the butter. The moisture present in the final droplets comes originally from milk serum, with a small contribution from water added in brine.

The by-product buttermilk is rich in milk fat globule membrane materials, including phospholipids, and is used as an ingredient in many food products; however, due to its high content of polyunsaturated fatty acids, it deteriorates quite rapidly due to oxidation (O'Connell and Fox 2000). The buttermilk may contain high levels of biologically active (e.g., anti-carcinogenic) membrane-derived lipids, including glycosphingolipids and gangliosides (Jensen 2002). Buttermilk isolates may also have potential application as emulsifying agents (Corrideg and Dalgleish 1997).

In some countries, a significant amount of butter is produced from cream that has been ripened with lactic acid bacteria (lactic or fermented butter). Bacterial acidification enhances the keeping quality of the butter and changes the flavor, through production of diacetyl. However, the production of lactic butter leads to the production of acidic buttermilk as an unwanted by-product, and in recent years alternative technologies have been developed for the production of lactic butter. One of the most successful is the NIZO method, in which a sweet (nonlactic) cream is used, and an aromatic starter and concentrated starter permeate are added to the butter grains midway through the process. This process leads to production of normal buttermilk and gives a well-flavored product, which is very resistant to autoxidation (Walstra et al. 1999).

The high cost of butter was the original reason that led, in the second half of the 19th century, to the development of alternatives to butter, such as margarine; consumer preference for reduced-fat (i.e., higher perceived "healthiness") products has strengthened this trend. A further significant disadvantage of butter for domestic applications is the fact that at typical refrigeration temperatures, butter behaves essentially as a solid and is not easily spreadable; moreover, at room temperature it oils off and exudes water. The spreadability of butter may be increased by blending with vegetable oils, or by modifications such as interesterification (Marangoni and Rousseau 1998).

Today, there are two principal classes of nonbutter, milk fat–based spreads: (1) full-fat products with partial replacement of milk fat by another (e.g., vegetable) fat, with physicochemical, rheological, economic, or dietary advantages, and (2) reduced-fat products containing varying levels of milk fat (Frede 2002b). Such products may be manufactured either in a modified continuous butter-making system (butter technology) or in a Votator scraped-surface cooler system (margarine technology). Generally, all products are made by preparation of aqueous (e.g., skim milk or cream) and lipid phases, each containing the appropriate ingredients, followed by mixing and phase inversion of the initial o/w emulsion to a final w/o emulsion. Reduction of the fat content of the mix requires increased attention to the structural characteristics of the aqueous phase, which must increasingly contribute to the texture and body of the product. Aqueous phase structuring agents, such as polysaccharides and proteins, may be added for this purpose (Keogh 1995, Frede 2002).

Butter is a relatively stable product; the small size and high salt content of the water droplets make them an inhospitable environment for microbial growth. Lower-fat spreads, while also being generally stable products, may have preservatives, such as sorbates, incorporated into their formulation to ensure shelf life and safety (Delamarre and Batt 1999).

Some new developments in spread technology include the production of triple-emulsion (o/w/o) products (where the droplets of aqueous phase in the product contain very small fat globules) (Frede 2002b).

LIPID OXIDATION

Lipids with double bonds (i.e., unsaturated fatty acids) are inherently susceptible to attack by active O_2, that is, lipid oxidation. Oxidation can give rise to a range of undesirable flavor compounds (such as aldehydes, ketones, and alcohols, which cause rancid off-flavors) and possibly result in toxic products. Oxidation is dependent on factors such as availability of oxygen, exposure to light, temperature, presence or pro- or antioxidants, and the exact nature of the fat (O'Brien and O'Connor 1995, 2002).

Milk fat contains a high level of monounsaturated fatty acids (although less than many other fats), but a low level of polyunsaturated fatty acids. It is susceptible to oxidation, but raw milk samples differ in susceptibility to oxidation, according to the following classifications:

- *Spontaneous:* Will develop oxidized flavor within 48 hours without the addition of prooxidant metals, such as iron and copper.
- *Susceptible:* Will not oxidize spontaneously, requires contamination by prooxidant metals.
- *Nonsusceptible:* Will not oxidize even in the presence of iron or copper.

The reasons for the differences between milk samples are not clear, but are likely to be related to lactational and dietary factors, and the enzyme xanthine oxidase in milk may be critical.

Of particular current interest is the oxidation of the unsaturated alcohol, cholesterol, in milk; consumption of cholesterol oxidation products (COPs) in the diet is tentatively linked to the incidence of artherosclerosis (Kumar and Singhal 1991), and hence these products are regarded as potential health hazards. COPs are not found in all dairy products, but they have been detected in whole milk powder, baby foods, butter, and certain cheese varieties, albeit at levels which probably do not pose a risk to consumers (Sieber et al. 1997, RoseSallin et al. 1997). The formation of COPs is influenced by factors such as temperature and exposure of the food to light (Angulo et al. 1997, RoseSallin et al. 1997, Hiesberger and Luf 2000).

ICE CREAM

Ice cream, probably the most popular dairy dessert, is a frozen aerated emulsion. The continuous phase consists of a syrup containing dissolved sugars and minerals, while the dispersed phase consists of air cells, milk fat (or other kinds of fat) globules, ice crystals, and insoluble proteins and hydrocolloids (Marshall 2002). The structure of ice cream is particularly complex, with several phases (e.g., ice crystals, fat globules and air bubbles, freeze-concentrated aqueous phase) coexisting in a single product.

The production technology for ice cream was reviewed by Goff (2002), and will be summarized hereafter. The base for production of ice cream is milk blended with sources of milk solids-not-fat (e.g., skim milk powder) and fat (e.g., cream) and added sugars or other sweeteners, emulsifying agents, and hydrocolloid stabilizers. The exact formulation depends on the characteristics of the final product, and once blended, the mix is pasteurized, in either batch (e.g., 69°C for 30 minutes) or continuous (e.g., 80°C for 25 seconds) processes, and homogenized at 15.5–18.9 MPa, first-stage, and 3.4 MPa, second stage. The mix is then cooled and stored at 2–4°C for at least 4 hours; this step is called ageing, and it facilitates the hydration of milk proteins and stabilizers and the crystallization of fat globules. During this period, emulsifiers generally displace milk proteins from the milk fat globule surface. Ageing improves the whipping quality of the mix and the melting and structural properties of the final ice cream.

After ageing, the ice cream is passed through a scraped-surface heat exchanger, cooled using a suitable refrigerant flowing in the jacket, under high shear conditions with the introduction of air into the mix. These conditions result in rapid ice crystal nucleation and freezing, yielding small ice crystals, and incorporation of air bubbles, resulting in a significant increase in volume (overrun) of the product. The partially crystalline fat phase at refrigeration temperatures undergoes partial coalescence during the whipping and freezing stage, and a network of agglomerated fat develops, which partially surrounds the air bubbles and produces a solid-like structure (Hartel 1996, Goff 1997).

Flavorings and colorings may be added either to the mix before freezing, or to the soft semifrozen mix exiting the heat exchanger. The mix typically exits the barrel of the freezer at −6°C, and is transferred immediately to a hardening chamber (−30°C or below) where the majority of the unfrozen water freezes.

Today, ice cream is available in a wide range of forms and shapes (e.g., stick, brick or tub, low- or full-fat varieties).

PROTEIN HYDROLYSATES

The bovine caseins contain several peptide sequences that have specific biological activities when released by enzymatic hydrolysis (Table 20.5). Such enzymatic hydrolysis can occur either in vivo, during the digestion of ingested food, or in vitro, by treating the parent protein with appropriate enzymes under closely controlled conditions.

Casein-derived bioactive peptides have been the subject of considerable research for several years,

Table 20.5. Range and Properties of Casein-Derived Peptides with Potential Biological Activity

Peptides	Putative Biological Activities
Phosphopeptides	Metal binding
Caseinomacropeptide	Anticancerogenic action; inhibition of viral and bacterial adhesion; bifidogenic action; immunomodulatory activity; suppression of gastric secretions
Casomorphins	Opioid agonist and ACE inhibitors (antihypertensive action)
Immunomodulating peptides	Immunomodulatory activity
Blood platelet-modifying (antithrombic) peptides (e.g., casoplatelin)	Inhibition of aggregation of platelets
Angiotensin converting enzyme (ACE) inhibitors (casokinins)	Antihypertension action; blood pressure regulation; effects on immune and nervous systems
Bacteriocidal peptides (casocidins)	Antibiotic-like activity

and the very extensive literature has been reviewed by Miesel (1998), Pihlanto-Lappälä (2002), Gobbetti et al. (2002), and FitzGerald and Meisel (2003). Several bioactive peptides are liberated during the digestion of bovine milk, as shown by studies of the intestinal contents of consumers, confirming that such peptides are liberated in vivo.

Laboratory-scale processes for the production and purification (e.g., using chromatography, salt fractionation, or UF) of many interesting peptides from the caseins have been developed; enzymes used for hydrolysis include chymotrypsin and pepsin (Pihlanto-Lappäla 2002).

Bioactive peptides may also be produced by enzymatic hydrolysis of whey proteins; α- and β-lactorphins, derived from α-lactalbumin and β-lactoglobulin, respectively, are opioid agonists and possess angiotensin (ACE) inhibitory activity. The whey proteins are also the source of lactokinins, which are probably ACE inhibitory.

Currently, few milk-derived biologically active peptides are produced commercially. Perhaps the peptides most likely to be commercially viable in the short term are the caseinophosphopeptides, which contain clusters of phosphoserine residues and are claimed to promote the absorption of metals (Ca, Fe, Zn), through chelation and by acting as passive transport carriers for the metals across the distal small intestine, although evidence for this is equivocal. Caseinophosphopeptides are currently used in some dietary and pharmaceutical supplements, for example, in the prevention of dental caries.

The caseinomacropeptide (CMP;]κ-CN f106–169) is a product of the hydrolysis of the Phe_{105}-Met_{106} bond of κ-casein by rennet; during cheese making, it diffuses into the whey, while the N-terminal portion of κ-casein remains with the cheese curd. CMP has several interesting biological properties; for example, it has no aromatic amino acids and is thus suitable for individuals suffering from phenylketonuria; however, it lacks several essential amino acids. It also inhibits viral and bacterial adhesion, acts as a bifidogenic factor, suppresses gastric secretions, modulates immune system responses, and inhibits the binding of bacterial toxins (e.g., toxins produced by cholera and *E. coli*). Of particular interest, from the viewpoint of the commercial exploitation of CMP, are the relatively high levels of this peptide present in whey (~ 4% of total casein, 15–20% of protein in cheese whey, an estimated 180 $\times 10^3$ mt/yr available globally in whey), and the fact that it can be quite easily recovered therefrom.

Overall, detailed information is lacking regarding the physiological efficacy and mechanism of action of many milk protein–derived peptides and their possible adverse effects. Technological barriers also remain in terms of methods for industrial-scale production and purification of desired products.

INFANT FORMULAS

Today, a high proportion of infants in the developed world receive some or all of their nutritional requirements during the first year of life from prepared infant formulas, as opposed to breast milk. The raw material for such formulas is usually bovine milk or ingredients derived therefrom, but there are significant differences between the composition of bovine and human milk. This fact has led to the development of specialized processing strategies for transforming its composition to a product more nutritionally acceptable to the human neonate. Today, most formulas are in fact prepared from isolated constituents of bovine milk (e.g., casein, whey proteins, lactose), blended with nonmilk components. This, combined with the requirement for high hygienic standards and the absence of potentially harmful agents, makes the manufacture of infant formula a highly specialized branch of the dairy processing industry, with almost pharmaceutical-grade quality control.

Most infant formulas are formulated by blending dairy proteins, vegetable (e.g., soy) proteins, lactose, and other sugars with vegetable oils and fats, minerals, vitamins, emulsifiers, and micronutrients. The mixture of ingredients is then homogenized and heat-treated to ensure microbiological safety. Subsequent processing steps differ in the case of dry or liquid formulas.

The dairy ingredients used are generally demineralized, as the mineral balance in bovine milk is very different from that in human milk (Burling 2002), and desired minerals are added back to the formula as required. Certain proteins (e.g., lactoferrin and α-lactalbumin) are present at higher levels in human than in bovine milk, and β-lg is absent from human milk. There is interest in fortifying infant formulas with α-lactalbumin and/or lactoferrin, although technological challenges exist in the economical production of such proteins at acceptable purity.

The exact formulations of infant formulas differ based on the age and special requirements of the infant. Formulas for very young (< 6 months) babies generally have a high proportion of whey protein (e.g., 60% of total protein), while follow-on formulas for older infants contain a higher level of casein. In cases of infants with allergies to milk proteins or, in some cases, proteins in general, formulas in which milk proteins have been substituted by soy proteins or in which the proteins have been hydrolyzed to small peptides and amino acids, may be used. Soy-based formulas are also used in cases of intolerance to lactose.

The exact composition and level of added vitamins and minerals also differ based on nutritional requirements (O'Callaghan and Wallingford 2002). Certain lipids are of interest as supplements for infant formulas, such as long-chain polyunsaturated fatty acids and conjugated linoleic acids. The final category of additives of interest for infant formulas is oligosaccharides, which are present at quite high levels (~ 1 g/L) in human milk. Such oligosaccharides may be produced enzymatically, be chemically synthesized, or be produced by fermentation.

Two main categories of infant formula are available in most countries: dry and liquid (UHT). For dry powder manufacture, the liquid mix is evaporated and spray-dried to yield a highly agglomerated powder that will disperse readily in warm water (dissolving infant formula is probably the most common dairy powder reconstitution operation practiced by consumers in the home). Some components may be dry-blended with the base powder, allowing flexibility in manufacture for different applications (e.g., infant age, dietary requirements, etc.). Powdered formulas are typically packaged in N_2/CO_2-flushed cans.

Liquid formulas (ready-to-feed) are generally subjected to far more severe thermal treatments than those intended for drying, e.g., UHT processing followed by aseptic packaging or retort sterilization of product in screw-capped glass jars. These products, which are stable at room temperature, have the obvious advantage of convenience over their dry counterparts.

NOVEL TECHNOLOGIES FOR PROCESSING MILK AND DAIRY PRODUCTS

In recent years, a number of novel processing techniques have been developed for applications in food processing; major reasons for this trend include consumer demand for minimally processed food products, and ongoing challenges in satisfactorily processing certain food products without significant loss of quality using existing technologies (e.g., deterioration of nutrient content on heating fruit juices, problems with the safety of shellfish).

One new process that has received particular attention during the last decade is high-pressure (HP) treatment. The principle of HP processing involves subjecting food products to a very high pressure (100–1000 MPa, or 1000–10,000 atm), typically at room temperature, for a fixed period (e.g., 1–30 minutes); under such conditions, microorganisms are killed, proteins denatured, and enzymes either activated (at low pressures) or inactivated (at higher pressures). The principal advantage of HP processing, which does not rupture covalent bonds, is that low molecular weight substances, such as vitamins, are not affected, and hence there are few detrimental effects on the nutritional or sensory characteristics of food.

Although first described for the inactivation of microorgansisms in milk in the 1890s by Bert Hite at the Agricultural Research Station in Morganstown, West Virginia, HP treatment remained unexploited in the food industry for most of the 20th century due to lack of available processing equipment. Only in the late 1990s were HP-processed foods such as shellfish in the United States, fruit juice in France, and meat in Spain launched.

To date, no HP-processed dairy products are available, probably due at least in part to the complexity of the effects of high pressure on dairy systems, which necessitates considerable fundamental research to underpin future commercial applications.

In short, high pressure affects the properties of milk in several, often unique, ways (for reviews see Huppertz et al. 2002, Trujillo et al. 2002). Key effects include

- Denaturation of the whey proteins, α-la and β-lg, at pressures ≥ 200 or ≥ 600 MPa, respectively, and interaction of denatured β-lg with the casein micelles.
- Increased casein micelle size (by ~ 25%) after treatment at 250 MPa for about 15 minutes (Huppertz et al. 2004a) and reductions in micelle size by about 50% on treatment at 300–800 MPa.
- Increased levels of nonmicellar α_{s1}-, α_{s2}-, β- and κ-caseins after HP treatment at ≥ 200 MPa.

- Inactivation of indigenous alkaline phosphatase and plasmin at pressures ≥ 400 MPa; lactoperoxidase is not inactivated by treatment at ≤ 700 MPa.
- Reduced rennet coagulation time of milk, reduced time required for the gel to become firm enough for cutting, and enhanced strength of the rennet gel, following treatment at 200 MPa for up to 60 minutes or 400 MPa for ≤ 15 minutes.
- Increased yield of rennet-coagulated cheese curd, in particular after treatment ≥ 400 MPa.
- Increased rate and level of creaming in milk, by up to 70%, after HP treatment at 100–250 MPa, whereas treatment at 400 or 600 MPa reduces the rate and level of creaming by up to 60% (Huppertz et al. 2004b)
- Reduced stability of milk to coagulation by ethanol (Johnston et al. 2002, Huppertz et al. 2003).

HP treatment can also be applied to cheese (for review see O'Reilly et al. 2000a). A patent issued in Japan in 1992 suggested that the ripening of Cheddar cheese could be significantly accelerated by treatment at 50 MPa for 3 days; however, subsequent research (O'Reilly et al. 2001) has failed to substantiate this claim. There have been studies on the effects of high pressure on the ripening of many other cheese varieties; varying degrees of acceleration have been reported [e.g., Saldo et al. (2002) reported effects of HP treatment on ripening of cheese made from caprine milk]. Of particular interest may be positive effects of HP treatment on the functionality of mozzarella cheese (Johnston et al. 2002, O'Reilly et al. 2002). HP may also be used to modify the microbial population (either starter or contaminant) of cheese (O'Reilly et al. 2000b, 2002).

Another technology which may be of potential interest for dairy processing in the future is high-pressure homogenization (HPH), which works like conventional homogenization, but at significantly higher pressures, up to 250 MPa; a related process, microfluidization, is based on the principle of collisions between high-speed liquid jets. As well as reducing the size of oil droplets in an emulsion, HPH may inactivate enzymes and microorganisms and denature proteins in food. Many effects of HPH are due to the extremely high shear forces encountered by a fluid being processed; however, there is also a significant heating effect during the process. Recent studies have indicated that HPH significantly inactivates bacteria in raw bovine milk and affects its rennet coagulation properties, fat globule size distribution, and enzyme profile (Hayes and Kelly 2003a,b).

Another novel process that may be applied to dairy products is pulsed electric field treatment (which inactivates microorganisms but has relatively few other effects; Deeth and Datta 2002). Some other processes, such as ultrasonication, irradiation, addition of antimicrobial peptides or enzymes, and addition of carbon dioxide, have been studied for potential application to milk (Datta and Deeth 2002). Arguably, HP processing is the most likely of the novel processes to be adopted by the dairy industry in the near future, due to the availability of equipment for commercial processing.

CONCLUSION AND SUMMARY

Milk is a very complex raw material, with constituents and properties that are sensitive to applied stresses such as heat and changes in pH or concentration. Changes that can occur include inactivation of enzymes, mineral distribution and equilibria, and denaturation of proteins. The extent of such changes and the consequences for the properties and stability of milk depend on the severity of the treatment.

As well as physicochemical changes that occur during processing, many dairy products involve complex enzymatic pathways, such as those involved in cheese ripening; exogenous enzymes may be added to milk to achieve a variety of end results. Milk can also be fractionated by any of a range of complex technological processes to yield a broad portfolio of food ingredients.

In conclusion, the processing of milk represents perhaps one of the most complex fields in food science and technology, and while many underpinning principles have been characterized, much research remains to be done in several areas.

REFERENCES

Andrews AT, Anderson M, Goodenough PW. 1987. A study of the heat stabilities of a number of indigenous milk enzymes. J Dairy Res 54:237–246.

Angulo AJ, Romera JM, Ramirez M, Gill A. 1997. Determination of cholesterol oxides in dairy products. Effect of storage conditions. J Agric Food Chem 45:4318–4323.

Balcão, VM, Malcata, FX. 1998. Lipase-catalysed modification of milkfat. Biotechnol Adv 16:309–341.

Becerra M, Baroli B, Fadda AM, Méndez JB, Gonzaléz-Siso MI. 2001. Lactose bioconversion by calcium-alginate immobilisation of *Kluyveromyces lactis* cells. Enz Microbiol Technol 29:506–512.

Birlouez-Aragon I, Sabat P, Gouti N. 2002. A new method of discriminating milk heat treatment. Int Dairy J 12:59–67.

Birsbach P. 1992. Pregastric lipases. In: Bulletin 269. Brussels: International Dairy Federation. Pp. 36–39.

Burling H. 2002. Whey processing. Demineralisation. In: H Roginski, JW Fuquay, PF Fox, editors. Encyclopedia of Dairy Science. London: Academic Press. Pp. 2745–2751.

Caríc M, Kalab M. 1987. Effects of drying techniques on milk powder quality and microstructure: A review. Food Microstructure 6:171–180.

Chen CS, Hsu CK, Chiang BH 2002. Optimization of the enzymic process for manufacturing low-lactose milk containing oligosaccharides. Process Biochem 38:801–808.

Claeys WL, Van Loey AM, Hendrickx ME. 2002. Intrinsic time temperature integrators for heat treatment of milk. Trends Food Sci Technol 13:293–311.

Coolbear KP, Elgar DF, Coolbear T, Ayers, JS. 1996. Comparative study of methods for the isolation and purification of bovine κ-casein and its hydrolysis by chymosin. J Dairy Res 63:61–71.

Corrideg M, Dalgleish DG. 1997. Isolates from industrial buttermilk: emulsifying properties of materials derived from the milk fat globule membrane. J Aric Food Chem 45:4595–4600.

Datta N, Deeth HC. 2001. Age gelation of UHT milk—A review. Trans Inst Chem Eng 71:197–210.

———. 2002.Other nonthermal technologies. In: H Roginski, JW Fuquay, PF Fox, editors. Encyclopedia of Dairy Science. London: Academic Press. Pp. 1339–1346.

Deeth HC, Datta N. 2002. Pulsed electric technologies. In: H Roginski, JW Fuquay, PF Fox, editors. Encyclopedia of Dairy Science. London: Academic Press. Pp. 1333–1339.

Delamarre S, Batt CA. 1999. The microbiology and historical safety of margarine. Food Microbiol 16: 327–333.

De Vuyst L, De Vin F, Vaningelgem F, Degeest B. 2001. Recent developments in the biosynthesis and applications of heteropolysaccharides from lactic acid bacteria. Int Dairy J 11:687–707.

De Vuyst L, Zamfir M, Mozzi F, Adriany T, Marshall V, Degeest B, Vaningelgem F. 2003. Exopolysaccharide-producing *Streptococcus thermophilus* strains as functional starter cultures in the production of fermented milks. Int Dairy J 13. (In press.)

Dickinson E, Yamamoto Y. 1996. Rheology of milk protein gels and protein-stabilised emulsion gels cross-linked with transglutaminase. J Agric Food Chem 44:1371–1377.

Early R. 1998. Milk concentrates and milk powders. In: R Early, editor. The Technology of Dairy Products, 2nd edition. London: Blackie Academic and Professional. Pp. 228–300.

Eck A, Gilles J-C. 2000. Cheesemaking—From Science to Quality Assurance, 2nd edition. Paris: Laviosier Technique and Documentation.

El-Gazzar FE, Marth EH. 1991. Ultrafiltration and reverse osmosis in dairy technology: A review. J Food Prot 54:801–809.

Færgamand M, Otte J, Qvist KB. 1998. Emulsifying properties of milk proteins cross-linked with microbial transglutaminase. Int Dairy J 8:715–723.

Færgamand M, Qvist KB. 1998 Transglutaminase: Effect on rheological properties, microstructure and permeability of set-style skim milk gel. Food Hydrocoll 11:287–292.

Farise J-F, Cayot P. 1998. New ultrarapid method for the separation of milk proteins by capillary electrophoresis. J Agric Food Chem 46:2628–2633.

Farkye NY, Imafidon GI. 1995. Thermal denaturation of indigenous milk enzymes. In: PF Fox, editor. Heat-Induced Changes in Milk. Special Issue 9501. Brussels: International Dairy Federation. Pp. 331–348.

FitzGerald R, Meisel H. 2003. Milk protein hydrolysates. In: Advanced Dairy Chemistry. PF Fox, PLH McSweeney, editors. Vol. 1, Proteins, 3rd edition. New York: Kluwer Academic-Plenum Publishers. Pp. 625–698.

Fizsman SM. Lluch MA, Salvador A. 1999. Effect of addition of gelatin on microstructure of acidic milk gels and yoghurt and on their rheological properties. Int Dairy J 9:895–901.

Fontes EAF, Passon FML, Passos FJV. 2001. A mechanistical mathematical model to predict lactose hydrolysis by β-galactosidase in a permeabilised cell mass of *Kluyveromyces lactis*: Variability and sensitivity analysis. Process Biochem 37:267–274.

Fox PF. 2001. Milk proteins as food ingredients. Int J Dairy Technol 54(2): 41–56.

Fox PF, Kelly AL. 2004. The caseins. In: R Yada, editor. Proteins in Food Processing, 28–71. New York: Woodhead Publishing.

Fox PF, McSweeney PLH. 1998. Dairy Chemistry and Biochemistry. London: Chapman and Hall.

Fox PF, McSweeney PLH, Cogan T, Guinee TP. 2004. Cheese: Chemistry, Physics and Microbiology. London: Elsevier.

Frede E. 2002a. Butter: The product and its manufacture. In: H Roginski, JW Fuquay, PF Fox, editors. Encyclopedia of Dairy Science. London: Academic Press. Pp. 220–226.

———. 2002b. Milk fat products: Milk fat based spreads. In: H Roginski, JW Fuquay, PF Fox, editors. Encyclopedia of Dairy Science. London: Academic Press. Pp. 1859–1868.

Frede E, Buchheim W. 1994. Buttermaking and the churning of blended fat emulsions. J Soc Dairy Technol 47:17–27.

Garem A, Schuk, P, Maubois, JL. 2000. Cheesemaking properties of a new dairy-based powder made by a combination of microfiltration and ultrafiltration. Lait 80:25–32.

Gobbetti M, Stepaniak L, Angelis M, Corsetti A, Di Cagni, R. 2002. Latent bioactive peptides in milk proteins: Proteolytic activation and significance in dairy products. CRC Crit Rev Food Sci Nutr 42:223–239.

Goff HD. 1997. Colloidal aspects of ice cream: A review. Int Dairy J. 7:363–373.

———. 2002. Ice cream and frozen desserts. Manufacture. In: H Roginski, JW Fuquay, PF Fox, editors. Encyclopedia of Dairy Science. London: Academic Press. Pp. 1374–1380.

González-Martínez C, Becerra M, Cháfer M, Albors A, Carot JM, Chiralt A. 2002. Influence of substituting milk powder for whey powder on yoghurt quality. Trends Food Sci Technol 13:334–340.

Grant IR, Rowe MT, Dundee L, Hitchings E. 2001. *Mycobacterium avium* ssp. *paratuberculosis*: Its incidence, heat resistance and detection in milk and dairy products. Int J Dairy Technol 54:2–13.

Hartel RW. 1996. Ice crystallisation during the manufacture of ice cream. Trends Food Sci Technol. 7: 315–321.

Hassan AN, Frank JF, El-Soda M. 2003. Observation of bacterial exopolysaccharide in dairy products using cryo-scanning electron microscopy. Int Dairy J. (In press.)

Hayes MG, Kelly AL. 2003a. High pressure homogenisation of raw whole bovine milk. (a) Effects on fat globule size and other properties. J Dairy Res 70: 297–305.

———. 2003b. High pressure homogenisation of milk. (b) Effects on indigenous enzymatic activity. J Dairy Res 70:307–313.

Hiesberger J, Luf W. 2000. Oxidation of cholesterol in butter during storage—Effects of light and temperature. European Food Res Technol 211:161–164.

Holsinger VH, Rajkowski KT, Stabel JR. 1997. Milk pasteurisation and safety: A brief history and update. Rev Sci Techn Office Int Epiz 16:441–451.

Horton BS. 1997. What ever happened to the ultrafiltration of milk? Aust J Dairy Technol 52:47–49.

Huppertz T, Kelly AL, Fox PF. 2002. Effects of high pressure on constituents and properties of milk: A review. Int Dairy J 12:561–572.

Huppertz T, Fox PF, Kelly AL. 2003. High pressure-induced changes in the creaming properties of bovine milk. Innov Food Sci Emerg Technol 4:349–359.

———. 2004. High pressure treatment of bovine milk: Effects on casein micelles and whey proteins. J Dairy Res 71:98–106.

Huppertz T, Grosman S, Fox PF, Kelly AL. 2004. Heat and ethanol stability of high pressure-treated bovine milk. Int Dairy J 14:125–133.

Ikura K, Komitani T, Yoshikawa M, Sasaki R, Chiba H. 1980. Crosslinking of casein components by transglutaminase. Agric Biol Chem. 44:1567–1573.

Jelen P. 2002. Whey processing: Utilisation and products. In: H Roginski, JW Fuquay, PF Fox, editors. Encyclopedia of Dairy Science. London: Academic Press. Pp. 2739–2745.

Jensen JD. 1975. Some recent advances in agglomeration, instantising and spray drying. Food Technol 29: 60–71.

Jensen, RJ. 2002. The composition of bovine milk lipids: January 1995 to December 2000. J Dairy Sci 85:295–350.

Ji T, Haque, ZU. 2003. Cheddar whey processing and source: I. Effect on composition and functional properties of whey protein concentrates. Int J Food Sci Technol 38:453–461.

Johnston, DE, O'Hagan, M, Balmer, DW. 2002. Effects of high pressure treatment on the texture and cooking performance of half-fat Cheddar cheese. Milchwissenschaft 57:198–201.

Johnston DE, Rutherford JA, McCreedy RW. 2002. Ethanol stability and chymosin-induced coagulation behaviour of high pressure treated milk. Milchwissenschaft 57:363–366.

Karleskind D, Laye I, Mei FI, Morr CV. 1995. Chemical pre-treatment and microfiltration for making delipidised whey-protein concentrate. J Food Sci 60: 221–226.

Kelly AL, O'Connell JE, Fox, PF. 2003. Manufacture and properties of milk powders. In: Advanced Dairy Chemistry. PF Fox, PLH McSweeney, editors. Vol.

1, Proteins, 3rd edition. New York: Kluwer Academic-Plenum Publishers. Pp. 1027–1062.

Kelly AL, O'Shea N. 2002. Pasteurisers: Design and operation. In: H Roginski, JW Fuquay, PF Fox, editors. Encyclopedia of Dairy Science. London: Academic Press. Pp. 2237–2244.

Kelly PM, Kelly J, Mehra R, Oldfield DJ, Raggett E, O'Kennedy, BT. 2000. Implementation of integrated membrane processes for pilot-scale development of fractionated milk components. Le Lait. 80:139–153.

Keogh KK. 1995. Chemistry and technology of milk fat spreads. In: Advanced Dairy Chemistry. PF Fox, editor. Vol. 2, Lipids. London: Chapman and Hall. Pp. 213–248.

Kilara A. 2002. Enzymes exogenous to milk. Lipases. In: H Roginski, JW Fuquay, PF Fox, editors. Encyclopedia of Dairy Science. London: Academic Press. Pp. 914–918.

Kilcawley KN, Wilkinson, MG, Fox, PF. 1998. Enzyme-modified cheese. Int Dairy J. 8:1–10.

Kumar N, Singhal OP. 1991. Cholesterol oxides and atherosclerosis—A review. J Sci Food Agric 55: 497–510.

Kyle WSA. 1993. Powdered milk. In: R McCrae, RK Robinson, MJ Sadler, editors. Encyclopedia of Food Science, Food Technology and Nutrition. New York: Academic Press. Pp. 3700–3713.

Lane R. 1998. Butter and mixed fat spreads. In: R Early, editor. The Technology of Dairy Products. London: Blackie Academic and Professional. Pp. 158–197.

Law BA. 1999. Technology of Cheesemaking. Boca Raton, Florida: CRC Press.

Le Graet Y, Brule G. 1982. Effect of concentration and drying on mineral equilibria in skim milk and retentates. Lait 62:113–125.

Lewis M, Heppell N. 2000. Continuous thermal processing of foods. Pasteurisation and UHT sterilization. Gaithersburg, Maryland; Aspen Publishers, Inc.

Lorenzen PC, Schlimme E. 1998. Properties and potential fields of application of transglutaminase preparations in dairying. Bulletin 332. Brussels: International Dairy Federation. Pp. 47–53.

Lorenzen PC. 2000. Renneting properties of transglutaminase-treated milk. Milchwissenschaft 55: 433–437.

Lu W, Piyasena P, Mittal GS. 2001. Modelling alkaline phosphatase inactivation in bovine milk during high-temperature short-time pasteurisation. Food Sci Technol Int 7:479–485.

Lucey JA, Singh H. 1998. Formation and physical properties of acid milk gels: A review. Food Res Int 30:529–542.

Maciunska J, Czyz, B, Syniwjecki J. 1998. Isolation and some properties of beta-galactosidase from the thermophilic bacterium *Thermus thermophilus*. Food Chem 63:441–445.

Mahoney RR. 1997. Lactose: Enzymatic modification. In: Advanced Dairy Chemistry. PF Fox, editor. Vol. 3, Lactose, Salts, Water and Vitamins. London: Chapman and Hall. Pp. 77–126.

———. 2002. Enzymes exogenous to milk in dairy technology: β-D-galactosidase. In: H Roginski, JW Fuquay, PF Fox, editors. Encyclopedia of Dairy Science. London: Academic Press. Pp. 907–914.

Marangoni AG, Rousseau D. 1998. Chemical and enzymatic modification of butter fat and butterfat-canola oil blends. Food Res Int 31:595–599.

Marshall RT. 2002. Ice cream and frozen desserts. Product types. In: H Roginski, JW Fuquay, PF Fox, editors. Encyclopedia of Dairy Science. London: Academic Press. Pp. 1367–1373.

Matthews ME. 1984. Whey protein recovery processes and products. J Dairy Sci 67:2680–2692.

Maubois J-L. 1997. Current uses and future perspectives of MF technology in the dairy industry. Bulletin 320. Brussels: International Dairy Federation. Pp. 154–159

McKellar RC, Modler HW, Couture H, Hughes A, Mayers P, Gleeson T, Ross WH. 1994. Predictive modelling of alkaline phosphatase inactivation in a high-temperature short-time pasteurizer. J Food Prot 57:424–430.

McKellar RC, Liou S, Modler HW. 1996. Predictive modelling of lactoperoxidase and γ-glutamyl transpeptidase inactivation in a high-temperature short-time pasteurizer. Int Dairy J 6:295–301.

Miesel H. 1998. Overview on milk protein-derived peptides. Int Dairy J 8:363–373.

Mistry VV, Pulgar JB. 1996. Physical and storage properties of high milk protein powder. Int Dairy J 6:195–203.

Modler HW, Gelda A, Yamaguchi M, Gelda, S. 1993. Production of fluid milk with a high degree of lactose hydrolysis. In: PF Fox, editor. Heat-induced changes in milk. Brussels: International Dairy Federation. Pp. 57–61.

Motoki M, Seguro K. 1998. Transglutaminase and its use for food processing. Trends Food Sci Technol 9:204–210.

Muir DD. 2002. Lactose; properties, production, applications. In: H Roginski, JW Fuquay, PF Fox, editors. Encyclopedia of Dairy Science. London: Academic Press. Pp. 1525–1529.

Mulvihill DM, Ennis MP. 2003. Functional milk proteins: Production and utilisation. In: Advanced Dairy Chemistry. PF Fox and PLH McSweeney, editors. Vol. 1, Proteins, 3rd edition. New York: Kluwer Academic-Plenum Publishers. Pp. 1175–1128.

Mulvihill DM, Fox PF. 1994. Developments in the production of milk proteins. In: S Damodaran, editor. New and Developing Sources of Food Proteins. New York: Plenum Press. Pp. 1–30.

Nakagawa T, Fujimoto Y, Uchino M, Miyaji T, Takano K, Tomizuka, N. 2003. Isolation and characterisation of psychrophiles producing cold-active beta-galactosidase. Lett Appl Microbiol 37:154–157.

Nieuwenhuijse JA. 1995. Changes in heat-treated milk products during storage. In: PF Fox, editor. Heat-Induced Changes in Milk. Special Issue 9501. Brussels: International Dairy Federation. Pp. 231–255.

Nieuwenhuijse JA, Timmermans W, Walstra P. 1988. Calcium and phosphate paritions during the manufacture of sterilised concentrated milk and their relations to heat stability. Neth Milk Dairy J 42:387–421.

Nieuwenhuijse JA, van Boekel, MAJS. 2003. Protein stability in sterilised milk and milk products. In: Advanced Dairy Chemistry. PF Fox, PLH McSweeney, editors. Vol. 1, Proteins, 3rd edition. New York: Kluwer Academic-Plenum Publishers. Pp. 947–974.

Obon JM, Castellar MR, Iborra JL, Manjøn A. 2000. β-galactosidase immobilisation for milk lactose hydrolysis: A simple experimental and modelling study of batch and continuous reactors. Biochem Edu 28: 164–168.

O'Brien J. 1995. Heat-induced changes in lactose: Isomerisation, degradation, Maillard browning. In: PF Fox, editor. Heat-Induced Changes in Milk. Special Issue 9501. Brussels: International Dairy Federation. Pp. 134–170.

O'Brien NM, O'Connor T. 1995. Lipid oxidation. In: Advanced Dairy Chemistry. PF Fox, editor. Vol. 2, Lipids. London: Chapman and Hall. Pp. 309–348.

———. 2002. Lipid oxidation. In: H Roginski, JW Fuquay, PF Fox, editors. Encyclopedia of Dairy Science. London: Academic Press. Pp. 1600–1604.

O'Connell JE, Fox PF. 2000. Heat stability of buttermilk. J Dairy Sci 83:1728–1732.

———. 2002. Heat stability of milk. In: H Roginski, JW Fuquay, PF Fox, editors. Encyclopedia of Dairy Science. London: Academic Press. Pp. 1321–1326.

———. 2003. Heat-induced coagulation of milk. In: Advanced Dairy Chemistry. PF Fox, PLH McSweeney, editors. Vol. 1, Proteins, 3rd edition. New York: Kluwer Academic-Plenum Publishers. Pp. 879–945.

Oldfield DJ. 1998. Heat Induced Whey Protein Reactions in Milk: Kinetics of Denaturation and Aggregation as Related to Milk Powder Manufacture. PhD Thesis, Massey University, Palmerstown North, New Zealand.

O'Reilly CE, Kelly AL, Murphy PM, Beresford TP. 2000a. Effect of high pressure on proteolysis during ripening of Cheddar cheese. Innov Food Sci Emerg Technol 1:109–117

———. 2001. High pressure treatment: Applications to cheese manufacture and ripening. Trends Food Sci Technol 12:51–59.

O'Reilly CE, O'Connor PM, Kelly AL, Beresford TP, Murphy PM. 2000b. Inactivation of a number of microbial species in cheese by high hydrostatic pressure. Applied Environ Microbiol 66:4890–4896.

O'Reilly CE, O'Connor PM, Murphy PM, Kelly AL, Beresford TP. 2002. The effect of high pressure on starter viability and autolysis and proteolysis in Cheddar cheese. Int Dairy J 12:915–922.

O'Reilly CE, Murphy PM, Kelly AL, Guinee TP, Beresford TP. 2002. The effect of high pressure treatment on the functional and rheological properties of Mozzarella cheese. Innov Food Sci Emerg Technol 3:3–9.

O'Sullivan MM, Kelly AL, Fox PF. 2002b. Influence of transglutaminase treatment on some physicochemical properties of milk. J Dairy Res 69:433–442.

———. 2002a. Effect of transglutaminase on the heat stability of milk: A possible mechanism. J Dairy Sci 85:1–7.

Pellegrino L, Resmini L, Luf W. 1995. Assessment (indices) of heat treatment of milk. In: PF Fox, editor. Heat-Induced Changes in Milk. Special Issue 9501. Brussels: International Dairy Federation. Pp. 419–453.

Pihlanto-Lappäla A. 2002. Bioactive peptides. In: H Roginski, JW Fuquay, PF Fox, editors. Encyclopedia of Dairy Science. London: Academic Press. Pp. 1960–1967.

Písecky J. 1997. Handbook of Milk Powder Manufacture. Copenhagen: Niro A/S.

Pouliot M, Pouliot Y, Britten M. 1996. On the conventional cross-flow microfiltration of skim milk for the production of native phosphocaseinate. Int Dairy J 6:105–111.

Puvanenthiran A, Williams RPW, Augustin MA. 2002. Structure and visco-elastic properties of set yoghurt with altered casein to whey protein ratios. Int Dairy J 12:383–391.

Ranjith HMP, Rajah KK. 2001. High fat content dairy products. In: AY Tamime, BA Law, editors. Mechanisation and Automation in Dairy Technology. Sheffield: Sheffield Academic Press. Pp. 119–151.

Remeuf F, Mohammed S, Sodini I, Tissier JP. 2003. Preliminary observations on the effects of milk fortification and heating on microstructure and physical properties of stirred yoghurt. Int Dairy J 13. (In press.)

Richardson BC. 1983. The proteinases of bovine milk and the effect of pasteurisation on their activity. NZ J Dairy Sci Technol 18:233–245.

Robinson RK. 2002a. Fermented milks. Yoghurt types and manufacture. In: H Roginski, JW Fuquay, PF Fox, editors. Encyclopedia of Dairy Science. London: Academic Press. Pp. 1055–1058.

———. 2002b. Fermented milks. Yoghurt, role of starter culture. In: H Roginski, JW Fuquay, PF Fox, editors. Encyclopedia of Dairy Science. London: Academic Press. Pp. 1059–1063.

Robinson RK, Wilbey RA. 1998. Cheesemaking Practice. 3rd edition. Gaithersburg, Maryland: Aspen Publishers.

RoseSallin C, Sieber R, Bosset, JO, Tabacchi R. 1997. Effects of storage or heat treatment on oxysterols formation in dairy products. Food Sci Technol—Lebens-Wissen Technol 30:170–177.

Ryser ET. 2002. Pasteurisation of liquid milk products. Principles, public health aspects. In: H Roginski, JW Fuquay, PF Fox, editors. Encyclopedia of Dairy Science. London: Academic Press. Pp. 2232–2237.

Saldo J, McSweeney, PLH, Sendra E, Kelly AL, Guamis B. 2002. Proteolysis in caprine milk cheese treated by high pressure to accelerate cheese ripening. Int Dairy J 12:35–44.

Sanderson WB. 1978. Instant milk powders: Manufacture and keeping quality. NZ J Dairy Sci Technol 13:137–143.

Sawyer L. 2003. β-lactoglobulin. In: Advanced Dairy Chemistry. PF Fox, PLH McSweeney, editors. Vol. 1, Proteins, 3rd edition. New York: Kluwer Academic-Plenum Publishers. Pp. 319–386.

Sharma R, Lorenzen PC, Qvist KB. 2001. Influence of transglutaminase treatment of skim milk on the formation of ε γ-glutamyl) lysine and the susceptibility of individual proteins towards crosslinking. Int Dairy J 11:785–793.

Shin, HS, Lee JH, Pestka, JJ, Ustonol, Z. 2000. Growth and viability of commercial *Bifidobacterium* spp. in skim milk containing oligosaccharides and inulin. J Food Sci 65(5): 884–887.

Sieber R, RoseSallin C, Bosset JO. 1997. Cholesterol and its oxidation products in dairy products: Analysis, effect of technological treatments, nutritional consequences. Sciences des Aliments. 17:243–252.

Sienkiewicz T, Riedel C-L. 1990. Whey and Whey Utilisation. Gelsenkirchen, Germany: Verlag Th. Mann.

Singh H, Creamer LK. 1991. Denaturation, aggregation and heat stability of milk protein during the manufacture of skim milk powder. J Dairy Res 58:269–283.

Stepaniak L, Sørhaug T. 1995. Thermal denaturation of bacterial enzymes in milk In: PF Fox, editor. Heat-induced changes in milk. Special Issue 9501. Brussels: International Dairy Federation. Pp. 349–363.

Surono I, Hosono A. 2002a. Fermented milks. Types and standards of identity. In: H Roginski, JW Fuquay, PF Fox, editors. Encyclopedia of Dairy Science. London: Academic Press. Pp. 1018–1023.

———. 2002b. Fermented milks. Starter cultures. In: H Roginski, JW Fuquay, PF Fox, editors. Encyclopedia of Dairy Science. London: Academic Press. Pp. 1023–1028.

Tamime AY, Robinson, RK. 1999. Yoghurt: Science and technology. Cambridge: Woodhead Publishers.

Tomasini A, Bustillo G, Lebeault J-M. 1995. Production of blue cheese flavor concentrates from different substrates supplemented with lipolysed cream. Int Dairy J. 5:247–257.

Trujillo AJ, Capellas M, Saldo J, Gervilla R, Guamis B. 2002. Applications of high-hydrostatic pressure on milk and dairy products: A review. Innov Food Sci Emerg Technol 3:295–307.

van Boekel MAJS. 1998. Effect of heating on Maillard reactions in milk. Food Chem 62:403–414.

Vanapalli SA, Coupland JN. 2001. Emulsions under shear—The formation and properties of partially coalesced lipid structures. Food Hydrocoll 51:507–512.

Vasiljevic T, Jelen P. 2001. Production of β-galactosidase for lactose hydrolysis in milk and dairy products using thermophilic lactic acid bacteria. Innov Food Sci Emerg Technol 2:75–85.

Walstra P, Guerts TJ, Noomen A, Jellema A, van Boekel, MAJS. 1999. Dairy Technology. Principles of Milk Properties and Processes. New York: Marcel Dekker, Inc.

Westhoff DC. 1978. Heating milk for microbial destruction: A historical outline and update. J Food Prot 41:122–130.

Whitaker JR. 1994. β-galactosidase. In: Principles of Enzymology for the Food Sciences. New York: Marcel Dekker Inc. Pp. 419–423.

Wilbey RA. 1996. Estimating the degree of heat treatment given to milk. J Soc Dairy Technol 49:109–112.

Zadow JG. 1993. Economic considerations related to the production of lactose and lactose by-products. In: Lactose Hydrolysis, IDF bulletin 289. Brussels: International Dairy Federation. Pp. 10–15.

Part V
Fruits, Vegetables, and Cereals

21
Biochemistry of Fruits

G. Paliyath and D. P. Murr

INTRODUCTION

Fruits and various products derived from fruits are at the center stage of human dietary choices in recent days, specifically because of various health benefits associated with the consumption of fruits and fruit products. The selection of trees that produce fruits with ideal edible quality has proceeded throughout human evolution. Fruits are developmental manifestations of the seed-bearing structures in plants, the ovaries. After fertilization, the hormonal changes induced in the ovary result in the development of the characteristic fruit, which may vary in ontogeny, form, structure, and quality. Pome fruits such as apple and pear are developed from the thalamus in the flower. In general, drupe fruits such as cherries, peaches, plums, and apricots are developed from the ovary wall (mesocarp) enclosing a single seed. Berry fruits such as tomato possess seeds embedded in a jelly-like pectinaceous matrix, with the ovary wall developing into the flesh of the fruit. Cucumbers and melons develop from an inferior ovary. Citrus fruits belong to the class hesperidium, where the ovary wall develops as a protective structure surrounding the juice-filled locules, which are the edible part of the fruit. In strawberry, the seeds are located outside the fruit, and it is the receptacle of the ovary (central portion) that develops into the edible part. The biological purpose of the fruit is to attract vectors that help in the dispersal of the seeds. For this, the fruits have developed various organoleptic (stimulatory to organs) characteristics that include attractive color, flavor, and taste. The biochemical characteristics and pathways in the fruits are structured to achieve these goals. The nutritional and food qualities of fruits arise as a result of the accumulation of components derived from these intricate biochemical pathways. In terms of production and volume, tomato, orange, banana, and grape are the major fruit crops used for consumption and processing around the world (Kays 1997).

BIOCHEMICAL COMPOSITION OF FRUITS

Fruits contain a large percentage of water, which can often exceed 95% by fresh weight. During ripening, activation of several metabolic pathways often leads to drastic changes in the biochemical composition of fruits. Fruits such as banana store starch during development, and hydrolyze the starch to sugars during ripening, which also results in fruit softening. Most fruits are capable of photosynthesis, store starch, and convert starches to sugars during ripening. Fruits such as apple, tomato, and grape have a high percentage of organic acids, which decreases during ripening. Fruits contain large amounts of fibrous materials such as cellulose and pectin. The degradation of these polymers into smaller water-soluble units during ripening leads to fruit softening, as exemplified by the breakdown of pectin in tomato and cellulose in avocado. Secondary plant products are major compositional ingredients in fruits. Anthocyanins are the major color components in grapes, blueberries, apples, and plums; carotenoids, specifically lycopene and carotene, are the major components that impart color in tomatoes. Aroma is derived from several types of compounds that include monoterpenes (as in lime, orange), ester volatiles (ethyl, methyl butyrate in apple, isoamyl acetate in banana), simple organic acids such as citric and malic acids (citrus fruits, apple), and small-chain aldehydes such as hexenal and hexanal (cucumber). Fruits are also rich in vitamin C. Lipid content is quite low in fruits, the exceptions being avocado and olives, in which triacylglycerols (oils) form the major storage components. The amounts of proteins are usually low in most fruits.

Carbohydrates, Storage, and Structural Components

As the name implies, carbohydrates are organic compounds containing carbon, hydrogen, and oxygen. Basically, all carbohydrates are derived by the photosynthetic reduction of CO_2, and the hexoses (glucose, fructose) and pentoses (ribose, ribulose) that are intermediates in the pathway are further converted to several sugar monomers. Polymerization of several sugar derivatives leads to various storage (starch) and structural components (cellulose, pectin).

During photosynthesis, the glucose formed is converted to starch and stored as starch granules. Glucose and its isomer fructose, along with phosphorylated forms (glucose-6-phosphate, glucose-1,6-diphosphate, fructose-6-phosphate, and fructose-1,6-diphosphate) can be considered to be the major metabolic hexose pool components that provide a carbon skeleton for the synthesis of carbohydrate polymers. Starch is the major storage carbohydrate in fruits. There are two molecular forms of starch, amylose and amylopectin, and both components are present in the starch grain. Starch is synthesized from glucose phosphate by the activities of a number of enzymes designated as ADP-glucose pyrophosphorylase, starch synthase, and a starch branching enzyme. ADP-glucose pyrophosphorylase catalyzes the reaction between glucose-1-phosphate and ATP that generates ADP-glucose and pyrophosphate. ADP-glucose is used by starch synthase to add glucose molecules to amylose or amylopectin chains, thus increasing their degree of polymerization. By contrast to cellulose, which is made up of glucose units in β-1,4-glycosidic linkages, the starch molecule contains glucose linked by α-1,4-glycosidic linkages. The starch branching enzyme introduces glucose molecules through α-1,6 linkages to a linear amylose molecule. These added glucose branch points serve as sites for further elongation by starch synthase, thus resulting in a branched starch molecule, also known as amylopectin.

The cell wall is a complex structure composed of cellulose and pectin, derived from hexoses such as glucose, galactose, rhamnose, and mannose, and pentoses such as xylose and arabinose, as well as some of their derivatives such as glucuronic and galacturonic acids. A model proposed by Keegstra et al. (1973) describes the cell wall as a polymeric structure constituted by cellulose microfibrils and hemicellulose embedded in the apoplastic matrix in association with pectic components and proteins. In combination, these components provide the structural rigidity that is characteristic of the plant cell. Most of the pectin is localized in the middle lamella. Cellulose is biosynthesized by the action of β-1,4-glucan synthase enzyme complexes that are localized on the plasma membrane. The enzyme uses uridine diphosphate glucose (UDP-glucose) as a substrate, and by adding UDP-glucose units to small cellulose units, it extends the length and polymerization of the cellulose chain. In addition to cellulose, there are polymers made of different hexoses and pentoses known as hemicelluloses, and based on their composition, they are categorized as xyloglucans, glucomannans and galactoglucomannans. The cellulose chains assemble into microfibrils through hydrogen bonds to form crystalline structures. In a similar manner, pectin is biosynthesized from UDP-galacturonic acid (galacturonic acid is derived from galactose, a six-carbon sugar), as well as other sugars and derivatives and includes galacturonans and rhamnogalacturonans that form the acidic fraction of pectin. As the name implies, rhamnogalacturonans are synthesized primarily from galacturonic acid and rhamnose. The acidic carboxylic groups complex with calcium, which provide the rigidity to the cell wall and the fruit. The neutral fraction of the pectin comprises polymers such as arabinans (polymers of arabinose), galactans (polymers of galactose) or arabinogalactans (containing both arabinose and galactose). All these polymeric components form a complex three-dimensional network stabilized by hydrogen bonds, ionic interactions involving calcium, phenolic components such as diferulic acid, and hydroxyproline-rich glycoproteins (Fry 1986). It is also important to visualize that these structures are not static, and the components of cell wall are constantly being turned over in response to growth conditions.

Lipids and Biomembranes

By structure, lipids can form both structural and storage components. The major forms of lipids include fatty acids, diacyl- and triacylglycerols, phospholipids, sterols, and waxes that provide an external barrier to the fruits. Fruits in general are not rich in lipids, with the exception of avocado and olives, which store large amounts of triacylglycerols or oil. As generally observed in plants, the major fatty acids in fruits include palmitic (16:0), stearic (18:0), oleic (18:1), linoleic (18:2), and linolenic (18:3) acids. Among these, oleic, linoleic and linolenic acids possess an increasing degree of unsaturation. Olive oil is rich in triacylglycerols containing the monounsaturated oleic acid and is considered as a healthy ingredient for human consumption.

Compartmentalization of cellular ingredients and ions is an essential characteristic of all life forms. The compartmentalization is achieved by biomembranes, formed by the assembly of phospholipids and several neutral lipids that include diacylglycerols and sterols, the major constituents of the biomembranes. Virtually all cellular structures include or are enclosed by biomembranes. The cytoplasm is surrounded by the plasma membrane; the biosynthetic and transport compartments such as the endoplasmic reticulum and the Golgi bodies form an integral network of membranes within the cell. Photosynthetic activity, which converts light energy into chemical energy, and respiration, which further converts chemical energy into more usable forms, occur on the thylakoid membrane matrix in the chloroplast and the cristae of mitochondria, respectively. All these membranes have their characteristic composition and enzyme complexes to perform their designated functions.

The major phospholipids that constitute the biomembranes include phosphatidylcholine, phosphatidylethanolamine, phosphatidylglycerol, and phosphatidylinositol. Their relative proportion may vary from tissue to tissue. In addition, metabolic intermediates of phospholipids such as phosphatidic acid, diacylglycerols, free fatty acids, and so on, are also present in the membrane in lower amounts. Phospholipids are integral functional components of hormonal and environmental signal transduction processes in the cell. Phosphorylated forms of phosphatidylinositol such as phosphatidylinositol-4-phosphate and phosphatidylinositol-4,5- bisphosphate are formed during signal transduction events, though their amounts can be very low. The membrane also contains sterols such as sitosterol, campesterol, and stigmasterol, as well as their glucosides, and they are extremely important for the regulation of membrane fluidity and function.

Biomembranes are bilamellar layers of phospholipids. The amphipathic nature of phospholipids, which have hydrophilic head groups (choline, ethanolamine, etc.) and hydrophobic fatty acyl chains, thermodynamically favors their assembly into bilamellar or micellar structures when they are exposed to an aqueous environment. In a biomembrane, the hydrophilic head groups are exposed to the external aqueous environment. The phospholipid composition between various fruits may differ, and within the same fruit, the inner and outer lamella of the membrane may have a different phospholipid composition. Such differences may cause changes in polarity between the outer and inner lamellae of the membrane and lead to the generation of a voltage across the membrane. These differences usually become operational during signal transduction events.

An essential characteristic of the membrane is its fluidity. The fluid-mosaic model of the membrane (Singer and Nicholson 1972) depicts the membrane as a planar matrix comprised of phospholipids and proteins. The proteins are embedded in the membrane bilayer (integral proteins) or are bound to the periphery (peripheral proteins). The nature of this interaction stems from the structure of the proteins. If the proteins have a much larger proportion of hydrophobic amino acids, they would tend to become embedded in the membrane bilayer. If the protein contains more hydrophilic amino acids, it may tend to prefer a more aqueous environment and thus remain as a peripheral protein. In addition, proteins may be covalently attached to phospholipids such as phosphatidylinositol. Proteins that remain in the cytosol may also become attached to the membrane in response to an increase in cytosolic calcium levels. The membrane is a highly dynamic entity. The semifluid nature of the membrane allows for the movement of phospholipids in the plane of the membrane and between the bilayers of the membrane. The proteins are also mobile within the plane of the membrane. However, this process is not always random, and it is regulated by the functional assembly of proteins into metabolons (photosynthetic units in thylakoid membrane, respiratory complexes in the mitochondria, cellulose synthase on plasma membrane, etc.), their interactions with the underlying cytoskeletal system (network of proteins such as actin and tubulin), and the fluidity of the membrane.

The maintenance of homeostasis (life processes) requires the maintenance of the integrity and function of discrete membrane compartments. This is essential for the compartmentalization of ions and metabolites, which may otherwise destroy the cell. For instance, calcium ions are highly compartmentalized within the cell. The concentration of calcium is maintained at millimolar levels within the cell wall compartment (apoplast), endoplasmic reticulum, and the tonoplast (vacuole). This is achieved by energy-dependent extrusion of calcium from the cytoplasm into these compartments by ATPases. As a result, the cytosolic calcium levels are maintained at low micromolar ($< 1\ \mu M$) levels. Maintenance of this concentration gradient across the membrane is a key requirement for the signal transduction events, as regulated entry of calcium into the cytosol can be achieved simply by opening calcium channels. Calcium can then activate several cellular biochemical reactions that mediate the response to the signal. Calcium is pumped back into the storage compartments when the signal diminishes in intensity. In a similar manner, cytosolic pH is highly regulated by the activity of proton ATPases. The pH of the apoplast and the vacuole is maintained near 4, whereas the pH of the cytosol is maintained in the range of 6–6.5. The pH gradient across the membrane is a key feature that regulates the absorption or extrusion of other ions and metabolites such as sugars. The cell could undergo senescence if this compartmentalization is lost.

There are several factors that affect the fluidity of the membrane. The major factor that affects the fluidity is the type and proportion of acyl chain fatty acids of the phospholipids. At a given temperature, a higher proportion of unsaturated fatty acyl chains (oleic, linoleic, linolenic) in the phospholipids can increase the fluidity of the membrane. An increase in saturated fatty acids such as palmitic and stearic acids can decrease the fluidity. Other membrane components such as sterols, and degradation products of fatty acids such as fatty aldehydes, alkanes and so on, can also decrease the fluidity. Based on the physiological status of the tissue, the membrane can exist in either a liquid crystalline state (where the phospholipids and their acyl chains are mobile) or a gel state (where they are packed as rigid ordered structures and their movements are much restricted). The membrane usually has coexisting domains of liquid crystalline and gel phase lipids, depending on growth conditions, temperature, ion concentration near the membrane surface, and so on. The tissue has the ability to adjust the fluidity of the membrane by altering the acyl lipid composition of the phospholipids. For instance, an increase in the gel phase lipid domains resulting from exposure to cold temperature could be counteracted by increasing the proportion of fatty acyl chains having a higher degree of unsaturation, and therefore a lower melting point. Thus, the membrane will tend to remain fluid even at a lower temperature. An increase in gel phase lipid domains can result in the loss of compartmentalization. The differences in the mobility properties of phospholipid acyl chains can cause packing imperfections at the interface between gel and liquid crystalline phases, and these regions can become leaky to calcium ions and protons that are highly compartmentalized. The membrane proteins are also excluded from the gel phase into the liquid crystalline phase. Thus, during examinations of membrane structure by freeze fracture electron microscopy, the gel phase domains can appear as regions devoid of proteins (Paliyath and Thompson 1990).

Proteins

Fruits in general are not very rich sources of proteins. During the early growth phase of fruits, the chloroplasts and mitochondria are the major organelles that contain structural proteins. The structural proteins include the light-harvesting complexes in chloroplast or the respiratory enzyme/protein complexes in mitochondria. Ribulosebisphosphate carboxylase/oxygenase (rubisco) is the most abundant enzyme in photosynthetic tissues. Fruits do not store proteins as an energy source. Green fruits such as bell peppers and tomato have a higher level of chloroplast proteins.

Organic Acids

Organic acids are major components of fruits. The acidity of fruits arises from the organic acids that are stored in the vacuole, and their composition can vary depending on the type of fruit. In general, young fruits contain more acids, which may decline during maturation and ripening due to their conversion to sugars (gluconeogenesis). Some fruit families are characterized by the presence of certain organic acids. For example, fruits of Oxalidaceae members (e.g., Starfruit, *Averrhoa carambola*) contain oxalic acid, and fruits of the citrus family, Rutaceae, are rich in citric acid. Apples contain malic acid, and grapes are characterized by the presence of tartaric acid. In general, citric and malic acids are the major organic acids of fruits. Grapes contain tartaric acid as the major organic acid. During ripening, these acids can enter the citric acid cycle and undergo further metabolic conversions.

L-(+)-tartaric acid is the optically active form of tartaric acid in grape berries. A peak in acid content is observed before the initiation of ripening, and the acid content declines on a fresh weight basis during ripening. Tartaric acid can be biosynthesized from carbohydrates and other organic acids. Radiolabeled glucose, glycolate, and ascorbate were all converted to tartarate in grape berries. Malate can be derived from the citric acid cycle or through carbon dioxide fixation of pyruvate by the malic enzyme (NADPH-dependent malate dehydrogenase). Malic acid, as the name implies, is also the major organic acid in apples.

FRUIT RIPENING AND SOFTENING

Fruit ripening is the physiological repercussion of very complex and interrelated biochemical changes that occur in the fruits. Ripening is the ultimate stage of the development of the fruit, which entails the development of ideal organoleptic characters such as taste, color, and aroma that are important features of attraction for the vectors (animals, birds, etc.) responsible for the dispersal of the fruit, and thus the seeds, in the ecosystem. Human beings have developed an agronomic system of cultivation, harvest, and storage of fruits with ideal food qualities. In most cases, the ripening process is very fast, and the fruits undergo senescence, resulting in the loss of desirable qualities. An understanding of the biochemistry and molecular biology of the fruit ripening process has resulted in development of biotechnological strategies for the preservation of postharvest shelf life and quality of fruits.

A key initiator of the ripening process is the gaseous plant hormone ethylene. In general, all plant tissues produce a low, basal level of ethylene. Based on the pattern of ethylene production and responsiveness to externally added ethylene, fruits are generally categorized into climacteric and nonclimacteric fruits. During ripening, the climacteric fruits show a burst in ethylene production and respiration (CO_2 production). Nonclimacteric fruits show a considerably low level of ethylene production. In climacteric fruits (apple, pear, banana, tomato, avocado, etc.), ethylene production can reach levels of 30–500 ppm (parts per million, microliters/liter), whereas in nonclimacteric fruits (orange, lemon, strawberry, pineapple, etc.) ethylene levels usually are in the range of 0.1–0.5 ppm. Ethylene can stimulate its own biosynthesis in climacteric fruits (autocatalytic ethylene production). As well, the respiratory carbon dioxide evolution increases in response to ethylene treatment (the respiratory climacteric). Climacteric fruits respond to external ethylene treatment by accelerating the respiratory climacteric and time required for ripening, in a concentration-dependent manner. Nonclimacteric fruits show increased respiration in response to increasing ethylene concentration without accelerating the time required for ripening.

Ethylene is biosynthesized through a common pathway that uses the amino acid methionine as the precursor (Yang 1981, Fluhr and Mattoo 1996; Fig. 21.1). The first reaction of the pathway involves the conversion of methionine to S-adenosyl methionine (SAM) mediated by the enzyme methionine adenosyl transferase. SAM is further converted into 1-aminocyclopropane-1-carboxylic acid (ACC) by the enzyme ACC synthase. The sulphur moiety of methylthioribose generated during this reaction is recycled back to methionine by the action of a number of enzymes. ACC is the immediate precursor of ethylene and is acted upon by ACC oxidase to generate ethylene. ACC synthase and ACC oxidase are the key control points in the biosynthesis of ethylene. ACC synthase is a soluble enzyme located in the cytoplasm, with a relative molecular mass of 50 kDa. ACC oxidase is found to be associated with the vacuolar or mitochondrial membrane. Using molecular biology tools, a cDNA (complementary DNA representing the coding sequences of a gene) for ACC oxidase was isolated from tomato (Hamilton et al. 1991) and is found to encode a protein with a relative molecular mass of 35 kDa. There are several isoforms of ACC synthase. These are differentially expressed in response to wounding, other stress factors, and the initiation of ripening. ACC oxidase reaction requires Fe^{2+}, ascorbate, and oxygen.

Regulation of the activities of ACC synthase and ACC oxidase is extremely important for the preservation of the shelf life and quality in fruits. Inhibition of the ACC synthase and ACC oxidase gene expression by the introduction of their respective antisense cDNAs resulted in delayed ripening and better preservation of the quality of tomato (Hamilton et al. 1990, Oeller et al. 1991) and apple (Hrazdina et. al. 2000) fruits. ACC synthase, which is the rate-limiting enzyme of the pathway, requires pyridoxal-5-phosphate as a cofactor and is inhibited by pyridoxal phosphate inhibitors such as aminoethoxyvinylglycine (AVG) and aminooxy acetic acid (AOA). Field application of AVG as a growth regulator (Retain,

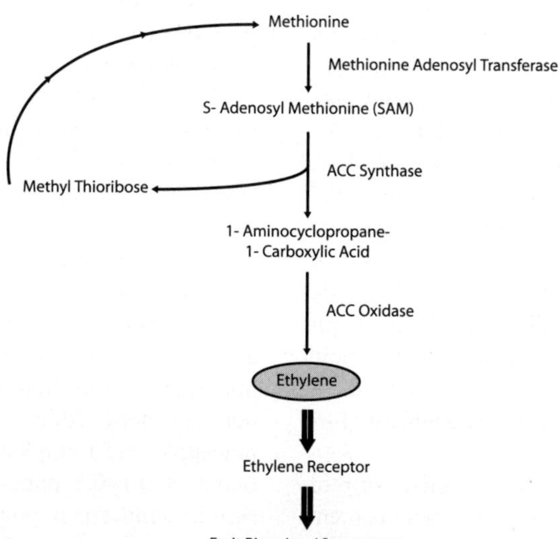

Figure 21.1. Ethylene biosynthetic pathway in plants. Key: ACC synthase, aminocyclopropane carboxylic acid synthase; ACC oxidase, aminocyclopropane carboxylic acid oxidase.

Abbot Laboratories) has been used to delay ripening in fruits such as apples, peaches, and pears. Also, commercial storage operations employ a controlled atmosphere with very low oxygen levels (1–3%) for long-term storage of fruits such as apples to reduce the production of ethylene, as oxygen is required for the conversion of ACC to ethylene.

In response to the initiation of ripening, several biochemical changes are induced in the fruit, which ultimately results in the development of ideal texture, taste, color, and flavor. Several biochemical pathways are involved in these processes, as described below.

Carbohydrate Metabolism

Cell Wall Degradation

Cell wall degradation is the major factor that causes softening of several fruits. This involves the degradation of cellulose components, pectin components, or both. Cellulose is degraded by the enzyme cellulase or β-1,4-glucanase. Pectin degradation involves the enzymes pectin methylesterase, polygalacturonase (pectinase), and β-galactosidase. The degradation of cell wall can be reduced by the application of calcium as a spray or drench in apple fruits. Calcium binds and cross-links the free carboxylic groups of polygalacturonic acid components in pectin. Calcium treatment therefore also enhances the firmness of the fruits.

The activities of both cellulase and pectinase have been observed to increase during ripening of avocado fruits and to result in their softening. Cellulase is an enzyme with a relative molecular mass of 54.2 kDa and is formed by extensive posttranslational processing of a native 54 kDa protein involving proteolytic cleavage of the signal peptide and glycosylation (Bennet and Christofferson 1986). Further studies have shown three isoforms of cellulose, ranging in molecular mass between 50 and 55 kDa. These forms are associated with the endoplasmic reticulum, the plasma membrane, and the cell wall (Dallman et al. 1989). The cellulase isoforms are initially synthesized at the style end of the fruit at the initiation of ripening, and the biosynthesis moves towards the stalk end of the fruit with the advancement of ripening. Degradation of hemicelluloses (xyloglucans, glucomannans, and galactoglucomannans) is also considered as an important feature that leads to fruit softening. Degradation of these polymers could be achieved by cellulases and galactosidases.

Loss of pectic polymers through the activity of polygalacturonases (PG) is a major factor involved in the softening of fruits such as tomato. There are three major isoforms of polygalacturonases responsible for pectin degradation in tomato, designated as PG1, PG2a, and PG2b (Fischer and Bennet 1991). PG1 has a relative molecular mass of 100 kDa and is the predominant form at the initiation of ripening. With the advancement of ripening, PG2a and PG2b isoforms increase, becoming the predominant isoforms in ripe fruit. The different molecular masses of the isozymes result from the posttranslational processing and glycosylation of the polypeptides. PG2a (43 kDa) and PG2b (45 kDa) appear to be the same polypeptide with different degrees of glycosylation. PG1 is a complex of three polypeptides, PG2a, PG2b, and a 38 kDa subunit known as the β-subunit. The 38 kDa subunit is believed to exist in the cell wall space, where it combines with PG2a and PG2b, forming the PG1 isoform. The increase in activity of PG1 is related to the rate of pectin solubilization and tomato fruit softening during the ripening process.

Research into the understanding of the regulation of biosynthesis and activity of PG using molecular biology tools has resulted in the development of strategies for enhancing the shelf life and quality of tomatoes. Polygalacturonase mRNA was one of the first ripening-related mRNAs isolated from tomato fruits. All the different PG isoforms are encoded by a single gene. The PG cDNA, which has an open reading frame of 1371 bases, encodes a polypeptide having 457 amino acids, including a 24 amino acid signal sequence (for targeting to the cell wall space) and a 47 amino acid prosequence at the N-terminal end, which are proteolytically removed during the formation of the active PG isoforms. A 13 amino acid C-terminal peptide is also removed, resulting in a 373 amino acid long polypeptide that undergoes different degrees of glycosylation to result in the PG2a and PG2b isozymes. Complex formation among PG2a, PG2b, and the 38 kDa subunit in the apoplast results in the PG1 isozyme (Grierson et al. 1986, Bird et al, 1988). In response to ethylene treatment of mature green tomato fruits that stimulates ripening, the levels of PG mRNA and PG are found to increase. These changes can be inhibited by treating tomatoes with silver ions, which interfere with the binding of ethylene to its receptor and with initiation of ethylene action (Davies et al. 1988). Thus, there is a link between ethylene, PG synthesis, and fruit softening.

Genetic engineering of tomato with the objective of regulating PG activity has yielded complex results. In the rin mutant of tomato, which lacks PG and does not soften, introduction of a PG gene resulted in the synthesis of an active enzyme; however, this did not cause fruit softening (Giovannoni et al. 1989). As a corollary to this, introduction of the PG gene in the antisense orientation resulted in near total inhibition of PG activity (Smith et al. 1988). In both these cases, there was very little effect on fruit softening, suggesting that factors other than pectin depolymerization may play an integral role in fruit softening. Further studies using tomato cultivars such as UC82B (Kramer et al. 1992) showed that antisense inhibition of ethylene biosynthesis or PG did indeed result in lowered PG activity, improved integrity of cell wall, and increased fruit firmness during fruit ripening. As well, increased activity of pectin methylesterase, which removes the methyl groups from esterified galacturonic acid moieties, may contribute to the fruit softening process.

The activities of pectin-degrading enzymes have been related to the incidence of physiological disorders such as "mealiness" or "wooliness" in mature, unripe peaches that are stored at a low temperature. The fruits with such a disorder show a lack of juice and a dry texture. Deesterification of pectin by the activity of pectin methyl esterase is thought to be responsible for the development of this disorder. Pectin methyl esterase isozymes with relative molecular masses in the range of 32 kDa have been observed in peaches, and their activity increases after 2 weeks of low temperature storage. Polygalacturonase activity increases as the fruit ripens. The ripening fruits, which possess both polygalacturonase and pectin methyl esterase do not develop mealy symptoms when stored at low temperature, implicating the potential role of pectin degradation in the development of mealiness in peaches.

There are two forms of polygalacturonase in peaches, the exo- and endopolygalacturonases. The endopolygalacturonases (endo-PG) are the predominant forms in the freestone type of peaches, whereas the exopolygalacturonases (exo-PG) are observed in the mesocarp of both freestone and clingstone varieties of peach. As the name implies, exopolygalacturonases remove galacturonic acid, moieties of pectin from the terminal reducing end of the chain, whereas the endopolygalacturonases can cleave the pectin chain at random within the chain. The activities of these enzymes increase during the ripening and softening of the fruit. Two exo-PG isozymes having a relative molecular mass of near 66 kDa have been identified in peach. The exo enzymes are activated by calcium. Peach endo-PG is observed to be similar to the tomato endo-PG. The peach endo-PG is inhibited by calcium. Freestone peaches possess enhanced activities of both exo-PG and endo-PG, leading to a high degree of fruit softening. However, the clingstone varieties with low levels of endo-PG activity do not soften as do the freestone varieties. In general, fruits such as peaches, tomatoes, strawberries, and pears that soften extensively possess high levels of endo-PG activity. Apple fruits that remain firm lack endo-PG activity.

Starch Degradation

Starch is the major storage form of carbohydrates. During ripening, starch is catabolized into glucose and fructose, which enter the metabolic pool, where they are used as respiratory substrates or further converted to other metabolites (Fig. 21.2). In fruits such as banana, the breakdown of starch into simple sugars is associated with fruit softening. There are several enzymes involved in the catabolism of starch. α-amylase hydrolyzes amylose molecules by cleaving the α-1,4 linkages between sugars to provide smaller chains of amylose, termed dextrins. β-amylase, another enzyme that acts upon the glucan chain, releases maltose, a diglucoside. The dextrins and maltose can be further catabolized to simple glucose units by the action of glucosidases. Another enzyme, starch phosphorylase, mediates the phosphorylytic cleavage of terminal glucose units at the nonreducing end of the starch molecule using inorganic phosphate, thus releasing glucose-1-phosphate. The amylopectin molecule is degraded in a similar manner to amylose, but its degradation also involves the action of debranching enzymes that cleave the α-1,6 linkages in amylopectin and release linear units of the glucan chain.

In general, starch is confined to the plastid compartments of fruit cells, where it exists as granules made up of both amylose and amylopectin molecules.

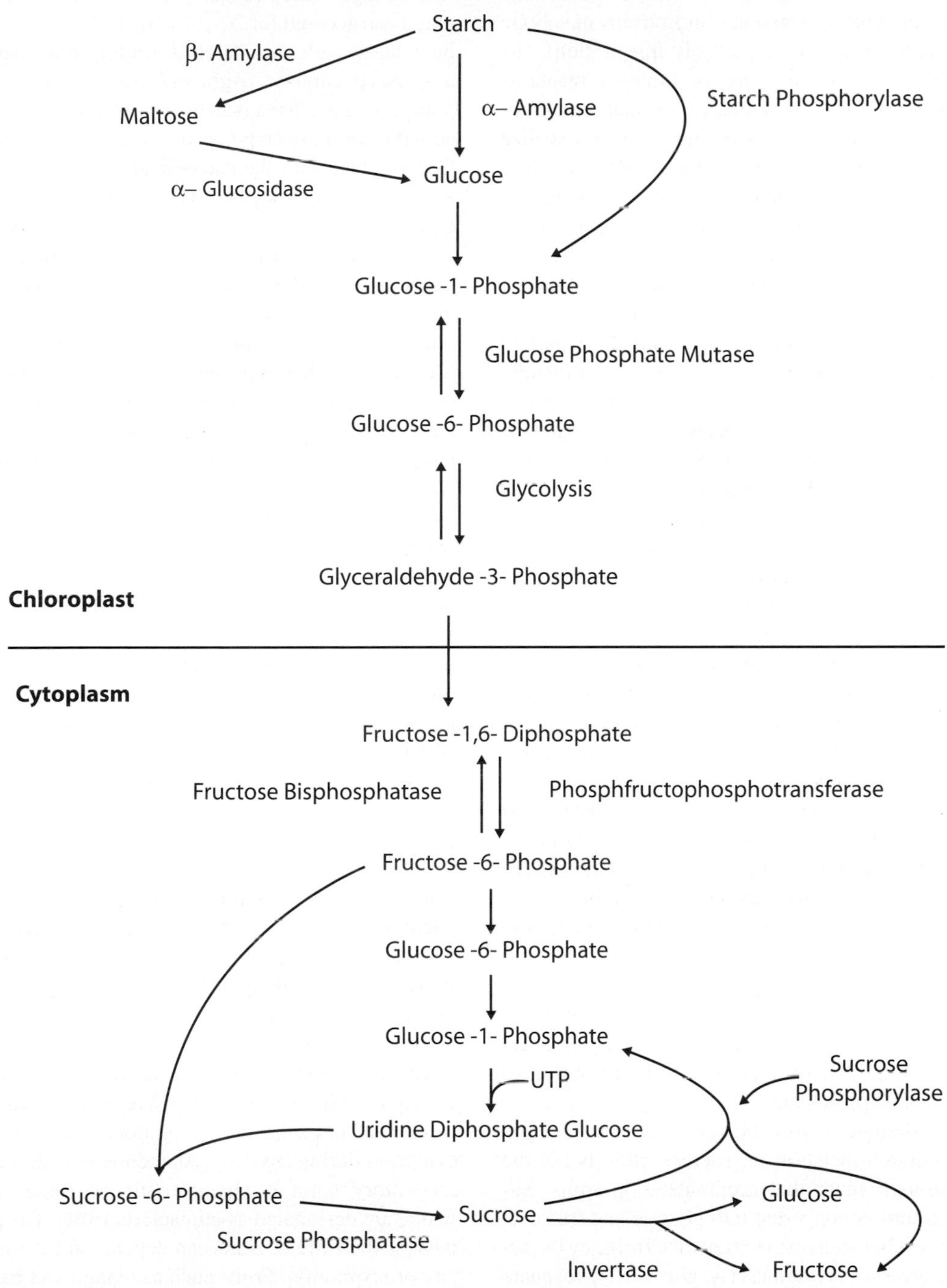

Figure 21.2. Starch-sugar interconversions in plants and metabolite transfer from chloroplast to the cytoplasm.

The enzymes that catabolize starch are also found in this compartment, and their activities increase during ripening. The glucose-1-phosphate generated by starch degradation (Fig. 21.2) is mobilized into the cytoplasm, where it can enter into various metabolic pools such as that of glycolysis (respiration), the pentose phosphate pathway, or turnover reactions that replenish lost or damaged cellular structures (cell wall components). It is important to visualize that the cell always tries to extend its life under regular developmental conditions (the exceptions are programmed cell death, which occurs during a hypersensitive response to kill invading pathogens, thus killing both the pathogen and the cell/tissue; formation of xylem vessels, secondary xylem tissues, etc.), and the turnover reactions are a part of maintaining the homeostasis. The cell ultimately succumbs to the catabolic reactions during senescence. The compartmentalization and storage of chemical energy in the form of metabolizable macromolecules are inherent properties of life, which is defined as a struggle against increasing entropy.

The biosynthesis and catabolism of sucrose is an important part of carbohydrate metabolism. Sucrose is the major form of transport sugar and is translocated through the phloem tissues to other parts of the plant. It is conceivable that photosynthetically fixed carbon from leaf tissues may be transported to the fruits as sucrose during fruit development. Sucrose is biosynthesized from glucose-1-phosphate by three major steps (Fig. 21.3). The first reaction involves the conversion of glucose-1-phosphate to UDP-glucose, by UDP-glucose pyrophosphorylase in the presence of UTP (uridine triphosphate). UDP-glucose is also an important substrate for the biosynthesis of cell wall components such as cellulose. UDP-glucose is converted to sucrose-6-phosphate by the enzyme sucrose phosphate synthase, which utilizes fructose-6-phosphate during this reaction. Finally, sucrose is formed from sucrose-6-phosphate by the action of phosphatase with the liberation of the inorganic phosphate.

Even though sucrose biosynthesis is an integral part of starch metabolism, sucrose often is not the predominant sugar that accumulates in fruits. Sucrose is further converted into glucose and fructose, which are characteristic to many ripe fruits, by the action of invertase. Alternatively, glucose-1-phosphate can be regenerated from sucrose by the actions of sucrose synthase and UDP-glucose pyrophosphorylase. As well, sugar alcohols such as sorbitol and mannitol are major transport and storage components in apple and olive, respectively.

Biosynthesis and catabolism of starch has been extensively studied in banana, where prior to ripening, it can account for 20–25% by fresh weight of the pulp tissue. All the starch-degrading enzymes, α-amylase, β-amylase, α-glucosidase, and starch phosphorylase, have been isolated from banana pulp. The activities of these enzymes increase during ripening. Concomitant with the catabolism of starch, there is an accumulation of the sugars, primarily sucrose, glucose, and fructose. At the initiation of ripening, sucrose appears to be the major sugar component, which declines during the advancement of ripening, with a simultaneous increase in glucose and fructose through the action of invertase (Beaudry et al. 1989). Mango is another fruit that stores large amounts of starch. The starch is degraded by the activities of amylases during the ripening process. In mango, glucose, fructose, and sucrose are the major forms of simple sugars (Selvaraj et al. 1989). The sugar content is generally very high in ripe mangoes and can reach levels in excess of 90% of the total soluble solids content. Fructose is the predominant sugar in mangoes. In contrast to the bananas, the sucrose levels increase with the advancement of ripening in mangoes, potentially due to gluconeogenesis from organic acids (Kumar and Selvaraj 1990). As well, the levels of pentose sugars increase during ripening, which could be related to an increase in the activity of the pentose phosphate pathway.

Glycolysis

The conversion of starch to sugars and their subsequent metabolism occur in different compartments. During the development of fruits, photosynthetically fixed carbon is utilized for both respiration and biosynthesis. During this phase, the biosynthetic processes dominate. As the fruit matures and begin to ripen, the pattern of sugar utilization changes. Ripening is a highly energy intensive process. And this is reflected in the burst of respiratory carbon dioxide evolution during ripening. As mentioned earlier, the respiratory burst is characteristic of some fruits, which are designated as climacteric fruits. The postharvest shelf life of fruits can depend on their intensity of respiration. Fruits such as mango and banana possess high levels of respiratory activity and are highly perishable. The application of controlled atmosphere conditions having low oxygen levels

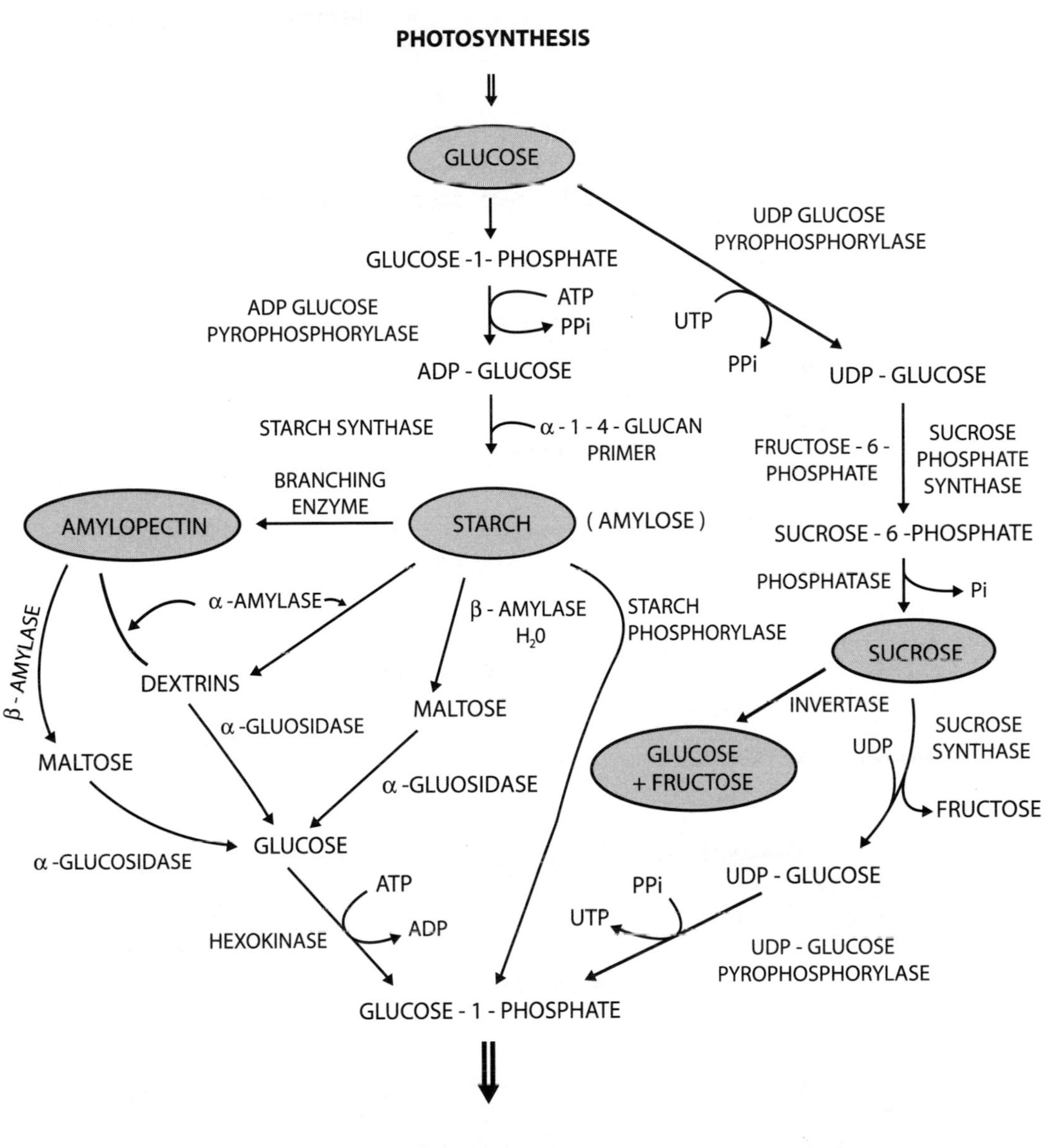

Figure 21.3. Carbohydrate metabolism in fruits.

and low temperature have thus become a routine technology for the long-term preservation of fruits.

The sugars and sugar phosphates generated during the catabolism of starch are metabolized through glycolysis and the citric acid cycle (Fig. 21.4). Sugar phosphates can also be channeled through the pentose phosphate pathway, a major metabolic cycle that provides reducing power for biosynthetic

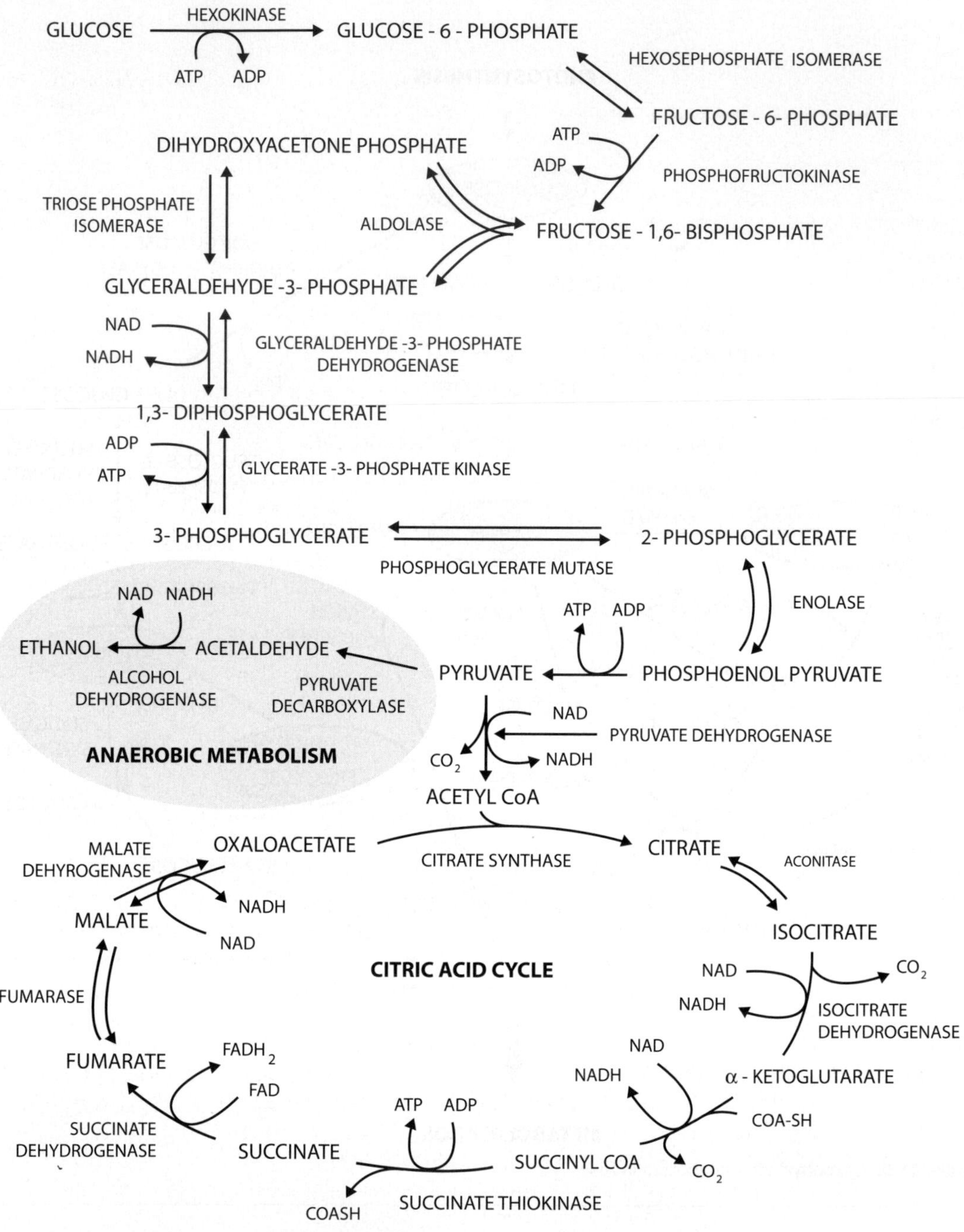

Figure 21.4. Catabolism of sugars through glycolytic pathway and citric acid cycle.

reactions in the form of NADPH (nicotinamide adenine dinucleotide phosphate, reduced form), and supplies carbon skeletons for the biosynthesis of several secondary plant products. The organic acids stored in the vacuole are metabolized through the functional reversal of the respiratory pathway; this process is termed gluconeogenesis. Altogether, sugar metabolism is a key biochemical characteristic of the fruits.

In the glycolytic steps of reactions (Fig. 21.4), glucose-6-phosphate is isomerized to fructose-6-phosphate by the enzyme hexose phosphate isomerase. Glucose-6-phosphate is derived from glucose-1-phosphate by the action of glucose phosphate mutase. Fructose-6-phosphate is phosphorylated at the C1 position, yielding fructose-1,6-bisphosphate. This reaction is catalyzed by the enzyme phosphofructokinase in the presence of ATP. Fructose-1, 6-bisphosphate is further cleaved into two, three-carbon intermediates, dihydroxyacetone phosphate and glyceraldehyde-3-phosphate, catalyzed by the enzyme aldolase. These two compounds are interconvertible through an isomerization reaction mediated by triose phosphate isomerase. Glyceraldehyde-3-phosphate is subsequently phosphorylated at the C1 position using orthophosphate, and oxidized using NAD, to generate 1,3-diphosphoglycerate and NADH. In the next reaction, 1,3-diphosphoglycerate is dephosphorylated by glycerate-3-phosphate kinase in the presence of ADP, along with the formation of ATP. Glycerate-3-phosphate formed during this reaction is further isomerized to 2-phosphoglycerate in the presence of phosphoglycerate mutase. In the presence of the enzyme enolase, 2-phosphoglycerate is converted to phosphoenol pyruvate (PEP). Dephosphorylation of PEP in the presence of ADP by pyruvate kinase yields pyruvate and ATP. The metabolic fate of pyruvate is highly regulated. Under normal conditions, it is converted to acetyl-CoA, which then enters the citric acid cycle. Under anaerobic conditions, pyruvate can be metabolized to ethanol, which is a by-product in several ripening fruits.

There are two key regulatory steps in glycolysis, one mediated by phosphofructokinase (PFK) and the other by pyruvate kinase. In addition, there are other types of modulation involving cofactors and enzyme structural changes reported to be involved in glycolytic control. ATP levels increase during ripening. However, in fruits, this does not cause a feedback inhibition of PFK like that observed in animal systems. There are two isozymes of PFK in plants, one localized in plastids and the other in the cytoplasm. These isozymes regulate the flow of carbon from the hexose phosphate pool to the pentose phosphate pool. PFK isozymes are strongly inhibited by phosphoenol pyruvate (PEP). Thus, any conditions that may cause the accumulation of PEP will tend to reduce the carbon flow through glycolysis. By contrast, inorganic phosphate is a strong activator of PFK. Thus, the ratio of PEP to inorganic phosphate would appear to be the major factor that regulates the activity of PFK and carbon flux through glycolysis. Structural alteration of PFK that increases the efficiency of utilization of fructose-6-phosphate is another means of regulation that can activate the carbon flow through the glycolytic pathway.

Other enzymes of the glycolytic pathway are involved in the regulation of starch/sucrose biosynthesis (see Figs. 21.2 and 21.3). Fructose-1,6-bisphosphate is converted back to fructose-6-phosphate by the enzyme fructose-1,6-bisphosphatase, also releasing inorganic phosphate. This enzyme is localized in the cytosol and the chloroplast. Fructose-6-phosphate is converted to fructose-2,6-bisphosphate by fructose-6-phosphate 2-kinase, which can be dephosphorylated at the 2 position by fructose-2,6-bisphosphatase. Fructose-6-phosphate is an intermediary in sucrose biosynthesis (Fig. 21.3). Sucrose phosphate synthase (SPS) is regulated by reversible phosphorylation (a form of posttranslational modification that involves addition of a phosphate moiety from ATP to an OH-amino acid residue in the protein, such as serine or threonine, mediated by a kinase; and dephosphorylation mediated by a phosphatase) by SPS kinase and SPS phosphatase. Phosphorylation of the enzyme makes it less active. Glucose-6-phosphate is an allosteric activator (a molecule that can bind to an enzyme and increase its activity through enzyme subunit association) of the active form of SPS (dephosphorylated). Glucose-6-phosphate is an inhibitor of SPS kinase, and inorganic phosphate is an inhibitor of SPS phosphatase. Thus, under conditions when glucose-6-phosphate/inorganic phosphate ratio is high, the active form of SPS will dominate, favoring sucrose phosphate biosynthesis. These regulations are highly complex and may be regulated by the flux of other sugars in several pathways.

The conversion of PEP to pyruvate, mediated by pyruvate kinase, is another key metabolic step in the glycolytic pathway and is irreversible. Pyruvate is used in several metabolic reactions. During respiration, pyruvate is further converted to acetyl coenzyme A (acetyl-CoA), which enters the citric acid cycle, through which it is completely oxidized to carbon dioxide (Fig. 21.3). The conversion of pyruvate to acetyl-CoA is mediated by the enzyme complex pyruvate dehydrogenase, in an oxidative step that involves the formation of NADH from NAD. Acetyl-CoA is a key metabolite and starting point for several biosynthetic reactions (fatty acids, isoprenoids, phenylpropanoids, etc.).

Citric Acid Cycle

The citric acid cycle involves the biosynthesis of several organic acids, many of which serve as precursors for the biosynthesis of several groups of amino acids. In the first reaction, oxaloacetate combines with acetyl-CoA, mediated by citrate synthase, to form citrate (Fig. 21.4). In the next step, citrate is converted to isocitrate by the action of aconitase. The next two steps in the cycle involve oxidative decarboxylation. The conversion of isocitrate to α-ketoglutarate involves the removal of a carbon dioxide molecule and reduction of NAD to NADH. This step is catalyzed by isocitrate dehydrogenase. α-ketoglutarate is converted to succinyl-CoA by α-ketoglutarate dehydrogenase, with the removal of another molecule of carbon dioxide and the conversion of NAD to NADH. Succinate, the next product, is formed from succinyl-CoA by the action of succinyl-CoA synthetase, which involves the removal of the CoA moiety and the conversion of ADP to ATP. Through these steps, the complete oxidation of the acetyl-CoA moiety has been achieved, with the removal of two molecules of carbon dioxide. Thus, succinate is a four-carbon organic acid. Succinate is further converted to fumarate and malate in the presence of succinate dehydrogenase and fumarase, respectively. Malate is oxidized to oxaloacetate by the enzyme malate dehydrogenase, with the conversion of NAD to NADH. Oxaloacetate, then can combine with another molecule of acetyl-CoA to repeat the cycle. The reducing power generated in the form of NADH and FADH (flavin adenine dinucleotide, reduced form; succinate dehydrogenation step) is used for the biosynthesis of ATP through the transport of electrons through the electron transport chain in the mitochondria.

Gluconeogenesis

Several fruits store large amounts of organic acids in their vacuoles, and these acids are converted back to sugars during ripening, a process called gluconeogenesis. Several irreversible steps in the glycolysis and citric acid cycle are bypassed during gluconeogenesis. Malate and citrate are the major organic acids present in fruits. In fruits such as grapes, where there is a transition from a sour to a sweet stage during ripening, organic acids content declines. Grape contains predominantly tartaric acid along with malate, citrate, succinate, fumarate, and several organic acid intermediates of metabolism. The content of organic acids in berries can affect their suitability for processing. High acid content coupled with low sugar content can result in poor quality wines. External warm growth conditions enhance the metabolism of malic acid in grapes during ripening and could result in a high tartarate/malate ratio, which is considered ideal for vinification.

The metabolism of malate during ripening is mediated by the malic enzyme, NADP-dependent malate dehydrogenase. With the decline in malate content, there is a concomitant increase in sugars, suggesting a possible metabolic precursor product relationship between these two events. Indeed, when grape berries were fed with radiolabeled malate, the radiolabel could be recovered in glucose. The metabolism of malate involves its conversion to oxaloacetate mediated by malate dehydrogenase, the decarboxylation of oxaloacetate to phosphoenol pyruvate catalyzed by PEP-carboxykinase, and a reversal of the glycolytic pathway, leading to sugar formation (Ruffner et al. 1983). The gluconeogenic pathway from malate may contribute only a small percentage (5%) of the sugars, and decrease in malate content could primarily result from reduced synthesis and increased catabolism through the citric acid cycle. The inhibition of malate synthesis by the inhibition of the glycolytic pathway could result in increased sugar accumulation. Metabolism of malate in apple fruits is catalyzed by the NADP-malic enzyme, which converts malate to pyruvate. In apples, malate appears to be primarily oxidized through the citric acid cycle. In citrus fruits, organic acids are important components; citric acid is the major form of acid, followed by malic acid and several less abundant acids such as acetate, pyruvate, oxalate, glutarate, fumarate, and others. In oranges, the acidity increases during maturation of the fruit and declines during the ripening phase. Lemon fruits, by contrast, increase their acid content through the accumulation of citrate. The citrate levels in various citrus fruits range from 75 to 88%, and malate levels range from 2 to 20%. Ascorbate is another major component of citrus fruits. Ascorbate levels can range from 20 to 60 mg/100 g juice in various citrus fruits. The orange skin may possess 150–340 mg/100 g fresh weight of ascorbate, which may not be extracted into the juice.

Anaerobic Respiration

Anaerobic respiration is a common event in the respiration of ripe fruits and especially becomes significant when fruits are exposed to low temperatures. Often, this may result from oxygen-depriving conditions induced inside the fruit. Under anoxia, ATP production through the citric acid cycle and the mitochondrial electron transport chain is inhibited. Anaerobic respiration is a means of regenerating NAD, which can drive the glycolyic pathway and produce minimal amounts of ATP (Fig. 21.4). Under anoxia, pyruvate formed through glycolysis is converted by lactate dehydrogenase to lactate, using NADH as the reducing factor and generating NAD. Accumulation of lactate in the cytosol could cause acidification, and under these low pH conditions, lactate dehydrogenase is inhibited. The formation of acetaldehyde by the decarboxylation of pyruvate is stimulated by the activation of pyruvate decarboxylase under low pH conditions in the cytosol. It is also likely that the increase in concentration of pyruvate in the cytoplasm may stimulate pyruvate decarboxylase directly. Acetaldehyde is reduced to ethanol by alcohol dehydrogenase using NADH as the reducing power. Thus, acetaldehyde and ethanol are common volatile components observed in the headspace of fruits, indicative of the occurrence of anaerobic respiration. Cytosolic acidification is a condition that stimulates deteriorative reactions. By removing lactate through efflux and converting pyruvate to ethanol, cytosolic acidification can be avoided.

Anaerobic respiration plays a significant role in the respiration of citrus fruits. During early stages of growth, respiratory activity predominantly occurs in the skin tissue. Oxygen uptake by the skin tissue was much higher than that of the juice vesicles (Purvis 1985). With advancing maturity, a decline in aerobic respiration and an increase in anaerobic respiration was observed in Hamlin orange skin (Bruemmer 1989). In parallel with this, the levels of ethanol and acetaldehyde increased. As well, a decrease in the organic acid substrates pyruvate and oxaloacetate was detectable in Hamlin orange juice. An increase in the activity levels of pyruvate decarboxylase, alcohol dehydrogenase, and malic enzyme was noticed in parallel with the decline in pyruvate and the accumulation of ethanol. In apple fruits, malic acid is converted to pyruvate by the action of NADP-malic enzyme, and pyruvate is subsequently converted to ethanol by the action of pyruvate decarboxylase and alcohol dehydrogenase. The alcohol dehydrogenase in apple can use NADPH as a cofactor, and NADP is regenerated during ethanol production, thus driving malate utilization. Ethanol is either released as a volatile or can be used for the biosynthesis of ethyl esters of volatiles.

Pentose Phosphate Pathway

Oxidative pentose phosphate pathway (PPP) is a key metabolic pathway that provides reducing power (NADPH) for biosynthetic reactions as well as carbon precursors for the biosynthesis of amino acids, nucleic acids, secondary plant products, and so forth. The PPP shares many of its sugar phosphate intermediates with the glycolytic pathway (Fig. 21.5). The PPP is characterized by the interconversion of sugar phosphates with three (glyceraldehyde-3-phosphate), four (erythrose-4-phosphate), five (ribulose, ribose, and xylulose phosphates), six (glucose-6-phosphate, fructose-6-phosphate), and seven (sedoheptulose-7-phosphate) carbons.

PPP involves the oxidation of glucose-6-phosphate, and the sugar phosphate intermediates formed are recycled. The first two reactions of PPP are oxidative reactions mediated by the enzymes glucose-6-phosphate dehydrogenase and 6-phosphogluconate dehydrogenase (Fig. 21.5). In the first step, glucose-6-phosphate is converted to 6-phosphogluconate by the removal of two hydrogen atoms by NADP to form NADPH. In the next step, 6-phosphogluconate, a six-carbon sugar acid phosphate, is converted to ribulose-5-phosphate, a five-carbon sugar phosphate. This reaction involves the removal of a carbon dioxide molecule along with the formation of NADPH. Ribulose-5-phosphate undergoes several metabolic conversions to yield fructose-6-phosphate. Fructose-6-phosphate can then be converted back to glucose-6-phosphate by the enzyme glucose-6-phosphate isomerase, and the cycle can be repeated. Thus, six complete turns of the cycle can result in the complete oxidation of a glucose molecule.

Despite the differences in the reaction sequences, the glycolytic pathway and the PPP intermediates

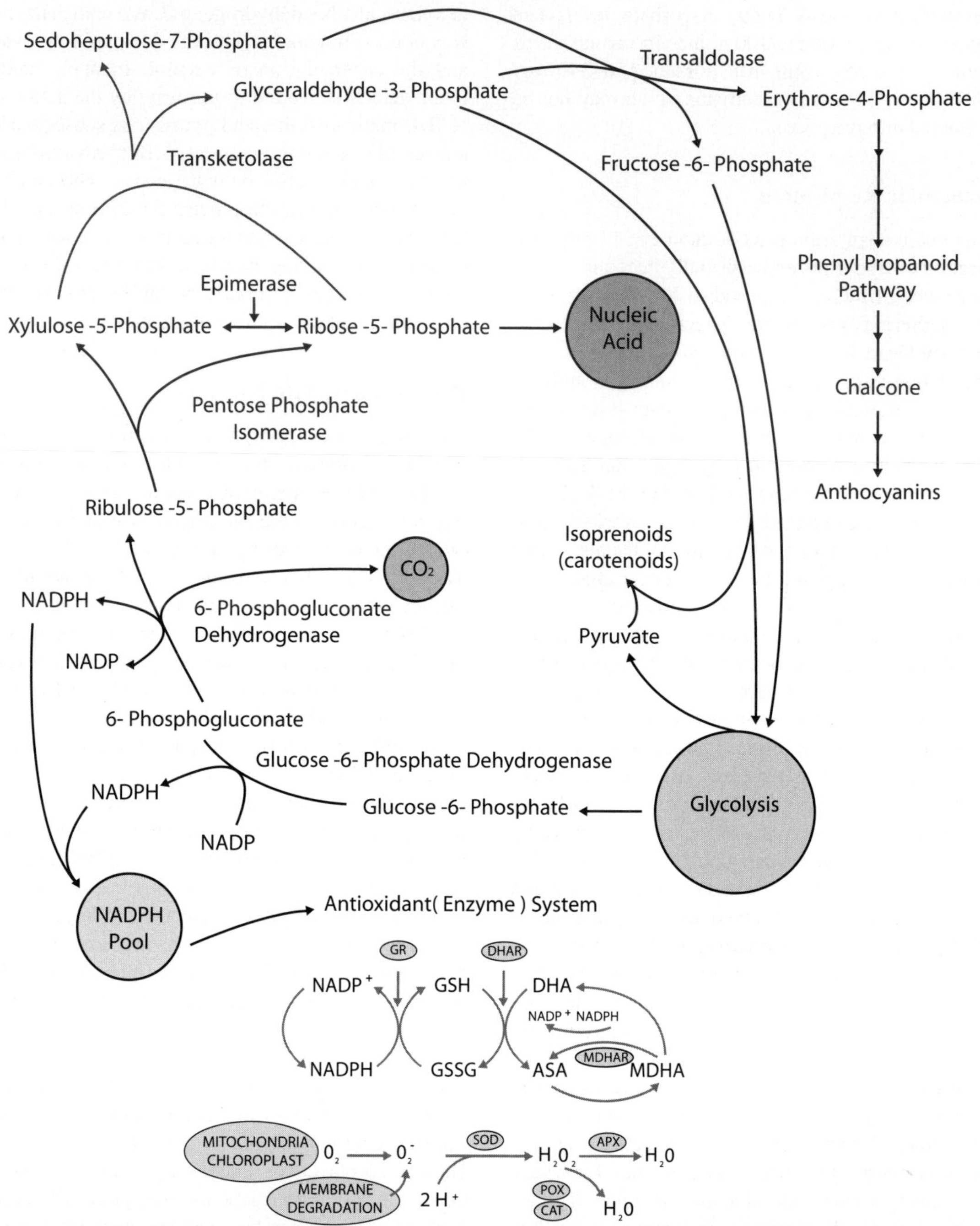

Figure 21.5. Oxidative pentose phosphate pathway in plants. NADPH generated from the pentose phosphate pathway is channeled into the antioxidant enzyme system, where the regeneration of oxidized intermediates requires NADPH. Key: GSH, reduced glutathione; GSSG, oxidized glutathione; ASA, reduced ascorbate; MDHA, monodehydroascorbate; DHA, dehydroascorbate; GR, glutathione reductase; DHAR, dehydroascorbate reductase; MDHAR, monodehydroascorbate reductase; SOD, superoxide dismutase; CAT, catalase; POX, peroxidase; APX, ascorbate peroxidase.

can interact with one another and share common intermediates. Intermediates of both the pathways are localized in plastids as well as the cytoplasm, and the intermediates can be transferred across the plastid membrane into the cytoplasm and back into the chloroplast. Glucose-6-phosphate dehydrogenase is localized both in the chloroplast and the cytoplasm. Cytosolic glucose-6-phosphate dehydrogenase activity is strongly inhibited by NADPH. Thus, the ratio of NADP to NADPH could be the regulatory control point for the enzyme function. The chloroplastic enzyme is regulated differently, through oxidation and reduction, and its regulation is related to the photosynthetic process. 6-Phosphogluconate dehydrogenase exists as distinct cytosol- and plastid-localized isozymes.

PPP is a key metabolic pathway related to the biosynthetic reactions, antioxidant enzyme function, and general stress tolerance of the fruits. Ribose-5-phosphate is used in the biosynthesis of nucleic acids, and erythrose-4-phosphate is channeled into phenyl propanoid pathway, leading to the biosynthesis of the amino acids phenylalanine and tryptophan. Phenylalanine is the metabolic starting point for the biosynthesis of flavonoids and anthocyanins in fruits. Glyceraldehyde-3-phosphate and pyruvate serve as the starting intermediates for the isoprenoid pathway localized in the chloroplast. Accumulation of sugars in fruits during ripening has been related to the function of PPP. In mangoes, the increase in the levels of pentose sugars observed during ripening has been related to increased activity of PPP. Increases in glucose-6-phosphate dehydrogenase and 6-phosphogluconate dehydrogenase activities were observed during ripening of mango.

NADPH is a key component required for the proper functioning of the antioxidant enzyme system (Fig. 21.5). During growth, stress conditions, fruit ripening, and senescence, free radicals are generated within the cell. Activated forms of oxygen, such as superoxide, hydroxyl, and peroxy radicals can attack enzymes and proteins, nucleic acids, lipids in the biomembrane, and so on, causing structural and functional alterations in these molecules. Under most conditions, these are deleterious changes, which are nullified by the action of antioxidants and antioxidant enzymes. Simple antioxidants such as ascorbate and vitamin E can scavenge the free radicals and protect the tissue. Anthocyanins and other polyphenols may also serve as simple antioxidants. In addition, the antioxidant enzyme system involves the integrated function of several enzymes. The key antioxidant enzymes are superoxide dismutase, catalase, ascorbate peroxidase, and peroxidase. Superoxide dismutase (SOD) converts superoxide into hydrogen peroxide. Hydrogen peroxide is immediately acted upon by catalase, generating water. Hydrogen peroxide can also be removed by the action of peroxidases. A peroxidase uses the oxidation of a substrate molecule (usually having a phenol structure, C—OH, which becomes a quinone, C=O, after the reaction) to react with hydrogen peroxide, converting it to water. Hydrogen peroxide can also be acted upon by ascorbate peroxidase, which uses ascorbate as the hydrogen donor for the reaction, resulting in water formation. The oxidized ascorbate is regenerated by the action of a series of enzymes (Fig. 21.5). These include monodehydroascorbate reductase (MDHAR), and dehydroascorbate reductase (DHAR). Dehydroascorbate is reduced to ascorbate using reduced glutathione (GSH) as a substrate, which itself gets oxidized (GSSG) during this reaction. The oxidized glutathione is reduced back to GSH by the activity of glutathione reductase using NADPH. Antioxidant enzymes exist as several functional isozymes with differing activities and kinetic properties in the same tissue. These enzymes are also compartmentalized in chloroplast, mitochondria, and cytoplasm. The functioning of the antioxidant enzyme system is crucial to the maintenance of fruit quality through preservation of cellular structure and function (Meir and Bramlage 1988, Ahn et al. 2002).

Lipid Metabolism

Among fruits, avocado and olive are the only fruits that significantly store reserves in the form of lipid triglycerides. In avocado, triglycerides form the major part of the neutral lipid fraction, which can account for nearly 95% of the total lipids. Palmitic (16:0), palmitoleic (16:1), oleic (18:1), and linoleic (18:2) acids are the major fatty acids of triglycerides. The oil content progressively increases during maturation of the fruit, and the oils are compartmentalized in oil bodies or oleosomes. The biosynthesis of fatty acids occurs in the plastids, and the fatty acids are exported into the endoplasmic reticulum, where they are esterified with glycerol-3-phosphate, by the action of a number of enzymes, to form the triglyceride. The triglyceride-enriched regions then are believed to bud off from the endoplasmic reticulum as the oil body. The oil body membranes are different from other cellular membranes since they are made up of only a single layer of phospholipids. The triglycerides are catabolized by the action of triacylglycerol lipases, which release the fatty acids. The fatty acids are then broken down into acetyl CoA units through β-oxidation.

Even though phospholipids constitute a small fraction of the lipids in fruits, the degradation of phospholipids is a key factor that controls the progression of senescence. As in several senescing systems, there is a decline in phospholipids as the fruit undergoes senescence. With the decline in phospholipids content, there is a progressive increase in the levels of neutral lipids, primarily diacylglycerols, free fatty acids, and fatty aldehydes. In addition, the levels of sterols may also increase. Thus, there is an increase in the sterol:phospholipid ratio. Such changes in the composition of membranes can cause the formation of gel phase or nonbilayer lipid structures (micelles). These changes can make the membranes leaky, resulting in the loss of compartmentalization, and ultimately, senescence (Paliyath and Droillard 1992).

Membrane lipid degradation occurs by the tandem action of several enzymes, one enzyme acting on the product released by the previous enzyme in the sequence. Phospholipase D (PLD) is the first enzyme of the pathway that initiates phospholipid catabolism, and it is a key enzyme of the pathway (Fig. 21.6). Phospholipase D acts on phospholipids, liberating phosphatidic acid and the respective headgroup (choline, ethanolamine, glycerol, inositol). Phosphatidic acid in turn is acted upon by phosphatidate phosphatase, which removes the phosphate group from phosphatidic acid, with the liberation of diacylglycerols (diglycerides). The acyl chains of diacylglycerols are then deesterified by the enzyme lipolytic acyl hydrolase, liberating free fatty acids. Unsaturated fatty acids with a *cis*-1,4-pentadiene structure (linoleic acid, linolenic acid) are acted upon by lipoxygenase causing the peroxidation of fatty acids. This step may also cause the production of activated oxygen species such as singlet oxygen, superoxide, and peroxy radicals. The peroxidation products of linolenic acid can be 9-hydroperoxy- or 13-hydroperoxylinoleic acid. The hydroperoxylinoleic acids undergo cleavage by hydroperoxide lyase, resulting in several products including hexanal, hexenal, and ω-keto fatty acids (keto group towards the methyl end of the molecule). For example, hydroperoxide lyase action on 13-hydroperoxylinolenic acid results in the formation of *cis*-3-hexenal and 12-keto-*cis*-9-dodecenoic acid. Hexanal and hexenal are important fruit volatiles. The short-chain fatty acids may feed into catabolic pathway (β-oxidation) that results in the formation of short-chain acyl-CoAs, ranging from acetyl-CoA to dodecanoyl-CoA. The short-chain acyl-CoAs and alcohols (ethanol, propanol, butanol, pentanol, hexanol, etc.) are esterified to form a variety of esters that constitute components of flavor volatiles that are characteristic to fruits. The free fatty acids and their catabolites (fatty aldehydes, fatty alcohols, alkanes, etc.) can accumulate in the membrane, causing membrane destabilization (formation of gel phase, nonbilayer structures, etc.). An interesting regulatory feature of this pathway is the very low substrate specifity of enzymes that act downstream from phospholipase D, for the phospholipids. Thus, phosphatidate phosphatase, lipolytic acyl hydrolase, and lipoxygenase do not directly act on phospholipids, though there are exceptions to this rule. Therefore, the degree of membrane lipid catabolism will be determined by the extent of activation of phospholipase D.

The membrane lipid catabolic pathway is considered to be an autocatalytic pathway. The destabilization of the cell membrane can cause the leakage of calcium and hydrogen ions from the cell wall space, as well as the inhibition of calcium and proton ATPases, the enzymes responsible for maintaining a physiological calcium and proton concentration within the cytoplasm (calcium concentration below micromolar range, pH in the 6–6.5 range). Under conditions of normal growth and development, these enzymes pump the extra calcium and hydrogen ions that enter the cytoplasm from storage areas such as the apoplast and the endoplasmic reticular lumen, in response to hormonal and environmental stimulation, using ATP as the energy source. The activities of calcium and proton ATPases localized on the plasma membrane, the endoplasmic reticulum, and the tonoplast are responsible for pumping the ions back into the storage areas. In fruits (and other senescing systems), with the advancement in ripening and senescence, there is a progressive increase in leakage of calcium and hydrogen ions. Phospholipase D is stimulated by low pH and calcium con-

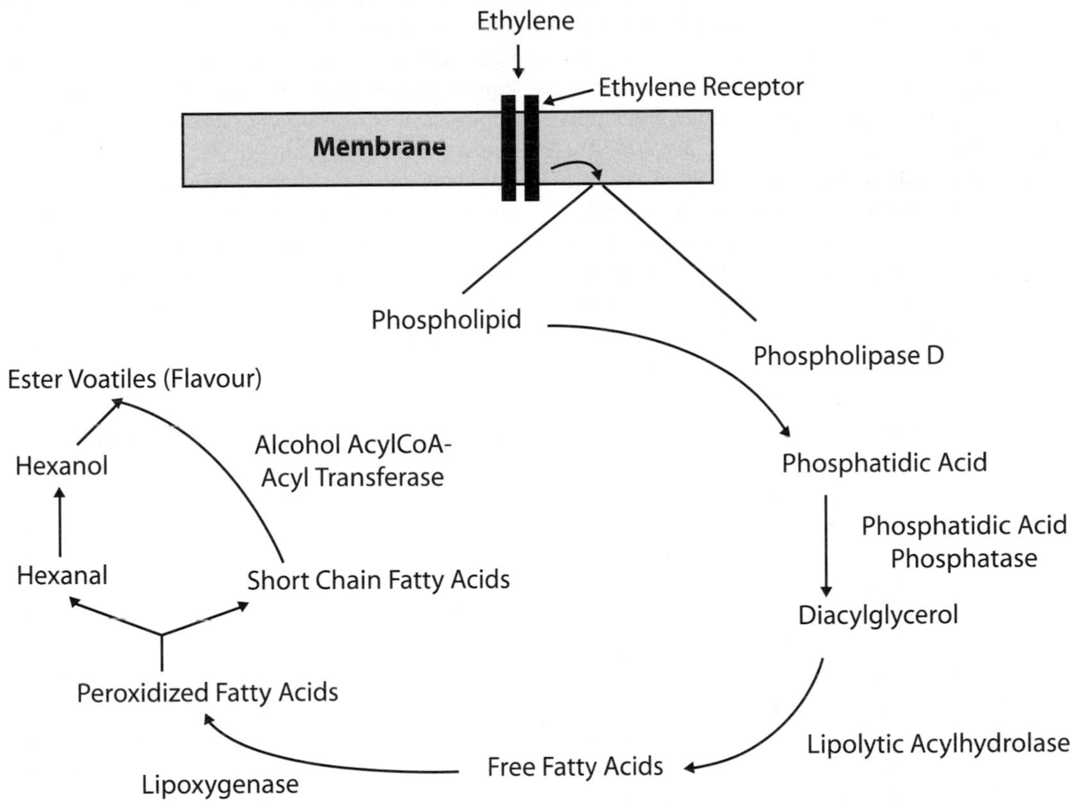

Figure 21.6. Phospholipid catabolic pathway and its relation to fruit ripening.

centrations over 10 μM. Thus, if the cytosolic concentrations of these ions progressively increase during ripening or senescence, the membranes are damaged as a consequence. However, this is an inherent feature of the ripening process in fruits and results in the development of ideal organoleptic qualities that makes them edible. However, uncontrolled membrane deterioration can result in the loss of shelf life and quality in fruits.

The properties and regulation of the membrane degradation pathway are increasingly becoming clear. Enzymes such as phospholipase D (PLD) and lipoxygenase (LOX) are very well studied. There are several isoforms of phospholipase D designated as PLD-alpha, PLD-beta, PLD-gamma, and so on. The expression and activity levels of PLD-alpha are much higher than those of the other PLD isoforms. Thus, PLD-alpha is considered a housekeeping enzyme. The regulation of PLD activity is an interesting feature. PLD is normally a soluble enzyme. The secondary structure of PLD shows the presence of a segment of around 130 amino acids at the N-terminal end, designated as the C2 domain. This domain is characteristic of several enzymes and proteins that are integral components of the hormone signal transduction system. In response to hormonal and environmental stimulation and the resulting increase in cytosolic calcium concentration, C2 domain binds calcium and transports PLD to the membrane where it can initiate membrane lipid degradation. The precise relation between the stimulation of the ethylene receptor and phospholipase D activation is not fully understood, but it could involve the release of calcium and migration of PLD to the membrane. PLD-alpha appears to be the key enzyme responsible for the initiation of membrane lipid degradation in tomato fruits. Antisense inhibition of PLD-alpha in tomato fruits resulted in the reduction of PLD activity and, consequently, an improvement in the shelf life, firmness, soluble solids, and lycopene content of the ripe fruits (Pinhero et al 2003, Oke et al. 2003). There are other phospholipid-degrading enzymes such as phospholipase C and phospholipase A_2. Several roles of these enzymes in signal transduction processes have been extensively reviewed (Wang 2001, Meijer and Munnik 2003).

Lipoxygenase exists in both soluble and membranous forms in tomato fruits (Todd et al. 1990). Very little information is available on phosphatidate phosphatase and lipolytic acyl hydrolase in fruits.

Proteolysis and Structure Breakdown in Chloroplasts

The major proteinaceous compartments in fruits are the chloroplasts, which are distributed in the epidermal and hypodermal layers of fruits. The chloroplasts are not very abundant in fruits. During senescence, the chloroplast structure is gradually disassembled, with a decline in chlorophyll levels due to degradation and disorganization of the grana lamellar stacks of the chloroplast. With the disorganization of the thylakoid, globular structures (plastoglobuli) that are rich in degraded lipids accumulate within the chloroplast stroma. The degradation of chloroplasts and chlorophyll results in the unmasking of other colored pigments and is a prelude to the state of ripening and the development of organoleptic qualities. Mitochondria, which are also rich in protein, are relatively stable and undergo disassembly during the latter part of ripening and senescence.

Chlorophyll degradation is initiated by the enzyme chlorophyllase, which splits chlorophyll into chlorophyllide and the phytol chain. The phytol chain is made up of isoprenoid units (methyl-1,3-butadiene), and its degradation products accumulate in the plastoglobuli. Flavor components such as 6-methyl-5-heptene-2-one, a characteristic component of tomato flavor, are also produced by the catabolism of the phytol chain. The removal of magnesium from chlorophyllide results in the formation of pheophorbide. Pheophorbide, which possesses a tetrapyrrole structure, is converted to a straight-chain colorless tetrapyrrole by the action of pheophorbide oxidase. Action of several other enzymes is necessary for the full catabolism of chlorophyll. The protein complexes that organize the chlorophyll, the light-harvesting complexes, are degraded by the action of several proteases. The enzyme ribulose-*bis*-phosphate carboxylase/oxygenase (rubisco), the key enzyme in photosynthetic carbon fixation, is the most abundant protein in chloroplast. Rubisco levels also decline during ripening/senescence due to proteolysis. The amino acids resulting from the catabolism of proteins may be translocated to regions where they are needed for biosynthesis. In fruits, they may just enrich the soluble fraction with amino acids.

SECONDARY PLANT PRODUCTS AND FLAVOR COMPONENTS

Secondary plant products are regarded as metabolites that are derived from primary metabolic intermediates through well-defined biosynthetic pathways. The importance of the secondary plant products to the plant or organ in question may not readily be obvious, but these compounds appear to have a role in the interaction of the plant with the environment. The secondary plant products may include nonprotein amino acids, alkaloids, isoprenoid components (terpenes, carotenoids, etc.), flavonoids and anthocyanins, ester volatiles, and several other organic compounds with diverse structures. The number and types of secondary plant products are enormous, but from the perspective of fruit quality, the important secondary plant products include isoprenoids, anthocyanins, and ester volatiles.

Isoprenoid Biosynthesis

In general, isoprenoids possess a basic five-carbon skeleton in the form of 2-methyl-1,3-butadiene (isoprene), which undergoes condensation to form larger molecules. There are two distinct pathways for the formation of isoprenoids, the acetate/mevalonate pathway (Bach et al. 1999) localized in the cytosol, and the DOXP pathway (Rohmer pathway; Rohmer et al. 1993), localized in the chloroplast (Fig. 21.7). The metabolic precursor for the acetate/mevalonate

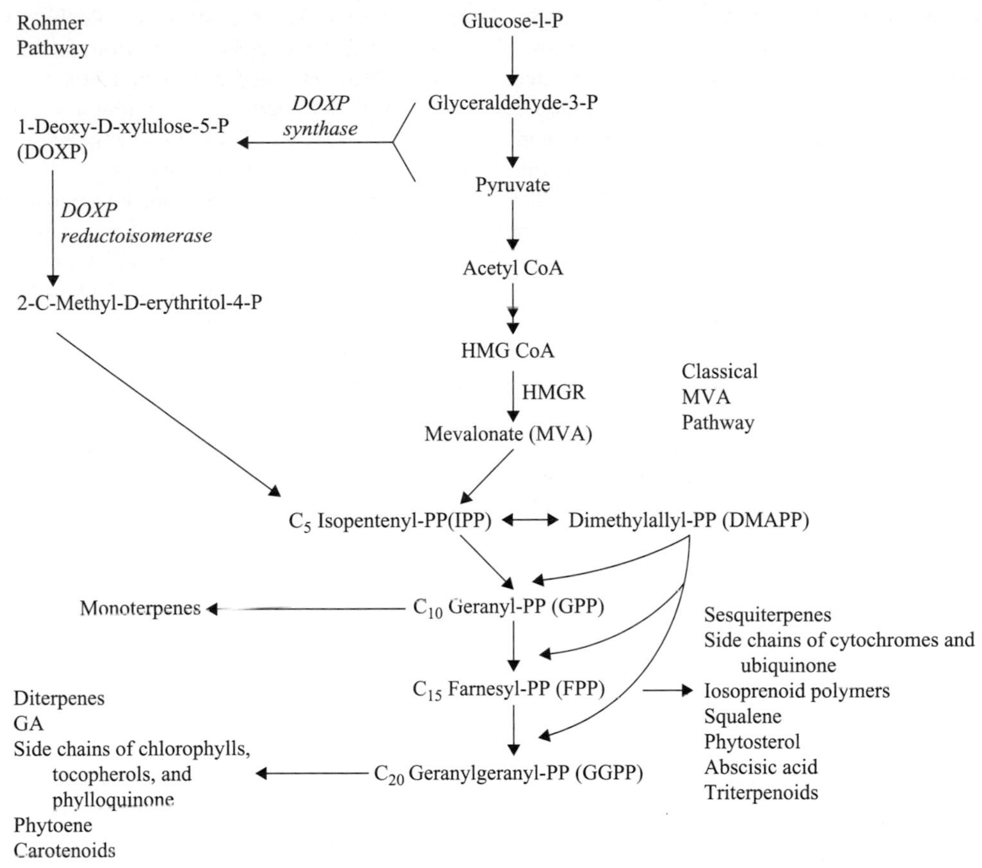

Figure 21.7. Isoprenoid biosynthetic pathway in plants.

pathway is acetyl coenzyme A (acetyl-CoA). Through the condensation of three acetyl-CoA molecules, a key component of the pathway, 3-hydroxy-3-methyl-glutaryl CoA (HMG-CoA) is generated. HMG-CoA undergoes reduction in the presence of NADPH, mediated by the key regulatory enzyme of the pathway, HMG-CoA reductase (HMGR), to form mevalonate. Mevalonate undergoes a two-step phosphorylation in the presence of ATP, mediated by kinases, to form isopentenyl pyrophosphate (IPP), the basic five-carbon condensational unit of several terpenes. IPP is isomerized to dimethylallylpyrophosphate (DMAPP), mediated by the enzyme IPP isomerase. Condensation of these two components results in the synthesis of C_{10} (geranyl), C_{15} (farnesyl), and C_{20} (geranylgeranyl) pyrophosphates. The C_{10} pyrophosphates give rise to monoterpenes, C_{15} pyrophosphates give rise to sesquiterpenes, and C_{20} pyrophosphates give rise to diterpenes. Monoterpenes are major volatile components of fruits. In citrus fruits, these include components such as limonene, myrcene, pinene, and others, occurring in various proportions. Derivatives of monoterpenes such as geranial, neral (aldehydes), geraniol, linalool, terpineol (alcohols), geranyl acetate, neryl acetate (esters), and so on, are also ingredients of the volatiles of citrus fruits. Citrus fruits are especially rich in monoterpenes and derivatives. Alpha-farnesene is a major sesquiterpene (C_{15}) component evolved by apples. The catabolism of alpha-farnesene in the presence of oxygen into oxidized forms has been implicated as a causative feature in the development of the physiological disorder superficial scald (a type of superficial browning) in certain varieties of apples such as red Delicious, McIntosh, Cortland, and others (Rupasinghe et al. 2000, 2003).

HMGR is a highly conserved enzyme in plants and is encoded by a multigene family (Lichtenthaler et al. 1997). The HMGR genes (*hmg1, hmg2, hmg3,* etc.) are nuclear encoded and can be differentiated from each other by the sequence differences at the 3'-untranslated regions of the cDNAs. There are three distinct genes for HMGR in tomato and two in apples. The different HMGR end products may be localized in different cellular compartments and are synthesized differentially in response to hormones, environmental signals, pathogen infections, and so on. In tomato fruits, the level of *hmg1* expression is high during the early stage of fruit development when cell division and expansion processes are rapid and require high levels of sterols for incorporation into the expanding membrane compartments. The expression of *hmg2,* which is not detectable in young fruits, increases during the latter part of fruit maturation and ripening.

HMGR activity can be detected in both membranous and cytosolic fractions of apple fruit skin tissue extract. HMGR is a membrane-localized enzyme, and the activity is detectable in the endoplasmic reticular, plastid, and mitochondrial membranes. It is likely that HMGR may have undergone proteolytic cleavage, releasing a fragment into the cytosol, which also possesses enzyme activity. There is a considerable degree of interaction between the different enzymes responsible for the biosynthesis of isoprenoids, which may exist as multienzyme complexes referred to as metabolons. The enzyme farnesyl pyrophosphate synthase, responsible for the synthesis of farnesyl pyrophosphate, is a cytosolic enzyme. Similarly, farnesene synthase, the enzyme that converts farnesyl pyrophosphate to alpha-farnesene in apples, is a cytosolic enzyme. Thus several enzymes may act in concert at the cytoplasm–endoplasmic reticulum boundary to synthesize isoprenoids.

HMG-CoA reductase expression and activities in apple fruits are hormonally regulated (Rupasinghe et al. 2001, 2003). There are two genes for HMGR in apples, designated as *hmg1* and *hmg2,* which are differentially expressed during storage. The expression of *hmg1* was constitutive, and the transcripts (mRNA) were present throughout the storage period. By contrast, the expression of *hmg2* increased during storage in parallel with the accumulation of alpha-farnesene. Ethylene production also increased during storage. Ethylene stimulates the biosynthesis of alpha-farnesene, as is evident from the inhibition of alpha-farnesene biosynthesis and the expression of *hmg2* by the ethylene action inhibitor 1-methylcyclopropene (MCP). Thus, biosynthesis of isoprenoids is a highly controlled process.

Carotenoids, which are major isoprenoid components of chloroplasts, are biosynthesized through the Rohmer pathway. The precursors of this pathway are pyruvate and glyceraldehyde-3-phosphate, and through a number of enzymatic steps, 1-deoxy-D-xylulose-5-phosphate (DOXP), a key metabolite of the pathway, is formed. NADPH-mediated reduction of DOXP leads ultimately to the formation of isopentenyl pyrophosphate (IPP). Subsequent condensation of IPP and DMAPP are similar as in the classical mevalonate pathway. Carotenoids have a stabilizing role in photosynthetic reactions. By virtue of their structure, they can accept and stabilize excess energy absorbed by the light-harvesting complex. During the early stages of fruit development, the carotenoids have a primarily photosynthetic function. As the fruit ripens, the composition of carotenoids changes to reveal the colored xanthophyll pigments. In tomato, lycopene is the major carotenoid pigment that accumulates during ripening. Lycopene is an intermediate of the carotene biosynthetic pathway. In young fruits, lycopene formed by the condensation of two geranylgeranyl pyrophosphate (C_{20}) moieties, mediated by the enzyme phytoene synthase, is converted to beta-carotene by the action of the enzyme sesquiterpene cyclase. However, as ripening proceeds, the levels and activity of sesquiterpene cyclase are reduced, leading to the accumulation of lycopene in the stroma. This leads to the development of red color in ripe tomato fruits. In yellow tomatoes, the carotene biosynthesis is not inhibited, and as the fruit ripens, the chlorophyll pigments are degraded, exposing the yellow carotenoids. Carotenoids are also major components that contribute to the color of melons. Beta-carotene is the major pigment in melons with an orange flesh. In addition, the contribution to color is also provided by alpha-carotene, delta-carotene, phytofluene, phytoene, lutein, and violaxanthin. In red-fleshed melons, lycopene is the major ingredient, whereas in yellow-fleshed melons, xanthophylls and beta-carotene predominate. Carotenoids not only provide a variety of color to the fruits, but also are important nutritional ingredients in the human diet. Beta-carotene is converted to vitamin A in the human body and thus serves as a precursor to vitamin A. Carotenoids are strong antioxidants. Lycopene is observed to provide protection from cardiovascular diseases and cancer (Giovanucci 1999). Lutein, a xanthophyll, has been proposed to play a protective role in the retina, maintaining vision.

Anthocyanin Biosynthesis

The development of color is a characteristic feature of the ripening process, and in several fruits, the color components are anthocyanins biosynthesized from metabolic precursors. The anthocyanins accumulate in the vacuole of the cell and are often abundant in the cells closer to the surface of the fruit. Anthocyanin biosynthesis starts with the condensation of three molecules of malonyl-CoA with p-coumaroyl-CoA to form tetrahydroxychalcone, mediated by the enzyme chalcone synthase (Fig. 21.8). Tetrahydroxychalcone has the basic flavonoid structure C_6-C_3-C_6, with two phenyl groups separated by a three-carbon link. Chalcone isomerase enables the ring closure of chalcone, leading to the formation of the flavanone naringenin, which possesses a flavonoid structure having two phenyl groups linked together by a heterocyclic ring (Fig. 21.9). The phenyl groups are designated as A and B, and the heterocyclic ring is designated as ring C. Subsequent conversions of naringenin by flavonol hydroxylases result in the formation of dihydrokaempferol, dihydromyricetin, and dihydroquercetin, which differ in their number of hydroxyl moieties. Dihydroflavonol reductase converts the dihydroflavonols into the colorless anthocyanidin compounds leucocyanidin, leucopelargonidin, and leucodelphinidin. Removal of hydrogens and the induction of unsaturation of the C ring at C2 and C3, mediated by anthocyanin synthase, result in the formation of cyanidin, pelargonidin, and delphinidin, the colored compounds (Fig. 21.9). Glycosylation, methylation, coumaroylation, and a variety of other additions of the anthocyanidins result in color stabilization of the diverse types of anthocyanins seen in fruits. Pelargonidins give orange, pink, and red color; cyanidins provide magenta and crimson coloration; and delphinidins provide the purple, mauve, and blue color characteristics to several fruits. The color characteristics of fruits may result from a combination of several forms of anthocyanins existing together, as well as the conditions of pH and ions present in the vacuole.

Anthocyanin pigments cause the diverse coloration of grape cultivars, resulting in skin colors varying from translucent to red and black. All the forms of anthocyanins including those with modifications of the hydroxyl groups are routinely present in the red and dark varieties of grapes. A glucose moiety is attached at the 3 or 5 position or at both in most grape anthocyanins. The glycosylation pattern can vary between the European *(Vitis vinifera)* and North American *(Vitis labrusca)* grape varieties. Anthocyanin accumulation occurs towards the end of ripening and is highly influenced by sugar levels, light, temperature, ethylene, and increased metabolite translocation from leaves to fruits. All these factors positively influence the anthocyanin levels. Most of the anthocyanin accumulation may be limited to epidermal cell layers and a few of the subepidermal cells. In certain high-anthocyanin-containing varieties, even the interior cells of the fruit may possess high levels of anthocyanins. In the red wine varieties such as merlot, pinot noir, and cabernet sauvignon, anthocyanin content may vary between 1500 and 3000 mg/kg fresh weight. In some high-anthocyanin-containing varieties such as Vincent, Lomanto, and Colobel, the anthocyanin levels can exceed 9000 mg/kg fresh weight. Anthocyanins are very strong antioxidants and are known to provide protection from the development of cardiovascular diseases and cancer.

Many fruits have a tart taste during the early stage of development, which is termed astringency; it is characteristic to fruits such as banana, kiwi, and grape. The astringency is due to the presence of tannins and several other phenolic components in fruits. Tannins are polymers of flavonoids such as catechin and epicatechin, phenolic acids (caffeoyl tartaric acid, coumaroyl tartaric acid, etc.). The contents of tannins decrease during ripening, making the fruit palatable.

Ester Volatile Biosynthesis

The sweet aroma characteristic to several ripe fruits is due to the evolution of several types of volatile components that include monoterpenes, esters, organic

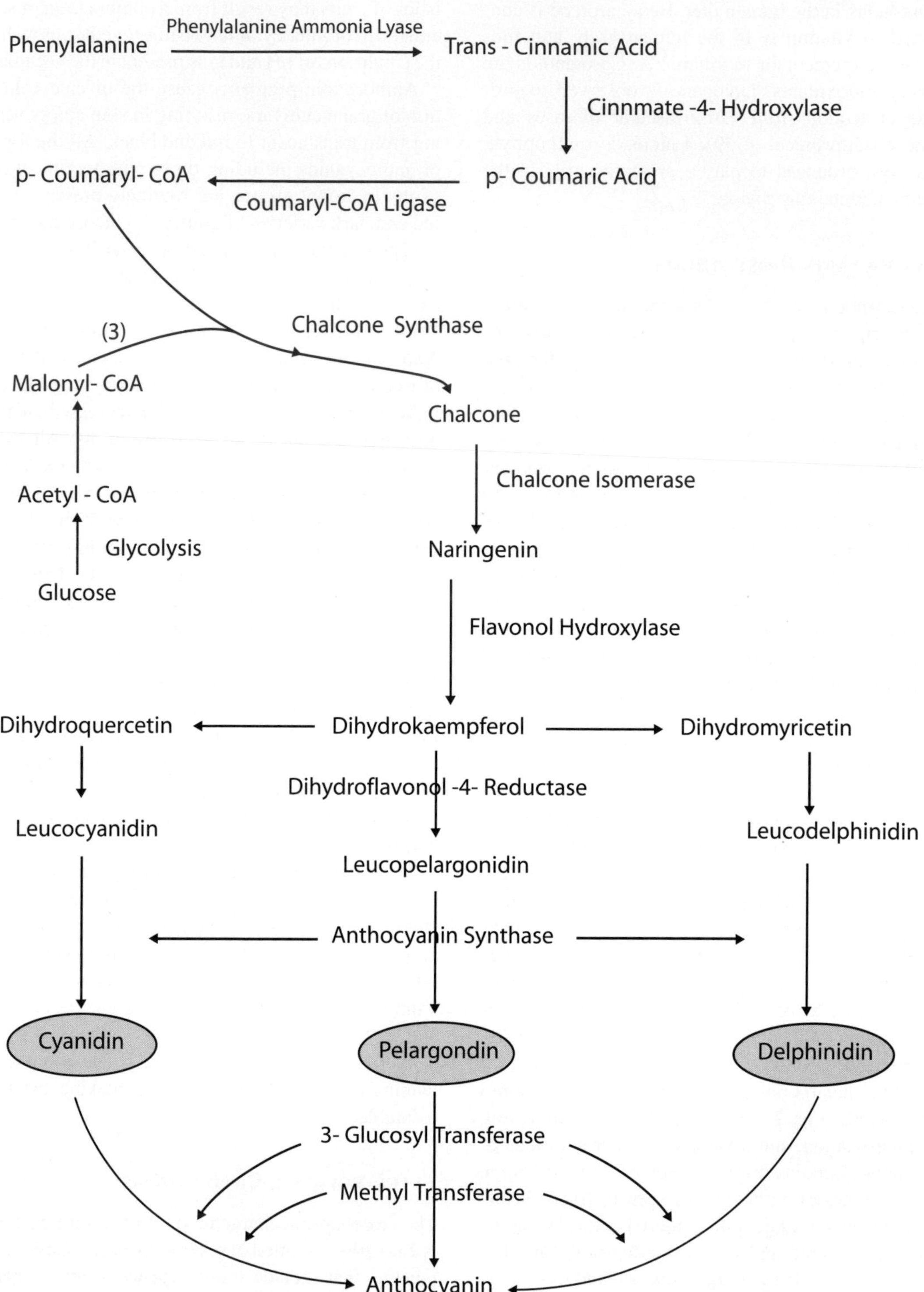

Figure 21.8. Anthocyanin biosynthetic pathway in plants.

ANTHO CYANIDINS

PELARGONIDIN

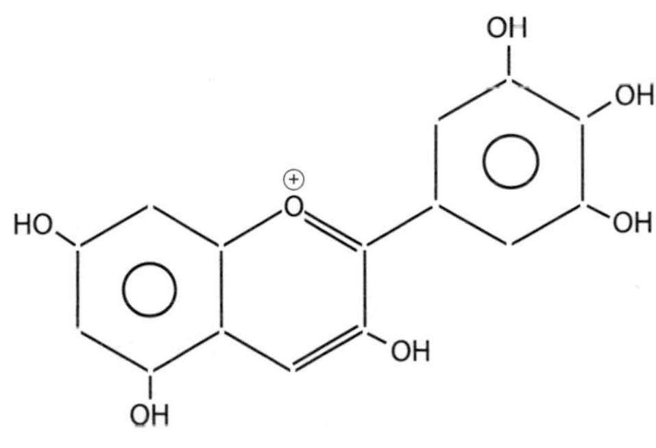

CYANIDIN

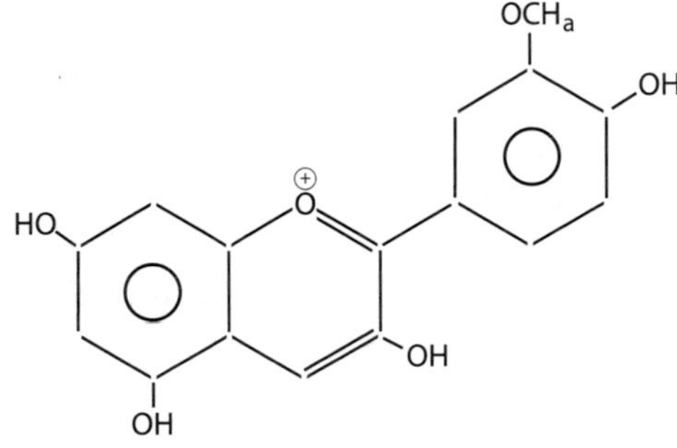

PEONIDIN

DELPHINIDIN

PETUNIDIN

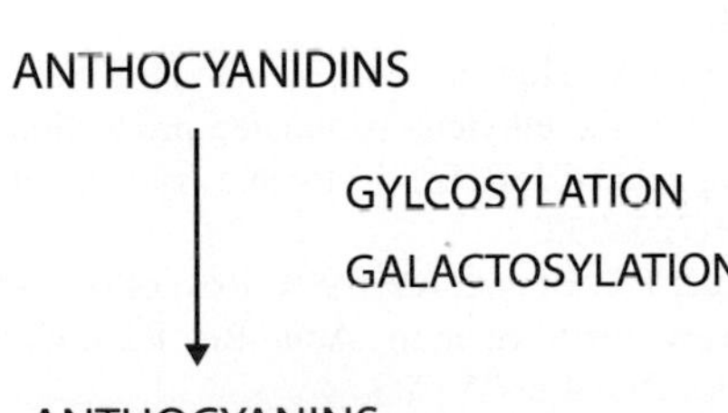

MALVIDIN

Figure 21.9. Some common anthocyanidins found in fruits and flowers.

acids, aldehydes, ketones, alkanes, and others. Some of these ingredients specifically provide the aroma characteristic to fruits and are referred to as character impact compounds. For instance, the banana flavor is predominantly from isoamyl acetate, apple flavor from ethyl-2-methyl butyrate, and the flavor of lime is primarily due to the monoterpene limonene. As the name implies, ester volatiles are formed from an alcohol and an organic acid through the formation of an ester linkage. The alcohols and acids are, in general, products of lipid catabolism. Several volatiles are esterified with ethanol, giving rise to ethyl derivatives of aliphatic acids (ethyl acetate, ethyl butyrate, etc.).

The ester volatiles are formed by the activity of the enzyme Acyl-CoA:alcohol acyl transferase, or as it is generally called, alcohol acyl transferase (AAT). In apple fruits, the major aroma components are ester volatiles (Paliyath et al. 1997). The alcohol can vary from ethanol to propanol, butanol, pentanol, hexanol, and so on. The organic acid moiety containing the CoA group can vary in chain length from C_2 (acetyl) to C_{12} (dodecanoyl). AAT activity has been identified in several fruits, including banana, strawberry, melon, apple, and others. In banana, esters are the predominant volatiles, enriched with esters such as acetates and butyrates. The flavor may result from the combined perception of amyl esters and butyl esters. Volatile production increases during ripening. The components for volatile biosynthesis may arise from amino acids and fatty acids. In melons, the volatile components comprise esters, aldehydes, alcohols, terpenes, and lactones. Hexyl acetate, isoamyl acetate and octyl acetate are the major aliphatic esters. Benzyl acetate, phenyl propyl acetate, and phenyl ethyl acetate are also observed. The aldehydes, alcohols, terpenes, and lactones are minor components in melons. In mango fruits, the characteristic aroma of each variety is based on the composition of volatiles. The variety "Baladi" is characterized by the presence of high levels of limonene, other monoterpenes and sesquiterpenes, and ethyl esters of even-numbered fatty acids. By contrast, the variety "Alphonso" is characterized by high levels of C_6 aldehydes and alcohols (hexanal, hexanol) that may indicate a high level of fatty acid peroxidation in ripe fruits. C_6 aldehydes are major flavor components of tomato fruits as well. In genetically transformed tomatoes (antisense phospholipase D), the evolution of pentanal and hexenal/hexanal was much higher after blending, suggesting the preservation of fatty acids in ripe fruits. Preserving the integrity of the membrane during ripening could help preserve the fatty acids that contribute to the flavor profile of the fruits, and this feature may provide a better flavor profile for fruits.

GENERAL READING

Biochemistry and Molecular Biology of Plants. 2000. BB Buchanan, W Gruissem, RL Jones, editors. American Society of Plant Physiologists, Bethesda, Maryland.

Biochemistry of Fruit Ripening. 1991. GB Seymour, JE Taylor, GA Tucker, editors. Chapman and Hall, London.

Postharvest Physiology of Perishable Plant Products. 1997. SJ Kays, editor. Exon Press, Athens, Georgia.

REFERENCES

Ahn T, Schofield A, Paliyath G. 2002. Changes in antioxidant enzyme activities during tomato fruit development. Physiol Mol Biol Plants 8:241–249.

Bach TJ, Boronat A, Campos N. 1999. Mevalonate biosynthesis in plants. Crit Rev Biochem Mol Biol 34: 107–122.

Beaudry RM, Severson RF, Black CC, Kays SJ. 1989. Banana ripening: implication of changes in glycolytic intermediate concentrations, glycolytic and gluconeogenic carbon flux, and fructose 2,6-bisphosphate concentration. Plant Physiol 91:1436–1444.

Bennet AB, Christofferson RE. 1986. Synthesis and processing of cellulose from ripening avocado fruit. Plant Physiol 81:830–835.

Bird CR, Smith CJS, Ray JA, et al. 1988. The tomato polygalacturonase gene and ripening specific expression in transgenic plants. Plant Mol Biol 11:651–662.

Bruemmer JH. 1989. Terminal oxidase activity during ripening of Hamlin orange. Phytochemistry 28: 2901–2902.

Dallman TF, Thomson WW, Eaks IL, Nothnagel EA. 1989. Expression and transport of cellulase in avocado mesocarp during ripening. Protoplasma 151:33–46.

Davies KM, Hobson GE, Grierson D. 1988. Silver ions inhibit the ethylene stimulated production of ripening related mRNAs in tomato fruit. Plant Cell Env 11:729–738.

Fischer RL, Bennet AB. 1991. Role of cell wall hydrolases in fruit ripening. Annu Rev Plant Physiol Plant Mol Biol 42:675–703.

Fluhr R, Mattoo AK. 1996. Ethylene—Biosynthesis and Perception. Crit Rev Plant Sci 15:479–523.

Fry SC. 1986. Cross linking of matrix polymers in the growing cell walls of angiosperms. Annu Rev Plant Physiol 37:165–186.

Giovannoni JJ, DellaPenna D, Bennet AB, Fischer RL. 1989. Expression of a chimeric PG gene in transgenic rin tomato fruit results in polyuronide degradation but not fruit softening. The Plant Cell 1:53–63.

Giovanucci EL. 1999. Tomatoes, tomato based products, lycopene andcancer: A review of the epidemiologic literature. J Natl Cancer Inst 91:317–329.

Grierson D, Tucker GA, Keen J, Ray J, Bird CR, Schuch W. 1986. Sequencing and identification of a cDNA clone for tomato polygalacturonase. Nucleic Acids Res 14:8595–8603.

Hamilton AJ, Bouzayen M, Grierson D. 1991. Identification of a tomato gene for the ethylene forming enzyme by expression in yeast. Proc Natl Acad Sci (USA) 88:7434–7437.

Hamilton AJ, Lycett GW, Grierson D. 1990. Antisense gene that inhibits synthesis of the hormone ethylene in transgenic plants. Nature 346:284–287.

Hrazdina G, Kiss E, Rosenfield CL, Norelli JL, Aldwinckle HS. 2000. Down regulation of ethylene production in apples. In: G Hrazdina, editor, Use of Agriculturally Important Genes in Biotechnology. IOS Press, Amsterdam. Pp. 26–32.

Kays, S. J. 1997. Postharvest physiology of perishable plant products. Van Nostrand Reinhold, New York.

Keegstra K, Talmadge KW, Bauer WD, Albersheim P. 1973. The structure of plant cell walls III. A model of the walls of suspension cultured sycamore cells based on the interaction of the macromolecular components. Plant Physiol51:188–196.

Kramer M, Sanders R, Bolkan H, Waters C, Sheehy RE, Hiatt WR. 1992. Postharvest evaluation of transgenic tomatoes with reduced levels of polygalacturonase: Processing, firmness and disease resistance. Postharv Biol Technol 1:241–255.

Kumar R, Selvaraj Y. 1990. Fructose-1,6-bisphosphatase in ripening mango fruit. Indian J Exp Biol 28:284–286.

Lichtenthaler HK, Rohmer M, Schwender J. 1997. Two independent biochemical pathways for isopentenyl diphosphate and isoprenoid biosynthesis in higher plants. Physiologia Plant 101:643–652.

Meijer HJG,Munnik T. 2003. Phospholipid-based signalling in plants. Annu Rev Plant Biol 54:265–306.

Meir S, Bramlage WJ. 1988. Antioxidant activity in "Cortland" apple peel and susceptibility to superficial scald after storage. J Amer Soc Hort Sci 113: 412–418.

Oeller PW, Min-Wong L, Taylor LP, Pike DA, Theologis A. 1991. Reversible inhibition of tomato fruit senescence by antisense RNA. Science 254:437–439.

Oke M, Pinhero RG, Paliyath G. 2003. The effects of genetic transformation of tomato with antisense phospholipase D cDNA on the quality characteristics of fruits and their processed products. Food Biotechnol 17:163–182.

Paliyath G, Droillard MJ. 1992. The mechanism of membrane deterioration and disassembly during senescence. Plant Physiol Biochem 30:789–812.

Paliyath G, Thompson JE. 1990. Evidence for early changes in membrane structure during post harvest development of cut carnation flowers. New Phytologist 114:555–562.

Paliyath G, Whiting MD, Stasiak MA, Murr DP, Clegg BS. 1997. Volatile production and fruit quality during development of superficial scald in Red Delicious apples. Food Res International 30:95–103.

Pinhero RG, Almquist KC, Novotna Z, Paliyath G. 2003. Developmental regulation of phospholipase D in tomato fruits. Plant Physiol Biochem 41:223–240.

Purvis AC. 1985. Low temperature induced azide-insensitive oxygen uptake in grapefruit flavedo tissue. J Amer Soc Hort Sci 110:782–785.

Rohmer M, Knani M, Simonin P, Sutter B, Sahm H. 1993. Isoprenoid biosynthesis in bacteria: A novel pathway for early steps leading to isopentenyl diphosphate. Biochem J 295:517–524.

Ruffner HP, Brem S, Malipiero U. 1983. The physiology of acid metabolism in grape berry ripening. Acta Horticulturae 139:123–128.

Rupasinghe HPV, Almquist KC, Paliyath G, Murr DP. 2001. Cloning of *hmg1* and *hmg2* cDNAs encoding 3-hydroxy-3-methylglutaryl coenzyme A reductase and their expression and activity in relation to alpha-farnesene synthesis in apple. Plant Physiol Biochem 39:933–947.

Rupasinghe HPV, Paliyath G, Murr DP. 2000. Sesquiterpene alpha-farnesene synthase: Partial purification, characterization and activity in relation to superficial scald development in "Delicious" apples. J Amer Soc Hort Sci 125:111–119.

———. 2003. Biosynthesis of isoprenoids in higher plants. Physiol Mol Biol Plants 9:19–28.

Selvaraj Y, Kumar R, Pal DK. 1989. Changes in sugars, organic acids, amino acids, lipid constituents and aroma characteristics of ripening mango fruit. J Food Sci Technol 26:308–313.

Singer SJ, Nicholson GL. 1972. The fluid mosaic model of the structure of cell membranes. Science 175: 720–731.

Smith CJS, Watson CF, Ray J, Bird CR, Morris PC, Schuch W, Grierson D. 1988. Antisense RNA inhibition of polygalacturonase gene expression in transgenic tomatoes. Nature 334:724–726.

Todd JF, Paliyath G, Thompson JE. 1990. Characteristics of a membrane-associated lipoxygenase in tomato fruit. Plant Physiol 94:1225–1232.

Wang X. 2001. Plant Phospholipases. Annu Rev Plant Physiol Plant Mol Biol 52:211–231.

Yang SF. 1981. Ethylene, the gaseous plant hormone and regulation of biosynthesis. Trends Biochem Sci 6:161–164.

22
Biochemistry of Fruit Processing

M. Oke and G. Paliyath

INTRODUCTION

Overall, the food and beverage processing industry is an important manufacturing sector all across the world. The United States is among the top producers and consumers of fruits and tree nuts in the world. Each year, fruit and tree nut production generates about 13% of U.S. farm cash receipts for all agricultural crops. Annual U.S. per capita use of fruit and tree nuts totals nearly 300 pounds (fresh weight equivalent). Oranges, apples, grapes, and bananas are the most popular fruits. The consumption of fruits and processed products has enjoyed an unprecedented growth during the past decade. Many factors motivate this increase, including consumers' awareness of the health benefits of fruit constituents such as the importance of dietary fiber, antioxidants, vitamins, minerals, and phytochemicals present in fruits. In food stores, one can buy fresh and processed exotic food items, in canned, frozen, dehydrated, fermented, or pickled form, or made into jams, jellies, and marmalades year-round. Several varieties of fruits are sold throughout the year in developed countries, and with the increase in international trade of fruits, even tropical fruits are available at a reasonable cost. The food processing industry uses fruits as ingredients in juice blends, snacks, baby foods, and many other processed food items. As a result, the world production of primary fruits has increased from 384 million metric tons in 1992 to 475 million metric tons in 2002 (FAO). The world production of fruits in 2002, by region, is shown in Table 22.1.

Table 22.1. World Production of Fruit in 2002 in Metric Tons

	Production (mt)					
Fruit	Africa	Asia	Europe	America	Australia and New Zealand	World
Primary fruits	61,934,001	204,640,809	73,435,847	129,642,621	4,438,090	475,503,880
Apples	1,696,109	30,870,497	15,820,454	7,875,880	831,999	57,094,939
Bananas	7,140,629	35,766,134	446,600	15,372,622	250,000	69,832,378
Grapes	3,113,940	14,761,974	28,739,848	12,529,900	1,872,588	61,018,250
Mangoes	2,535,781	19,841,469	–	3,343,697	28,000	25,748,947
Oranges	4,920,544	14,177,590	6,180,846	38,402,953	444,500	64,128,523
Pears	530,547	10,996,503	3,626,020	1,759,538	202,597	17,115,205
Plums	173,946	5,385,634	2,661,194	1,069,953	24,000	9,314,727
Strawberries	171,132	325,670	1,276,192	1,157,540	21,300	3,237,533

Source: Calculated from FAOSTAT database. http://apps.fao.org/default.htm. Accessed on November 08, 2003.

Advances in fruit processing technologies mostly occur in response to consumer demands or improvement in the efficiency of technology. Traditional methods of canning, freezing, and dehydration are progressively being replaced by fresh cut, ready-to-eat fruit preparations. The use of modified atmospheres and irradiation also helps in extending the shelf life of produce.

Fruits are essential sources of minerals, vitamins, and dietary fiber. In addition to these components, they provide carbohydrates and, to a lesser extent, proteins. Fruits play an important role in the digestion of meat, cheese, and other high-energy foods by neutralizing the acids produced by the hydrolysis of lipids.

FRUIT CLASSIFICATION

Fruits can be broadly grouped into three categories according to their growth latitude: temperate, subtropical, or tropical regions.

TEMPERATE ZONE FRUITS

Four subgroups can be distinguished among temperate zone fruits. These subgroups include small fruits and berries (e.g., grape, strawberry, blueberry, blackberry, cranberry), stone fruits (e.g., peach, plum, cherry, apricot, nectarine) and pome fruits, (e.g., apple, pear, Asian pear or Nashi, and European pear or quince).

SUBTROPICAL FRUITS

Two subgroups can be differentiated among subtropical fruits: citrus fruits (e.g., orange, grapefruit, lemon, tangerine, mandarin, lime, and pomelo), and noncitrus fruits (e.g., avocado, fig, kiwifruit, olive, pomegranate, and cherimoya).

TROPICAL FRUITS

Major tropical fruits include banana, mango, papaya, and pineapple. Minor tropical fruit include passion fruit, cashew apple, guava, longan, lychee, mangosteen, carambola, rambutan, sapota, and others.

CHEMICAL COMPOSITION OF FRUITS

CARBOHYDRATES

Fruits typically contain between 10 and 25% carbohydrates, less than 1% protein and less than 0.5% fat (Somogyi et al. 1996b). Carbohydrates, sugars, and starches are broken down into CO_2, water, and energy during catabolism. The major sources of carbohydrates are banana, plantain, date, raisin, breadfruit, and jackfruit. Proteins and amino acids are contained in dried apricot and fig, whereas fats are the major components of avocado and olives. Sugars are the major carbohydrate components of fruits, and their composition varies from fruit to fruit. In general, sucrose, glucose, and fructose are the major sugar components present in fruits (Table 22.2). Several fruits also contain sugar alcohols such as sorbitol.

VITAMINS

Fruits and vegetables are major contributors to our daily vitamin requirements. The nutrient contribution from a specific fruit or vegetable is dependent on the amount of vitamins present in the fruit or vegetable, as well as the amount consumed. The approximate percentage contribution to daily vitamin intake from fruits and vegetables is vitamin A, 50%; thiamine, 60%; riboflavin, 30%; niacin, 50%; and vitamin C, 100% (Somogyi et al. 1996b). Vitamins are sensitive to different processing conditions such as exposure to heat, cold, reduced or high levels of oxygen, light, free water, and mineral ions. Trimming, washing, blanching, and canning can also cause loss in the vitamin content of fruits and vegetables.

MINERALS

Minerals found in fruits in general may not be fully nutritionally available (e.g. calcium, found as calcium oxalate in certain produce is not nutritionally available). Major minerals in fruits are base-forming elements (K, Ca, Mg, Na) and acid-forming elements (P, Cl, S). Minerals often present in microquantities are Mn, Zn, Fe, Cu, Co, Mo, and I. Potassium is the most abundant mineral in fruits, followed by calcium and magnesium. High potassium content is often associated with increased acidity

Table 22.2. Sugar Composition of Selected Fruits

Fruit	Sugar (g/100 mL of juice)			
	Sucrose	Glucose	Fructose	Sorbitol
Apple	0.82±0.13	2.14±0.43	5.31±0.94	0.20±0.04
Cherry	0.08±0.02	7.50±0.81	6.83±0.74	2.95±0.33
Grape	0.29±0.08	9.59±1.03	10.53±1.04	ND
Nectarine	8.38±0.73	0.85±0.04	0.59±0.02	0.27±0.04
Peach	5.68±0.52	0.67±0.06	0.49±0.01	0.09±0.02
Pear	0.55±0.12	1.68±0.36	8.12±1.56	4.08±0.79
Plum	0.51±0.36	4.28±1.18	4.86±1.30	6.29±1.97
Kiwifruit	1.81±0.72	6.94±2.85	8.24±3.43	ND
Strawberry	0.17±0.06	1.80±0.16	2.18±0.19	ND

Source: Reprinted with permission from Van Gorsel et al., J Agr Food Chem, 1992, 40:784–789. Copyright, American Chemical Society.

ND = Not detected (less than 0.05 g/100 ml).

and improved color, whereas high calcium content reduces the incidence of physiological disorders and improves the quality of fruits. Phosphorus is a constituent of several metabolites, and nucleic acids and plays an important role in carbohydrate metabolism and energy transfer.

Dietary Fiber

Dietary fiber consists of cellulose, hemicellulose, lignin, and pectic substances, which are derived from fruit cell walls and skins. The dietary fiber content of fruits ranges from 0.5 to 1.5% by fresh weight. Because of its properties, which include a high water-holding capacity, dietary fiber plays a major role in the movement of digested food and in reducing the incidence of colon cancer and cardiovascular disease.

Proteins

Fruits contain less than 1% protein (as opposed to 9–20% protein in nuts). In general, plant protein sources provide a significant portion of the dietary protein requirement in countries where animal proteins are in short supply. Plant proteins, unlike animal proteins are often deficient or limiting in one or more essential amino acids. The green fruits are important sources of proteins as they contain enzymes and proteins associated with the photosynthetic apparatus. Enzymes, which catalyze metabolic processes in fruits, are important in the reactions involved in fruit ripening and senescence. Some of the enzymes important to fruit quality are

- *Ascorbic acid oxidase:* Catalyzes oxidation of ascorbic acid and results in loss of nutritional quality.
- *Chlorophyllase:* Catalyzes removal of the phytol ring from chlorophyll; results in loss of green color.
- *Polyphenol oxidase:* Catalyzes oxidation of phenolics, resulting in the formation of brown-colored polymers.
- *Lipoxygenase:* Catalyzes oxidation of unsaturated fatty acids; results in off-odor and off-flavor production.
- *Polygalacturonase:* Catalyzes hydrolysis of glycosidic bonds between adjacent polygalacturonic acid residues in pectin; results in tissue softening.
- *Pectin esterase:* Catalyzes deesterification of methyl groups in pectin; acts in conjunction with polygalacturonases, leading to tissue softening.
- *Cellulase:* Catalyzes hydrolysis of cellulose polymers in the cell wall and therefore is involved in fruit softening.
- *Phospholipase D:* Initiates the degradation of cell membrane.

Lipids

In general the lipid content of fruits is very small, amounting to 0.1–0.2%. Avocado, olive, and nuts are exceptions. Despite the relatively small amount of storage lipids in fresh fruits, they still possess cell membrane lipids. Lipids make up the surface wax, which contributes to the shiny appearance of fruits, and cuticle, which protects fruits against water loss and pathogens. The degree of fatty acid saturation establishes membrane fluidity, with greater saturation resulting in less fluidity. Lipids play a significant role in the characteristic aroma and flavor of fruits. For example, the characteristic aromas and flavors of cut fruits result from the action of the enzyme lipoxygenase on fatty acids (linoleic and linolenic acids), which eventually leads to the production of volatile compounds. The action of lipoxygenase increases after fruit is cut because there is a greater chance for the enzyme and substrate to mix together. Lipoxygenase can also be responsible for the off-flavor and off-aroma in certain plant products (soybean oil).

VOLATILES

The specific aroma of fruits is due to the amount and diversity of volatiles they contain. Volatiles are present in extremely small quantities (< 100 μg/g fresh weight). Ethylene is the most abundant volatile in fruit, but it has no typical fruit aroma. Characteristic flavors and aromas are a combination of various character-impact compounds, mainly esters, terpenes, short-chain aldehydes, ketones, organic acids, and so on. Their relative importance depends upon the threshold concentration (sometimes < 1 ppb), and interaction with other compounds. For some fruits such as apple and banana, the specific aroma is determined by the presence of a single compound, ethyl-2-methylbutyrate in apples, and isoamylacetate in bananas.

WATER

Water is the most abundant single component of fruits and vegetables (up to 90% of total weight). The maximum water content of fruits varies due to structural differences. Agricultural conditions also influence the water content of plants. As a major component of fruits, water has an impact on both their quality and their deterioration. Turgidity is a major quality determinant factor in fruits. Loss of turgor is associated with loss of quality and is a major problem during postharvest storage and transportation. Harvest should be done during the cool part of the day in order to keep turgidity at its optimum.

ORGANIC ACIDS

The role of organic acids in fruits is twofold:

1. *Organic acids are an integral part of many metabolic pathways, especially the Krebs (TCA) cycle.* The tricarboxylic acid cycle (respiration) is the main channel for the oxidation of organic acids and serves as an important energy source for the cells. Organic acids are metabolized into many constituents, including amino acids. Major organic acids present in fruits include citric acid, malic acid, and quinic acid.
2. *Organic acids are important contributors to the taste and flavor of many fruits and vegetables.* Total titratable acidity, the quantity and specificity of organic acids present in fruits, and other factors influence the buffering system and the pH. Acid content decreases during ripening, because part of it is used for respiration, and another part is transformed into sugars (gluconeogenesis). The composition of organic acids in some fruits is given in (Table 22.3).

PIGMENTS

Pigments are mainly responsible for the skin and flesh colors in fruits and vegetables. They undergo changes during maturation and ripening of the fruits, including loss of chlorophyll (green color), synthesis and/or revelation of carotenoids (yellow and orange), and biosynthesis of anthocyanins (red, blue, and purple) (see Table 22.6, below). Anthocyanins occur as glycosides and are water soluble, unstable, and easily hydrolyzed by enzymes to free anthocyanins, which may be oxidized to brown products by phenol oxidases after wounding.

PHENOLICS

Fruit phenolics include chlorogenic acid, catechin, epicatechin, leucoanthocyanidins and anthocyanins, flavonoids, cinnamic acid derivatives, and simple phenols. The total phenolic content of fruits varies from 1 to 2 g/100 g fresh weight. Chlorogenic acid is the main substrate responsible for the enzymatic browning of damaged fruit tissue. Phenolic compo-

Table 22.3. Organic Acids of Selected Fruits

Fruit	Organic Acid (mg/100 mL of juice)				
	Citric	Ascorbic	Malic	Quinic	Tartaric
Apple	ND	tr	518±32	ND	ND
Cherry	ND	tr	727±20	ND	ND
Grape	tr	tr	285±58	ND	162±24
Nectarine	730±92	114±6	501±42	774±57	tr
Peach	140±39	tr	383±67	136±28	ND
Pear	109±16	tr	358±72	121±11	tr
Plum	ND	tr	371±16	220±2	ND
Kiwifruit	ND	tr	294±24	214±68	ND
Strawberry	207±35	56±4	199±26	ND	ND

Source: Reprinted with permission from Van Gorsel et al., J Agr Food Chem, 1992, 40:784–789. Copyright, American Chemical Society.
ND = Not detected, tr = (< 10 mg/100 mL).

nents are oxidized by polyphenol oxidase enzyme in the presence of O_2 to brown products according to the following reactions:

$$\text{Monophenol} + O_2 \xrightarrow{\text{Polyphenol oxidase}} \text{quinone} + H_2O$$

$$\text{1,4,diphenol} + O_2 \xrightarrow{\text{Polyphenol oxidase}} \text{1,4 quinoine} + H_2O$$

Phenolic components are responsible for the astringency in fruits and decrease with maturity, because of their conversion from soluble to insoluble forms, and metabolic conversions.

Cell Structure

The fruit is composed of three kinds of cells: parenchyma, collenchyma, and sclerenchyma. *Parenchyma cells* are the most abundant cells found in fruits. These cells are mostly responsible for the texture resulting from turgor pressure. *Collenchyma cells* contribute to the plasticity of fruit material. The collenchyma cells usually are located closer to the periphery of fruits. Collenchyma cells have thickened lignified cell walls and are elongated in their form. *Sclerenchyma cells,* also called stone cells (e.g., in pears), are responsible for the gritty/sandy texture of pears. They have very thick lignified cell walls and are also responsible for the stringy texture in such commodities as asparagus. For some commodities such as asparagus, the number of sclerenchyma cells increases after harvest and during handling and storage.

FRUIT PROCESSING

Harvesting and Processing of Fruits

Harvesting is one of the most important of fruit-growing activities. It represents about half of the cost involved in fruit production because most fruit crops are still harvested by hand. Improvements are needed for the development of mechanical harvesters. Production of fruits of nearly equal size that are resistant to mechanical damage during harvest is one area addressed by breeding technology. Certain fruits that are not to be processed are harvested with stems in order to avoid microbial invasion. Examples of this include cherries and apples.

Harvested fruit is washed to remove soil, microorganisms, and pesticide residues and then sorted according to size and quality. Fruit sorting can be done by hand or mechanically. There are two ways of mechanically sorting fruits: (1) sorting with water, which takes advantage of changes in density with ripening, and (2) automatic high-speed sorting, in which compressed air jets separate fruit in response to differences in color and ripeness as measured by light reflectance or transmittance. When not marketed as fresh, fruits are processed in many ways including canning, freezing, concentration, and drying.

Freezing and Canning of Fruits

For freezing and canning of fruits, the general sequence of operations include (1) washing; (2) peeling; (3) cutting, trimming, and coring; (4) blanching; (5) packaging and sealing; and (6) heat processing or freezing

Freezing is generally superior to canning for preserving the firmness of fruits. In order to minimize postharvest changes, the freezing process is done at the harvest site for certain fruits. Fruit to be frozen must be stabilized against enzymatic changes during frozen storage and on thawing. The principal enzymatic changes are oxidations, causing darkening of color and the production of off-flavor. An important color change is enzymatic browning in apples, peaches, and bananas. This is due to the oxidation of pigment precursors such as O-diphenols and tannins by enzymes of the group known as phenol oxidases and polyphenol oxidases. Frozen fruits are produced for home use, restaurants, baking manufacturers, and other food industries. Depending on the intended end use, various techniques are employed to prevent oxidation.

The objective of blanching is to inhibit various enzyme reactions that reduce the quality of fruits. Fruits generally are not heat blanched because the heat causes loss of turgor, resulting in sogginess and juice drainage after thawing. An exception to this is when the frozen fruit is to receive heating during the baking operation, in which case calcium salts are added to the blanching water or after blanching. Calcium salts increase the firmness of the fruit by forming calcium pectates. For the same purpose, pectin, carboxymethyl cellulose, alginates, and other colloidal thickeners can also be added to fruit prior to freezing. Instead of heat treatments, certain chemicals can be used to inactivate oxidative enzymes or to act as antioxidants. The major antioxidant treatments include ascorbic acid dip and treatment with sulphites. Other methods for the prevention of contact with oxygen using physical barriers (sugar syrup) can be used.

Commonly dissolved in sugar syrup, vitamin C acts as an antioxidant and protects the fruit from darkening by subjecting itself to oxidation. A concentration of 0.05–0.2% ascorbic acid in syrup is effective for apple and peach. Peaches subjected to this treatment may not darken during frozen storage at −18°C for up to 2 years. Because low pH also helps in delaying oxidative processes, vitamin C and citric acid may be used in combination. Furthermore, citric acid removes cofactors of oxidative enzymes such as copper ions in polyphenol oxidase.

Sulphite is a multiple function antioxidant. Bisulphites of sodium or calcium are used to stabilize the color of fresh fruits. Like vitamin C, sulphite is an oxygen acceptor and inhibits the activity of oxidizing enzymes. Furthermore, sulphite reduces the nonenzymatic Maillard-type browning reactions by reacting with aldehyde groups of sugars so that they are no longer free to combine with amino acids. Inhibition of browning is especially important in dried fruits such as apples, apricots, and pears. Sulphite also exhibits antimicrobial properties. The disadvantage of sulphite is that some people are allergic to this chemical, which makes its use restricted; for example, FDA prohibited the use of sulphite in fresh produce and limited the residual sulphite in processed products to 10 ppm, with appropriate labeling.

Sugar syrup had long been used to minimize oxidation before the mechanisms of browning reactions were understood, and still remains as a common practice today. Application of sugar syrup provides a coating to the fruit and thus prevents contact of the cut surface and oxidizable components with atmospheric oxygen. Sugar syrup also increases the sensory attributes of fruits by reducing the loss of volatiles and improving the taste.

Vacuum treatment is used in combination with chemical dips in order to improve their effects. While fruits are submerged in the dip, vacuum is applied to draw air from the fruit tissue, allowing better penetration.

In order to reduce the cost of handling and shipping, some high-moisture fruits are pureed and concentrated to two or three times their natural solids content. Many others are dried for various purposes to different moisture levels. The majority of fruits including apricot, apple, figs, pears, prunes, and raisins are sun dried. Sulphite is commonly used to preserve the color when fruits are dried under high temperatures that do not inactivate the oxidative enzymes.

Nonenzymatic Browning

Also called Maillard reaction, nonenzymatic browning is of great importance in fruit processing. During the Maillard reaction, the amino groups of amino acids, peptides, or proteins, react with aldehyde groups of sugars, resulting in the formation of brown nitrogenous polymers called melanoidins (Ellis 1959, deMan 1999). The velocity and pattern of the reaction depend on the nature of the reacting compounds, the pH of the medium, and the temperature. Each kind of food shows a different browning pattern. Lysine is the most reactive amino acid because it contains a free amino group. The destruction of lysine reduces the nutritional value of a food since it is a limiting essential amino acid. The major steps involved in the Maillard reaction are as follows:

- An aldose or ketose reacts with a primary amino group of an amino acid, peptide, or protein to produce an N-substituted glycosylamine;
- The Amadori reaction, which rearranges the glycosylamine to yield ketoseamine or aldoseamine;
- Rearrangement of the ketoseamine with a second molecule of aldose to yield diketoseamine, or the rearrangement of the aldoseamine with a second molecule of amino acid to yield a diamino sugar;
- Amino sugars are degraded by losing water to give rise to amino or nonamino compounds; and
- The condensation of the obtained products with amino acid or with each other.

In the Maillard reaction, the basic amino group is the reactive component, and therefore the browning is dependant on initial pH, or the presence of a buffer system. Low pH results in the protonation of the basic amino group; it therefore inhibits the reaction. The effect of pH is also heavily dependent on the moisture content of the product. When the moisture is high, browning is caused by caramelization, whereas at low water content and pH of about 6, the Maillard reaction prevails. The flavors produced by the Maillard reaction in some cases are reminiscent of caramelization.

The nonenzymic browning can be prevented by closely monitoring the factors leading to its occurrence, including temperature, pH, and moisture content. The use of inhibitors such as sulphite is an effective way of controlling nonenzymic browning. It is believed that sulphite reacts with the degradation products of the amino sugars to prevent the condensation of these products into melanoidins. However, sulphite can react with thiamine, which prohibits its use in foods containing this vitamin.

FRUIT JUICE PROCESSING

The quality of the juice depends on the quality of the raw material, regardless of the process. Often the

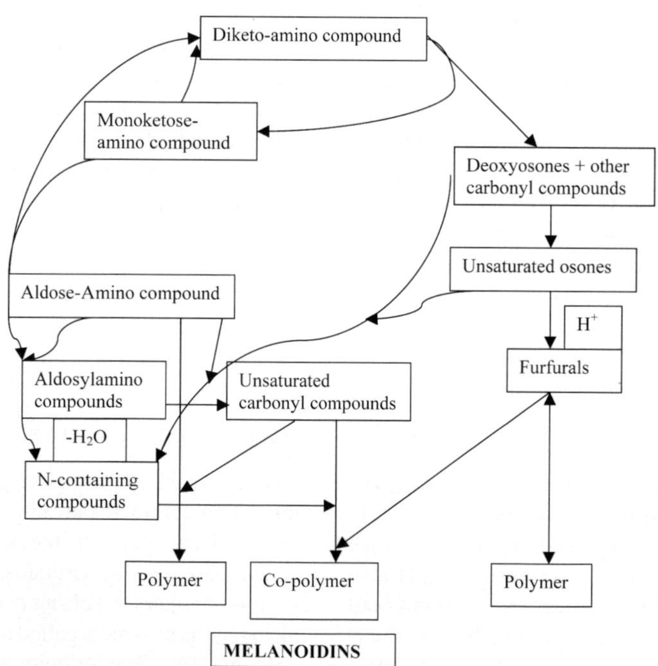

Figure 22.1. Mechanism of browning reaction.

quality of the fruit is dependent on the stage of maturity or the stage of ripening. The major physicochemical parameters used in assessing fruit ripening are sugar content, acidity, starch content, and firmness (Somogyi 1996a,b).

The main steps involved in the processing of most type of juice include the extraction of the juice, clarification, juice deaeration, pasteurization, concentration, essence add-back, canning or bottling, and freezing (less frequent). Juice extractors for oranges and grapefruits, whose peels contain bitter oils, are designed to cause the peel oil to run down the outside of the fruit and not enter the juice stream. Since apples do not contain bitter oil, the whole apples are pressed. Juice extraction should be done as quickly as possible in order to minimize oxidation of phenolic

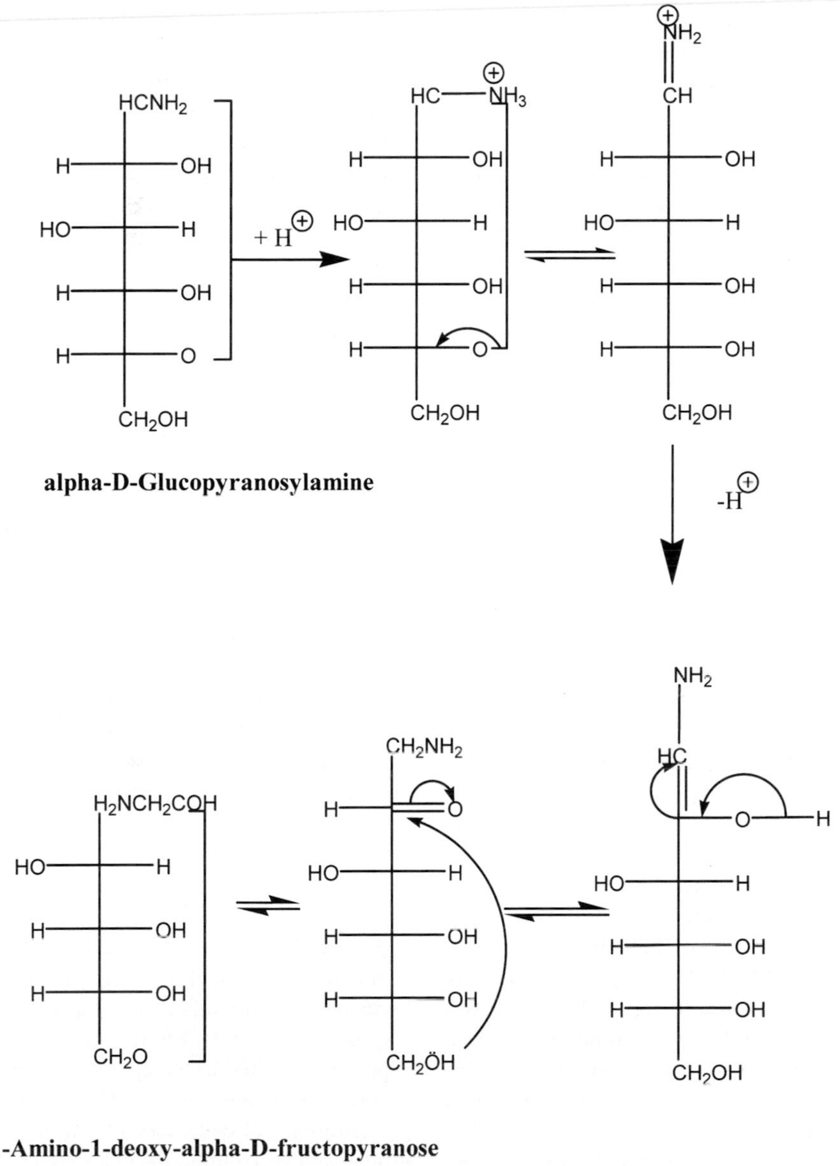

Figure 22.2. Amadori rearrangement. Reactions of free sugars with aqueous ammonia.

compounds in fruit juice by naturally present enzymes.

Apple Juice Processing

Apple juice is processed and sold in many forms including American apple cider, European apple cider, and shelf-stable apple juice. American apple cider is the product of sound, ripe fruit that has been pressed and bottled or packaged with no form of preservatives added and stored under refrigeration. This is a sweet type of cider, which has not been fermented. Because of an increased number of foodborne outbreaks of *E. coli* 0157:H7 and *Cryptosporidium parvum* in apple cider, many jurisdictions are working toward implementation of a kill step in the processing of apple cider. The recent outbreaks make apple cider not suitable for children, the elderly, and immuno (1) compromised people.

European-type apple cider is a naturally fermented apple juice, usually fermented to a specific gravity of 1 or less (Anonymous 1980). Shelf-stable apple juice includes clarified and crushed apple juice and concentrates. The most popular type of apple juice is the one that has been clarified and filtered before pasteurization and bottling. Natural juice is a juice that comes from the press. Up to 2% of ascorbic acid can be added before pasteurization and bottling in order to preserve the color. Crushed apple juice is produced by passing coarsely ground apples through a pulper and a juice extraction device before pasteurization.

Juice concentrates can be of two types: (1) frozen, when the Brix is about 42°, or (2) regular (commercial) concentrate, with a Brix of 70° or more. Regardless of the type of apple product that is to be produced, the quality of the apple juice is directly related to the quality of the raw material used in its production. Only sound ripe fruits should be used in the processing of apple juice. Windfalls, apples that are fallen and picked from the ground, should not be used for the juice due to the risk of contamination by pathogens such as *E. coli* O157:H7, *Cryptosporidium parvum*, *Penicillium expansium* (producing the mycotoxin patulin) and also the risk of pronounced musty or earthy off-flavor in the juice. Overmature apples are very difficult to process, whereas immature apples give a starchy and astringent juice with poor flavor.

The processing of apple juice starts with washing and sorting of the fruit in order to remove soil and other foreign material as well as decayed fruits. Any damaged or decayed fruit should be removed or trimmed in order to keep down the level of patulin in the finished juice. Patulin is an indicator that tells if the juice was produced from windfalls or spoiled apples. The acceptable level of patulin in most countries is less than 50 ppb. Patulin is carcinogenic and teratogenic. Various methods are currently used to reduce the levels of patulin in apple juice, namely, charcoal treatment, chemical preservation (with sulfur dioxide), gamma irradiation, fermentation, and trimming of fungus-infected apples. A recent study shows that pressing followed by centrifugation resulted in an average toxin reduction of 89%. Total toxin reduction using filtration, enzyme treatment, and fining was 70, 73, and 77%, respectively, in the finished juice (Bissessur et al. 2001). Patulin reduction was due to the binding of the toxin to solid substrates such as the filter cake, pellet, and sediment.

Prior to pressing, apples are ground using disintegrators, hammers, or grating mills. The effectiveness of the pressing operation depends on the maturity level of the fruits, as more mature fruits are often difficult to press. A wide range of presses are used in juice extraction including hydraulic, pneumatic, and screw/basket types. The vertical hydraulic press is a batch-type press that requires no press aid. Its main disadvantage is that the press is labor intensive and produces juice with low solids content. Hydraulic presses are the oldest type of press and are still used worldwide. The newer versions are automated and require press aids such as up to 1–2% paper pulp, rice hulls, or both in order to reduce spillage and increase juice channel in the mash.

Although the apple mash has many natural enzymes, enzymatic mash treatment has been developed to improve the pressability of the mash and, therefore, the throughput and yield. Pectinolytic enzyme products contain the primary types of pectinases, pectin methylesterase (PME), polygalacturonase (PG), pectin lyase, and pectin transeliminase (PTE). PME deesterifies methylated carboxylic acid moieties of pectin, liberating methanol from the side chain, after which PG can hydrolyze the long pectin chains (Fig. 22.3). From 80 to 120 mL of enzyme per ton is added to the apple mash in order to break down the cell structure. (High molecular weight constituents of cell walls, like protopectin, are insoluble, and they inhibit the extraction of the juice from the fruit and keep solid particles suspended in the juice.) Pectinase used in apple juice processing is extracted from the fungus *Aspergillus niger*. Pectinase developed for apple mash pretreatment acts mainly on the cell wall, breaking the structure and freeing the juice. Also, the viscosity of the juice is lowered, and it can emerge more easily from the mash. The high content of pectin esterase (PE) in the mash causes the formation of deesterified pectin fragments, which have a low water-binding capacity and reduce the slipperiness of the mash. These pectins consist of chains of galacturonic acid joined by alpha-glycoside linkage. In addition, polymers of xylose, galactose, and arabinose (hemicelluloses) form a link with the cellulose. The entire system forms a gel that retains the juice in the mash. The extraction is easier, even if the pectins are partially broken by pectin esterase. The pomace acts like a pressing aid, when used with mash predraining. The application of enzyme treatment can increase the press throughput by 30–40% and the juice yield by over 20%. Mash pretreatment also increases the flux rate of ultrafiltered apple juice by up to 50%.

An important by-product of apple juice industry is pectin, so overtreatment of mash with pectinolytic enzymes could render the pomace unsuitable for the production of pectin. Inactivation of the enzyme after reaching the appropriate level of pectin degradation is therefore an important step in the production of apple juice. Also, residual pectic enzymes in apple juice concentrate could cause setup problems, for example, when concentrate is used for making apple jelly.

One recent development in apple juice extraction is the process of liquefaction. Liquefaction is a process of completely breaking down the mash by using an enzyme preparation, temperature, and time combination. The liquefied juice is extracted from the residual solid by the use of decanter centrifuges and rotary vacuum filters. It is a common practice, with the addition of cellulose enzyme to the mash, to further degrade the cellulose to soluble solids, increasing the juice Brix by nearly 5°. Commercially available enzyme preparations contain more than 120 substrate-specific enzyme components.

Another popular extraction method is the countercurrent extraction method, developed in South Africa and refined in Europe in the 1970s. The principle of the system is as follows: the mash is heated, predrained, and counter-washed with water and recycled hot juice. A 90–95% recovery is possible when the throughput is about 3 tons/hour. The main disadvantage of the system is the lower soluble solids content of the juice obtained (6–8 versus 11–12 from other methods). Juice yield from different types of extraction varies from 75 to 95% and depends on many factors, including the cultivar and maturity of the fruit, the type of extraction, the equipment and press aids, the time, the temperature, the addition of enzyme to the apple mash, and the concentration of the added enzyme.

Enzyme Applications

Enzyme applications in apple juice processing follow extraction, especially when producing a clarified type of juice. The main objective is to remove the suspended particles from the juice (Smock and Neubert 1950). The soluble pectin in the juice has colloidal properties and inhibits the separation of the undissolved cloud particles from the clear juice. Pectinase hydrolyzes the pectin molecule so it can no longer hold juice. The treatment dosage of pectinase depends on the enzyme strength and varies from one manufacturer to another. A typical "3X" enzyme dosage is about 100 mL/4000 L of raw juice. Depectinization is important for viscosity reduction and the formation of galacturonic acid groups that help flocculate the suspended matter. This material, if not removed, binds to the filters, reduces production, and can result in haze formation in the final product. There are two methods of enzyme treatment commonly used in the juice industry: (1) hot treatment, where the enzymes are added to 54°C juice, mixed, and held for 1–2 hours; and (2) cold treatment, where the enzymes are added to the juice at reduced temperature (20°C) and held for 6–8 hours. The complete breakdown of the pectin is monitored by means of an acidified alcohol test. Five milliliters of juice are added to 15 mL of HCl-acidified ethyl alcohol. Pectin is present if a gel develops in 3–5 minutes after mixing the juice with the ethanol solution. The absence of gel formation means the juice depectination is complete. In the cloud of the postprocessed juice, other polymers such as starch and arabinans can be present; therefore, the clear juice is treated with alpha-amylase and hemicellulase enzymes, in order to partially or completely degrade the polymers. Gelatin can be used to remove fragmented pectin chains and tannins from the juice. Best results are obtained when

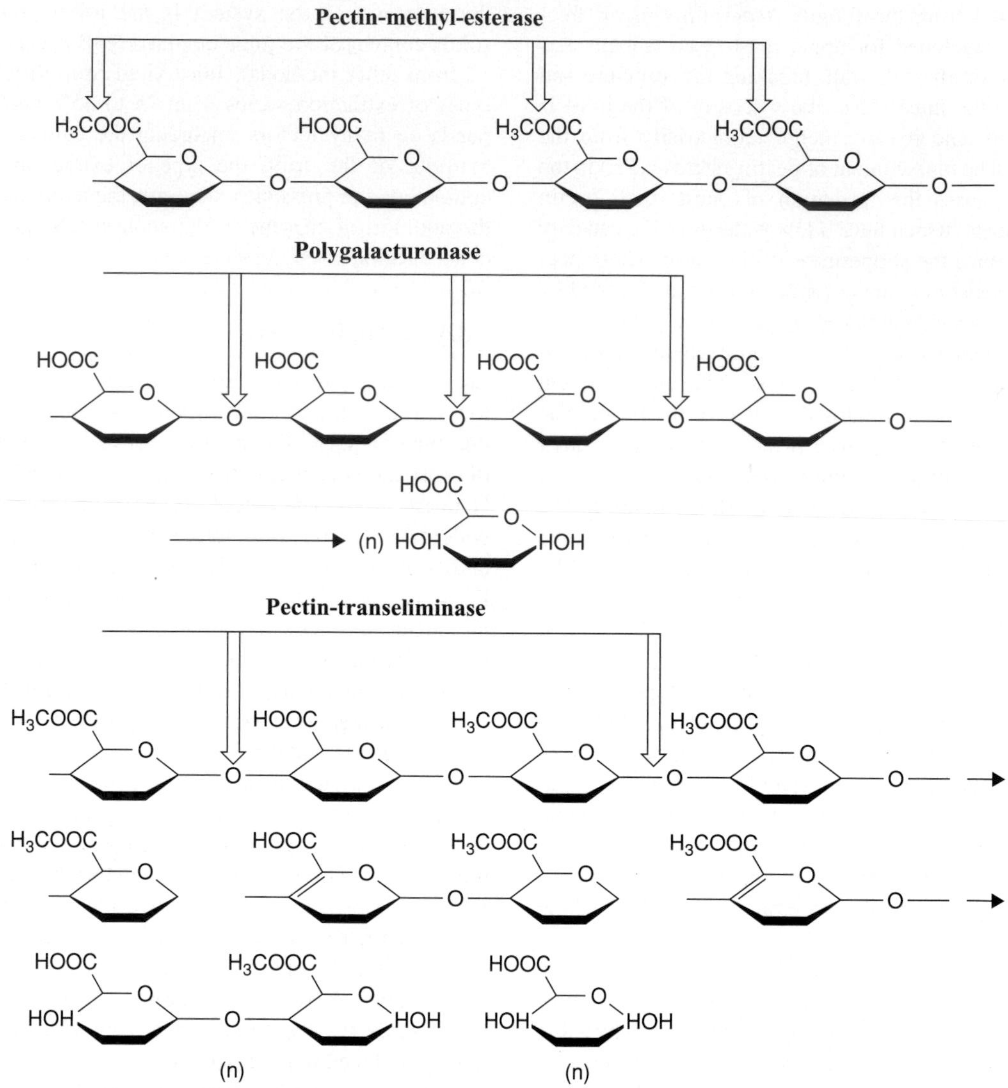

Figure 22.3. Pectin degradation by pectin methylesterase, polygalacturonase, and pectin transeliminase.

hydrating 1% gelatin in warm 60°C water. Gelatin can be added either in combination with the enzyme treatment, bentonite, or midway through the enzyme treatment period. The positively charged gelatin facilitates the removal of the negatively charged suspended colloidal material from the juice. Bentonite can be added in the range of 1.25–2.5 kg of rehydrated bentonite per 4000 L of juice. Bentonite is also added to increase the efficiency of settling, protein removal, and prevention of cloudiness caused by metal ions. For cloudy and natural apple juice, enzymes are usually not used. The enzyme-treated, refined, and settled apple juice is then pumped from the settled material (lees) and further clarified by filtration.

The filtration of apple juice is done with or without a filter aid. The major types of filters are pressure leaf, rotary vacuum, frame, belt, and Millipore filters. To obtain the desired product color and clarity, most juice manufacturers use a filter medium or

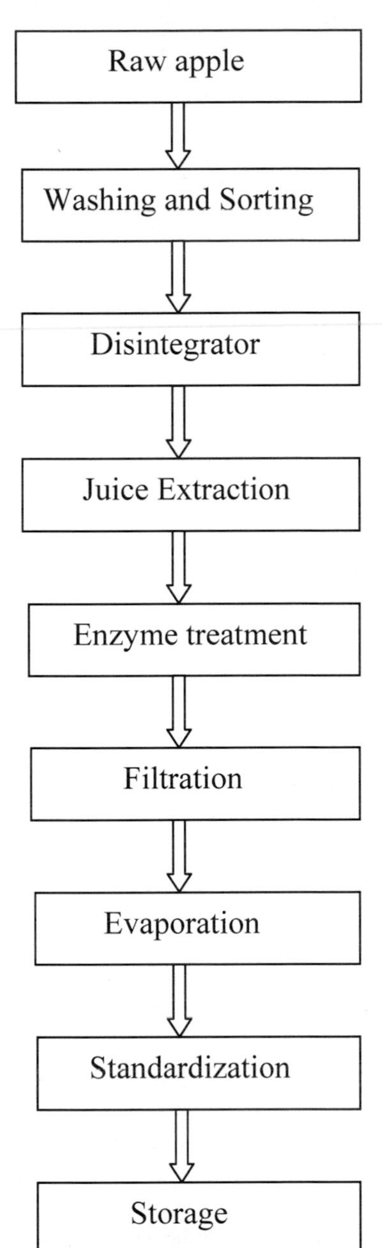

Figure 22.4. Apple juice and concentrate flowchart.

filter aid in the filtration process. The filter media include diatomaceous earth, paper pulp pads, cloth pads or socks, and ceramic membranes, to name a few. The filter aid helps prevent the blinding of the filters and increases throughput. As the fruit matures, more filter aid will be required. Several types of filter aids are available; the most commonly used are diatomaceous earth or cellulose-type materials. Additional juice can be recovered from the tank bottoms or "lees" by centrifugation or filtration. This recovered juice can be added to the raw juice before filtration.

Diatomaceous earth, or kieselguhr, is a form of hydrated silica. It has also been called fossil silica or infusorial earth. Diatomaceous earth is made up of the skeletal remains of prehistoric diatoms, single-celled plants that are related to the algae that grow in lakes and oceans. Diatomaceous earth filtration is a three-step operation:(1) building up a firm, thin protective precoat layer of filter aid, usually of cellulose, on the filter septum (which is usually a fine wire screen, synthetic cloth, or felt); (2) use of the correct amount of a diatomite body feed or admix (about 4.54 kg/3000 cm^2 of filter screen); and (3) the separation of the spent filter cake from the septum prior to the next filter cycle.

Prior to the filtration, centrifugation may be used to remove high molecular weight suspended solids. In some juice plants, high-speed centrifugation is used prior to the filtration. This centrifugation step reduces the solids by about 50%, thus minimizing the amount of filter aid required. The filtration process is critical, not only from the point of view of production, but also for the quality of the end product. Both pressure and vacuum filters have been used with success in juice production (Nelson and Tresler 1980).

A recent development in the juice industry is membrane ultrafiltration. Ultrafiltration, based on membrane separation, has been used with good results to separate, clarify, and concentrate various food products. Ultrafiltration of apple juice not only will clarify the product, but also, depending on the size of the membrane, will remove the yeast and mold microorganisms common in apple juice (Fig. 22.5).

Apple Juice Preservation

Preservation of apple juice can be achieved by refrigeration, pasteurization, concentration, chemical treatment, membrane filtration, or irradiation. The most common method is pasteurization based on temperature and time of exposure. The juice is heated

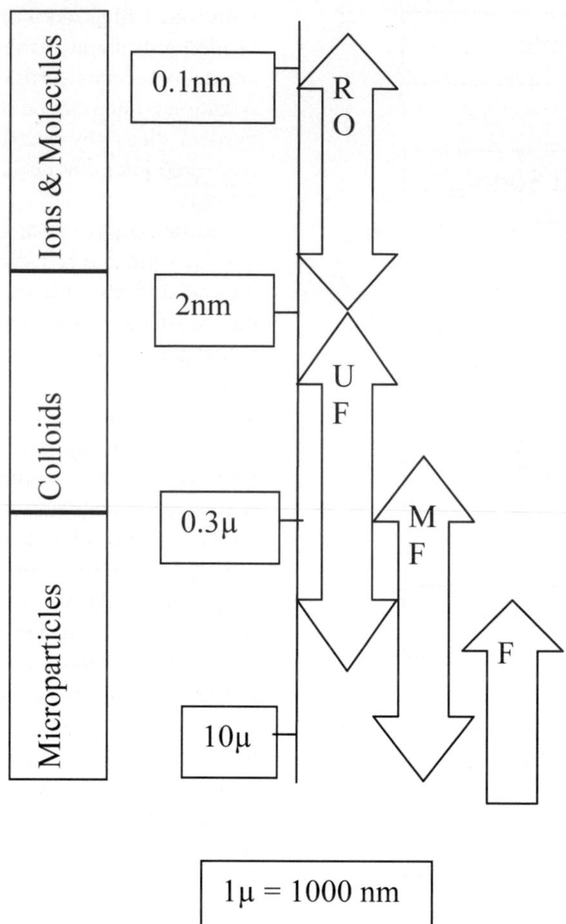

Figure 22.5. Membrane filtration. Key: RO, reverse osmosis; UF, ultrafiltration; MF, microfiltration; F, filtration.

to over 83°C, held for 3 minutes, filled hot into the container (cans or bottles), and hermetically sealed. The apple juice is held for 1 minute at 83°C and then cooled to less than 37°C. When containers are closed hot and then cooled, a vacuum develops, reducing the available oxygen, which also aids in the prevention of microbial growth. After heat treatment, the juice may also be stored in bulk containers. Aseptic packaging is another common process in which, after pasteurization, the juice is cooled and packed in a closed, commercially sterile system under aseptic conditions. This process provides shelf-stable juice in laminated, soft-sided consumer cartons, bag-in-box cartons, or aseptic bags in 200–250 L drums.

Apple juice concentration is another common method of preservation. The single-strength apple juice is concentrated by evaporation, preferably to 70° or 71° Brix. By an alternate method, the single-strength juice is preconcentrated by reverse osmosis to about 40° Brix, then further concentrated by evaporation methods. The reduced water content and natural acidity make the final concentrated apple juice shelf stable at room temperature. There are several evaporators used in apple juice production including rising-film evaporators, falling-film evaporators, and multieffect tubular and plate evaporators. Due to the heat sensitivity of the apple juice, the multieffect evaporator with aroma recovery is the most commonly used. The general method in a multieffect evaporator is to heat the juice to about 90°C and capture the volatile (aroma) components by cooling and condensation. This is followed by reheating the 20–25° Brix juice concentrate from the first stage to about 100°C and concentrating it to about 40–45° Brix. Another stage of heating and evaporation at about 50°C to 60° Brix, and final heating and concentration in the fourth stage to 71° Brix, provides fully concentrated juice. The warm concentrate is then chilled to 4–5°C prior to adjusting the Brix to 70° before barreling or bulk storage.

Chemicals such as benzoic acid, sorbic acid, and sulphite are sometimes used to reduce spoilage of unpasteurized apple juice, either in bulk or as an aid in helping to preserve refrigerated products. Irradiation and ultrasonication are new, emerging methods of preservation with high potential, even though they are not fully accepted by consumers at this time.

Apple essence is recovered during the concentration of apple juice. The identification of volatile apple constituents, commonly known as essence or aroma, has been the subject of considerable research. In 1967, researchers at the U.S. Department of Agriculture (USDA) identified 56 separate compounds from apple essence. These compounds were further refined by organoleptic identification using a trained panel of sensory specialists. These laboratory evaluations revealed 18 threshold compounds identified as "Delicious" apple compounds consisting of alcohols, aldehydes, and esters. Three of the 18 compounds had "apple-like aromas" according to the taste panel. These were 1-hexanal, trans-2-hexenal, and ethyl-2-methyl butyrate (Flath et al. 1967, Somogyi et al. 1996b).

Processed Apple Products

The popularity of blended traditional and tropical juice products has pushed the consumption of traditional fruit juices such as apple, orange, and grape juices to 25 L in 2002, an increase of more than 24% from 1992 (Statistics Canada 2003). Apples are normally processed into a variety of products, although apple juice is the most popular processed apple product. With a production of 465,418 metric tons (mt) of apple in 2001, Canada exported 18,538 mt of concentrated apple juice. For the same year, the United States exported 25,170 mt of concentrated apple juice (FAO 2003). Apples for processing should be of high quality, proper maturity, of medium size, and uniform in shape. Apples are processed into, frozen, canned, dehydrated apple slices and dices, and different kinds of applesauce. Apples that are unsuitable for peeling are diverted to juice processing.

Applesauce

Diced or chopped apples with added sugar, preferably a sugar concentrate, are cooked at 93–98°C for 4–5 minutes in order to soften the fruit and inactivate polyphenol oxidase. Sauce with a good texture, color, and consistency is produced with high quality raw apples and a good combination of time and temperature treatment. Cooked applesauce is passed through a pulper with a 1.65–3.2 mm finishing screen to remove unwanted debris and improve the texture. Applesauce is then heated to 90°C and immediately filled in glass jars or metal cans. The filled containers are seamed or capped at 88°C and cooled to 35–40°C after 1–2 minutes (Fig. 22.6). There are various types of applesauce; these include natural, no sugar added, "chunky," and cinnamon applesauces, and mixtures of applesauce and other fruits such as apricot, peach, or cherry.

Sliced Apples

Sliced apples have multiple uses and are preserved by many different states including canned, refrigerated, frozen, or dehydrated states. About 85% of sliced apples are processed, whereas only 15% are refrigerated, frozen, or dehydrated and frozen (dehydrofrozen). Apple slice texture and the consistency of the slice size are very important; therefore, apples with firm flesh and high quality, falling within a specified size range, are desirable. Apples are sliced into 12 to 16 slices and blanched after inspection for eventual defects. The blanched apple slices are hot-filled into cans and closed under steam vacuum after the addition of hot water or sugar syrup. The canned apples are heated at 82.2°C and immediately cooled to 37–40°C.

When bulk frozen, apple slices are vacuum treated, blanched, and filled into 13–15 kg tins or poly-lined boxes. The tins or boxes are then sealed, frozen, and stored at −17°C.

Individual quick frozen apple slices (IQF) are usually treated with sodium bisulphite after inspection. Nitrogen (N_2) and carbon dioxide (CO_2) are the

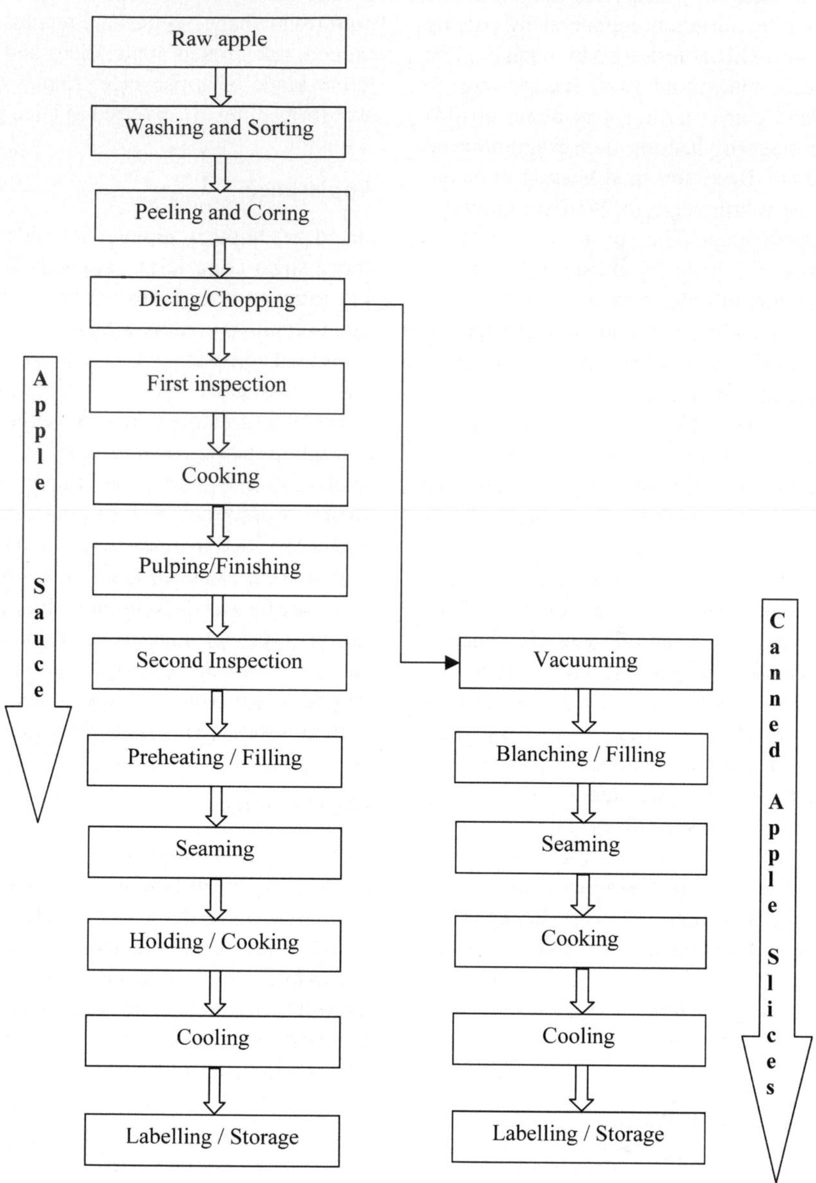

Figure 22.6. Apple processing flowchart.

most popular freezing media. From the vacuum tank, the apple slices pass through an IQF freezing unit where the slices are individually frozen. From the freezing unit the slices are filled into tins or poly-lined boxes and stored frozen at −17°C or below.

Dehydrofrozen apple slices are dehydrated to less than 50% of their original weight and then frozen. The dehydrofrozen slices are packed in cardboard containers or large metal cans with polyethylene liners and rapidly frozen before storing. Frozen slices are thawed and then soaked in a combined solution of sugar, $CaCl_2$, and ascorbic acid or bisulphite. For fresh and refrigerated apple slices the use of 0.1–0.2% calcium chloride and ascorbate protects apple slices from browning and microbial spoilage. Blanched slices resist browning for approximately 2 days, but lose flavor, sugar, and acid. This type of product has a very short shelf life. Recently, the use of natural protein polymer coatings (NatureSeal) has shown promise in enhancing the shelf life and quality of fresh cut apple products.

Dried Apple Products

Most processing cultivars are used for drying, but the best quality dried apple slices are obtained from "Red Delicious" and "Golden Delicious" apples. A desirable quality attribute of apples used for drying is a high sugar/water ratio. Color preservation and reduction in undesirable enzyme activities are achieved by the use of bisulphite. There are two types of dried apple products including evaporated and dehydrated apples. Evaporated apples are cut into rings, pie pieces or dices and then dried to less than 24% moisture by weight. The dehydrated apples on the other hand, are cut into dices, pie pieces, granules and flakes prior to drying to 3–0.5% moisture content. Up to 300 ppm of bisulphite is used to prevent color deterioration. The maximum allowed SO_2 in dried apples in Europe is 500 ppm, whereas in the United States, the limit is 1000 ppm.

Quality Control

Apples contain several organic acids, and therefore only a limited number of microorganisms can grow in them. The most common microbes are molds, yeasts, aciduric bacteria, and certain pathogens such as *E. coli* O157:H7 that are capable of growing at low pH (Swanson 1989). Several approaches to quality management are available today, including total quality management (TQM), statistical quality control (SQC), and hazard analysis critical control points (HACCP).

Biochemical Composition and Nutritional Value of Processed Apples

The nutritive value of most processed apple products is similar to that of the fresh raw product. Apple products are sources of potassium, phosphorus, calcium, vitamin A, and ascorbic acid. Glucose, sucrose, and fructose are the most abundant sugars. Dried or dehydrated apples have a higher energy value per gram tissue due to the concentration of sugars (Tables 22.4 and 22.5).

The nutritional value of apples, and fruit in general, is enhanced by the presence of flavonoids. More than 4000 flavonoids have been identified to date. There are many classes of flavonoids, of which flavanones, flavones, flavonols, isoflavonoids, anthocyanins, and flavans are of interest. Flavanones occur predominantly in citrus fruits; anthocyanins, catechins, and flavonols are widely distributed in fruits; and isoflavonoids are present in legumes. Structurally, flavonoids are characterized by a C_6—C_3—C_6 carbon skeleton (Fig. 22.7). They occur as aglycones (without sugar moieties) and glycosides (with sugar moieties). Processing may decrease the content of flavonoids by up to 50%, due to leaching into water or removal of portions of the fruit, such as the skin, that are rich in flavonoids and anthocyanins. It is estimated that the dietary intake of flavonoids may vary from 23 to 1000 mg/day in various populations (Peterson

Table 22.4. Protein, Fat, and Carbohydrate Nutrients in Fruits and Processed Products (454 g)

Apples	Food Energy (kcal)	Protein (g)	Fat (g)	Carbohydrate (g)
Raw fresh	242	0.8	2.5	60.5
Applesauce[a]	413	0.9	0.5	108.0
Unsweetened Juice	186	0.9	0.9	49.0
Apple juice	213	0.5	0.1	54.0
Froz. Sliced[a]	422	0.9	0.5	110.2
Apple butter[a]	844	1.8	3.6	212.3
Dried, 24%	1247	4.5	7.3	325.7
Dehyd. 2%	1601	6.4	9.1	417.8

Source: www.nal.usda.gov/fnic.
[a]With sugar.

Table 22.5. Vitamin and Mineral Nutrients in Fruit and Processed Products (454 g)

Apples	Ca (mg)	P (mg)	Fe (mg)	Na (mg)	K (mg)	Vit A (IU)	Thiamin (mg)	Riboflavin (mg)	Niacin (mg)	Vit C (mg)
Raw fresh	29	42	1.3	4	459	380	0.12	0.08	0.3	16
Applesauce[a]	18	23	2.3	9	295	180	0.08	0.05	0.2	5
Unsweetened juice	18	23	2.3	9	354	180	0.08	0.05	0.2	5
Apple juice	27	41	2.7	5	458	–	0.03	0.07	0.4	4
Froz. sliced[a]	23	27	2.3	64	308	80	0.05	0.014	1.0	33
Apple butter[a]	64	163	3.2	9	1143	0	0.05	0.09	0.7	9
Dried, 24%	141	236	7.3	23	22,581	–	0.26	0.053	2.3	48
Dehyd. 2%	181	299	9.1	32	3311	–	0.02	0.026	2.9	47

Source: www.nal.usda.gov/fnic.
[a]With sugar.

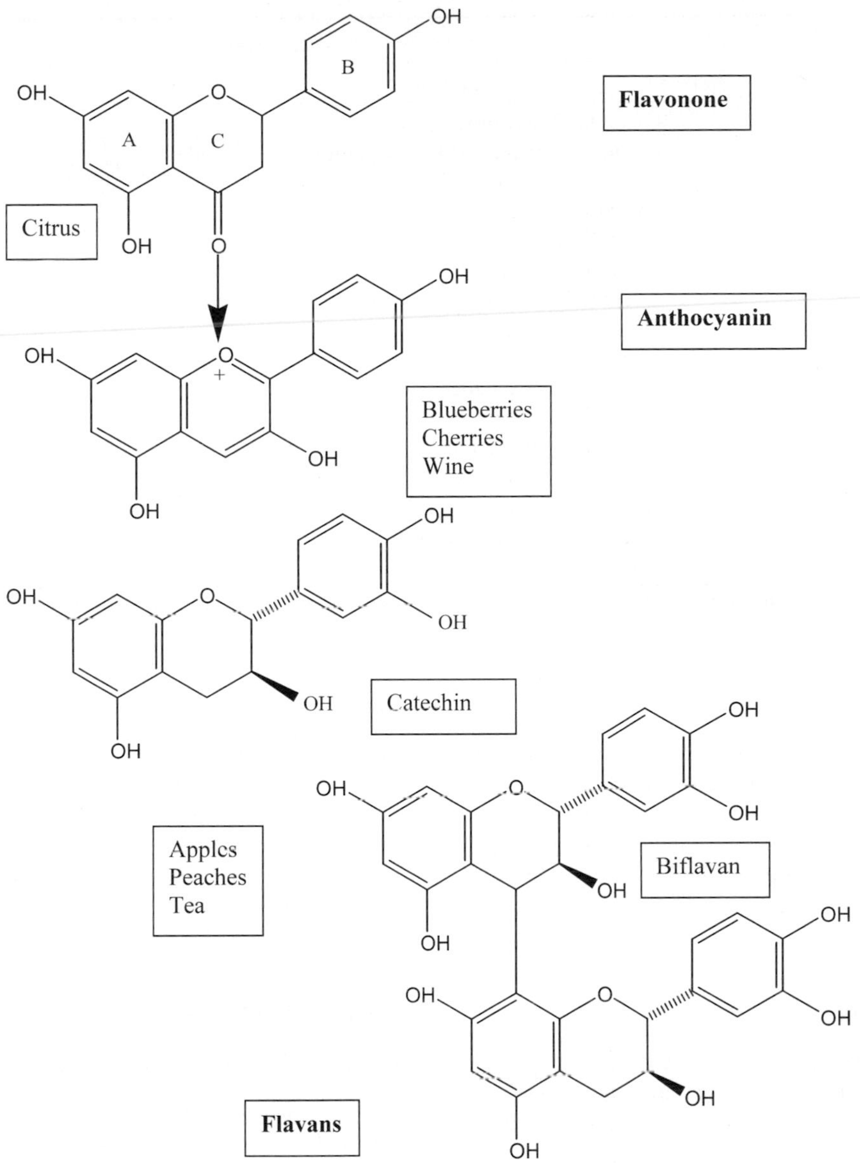

Flavonone

Citrus

Anthocyanin

Blueberries
Cherries
Wine

Catechin

Biflavan

Apples
Peaches
Tea

Flavans

Figure 22.7. Chemical structures of fruit flavonoids. The propanoid structure consists of two fused rings. "A" is the aromatic ring, and "C" is a heterocyclic ring attached by carbon-carbon bond to the "B" ring. The flavonoids differ in their structure at the "B" and "C" rings.

and Dwyer 1998). Flavonoids are increasingly recognized as playing potentially important roles in health, including but not limited to their roles as antioxidants. Epidemiological studies suggest that flavonoids may reduce the risk of developing cardiovascular diseases and stroke.

Flavanones

The major source of flavanones is citrus fruits and juices. Flavanones contribute to the flavor of citrus; for example, naringin, found in grapefruit, provides the bitter taste, whereas hesperidin, found in oranges, is tasteless.

Table 22.6. Anthocyanins of Selected Fruits

Fruit	Anthocyanin	
	Identification	Peak Area
Apple	Cyanidin 3-arabinoside	22±6
	Cyanidin 7-arabinoside or Cyanidin 3-glucoside	27±15
	Cyanidin 3-galactoside	260±69
Cherry	Cyanidin 3-glucoside	189±40
	Cyanidin 3-rutinoside	1320±109
	Peonidin 3 rutinoside	47±36
Grape	Cyanidin 3-glucoside	121±33
	Delphinidin 3-glucoside	586±110
	Malvidin 3-glucoside	2157±375
	Peonidin 3-glucoside	478±92
Nectarine	Cyanidin 3-glucoside	322±51
Peach	Cyanidin 3-glucoside	180±43
Plum	Cyanidin 3-glucoside	42±5
Strawberry	Cyanidin 3-glucoside	70±18
	Pelargonidin 3-glucoside	1302±29
	Pelargonidin 3-glycoside	78±9

Source: Reprinted with permission from Van Gorsel et al., J Agr Food Chem, 1992, 40:784–789. Copyright, American Chemical Society.

Flavonols

The best-known flavonols are quercetin and kaempferol. Quercetin is ubiquitous in fruits and vegetables. Kaempferol is most common among fruits and leafy vegetables. In fruits, flavonols and their glycosides are found predominantly in the skin. Myricetin is found most often in berries.

Anthocyanins

Anthocyanins (Table 22.6) often occur as a complex mixture. Grape extracts can have glucosides, acetyl glucosides, and coumaryl glucosides of delphinidin, cyanidin, petunidin, peonidin, and malvidin. The color of anthocyanins is pH dependent. An anthocyanin is usually red at a pH of 3.5, becoming colorless and then shifting to blue as the pH increases. Fruit anthocyanin content increases with maturity. In stone fruits (peaches and plums) and pome fruits (apple and pears), anthocyanins are restricted to the skin, whereas in soft fruits (berries), they are present both in skin and flesh. Anthocyanins are used as coloring agents for beverages and other food products.

Flavans

Flavans are what was once called catechins, leucoanthocyanins, proanthocyanins, and tannins. They occur as monoflavans, biflavans, and triflavans. Monoflavans are found in ripe fruits and fresh leaves. Biflavans and triflavans are found in fruits such as apples, blackberries, black currants, cranberries, grapes, peaches, and strawberries.

REFERENCES

Further Reading Materials

Arthey D, Ashwurst PR. 2001. Fruit Processing: Nutrition, Products and Quality Management. Aspen Publishers, Gaithersburg, Maryland. 312 pages.

Enachescu DM. 1995. Fruit and Vegetable Processing. FAO Agricultural Services Bulletin 119. FAO, 382 pages.

Jongen WMF. 2002. Fruit and Vegetable Processing: Improving Quality (Electronic Resource). CRC, Boca Raton, Florida.

Salunkhe DK, Kadam SS. 1998. Handbook of Vegetable Science and Technology: Production, Composition, Storage and Processing. Marcel Dekker Inc., New York. 721 pages.

Cited References

Anonymous. 1980. Cider. National Association of Cider Makers, NACM, Dorchester, Dorset, England.

Bissessur J, Permaul K, Odhav B. 2001. Reduction of patulin during apple juice clarification. J Food Protection 64:1216–1222.

deMan JM. 1999. Principles of Food Chemistry, 3rd edition. Aspen Publishers, Inc., Gaithersburg, Maryland. Pp. 120–137.

Ellis G. P.1959. The Maillard reaction. In: ML Wolfram, RS Tipson, editors. Advances in Carbohydrate Chemistry, vol. 14. Academic Press. New York.

Flath RA, Black DR, Guadagni DG, McFadden WH, Schultz TR. 1967. J Agric Food Chem 15:29–38.

http://atn-riae.agr.ca/applecanada/production-e.htm, accessed on Dec. 03, 2003

http://apps.fao.org/page/collections?subset=agriculture, accessed on October 27, 2003.

http://www.statcan.ca/english/ads/23F0001XCB/highlight.htm, accessed on Dec. 03, 2003.

Nelson PE, Tresler DK. 1980. Fruit and Vegetable Juice Production Technology, 3rd edition. AVI Co., Westport, Connecticut.

Peterson J, Dwyer J. 1998. Flavonoids: Dietary occurrence and biochemical activity. Nutrition Research 18:1995–2018.

Smock RM, Neubert AM. 1950. Apples and Apple Products. Interscience Publishers, New York.

Somogyi LP, Barrett DM, Hui YH. 1996a. Processing Fruits: Science and Technology. Vol. 2, Major Processed Products. Technomic Publishing, Inc., Lancaster, Pennsylvania. Pp. 9–35.

Somogyi LP, Ramaswamy HS, Hui YH. 1996b. Processing Fruits: Science and Technology. Vol. 1, Biology, Principles, and Applications. Technomic publishing, Inc., Lancaster, Pennsylvania. Pp. 1–24.

Swanson KMJ. 1989. Microbiology and Preservation. Processed Apple Products. Van Nostrand Reinhold, New York.

Van Gorsel H, Li C, Kerbel EL, Smits M, Kader AA. 1992. Compositional characterisation of prune juice. J Agric Food Chem, 40:784–789.

23
Biochemistry of Vegetable Processing

M. Oke and G. Paliyath

INTRODUCTION

Vegetables and fruits have many similarities with respect to their composition, harvesting, storage properties, and processing. In the true botanical sense, many vegetables are considered fruits. Thus, tomatoes, cucumbers, eggplant, peppers, and others could be considered as fruits, since fruits are those portions of a plant that house seeds. However, the important distinction between fruits and vegetables is based on their use. Vegetables are considered as plant items that are eaten with the main course of a meal, whereas fruits are generally eaten alone or as a dessert. The United States is one of the world's leading producers and consumers of vegetables and melons. In 2002, the farmgate value of vegetables and melons (including mushrooms) sold in the United States reached $17.7 billion. Annual per capita use

Table 23.1. World Production of Selected Vegetables in 2002

Countries	Production in 2002* (metric tons)	
	Fresh Vegetables	Tomatoes
World	233,223,758	108,499,056
Africa	12,387,390	12,428,174
Asia	199,192,449	53,290,273
Australia	80,000	400,000
Canada	124,000	690,000
European Union (15)	8,462,000	14,534,582
New Zealand	120,000	87,000
North and Central America	2,023,558	15,837,877
South America	3,503,611	6,481,410
United States of America	1,060,000	12,266,810

Countries	Tomato Production in 2002 (metric tons)
World	108,499,056
Canada	690,000
United States of America	12,266,810
Mexico	2,083,558

**Source:* FAOSTAT: http://apps.fao.org/default.htm. Accessed on November 08, 2003.

of fresh vegetables and melons rose 7% between 1990–92 and 2000–02, reaching 442 pounds, as fresh consumption increased and consumption of processed products decreased. According to FAO data, the world production of vegetables has increased in the last decade by 58.4%, from 147 million metric tons (Mt) in 1992 to 233 million Mt in 2002. Vegetables remain a popular choice for consumers worldwide. In 2002, each individual ate an average of 110 kg of vegetables (including potatoes), up from 106 kg a decade earlier. Potatoes represented 35% of all vegetables consumed. In 2003, per capita consumption of fresh or processed potatoes in Canada was 38 kg on the average and included products such as French fries, potato chips, stuffed baked potatoes, or frozen mashed potatoes. That compares with an average intake of 33 kg a decade ago (Statistics Canada 2002). The world production of selected vegetables is shown in Table 23.1.

CLASSIFICATION OF VEGETABLES

Vegetables can be classified according to the part of the plant from which they are derived: leaves, roots, stems, or buds (Table 23.2). They also can be classified into "wet" or "dry" crops. Wet crops such as celery or lettuce have water as their major component, whereas dry crops such as soybeans have carbohy-

Table 23.2. Classification of Vegetables

Types of Vegetables	Examples
Earth vegetables	
Roots	Sweet potatoes, carrots
Modified stems	
Corms	Taro
Tubers	Potatoes
Modified buds	
Bulbs	Onions, garlic
Herbage vegetables	
Leaves	Cabbage, spinach, lettuce
Petioles (leaf stalk)	Celery, rhubarb
Flower buds	Cauliflower, artichokes
Sprouts, shoots (young stems)	Asparagus, bamboo shoots
Fruit vegetables	
Legumes	Peas, green beans
Cereal	Sweet corn
Vine fruits	Squash, cucumber
Berry fruits	Tomato, eggplant
Tree fruits	Avocado, breadfruit

drates, protein, and fat as their major constituents, and water as a minor constituent. Soybeans, for example, are composed of 36% protein, 35% carbohydrate, 19% fat, and 10% water. Wet crops tend to perish more rapidly than dry crops.

CHEMICAL COMPOSITION OF VEGETABLES

Fresh vegetables contain more than 70% water, and very frequently greater than 85%. Beans and other dry crops are exceptions. The protein content is often less than 3.5% and the fat content less than 0.5%. Vegetables are also important sources of digestible and indigestible carbohydrate, as well as of minerals and vitamins. They contain the precursor of vitamin A, beta-carotene, and other carotenoids. Carrots are one of the richest sources of beta-carotene (provitamin A).

VITAMINS

Vegetables are major contributors to our daily vitamin requirements. The nutrient contribution from a specific vegetable is dependent on the amount of vitamins present in the vegetable, as well as the amount consumed. The approximate percentage that vegetables contribute to daily vitamin intake is vitamin A, 50%; thiamine, 60%; riboflavin, 30%; niacin, 50%; vitamin C, 100%. Vitamins are sensitive to different processing conditions including exposure to heat, oxygen, light, free water, or traces of certain minerals. Trimming, washing, blanching, and canning can cause loss in the vitamin content of fruits and vegetables.

MINERALS

The amount and types of minerals depends on the specific vegetable. Not all minerals in plant materials are readily available, and they are mostly in the form of complexes. An example is calcium found in vegetables as calcium oxalate.

DIETARY FIBER

The major polysaccharides found in vegetables include starch and dietary fiber, cellulose, hemicellulose, pectic substances, and lignin. Cell walls in young vegetables are composed of cellulose. As the produce ages, cell walls become higher in hemicellulose and lignin. These materials are tough and fibrous, and their consistency is not affected by processing.

PROTEINS

Most vegetables contain less than 3.5% protein. Soybeans are an exception. In general, plant proteins are major sources of dietary protein in places where animal protein is in short supply. Plant proteins, however, are often deficient or limiting in one or more essential amino acids. Wheat protein is limiting in lysine, while soybean protein is limiting in methionine. That is why multiple sources of plant proteins are recommended in the diet. Enzymes, which catalyze metabolic processes, are important in the reactions involved in fruit ripening and senescence.

LIPIDS

The lipid content of vegetables is less than 0.5% and is primarily found in the cuticles, as constituents of the cell membrane, and in some cases, as a part of the internal cell structure (oleosomes). Even though lipids are a minor component of vegetables, they play an important role in the characteristic aroma and flavor of the vegetable. The characteristic aromas of cut tomato and cucumber result from components released from the lipoxygenase pathway, through the action of lipoxygenase upon linoleic and linolenic acids and the action of hydroperoxide lyase on the peroxidized fatty acids to produce volatile compounds. This action is accentuated when the tissue is damaged. The results of the action of lipoxygenase are sometimes deleterious to quality; for example, the action of lipoxygenase on soybean oil leads to rancid flavors and aromas.

VOLATILES

The specific aroma of vegetables is due to the amount and diversity of volatiles they contain. Volatiles are present in extremely small quantities (< 100 μg/g fresh weight). Characteristic flavors and aromas are a combination of various compounds, mainly short-chain aldehydes, ketones, and organic acids. Their relative importance depends upon threshold concentration (sometimes < 1 ppb) and interaction with other compounds. Of more than 400 volatiles identified in tomato, the following have been reported to play important roles in fresh tomato flavor: hexanal, trans-2-hexenal, *cis*-3-hexenal, *cis*-3-hexenol, trans-2-trans-4 decadienal, 2-isobutylthiazole, 6-methyl-5-hepten-2-one, 1-penten-3-one, and β-ionone (Petro-Turza 1986–87).

Water

In general, water is the most abundant single component of vegetables (up to 90% of total weight). The maximum water content of vegetables varies between individuals due to structural differences. Agricultural conditions also influence the water content of plants. As a major component of vegetables, water impacts both quality and rate of deterioration. Harvest should be done during the cool part of the day in order to keep turgidity at its optimum. Loss of turgor under postharvest storage is a major quality-reducing factor in vegetables (wilting of leaves such as spinach).

Organic Acids

Organic acids are important contributors to the taste and flavor of many vegetables (tomato). Total titratable acidity and the quantity and specificity of organic acids present in vegetables influence the buffering system and the pH. Acid content decreases during maturation, because part of it is used for respiration and another part is transformed into sugars (gluconeogenesis).

Pigments

Pigments are mainly responsible for the skin and flesh colors in vegetables. They undergo changes during maturation and ripening of vegetables, including loss of chlorophyll (green color), synthesis and/or revelation of carotenoids (yellow and orange), and development of anthocyanins as in eggplant (red, blue, and purple). Vegetables such as carrots are especially rich in carotene, and red beets owe their red color to betacyanins. Anthocyanins belong to the group of flavonoids, occur as glycosides, and are water soluble. They are unstable and easily hydrolyzed by enzymes to free anthocyanins. The latter are oxidized to brown products by phenol oxidases. The colors of anthocyanins are pH dependent. In a basic medium they are mostly violet or blue, whereas in an acidic medium they tend to be red.

Phenolics

Phenolic compounds found in vegetables vary in structure from simple monomers to complex tannins. Under most circumstances they are colorless, but after reaction with metal ions, they assume red, brown, green, gray, or black coloration. The various shades of color depend on the particular tannins, the specific metal ion, pH, and the concentration of the phenolic complex. Phenolics, which are responsible for astringency in vegetables, decrease with maturity, because of the conversion of astringent phenolics from soluble to insoluble form (eggplant, plantain).

Carbohydrates

The carbohydrate content of vegetables is between 3 and 27% (Table 23.3). Carbohydrates vary from low molecular weight sugars to polysaccharides such as starch, cellulose, hemicellulose, pectin, and lignin. The main sugars found in vegetables are glucose, fructose, and sucrose. Vegetables contain higher amounts of polysaccharides (starch), whereas fruits contain more mono- and disaccharides instead.

Turgor and Texture

The predominant structural feature of vegetables is the parenchyma cell. The texture of vegetables is largely related to the elasticity and permeability of the parenchyma cells. Cells with a high content of water exhibit a crisp texture. The cell vacuoles contain most of the water of plant cells. The vacuolar solution contains dissolved sugars, acids, salts, amino acids, pigments, and vitamins, as well as several other low molecular weight constituents. The osmotic pressure within the cell vacuole, and within the protoplast against the cell walls, causes the cell walls to stretch slightly in accordance with their elastic properties. These processes determine the specific appearance and crispness of vegetables. Damaged vegetables, after processing, lose their turgor and tend to become soft unless precautions are taken in packaging and storage.

Table 23.3. Percentage Composition of Vegetables

	Component				
Food	Carbohydrate	Protein	Fat	Ash	Water
Cereals					
Maize (corn) whole grain	72.9	9.5	4.3	1.3	12
Earth vegetables					
Potatoes, white	18.9	2.0	0.1	1.0	78
Sweet potatoes	27.3	1.3	0.4	1.0	70
Vegetables					
Carrots	9.1	1.1	0.2	1.0	88.6
Radishes	4.2	1.1	0.1	0.9	93.7
Asparagus	4.1	2.1	0.2	0.7	92.9
Beans, snap, green	7.6	2.4	0.2	0.7	89.1
Peas, fresh	17.0	6.7	0.4	0.9	75.0
Lettuce	2.8	1.3	0.2	0.9	94.8

Source: Food and Agriculture Organization (FAO).

VEGETABLE PROCESSING

Harvesting and Processing of Vegetables

The quality of the vegetables constantly varies depending on the growth conditions. Since optimum quality is transient, harvesting and processing of several vegetables such as corn, peas, and tomatoes are strictly planned. Serious losses can occur when harvesting is done before or after the peak quality stage. The time of harvesting vegetables is very important to the quality of the raw produce, and the manner of harvesting and handling is critical economically. A study on sweet corn showed that 26% of total sugars were lost in just 24 hours by storing the harvested corn at room temperature. Even when stored at low temperatures, the sugar loss could reach 22% in 4 days. Peas and lima beans can lose up to 50% of their sugars in just 1 day. Losses are slower under refrigeration, but the deterioration of the sweetness and freshness of the produce is an irreversible process. It is assumed that a part of the sugars is used for respiration and starch formation in commodities such as corn, whereas the sugars are converted to cellulose, hemicellulose, and lignin in the case of asparagus. Each type of vegetable has its optimum cold storage temperature, which may vary between 0 and 10°C. Water loss is another problem that reduces the quality of the produce. Continued water loss due to transpiration and drying of cut surfaces results in wilting of leafy vegetables. Hermetic packaging (anaerobic packaging) does not prevent water loss, but instead creates conditions that prevent deterioration due to an increase in the level of carbon dioxide and a decrease in the oxygen level. In order to keep consistent produce quality, many processors monitor growing practices so that harvest and processing are programmed according to the capacity of the processing plant.

Preprocessing Operations

Vegetables can be processed in different ways including canning, freezing, freeze-drying, pickling, and dehydration. The operations involved in the processing depend on the type of vegetable and the method to be used. After harvest, the processing steps involved in canning are washing, sorting and grading, peeling, cutting and sizing, blanching, filling and brining (brining is very important for filling weight and heat transfer), exhausting (helps maintain high vacuum; exhaust temperature in the center of the can should be about 71°C), sealing, processing (heating cycle), cooling, labeling, and storage.

Harvesting

The decision to harvest should be based on experience and on objective testing methods. It is recommended that the vegetables be harvested at optimum maturity and processed promptly (Luh and Kean 1988, Woodroof 1988).

Sorting and Grading

This operation is done using a roller grader, air blower, or any mechanical device, followed by sorting on conveyor belts. Electronic sorting is commonly used recently to remove vegetables affected by diseases and insects.

Washing

Vegetables are washed to remove not only field soil and surface microorganisms, but also fungicides, insecticides, and other pesticides. There are laws specifying the maximum allowed level of contaminant. In order to remove dirt, insects, and small debris, vegetables are rinsed with water, or with detergent in some cases. Mechanically harvested tomatoes, potatoes, red beets, and leafy vegetables are washed with fruit grade detergents. The choice of washing equipment depends on the size, shape, and fragility of the particular type of vegetable. Flotation cleaners can be used for peas and other small vegetables, whereas fragile vegetables such as asparagus may be washed by gentle spraying while being transported on conveyer belts.

Peeling

Several methods are used to remove skins from vegetables including lye, steam, and direct flame. Lye peeling of mechanically harvested tomatoes and potatoes is a common practice. Vegetables with loosened skins are jet washed with water to remove skins and residual sodium hydroxide. Steam is used to peel vegetables with thick skins such as red beets and sweet potatoes, whereas for onions and peppers direct flame or hot gases in rotary tube flame peelers are used.

Cutting and Trimming

Cutting, stemming, pitting, or coring depends on the type of vegetable. Asparagus spears are cut to precise lengths. The most fibrous part is used for soup and other heated products where heat tenderizes them. Green beans are cut by machine into several different shapes along the length of the vegetable. Brussels sprouts are trimmed by hand by pressing the base against a rapidly rotating knife. Olives are pitted by aligning them in small cups and mechanically pushing plungers through the olives.

Blanching

The purpose of blanching is to inactivate the enzymes present in the vegetables. Since many vegetables don't receive a high temperature heat treatment, heating to a minimal temperature before processing or storing inactivates the activity of enzymes responsible for changes in the texture, color, flavor, and nutritional quality of the produce. Several enzymes are responsible for the loss of quality in vegetables. The deterioration of the cell membrane caused by the action of phospholipase D and lipoxygenases account for the flavor development in vegetables (Pinhero et al. 2003, Oke et al. 2003, Kruger et al. 1991). Proteases and chlorophyllases contribute to the destruction of chloroplast and chlorophyll. Changes in texture occur due to the activity of pectic enzymes and cellulases. Color deterioration occurs due to the activity of polyphenol oxidase, chlorophyllase, and peroxidase (Robinson 1991). Changes in nutritional quality can occur by the activity of enzymes that destroy the vitamins. Ascorbic acid oxidase can cause a decline in the level of vitamin C.

The blanching process also reduces the microbial load of vegetables and renders packaging into containers easier. To evaluate the effectiveness of blanching, indicator enzymes such as catalase and peroxidase are traditionally used. The reason for using indicators is that blanching is not a process of indiscriminate heating: too little heating is ineffective, whereas heating too much negatively impacts the freshness of certain vegetables. The choice of an indicator depends on the vegetable being processed. For example lipoxygenase may be an ideal indicator for peas and beans. The problem with using peroxidase as a universal indicator is that it sometimes overestimates heat requirements, which may vary from one product to another. Blanching prior to freezing has the advantage of stabilizing color, texture, flavor, and nutritional quality, as well as helping destroy microorganisms. Blanching, however, can cause deterioration of taste, color, texture, flavor, and nutritional quality, because of heating (Table 23.4.). There are three ways of blanching produce: water, steam, or microwave. Blanchers need to be energy efficient, give a uniform heat distribution and time, and have the ability to maintain the quality of the produce while destroying enzymes and reducing microbial load. Blanching using water is done at 70–100°C for a specific time frame, giving a thermal energy transfer efficiency of about 60%, versus 5% for steam blanchers. The time-temperature combination is very important in order to inactivate enzymes while maintaining the quality of the vegetables. Effects of blanching on plant tissues include alteration of membranes, pectin demethylation, protein denaturation, and starch gelatinization. Microwave blanching gives a result similar to that of water blanching, but the loss of vitamins is higher than in steam and water methods (Table 23.5.)

Canning Procedures

Depending on their pH, vegetables can be grouped into four categories (Banwart 1989): (1) high-acid vegetables with pH < 3.7, (2) acid vegetables with pH 3.7–4.6, (3) medium-acid vegetables with pH 4.6–5.3, and (4) low-acid vegetables with pH > 5.3.

The purpose of canning is to ensure food safety and high quality of the product, as the growth of several pathogens may compromise these parameters. *Clostridium botulinum (C. botulinum)*, an anaerobic and neuroparalytic toxin-forming bacteria, is a primary safety concern in hermetically sealed, canned vegetables. Other important spoilage organisms include the *C. sporogenes* group including putrefactive anaerobe 3679, *C. thermosaccharolyticum*, *Bacillus stearothermophillus*, and related species (Stumbo et al. 1983). Although *C. botulinum* spores do not produce toxin, the vegetative cells formed after germination produce a deadly neurotoxin in canned low-acid vegetables. Botulism, the poisoning caused by the *C. botulinum* is mostly associated with canned products including vegetables. Canned vegetables are commercially sterile, meaning that all pathogens and spoilage organisms of concern have been destroyed. The product may still contain a few microbial spores that could be viable under better conditions. During the processing of low-acid vegetables, it is necessary to provide a margin of safety in the methods schedule. This is achieved according to a "12D" process. D-value for a given temperature is taken from a thermal death time (TDT) curve. D-value is the time in minutes required to kill 90% of a bacterial population. The assumption is that by increasing this time by 12, any population of *C. botulinum* present in the canned product will decrease by 12 log cycles. This process time allows for adequate reduction of bacterial load to achieve a commercially sterile product.

$$t = D(\log a - \log b) \quad (2.1)$$

where t is heating time in minutes at a constant lethal temperature, D is the time (in minutes) required to kill 90% of a bacterial population, log a is the log of the initial number of viable cells, and log b is the log of the number of viable cells after time t.

For *C. botulinum*, with D = 0.21 at 121°C, 12D is equal to 2.52 min (12 × 0.21 minutes), which means that if a can contains one spore of *C. botulinum* with this D value, then from the above equation, $2.52 = 0.21(\log 1 - \log b)$, or $\log b = -2.52/0.21 = -12$; therefore $b = 10^{-12}$. The probability of survival of a single *C. botulinum* spore in the can is one in 10^{12} (Hersom and Hulland 1980). In canning low- and medium-acid vegetables, where the destruction of spores of *C. botulinum* is the major concern, a 12D process (2.52 or 3 minutes at 121°C) is the minimum safe standard for the "botulinum cook" (Banwart 1989).

Table 23.4. Relative Content of Vitamins in Peas during Processing

Processing	Vitamins, % of original			
	C	B1	B2	Niacin
Fresh	100	100	100	100
Blanched	67	95	81	90
Blanched/ frozen	55	94	78	76
Blanched, frozen, Cooked	38	63	72	79

Table 23.5. Effect of Blanching Methods on Vitamin C

	Vitamin C (mg/100g fresh weight)			
	Asparagus	Beans	Peas	Corn
Water	35.7	22.5	15.6	15.8
Steam	35.3	23.3	11.0	13.6
Microwave	18.9	13.1	9.3	12.9

Canned Tomatoes

Tomato, *Lycopersicon esculentum* Mill, belongs to the family of Solanaceae. Tomato is a major vegetable crop in North America. In North America alone, over 15 million metric tons of processing tomatoes are produced. Canned tomatoes are prepared from red ripe tomatoes as whole, diced, sliced, or in wedges. The fruit may or may not be peeled, but stems and calices should be removed. Canned tomatoes may be packed with or without added liquid. Calcium salts, varying from 0.045 to 0.08% by weight of the finished products, can also be added. Other ingredients such as organic acids, spices, oil, and flavorings can be added up to 10%. There are three categories of canned tomatoes. The label tomatoes are valid only for peeled and canned tomato. Unpeeled tomatoes are labeled accordingly. Stewed tomatoes are canned tomatoes containing onion, celery, and peppers (Anon. 1993). The flowchart for the manufacture of canned tomatoes is as follows: fresh tomatoes → sorting → washing → re-sorting, trimming → peeling → final inspection → cutting (except for whole tomatoes) → filling → exhausting → steaming and thermal processing. Most of the operations are similar to the ones described for canning in general.

Peeling

Tomatoes for canning are peeled with hot water, steam, or lye. Tomatoes are passed through a boiling water bath or live steam chamber for 15–60 seconds, depending on the variety, size, and ripeness of the fruit. A temperature of 98°C or above is recommended. The hotter the water, the shorter is the scalding time. Boiled tomatoes are immediately cooled in cold water and then peeled by hand or with a machine. Hand peeling is labor intensive and is completed by removing the core with a knife. Machine peeling is faster and is done by scrubbing or cutting and squeezing. For lye treatment, a solution of hot 14–20% soda caustic (sodium hydroxide, NaOH) is normally used in two ways. Tomatoes could be immersed in the solution or sprayed. Then, a scrubber removes the disintegrated skins. Lye peeling is less labor intensive but uses more water than hot water and steam peeling. Since the peeled tomatoes are thoroughly rinsed to remove the residual lye, treatment with up to 10% of citric acid is recommended. Lye treatment creates more pollutants and loss of soluble solids that either of the other two methods.

Inspection

The final inspection after peeling is very important for the grading of the finished product. Canned tomatoes are graded in categories: A, B, C, or first-, second-, or third-class product. The purpose of the final inspection is to remove any visible defects in the products including residual peel fragments and extraneous vegetable materials, and to check the wholeness of the product (when necessary).

Cutting

Peeled tomato may be cut into halves, slices, dices, or wedges when necessary, using cutter, slicer, or dicer before filling in the can.

Filling

Filling of tomato in cans can be done by hand or by machine. The best quality whole tomatoes are filled by hand and filled up with tomato juice. Softening during heating can be avoided by addition of calcium chloride or calcium sulfate, in the form of tablets or mixed in tomato juice and dispensed in each can. In order to inhibit the action of *C. botulinum*, a required pH less than 4.6 is secured by the addition of citric acid in the form of tablets in each can. For stewed tomato, three-quarter filled cans with peeled tomatoes are spiced with dehydrated onion, garlic, chopped celery, and green bell pepper dices, as well as tablets made of citric acid and a mixture of salt, sugar, and calcium chloride. The can is then filled up with tomato juice until an acceptable level of headspace is reached.

Exhausting

The purpose of exhausting is to create enough vacuum in the can to avoid fast deterioration of the canned product during the summer season. A minimum temperature of 71°C at the center of a can after the completion of exhaust is recommended (Luh and Kean 1988). Exhaust is normally done in steam chambers. The exhaust time in the chamber depends on the size of the can and varies from 3 minutes for

300 mm × 407 mm cans to 10 minutes for 603 mm × 700 mm cans. Exhaust is also done by mechanical vacuum closing machines (Lopez 1987, Gould 1992).

Processing

The processing time of canned tomato depends on many factors including the pH of the canned tomato, the major spoilage microorganisms of concern, the size of the can, and the type of retort (sterilizer). Organisms of concern in canned tomatoes are *C. pasteurianum* and *C. butyricum*, as well as *Bacillus coagulans (B. thermoacidurans)*. Spores of butyric acid anaerobes are destroyed at 93.3°C for 10 minutes when the pH is higher than 4.3, and after 5 minutes when the pH is between 4 and 4.3. Canned tomatoes can be processed with a rotary sterilizer, with a conventional stationary retort, or in an open nonagitating cooker. In general, the processing time is longer if the cooling is done by air instead of with water. For a rotary sterilizer the processing time at 100°C for a can of 307 mm × 409 mm may be reduced to 9 minutes or less if air cooled, and to 13 minutes or less if water cooled. For the same size of can at 100°C an open nonagitating cooker or a conventional retort would require a processing time of 35–55 minutes. The can center temperature should reach 82°C when air cooled, and 90°C when water cooled (Lopez 1987). The retort time depends on the size of the can, the fill weight, and the initial temperature of the canned product.

TOMATO JUICE PROCESSING

Processing

Tomato juice is defined as the unconcentrated, pasteurized liquid containing a substantial portion of fine tomato pulp extracted from good quality, ripe, whole tomatoes from which all stems and unnecessary portions have been removed by any method that does not increase the water content; it may contain salt and a sweetening ingredient in dry form (Canadian Food and Drugs Act). Only high quality tomatoes should be used for juice production. Tomatoes are an important source of vitamins A and C and antioxidants such as lycopene. In tomato and tomato products, color serves as a measure of total quality. Consumers notice color first, and their observation often supplements preconceived ideas about other quality attributes such as aroma and flavor. Color in tomato is due to carotenoids, a class of isoprenoid compounds varying from yellow to red color. Most carotenoids are tetraterpenes (C_{40}), derived from two C_{20} isoprene units (geranylgeranyl pyrophosphate). The most isolated and quantified carotenoids in tomato and tomato products include lycopene, lycope-5-6-diol, α-carotene, β-carotene, γ-carotene, δ-carotene, lutein, xanthophylls (carotenol), neurosporene, phytoene, and phytofluene. Lycopene is the major carotenoid of tomato and comprises about 83% of the total pigments present in the ripe fruit (Thakur et al. 1996). Therefore the levels of lycopene are very important in determining the quality of processed tomato products. It not only determines the color of tomato products, but also provides antioxidant properties to them. Lycopene is considered as a preventive agent against coronary heart disease and cancers (Gerster 1991, Clinton 1998). The flowchart for making tomato juice is as follows: fresh tomatoes → washing → sorting and trimming → comminution → extraction → deaeration → homogenization → salting and acidification → thermal processing → tomato juice.

Comminution Comminution is a process of chopping or crushing tomatoes into small particles prior to extraction. The comminuted tomatoes are subjected to either cold break or hot break processing. The cold break processing produces tomato juice with a more natural color, fresh flavor, and higher vitamin C content than does the hot break process. The hot break process, on the other hand, produces tomato juice with higher consistency and less tendency to separate, but with a cooked flavor. During the cold break process, the comminuted tomato is heated below 65°C (to introduce rapid enzyme inactivation) and held at room temperature for a short time (a few seconds to many minutes) prior to extraction. In the hot break process, the tomatoes are rapidly heated to above 82°C, immediately following comminution, in order to inactivate pectin esterase and enhance pectin extraction. Hot break process can be done either in a rotary heat exchanger or in a rotary coil tank. The latter not only inactivates the enzymes fast enough to retain most of the serum viscosity, but also deaerates the juice. Low pH inhibits pectic enzymes and enhances the extraction of macromolecules such as pectin. Therefore, the addition of citric acid to the juice during comminution improves the consistency of the juice (Miers et al. 1970, Becker et al. 1972).

Extraction Extraction is a process of separating the seeds and skins from the juice. There are two types of extractors: screw type and paddle type. The screw-type presses tomatoes between a screw and a screen with openings of about 0.5–0.8 mm (Lopez 1987). Paddle-type extractors beat the tomato against screens. The extraction yields 3% seeds and skins and 97% juice. Aiming for lower yield of 70–80% results in a better quality juice and a high quality residue for other purposes.

Deaeration The deaeration process improves the color, flavor, and ascorbic acid content of the juice. It is done immediately after the extraction by a vacuum deaerator. Removal of oxygen inhibits oxidative processes.

Homogenization A stable juice is one in which the solid and liquid phases do not separate during storage for a long period of time. The ability of a juice to separate depends on many factors, including the serum viscosity, the gross viscosity, the pH of the medium, the size and shape of the suspended solids, and others. Homogenization is a process of forcing the juice through narrow orifices at a pressure of 6.9–9.7 MPa and a temperature of about 65°C to finely break up the suspended particles (Lopez 1987).

Salting Sodium chloride is sometimes added to the juice at a rate of 0.5–1.25% by weight in order to improve the taste of the tomato juice. Citric acid is added to improve the color, flavor, and taste. As well, citric acid inhibits polyphenol oxidase by removing copper, reducing browning reactions, and improving color.

Quality Attributes of Tomatoes

The major quality attribute of ripe tomato is its red color, which is due to the lycopene content of the fruit. Other important physicochemical parameters, which determine the quality of tomato are Brix, acidity, pH, vitamin C, ash, dry matter, firmness, fruit weight, and flavor volatiles. For processed tomato products, the required quality attributes are precipitate weight ratio, serum viscosity, total viscosity (Brookfield), and lycopene content. Several quality attributes of tomato and tomato products can be improved by genetic modification of tomatoes (Oke et al. 2003). These comparisons were made between fruits obtained from untransformed and genetically transformed tomato plants carrying antisense phospholipase D cDNA, and between juices prepared from these fruits. The levels and activity of phospholipase D, the key enzyme involved in membrane lipid degradation, was considerably reduced by antisense transformation. These changes potentially resulted in increased membrane stability and function that also improved several quality parameters. The flavor profiles of blended antisense tomato fruits were different from those of the controls, being enriched in volatile aldehydes such as pentenal and hexenal. Increased membrane stability in transgenic fruits potentially resulted in lowered degradation of unsaturated fatty acids such as linoleic and linolenic acids and may have contributed to increased substrate availability for lipoxygenase pathway enzymes (lipoxygenase, hydroperoxide lyase, etc.) during blending, with an increased evolution of volatile aldehydes. Table 23.6 provides a comparison of various quality parameters between a genetically modified tomato and a control; it shows improvements in several quality parameters due to the transformation.

Physicochemical Stability of Juices

Kinetic Stability There are two categories of juices: clear and comminuted. Clear juice such as apple juice contains no visible vegetal particles, whereas a comminuted juice such as tomato juice contains mostly vegetal particles suspended in a liquid. To be stable, a clear juice needs to remain clear (without sediment) during its shelf life. On the other hand, a comminuted juice may not separate into distinct phases during the shelf life of the product. An important quality attribute of juices such as tomato juice is the stability of their dispersion system. In order to be stable, a juice needs to be kinetically and physically stable.

The ability of a polydisperse system containing suspended particles to maintain its homogenous distribution without agglomeration is called **kinetic stability** or sedimentation stability. Kinetic stability depends on many factors, the most important of which include the size of suspended vegetable parti-

Table 23.6. Physicochemical Parameters of Tomato Fruits and Processed Juice from Transgenic and Control Tomato

Properties	Control	Transgenic
Fruit weight (g)	95.14±36.56[a]	34.87±21.57[b]
Firmness (N)	4.97±0.59[a]	5.99±1.03[b]
Redness (a+)	29.6±0.70[a]	31.70±1.67[a]
Acidity (%)	0.36±0.00[a]	0.38±0.05[a]
°Brix	4.55±0.71[a]	4.75±0.71[b]
Dry matter—NSS (%)	3.54±0.05[a]	5.15±0.03[b]
Ash (%)	0,63±0.03[a]	0.089±0.13[b]
Vitamin C (mg/100g)	3.9±0.00[a]	10.4±0.016[b]
PPT, %	15.70±0.33[a]	16.17±0.48[a]
Serum viscosity (mPa·s)	1.0919±0.04[a]	1.2503±0.010[b]
Brookfield viscosity (mPa·s)	1075±35[a]	1400±35[b]
Lycopene (mg/100g)	11.73±2.10[a]	17.47±0.58[b]

Note: The values showing different superscripts are significantly different at $P < 0.05$.

cles, the viscosity of the disperse medium, and the intensity of the Brownian motion. In a disperse medium, heavier or larger particles sediment faster than the lighter ones (in response to gravity). The sedimentation velocity of any particle is described by the Stokes equation:

$$V = \frac{2}{9r^2}(\rho_1 - \rho_2)\left(\frac{1}{\eta}\right)g \tag{2.2}$$

where V is the sedimentation velocity (m/s); r is the radius of the suspended particles (m); ρ_1 and ρ_2 are the density of the particles and serum, respectively (kg/m^3; serum = liquid medium in which particles are suspended); η is the viscosity of the juice (Pascal second, Pa.s); and g is gravitational force (g = 9.81 m/s^2).

The sedimentation time of a given particle is about four times longer in a juice than in water (Table 23.7).

Particles larger than 10 μ will sediment in a few seconds. This is the reason why the sedimentation stability of juices, especially comminuted juices such as tomato juice, is a serious processing issue.

The physical force that affects the sedimentation of particles in a juice is called normal force and can be calculated by

$$f' = mg \tag{2.3}$$

where m is the mass (kg) of the suspended particle. In general, a particle with a spherical shape has

$$m = \frac{4}{3}\pi r^3 \rho_1 \tag{2.4}$$

Where $\pi = 3.14$, r is the radius of the particle, and ρ_1 is the density of the particle.

For particles with nonspherical shape (most of suspended particles), r is equal to the nearest equivalent value of a spherical particle with an identical mass and an identical density.

During particle sedimentation, another important force, friction, also comes into play. Friction between particles results in a reduction in their movement.

The frictional force $f'' = 6\pi\eta$ (2.5)

Table 23.7. Sedimentation of Spherical Mineral Particles in Water and in Juice with a Depth of 1 cm

Particle Size (μ)	Velocity (m/s)	Sedimentation Time in Water	Sedimentation Time in Juice
10	3.223×10^{-4}	31.03 seconds	2.29 minutes
0.1	3.223×10^{-8}	86.2 hours	16 days
0.001	3.223×10^{-12}	100 years	436 years

Where η = viscosity of the medium, r = radius of the particle in meters, and v = velocity of the particle (m/s).

Friction between the particles increases depending on the density of the medium: the higher the density, the higher the friction. When $f' = f''$ sedimentation of the particles occur.

Kinetic stability of a heterogeneous dispersion system also depends on Brownian motion. The most dispersed particles have a very complex motion due to collisions from molecules in the dispersion medium. Because of this, the suspended particles are subjected to constant changes in their velocity and trajectory. Molecular kinetics shows that dispersed particles with colloidal size change their path 10^{20} times/s. These particles may also acquire a rotational Brownian motion. This is why colloidal particles have a higher sedimentation stability than larger particles. With an increase in the mass of the suspended particles, their momentum also increases. Particles smaller than 5×10^{-4} cm in diameter that oscillate around a point do not sediment, whereas larger particles that do not experience as much Brownian motion as smaller particles easily sediment. Thus, when particles have reduced motion, they tend to aggregate and enhance sedimentation. Under ideal conditions, obtaining a colloidal particulate size will enhance the stability of a juice preparation.

Physical Stability **Physical stability** is a result of the ability of a polydisperse system to inhibit the agglomeration of suspended particles. In a system with low physical stability, suspended particles agglomerate to form heavier particles (> 5×10^{-4} cm in diameter), which easily sediment. Physical stability depends on two opposing forces, attracting and repulsing.

Attracting forces between molecules are referred to as VanDerVaals forces, and they reduce the physical stability of a heterogeneous colloidal system. The intensity of attractive forces increases as the distance between suspended particles decreases.

Repulsing forces between particles are caused by the charges surrounding the particles designated by their ζ-potential. When ζ-potential is zero, the net charge surrounding the particle is also zero, and the suspended particles are said to be at their isoelectric point. Agglomeration of particles begins at a given value of ζ-potential called the critical potential. This is the point at which equilibrium is reached between VanDerVaals (attracting) forces and repulsing forces. Different heterogeneous colloidal dispersion systems have different values of critical potential. When the ζ-potential (repulsive forces) of a particle is higher than the critical potential, hydrophilic colloids are stable due to the repulsion between particles, whereas, at a lower potential than the critical potential ζ, the particles tend to aggregate. The effective energy of interaction between particles of a heterogeneous colloidal system is expressed by

$$E = E_A + E_R, \tag{2.6}$$

where E_A is attracting energy, and E_R is repulsing energy.

Attracting energy is actually the integration of the sum of all attracting forces between molecules of two colloidal particles. For two particles with radius r the potential energy is expressed by

$$E_A = \frac{-Ar}{12h} \tag{2.7}$$

where h is the distance between surfaces of the two particles, and A is the Hamaker constant (10^{-1} to 10^{-21} J). Consequently, attracting forces decrease as the distance between particles increases.

On the other hand, particles can come closer to one another up to a certain distance, after which they start repulsing each other because of their ζ-potential. When two particles are very close, their similar ionic charge (positive or negative) layers create a repulsive force, which keeps them apart. The repulsing energy between two particles with the same radius r and the same surface potential ψ_0 is expressed by

$$E_R = -2\int \Delta P \, d\delta, \tag{2.8}$$

where ΔP is the increase in osmotic pressure, δ is thickness of the double ionic layer, and the number 2 relates to energy changes between two particles.

$$\Delta P = \Delta n \mathrm{K} T \tag{2.9}$$

where Δn is the increase in the ion concentration between the two particles, K is Boltzman's constant, and T is temperature in degrees Kelvin (K).

Assuming that the two particles are spherical, Equation 2.8 can be expressed by

$$E_R \approx \frac{[(\varepsilon r \psi_0^2 \exp\{-\mathrm{K}ih\})]}{2}, \tag{2.10}$$

Where ε is the dielectric constant, r is the radius of particles, h is the distance between the two particles, ψ_0 is the surface potential, and Ki is a constant that characterizes the double ionic layer.

The inferences from the Equation 2.10 are that

- Repulsing energy exponentially decreases as the distance between the particles increases, and
- Repulsing energy quadratically increases as the surface potential increases and as the radius linearly increases.

For particles with radius r and a constant surface potential, E_R depends on Ki. Figure 23.1 shows the interaction between two particles (1 and 2) under constantly increasing Ki. As a result of decreasing the thickness of the double ionic layer and a decrease in ζ-potential, the distance between the two particles decreases, leading to an increase in the repulsing energy according to Equation 2.10. For a given h, the residual energy reaches the maximum value (E_{max}), which must be overcome by particle 1 in order to aggregate with particle 2. Continual increase of Ki and a decrease in the distance between particles h, reduces the effect of repulsing energy E_R over the residual energy E ($E_R - E_A$ or E_{max}), which reaches its minimal value M. At this stage, the attracting energy E_A dominates, and the particles aggregate. The speed of aggregation is an important physicochemical parameter for the stability of a microheterogeneous system. E_{max} is called coagulation energy and determines the speed of aggregation. When $E_{max} > 0$ (Fig. 23.1A), the aggregation process is slow, whereas when $E_{max} < 0$ (Fig. 23.1B), the process is quick, since the particles of the system do not have to overcome E_{max} (when $E_A > E_R$, $E_R - E_A$ is < 0; Fig. 23.1A,B). The critical concentration of coagulation is reached at $E_{max} = 0$, when the attractive and repulsive energies are equal.

In the real situation of a polydisperse colloidal system such as juice, suspended particles have different sizes and different ψ_0 and ζ-potential. When two particles with different sizes and different surface ζ-potential approach each other, an opposite charge is induced upon the particle with lower ζ-potential, which activates the agglomeration. Another important factor in the stability of juices is the hydrated layer of suspended colloidal particles. The formation of the hydrated layer is due to the orientation of water molecules towards the hydrophilic groups, such as —COOH and —OH, situated on the surface of the particles. A decrease in ζ-potential leads to a decrease in the hydrated layer of suspended particles, and therefore increases the susceptibility to aggregation. The ability of the hydrated layer to increase stability is explained by Deryagin theory, which stipulates that when suspended particles are close enough, their hydrated layers are reduced, leading to an increase in repulsing forces between them. Only strong hydrated layers can affect the aggregative stability of a polydisperse system. The probability of the formation of such hydrated layers around vegetal particles in juices is very low, because of the weak energetic relationship between the particles and the disperse phase. In this case, hydrated

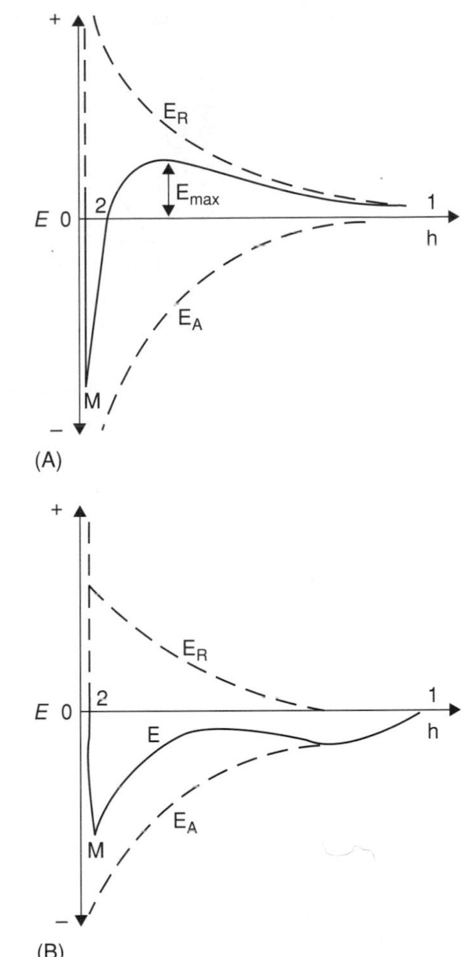

Figure 23.1. Interactive energy between two particles.

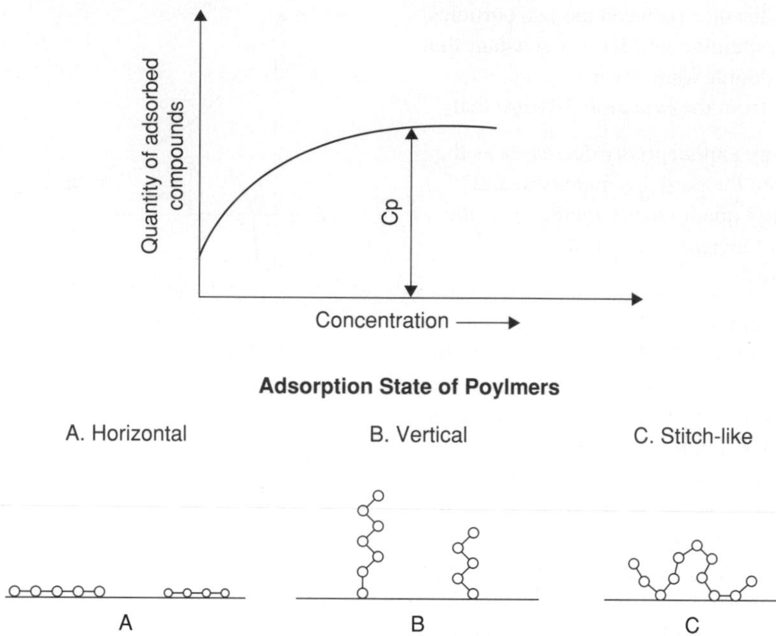

Figure 23.2. Isotherm of adsorption of polymers; polymer adsorption states.

layers play a secondary role to supplement the action of the double ionic layer.

Stability from particular aggregation in juice is also influenced by the viscosity of the dispersion medium. The viscosity of the dispersion phase or serum of juices is mainly due to high molecular mass compounds such as pectin, protein, and starch. Practically, this is achieved by the addition of such compounds to the juice. The mechanism of action is as follows. Upon addition of high molecular mass compounds (pectin), these compounds adsorb on the surface of the hydrophobic vegetal particles to make a layer of hydrophilic molecules, circled by a thick hydrated layer, which gradually blends into the dispersion medium. Pectin is the stabilizer of choice in the juice processing industry, over protein and starch. With an increase in the concentration of high molecular mass compounds, the sedimentation stability of the juice increases. It has been established that with the increase in the concentration of pectin in the juice, the stability increases to reach a plateau, after which a further increase does not provide any added beneficial effect (Idrissou 1992). The mechanism of the sedimentation stability of the juice is as follows:

- Low adsorption velocity: Lower adsorption velocity is observed under lower concentration of stabilizers.
- Irreversible adsorption: Stabilizers in most cases are bonded to the particular phase surface with strong adsorption bonds.
- Adsorption is described by the curve in Figure 23.2, following a typical saturation curve.
- Different configurations of adsorption by particles are shown in Figure 23.2. Molecules of linear polymers can be adsorbed in three different ways, depending on their affinity to the liquid phase and the surface of the suspended particles, adopting horizontal, vertical, and stitch-like configurations.

Minimally Processed Vegetables

Definitions

Rolle and Chism (1987) defined minimally processed refrigerated fruits and vegetables (MPR F&V) as produce which has undergone minimal processing such as washing, sorting, peeling, slicing, or shredding prior to packaging and sale to consumers.

Odumeru et al. (1997, 1999) defined ready-to-use vegetables as fresh-cut, packaged vegetables requiring minimal or no processing prior to consumption. It is generally accepted that MPR F&V are products that contain live tissues or tissues that have been only slightly modified from the fresh condition and are fresh-like in character and quality.

The rapid growth of fast-food restaurants in the 1980s encouraged the development and use of fresh-cut products. As the demand for these products increased, new technology and product innovation were developed that helped make fresh-cut produce one of the fastest growing segments of the food industry. It is estimated that in the United States fresh-cut produce market, up to 8% of all produce is sold in retail grocery outlets and 20% in the food service industry. This market is expected to continue growing in size and popularity as more products are introduced and consumers change their buying habits. In the United States over the next 5 years, fresh-cut produce sales are expected to reach $19 billion, compared with $5 billion in 1994. Freshly prepared, ready-to-eat salads, now with cut tomatoes, will have a shelf life of up to 14 days utilizing unique modified atmosphere packaging technology.

Consumption

In the United States, consumption of fresh vegetables (excluding potatoes) has increased from 60 kg in 1986 to 72 kg in 2003. In 2003 the most consumed fresh vegetable was head lettuce at 12 kg, followed by onions at 9 kg, and fresh tomatoes at 8 kg. In the same year, the consumption of processed tomatoes per capita was 35 kg and processed sweet corn 9.5 kg. In Canada, the consumption of fresh vegetables (excluding potatoes) has been increasing steadily, reaching 70.2 kg/capita in 1997 from 41 kg/capita in 1971. Also in 1997, the consumption of fresh vegetables had declined 2.4% from the amount of fresh vegetables consumed in 1996 (71.9 kg/capita). Lettuce (16%) is the most consumed fresh vegetable, followed by onions (12.1%), carrots (12.0%), tomatoes (11.6%), and cabbage (8.1%). Brussels sprouts, parsnips, asparagus, beets, and peas each represent less than 1.0% of Canadian fresh vegetable diet. On a per capita basis, Canada has one of the highest consumption rates of fresh vegetables in the world. Frozen vegetable consumption has declined 3.4% to 5.6 kg/capita, while canned vegetables and vegetable juices have increased 2.4% to 12.9 kg/capita for the same period. Vegetable purchases by consumers represent 6.6% of total food expenditures, virtually unchanged in the past 10 years.

Processing

The major operations in processing MPR produce are sorting, sizing, and grading; cleaning, washing, and disinfection; centrifugation; and packaging and distribution. For certain vegetables such as eggplant, the peeling operation follows the sorting operation, and then there is a size reduction, mixing whenever necessary, packaging, and distribution. Washing of vegetables for minimal processing is done with a lot of water (5–10 L/kg) in order to reduce the bacterial load. The main pathogen of concern in this type of product is *Listeria monocytogenes*. A water temperature of about 4°C is recommended. The safety of the product is best secured through disinfection with chlorine in the concentration of about 100 mg/L. Two forms of chlorine are commonly used: gaseous chlorine, which is 100% active chlorine, and calcium or sodium hypochlorites. Calcium and sodium hypochlorites are most popular in MPR F&V industry.

Centrifugation is one of the major operations in MPR processing. The purpose is to dry the vegetables rapidly. The efficacy of the centrifugation depends on the speed and time of rotation of the centrifuge. Certain antioxidants such as ascorbic acid and citric acid, at a concentration of about 300 ppm in the wash solution, would enhance the quality of the products. The processing techniques employed usually wound the tissue and may cause the liberation of "wound ethylene," which can activate deteriorative reactions within the tissue. Vegetables of high quality may also have a good complement of the antioxidants (vitamins C and E, reduced glutathione) and the antioxidant enzyme system that comprises enzymes such as superoxide dismutase, catalase, peroxidase, and ascorbate peroxidase. An active pentose phosphate pathway is required for the supply of reduced nicotinamide adenine dinucleotide phosphate (NADPH) for the efficient functioning of the antioxidant enzyme system.

Membrane phospholipid degradation mediated by phospholipase D is activated in response to wounding. Damage to the tissue during processing can cause the leakage of ions such as Ca^{2+} and H^+, which can activate phospholipase D. Naturally occurring phospholipase D inhibitors such as hexanal can be used to reduce the activity of phospholipase D in processed tissues and enhance their shelf life and quality (Paliyath et al. 2003). The quality of MPR F&V can be increased by following good manufacturing practices (GMP), the main points of which are minimizing handling frequency and providing continued control of temperature, relative humidity (% RH), and modified and controlled atmosphere (MA/CA storage, etc.). Always, the product is transferred from truck to refrigerated storage immediately to minimize degradative and oxidative reactions. The products are replaced on a first-in–first-out basis, and the inventory is replaced on a weekly basis. Currently MPR produce includes the following items: ready-to-eat fruits and vegetables, ready-to-cook fruits and vegetables, ready-to-cook mixed meals, and fresh ready-to-use herbs and sprouts.

Quality of MPR

Consumers expect fresh-cut products to be of optimum maturity, without defects, and in fresh condition. The most important physicochemical quality parameters targeted for preservation include good appearance, nutrients, and excellent sensory attributes such as texture, firmness, and taste. Minimally processed products are vulnerable to discoloration because of damaged cells and tissues, which become dehydrated. Cutting and slicing of carrots with a very sharp blade reduces the amount of damaged cells and dehydration, when compared with those sliced with a regular culinary knife. To overcome the problem of dehydration, fresh-cut produce can be treated with calcium chloride and kept in a high-humidity atmosphere. Enzymatic browning is a serious problem with minimally processed produce. During the processing of produce, several types of oxidative reactions may occur, leading to the formation of oxidized products. These reactions cause browning reactions, resulting in the loss of nutritional value by the destruction of vitamins and essential fatty acids. In lipid-rich vegetables, oxidative processes lead to the development of rancid off-flavors and sometimes to toxic oxidative products (Dziezak 1986). There are a number of chemicals used to stabilize MPR produce, including (1) free radical scavengers such as tocopherols, (2) reducing agents and oxygen scavengers such as ascorbic acid and erythorbic acid, (3) chelating agents such as citric acid, and (4) other secondary antioxidants such as carotenoids. Ascorbic acid is commonly used either alone or in combination with other organic acids. Among the factors influencing the general quality of MPR produce, temperature is the most important. When temperature increases from 0 to 10°C, respiration rate increases substantially, with Q_{10} ranging from 3.4 to 8.3 among various fresh-cut products. With the increase in respiration rate, deterioration rate also increases at a comparable rate; therefore, low temperature storage is essential for maintaining good quality (Watada et al. 1996). Modified atmosphere (MA) within MPR containers or bags is useful in maintaining the quality of the produce (Gorny 1997). Gas mixtures suitable for MA storage have been shown to be the same as those recommended for the whole commodity (Saltveit 1997). A controlled atmosphere (CA) system is used to simulate the MA, with similar gas composition. A mixture of 10% O_2 + 10% CO_2 has been shown to retard chlorophyll degradation in parsley (Yamauchi 1993). An atmosphere of 3% O_2 + 10% CO_2 was beneficial for fresh-cut iceberg lettuce, slightly beneficial for romaine lettuce, and not beneficial for butterhead lettuce (Lopez-Galvez et al. 1996). The oxygen level can be allowed to drop to the level of the respiratory quotient breakpoint. The O_2 level could be dropped to 0.25% for zucchini slices (Izumi et al. 1996) and 0.8% for spinach (Ko et al. 1996).

REFERENCES

Further Reading Materials

Arthey D, Ashwurst PR. 2001. Fruit Processing: Nutrition, Products and Quality Management. Aspen Publishers, Gaithersburg, Maryland. 312 pages.

Enachescu DM. 1995. Fruit and Vegetable Processing. FAO Agricultural Services Bulletin 119. FAO. 382 pages.

Jongen WMF. 2002. Fruit and Vegetable Processing: Improving Quality (Electronic Resource). CRC, Boca Raton, Florida.

Salunkhe DK, Kadam SS. 1998. Handbook of Vegetable Science and Technology: Production, Composition, Storage and Processing. Marcel Dekker, Inc., New York. 721 pages.

Cited References

Anonymous. 1993. Canned tomatoes. Tomato concentrates. Catsup. In: Code of Federal Regulations, Title 21, Sections 155.190–155.194.

Banwart GJ. 1989. Control of microorganisms by destruction. In: Basic Food Microbiology, 2nd edition. AVI Press, New York.

Becker R, Miers JC, Nutting MD, Dietrich WC, Wagner JR. 1972. Consistency of tomato products. Effects of acidification on cell walls and cell breakage. J Food Sci 37:118–125.

Clinton SK. 1998. Lycopene: Chemistry, biology and implications for human health and disease. Nutr Rev 56:35–51.

Dziezak JD. 1986. Preservative systems in foods, antioxidants and antimicrobial agents. Food Technol 40: 94–136.

Gerster H. 1991. Potential role of β-carotene in the prevention of cardiovascular disease. Int J Vitamin Nutr Res 61:277–291.

Gorny JR. 1997. Summary of CA and MA requirements and recommendations for fresh-cut (minimally processed) fruits and vegetables. In: JR Gorny, editor. Proceedings of the Seventh International Controlled Atmosphere Conference, vol. 5. Postharvest Outreach Program, University of California, Davis. Pp. 30–66.

Gould WA. 1992. Tomato production. Processing and Technology, 3rd edition. CTI Publications, Baltimore, Maryland.

Hersom AC, Hulland ED. 1980. Principles of thermal processing. In: Canned Foods, 7th edition. Churchill Livingstone, New York.

http://apps.fao.org/page/collections?subset=agriculture, accessed on October 27, 2003.

http://www.hc-sc.gc.ca/food-aliment/friia-raaii/food_drugs-aliments_drogues/act-loi/pdf/e_b-text-1.pdf, accessed on December 20, 2003.

http://www.statscan.ca/english/ads/23F0001XCB/highlight.htm, accessed on Dec. 03, 2003.

Idrissou M. 1992. Investigation and Improvement of the Technology of Processing Citrus Bases and Concentrates. Ph.D. thesis. Higher Institute for Food and Flavour Industries. Plovdiv, Bulgaria (in Bulgarian).

Izumi H, Watada AE, Douglas W. 1996. Low O_2 atmospheres affect storage quality of zucchini squash slices treated with calcium. J Food Sci 41:317–321.

Ko NP, Watada AE, Schlimme DV, Bouwkamp JC. 1996. Storage of spinach under low oxygen atmosphere above the extinction point. J Food Sci 61:398–400.

Kruger JE, MacGregor AW, Marchylo BA. 1991. Chapter 17. Endogenous cereal enzymes. In: PF Fox, editor. Food Enzymology, vol. 2. Elsevier Applied Science, London and New York.

Lopez A. 1987. Canning of vegetables. In: A complete Course in Canning, Book 3, 12th edition. The Canning Trade Inc., Maryland.

Lopez-Galvez G, Saltveit M, Cantwell M. 1996. The visual quality of minimally processed lettuce stored in air or controlled atmosphere with emphasis on romaine and iceberg types. Postharvest Biol Technol 8:179–190.

Luh BS, Kean CE. 1988. Canning of vegetables. In: BS Luh, JG Woodroof, editors. Commercial Vegetable Processing, 2nd edition. Van Nostrand Reinhold, New York.

Luh BS, Woodroof JG. 1988. Commercial Vegetable Processing, 2nd edition. Van Nostrand Reinhold, New York.

Miers JC, Sanshuck DW, Nutting MD, Wagner JR. 1970. Consistency of tomato products. 6. Effects of holding temperature and pH. Food Technol 24:1399–1403.

Odumeru JA. 1999. Microbiology of fresh-cut ready-to-use vegetables. Recent Res Developments Microbiol 3:113–124.

Odumeru JA, Mitchell SJ, Alves DM, Lynch JA, Yee AJ, Wang SL, Styliadis S, Farber JM. 1997. Assessment of the microbiological quality of ready-to-use vegetables for health-care food services. J Food Protection, 60:954–960.

Oke M, Pinhero RG, Paliyath G. 2003. The effects of genetic transformation of tomato with antisense phospholipase D cDNA on the quality characteristics of fruits and their processed products. Food Biotechology 17:163–182.

Paliyath G, Yada RY, Murr DP, Pinhero RG. 2003. Inhibition of phospholipase D. U.S. patent no. 6,514,914.

Pinhero RP, Almquist KC, Novotna Z, Paliyath G. 2003. Developmental regulation of phospholipase D in tomato fruits. Plant Physiol Biochem 41:223–240.

Petro-Turza M. 1986–87. Flavour of tomato and tomato products. Food Rev International 2:309–351.

Potter NN, Hotchkiss JH. 1995. Food Science, 5th edition. Chapman and Hall International Thompson Publishing. P. 410.

Robinson DS. 1991. Chapter 10. Peroxidases and their significance in fruits and vegetables. In: PF Fox, editor. Food Enzymology, vol. 1. Elsevier Applied Science, London and New York. Pp. 399–426.

Rodrigo M, Martinez A, Sanchis J, Trama J, Giver V. 1990. Determination of hot-fill-hold-cool process specifications for crushed tomatoes. J Food Sci 55: 1029–1032.

Rolle RS, Chism GW III. 1987. Physiological consequences of minimally processed fruits and vegetables. J Food Quality 10:157–177.

Saltveit ME Jr. 1997. A summary of CA and MA requirements and recommendations for harvested vegetables. In: ME Saltveit, editor. Proceedings of the Seventh International Controlled Atmosphere Conference, vol. 4. Postharvest Outreach Program, University of California, Davis. Pp. 98–117.

Singh RK, Nelson PE. 1996. Quality attributes of processed tomato products: A review. Food Rev International, 12:375–401.

Smith DS, Cash JN, Nip W-K Hui YH (editors). 1997. Processing vegetables: Science and Technology. Technomic Publishing Company, Lancaster, Pennsylvania. Pp. 400–407.

Stumbo CR, Purohit RS, Ramakrishnan TV, Evans DA, Francis FJ. 1983. Introduction to thermal processing of low acid foods. In: Handbook of Lethality Guides for Low-Acid Foods. Vol. 1, Conduction Heating. CRC Press, Inc., Florida.

Thakur BR, Singh RK, Nelson PE. 1996. Quality attributes of processed tomato products: A review. Food Rev International, 12:375–401.

Watada AE, Ko NP, Minott DA. 1996. Factors affecting quality of fresh-cut horticultural products. Postharv Biol Technol 9:115–125.

Woodroof JG. 1988. Harvesting, handling, and storing vegetables. In: Commercial Vegetable Processing. Van Nostrand Reinhold, New York.

Yamaguchi N, Watada AE. 1993. Pigment changes in parsley leaves during storage in controlled or ethylene containing atmosphere. J Food Sci 616–618.

24
Nonenzymatic Browning of Cookies, Crackers, and Breakfast Cereals

M. Villamiel

INTRODUCTION

Cereal-based products such as cookies, crackers, and breakfast cereals represent a predominant source of energy in the human diet, especially for children consuming cereal derivatives for breakfast meals.

Cookies, crackers, and breakfast cereals can be manufactured by means of traditional processes or by extrusion cooking. In general, during conventional treatment of flour products (generally, temperatures close to 200°C or higher for several minutes), more intense processing conditions are applied as compared with the extrusion process (González-Galán et al. 1991, Manzaneque Ramos 1994, Huang 1998). Extrusion cooking is a well-established industrial technology with a number of food applications since, in addition to the usual benefits of heat processing, extrusion has the possibility of changing the functional properties of food ingredients and/or of texturizing them (Cheftel 1986). In the extruder, the mixture of ingredients is subjected to intense mechanical shear through the action of one or two rotating screws. The cooking can occur at high temperatures (up to 250°C), relatively short residence times (1–2 minutes), high pressures (up to 25 MPa), intense shear forces (100 rpm), and low moisture conditions (below 30%). In addition to the cooking step, cookie, cracker, and breakfast cereal manufacture involves toasting and/or drying operations (Cheftel 1986, Camire and Belbez 1996, Huang 1998).

During these technological treatments, due to the elevated temperatures and low moisture conditions used, different chemical reactions such as the nonenzymatic browning can take place. Nonenzymatic browning includes the Maillard reaction and caramelization. The products resulting from both reactions depend on food composition, temperature, water activity, and pH and both reactions can occur simultaneously (Zanoni 1995).

The Maillard reaction that occurs between reducing sugars such as glucose, fructose, lactose, and maltose and free amino groups of amino acids or proteins (usually the epsilon-amino group of lysine) is favored in foods with high protein and reducing carbohydrate content at intermediate moisture content, temperatures above 50°C, and a pH in the range from 4 to 7. Caramelization depends on direct degradation of carbohydrates due to heat, and it needs more drastic conditions than the Maillard reaction; thus, at temperatures higher than 120°C, pH lower than 3 or higher than 9, and very low moisture content, caramelization is favored (Kroh 1994).

In addition, other chemical reactions that can occur during processing of these products may affect the extent of the nonenzymatic browning. Thus, starch and nonreducing sugars such as sucrose can be hydrolyzed into reducing sugars that can later be involved in other reactions, for instance, in the Maillard reaction (Linko et al. 1981, Camire et al. 1990). Noguchi et al. (1982) observed that 10% of the initial sucrose molecules in a severely processed biscuit mix were hydrolyzed into glucose and fructose, thus permitting the Maillard reaction.

The chemical changes that take place during the technological processes used in the elaboration of these types of cereal-based foods, contribute, to some extent, to their typical organoleptic characteristics. However, important losses of lysine due to the formation of chemically stable and nutritionally unavailable derivatives of protein-bound lysine can be observed under Maillard reaction conditions (McAuley et al. 1987).

The amount of lysine and its biological availability are meaningful criteria for the protein nutritive quality of cereals. Foods processed from cereal grains are low in essential amino acids such as lysine and methionine (Meredith and Caster, 1984). As this deficiency can be further impaired by losses from browning reactions during processing, a compromise must be found in which the objectives of heat treatment are reached with a minimal decrease in the nutritional quality of the food. For that purpose, indicators of heat-treated foods have proved to be useful for the control of processes, allowing the possibility of optimizing conditions. Thus, many indicators are available for evaluating the extent of nonenzymatic browning in cereal-based foods. Although not all the studies on cereal-based products reported in this review are carried out directly on cookies, crackers, or breakfast cereals, they have been considered since both the manufacture and composition could be similar, and consequently the same chemical reactions may be involved.

AVAILABLE LYSINE

The determination of available lysine has been used to evaluate the effect of heating on the protein quality of the following cereal-based products: pasta (Nepal-Sing and Chauhan 1989, Acquistucci and Quattrucci 1993), breads (Tsen et al. 1983, Ramírez-Jiménez et al. 2001), infant cereals (Fernández-Artigas et al. 1999a), cookies fortified with oilseed flours (Martinkus et al. 1977), and biscuits (Singh et al. 2000).

Since extrusion cooking is a well-established technology for the industrial elaboration of cereal-based foods, several authors have studied the effect of the initial composition and the different operating conditions on lysine loss in extruded materials. Noguchi et al. (1982), in samples of protein-enriched biscuits, found that the loss of reactive lysine is significant (up to 40% of the initial value) when the extrusion cooking is carried out at a high temperature range (190–210°C) and relatively low water content (13%). When the water content is increased to 18%, the lysine loss is much less pronounced or even negligible.

Noguchi et al. (1982) also studied the effect of the decrease of pH on lysine loss in biscuits obtained by extrusion; they observed that lysine loss increases with low pH, since strong acidification markedly increases starch or sucrose hydrolysis and consequently the formation of reducing carbohydrates. The formation of reducing sugars by starch hydrolysis was proposed to be the main cause of lysine loss in extruded wheat flours (Bjorck et al. 1984).

Bjorck et al. (1983) measured the effect of extrusion cooking on available lysine in extruded biscuits and found that the decrease in lysine was about 11%, and lysine retention was negatively influenced by increasing the process temperature and positively influenced by increasing the moisture content of the mix of ingredients.

McAuley et al. (1987) observed an important decrease of available lysine content (40.9–69.2%) in wheat grain processed by flaking and toasting, and they attributed these results to the high temperature reached during toasting.

In general, extrusion cooking of cereal-based foods appears to cause lysine losses that do not exceed those for other methods of food processing. In order to keep lysine losses low (10–15%), it is necessary to avoid operating conditions above 180°C at water contents below 15% (even if a subsequent drying step is then necessary; Cheftel 1986). Phillips (1988) suggested that if the processing conditions are controlled, the Maillard reaction is more likely to occur in expanded snack foods in which nutritional quality is not a major factor than in other extruded foods with higher moisture content.

Horvatić and Guterman (1997), in a study on available lysine content during industrial cereal (wheat, rye, barley, and oat) flake production, found that the effects of particular processing phases can result in a significant decrease of available lysine in rye and oat flakes, whereas less influence can be observed in the case of wheat and barley flakes. Apart from the importance of processing conditions, the results obtained by these authors seem to indicate that the decrease of lysine availability is higher for cereals with greater available lysine content in total proteins.

Another important consideration is to avoid the presence of reducing sugars during extrusion. Lysine loss and browning are more intense when reducing carbohydrates such as glucose, fructose, and lactose (added as skim milk) are added to the food mix at levels above 2–5% (Cheftel et al. 1981). Singh et al. (2000) and Awasthi and Yadav (2000) also found higher lysine loss and browning in traditionally elaborated biscuits when whey or skim milk was added to the initial mixture.

Horvatić and Eres (2002) performed an investigation of the changes in available lysine content during industrial production of dietetic biscuits. They found that dough preparation did not significantly affect the available lysine content. However, after baking, a significant loss (27–47%) of available lysine was observed in the studied biscuits. The loss of available lysine was found to be significantly correlated with technological parameters, mainly baking temperature and time conditions.

The influence of storage on the lysine loss in protein-enriched biscuits was studied by Noguchi et al. (1982). Lysine loss was observed to increase when samples were stored at room temperature for long periods of time. By measurement of lysine, Hozova et al. (1997) estimated the nutritional quality of amaranth biscuits and crackers stored during 4 months under laboratory conditions (20°C and 62% relative humidity). Although a slight decrease in the level of lysine was detected, this was not significant. However, these authors suggested that lysine degradation can continue with prolonged storage, and it is necessary to consider this fact in relation to consumers and the extension of storage time.

FUROSINE

The determination of furosine (ε-N-2-furoylmethyl-lysine), generated from the acid hydrolysis of Amadori compounds formed during the early stages of the Maillard reaction (Erbersdobler and Hupe 1991), has been used for the assessment of lysine loss in malt (Molnár-Perl et al. 1986), pasta (Resmini and Pellegrino 1991, García-Baños et al. 2004), baby cereals (Guerra-Hernández and Corzo 1996, Guerra-Hernández et al. 1999), baby biscuits (Carratú et al. 1993), and bread (Ramírez-Jiménez et al. 2001, Cárdenas-Ruiz 2004). Other 2-furoylmethyl derivatives such as that corresponding to GABA (γ-aminobutyric acid), have been, among others, suggested as indicators of the extent of the Maillard reaction in vegetable products (Del Castillo et al. 2000; Sanz et al. 2000, 2001).

In commercial cookies, crackers, and breakfast cereals the initial steps of the Maillard reaction also have been evaluated by furosine determination (Rada-Mendoza et al. 2004). Furosine was detected in all analyzed samples, but showed a wide variation. Figure 24.1 illustrates, as an example, the HPLC (high performance liquid chromatography) chromatogram of the 2-furoylmethyl-amino acids after acid hydrolysis of a breakfast cereal sample.

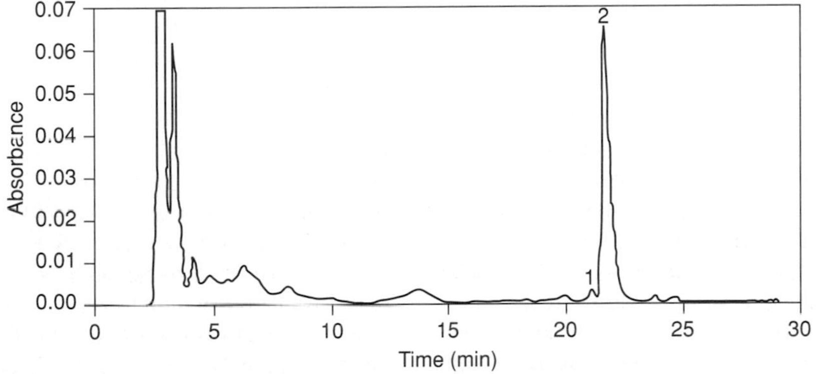

Figure 24.1. HPLC chromatogram of the 2-furoylmethyl-GABA (1) and furosine (2) in an acid hydrolyzate of a breakfast cereal sample. (Taken from Rada-Mendoza et al. 2004 with permission from Elsevier.)

Since considerable amounts of furosine have been detected in dried milk (Corzo et al. 1994), part of the furosine found in cereal products containing dried milk might already be present in dried milk ingredients. Besides furosine, Rada-Mendoza et al. (2004) found the 2-furoylmethyl derivative corresponding to GABA in samples of crackers and breakfast cereals; however, it was absent in most cookie samples. The presence of this compound in breakfast cereals and crackers may be attributed to the considerable amount of free GABA present in the rice and corn used in their manufacture. The presence of a substantial amount of free GABA in cornflakes has been previously reported (Marchenko et al. 1973). The variable amounts of dried milk used in the manufacture of cereal-based products and the different levels of free GABA in unprocessed cereals appears to be a major drawback for the use of furosine and 2-furoylmethyl-GABA as suitable indicators for differentiating among commercial cereal-based products. However, in the cereal industry, where exact ingredient composition is known, measurement of the 2-furoylmethyl-GABA and furosine formed might be used as indicators to monitor processing conditions during the manufacture of cereal products.

HYDROXYMETHYLFURFURAL

Hydroxymethylfurfural is formed by the degradation of hexoses and is also an intermediate product in the Maillard reaction (Hodge 1953, Kroh 1994). Hydroxymethylfurfural is a classic indicator of browning in several foods such as milk (van Boekel and Zia-Ur-Rehman 1987, Morales et al. 1997), juices (Lee and Nagy 1988), and honey (Jeuring and Kuppers 1980, Sanz et al. 2003).

In the case of cereal-based foods hydroxymethylfurfural has been also used as chemical indicator. Thus, this compound has been detected in dried pasta (Acquistucci and Bassotti 1992, Resmini et al. 1993), baby cereals (Guerra-Hernández et al. 1992, Fernández-Artigas et al. 1999b), and bread (Ramírez-Jiménez et al. 2000, Cárdenas-Ruiz et al. 2004). In a study on the effect of various sugars on the quality of baked cookies, furfural (in cookies elaborated with pentoses) and hydroxymethylfurfural (in cookies elaborated with hexoses) were detected (Nishibori and Kawakishi 1992). Hydroxymethylfurfural has also been found in model systems of cookies baked at 150°C for 10 minutes (Nishibori and Kawakishi 1995, Nishibori et al. 1998).

García-Villanova et al. (1993) proposed the determination of hydroxymethylfurfural to control the heating procedure in breakfast cereals as well as in other cereal derivatives. Birlouez-Aragon et al. (2001) detected hydroxymethylfurfural in commercial samples of breakfast cereals (cornflakes), in samples (before and after processing) using traditional cooking and roasting, and in a model system of wheat, oats, and rice with two levels of sugars subjected to extrusion. Hydroxymethylfurfural was formed during the manufacture of cornflakes; however, only traces were detected in the raw material. Hydroxymethylfurfural formation during extrusion increased proportionally as sugar concentration increased. Among all analyzed samples, the highest content of hydroxymethylfurfural was found in commercial samples.

COLOR

Color is an important characteristic of cereal-based foods and, together with texture and aroma, contributes to consumer preference. Color is another indication of the extent of Maillard reaction and caramelization. The kinetic parameters of these reactions are extremely complex for cereal products. As a consequence, the coloring reaction is always studied globally, without taking into account individual reaction mechanisms (Chevallier et al. 2002). Color depends both on the physicochemical characteristics of the raw dough (water content, pH, reducing sugars, and amino acid content) and on the operating conditions during processing (Zanoni et al. 1995).

Browning pigment (melanoidins) formation occurs at the advanced stages of browning reactions, and although is undesirable in milk, fruit juices, and tomatoes, among other foods (De Man, 1980), it is desirable during the manufacture of cookies, crackers, and breakfast cereals. Color development has been studied in cereal-based foods such as baby cereals (Fernández-Artigas et al. 1999b), bread (Ramírez-Jiménez et al. 2001), cookies (Kane et al. 2003), and gluten-free biscuits (Schober et al. 2003).

The formation of melanoidins and other products of the Maillard reaction darkens a food, but a darker color is not always attributed to the presence of these compounds, since the initial composition of the mixture can also afford color. In a study on the protein nutritional value of extrusion-cooked wheat flours, Bjorck et al. (1984) determined the color of the extruded material and found a correlation between reflectance values and the total lysine content of extruded wheat. The reflectance of the raw flour was very different, and comparison could not be made to determine the effect of extrusion.

Hunter "L" values determined for wheat flours, commercial flaked-toasted, extruded-toasted, and extruded-puffed breakfast cereals were found to be positively correlated with available lysine (McAuley et al. 1987).

During industrial baking of cookies, the effect of time on color development and other parameters (volume, structure, weight, crispness) was studied by Piazza and Masi (1997). The development of crispness increased with time and was found to be related to the other physical processes that occur during baking.

Bernussi et al. (1998) studied the effects of microwave baking on the moisture gradient and overall quality of cookies, and they observed that color did not differ significantly from that of the control samples (cookies baked using the traditional process).

Broyart et al. (1998) carried out a study on the kinetics of color formation during the baking of crackers in a static electrically heated oven. These authors observed that the darkening step starts when the product temperature reaches a critical value in the range of 105–115°C. A kinetic model was developed in order to predict the lightness variation of the cracker surface using the product temperature and moisture content variations during baking. The evolution of lightness appears to follow a first-order kinetic influenced by these two parameters.

Color development has also been included, together with other parameters (temperature, water loss, etc.), in a mathematical model that simulates the functioning of a continuous industrial-scale biscuit oven (Broyart and Trystram 2003).

Gallagher et al. (2003) observed different color development in the production of a functional low-fat, low-sugar biscuit depending upon the quantities of sugar and protein present.

The manufacture of many breakfast cereals starts with the cooking of whole cereal grains in a rotary pressure cooker. During this operation, the grains absorb heat and moisture and undergo chemical (browning reactions) and physicochemical changes as a consequence. The cooking stage is thought to have a key influence on the properties, such as color, flavor, and texture, of the final product. Horrobin et al. (2003) studied the interior and surface color development during wheat grain steaming and the results obtained indicated a possible relationship between color development and moisture uptake during the cooking process.

In a study on the effects of oven humidity on foods (bread, cakes, and cookies) baked in gas convection ovens, Xue et al. (2004) observed that increased oven humidity results in products with lighter color and reduced firmness.

As mentioned above, the formation of Maillard reaction products and intense color is responsible for the organoleptic properties of this type of product. Moreover, it is also important to consider that the brown pigments formed could present some biological activities. Thus, Bressa et al. (1996) observed a considerable antioxidant capacity in cookies during the first 20–30 minutes of cooking (when browning takes place). Whole grain breakfast cereals also have been proved to be an important dietary source of antioxidants (Miller et al. 2000). Borreli et al. (2003) studied the formation of colored compounds in bread and biscuits, and they examined the antioxidant activity and the potential cytotoxic effects of the formed products.

FLUORESCENCE

During the advanced stages of the nonenzymatic browning, compounds with fluorescence are also produced. Recently, some analytical methods based on fluorescence measurements have been used to evaluate the extent of this reaction. For instance, the FAST (fluorescence of advanced Maillard products and soluble tryptophan) method proposed by Birlouez-Aragon et al. (1998) is based on the determination of maximal fluorescence emission at an excitation wavelength of 330–350 nm, which corresponds to molecular structures formed between reducing sugars and the lysine residues of proteins. This fluorescence is dependent on heat treatment and is related to protein nutritional loss. Thus, this method, firstly validated on milk samples, has been used in other foods modified by the Maillard reaction, such as breakfast cereals. Birlouez-Aragon et al. (2001) studied the correlation between the FAST index, lysine loss, and hydroxymethylfurfural formation during the manufacture of breakfast cereals by extrusion and in commercial samples. The FAST index was in good agreement with hydroxymethylfurfural formation. These authors also found that the relationship between the FAST index and lysine loss indicates that, for severe treatments inducing lysine blockage higher than 30%, lysine loss is less rapid than the increase in the FAST index. Thus, the fluorimetric FAST method appears to be an interesting alternative for evaluating nutritional damage in a great variety of cereal-based products submitted to heat treatment.

ACRYLAMIDE

Recent studies have reported the presence of acrylamide (2-propenamide), "a probably carcinogenic to humans" compound (IARC 1994), in a number of fried and oven-cooked foods (Riediker and Stadler 2003). One of the possible mechanisms involved in acrylamide formation is the reaction between asparagine and reducing sugars such as glucose or fructose via the Maillard reaction (Mottram et al. 2002; Weisshaar and Gutsche 2002). Since asparagine is a major amino acid in cereals, the possible formation of acrylamide in cereal-based foods should be considered.

Variable amounts of acrylamide have been found in cereal-based foods such as bread, cookies, crackers, biscuits, and breakfast cereals (Yoshida et al. 2002, Ono et al. 2003, Riediker and Stadler 2003, Svensson et al. 2003, Delatour et al. 2004, Wenzl et al. 2004, Taeymans et al. 2004).

MALTULOSE

Besides the Maillard reaction, isomerization of reducing carbohydrates may take place during the

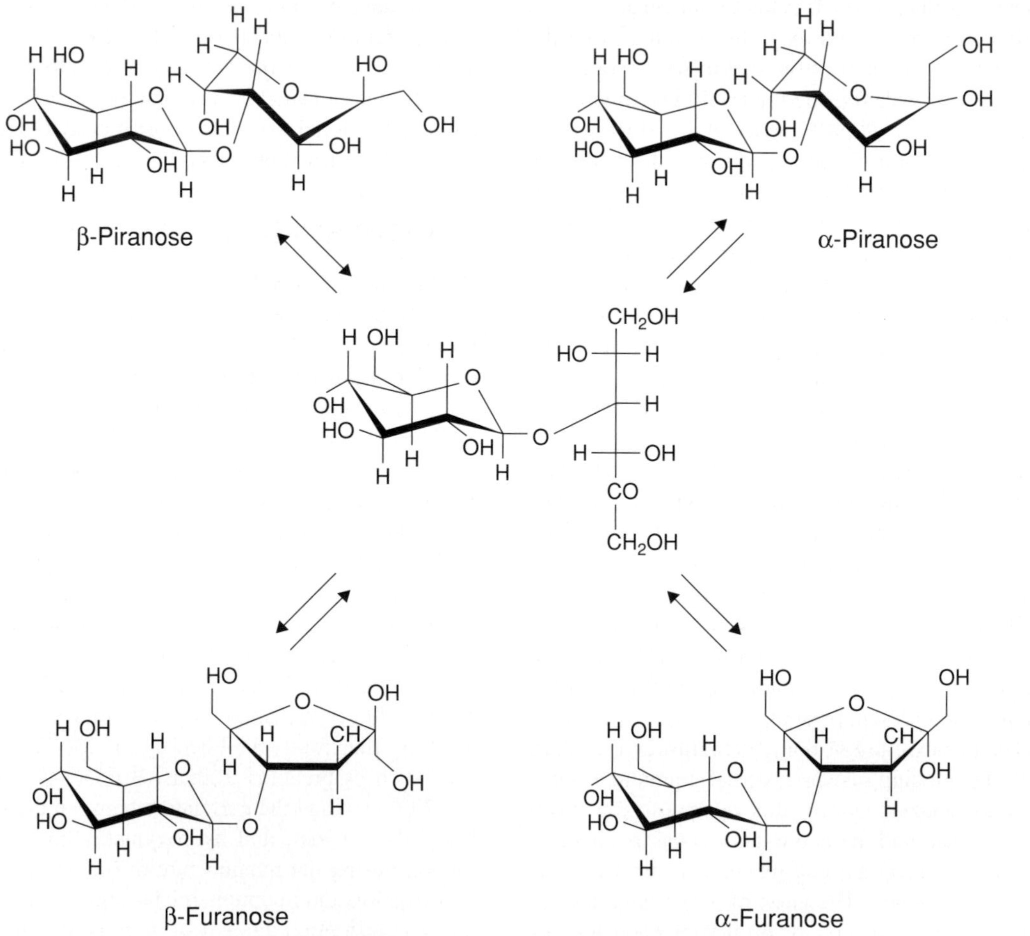

β-Piranose

α-Piranose

β-Furanose

α-Furanose

Figure 24.2. Anomeric forms of maltulose.

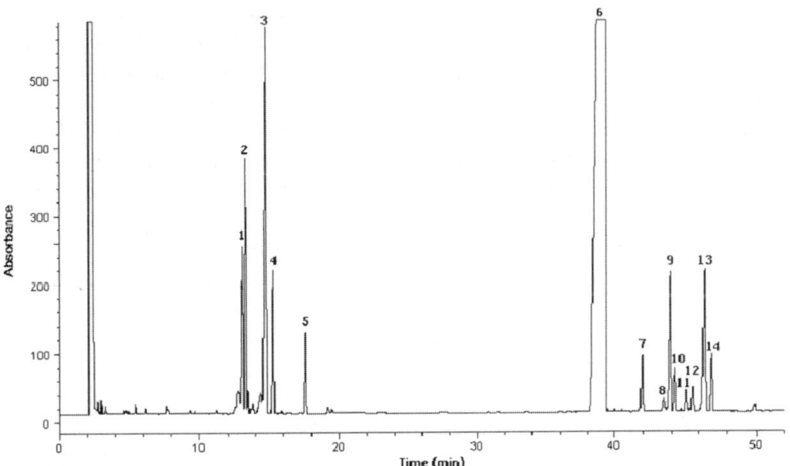

Figure 24.3. Gas chromatographic profile of the oxime-trimethylsilyl derivatives of fructose (1,2), glucose (3,4), sucrose (6), lactulose (8), lactose (9,10), maltulose (11,12), and maltose (13,14) of a cookie sample. Peaks 5 and 7 were the internal standards *myo*-inositol and trehalose, respectively. (Taken from Rada-Mendoza et al. 2004 with permission from Elsevier.)

processing of cookies, crackers, and breakfast cereals. Maltulose (Fig. 24.2), an epimerization product of maltose, has been found in the crust of bread (Westerlund et al. 1989). Maltulose formation is also observed during the heating of maltodextrin solution at high temperatures (180°C) (Kroh et al. 1996). García Baños et al. (2000, 2002), detected maltulose in commercial enteral products and proposed the maltose:maltulose ratio as a heat treatment and storage indicator of enteral formula.

In cookies, crackers, and breakfast cereals, Rada-Mendoza et al. (2004) detected maltulose (from traces to 842 mg/100 g) in all commercial samples analyzed. Figure 24.3 shows the gas chromatographic profile of the oxime-trimethylsilyl derivatives of mono- and disaccharide fractions of a cookie sample. Similar patterns were obtained for crackers and breakfast cereals. The formation of maltulose depends mainly on initial maltose content, pH, and the heat treatment intensity of the processes. Since cookies, crackers, and breakfast cereal samples may contain variable amounts of maltose, the usefulness of maltulose as an indicator of heat treatment may be questionable. Previous studies on the formation of maltulose during heating of enteral formula (García-Baños et al. 2002) pointed out that values of the maltose:maltulose ratio were similar in samples with different maltose content submitted to the same heat treatment. Therefore, the maltose:maltulose ratio is an adequate parameter for comparing samples with different initial maltose contents. Because maltose isomerization increases with pH, differences in the maltose:maltulose ratio can be due to different heat processing conditions only in samples with a similar pH. These results, shown by Rada-Mendoza et al. (2004), seem to indicate that the maltose:maltulose ratio can allow differentiation among commercial cereal-based products and may serve as an indicator of the heat load during its manufacture.

REFERENCES

Acquistucci R, Bassotti G. 1992. Effects of Maillard reaction on protein in spaghetti samples. Chemical reaction in Foods II. Abstract Papers, Symposium Prague, September. P.94.

Acquistucci R, Quattrucci E. 1993. In vivo protein digestibility and lysine availability in pasta samples dried under different conditions. Nutritional, chemical and food processing implications of nutrient availability. Bioavailability 1:23–27.

Awasthi P, Yadav MC. 2000. Effect of incorporation of liquid dairy by-products on chemical characteristics of soy-fortified biscuits. Journal of Food Science and Technology-Mysore 37(2): 158–161.

Bernussi ALM, Chang YK, Martínez-Bustos E. 1998. Effects of production by microwave heating after conventional baking on moisture gradient and product quality of biscuits (cookies). Cereal chemistry 75(5): 606–611.

Birlouez-Aragon I, Nicolas M, Métais A, Marchon N, Grenier J, Calvo D. 1998. A rapid fluorimetric meted to estimate the heat treatment of liquid milk. International Dairy Journal 8(9): 771–777.

Birlouez-Aragon I, Leclere J, Quedraogo CL, Birlouez E, Grongnet J-F. 2001. The FAST method, a rapid approach of the nutritional quality of heat-treated foods. Nahrung/Food 45(3): 201–205.

Bjorck I, Noguchi A, Asp N-G, Cheftel JC, Dahlqvist A. 1983. Protein nutritional value of a biscuit processed by extrusion cooking: Effects on available lysine. Journal of Agricultural and Food Chemistry 31(3): 488–492.

Bjorck I, Asp N-G, Dahlqvist A. 1984.Protein nutritional value of extrusion-cooked wheat flours. Food Chemistry 15(8): 203–214.

Borrelli RC, Mennella C, Barba F, Russo M, Krome K, Erbersdobler HF, Faist V, Fogliano V. 2003. Characterization of coloured compounds obtained by enzymatic extraction of bakery products. Food and Chemical Toxicology 41(10): 1367–1374.

Bressa F, Tesson N, DallaRosa M, Sensidoni A, Tubaru F. 1996. Antioxidant effect of Maillard reaction products: Application to a butter cookie of a competition kinetics analysis. Journal of Agricultural and Food Chemistry 44(3): 692–695.

Broyart B, Trystram G. 2003. Modelling of heat and mass transfer phenomena and quality changes during continuous biscuit baking using both deductive and inductive (neural network) modelling principles. Food and Bioproducts Processing 81(C4): 316–326.

Broyart B, Trystram G, Duquenoy A. 1998. Predicting colour kinetics during cracker baking. Journal of Food Engineering 35(3): 351–368.

Camire ME, Belbez EO. 1996. Flavor formation during extrusion cooking. Cereal Foods World 41(9): 734–736.

Camire ME, Camire A, Krumhar K. 1990. Chemical and nutritional changes in foods during extrusion. Critical Review in Food Science and Nutrition 29(1): 35–57.

Cárdenas-Ruiz, J., Guerra-Hernández, E., García-Villanova, B. 2004. Furosine is a useful indicator in pre-baked bread. Journal of the Science of Food and Agriculture 84(4): 366–370.

Carratú B, Boniglia C, Filesi C, Bellomonte G. 1993. Influenza del trattamento tecnologico sugli alimenti: Determinazione per cromatografía ionica di piridosina e furosina in prodotti dietetici ed alimentari. La Rivista di Scienza dell'Alimentazione, anno 22(4): 455–457.

Cheftel JC. 1986. Nutritional effects of extrusion cooking. Food Chemistry 20:263–283.

Cheftel JC, Li S-F, Mosso JC, Arnauld J. 1981. In: Progress Food Nutrition Science, vol. 5. Pergamon Press, London. P.487.

Chevallier S, Della Valle G, Colonia P, Broyart B, Trystram G. 2002. Structural and chemical modifications of short dough during baking. Journal of Cereal Science 35(1): 1–10.

Corzo N, Delgado T, Troyano E, Olano A. 1994. Ratio of lactulose to furosine as indicator of quality of commercial milks. Journal of Food Protection 57(8): 737–739.

Delatour Y, Perisset A, Goldmann T, Riediker S, Stadler RH. 2004. Improved sample preparation to determine acrylamide in difficult matrixes such as chocolate powder, cocoa, and coffee by liquid chromatography tandem mass spectroscopy. Journal of Agricultural and Food Chemistry 52(15): 4625–4631.

Del Castillo MD, Villamiel M, Olano A, Corzo N. 2000. Use of 2-furoylmethyl derivatives of GABA and arginine as indicators of the initial steps of Maillard reaction in orange juice. Journal of Agricultural and Food Chemistry 48(9): 4217–4220.

De Man JM. 1980. Nonenzymatic browning. In: Principles of Food Chemistry. AVI Publishing, Westport, Connecticut. Pp. 98–112.

Erbersdobler HF, Hupe A. 1991. Determination of the lysine damage and calculation of lysine bio-availability in several processed foods. Zeitschrift für Ernaührungswissenchaft 30(1): 46–49.

Fernández-Artigas P, García-Villanova B, Guerra-Hernández E. 1999a. Blockage of available lysine at different stages of infant cereal production. Journal of the Science of Food and Agriculture 79(6): 851–854.

Fernández-Artigas P, Guerra-Hernández E, García-Villanova B. 1999b. Browning indicators in model systems and baby cereals. Journal of Agricultural and Food Chemistry 47(7): 2872–2878.

Gallagher E, O'Brien CM, Scannell AGM, Arendt EK. 2003. Use of response surface methodology to produce functional short dough biscuits. Journal of Food Engineering 56(2–3): 269–271.

García-Baños JL, Olano A, Corzo N. 2000. Determination of mono and disaccharide content of enteral formulations by gas chromatography. Chromatographia 52(3/4): 221–224.

———. 2002. Changes in carbohydrate fraction during manufacture and storage of enteral formulas. Journal of Food Science 67(9): 3232–3235.

García-Baños JL, Corzo N, Sanz ML, Olano A. 2004. Maltulose and furosine as indicators of quality of pasta products. Food Chemistry 88(1): 35–38.

García-Villanova B, Guerra-Hernández E, Martínez-Gómez E, Montilla J. 1993. Liquid Chromatography for the determination of 5-hydroxymethyl-2-furaldehyde in breakfast cereals. Journal of Agricultural and Food Chemistry 41(8): 1254–1255.

González-Galán A, Wang SH, Sgarbieri VC, Moraes MAC. 1991. Sensory and nutritional properties of cookies based on wheat-rice-soybean flours baked in a microwave oven. Journal of Food Science 56(6): 1699–1701, 1706.

Guerra-Hernández E, Corzo N. 1996. Furosine determination in baby cereals by ion-pair reversed phase liquid chromatography. Cereal Chemistry 73(6): 729–731.

Guerra-Hernández E, Corzo N, García-Villanova B. 1999. Maillard reaction evaluation by furosine determination during infant cereal processing. Journal of Cereal Science 29(2): 171–176.

Guerra-Hernández E, García-Villanova B, Montilla Gómez J. 1992. Determination of hydroxymethylfurfural in baby cereals by high-performance liquid chromatography. Journal of Liquid Chromatography 15(14): 2551–2559.

Hodge JE. 1953. Dehydrated foods: Chemistry of browning reactions in model systems. Journal of Agricultural and Food Chemistry 1(15): 928–943.

Horrobin DJ, Landman KA, Ryder L. 2003. Interior and surface color development during wheat grain steaming. Journal of Food Engineering 57(1): 33–43.

Horvatić M, Eres M. 2002. Protein nutritive quality during production and storage of dietetic biscuits. Journal of the Science of Food and Agriculture 82(14): 1617–1620.

Horvatić M, Guterman M. 1997. Available lysine content during cereal flake production. Journal of the Science of Food and Agriculture 74(3): 354–358.

Hozova B, Buchtova V, Dodok L, Zemanovic J. 1997. Microbiological, nutritional and sensory aspects of stored amaranth biscuits and amaranth crackers. Nahrung/Food 41(3): 155–158.

Huang WN. 1998. Comparing cornflake manufacturing processes. Cereal Foods World 43(8): 641–643.

IARC. 1994. International Agency for Research on Cancer. Monographs on the Evaluation of Carcinogenic Risks to Humans: Some Industrial Chemicals No. 60. IARC, Lyon, France. P.389.

Jeuring HJ, Kuppers FJEM. 1980. High performance liquid chromatography of furfural and hydroxymethylfurfural in spirits and honey. Journal of the Association of Official Analytical Chemists 63(6): 1215–1218.

Kane AM, Lyon BG, Swanson RB, Savage EM. 2003. Comparison of two sensory and two instrumental methods to evaluate cookie color. Journal of Food Science 68(5): 1831–1837.

Kroh LW. 1994. Caramelisation in Food and Beverages. Food Chemistry 51(4): 373–379.

Kroh LW, Jalyscho W, Häseler J. 1996. Nonvolatile reaction products by heat-induced degradation of α-glucans. Part I: Analysis of oligomeric maltodextrins and anhydrosugars. Starch 48(11–12): 426–433.

Lee HS, Nagy S. 1988. Quality changes and nonenzymatic browning intermediates in grapefruit juice during storage. Journal of Food Science 53(1): 168–172.

Linko O, Colonna P, Mercier C. 1981. High temperature, short time extrusion cooking. Advances in Cereal Science and Technology 4:145–235.

Manzaneque Ramos A. 1994. Snacks y preparados especiales para desayuno. Alimentación, Equipos y Tecnología num. 2:61–65.

Marchenko NV, Pankratova TS, Lyubushkin VT. 1973. Study of the free amino acids in cornflakes. Konservnaya I Ovoshchesushil'naya Promyshennost 3:40–41.

Martinkus VB, Alford BB, Pyke RE. 1977. Effect of heat-treatment on available lysine and other amino-acids in a cookie fortified with oilseed flours. Federation Proceedings 36(3): 1111–1111.

McAuley JA, Kunkel ME, Acton JC. 1987. Relationships of available lysine to lignin, color and protein digestibility of selected wheat-based breakfast cereals. Journal of Food Science 52(6): 1580–1582.

Meredith FI, Caster WO. 1984. Amino acid content in selected breakfast cereals. Journal of Food Science 49(6): 1624–1625.

Miller HE, Rigelhof F, Marquart L, Prakash A, Kanter M. 2000. Antioxidant content of whole grain breakfast cereals, fruits and vegetables. Journal of the American College of Nutrition 19(3): 312S–319S. Suppl. S.

Molnár-Perl I, Pinter-Szakács M, Wittmann R, Reutter M, Eichner K. 1986. Optimum yield of pyridosine and furosine originating from Maillard reactions monitored by ion-exchange chromatography. Journal of Chromatography 361:311–320.

Mottram DS, Wedzicha BL, Dodson AT. 2002. Acrylamide is formed in the Maillard reaction. Nature 419(6906): 448–449.

Morales FJ, Romero C, Jiménez-Pérez S. 1997. Chromatographic determination of bound hydroxymethylfurfural as an index of milk protein glycosylation. Journal of Agricultural and Food Chemistry 45(5): 1570–1573.

Nepal-Sing S, Chauhan GS. 1989. Some physicochemical characteristics of defatted soy flour fortified noodles. Journal of Food Science and Technology 26:210–212.

Nishibori S, Berhnard RA, Osawa T, Kawakishi S. 1998. Volatile components formed from reaction of sugar and beta-alanine as a model system of cookie processing. Process-Induced Chemical Changes in Food Advances in Experimental Medicine and Biology 434:255–267.

Nishibori S, Kawakishi S. 1992. Effect of various sugars on the quality of baked cookies. Cereal Chemistry 69(2): 160–163.

———. 1995. Effect of various amino-acids on formation of volatile compounds during baking in a low moisture food system. Journal of the Japanese Society for Food Science and Technology-Nippon Shokuhin Kagaku Kogaku Kaishi 42(1): 20–25.

Noguchi A, Mosso K, Aymard C, Jeunink J, Cheftel JC. 1982. Maillard reactions during extrusion-cooking of protein-enriched biscuits. Lebensmittel Wissenschaft und Technologie 15(2): 105–110.

Ono H, Chuda Y, Ohnishi-Kameyama M, Yada H, Ishizaka M, Kobayashi H, Yoshida M. 2003. Analysis of acrylamide by LC-MS/MS and GC-MS in processed Japanese foods. Food Additives and Contaminants 20(3): 215–220.

Piazza L, Masi P. 1997. Development of crispness in cookies during baking in an industrial oven. Cereal Chemistry 74(2): 135–140.

Phillips RD. 1988. Effect of extrusion cooking on the nutritional quality of plant proteins. In: RD Phillips, JW Fionley, Editors. Protein Quality and the Effects of Processing. Marcel Dekker, New York.

Rada-Mendoza M, García-Baños JL, Villamiel M, Olano A. 2004. Study on nonenzymatic browning in cookies, crackers and breakfast cereals by maltulose and furosine determination. Journal of Cereal Science 39(2): 167–173.

Ramírez-Jiménez A, García-Villanova B, Guerra-Hernández E. 2001. Effect of toasting time on browning of sliced bread. Journal of the Science of Food and Agriculture 81(5): 513–518.

Ramírez-Jiménez, A., Guerra-Hernández, E., García-Villanova, B. 2000. Browning indicators in bread. Journal of Agricultural and Food Chemistry 48(9): 4176–4181.

Resmini P, Pellegrino L. 1991. Analysis of food heat damage by direct HPLC of furosine. International Chromatography Laboratory 6:7–11.

Resmini P, Pellegrino L, Pagani MA, De Noni I. 1993. Formation of 2-acetyl-3-D-glucopyranosylfuran (glucosylisomaltol) from nonenzymatic browning in pasta drying. Italian Journal Food Science 5(4): 341–353.

Riediker S, Stadler RH. 2003. Analysis of acrylamide in food by isotope-dilution liquid chromatography coupled with electrospray ionisation tandem mass spectrometry. Journal of chromatography A 1020(1): 121–130.

Sanz ML, Del Castillo MD, Corzo N, Olano A. 2000. Presence of 2-furoylmethyl derivatives in hydrolysates of processed tomato products. Journal of Agricultural and Food Chemistry 48(2): 468–471.

———. 2001. Formation of Amadori compounds in dehydrated fruits. Journal of Agricultural and Food Chemistry 49(11): 5228–5231.

Sanz ML, Corzo N, Olano A. 2003. 2-Furoylmethyl amino acids and hydroxymethylfurfural as indicators of honey quality. Journal of Agricultural and Food Chemistry 51(15): 4278–4283.

Schober TJ, O'Brien CM, McCarthy D, Darnedde A, Arendt EK. 2003. Influence of gluten-free flour mixes and fat powders on the quality of gluten-free biscuits. European Food Research and Technology 216(5): 369–376.

Singh R, Singh G, Chauhan GS. 2000. Nutritional evaluation of soy fortified biscuits. Journal of Food Science and Technology-Mysore 37(2): 162–164.

Svensson, K, Abramsson L, Becker W, Glynn A, Hellenäs K-E, Lind Y, Rosén J. 2003. Dietary intake of acrylamide in Sweden. Food and Chemical Toxicology 41(11): 1581–1586.

Taeymans D, Wood J, Ashby P, Blank I, Studer A, Stadler RH, Gonde P, van Eijck P, Lalljie S, Lingnert H, Lindblom M, Matissek R, Muller D, Tallmadge D, O'Brien J, Thompson S, Silvani D, Whitmore T. 2004. A review of acrylamide: An industry perspective on research, analysis, formation and control. Critical Reviews in Food Science and Nutrition 44(5): 323–347.

Tsen CC, Reddy PRK, El-Samahy SK, Gehrke CW. 1983. Effects of the Maillard browning reaction on the nutritive value of breads and pizza crusts. In: GR Waller, MS Feather, editors. The Maillard Reaction in Food and Nutrition. ACS, Washington, D.C. Pp. 379–394.

van Boekel MAJS, Zia-Ur-Rehman. 1987. Determination of HMF in heated milk by HPLC. Netherlands Milk Dairy Journal 41(4): 297–306.

Weisshaar R, Gutsche B. 2002. Formation of acrylamide in heated potato products. Model experiments pointing to asparagines as precursor. Deutsche Lebbensmittel 98 Jahrgang (11): 397–400.

Wenzl T, de la Calle B, Gatermann R, Hoenicke K, Ulberth F, Anklam E. 2004. Evaluation of the results from an Inter-laboratory comparison study of the determination of acrylamide in crispbread and butter cookies. Analytical and Bioanalytical Chemistry 379(3): 449–457.

Westerlund E, Theander O, Aman P. 1989. Effects of baking on protein and ethanol-extractable carbohydrate in white bread fractions. A review of acrylamide: An industry perspective Journal of Cereal Science 10(2): 139–147.

Xue J, Lefort G, Walker CE. 2004. Effects of oven humidity on foods baked in gas convection ovens. Journal of Food Processing and Preservation 28(3): 179–200.

Yoshida M, Ono H, Ohnishi-Kameyama M, Chuda Y, Yada H, Kobayashi H, Ishizaka M. 2002. Determination of acrylamide in processed foodstuffs in Japan. Journal of the Japancse Society for Food Science and Technology-Nippon Shokuhin Kagaku Kogaku Kaishi 49(12): 822–825.

Zanoni B, Peri C, Bruno D. 1995. Modelling of browning kinetics of bread crust during baking. Lebensmittel Wissenschaft und Technologie 28(3): 604–609.

25
Rye Constituents and Their Impact on Rye Processing

*T. Verwimp, C. M. Courtin, and J. A. Delcour**

*Corresponding author.

CLASSIFICATION

Rye (*Secale cereale* L.), member of the grass family (Gramineae) and typically classified into winter and spring varieties, is a diploid, seven pair chromosome cereal, with tetraploid varieties sometimes being produced artificially. While cultivated rye varieties are less numerous than those of other cereal crops, most varieties are mixed populations because of cross-pollination.

PRODUCTION

Although rye production (15 million tons) is low compared with the 2003 total world production of the three major cereals, wheat, corn, and rice (556, 638, and 589 million tons, respectively) (FAO 2004), rye's low soil and fertilization requirements as well as its relatively good overwintering ability guarantee continuing interest in the cereal, as it can be cultivated in areas generally not suited for other cereal crops. In 2003, high amounts of rye were produced by Russia (4.1 million tons), Poland (3.2 million tons), Germany (2.3 million tons), and Belarus (1.4 million tons) (FAO 2004), making this cereal crop of major importance in parts of Europe and Asia, but of less importance in America. Production has remained fairly constant, but with a downward trend, during the past 10 years.

THE RYE KERNEL

The main components of the rye kernel are, from the outside to the inside, the hull, pericarp, testa, nucellar epidermis, aleurone and starchy endosperm, and the germ. As the hull threshes free, the pericarp is the actual outer part of the grain, surrounding the seed and consisting of different layers, ranging from the epidermis, hypodermis, and cross cells to tube cells. The latter adhere to the testa or seed coat, which is in turn lined with the nucellar epidermis. The aleurone cells are located beneath the nucellar epidermis, forming the outermost layer of the endosperm tissue, and merge with the scutellum of the germ. The starchy endosperm is composed of three main cell types, that is, the subaleurone or peripheral cells, the prismatic cells, and the central endosperm cells, and contains two major storage reserves. The subaleurone cells contain high levels of storage protein, while the prismatic cells contain high levels of starch granules (Shewry and Bechtel 2001). The pericarp, testa, and nucellar epidermis form the botanical bran and are separated from the starchy endosperm together with the aleurone layer and the germ in the milling process, forming the technical or miller's bran.

RYE CONSTITUENTS

The major rye constituents are starch and nonstarch polysaccharides and protein. Its lipids, vitamins, and minerals are minor constituents (Table 25.1).

STARCH

Starch Composition

Starch granules are mainly composed of two types of polymers, amylose and amylopectin. In general, amylose is an essentially linear polymer consisting of α-D-glucopyranose residues linked by α-1,4 bonds with few (< 1%; Ball et al. 1996) α-1,6 bonds. Amylose occurs in both free and lipid-complexed forms. Lipid-complexed amylose may be present in native starch, but is possibly also formed during gelatinization of starch (Andreev et al. 1999, Morrison 1995). Amylose contents in rye starch vary from 12 to 30% (Andreev et al. 1999, Fredriksson et al. 1998, Mohammadkhani et al. 1998, Wasserman et al. 2001). Berry et al. (1971) reported the rye amylose fraction to have a molecular weight of about 220 k.

Amylopectin is a highly branched polymer consisting of α-D-glucopyranose residues linked by α-1,4 linkages with 4–5% (according to Keetels et al. 1996) α-1,6 linkages. Berry et al. (1971) showed that rye amylopectin has 4.8% branching, with branch points about every 21 glucose units on aver-

Table 25.1. Composition of the Rye Kernel

Constituent	Content (%)	References
Starch	57.0–65.6	Aman et al. 1997, Hansen et al. 2004
Nonstarch polysaccharide	13.0–15.0	Härkönen et al. 1997
Arabinoxylan	6.5–12.2	Aman et al. 1997; Hansen et al. 2003, 2004; Henry 1987; Vinkx and Delcour 1996
Mixed-linkage β-glucan	1.5–2.6	Aman et al. 1997, Hansen et al. 2003, Henry 1987
Cellulose	2.1–2.6	Aman et al. 1997, Vinkx and Delcour 1996
Arabinogalactan peptide	0.2	Van den Bulck et al. 2004
Fructan	4.6–6.6	Hansen et al. 2003, Karpinnen et al. 2003
Protein	8.0–17.7	Fowler et al. 1990, Hansen et al. 2004
Lipid	2.0–2.5	Härkönen et al. 1997, Vinkx and Delcour 1996
Mineral	1.7–2.2	Aman et al. 1997, Hansen et al. 2004

age. The chains of amylopectin can be classified into three types: A-, B-, and C-chains. The A-chains are unbranched outer chains and are attached to the inner B-chains through their potential reducing end. In their turn, the B-chains are attached to other B-chains or to the single C-chain through their potential reducing end. The C-chain is the only chain carrying a reducing end group. Lii and Lineback (1977) reported an A-chain to B-chain ratio of 1.71–1.81 and an average unit chain length of 26 for rye amylopectin.

Minor components of starch granules are lipids, proteins, phosphorus, and ash.

Starch Structure

Rye starch has a bimodal particle size distribution comprising a major population of large, lenticular A-type granules (> 10 μm, 85%), and a minor population of small, spherical B-type granules (≤ 10 μm, 15%) (Schierbaum et al. 1991, Verwimp et al. 2004). Rye A-type starch granules have larger average particle sizes and broader particle size distribution profiles than their wheat counterparts (Fredriksson et al. 1998, Verwimp et al. 2004).

In starch granules, the amylose and amylopectin molecules are radially ordered, with their single reducing end groups towards the center, or hilum. Ordering of crystallites within the starch granule causes optical birefringence, and a Maltese cross can be observed under polarized light. Starch granules are made up of alternating semicrystalline and amorphous growth rings, arranged concentrically around the hilum. The amorphous growth ring contains amylose and probably less-ordered amylopectin (Morrison 1995). The semicrystalline growth ring consists of alternating crystalline and amorphous lamellae. The crystalline lamellae are built from amylopectin double helices, whereas the amorphous lamellae contain the branch points of the amylopectin side chains (Andreev et al. 1999, Gallant et al. 1997). Evidence has been provided that the crystalline and amorphous lamellae of the amylopectin are organized into larger, more or less spherical blocklets (Gallant et al. 1997).

The structure of starch can be studied at different levels by several techniques such as optical microscopy, scanning electron microscopy, transmission electron microscopy, and X-ray diffraction. X-ray diffraction provides information about the crystal structure of the starch polymers, but also about the relative amounts of the crystalline and amorphous phases. Cereal starches have an A-type X-ray diffraction pattern, while B-type and C-type X-ray diffraction patterns are characteristic for tuber and root starches and for legume starches, respectively. C-type starches are considered to be mixtures of A-type and B-type starches. In general, rye starches show A-type X-ray diffraction patterns, but some researchers (Gernat et al. 1993, Schierbaum et al. 1991) also reported a significant B-type portion.

Starch Physicochemical Properties

The physicochemical gelatinization, pasting, gelation, and retrogradation properties of starches are important parameters in determining the behavior of cereals in food systems.

When a starch suspension is heated above a characteristic temperature (the gelatinization temperature) in the presence of a sufficient amount of water, the starch granules undergo an order-disorder transition known as gelatinization. The phenomenon is accompanied by melting of the crystallites (loss of X-ray pattern), irreversible swelling of the granules, and solubilization of the amylose. Gelatinization can be studied by measuring the loss of birefringence of the starch granules when observed under polarized light. The end point and the range of temperatures at which gelatinization occurs can be determined. For rye starch, birefringence end point temperatures of 59.6°C (Klassen and Hill 1971) and 58.0°C (Lii and Lineback 1977) were reported. These birefringence end point temperatures were somewhat lower than those observed for wheat starches by the same authors. Another technique often used to study gelatinization-associated phenomena is differential scanning calorimetry (DSC). DSC enables determination of onset, peak, and conclusion gelatinization temperatures, gelatinization temperature ranges, and gelatinization enthalpy. The gelatinization temperature is a qualitative index of crystal structure, whereas gelatinization enthalpy is a quantitative measure of order. The DSC technique also measures the temperature and enthalpy of the dissociation of the amylose-lipid complexes. Rye starches show an onset of gelatinization at rather low temperatures. Gudmundsson and Eliasson (1991) reported onset gelatinization temperatures of 54.8–60.3°C, Radosta et al. (1992) of 53.5°C, and Verwimp et al. (2004) of 49.2–51.4°C. Rye starch has somewhat lower gelatinization temperatures and enthalpies than wheat starch (Gudmundsson and Eliasson 1991, Radosta et al. 1992, Verwimp et al. 2004). Lower dissociation enthalpies of the amylose-lipid complexes for rye starch than for wheat starch were observed by Gudmundsson and Eliasson (1991) and Verwimp et al. (2004), indicating lower levels of lipids in rye starch.

When starch suspensions are heated beyond their gelatinization temperature, granule swelling and amylose and/or amylopectin leaching continue. Eventually, total disruption of the granules occurs, and a starch paste with viscoelastic properties is formed. This process is called pasting. Upon cooling, gelation occurs. During gelation, at sufficiently high starch concentrations (> 6%, w/w), the starch paste is converted into a gel. The gel consists of an amylose matrix enriched with swollen granules. During gelation and storage of starch gels, retrogradation occurs. This is a process whereby the starch molecules begin to reassociate in an ordered structure, without regaining the original molecular order. Initially, starch retrogradation is dominated by gelation and crystallization of the amylose molecules. In a longer time frame, amylopectin molecules recrystallize. Rye starch exhibits a unique, typical swelling behavior with a high pasting temperature (79–87°C according to Schierbaum and Kettlitz 1994; 75°C according to Verwimp et al. 2004), a stable consistency during cooking, and a high viscosity increase during cooling (Schierbaum and Kettlitz 1994, Verwimp et al. 2004). It has lower pasting temperatures than wheat starch (Schierbaum and Kettlitz 1994, Verwimp et al. 2004). Fredriksson et al. (1998) and Gudmundsson and Eliasson (1991) showed that rye starches retrograde to a lesser extent than starches from wheat or corn and ascribed this to differences in the structure of the amylopectin polymers.

Nonstarch Polysaccharides

Rye contains on average 13.0–15.0% nonstarch polysaccharides, which serve mainly as cell wall components and the majority of which are arabinoxylans (Table 25.1). Other nonstarch polysaccharide constituents of rye are mixed-linked β-glucan, cellulose, arabinogalactan peptides, and fructans (Table 25.1). The variation in nonstarch polysaccharide content and composition is significantly influenced by both harvest year and rye genotype (Hansen et al. 2003, 2004). Most of the nonstarch polysaccharides are classified as dietary fiber because of their resistance to digestion by human digestive enzymes. In view of the major importance of arabinoxylans in rye, these components will be further discussed in detail.

Arabinoxylan Occurrence and Structure

Arabinoxylans consist of a backbone of 1,4-linked β-D-xylopyranose residues, unsubstituted or mono- or disubstituted with single α-L-arabinofuranose residues at the O-2- and/or O-3-position. To some of these arabinose residues, a ferulic acid moiety is ester bound at the O-5 position. In rye, as in other cereals, two types of arabinoxylans can be distinguished: water extractable and water unextractable. Water-unextractable arabinoxylans, representing approximately 75% of the total arabinoxylan population of the rye kernel, are unextractable because they are retained in the cell walls by covalent and noncovalent interactions among arabinoxylans and between arabinoxylans and other cell wall constituents. Water-extractable arabinoxylans are thought to be loosely bound at the cell wall surface. Their content in the rye kernel ranges from 1.5 to 3.0% (Figueroa-Espinoza et al. 2002; Hansen et al. 2003, 2004) and is influenced by environmental and genetic factors but also by the conditions under which extraction occurs (Cyran et al. 2003, Härkönen et al. 1995). While the rye bran and shorts are high in arabinoxylan (14.9–39.5% and 13.5–18.7%, respectively), the endosperm contain lower levels (2.1–4.9%) (Fengler and Marquardt 1988a, Glitso and Bach Knudsen 1999, Härkönen et al. 1997, Nilsson et al. 1997a).

A considerable variation in substitution degree and pattern of xylose residues, arabinoxylan molecular weight, and ferulic acid content has been reported for both water-extractable and water-unextractable arabinoxylan. This variability in structure can, in part, account for the differences in extractability and solubility of the arabinoxylans.

Substitution Degree and Pattern of Xylose Residues Bengtsson et al. (1992a) reported on what could either be two different types of polymers or two different regions in the same molecule, that is, arabinoxylan I and arabinoxylan II. Arabinoxylan I contains mainly un- and monosubstituted xylose residues with on average 50% of the xylose residues substituted at O-3 with an arabinose residue. Only 2% of the xylose residues are disubstituted at O-2 and O-3 with arabinose residues (Bengtsson and Aman 1990). The xylose residues carrying an arabinose side chain occur predominantly as isolated residues (36%) or small blocks of two residues (62%) (Aman and Bengtsson 1991). Arabinoxylan II contains mainly un- and disubstituted xylose residues with on average 60–70% of the xylose residues substituted at O-2 and O-3 with arabinose residues (Bengtsson et al. 1992a). The levels of arabinoxylan I and II in different rye varieties ranges from 1.4 to 1.7% and from 0.6 to 1.0%, respectively (Bengtsson et al. 1992b). In contrast to the two classes of arabinoxylans described by Bengtsson et al. (1992a), Vinkx et al. (1993) concluded that there were a range of rye water-extractable arabinoxylans. The latter authors fractionated rye water-extractable arabinoxylans by graded ammonium sulfate precipitation into several fractions with arabinose:xylose (A/X) ratios of 0.5–1.4, with among them a major fraction containing almost purely O-3 monosubstituted arabinoxylans and a minor fraction consisting of almost purely disubstituted arabinoxylans. All arabinoxylan fractions contained a small amount of xylose residues substituted at O-2 with arabinose (Vinkx et al. 1995a).

Cyran et al. (2003) obtained rye flour arabinoxylans with different structural features after sequential extraction with water at different temperatures. A gradual increase in the degree of substitution in general and disubstitution in particular and a decrease in O-3 monosubstitution were observed from cold to hot water-extractable fractions. Within a water-extractable fraction, subfractions obtained after ammonium sulfate precipitation showed structures analogous to those reported by Bengtsson et al. (1992a) and Vinkx et al. (1993).

Rye water-unextractable arabinoxylan molecules also consist of a range of structures, which can only be studied after alkaline solubilization of the arabinoxylans. This treatment results in the saponification of the ester bonds linking ferulic acid to arabinose, releasing individual arabinoxylan molecules from the cell wall structure. Based on the studies of Hromadkova et al. (1987), Nilsson et al. (1996, 1999) and Vinkx et al. (1995b) (Table 25.2), three different groups of alkali-extractable arabinoxylans can be distinguished in rye bran. A first group shows an intermediate arabinose:xylose ratio (0.54–0.65) and contains mainly unsubstituted (57–64%) and monosubstituted (24–29%) xylose residues. A second group consists of almost pure unsubstituted (89%) arabinoxylans with a low arabinose:xylose ratio (0.20–0.27). A third group is characterized by a high arabinose:xylose ratio (1.08–1.10) and contains approximately 46% monosubstituted and 33% disubstituted xylose residues. In contrast to rye bran water-unextractable arabinoxylans, the rye flour alkali-solubilized arabinoxylans all show similar xylose substitution levels (Cyran et al. 2004, Vinkx 1994) (Table 25.2). However, further fractionation of the alkali-extracted arabinoxylans from rye flour by ammonium sulfate precipitation yields subfractions that differ in structure (Cyran et al. 2004). The arabinoxylans sequentially extracted with alkali from rye bran show lower degrees of branching than those extracted from rye flour by the same extraction solvent (Table 25.2).

In general, for rye bran as well as for rye flour, low yields of alkali-extractable arabinoxylan fractions with low (0.2) and high (1.1) arabinose: xylose ratios are obtained. Glitso and Bach Knudsen (1999) found lower substitution degrees for water-unextractable arabinoxylans in the aleurone layer (0.35) than for those in the starchy endosperm (0.83) and pericarp/testa fraction (1.02). From these results and because in rye milling the endosperm can be contaminated with bran, one can assume that the arabinoxylan fractions with high substitution degrees isolated from rye flour originate from contamination with bran fractions. Similarly, the arabinoxylan fractions isolated from rye bran might be contaminated with endosperm fractions.

Water-unextractable arabinoxylans in bran have a lower degree of branching than their water-extractable counterparts (Glitso and Bach Knudsen 1999), whereas the opposite is the case in the endosperm (Cyran and Cygankiewicz 2004, Glitso and Bach Knudsen 1999).

Molecular Weight Large differences in molecular weight of rye arabinoxylans exist. For rye whole meal water-extractable arabinoxylans, average molecular weights of 770 k (Girhammar and Nair 1992a) and more than 1000 k (Härkönen et al. 1995), as determined by gel permeation chromatography, have been

Table 25.2. Substitution Degrees (A/X Ratios) and Xylose Substitution Levels (%) of Alkali-Solubilized Water-Unextractable Arabinoxylans (AS-AX) from Rye Bran and Flour

AS-AX Source	Extraction Solvent[a]	A/X	Un[b]	Mono[b]	Di[b]	Reference
Rye bran	1.1 M NaOH	0.14	88	12	0	Hromadkova et al. 1987
Rye bran	1% NH_4OH	0.78	41	33	26	Ebringerova et al. 1990
Rye bran	Saturated $Ba(OH)_2$	0.65	61	29	10	Vinkx et al. 1995b
	H_2O	0.55	64	24	12	
	1.0 M KOH	0.21	89	8	3	
	Delignification followed by 1.0 M KOH	1.10	21	46	33	
Rye bran	Saturated $Ba(OH)_2$	0.54	ND[c]	ND	ND	Nilsson et al. 1996
	H_2O	0.60	ND	ND	ND	
	4.0 M KOH	0.27	ND	ND	ND	
	2.0 M KOH	1.08	ND	ND	ND	
Rye bran	Saturated $Ba(OH)_2$	0.56	57	29	14	Nilsson et al. 1999
Rye flour	Saturated $Ba(OH)_2$	0.68	46	40	14	Vinkx 1994
	H_2O	0.78	44	34	22	
	1.0 M KOH	0.72	ND	ND	ND	
Rye flour	Saturated $Ba(OH)_2$	0.67	ND	ND	ND	Nilsson et al. 1996
	H_2O	0.93	ND	ND	ND	
	4.0 M KOH	0.63	ND	ND	ND	
	2.0 M KOH	1.10				
Rye flour	Saturated $Ba(OH)_2$	0.69–0.70	51	28–29	20–21	Cyran et al. 2004
	H_2O	0.79–0.84	42–44	32–34	22–26	
	1.0 M NaOH	0.70–0.77	45–48	33	19–22	

[a]Consecutive extraction solvents in one reference indicate sequential alkaline extraction.

[b]Un, mono, and di: percentages of total xylose occurring as unsubstituted, O-2 and/or O-3 monosubstituted, and O-2, O-3 disubstituted xylose residues.

[c]ND = Not determined.

reported. Whether the water-extractable arabinoxylans from rye bran have a higher (Meuser et al. 1986) or lower (Härkönen et al. 1997) mean molecular weight than those from rye flour is not clear. It is also not clear whether extraction temperature affects the molecular size distribution of rye flour water-extractable arabinoxylans (Cyran et al. 2003) or not (Härkönen et al. 1995, Meuser et al. 1986). For water-extractable arabinoxylans isolated from rye varieties differing in extract viscosity, increasing molecular weights were correlated with increasing rye extract viscosity (Ragaee et al. 2001). Rye water-extractable arabinoxylans have higher average molecular weights than those of wheat (Dervilly-Pinel et al. 2001, Girhammar and Nair 1992a).

The water-unextractable arabinoxylans from rye flour have a higher molecular weight (1500–3000 k) than their water-extractable counterparts (620–1200 k) (Meuser et al. 1986). For rye bran water-unextractable arabinoxylans, however, Vinkx et al. (1995b) found that the molecular weight of arabinoxylans solubilized with alkali were in the same range as those reported for water-extractable arabinoxylans (Vinkx et al. 1993).

Rye arabinoxylans were reported to have an extended, rodlike conformation (Anger et al. 1986, Girhammar and Nair 1992a). In contrast, Dervilly-Pinel et al. (2001) stated that the water-extractable arabinoxylans from rye have a random coil conformation.

Ferulic Acid Content Ferulic acid residues can link adjacent arabinoxylan chains through formation of diferulic acid bridges, thereby affecting extractability of arabinoxylans.

Ferulic acid and ferulic acid dehydrodimer contents in rye vary from 0.90 to 1.17% and from 0.24 to 0.41%, respectively, and are significantly influenced by both rye genotype and harvest year (Andreasen et al. 2000b). These phenolic compounds are concentrated in the bran of the rye kernel (Andreasen et al. 2000a, Glitso and Bach Knudsen 1999).

The water-extractable arabinoxylans from rye flour contain low levels of ferulic acid (0.03–0.15%) (Figueroa-Espinoza et al. 2002, Vinkx et al. 1993) and ferulic acid dehydrodimers (0.03×10^{-2}%) (Dervilly-Pinel et al. 2001, Figueroa-Espinoza et al. 2002).

The water-unextractable arabinoxylans from rye flour contain higher amounts of ferulic acid (0.20–0.36%) and ferulic acid dehydrodimers (0.35%) than their water-extractable counterparts (Figueroa-Espinoza et al. 2002). After alkaline extraction with saturated barium hydroxide and 1.0 M sodium hydroxide, rye flour arabinoxylans still contain a substantial level of ferulic acid. Further fractionation of these arabinoxylans by ammonium sulfate precipitation reveals increasing levels of ferulic acid with increasing ammonium sulfate concentration (Cyran et al. 2004). This observation might corroborate the above-mentioned hypothesis that the different arabinoxylan fractions with varying structures obtained from rye flour may reflect contamination of the rye flour with bran fractions.

Arabinoxylan Physicochemical Properties

Solubility of a large part of the individual arabinoxylan molecules in water is mainly associated with the presence of arabinose substituents attached to the xylose residues, which prevent intermolecular aggregation of unsubstituted xylose residues. High molecular weight arabinoxylans with arabinose: xylose ratios below 0.3 tend to be insoluble.

Rye arabinoxylans are said to have high water-holding capacities, with water-extractable and water-unextractable arabinoxylans able to absorb 11 and 10 times their weight of water, respectively (Girhammar and Nair 1992a). These values were typically determined by addition of arabinoxylan to flour and measurement of the consecutive rise in Farinograph absorption. However, it is not clear whether the results obtained by this technique may be simply defined as "water-holding capacity." The mechanism behind water binding of discrete cell wall fragments that contain water-unextractable arabinoxylans is bound to be different from that of water-extractable arabinoxylans in solution. It is also unclear whether the increased Farinograph absorption is solely caused by the arabinoxylans. That water-holding values obtained with the Farinograph method have to be interpreted with care is clearly demonstrated by Girhammar and Nair (1992b), who reported a water-holding capacity of 0.47 g/g dry matter for rye water-extractable arabinoxylan, when determined by measuring the level of unfreezable water associated with the water-extractable arabinoxylan using differential scanning calorimetry. These authors stated that the water-holding capacity was not related to the molecular weight of the arabinoxylan. Different milling fractions of rye showed different water absorption capacities in relation to this total arabinoxylan content (Härkönen et al. 1997).

Even at relatively low concentrations, water-extractable arabinoxylans are able to form highly viscous solutions in water (Girhammar and Nair 1992b). Highly positive correlations were found between the viscosity of a rye extract and its content of water-extractable arabinoxylans (Boros et al. 1993; Fengler and Marquardt 1988a; Härkönen et al. 1995, 1997; Ragaee et al. 2001). For different rye milling fractions, extract viscosity correlates with the content of water-extractable arabinoxylans (Fengler and Marquardt 1988a, Glitso and Bach Knudsen 1999, Härkönen et al. 1997). Water extracts from rye are more viscous than those from other cereals (Boros et al. 1993, Fengler and Marquardt 1988a), which can be ascribed to the higher concentration of water-extractable arabinoxylans in rye than in other cereals. Differences in viscosity are related to differences in the structural features and molecular weight of water-extractable arabinoxylans. Although the impact of un-, mono-, and disubstitution is unclear (Bengtsson et al. 1992b, Dervilly-Pinel et al. 2001, Ragaee et al. 2001), molecular weight seems a logical determining parameter (Girhammar and Nair 1992b, Nilsson et al. 2000, Ragaee et al. 2001, Vinkx et al. 1993). Arabinoxylan conformation was also found to have a bearing on viscosity, with a higher viscosity being associated with a larger radius of gyration (Dervilly-Pinel et al. 2001, Ragaee et al. 2001). Viscosity measurement conditions such as shear rate (Bengtsson et al. 1992b, Härkönen et al. 1997, Nilsson et al. 2000) and temperature, pH, and salt concentration (Girhammar and Nair 1992b) also influence the viscosity of rye water-extractable arabinoxylan solutions.

Water-extractable arabinoxylans undergo oxidative gelation upon addition of hydrogen peroxide and peroxidase, resulting in an increase in viscosity and eventually the formation of a gel (Dervilly-Pinel et al. 2001, Vinkx et al. 1991). The phenomenon is caused by the formation of covalent linkages through oxidative coupling of ferulic acid residues esterified to arabinoxylan. The aromatic ring and not the propenoic moiety of ferulic acid is involved in the oxidative gelation (Vinkx et al. 1991). The oxidative gelation capacity of arabinoxylans depends on several parameters including the ferulic acid content, the molecular weight, and the structure of the arabinoxylans. Arabinoxylan fractions with higher ferulic acid content, higher molecular weights, and fewer disubstituted xylose residues have a higher gelation potential (Dervilly-Pinel et al. 2001, Vinkx et al. 1993). Dervilly-Pinel et al. (2001) stated that the intrinsic viscosity of arabinoxylans is the main parameter governing gel rigidity.

PROTEINS

The total protein content in rye is influenced by both environmental and genotypic factors and varies from 8.0 to 17.7% (see Table 25.1). The protein content is higher in the outer layers than in the inner layers of the rye kernel (Glitso and Bach Knudsen 1999; Nilsson et al. 1996, 1997a).

Protein Classification

Rye proteins can be separated by the classical Osborne fractionation procedure. The albumins are extractable with water, the globulins with dilute salt solutions, the prolamins with alcohol, and the glutelins with dilute acid or alkali. Chen and Bushuk (1970) found that the albumin, globulin, prolamin, and glutelin fractions accounted for 34, 11, 19, and 9%, respectively, of the total nitrogen content in a rye flour, with 21% of the protein remaining in the residue. Similar results were obtained by Preston and Woodbury (1975) and Jahn-Deesbach and Schipper (1980), except that the latter authors extracted a much higher proportion of glutelins. Gellrich et al. (2003), however, extracted fewer albumins and globulins and many more prolamins from rye flour. In all cases, rye flour contained higher levels of albumins and globulins and lower levels of prolamins and glutelins than wheat (Chen and Bushuk 1970, Gellrich et al. 2003, Jahn-Deesbach and Schipper 1980).

From a physiological point of view, rye proteins can be divided into two main groups: the storage proteins and the nonstorage proteins. The storage proteins are deposited in protein bodies in the developing starchy endosperm and function solely as a supply of amino acids for use during germination and seedling growth. They include the prolamins, glutelins, and the proteins in the Osborne residue. The nonstorage proteins include, amongst others, enzymes and (in some cases) their inhibitors and correspond with the Osborne albumin and globulin fractions.

Major differences are found in the amino acid compositions of the proteins in the two groups. The nonstorage proteins contain high proportions of lysine and arginine and low proportions of glutamine/glutamate and proline, whereas the opposite was observed for the storage proteins (Dexter and Dronzek 1975, Gellrich et al. 2003, Preston and Woodbury 1975). No significant differences exist in the amino acid composition of the proteins from different rye varieties (Dembinski and Bany 1991, Morey and Evans 1983). The higher proportions of lysine make rye proteins nutritionally superior to wheat proteins (Chen and Bushuk 1970, Dexter and Dronzek 1975, Jahn-Deesbach and Schipper 1980, Morey and Evans 1983).

Table 25.3 gives an overview of the classification of the rye proteins, which are of technological importance and are further discussed in detail below.

Storage Proteins

Secalins The rye alcohol-soluble prolamins, also called "secalins," form the major storage proteins in rye (Gellrich et al. 2003). They have unusual amino acid compositions with high glutamine/glutamate and proline contents and low levels of basic and acidic amino acids (Dexter and Dronzek 1975, Gellrich et al. 2003, Preston and Woodbury 1975). The secalins are highly polymorphic and can be classified into three groups on the basis of their amino acid compositions: S-rich, S-poor, and high molecular weight (HMW) secalins (Fig. 25.1). Some

Table 25.3. Classification of Rye Proteins

Physiological Classification	Osborne Classification	Technologically Important Proteins
Nonstorage proteins	Albumins Globulins	Enzymes → α-amylases, Endoxylanases, Arabinofuranosidases, Xylosidases, Ferulic acid esterases, Proteases → (Proteases) Serine proteases, Metalloproteases, Aspartic proteases, Cysteine proteases Enzyme inhibitors → α-amylase inhibitors, Endoxylanase inhibitors, Protease inhibitors
Storage proteins	Secalins	γ-secalins → 40k γ-secalins, 75k γ-secalins ω-secalins HMW secalins
	Glutelins	γ-secalins → 40k γ-secalins, 75k γ-secalins HMW secalins

researchers also purified secalins of low molecular weight (LMW) (10–16 k) (Charbonnier et al. 1981; Preston and Woodbury 1975), which is in the same range as that of wheat low molecular weight gliadin and barley A hordein.

The S-rich or γ-secalins are the most abundant secalins and consist of two major groups of polypeptides: 40k γ-secalins (26% of total secalin fraction) and 75k γ-secalins (45% of total secalin fraction) (Gellrich et al. 2003, Shewry et al. 1982). The molecular weights of 40k γ- and 75k γ-secalins are 40 k and 75 k, respectively, as determined by SDS-PAGE (sodium dodecyl sulfate–polyacrylamide gel electrophoresis), hence their names (Field et al. 1983, Gellrich et al. 2003, Shewry et al. 1982), and 33 k and 54 k, respectively, as determined by sedimentation equilibrium ultracentrifugation (Shewry et al. 1982) or mass spectrometry (Gellrich et al. 2001, 2003). The two groups differ in amino acid composition, with 75k γ-secalins having higher contents of glutamine/glutamate and proline and lower contents of cysteine than the 40k γ-secalins (Gellrich et al. 2003, Shewry et al. 1982, Tatham and Shewry 1991).

The γ-secalins contain two structural domains: an N-terminal domain rich in glutamine and proline and consisting of repetitive sequences (Gellrich et al. 2001, 2004a; Kreis et al. 1985; Shewry et al. 1982), and a C-terminal domain with a nonrepetitive structure, which has lower glutamine and proline content and is rich in cysteine (Kreis et al. 1985). The N-terminal sequences of the 40k γ- and 75k γ-secalins are identical at 16 or 17 of the first 20 positions (Gellrich et al. 2003, 2004a; Shewry et al. 1982). The N-terminal repetitive domains probably have a rod-like conformation (Shewry and Tatham 1990), whereas the C-terminal nonrepetitive domains probably have a compact globular conformation (Shewry and Tatham 1990) rich in α-helical structures (Tatham and Shewry 1991). The 75k γ-secalins have a lower content of α-helical structures than the 40k γ-secalins (Tatham and Shewry 1991),

Prolamins (mainly monomeric, some smaller polymers) | **Glutelins** (polymeric)

Rye proteins

Prolamins		Classification		Glutelins
• Secalins				• Glutelins
- 40k γ-, 75k γ-secalins	→	S-rich prolamins	←	- 40k γ-, 75k γ-secalins
- ω-secalins	→	S-poor prolamins		
- HMW secalins	→	HMW prolamins	←	- HMW secalins

Wheat proteins

Prolamins		Classification		Glutelins
• Gliadins				• Glutenins
- α-, β-, γ-gliadins	→	S-rich prolamins	←	- B and C LMW-GS*
- ω-gliadins	→	S-poor prolamins	←	- D LMW-GS*
	→	HMW prolamins	←	- HMW-GS

*LMW-GS show strong similarities with gliadins, differing in one very important characteristic, i.e. the occurrence of inter-chain disulphide bonds in LMW-GS

Figure 25.1. Schematic representation of the classification of rye and wheat proteins.

consistent with a higher proportion of repetitive sequences in the former group (Shewry et al. 1982).

The two γ-secalin types differ in aggregation behavior. The 40k γ-secalins predominantly occur as monomers containing intramolecular disulphide bonds, while the 75k γ-secalins are present predominantly as disulphide-linked aggregates (Field et al. 1983, Gellrich et al. 2003, Shewry et al. 1983). A possible explanation for this different aggregation behavior is the presence or absence of cysteine residues in positions that are favorable or unfavorable for formation of intermolecular disulphide bonds. Alternatively, it might result from the specific action of enzymes responsible for the formation of such bonds (Field et al. 1983, Shewry et al. 1983). Gellrich et al. (2001, 2004b) showed that the eight cysteine residues of the C-terminal domain of 75k γ-secalin are linked by intramolecular disulphide bonds and that the cysteine residue located in position 12 of the N-terminal domain of the 75k γ-secalin forms an intermolecular disulphide bond with 75k γ-secalins. The cysteine residue in the N-terminal domain is unique to the 75k γ-secalins and is likely to be responsible for their aggregative nature (Gellrich et al. 2003, 2004b).

The 40k γ-secalins are homologous to the wheat γ-gliadins and to the B hordeins of barley (Field et al. 1983; Gellrich et al. 2001, 2003; Shewry et al. 1982), whereas the 75k γ-secalins differ from them due to the presence of additional repetitive sequences rich in glutamine and proline and their occurrence as aggregates. The latter are suggested to be homologous to the LMW glutenin subunits (LMW-GS) of wheat (Shewry et al. 1982) (Fig. 25.1).

The S-poor or ω-secalins are quantitatively minor components, accounting for about 19% of the total secalin fraction (Gellrich et al. 2003). They have SDS-PAGE molecular weights of about 48 k to 53 k (Field et al. 1983, Gellrich et al. 2003, Shewry et al. 1983). The amino acid compositions of the ω-secalins are characterized by high levels of glutamine/glutamate, proline, and phenylalanine and the absence of cysteine and methionine (Tatham and Shewry 1991). The ω-secalins consist almost entirely of repetitive sequences (Tatham and Shewry 1991) forming a stiff, wormlike coil (Tatham and Shewry 1995). They are present predominantly as monomers (Field et al. 1983, Gellrich et al. 2003) and are homologous to the ω-gliadins of wheat and to the C hordeins of barley (Field et al. 1983, Gellrich et al. 2003, Tatham and Shewry 1991) (Fig. 25.1).

HMW secalins are also quantitatively minor components, accounting for about 5% of the total secalin fraction (Gellrich et al. 2003). SDS-PAGE molecular weights above 100 k were observed (Field et al. 1982, Gellrich et al. 2003, Kipp et al. 1996, Shewry et al. 1983), whereas sedimentation equilibrium ultracentrifugation showed a molecular weight of 68 k (Field et al. 1982). The HMW secalins have a high content of glycine and glutamine/glutamate and contain lower levels of proline than the other secalin groups (Field et al. 1982, Tatham and Shewry 1991). The conformation of HMW secalins has not been reported in detail, but based on studies with HMW glutenin subunits (HMW-GS) of wheat, it was suggested that they contain central repetitive sequences with a rodlike shape, flanked by nonrepetitive domains having a globular conformation rich in α-helical structure. The nonrepetitive domains contain most of the cysteine residues involved in intermolecular cross-linking (Shewry et al. 1995). The HMW secalins are present only in an aggregated state stabilized by disulphide bonds (Field et al. 1983, Gellrich et al. 2003, Shewry et al. 1983). They are homologous to the HMW-GS of wheat and the D hordeins of barley (Field et al. 1982, 1983; Gellrich et al. 2003; Shewry et al. 1982, 1988; Fig. 25.1).

Glutelins In contrast to the wheat glutelins, which are called "glutenins," no alternative name has been given to the rye glutelin fraction. The glutelins are not soluble in aqueous alcohols. They form HMW polymers stabilized by interchain disulphide bonds. However, the individual subunits of the polymers obtained in the presence of a reducing agent are soluble in alcohol media and rich in proline and glutamine (Field et al. 1982, 1983; Gellrich et al. 2003; Kipp et al. 1996; Shewry et al. 1983). The building blocks of the glutelins are therefore considered to be secalins. The glutelins soluble in alcohol after reduction contain mainly HMW-secalins (26%) and 75k γ-secalins (52%) and small levels of 40k γ-secalins (12%) (Gellrich et al. 2003; Fig. 25.1). When the distribution of the 75k γ-, 40k γ-, ω- and HMW secalins over the storage proteins (secalins + glutelins) is considered, it can be concluded that the major portion of 75k γ-secalins is present in the prolamin fraction (40%) and only 7% in the glutelin fraction. The 40k γ-secalins (23%) and ω-secalins (16%) appear mainly in the prolamin fraction due to their monomeric state. A larger portion of HMW secalins is present in the prolamin (4%) than in the glutelin fraction (3%) (Gellrich et al. 2003).

Gluten Formation The wheat gliadin and glutenin fractions can form a polymeric gluten network by intermolecular noncovalent interactions and disulphide bonds. In contrast to wheat, rye storage proteins do not form gluten. Despite partial homology, storage proteins of rye differ significantly from those of wheat with respect to quantitative and structural parameters that are important for the formation and properties of gluten. Typical for rye are the low content of storage proteins and the high ratio of alcohol-soluble proteins to insoluble proteins (Chen and Bushuk 1970, Gellrich et al. 2003, Preston and Woodbury 1975). The 75k γ-secalins and HMW secalins differ in aggregation behavior from the LMW-GS and HMW-GS of wheat, respectively, due to the absence (Gellrich et al. 2004b) or presence (Köhler and Wieser 2000) of cysteine residues in the C-terminal domain in positions favorable for formation of intramolecular disulphide bonds and unfavorable for formation of intermolecular disulphide bonds. It is likely that these structural differences contribute to the lack of ability of rye to form gluten. Another factor hampering the formation of a protein network from rye secalin and glutelin fractions may be a higher degree of glycosylation in rye subunits than in wheat subunits (Kipp et al. 1996).

Functional Proteins

Among the many enzymes and structural proteins in cereals in general, the most important enzymes present in rye grain that break down its major constituents (starch, arabinoxylan, and protein) are the α-amylases, endoxylanases, and proteases, respectively. Such enzymes can be of major importance in rye processing. Rye also contains specific proteins with inhibition activity against these enzymes. Presumably, these enzyme inhibitors contribute to plant defense mechanisms and/or possibly intervene in the complex regulation of plant metabolic processes.

Starch-Degrading Enzymes and Their Inhibitors α-amylases or, in full, 1,4-α-D-glucan glucanohydrolases (E.C. 3.2.1.1) catalyze the hydrolysis of the internal 1,4-α-D-glucan linkages in starch components.

Rye synthesizes two groups of α-amylases: low pI (≤ pI 5.5) and high pI (≤ pI 5.8). The low-pI α-amylases are synthesized during grain development, particularly in the pericarp (Dedio et al. 1975). Their activity decreases during the later stages of grain development and is low at maturity. The high-pI α-amylases are synthesized during germination. In germinated rye, the high-pI groups represent a high proportion of the total amylase activity (MacGregor et al. 1988). Gabor et al. (1991) isolated a high-pI α-amylase from germinated rye, also called "germination-specific" α-amylase, which showed optimal activity at pH 5.5 and is stable at pH 6.0–10.0 after storage at 4°C. The temperature optimum over a 3 minute incubation period was 65°C, but 40% of the activity was lost when the enzyme was incubated without substrate for 4 hours at 55°C. Täufel et al. (1991) also isolated a germination-specific α-amylase from rye and reported a pH optimum of 5.0, pH stability between pH 5.0 and 7.0 and thermal stability up to 55°C.

Variation in α-amylase levels exists between different rye lines (Masojc and Larssonraznikiewicz 1991a). Masojc and Larssonraznikiewicz (1991b) showed the existence of low and high α-amylase genotypes. In contrast, Hansen et al. (2004) stated that the α-amylase activity is mostly influenced by harvest year, attributing only a small effect to genotype.

α-amylase inhibitors can reduce the activity of one or more α-amylases and are mainly found in two major families, that is, the "cereal trypsin/α-amylase inhibitor" and the "Kunitz" or "α-amylase/subtilisin inhibitor" families (Garcia-Olmedo et al. 1987). The cereal trypsin/α-amylase inhibitor family represents the major part of the albumin and globulin fraction of the endosperm and consists of monofunctional inhibitors, active against either trypsin or exogenous α-amylase. The Kunitz-type inhibitors are bifunctional inhibitors of about 20 k, containing two intramolecular disulphide bridges. They can simultaneously inhibit both subtilisin and endogenous α-amylases, in particular the germination specific, high-pI α-amylase.

Täufel et al. (1991, 1997) purified two proteinaceous inhibitors of α-amylase from rye: a regulation or R-type inhibitor, which only inhibits the endogenous germination-specific α-amylase, and a defense or D-type inhibitor, which only inhibits exogenous α-amylases of animal and human origin. The inhibitors are present in different tissues of the rye kernel (Täufel et al. 1997). The R-type inhibitor predominantly occurs in the bran, whereas the D-type inhibitor is enriched in the germ and the endosperm. The high D-type inhibitor activity in these kernel sections is consistent with its assumed defense function, as it can protect the starch against exogenous α-amylases from insects and rodents. The levels of both types of inhibitor vary with growing conditions (Täufel et al. 1997). Rye varieties cultivated under drought conditions show high D-type inhibitor activities and low R-type inhibitor activities, whereas the opposite is true when they are cultivated under wet conditions. Small variations in α-amylase inhibitor activities exist between sprout-stable and sprout-sensitive rye genotypes (Täufel et al. 1997). Whereas α-amylase activity increases rapidly after 24 hours of germination, the R- and D-type inhibitor activities are stable during the 72 hours of germination (Täufel et al. 1997). The α-amylase inhibitors obviously have no importance at the early stages of growth.

Arabinoxylan-Degrading Enzymes and Their Inhibitors Arabinoxylan structural and physicochemical properties can be impacted by enzymes such as arabinofuranosidases, xylosidases, esterases, and endoxylanases, with the latter being by far the most relevant in cereal processing.

Endoxylanases or, in full, endo-1,4-β-D-xylan xylanohydrolases(E.C. 3.2.1.8) hydrolyze internal 1,4 linkages between β-D-xylopyranose residues of (arabino-) xylan molecules, generating (arabino-) xylan fragments with lower molecular weight and (un)substituted xylooligosaccharides. Their action can thus lead to (partial) solubilization of water-unextractable arabinoxylans and to a decrease in molecular weight of water-extractable and/or solubilized arabinoxylans. The susceptibility of arabinoxylans to endoxylanase attack probably depends on their substitution degree and their substitution pattern as well as on their linkages to other cell wall components. Enzymic hydrolysis also depends on the substrate selectivity of the endoxylanases. Certain endoxylanases preferentially release high molecular weight arabinoxylans from water-unextractable arabinoxylans and slowly degrade water-extractable arabinoxylans; other endoxylanases solubilize arabinoxylans of low molecular weight and/or extensively degrade water-extractable arabinoxylans.

Based on their primary sequence and structure, endoxylanases are classified into two main groups: glycosyl hydrolase family 10 and family 11 (Henrissat 1991). Plant endoxylanases to date have been exclusively classified in glycosyl hydrolase family 10 in contrast to fungal and bacterial endoxylanases, which are present in both glycosyl hydrolase families 10 and 11 (Coutinho and Henrissat 1999).

Arabinofuranosidases (α-L-arabinofuranosidases, E.C. 3.2.1.55) hydrolyze the linkage between arabinose residues and the xylan backbone, rendering the latter less branched and more accessible to depolymerization by endoxylanases. Ferulic acid esterase (feruloyl esterase, E.C. 3.1.1.73) can hydrolyze the ester linkage between ferulic acid and arabinose. Xylosidases (exo-1,4-β-xylosidase, E.C. 3.2.1.37) release single xylose residues from the nonreducing end of arabinoxylan fragments, thereby increasing the level of reducing end sugars. While synergy between these classes of enzymes and endoxylanases can be expected, it is not always observed (Figueroa-Espinoza et al. 2002, 2004).

Endogenous xylanolytic activity, although low, has been reported to be present in ungerminated rye (Rasmussen et al. 2001). The rye endogenous endoxylanase showed optimal activity at pH 4.5 and 40°C and was pH stable but heat labile. Xylosidase and arabinofuranosidase enzymes are also present in ungerminated rye grain (Rasmussen et al. 2001). During germination of rye, endoxylanase activity increases significantly after the first day (Autio et al. 2001).

Endoxylanase inhibitors can affect the enzymic hydrolysis of arabinoxylans by endoxylanases. Two different types of such inhibitors have been found to be present in rye (Goesaert et al. 2002, 2003).

The first type of inhibitor found in rye is the *Secale cereale* L. xylanase inhibitor (SCXI) (Goesaert et al. 2002), representing a family of isoinhibitors (SCXI I–IV) with similar structures and specificities. These inhibitors are basic proteins with isoelectric points of at least 9.0 and have highly homologous N-terminal amino acid sequences. They occur in two molecular forms, that is, a monomeric 40 k protein with at least one intramolecular disulphide bridge and, presumably following proteolytic cleaving of this form, a heterodimer consisting of two disulphide linked subunits (30 and 10 k). They specifically inhibit family 11 endoxylanases, while fungal family 10 endoxylanase is not affected (Goesaert et al. 2002). These inhibitors are homologous with wheat *Triticum aestivum* L. xylanase inhibitor I (TAXI I) (Gebruers et al. 2001).

Another family of isoinhibitors in rye, is the XIP-type (xylanase inhibiting protein) endoxylanase inhibitor family (Elliott et al. 2003, Goesaert et al. 2003). These inhibitors are basic, monomeric proteins with a molecular weight of 30 k with pI values of at least 8.5. They show inhibitory activity against fungal glycosyl hydrolase family 10 and 11 endoxylanases (Goesaert et al. 2003). Structural characteristics and inhibition specificities from the rye XIP-type inhibitors are similar to those of wheat XIP-type inhibitors (Gebruers et al. 2002).

Protein-Degrading Enzymes and Their Inhibitors Literature on proteins from different botanical sources (e.g., soy, wheat, rice) reveals that their enzymic hydrolysis can significantly alter their properties (Caldéron de la Barca et al. 2000, Popineau et al. 2002, Shih and Daigle 1997, Wu et al. 1998). This, however, has not been reported for rye protein.

Proteases hydrolyze the peptide bonds between the amino end of one amino acid residue and the carboxyl end of the adjacent amino acid residue in a protein. Endoproteases hydrolyze peptide bonds somewhere along the protein chain, whereas exoproteases attack the ends of the protein chain and remove one amino acid at a time. The latter are called carboxypeptidases when acting from the carboxy terminus or aminopeptidases when acting from the amino terminus. Based on the chemistry of their catalytic mechanism, proteases have been classified into four groups: serine (E.C. 3.4.21), metallo- (E.C. 3.4.24), aspartic (E.C. 3.4.23), and cysteine (E.C. 3.4.22) proteases.

Nowak and Mierzwinska (1978) reported the presence of endopeptidase, carboxypeptidase, and aminopeptidase activities in the embryo and endosperm of rye seeds. Endoproteolytic, exoproteolytic, carboxypeptidase, aminopeptidase, and N-α-benzoylarginine-*p*-nitroanilide (BAPA) hydrolyzing activities have also been found in whole meal extracts from ungerminated rye seeds by Brijs et al. (1999). During germination, proteolytic activity increases during the first 3 days to then remain constant (Brijs et al. 2002). Milling fractions of germinated rye (Brijs et al. 2002) have much higher proteolytic activities than those of ungerminated rye (Brijs et al. 1999). The enzymes are mainly present in the bran and shorts fractions. Ungerminated rye mainly contains two classes of proteases (serine and aspartic proteases) (Brijs et al. 1999), whereas all four protease classes are present during germination (Brijs et al. 2002). Germinated rye contains mainly aspartic and cysteine protease activities; serine and metalloproteases are less abundant.

Aspartic proteases have been purified from bran of ungerminated rye (Brijs 2001). Two heterodimeric aspartic proteases were found: a larger 48 k enzyme, consisting of 32 k and 16 k subunits which are probably linked by disulphide bridges and a smaller one of 40 k, consisting of two noncovalently bound 29 k and 11 k subunits. Both proteins have isoelectric points close to pI 4.6. Rye aspartic proteases show optimal activity around pH 3.0 and 45°C and can be completely inhibited by pepstatin A (Brijs 2001). Amino acid sequence alignment showed that the rye aspartic proteases resemble those from wheat (Bleukx et al. 1998) and barley (Sarkkinen et al. 1992). Cysteine proteases were isolated from the starchy endosperm of 3-day germinated rye (Brijs 2001). SDS-PAGE revealed three protein bands with apparent molecular weights of 67 k, 43 k, and 30 k, although it was not established that all three protein bands are cysteine proteases. The rye cysteine proteases are optimally active around pH 4.0 and at 45°C and can be totally inhibited by L-transepoxysuccinyl-L leucylamido-(4-guanidino)butane (or E-64).

Protease inhibitors can reduce the activity of proteases and have also been found in rye. Boisen (1983) and Lyons et al. (1987) isolated a trypsin inhibitor from rye endosperm and rye seed, respectively. The inhibitor had a molecular weight of about 12 to 14 k, contained four disulphide bridges, and was resistant to trypsin and elastase but was completely inactivated by chymotrypsin. The amino acid composition was very similar to that of the barley and wheat trypsin inhibitors.

Mosolov and Shul'gin (1986) purified a subtilisin inhibitor from rye that belongs to the Kunitz-type inhibitor family. The inhibitor consisted of two forms with pI values of 6.8 and 7.1 and had a molecular weight of about 20 k. It inhibited subtilisin (at a molar ration of 1:1) and fungal proteinases from the genus *Aspergillus,* but was inactive against trypsin, chymotrypsin, and pancreatic elastase. The amino acid composition of the rye subtilisin inhibitor was similar to that from its wheat and barley counterparts.

Hejgaard (2001) purified and characterized six serine protease inhibitors, also called "serpins," from rye grain. The serpins each have different regulatory functions based on different reactive center loop sequences, allowing for specific associations with the substrate-binding subsites of the proteases. The reactive center loops of the rye serpins were similar to those of wheat serpins. Five of the six serpins contained one or two glutamine residues at the reactive center bond, which were differently positioned than in the wheat reactive center loops. The rye serpins have a molecular weight of 43 k and inhibit proteases with chymotrypsin-like specificity, but not pancreatic elastase or proteases with trypsin-like specificity.

Protein Physicochemical Properties

The properties of rye proteins, for example, their foam-forming and emulsifying capacities, have hitherto not been studied well. Ludwig and Ludwig (1989) isolated a protein mixture from rye flour that exhibited low foam stability but excellent emulsifying properties at pH 3.5. Meuser et al. (2001) reported on a rye water-soluble protein with foam-forming capacity. It had a molecular weight of 12.3 k and an isoelectric point of 5.97. The optimum pH value for the water-soluble rye protein to form stable foam was pH 6.0. The protein did not appear to be located in specific morphological parts of the rye kernel. Wannerberger et al. (1997) found that secalins are more surface-active than gliadins. The lower molecular weight secalins spread more rapidly at the air-water interface, resulting in a higher surface pressure, than did the higher molecular weight gliadins. The surface pressure developed by both secalin and gliadin decreased with decreasing pH. Different rye milling fractions showed differences in surface behavior in relation to their protein content. The milling fraction with the highest protein content spread fastest and reached the highest surface pressure value.

RYE PROCESSING

While rye processing has gained interest because of its perceived positive health effects (Poutanen 1997), the functional properties of the rye major constituents limit the processing performance of rye in industrial processes (Weipert 1997). Milling and bread making are the most common processes in rye grain processing for human consumption.

RYE MILLING

The objective of milling is to separate the bran and the germ from the starchy endosperm and to reduce the starchy endosperm in size, in order to obtain a refined product, that is, flour. Rye milling differs some what from wheat milling. Rye kernels are smaller than wheat kernels, and the separation of the endosperm from the bran is poorer in rye than in wheat (Nyman et al. 1984). The extraction rate of rye flour is lower than that of wheat flour, and at any given extraction rate, the ash content of rye flour is always higher than that of wheat flour (Weipert 1997).

Cleaning and Tempering of Rye

Prior to milling, rye grain is cleaned to remove stones, magnetic objects, broken rye kernels, cockle, oats, and ergot. Cleaning is based on a separation between rye and the impurities on the basis of size, density, and shape.

The second step involves tempering of the rye grain by adding a required quantity of moisture to bring it to an optimum condition for milling. Rye is, in most cases, milled at a moisture level of 15%, (which is 1% less than the preferred moisture level for wheat) and, in general, is tempered for a shorter time (2–4 hours with an absolute maximum of 6 hours) than wheat. This is due to the fact that water penetrates the rye kernel faster than the wheat kernel due to rye's weaker cell structure. The tempering process can be eliminated to reduce bacterial counts. However, if rye is milled too dry, the risk of shredding the bran into smaller particle sizes and thus contaminating the rye flour exists.

After tempering, a second cleaning step removes surface dirt, loose hairs, and outer bran layers. Just before milling, an additional 0.5% moisture is added (Zwingelberg and Sarkar 2001).

Milling of the Rye

In Europe, "roller" milling and "pin" milling are the most important rye milling processes. In roller milling, each passage through rolls involves particle size reduction by pressure and shear forces, followed by flour separation by particle size using sieves. Rolls are matched to the product needed. Their size, surface flutes, rotation velocity, and the gap between pairs of rolls rotating in opposite directions at dissimilar speeds, all can be selected or adjusted. Typically, milling starts with prebreak or flatting rolls, followed by six to seven break rolls, six bran finishers, and six reduction rolls. All rolls run at a differential speed of 3:1, except the flatting rolls (1:1 ratio). The first four break and reduction rolls are run dull to dull and the following breaks, sharp to sharp. A complex sieving system with plansifters and centrifugal sifters is used. In this process, rye is fractionated into different milling streams, a number of which are combined to give the final flour. When practically all streams are used, an extraction rate of approximately 100% is reached and whole meal rye flour is obtained. When an increasing number of the fractions containing outer layer materials are left out of the reconstitution process, extraction rates decrease. For straight-run flour, a typical extraction rate is 58–68%.

In pin milling, rye is milled in impact mills, which work by the principle of impact disintegration. As this process requires a moisture content of 17–18%, rye is tempered a second time. In impact milling, five to seven milling steps are used. The extraction rate, ash content, and flour quality are similar for impact mills and roller mills. The starch damage, however, is lower in rye flour from impact mills than in that from roller mills because of the number of milling steps required.

In North America, the main objective in rye milling is to produce as much flour of the desired granulation as possible. The ash content and color of rye flours are not as critical as for common wheat flours. Therefore, the flow sheet of a rye mill is simple relative to that of a common wheat mill. A typical rye mill contains three to five break rolls, up to two sizings, and four to six reduction rolls. All grinding rolls are corrugated and run at a differential rate of 3:1. To properly remove the bran, bran dusters precede the last two or three break passages rather than coming just at the end. Because rye flour is sticky and difficult to sift, generous sifting surfaces are required (Zwingelberg and Sarkar 2001).

Impact of Rye Constituents on Rye Milling

Rye contains higher levels of arabinoxylans than wheat (Aman et al. 1997) and is accordingly more difficult to mill. Because the arabinoxylans are hygroscopic, the flour absorbs ambient humidity and tends to clump. This interferes with sifting. For this reason, rye is milled at a lower moisture content and requires more sifting surface than wheat (Zwingelberg and Sarkar 2001).

Protein content also influences the performance of rye in the mill. When it exceeds 12%, rye becomes more difficult to grind and has a lower extraction rate (Weipert 1993a).

Rye Bread Making

The bread-making behavior and performance of rye differ considerably from those of wheat (Weipert 1997). The principal reason for this is that rye proteins cannot form a gluten network, which lies at the basis of wheat bread quality. In addition, rye starch gelatinizes at a lower temperature than wheat starch and is therefore more prone to enzymic degradation during the oven phase than wheat starch. The main differences between rye and wheat breads are observed in terms of volume yield, crumb texture, and shelf life. Baking volume of rye breads is normally only about half that of wheat. However, rye bread has a longer shelf life and is richer in taste and aroma (Weipert 1997).

A wide assortment of rye-containing breads exists. They vary in size, shape, formulation, bread-making process, and sensory properties. Rye flour and whole meal rye can be incorporated into rye bread (minimum 90% rye), mixed rye/wheat bread (minimum 50% rye), or mixed wheat/rye bread (minimum 50% wheat). Other ingredients can be crushed rye grains, malted and crushed rye grains, malted rye kernels, or precooked rye kernels (Kujala 2004). Examples of rye bread types are pumpernickel (containing whole kernels), soft sourdough bread, crisp bread, rolls, and buns.

Rye Bread-Making Process

For rye bread production, acidification is a prerequisite to obtain a good quality product. The traditional method of preparing rye bread uses sourdough, although sometimes, direct acidification by lactic acid or other organic acids is used.

Preparation of Sourdough Sourdough is classically made by mixing rye flour or whole meal rye with water, yeast, and a starter and allowing the mixture to ferment (Kariluoto et al. 2004, Meuser et al. 1994, Seibel and Brümmer 1991). As starter, a commercial culture containing lactic acid bacteria or a portion of previous sourdough, equally containing lactic acid bacteria, is used (Seibel and Brümmer 1991). In homofermentative lactic acid fermentation, the lactic acid bacteria ferment glucose mainly into lactic acid; in heterofermentative lactic acid fermentation, considerable levels of acetic acid are formed in addition to lactic acid. The formation of lactic acid and acetic acid during lactic acid fermentation can and must be controlled, as the ratio of lactic acid to acetic acid is quite important for optimal bread quality. Sourdoughs with a high lactic acid content are used mainly for rye and mixed rye breads, while sourdoughs with a high acetic acid content are ideally suited for mixed wheat bread (Seibel and Brümmer 1991). One-, two-, and three-stage sourdough processes are used (Meuser et al. 1994, Seibel and Brümmer 1991). These processes require different fermentation times, dough temperatures, and water to flour ratios, and the resulting doughs have different lactic acid to acetic acid ratios (Seibel and Brümmer 1991). Depending on the proportion of rye flour in the sourdough, various degrees of acidity are required (Seibel and Brümmer 1991).

There are different reasons for using sourdough. Acid conditions have a positive influence on the swelling of rye flour constituents and control the enzyme activity in the dough, thereby preventing early staling (Meuser et al. 1994, Seibel and Brümmer 1991). Rye flours with high enzyme activity require a higher degree of acidification than flours with less enzyme activity (Meuser et al. 1994). Bread sensory properties, such as taste and aroma, can be improved by optimal use of sourdough (Heiniö et al. 2003a, Meuser et al. 1994, Seibel and Brümmer 1991). It contributes to aroma in the bread due to the production of acids, alcohols, and other volatile compounds (Heiniö et al. 2003a, Poutanen 1997). Sourdough bread has good crumb characteristics (Seibel and Brümmer 1991) and nutritional properties (Kariluoto et al. 2004, Meuser et al. 1994, Poutanen 1997). Acidification of rye dough also protects against spoilage by mold growth, thereby improving product shelf life (Meuser et al. 1994, Poutanen 1997, Seibel and Brümmer 1991). When sourdough is used in wheat bread making, its main function is to improve sensory properties and to prolong shelf life (Brümmer and Lorenz 1991).

Mixing and Kneading Rye flour, yeast, salt, and water are optimally mixed and kneaded with sourdough, lactic acid, or lactic and acetic acid to form a viscoelastic dough. The level of water is that yielding the required dough consistency. Rye doughs have to be mixed slowly because of their high arabinoxylan content, which leads to tough doughs when the energy input is excessive. The optimal mixing time depends on the composition and enzymic activities of the rye flour (Seibel and Weipert 2001a).

Important dough quality characteristics are dough yield, that is, the amount of optimally developed dough obtained from 100 g flour, and dough structure.

Fermentation Mixing and kneading of the dough is followed by a floor time. Thereafter, the dough is divided into pieces, which are molded by hand into the desired shape. The doughs are then proofed to their best potential. Fermentation takes place at 30–36°C and 70–85% relative humidity. Fermentation time depends on the composition of the flour, the activity of the flour enzymes, the dough formulation, and the fermentation temperatures.

During fermentation, carbon dioxide is produced, and the gas is retained by the dough, resulting in a volume increase of the dough. Although the rate of gas production in rye flour dough is high, the gas-retaining capacity is low, resulting in dense, compact loaves (He and Hoseney 1991).

Baking After fermentation, the dough is placed in an oven and transformed into bread. Bread quality depends on baking temperature, uniformity of temperature in the oven, and baking time. Optimum baking temperature and time depend on the weight of the dough pieces and the shape of the loaves, which determine the rate of heat transfer. In rye bread making, at the beginning of the baking stage, steam is often used in the oven to develop good crust characteristics (Seibel and Weipert 2001a).

Important bread quality characteristics are specific volume, crumb structure, crumb elasticity, crumb firmness during storage; and sensory properties such as taste, flavor, and odor.

Impact of Rye Constituents on Dough and Bread Quality Characteristics

Different rye varieties show highly different bread-making quality as influenced by harvest year and genotype (Hansen et al. 2004, Nilsson et al. 1997b, Rattunde et al. 1994). This can be attributed to significant differences in the content and structure of the major rye constituents, which are influenced by both factors (Hansen et al. 2003).

Starch Starch gelatinization and α-amylase activity are very important parameters for the bread-making quality of rye flour. Both properties can, for example, be measured by the falling number method or the amylograph test. The falling number is inversely related to α-amylase activity. An important parameter in the amylograph test is amylograph peak viscosity, which is correlated positively with starch content and negatively with α-amylase activity. It should be noted, however, that arabinoxylans can influence falling number and amylograph characteristics of rye flours. At similar α-amylase activity levels, higher falling numbers (Weipert 1993a, 1993b, 1994) and higher amylograph peak viscosities (Nilsson et al. 1997b) can be measured due to higher arabinoxylan contents. The arabinoxylans increase the viscosity, but also protect the starch granules against enzymic breakdown by hindering and retarding the activity of α-amylase (Weipert 1993b, 1994).

For rye bread-making suitability, a minimum starch gelatinization temperature of 63°C and, under specified experimental conditions, a minumum amylograph peak viscosity of 200 amylograph units (AU) are required. Rye shows optimal bread-making quality at a starch gelatinization temperature of 65–69°C and an amylograph peak viscosity of 400–600 AU (Weipert 1993a). Maximum bread-making quality is associated with intermediate values of falling number (Rattunde et al. 1994). Rye flours with high α-amylase activity give a soft, sticky dough, which may be too fluid to process into bread or results in a doughlike crumb. Therefore, preharvest sprouting, which goes hand in hand with high α-amylase activity, is disadvantageous for rye flour quality. Rye flours with low α-amylase activity yield a rigid, stable dough but lead to breads of low volume having a dry and firm crumb, resulting in a short shelf life (Autio and Salmenkallio-Marttila 2001, Fabritius et al. 1997). However, when rye flours with low α-amylase activity contain high levels of total and water-extractable arabinoxylans, better dough and bread characteristics can be observed (Weipert 1993a). It is believed that optimum bread-making performance is reached at an optimal arabinoxylan to starch ratio of 1:16 (Seibel and Brümmer 1991). This ratio provides balanced water-binding capacity and water distribution in the dough and the bread crumb. A shift in favor of more arabinoxylan causes stiffening of the dough (Kühn and Grosch 1989), due to the high water-holding capacity of the arabinoxylans. It also results in breads of greater specific volume and a firmer crumb (Kühn and Grosch 1989). A shift in favor of starch results in a low dough yield, soft dough, a flat bread form, a split and tight crumb, and a low shelf life (Vinkx and Delcour 1996).

Arabinoxylans Although many studies have reported conflicting results concerning the functional role of arabinoxylans in the bread-making process and during storage of breads, they are of considerable importance for rye bread-making quality.

Rye cultivars with high total arabinoxylan content, particularly when they also have a high water-extractable arabinoxylan fraction, perform much better in bread making than those with low arabinoxylan content. This is due to the high water-holding capacity of the arabinoxylans, which results in a delay in starch gelatinization and protects starch from α-amylase degradation, resulting in a higher volume yield, a better crumb elasticity, and a longer bread shelf life (Weipert 1993a, 1994, 1997). Small differences in arabinoxylan content cause differences in water absorption and significant differences in bread-making performance (Hansen et al. 2004, Weipert 1997).

Besides the content of arabinoxylans, their structure also significantly influences rye bread-making performance. Differences in arabinoxylan structure, such as degree of branching and ferulic acid cross-linking, as well as molecular size, may have influence on water absorption (Hansen et al. 2004). A high degree of branching and a high degree of ferulic acid cross-linking of arabinoxylans are related to low water absorption in rye flour (Hansen et al. 2004).

The dough and bread-making properties of rye flour also depend on the type of arabinoxylans present. In general, dough consistency is negatively related to dough water content. Because of their high (and similar) water-holding capacities, both water-extractable and water-unextractable arabinoxylans affect dough consistency (Kühn and Grosch 1989). However, the water-extractable and -unextractable arabinoxylans affect the bread-making properties of rye flour differently. The water-extractable arabinoxylans increase bread specific volume and decrease crumb firmness, whereas the water-unextractable arabinoxylans decrease bread specific volume and form ratio and slightly increase crumb firmness (Kühn and Grosch 1989). Increased water-extractable to water-unextractable arabinoxylan ratios result in flatter breads with higher specific volume, softer crumb, and darker crust (Cyran and Cygankiewicz 2004, Kühn and Grosch 1989).

Enzymic breakdown of the water-unextractable arabinoxylans by endoxylanases to solubilized arabinoxylans with medium to high molecular weight and with limited depolymerization of the water-extractable fraction has been reported to improve both rye (Kühn and Grosch 1988) and wheat (Courtin et al. 1999) bread-making performance. The increase in the level of high molecular weight arabinoxylans in the dough aqueous phase results in an improved water distribution and facilitation of molecular interactions between the dough constituent macromolecules. The subsequent increase in viscosity of the dough aqueous phase is likely to play a positive role in dough rheology and gas retention (Courtin et al. 1999, Rouau et al. 1994). Kühn and Grosch (1988) showed an improved crumb structure, decreased crumb firmness, increased width to height ratio, and increased loaf volume of rye bread after enzymic treatment of the water-unextractable arabinoxylans. Shelf life is also increased, resulting from an initial decrease in crumb firmness (Kühn and Grosch 1988). The endogenous arabinoxylan-degrading enzymes of rye might be active under the conditions prevailing in rye dough (Hansen et al. 2002, Rasmussen et al. 2001) and in the first part of the oven phase, thus prior to enzyme inactivation (Rasmussen et al. 2001). Endogenous enzymes in flour may thus affect crumb properties such as firmness, elasticity, and stickiness (Nilsson et al. 1997b).

According to Figueroa-Espinoza et al. (2004), the enzymic release of high molecular weight arabinoxylans may be positive due to the capacity of the latter to gel and form a polysaccharide network in the dough, thereby increasing water absorption, gas retention, bread volume, and shelf life.

Proteins Proteins are only of secondary importance for rye bread-making quality because of their inability to form a gluten network (Gellrich et al. 2003, 2004b; Kipp et al. 1996; Köhler and Wieser 2000). However, they should not be neglected, as they seem to be important at the dough-mixing step, at least for certain cultivars (Parkkonen et al. 1994). Indeed, although rye proteins do not form a regular gluten network, they have some ability to aggregate (Field et al. 1983) and are surface active (Wannerberger et al. 1997).

Sensory Properties of Rye Bread

Rye bread has a typical strong and slightly bitter flavor. The most dominant sensory attribute of rye bread is its rye-like flavor, but perceptions of sourness and saltiness are also substantially associated with rye bread (Heiniö et al. 1997). The aroma results from a mixture of many different volatile compounds formed by enzymic reactions during fermentation and thermal reactions during baking (Heiniö et al. 2003a, 2003b; Kirchhoff and Schieberle 2001). Besides the generation of aromas during these steps, the flour itself is an important source of dough aroma (Kirchhoff and Schieberle 2002). The specific rye-like flavor can be slightly modified to have a milder taste, without decreasing the high dietary fiber content and other positive health effects of rye, for example, by using the processing techniques of sourdough fermentation, germination prior to use, and extrusion cooking (Heiniö et al. 2003a), or by milling fractionation (Heiniö et al. 2003b).

Rye in Other Food Products

In addition to bread, many other food products such as ginger cake, porridge, pastries, breakfeast cereals, and pasta products are based on rye (Kujala 2004, Poutanen 1997). Rye can also be fermented to produce alcoholic beverages (Seibel and Weipert 2001a).

Rye in Feed

Although the majority of rye is fed to animals, rye is considered to be of inferior quality to other feed grains. The soluble, highly viscous rye arabinoxylans cause reductions in both the rate of digestion and the retention of nutrients in the gastrointestinal tract (Boros et al. 2002, Fengler and Marquardt 1988b). Therefore, rye is generally used in small proportions in mixtures with other grains. The use of enzymes (Boros and Bedford 1999, Boros et al. 2002, He et al. 2003, Lazaro et al. 2003), however, has recently substantially increased the proportion of rye included in animal feed.

Rye Fractionation

Wheat is an important raw material in the industrial separation of gluten and starch because of the abundant applications for gluten and starch in both the food and nonfood sectors. Rye would be an interesting alternative raw material because of the specific properties of the rye constituents. However, in spite of the development of some methods on the laboratory scale to separate starch from the arabinoxylans and proteins in rye (Schierbaum et al. 1991), no such industrial process exists for rye, presumably because of its relatively high content of arabinoxylans with high water-holding capacity and the inability of the rye proteins to form gluten.

Rye in Industrial Uses

Rye and rye products are also used in industrial applications such as production of adhesives and glue, film coating, sludge and oil-well drilling materials, and textiles and paper (Seibel and Weipert 2001b).

RYE AND NUTRITION

Rye products, especially when they contain whole meal rye flour, provide considerable levels of dietary fiber, vitamins, minerals, and phytoestrogens, all of which are considered to have positive health effects (Poutanen 1997). High rye consumption leads to positive effects on digestion and decreased risk of heart disease, hypercholesterolemia, obesity, and non-inulin-dependent diabetes (Hallmans et al. 2003, Smith et al. 2003). It probably also protects against some hormone-dependent cancer types (Hallmans et al. 2003, Smith et al. 2003). Liukkonen et al. (2003) showed that many of the bioactive compounds in whole meal rye are stable during processing and that their levels can even be increased through suitable processing (e.g., sourdough bread making). Rye milling, however, is accompanied by losses in dietary fiber and other phytochemicals, since these compounds are concentrated in the outer layers of the rye kernel (Glitso and Bach Knudsen 1999, Liukkonen et al. 2003, Nilsson et al. 1997a). In overall nutritional value, rye has some advantages over wheat because of the higher content of dietary fiber, vitamins, minerals, phytoestrogens, and lysine in rye.

REFERENCES

Aman P, Bengtsson S. 1991. Periodate oxidation and degradation studies on the major water-soluble arabinoxylan in rye grain. Carbohydrate Polymers 15: 405-414.

Aman P, Nilsson M, Andersson R. 1997. Positive health effects of rye. Cereal Foods World 42: 684-688.

Andreasen MF, Christensen LP, Meyer AS, Hansen A. 2000a. Ferulic acid dehydrodimers in rye (*Secale cereale* L.). Journal of Cereal Science 31: 303-307.

Andreasen MF, Christensen LP, Meyer AS, Hansen A. 2000b. Content of phenolic acids and ferulic acid dehydrodimers in 17 rye (*Secale cereale* L.) varieties. Journal of Agricultural and Food Chemistry 48: 2837-2842.

Andreev NR, Kalistratova EN, Wasserman LA, Yuryev VP. 1999. The influence of heating rate and annealing on the melting thermodynamic parameters of some cereal starches in excess water. Starch 51: 422-429.

Anger H, Dorfer J, Berth G. 1986. Untersuchungen zur Molmasse und Grenzviskositat von Arabinoxylan (Pentosan) aus Roggen (*Secale cereale*) zur Aufstellung der Mark-Houwink-Beziehung. Die Nahrung 30: 205-208.

Autio K, Salmenkallio-Martilla M. 2001. Light microscopic investigations of cereal grains, doughs and breads. Lebensmittel Wissenschaft und Technologie 34: 18-22.

Autio K, Simoinen T, Suortti T, Salmenkallio-Marttila M, Lassila K, Wilhelmson A. 2001. Structural and enzymic changes in germinated barley and rye. Journal of the Institute of Brewing 107: 19-25.

Ball S, Guan HP, James M, Myers A, Keeling P, Mouille G, Buleons A, Colonna P, Preiss J. 1996. From glycogen to amylopectin: a model for the biogenesis of the plant starch granule. Cell 86: 349-352.

Bengtsson S, Aman P. 1990. Isolation and chemical characterization of water-soluble arabinoxylans in rye grain. Carbohydrate Polymers 12: 267-277.

Bengtsson S, Aman P, Andersson RE. 1992a. Structural studies on water-soluble arabinoxylans in rye grain using enzymatic hydrolysis. Carbohydrate Polymers 17: 277-284.

Bengtsson S, Andersson R, Westerlund E, Aman P. 1992b. Content, structure and viscosity of soluble arabinoxylans in rye grain from several countries. Journal of Scientific Agriculture 58: 331-337.

Berry CP, D'Appolonia BL, Gilles KA. 1971. The characterization of triticale starch and its comparison with starches of rye, durum and HRS wheat. Cereal Chemistry 48: 415-427.

Bleukx W, Torrekens S, Van Leuven F, Delcour JA. 1998. Purification, properties and N-terminal amino acid sequence of a wheat gluten aspartic proteinase. Journal of Cereal Science 28: 223-232.

Boisen S. 1983. Comparative physico-chemical studies on purified trypsin inhibitors from the endosperm of barley, rye and wheat. Zeitschrift fur Lebensmittel-Untersuchung und -Forschung 176: 434-439.

Boros D, Marquardt RR, Slominski BA, Guenter W. 1993. Extract viscosity as an indirect assay for water-soluble pentosan content in rye. Cereal Chemistry 70: 575-580.

Boros D, Bedford MR. 1999. Influence of water extract viscosity and exogenous enzymes on nutritive value of rye hybrids in broiler diets. Journal of Animal and Feed Sciences 8: 579-588.

Boros D, Marquardt RR, Guenter W, Brufau J. 2002. Chick adaptation to diets based on milling fractions of rye varying in arabinoxylans content. Animal Feed Science and Technology 101: 135-149.

Brijs K, Bleukx W, Delcour JA. 1999. Proteolytic activities in dormant rye (*Secale cereale* L.) grain. Journal of Agricultural and Food Chemistry 47: 3572-3578.

Brijs K. 2001. Proteolytic enzymes in rye: isolation, characterization and specificities [Ph. D. dissertation]. Leuven, Belgium: Katholieke Universiteit Leuven. 141 p.

Brijs K, Trogh I, Jones BL, Delcour JA. 2002. Proteolytic enzymes in germinating rye grains. Cereal Chemistry 79: 423-428.

Brümmer JM, Lorenz K. 1991. European developments in wheat sourdoughs. Cereal Foods World 36: 310-314.

Caldéron de la Barca AM, Ruiz-Salazar RA, Jara-Marini ME. 2000. Enzymatic hydrolysis and synthesis of soy protein to improve its amino acid composition and functional properties. Journal of Food Science 65: 246-253.

Charbonnier L, Tercé-Laforgue T, Mossé J. 1981. Rye prolamins: extractability, separation and characterization. Journal of Agricultural and Food Chemistry 29: 968-973.

Chen CH, Bushuk W. 1970. Nature of proteins in triticale and its parental species. I. Solubility characteristics and amino acid composition of endosperm proteins. Canadian Journal of Plant Science 50: 9-14.

Courtin CM, Roelants A, Delcour JA. 1999. Fractionation-reconstitution experiments provide insight into the role of endoxylanases in bread-making. Journal of Agricultural and Food Chemistry 47: 1870-1877.

Coutinho PM, Henrissat B. 1999. Carbohydrate-active enzymes server at URL: http://afmb.cnrs-mrs.fr/CAZY/.

Cyran M, Courtin CM, Delcour JA. 2003. Structural features of arabinoxylans extracted with water at different temperatures from two rye flours of diverse breadmaking quality. Journal of Agricultural and Food Chemistry 51: 4404-4416.

Cyran M, Courtin CM, Delcour JA. 2004. Heterogeneity in the fine structure of alkali-extractable arabinoxylans isolated from two rye flours with high and low breadmaking quality and their coexistence with other cell wall components. Journal of Agricultural and Food Chemistry 52: 2671-2680.

Cyran M, Cygankiewicz A. 2004. Variability in the content of water-extractable and water-unextractable non-starch polysaccharides in rye flour and their relationship to baking quality parameters. Cereal Research Communications 32: 143-150.

Dedio W, Simmonds DH, Hill RD, Shealy H. 1975. Distribution of alpha-amylase in triticale kernel during development. Canadian Journal of Plant Science 55: 29-36.

Delcour JA, Vanhamel S, Hoseney RC. 1991. Physicochemical and functional properties of rye nonstarch polysaccharides. II. Impact of a fraction containing water-soluble pentosans and proteins on gluten-starch loaf volumes. Cereal Chemistry 68: 72-76.

Dembinski E, Bany S. 1991. The amino acid pool of high and low protein rye inbred lines (*Secale cereale* L.). Journal of Plant Physiology 138: 494-496.

Denli E, Ercan R. 2001. Effect of added pentosans isolated from wheat and rye grain on some properties of bread. European Food Research and Technology 212: 374-376.

Dervilly-Pinel G, Rimsten L, Saulnier L, Andersson R, Aman P. 2001. Water-extractable arabinoxylan from pearled flours of wheat, barley, rye and triticale. Evidence for the presence of ferulic acid dimers and their involvement in gel formation. Journal of Cereal Science 34: 207-214.

Dexter JE, Dronzek BL. 1975. Note on the amino acid composition of protein fractions from a developing triticale and its rye and durum wheat parents. Cereal Chemistry 52: 587-596.

Ebringerova A, Hromadkova Z, Petrakova E, Hricovini M. 1990. Structural features of a water-soluble L-arabino-D-xylan from rye bran. Carbohydrate Research 198: 57-66.

Elliott, GO, McLauchlan WR, Williamson G, Kroon PA. 2003. A wheat xylanase inhibitor protein (XIP-I) accumulates in the grain and has homologues in other cereals. Journal of Cereal Science 37: 187-194.

Fabritius M, Autio K, Gates F, Salovaara H. 1997. Structural changes in insoluble cell walls in wholemeal rye doughs. Lebensmittel Wissenschaft und Technologie 30: 367-372.

FAOSTAT data. 2004.

Fengler AI, Marquardt RR. 1988a. Water-soluble pentosans from rye: I. Isolation, partial purification and characterization. Cereal Chemistry 65: 291-297.

Fengler AI, Marquardt RR. 1988b. Water-soluble pentosans from rye: II. Effects on rate of dialysis and on the retention of nutrients by the chick. Cereal Chemistry 65: 298-302.

Field JM, Shewry PR, Miflin BJ, March JF. 1982. The purification and characterization of homologous high molecular weight storage proteins from grain of wheat, rye and barley. Theoretical and Applied Genetics 62: 329-336.

Field JM, Shewry PR, Miflin BJ. 1983. Aggregation states of alcohol-soluble storage proteins of barley, rye, wheat and maize. Journal of the Science of Food and Agriculture 34: 362-369.

Figueroa-Espinoza MC, Poulsen C, Borch Soe J, Zargahi MR, Rouau X. 2002. Enzymatic solubilization of arabinoxylans from isolated rye pentosans and rye flour by different endo-xylanases and other hydrolyzing enzymes. Effect of a fungal laccase on the flour extracts oxidative gelation. Journal of Agricultural and Food Chemistry 50: 6473-6484.

Figueroa-Espinoza MC, Poulsen C, Borch Soe J, Zargahi MR, Rouau X. 2004. Enzymatic solubilization of arabinoxylans from native, extruded, and high-shear-treated rye bran by different endo-xylanases and other hydrolyzing enzymes. Journal of Agricultural and Food Chemistry 52: 4240-4249.

Fowler DB, Brydon J, Darroch BA, Entz MH, Johnston AM. 1990. Environment and genotype influence on grain protein concentration of wheat and rye. Agronomy Journal 82: 655-664.

Fredriksson H, Silverio J, Andersson R, Eliasson AC, Aman P. 1998. The influence of amylose and amylopectin characteristics on gelatinization and retrogradation properties of different starches. Carbohydrate Polymers 35: 119-134.

Gabor R, Täufel A, Behnke U, Heckel J. 1991. Studies on the germination specific alpha-amylase and its inhibitor of rye (*Secale cereale*). 1. Isolation and characterization of the enzyme. Zeitschrift fur Lebensmittel-Untersuchung und -Forschung 192: 230-233.

Gallant DJ, Bouchet B, Baldwin PM. 1997. Microscopy of starch: evidence of a new level of granule organization. Carbohydrate Polymers 32: 177-191.

Garcia-Olmedo F, Salcedo G, Sanchez-Monge R, Gomez L, Royo J, Carbonero P. 1987. Plant proteinaceous inhibitors of proteinases and α-amylases. Oxford Surveys of Plant Molecular and Cell Biology 4: 275-334.

Gebruers K, Debyser W, Goesaert H, Proost P, Van Damme J, Delcour JA. 2001. *Triticum aestivum* L. endoxylanase inhibitor (TAXI) consists of two inhibitors, TAXI I and TAXI II, with different specificities. Biochemical Journal 353: 239-244.

Gebruers K, Brijs K, Courtin CM, Goesaert H, Proost P, Van Damme J, Delcour JA. 2002. Affinity chromatography with immobilised endoxylanases separates TAXI- and XIP-type endoxylanase inhibitors from wheat (*Triticum aestivum* L.). Journal of Cereal Science 36: 367-375.

Gellrich C, Schieberle P, Wieser H. 2001. Isolierung und Charakterisierung der γ-Secaline von Roggen. Getreide Mehl und Brot 55: 275-277.

Gellrich C, Schieberle P, Wieser H. 2003. Biochemical characterization and quantification of the storage protein (secalin) types in rye flour. Cereal Chemistry 80: 102-109.

Gellrich C, Schieberle P, Wieser H. 2004a. Biochemical characterization of γ-75k secalins of rye. I. Amino acid sequences. Cereal Chemistry 81: 290-295.

Gellrich C, Schieberle P, Wieser H. 2004b. Biochemical characterization of γ-75k secalins of rye. II. Disulfide bonds. Cereal Chemistry 81: 296-299.

Gernat C, Radosta S, Anger H, Damaschun G. 1993. Crystalline parts of three different conformations detected in native and enzymatically degraded starches. Starch 45: 309-314.

Girhammar U, Nair BM. 1992a. Isolation, separation and characterization of water soluble non-starch polysaccharides from wheat and rye. Food Hydrocolloids 6: 285-299.

Girhammar U, Nair BM. 1992b. Certain physical properties of water soluble non-starch polysaccharides from wheat, rye, triticale, barley and oats. Food Hydrocolloids 6: 329-343.

Glitso LV, Bach Knudsen KE. 1999. Milling of whole grain rye to obtain fractions with different dietary fibre characteristics. Journal of Cereal Science 29: 89-97.

Goesaert H, Gebruers K, Courtin CM, Proost P, Van Damme J, Delcour JA. 2002. A family of 'TAXI'-like endoxylanase inhibitors in rye. Journal of Cereal Science 36: 177-185.

Goesaert H, Gebruers K, Brijs K, Courtin CM, Delcour JA. 2003. XIP-type endoxylanase inhibitors in different cereals. Journal of Cereal Science 38: 317-324.

Gudmundsson M, Eliasson AC. 1991. Thermal and viscous properties of rye starch extracted from different varieties. Cereal Chemistry 68: 172-177.

Gudmundsson M, Eliasson AC, Bengtsson S, Aman P. 1991. The effects of water soluble arabinoxylan on gelatinization and retrogradation of starch. Starch 43: 5-10.

Hallmans G, Zhang, JX, Lundin E, Stattin P, Johansson A, Johansson I, Hulten K, Winkvist A, Lenner P, Aman P, Adlercreutz H. 2003. Rye, lignans and human health. Proceedings of the Nutrition Society 62: 193-199.

Hansen HB, Andreasen MF, Nielsen MM, Larsen LM, Bach Knudsen KE, Meyer AS, Christensen LP, Hansen A. 2002. Changes in dietary fibre, phenolic acids and activity of endogenous enzymes during rye breadmaking. European Food Research and Technology 214: 33-42.

Hansen HB, Rasmussen CV, Bach Knudsen KE, Hansen A. 2003. Effects of genotype and harvest year on content and composition of dietary fibre in rye (*Secale cereale* L.) grain. Journal of the Science of Food and Agriculture 83: 76-85.

Hansen HB, Moller B, Andersen SB, Jorgensen JR, Hansen A. 2004. Grain characteristics, chemical composition and functional properties of rye (*Secale cereale* L.) as influenced by genotype and harvest year. Journal of Agricultural and Food Chemistry 52: 2282-2291.

Härkönen H, Lehtinen P, Suortti T, Parkkonen T, Siikaaho M, Poutanen K. 1995. The effects of a xylanase and a β-glucanase from *Trichoderma reesei* on the non-starch polysaccharides of whole meal rye slurry. Journal of Cereal Science 21: 173-183.

Härkönen H, Pessa E, Suortti T, Poutanen K. 1997. Distribution and some properties of cell wall polysaccharides in rye milling fractions. Journal of Cereal Science 26: 95-104.

He H, Hoseney RC. 1991. Gas retention of different cereal flours. Cereal Chemistry 68: 334-336.

He T, Thacker PA, McLeod JG, Campbell GL. 2003. Performance of broiler chicks fed normal and low viscosity rye or barley with or without enzyme supplementation. Asian Australian Journal of Animal Sciences 16: 234-238.

Heiniö RL, Urala N, Vainionpää J, Poutanen K, Tuorila H. 1997. Identity and overall acceptance of two types of sour rye bread. International Journal of Food Science and Technology 32: 169-178.

Heiniö RL, Katina K, Wilhelmson A, Myllymäki O, Rajamäki T, Latva-Kala K, Liukkonen KH, Poutanen K. 2003a. Relationship between sensory perception and flavour-active volatile compounds of germinated, sourdough fermented and native rye following the extrusion process. Lebensmittel Wissenschaft und Technologie 36: 533-545.

Heiniö RL, Liukkonen KH, Katina K, Myllymäki O, Poutanen K. 2003b. Milling fractionation of rye produces different sensory profiles of both flour and bread. Lebensmittel Wissenschaft und Technologie 36: 577-583.

Hejgaard J. 2001. Inhibitory serpins from rye grain with glutamine as P_1 and P_2 residues in the reactive center. FEBS Letters 488: 149-153.

Henrissat B. 1991. A classification of glycosyl hydrolases based on amino acid sequence similarities. Biochemical Journal 280: 309-316.

Henry RJ. 1987. Pentosan and (1→3, 1→4)-β-glucan concentrations in endosperm and wholegrain of wheat, barley, oats and rye. Journal of Cereal Science 6: 253-258.

Hromadkova Z, Ebringerova A, Petrakova E. 1987. Structural features of a rye bran arabinoxylan with a low degree of branching. Carbohydrate Research 163: 73-79.

Jahn-Deesbach W, Schipper A. 1980. Protein-Fraktionen und Aminosäuren in ungekeimten und gekeimten Körnern von Weizen, Gerste, Roggen und Hafer. Getreide Mehl und Brot 34: 281-287.

Kariluoto S, Vahteristo L, Salovaara H, Katina K, Liukkonen KH, Piironen V. 2004. Effect of baking method and fermentation on folate content of rye and wheat breads. Cereal Chemistry 81: 134-139.

Karppinen S, Myllymaki O, Forssell P, Poutanen K. 2003. Fructan content of rye and rye products. Cereal Chemistry 80: 168-171.

Keetels CJAM, Oostergetel GT, van Vliet T. 1996. Recrystallization of amylopectin in concentrated starch gels. Carbohydrate Polymers 30: 61-64.

Kipp B, Belitz H-D, Seilmeier W, Wieser H. 1996. Comparative studies of high M_r subunits of rye and wheat. I. Isolation and biochemical characterization and effects on gluten extensibility. Journal of Cereal Science 23: 227-234.

Kirchhoff E, Schieberle P. 2001. Determination of key aroma compounds in the crumb of a three-stage sourdough rye bread by stable isotope dilution assays and sensory studies. Journal of Agricultural and Food Chemistry 49: 4304-4311.

Kirchhoff E, Schieberle P. 2002. Quantitation of odor-active compounds in rye flour and rye sourdough using stable isotope dilution assays. Journal of Agricultural and Food Chemistry 50: 5378-5385.

Klassen AJ, Hill RD. 1971. Comparison of starch from triticale and its parental species. Cereal Chemistry 48, 647-654.

Köhler P, Wieser H. 2000. Nachweis von Cysteinresten in hochmolekularen Glutelin-Untereinheiten aus Roggen. Getreide Mehl und Brot 54: 283-289.

Kreis M, Forde BG, Miflin BJ, Shewry PR. 1985. Molecular evolution of the seed storage proteins of barley, rye and wheat. Journal of Molecular Biology 183: 499-502.

Kühn MC, Grosch W. 1988. Influence of the enzymic modification of the nonstarchy polysaccharide fractions on the baking properties of reconstituted rye flour. Journal of Food Science 53: 889-895.

Kühn MC, Grosch W. 1989. Baking functionality of reconstituted rye flours having different nonstarchy polysaccharide and starch contents. Cereal Chemistry 66: 149-154.

Kujala T. 2004. Rye and health at URL: http://rye.vtt.fi/.

Lazaro R, Garcia M, Medel P, Mateos GG. 2003. Influence of enzymes on performance and digestive parameters of broilers fed rye-based diets. Poultry Science 82: 132-140.

Lii CY, Lineback DR. 1977. Characterization and comparison of cereal starches. Cereal Chemistry 54: 138-149.

Liukkonen KH, Katina K, Wilhelmson A, Myllymäki O, Lampi AM, Kariluoto S, Piironen V, Heinonen SM, Nurmi T, Adlercreutz H, Peltoketo A, Pihlava JM, Hietaniemi V, Poutanen K. 2003. Process-induced changes on bioactive compounds in whole grain rye. Proceedings of the Nutrition Society 62: 117-122.

Ludwig I, Ludwig E. 1989. Untersuchungen zur Extraktion von Roggenproteinen mit grenzflächenaktiven Verbindungen. Die Nahrung 33: 761-765.

Lyons A, Richardson M, Tatham AS, Shewry PR. 1987. Characterization of homologous inhibitors of trypsin and α-amylase from seeds of rye (*Secale cereale* L.). Biochimica et Biophysica Acta 915: 305-313.

MacGregor AW, Marchylo BA, Kruger JE. 1988. Multiple α-amylase components in germinated cereal grains determined by isoelectric focusing and chromatofocusing. Cereal Chemistry 65: 326-333.

Masojc P, Larssonraznikiewicz M. 1991a. Variations of the levels of alpha-amylase and endogenous alpha-amylase inhibitor in rye and triticale grain. Swedish Journal of Agricultural Research 21: 3-9.

Masojc P, Larssonraznikiewicz M. 1991b. Genetic variation of alpha-amylase levels among rye (*Secale cereale* L.) kernels tested by gel-diffusion technique. Swedish Journal of Agricultural Research 21: 141-145.

Meuser F, Suckow P, Abdel-Gawad A. 1986. Chemisch-physikalische Charakterisierung von Roggenpentosanen. Getreide Mehl und Brot 40: 198-204.

Meuser F, Brümmer JM, Seibel W. 1994. Bread varieties in Central Europe. Cereal Foods World 39: 222-230.

Meuser F, Busch KG, Fuhrmeister H, Rubach K. 2001. Foam-forming capacity of substances present in rye. Cereal Chemistry 78: 50-54.

Michniewicz J, Biliaderis CG, Bushuk W. 1992. Effect of added pentosans on some properties of wheat bread. Food Chemistry 43: 251-257.

Mohammadkhani A, Stoddard FL, Marshall DR. 1998. Survey of amylose content in *Secale cereale*, *Triticum monococcum*, *T. turgidum* and *T. tauschii*. Journal of Cereal Science 28: 273-280.

Morey DD, Evans JJ. 1983. Amino acid composition of six grains and winter wheat forage. Cereal Chemistry 60: 461-464.

Morrison WR. 1995. Starch lipids and how they relate to starch granule structure and functionality. Cereal Foods World 40: 437-445.

Mosolov VV, Shul'gin MN. 1986. Protein inhibitors of microbial proteinases from wheat, rye and triticale. Planta 167: 595-600.

Nilsson M, Saulnier L, Andersson R, Aman P. 1996. Water unextractable polysaccharides from three milling fractions of rye grain. Carbohydrate Polymers 30: 229-237.

Nilsson M, Aman P, Härkönen H, Hallmans G, Bach Knudsen KE, Mazur W, Adlercreutz H. 1997a. Content of nutrients and lignans in roller milled fractions of rye. Journal of the Science of Food and Agriculture 73: 143-148.

Nilsson M, Aman P, Härkönen H, Hallmans G, Bach Knudsen KE, Mazur W, Adlercreutz H. 1997b. Nutrient and lignan content, dough properties and baking performance of rye samples used in Scandinavia. Acta Agriculturae Scandinavica 47: 26-34.

Nilsson M, Andersson R, Aman P. 1999. Arabinoxylan fractionation on DEAE-cellulose chromatography influenced by protease pre-treatment. Carbohydrate Polymers 39: 321-326.

Nilsson M, Andersson R, Andersson RE, Autio K, Aman P. 2000. Heterogeneity in a water-extractable rye arabinoxylan with a low degree of disubstitution. Carbohydrate Polymers 41: 397-405.

Nowak J, Mierzwinska T. 1978. Activity of proteolytic enzymes in rye seeds of different ages. Zeitschrift fur Pflanzenphysiologie 86: 15-22.

Nyman M, Siljeström M, Pedersen B, Bach Knudsen KE, Asp NG, Johansson CG, Eggum BO. 1984. Dietary fiber content and composition in six cereals at different extraction rates. Cereal Chemistry 61: 14-19.

Parkkonen T, Härkönen H, Autio K. 1994. Effect of baking on the microstructure of rye cell walls and protein. Cereal Chemistry 71: 58-63.

Popineau Y, Huchet B, Larré C, Bérot S. 2002. Foaming and emulsifying properties of fractions of gluten peptides obtained by limited enzymatic hydrolysis and ultrafiltration. Journal of Cereal Science 35: 327-335.

Poutanen K. 1997. Rye bread: added value in the world's bread basket. Cereal Foods World 42: 682-683.

Preston KR, Woodbury W. 1975. Amino acid composition and subunit structure of rye gliadin proteins fractionated by gel filtration. Cereal Chemistry 52: 719-726.

Radosta S, Kettlitz B, Schierbaum F, Gernat C. 1992. Studies on rye starch properties and modification. Part II: Swelling and solubility behaviour of rye starch granules. Starch 44: 8-14.

Ragaee SM, Campbell GL, Scoles GJ, McLeod JG, Tyler RT. 2001. Studies on rye (*Secale cereale*) lines exhibiting a range of extract viscosities. 1. Composition, molecular weight distribution of water extracts, and biochemical characteristics of purified water-extractable arabinoxylan. Journal of Agricultural and Food Chemistry 49: 2437-2445.

Rasmussen CV, Hansen HB, Hansen A, Larsen LM. 2001. pH-, temperature- and time-dependent activities of endogenous endo-β-D-xylanase, β-D-xylosidase and α-L-arabinofuranosidase in extracts from ungerminated rye (*Secale cereale* L.) grain. Journal of Cereal Science 34: 49-60.

Rattunde HF, Geiger HH, Weipert D. 1994. Variation and covariation of milling and baking quality characteristics among winter rye single-cross hybrids. Plant Breeding 113: 287-293.

Rouau X, El-Hayek ML, Moreau D. 1994. Effect of an enzyme preparation containing pentosanases on the bread-making quality of flours in relation to changes in pentosan properties. Journal of Cereal Science 19: 259-272.

Sarkkinen P, Kalkkinen N, Tilgmann C, Siuro J, Kervinen J, Mikola L. 1992. Aspartic proteinase from barley grains is related to mammalian lysosomal cathepsin-D. Planta 186: 317-323.

Schierbaum F, Radosta S, Richter M, Kettlitz B, Gernat C. 1991. Studies on rye starch properties and modification. Part I: Composition and properties of rye starch granules. Starch 43: 331-339.

Schierbaum F, Kettlitz B. 1994. Studies on rye starch properties and modification. Part III: Viscograph pasting characteristics of rye starches. Starch 46: 2-8.

Seibel W, Brümmer JM. 1991. The sourdough process for bread in Germany. Cereal Foods World 36: 299-304.

Seibel W, Weipert D. 2001a. Bread baking and other food uses around the world. In: Busuk W, editor. Rye: production, chemistry and technology. 2nd ed. St. Paul, Minnesota, USA: American Association of Cereal Chemists. p 147-211.

Seibel W, Weipert D. 2001b. Animal feed and industrial uses. In: Busuk W, editor. Rye: production, chemistry and technology. 2nd ed. St. Paul, Minnesota, USA: American Association of Cereal Chemists. p 213-233.

Shewry PR, Field JM, Lew EJ-L, Kasarda DD. 1982. The purification and characterization of two groups of storage proteins (secalins) from rye (*Secale cereale* L.). Journal of Experimental Botany 33: 261-268.

Shewry PR, Parmar S, Miflin BJ. 1983. Extraction, separation and polymorphism of the prolamin storage proteins (secalins) of rye. Cereal Chemistry 60: 1-6.

Shewry PR, Tatham AS, Pappin DJ, Keen J. 1988. N-terminal amino acid sequences show that D hordein of barley and high molecular weight (HMW) secalins of rye are homologous with HMW glutenin subunits of wheat. Cereal Chemistry 65: 510-511.

Shewry PR, Tatham AS. 1990. The prolamin storage proteins of cereal seeds: structure and evolution. Biochemical Journal 267: 1-12.

Shewry PR, Napier JA, Tatham AS. 1995. Seed storage proteins: structures and biosynthesis. The Plant Cell 7: 945-956.

Shewry PR, Bechtel DB. 2001. Morphology and chemistry of the rye grain. In: Busuk W, editor. Rye: production, chemistry and technology. 2nd ed. St. Paul, Minnesota, USA: American Association of Cereal Chemists. p 69-127.

Shih FF, Daigle K. 1997. Use of enzymes for the separation of protein from rice flour. Cereal Chemistry 74: 437-441.

Smith AT, Kuznesof S, Richardson DP, Seal CJ. 2003. Behavioural, attitudinal and dietary responses to the consumption of wholegrain foods. Proceedings of the Nutrition Society 62: 455-467.

Smulikowska S, Nguyen VC. 2001. A note on variability of water extract viscosity of rye grain from northeast regions of Poland. Journal of Animal and Feed Sciences 10: 687-693.

Tatham AS, Shewry PR. 1991. Conformational analysis of the secalin storage proteins of rye (*Secale cereale* L.). Journal of Cereal Science 14: 15-23.

Tatham AS, Shewry PR. 1995. The S-poor prolamins of wheat, barley and rye. Journal of Cereal Science 22: 1-16.

Täufel A, Behnke U, Emmer I., Gabor R. 1991. Studies on the germination specific α-amylase and its inhibitor of rye (*Secale cereale*). 2. Isolation and characterization of the inhibitor. Zeitschrift fur Lebensmittel-Untersuchung und -Forschung 193: 9-14.

Täufel A, Böhm H, Flamme W. 1997. Protein inhibitors of alpha-amylase in mature and germinating grain of rye (*Secale cereale*). Journal of Cereal Science 25: 267-273.

Van den Bulck K, Swennen K, Loosveld A-M A, Courtin CM, Brijs K, Proost P, Van Damme J, Van Campenhout S, Mort A, Delcour JA. 2005. Isolation of cereal arabinogalactan-peptides and structural comparison of their carbohydrate and peptide moieties. Journal of Cereal Science 41: 59-67.

Vanhamel S, Cleemput G, Delcour JA, Nys M, Darius PL. 1993. Physicochemical and functional properties of rye nonstarch polysaccharides. IV. The effect of high molecular weight water-soluble pentosans on wheat-bread quality in a straight-dough procedure. Cereal Chemistry 70: 306-311.

Verwimp T, Vandeputte GE, Marrant K, Delcour JA. 2004. Isolation and characterization of rye starch. Journal of Cereal Science 39: 85-90.

Vinkx CJA, Van Nieuwenhove CG, Delcour JA. 1991. Physicochemical and functional properties of rye nonstarch polysaccharides. III. Oxidative gelation of a fraction containing water-soluble pentosans and proteins. Cereal Chemistry 68: 617-622.

Vinkx CJA, Reynaert HR, Grobet PJ, Delcour JA. 1993. Physicochemical and functional properties of rye nonstarch polysaccharides. V. Variability in the structure of water-soluble arabinoxylans. Cereal Chemistry 70: 311-317.

Vinkx CJA. 1994. Structure and properties of arabinoxylans from rye [Ph. D. dissertation]. Leuven, Belgium: Katholieke Universiteit Leuven. 127 p.

Vinkx CJA, Delcour JA, Verbruggen MA, Gruppen H. 1995a. Rye water-soluble arabinoxylans also vary in their 2-monosubstituted xylose content. Cereal Chemistry 72: 227-228.

Vinkx CJA, Stevens I, Gruppen H, Grobet PJ, Delcour JA. 1995b. Physicochemical and functional properties of rye nonstarch polysaccharides. VI. Variability in the structure of water-unextractable arabinoxylans. Cereal Chemistry 72: 411-418.

Vinkx CJA, Delcour JA. 1996. Rye (*Secale cereale*) arabinoxylans: a critical review. Journal of Cereal Science 23: 1-14.

Wannerberger L, Eliasson A-C, Sindberg A. 1997. Interfacial behaviour of secalin and rye flour-milling streams in comparison with gliadin. Journal of Cereal Science 25: 243-252.

Wasserman LA, Eiges NS, Koltysheva GI, Andreev NR, Karpov VG, Yuryev VP. 2001. The application of different physical approaches for the description of structural features in wheat and rye starches. A DSC study. Starch 53: 629-634.

Weipert D. 1993a. Verarbeitungswert von deutschen Roggensorten. Getreide Mehl und Brot 47 (2): 6-12.

Weipert D. 1993b. Messung der Auswuchsfestigkeit bei Roggen. Getreide Mehl und Brot 47 (4): 3-9.

Weipert D. 1994. Beschreibung des Verarbeitungswertes neuer Roggensorten. Getreide Mehl und Brot 4: 3-5.

Weipert D. 1997. Processing performance of rye as compared to wheat. Cereal Foods World 42: 706-712.

Wu WU, Hettiarachchy NS, Qi M. 1998. Hydrophobicity, solubility and emulsifying properties of soy protein peptides prepared by papain modification and ultrafiltration. Journal of the American Oil Chemists Society 75: 845-850.

Zwingelberg H, Sarkar A. 2001. Milling of rye. In: Busuk W, editor. Rye: production, chemistry and technology. 2nd ed. St. Paul, Minnesota, USA: American Association of Cereal Chemists. p 129-145.

Part VI
Fermented Foods

26
Dairy Products

T. D. Boylston

INTRODUCTION

Fermented dairy products were originally developed as a means to preserve milk through the production of lactic acid. However, these products are also recognized to have desirable sensory characteristics and nutritional value. The term, cultured dairy products, is currently used to indicate that these products are prepared using lactic acid bacteria (LAB) starter cultures and controlled fermentation. LAB utilize the nutrients in milk to support their growth. The production of lactic acid reduces the pH of these products to inhibit the growth of many pathogenic and spoilage microorganisms. In addition, the fermentation process develops a wide range of dairy products with a diversity of flavor and textural attributes, including cheese, yogurt, buttermilk, butter, acidophilus milk, and sour cream.

Chapters 19 and 20 provided detailed information on the biochemistry and processing of milk and milk products. This chapter will focus on the biochemistry and processing involved in the formation of cultured dairy products. This chapter will include a general discussion of the biochemistry and processing of cultured dairy products, followed by a discussion of the processing of the individual dairy products.

BIOCHEMISTRY OF CULTURED DAIRY PRODUCTS

Composition of Milk

The quality of the cultured dairy products is influenced by the composition and quality of the raw milk. Species, breed, nutritional status, health, and stage of lactation of the cow can have an impact on the fat, protein, and calcium content and overall composition and quality of the milk. On the average, cow milk consists of 3.7% fat, 3.4% protein, 4.8% lactose, and 0.7% ash and has a pH of 6.6 (Fox et al. 2000). Good microbiological quality of the raw milk is also essential to the production of safe, high quality, cultured dairy products.

The predominant sugar in milk is lactose, a disaccharide of glucose and galactose. The fermentation of lactose by lactic acid bacteria in cultured dairy products provides the flavor and textural attributes that are desirable in cultured dairy products.

Triacylglycerols are the predominant lipid fraction in milk, accounting for 98% of the total lipids. Diacylglycerols, monoacylglycerols, fatty acids, phospholipids, and sterols account for the remaining lipid fraction. Approximately 65% of the fatty acids in milk fat are saturated, including 26% palmitic acid and 15% stearic acid. A significant amount of short- and medium-chain fatty acids, including 3.3% butyric acid are present. These fatty acids and the breakdown products of these fatty acids are important contributors to the flavor of many cultured dairy products.

Two major classes of proteins, the caseins and the whey proteins, are present in milk. The caseins, which make up 80% of the total protein in cow milk, are insoluble at a pH of 4.6, but are stable to heating. The whey proteins remain soluble at pH 4.6 and are heat sensitive. The caseins are the major protein present in cultured dairy products.

The casein micelles exist in milk as a colloidal dispersion, with a diameter ranging from 40 to 300 nm and containing approximately 10,000 casein molecules. The principal casein proteins, α_{s1}-, α_{s2}-, β-, and κ-casein, present in the ratio 40:10:35:12, vary in the number of phosphate residues, calcium sensitivity, and hydrophobicity. Within the casein micelle, the more hydrophobic proteins, such as β-casein are located on the interior of the micelle, while the more hydrophilic proteins, such as κ-casein, are located on the surface of the micelle. The carboxyl end of κ-casein is dominated by glutamic acid residues and glycoside groups. These hydrophilic carboxyl ends are represented as "hairy" regions in the model of the casein micelle and promote the stability of the casein micelle in solution (Walstra and Jenness 1984). Calcium phosphate further facilitates the association of individual calcium-sensitive casein proteins (α_{s1}-, α_{s2}-, and β-casein), within the casein micelle. Processing treatments applied during the formation of fermented dairy products, such as the addition of acid or enzymes, destabilize the casein micelle causing the casein proteins to precipitate.

The whey proteins consist of four major proteins, β-lactoglobulin (50%), α-lactalbumin (20%), blood serum albumin (10%), and immunoglobulins (10%). These proteins have a significant number of cysteine and cystine residues and are able to form disulfide linkages with other proteins following heat treatment.

Lactic Acid Bacteria

The lactic acid bacteria used in the development of fermented dairy products include *Streptococcus, Lactococcus, Leuconostoc*, and *Lactobacillus* genera. These bacteria are gram-positive bacteria and belong to either the Streptococcaceae or Lactobacillaceae families, depending on the morphology of the bacteria as cocci or rods, respectively. These bacteria also differ in their optimal temperature for growth: 20–30°C is the optimal temperature for mesophilic bacteria, and 35–45°C is the optimal temperature for thermophilic bacteria. Although the lactic acid bacteria are quite diverse in growth requirements, morphology, and physiology, they all have the ability to metabolize lactose to lactic acid and reduce the pH of the milk to produce specific cultured dairy products. The choice of lactic acid bacteria is in part dictated by the heat treatment the cultured dairy products undergo following inoculation. Table 26.1 summarizes the growth characteristics of common lactic acid bacteria.

KEY PROCESSING STEPS IN CULTURED DAIRY PRODUCTS

Many of the processing steps important in the production of cultured dairy products are not unique to a specific product. Therefore, the following discussion will provide an overview of the key processing steps that are used in the production of several cultured dairy products. Specific processing treatments and concerns will be highlighted within the discussion of the processing of the specific cultured dairy products.

Lactose Fermentation

Lactic acid bacteria use the lactose in milk to produce lactic acid and other important flavor com-

Table 26.1. Characteristics of Lactic Acid Bacteria Used in Cultured Dairy Products

	Morphology	Lactose Fermentation	Temperature for Growth (°C)			Lactic Acid Production[a]	Primary Metabolic Products
			Minimum	Optimal	Maximum		
Lactococcus lactis							
Lc. lactis ssp. *lactis*	Cocci	Homofermentative	8–10	28–32	40	~ 0.9	Lactate
Lc. lactis ssp. *cremoris*	Cocci	Homofermentative	8–10	22	37–39	~ 0.9	Lactate
Leuconostoc							
Ln. mesenteroides ssp. *cremoris*	Cocci	Heterofermentative	4–10	20–25	~37	tr	Lactate, diacetyl, carbon dioxide
Ln. lactis	Cocci	Heterofermentative	4–10	20–25	~37	~ 0.8	Lactate, diacetyl, carbon dioxide
Streptococcus							
Streptococcus thermophilus	Cocci	Homofermentative	20	40	50	~ 0.9	Lactate, acetaldehyde
Lactobacillus							
Lb. delbrueckii ssp. *bulgaricus*	Rods	Homofermentative	22	40–45	52	~ 2.5	Lactate, acetaldehyde
Lb. helveticus	Rods	Homofermentative	20–22	42	54	~ 2.5	Lactate
Lb. delbrueckii ssp. *lactis*	Rods	Homofermentative	18	40	50	~ 1.2	Lactate
Lb. acidophilus	Rods	Homofermentative	20–22	37	45–48	~ 1.0	

Sources: Walstra et al. 1999, Stanley 1998.

[a]Percent in milk in 24 hours at optimum temperature.

pounds in cultured dairy products. Many of these bacteria have lactase activity and hydrolyze lactose to its monosaccharide units, glucose and galactose, prior to further metabolism. The hydrolysis of lactose in most cultured dairy products is significant for individuals who are lactose intolerant, allowing them to consume dairy products without the undesirable effects of the inability to hydrolyze lactose.

Homofermentative lactic acid bacteria will produce lactic acid from these monosaccharides, according to the reaction

Lactose + 4 ADP + 4 $H_3PO_4 \rightarrow$ 4 Lactic Acid + 4 ATP + $3H_2O$

The glucose and galactose molecules are metabolized through the glycolytic and tagatose pathways, respectively. Key steps in the formation of lactic acid from the monosaccharides include the hydrolysis of the hexose diphosphates to glyceraldehyde-3-phosphate by aldolases, the formation of pyruvate from phosphoenol pyruvate by pyruvate kinase, and the reduction of pyruvate to lactate by lactate dehydrogenase.

Heterofermentative lactic acid bacteria produce carbon dioxide, acetic acid, and ethanol, in addition to lactic acid during the metabolism of the monosaccharides, according to the reaction

Lactose + 2 H_3PO_4 + 2 ADP $\rightarrow$ 2 Lactic Acid + 2 Ethanol + 2 CO_2 + 2 ATP + H_2O

The heterofermentative lactic acid bacteria lack aldolases; therefore, the sugars are metabolized through 6-P-gluconate pathway rather than the glycolytic pathway. In this pathway, glucose-6-phosphate is oxidized to 6-phosphogluconate, which is decarboxylated to a pentose-5-phosphate and carbon dioxide by phosphoketolase. The pentose-5-phosphate is converted to glyceraldehyde-3-phosphate, which enters the glycolytic pathway to form lactic acid and acetyl phosphate, which is metabolized to acetaldehyde. The acetaldehyde is reduced to ethanol by alcohol dehydrogenase. However, several bacteria, including *Lactococcus lactis* ssp. *lactis* biovar. *diacetylactis, Streptococcus thermophilus,* and *Lactobacillus delbrueckii* ssp. *bulgaricus* lack dehydrogenase and accumulate acetaldehyde.

The production of acid by the lactic acid bacteria has a significant impact on the safety and quality of the cultured dairy products. The reduction in pH increases the shelf life and safety of the fermented dairy products through the inhibition of spoilage and pathogenic microorganisms. Acid production by these lactic acid bacteria is critical for the precipitation of the casein proteins in the formation of several cultured dairy products, such as yogurt, sour cream, and unripened cheeses. These bacteria may also contribute to the degradation of proteins and lipids through proteolytic and lipolytic reactions to further develop the unique texture and flavor characteristics of the cultured dairy products. These reactions are especially important in ripened cheeses. The selection of the appropriate type and level of starter culture is imperative for the development of the appropriate flavor characteristics.

Many commercial dairy processors now use the direct vat inoculation (DVI) process for frozen or freeze-dried cultures (up to 10^{12} bacteria per gram of starter) in the processing of fermented dairy products. The use of DVI cultures allows the dairy processor to directly add the cultures to the milk and bypass on-site culture preparation. This recent progress in the development of starter cultures has also increased phage resistance, minimized the formation of mutants that may alter the characteristics of the starter cultures, enhanced the ability to characterize the composition of the cultures, and improved the quality consistency of cultured dairy products. However, the DVI process is limited by the additional cost of these cultures, the dependence of the cheese plants on the starter suppliers for the selection and production of the starters, and the increased lag phase of these cultures in comparison with that of on-site culture preparation (Stanley 1998, Tamime and Robinson 1999a, Canteri 2000).

Coagulation of Milk Proteins

The production of lactic acid by LAB decreases the pH of the milk to cause coagulation of the casein. As the pH decreases to less than 5.3, colloidal calcium phosphate is solubilized from the casein micelle, causing the micelles to dissociate and the casein proteins to aggregate and precipitate at the isoelectric point of casein (pH 4.6). The resulting gel, which is somewhat fragile in nature, provides the structure for sour cream, yogurt, and acid-precipitated cheeses, such as cream cheese and cottage cheese.

The casein micelles are also susceptible to coagulation through enzymatic activity. Rennet, a mixture of chymosin and pepsin, obtained from calf stomach, is most commonly recognized as the enzyme for coaguluating casein. However, proteases from microorganisms and those produced through recombinant DNA technologies have been successfully adapted as alternatives to calf rennet (Banks 1998). Chymosin, the major enzyme present in rennet, cleaves the peptide bond between Phe_{105} and Met_{106} of κ-casein, releasing the hydrophilic, charged casein macropeptide, while the para-κ-casein remains associated with the casein micelle. The loss of the charged macropeptide reduces the surface charge of the casein micelle and results in the aggregation of the casein micelles to form a gel network stabilized by hydrophobic interactions. Temperature influences both the rate of the enzymatic reaction and the aggregation of the casein proteins; the optimal temperature for casein coagulation is 40–42°C. Rennet is used in the manufacture of most ripened cheeses to hydrolyze the peptide bond and cause aggregation of the casein micelles.

Homogenization

Milk fat globules have a tendency to coalesce and separate upon standing. Homogenization reduces the diameter of the fat globules from 1–10 μm to less than 2 μm and increases the total fat globule surface area. The physical change in the fat globule occurs through forcing the milk through a small orifice under high pressure. The decrease in the size of the milk fat globules reduces the tendency of the fat globules to aggregate during the gelation period. In addition, denaturation of the whey proteins and interactions of the whey proteins with casein or the fat globules can alter the physical and chemical properties of the milk proteins to result in a firmer gel with reduced syneresis (Tamime and Robinson 1999b, Fox et al. 2000). Milk to be used to process yogurt, cultured buttermilk, and unripened cheeses is commonly homogenized to improve the quality of the final product.

Pasteurization

The original fermented dairy products relied on the native microorganisms in the milk for the fermentation process. Current commercial methods for all cultured dairy products include a pasteurization treatment to kill the native microorganisms, followed by inoculation with starter cultures to produce the desired product. The heat process, which must be sufficient to inactivate alkaline phosphatase, also destroys many pathogenic and spoilage microorganisms and enzymes that may have a negative impact on the quality of the finished products. The time-temperature treatments for fluid milk pasteurization have been adapted for milk to be used in the processing of cultured dairy products (62.8°C for 30 minutes or 71.1°C for 15 seconds). More severe heat treatments than are characteristic of pasteurization cause denaturation of whey proteins and interactions between β-lactoglobulin and κ-casein. In cheeses, this interaction decreases the ability of chymosin to hydrolyze the casein molecule and initiate curd precipitation and formation.

Cooling

The processing of cultured dairy products relies on the metabolic activity of the starter cultures to contribute to acid formation and flavor and texture development. Once the desired pH or titratable acidity is reached for these products, the products are cooled to 5–10°C to slow the growth of the bacteria and limit further acid production and other biological reactions.

PROCESSING OF CULTURED DAIRY PRODUCTS

The following discussion highlights the unique processing steps that are involved in the production of cultured dairy products. These processing steps contribute to the unique flavor, texture, and overall sensory characteristics of these products.

Cheese

Over 400 different varieties of cheese have been recognized throughout the world. The wide diversity in the flavor, texture, and appearance of these cheeses is attributed to differences in the milk source, starter cultures, ripening conditions, and chemical composition. Cheeses are frequently classified based on moisture content, method of precipitation of the cheese proteins, and the ripening process. Table 26.2 compares processing methods and compositions of selected cheeses.

The coagulation of the casein proteins, separation of the curds from the whey, and ripening of the curd

Table 26.2. Comparison of Processing Characteristics of Different Types of Ripened, Natural Cheeses

Cheese	Starter Cultures	Ripening	Scald Temperature (°C)	Other Treatments	Composition (%) Moisture	Fat	Salt
Very hard							
Romano	Thermophilic cultures	18–24 months at 16–18°C	45–48	–	23.0	24.0	5.5
Hard—Ripened by bacteria, without eyes							
Cheddar	Mesophilic D cultures	0.5–2 years at 6–8°C	37–39	Curds are cheddared prior to salting, dry salting	37.0	32.0	1.5
Hard—Ripened by bacteria, with eyes							
Emmental	Thermophilic cultures and propionibacteria	10–14 days at 10–15°C; 3 weeks–2 months at 20–24°C; 1–2 months at 7°C	55	Brine-salted	35.5	30.5	1.2
Swiss	Thermophilic cultures, *Pr. freudenreichii* ssp. *shermanii*	7–14 days at 21–25°C; 6–9 months at 2–5°C	50–53C	Brine-salted	37.1	27.8	0.5
Semi-soft—Ripened principally by bacteria, without eyes							
Colby	Mesophilic D cultures	2–3 months at 3–4°C	37–39C	Curds and whey stirred after cooking, as whey is drained	38.2	32.1	1.5
Provalone	*L. bulgaricus*	1–12 months at 15–16°C	44–50	Curd—knead and stretch in 85–90°C water	42.5	27.0	3.0
Mozzarella	Thermophilic cultures	< 1 month	41	Curd—kneaded and stretched in 70°C water	54.0	18.0	0.7

Feta	Thermophilic or mesophilic starter culture	7 days in salt brine at 14–16°C; 2 months at 3–4°C	Not cooked	From sheep or sheep/goat milk	59.7	20.3	2.2
Semi-soft—Ripened principally by bacteria, with eyes							
Gouda	Mesophilic DL culture	2–3 months at 15°C	36–38	Curd washed with hot water, stirred for 30 minutes at 36–38°C, pressed under whey, brine salted	41.0	28.5	2.0
Edam	Mesophilic DL culture	6 weeks to 6 months	36	Similar to Gouda, made from semi-skimmed (2.5% fat) milk	43.0	24.0	2.0
Semi-soft—Ripened by bacteria and surface microorganisms							
Limburger	Mesophilic DL culture	2–3 weeks at 10–15°C; 3–8 weeks at 4°C (wrapped in foil)	37	Whey drained and replaced with salt brine, Surface smeared with microflora during initial ripening	45.0	28.0	2.0
Semi-soft—Ripened principally by blue mold in the interior							
Roquefort	Indigenous microflora	3–5 months in limestone caves at 5–10°C and 95% RH	Not cooked	Ewe's milk. Curds mixed with *Penicillium roqueforti* spores, Cheeses pierced during ripening	40.0	31.0	3.5
Soft—Ripened principally by surface molds							
Camembert	Mesophilic DL culture	10–12 days at 12°C, 7–10 days at 7°C	Not cooked	Surface mold growth by *Penicillium camemberti*	52.5	23.0	2.5

Sources: Kosikowski 1977, Fox et al. 2000.

are the primary steps involved in the processing of cheese. The resulting product is a highly nutritious product in which the casein and fat from the milk are concentrated. The fat plays a critical role in the texture of the cheese by preventing the casein molecules from associating to form a tough structure. In general, most cheeses can be classified as natural or processed cheeses. The natural cheeses include both ripened and unripened cheeses. The stages involved in processing these different types of cheeses and their unique characteristics will be discussed further.

Natural Cheeses

A simplified overview of the steps involved in processing fresh and ripened natural cheeses is presented in Figure 26.1. Fresh cheeses are consumed immediately after processing and are characterized as having a high moisture content and mild flavor. In most cases, lactic acid produced by the starter cultures cause the precipitation of the caseins. The final pH of the acid-coagulated cheeses is 4.6. Ripened cheeses undergo a ripening period, ranging from 3 weeks to more than 2 years, following processing, which contributes to the development of the flavor and texture of the cheese. The moisture content of these cheeses ranges from 30 to 55% and the pH ranges from 5.0 to 5.3. Rennet is primarily used for the coagulation of the casein proteins and curd formation. Starter cultures are added to produce acid and contribute enzymes for flavor and texture development during ripening.

Standardization of the Milk The casein and fat content of the milk are standardized to minimize variations in the quality of the cheese due to seasonal effects and variation in the milk supply. The casein-to-fat ratio can be adjusted by the addition of skim milk, cream, milk powder, or evaporated milk or the removal of fat. Calcium chloride (0.1%) may also be added to improve coagulation of the milk by rennet and further processing of the cheese. The actual casein and fat content of the milk will vary for each cheese type and influence the curd formation, cheese yield, fat content, and texture of the cheese (Banks 1998).

Coagulation of the Milk Proteins Aggregation of the casein micelles to form a three-dimensional gel protein network is initiated through the addition of rennet or other proteolytic enzymes or the addition of acid. Fat and water molecules are also entrapped within this protein network. Enzymes and starter bacteria also tend to associate with the curds, and thus contribute to a number of biochemical changes that occur during the ripening process. The whey, which includes water, salts, lactose, and the soluble whey proteins, is expelled from the gel. The aggregation of the casein micelles by either enzyme or acid treatment results in gels with different characteristics.

In most natural, aged cheeses, coagulation of the casein proteins by the addition of rennet is most common. This process is temperature dependent, with no coagulation occurring below 10°C, and an increase in coagulation rate accompanying an increase in temperature until the optimal temperature for coagulation (40–45°C) is reached. Above 65°C, the enzyme is inactivated (Fox 1969, Brulé et al. 2000). The aggregation of the casein micelles is influenced by temperature ($Q_{10} \sim 12$) to a greater degree than the enzymatic hydrolysis of κ-casein ($Q_{10} \sim 2$) (Cheryan et al. 1975). Aggregation of the micelles begins when approximately 70–85% of the κ-casein molecules are hydrolyzed, which reduces the steric hindrance between the micelles. Reducing the pH or increasing the temperature reduces the degree of casein hydrolysis necessary for coagulation (Fox and McSweeney 1997). The presence of Ca^{2+} ions further facilitates the aggregation of the casein micelles through the neutralization of the negative charge on the micelle and the formation of ionic bonds. The resulting gel has an irregular network, is highly elastic and porous, and exhibits a high degree of syneresis.

The production of acid by lactic acid bacteria or the direct addition of hydrochloric or lactic acid can also result in aggregation of the casein micelles and formation of clots. As the pH of the milk is reduced, the casein micelles become insoluble and begin to aggregate. Because calcium phosphate is solubilized as the pH of the milk drops, the gel formed during acid coagulation is not stabilized by calcium ions. The acid-coagulated gels are less cohesive and exhibit less syneresis than enzyme-coagulated cheeses. These cheeses generally have a high moisture content and a low mineral content. Acid coagulation is most frequently used in the manufacture of cottage cheese and other unripened cheeses.

A few unique types of cheese are prepared through acid coagulation of whey or a blend of whey and

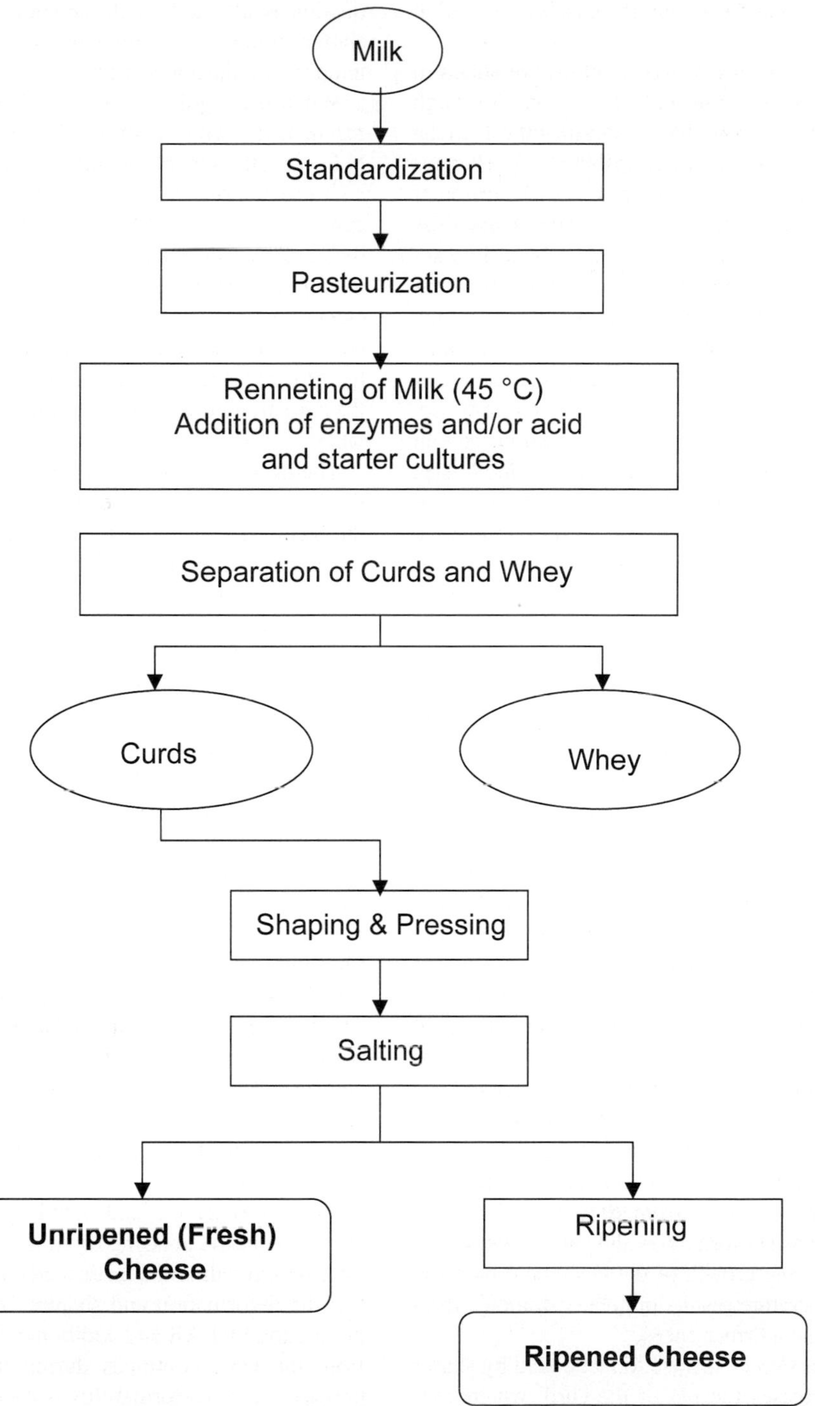

Figure 26.1. Processing scheme for natural cheeses.

skim milk in conjunction with heat treatment. Ricotta cheese is the most common cheese prepared in this manner.

Lactic acid bacteria (LAB) cultures are added to the milk in conjunction with the rennet. Although the LAB cultures do not have a significant role in the coagulation of casein, they contribute to the changes that occur during the ripening process. The different strains of starter cultures differ in such characteristics as growth rate, metabolic rate, phage interactions, proteolytic activity, and flavor promotion (Stanley 1998). Frequently, mixed-strain cultures, which contain unknown numbers of strains of the same species, are used. Mesophilic L cultures contain *Leuconostoc* species, including *Ln. mesenteroides* ssp. *cremoris* and *Ln. lactis,* while the main species in mesophilic D cultures are *Lactococcus lactis* ssp. *cremoris* with lesser amounts of *Lc. lactis* ssp. *lactis.* Mesophilic DL cultures would consist of both lactococci and leuconostocs. The thermophilic cultures consist of *Streptococcus thermophilus* and one of the following: *Lactobacillus helveticus, Lb. delbrueckii* ssp. *lactis* or *Lb. delbrueckii* ssp. *bulgaricus.*

Cutting the Coagulum Cutting the coagulum increases the drainage of the whey from the curds and contributes to a sharp decrease in the moisture content of the curds. Acid-coagulated and rennet-coagulated cheeses differ in the pH at which curd formation occurs and, subsequently, the pH at which the coagulum is cut. For acid-coagulated cheeses, curd formation occurs at pH 4.6, the isoelectric point of casein. Curd formation of rennet-coagulated cheeses occurs at higher pH, ranging from pH 6.3 to 6.6. Following the cutting of the coagulum, the curd and whey mixture is heated and agitated in a process called "scalding." The agitation is necessary to keep the curds suspended in the whey and to promote drainage of whey from the curds. The temperature during the scalding process is dependent on the type of the cheese and ranges from 20 to 55°C. The temperature affects gel formation and gel viscoelasticity and regulates the growth of the lactic acid bacteria. A high temperature results in greater drainage from the cheese and a firmer cheese.

The conversion of lactose to lactic acid by starter cultures decreases the pH of the curd, which contributes to the loss of whey from the curd and a decrease in moisture content. While the curds are in the whey, diffusion of lactic acid into the whey and lactose into the curds occurs. The rate of acid production is affected by the amount and type of the starter culture, the composition of the milk, and the temperature during acid production.

When the required acidity of the cheese curds is reached, the whey is drained to recover the curds. Following the separation of the curds from the whey, acid production continues at an increased rate, because of the lack of diffusion of the lactic acid from the curd. To minimize excess acid development following the drainage of the whey, some types of cheese include a washing step to reduce the lactose content. The moisture content of the curd is affected by the extent the coagulum is cut, the temperature of the curd after cutting, and agitation of the curd in the whey.

The handling of the curd following the cutting of the coagulum greatly affects the characteristics of the cheese. The cheddaring process, used in the manufacture of Cheddar, Colby, Monterey, and mozzarella cheeses, involves piling blocks of curd on top of each other, with regular turning to allow the curds to fuse together. In Colby and Monterey cheeses, vigorous stirring during the drainage of the whey inhibits the development of a curd structure and results in a softer cheese with a higher moisture content. Frequently, the whey may be partially drained off and replaced with water or salt brine to remove lactose and reduce the development of acidity, as is done in the processing of Gouda and Limburger cheeses.

Shaping and Pressing The resulting curds are shaped to form a coherent mass that is easy to handle. The curds are placed in a mold and are often pressed with an external force to cause the curds to deform and fuse. The pressure and time of pressing ranges from a few grams per square centimeter for a few minutes for moist cheeses to 200–500 g/cm^2 for up to 16–48 hours for cooked and hard cheeses. The temperature (20–27°C) and humidity (95% relative humidity) of the pressing room is controlled to optimize the growth of the lactic acid bacteria and facilitate the deformation and shaping of the curd. Acid production by LAB and additional drainage of whey from the curd continues during the shaping and pressing stage. Deformability is affected by the composition of the cheese, and it increases until the curds reach a pH of 5.2–5.3. Deformability also increases with an increase in moisture content and temperature.

Salting The salting step reduces the moisture content of the curd, inhibits the growth of starter bacteria, and affects the flavor, preservation, texture, and rate of ripening of the cheese. The final salt content of cheese ranges from 0.7% to 4% (2–10% salt in moisture content). The amount of salt, the method of application of the salt, and the timing of the salting is dependent on the specific type of cheese. The salt may be incorporated through (1) mixing with dry-milled curd pieces, (2) rubbing onto the surface of the molded cheese, or (3) immersing the cheese in a salt brine. Following the salting step, the salt diffuses into the interior of the cheese, with the subsequent displacement of whey. Depending on the size of the cheese block and the composition of the cheese, it may take from 7 days to over 4 months for the salt to equilibrate within the cheese.

Ripening Fresh, green cheese has a bland flavor and a smooth, rubbery texture. During the ripening process, the characteristic texture and flavor of the cheese develop through a complex series of biochemical reactions. Ripening starter cultures are selected to develop the texture and flavor characteristics of the specific cheese type. Enzymes released following lysis of the microorganisms catalyze the degradation of proteins, lipids, and lactose in the cheese. As the ripening time increases, the moisture content of the cheese decreases, and the intensity of the flavor increases. The resulting quality attributes of the finished cheese depend not only on the initial composition of the milk and the starter cultures used, but also on the water activity of the cheeses and the temperature, time, and humidity during the ripening period. Depending on the type of cheese, the ripening period can range from 3 weeks to more than 2 years.

The ripening temperature influences the rate of the microbial growth and enzyme activity during the process and the equilibrium between the biochemical reactions that occurs during ripening. Ripening temperatures generally range from 5 to 20°C, which is well below the optimum temperatures for microbial growth and enzyme activity. Soft cheeses are often ripened at 4°C to slow the biochemical processes. An increase in ripening temperature for hard cheeses reduces the ripening time necessary for flavor development, with a 5°C increase in ripening temperature reducing the ripening time 2 to 3 months. However, caution must be exercised in altering ripening temperatures since not all microorganisms and enzymes respond to temperature changes in the same manner, resulting in an imbalance in flavor characteristics (Choisy et al. 2000).

The growth of most of the starter bacteria added to the milk in the initial stages of cheese making is slowed as the pH of the cheese approaches 5.7 and following the addition of salt, but fermentation and the decrease in pH continue. The fermentation of lactose to lactic acid by the starter cultures provides an environment that prevents the growth of undesirable microorganisms through reduced pH and the formation of an anaerobic environment. The optimal activity for proteases is between pH 5.5 and 6.5 and that for lipases is between pH 6.5 and 7.5.

Protease and lipase activity during the ripening is probably most important to the development of the flavor and texture of the cheese. Enzymes of the starter bacteria, nonstarter lactic acid bacteria, and secondary cultures added during cheese making are most important in the development of the flavor and texture of the cheese during ripening. These enzymes are released by the lysis of the cell wall of the bacteria. Rennet enzymes and endogenous milk enzymes, such as plasmin, also contribute to these hydrolytic reactions during ripening. The extent of these enzymatic reactions depends on the activity and specificity of the enzymes, the concentration of the substrates, pH, water activity, salt concentration, and ripening temperature and duration. The degradation of the amino acids and fatty acids, through enzymatic and nonenzymatic reactions, result in the formation of several important volatile flavor compounds, including sulfur-containing compounds, amines, aldehydes, alcohols, esters, and lactones.

Rennet and plasmin are associated with the primary phase of proteolysis and hydrolyze the caseins to large polypeptides. This proteolysis alters the three-dimensional protein network of the cheese to form a less firm and less elastic cheese. Although these polypeptides do not have a direct impact on flavor, they do function as a substrate for the proteases associated with the starter and nonstarter bacteria. However, if the primary proteolysis is extensive, bitter peptides, with a high percentage of hydrophobic amino acids, predominate. Free amino acids and short-chain peptides contribute sweet, bitter, and brothy-like taste characteristics to the cheese. Further degradation and chemical reactions of these peptides and amino acids through the action of decarboxylases, transaminases, or deaminases contribute to the formation of amines, acids, ammonia, and thiols, which contribute to cheese flavor.

Lipases break down triglycerides into free fatty acids and mono- and diglycerides. The short-chain free fatty acids contribute to the sharp, pungent flavor characteristics of the cheese. The degree of lipolysis that is acceptable without producing soapy and rancid flavors depends on the type of cheese. Several *Penicillium* strains form methyl ketones, lactones, and unsaturated alcohols through their enzymatic systems associated with β-oxidation and decarboxylation, β-oxidation and lactonization, and lipoxygenase activity. Aliphatic and aromatic esters are synthesized by esterases present in a range of microorganisms, including mesophilic and thermophilic LAB (Choisy et al. 2000).

The texture of cheese is attributed to the three-dimensional protein network that entraps fat and whey. This structure is altered through proteolysis during ripening to form a less firm and less elastic cheese.

Carbon dioxide produced by the metabolism of the bacteria and entrapped within the curd results in the formation of eyes in several types of cheeses. The small eyes characteristics of Edam, Gouda, and related cheese varieties are formed by carbon dioxide produced from citrate by the *Leuconostoc* ssp. In Swiss type cheeses, *Propionibacterium freudenreichii* ssp. *shermanii* metabolize lactate to form the carbon dioxide responsible for the eye formation. These cheeses are ripened in hot rooms (20–22°C) to optimize the growth of the propionibacteria and allow eye formation.

Ripening of cheese is a time-intensive process, thus, there is much interest in accelerating the ripening process without an adverse effect on cheese quality. The addition of attenuated starters or a mixture of enzymes has been explored as a possible means to shorten the ripening processing. An important consideration is that increased proteolysis or an imbalance of enzyme activity can result in higher contents of hydrophobic amino acids and peptides that contribute to bitterness (Choisy et al. 2000).

Processed Cheese

Processed cheeses are produced by blending and heating several natural cheeses with emulsifiers, water, butterfat, whey powder, and/or caseinates to form a homogeneous mixture. The proportion of cheese used in the formulation ranges from 51% for cheese spreads and cheese foods to 98% in processed cheeses. Different types of natural cheeses will produce processed cheeses with different flavor and textural characteristics. The formulation of these ingredients affects the consistency, texture, flavor, and melting characteristics of the processed cheese. The heating process inactivates the microorganisms and denatures enzymes to produce a stable product. The flavor of the processed cheese is generally milder than the natural cheeses, due to the effects of the heat. However, the melting properties of the processed cheese are much improved due to the addition of the emulsifiers. Processed cheeses range in consistency from block cheese to sliced cheese to cheese for spreading (Banks 1998, Fox et al. 2000).

Butter

Flavor and consistency are the most important quality attributes to consider during the processing of butter. The characteristic flavor of butter is a result of the formation of diacetyl by heterofermentative LAB. Off-flavors due to lipolysis, oxidation, or other contaminants should be avoided. The butter should be firm enough to hold its shape, yet soft enough to spread easily. Figure 26.2 outlines the process involved for making butter and the by-product buttermilk.

The first step in the processing of butter is skimming the milk to obtain cream with a fat content of at least 35%, in which the fat is dispersed as droplets within the aqueous phase. This step increases the efficiency of the process through increasing the yield of butter, reducing the yield of buttermilk, and reducing the size of machinery needed. The function of the pasteurization step is to kill microorganisms, inactivate enzymes, and make the butter less susceptible to oxidative degradation. Following pasteurization, the cream is inoculated with a starter culture mixture (1–2%) of *Lactococcus* ssp. and *Leuconostoc* ssp., which contributes to the development of the characteristic butter flavor. It is desirable that the starter cultures grow rapidly at low temperatures. During ripening, the cream begins to ferment and the fat begins to crystallize. The formation of lactic acid and diacetyl by the starter cultures contributes to flavor development in the butter and buttermilk.

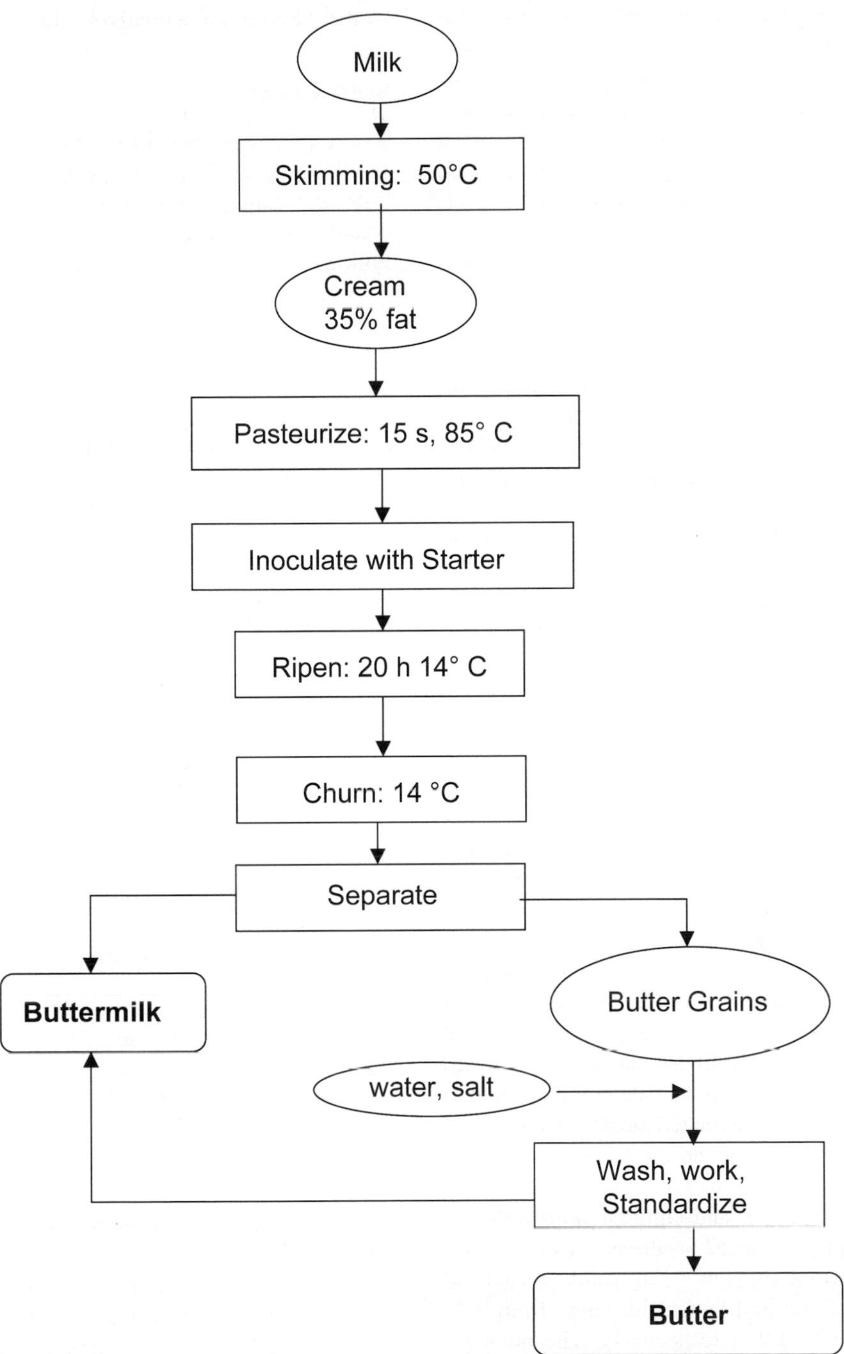

Figure 26.2. Processing scheme for butter processed from ripened cream.

Crystallization of the fat is important to maximize the yield of the butter and minimize loss of fat into the buttermilk during the churning process. During the churning process, air is incorporated into the cream using dashboards or rotary agitators, contributing to the aggregation of fat globules to form butter granules. The resulting butter grains may be washed with water to remove nonfat solids, followed by working or kneading the butter grains. During the working step, a water-in-oil emulsion is formed as the small water droplets are dispersed into the fat matrix.

The churning process is probably most critical to the textural quality of the butter. As air is incorporated into the cream, the fat globules surround the air bubbles and coalesce with other fat-coated air bubbles to form clumps. These clumps continue to coalesce during the churning process, and the volume of air bubbles decreases. The proportion of solid fat, which is influenced by temperature, is critical to the aggregation of fat clumps to form butter. If the rate of churning is too fast, the fat globules are less stable and less likely to coalesce, resulting in a greater loss of fat into the buttermilk.

Buttermilk

Natural buttermilk is a by-product of butter manufacture (Fig. 26.3) and is categorized as sweet (nonfermented) or acidic (fermented). Sweet and acid buttermilk are obtained from the manufacture of butter from nonfermented, sweet cream and sour or cultured cream, respectively. The fat content of natural buttermilk is approximately 0.4% and consists primarily of the membrane components of the fat globules. Because of the high content of unsaturated fatty acids associated with the membrane lipids, natural buttermilk has a characteristic flavor and is also more susceptible to the development of oxidized off-flavors.

Cultured buttermilk and cultured skim milk are more frequently produced as alternatives to the traditional buttermilk (Fig. 26.3). The solids-not-fat and fat content of cultured buttermilk range from 7.4–11.4% and 0.25–1.9%, respectively. The homogenized and pasteurized milk with the desired fat content is inoculated with a 1% starter culture of *Lc. lactis* ssp. *lactis* and *Lc. lactis* biovar. *diacetylactis* and *Ln. mesenteroides* ssp. *cremoris*. These bacteria metabolize citrate to diacetyl to contribute to the development of "buttery" flavor. Acid is produced by these bacteria at a relatively slow rate.

Sour Cream

Sour cream is produced from pasteurized, homogenized cream (20–30% fat) and has a pleasant acidic taste and buttery aroma (Fig. 26.4). Low-fat sour cream has a fat content ranging from 10 to 12%. Following standardization of the milk and cream to

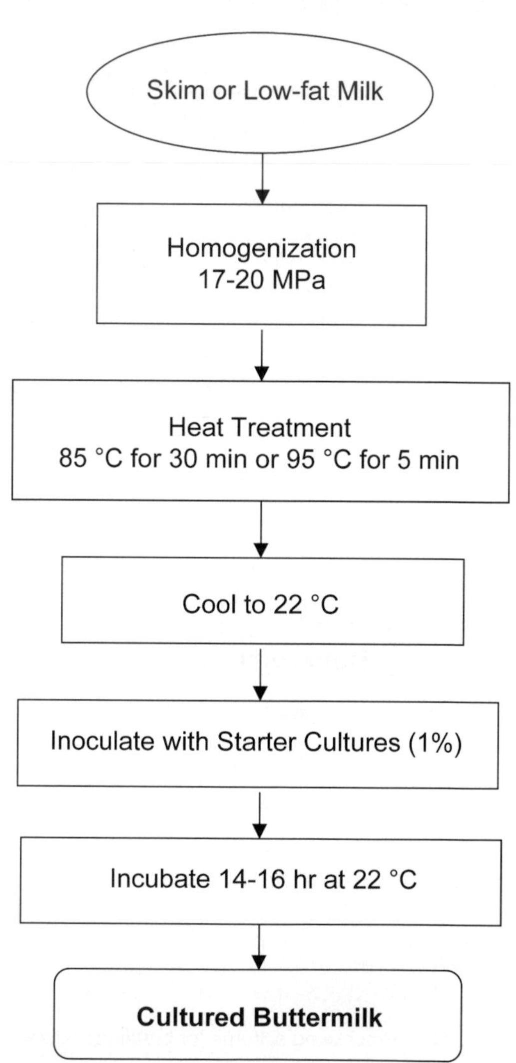

Figure 26.3. Processing scheme for cultured buttermilk.

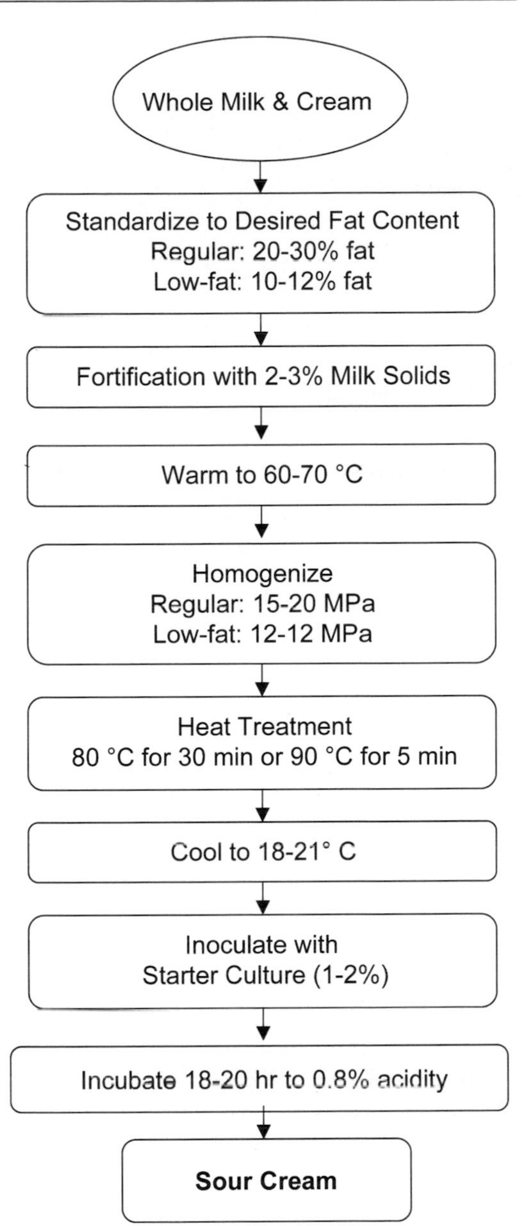

Figure 26.4. Processing scheme for sour cream.

the desired fat content, the mixture is warmed and homogenized to improve the consistency of the final product. The starter cultures include *Lc. lactis* ssp. *lactis* and *Lc. lactis* biovar. *diacetylactis* and *Ln. mesenteroides* ssp. *cremoris*. As in buttermilk, these cultures contribute to the acid and buttery flavor of the sour cream. The sour cream may be packaged prior to or following fermentation at 20°C until the pH is reduced to 4.5. Packaging prior to fermentation results in a thicker product because the gel is not disturbed. As the pH decreases, the fat clusters aggregate to form a viscous cream. Rennet or thickening agents are sometimes added to increase the firmness of the sour cream.

Yogurt

The tremendous increase in the popularity of yogurt in recent decades has been attributed to its health food image and the wide diversity of flavors, compositions, and viscosities available to consumers. The manufacturing methods, raw materials, and formulations vary widely from country to country, resulting in products with a diversity of flavor and texture characteristics. While in many Western societies, yogurt is produced from cows' milk, other mammalian milks can be used to produce yogurt. Figure 26.5 outlines the steps involved in the processing of stirred- and set-style yogurts.

The milk is initially standardized to the desired fat (0.5–3.5% fat) and milk solids-not-fat (12.5%) content. Increase in the protein content is most commonly achieved through the addition of nonfat milk powder, which improves the body and decreases syneresis of the final product. In addition to decreasing the size of the fat globule, homogenization of the milk alters the milk proteins to reduce syneresis and increase firmness (Tamime and Robinson 1999b). The heat treatment eliminates pathogenic microorganisms and reduces the oxygen in the milk to provide a good growth medium for the starter cultures. Enzymes and the major whey proteins, including β-lactoglobulin and α-lactalbumin, but not the casein proteins, are also denatured by the heat treatment. The denaturation of the whey proteins and subsequent interactions between the whey proteins and casein and/or fat globules improves the stability of the gel and decreases syneresis (Tamime and Robinson 1999b).

Following heat treatment, the milk is cooled to 43–45°C for inoculation of the starter cultures. The fast acid-producing thermophilic LAB *Streptococcus salivarius* ssp. *thermophilus* and *Lactobacillus delbrueckii* ssp. *bulgaricus* are the primary microorganisms used in the production of yogurt. These bacteria have a synergistic effect on each others'

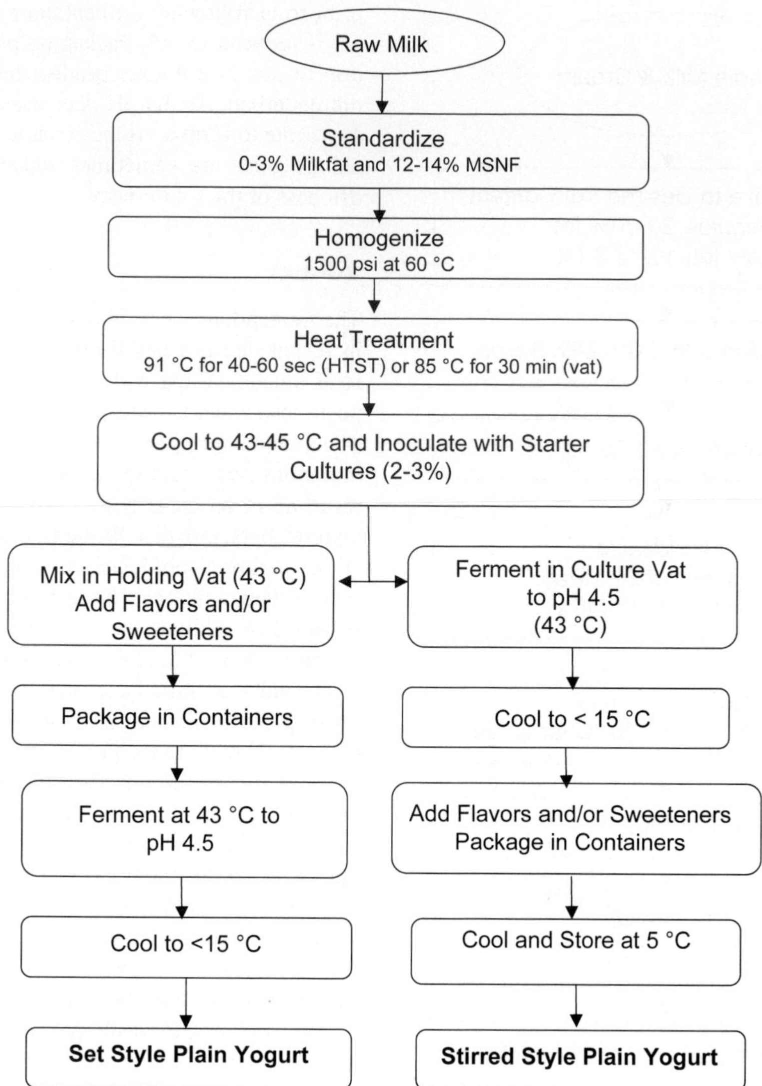

Figure 26.5. Processing scheme for set- and stirred-style yogurts.

growth and should be present in approximately equal numbers for optimal flavor development. *Lb. delbrueckii* ssp. *bulgaricus* has important protease activity and hydrolyzes the milk proteins to small peptides and amino acids. These peptides and amino acids enhance the growth of *S. thermophilus,* which has limited proteolytic activity. *S. thermophilus* metabolizes pyruvic acid to formic acid and carbon dioxoide, which in turn, stimulates the growth of *Lb. delbrueckii.* Initially, *S. thermophilus* grows faster than *Lb. delbrueckii*; however, at the later stages of the fermentation process, the growth of the *S. thermophilus* is inhibited by the reduced pH of the yogurt. The mutual stimulation of the yogurt cultures through their metabolic activity significantly increases the formation of lactic acid to a rate greater than would be possible by the individual cultures.

The acid production by the yogurt cultures results in the aggregation of casein micelles and gel formation. Two types of yogurt, set and stirred, are produced. Set-style yogurt is fermented following packaging, with color and flavors added to the container prior to the addition of the inoculated milk, resulting in a gel that forms a firm, unbroken coagulum. Stirred yogurt is fermented in a vat prior to packaging, with the gel structure broken following fermentation during the addition of flavors and colors and during the cooling and packaging stages. For both types of yogurt, either a short incubation period at 40–45°C for 2–3 hours or a long incubation period at 30°C for 16–18 hours may be used to attain a pH of 4.5 or titratable acidity of 0.9% lactic acid prior to cooling.

The metabolism of the yogurt cultures contributes to the characteristic flavor of yogurt. Both bacteria are heterofermentative and produce lactic acid from glucose, yet are unable to metabolize galactose. Acetaldehyde, a key flavor component of yogurt, is produced by the degradation of threonine to acetaldehyde and glycine by the enzyme threonine aldolase. Although *S. thermophilus* forms a majority of the acetaldehyde produced, the proteolytic activity of *Lb. delbrueckii* ssp. *bulgaricus* generates the precursors for the formation of acetaldehyde.

Syneresis, the expelling of interstitial liquid due to association of the protein molecules and shrinkage of a gel network, is undesirable in yogurt. Syneresis increases with an increase in incubation temperature. In stirred yogurt, extensive syneresis results in a thin product. Therefore, incubation of the yogurt at a lower temperature, such as 32°C, is recommended to maintain the desirable viscosity.

Acidophilus Milk

Acidophilus milk was developed as a fluid milk product that provides therapeutic benefits through the alleviation of intestinal disorders. Following homogenization and heat treatment of the milk to minimize contamination by other microorganisms, the milk is cooled to 37°C and inoculated with a 1% culture of *Lactobacillus acidophilus* (Fig. 26.6). The milk is incubated at this temperature for 18–24 hours until the titratable acidity reaches 0.63–0.72%. *Lb. acidophilus* is a thermophilic starter that grows slowly in milk and has a relatively high acid tolerance. To provide the desired therapeutic effects, the milk should contain 5×10^8 cfu/mL at the time of consumption (Tamime and Marshall 1997).

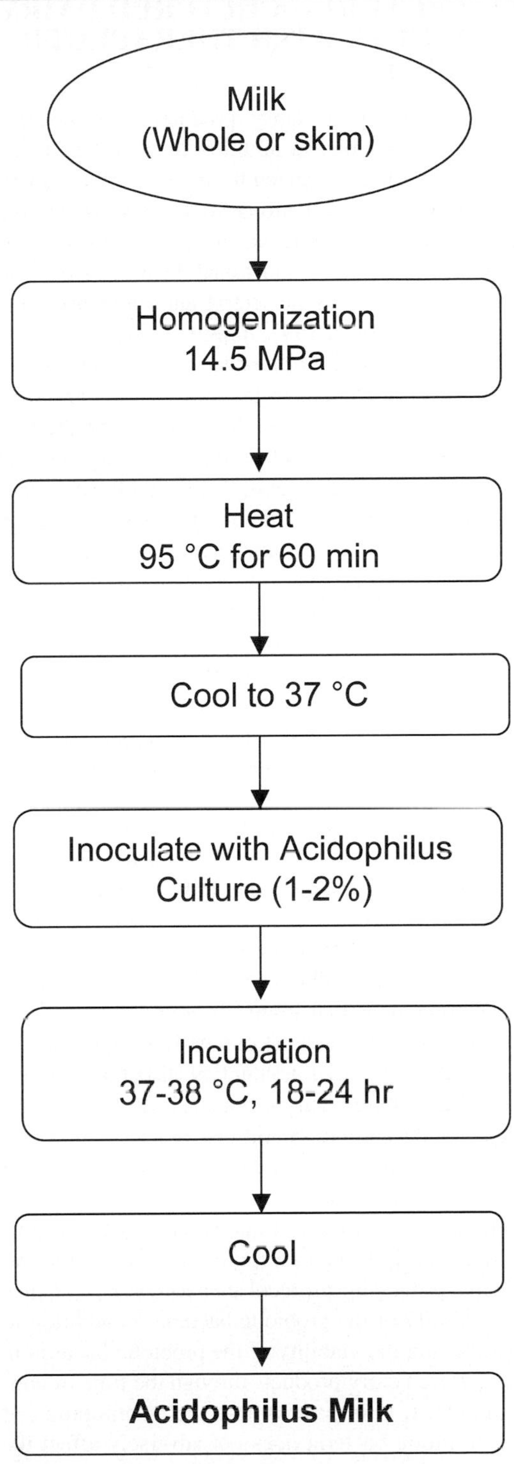

Figure 26.6. Processing scheme for acidophilus milk.

THE FUTURE—CULTURED DAIRY PRODUCTS WITH THERAPEUTIC BENEFITS

The availability of dairy products with positive nutritional benefits, in addition to yogurt and acidophilus milk, is expected to increase in the future. The growth of the functional foods area has led to a growing consumer interest in cultured dairy products with enhanced nutritional benefits from the incorporation of bifidobacteria and other probiotic bacteria (Robinson 1991, Tamime 2002). Probiotic bacteria, specifically bifidobacteria and lactobacilli, are normal inhabitants of the human colon and beneficially affect human health by improving the balance of intestinal microflora and improving mucosal defenses against pathogens. Additional health benefits include enhanced immune response, reduction of serum cholesterol, vitamin synthesis, anticarcinogenic activity, and antibacterial activity (Arunachalam 1999, Lourens-Hattingh and Viljoen 2001). A daily intake of at least 10^8 to 10^9 viable cells, which could be achieved with a daily consumption of at least 100 g of a product containing between 10^6 and 10^7 viable cells per gram, has been suggested as the minimum intake to provide a therapeutic effect (Gomes and Malcata 1999).

The most popular food delivery systems for probiotic cultures have been freshly fermented or unfermented dairy foods, including milk, yogurt, cheeses, ice cream, and desserts. Successful incorporation of these probiotic bacteria into cultured dairy products must recognize the challenges related to the instability of some intestinal strains of probiotic bacteria in cultured milk products (Lourens-Hattingh and Viljoen 2001). The environment typical of many yogurt and other cultured dairy products, including the low pH and the aerobic conditions of production and packaging, can result in decreases in the count of bifidobacteria in these dairy products to below the therapeutic minimum (Lourens-Hattingh and Viljoen 2001). In some cases, modifications of the traditional processing protocol are necessary to enhance the viability of the probiotic bacteria. In addition to maintaining the viability of the probiotic bacteria in the cultured dairy products through the time of consumption, it is imperative that the incorporation of the probiotic bacteria does not adversely affect the flavour, texture, and other quality attributes of the cultured dairy products. The development of cultured dairy products with improved therapeutic benefits will provide numerous challenges and rewards to food scientists.

ACKNOWLEDGMENT

This chapter of the Iowa Agriculture and Home Economics Experiment Station, Ames, Iowa, Project No. 3574, was supported by Hatch Act and State of Iowa funds.

REFERENCES

General References

Early R. 1998. The Technology of Dairy Products, 2nd edition. London: Blackie Academic and Professional. 446 pp.

Fox PF, Guinee TP, Cogan TM, McSweeney PLH. 2000. Fundamentals of Cheese Science. Gaithersburg, Maryland: Aspen Publishers, Inc.

Kurmann JA, Rasic JL, Kroger M. 1992. Encyclopedia of Fermented Fresh Milk Products: An International Inventory of Fermented Milk, Cream, Buttermilk, Wheyand Related Products. New York: Van Nostrand Reihnold. 368 pp.

Law BA. 1997. Microbiology and Biochemistry of Cheese and Fermented Milk, 2nd edition. London: Blackie Academic and Professional. 365 pp.

Tamime AY, Robinson RK. 1999. Yoghurt—Science and Technology, 2nd edition. Boca Raton, Florida: CRC Press. 619 pp.

Walstra P, Geurts TJ, Noomen A, Jellema A, van Boekel, MAJS. 1999. Dairy Technology: Principles of Milk Properties and Processes. New York: Marcel Dekker, Inc.

Cited References

Arunachalam KD. 1999. Role of bifidobacteria in nutrition, medicine and technology. Nutr Res 19:1559–1597.

Banks JM. 1998. Cheese. Chapter 3. In: R Early, editor. The Technology of Dairy Products, 2nd edition. London: Blackie Academic and Professional. Pp. 81–122.

Brulé G, Lenoir J, Remeuf F. 2000. The casein micelle and milk coagulation. Chapter 1. In: A Eck, J-C Gillis, editors. Cheesemaking—From Science to Quality Assurance, 2nd edition. Paris: Lavoisier Publishing. Pp. 7–40.

Canteri G. 2000. Lactic starters. Chapter 5.2 In: A Eck, J-C Gillis, editors. Cheesemaking—From Science to Quality Assurance, 2nd edition. Paris: Lavoisier Publishing. Pp. 164–182.

Cheryan M, Van Wik PJ, Olson NF, Richardson T. 1975. J Dairy Sci 58:477–481.

Choisy C, Desmazeaud M, Gripon JC, Lamberet G, Lenoir J. 2000. Chapter 4. The biochemistry of ripening. In: A Eck, J-C Gillis, editors. Cheesemaking—From Science to Quality Assurance, 2nd edition. Paris: Lavoisier Publishing. Pp. 82–151.

Fox PF. 1969. Milk-clotting and proteolytic activities of rennet, bovine pepsin, and porcine pepsin. J Dairy Res 36:427–433.

Fox PF, Guinee TP, Cogan TM, McSweeney PLH. 2000. Fundamentals of cheese science. Gaithersburg, Maryland: Aspen Publishers, Inc.

Fox PF, McSweeney PLH. 1997. Chapter 1. Rennets: Their role in milk coagulation and cheese ripening. In: BA Law, editor. Microbiology and Biochemistry of Cheese and Fermented Milk, 2nd edition. London: Blackie Academic and Professional. Pp. 1–49.

Gomes AMP, Malcata FX. 1999. *Bifidobacterium* spp. and *Lactobacillus acidophilus*: Biological, biochemical, technological, and therapeutical properties relevant for use as probiotics. Trends Food Sci Technol 10:139–157.

Kosikowski F. 1977. Cheese and Fermented Milk Foods, 2nd edition. Brooktondale, New York: F.V. Kosikowski and Associates. 711 pp.

Lourens-Hattingh A, Viljoen BC. 2001. Yogurt as probiotic carrier food. Int Dairy J 11:1–17.

Robinson RK. 1991. Microorganisms of fermented milks. In: RK Robinson, editor. Therapeutic Properties of Fermented Milks. London: Elsevier Applied Science. Pp. 23–43.

Stanley G. 1998. Chapter 2. Microbiology of fermented milk products. In: R Early, editor. The Technology of Dairy Products, 2nd edition. London: Blackie Academic and Professional. Pp. 50–80.

Tamime AY. 2002. Microbiology of starter cultures. In: RK Robinson, editor. Dairy Microbiology Handbook, 3rd edition. New York: John Wiley and Sons, Inc. Pp. 261–366.

Tamime AY, Marshall VME. 1997. Chapter 3. Microbiology and technology of fermented milks. In: BA Law, editor, Microbiology and Biochemistry of Cheese and Fermented Milk, 2nd ed. London: Blackie Academic and Professional. Pp. 57–152.

Tamime AY, Robinson RK. 1999a. Chapter 8. Preservation and production of starter cultures. In: Yoghurt—Science and Technology, 2nd edition. Boca Raton, Florida: CRC Press. Pp. 486–534.

———. 1999b. Chapter 2. Background to manufacturing practice. In: Yoghurt—Science and Technology, 2nd edition. Boca Raton, Florida: CRC Press. Pp. 11–128.

Walstra P. 1990. On the stability of casein micelles. J Dairy Sci 73:1965–79.

Walstra P, Jenness R. 1984. Dairy Chemistry and Physics. New York: Wiley and Sons.

27
Bakery and Cereal Products

J. A. Narvhus and T. Sørhaug

INTRODUCTION

Cereals are the edible seeds of plants of the grass family. They can be grown in a large part of the world and provide the staple food for most of mankind. Maize, wheat, and rice contribute about equally to 85% of world cereal production, which is at present about 2000 million tons (FAO 1999).

Cereals in their dry state are not subject to fermentation due to their low water content. Properly dried cereals contain less than 14% water, and this limits microbial growth and chemical changes during storage. However, on mixing grains or cereal flour with water or other water-based fluids, enzymatic changes occur that may be attributed to the enzymes inherent in the grain itself and/or to microorganisms. These microorganisms can either be those present as the natural contaminating flora of the cereal, or they can be added as a starter culture.

This chapter will be mainly devoted to fermented bakery products made from wheat. However, on a global basis, many fermented cereal products are derived wholly or in part from other grains such as rice, maize, sorghum, millet, barley, and rye. Different cereals differ not only in nutrient content, but also in the composition of the protein and carbohydrate polymers. The functional and sensory characteristics of products made from different cereals will therefore vary at the outset due to these factors. In addition to this, the opportunity to vary technological procedures and microbiological content and activity provides us with the vast range of fermented cereal products that are prepared and consumed in the world today.

CEREAL COMPOSITION

Carbohydrates are quantitatively the most important constituents of cereal grains, contributing 77–87% of the total dry matter. In wheat, the carbohydrate in the endosperm is mainly starch, whereas the pericarp, testa, and aleurone contain most of the crude and dietary fiber present in the grain. The pericarp, testa, and aleurone also contain over half the total mineral matter. Whole meal flour is derived, by definition, from the whole grain and contains all its nutrients. When wheat is milled into flour, the yield of flour from the grain (extraction rate) reflects the extent to which the bran and the germ are removed and thereby determines not only the whiteness of the flour, but also its nutritive value and baking properties. Decreasing extraction rate results in a marked, and nutritionally important, decrease in fiber, fat, vitamins, and minerals (Kent 1983). The protein content of different cereal grains varies between 7 and 20%, governed not only by cereal genus, species, or variety (i.e., genetically regulated), but also by plant growth conditions such as temperature, availability during plant growth of water and also of nitrogen and other minerals in the soil. There is an uneven distribution of different protein types in the different parts of the grain, so that although the protein concentration is not radically affected by milling, the proteins present in different milling fractions will vary.

STARCH

The starch in cereals is contained in granules that vary in size, from 2–3 μm to about 30 μm according to grain species. Barley, rye, and wheat have starch granules with a bimodal size distribution, with large lenticular and small spherical grains. Almost 100% of the starch granule is composed of the polysaccharides amylose and amylopectin, and the relative proportions of these polymers vary according not only to species of cereal, but also to variety within a species. However, both wheat and maize contain about 28% amylose, but in wheat the ratio of amylose to amylopectin does not vary (Fenema 1996; Hoseney 1998). The amylose molecule is essentially linear, with up to 5000 glucose molecules polymerized by α-1,4 linkages and only occasional α-1,6 linkages. Amylopectin is a much larger (up to 10^6 glucose units) and more highly branched molecule with approximately 4% of α-1,6 linkages which cause branching in the α-1,4 glucosidic chain.

During milling of grains, some of the starch granules become damaged, particularly in hard wheat. The starch exposed in these broken granules is more susceptible to attack by amylases and also absorbs water much more readily. The degree of damage to the starch grains therefore dictates the functionality of the flour in various baking processes. When water is added to starch grains, they absorb water, and soluble starch leaks out of damaged granules. Heating of this mixture results in an increase in viscosity and a pasting of the starch, which on further heating leads to gelatinization as the ordered crystalline structure is disrupted and water forms hydration layers around the separated molecules. The gelatinization temperature of starch from different cereals varies from 55 to 78°C, partly due to the ratio of amylose to amylopectin. The gelatinization of wheat starts at about 60°C. Despite the fact that there is not sufficient water to totally hydrate the starch in most bakery foods, the heat causes irreversible changes to the starch. On cooling a heated cereal product, some starch molecules reassociate, causing a firming of the product.

Nonstarch polysaccharides, the pentosans, which are principally arabinoxylans, comprise approximately 2–3% of the weight of flour. They are derived from the grain cell walls and are polymers that may contain both pentoses and hexoses. They are able to absorb many times their own weight in water and contribute in baking by increasing the viscosity of the aqueous phase, but they may also compete with the gluten proteins for available water. Cereals also contain small amounts (1–3%) of mono-, di- and oligosaccharides, and these are important as an energy source for yeast at the start of dough fermentation.

PROTEIN

Cereal proteins contribute to the nutritional value of the diet, and therefore the composition and amount of protein present are inherently important. However, the protein content in cereals also has several important aspects in fermented bakery products. The amount and type of some of the proteins is important for the formation of an elastic dough and for its gas-retaining properties. Other proteins in cereals are enzymes with specific functions, not only for the developing germ, but also for various changes that take place from the processing of flour to bakery products.

Gluten Proteins

The unique storage proteins of wheat are also the functional proteins in baking. The gliadins and glutenins, collectively called gluten proteins, make up about 80% of the total protein in the grain and are mostly found in the endosperm. These proteins have very limited solubility in water or salt solutions, unlike albumins and globulins (Fig. 27.1). A good bread flour (known as "strong" flour) must contain adequate amounts of gluten proteins to give the desired dough characteristics, and extra gluten may be added to the bread formulation.

Enzyme Proteins

The albumin and globulin proteins are concentrated in the bran, germ, and aleurone.

Amylases The primary function of starch-hydrolyzing enzymes is to mobilize the storage polysaccharides to readily metabolized carbohydrates when the grain germinates (Hoseney 1998).

α-amylase hydrolyzes α-1,4 glycosidic bonds at random in the starch molecule chain but is unable to attack the α-1,6 linkages at the branching points on the amylopectin molecule. The activity of α-amylase causes a rapid reduction in size of the large starch molecule, and the viscosity of a heated solution or slurry of starch is greatly decreased. It is most active on gelatinized starch, but granular starch is also slowly degraded.

β-amylase splits off two glucose units (maltose) at a time from the nonreducing end of the starch chain, thus providing a large amount of fermentable carbohydrate. β-amylase is also called a saccharifying enzyme since its action causes a marked increase in sweetness of the hydrated cereal. Neither the hydrolysis of amylopectin nor of amylose is completed by β-amylase, since the enzyme is not able to move past the branching points. The presence of both α- and β-amylases, however, leads to a much more comprehensive hydrolysis, since α-amylase produces several new reducing ends in each starch molecule.

The level of α-amylase is very low in intact grain but increases markedly on germination, whereas β-amylase levels in intact and germinated grain are similar.

Flour containing too much α-amylase absorbs less water and therefore results in heavy bread. In addition, the dough is sticky and hard to handle, and the texture of the loaf is usually faulty, having large open holes and a sticky crumb texture. However, some activity is required, and bakers may add amylase either as an enzyme preparation or as wheat or barley malt in order to slightly increase loaf volume and improve crumb texture. The thermal stability of amylases from different sources dictates their activity during the baking process. Microbial amylases with greater thermal stability have been used in bread to

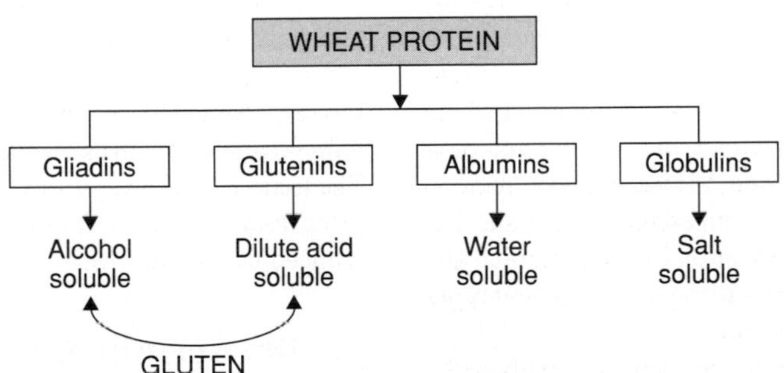

Figure 27.1. Wheat proteins.

decrease firming (retrogradation) upon storage since these enzymes are not fully denatured during baking.

Proteases Proteinases and peptidases are found in cereals, and their primary function is to make small amino nitrogen compounds available for the developing seed embryo during germination, when the levels of these enzymes also increase. However, whether these enzymes have a role in bread baking is not certain. Peptidases may furnish the yeast with soluble nitrogen during fermentation, and a proteinase in wheat that is active at low pH may be important in acidic fermentations such as sourdough bread.

Lipases Lipases are present in all grains, but oats and pearl millet have a relatively high activity of lipase compared with wheat or barley (Linko et al. 1997). In flour of the former grain types, hydrolytic rancidity of the grain lipids and added baking fat may be a problem.

Lipids

Lipids are present in grains as a large number of different compounds, and they vary from species to species and also within each cereal grain. Most lipids are found within the germ. Wheat flour contains about 2.5% lipids, of which about 1% are polar lipids (tri- and diglycerides, free fatty acids, and sterol esters) and 1.5% are nonpolar lipids (phospholipids and galactosyl glycerides). During dough mixing, much of the lipid forms hydrophobic bonds to the gluten protein (Hoseney 1998).

BREAD

Many different types of bread are produced in the world. Bread formulations and technologies differ both within and between countries due to both traditional and technological factors including (1) which cereals are traditionally grown in a country and their suitability for bread baking, (2) the status of bread in the traditional diet, (3) changes in lifestyle and living standards, (4) globalization of eating habits, and (5) economic possibilities for investing in new types of bread-making equipment.

The basic production of most bread involves the addition of water to wheat flour, yeast, and salt. Other cereal flours may be blended into the mixture, and other optional ingredients include sugar, fat, malt flour, milk and milk products, emulsifiers, and gluten (for further ingredients and their roles in bread, see Table 27.1). The mixture is worked into an elastic dough that is then leavened by the yeast to a soft and spongy dough that retains its shape and porosity when baked. An exception to this is the production of bread containing 20–100% rye flour, where the application of sourdough and low pH are required.

Bread Formulation

The formulation of bread is determined by several factors. In a simple bread, the baking properties of the flour are of vital importance in determining the characteristics of the loaf using a given technology. In addition, the bread obtained from using poor bread flour or suboptimum technology may be improved by using certain additives (Table 27.1).

The major methods used to prepare bread are summarized in Figure 27.2. In the straight dough method, all the ingredients are added together at the start of the process, which includes two fermentation steps and then two proofing steps. In the sponge and dough method, only part of the dry ingredients are added to the water, and this soft dough undergoes a fermentation of about 5 hours before the remainder of the ingredients are added and the dough is kneaded to develop the structure. Although these processes are time consuming, their advantages are that they develop a good flavor in the bread and that the timing and technology of the processes are less critical (Hoseney 1998). Mechanical dough development processes, such as the Chorleywood bread process developed in the United Kingdom in the 1960s, radically cut down the total bread-making time. The fermentation step is virtually eliminated, and dough formation is achieved by intense mechanical mixing and by various additives that hasten the process (Kent 1983). The resulting loaf has a high volume and a thin crust but lacks flavor and aroma. The trend is now away from this kind of process due to customer demand for more flavorful bread and reduced use of additives.

The Development of Dough Structure

When wheat flour and water are mixed together in an approximately 3:1 ratio and kneaded, a viscoelas-

Table 27.1. Bread Additives

Dough Additive	Role
Cysteine. Sodium sulphite and metabisulphite[a]	Reducing agent. Aids optimal dough development during mixing by disrupting disulphide (–S–S–) bonds. A "dough relaxer."
Amylase	Releases soluble carbohydrate for yeast fermentation and Maillard browning reaction. Reduces starch retrogradation.
Ascorbic acid	Oxidizing agent. Strengthens gluten and increases bread volume by improving gas retention.
Potassium iodate, calcium iodate; calcium peroxide; azodicarbonamide[a]	Fast-acting oxidants; oxidizes flour lipids, carotene and converts sulphydryl (–SH) groups to disulphide (–S–S–) bonds.
Potassium and calcium bromates[a]	Delayed-acting oxidants: Develops dough consistency, reduces proofing stage.
Emulsifiers, strengtheners/ conditioners and crumb softeners	Dispersion of fat in the dough. Increase dough extensibility. Interact with the gluten-starch complex and thereby retard staling.
Soy flour	Increases nutritional value, bleaches flour pigments, increases in loaf volume, increases crumb firmness and crust appearance, promotes a longer shelf life.
Vital wheat gluten and its derivatives	Increases gluten content, used especially when mixing time or fermentation time is reduced. Water adsorbant. Improves dough and loaf properties.
Hydrocolloids: Starch-based products from various plants	Regulates water distribution and water-holding capacity and thereby improves yield. Strengthens bread crumb structure and improves digestibility.
Cellulose and cellulose-based derivatives	Source of dietary fiber.
Salt	Enhances flavor (ca. 2% based on flour weight) and modifies mixing time for bread and rolls. Increases dough stability, firmness and gas retention properties. Raises starch gelatinization temperature

Source: Compiled from Stear 1990, Williams and Pullen 1998.
[a]Not allowed in all countries.

tic dough is formed that can entrap the gas formed during the subsequent fermentation. The amount of water absorbed by the flour is dependent upon, and therefore must be adjusted to, the integrity of the starch granules and the amount of protein present. A high proportion of damaged granules, as found in hard wheat flour, results in greater water absorption. The unique elastic property of the dough is due to the nature of the gluten proteins. Hydrated gliadin is sticky and extensible, whereas glutenin is cohesive and plastic. When hydrated during the mixing process, the gluten proteins unfold and bond to each other by forming a complex (gluten) as kneading proceeds, with an increasing number of cross-linkages between the protein molecules as they become aligned. Disulphide bonds (-S-S-) break and re-form within and between the protein molecules during mixing.

Gluten does not form spontaneously when flour and water are mixed; energy must be provided (i.e., in the actual mixing process) in order for the molecular bonds to break and re-form as the gluten structure. At this point, the dough stiffens and becomes smooth and shiny. The gluten is now composed of protein sheets in which the starch granules are embedded. In addition, free polar lipids and glycolipids are incorporated in the complex by hydrophobic and hydrogen bonds. The properties of the dough are determined by the amount of protein present and by the relative proportions of the gluten proteins.

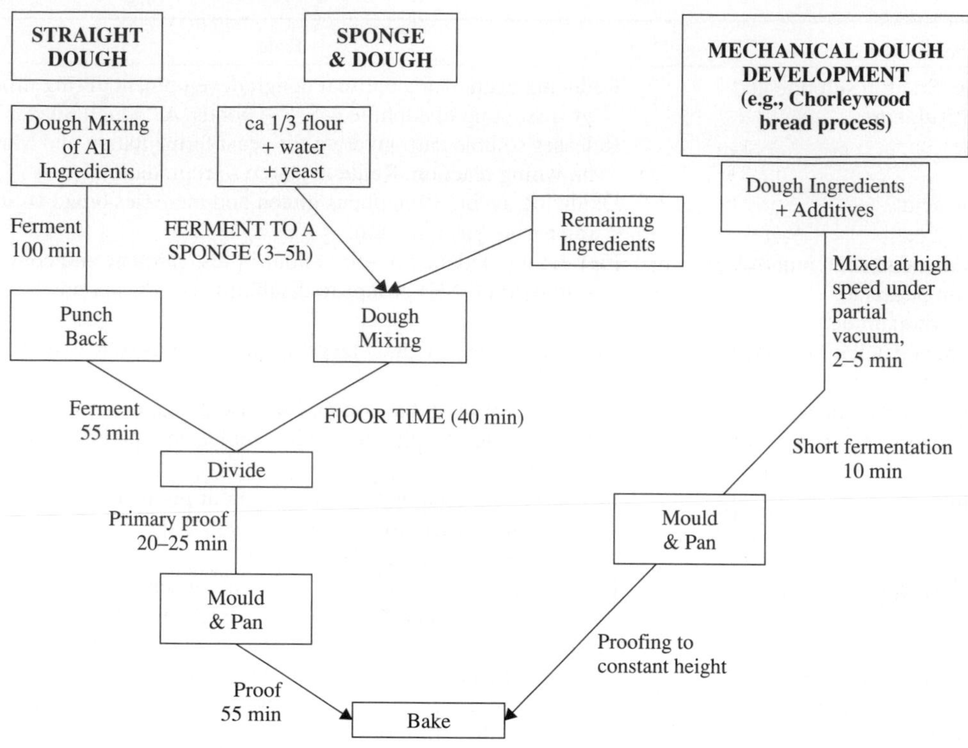

Figure 27.2. Bread-processing methods. (Adapted from Hoseney 1994.)

Another important part of the dough formation is the incorporation of air, in particular nitrogen. This forms insoluble bubbles in the dough that become weak points where carbon dioxide collects during the subsequent fermentation step. In the Chorleywood bread process, the dough is mixed under partial vacuum so that the incorporated bubbles expand and are then split into many small ones as mixing continues, thus giving a fine-pored loaf crumb after baking.

Dough Fermentation

During the fermentation step, several processes happen simultaneously, and in order to produce a bread of the required quality characteristics, each of these processes must be optimized to that end.

Yeasts have been used to leaven bread for thousands of years, but only in comparatively recent times have pure cultures of the yeast *Saccharomyces (S.) cerevisiae* been added to the bread dough as a leavening agent. The commercial production of baker's yeast follows procedures similar to those used in the production of brewing, wine-making, and distilling strains of this same species. Indeed, the baking industry was originally supplied with yeast waste from the brewing industry until about 1860 (Ponte and Tsen 1987). However, commercial production of yeast biomass specifically for the baking industry developed alongside an increasingly expanding manufacture of bread in commercial bakeries and the development of the technology that provided the great volumes required by the industry.

Commercial Production of Baker's Yeast

Saccharomyces cerevisiae was originally produced commercially using grain mash as a growth substrate, but for economic reasons, it is now grown on sucrose-rich molasses, a by-product from the sugar cane or sugar beet refining industry. Nitrogen, phosphorous, and essential mineral ions such as magnesium are added to promote growth. The production of the yeast biomass for the baking industry is multistage and takes about 10–13 hours at 30°C. *S. cerevisiae* shows the Crabtree effect, as its metabolism favors fermentative metabolism at high levels of energy-giving substrate, thus resulting in a low production of biomass (Walker 1998a). To avoid this, molasses is added incrementally towards the end of the production of yeast biomass, and the mixture is vigorously aerated in order to promote respiration and avoid fermentative metabolism. At the end of the production, the yeast is allowed to "ripen" by aeration in the absence of nutrients. This step synchronizes the yeast cells into the stationary growth phase and also promotes an increase in the storage sugar trehalose in the cells, thus improving their viability and activity.

When the fermentation is complete, the amount of yeast is about 3.5–4% w/v. The biomass is separated and concentrated by centrifugation and filtering. The yeast cream is then processed into pressed yeast or is dried. The most usual types of commercial yeast preparations are (Stear 1990) the following:

- *Cream yeast* is a near liquid form of baker's yeast that must be kept at refrigerated temperatures. It may be added directly to the bakery product being made.
- *Compressed yeast* is formed by filtering cream yeast under pressure to give approximately 30% solids. It has a refrigerated shelf life of 3 weeks.
- *Active dry yeast* (ADY) is produced by extruding compressed yeast through a perforated steel plate. The resulting thin strands are dried and then broken into short lengths to give a free-flowing granular product after further drying. Depending on the subsequent treatment and packaging, ADY may have a shelf life of over a year. However, ADY requires rehydration before application in dough, and this can be a labor-intensive operation in a large bakery. The product rehydrates best using steam or in water with added sugar at 40°C. Rehydration in pure water promotes leaching of cell contents and a reduction in the activity of the yeast.
- *High activity dry yeast* (HADY) (instant active dry yeast, IADY) is a similar product, where improved drying techniques are used to give a product with smaller particle size that does not need to be rehydrated before use and can therefore be incorporated directly into bread dough without prior treatment.

Desirable Properties of Baker's Yeast

Yeast plays a critically important role in leavened bread production, and over the decades of commercial production, strains have been selected that give improved performance. Desirable characteristics include

- High CO_2 production during the dough fermentation due to high glycolytic rate;
- The ability to quickly commence maltose utilization when the glucose in the flour is depleted;
- The ability to store high concentrations of trehalose, which gives tolerance to freezing and to high sugar and salt concentrations;
- Tolerance to bread preservatives such as propionate; and
- Viability and retained activity during various storage conditions.

In the future, strains will probably be developed with even more useful properties. In particular, the flavor-forming properties will receive special attention (Walker 1998b).

The Role of Yeast in Leavened Bread

When yeast is incorporated into the dough, conditions allow a resumption of metabolic activity, although there is little actual multiplication of the yeast during shorter bread-making processes such as the straight dough and Chorleywood processes (Fig. 27.2). The yeast has been produced under aerobic (respiratory) conditions and is therefore adapted to this metabolism, but conditions very quickly become anaerobic in bread dough since the oxygen incorporated in the dough is soon depleted. The sugars are metabolized to pyruvate by glycolysis; pyruvate is then decarboxylated to acetaldehyde, thus producing carbon dioxide; and then ethanol is formed by reduction of acetaldehyde by $NADH_2$ (Fig. 27.3). For each molecule of glucose (or half molecule of maltose) that is metabolized, two molecules each of ethanol and carbon dioxide are produced. This fermentative metabolism is the prevalent pathway in *S. cerevisiae* in dough due both to

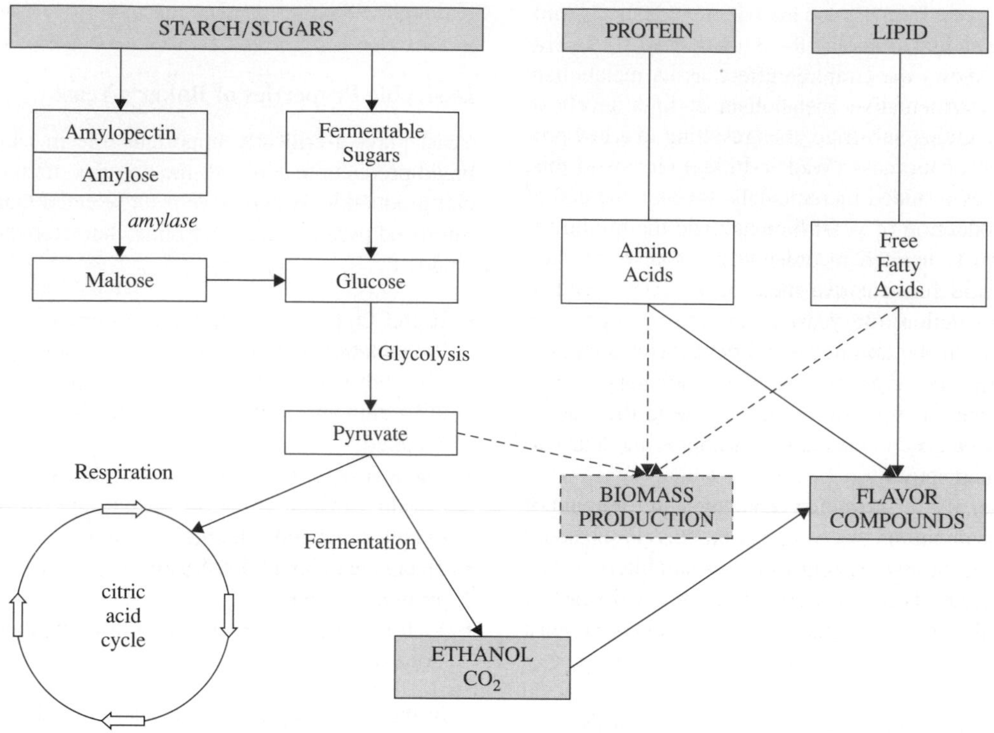

Figure 27.3. Biochemical changes during yeast fermentation of bread.

the absence of oxygen and to the nonlimiting supply of fermentable sugars (Maloney and Foy 2003).

The amount of maltose available is a complex interaction between the amount of damaged starch, the level of amylases in the flour and the stage and length of the fermentation process. Maltose accumulates during the early stages of the fermentation because it is generated by amylase but is not metabolized by the yeast because the presence of glucose represses maltose utilization.

When readily fermentable sugars (glucose and fructose) are exhausted, the yeast shows a lag in fermentation and then turns its metabolism to the maltose produced from the action of β-amylase on starch. If sucrose has been added in the bread formulation (e.g., 4%), this is fermented in preference to maltose, and the lag in the fermentation may not be observed. High amounts of added sucrose (e.g., 20%) significantly retard fermentation due to the high osmotic stress on the yeast (Maloney and Foy 2003).

The products of yeast metabolism in dough fermentation vary considerably with pH. In bread, the pH is usually below 6.0, but above this, end products in addition to ethanol and CO_2 are formed, such as succinate, acetic acid, and glycerol, and less ethanol and CO_2 are formed. *S. cerevisiae* is also able to degrade proteins and lipids, and several flavor compounds are produced (Fig. 27.3).

It is generally not considered that the yeast fermentation is important for bread flavor and aroma development in traditional bread processes. However, the modern mechanical dough development processes, where the fermentation stage has been radically reduced, produce bread with a flavor that is inferior to that produced by the traditional straight dough process. This indicates that the yeast fermentation does make a positive contribution to bread flavor (Stear 1990). Zehentbauer and Grosch (1998) showed that yeast level and fermentation time and temperature affected aroma in the crust of baguettes, and they identified the flavor compounds 2-acetyl-1-

pyrroline (roasty), methyl propanal and 2- and 3-methylbutanal (malty), and 1-octene-3-ol and (E)-2-nonenal (fatty). An increase in fermentation time allows for a development of flavor, but this trend is not really noticeable until much longer fermentation times are used, as in sourdough breads. There is not a clear borderline between regular bread and sourdough bread.

The production of ethanol and CO_2 is essential for the development of the desired bread crumb structure, and several factors affect both the development of the dough and its leavening (Fig. 27.4). During fermentation, some of the CO_2 is lost to the atmosphere, but most either collects in the small pockets of air incorporated during dough mixing or is dissolved in the dough's aqueous phase. The amount that can be dissolved in the aqueous phase is dependent on temperature, and is greater at lower temperatures. As the aqueous phase is already saturated with CO_2, it cannot escape from the bubbles by diffusion into the dough, so the bread begins to increase in volume. As the gas collects, the rheological properties of the dough allow it to expand in order to equalize the pressure that builds up. Ethanol reacts with the gluten to slightly soften it, allowing for easier expansion of the dough. It is important that CO_2 develop immediately after dough preparation and proceed at an adequate intensity. In addition, the dough must have the physical properties necessary to withstand dough manipulation and allow for gas retention, so that the optimal structure has been obtained for the final proof and baking (Stear 1990).

The Bread-Baking Process

When the bread has undergone the final proofing and is put in the oven, the outer surface rapidly starts to form the crust. A temperature gradient develops due to transfer of heat from the pan to the loaf, and if the loaf is to achieve optimal properties, then the heat of the oven and the state of the bread proof need to be synchronized (Stear 1990). Apart from the outer crust, no part of the bread ever becomes dry; therefore, despite oven temperatures of well over 200°C, the temperature in most of the loaf will not exceed 100°C. The primary rise in temperature increases the activity of the yeast, and its production of CO_2. At the same time, the solubility of CO_2

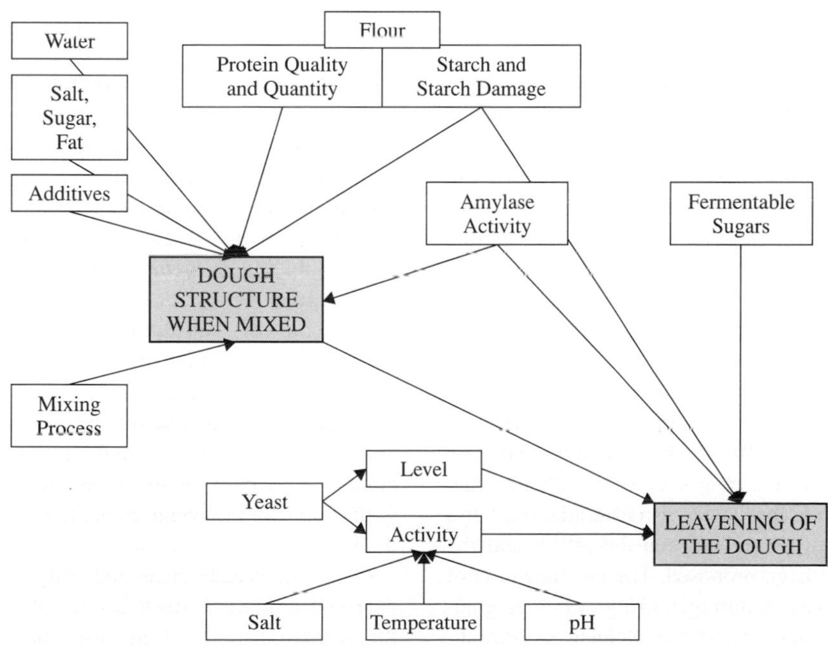

Figure 27.4. Important factors for bread leavening.

decreases, ethanol and water evaporate, and the gases increase in volume. This results in a marked increase in the volume of the dough, called "oven spring."

As the temperature in the loaf continues to rise, several other changes take place. The yeast is increasingly inhibited, and its enzymes are inactivated at about 65°C. The amylases in the dough are active until about 65–70°C is reached and a rapid increase in the amount of soluble carbohydrate takes place. Gelatinization of starch occurs at 55–65°C, and the water that this requires is taken from the gluten protein network, which then becomes more rigid, viscous, and elastic until a temperature is reached at which the protein begins to coagulate. At this stage, the structure of the dough has changed to a more rigid structure due to denatured protein and gelatinized starch. These changes first occur near the crust and gradually move into the crumb as the heat is transferred inwards.

Towards the end of baking, the temperature at the crust is much higher than 100°C. The crust becomes brown, and aroma compounds, predominantly aldehydes and ketones, are formed, mainly from Maillard reactions. The formation of flavor compounds is two staged. First, compounds are formed from the fermentation itself, and then during baking some of these compounds may react with each other or with the bread components to form other flavor compounds. Some other flavor compounds formed during the fermentation may be lost due to the high temperature, but those that remain gradually diffuse into the crumb after cooling. On further storage, the levels of flavor compounds decrease due to volatilization. Over 200 different flavor and aroma compounds have been identified in bread (Stear 1990).

Staling

The two main components of staling (firming of the bread crumb) are loss of moisture, mainly due to migration of moisture from the crumb to the crust (which becomes soft and leathery), and the retrogradation of starch. The major chemical change that occurs during staling is starch retrogradation, but a redistribution of water between the starch and the gluten also has been proposed. The gelatinization of the starch that occurs during cooking or baking gradually reverses, and the starch molecules form intermolecular bonds and crystallize, expelling water molecules and resulting in the firming of crumb texture. Starch retrogradation is a time- and temperature-dependent process and proceeds fastest at low temperatures, just above freezing point. Since the rearrangement of the starch molecules is facilitated by a high water activity, staling is retarded by the addition of ingredients that lower water activity (e.g., salt and sugar) or bind water (e.g., hydrocolloids and proteins). The staling rate can also be slowed by the incorporation of surfactants, shortening, or heat-stable α-amylase. Freezing of baked goods also retards staling since the water activity is drastically lowered. Much of the firming of the loaf during cooling is due to retrogradation of the amylose whereas the slower reaction of staling is due, in addition, to retrogradation of amylopectin (Hoseney 1998, Stear 1990).

SOURDOUGH BREAD

When cereal flour does not contain gluten, it is not suitable for production of leavened bread in the manner described above. However, if rye flour, which is very low in gluten proteins, is mixed with water and incubated at 25–30°C for a day or two, there is a good possibility that first step of sourdough production will be started. This mixture will regularly develop fermentation with lactic acid bacteria (LAB) and yeasts. This forms the basis of sourdough production, and this low-pH dough is able to leaven.

The use of cereal flour and water as a basis for spontaneous or directed fermentation products is common in many countries. In Africa, fermented porridge and gruel as well as their diluted thirst quenchers, are the main products of these natural fermentations, whereas Europeans and Americans and their descendants enjoy a variety of sourdough breads. In all these areas beers are also produced.

This great variety of fermented products has an historic prototype in the earliest reported leavened breads in Egypt about 1500 BC. Considering the simplicity of the process and the ease with which it succeeds, it has been suggested that peoples in several places must have shared this experience independently. It may be surmised that the experience with gruels and porridge preceded the idea of making bread.

Common bread fermented only with yeast appeared later in our history, and it was a staple food in the Roman Empire. This also indicates that the Romans had wheat with sufficient gluten potential.

It is possible to make leavened bread without gluten using sourdough, and this bread has become a favorite among many peoples (Hammes and Gänzle 1998, Wood 2000).

Advantages of Making Sourdough Bread

- Sourdough bread does not have to contain high levels of gluten for successful leavening.
- Low pH inhibits amylase, and thereby, degradation of starch is avoided.
- Sourdough improves the water-binding capacity of starch and the swelling and solubility of pentosans.
- Sourdough bread has very good keeping quality and an excellent safety potential.
- Less costly cereal flours can be used.
- A different variety of flavor and taste attributes can be offered.
- Sourdough bread can nutritionally compete with regular bread.
- Phytic acid is degraded by phytase in flour and from lactic acid bacteria. This improves the availability of iron and other minerals.
- Bread volume is increased, crumb quality is improved, and staling is delayed.

Rye flour is very low in gluten proteins, and instead, starch and pentosans make an important contribution to bread structure. The swelling and solubility of pentosans increase when LAB fermentation lowers pH. Gelatinization of starch occurs at about 55–58°C. Considering that the flour amylase has a temperature optimum around 50–52°C, it is crucial that the amylase is actually inactivated in the pH range that is obtained during sourdough fermentation. When mixtures of wheat and rye flour are used for bread making, a sourdough process is necessary if the content of rye flour exceeds 20%.

Rye and wheat flour contain phytic acid that binds minerals, particularly iron, that then become nutritionally unavailable. However, these cereals also contain phytases with pH optima around 5.0–5.5; thus phytate degradation is very good in fermented flour, where these phytate complexes are also more soluble. Lactic acid bacteria also appear to have some phytase activity.

As wheat flour is able to form gluten, some of the considerations about amylases and starch are not equally relevant when baking wheat bread. Nevertheless, wheat flour is often used in sourdough bread.

However, preferred qualities like improved keeping and safety potential as well as the increased variety of flavors appeal to many consumers. These desirable qualities are also praised because they represent an alternative natural preservation method (Gobbetti 1998, Hammes and Gänzle 1998, Wood 2000).

Microbiology of Sourdough

An established, "natural" sourdough is dominated by a few representatives of some bacteria and yeast species. This results from the selective ecological pressures exerted in the (rye) flour-water environment. Rye flour is an appropriate choice for this mixture because leavening of the dough is dependent on sourdough development. At the start of fermentation, a 50:50 (w/w) rye flour–water mixture at 25–30°C will harbor approximately the following:

Mesophilic microorganisms, aerobes	10^3–10^7 cfu/g
Lactic acid bacteria	< 10 to 5×10^2 cfu/g
Yeast	10–10^3 cfu/g
Molds	10^2 to 5×10^4 cfu/g

Among the mesophiles at the start, members of the Enterobacteriaceae dominate. Microorganisms dominating in the sourdough after 1–2 days at 25–30°C are as follows:

Lactic acid bacteria	10^9 cfu/g
Yeast	10^6 to 5×10^7 cfu/g

Some important properties of the rye flour–water environment determine that certain LAB and yeasts will compete most favorably. Lactobacilli have a superior ability to ferment maltose, they thrive despite limited iron due to the presence of phytic acid, and they are able to grow at about pH 5.0 and lower.

Reports show that different LAB may be isolated from sourdoughs; however, *Lactobacillus sanfranciscencis* has been found most often. Table 27.2 presents some of the other *Lactobacillus* species that have been isolated from sourdough. The selection of yeasts may be even narrower, with *Candida milleri* often cited (Table 27.2). Several other species are isolated occasionally (Spicher 1983, Gobbetti and Corsetti 1997, Hammes and Gänzle 1998, Martinez-Anaya 2003, Stolz 2003).

Starters

It is traditional bakers' practice to maintain a good sourdough over time by regular transfer, for example,

Table 27.2. Common Representatives of Lactic Acid Bacteria and Yeasts Isolated from Mature Sourdoughs

	Heterofermentative	Homofermentative
Lactic acid bacteria	*Lactobacillus sanfranciscencis* (formerly *Lb. sanfrancisco, Lb. brevis* subsp. *lindneri*) *Lb. brevis, Lb. fructivorans, Lb. fermentum, Lb. pontis, Lb. sakei* (formerly *Lb. bavaricus, Lb. reuteri*)	*Lb. plantarum* *Lb. delbrueckii*
Yeasts	*Candida milleri, C. krusei, Saccharomyces cerevisiae, S. exiguus Torulopsis holmii, T. candida*	

every 8 hours. This is called "rebuilding." Such established cultures are referred to as Type I sourdoughs (Type I process), and a three-stage fermentation procedure is considered necessary to obtain an optimal sourdough. Each step is defined by specific dough yield, temperature, and incubation time.

Dough yield is defined as:

$$\frac{\text{(Flour + Water) by weight}}{\text{Flour by weight}} = \times 100 \text{ Dough yield}$$

A high dough yield implies that a relatively large amount of water is used to make the dough; such a dough would conform with certain requirements in industrial production when there is a need to pump the dough.

The lactobacilli and yeasts in Table 27.2 are all common in sourdoughs; however, the composition of starter cultures for Type I processes have been continuously stably maintained for many years, and may be compared to certain mixed cultures in dairy technology. The LAB and yeasts in such cultures will be particularly well adjusted and adapted for the conditions in sourdoughs. In Germany, established natural sourdough starter cultures with a stable composition of *Lb. sanfranciscensis* and *Candida milleri* have been propagated for decades; they are marketed as "Reinzuchtsauerteig." When the sourdough process has been started through a one-stage or a three-stage procedure, a part of the optimized dough is withdrawn to start up the sourdough production for the next day. These sourdoughs are mentioned in different languages as Anstellgut (German), mother sponge (English), chef (French), masa madre (Spanish), madre (Italian).

Sourdoughs in Type II processes are used mainly for enhancing the flavor and taste of the regular bread. Addition of baker's yeast is required for efficient leavening. Type I sourdoughs are good alternatives as starters for the production of Type II sourdoughs. At the start of a production period, industrially large quantities of Type II sourdoughs may be stocked for portion-wise use over time.

Sourdoughs for Type III processes are dried sourdough preparations.

Defined cultures have also been marketed; however, the suitability of Type I and Type II cultures appears to outcompete the alternatives offered. In addition to baker's yeast the collection of defined cultures comprise at least pure cultures of *Lb. brevis* (heterofermentative) and *Lb. delbrueckii* and *Lb. plantarum* (homofermentative). The homofermentative cultures produce mainly lactic acid under anaerobic conditions, whereas the heterofermentative cultures will also produce acetic acid or ethanol and carbon dioxide. By controlling, if possible, the contributions from these different cultures, the relative amounts of acetic and lactic acids may be regulated. This important relationship

$$\frac{\text{Lactic acid (Mole)}}{\text{Acetic acid (Mole)}}$$

is called the fermentation quotient (FQ).

Relatively mild acidity will have an FQ about 4–9, whereas a more strongly flavored rye bread, as produced in Germany, requires a much lower FQ, for example, 1.5–4.0 (Spicher 1983, Gobbetti and Corsetti 1997, Hammes and Gänzle 1998, Martinez-Anaya 2003, Stolz 2003).

Sourdough Processes

Several more or less traditional sourdough processes are practiced on a large scale in present-day bakery industry. One line comprises processes designed for baking with rye flour. They may be the Type I processes mentioned above. The Berliner short-sour process, the Detmolder one-stage process, and the Lönner one-stage process are typical for central and northern Europe. In every case, the process is initiated by a starter culture, 2–20% (often 9–10%), in a rye flour–water (close to 50:50 w/w) mixture that is incubated for 3–24 hours at 20–35°C, depending on the process. When the resulting sourdough is ready, bread making starts with an "inoculation" of about 30% sourdough together with rye flour, wheat flour, baker's yeast, and salt. The final rye:wheat ratio is regularly 70:30 w/w. Dough yield is adjusted to satisfy handling (e.g., pumping) and microbial nutrition requirements.

Another line of sourdough processes comprises those for baking with wheat only. The San Francisco sourdough process is often mentioned in this connection; however, Italian wheat sourdough products, for example, Panettone, Colomba, and Pandoro, are also very important (Cauvain 1998, Gobbetti and Corsetti 1997, Spicher 1983, Wood 2000).

The starter for San Francisco sourdough is ideally rebuilt every 8 hours to maintain maximum activity. However, a mature sponge may be kept refrigerated for days with acceptable performance. For rebuilding the starter, sponge (40%) is mixed with high-gluten wheat flour (40%) and water (about 20%) to ferment at bakery temperature. Final bread making requires a dough consisting of the ripe sourdough (9.2%), regular wheat flour (45.7%), water (44.2%), and salt (0.9%). Proofing for 7 hours follows, during which the pH decreases from about 5.3 to about 3.9. A sourdough starter culture for San Francisco French bread production commonly contains *Lb. sanfranciscensis* and *Candida holmii*.

Selection and Biochemistry of Microorganisms in Sourdough

Sourdough breads based on rye or wheat flour or their mixtures enjoys a remarkable standing in many societies, either as established, traditional products or as "innovative" developments for more natural products and a wider choice of flavors. Well-functioning and popular sourdough starters that have been maintained by simple rebuilding for decades, and the reestablishment of the "same" stable starter over and

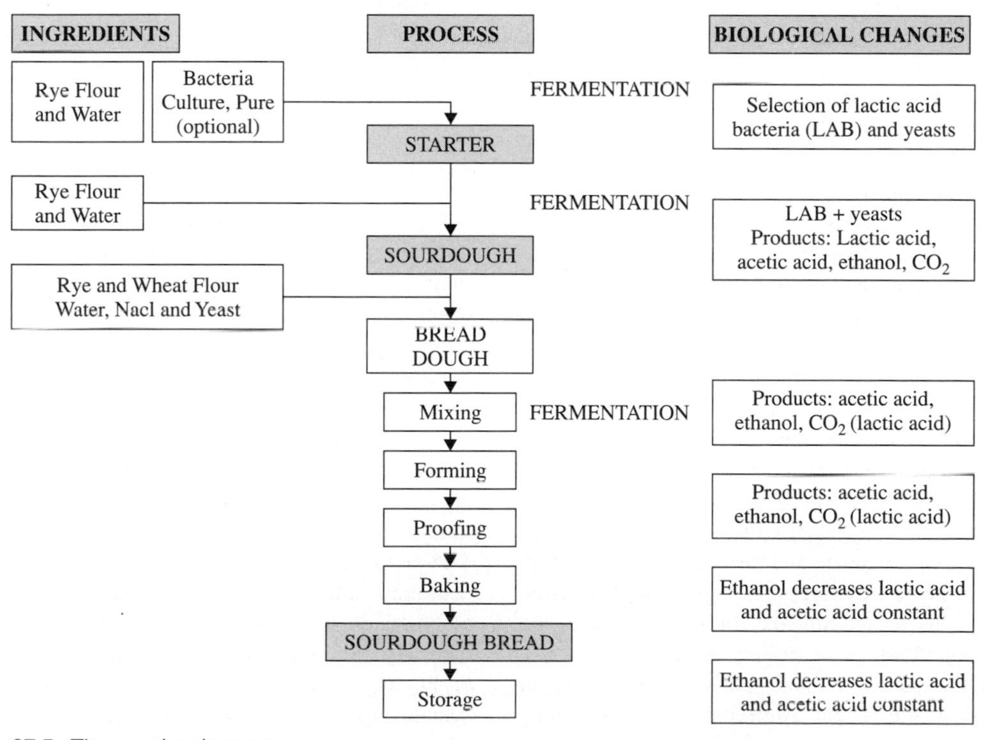

Figure 27.5. The sourdough process.

over again from a constant quality flour, are both expressions of stable ecological conditions. The simplicity of the procedures may be somewhat deceiving with respect to the actual complexity of these biological systems. In the following sections, metabolic events and biochemical aspects of sourdough fermentation will be discussed. Attention will be drawn to some of the more clear points about the metabolic events and other biochemical facts. The near future should bring us closer to a comprehensive understanding.

Carbohydrate Metabolism

The development of a mixture (1:1) of rye flour or even wheat flour with water incubated some hours at 25–30°C will almost inevitably lead to a microbiological population consisting of lactic acid bacteria (LAB) and yeasts. It may need rebuilding several times in order to stabilize it, but from then on the composition of the microflora may be constant for years, provided the composition of the flour and the conditions for growth are not changed much. Representative LAB and yeasts have been presented in Table 27.2. These microorganisms have certain characteristics in common. First, the selected LAB are very efficient maltose fermenters, a prime reason why they competed so well in the first place. Several lactobacilli in sourdoughs, e.g. *Lb. sanfranciscensis*, *Lb. pontis*, *Lb. reuteri,* and *Lb. fermentum*, harbor a key enzyme, maltose phophorylase, which cleaves maltose (the phosphorolytic reaction) to glucose-1-phosphate and glucose. Glucose-1-phosphate is metabolized heterofermentatively via the phosphogluconate pathway, while glucose is excreted into the growth medium. Glucose repression has not been observed with these lactobacilli. Most of the yeast species identified in sourdoughs are, per se, maltose negative, and will thus prefer to take up glucose when it is available. Other microorganisms may experience glucose repression of the maltose enzymes, to the benefit of the sourdough lactobacilli. Among the yeasts, *S. cerevisiae*, which is maltose positive and transports maltose and hexoses very efficiently, cannot take up maltose due to glucose repression and will, as a consequence, be defeated from the sourdough flora. *S. cerevisiae* as baker's yeast is, however, used at the bread-making stage, but as an addition in the recipe. Additional yeast cells may also be necessary for fast and efficient CO_2 production, because the yeasts are relatively sensitive to acids, particularly to acetic acid, which is excreted by the heterofermentative lactobacilli that often dominate the LAB flora of the sourdough. *Candida milleri* (syn. *S. exiguus*, *Torulopsis holmii*) is common in sourdoughs for San Francisco French bread. This yeast tolerates the acetic acid from heterolactic fermentation and thrives on glucose and sucrose in preference to maltose; it thus appears to be a near ideal partner for *Lb. sanfranciscensis* (Gobbetti and Corsetti 1997, Gobbetti 1998, Wood 2000, Hammes and Gänzle 1998).

Wheat and rye flour contain mainly maltose as a readily available carbohydrate, although rye flour has greater amylase activity and therefore has a greater potential for release of maltose. Early work in the United States on *Lb. sanfranciscensis* indicated that this organism would only ferment maltose (Kline and Sugihara 1971). However, strains isolated in Europe appeared more diversified, and some of them would ferment up to eight different sugars (Hammes and Gänzle 1998). Utilization of maltose by *Lb. sanfranciscensis*, *Lb. pontis*, *Lb. reuteri,* and *Lb. fermentum* through phosphorolytic cleavage with maltose phophorylase is energetically very favorable (Stolz et al. 1993) and shows increased cell yield and excretion of glucose when maltose is available. In these conditions the cells have very low levels of hexokinase.

Co-metabolism

Lactobacilli in sourdough production are not only specialized for maltose fermentation they also exploit co-fermentations for optimized energy yield (Gobbetti and Corsetti 1997, Hammes and Gänzle 1998, Stolz et al. 1995, Romano et al. 1987). *Lactobacillus sanfranciscensis*, *Lb. pontis* and *Lb. fermentum* all have mannitol dehydrogenase. Thus fructose may be used as an electron acceptor for the reoxidation of NADH in maltose or glucose metabolism, and then acetylphosphate may react on acetate kinase to yield ATP and acetate (Axelsson 1993). The lactobacilli gain energetically and more acetic acid may contribute to the desirable taste and flavor of bread. In practical terms addition of fructose is used to increase acetate in the products, that is, lower FQ (Spicher 1983). Comparable regulation of acetate production may be achieved by providing citrate, malate or oxygen as electron acceptors, resulting in products like succinate, glycerol and acetate (Gobbetti and Corsetti 1996, Condon 1987, Stolz et al. 1993).

Proteolysis and Amino Compounds

In a sourdough, the flour contributes considerable amounts of amino acids and peptides; however, in order to satisfy nutritional requirements of growing LAB and provide sufficient amino compounds, precursors, for flavor development, some proteolytic action is necessary. The LAB have been suspected as the main contributors of proteinase and peptidase activities for release of amino acids in sourdoughs (Spicher and Nierle 1984, Spicher and Nierle 1988, Gobbetti et al. 1996), although the flour enzymes may also have considerable input (Hammes and Gänzle 1998). In addition, lysis of microbial cells, particularly yeast cells, add to the pool of amino acids; a stimulant peptide containing aspartic acid, cysteine, glutamic acid, glycine, and lysine that appears in the autolytic process of *C. milleri* has also been identified (Berg et al. 1981). *Lactobacillus sanfranciscensis* has been found to have a regime of intracellular peptidases, endopeptidase, and proteinase, as well as a dipeptidase and proteinase in the cell envelope (Gobbetti et al. 1996). Limited autolysis of lactobacillus populations in sourdoughs may add to the repertoire of enzymes that will release amino acids from flour proteins, including those from proline-rich gluten in wheat. Some of the enzymes have been purified for further characterization (Gobbetti et al. 1996), and they express interesting activity levels at sourdough pH and temperatures.

The addition of exogenous microbial glucose oxidase, lipase, endoxylanase, α-amylase, or protease in the production of sourdough with 11 different LAB cultures showed positive effects on acidification rate and level for only three cultures, one *Leuconostoc citreum*, one *Lactococcus lactis* subsp. *lactis* and one *Lb. hilgardii*. *Lactobacillus hilgardii* with lipase, endoxylanase or α-amylase showed increased production of acetic acid. *Lactobacillus hilgardii* interacted with the different enzymes for higher stability and softening of doughs (Di Cagno et al. 2003).

Recent work with *Lb. sanfranciscensis*, *Lb. brevis*, and *Lb. alimentarius* in model sourdough fermentations showed, by using two-dimensional electrophoresis, that 37–42 polypeptides had been hydrolyzed. The polypeptides varied over wide ranges of pIs and molecular masses, and they originated from albumin, globulin, and gliadin, but not from glutenin. Free amino acid concentrations increased, in particular those of proline and glutamic and aspartic acid. Proteolysis by the lactobacilli had a positive effect on the softening of the dough. A toxic peptide for celiac patients, A-gliadin fragment 31–43, was degraded by enzymes from lactobacilli. The agglutination of human myelogenous leukemia–derived cells (K562) by toxic peptic-tryptic digest of gliadins was abolished by enzymes from lactobacilli (Di Cagno et al. 2002).

Volatile Compounds and Carbon Dioxide

Both yeasts and LAB contribute to CO_2 production in sourdough products, but the importance of the two varies. In bread production with only the (natural) sourdough microflora, the input from LAB may even be decisive for leavening because the counts and kinds of yeast may not be optimal for gas production. Relatively low temperature (e.g., 25°C) and low dough yield (e.g., 135) would select for LAB activities and less yeast metabolism. More complete volatile profiles were obtained at higher temperatures (e.g., 30°C) and with a more fluid dough. Of course increasing the leavening time may give substantially richer volatile profiles (Gobbetti et al. 1995). If baker's yeast, *S. cerevisiae*, is added to optimize and speed up the production process, the contribution from yeasts will dominate (Gobbetti 1998, Hammes and Gänzle 1998).

Bread made with chemical acidification without fermentation starter failed in sensory analysis. This indicates that fermentation with yeasts and LAB is important for good flavor, although high quality raw materials and proofing and baking are also decisive factors. Flavor compounds distinguishing the different metabolic contributions in sourdough are as follows (Gobbetti 1998):

- *Yeast fermentation (alcoholic):* 2-methyl-1-propanol, 2,3-methyl-1-butanol
- *LAB homofermentative:* diacetyl, other carbonyls
- *LAB heterofermentative:* ethyl acetate, other alcohols and carbonyls.

Antimicrobial Compounds from Sourdough LAB

The primary antimicrobial compounds produced by sourdough LAB are lactic and acetic acid, diacetyl, hydrogen peroxide, carbon dioxide,and ethanol, and among these, the two organic acids continue to be the most important contributions for beneficial effects in fermentations.

Researchers in the field, of course, also consider and test possibilities that LAB may produce bacteriocins and other antimicrobials. Thus antifungal compounds from *Lb. plantarum* 21B have been identified, for example, phenyl lactic acid and 4-hydroxyphenyl lactic acid (Lavermicocca et al. 2000). Caproic acid from *Lb. sanfranciscensis* also has some antifungal activity (Corsetti et al. 1998).

A real broad-spectrum antimicrobial from *Lb. reuteri* is reuterin (β-hydroxypropionic aldehyde), which comes as a monomer and a cyclic dimer (El-Ziney et al. 2000). Reutericyclin, which was isolated from *Lb. reuteri* LTH2584 after the screening of 65 lactobacilli, is a tetramic acid derivative. Reutericyclin inhibited Gram-positive bacteria (e.g., *Lactobacillus* spp., *Bacillus subtilis*, *B. cereus*, *Enterococcus faecalis*, *Staphylococcus aureus,* and *Listeria innocua*), and it was bactericidal towards *B. subtilis*, *S. aureus,* and *Lb. sanfranciscensis*. The ability to produce reutericyclin was stable in sourdough fermentations over a period of several years. Reutericyclin produced in sourdough was also active in the dough (Gänzle et al. 2000; Höltzel et al. 2000; Gänzle and Vogel 2003).

A few bacteriocins or bacteriocin-like compounds have also been identified, isolated, and characterized (Messens and De Vuyst 2002). Bavaricin A from *Lb. sakei* MI401 was selected by screening 335 LAB strains, including 58 positive strains (Larsen et al. 1993). Bavaricin A (and Bavaricin MN from *Lb. sakei* MN) have the N-terminal consensus motif of bacteriocin class IIA in common, comprise 41 and 42 amino acids, respectively, and have interesting sequence homologies and similar hydrophobic regions. Bavaricin A inhibits *Listeria* strains and some other Gram-positive bacteria but not *Bacillus* or *Staphylococcus* (Larsen et al. 1993, Kaiser and Montville 1996). Plantaricin ST31 is produced by *Lb. plantarum* ST; it contains 20 amino acids, and the activity spectrum includes several Gram-positive bacteria but not *Listeria* (Todorov et al. 1999). A bacteriocin-like compound, BLIS C57 from *Lb. sanfranciscensis* C57, was detected after screening 232 *Lactobacillus* isolates, including 52 strains expressing antimicrobial activity. BLIS C57 inhibits Gram-positive bacteria including bacilli and *Listeria* strains (Corsetti et al. 1996).

TRADITIONAL FERMENTED CEREAL PRODUCTS

Only two cereals, wheat and rye, contain gluten and are thereby suitable for the production of leavened bread, but many other food cereals are grown in the world. On a global basis, a great proportion of cereals are consumed as spontaneously fermented products, in particular in Africa, Asia, and Latin America. Most fermented cereals are dominated by lactic acid bacteria (LAB), and the microflora associated with the grains, flour, or any other ingredient, together with contamination from water, food-making equipment, and the producers themselves, represent the initial fermentation flora. Malted flour is also an important source of microorganisms. The changes that take place during the fermentation are due to both the metabolism of the microorganisms present and the activity of enzymes in the cereal, and these are in turn affected by the great variety of technologies that are used. The technology may be simple, involving little more than a mixing of flour with water and allowing it to ferment, or it may be extremely complex and involve many steps with obscure roles. Indigenous fermented foods are usually based upon raw materials that have a sustainable production in their country of origin and are therefore attracting increasing interest from researchers—both within pure and applied food science and also in anthropology. These ancient technologies often have deep roots in the culture of a country, and there is increasing awareness of the importance of preserving these traditional foods. Many products are not yet described in the literature, and knowledge of them is in danger of disappearing. It is therefore necessary to document the technologies used and to identify the fermenting organisms and the metabolic changes that are essential for the characteristics of the product. It is, however, often difficult to describe the sensory attributes of a product that is inherently variable.

In Africa, as much as 77% of the total caloric consumption is provided by cereals, of which rice, maize, sorghum, and millet are most important. Cereals are also significant sources of protein. Most of the cereal foods consumed in Africa are traditional fermented products and are very important both as weaning foods and as staple foods and beverages for adults. In Asia, many products are based on rice, and maize is most widely utilized in Latin America (FAO 1999).

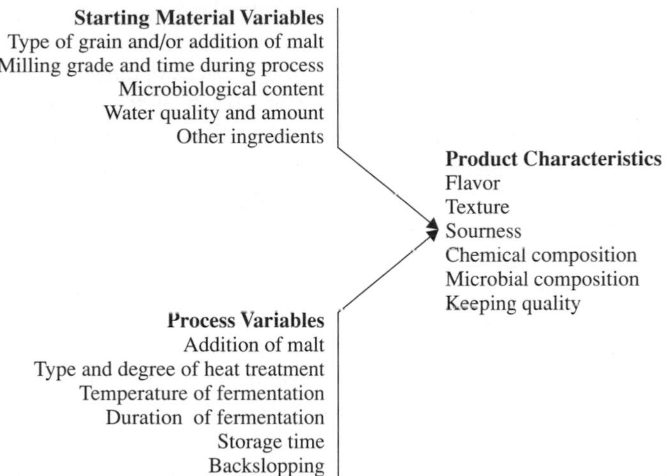

Figure 27.6. Important factors determining the characteristics of spontaneously fermented cereal products.

Indigenous fermented cereals can be classified according to raw material, type of fermentation, technology used, product usage, or geographical location. They can range from quite solid products such as baked flat breads to sour, sometimes mildly alcoholic, refreshing beverages.

Many factors have an influence on the characteristics of an indigenous product (Fig. 27.6). The choice of raw material may be primarily influenced by price and availability rather than by preference.

For instance Togwa, a Tanzanian fermented beverage, may be made from maize in the inland areas of Morogoro and Iringa, but from sorghum in the coastal areas of Dar es Salaam and Zanzibar (Mugula 2001). Similarly, the Ethiopian product borde may be made from several different grains according to availability—sorghum, maize, millet, barley and also the Ethiopian cereal tef (Abegaz et al. 2002). The use of different grains obviously affects the sensory characteristics of a product, and yet it may have the same name throughout the country. Some fermented cereal products also contain other ingredients. Idli is a leavened steamed cake made primarily from rice to which black gram dahl is added. This not only improves the nutritional quality, but in addition, the black gram imparts a viscosity, apparently specific for this legume, which may aid air entrapment during fermentation and thereby lighten the texture of the product (Soni and Sandu 1990). However, on a broader basis, the addition of legumes such as soya bean flour to fermented cereals has been suggested as an economically feasible way to generally improve the nutritional quality of cereal foods.

Some fermented cereal products are made using unmalted grain, with no extra addition of amylase, but they tend to either be very thick or of low nutritional density. Malted flour is added to many indigenous fermented cereals, a traditional technology that has far-reaching effects on several product characteristics. The addition of malt provides amylases (in particular α-amylase) that hydrolyze the starch, sweeten the product, and also cause a considerable decrease in viscosity of the product after heat treatment. The malting process, the germination of grain following steeping in water, is associated with colossal microbiological proliferation, and the organisms that develop during malting are a source of fermenting organisms. Many Asian products, for example koji, a Japanese fermented cereal or soya bean product, are first inoculated with a fungus, as a source of amylase, in order to liberate fermentable sugars from the cereal starch (Lotong 1998).

Many fermented cereals are multipurpose. A single product may be prepared in varying thicknesses and used as a fermented gruel for both adults and children, or it may be watered down and used as a fermented thirst-quenching beverage. As Wood (1994) remarked, the latter type of product makes a meaningful contribution to nutrition; the potential of their replacement by cola-type beverages would result in a serious negative impact on the nutrition of people in developing countries.

The use of fermented cereals as weaning foods in developing countries raises several important issues. Unfermented gruels deteriorate very rapidly in unhygienic conditions, especially if refrigeration is not available. They then represent a significant source of foodborne infections that annually claim the lives of millions of young children (Adams 1998). Fermented malted cereal gruels have been shown on the whole to contain low numbers of pathogenic organisms since these are inhibited and killed by the low pH that rapidly develops in the product. Fermented cereals are therefore usually regarded as safer than their unfermented counterparts (Nout and Motarjemi 1997). A weaning food made from unmalted cereals may be a cause of malnutrition because its thick viscosity limits the nutritional intake of a small child. Addition of malted flour decreases the viscosity so that more food can be ingested. If the fermentation flora includes yeasts in addition to LAB, a measurable reduction of carbohydrate will occur due to the production of CO_2 and other volatile compounds (Muyanja 2001). Analysis of fermented cereal products therefore shows that the protein:carbohydrate ratio is improved during the fermentation, and this obviously has nutritional benefits.

Milling of cereals into flour is usually done prior to fermentation, but in some products, for example, borde (Ethiopia), wet milling is used. This technique can be used when mechanical grain mills are not available and if the product is required to be smooth and without bits of suspended bran. The starch is also liberated from the grain more thoroughly when slurried with water and sieved than if it has been previously dry milled (Abegaz 2002).

A heat treatment step is found at some point in the production technology of most fermented cereal products and may involve boiling, steaming, or roasting. The type of heating employed is likely to have an effect on the flavor of the product, certainly if the temperature attained is sufficient to promote Maillard reactions. The heat also gelatinizes the starch, making it more susceptible to amylolytic enzymes, thus providing greater amounts of fermentable carbohydrates. However, at the same time, most of the natural contaminating (and potentially fermenting) flora and cereal enzymes are destroyed. Such products are also prone to contamination after the heat treatment step, and are thereby potentially unsafe should pathogenic organisms grow during the subsequent fermentation. The traditional solution to this is to use "backslopping," the addition of some of a previous batch of the product, and/or the addition of malted flour. Regular backslopping results in a selection of acid-tolerant organisms and functions as an empirical starter culture.

Fermentation usually takes place at ambient temperatures, and this may cause seasonal variations in products due to selection of different microorganisms at different temperatures. The duration of fermentation is largely a matter of personal choice, based on expected sensory attributes. Heat treatment after fermentation makes for a safer product, but it has the disadvantage of change of taste or loss of volatile flavor and aroma compounds.

The Microflora of Spontaneously Fermented Cereals

Spontaneously acid-fermented cereal products may contain a variety of microorganisms, but the flora in the final product is generally dominated by acid-tolerant LAB. Yeasts are also invariably present in large numbers when the fermentation is prolonged. A typical fermented cereal product contains approximately 10^9 and 10^7 cfu/g of product, of LAB and yeasts, respectively. However, since yeast cells are considerably larger than bacteria cells, their metabolic contribution to product characteristics is likely to be just as important as that of the LAB. The buffer capacity of cereal slurries is low, and the pH therefore drops quickly as acid is produced. Pathogenic organisms are inhibited by a fast acid production, so the addition of starter cultures, either as a pure culture or by "backslopping" promotes acid production and contributes to the safety of the fermented product (Nout et al. 1989).

The potential and the need for upgrading traditional fermentation technologies have initiated considerable research (Holzapfel 2002). In some recent studies of spontaneously fermented cereals, the LAB and yeasts have been isolated and identified as a first stage towards developing starter cultures for small-scale production of traditional fermented cereals. Muyanja et al. (2003) recorded that bushera, a traditional Ugandan fermented sorghum beverage that contains high numbers of LAB, was usually consumed by children after one day of fermentation as "sweet bushera." After 2–4 days, the product became sour and alcoholic and was consumed by adults. However, the sweet bushera showed very high counts of coliforms and had a reputation for causing diarrhea (Muyanja 2001). Clearly, the development of defined starter cultures would improve the safety of this and similar products.

Some recent examples of studies on the microbial flora of spontaneously fermented cereals are shown in Table 27.3. For each product, several different types of organisms have been isolated. In other words, a specific product is not produced from fermentation by a specific organism or organisms. *Lb. plantarum* seems to be the most commonly isolated *Lactobacillus* species in fermented cereals. In addition, heterofermentative LAB such as leuconostocs, *Lb. brevis,* and *Lb. fermentum* frequently occur. Yeasts are always present in spontaneously fermented products, but few studies have characterized the predominating species. However, Jespersen (2003) reported that *S. cerevisciae* is the predominant yeast in many African fermented foods and beverages.

All spontaneously fermented products contain, or have contained, many different types of microorganisms. These have grown in the product and will have metabolized some of the cereal components, thereby making a contribution (positive or negative) with their metabolites to the overall sensory characteristics of that product. However, studies on spontaneously fermented products have focused on LAB and yeasts since these organisms are often associated with other, better known, fermented products and have a history of safe use in food. Stanton (1998) proposed that the nature of the substrate (raw material) and the technology used to produce fermented foods are the predominating factors that determine the development of microorganisms and, thereby, the properties of a product.

Desirable Properties of the Fermenting Microflora

The most important property of a starter culture for a fermented cereal is the ability to quickly produce copious amounts of lactic acid in order to achieve a rapid decline in pH and retard the growth of pathogens and other undesirable organisms. Some workers (Sanni et al. 2002) have sought amylolytic LAB strains, as this could remove the need for using the highly contaminated malted flour in a product. The starter should also be able to hydrolyze the cereal protein in order to obtain the amino acids sufficient for rapid growth, and it should produce desirable and product-typical aroma and flavor compounds, but not off-flavors. Some products are characterized by a foaming consistency, and heterofermentative organisms (LAB or yeasts) are required for this property. Bacteriocin-producing strains have also been sought (Holzapfel 2002) in an attempt to increase the microbiological safety of the products. Starter cultures must also be commercially propagable and be able to survive preservation methods without loss of viability, activity, or metabolic traits.

Microbiological and Biochemical Changes in Traditional Fermented Cereals

Few studies have been made on the biochemical changes that take place in traditional fermented cereals. Mugula et al. (2003) analyzed samples of naturally fermented togwa made from sorghum and maize, to which togwa was backslopped and malt was added. The development of groups of microorganisms, organic acids, soluble carbohydrates, and volatile components was studied during the 24-hour fermentation. Maltose and glucose increased during the first part of the fermentation due to the action of cereal amylases, but later were reduced as the growth of LAB and yeasts increased. The pH dropped from around 5.0 to 3.2 in 24 hours, and this was mirrored by a rise in lactic acid to about 0.5%. Ethanol and secondary alcohols and aldehydes increased during the secondary part of the fermentation. Malty flavors are typical for fermented cereal products and may be produced during grain malting. Secondary aldehydes and alcohols are responsible for these flavors and may also originate from microbial metabolism of the branched-chain amino acids leucine, isoleucine, and valine. These compounds are produced by yeasts, some LAB, and probably also by other microorganisms in the product.

Many spontaneously fermented cereals also have a very short shelf life, since fermentation continues in the absence of refrigeration. Off-flavors, in particular vinegary notes, are a common problem. The very low pH in fermented cereal products may be sensorially compensated for by saccharification by β-amylase.

Table 27.3. Lactic Acid Bacteria and Yeasts Isolated from Some Traditional Fermented Cereal Products

Product (Country of Origin)	Cereal Basis	Most Prevalent		Reference
		Species of LAB	Species of Yeast	
Togwa (Tanzania)	Various	*Lb. brevis, Lb. cellobiosus, Lb. Plantarum* *W. confusa* *P. pentosaceus*	*P. orientalis* *S. cerevisiae* *C. pelliculosa* *C. tropicalis*	Mugula et al. 2003
Bushera (Uganda)	Sorghum, millet	*Lb. brevis; E. faecium* *Ln. mesenteroides* subsp. *mesenteroides*	NR	Muyanja et al. 2003
Ogi (Nigeria)	Maize	*Lb. reuteri, Lb. Leichmanii, Lb. plantarum, Lb. casei, Lb. fermentum, Lb. brevis, Lb. alimentarius, Lb. buchneri, Lb. jensenii.*	*S. cereviseae* and *Candida mycoderma*	Ogunbanwo et al. 2003 Odunfa 1985
Pozol (Mexico)	Maize	*Streptococcus* spp., *Lb. plantarum, Lb.fermentum*		Omar and Ampe 2000
		Lb. brevis *Ln. mesenteroides* *Lc. lactis, Lc. raffinolactis*	*C. mycoderma, S. cerevisiae, Rhodotorula* spp.	Nuraida and others 1995
Borde (Ethiopia)	Various	*Lb. brevis* *W. confusa* *P. pentosaceus*	NR	Abegaz 2002
Idli (India)	Rice and blackgram beans	*Leuconostocs* spp. *Enterococcus faecalis*	*Saccharomyces* spp.	Soni and Sandu 1990

Note: Lb., Lactobacillus; Ln., Leuconostoc; L., Lactococcus; P., Pediococcus, E., Enterococcus; W., Weisella; S., Saccharomyces; I., Isa; NR, Not recorded.

FERMENTED PROBIOTIC CEREAL FOODS

A probiotic food is a live bacterial food supplement, which when ingested, may improve the well-being of the host in a variety of ways by influencing the balance of the host's intestinal flora (Fuller 1989). Most probiotic bacteria have been isolated from the healthy human intestine and are members of the genus *Lactobacillus*, but some products may contain *Bifidobacterium* spp. or the yeast *Saccharomyces boulardii*. While the potential benefits of probiotic bacteria have been generally accepted for many decades, it is only in comparatively recent years that research has been able to scientifically document the beneficial medical effect due to some specific strains (Gorbach 2002). There is now strong scientific evidence that specific strains of probiotic microorganisms are able to

- Show a prophylactic action against and alleviate diarrhea caused by bacterial and viral infections, radiation therapy or the use of antibiotics,
- Suppress undesirable bacteria in the gut with beneficial results for patients with conditions such as irritable bowel syndrome and ulcerative colitis, and
- Influence the immune system, showing positive results for infant atopic eczema and other allergies.

Indeed, in addition to the list above, other effects have been proposed: the lowering of blood cholesterol; the prevention of acute respiratory infections, *Helicobacter pylori* infections, and colonization by potential pathogens in intensive care units in hospitals; relief of constipation; and a protection against the development of various forms of cancer. However, so far, convincing proof for the efficacy of probiotics against these problems has not been obtained. The positive effects that have been documented have led to a great interest from food manufacturers and consumers alike. The main motivation for consuming probiotic products is said to be the developing consumer trend towards healthy living though natural foods and medicines and a trend away from the use of antibiotics and the incorporation of chemical additives in food. As the beneficial effects of probiotic foods become scientifically accepted, there will be increasing pressure from food manufacturers on the authorities to allow health claims to be used in product advertising. Probiotic fermented milks were the first probiotic products to be produced commercially and are available in many countries (Tamime and Marshall 1997).

Some fermented probiotic cereal products are now being prepared and marketed (Table 27.4) and may have an appeal for those who do not consume dairy products.

Oats are a popular basis for probiotic cereal foods. This choice is due to the healthy image of oats with respect to soluble and insoluble fiber content and the potential to reduce blood cholesterol due to β-glucans. A prebiotic is a compound, usually an oligosaccharide, that reaches the colon undigested by the host's enzymes and selectively favors the growth of probiotic bacteria. Such compounds

Table 27.4. Fermented Probiotic Cereal Foods

Type of Product (Commercial Name)	Cereal Constituent	Probiotic Constituent	Reference
Fermented fruit flavored cereal drink (Pro Viva)	Oat + malted barley flour	*Lb. plantarum* 299v	Molin 2001
Fermented cereal drink	Oat 'milk'	*LB. reuteri; lb. acidophilus;* *B. bifidus*	Mårtensson, Öste and Holst 2002
Fruit flavored cereal pudding (Yosa)	Oat flour	*Lb. acidophilus; B. bifidus*	Blandino et al. 2003
Cereal-based weaning food	Maize + malted barley flour	*Lb. acidophilus* *Lb. rhamnosus 'GG'* *Lb. reuteri*	Helland et al. 2004

Note: *LB = Lactobacillus; B = Bifidobacterium.*

include lactulose, fructooligosaccharides, and inulin. It has been suggested that the best probiotic results may be obtained by using a combination of a prebiotic (such as oats) and a probiotic organism (Charalampopoulos et al. 2002). In this way, the total physiological effect of the food could be increased.

In order for a probiotic product to have a physiological effect, it has been suggested that it should contain at least 10^6 cfu/g product, and that daily intake should be at least 100 g (Sanders and Huis in't Veld 1999). The final acidity in the product has been shown to be of critical importance for the survival of probiotic bacteria during storage (Mårtensson, Öste and Holst 2002). Many probiotic bacteria do not tolerate a pH below 4.0, and fermented cereals frequently reach this pH due to the poor buffering capacity of the substrate. In addition, the physiological state of the probiotic organisms at the time of storage also determines their survival. Organisms that show poor growth during a fermentation period are more likely to die out during cold storage. This necessitates careful formulation of the product as well as selection of the right probiotic culture.

The choice of a substrate for a probiotic food is partially governed by the tolerance of the food towards heat pasteurization or even sterilization before fermentation, and cereal mixtures lend themselves well to this treatment. Probiotic products require fermentation at around 37°C for 8 to 18 hours, depending on substrate. The suitability of such conditions for the growth of pathogenic organisms necessitates strict adherence to hygiene both before and during fermentation. A fast lactic acid development in the product during fermentation is a critical step, and the growth of probiotic organisms in cereal products is greatly stimulated by the addition of malted flour (either of the same grain or of barley malt) or milk, due to the increased availability of fermentable sugars, peptides, and amino acids. For probiotic weaning foods, the use of malt has a further advantage since at a given viscosity, the product has a higher nutritional density.

Probiotic cereal foods are in their infancy, and the future will probably see further development in this type of product. New strains with proven probiotic efficacy and good flavor-forming abilities will increase the range of probiotic products available.

REFERENCES

Abegaz K. 2002. Traditional and Improved Fermentation of *Borde*, a Cereal-Based Ethiopian Beverage. [PhD thesis] Ås, Norway. Agric Univ of Norway. ISBN: 82-575-0508-0.

Abegaz K, Beyene F, Langsrud T, Narvhus JA. 2002. Indigenous processing methods and raw materials of *borde*, an Ethiopian traditional fermented beverage. J Food Technol Africa 7:59–64.

Adams MR. 1998. Fermented weaning foods. In: BJB Wood, editor. Microbiology of Fermented Foods. London: Blackie Academic and Professional.

Axelsson LT. 1993. Lactic acid bacteria: classification and physiology. In: S Salminen, A von Wright, editors. Lactic Acid Bacteria. New York: Marcel Dekker, Inc. Pp. 1–63.

Berg RV, Sandine WE, Anderson AW. 1981. Identification of a growth stimulant for *Lactobacillus sanfrancisco*. Appl Env Microbiol 42:786–788.

Blandino A, Al-Aseeri ME, Pandiella SS, Cantero D, Webb C. 2003. Cereal-based fermented foods and beverages. Food Res Int 37:527–543.

Cauvain S P. 1998. Other cereals in breadmaking. In: Technology of Breadmaking. Cauvain SP, Young LS, editors, London: Blackie Academic and Professional. Pp. 330–346.

Charalampopoulos D, Wang R, Pandiella SS, Webb C. 2002. Application of cereal components in functional foods: a review. Int J Food Microbiol 79:131–141.

Condon S. 1987. Responses of lactic acid bacteria to oxygen. FEMS Microbiol Rev 46:269–280.

Corsetti A, Gobbetti M, Rossi J, Damiani P. 1998. Antimould activity of sourdough lactic acid bacteria: Identification of a mixture of organic acids produced by *Lactobacillus sanfrancisco* CB1. Appl Microbiol Biotechnol 50:253–256.

Corsetti A, Gobbetti M, Smacchi E. 1996. Antimicrobial activity of sourdough lactic acid bacteria: Isolation of a bacteriocin-like inhibitory substance from *Lactobacillus sanfrancisco* C57. Food Microbiol 13: 447–456.

Di Cagno R, De Angelis M, Corsetti A, Lavermicocca P, Arnault P, Tossut P, Gallo G, Gobbetti M. 2003. Interactions between sourdough lactic acid bacteria and exogenous enzymes: Effects on the microbial kinetics of acidification and dough textural properties. Food Microbiol 20:67–75.

Di Cagno R, De Angelis M, Lavermicocca P, De Vincenzi M, Giovannini C, Faccia M, Gobbetti M. 2002. Proteolysis by lactic acid bacteria: Effects on wheat flour protein fractions and gliadin peptides involved in human cereal tolerance. Appl Env Microbiol 68:623–633.

El-Ziney MG, Debevere J, Jakobsen M. 2000. Reuterin. In: AS Naidu, editor. Natural Food Antimicrobial Systems. London: CRC Press. Pp. 567–587.

FAO. 1999. Fermented cereals. A Global Perspective. Food and Agricultural Services Bulletin No. 138. Rome, Italy: Food and Agriculture Organization of the United Nations.

Fenema OR. 1996. Food Chemistry. New York: Marcel Dekker.

Fuller R. 1989. Probiotics in man and animals. J Appl Bact 66:365–378.

Gänzle MG, Höltzel A, Walter J, Jung G, Hammes WP. 2000. Characterization of reutericyclin produced by *Lactobacillus reuteri* LTH2584. Appl Env Microbiol 66:4325–4333.

Gänzle MG, Vogel R. 2003. Contribution of reutericyclin production to the stable persistence of *Lactobacillus reuteri* in an industrial sourdough fermentation. Int J Food Microbiol 80:31–45.

Gobbetti M. 1998. The sourdough microflora: Interactions of lactic acid bacteria and yeasts. Trends Food Sci Technol 9:267–274.

Gobbetti M, Corsetti A. 1996. Co-metabolism of citrate and maltose by *Lactobacillus brevis* subsp. *lindneri* CB1 citrate-negative strain: Effect on growth, end-products and sourdough fermentation. Z Lebensmittel-Untersuchung und -Forschung 203:82–87.

———. 1997. *Lactobacillus sanfrancisco* a key sourdough lactic acid bacterium: A review. Food Microbiol 14:175–187.

Gobbetti M, Simonetti MS, Corsetti A, Santinelli F, Rossi J, Damiani P. 1995. Volatile compound and organic acid productions by mixed wheat sour dough starters: Influence of fermentation parameters and dynamics during baking. Food Microbiol 12:497–507.

Gobbetti M, Smacchi E, Corsetti A. 1996. The proteolytic system of *Lactobacillus sanfrancisco* CB1: Purification and characterization of a proteinase, dipeptidase and aminopeptidase. Appl Env Microbiol 62:3220–3226.

Gobbetti M, Smacchi E, Fox P, Stepaniak L, Corsetti A. 1996. The sourdough microflora. Cellular localization and characterization of proteolytic enzymes in lactic acid bacteria. Lebensmittel-Wissenschaft und -Technologie 29:561–569.

Gorbach SL. 2002. Probiotics in the third millenium. Digest Liver Dis 34:S2–7.

Hammes WP, Gänzle MG. 1998. Sourdough breads and related products. In: BJB Wood, editor. Microbiology of Fermented Foods. London: Blackie Academic and Professional. Pp. 199–216.

Helland M, Wicklund T, Narvhus JA. 2004. Growth and metabolism of selected strains of probiotic bacteria, in maize porridge with added malted barley. Int J Food Microbiol 44:957–965.

Höltzel A, Gänzle MG, Nicholson GJ, Hammes WP, Jung G. 2000. The first low-molecular-weight antibiotic from lactic acid bacteria: Reutericyclin, a new tetramic acid. Angewandte Chemie Int Ed 39:2766–2768.

Holzapfel WH. 2002. Appropriate starter culture technologies for small-scale fermentation in developing countries. Int J Food Microbiol 75:197–212.

Hoseney RC. 1998. Cereal Science and Technology, 2nd edition. St. Paul, Minnesota: American Association of Cereal Chemists.

Jespersen L. 2003. Occurrence and taxonomic characteristics of strains of *Saccharomyces cerevisiae* predominant in African indigenous fermented foods and beverages. FEMS Yeast Res 3:191–200.

Kaiser AL, Montville TJ. 1996. Purification of the bacteriocin bavaricin MN and characterization of its mode of action against *Listeria monocytogenes* Scott A cells and lipid vesicles. Appl Env Microbiol 62: 4529–4535.

Kent NL. 1983. Technology of Cereals, 2nd edition. Oxford: Pergamon Press, Ltd.

Kline L, Sugihara TF. 1971. Microorganisms of the San Francisco sourdough bread process. II. Isolation and characterization of undescribed bacterial species responsible for souring activity. Appl Microbiol 21: 459–465.

Kunene NF, Geornaras I, von Holy A, Hastings JW. 2000. Characterization and determination of origin of lactic acid bacteria from a sorghum-based fermented weaning food by analysis of soluble proteins and amplified fragment length polymorphism fingerprinting. Appl Environ Microbiol. 66:1084-1092.

Larsen AG, Vogensen FK, Josephsen J. 1993. Antimicrobial activity of lactic acid bacteria isolated from sour doughs: Purification and characterization of bavaricin A, a bacteriocin produced by *Lactobacillus bavaricus* MI401. J Appl Bacteriol 75:113–122.

Lavermicocca P, Valerio F, Evidente A, Lazzaroni S, Corsetti A, Gobbetti M. 2000. Purification and characterization of novel antifungal compounds from the sourdough *Lactobacillus plantarum* strain 21B. Appl Env Microbiol 66:4084–4090.

Linko Y-Y, Javanainen P, Linko S. 1997. Biotechnology of bread baking. Trends Food Sci Technol 8:339–344.

Lotong N. 1998. Koji. In: BJB Wood, editor. Microbiology of Fermented Foods. London: Blackie Academic and Professional.

Maloney DH, Foy JJ. 2003. Yeast Fermentations. In: K Kulp, K Lorenz, editors. Handbook of Dough Fermentations. New York: Marcel Dekker, Inc.

Mårtensson O, Öste R, Holst O. 2002. The effect of yoghurt culture on the survival of probiotic bacteria in oat-based, non-dairy products. Food Res Int 35: 775–784.

Martinez-Anaya MA. 2003. Associations and interactions of microorganisms in dough fermentations: Effects on dough and bread characteristics. In: K Kulp, K Lorenz, editors. Handbook of Dough Fermentations. New York: Marcel Dekker, Inc. Pp. 63–95.

Messens W, De Vuyst L. 2002. Inhibitory substances produced by *Lactobacilli* isolated from sourdoughs - a review. Int J Food Microbiol 72:31–43.

Molin G. 2001. Probiotics in food not containing milk or milk constituents, with special reference to *Lactobacillus plantarum* 299v. Am J Clinical Nutr 73: 380S–385S.

Mugula JK. 2001. Microbiology, fermentation and shelf life extension of togwa, a Tanzanian indigenous food. [PhD thesis]. Ås, Norway. Agric Univ of Norway. ISBN 82-575-0457-2.

Mugula JK, Narvhus JA, Sorhaug T. 2003. Use of starter cultures of lactic acid bacteria and yeasts in the preparation of *togwa*, a Tanzanian fermented food. Int J Food Microbiol 83:307–318.

Muyanja CBK. 2001. Studies on the Fermentation of *Bushera* : A Ugandan Traditional Fermented Cereal-Based Beverage. [PhD thesis]. Ås, Norway, Agric Univ Norway. ISBN 82-575-0486-8.

Muyanja CBK, Kikafunda JK, Narvhus JA, Helgetun K, Langsrud T. 2003. Production methods and composition of *bushera*, a Ugandan Traditional Fermented Cereal Beverage. African J Food, Agric, Nutr Development 3:10–19.

Muyanja C, Narvhus JA, Langsrud T. 2004. Chemical changes during spontaneous and lactic acid starter bacteria starter culture fermentation of bushera. MURARIK Bulletin 7:606–616.

Nuraida L, Wacher MC, Owens JD. 1995. Microbiology of *pozol*, a Mexican fermented maize dough. World J Microbiol Biotechnol 11:567–571.

Nout MJR, Motarjemi Y. 1997. Assessment of fermentation as a household technology for improving food safety: A joint FAO/WHO workshop. Food Control 8:221–226.

Nout MJR, Rombouts FM, Havelaar A. 1989. Effect of accelerated natural lactic fermentation of infant food ingredients on some pathogenic microorganisms. Int J Food Microbiol 8:351–361.

Odunfa SA. 1985. African Fermented Foods. In: BJB Wood, editor. Microbiology of Fermented Foods, Vol, 2. London and New York: Elsevier Applied Science Publishers.

Ogunbanwo ST, Sanni AI, Onilude AA. 2003. Characterization of bacteriocin produced by *Lactobacillus plantarum* F1 and *Lactobacillus brevis* OG1. African J Biotechnol 2:219–227.

Omar ben N, Ampe F. 2000. Microbial community dynamics during production of the Mexican fermented maize dough Pozol. Appl and Env Microbiol 66:3644–3673.

Ponte JG, Tsen CC. 1987. Bakery Products. In: LR Beuchat, editor. Food and Beverage Mycology, 2nd edition. New York: Van Nostrand Reinhold Company, Inc.

Romano AE, Brino G, Peterkofsky A, Reizer J. 1987. Regulation of β-galactoside transport and accumulation in heterofermentative lactic acid bacteria. J Appl Bacteriol 169:5589–5596.

Sanders ME, Huis in't Veld JHJ. 1999. Bringing a probiotic-containing functional food to the market: Microbiological, product, regulatory and labelling issues. In: WN Konings, OP Kuipers, Hui in T'Veld JHJ, editors. Lactic Acid Bacteria: Genetics, Metabolism and Applications. Dordrecht, The Netherlands: Kluwer Academic Publishing. Pp. 293–315.

Sanni AI, Morlon-Guyot J, Guyot JP. 2002. New efficient amylase-producing strains of *Lactobacillus plantarum* and *L. fermentum* isolated from different Nigerian fermented foods. Int J Food Microbiol 72:53–62.

Soni SK, Sandu DK. 1990. Indian fermented foods. Microbiological and biochemical aspects. Indian J Microbiol. 30:135–157.

Spicher G. 1983. Baked goods. In: HJ Rehm, G Reed, editors. Biotechnology, Vol. 5. Weinheim: Verlag Chemie. Pp. 1–80.

Spicher G, Nierle W. 1984. The microflora of sourdough. XVIII. Communication: The protein degrading capabilities of the lactic acid bacteria of sourdough. Z für Lebensmittel-Untersuchung und-Forschung 178:389–392.

———. 1988. Proteolytic activity of sourdough bacteria. Appl Microbiol Biotechnol 28:487–492.

Stanton WR. 1998. Food fermentation in the tropics. In: BJB Wood, editor. Microbiology of Fermented Foods. London: Blackie Academic and Professional. Pp. 696–712.

Stear CA. 1990. Handbook of Breadmaking Technology. Barking, United Kingdom: Elsevier Applied Science.

Stolz P. 2003. Biological fundamentals of yeast and lactobacilli fermentation in bread dough. In: K Kulp, K Lorenz, editors. Handbook of Dough Fermentations. New York: Marcel Dekker, Inc. Pp. 23–42.

Stolz P, Böcker G, Vogel RF, Hammes WP. 1993. Utilization of maltose and glucose by lactobacilli isolated from sourdough. FEMS Microbiol Lett 109:237–242.

Stolz P, Vogel RF, Hammes WP. 1995. Utilization of electron acceptors by lactobacilli isolated from sourdough. II. *Lactobacillus pontis, L. reuteri, L. amylovorus,* and *L fermentum.* Z für Lebensmittel-Untersuchung und -Forschung 201:402–410.

Todorov S, Onno B, Sorokine O, Chobert JM, Ivanova I, Dousset X. 1999. Detection and characterization of a novel antibacterial substance produced by *Lactobacillus plantarum* ST31 isolated from sourdough. Int J Food Microbiol 48:167–177.

Tamime AY, Marshall VME.1997. Microbiology and technology of fermented milks. In: BA Law, editor. Microbiology and Biochemistry of Cheese and Fermented Milk. London: Blackie Academic and Professional.

Walker GM. 1998a. Yeast technology. In: Yeast Physiology and Technology. New York: John Wiley and Sons, Inc.

———. 1998b. Yeast metabolism. In: Yeast Physiology and Technology. New York: John Wiley and Sons, Inc.

Williams T, Pullen G. 1998. Functional ingredients. In: SP Cauvain, LS Young, editors. Technology of Breadmaking. London: Blackie Academic and Professional.

Wood BJB. 1994. Technology transfer and indigenous fermented foods. Food Res Int 27:269.

———. 2000. Sourdough bread. In: RK Robinson, CA Blatt, PD Patel, editors. Encyclopedia of Food Microbiology.San Diego: Academic Press. Pp. 295–301.

Zehentbauer G, Grosch W. 1998. Crust aroma of baguettes II. Dependence of the concentrations of key odorants on yeast level and dough processing. J Cereal Sci 28:93–96.

28
Biochemistry of Fermented Meat

F. Toldrá

BACKGROUND INFORMATION

The origin of fermented meats is quite far away in time. Ancient Romans and Greeks already manufactured fermented sausages, and in fact, the origin of words like sausage and salami may proceed from the Latin expressions "salsicia" and "salumen," respectively (Toldrá 2002). The production and consumption of fermented meats expanded throughout Europe in the Middle Ages, being adapted to climatic conditions (i.e., smoked in northern Europe and dried in Mediterranean countries). The experience in manufacturing these meats went to America with settlers (e.g., states like Wisconsin still have a good number of typical northern European sausages like Norwegian and German sausages).

Today, a wide variety of fermented sausages are produced, depending on the raw materials, microbial population, and processing conditions. For instance, northern European type sausages contain beef and pork as raw meats, are ripened for short periods (up to 3 weeks), and are usually subjected to smoking. In these sausages, shelf life is mainly due to acid pH and smoking rather than drying. On the other hand, Mediterranean sausages mostly use only pork and are ripened for longer periods (several weeks or even months), and smoke is not so typically applied (Flores and Toldrá 1993). Examples for different types of fermented sausages, according to the intensity of drying, are shown in Table 28.1. Undry and semidry sausages are fermented, to reach low pH values, and are usually smoked and cooked before consumption. Shelf life and safety is mostly determined by pH drop and reduced water activity as a consequence of fermentation and drying, respectively. The product may be considered stable at room temperature when $pH < 5.0$ and the moisture:protein ratio is below 3.1:1 (Sebranek 2004). Moisture:protein ratios are defined for the different

Table 28.1. Examples of Fermented Meats with Different Dryness Degree

Product	Type	Examples	Weight Loss (%)	Drying/ Ripening
Undry fermented sausages	Spreadable	German teewurst	< 10	No drying
		Frische mettwurst	< 10	No drying
Semidry fermented sausages	Sliceable	Summer sausage	< 20	Short
		Lebanon Bologna	< 20	Short
		Saucisson d´Alsace	< 20	Short
		Chinese Laap ch´eung	< 20	Short
		Chinese Xunchang	< 20	Short
Dry fermented sausages	Sliceable	Hungarian and Italian salami	> 30	Long
		Pepperoni	> 30	Long
		Spanish chorizo	> 30	Long
		Spanish salchichón	> 30	Long
		French saucisson	> 30	Long

Sources: Lücke 1985, Campbell-Platt 1995, Roca and Incze 1990, Toldrá 2002.

dry and semidry fermented sausages in the United States, while water activity values are preferred in Europe.

RAW MATERIAL PREPARATION

There are several considerations (listed in Table 28.2) that need to be taken into account when producing fermented meats. The selection of the different options, which will be discussed in the following sections, facilitates the choice of the most adequate conditions for the correct processing, safety, and optimal final quality.

INGREDIENTS

Lean meats from pork and beef, in equal amounts, or only pork are generally used. Quality characteristics such as color, pH (preferably < 5.8), and water-holding capacity are very important. When the pH of pork meat is > 6.0, the meat is known as DFD (dark, firm, and dry). This type of meat binds water tightly and spoils easily. Pork meat with another defect, known as PSE (pale, soft, and exudative), is not recommended because the color is pale, and the sausage would release water too fast, which could cause casings to wrinkle. Meat from older animals is preferred because of its more intense color, which is due to the accumulation of myoglobin, a sarcoplasmic protein that is the natural pigment responsible for color in meat.

Pork back and belly fats constitute the main source for fats. Special care must be taken for the polyunsaturated fatty acid profile, which should be lower than 12%; and the level of oxidation, measured as peroxide value, should be as low as possible (Demeyer 1992). Some rancidity may develop after long-term frozen storage since lipases present in adipose tissue are active even at temperatures as low as −18°C and are responsible for the continuous release of free fatty acids that are susceptible to oxidation (Hernández et al. 1999). So, extreme caution must be taken with fats stored for several months as they may develop a rancid flavor.

OTHER INGREDIENTS AND ADDITIVES

Salt is the oldest additive used in cured meat products since ancient times. Salt, at about 2–4%, serves several functions, including (1) an initial reduction in water activity, (2) providing a characteristic salty taste, and (3) contributing to increased solubility of myofibrillar proteins. Nitrite is a typical curing agent used as a preservative against pathogens, especially *Clostridium botulinum*. Nitrite is also responsible for the development of the typical cured meat color, prevention of oxidation, and contribution to the cured meat flavor (Gray and Pearson 1984). The

Table 28.2. Some Decisions to Adopt and Options to Choose in the Processing of Fermented Meats

Aspects	Options
Type of meat	Pork, beef, etc.
Quality of meat	Choose good quality. Reject defective meats (pork PSE and DFD), abnormal colors, exudation, etc.
Origin of fat	Choose either chilled or frozen (how long?) fats. Reject oxidized fats.
Type of fat	Control of fatty acids profile (excess of PUFA?).
Ratio	Choose desired meat:fat ratio.
Particle size	Choose adequate plate (grinder) or speeds (cutter).
Additives: salt	Decide concentration.
Additives: curing agent	Nitrite or nitrate depending on type and length of process.
Additives: carbohydrates	Type and concentration depending on type of process and required pH drop.
Spices	Choose according to required specific flavor.
Microflora	Natural or added as starter?
Starters	Choose microorganisms depending on type of process and product.
Casing	Material and diameter depending on type of product.
Fermentation	Conditions depending on type of starter used and product.
Ripening/drying	Conditions depending on type of product.
Smoking	Optional application. Conditions depending on type of product and specific flavor.
Color	Depends on raw meat, nitrite, and processing conditions.
Texture	Depends on meat:fat ratio, stuffing pressure, and extent of drying.
Flavor	Choose adequate starter and process conditions.
Water activity	Depends on drying conditions and length of process.

reduction of nitrite to nitric oxide is favored by the presence of ascorbic and erythorbic acids or their sodium salts. They also exert antioxidative action and inhibit the formation of nitrosamines.

Carbohydrates like glucose and lactose are used quite often as substrates for microbial growth and development. Disaccharides, and especially polysaccharides, may delay the growth and pH drop rate because they have to be hydrolyzed to monosaccharides by microorganisms.

Sometimes, additional substances may be used for specific purposes (Demeyer and Toldrá 2004). This is the case of glucono-delta-lactone, added at 0.5%, which may simulate bacterial acidulation. In the presence of water, glucono-delta-lactone is hydrolyzed to gluconic acid and produces a rapid decrease in pH. The quality is rather poor because the rapid pH drop drastically reduces the activity of flavor-related enzymes such as exopeptidases and lipases. Phosphates may be added to improve stability against oxidation; vegetable proteins, such as soy isolates, to replace meat proteins; and manganese sulphate as a cofactor for lactic acid bacteria.

Spices, either in natural form or as extracts, are added to give a characteristic aroma or color to the fermented sausage. There are a wide variety of spices (pepper, paprika, oregano, rosemary, garlic, onion, etc.), each one giving a particular aroma to the product. Some spices also contain powerful antioxidants. The most important aromatic volatile compounds may vary depending on the geographical and/or plant origin. For instance, garlic, which gives a pungent and penetrating smell, is typically used in chorizo, and pepper is used in salchichón and salami. Paprika gives a characteristic flavor and color due to its high content of carotenoids (Ordoñez et al. 1999). The presence of manganese in some spices, like red pepper and mustard, is necessary for the activity of several enzymes involved in glycolysis and thus enhances the generation of lactic acid (Lücke 1985).

Starters

Typical fermented products were initially based on the development and growth of desirable indigenous flora, sometimes reinforced with backslopping, which consists in the addition of a previous ripened, fermented sausage with adequate sensory properties. However, this practice usually gave a high heterogeneity in the product quality. The use of microbial starters, as a way to standardize processing as well as quality and safety, is relatively new. In fact, the first commercial use was in the United States in the 1950s, followed by its use in Europe in the 1960s; since then, starters have seen extensive use. Today, most of the fermented sausages are produced with a combination of lactic acid bacteria, to get an adequate acidulation, and two or more cultures to develop flavor and facilitate other reactions, such as nitrate reduction.

In general, microorganisms used as starter cultures must satisfy several requirements according with the purposes of their use: nontoxicity for humans, good stability under the processing conditions (resistance to acid pH, low water activity, tolerance to salt, resistance to phage infections), intense growth at the fermentation temperature (i.e., 18–25°C in Europe or 35–40°C in the United States), generation of products with technological interest (i.e., lactic acid for pH drop, volatile compounds for aroma, nitrate reduction, secretion of bacteriocins, etc.), and lack of undesirable enzymes (e.g., decarboxylases responsible for amine generation). Thus, the most adequate strains must be carefully selected and controlled as they will have a very important role in the process and will be decisive for the final quality. The most important microorganisms used as starters belong to one of the following groups: lactic acid bacteria, Micrococcaceae, yeasts, or molds (Leistner 1992). The main roles and functions for each group are shown in Table 28.3.

Lactic Acid Bacteria (LAB)

The most important function of lactic acid bacteria consist in the generation of lactic acid from glucose or other carbohydrates through either homo or heterofermentative pathway. The accumulation of lactic acid produces a pH drop in the sausage. However, some undesirable secondary products like acetic acid, hydrogen peroxide, acetoin, . . . may be generated in the case of certain species having heterofermentative pathways. *L. sakei* and *L. curvatus* grow at mild temperatures being usual in the processing of European sausages while *L. plantarum* and *P. acidilactici* grow well at higher temperatures (30–35°C) closer to the fermentation conditions in the sausages produced in the United States. Lactic acid bacteria also have a proteolytic system, consisting in endo- and exopeptidases, that contributes to the generation of free amino acids during processing and, most of them, are also able to generate different types of bacteriocins with antimicrobial properties.

Micrococcaceae

This group consists of *Staphylococcus* and *Kocuria* (formerly *Micrococcus*), which are major contributors to flavor due to their proteolytic and lipolytic activity. Another important function consists in the nitrate reductase activity, necessary to reduce nitrate to nitrite and contribute to color formation and safety. However, these microorganisms must be added in high amounts because they grow little or even die just at the onset of fermentation, when low pH conditions prevail. Preferably, low pH–tolerant strains should be carefully selected. The species from this family also have an important catalase activity that contributes to color stability and, somehow, prevention of lipid oxidation.

Yeasts

Debaryomyces hansenii is the predominant yeast in fermented meats, mainly growing in the outer area of the sausage due to its aerobic metabolism. *D. hansenii* has a good lipolytic activity and is able to degrade lactic acid. In addition, it also exhibits an important deaminase/deamidase activity, using free amino acids as substrates and producing ammonia as a subproduct that raises the pH in the sausage (Durá et al. 2002).

Molds

Some typical Mediterranean dry fermented sausages have molds on the surface. The most usual are *Penicillium nalgiovense* and *P. chrysogenum*. They contribute to flavor, through their proteolytic and lipolytic activity, and to appearance, in the form of a white coating on the surface (Sunesen and Stahnke 2003). They also generate ammonia through their deaminase and deamidase activity and contribute to pH rise. Inoculation of sausages with natural molds present in the fermentation room is dangerous because

Table 28.3. Main Roles and Effects of Microorganisms in Fermented Meats

Group	Microorganisms	Primary Role	Secondary Role	Chemical Contribution	Effects
Lactic acid bacteria	*L. sakei, L. curvatus, L. plantarum, L. pentosus, P. acidilactici, P. pentosaceus*	Glycolysis	Proteolysis (endo and exo)	Lactic acid generation Generation of free amino acids	pH drop Safety Firmness Taste
Micrococcaceae	*K. varians, S. Xylosus, S. carnosus*	Nitrate reductase, lipolysis, catalase	Proteolysis (exo)	Reduction of nitrate to nitrite Generation of free fatty acids, ready for oxidation Generation of free amino acids Degradation of hydrogen peroxide	Safety Aroma Taste
Yeasts	Debaryomyces hansenii	Lipolysis	Deamination/ deamidation Transamination	Generation of free fatty acids Transformation of amino acids Lactic acid consumption and generation of ammonia	Aroma Taste pH rise
Molds	*Penicillium nalgiovense, P. chrysogenum*	Lipolysis, proteolysis (exo)	Deamination/ deamidation Transamination	Generation of free fatty acids, ready for oxidation Generation of free amino acids and its transformation Generation of ammonia	Aroma Taste pH rise External aspect

toxigenic molds might grow. So fungal starter cultures are mainly used as a preventive measure against the growth of other mycotoxin-producing molds and give a typical white color on the surface as demanded in certain Mediterranean areas.

Casings

Casings may be natural, semisynthetic, or synthetic, but a common required characteristic is its permeability to water and air. Natural casings are natural portions of the gastrointestinal tract of pigs, sheep, and cattle, and although irregular in shape, they have good elasticity, tensile strength, and permeability. Natural casings are typically used for traditional sausages because they give a homemade aspect to the product. Semisynthetic casings are based on collagen that shrinks with the product and is permeable, but cannot be overstuffed (Toldrá et al. 2004). Synthetic cellulose-based casings are nonedible but are preferred for industrial processes due to important advantages such as controlled and regular pore size, uniformity for standard products, and hygiene. These casings are easily peeled off.

A wide range of sizes, between 2 and 15 cm, may be used, depending on the type of product. Of course, the diameter strongly affects fermentation and drying conditions. So pH drop is more important in large diameter sausages, where drying is more difficult to achieve.

PROCESSING STAGE 1: COMMINUTION

An example of a flow diagram for the processing of fermented sausages is shown in Figure 28.1. Chilled meats, pork alone or mixtures of pork and beef, and porcine fats are submitted to comminution in a grinder (Fig. 28.2). There are several plates with different hole sizes depending on the desired particle size. Previous trimming for removal of connective tissue is recommended, especially when processing undry or semidry fermented sausages where no further hydrolysis of collagen will occur. Salt, nitrate and/or nitrite, carbohydrates, microbial starters, spices, sodium ascorbate, and optionally, other nonmeat proteins are added to the ground mass, and the whole mix is homogenized under vacuum to avoid bubbles and undesirable oxidations that affect color and flavor (Fig. 28.3). Grinding and mixing take several minutes, depending on the amount. Industrial processes may use a cutter, as an alternative to grinding and mixing, when the required particle sizes are small. The cutter consists in a slowly moving bowl, containing the meats, fat, and additives, which rotates against a set of knives operating with rapid rotation. The fat and meat must be prefrozen (−6 to −7°C) to avoid smearing of fat particles during chopping. This phenomenon consists in a fine film of fat formed over the lean parts that may reduce the release of water during drying (Roca and Incze 1990). The cutter operates under vacuum to avoid any damage by oxygen, although the opera-

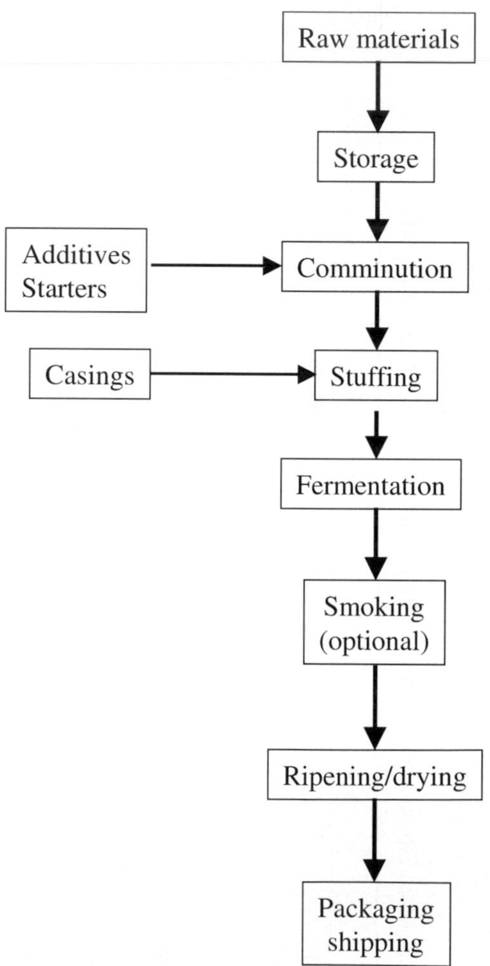

Figure 28.1. Flow diagram showing the most important stages in the processing of fermented sausages.

Figure 28.2. Grinding of meats and fats. There are many sizes of grinder plates in accordance to the required particle size.

Figure 28.3. Detail of the batter after mixing in a vacuum mixer massager.

tion takes only a short time, just a few minutes; and the ratio of bowl speed to knife speed determines the desired particle size.

PROCESSING STAGE 2: STUFFING

The mixture is stuffed under vacuum into casings, natural, collagen-based, or synthetic, with both extremes clipped. The vacuum avoids the presence of bubbles within the sausage and disruptions in the casing. The stuffing must be adequate in order to avoid smearing of the batter, and temperature must be kept below 2°C to avoid this problem. Once stuffed (Fig. 28.4), the sausages are hung in racks and placed in natural or air-conditioned drying chambers.

PROCESSING STAGE 3: FERMENTATION

FERMENTATION TECHNOLOGY

Once sausages are stuffed, they are placed in computer-controlled air-conditioned chambers and left to ferment for microbial growth and development. A typical chamber is shown in Figure 28.5. Temperature, relative humidity, and air speed must be carefully controlled in order to have correct microbial growth and enzyme action. The whole process can be considered as a lactic acid solid-state fermentation in which several simultaneous processes take place: (1) microbial growth and development, (2) biochemical changes, mainly enzymatic breakdown of carbohydrates, proteins, and lipids, and (3) physical changes, mainly acid gelation of meat proteins and drying.

Meat fermentation technology differs between the United States and Europe. High fermentation temperatures (35–40°C) are typical in U.S. sausages, followed by a mild heating process, as a kind of pasteurization, instead of drying, to kill any trichinellae.

Figure 28.4. Sausages stuffed into a collagen casing, 80 mm diameter, and clipped on both extremes.

Figure 28.5. Example of a fermentation/drying chamber with computer control of temperature, relative humidity, and air flow rate. (By courtesy of Embutidos y Conservas Tabanera, Segovia, Spain.)

Thus, starters such as *Lactobacillus plantarum* or *Pediococcus acidilactici,* which grow well at those temperatures, are typically used. In Europe, different technologies may be found, depending on the location and climate. There is a historical trend towards short-processed, smoked sausages in cold and humid countries, as in northern Europe, and long-processed, dried sausages in warmer and drier countries, as in the Mediterranean area. In the case of northern European (NES) countries, sausages are fermented for about 3 days at intermediate temperatures (25–30°C), followed by short ripening periods (up to 3 weeks). These sausages are subjected to a rapid pH drop and are usually smoked for a specific flavor (Demeyer and Stanhke 2002). On the other hand, Mediterranean sausages require longer processing times. Fermentation takes place at milder temperatures (18–24°C) for about 4 days, followed by mild drying conditions for a longer time, usually several weeks or months. *L. sakei* or *L. curvatus* are the LAB most often used as starter cultures (Toldrá et al. 2001). The time required for the fermentation stage is a function of the temperature and the type of microorganisms used as starters.

The technology is quite different in China and other Asian countries. Sausages are first dried over charcoal at 48°C and 65% relative humidity for 36 hours and then at 20°C and 75% relative humidity for 3 days. Water activity rapidly drops below 0.80, although pH remains about 5.9, which is a relatively high value. Fermentation is relatively poor, and the sour taste, which is considered undesirable, is reduced. Chinese raw sausage is consumed after heating (Leistner 1992).

Microbial Metabolism of Carbohydrates

The added carbohydrate is converted, during the fermentation, into lactic acid of either the D(-) or L(+) configuration, or a mixture of both, depending on the species of lactic acid bacteria used as starter. The ratio between the L and D enantiomers depends on the action of L and D lactate dehydrogenase, respectively, and the presence of lactate racemase. The rate of generation and the final amount of lactic acid depend on the type of LAB species used as starter, the type and content of carbohydrates, the fermentation temperature, and other processing parameters. The accumulation of lactic acid produces a pH drop more or less intense depending on its generation rate. Some secondary products such as acetic acid, acetoin, and others may be formed through heterofermentative pathways (Demeyer and Stahnke 2002). Acid pH favors protein coagulation, as pH approaches its isoelectric point, and thus also favors water release. Acid pH also contributes to safety by contributing to the inhibition of undesirable pathogenic or spoilage bacteria. The pH drop favors initial proteolysis and lipolysis by stimulating the activity of muscle cathepsin D and lysosomal acid lipase, both active at acid pH, but an excessive pH drop does not favor later enzymatic reactions involved in the generation of flavor compounds (Toldrá and Verplaetse 1995).

PROCESSING STAGE 4: RIPENING AND DRYING

Temperature, relative humidity, and air flow have to be carefully controlled during fermentation and ripening to allow correct microbial growth and enzyme action while maintaining adequate drying progress. The air velocity is kept at around 0.1 m/s, which is enough for a good homogenization of the environment. Ripening and drying are important for enzymatic reactions related to flavor development and obtaining the required water loss and thus reduction in water activity. The length of the ripening/drying period takes from 7 to 90 days, depending on many factors, including the kind of product, its diameter, dryness degree, fat content, desired flavor intensity, and so on. The reduction in a_w is slower in beef-containing sausages. The casing must remain attached to the sausage when it shrinks during drying. In general, long-ripened products tend to be drier and more flavorful.

Physical Changes

The most important physical changes during fermentation and ripening/drying are summarized in Figure 28.6. The acidulation produced during the fermentation stage induces protein coagulation and thus some water release. The acidulation also reduces the solubility of sarcoplasmic and myofibrillar proteins, and the sausage starts to develop consistency. The drying process is a delicate operation that must achieve an equilibrium between two different mass transfer processes, diffusion and evaporation (Baldini et al. 2000). Water inside the sausage must diffuse to the outer surface and then evaporate to the environment. Both rates must be in equilibrium because a very fast reduction in the relative humidity of the chamber would cause excessive evaporation from the sausage surface that would reduce the water content on the outer parts of the sausage, causing hardening. This is typical of sausages of large diameter because of the slow water diffusion rate. The cross section of these sausages shows a darker, dry, hard outer ring. On the other hand, when the water diffusion rate is much higher than the evaporation rate, water accumulates on the surface of the sausage, causing a wrinkled casing. This situation may happen in small diameter sausages being ripened in a chamber with high relative humidity. The progress in drying reduces the water content, up to 20% weight loss in semidry sausages and 30% in dry

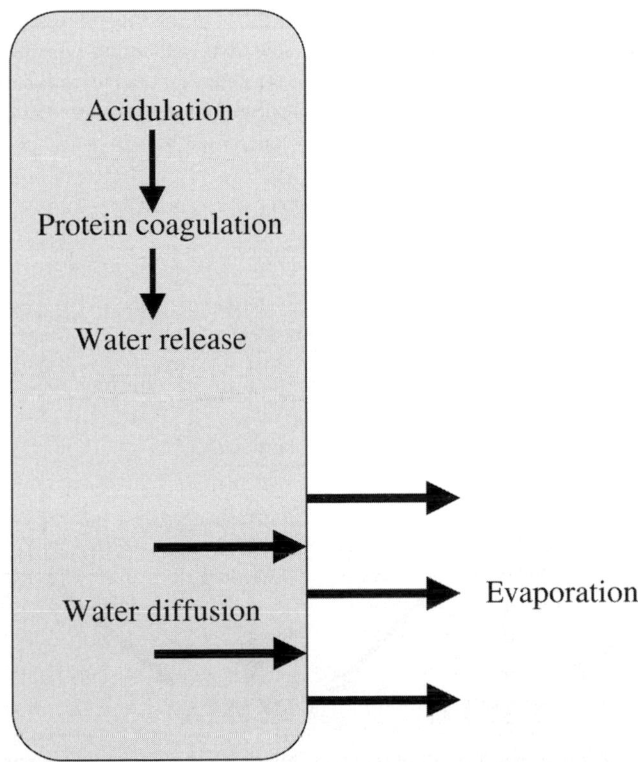

Figure 28.6. Scheme showing important physical changes during the processing of fermented meats.

sausages (Table 28.1). The water activity decreases according to the drying rate, reaching values below 0.90 for long-ripened sausages.

Chemical Changes

There are different enzymes, of both muscle and microbial origin, involved in reactions related to color, texture, and flavor generation. These reactions, which are summarized in Figure 28.7, are very important for the final sensory quality of the product. One of the most important group of reactions, mainly affecting myofibrillar proteins and yielding small peptides and free amino acids as final products, is known as proteolysis (Toldrá 1998). An intense proteolysis during fermentation and ripening is mainly carried out by endogenous cathepsin D, an acid muscle proteinase that is very active at acid pH. This enzyme hydrolyses myosin and actin, producing an accumulation of polypeptides that are further hydrolyzed to small peptides by muscle and microbial peptidylpeptidases and to free amino acids by muscle and microbial aminopeptidases (Sanz et al. 2002). The generation of small peptides and free amino acids increases with the length of processing, although the generation rate is reduced at acid pH values because the conditions are far from optimal for the enzyme activity. Free amino acids may be further transformed into other products, for example, volatile compounds through Strecker degradations and Maillard reactions; ammonia through deamination and/or deamidation reactions by deaminases and deamidases, respectively, present in yeasts and molds; or amines by microbial decarboxylases.

Another important group of enzymatic reactions, affecting muscle and adipose tissue lipids, is known as lipolysis (Toldrá 1998). Thus, a large amount of free fatty acids (between 0.5 and 7%) is generated through the enzymatic hydrolysis of triacylglycerols and phospholipids. Most of the observed lipolysis is attributed, after extensive studies on model sterile systems and sausages with added antibiotics, to endogenous lipases present in muscle and adipose tissue (e.g., lysosomal acid lipase, present in the lysosomes and very active at acid pH; Toldrá 1992, Hierro et al. 1997, Molly et al. 1997).

Catalases are mainly present in microorganisms such as *Kocuria* and *Staphylococcus;* they are responsible for peroxide reduction and thus contribute to color and flavor stabilization. Nitrate reductase, also present in these microorganisms, is also important for reducing nitrate to nitrite in slow-ripened sausages with an initial addition of nitrate. Recently, two strains of *Lactobacillus fermentum* have proved to be able to generate nitric oxide and give an acceptable color in sausages without nitrate/nitrite.

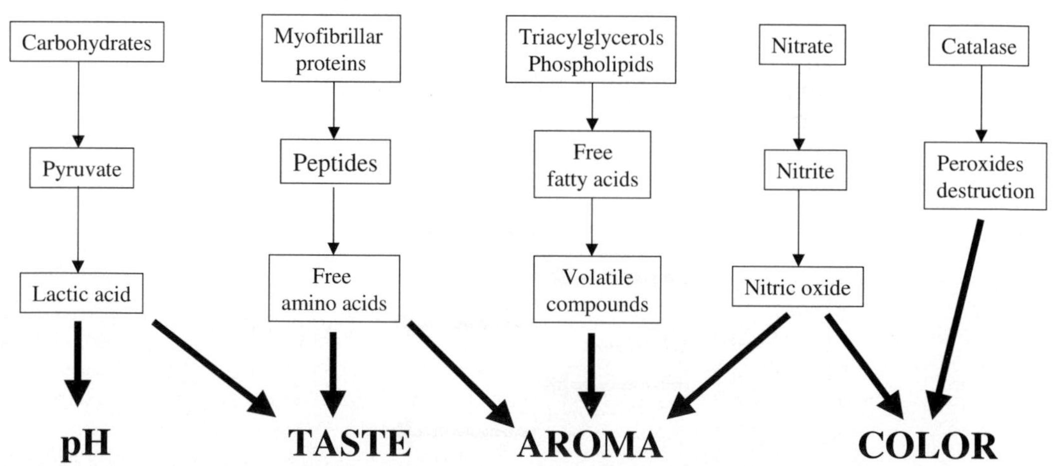

Figure 28.7. Scheme showing the most important reactions by muscle and microbial enzymes involved in chemical and biochemical changes affecting the sensory quality of fermented meats.

This could be used to produce cured meats free of nitrate and nitrite (Moller et al. 2003).

PROCESSING STAGE 5: SMOKING

Smoking is mostly applied in northern countries with cold and/or humid climates. Initially, it was used for preservation purposes, but today its contribution to flavor and color is more important (Ellis 2001). In some cases, smoking can be applied just after fermentation or even at the start of the fermentation. Smoking can be accompanied by heating at 60°C and has a strong impact on the final sensory properties. It has a strong antioxidative effect and gives a characteristic color and flavor to the product, which is the primary role of smoking. The bacteriostatic effect of smoking compounds inhibits the growth of yeasts, molds, and certain bacteria.

SAFETY

The stability of the sausage against pathogen and/or spoilage microorganisms is the result of successive hurdles (Leistner 1992). Initially, the added nitrite curing salt is very important for the microbial stability of the mix. During mixing under vacuum, oxygen is gradually removed, and redox potential is reduced. This effect is enhanced when ascorbic acid or ascorbate is added. Low redox potential values inhibit aerobic bacteria and make nitrite more effective as bactericide. During the fermentation, lactic acid bacteria can inhibit other bacteria, not only by the generation of lactic acid (and the subsequent pH drop), but also by generation of other metabolic products such as acetic acid and hydrogen peroxide and, especially, bacteriocins (low molecular mass peptides synthetized in bacteriocin-positive strains; Lücke 1992). The drying of the sausage continues the reduction in water activity to low values (a_w below 0.92) that inhibit growth of spoilage and/or pathogenic microorganisms. Thus, the correct interaction of all these factors assures the stability of the product.

Some foodborne pathogens that might be found in fermented meats are briefly described. *Salmonella* is more usual in fresh, spreadable sausages (Lücke 1985) but can be inhibited by acidification to pH 5.0 and/or drying to $a_w < 0.95$ (Talon et al. 2002). Lactic acid bacteria exert an antagonistic effect against *Salmonella* (Roca and Incze 1990). *Staphylococcus aureus* may grow under aerobic or anaerobic conditions and requires $a_w < 0.91$ for inhibition, but is sensitive to acid pH. So, it is important to control the elapsed time before reaching the pH drop in order to avoid toxin production. Furthermore, this toxin is produced only in aerobic conditions (Roca and Incze 1990). *Clostridium botulinum* and its toxin production capability are affected by a rapid pH drop and low a_w even more than by the addition of lactic acid bacteria and nitrite (Lücke 1985). *Listeria monocytogenes* is limited in growth at $a_w < 0.90$ combined with low pH values and specific starter cultures (Hugas et al. 2002). *Escherichia coli* is rather resistant to low pH and a_w but is reduced when exposed to $a_w < 0.91$ (Nissen and Holck 1998). Adequate prevention measures consist in correct cooling and a hazard analysis critical control point (HACCP) plan with application of good manufacturing practices (GMP), sanitation, and strict hygiene control of personnel and raw materials.

In recent years, most attention has been paid to biopreservation as a way to enhance preservation against spoilage bacteria and foodborne pathogens. The bioprotective culture consists in a competitive bacterial strain that grows very fast or produces antagonistic substances like bacteriocins. Another precise way consists in the direct addition of purified bacteriocins. Those bacteriocins belonging to group IIa (also called pediocin-like) that display inhibition against *Listeria* have been reported to be the most interesting for the meat industry (Hugas et al. 2002).

Parasites like *Trichinella spiralis* are almost eliminated through modern breeding systems. Pork meat free of trichinae must be used as raw material for fermented sausages; otherwise, heat treatments of the sausage to reach internal temperatures above 62.2°C are required to inactivate them (Sebranek 2004).

The generation of undesirable compounds, listed in Table 28.4, depends on several factors. The most important factor is the hygienic quality of the raw materials. For instance, the presence of cadaverine and/or putrescine may be indicative of the presence of contaminating meat flora. Another factor is processing conditions, which may favor the generation of biogenic amines; however, the type of natural flora or microbial starters used for the process is the most important issue, because the presence of microorganisms with decarboxylase activity can induce the generation of biogenic amines. In general, tyramine is the amine generated in higher amounts, and it is formed by certain lactic acid bacteria that exhibit enzymatic activity for decarboxylation of tyrosine (Eerola et al. 1996). Tyramine releases noradrenaline from the sympathetic nervous system, and the peripheral vasoconstriction and increase in cardiac output result in higher blood pressure and risk for hypertensive crisis (Shalaby 1996). However, the estimated tolerance level for tyramine (100–800 mg/kg) is higher than for other amines (Nout 1994). The amines derived from foods are generally degraded in humans by the enzyme monoamine oxidase (MAO) through oxidative deamination reactions. Those consumers using MAO inhibitors are less protected against amines and are thus susceptible for risk situations such as hypertensive crisis when ingesting significant amounts of amines. Other amines may also cause problems; for example, phenylethylamine, which may cause migraine and an increase in blood pressure; or histamine, which excites the smooth muscles of the uterus, the intestine, and the respiratory tract. One way to reduce health risks from amines consists in the use of starter cultures that are unable to produce amines but are competitive against amine-producing microorganisms. Additionally, the use of microorganisms that exhibit amine oxidase activity and are able to degrade amines, the selection of raw materials of high quality, and the use of GMP assures products of high quality and reduced risks (Talon et al. 2002). Finally, the generation of nitrosamines during the process is almost negligible due to the restricted amount of nitrate and/or nitrite that can be initially added and the low amount of residual nitrite remaining by the end of the process (Cassens 1997).

The processing conditions may favor the oxidation of cholesterol. Some generated oxides can be involved in cardiovascular-related diseases (e.g., 7-ketocholesterol and 5,6-α-epoxycholesterol), but in general, the reported levels of all cholesterol oxides is very low, less than 0.15 mg/100g, for exerting any toxic effect (Demeyer et al. 2000).

FINISHED PRODUCT

Once the product is finished, it is packaged and distributed. Fermented sausages can be sold as either entire or as thin slices (Fig. 28.8). The developed color, texture, and flavor depend on the processing and type of product. Main sensory properties are described below.

Color

The color of the sausage depends on the moisture and fat content as well as its content of hemoprotein, particularly myoglobin. Color is also influenced by pH drop rate and the ultimate pH, but it may be also affected by the presence of spices like red pepper. An excess of acid generation by lactobacilli may also affect color.

The characteristic color is due to the action of nitrite with myoglobin. Nitrite is reduced to nitric oxide, favored by the presence of ascorbate/erythorbate. Myoglobin and nitric oxide may then interact to form nitric oxide myoglobin, which gives the characteristic cured pinkish-red color (Pegg and Shahidi 1996). This reaction is favored at low pH. Long-processing sausages using nitrate need some time for the growth of Micrococcaceae before pH

Table 28.4. Safety Aspects: Generation of Undesirable Compounds in Dry Fermented Meats

Compounds	Route of Formation	Origin	Concentrations (mg/100 g)
Tyramine	Microbial decarboxylation	Tyrosine	< 16.0
Tryptamine	Microbial decarboxylation	Trytophane	< 6.0
Phenylethylamine	Microbial decarboxylation	Phenylalanine	< 3.5
Cadaverine	Microbial decarboxylation	Lysine	< 0.6
Histamine	Microbial decarboxylation	Histidine	< 3.6
Putrescine	Microbial decarboxylation	Ornithine	< 10.0
Spermine	Microbial decarboxylation	Methionine	< 3.0
Spermidine	Microbial decarboxylation	Methionine	< 0.5
Cholesterol oxides	Oxidation	Cholesterol	< 0.15

Sources: Adapted from Maijala et al. 1995, Shalaby 1996, Hernández-Jover et al. 1997, Demeyer et al. 2000.

Figure 28.8. Picture of a typical small-diameter salchichón, showing its cross section.

drops. Nitrate reductase, which is present in Micrococcaceae, reduces nitrate to nitrite, which is afterwards further reduced to nitric oxide, which can react with myoglobin. Oxidative discoloration consists in the conversion of nitrosylmyoglobin to nitrate and metmyoglobin, which affects the oxidative stability because of the prooxidant effect of ferric heme.

Texture

The consistency of fermented meats is initiated with the salt addition and pH reduction. The water-binding capacity of myofibrillar proteins decreases as pH approaches the proteins' isoelectric point and releases water. The solubility of myofibrillar proteins is also reduced, with a trend towards aggregation and coagulation, forming a gel. The consistency of this gel increases with water loss during drying. So, there is a continuous development of textural characteristics such as firmness, hardness, and cohesiveness of meat particles during drying (Toldrá 2002). The meat:fat ratio may affect some of these textural characteristics, but in general, the final texture of the sausage will mainly depend on the extent of drying (Toldrá et al. 2004).

Flavor

Little or no flavor is usually detected before meat fermentation, although a large number of flavor precursors are present. As fermentation and further ripening/drying progresses, the combined effect of endogenous muscle enzyme and microbial activity produces a high number of nonvolatile and volatile compounds with sensory impact. The longer the process, the more the accumulation of these compounds is increased and their sensory impact enhanced. Although not so important as in meat cooking, some compounds with sensory impact may be produced through further chemical reactions. The addition of spices also has an intense contribution to specific flavors.

Taste

The main nonvolatile compounds contributing to taste of fermented meats are summarized in Table 28.5. Sour taste, mainly resulting from lactic acid generation through microbial glycolysis, is the most relevant taste in fermented meats. Sourness is also correlated with other microbial metabolites such as acetic acid. Ammonia may be generated through deaminase and deamidase activity, usually present in yeasts and molds, reducing the intensity of the acid taste. Salty taste is usually perceived as a direct taste from salt addition. ATP-derived compounds such as inosine monophosphate and guanosine monophosphate exert some taste enhancement, while hypoxanthine contributes to bitterness. Other taste contributors are those compounds resulting from protein hydrolysis. The generation and accumulation of small peptides and free amino acids contribute to taste perception, which increases with the length of process. Some of these small peptides (e.g., leucine, isoleucine, and valine) also act as aroma precursors, as described below.

Aroma

The origin of aroma mainly depends on the ingredients and processing conditions. Different pathways are responsible for the formation of volatile compounds with aroma impact (Table 28.6). As mentioned above, proteolysis originates a large amount of small peptides and free amino acids. Microorganisms can convert the amino acids leucine, isoleucine, valine, phenylalanine, and methionine to important sensory compounds with low threshold values. Some of the most important are branched aldehydes such as 2- and 3-methylbutanal and 2-methylpropanal, branched alcohols, acids such as

Table 28.5. Quality Aspects: Generation or Presence of Desirable Nonvolatile Compounds Contributing to Taste in Fermented Meats

Group of Compounds	Main Representative Compounds	Routes of Generation	Presence in Final Product	Main Contribution	Expected Intensity
Peptides	Tri- and dipeptides	Proteolysis	Increases with length of process	Taste	High
Free amino acids	Glutamic acid, aspartic acid, alanine, lysine, threonine	Proteolysis	Increases with length of process	Taste	High
Nucleotides and nucleosides	Inosine monophosphate, guanosine monophosphate, inosine, hypoxanthine	ATP degradation	Around 100 mg/100g	Taste enhancement	Low
Long-chain free fatty acids	Oleic acid, linoleic acid, linolenic acid, arachidonic acid, palmitic acid	Lipolysis	Increases with length of process	Taste	Low
Short-chain fatty acids	Acetic acid, propionic acid	Microbial metabolism	Depends on microflora	Taste	Medium
Acids	Lactic acid	Glycolysis	Depends on initial amount of sugar and fermentation	Sour taste	High
Carbohydrates	Glucose, lactose	Remaining (not consumed through glycolysis)	Depends on initial amount of sugar and microflora	Sweet taste	Low
Inorganic compounds	Salt	Addition	Depends on initial amount	Salty taste	High

Table 28.6. Quality Aspects: Generation of Desirable Volatile Compounds Contributing to Aroma in Fermented Meats

Group of Compounds	Main Representative Compounds	Routes of Generation	Main Aroma	Expected Contribution
Aliphatic aldehydes	Hexanal, pentanal, octanal, etc.	Oxidation of unsaturated fatty acids	Green	High
Strecker aldehydes	2- and 3-methylbutanal, etc.	Strecker degradation of free amino acids	Roasted cocoa, cheesy-green	High
Branched-chain acids	2- and 3-methylbutanoic acid	Secondary products of previous Strecker degradation	Sweaty	Medium
Alcohols	Ethanol, butanol, etc.	Oxidative decomposition of lipids	Sweet, alcohol, etc.	Low
Ketones	2-pentanone, 2-heptanone, 2-octanone, etc.	Lipid oxidation	Ethereal, soapy	Medium
Sulfides	Dimethyldisulfide	Strecker degradation of sulfur-containing amino acids (methionine)	Dirty socks	Low
Esters	Ethyl acetate, ethyl 2-methylbutanoate	Interaction of carboxylic acids and alcohols	Pineapple, fruity	High
Hydrocarbons	Pentane, heptane, etc.	Lipids autoxidation	Alkane	Very low
Dicarbonyl products	Diacetyl, acetoin, acetaldehyde	Pyruvate microbial metabolism	Butter	Low
Nitrogen compounds	Ammonia	Deamination, deamidation	Ammonia	Variable, depends on growth of yeasts and molds

Sources: Adapted from Flores et al. 1997, Viallon et al. 1996, Stahnke 2002, Toldrá 2002, and Talon et al. 2002.

2- and 3-methylbutanoic and 2-methylpropanoic acids, and esters such as ethyl 2- and 3-methylbutanoate (Stahnke 2002). Some of these branched-chain aldehydes may also be formed through the Strecker degradation, consisting in the reaction of amino acids with diketones. However, conditions found in sausages are far from those optimal for this kind of reaction, which needs high temperature and low water activity (Talon et al. 2002).

Methyl ketones may be formed either by β-oxidation of free fatty acids or decarboxylation of free β-keto acids. Other nonbranched aliphatic compounds generated by lipid oxidation are alkanes, alkenes, aldehydes, alcohols, and several furanic cycles.

A large number of volatile compounds are generated by chemical oxidation of the unsaturated fatty acids. These volatile compounds are mainly generated during ripening and further storage. Other low molecular weight volatile compounds are generated by microorganisms from carbohydrate catabolism. The most usual compounds are diacetyl, acetoin, butanediol, acetaldehyde, ethanol, and acetic propionic and butyric acids. However, some of these compounds may be derived from pyruvate originated through other metabolic pathways than carbohydrate glycolysis (Demeyer and Stahnke 2002, Demeyer and Toldrá 2004). The flavor profile may have important variations dependant on the type of microorganisms used as starters (Berdagué et al. 1993).

GLOSSARY

ATP—Adenosine triphosphate.

Aminopeptidases—Exopeptidases that catalyze the release of an amino acid from the amino terminus of a peptide.

Backslopping—Traditional practice consisting in the addition of previously fermented sausage with successful sensory properties.

Bacteriocin—Peptides of low molecular mass produced by lactic acid bacteria with inhibitory action against certain spoilage bacteria and foodborne pathogens.

Catalase—Enzyme able to catalyze the decomposition of hydrogen peroxide into molecular oxygen and water.

Cathepsins—Enzymes located in lysosomes and able to hydrolyze myofibrillar proteins to polypeptides.

Decarboxylases—Enzymes able to transform an amino acid into an amine.

DFD—Pork meat with dark, firm, and dry characteristics due to a lack of carbohydrate in muscle and thus poor glycolysis and reduced latic acid generation. These meats have pH values above 6.0 after 24 hours postmortem and are typical of exhausted, stressed pigs before slaughtering.

Glycolysis—Enzymatic breakdown of carbohydrates with the formation of pyruvic acid and lactic acid and the release of energy in the form of ATP.

Homofermentative—Bacteria generating a single end product (lactic acid) from fermentation of carbohydrates.

Heterofermentative—Bacteria generating several end products (lactic acid, acetoin, ethanol, CO_2, etc.) from fermentation of carbohydrates.

Lactate dehydrogenase—Enzyme that catalyzes the oxidation of pyruvic acid to lactic acid.

Lactate racemase—Enzyme that catalyzes lactic acid racemization reactions.

Lipolysis—Enzymatic breakdown of lipids with the formation of free fatty acids.

Lysosomal acid lipase—Enzyme that catalyzes the release of fatty acids by hydrolysis of triacylglycerols at positions 1 and 3.

Peroxide value—Term used to measure rancidity and expressed as millimoles of peroxide taken up by 1000 g of fat.

Proteolysis—Enzymatic breakdown of proteins with the formation of peptides and free amino acids.

PSE—Pork meat with pale, soft, and exudative characteristics resulting from an accelerated glycolysis and thus rapid lactic acid generation. The pH drop is very fast, reaching values as low as 5.6 in just 1 hour postmortem.

Water activity (a_w)—It indicates the availability of water in a food and is defined as the ratio of the equilibrium water vapor pressure over the system to the vapor pressure of pure water at the same temperature.

REFERENCES

Baldini P, Cantoni E, Colla F, Diaferia C, Gabba L, Spotti E, Marchelli R, Dossena A, Virgili R, Sforza S, Tenca P, Mangia A, Jordano R, Lopez MC, Medina L, Coudurier S, Oddou S, Solignat G. 2000. Dry sausages ripening: Influence of thermohygrometric conditions on microbiological, chemical and physico-chemical characteristics. Food Research Int 33:161–170.

Berdagué JL, Monteil, P, Montel, MC, Talon R. 1993. Effects of starter cultures on the formation of flavour compounds in dry sausages. Meat Sci 35:275–287.

Campbell-Platt G. 1995. Fermented meats—A world perspective. In: G Campbell-Platt, PE Cook, editors. Fermented Meats, pp. 39–51. London: Blackie Academic and Professional.

Cassens RG. 1997. Composition and safety of cured meats in the USA. Food Chem 59:561–566.

Demeyer D. 1992. Meat fermentation as an integrated process. In: FJM Smulders, F Toldrá, J Flores, M Prieto, editors. New Technologies for Meat and Meat Products, pp. 21–36. Nijmegen (The Netherlands): Audet.

Demeyer D, Stahnke L. 2002. Quality control of fermented meat products. In: J Kerry, J Kerry, D Ledward, editors. Meat processing: Improving quality, pp. 359–393. Cambridge: Woodhead Publishing. Co.

Demeyer D, Toldrá F. 2004. Fermentation. In: W Jensen, C Devine, M Dikemann, editors. Encyclopedia of Meat Sciences, pp. 467–474. London: Elsevier Science.

Demeyer DI, Raemakers M, Rizzo A, Holck A, De Smedt A, Ten Brink B, Hagen B, Montel C, Zanardi E, Murbrek E, Leroy F, Vanderdriessche F, Lorentsen K, Venema K, Sunesen L, Stahnke L, De Vuyst L, Talon R, Chizzolini R, Eerola S. 2000. Control of bioflavor and safety in fermented sausages: First results of a European project. Food Research Int 33: 171–180.

Durá A, Flores M, Toldrá F. 2002. Purification and characterization of a glutaminase from *Debaryomices* spp. Int J Food Microbiol 76:117–126.

Eerola S, Maijala R, Roig-Sangués AX, Salminen M, Hirvi T. 1996. Biogenic amines in dry sausages as affected by starter culture and contaminant amine-positive *Lactobacillus*. J Food Sci 61:1243–1246.

Ellis DF. 2001. Meat smoking technology. In: YH Hui, WK Nip, RW Rogers, OA Young, editors. Meat Science and Applications, pp. 509–519. New York: Marcel Dekker, Inc.

Flores J, Toldrá F. 1993. Curing: Processes and applications. In: R MacCrae, R Robinson, M Sadle, G Fullerlove,editors. Encyclopedia of Food Science, Food Technology and Nutrition, pp.1277–1282. London: Academic Press.

Gray JY, Pearson AM. 1984. Cured meat flavor. In: CO Chichester, EM Mrak, BS Schweigert, editors. Advances in Food Research, pp. 2–70, Orlando, Florida: Academic Press.

Hernández P, Navarro JL, Toldrá F. 1999. Effect of frozen storage on lipids and lipolytic activities in the *Longissimus dorsi* muscle of the pig. Z Lebensm Unters Forsch. A 208:110–115.

Hierro E, De la Hoz L, Ordoñez JA. 1997. Contribution of microbial and meat endogenous enzymes to the lipolysis of dry fermented sausages. J Agric Food Chem 45:2989–2995.

Hugas M, Garriga M, Aymerich MT, Monfort JM. 2002. Bacterial cultures and metabolites for the enhancement of safety and quality of meat products. In: F Toldrá, editor. Research Advances in the Quality of Meat and Meat Products, pp. 225–247. Trivandrum (India): Research Signpost.

Leistner L. 1992. The essentials of producing stable and safe raw fermented sausages. In: FJM Smulders, F Toldrá, J Flores, M Prieto, editors. New Technologies for Meat and Meat Products, pp. 1–19. Nijmegen (The Netherlands): Audet.

Lücke FK. 1985. Fermented sausages. In: BJB Wood, editor. Microbiology of Fermented Foods, pp. 41–83. London: Elsevier Applied Science.

———. 1992. Prospects for the use of bacteriocins against meat-borne pathogens. In: FJM Smulders, F Toldrá, J Flores, M Prieto, editor. New Technologies for Meat and Meat Products, pp. 37–52. Nijmegen (The Netherlands): Audet.

Moller JKS, Jensen JS, Skibsted LH, Knöchel S. 2003. Microbial formation of nitrite-cured pigment, nitrosylmyoglobin, from metmyoglobin in model systems and smoked fermented sausages by Lactobacillus fermentum strains and a commercial starter culture. Eur Food Res Technol 216:463–469.

Molly K, Demeyer DI, Johansson G, Raemaekers M, Ghistelinck M, Geenen I. 1997. The importance of meat enzymes in ripening and flavor generation in dry fermented sausages. First results of a European project. Food Chem 54:539–545.

Nissen H, Holck AL. 1998. Survival of *Escherichia coli* O157:H7, *Listeria monocytogenes* and *Salmonella kentucky* in Norwegian fermeted dry sausage. Food Microbiol 15:273–279.

Nout MJR. 1994. Fermented foods and food safety. Food Res Int 27:291–296.

Ordoñez JA, Hierro EM, Bruna JM, de la Hoz L. 1999. Changes in the components of dry-fermented sausages during ripening. Crit Rev Food Sci Nutr 39: 329–367.

Pegg BR, Shahidi F. 1996. A novel titration methodology for elucidation of the structure of preformed cooked cured-meat pigment by visible spectroscopy. Food Chem 56:105–110.

Roca M, Incze K. 1990. Fermented sausages. Food Reviews Int 6:91–118.

Sanz Y, Sentandreu MA, Toldrá F. 2002. Role of muscle and bacterial exopeptidases in meat fermentation. In: F Toldrá, editor. Research Advances in the Quality of Meat and Meat Products, pp. 143–155. Trivandrum (India): Research Signpost.

Sebranek JG 2004. Semi-dry fermented sausages. In: YH Hui, LM Goddik, J Josephsen, PS Stanfield, AS Hansen, WK Nip, F Toldrá, editors. Handbook of Food and Beverage Fermentation Technology. New York: Marcel Dekker, Inc. (In press.)

Shalaby AR. 1996. Significance of biogenic amines to food safety and human health. Food Res Int 29:675–690.

Stahnke L. 2002. Flavour formation in fermented sausage. In: F Toldrá, editor. Research Advances in the Quality of Meat and Meat Products, pp. 193–223. Trivandrum (India): Research Signpost.

Sunesen LO, Stahnke LH. 2003. Mould starter cultures for dry sausages-selection, application and effects. Meat Sci 65:935–948.

Talon R, Leroy-Sétrin S, Fadda S. 2002. Bacterial starters involved in the quality of fermented meat products. In: F Toldrá, editor. Research advances in the quality of meat and meat products, pp. 175–191. Trivandrum (India): Research Signpost.

Toldrá F. 1992. The enzymology of dry-curing of meat products. In: FJM Smulders, F Toldrá, J Flores, M Prieto, editors. New Technologies for Meat and Meat Products, pp. 209–231. Nijmegen (The Netherlands): Audet.

———. 1998. Proteolysis and lipolysis in flavour development of dry-cured meat products. Meat Sci 49:s101–s110.

———. 2002. Dry-Cured Meat Products. Pp. 1–238. Trumbull, Connecticut: Food and Nutrition Press.

Toldrá F, Gavara G, Lagarón JM. 2004. Packaging and quality control. In: YH Hui, LM Goddik, J Josephsen, PS Stanfield, AS Hansen, WK Nip, F Toldrá, editors. Handbook of Food and Beverage Fermentation Technology, pp. 445–458. New York: Marcel Dekker, Inc.

Toldrá F, Sanz Y, Flores M. 2001. Meat fermentation technology. In: YH Hui, WK Nip, RW Rogers, OA Young, editors. Meat Science and Applications, pp. 537–561. New York: Marcel Dekker, Inc.

Toldrá F, Verplaetse A. 1995. Endogenous enzyme activity and quality for raw product processing. In: K Lündstrom, I Hansson, E Winklund, Composition of Meat in Relation to Processing, Nutritional and Sensory Quality, pp. 41–55. Uppsala (Sweden): Ecceamst.

Viallon C, Berdagué JL, Montel MC, Talon R, Martin JF, Kondjoyan N, Denoyer C. 1996. The effect of stage of ripening and packaging on volatile content and flavour of dry sausage. Food Res Int 29, 667–674.

29
Biochemistry and Fermentation of Beer*

R. Willaert

INTRODUCTION

The production of alcoholic beverages is as old has history. Wine may have an archeological record going back more than 7500 years, with the early suspected wine residues dating from early to mid-fifth millennium BC (McGovern et al. 1996). Clear evidence of intentional wine making first appears in the representations of wine presses that date back to the reign of Udimu in Egypt, some 5000 years ago. The direct fermentation of fruit juices, such as that of grape, had doubtlessly taken place for many thousands of years before early thinking man developed beer brewing and, probably coincidentally, bread baking (Hardwick 1995). The oldest historical evidence of formal brewing dates back to about 6000 BC in ancient Babylonia: a piece of pottery found there shows workers either stirring or skimming a brewing vat.

Nowadays, alcoholic beverage production represents a significant contribution to the economies of many countries. The most important beverages today are beer, wine, distilled spirits, cider, sake, and liqueurs (Lea and Piggott 1995). In Belgium ("the beer paradise"), beer is the most important alcoholic beverage, although the beer consumption has declined in the last 40 years: from 11,096,717 hL in 1965 to 10,059,513 hL in 2001 (N.N. 2002). In this time frame, wine consumption doubled from 1,059,964 to 2,215,579 hL. Another trend is the spectacular increase in water and soft drink consumption (from 5,215,056 hL to 24,628,781 hL).

In this chapter, the biochemistry and fermentation of beer is reviewed. Firstly, the carbohydrate metabolism in brewer's yeast is discussed. The maltose metabolism is of major importance in beer brewing since this sugar is present in a high concentration in

*The information in this chapter has been derived from "Beer," in *Alcoholic Beverages Manual,* ©2004 by Ronnie Willaert. Used with permission.

wort. For the production of a high quality beer, a well-controlled fermentation needs to be performed. During fermentation, major flavor-active compounds are produced (and some of them are again degraded) by the yeast cells. The metabolism of the most important fermentation by-products during main and secondary fermentation is discussed in detail. The latest trend in beer fermentation technology is process intensification using immobilized cell technology. This new technology is explained, and some illustrative applications—small and large scale—are discussed.

THE BEER BREWING PROCESS

The principal raw materials used to brew beer are water, malted barley, hops, and yeast. The brewing process involves extracting and breaking down the carbohydrate from the malted barley to make a sugar solution (called "wort"), which also contains essential nutrients for yeast growth, and using this as a source of nutrients for "anaerobic" yeast growth. During yeast fermentation, simple sugars are consumed, releasing energy and producing ethanol and other metabolic flavoring by-products. The major biological changes that occur in the brewing process are produced by naturally produced enzymes from barley (during malting) and yeast. The rest of the brewing process largely involves heat exchange, separation, and clarification, which only produce minor changes in chemical composition when compared with the enzyme-catalyzed reactions. Barley is able to produce all the enzymes that are needed to degrade starch, β-glucan, pentosans, lipids, and proteins, which are the major compounds of interest to the brewer. An overview of the brewing process is shown in Figure 29.1, where the input and output flows also are indicated. Table 29.1 gives a more detailed explanation of each step in the process.

CARBOHYDRATE METABOLISM—ETHANOL PRODUCTION

Carbohydrate Uptake

Carbohydrates in wort make up 90–92% of wort solids. Wort from barley malt contains the fermentable sugars sucrose, fructose, glucose, maltose, and maltotriose, and some dextrin material (Table 29.2). The fermentable sugars typically make up 70–80% of the total carbohydrate (MacWilliam 1968). The three major fermentable sugars are glucose and the α-glucosides maltose and maltotriose. Maltose is by far the most abundant of these sugars, typically accounting for 50–70% of the total fermentable sugars in an all-malt wort. Sucrose and fructose are present in low concentrations. The unfermentable dextrins play little part in brewing. Wort fermentability may be reduced or increased by using solid or liquid adjuncts.

Brewing strains consume the wort sugars in a specific sequence: glucose is consumed first, followed by fructose, maltose, and finally maltotriose. The uptake and consumption of maltose and maltotriose is repressed or inactivated at elevated glucose concentrations. Only when 60% of the wort glucose has been taken up by the yeast, will the uptake and consumption of maltose start. Maltotriose uptake is inhibited by high glucose and maltose concentrations. When high amounts of carbohydrate adjuncts (e.g., glucose) or high-gravity wort are employed, the glucose repression is even more pronounced, resulting in fermentation delays (Stewart and Russell 1993).

The efficiency of brewer's yeast strains to effect alcoholic fermentation is dependent upon their ability to utilize the sugars present in wort. This ability very largely determines the fermentation rate as well as the final quality of the beer produced. In order to optimize the fermentation efficiency of the primary fermentation, a detailed knowledge of the sugar consumption kinetics, which is linked to the yeast growth kinetics, is required (Willaert 2001).

Maltose and Maltotriose Metabolism

The yeast *Saccharomyces cerevisiae* transports the monosaccharides across the cell membrane by the hexose transporters. There are 19, or possibly 20, genes encoding hexose transporters (Dickinson 1999). The disaccharide maltose and the trisaccharide maltotriose are transported by specific transporters into the cytoplasm, where these molecules are hydrolyzed by the same α-glucosidase, yielding two or three molecules of glucose, respectively (Panchal and Stewart 1979, Zheng et al. 1994).

Maltose utilization in yeast is conferred by any one of five *MAL* loci: *MAL1* to *MAL4* and *MAL6* (Bisson et al. 1993, Dickinson 1999). Each locus consists of three genes: gene 1 encodes a maltose

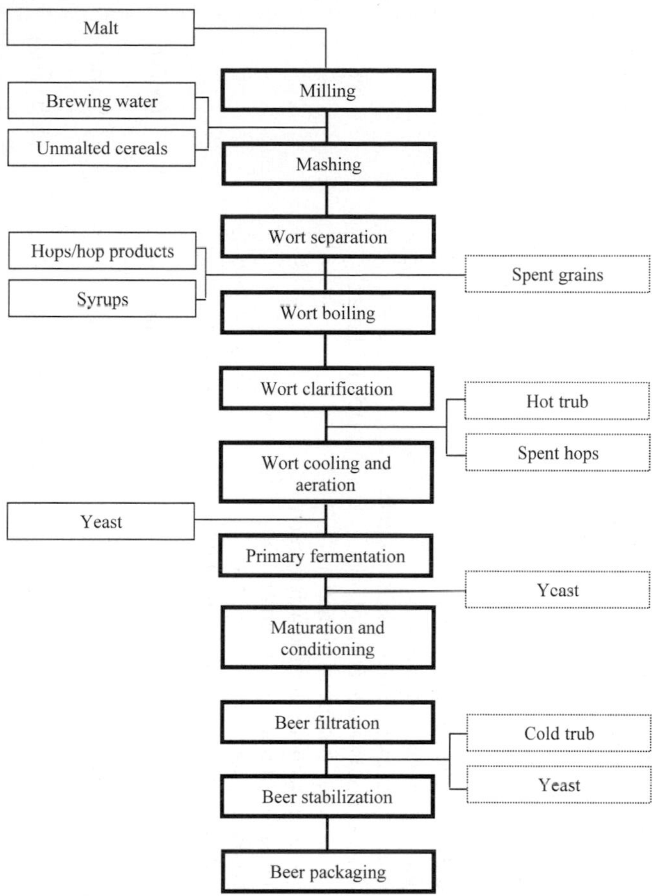

Figure 29.1. Schematic overview of the brewing process (input flows are indicated on the left side and output flows on the right side).

transporter (permease), gene 2 encodes a maltase (α-glucosidase), and gene 3 encodes a transcriptional activator of the other two genes. Thus, for example, the maltose transporter gene at the *MAL1* locus is designated *MAL61*. The three genes of a *MAL* locus are all required to allow fermentation. Some authors persist in using gene designations such as for the *MAL1* locus: *MAL1T* (transporter = permease), *MAL1R* (regulator) and *MAL1S* (maltase). The five *MAL* loci each map to a different chromosome. The *MAL* loci exhibit a very high degree of homology and are telomere linked, suggesting that they evolved by translocation from telomeric regions of different chromosomes (Michels et al. 1992). Since a fully functional or partial allele of the *MAL1* locus is found in all strains of *S. cerevisiae*, this locus is proposed as the progenitor of the other *MAL* loci (Chow et al. 1983). Gene dosage studies performed with laboratory strains of yeast have shown that the transport of maltose in the cell may be the rate-limiting step in the utilization of this sugar (Goldenthal et al. 1987). Constitutive expression of the maltose transporter gene (*MALT*) with high-copy-number plasmids in a lager strain of yeast has been found to accelerate the fermentation of maltose during high-gravity (24°P) brewing (Kodama et al. 1995). The constitutive expression of *MALS* and *MALR* had no effect on maltose fermentability.

The control over *MAL* gene expression is exerted at three levels. The presence of maltose induces,

Table 29.1. Overview of the Brewing Processing Steps: From Barley to Beer

Process	Action	Objectives	Time	Temperature (°C)
Malting				
Steeping	Moistening and aeration of barley	Preparation for the germination process	48 hours	12–22
Germination	Barley germination	Enzyme production, chemical structure modification	3–5 days	22
Kilning	Kilning of the green malt	Ending of germination and modification, production of flavoring and coloring substances	24–48 hours	22–110
Milling	Grain crushing without disintegrating the husks	Enzyme release and increase of surface area	1–2 hours	22
Mashing + wort separation	Addition of warm/hot water	Stimulation of enzyme action, extraction and dissolution of compounds, wort filtration, to obtain the desired fermentable extract as quick as possible	1–2 hours	30–72
Wort boiling	Boiling of wort and hops	Extraction and isomerization of hop components, hot break formation, wort sterilization, enzyme inactivation, formation of reducing, aromatic and coloring compounds, removal of undesired volatile aroma compounds, wort acidification, evaporation of water	0.5–1.5 hours	> 98
Wort clarification	Sedimentation or centrifugation	Removal of spent hops, clarification (whirlpool, centrifuge, settling tank)	< 1 hour	100–80
Wort cooling and aeration	Use of heat exchanger, injection of air bubbles	Preparing the wort for yeast growth	< 1 hour	12–18
Fermentation	Adding yeast, controlling the specific gravity, removal of yeast	Production of green beer, to obtain yeast for subsequent fermentations, carbon dioxide recovery	2–7 days	12–22 (ale) 4–15 (lager)

Maturation and conditioning	Beer storage in oxygen free tank, beer cooling, adding processing aids	Beer maturation, adjustment of the taste, adjustment of CO_2 content, sedimentation of yeast and cold trub, beer stabilization	7–21 hours	−1–0
Beer clarification	Centrifugation, filtration	Removal of yeast and cold trub	1–2 hours	−1–0
Biological stabilization	Pasteurization of sterile filtration	Killing or removing of microorganisms	1–2 hours	62–72 (pasteurization) −1–0 (filtration)
Packaging	Filling of bottles, cans, casks and kegs; pasteurization of small volumes in packings	Production of packaged beer according to specifications	0.5–1.5 hours	−1–room temperature

Table 29.2. Carbohydrate Composition of Worts

		Wort Carbohydrate Content		
Origin		Danish	Canadian	British
Type of Wort		Lager	Lager	Pale Ale
Original Gravity		1043.0	1054.0	1040.0
Fructose	(g/L)	2.1	1.5	3.3
	(%)[a]	2.7	1.6	4.8
Glucose	(g/L)	9.1	10.3	10.0
	(%)[a]	11.6	10.9	14.5
Sucrose	(g/L)	2.3	4.2	5.3
	(%)[a]	2.9	4.5	7.7
Maltose	(g/L)	52.4	60.4	38.9
	(%)[a]	66.6	64.2	56.5
Maltotriose	(g/L)	12.8	17.7	11.4
	(%)[a]	16.3	18.8	16.5
Total ferm. sugars	(g/L)	78.7	94.1	68.9
Maltotetraose	(g/L)	2.6	7.2	2.0
Higher sugars	(g/L)	21.3	26.8	25.2
Total dextrins	(g/L)	23.9	34.0	25.2
Total sugars	(g/L)	102.6	128.1	94.1

Source: Adapted from Hough et al. 1982.
[a]Percent of the total fermentable sugars.

whereas glucose represses, the transcription of *MALS* and *MALT* genes (Federoff et al. 1983a,b; Needleman et al. 1984). The constitutively expressed regulatory protein (*MALR*) binds near the *MALS* and *MALT* promotors and mediates the induction of *MALS* and *MALT* transcription (Cohen et al. 1984, Chang et al. 1988, Ni and Needleman 1990). Experiments with *MALR*-disrupted strains led to the conclusion that MalRp is involved in glucose repression (Goldenthal and Vanoni 1990, Yao et al. 1994). Relatively little attention has been paid to posttranscriptional control (i.e., the control of translational efficiency) or mRNA turnover as mechanisms complementing glucose repression (Soler et al. 1987). The addition of glucose to induced cells has been reported to cause a 70% increase in the lability of a mRNA population containing a fragment of *MALS* (Federoff et al. 1983a). The third level of control is posttranslational modification. In the presence of glucose, maltose permease is either reversibly converted to a conformational variant with decreased affinity (Siro and Lövgren 1979, Peinado and Loureiro-Dias 1986) or irreversibly proteolytically degraded, depending on the physiological conditions (Lucero et al. 1993, Riballo et al. 1995). The latter phenomenon is called catabolite inactivation.

Glucose repression is accomplished by the Mig1p repressor protein, which is encoded by the *MIG1* gene (Nehlin and Ronne 1990). It has been shown that Mig1p represses the transcription of all three *MAL* genes by binding upstream of them (Hu et al. 1995). The *MIG1* gene has been disrupted in a haploid laboratory strain and in an industrial polyploid strain of *S. cerevisiae* (Klein *et al.* 1996). In the *MIG1*-disrupted haploid strain, glucose repression was partly alleviated; that is, maltose metabolism was initiated at higher glucose concentrations than in the corresponding wild-type strain. In contrast, the polyploid Δ*mig1* strain exhibited an even more stringent glucose control of maltose metabolism than the corresponding wild-type strain, which could be explained by a more rigid catabolite inactivation of maltose permease, affecting the uptake of maltose.

Recently, the gene A*GT1*, which codes for a α-glucoside transporter, has been characterized (Han et al. 1995). *AGT1* is found in many *S. cerevisiae* laboratory strains and maps to a naturally occurring, partially functional allele of the *MAL1* locus. Agt1p is a highly hydrophobic, postulated integral membrane protein. It is 57% identical to Mal61p (the maltose permease encoded at *MAL6*) and is also a member of the 12 transmembrane domain superfamily of sugar transporters (Nelissen et al. 1995). Like Mal61p, Agt1p is a high-affinity, maltose/proton symporter, but Mal61p is capable of transporting only maltose and turanose, while Agt1p transports these two α-glucosides as well as several others, including isomaltose, α-methylglucoside, maltotriose, palatinose, trehalose, and melezitose. *AGT1* expression is maltose inducible, and induction is mediated by the Mal activator.

Brewing strains of yeast are polyploid, aneuploid, or, in the case of lager strains, alloploid. Recently, Jespersen et al. (1999) examined 30 brewing strains of yeast (five ale strains and 25 lager strains) with the aim of examining the alleles of maltose and maltotriose transporter genes contained by them. All the strains of brewer's yeast examined, except two, were found to contain *MAL11* and *MAL31* sequences, and only one of these strains lacked *MAL41*. *MAL21* was not present in the five ale strains and 12 of the lager strains. *MAL61* was not found in any of the yeast chromosomes other than those known to carry *MAL* loci. Sequences corresponding to the *AGT1* gene (transport of maltose and maltotriose) were detected in all but one of the yeast strains.

Wort maltotriose has the lowest priority for uptake by brewer's yeast cells and incomplete maltotriose uptake results in yeast-fermentable extract in beer, material loss, greater potential for microbiological stability, and sometimes atypical beer flavor profiles (Stewart and Russell 1993). Maltotriose uptake from wort is always slower with ale strains than with lager strains under similar fermentation conditions. However, the initial transport rates are similar to those of maltose in a number of ale and lager strains. Elevated osmotic pressure inhibits the transport and uptake of glucose, maltose, and maltotriose, with maltose and maltotriose being more sensitive to osmotic pressure than glucose in both lager and ale strains. Ethanol (5% w/v) stimulated the transport of maltose and maltotriose, due in all probability to an ethanol-induced change in the plasma membrane configuration, but had no effect on glucose transport. Higher ethanol concentrations inhibited the transport of all three sugars.

Wort Fermentation

Before the fermentation process starts, wort is aerated. This is a necessary step since oxygen is required for the synthesis of sterols and unsaturated fatty acids, which are incorporated in the yeast cell membrane (Rogers and Stewart 1973). It has been shown that both ergosterol and unsaturated fatty acids increase in concentration as long as oxygen is present in the wort (e.g., Haukeli and Lie 1979). A maximum concentration is obtained in 5–6 hours after pitching, but the formation rate is dependent upon pitching rate and temperature. Unsaturated fatty acids can also be taken up from the wort, but not all malt wort contains sufficient unsaturated lipids to support a normal yeast growth rate. Adding lipids to wort, especially unsaturated fatty acids, might be an interesting alternative (Moonjai et al. 2000, 2002). The oxygen required for lipid biosynthesis can also be introduced by oxygenation of the separated yeast cells.

Different devices are used to aerate the cold wort: ceramic or sintered metal candles, aeration plants employing Venturi pipes, two-component jets, static mixers, or centrifugal mixers (Kunze 1999). The principle of these devices is that very small air (oxygen) bubbles are produced and quickly dissolve during turbulent mixing.

As a result of this aeration step, carbohydrates are degraded aerobically during the first few hours of the "fermentation" process. The aerobic carbohydrate catabolism for a lager fermentation typically takes 12 hours.

During the first hours of the fermentation process, oxidative degradation of carbohydrates occurs through glycolysis and the Krebs (tricarboxylic acid, TCA) cycle. The energy efficiency of glucose oxidation is derived from the large number of $NADH_2^+$ produced for each mole of glucose oxidized to CO_2. The actual wort fermentation gives alcohol and carbon dioxide via the Embden-Meyerhof-Parnas (glycolytic) pathway. The reductive pathway from pyruvate to ethanol is important since it regenerates NAD^+. Energy is obtained solely from the ATP-producing steps of the Embden-Meyerhof-Parnas pathway. During fermentation, the activity of the TCA cycle is greatly reduced, although it still serves as a source of intermediates for biosynthesis (Lievense and Lim 1982).

Lagunas (1979) observed that during aerobic growth of *S. cerevisiae*, respiration accounts for less than 10% of glucose catabolism, the remainder being fermented. Increasing sugar concentrations resulting in a decreased oxidative metabolism is known as the Crabtree effect. This was traditionally explained as an inhibition of the oxidative system by high concentrations of glucose. Nowadays, it is generally accepted that the formation of ethanol at aerobic conditions is a consequence of a bottleneck in the oxidation of pyruvate (e.g., in the respiratory system; Petrik et al. 1983, Rieger et al. 1983, Käppeli et al. 1985, Fraleigh et al. 1989, Alexander and Jeffries 1990).

A reduction of ethanol production can be achieved by metabolic engineering of the carbon flux in yeast, resulting in an increased formation of other fermentation products. A shift of the carbon flux towards glycerol at the expense of ethanol formation in yeast was achieved by simply increasing the level of glycerol-3-phosphate dehydrogenase (Michnick et al. 1997, Nevoigt and Stahl 1997, Remize et al. 1999, Dequin 2001). The *GDP1* gene, which encodes glycerol-3-phosphate dehydrogenase, has been overexpressed in an industrial lager brewing yeast to reduce the ethanol content in beer (Nevoigt et al. 2002). The amount of glycerol produced by the *GDP1*-overexpressing yeast in fermentation experiments—simulating brewing conditions—was increased 5.6 times, and ethanol was decreased by 18% compared with the wild-type strain. Overexpression did not affect the consumption of wort sugars, and only minor changes in the concentration of higher alcohols, esters, and fatty acids could be observed. However, the concentrations of several other by-products, particularly acetoin, diacetyl, and acetaldehyde, were considerably increased.

METABOLISM OF BIOFLAVORING BY-PRODUCTS

Yeast is an important contributor to flavor development in fermented beverages. The compounds that are produced during fermentation are many and varied, depending on both the raw materials and the microorganisms used. The interrelation between yeast metabolism and the production of bioflavoring by-products is illustrated in Figure 29.2.

BIOSYNTHESIS OF HIGHER ALCOHOLS

During beer fermentation, higher alcohols (also called "fusel alcohols") are produced by yeast cells as by-products and represent the major fraction of the volatile compounds. More than 35 higher alcohols in beer have been described. Table 29.3 gives the most important compounds, which can be classified into aliphatic [n-propanol, isobutanol, 2-methylbutanol (or active amyl alcohol), and 3-methylbutanol (or

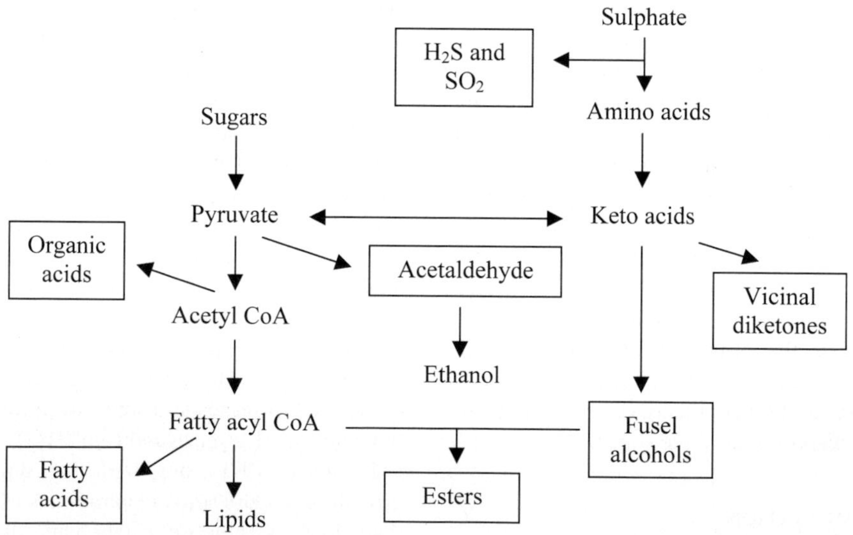

Figure 29.2. Interrelationships between yeast metabolism and the production of flavor-active compounds. (Hammond 1993.)

Table 29.3. Major Higher Alcohols in Beer (N.N. 2000)

Compound	Flavor Threshold (mg/l)	Aroma or Taste[b]	Concentration Range (mg/L), Bottom Fermentation	Concentration Range (mg/L), Top Fermentation
n-Propanol	600[c], 800[b]	Alcohol	7–19 (12)[*,f]	20–45[i]
Isobutanol	100[c],80–100[g],200[b]	Alcohol	4–20 (12)[f]	10–24[i]
2-Methylbutanol	50[c], 50–60[g], 70[b]	Alcohol	9–25 (15)[a]	80–140[i]
3-Methylbutanol	50[c], 50–60[g], 65[b]	Fusely, pungent	25–75 (46)[a]	80–140[i]
2-Phenylethanol	5[a], 40[c], 45–50[g], 75[d],125[b]	Roses, sweetish	11–51 (28)[f], 4–22[g], 16–42[h]	35–50[g], 8–25[a], 18–45[i]
Tyrosol	10[a], 10–20[e], 20[c], 100[d,g], 200[b]	Bitter chemical	6–9[a], 6–15[a]	8–12[g], 7–22[g]
Tryptophol	10[a], 10–20[e], 200[d]	Almonds, solvent	0.5–14[a]	2–12[g]

Source: N.N. 2000.
References: [a]Szlavko 1973; [b]Meilgaard 1975a; [c]Engan 1972; [d]Rosculet 1971;[e] Charalambous *et al.* 1972; [f]Values in 48 European lagers, Dufour (unpublished data); [g]Reed and Nogodawithana 1991; [h]Iverson 1994; [i]Derdelinckx (unpublished data).
[*]Mean value.

isoamyl alcohol)] and aromatic (2-phenylethanol, tyrosol, tryptophol) higher alcohols. Aliphatic higher alcohols contribute to the "alcoholic" or "solvent" aroma of beer, and produce a warm mouthfeel. The aromatic alcohol 2-phenylethanol has a sweet aroma and makes a positive contribution to the beer aroma, whereas the aroma of tyrosol and tryptophol are undesirable.

Higher alcohols are synthesized by yeast during fermentation via the catabolic (Ehrlich 1904) and anabolic pathways (amino acid metabolism) (Ehrlich 1904, Chen 1978, Oshita et al. 1995). In the catabolic pathway, the yeast uses the amino acids of the wort to produce the corresponding α-keto acid via a transamination reaction. The excess oxoacids are subsequently decarboxylated into aldehydes and further reduced (alcohol dehydrogenase) to higher alcohols. This last reduction step also regenerates NAD^+.

Dickinson et al. looked at the genes and enzymes that are used by *S. cerevisiae* in the catabolism of leucine to isoamyl alcohol (Dickinson et al. 1997), valine to isobutanol (Dickinson et al. 1998), and isoleucine to active amyl alcohol (Dickinson et al. 2000). In all cases, the general sequence of biochemical reactions is similar, but the details for the formation of the individual alcohols are surprisingly different. The branched-chain amino acids are first deaminated to the corresponding α-keto acids (α-ketoisocapric acid from leucine, α-ketoisovaleric acid from valine, and α-keto-β-methylvaleric acid from leucine). There are significant differences in the way each α-keto acid is subsequently decarboxylated. Recently, the catabolism of phenylalanine to 2-phenylethanol and of tryptophan were also studied (Dickinson et al. 2003). Phenylalanine and tryptophan are first deaminated to 3-phenylpyruvate and 3-indolepyruvate, respectively, and then decarboxylated. These studies revealed that all amino acid catabolic pathways studied to date use a subtly different spectrum of decarboxylases from the five-member family that comprises Pdc1p, Pdc5p, Pdc6p, Ydl080cp, and Ydr380wp. Using strains containing all possible combinations of mutations affecting the seven *AAD* genes (putative aryl alcohol dehydrogenases), five *ADH*, and *SFA1* (other alcohol dehydrogenase genes) showed that the final step of amino acid catabolism can be accomplished by any one of the ethanol dehydrogenases (Ahd1p, Ahd2p, Ahd3p, Ahd4p, Ahd5p) or Sfa1p (formaldehyde dehydrogenase).

In the anabolic pathway, the higher alcohols are synthesized from α-keto acids during the synthesis of amino acids from the carbohydrate source. The pathway choice depends on the individual higher alcohol and on the level of available amino acids available. The importance of the anabolic pathway decreases as the number of carbon atoms in the alcohol increases (Chen 1978), and increases in the later stage of fermentation as wort amino acids are depleted (MacDonald et al. 1984). Yeast strain, fermentation conditions, and wort composition all have significant effects on the combination and levels of higher alcohols that are formed (MacDonald et al. 1984).

Conditions that promote yeast cell growth—such as high levels of nutrients (amino acids, oxygen, lipids, zinc, etc.), increased temperature, and agitation—stimulate the production of higher alcohols. The synthesis of aromatic alcohols is especially sensitive to temperature changes. On the other hand, conditions that restrict yeast growth—such as lower temperature and higher pressure—reduce the extent of higher alcohol production.

Biosynthesis of Esters

Esters are very important flavor compounds in beer. They have an effect on the fruity/flowery aromas. Table 29.4 shows the most important esters with their threshold values, which are considerably lower than those for higher alcohols. The major esters can be subdivided into acetate esters and C_6–C_{10} medium-chain fatty acid ethyl esters. They are desirable components of beer when present in appropriate quantities and proportions but can become unpleasant when in excess. Ester formation is highly dependent on the yeast strain used (Nykänen and Nykänen 1977, Peddie 1990) and on certain fermentation parameters such as temperature (Engan and Aubert 1977, Gee and Ramirez 1994, Sablayrolles and Ball 1995), specific growth rate (Gee and Ramirez 1994), pitching rate (Maule 1967, D'Amore et al. 1991, Gee and Ramirez 1994), and top pressure. Additionally, the concentrations of assimilable nitrogen compounds (Hammond 1993, Calderbank and Hammond 1994, Sablayrolles and Ball 1995), carbon sources (Pfisterer and Stewart 1975; White and Portno 1979; Younis and Stewart 1998, 2000), dissolved oxygen (Anderson and Kirsop 1975a,b; Avhenainen and Mäkinen 1989; Sablayrolles and Ball 1995), and fatty acids (Thurston et al. 1981, 1982) can influence the ester production rate.

Esters are produced by yeast both during the growth phase (60%) and during the stationary phase (40%). They are formed by the intracellular reaction between a fatty acyl-coenzyme A (acyl-CoA) and an alcohol:

$$R'OH + RCO\text{-}SCoA \rightarrow RCOOR' + CoASH \quad (1)$$

This reaction is catalyzed by an alcohol acyltransferase (or ester synthetase). Since acetyl-CoA is also a central molecule in the synthesis of lipids and sterols, ester synthesis is linked to fatty acid metabolism (see also Fig. 29.2).

Alcohol acetyltransferase (AAT) has been localized in the plasma membrane (Malcorps and Dufour 1987) and found to be strongly inhibited by unsaturated fatty acids, ergosterol, heavy metal ions, and sulphydryl reagents (Minetoki et al. 1993). Subcellular fractionation studies conducted during the batch fermentation cycle demonstrated the existence of both cytosolic and membrane-bound AAT (Ramos-Jeunehomme et al. 1989, Ramos-Jeunehomme et al. 1991). In terms of controlling ester formation on a metabolic basis, it has further been shown that the ester-synthesizing activity of AAT is dependent on its positioning within the yeast cell. An interesting fea-

Table 29.4. Major Esters in Beer

Compound	Flavor Threshold (mg/L)	Aroma	Concentration Range (mg/L) in 48 Lagers
Ethyl actetate	20–30, 30[a]	Fruity, solventlike	8–32 (18.4)*
Isoamyl acetate	0.6–1.2, 1.2[a]	Banana, peardrop	0.3-3.8 (1.72)
Ethyl caproate (ethyl hexanoate)	0.17–0.21, 0.21[a]	Applelike with note of aniseed	0.05–0.3 (0.14)
Ethyl caprylate (ethyl octanoate)	0.3–0.9, 0.9[a]	Applelike	0.04–0.53 (0.17)
2-Phenylethyl acetate	3.8[a]	Roses, honey, apple, sweetish	0.10–0.73 (0.54)

Source: Dufour and Malcorps 1994.
Reference: [a]Meilgaard 1975b.
*Mean value.

ture of this distribution pattern is that specific rates of acetate ester formation varied directly with the level of cytosolic AAT activity (Masschelein 1997).

The *ATF1* gene, which encodes alcohol acetyltransferase, has been cloned from *S. cerevisiae* and brewery lager yeast (*S. cerevisiae uvarum*) (Fujii et al. 1994). A hydrophobicity analysis suggested that alcohol acetyltransferase does not have a membrane-spanning region that is significantly hydrophobic, which contradicts the membrane-bound assumption. A Southern analysis of the yeast genomes in which the *ATF1* gene was used as a probe, revealed that *S. cerevisiae* has one *ATF1* gene, while brewery lager yeast has one *ATF1* gene and another, homologous gene (*Lg-ATF1*). The AAT activity of *S. cerevisiae* has been compared in vivo and in vitro under different fermentation conditions (Malcorps et al. 1991). This study suggested that ester synthesis is modulated by a repression-induction of enzyme synthesis or processing, the regulator of which is presumably linked to yeast metabolism.

The ester production can be altered by changing the synthesis rate of certain fusel alcohols. Hirata et al. (1992) increased the isoamyl acetate levels by introducing extra copies of the *LEU4* gene in the *S. cerevisiae* genome. A comparable *S. cerevisiae uvarum* mutant has been isolated (Lee et al. 1995). The mutants have an altered regulation pattern of amino acid metabolism and produce more isoamyl acetate and phenylethyl acetate.

Isoamyl acetate is synthesized from isoamyl alcohol and acetyl-CoA by AAT and is hydrolyzed by esterases at the same time in *S. cerevisiae*. To study the effect of balancing both enzyme activities, yeast strains with different numbers of copies of the *ATF1* gene and isoamyl acetate–hydrolyzing esterase gene (*IAH1*) have been constructed and used in small-scale sake brewing (Fukuda et al. 1998). Fermentation profiles as well as components of the resulting sake were largely alike. However, the amount of isoamyl acetate in the sake increased with increasing ratio of AAT:Iah1p esterase activity. Therefore, it was concluded that the balance of these two enzyme activities are important for isoamyl acetate accumulation in sake mash.

The synthesis of acetate esters by *S. cerevisiae* during fermentation is ascribed to at least three acetyltransferase activities, namely alcohol acetyltransferase (AAT), ethanol acetyltransferase, and isoamyl AAT (Lilly et al. 2000). To investigate the effect of increased AAT activity on the sensory quality of Chenin blanc wines and distillates from Colombar base wines, the *ATF1* gene of *S. cerevisiae* was overexpressed. Northern blot analysis indicated constitutive expression of *ATF1* at high levels in these transformants. The levels of ethyl acetate, isoamyl acetate, and 2-phenylethyl acetate increased 3- to 10-fold, 3.8- to 12-fold, and 2- to 10-fold, respectively, depending on the fermentation temperature, cultivar, and yeast used. The concentrations of ethyl caprate, ethyl caprylate, and hexyl acetate only showed minor changes, whereas the acetic acid concentration decreased by more than half. This study established the concept that the overexpression of acetyltransferase genes such as *ATF1* could profoundly affect the flavor profiles of wines and distillates deficient in aroma.

In order to investigate and compare the roles of the known *S. cerevisiae* alcohol acetyltransferases Atf1p, Atf2p, and Lg-Atf1p in volatile ester production, the respective genes were either deleted or overexpressed in a laboratory strain and a commercial brewing strain (Verstrepen et al. 2003). Analysis of the fermentation products confirmed that the expression levels of *ATF1* and *ATF2* greatly affect the production of ethyl acetate and isoamyl acetate. Gas chromatography/mass spectrometry (GC/MS) analysis revealed that Atf1p and Atf2p are also responsible for the formation of a broad range of less volatile esters, such as propyl acetate, isobutyl acetate, pentyl acetate, hexyl acetate, heptyl acetate, octyl acetate, and phenylethyl acetate. With respect to the esters analyzed in this study, Atf2p seemed to play only a minor role compared to Atf1p. The *atf1 Δatf2Δ* double deletion strain did not form any isoamyl acetate, showing that together, Atf1p and Atf2p are responsible for the total cellular isoamyl alcohol acetyltransferase activity. However, the double deletion strain still produced considerable amounts of certain other esters, such as ethyl acetate (50% of the wild-type strain), propyl acetate (50%), and isobutyl acetate (40%), which provides evidence for the existence of additional, as yet unknown, ester synthases in the yeast proteome. Interestingly, overexpression of different alleles of *ATF1* and *ATF2* led to different ester production rates, indicating that differences in the aroma profiles of yeast strains may be partially due to mutations in their *ATF* genes.

Recently, it has been discovered that the Atf1 enzyme is localized inside lipid vesicles in the cytoplasm of the yeast cell (Verstrepen 2003). Lipid vesicles are small organelles in which certain neutral lipids are metabolized or stored. This indicates that fruity esters are possibly by-products of these processes.

Biosynthesis of Organic Acids

Important organic acids detected in beer include acetate, lactate, succinate, pyroglutamate, malate, citrate, α-ketoglutarate and α-hydroxyglutarate (Coote and Kirsop 1974). They influence flavor directly, when present above their taste thresholds, and by their influence on beer pH. These components have their origin in raw materials (malt, hops) and are produced during the beer fermentation. Organic acids that are excreted by yeast cells are synthesized via amino acid biosynthesis pathways and carbohydrate metabolism. Especially, they are overflow products of the incomplete Krebs cycle during beer fermentation. Excretion of organic acids is influenced by yeast strain and fermentation vigor. Sluggish fermentations lead to lower levels of excretion. Pyruvate excretion follows the yeast growth: maximal concentration is reached just before the maximal yeast growth, and the pyruvate is next taken up by the yeast and converted to acetate. Acetate is synthesized quickly during early fermentation and is later partially re-used by the yeast during yeast growth. At the end of the fermentation, acetate is accumulated. The reduction of pyruvate results in the production of D-lactate or L-lactate (most yeast strains produce preferentially D-lactate). The highest amount of lactate is produced during the most active fermentation period.

The change in organic acid productivity by disruption of the gene encoding fumarase *(FUM1)* has been investigated, and it has been suggested that malate and succinate are produced via the oxidative pathway of the TCA cycle under static and sake brewing conditions (Magarifuchi et al. 1995). Using a NAD^+-dependent isocitrate dehydrogenase gene *(IDH1, IDH2)* disruptant, approximately half of the succinate in sake mash was found to be synthesized via the oxidative pathway of the TCA cycle in sake yeast (Asano et al. 1999).

Sake yeast strains possessing various organic acid productivities were isolated by gene disruption (Arikawa et al. 1999). Sake fermented using the aconitase gene *(ACO1)* disruptant contained a two-fold higher concentration of malate and a two-fold lower concentration of succinate than that made using the wild-type strain. The fumarate reductase gene *(OSM1)* disruptant produced sake containing a 1.5-fold higher concentration of succinate, whereas the α-ketoglutarate dehydrogenase gene (*KGD1*) and fumarase gene (*FUM1*) disruptants gave lower succinate concentrations. In *S. cerevisiae*, there are two isoenzymes of fumarate reductase (FRDS1 and FDRS2), encoded by the *FRDS* and *OSM1* genes, respectively (Arikawa et al. 1998). Recent results suggest that these isoenzymes are required for the reoxidation of intracellular NADH under anaerobic conditions, but not under aerobic conditions (Enomoto et al. 2002).

Succinate dehydrogenase is an enzyme of the TCA cycle and is thus essential for respiration. In *S. cerevisiae*, this enzyme is composed of four non-identical subunits, that is, the flavoprotein, the iron-sulfur protein, the cytochrome b_{560}, and the ubiquinone reduction protein encoded by the *SDH1*, *SDH2*, *SDH3*, and *SDH4* genes, respectively (Lombardo et al. 1990, Chapman et al. 1992, Bullis and Lamire 1994, Daignan-Fournier et al. 1994). Sdh1p and Sdh2p comprise the catalytic domain involved in succinate oxidation. These proteins are anchored to the inner mitochondrial membrane by Sdh3p and Sdh4p, which are necessary for electron transfer and ubiquinone reduction, and constitute the succinate: ubiquinone oxidoreductase (complex II) of the electron transport chain. Single or double disruptants of the *SDH1*, *SDH1b* (which is a homologue of the *SDH1* gene), *SDH2*, *SDH3,* and *SDH4* genes have been constructed, and it has been shown that the succinate dehydrogenase activity was retained in the *SDH2* disruptant and that double disruption of *SDH1* and *SDH2* or *SDH1b* genes is necessary to cause deficiency of succinate dehydrogenase activity in sake yeast (Kubo et al. 2000). The role of each subunit in succinate dehydrogenase activity and the effect of succinate dehydrogenase on succinate production, using strains which were deficient in succinate dehydrogenase, have also been determined. The results suggested that succinate dehydrogenase activity contributes to succinate production under shaking conditions, but not under static and sake brewing conditions.

Biosynthesis of Vicinal Diketones

Vicinal diketones are ketones with two adjacent carbonyl groups. During fermentations, these flavor-active compounds are produced as by-products of the synthesis pathway of isoleucine, leucine, and valine (ILV pathway) (see Fig. 29.3) and are thus also linked to amino acid metabolism (Nakatani et al. 1984) and the synthesis of higher alcohols. They impart a "buttery," "butterscotch" aroma to alcoholic drinks. Two of these compounds are important in beer: diacetyl (2,3-butanedione) and 2,3-pentanedione. Diacetyl is quantitatively more important than 2,3-pentanedione. It has a taste threshold around 0.10–0.15 mg/L in lager beer, approximately 10 times lower than that of pentanedione (Wainwright 1973).

The excreted α-acetohydroxy acids are overflow products of the ILV pathway that are nonenzymatically degraded to the corresponding vicinal diketones (Inoue et al. 1968). Tetraploid gene dosage series for various *ILV* genes have been constructed, and the obtained yeast strains were used to study the influence of the copy number of *ILV* genes on the production of vicinal diketones (Debourg et al. 1990, Debourg 2002). It was shown that the *ILV5* activity is the rate-limiting step in the ILV pathway and is responsible for the overflow (Fig. 29.3). The nonenzymatic oxidative decarboxylation step is the rate-limiting step in the conversion of α-acetolactate to 2,3-butanediol and proceeds faster at a higher temperature and a lower pH (Inoue and Yamamoto 1970, Haukeli and Lie 1978). The produced amount of α-acetolactate is very dependent on the yeast strain used. The production increases with increasing yeast growth. For a classical fermentation, 0.6 ppm α-acetolactate is formed (Delvaux 1998). At high aeration, this value can be increased to 0.9 ppm, and even to 1.2–1.5 ppm in cylindroconic fermentations tanks.

Yeast cells posses the necessary enzymes (reductases) to reduce diacetyl to acetoin and further to 2,3-butanediol, and 2,3-pentanedione to 2,3-pentanediol. These reduced compounds have much higher taste thresholds than the diacetyl diketones and have no impact on the beer flavor (Van Den Berg et al. 1983). The reduction reactions are yeast-strain dependent. The reduction occurs at the end of the main fermentation and during the maturation. Sufficient yeast cells in suspension are necessary to obtain an efficient reduction. Yeast strains that flocculate early during the main fermentation need a long maturation time to reduce the vicinal diketones. Diacetyl can be complexed using SO_2. These complexes cannot be reduced, but diacetyl can again be liberated at a later stage by aldehydes. This situation is especially applicable to yeast strains that produce a lot of SO_2. Worts, which are produced using many adjuncts, can be low in free amino acid content. These worts can give rise to a high diacetyl peak at the end of the fermentation.

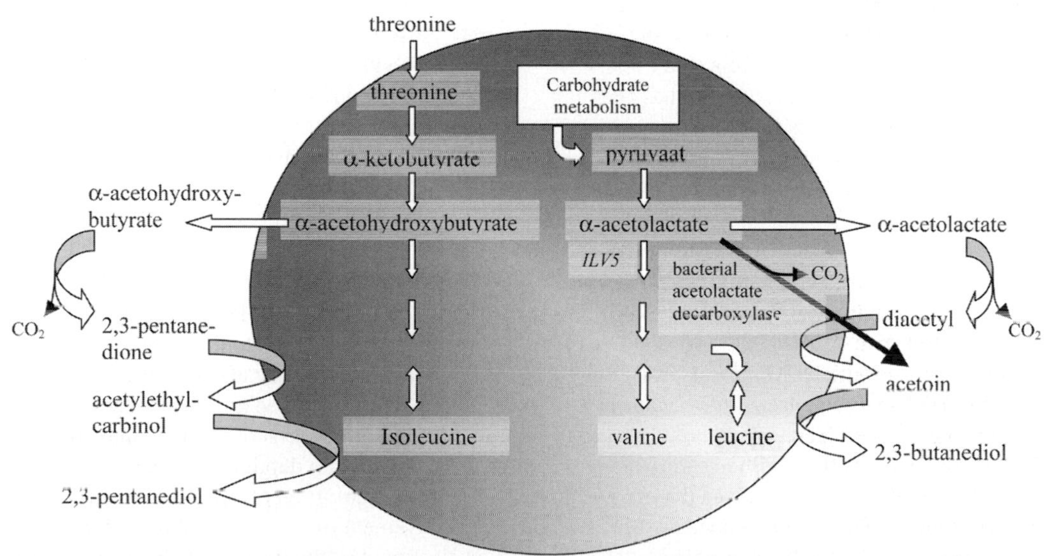

Figure 29.3. The synthesis and reduction of vicinal diketones in *S. cerevisiae*.

There are several strategies that can be chosen to reduce the vicinal diketones during fermentation:

1. Since the temperature has a positive effect on the reduction efficiency of the α-acetohydroxy acids, a warm rest period at the end of the main fermentation, and a warm maturation are applied in many breweries. In this case, temperature should be well controlled to avoid yeast autolysis.
2. Since the rapid removal of vicinal diketones requires yeast cells in an active metabolic condition, the addition of 5–10% Krausen (containing active, growing yeast) is a procedure that gives enhanced transformation of vicinal diketones (N.N. 2000). However, this procedure can lead to overproduction of hydrogen sulphide, depending upon the proportions of threonine and methionine carried forward from primary fermentation.
3. Heating up the green beer, beer obtained after the primary fermentation, to a high temperature (90°C) and holding it there for a short period (ca. 7–10 minutes) to decarboxylate all excreted α-acetohydroxy acids. To avoid cell autolysis, yeast cells are removed by centrifugation prior to heating. The vicinal diketones can be further reduced by immobilized yeast cells in a few hours (typically at 4°C) (see further).
4. Adding the enzyme α-acetolactate decarboxylase (Godtfredsen et al. 1984, Rostgaard-Jensen et al. 1987). This enzyme decarboxylates α-acetolactate directly into acetoin (see Fig. 29.3). It is not present in *S. cerevisiae*, but has been isolated from various bacteria such as *Enterobacter aerogenes*, *Aerobacter aerogenes*, *Streptococcus lactis*, *Lactobacillus casei*, *Acetobacter aceti*, and *Acetobacter pasteurianus*. It has been shown that the addition of α-acetolactate decarboxylase from *Lactobacillus casei* can reduce the maturation time to 22 hours (Godtfredsen et al. 1983, 1984). An example of a commercial product is Maturex L from Novo Nordisk (Denmark) (Jensen 1993). Maturex L is a purified α-acetolactate decarboxylase produced by a genetically modified strain of *Bacillus subtilis* that has received the gene from *Bacillus brevis*. The recommended dosage is 1–2 kg per 1000 hL wort, to be added to the cold wort at the beginning of fermentation.
5. Using genetic modified yeast strains:
    a. Introducing the bacterial α-acetolactate decarboxylase gene into yeast chromosomes (Fujii et al. 1990, Suihko et al. 1990, Blomqvist et al. 1991, Enari et al. 1992, Linko et al. 1993, Yamano et al. 1994, Tada et al. 1995, Onnela et al. 1996). Transformants possessed a very high α-acetolactate decarboxylase activity, which reduced the diacetyl concentration considerably during beer fermentations.
    b. Modifying the biosynthetic flux through the ILV pathway. Spontaneous mutants resistant to the herbicide sulfometuron methyl have been selected. These strains showed a partial inactivation of the α-acetolactate synthase activity, and some mutants produced 50% less diacetyl compared with the parental strain (Gjermansen et al. 1988).
    c. Increasing the flux of α-acetolactate acid isomeroreductase activity encoded by the *ILV5* gene (Dillemans et al. 1987). Since α-acetolactate acid isomeroreductase activity is responsible for the rate-limiting step, increasing its activity reduces the overflow of α-acetolactate. A multicopy transformant resulted in a 70% decreased production of vicinal diketones (Villaneuba et al. 1990), whereas an integrative transformant gave a 50% reduction (Goossens et al. 1993). A tandem integration of multiple *ILV5* copies also resulted in elevated transciption in a polyploid industrial yeast strain (Mithieux and Weiss 1995).

SECONDARY FERMENTATION

During the secondary fermentation or maturation of beer, several objectives should be realized:

- Sedimentation of yeast cells,
- Improvement of the colloidal stability by sedimentation of the tannin-protein complexes,
- Beer saturation with carbon dioxide,
- Removal of unwanted aroma compounds,
- Excretion of flavor-active compounds from yeast to give body and depth to the beer,
- Fermentation of the remaining extract,
- Improvement of the foam stability of the beer,
- Adjustment of the beer color (if necessary) by adding coloring substances (e.g., caramel),

- Adjustment of the bitterness of beer (if necessary) by adding hop products.

In the presence of yeast, the principal changes that occur are the elimination of undesirable flavor compounds—such as vicinal diketones, hydrogen sulphide, and acetaldehyde—and the excretion of compounds enhancing the flavor fullness (body) of the beer.

Vicinal Diketones

In traditional fermentation lagering processes, the elimination of vicinal diketones required several weeks and determined the length of the maturation process. Currently, the maturation phase is much shorter since strategies are used to accelerate the vicinal diketone removal (see above). Diacetyl is used as a marker molecule. The objective during lagering is to reduce the diacetyl concentration below its taste threshold (< 0.10 mg/mL).

Hydrogen Sulphide

Hydrogen sulphide plays an important role during maturation. Hydrogen sulphide, which is not incorporated into S-containing amino acids, is excreted by the yeast cell during the growth phase (see Fig. 29.4). The excreted amount depends on the used yeast strain, the sulphate content of the wort, and the growth conditions (Romano and Suzzi 1992). At the end of the primary fermentation and during the maturation, the excess H_2S is reutilized by the yeast. A warm conditioning period at 10–12°C may be used to remove excessive levels of H_2S.

Brewing yeasts produce H_2S when they are deficient in the vitamin pantothenate (Walker 1998). This vitamin is a precursor of coenzyme A, which is required for metabolism of sulphate into methionine. Panthothenate deficiency may therefore result in an imbalance in sulphur–amino acid biosynthesis, leading to excess sulphate uptake and excretion of H_2S (Slaughter and Jordan 1986).

Sulphite is a versatile food additive used to preserve a large range of beverages and foodstuffs. In beer, sulphite has a dual purpose, acting both as an antioxidant and as an agent for masking certain off-flavors. Some of the flavor-stabilizing properties of sulphite are suggested to be due to complex formation of bisulphate with varying carbonyl compounds, of which some would give rise to off-flavors in bottled beer (Dufour 1991). Especially, the unwanted carbonyl *trans*-2-nonenal has received particular attention since it is responsible for the "cardboard" flavor of some types of stale beer. It has been suggested that it would be better to use a yeast strain with reduced sulphite excretion during fermentation and to add sulphite at the point of bottling to ensure good flavor stability (Francke Johannesen et al. 1999). Therefore, a brewer's yeast disabled in the production of sulphite has been constructed by inactivating both copies of the two alleles of the *MET14* gene (which encodes for adenylylsulphate kinase). Fermentation experiments showed that there was no qualitative difference between yeast-derived and artificially added sulphite, with respect to *trans*-2-nonenal content and flavor stability of the final beer.

The elimination of the gene encoding sulphite reductase *(MET10)* in brewing strains of *Saccharomyces* results in increased accumulation of SO_2 in beer (Hansen and Kielbrandt 1996a). The inactivation of *MET2* resulted in elevated sulphite concentrations in beer (Hansen and Kielbrandt 1996b). Beers produced with increased levels of sulphite showed improved flavor stability.

The production of H_2S could be reduced by the expression of cystathione synthase genes from *S. cerevisiae* in a brewing yeast strain (Tezuka et al. 1992).

Acetaldehyde

Aldehydes—in particular acetaldehyde (green apple–like flavor)—have an impact on the flavor of green beer. Acetaldehyde synthesis is linked to yeast growth. Its concentration is maximal at the end of the growth phase, and is reduced at the end of the primary fermentation and during maturation by the yeast cells. As with diacetyl, levels may be enhanced if yeast metabolism is stimulated during transfer, especially by oxygen ingress. Removal also requires the presence of enough active yeast. Fermentations with early flocculating yeast cells can result in too high acetaldehyde concentrations at the end.

Development of Flavor Fullness

During maturation, the residual yeast will excrete compounds (i.e., amino acids, phosphates, peptides, nucleic acids, etc.) into the beer. The amount and

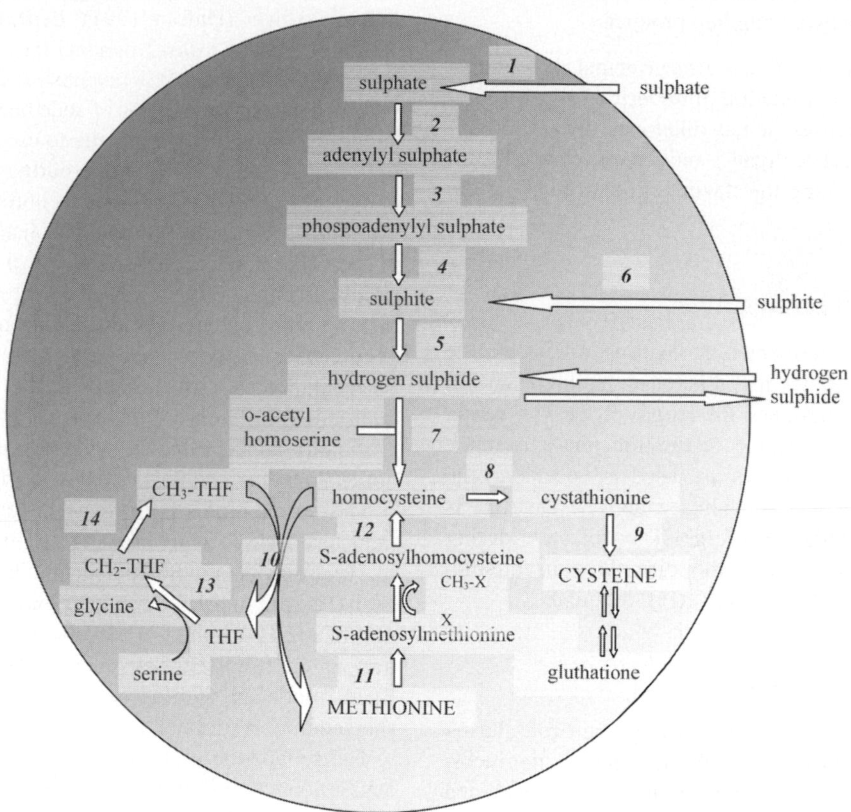

Figure 29.4. The remethylation, transulfuration, and sulfur assimilation pathways. Genes and enzymes catalyzing individual reactions are *1,* sulphate permease; *2,* ATP sulphurylase; *3, MET1; 4,* adenylylsulfate kinase (EC 2.7.1.25); *5, MET10*: sulfite reductase (EC 1.8.1.2); *6,* sulphite permease; *7, MET17*: *O*-acetylhomoserine (thiol)-lyase (EC 2.5.1.49); *8, CYS4*: cystathionine β-synthase (CBS, EC 4.2.1.22); *9, CYS3*: cystathionine γ-lyase (EC 4.4.1.1); *10, MET6*: methionine synthase (EC 2.1.1.14); *11, SAM1* and *SAM2*: *S*-adenosylmethionine synthetase (EC 2.5.1.6); *12, SAH1*: *S*-adenosylhomocysteine hydrolase (EC 3.3.1.1); *13, SHM1* and *SHM2*: serine hydroxymethyltransferase (SHMT, EC 2.1.2.1); *14, MET12* and *MET13*: methylenetetrahydrofolate reductase (MTHFR, EC 1.5.1.20). "X" represents any methyl group acceptor; THF, tetrahydrofolate; CH_2-THF, 5,10-methylenetetrahydrofolate; CH_3-THF, 5-methyltetrahydrofolate. (Partly adapted from Chan and Appling 2003.)

"quality" of these excreted materials depend on the yeast concentration, yeast strain, yeast metabolic state, and the temperature (N.N. 2000). Rapid excretion of material is best achieved at a temperature of 5–7°C during 10 days (Van de Meersche et al. 1977).

When the conditioning period is too long or when the temperature is too high, yeast cell autolysis will occur. Some enzymes are liberated (e.g., α-glucosidase) that will produce glucose from traces of residual maltose (N.N. 2000). At the bottom of a fermentation tank, the amount of α-amino-nitrogen can rise to 40–10,000 mg/L, which accounts for an increase of 30 mg/L for the total beer volume. The increase in amino acid concentration in the beer has a positive effect on the flavor fullness of the beer. Undesirable medium-chain fatty acids can also be produced in significant amounts if the maturation temperature is too high (Masschelein 1981). Measurement of these compounds indicates the level of autolysis and permits the determination of the most appropriate conditioning period and temperature.

BEER FERMENTATION USING IMMOBILIZED CELL TECHNOLOGY

The advantages of continuous fermentation—such as greater efficiency in utilization of carbohydrates and better use of equipment—led also to the development of continuous beer fermentation processes. Since the beginning of the 20th century, many different systems using suspended yeast cells have been developed. The excitement over continuous beer fermentation led—especially during the 1950s and 1960s—to the development of various interesting systems. These systems can be classified as (1) stirred versus unstirred tank reactors, (2) single-vessel systems versus a number of vessels connected in series, and (3) vessels that allow yeast to overflow freely with the beer ("open system") versus vessels that have abnormally high yeast concentrations ("closed" or "semiclosed system") (Hough et al. 1982, Wellhoener 1954, Coutts 1957, Bishop 1970). However, these continuous beer fermentation processes were not commercially successful due to many practical problems, such as the increased danger of contamination (not only during fermentation but also during storage of wort in supplementary holdings tanks, which are required since the upstream and downstream brewing processes are usually not continuous), changes in beer flavor (Thorne 1968), and a poor understanding of the beer fermentation kinetics under continuous conditions. One of the well-known exceptions is the successful implementation of a continuous beer production process in New Zealand by Morton Coutts (Dominion Breweries) that is still in use today (Hough et al. 1982, Coutts 1957).

In the 1970s, there was a revival of interest in developing continuous beer fermentation systems due to the progress in research on immobilization bioprocesses using living cells. Immobilization allows fermentation processes with high cell densities, resulting in a drastic increase in fermentation productivities compared with the traditional time-consuming batch fermentation processes.

In the last 30 years, immobilized cell technology (ICT) has been extensively examined, and some designs have already reached commercial exploitation. Immobilized cell systems are heterogeneous systems in which considerable mass transfer limitations can occur, resulting in a changed yeast cell metabolism. Therefore, successful exploitation of ICT requires a thorough understanding of mass transfer and of the intrinsic yeast kinetic behavior of these systems.

Carrier Materials

Various cell immobilization carrier materials have been tested and used for beer production/bioflavoring. Selection criteria are summarized in Table 29.5. Depending on the particular application, reactor type, and operational conditions, some selection criteria will be more appropriate. Examples of selected carrier materials for particular applications are tabulated in Table 29.6.

Applications of Immobilized Cell Technology (ICT) in the Brewing Industry

Flavor Maturation of Green Beer

The objective of flavor maturation is the removal of diacetyl and 2,3-pentanedione and their precursors, α-acetolactate and α-acetohydroxybutyrate, which are produced during the main fermentation (see above). The conversion of α-acetohydroxy acids to the vicinal diketones is the rate-limiting step. This reaction step can be accelerated by heating the beer—after yeast removal—to 80–90°C during a couple of minutes. The resulting vicinal diketones are subsequently reduced by immobilized cells into their less flavor-active compounds.

The traditional maturation process is characterized by a near-zero temperature, low pH, and low yeast concentration, resulting in a very long maturation period of 3–4 weeks. Using immobilized cell technology, this long period can be reduced to 2 hours. An ICT maturation process using a packed-bed bioreactor with diethylaminoethyl (DEAE) cellulose beads has been successfully integrated in Synebrychoff Brewery (Finland) for the treatment of 1 million hL/year (Pajunen 1995). Alfa Laval and Schott Engineering developed a maturation system based on porous glass beads (Dillenhöfer and Rönn 1996). This system has been implemented in several breweries in Finland (Hyttinen et al. 1995), Belgium, Germany, and elsewhere. The German company Brau and Brunnen has also shown an interest in the Alfa Laval maturation technology. In 1996, a

Table 29.5. Selection Criteria for Yeast Cell Immobilization Carrier Materials

- High cell mass loading capacity
- Easy access to nutrient media
- Simple and gentle immobilization procedure
- Immobilization compounds approved for food applications
- High surface area-to-volume ratio
- Optimum mass transfer distance from flowing media to center of support
- Mechanical stability (compression, abrasion)
- Chemical stability
- Highly flexible: rapid start-up after shutdown
- Sterilizable and reusable
- Suitable for conventional reactor systems
- Low shear experienced by cells
- Easy separation of cells and carrier from media
- Readily upscalable
- Economically feasible (low capital and operating costs)
- Desired flavor profile and consistent product
- Complete attenuation
- Controlled oxygenation
- Control of contamination
- Controlled yeast growth
- Wide choice of yeast

Source: Nedovic and Willaert 2004.

30,000 hL/year pilot-scale system was purchased and installed in their plant (Mensour et al. 1997). The Alfa Laval maturation system has been implemented in a medium-sized German brewery (Schäff/Treuchtlingen) (Back et al. 1998). The obtained beers yielded overall good analytical and sensorial results.

Production of Alcohol-Free or Low-Alcohol Beer

The classical technology to produce alcohol-free or low-alcohol beer is based on the suppression of alcohol formation by arrested batch fermentation (Narziss et al. 1992). However, the resulting beers are characterized by an undesirable wort aroma, since the wort aldehydes have only been reduced to a limited degree (Collin et al. 1991, Debourg et al. 1994, van Iersel et al. 1998). The reduction of these wort aldehydes can be quickly achieved by a short contact time with the immobilized yeast cells at a low temperature, without undesirable cell growth and ethanol production. A disadvantage of this short-contact process is the production of only a small amount of desirable esters.

Controlled ethanol production for low-alcohol and alcohol-free beers has been successfully achieved by partial fermentation using DEAE cellulose as the carrier material in a column reactor (Collin et al. 1991, Van Dieren 1995). This technology has been successfully implemented by Bavaria Brewery (The Netherlands) to produce malt beer on an industrial scale (150,000 hL/year) (Pittner et al. 1993). Several other companies—that is, Faxe (Denmark), Ottakringer (Austria) and a Spanish brewery—have also implemented this technology (Mensour et al. 1997). In Brewery Beck (Germany), a fluidized-bed pilot-scale reactor (8 hL/day) filled with porous glass beads was used for the continuous production of nonalcohol beer (Aivasidis et al. 1991, Breitenbücher and Mistler 1995, Aivasidis 1996). Yeast cells immobilized in silicon carbide rods and arranged in a multichannel loop reactor (Meura, Belgium) have been used by Grolsch Brewery (The Netherlands) and Guinness Brewery (Ireland) to produce alcohol-free beer at the pilot scale (Van De Winkel 1995).

Nuclear mutants of *S. cerevisiae* that are defective in the synthesis of tricarboxylic acid cycle enzymes—that is, fumarase (Kaclíková et al. 1992)

Table 29.6. Some Selected Applications of Cell Immobilization Systems Used for Beer Production

Carrier Material	Reactor Type	Reference
Flavor maturation		
Calcium alginate beads	Fixed bed	Shindo et al. 1994
DEAE cellulose	Fixed bed	Pajunen and Grönqvist 1994
Polyvinyl alcohol beads	Fixed bed	Smogrovicová et al. 2001
Porous glass beads	Fixed bed	Linko et al. 1993, Aivasidis 1996
Alcohol-free beer		
DEAE-cellulose beads	Fixed bed	Collin et al. 1991, Lomni et al. 1990
Porous glass beads	Fixed bed	Aivasidis et al. 1991
Silicon carbide rods	Monolith reactor	Van De Winkel et al. 1991
Acidified wort		
DEAE-cellulose beads	Fixed bed	Pittner et al. 1993
Main fermentation		
Calcium alginate beads	Gas lift	Nedovic et al. 1997
Calcium pectate beads	Gas lift	Smogrovicová et al. 1997
κ-carrageenan beads	Gas lift	Mensour et al. 1996
Ceramic beads	Fixed bed	Inoue 1995
Gluten pellets	Fixed bed	Bardi et al. 1997
Polyvinyl alcohol beads	Gas lift	Smogrovicová et al. 2001
Porous glass beads	Fixed bed	Virkajärvi and Krönlof 1998
Porous chitosan beads	Fluidized bed	Unemoto et al. 1998, Maeba et al. 2000
Silicon carbide rods	Monolith reactor	Andries et al. 1996
Spent grains	Gas lift	Brányik et al. 2002
Wood chips	Fixed bed	Linko et al. 1997, Kronlöf and Virkajärvi 1999

or 2-oxoglutarate dehydrogenase (Mockovciaková et al. 1993)—have been immobilized in calcium pectate gel beads and used in a continuous process for the production of nonalcoholic beer (Navrátil et al. 2000). These strains produced minimal amounts of ethanol, and they were also able to produce much lactic acid (up to 0.64 g/dm^3).

Production of Acidified Wort Using Immobilized Lactic Acid Bacteria

The objective of this technology is the acidification of the wort according to the "Reinheidsgebot," before the start of the boiling process in the brewhouse. An increased productivity of acidified wort has been obtained using immobilized *Lactobacillus amylovorus* on DEAE cellulose beads (Pittner et al. 1993, Meersman 1994). The pH of wort was reduced below a value of 4.0 after contact times of 7–12 minutes using a packed-bed reactor in downflow mode. The produced acidified wort was stored in a holding tank and used during wort production to adjust the pH.

Continuous Main Fermentation

The Japanese brewery Kirin developed a multistage continuous fermentation process (Inoue 1995, Yamauchi et al. 1994, Yamauchi and Kasahira 1995). The first stage is a stirred-tank reactor for yeast growth, followed by packed-bed fermenters, and the final step is a packed-bed maturation column. The first stage ensures adequate yeast cell growth with the desirable free amino nitrogen consumption. Ca-alginate was initially selected as carrier material to immobilize the yeast cells. These alginate beads were later replaced by ceramic beads ("Bioceramic®"). This system allowed production of beer within 3 to 5 days.

The engineering company Meura (Belgium) developed a reactor configuration with a first stage with immobilized yeast cells, where partial attenuation and yeast growth occurs, followed by a stirred-tank reactor (with free yeast cells) for complete attenuation, ester formation, and flavor maturation (Andries et al. 1996, Masschelein and Andries 1995). Silicon carbide rods are used in the first reactor as the immobilization carrier material. The stirred tank (second reactor) is continuously inoculated by free cells that escape from the first immobilized yeast cell reactor.

Labatt Breweries (Interbrew, Canada), in collaboration with the Department of Chemical and Biochemical Engineering at the University of Western Ontario (Canada), developed a continuous system using κ-carrageenan-immobilized yeast cells in an airlift reactor (Mensour et al. 1995, 1996, 1997). Pilot-scale research showed that full attenuation was reached in 20–24 hours with this system compared with 5–7 days for the traditional batch fermentation. The flavor profile of the beer produced using ICT was similar to that for the batch-fermented beer.

Hartwell Lahti and VTT Research Institute (Finland) developed a primary fermentation system using ICT on a pilot scale of 600 L/day (Kronlöf and Virkajärvi 1999). Woodchips were used as the carrier material, which reduced the total investment cost by one-third compared with more expensive carriers. The results showed that fermentation and flavor formation were very similar to those of a traditional batch process, although the process time was reduced to 40 hours.

Andersen et al. (1999) developed a new ICT process in which the concentration of carbon dioxide is controlled in a fixed-bed reactor in such a way that the CO_2 formed is kept dissolved, and is removed from the beer without foaming problems. DEAE cellulose was used as carrier material. High-gravity beer of acceptable quality has been fermented in 20 hours at a capacity of 50 L/hour.

REFERENCES

Aivasidis A. 1996. Another look at immobilized yeast systems. Cerevisia 21(1): 27–32.

Aivasidis A, Wandrey C, Eils H-G, Katzke M. 1991. Continuous fermentation of alcohol-free beer with immobilized yeast cells in fluidized bed reactors. Proceedings European Brewery Convention Congress, pp. 569–576.

Alexander MA, Jeffries TW. 1990. Respiratory efficiency and metabolite partitioning as regulatory phenomena in yeast. Enzyme Microb Technol 12:2–19.

Andersen K, Bergin J, Ranta B, Viljava T. 1999. New process for the continuous fermentation of beer. Proceedings European Brewery Convention Congress, pp. 771–778.

Anderson RG, Kirsop BH. 1975a. Oxygen as regulator of ester accumulation during the fermentation of worts of high gravity. J Inst Brew 80:111–115.

———. 1975b. Quantitative aspects of the control by oxygenation of acetate ester formation of worts of high specific gravity. J Inst Brew 81:296–301.

Andries M, Van Beveren PC, Goffin O, Masschelein CA. 1996. Design and application of an immobilized loop bioreactor for continuous beer fermentation. In: RH Wijffels, RM Buitelaar, C Bucke, J Tramper, editors. Immobilized Cells: Basics and Applications. Amsterdam: Elsevier. Pp. 672–678.

Arikawa Y, Enomoto K, Muratsubaki H, Okazaki M. 1998. Soluble fumarate reductase isoenzymes from *Saccharomyces cerevisiae* are required for anaerobic growth. FEMS Microb Lett 165:111–116.

Arikawa Y, Kobayashi M, Kodaira R, Shimosaka M, Muratsubaki H, Enomoto K, Okazaki M. 1999. Isolation of sake yeast strains possessing various levels of succinate- and/or malate-producing abilities by gene disruption or mutation. J Biosci Bioeng 87:333–339.

Asano T, Kurose N, Hiraoka N, Kawakita S. 1999. Effect of NAD^+-dependent isocitrate dehydrogenase gene (*IDH1*, *IDH2*) disruption of sake yeast on organic acid composition in sake mash. J Biosc Bioeng 88:258–263.

Avhenainen J, Mäkinen V. 1989. The effect of pitching yeast aeration on fermentation and beer flavor. Proceedings European Brewery Convention Congress, pp. 517–519.

Back W, Krottenthaler M, Braun T. 1998. Investigations into continuous beer maturation. Brauwelt Int 3:222–226.

Bardi E, Koutinas AA, Kanellaki M. 1997. Room and low temperature brewing with yeast immobilized on gluten pellets. Process Biochem 32:691–696.

Bishop LR. 1970. A system of continuous fermentations. J Inst Brew 76:172–181.

Bisson LF, Coons DM, Kruckeberg AL, Lewis DA. 1993. Yeast sugar transporters. Crit Rev Biochem Mol Biol 28:259–308.

Blomqvist K, Suihko M-L, Knowles J, Penttilä M. 1991. Chromosomal integration and expression of two bacterial α-acetolactate decarboxylase genes in brewer's yeast. Appl Environ Microbiol 57:2796–2803.

Brányik T, Vicente A, Cruz JM, Teixeira J. 2002. Continuous primary beer fermentation with brewing yeast immobilized on spent grains. J Inst Brew 108: 410–415.

Breitenbücher K, Mistler M. 1995. Fluidized-bed fermenters for the continuous production of non-alcoholic beer with open-pore sintered glass carriers. In: EBC Monograph XXIV, EBC Symposium on Immobilized Yeast Applications in the Brewing Industry, pp. 77–89.

Bullis BL, Lamire BD. 1994. Isolation and characterization of the *Saccharomyces cerevisiae SDH4* gene encoding a membrane anchor subunit of succinate dehydrogenase. J Biol Chem 269:6543–6549.

Calderbank J, Hammond JRM. 1994. Influence of higher alcohol availability on ester formation by yeast. J Am Soc Brew Chem 52:84–90.

Chan SY, Appling DR. 2003. Regulation of S-adenosylmethionine levels in *Saccharomyces cerevisiae*. J Biol Chem 22. 278:43051–43059.

Chang YS, Dubin RA, Perkins E, Forrest D, Michels CA, Needleman RB. 1988. *MAL63* codes for a positive regulator of maltose fermentation in *Saccharomyces cerevisiae*. Curr Genet 14:201–209.

Chapman KB, Solomon SD, Boeke JD. 1992. *SDH1*, the gene encoding the succinate dehydrogenase flavoprotein subunit from *Saccharomyces cerevisiae*. Gene 118:131–136.

Charalambous G, Bruckner KJ, Hardwick WA, Weatherby TJ. 1972. Effect of structurally modified polyphenols on beer flavour and flavour stability, part II. Tech Q Master Brew Assoc Am 9:131–135.

Chen EC-H. 1978. Relative contribution of Ehrlich and biosynthetic pathways to the formation of fusel alcohols. J Am Soc Brew Chem 36:39–43.

Chow T, Goldenthal MJ, Cohen JD, Hegde M, Marmur J. 1983. Identification and physical characterization of the maltose of yeast maltase structural genes. Mol Gen Genet 191:366–371.

Cohen JD, Goldenthal MJ, Buchferer B, Marmur J. 1984. Mutational analysis of *MAL1* locus of *Saccharomyces*: Identification and functional characterization of three genes. Mol Gen Genet 196:208–216.

Collin S, Montesinos M, Meersman E, Swinkels W, Dufour JP. 1991. Yeast dehydrogenase activities in relation to carbonyl compounds removal from wort and beer. Proceedings European Brewery Convention Congress, pp. 409–416.

Coote N, Kirsop BH. 1974. The content of some organic acids in beer and in other fermented media. J Inst Brew 80:474–483.

Coutts MW. 1957. A continuous process for the production of beer. UK Patents 872,391–400.

Daignan-Fournier B, Valens M, Lemire BD, Bolotin-Fukuhara M. 1994. Structure and regulation of *SDH3*, the yeast gene encoding the cytochrome b_{560} subunit by respiratory complex II. J Biol Chem 269:15469–15472.

D'Amore T, Celotto G, Stewart GG. 1991. Advances in the fermentation of high gravity wort. Proceedings European Brewery Convention Congress, pp. 337–344.

Debourg A. 2002. Yeast in action: From wort to beer. Cerevisia 27(3): 144–154.

Debourg A, Laurent M, Goossens E, Van De Winkel L, Masschelein CA. 1994. Wort aldehyde reduction potential in free and immobilized yeast systems. J Am Soc Brew Chem 52:100–106.

Debourg A, Masschelein CA, Piérard A. 1990. Effects of *ILV* genes doses on the flux through the isoleucine-valine pathway in *Saccharomyces cerevisiae*. Abstract C31, 6th International Symposium on Genetics of Industrial Microorganisms.

Delvaux F. 1998. Beheersing van gistingsproducten met belangrijke sensorische eigenschappen. Cerevisia 23(4): 36–45.

Dequin S. 2001. The potential of genetic engineering for improved brewing wine-making and baking yeasts. Appl Microbiol Biotechnol 56:577–588.

Dickinson JR. 1999 Carbon metabolism. In: JR Dickinson, M Schweizer, editors. The Metabolism and Molecular Physiology of *Saccharomyces cerevisiae*. London: Taylor and Francis, Ltd. Pp. 23–55.

Dickinson JR, Harrison SJ, Dickinson JA, Hewlins MJE. 2000. An investigation of the metabolism of isoleucine to active amyl alcohol in *Saccharomyces cerevisiae*. J Biol Chem 275:10937–10942.

Dickinson JR, Harrison SJ, Hewlins MJE. 1998. An investigation of the metabolism of valine to isobutyl alcohol in *Saccharomyces cerevisiae*. J Biol Chem 273:25751–25756.

Dickinson JR, Lanterman MM, Danner DJ, Pearson BM, Sanz P, Harrison SJ, Hewlins MJE. 1997. A ^{13}C nuclear magnetic resonance investigation of the metabolism of leucine to isoamyl alcohol in *Saccharomyces cerevisiae*. J Biol Chem 272:26871–26878.

Dickinson JR, Salgado LEJ, Hewlins MJE. 2003. The catabolism of amino acids to long chain and complex alcohols in *Saccharomyces cerevisiae*. J Biol Chem 278:8028–8034.

Dillemans M, Goossens E, Goffin O, Masschelein CA. 1987. The amplification effect of the *ILV5* gene on the production of vicinal diketones in *Saccharomyces cerevisiae*. J Am Soc Brew Chem 45:81–85.

Dillenhöfer W, Rönn D. 1996. Secondary fermentation of beer with immobilized yeast. Brauwelt Int 14: 344–346.

Dufour J-P. 1991. Influence of industrial brewing and fermentation working conditions on beer SO_2 levels and flavour stability. Proceedings European Brewery Convention Congress, 209–216.

Dufour J-P, Malcorps P. 1994. Ester synthesis during fermentation: Enzyme characterization and modulation mechanism. Proceedings 4th Inst Brew Aviemore Conference, pp. 137–151.

Ehrlich F. 1904. Uber das natürliche isomere des leucins. Berichte der Deutschen Chemisten Gesellschaft 37:1809–1840.

Enari T-M, Nikkola M, Suihko M-L, Penttilä M, Knowles J, Lehtovaara-Helenius P. 1992. Process for accelerated beer production by integrative expression in the *PGK1* or *ADC1* genes. US Patent 5,108,925.

Engan S. 1972. Organoleptic threshold values of some alcohols and esters in beer. J Inst Brew 78:33–36.

Engan S, Aubert O. 1977. Relations between fermentation temperature and the formation of some flavour components. Proceedings European Brewery Convention Congress, pp. 591–607.

Enomoto K, Arikawa Y, Muratsubaki H. 2002. Physiological role of soluble fumarate reductase in redox balancing during anaerobiosis in *Saccharomyces cerevisiae*. FEMS Microbiol Lett 24:103–108.

Federoff HJ, Ecclesall TR, Marmur J. 1983a. Carbon catabolite repression of maltase synthesis in *Saccharomyces cerevisiae*. J Bacteriol 156:301–307.

———. 1983b. Regulation of maltase synthesis in *Saccharomyces carlsbergensis*. J Bacteriol 154:1301–1308.

Fraleigh S, Bungay H, Fiechter A. 1989. Regulation of oxidoreductive yeast metabolism by extracellular factors. J Biotechnol 12:185–198.

Francke Johannesen P, Nyborg M, Hansen J. 1999. Construction of *S. carlsbergensis* brewer's yeast without production of sulfite. Proceedings European Brewery Convention Congress, pp. 655–662.

Fujii T, Kondo K, Shimizu F, Sone H, Tanaka JI, Inoue T. 1990. Application of a ribosomial DNA integration vector in the construction of a brewers' yeast having α-acetolactate decarboxylase activity. Appl Environ Microbiol 56:997–1003.

Fujii T, Nagasawa N, Iwamatsu A, Bogaki T, Tamai Y, Hamashi M. 1994. Molecular cloning, sequence analysis, and expression of the yeast alcohol acetyltransferase gene. Appl Environ Microbiol 60:2786–2792.

Fukuda K, Yamamoto N, Kiyokawa Y, Yanagiuchi T, Wakai Y, Kitamoto K, Inoue Y, Kimura A. 1998. Balance of activities of alcohol acetyltransferase and esterase in *Saccharomyces cerevisiae* is important for production of isoamyl acetate. Appl Environ Microbiol 64:4076–4078.

Gee DA, Ramirez WF. 1994. A flavour model for beer fermentation. J Inst Brew 100:321–329.

Gjermansen C, Nilsson-Tillgren T, Petersen JGL, Kielland-Bandt MC, Sisgaard P, Holmberg S. 1988. Towards diacetyl-less brewers' yeast. Influence of *ilv2* and *ilv5* mutations. J Basic Microbiol 28:175–183.

Godtfredsen SE, Ottesen M, Sisgaard P, Erdal K, Mathiasen T, Ahrenst-Larsen B. 1983. Application of the acetolactate decarboxylase from *Lactobacillus casei* for accelerated maturation of beer. Proceedings European Brewery Convention Congress, pp. 161–168.

Godtfredsen SE, Rasmussen AM, Ottesen M, Rafin P Petersen N. 1984. Occurrence of α-acetolactate decarboxylases among lactic acid bacteria and their utilization for maturation of beer. Appl Microbiol Biotechnol 20:23–28.

Goldenthal MJ, Vanoni M. 1990. Genetic mapping and biochemical analysis of mutants in the maltose regulatory gene of the *MAL1* locus of *Saccharomyces cerevisiae*. Arch Microbiol 154:544–549.

Goldenthal MJ, Vanoni M, Buchferer B, Marmur J. 1987. Regulation of *MAL* gene expression in yeast: Gene dosage effects. Mol Gen Genet 209:508–517.

Goossens E, Debourg A, Villaneuba KD, Masschelein CA. 1993. Decreased diacetyl production in lager brewing yeast by integration of the *ILV5* gene. Proceedings European Brewery Convention Congress, pp. 251–258.

Hammond JRM. 1993. Brewer's yeast. In: HA Rose, JS Harrison, editor. The Yeasts, vol. 5. London: Academic Press. Pp. 7–67.

Han E-K, Cotty F, Sottas C, Jiang H, Michels CA. 1995. Characterization of *AGT1* encoding a general α-glucoside transporter from *Saccharomyces*. Mol Microbiol 17:1093–1107.

Hammond JRM. 1993. Brewers' yeast. In: AH Rose, JS Harrison, editors. The Yeasts. London: Academic Press. Pp. 7–67.

Hansen J, Kielbrandt MC. 1996a. Inactivation of *MET10* in brewers' yeast specifically increases SO_2 formation during beer production. Nat Biotechnol 14:1587–1591.

———. 1996b. Inactivation of *MET2* in brewers' yeast increases the level of sulphite in beer. J Biotechnol 50:75–87.

Hardwick WA. 1995. History and antecedents of brewing. In: WA Hardwick, editor. Handbook of Brewing. New York: Marcel Dekker. Pp. 37–51.

Haukeli AD, Lie S. 1978. Conversion of α-acetolactate and removal of diacetyl: A kinetic study. J Inst Brew 84:85–89.

———. 1979. Yeast growth and metabolic changes during brewery fermentation. Proceedings European Brewery Convention Congress, pp. 461–473.

Hirata D, Aoki S, Watanabe K, Tsukioka M, Suzuki T. 1992. Stable overproduction of isoamylalcohol by *S. cerevisiae* with chromosome-integrated copies of the *LEU4* genes. Biosc Biotechnol Biochem 56:1682–1683.

Hough JS, Briggs DE, Stevens R, Young TW. 1982. Malting and Brewing Science.Vol. 2, Hopped Wort and Beer. London: Chapman and Hall.

Hu Z, Nehlin JO, Ronne H, Michels CA. 1995. *MIG1*-dependent and *MIG1*-independent glucose regulation of *MAL* gene expression in *Saccharomyces cerevisiae*. Curr Genet 28:258–266.

Hyttinen I, Kronlöf J, Hartwall P. 1995. Use of porous glass at Hartwall brewery in the maturation of beer with immobilized yeast. European Brewery Convention, Monograph XXIV, EBC Symposium Immobilized Yeast Applications in the Brewery Industry, Hans Carl Getränke-Fachverlag. Pp. 55–65.

Inoue T. 1995. Development of a two-stage immobilized yeast fermentation system for continuous beer brewing. Proceedings European Brewery Convention Congress, pp. 25–36.

Inoue T, Masuyama K, Yamamoto Y, Okada K, Kuroiwa Y. 1968. Mechanism of diacetyl formation in beer. Proc Am Soc Brew Chem, pp. 158–165.

Inoue T, Yamamoto Y. 1970. Diacetyl and beer fermentation. Proc Am Soc Brew Chem, pp. 198–208.

Iverson WG. 1994. Ethyl acetate extraction of beer: Quantitative determination of additional fermentation by-products. J Am Soc Brew Chem 52:91–95.

Jensen S. 1993. Using ALDC to speed up fermentation. Brewers' Gardian (September): 55–56.

Jespersen, L., Cesar, L.B., Meaden, P.G., Jakobsen, M. 1999. Multiple α-glucoside transporter genes in brewer's yeast. Appl Environ Microbiol 65:450–456.

Kaclíková E, Lachovicz TM, Gbelská Y, Subík J. 1992. Fumaric acid overproduction in yeast mutants deficient in fumarase. FEMS Microbiol Lett 91:101–106.

Käppeli O, Gschwend-Petrik M, Fiechter A. 1985. Transient responses of *Saccharomyces uvarum* to a change of growth-limiting nutrient in continuous culture. J Gen Microbiol 131:47–52.

Klein CJL, Olsson L, Ronnow B, Mikkelsen JD, Nielsen J. 1996. Alleviation of glucose repression of maltose metabolism by *MIG1* disruption in *Saccharomyces cerevisiae*. Appl Environ Microbiol 62: 4441–4449.

Kodama Y, Fukui N, Ashikari T, Shibano Y, Morioka-Fujimoto K, Hiraki Y, Kazuo N. 1995. Improvement of maltose fermentation efficiency: Constitutive expression of *MAL* genes in brewing yeasts. J Am Soc Brew Chem 53:24–29.

Kronlöf J, Virkajärvi I. 1999. Primary fermentation with immobilized yeast. Proceedings European Brewery Convention Congress, pp. 761–770.

Kubo Y, Takagi H, Nakamori S. 2000. Effect of gene disruption of succinate dehydrogenase on succinate production in sake yeast strain. J Biosci Bioeng 90: 619–624.

Kunze W. 1999. Technology of Brewing and Malting. Berlin: VLB.

Lagunas R. 1979. Energetic irrelevance of aerobiosis for *S. cerevisiae* growing on sugars. Mol Cell Biochem 27:139–146.

Lea AH, Piggott JRP. 1995. Fermented Beverage Production. Glasgow, United Kingdom: Blackie Academic and Professional.

Lee S, Villa K, Patino H. 1995. Yeast strain development for enhanced production of desirable alcohols/esters in beer. J Am Soc Brew Chem 39:153–156.

Lievense JC, Lim HC. 1982. The growth and dynamics of *Saccharomyces cerevisiae*. In: GT Tsao, MC Flickinger, RK Finn, editors. Annual Reports on Fermentation Processes. New York: Academic Press.

Lilly M, Lambrechts MG, Pretorius IS. 2000. Effect of increased yeast alcohol acetyltransferase activity on flavor profiles of wine and distillates. Appl Environ Microbiol 66:744–753.

Linko M, Suihko M-L, Kronlöf J, Home S. 1993. Use of brewer's yeast expressing α-acetolactate decarboxylase in conventional and immobilized fermentations. MBAA Tech Q 30:93–97.

Linko M, Virkajärvi I, Pohjala N, Lindborg K, Kronlöf J, Pajunen E. 1997. Main fermentation with immobilized yeast—A breakthrough? Proceedings European Brewery Convention Congress, pp. 385–394.

Lombardo A, Carine K, Scheffler IE. 1990. Cloning and characterization of the iron-sulfur subunit gene succinate dehydrogenase from *Saccharomyces cerevisiae*. J Biol Chem 265:10419–10423.

Lomni H, Grönqvist A, Pajunen E. 1990. Immobilized yeast reactor speeds beer production. Food Techn 5: 128–133.

Lucero P, Herweijer M, Lagunas R. 1993. Catabolite inactivation of the yeast maltose transporter is due to proteolysis. FEBS Lett 333:165–168.

MacDonald J, Reeve PTV, Ruddlesden JD, White FH. 1984. Current approaches to brewery fermentations. Progr Ind Microbiol 19:47–198.

MacWilliam IC. 1968. Wort composition. J Inst Brew 74:38–54.

Maeba H, Unemoto S, Sato M, Shinotsuka K. 2000. Primary fermenation with immobilized yeast in porous chitosan beads. Pilot scale trial. Proceedings 26th Convention Inst Brew Australia and New Zealand Section, Singapore, pp. 82–86.

Magarifuchi T, Goto K, Iimura Y, Tademuma M, Tamura G. 1995. Effect of yeast fumarase gene (*FUM1*) disruption on the production of malic, fumaric and succinic acids in sake mash. J Ferment Bioeng 80:355–361.

Malcorps P, Cheval JM, Jamil S, Dufour J-P. 1991. A new model for the regulation of ester synthesis by alcohol acetyltransferase in *Saccharomyces cerevisiae*. J Am Soc Brew Chem 49:47–53.

Malcorps P, Dufour J-P. 1987. Proceedings European Brewery Convention Congress, p. 377.

Masschelein CA. 1981. Role of fatty acids in beer flavour. EBC Flavour Symposium, Copenhagen, Denmark. EBC Monograph VII, pp. 211–221, 237–238.

———. 1997. A realistic view on the role of research in the brewing industry today. J Inst Brew 103:103–113.

Masschelein CA, Andries M. 1995. Future scenario of immobilized systems: Promises and limitations. In: EBC Monograph XXIV, EBC Symposium on Immobilized Yeast Applications in the Brewing Industry, pp. 223–241.

Maule DRJ. 1967. Rapid gas chromatographic examination of beer flavour. J Inst Brew 73:351–361.

McGovern PE, Glusker DL, Exner LJ, Voigt MM. 1996. Neolithic resinated wine. Nature 381:480–481.

Meersman E. 1994. Biologische aanzuring met geïmmobiliseerde melkzuurbacteriën. Cerevisia Biotechnol 19(4): 42–46.

Meilgaard MC. 1975a. Flavour chemistry of beer. Part 1. Flavour interactions between principle volatiles. Tech Q Master Brew Assoc Am 12:107–117.

———. 1975b. Flavour chemistry of beer. Part 2. Flavour and threshold of 239 aroma volatiles. Tech Q Master Brew Assoc Am 12:151–173.

Mensour N, Margaritis A, Briens CL, Pilkington H, Russell I. 1995. Gas lift systems for immobilized cell systems. In EBC Monograph XXIV, EBC Symposium on Immobilized Yeast Applications in the Brewing Industry, pp. 125–133.

———. 1996. Applications of immobilized yeast cells in the brewing industry. In: RH Wijffels, RM Buitelaar, C Bucke, J Tramper, editors. Immobilized Cells: Basics and Applications. Amsterdam: Elsevier. Pp. 661–671.

———. 1997. New developments in the brewing industry using immobilized yeast cell bioreactor systems. J Inst Brew 103:363–370.

Michels CA, Read E, Nat K, Charron MJ. 1992. The telomere-associated *MAL3* locus of *Saccharomyces* is a tandem array of repeated genes. Yeast 8:655–665.

Michnick S, Roustan JL, Remize F, Barre P, Dequin S. 1997. Modulation of glycerol and ethanol yields during alcoholic fermentation in *Saccharomyces cerevisiae* strains overexpressed or disrupted for *GPD1* encoding glycerol-3-phosphate dehydrogenase. Yeast 13:783–793.

Minetoki T, Bogaki T, Iwamatsu A, Fujii T, Hama H. 1993. The purification, properties and internal peptide sequences of alcohol acetyltransferase isolated from *Saccharomyces cerevisiae*. Biosci Biotechnol Biochem 57:2094–2098.

Mithieux SM, Weiss AS. 1995. Tandem integration of multiple *ILV5* copies and elevated transcription in polyploid yeast. Yeast 11:311–316.

Mockovciaková D, Janitorová V, Kaclíková E, Zagulski M, Subík J. 1993. The *ogd1* and *kgd1* mutants lacking 2-oxoglutarate dehydrogenase activity in yeast are allelic and can be differentiated by the cloned amber suppressor. Curr Genet 24:377–381.

Moonjai N, Delvaux F, Derdelinckx G, Verachtert H. 2000. Unsaturated fatty acid supplementation of stationary phase brewing yeast. Cerevisia 25(3): 37–50.

Moonjai N, Verstrepen KJ, Delvaux FR, Derdelinckx G, Verachtert H. 2002. The effects of linoleic acid supplementation of cropped yeast on its performance and acetate ester synthesis. J Inst Brew 108:227–235.

Nakatani K, Takahashi T, Nagami K, Kumada J. 1984. Kinetic studies of vicinal diketones in brewing. 2. Theoretical aspects for the formation of total vicinal diketones. Tech Q Master Brew Assoc Am 21:175–183.

Narziss L, Miedaner H, Kern E, Leibhard M. 1992. Technology and composition of non-alcoholic beers. Brauwelt Int 4:396.

Navrátil M, Gemeiner P, Sturdík E, Dömény Z, Smogrovicová D, Antalova Z. 2000. Fermented beverages produced by yeast cells entrapped in ionotropic hydrogels of polysaccharide nature. Minerva Biotec 12:337–344.

Nedovic V, Willaert R. 2005. Beer production using immobilised cells. In: V Nedovic, R Willaert, editors. Applications of Cell Immobilisation Biotechnology. Dordrecht (The Netherlands): Kluwer Academic Publishers. (In press.)

Nedovic VA, Pesic R, Leskosek-Cukalovic I, Laketic D, Vunjal-Novakovic G. 1997. Analysis of liquid axial dispersion in an internal loop gas-lift bioreactor for beer fermentation with immobilized yeast cells. Proceedings 2nd European Conference on Fluidization, Bilbao, pp. 627–635.

Needleman RB, Kaback DB, Dubin RA, Perkins EL, Rosenberg NG, Sutherland KA, Forrest DB, Michels CA. 1984. *MAL6* of *Saccharomyces*: A complex genetic locus containing three genes required for maltose fermentation. Proc Natl Acad Sci USA 81:2811–2815.

Nehlin JO, Ronne H. 1990. Yeast *MIG1* repressor is related to the mammalian early growth response and Wilms' tumour finger proteins. EMBO J 9:2891–2898.

Nelissen B, Mordant P, Jonniaux J-J, De Wachter R, Goffeau A. 1995. Phylogenetic classification of the major superfamily of membrane transport facilitators, as deduced from yeast genome sequencing. FEBS Lett 377:232–236.

Nevoigt E, Pilger R, Mast-Gerlach E, Schmidt U, Freihammer S, Eschenbrenner M, Garbe L, Stahl U. 2002. Genetic engineering of brewing yeast to reduce the content of ethanol in beer. FEMS Yeast Res 2:225–232.

Nevoigt E, Stahl U. 1997. Reduced pyruvate decarboxylase and increased glycerol-3-phosphate degydrogenase [NAD^+] levels enhance glycerol production in *Saccharomyces cerevisiae*. Yeast 12:1331–1337.

Ni B, Needleman RB. 1990. Identification of the upstream activating sequence of *MAL* and the binding sites for MAL63 activator of *Saccharomyces cerevisiae*. Mol Cell Biol 10:3797–3800.

N.N. 2000. Fermentation and Maturation, European Brewery Convention Manual of Good Practice. Nürnberg (Germany): Getränke-Fachverlag Hans Carl.

———. 2002. België (bijna) tweede bier-exporteur van Europa. Het Brouwersblad, June, pp. 9–29.

Nykänen L, Nykänen I. 1977. Production of esters by different yeast strains in sugar fermentations. J Inst Brew 83:30–31.

Onnela M-L, Suihko M-L, Penttilä M, Keränen S. 1996. Use of a modified alcohol dehydrogenase, *ADH1*, promotor in construction of diacetyl non-producing brewer's yeast. J Biotechnol 49:101–109.

Oshita K, Kubota M, Uchida M, Ono M. 1995. Clarification of the relationship between fusel alcohol formation and amino acid assimilation by brewing yeast using ^{13}C-labeled amino acid. Proceedings of the European Brewery Convention Congress, pp. 387–402.

Pajunen E. 1995. Immobilized yeast lager beer maturation: DEAE-cellulose at Synebrychoff. In: EBC Monograph XXIV, EBC Symposium on Immobilized Yeast Applications in the Brewing Industry, pp. 24–40.

Pajunen, E., Grönqvist, A. 1994. Immobilized yeast fermenters for continuous lager beer maturation. Proceedings 23rd Convention Inst of Brew Australia and New Zealand Section, Sydney, pp. 101–103.

Panchal CJ, Stewart GG. 1979. Utilization of wort carbohydrates. Brewers Digest (June): 36–46.

Peddie, H.A.B. 1990. Ester formation in brewery fermentations. J Inst Brew 96:327–331.

Peinado JM, Loureiro-Dias MC. 1986. Reversible loss of affinity induced by glucose in the maltose-H^+ symport of *Saccharomyces cerevisiae*. Biochim Biophys Acta 856:189–192.

Petrik M, Käppeli O, Fiechter A. 1983. An expanded concept for the glucose effect in the yeast *Saccharomyces uvarum*: involvement of short- and long-term regulation. J Gen Microbiol 129:43–49.

Pfisterer E, Stewart GG. 1975. Some aspects on the fermentation of high gravity worts. Proceedings European Brewery Convention Congress, pp. 255–267.

Pittner H, Back W, Swinkels W, Meersman E, Van Dieren B, Lomni H. 1993. Continuous production of acidified wort for alcohol-free-beer with immobilized lactic acid bacteria. Proceedings European Brewery Convention Congress, pp. 323–329.

Ramos-Jeunehomme C, Laub R, Masschelein CA. 1989. Proceedings European Brewery Convention Congress, pp. 513–519.

———. 1991. Why is ester formation in brewery fermentations yeast strain dependent? Proceedings European Brewery Convention Congress, pp. 257–264.

Reed G, Nogodawithana TW. 1991. Chapter 3. In: Yeast Technology. New York: Van Nostrand Reinhold.

Remize F, Roustan JL, Sablayrolles JM, Barre P, Dequin S. 1999. Glycerol overproduction by engineered *Saccharomyces cerevisiae* wine yeast strains leads to substantial changes in by-product formation and to stimulation of fermentation rate in stationary phase. Appl Environ Microbiol 65:143–149.

Riballo E, Herweijer M, Wolf DH, Lagunas R. 1995. Catabolite inactivation of the yeast maltose transporter occurs in the vacuole after internalization by endocytosis. J Bacteriol 177:5622–5627.

Rieger M, Käppeli O, Fiechter A. 1983. The role of limited respiration in the incomplete oxidation of glucose by *Saccharomyces cerevisiae*. J Gen Microbiol 129:653–661.

Rogers PJ, Stewart PR. 1973. Mitochondrial and peroxisomal contributions to the energy metabolism of *Saccharomyces cerevisiae* in continuous culture. J Gen Microbiol 79:205–217.

Romano P, Suzzi G. 1992. Production of H_2S by different yeast strains during fermentation. Proceedings 22nd Convention of the Institute of Brewing, Melbourne, Institute of Brewing, Adelaide, Australia, pp. 96–98.

Rosculet G. 1971. Aroma and flavour of beer. Part II. The origin and nature of less-volatile and non-volatile components of beer. Brew Dig 46(6): 96–98.

Rostgaard-Jensen B, Svendsen I, Ottesen M. 1987. Isolation and characterization of an α-acetolactate decarboxylase useful for accerelated beer maturation. Proceedings European Brewery Convention Congress, pp. 393–400.

Sablayrolles JM, Ball CB. 1995. Fermentation kinetics and the production of volatiles during alcoholic fermentation. J Am Soc Brew Chem 53:71–78.

Shindo S, Sahara H, Koshino S. 1994. Suppression of α-acetolactate formation in brewing with immobilized yeast. J Inst Brew 100:69–72.

Siro M-R, Lövgren T. 1979. Influence of glucose on the α-glucoside permease activity of yeast. Eur J Appl Microbiol 7:59–66.

Slaughter JC, Jordan B. 1986. The production of hydrogen sulphide by yeast. In: I Campbell, FG Priest, editors. Proceedings of the Second Aviemore Conference on Malting, Brewing and Distilling, London: Institute of Brewing. Pp. 308–310.

Smogrovicová D, Döměny Z, Gemeiner P, Malovíková A, Sturdík E. 1997. Reactors for the continuous primary beer fermentation using immobilised yeast. Biotechn Techn 11:261–264.

Smogrovicová D, Döměny Z, Navrátil M, Dvorák P. 2001. Continuous beer fermentation using polyvinyl alcohol entrapped yeast. Proceedings European Brewery Convention Congress 50:1–9.

Soler AP, Casanova M, Gozalbo D, Sentandreu R. 1987. Differential translational efficiency of the mRNA isolated from derepressed and glucose repressed *Saccharomyces cerevisiae*. J Gen Microbiol 133:1471–1480.

Stewart, G.G., Russell, I. 1993. Fermentation—The "black box" of the brewing process. MBAA Tech Quart 30:159–4168.

Suihko M, Blomqvist K, Penttilä M, Gisler R, Knowles J. 1990. Recombinant brewer's yeast strains suitable for accelerated brewing. J Biotechnol 14:285–300.

Szlavko CM. 1973. Tryptophol, tyrosol and phenylethanol—The aromatic higher alcohols in beer. J Inst Brew 83:283–243.

Tada S, Takeuchi T, Sone H, Yamano S, Schofiels MA, Hammond JRM, Inoue T. 1995. Pilot scale brewing with industrial yeasts which produce the α-acetolactate decarboxylase of *Acetobacter Aceti* ssp. *xylium*. Proceedings European Brewery Convention Congress, pp. 369–376.

Tezuka H, Mori T, Okumura Y, Kitabatake K, Tsumura Y. 1992. Cloning of a gene suppressing hydrogen sulphide production by *Saccharomyces cerevisiae* and its expression in a brewing yeast. J Am Soc Brew Chem 50:130–133.

Thorne RSW. 1968. Continuous fermentation in retrospect. Brew Dig 43(2): 50–55.

Thurston PA, Quain DE, Tubb RS. 1982. Lipid metabolism and the regulation of volatile synthesis in *Saccharomyces cerevisiae*. J Inst Brew 88:90–94.

Thurston PA, Taylor R, Avhenainen J. 1981. Effects of linoleic acid supplements on the synthesis by yeast of lipid and acetate esters. J Inst Brew 87:92–95.

Unemoto S, Mitani Y, Shinotsuka K. 1998. Primary fermentation with immobilized yeast in a fluidized bed reactor. MBAA Tech Q 35:58–61.

Van de Meersche J, Devreux A, Masschelein CA. 1977. The role of yeasts in the maturation of beer flavour. Proceedings European Brewery Convention Congress, pp.561–575.

Van Den Berg R, Harteveld R, Martens FB. 1983. Diacetyl reducing activity in brewer's yeast. Proceedings European Brewery Convention Congress, pp. 497–504.

Van De Winkel L. 1995. Design and optimization of a multipurpose immobilized yeast bioreactor system for brewery fermentations. Cerevisia 20(1): 77–80.

Van De Winkel L, Van Beveren PC, Masschelein CA. 1991. The application of an immobilized yeast loop reactor to the continuous production of alcohol-free beer. Proceedings European Brewery Convention Congress, pp. 307–314.

Van Dieren D. 1995. Yeast metabolism and the production of alcohol-free beer. In: EBC Monograph XXIV, EBC Symposium on Immobilized Yeast Applications in the Brewing Industry, pp. 66–76.

van Iersel MFM, Meersman E, Arntz M, Ronbouts FM., Abee, T. 1998. Effect of environmental conditions on flocculation and immobilization of brewer's yeast during production of alcohol-free beer. J Inst Brew 104:131–136.

Verstrepen KJ. 2003. Flavour-Active Ester Synthesis in *Saccharomyces cerevisiae*. PhD thesis, Katholieke Universiteit Leuven, Belgium.

Verstrepen KJ, Van Laere SDM, Vanderhaegen BMP, Derdelinckx G, Dufour J-P, Pretorius IS, Winderickx J, Thevelein JM, Delvaux FR. 2003. Expression levels of the yeast alcohol acetyltransferase genes *ATF1*, *Lg-ATF1*, and *ATF2* control the formation of a broad range of volatile esters. Appl Environ Microbiol 69:5228–5237.

Villaneuba KD, Goossens E, Masschelein CA. 1990. Subthreshold vicinal diketone levels in lager brewing yeast fermentations by means of *ILV5* gene amplification. J Am Soc Brew Chem 48:111–114.

Virkajärvi I, Krönlof J. 1998. Long-term stability of immobilized yeast columns in primary fermentation. J Am Soc Brew Chem 56:70–75.

Wainwright T. 1973. Diacetyl—A review. J Inst Brew 79:451–470.

Walker GM. 1998. Yeast—Physiology and Biotechnology. Chichester United Kingdom: John Wiley and Sons.

Wellhoener HJ. 1954. Ein Kontinuierliches Gär- und Reifungsverfahren für Bier Brauwelt 94(1): 624–626.

White FH, Portno AD. 1979. The influence of wort composition on beer ester levels. Proceedings European Brewery Convention Congress, pp. 447–460.

Willaert R. 2001. Sugar consumption kinetics by brewer's yeast during the primary beer fermentation. Cerevisia 26(1): 43–49.

Yamano S, Kondo K, Tanaka J, Inoue T. 1994. Construction of a brewer's yeast having α-acetolactate decarboxylase gene from *Acetobacter acetii* ssp. *xylinum* integrated in the genome. J Biotechnol 32: 173–178.

Yamauchi Y, Kashihara T. 1995. Kirin immobilized system. In: EBC Monograph XXIV, EBC Symposium on Immobilized Yeast Applications in the Brewing Industry, pp. 99–117.

Yamauchi Y, Okamato T, Murayama H, Nagara A, Kashihara T, Nakanishi K. 1994. Beer brewing using an immobilized yeast bioreactor design of an immobilized yeast bioreactor for rapid beer brewing system. J Ferm Bioeng 78:443–449.

Yao B, Sollitti P, Zhang X, Marmur J. 1994. Shared control of maltose induction and catabolite repression of the *MAL* structural genes in *Saccharomyces*. Mol Gen Genet 243:622–630.

Younis OS, Stewart GG. 1998. Sugar uptake and subsequent ester and higher alcohol production by *Saccharomyces cerevisiae*. J Inst Brew 104:255–264.

———. 2000. The effect of wort maltose content on volatile production and fermentation performance in brewing yeast. In: K Smart, editor. Brewing Yeast Fermentation and Performance, vol. 1. Oxford: Blackwell Science. Pp. 170–176.

Zheng X, D'Amore T, Russell I, Stewart GG. 1994. J Ind Microbiol 13:159–166.

Part VII
Food Safety

30
Microbial Safety of Food and Food Products

J. A. Odumeru

PREFACE

Globally, food safety issues are of top priorities to the food industry, government food safety regulators, and consumers as a result of a significant increase in the number of foodborne disease cases and outbreaks reported worldwide in the 20th century. These issues led to the proliferation of several food safety programs designed to reduce the incidence of foodborne illness. Although a number of producers and processors have implemented a variety of food safety programs, the occurrence of foodborne illness from emerging and existing pathogens remains a challenge to the food industry and food safety regulators. Food safety begins on the farm and continues through processing, transportation, and storage until the food is consumed. Food safety programs such as Good Manufacturing Practices (GMP), Sanitation, Food Quality, and Safety Tests, and Hazard Analysis Critical Control Points (HACCP) are examples of food safety programs that are commonly used to control and monitor microbial contamination of food.

The three main categories of food safety concerns in the food industry include microbiological, chemical, and physical hazards. The microbiological hazards are those involving foodborne pathogens; chemical hazards include concerns related to antibiotics, pesticides, and herbicides; and physical hazards are those related to foreign objects in foods that can result in injury or illness when consumed with foods. Although this chapter addresses issues related to microbial hazards, food safety programs, which provide protection against these three types of hazards, especially during food processing, will be discussed. Foodborne organisms, sources of microbial contamination of foods and emerging pathogens, will be reviewed in relation to food safety issues.

INTRODUCTION

Food safety concerns are currently at an all time high due to worldwide publicity about cases and outbreaks of foodborne illness. These concerns are now of top priority in the political and economic agendas of governments at various levels. One of the worst nightmares for food producers or processors is to have the name of their company show up in a news report as the source of a foodborne illness. Apart from the loss of consumer confidence and loss of sales, there are also legal aspects about which food companies must be concerned (Odumeru 2002). An estimated 76 million cases of foodborne illness per year occur in the United States, resulting in 325,000 hospitalizations and 5000 deaths (Mead et al. 1999). The economic impact of these illnesses is estimated at $5 billion or more. A number of food safety programs are currently in place in the food industry in an attempt to reduce the incidence of foodborne illness, which has been on the rise in the last two decades (Maurice 1994). The increase in the incidence of foodborne disease has been attributed to a combination of factors. These include changes in food production and processing practices, changes in retail distribution, social changes including consumer preferences and eating habits, lack of experience of mass kitchen personnel, changes in population demographics, and increases in population mobility worldwide as a result of increases in international trade and travel (McMeekin and Olley 1995, Baird Parker 1994). Furthermore, advances in sciences in the area of analytical methods development has led to the availability of better detection methods for the diagnosis of foodborne illness and a subsequent increase in the number of cases reported. Other factors such as better reporting systems and increases in the occurrence of emerging pathogens have contributed to the increase in cases of foodborne diseases reported. Issues related to emerging pathogens will be discussed in a subsection of this chapter. A number of scientific tools are now available to the food industry and food safety regulators for implementing food safety programs designed to reduce the incidence of foodborne illness. These include the use of risk assessment of foods and food ingredients, to determine the risks associated with various types of foods under certain processing conditions; predictive modeling, which estimates the growth and survival of pathogens and spoilage organisms under specified conditions; and rapid methods for screening foods for quality and safety during and after production. This chapter will provide an overview of issues related to microbial safety of food and food products, food preservation technologies, emerging pathogens, food safety programs to control microbial contamination, and future perspectives on food safety.

SHELF LIFE OF FOODS AND FOOD INGREDIENTS AND FOOD SAFETY

The shelf life of a food product generally refers to the keeping quality of the food. An estimated 25% of the food supplies worldwide are lost as a result of spoilage; hence, it is economically beneficial to maintain the quality of food products at various stages of food production and storage. There are two categories of foods in relation to shelf life: shelf stable and perishable. Whether a particular food product is shelf stable or perishable depends on the intrinsic properties of the food (e.g., pH, water activity, and structure). Shelf-stable foods usually have low water activity, low pH, or a combination of both, while perishable foods tend to have high water activity and high pH. The structure or texture of the food is also an important factor in shelf stability. Extrinsic factors such as storage temperature, gaseous atmosphere, and relative humidity also determine the shelf stability of food products (McMeekin and Ross 1996). These intrinsic and extrinsic factors influence the survival and growth not only of spoilage organisms but also of pathogenic organisms in foods. Food spoilage occurs as a result of physical or chemical changes in the food or of the by-products of spoilage microorganisms growing in the food product. Pathogens present in low levels may not produce identifiable changes in the food; hence, the presence of pathogens cannot be determined using noticeable changes in the food as an indicator.

Although shelf-stable foods are less likely to be implicated in foodborne illness than perishable foods, cross-contamination of shelf-stable or perishable foods by pathogens can be a source of foodborne illness. A number of preservation applications used in the food industry are designed to extend the shelf life of the food product by reducing microbial growth; however, pathogens that are able to survive or even grow under preservation techniques such as refrigeration can cause foodborne illness. Effective strategies for controlling the presence of spoilage and foodborne pathogens in foods should include elimination of sources of contamination combined with food preservation technologies such as drying, freezing, smoking, curing, fermenting, refrigeration (Baird-Parker 2000) and modified-atmosphere packaging (Farber 1991)

CATEGORIES OF FOODBORNE ORGANISMS

Microorganisms that can be transmitted to humans or animals through food are referred to as foodborne organisms. There are three main categories of foodborne organisms: spoilage, pathogenic, and beneficial. Spoilage organisms can grow and produce physical and chemical changes in foods, resulting in unacceptable flavor, odor, formation of slime, gas accumulation, release of liquid exudates or purge, and changes in consistency, color, and appearance. Also, extracellular or intracellular enzymes released by spoilage organisms can result in deterioration of food quality. Growth of microorganisms to high numbers is usually required before spoilage becomes noticeable. Hence, control of growth of spoilage organisms is required to impede microbial spoilage. The presence of foodborne pathogens in foods in low concentrations can render foods harmful to humans if consumed. Because pathogenic organisms at low levels may not produce noticeable changes in foods, consumers may not have advance warning signals of the danger associated with consumption of contaminated foods. "Beneficial" or "useful" organisms include microorganisms used in various food fermentation processes. These organisms are either naturally present in such foods or added to produce the desired by-product of fermentation. Various types of foods such as fruits and vegetables, pickles, dairy products, meats, sausages, cheeses, and yogurt are common types of fermented products involving the use of beneficial organisms. Beneficial organisms include organisms in the group of lactic acid bacteria, yeasts, and molds. Bacteria species from 10 genera are included in the group of lactic acid bacteria. These include *Lactococcus, Leuconostoc, Streptococcus, Lactobacillus, Pediococcus, Carnobacterium, Tetragenococcus, Aerococcus, Vagococcus,* and *Enterococcus*. The most important type of yeast used for fermentation of food and alcohol is *Saccharomyces cerevisiae.* This yeast is used for leavening bread and production of beer, wine, and liquors. It is also used for food flavor. Non-mycotoxin-producing molds from the genera *Penicillium* and *Aspergillus,* and some in the *Rhizopus* and *Mucor* genera, have been used for beneficial purposes in food preparation (Bibek 1996).

SOURCES OF FOODBORNE PATHOGENS

In order to determine and implement effective control measures for pathogens in foods, it is important to identify potential sources of contamination. Plants and animals are the main source of human food supply. The exterior and, in some cases, the interior of plants and animals harbor microorganisms from external sources such as soil, water, and air. These environmental sources contain a wide variety of microorganisms, some of which are pathogenic to man. Contamination of the food supply can occur at various stages of production, processing, transportation, and storage. For example, fruits and vegetables can be contaminated at the farm level as well as during harvesting, transportation, and processing. Potential sources of on-farm contamination of fruits and vegetables are summarized in Figure 30.1. Meats can be contaminated at the time of slaughter, processing, and storage. Microbial contamination can come from slaughtered animals, water, equipment, utensils, the slaughterhouse environment, and workers. Thus intervention strategies to control microbial contamination of the food supply must be implemented at various stages of food production, processing, transportation, and storage, and also at the consumer end of food preparation.

FOODBORNE DISEASE CASES AND OUTBREAKS

Foodborne disease in a susceptible host can result from consumption of food or water contaminated with pathogenic organisms. A single or sporadic case of foodborne illness refers to an instance when an illness that is unrelated to other cases occurs as a result of consumption of contaminated food or water. An outbreak, on the other hand, refers to an incident in which two or more persons become ill after consuming the same food or water from the same source. The occurrence of a foodborne illness depends on a number of risk factors such as (1) type and number of pathogenic microorganisms in the food, (2) effect of food product formulation or processing on the viability of the pathogen, (3) storage conditions of the food that may promote contamination, growth, and survival of the pathogen, and (4) the susceptibility of the individual to foodborne

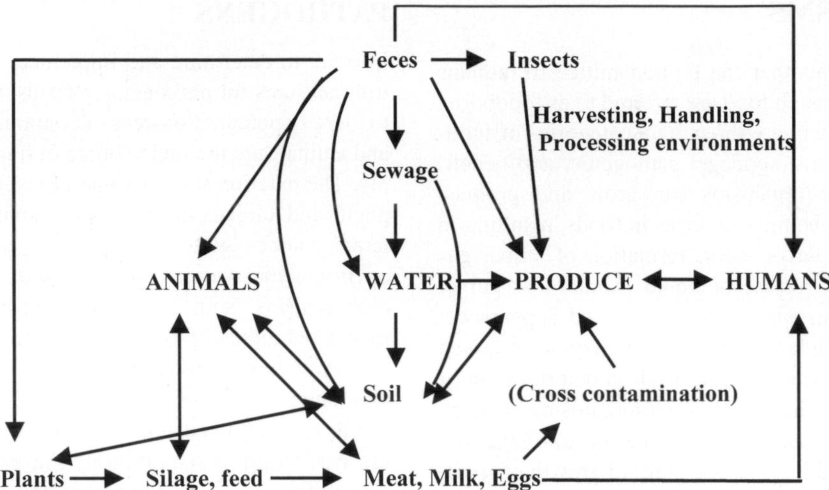

Figure 30.1. On-farm sources of microbial contamination of water and food products of plant and animal origin. (Adapted from Beuchat 1995.)

illness. It is believed that the number of reported outbreaks represents only 10% of the real incidence of foodborne disease, even in countries with well-established surveillance systems (Baird-Parker 2000). Foodborne illness resulting from severe infections such as hemolytic uremic syndrome, botulism, and listeriosis often require hospitalization and are more likely to be reported, while self-limiting foodborne illness such as salmonellosis, campylobacteriosis, and *S. aureus* enterotoxin–related infections are less likely to be reported. It is believed that viral foodborne infections account for a large portion of cases of foodborne illness (Caul 2000). However, the extent of the problem on a global scale is difficult to assess as a result of lack of surveillance data in most parts of the world, coupled with the fact that most viral infections are self-limiting. Although viruses and bacteria-related infections account for the majority of foodborne diseases, certain groups of parasites and fungi are also etiologic agents of foodborne diseases.

Foodborne Bacterial Infections

Several types of foodborne bacterial pathogens are implicated in foodborne diseases. Examples of bacterial genera most commonly implicated in foodborne infections, onset and duration of the symptoms of the disease, types of foods that are likely to be contaminated by these groups of bacteria, and potential sources of contamination are summarized in Table 30.1. Bacterial pathogens including *Campylobacter, Salmonella, Listeria monocytogenes* and verocytotoxigenic *E. coli* are the top causes of bacterial foodborne infections in westernized countries, based on the number of reported cases (Sharp and Reilly 2000). The number of reported cases of *Salmonella, Campylobacter* and *E. coli* O157 increased from 2- to 40-fold between 1982 and 1994 (Sharp and Reilly 2000). The increase in the number of cases was attributed to heightened awareness of foodborne related illness, improved diagnostic methods, and improved reporting and surveillance procedures. *Campylobacter* is currently the leading cause of foodborne disease in the United States and Canada, followed by *Salmonella*-related infections. The numbers of estimated cases of foodborne illness per year in the United States and Canada countries are listed in Table 30.2. Although the number of *Listeria monocytogenes*–related infections is considerably lower than those of other pathogens, foodborne infections caused by this organism are characterized by a high case mortality ratio, especially in certain high risk groups such as the elderly, the young, immunocompromised individuals, and pregnant women. Infections caused by *E. coli* O157:H7 often

Table 30.1. Foodborne Bacterial Pathogens, Symptoms of Disease, Target Populations and Potential Sources of Food Contamination

Organism	Symptoms	Onset	Duration	Target Population	Source	Suspect Foods
Bacillus cereus (B. cereus)	Two forms: 1. Diarrheal illness including abdominal cramps, nausea (vomiting rare) 2. Emetic illness including nausea and vomiting	Diarrheal: 6–15 hours Emetic: 30 minutes to 5 hours	Diarrheal: 12–24 hours Emetic: 6–24 hours	Everyone	Soil, Dust	Diarrheal: meat, milk, vegetables, fish, soups Emetic: rice products, potatoes, pasta, cheese products
Campylobacter jejuni	Diarrhea (may contain blood), fever, nausea, vomiting, abdominal pain, headache, muscle pain	2–5 days	2–10 days	Children under five, young adults 15–29	Cattle, chicken, birds, flies, stream, pond water	Raw chicken, turkey, raw milk, beef, shellfish, water
Clostridium botulinum	Fatigue, weakness, double vision, respiratory failure	18–36 hours, varies 4 hours to 8 days	Months	Everyone	Soil, sediment, intestinal tracts of fish and mammals	Canned foods, smoked and salted fish, chopped bottled garlic, sautéed onions, honey
Clostridium perfringens	Diarrhea, cramps, nausea (vomiting rare)	6–22 hours	Under 24 hours, may persist 1–2 weeks	Everyone, but young and elderly are most affected	Soil, feces	Meats, meat products, poultry, gravy
Escherichia coli (Traveller's diarrhea)	Watery diarrhea, abdominal cramps, low-grade fever, nausea, malaise	1–3 days	Days	Everyone	Water, Human sewage	Dairy products

(Continues)

Table 30.1. (*Continued*)

Organism	Symptoms	Onset	Duration	Target Population	Source	Suspect Foods
E. coli O157:H7	Severe cramping, watery diarrhea becoming bloody, low-grade fever	12–60 hours	2–9 days to weeks	Children	Cattle, deer	Undercooked or raw hamburger, raw milk, unpasteurized apple cider
Listeria monocytogenes	Septicemia, maningoen-cephalitis, spontaneous abortions, stillbirths, influenza-like symptoms	A few days to 6 weeks	Days to weeks	Pregnant woman, immunocom-promised persons, cancer and AIDS patients and those with chronic diseases, elderly, (fatality rate is as high as 70%)	Soil, improperly made silage	Raw milk, soft ripened cheeses, ice cream, raw vegetables, fermented raw meat, sausages, hot dogs, luncheon meats, raw and cooked poultry, raw and smoked fish
Salmonella	Nausea, chills, vomiting, cramps, fever, headache, diarrhea, dehydration	6–48 hours	1–4 days	Everyone, but more severe symptoms in infants, elderly, infirm and AIDS patients	Water, soil, insects, animal feces, raw meat, raw poultry, raw seafood	Raw meats, poultry, eggs, milk, dairy products, fish, shrimp, frogs legs, yeast, coconut, sauces and salad dressings, cake mixes, cream-filled desserts and toppings, dried gelatin, peanut butter, cocoa, chocolate

Shigella	Abdominal pain, diarrhea, fever, vomiting, sometimes cramps, nausea	1–7 days	4–7 days	Infants, elderly, infirm, AIDS patients	N/A	Salads (Potato, tuna, shrimp, macaroni and chicken), raw vegetables, milk and dairy products, poultry
Staphylococcus aureus	Nausea, vomiting, cramps, diarrhea, prostrations	1–6 hours	1–2 days	Everyone	Air, dust, sewage, water, milk, food equipment, humans, animals, nasal passages and throats, hair, skin, boils	Meat and meat products, poultry and poultry products, salads, (egg, tuna, chicken, potato, macaroni) cream-filled pastries, cream pies, milk and dairy products
Vibrio cholerae (Cholera)	Mild, watery diarrhea, abdominal cramps, nausea, vomiting, dehydration, shock	6 hours to 5 days	Days	Everyone, but especially those with immunocom-promised systems, reduced gastric acidity, malnutrition	Raw shellfish, water, poor sanitation	Seafood, water

Source: Adapted from American Academy of Microbiology: Food Safety Current Status and Future Needs 1999, research paper by Stephanie Doores.

result in a high morbidity and mortality rate. For example, complications such as hemolytic uremic syndrome can result in kidney failure and death (Riley et al. 1983, Karmali 1989). Sources of bacterial infections are not limited to contaminated foods; contaminated water and person-to-person transmission have also been implicated. For example, in a waterborne outbreak of *E. coli* 0157:H7 in Walkerton, Ontario, 1700 cases and seven deaths were reported. This led to changes in the Ontario Drinking Water Regulations to include strict water treatment procedures and zero tolerance for the presence of coliforms and *E. coli* in drinking water. Current food safety programs in the food industry must now document the use of pathogen-free water in food production and processing facilities.

Campylobacter Enteritis

Campylobacter jejuni and *C. coli* are the species of *Campylobacter* implicated in most *Campylobacter* enteritis infections. Other species of *Campylobacter* such as *C. lari* and *C. upsaleusis* may also cause *Campylobacter* enteritis. Symptoms of *Campylobacter* infection in humans are watery and/or bloody diarrhea, abdominal pain, fever, and general malaise. The disease is usually self-limiting, and antibiotics are required only when complications occur. It is estimated that *Campylobacter* enteritis accounts for 10% of cases of foodborne illness, and death is rarely reported. Most cases of this disease are sporadic, and foods of animal origin, particularly poultry, are largely responsible for most infections. Contaminated fruits and vegetables have also been implicated in *Campylobacter* enteritis cases.

Salmonellosis

Salmonellosis infection ranks second in incidence to *Campylobacter* enteritis. This organism is ubiquitous (present everywhere) in the environment, especially in the feces of most food-producing animals; hence, a variety of foods are readily contaminated by this organism. *Salmonella* are present in animals without causing apparent illness. However, certain serotypes of *Salmonella* such as *S. enteritidis*, can penetrate poultry reproductive organs, resulting in contamination of egg content and posing a health risk to consumers. Salmonellosis symptoms include watery diarrhea, abdominal pain, nausea, fever, headache, and occasional constipation. Hospitalization may be required in cases of severe infections. Foods that can become contaminated with *Salmonella* include meat, raw milk, poultry, eggs, dairy products, and other types of foods that can become contaminated with fecal material. An increase in the number of cases of Salmonellosis linked to consumption of contaminated fruits and vegetables such as bean sprouts, raw tomatoes, melons, and cantaloupes has been reported. In addition to fecal contamination, cross-contamination of foods by *Salmonella* during food preparation can be a source of foodborne illness.

Listeriosis

Listeriosis is caused by *Listeria monocytogenes*. The ubiquitous nature of this organism contributes to the widespread incidence of the organism in foods. It is psychrotropic in nature and is able to grow at refrigeration temperatures. Hence, this organism is of concern in refrigerated foods with ex-

Table 30.2. The Incidence of the Four Most Frequent Foodborne Disease–Causing Pathogens in Canada and the United States

	Estimated Cases/Year (Based on Population Ratio)	
Bacteria	Canada	United States[a]
Campylobacter	110,000–700,000	1,100,000–7,000,000
Listeria	93–177	928–1,767
Salmonella	69,600–384,000	696,000–3,840,000
Verotoxigenic *E. coli* (Including *E.coli* O157:H7)	1,600–3,200	16,000–32,000
Total	181,3000–1,087,400	1,812,900–10,873,800

[a]Adapted from Buzby and Roberts 1997.

tended shelf life. Listeriosis infections often result in septicemia and/or meningitis. The case mortality ratio for Listeriosis infections is much higher than those of *Salmonella*, *Campylobacter,* and *E. coli* 0157:H7 infections. An estimated 500 deaths associated with Listeriosis are reported annually in the United States. Individuals most susceptible to Listeriosis include the elderly, the young, immunocompromised patients, and pregnant woman. A wide variety of foods including processed meats, milk and dairy products, fruits,and vegetables have been implicated in a number of Listerioses outbreaks. In 1999, a multistate outbreak involving 1000 cases of Listeriosis in the United States was linked to consumption of contaminated hot dogs and delicatessen meats, resulting in a recall of 35 million pounds of these food products (MMWR 1999).

Verotoxigenic *Escherichia coli* (VTEC) Infections

Verotoxigenic *Escherichia coli* (VTEC) belonging to the O157 serogroup is the most common VTEC serogroup implicated in foodborne disease outbreaks. This organism was first recognized as a cause of hemorrhagic colitis in 1982. VTEC O157 and serogroups O5, O26, and O111 are of high prevalence in feces of healthy cattle. Thus, foods of animal origin such as meat, milk, and dairy products can be contaminated with these organisms. Other types of foods such as salads, vegetables, fruits, sandwiches, and cooked meat can be contaminated during preparation. Contaminated water can also be a source of infection by this organism. The majority of food- and waterborne outbreaks had links with contamination of fecal origin. Infection by VTEC can result in hemorrhagic colitis (HC) and hemolytic uremic syndrome (HUS), abdominal pain, and watery diarrhea (bloody or nonbloody). The HC symptoms may lead to complications of the HUS and subsequent renal failure. Although HUS can occur in any age group, children are more susceptible (Karmali 1989). In adults, 50% of thrombocytopenic purpura (TP) cases are caused by HUS.

Other bacterial foodborne infections can occur as a result of ingestion of food contaminated with *Staphylococcus aureus* toxin, *Clostridium perfringens, Clostridium botulinum* toxin, *Shigella* spp., *Yersinia enterocolitica, Brucella* spp., *Vibrio cholera, Vibrio paraheamolylicus*, and *Bacillus cereus* enterotoxin.

Foodborne Parasite Infections

Foodborne parasite infections are caused by certain groups of protozoa. It is believed that foodborne infections related to parasites are underreported by an estimated factor of 10 or more (Casemore 1991), and the etiological agent of foodborne outbreaks is identified in less than 50% of the cases (Bean et al. 1990, Cliver 1987). Parasites commonly associated with foodborne disease and possible food and water sources are listed in Table 30.3.

Cryptosporidium parvum is the species of *Cryptosporidium* associated with infection in humans (Current and Blagburn 1990). It is a common cause of gastrointestinal infections in immunocompromised individuals and is an opportunistic pathogen associated with infections in patients with acquired immune deficiency syndrome (AIDS), but it affects healthy people as well. Although drinking water contaminated with human or animal feces is the usual source of transmission of this organism to humans, epidemiological links between the consumption of contaminated foods such as raw sausage, offal, and raw milk and cryptosporidiosis have been reported (Casemore 1990, Casemore et al. 1986). The largest outbreak of waterborne cryptosporidiosis was reported in Milwaukee, Wisconsin, with 403,000 persons ill. Reported outbreaks in most countries are associated with contaminated drinking water or contaminated swimming pools.

Cyclospora is an emerging pathogen, which causes diarrheal infections in humans. The first reported outbreak of *Cyclospora* infection occurred in 20 U.S. states and two provinces in Canada, Ontario and Quebec, during the months of May and June, 1996, and April and May, 1997. Consumption of contaminated imported raspberries was implicated in many of these outbreaks (Herwaldt and Ackers 1996, Herwaldt and Beach 1997). Contaminated mesculun lettuce and basil have also been linked to *Cyclospora* outbreaks (MMWR 1997).

Sarcocytis is another coccidian parasite, which is prevalent in livestock. Human infection occurs as a result of ingestion of raw or undercooked meat containing mature sarcocysts (Tenter 1995).

Toxoplasma gondii is an intracellular parasite, which is prevalent in man and animals (Fayer 1981). Ingestion of raw or undercooked meat from livestock and game animals is often implicated in human toxoplasmosis infections. Another parasite

Table 30.3. Foodborne Parasites, Fungi, and Viruses Associated with Food- or Waterborne Illness

Pathogen	Sources of Pathogen or Toxin
Parasites	
Cryptosporidium	Contaminated surface water or foods in contact with contaminated water
Cyclospora	Raspberry, mesculun lettuce, and basil
Sarcocystis	Raw or undercooked meat
Toxoplasma	Raw or undercooked meat
Giardia	Contaminated surface water
Fungi	
Penicillium	Mycotoxins in apple juice, walnuts, corn and cereals
Aspergillus	Mycotoxins in groundnuts, corn, figs and tree nuts
Fusarium	Mycotoxins in cereals, corn, wheat and barley
Viruses	
Hepatitis A and E	Contaminated water, fruits and vegetables, raw shellfish, raw oysters
Norwalk/Norwalk–like viruses	Variety of foods and water
Calcivirus	Water and contaminated foods
Astrovirus	Water and contaminated foods

commonly implicated in food and waterborne illness is *Girdia.* Outbreaks of food and waterborne illness have been reported in the United States and the United Kingdom (Craun 1990, Porter et al. 1990). Infection with this parasite is associated with unsanitary conditions, and contaminated water is the most common source of infection.

Foodborne Fungi

Fungi most commonly associated with foodborne intoxications are *Penicillium, Aspergillus,* and *Fusarium*. The types of foods that may contain mycotoxin contamination are listed in Table 30.3. A foodborne fungi outbreak associated with consumption of corn contaminated with *Aspergillus flavus* was reported in India. The outbreak involved 1000 cases and 100 deaths (Krishnamachari et al. 1975). Mold growth can occur in foods stored under high temperature or humidity, resulting in the production of mycotoxin, which can be harmful to humans.

Foodborne Virus Infection

Viruses from human fecal origin can result in illness if ingested with food or water. Food or waterborne viruses commonly associated with human illness are listed in Table 30.3. Viral gastroenteritis is caused by Norwalk-like viruses and in some cases by the calicivirus and astrovirus groups. The symptoms often include acute but short self-limiting episode of diarrhea and vomiting. The viruses can be transmitted from person to person via contaminated utensils and foods. Foodborne viral hepatitis in humans is caused by the hepatitis A or hepatitis E virus. The onset of hepatitis may be preceded by anorexia, fever, fatigue, nausea, and vomiting. Infected individuals shed the organism in feces, which if allowed to contaminate food or water, can result in person-to-person transmission (Caul 2000).

EMERGING PATHOGENS AND FOOD SAFETY

Emerging infectious diseases have increased in the last two decades and are likely to increase in the future. Hence, the characteristics of the etiologic agents of these diseases must be considered when designing control measures to ensure food safety. There are two types of emergence: true emergence and reemergence. A true emergence involves the occurrence of microbial agent not previously identified as a public health threat. A reemergence involves the occurrence of a microbial agent causing disease in a new way not previously reported or the reemergence of a human disease after a decline in incidence. Emergence may be due to the introduction of a new agent, to recognition of an existing disease that was not previously detected, or to environmental pressures resulting in occurrence of a pathogen that can cause human disease. For example the occurrence of

enterohemorrhagic *E. coli* O157:H7 as a pathogen in 1982 is believed to be due to introduction of a new agent. The organism was involved in two outbreaks that year in the United States, and these outbreaks were associated with consumption of undercooked hamburgers from a fast-food restaurant chain. Several countries worldwide have reported outbreaks of infection caused by this organism. The appearance of *Salmonella typhimurium* phage type 104 is another example of a new agent. Increases in the use of antibiotics in humans and animals may have provided the environmental pressures resulting in the occurrence of *Salmonella typhimurium* phage type 104 with multiresistance to five antibiotics: ampicillin, chloramphenicol, streptomycin, sulphonamides, and tetracycline. This multiresistant characteristic leaves little choice of antibiotics for treatment of disease caused by this organism. The occurrence of *Listeria monocytogenes* as a foodborne agent is an example of reemergence. The organism is a well-known infectious agent; however, its role as a foodborne organism was not detected until the early 1980s. Examples of emerging foodborne pathogens and possible causes of emergence are provided in Table 30.4. The challenge posed by the appearance of emerging pathogens is that these organisms may not always behave as traditional pathogens. Therefore, new control measures to ensure that foods are free of these pathogens may be required.

CONTROL MEASURES FOR MICROBIAL CONTAMINANTS

A combination of factors is normally responsible for occurrence of an incident of foodborne illness. The pathogen must first reach the food involved; the organism must survive until food is ingested; in many cases, the organism must multiply to an infectious level or produce toxins; and lastly, the host must be susceptible to the level of organisms ingested with the food. Control measures to ensure food safety include (1) prevention of contamination of foods by pathogenic organisms (2) inhibition of growth or elimination of pathogens in foods and food products. The first stage of control measures is to prevent contamination of food animals and plants during the production stage. Production practices such as the use of manure and other organic fertilizer materials can provide the vehicle for contamination of food crops; hence, it is important to ensure that this type of fertilizer is pathogen free prior to use. The second stage of control measure is to prevent contamination and growth of pathogenic organisms during harvesting and transportation of food products. It is generally believed that control of microbial contamination early in production is more effective than control measures applied at a later stage of production. However, in cases when the presence of pathogens cannot be eliminated during the production stage, processing procedures that are designed to control the presence of pathogens must be implemented. A combination of intrinsic factors, (i.e., factors associated with the properties of the food) and extrinsic factors (i.e., factors associated with external conditions) is often used during processing and storage to control microbial growth or survival in foods (Bibek 1996). Intrinsic factors such as the pH, water activity, and food components can promote or inhibit microbial growth or survival. These properties can be used in combination with extrinsic factors such as

Table 30.4. Emerging Pathogens and Suspected Causes of Emergence

Pathogens	Cause of Emergence
Salmonella DT104	Resistance to antibiotics
E. coli O157:H7	Development of a new pathogen
Cyclospora cayetanensis	Development of a new pathogen
Cryptosporidium parvum	Development of a new watershed areas
Hepatitis E virus	Newly recognized
Norwalk Virus	Increased recognition
Aeromonas Spp.	Immunosuppression and improved detection
C. jejuni	Increased recognition, consumption of uncooked poultry
L. monocytogenes	Increased awareness
Helicobacter pylori	Increased recognition
Vibrio vulnificus	Increased recognition

heat, preservatives, irradiation, and storage conditions (e.g., refrigeration, relative humidity, and gaseous atmosphere).

Currently there are several food preservation technologies available for controlling microbial growth and survival. The most commonly used technologies for controlling microbial growth include application of low temperatures such as refrigeration or freezing, reduction of water activity of the food by drying, curing with salt or increased sugar level, pH reduction by acidification or fermentation, use of food preservatives, and modified atmosphere techniques. Other preservation technologies designed to inactivate foodborne organism include heat treatment such as sterilization, cooking, retorting, pasteurization, irradiation, and the use of hydrostatic pressure and pulse light. Also, use of packaging materials and adequate sanitation of the food production or processing environment can be used to limit entry of microorganisms into the food product and prevent recontamination of processed foods (Gould 1989, Hall 1997). The use of a combination of preservation techniques, often referred to as the hurdle concept, can be very effective in controlling microbial growth. A combination of suboptimal levels of the growth-limiting factors can be very useful where higher levels of one of the factors can be detrimental to the quality of the product. The most important hurdles used in food preservation include high or low temperature, water activity (a_w), redox potential (Eh), acidity (pH), preservatives such as sulfite, nitrite, and sorbate, and the use of competing microorganisms such as lactic acid bacteria. The application of an appropriate combination of these hurdles can improve the microbial stability, sensory and nutritional qualities, and safety of the foods.

FOOD SAFETY PROGRAMS

Traditional approaches to controlling the safety and quality of foods involve inspection of foods after production or processing for compliance with general hygienic practice, and where appropriate, foods are sampled for laboratory testing. This approach does not ensure food safety since reliance on visual inspection and testing of finished products cannot guarantee the absence of harmful pathogens in food. A more effective food safety control program, called the hazard analysis critical control point (HACCP) program, was introduced to the food industry in the early 1970s, and various food safety regulations and trade bodies worldwide endorsed its use as an effective and rational approach to assurance of food safety. HACCP is a systematic approach to hazard identification, assessment, and control. The benefits of the HACCP approach to ensuring food safety include the following:

- The food industry has a better proactive tool for ensuring the safety of foods produced,
- Potential food safety problems can be detected early,
- Food inspectors can focus more on verifying plant controls,
- More effective use is made of resources by directing attention to where the need is greatest, and
- A significant reduction in the cost of end product testing is achieved.

There are seven key principles of HACCP. Principle number one includes hazard analysis to identify the microbiological, chemical, and physical hazards of public health concern. Raw materials processing procedures, including packaging and storage, are assessed for microbiological hazards. The second principle is to determine the procedures or points in the food operation where hazards can be controlled effectively. These are called the critical control points (CCPs). The third principle includes the establishment of critical limits that separate acceptable from nonacceptable limits. The fourth principle involves the development of a system to monitor the CCPs so that the limits are not exceeded. The fifth principle is to determine what corrective actions to take when the CCPs are exceeded. The sixth principle includes the establishment of procedures for verifying that the HACCP program is working as expected. The seventh principle involves the documentation procedures and records for all aspects of these six principles. These HACCP principles have received worldwide recognition, by governments and the food industry. Many industrial organizations have adopted this food safety program as a means of controlling food safety hazards.

One of the most successful food safety programs at the consumer level is the FightBac program developed by the Partnership for Food Safety Education (PFSE), formed in 1997. This nonprofit organization provides education on safe handling of food to the public. The key features of the information on safe

Table 30.5. Four Simple Steps for Consumers to Ensure Food Safety (FightBac!™)

Clean: Wash hands and surfaces often
- Wash your cutting boards, dishes, utensils, and counter tops with hot soapy water after preparing each food item and before you go onto the next food
- Use plastic or other nonporous cutting boards
- Consider using paper towels to clean up kitchen surfaces

Separate: Don't cross-contaminate
- This is especially true when handling raw meat, poultry, and seafood
- Never place cooked food on a plate that previously held raw meat, poultry, or seafood

Cook: Cook to proper temperatures
- Use a clean thermometer that measures the internal temperatures of cooked foods to make sure meat, poultry, casseroles, and other foods are cooked all the way through
- Cook roasts and steaks to at least 145°F. Whole poultry should be cooked to 180°F for doneness
- Cook ground beef, where bacteria can spread during processing, to at least 160°F
- Don't use recipes in which eggs remain raw or only partially cooked

Chill: Refrigerate promptly
- Refrigerate or freeze perishables, prepared foods, and leftovers within 2 hours or sooner
- Never defrost at room temperatures. Thaw food in the refrigerator, under cold running water, or in the microwave. Marine foods in the refrigerator
- Divide large amounts of leftovers into small, shallow containers for quick cooling in the refrigerator

Source: FightBac!, Partnership for Food Safety Education. Available at www.fightbac.org.

handling of food in the FightBac program are summarized in Table 30.5. Further information on this program can be found in the FightBac website (www.fightbac.org). Implementation of these safe food-handling procedures by consumers can prevent the occurrence of foodborne illness.

FUTURE PERSPECTIVES ON FOOD SAFETY

Foodborne disease is preventable, provided control measures to prevent contamination of the food supply are implemented at various stages of the food chain, from the farm to the point of consumption. On-farm food safety programs include control measures for preventing contamination of plants, animals, and plant and animal products that are produced for human consumption. Implementation of effective environmental sanitation, good manufacturing practices (GMPs) and hazard analysis critical control points (HACCP) programs are key control measures for microbial, chemical, and physical hazards, not only at the farm level, but also during food processing, transportation, storage, and final preparation of foods prior to consumption. Furthermore, education and training of producers, processors, retailers, restaurant personnel, and consumers are important in the implementation of control measures to ensure the safety of foods consumed. Also, adequate inspection of food processing facilities and operations is very important to ensure that sanitation, GMP and HACCP protocols are being followed on a consistent basis. In order to implement a HACCP program successfully, the management of food operations must be committed to the program, by being aware of the benefits the program offers, and be prepared to invest time and money into the program. The use of microbiological testing as a verification tool for food safety programs including HACCP is important in providing evidence that indeed the control measures in place are effective. Finished product testing alone must not be relied on as means of ensuring food safety. The application of rapid methods for microbiological tests provides a useful tool for ensuring production of a safe food product at various points in the HACCP implementation. Microbiological tests can be performed for (1) the raw materials required for processing, (2) monitoring critical control points, and (3) HACCP verification. The reliance on testing of finished food products alone is a thing of the past. The challenge to food microbiologists now is to develop on-line food safety testing tools that can be used during food production. The limitation of current rapid tests for foodborne pathogens is that they require 24 hours or more to complete because of the requirement for amplification of pathogen numbers to detectable levels. The ability to detect pathogens at very low levels within seconds, minutes, or in less than 24 hours is an important future development for food microbiologists.

The likelihood of future increases in the incidence of foodborne illness is high as a result of increases in the number of susceptible aging populations, the treatment of disease with immunosuppressing drugs, and increases in the availability of ready-to-eat foods requiring no further processing prior to consumption. Unless steps are taken to implement food safety programs, which can reduce microbial contamination of foods at all levels of food production, processing, and retailing, and to educate restaurant personnel and consumers about the safe handling of foods, outbreaks of foodborne illness will continue to occur.

REFERENCES

Baird-Parker AC. 1994. Food and microbiological risks. Microbiology 140:687–695.

Baird-Parker TC. 2000. Chapter 1. The production of microbiologically safe and stable foods. In: BM Lund, TC Baird-Parker, GW Gould, editors. The Microbiological Safety and Quality of Food, vol. 1, pp. 3–18. Aspen Publishers, Inc., Gaithersburg, Maryland.

Bean NH, Griffin P.M, Goulding JS, Ivey CB. 1990. Foodborne disease outbreaks, five-year summary, 1983–1987. Morbility and Mortality Weekly Report 39:15–57.

Bibek R. 1996. Chapter 6. Fundamental Food Microbiology, pp. 61–72. CRC Press, Boca Raton, Florida.

Beuchat LR. 1995. Pathogenic microorganisms associated with fresh produce. Journal of Food Protection 59:204–216.

Buzby FL, Roberts T. (1997. Guillain-Barre Syndrome increases foodborne disease costs. Food Review 20(3): 36–42.

Casemore DP. 1990. Epidemiological aspects of human cryptosporidiosis. Epidemiology and Infection 104: 1–28.

———. 1991. Foodborne protozoal infection. In: WM Waites, JP Arbuthnott, editors. Foodborne Illness. A Lancet Review, pp. 108–119. Edward Arnold, London.

Casemore DP, Jessop EG, Douce D, Jackson FB. 1986. *Cryptosporidium* plus *Campylobacter*: An outbreak in a semi-rural population. Journal of Hygiene 96: 95–105.

Caul EO. 2000. Chapter 52. Foodborne viruses. In: The Microbiological Safety and Quality of Food, vol. 1, pp. 1457–1489. Aspen Publishers, Inc., Gaithersburg, Maryland.

Cliver DO. 1987. Foodborne disease in the United States, 1946–1986. International Journal of Food Microbiology 4:260–277.

Craun GF. 1990. Waterborne giardiasis. In: EA Meyer, editor. *Giardiasis*, pp. 305–313. Elsevier, New York.

Current WL, Blagburn BL. 1990. Cryptosporidium: infections in man and domestic animals. In: PL Long, editor. Coccidiosis of Man and Domestic Animals, pp. 155–185. CRC Press, Boca Raton, Florida..

Farber JM. 1991. Microbiological aspects of modified atmosphere packaging technology: A review. Journal of Food Protection 54:58–70.

Fayer R. 1981. Toxoplasmosis and public health implications. Canadian Veterinary Journal 22:344–352.

Gould GW. 1989. Introduction. In: Mechanisms of Action of Food Preservation Procedures, pp. 1–10. Elsevier Applied Science, London.

Hall RL. 1997. Foodborne illness: Implications for the future. Emerging Infectious Diseases 3:555–559.

Herwaldt BL, Ackers M-L. 1996. Cyclospora Working Group. An outbreak in 1996 of cyclosporiasis associated with imported raspberries. N England Journal of Medicine 336:1548–1556.

Herwaldt BL, Beach MJ. 1997. Cyclospora Working Group. The return of *Cyclospora* in 1997: Another outbreak of cyclosporiasis in North America associated with imported raspberries. Annals of Internal Medicine 130:210–220.

Kaferstein FK, Motarjemi Y, Bettcher DW. 1997. Foodborne disease control: A transnational challenge. Emerging Infectious Diseases 3:503–510.

Karmali MA. 1989. Infection by verocytotoxin-producing *Echerichia coli*. Clinical Microbiological Review 2:15–38.

Krishnamachari KAVR, Bhat RV, Nagarajon V, Tilek TBG. 1975. Hepatitis due to aflatoxicosis. Lancet 1:1061–1063.

Maurice J. 1994. The rise and rise of food poisoning. New Scientist 144:28–33.

Mead PS, Slutsker L, Dietz V, et al. 1999. Food-related illness and death in the United States. Emerging Infectious Disease 5:607–624.

McMeekin TA, Olley J. 1995. Predictive microbiology and the rise and fall of food poisoning. ATS Focus, pp. 14–20.

McMeekin TA, Ross T. 1996. Shelf life prediction: Status and future possibilities. International Journal of Food Microbiology 28:65–83.

Morbidity and Mortality Weekly Report. 1997. Update: Outbreaks of Cyclosporiasis—United States and Canada. MMWR 46:521–523.

———. 1999. Centre of Disease Control and Prevention. Update: Multistate outbreak of listeriosis—United States, 1998–1999. MMWR 47:1117–1118.

Odumeru J. 2002. Inside Microbiology: Current microbial concerns in the dairy industry. Food Safety Magazine 8(1): 20–23.

Porter JDH, Gaffney C, Heymann D,Parkin W. 1990. Foodborne outbreak of *Giardia lamblia*. American Journal of Public Health 80:1259–1260.

Riley LW, Remis RS, Helegerson SD, et al. 1983. Hemorrhagic colitis associated with a rare *Echerichia coli* serotype. New England Journal of Medicine 308:681–685.

Sharp JCM, Reilly W. 2000. Chapter 37. Surveillance of foodborne disease. In: BM Lund, TC Baird-Parker, GW Gould, editors. The Microbiological Safety and Quality of Food, vol. 2., pp. 975–1010. Aspen Publishers, Inc., Gaithersburg, Maryland.

Tenter AM. 1995. Current research on *Sarcocystis* species of domestic animals. International Journal of Parasitology 25:1311–1330.

31
Emerging Bacterial Foodborne Pathogens and Methods of Detection

*R. L. T. Churchill, H. Lee, and J. C. Hall**

*Corresponding author.

INTRODUCTION

Bacterial foodborne pathogens are bacteria that contaminate food and cause illnesses in humans when that food is ingested (Tauxe 2002). In 1999, the estimated number of foodborne illnesses in the United States was 76 million, that is, one in four people became ill, with approximately 325,000 hospitalizations in the year and 5000 deaths (Mead et al. 1999). Despite efforts made to prevent food contamination, new pathogens often emerge as a result of changing demography, food consumption habits, food technology, commerce, changes in water sources and environmental factors (Gugnani 1999), and microbial adaptation (Altekruse et al. 1997). Some of these bacterial foodborne pathogens are listed in Table 31.1.

There is no accepted general definition of an "emerging foodborne pathogen," but disease-causing organisms are considered to be emerging if they have one or more of the following properties. First, the microorganism has not been previously recognized, or it has not been known to cause problems in food. Second, the microorganism was known to be pathogenic, but has significantly increased in incidence over the past decades (Schlundt 2001). Third, it may be a pathogen that is predicted to increase in

Table 31.1. Some Characteristics of Selected Bacterial Foodborne Pathogens and the Estimated Number of Illnesses They Cause per Year in the United States

Bacterial Strains Causing Illness	Gram Staining Properties	Estimated Number of Illnesses Caused per Year in the United States[a,b]	Infectious Dose (cfu)[c]	Symptoms Associated with Infection[c,d]	Foods Commonly Associated with the Pathogen
Campylobacter	Negative	1,963,000–2,453,926	$\geq 10^3$	ABDF	Poultry, eggs (Park 2002)
Clostridium botulinum	Positive	56–58	Toxin causes disease symptoms	Neurological	Any food that is above pH 4.6, improperly canned (allowing spore survival), and eaten without heating—e.g., corn, asparagus, etc. (Anonymous, 1992)
Clostridium perfringens	Positive	248,520–249,000	10^7–10^8	ADFN (type A)	Pork and protein-rich foods (Granum and Brynstad 1999)
Escherichia coli O157:H7	Negative	73,480–92,000	10	DABH	Unpasteurized cider, salami, dairy products, mayonnaise, ground beef, lettuce, untreated water (Altekruse et al. 1997)
Listeria monocytogenes	Positive	2000–2,518	10^7–10^7	Systemic[e]	Dairy products, coleslaw, soft cheeses, deli meats
Salmonella (non-typhoid)	Negative	1,342,000–1,412,498	10^3–10^6	ADFHV	Poultry, raw meats, eggs, milk (Tauxe 1998)
Salmonella typhi	Negative	659–824	1–10^2	Systemic[e]	Raw meats, poultry eggs, milk
Staphylococcus	Positive	185,000–185,060	Toxin causes disease symptoms	ADFNV	Poultry and meat products, seafood, milk (Balaban and Rasooly 2000)
Yersinia enterocolitica	Negative	90,000–96,368	10^6–10^7	ADFHV	Pork products, raw milk (Granum and Brynstad 1999)

[a]Adapted from Mead et al. 1999, Tauxe 2002, Woteki and Kineman 2003.
[b]Numbers reported are from 1999–2003.
[c]From Granum and Brynstad 1999.
[d]A, abdominal pain; B, bloody diarrhea; D, diarrhea; F, fever; H, headache; N, nausea; V, vomiting.
[e]Systemic infections involve septicaemia, meningitis, and flu like symptoms that affect the whole body.

incidence in the next two decades (Gugnani 1999). Fourth, the pathogen may already be a recognized problem in some parts of the world, but is increasing significantly in its geographic range (Schlundt 2001). These definitions are illustrated using a few pathogens that have been classified as "emerging." For example, *Escherichia coli* O157:H7 evolved relatively recently and may be a new pathogen. *Vibrio vulnificus* has existed for centuries but has only recently been recognized as causing illness. Finally, *Listeria monocytogenes* and *Vibrio cholerae* are well-known bacterial pathogens, but their transmission in foods has only recently been recognized (Mead et al. 1999).

In 2002, Tauxe (2002) examined the shifting spectrum of foodborne pathogens over time. Many of the significant bacterial pathogens at the turn of the previous century are now extremely uncommon in the clinical setting. These include *Salmonella typhi*, *Brucella*, and *Clostridium botulinum*. Furthermore, of the 27 main microbial pathogens known to cause foodborne illnesses, 13 (including bacteria, viruses, and protozoa) have emerged in the last 25 years. These 13 account for 82% of the foodborne illness (or 13.8 million illnesses in the United States), and 61% of the 1800 deaths caused by known pathogens (Mead et al. 1999, Tauxe 2002).

We have an appalling lack of knowledge about foodborne pathogens. For example, if all of the 27 known foodborne pathogens are taken into account, only 19% of the estimated number of foodborne illnesses and 36% of the deaths are accounted for, meaning that there must be a number of foodborne pathogens that have not been identified (Tauxe 2002). Of those illnesses caused by known pathogens, three pathogens, *Salmonella*, *Listeria*, and *Toxoplasma* (a parasite), cause more than 75% of the deaths (Mead et al. 1999).

In this chapter a few of the more common emerging bacterial foodborne pathogens will be described and some of the reasons for the pathogens to emerge and the reasons why these bacteria are of concern will be examined. The chapter will cover the toxins that these bacteria produce, how the toxins mediate some of the pathogenic properties of the bacteria, and how these toxins can be used to detect the pathogens. The final section of the chapter will focus on methodologies used to detect pathogens in an attempt to keep them out of the food chain.

Factors Affecting the Emergence of New Pathogens

A number of factors contribute to the emergence of foodborne pathogens. One of the most important is the changing demographics of the Western world. The median age of the population has been increasing steadily over the past few decades. As well, with the advent of HIV, and advances in treatments for chronic diseases, there are an increasing number of people who are immunodeficient. These factors have led to an increase in the proportion of the population that is susceptible to a number of foodborne diseases (Altekruse et al. 1997). Another factor that contributes to the emergence of pathogens is changes in human behavior. Fresh fruit and vegetable consumption increased by 50% between 1970 and 1994 in the United States. In addition to this, people eat outside of the home much more regularly than at any other time in history, and these foods cause about 80% of the outbreaks seen in the United States (Altekruse et al. 1997).

Changing practices in industry and differences in technology can affect pathogen distribution. With the consolidation of many processing facilities, contamination is likely to affect more people. Increased consolidation also means longer transport distances, providing more time for growth of pathogenic organisms. Related to this increase in transportation are the changes in travel and commerce. With global trade, more people are travelling, exposing more people to different foods, with a higher probability of acquiring infections. Increasing global trade also means more international transport of foods. These foods may not be exposed to the same inspection standards. Increased immigration and the foods made popular by immigrants have also contributed to exposure to different pathogens (Altekruse et al. 1997).

Other factors that contribute to the emergence of new pathogens are changing land use and a breakdown of the public health infrastructure in North America (Altekruse et al. 1997). Changing land use involves problems such as the disposal of waste from feedlots and other commercial animal-raising facilities. The breakdown of the public health infrastructure relates to the decreasing budgets of many of the agencies responsible for looking into food and water safety. For instance, one of the contributing factors to the *E. coli* O157:H7 outbreak in Walkerton,

Ontario, Canada, in 2000 was related to government cutbacks in the number of inspections and uncertainty over reporting procedures (MacKay 2002).

Finally, microbial adaptation is an important reason for the emergence of new pathogens (Altekruse et al. 1997). Natural selection and the overuse of antibiotics, both in humans and in animal feed, have led to the emergence of a number of new hardy strains of pathogenic bacteria. Chemical treatments have also had an effect. For instance, the virulence plasmid of *Yersinia enterocolitica* has been associated with resistance to arsenate and arsenite (Neyt et al. 1997). The spread of this pathogen has been linked to the previous practice on pig farms of spreading arsenic to prevent spirochetal infections (Tauxe 2002).

COMMON FOODBORNE PATHOGENS

Food recalls are expensive for the food industry. In order to avoid unnecessary recalls, the food industry is only interested in detecting bacteria that may potentially cause illness. Many bacterial pathogens are closely related to nonpathogenic bacteria that may inhabit the same environments. These related bacteria may be members of the same genus or even different strains of the same species. A good detection method for the pathogen must be able to distinguish between the pathogens and closely related nonpathogens. Toxins produced by pathogenic bacteria often provide a convenient way to detect the organisms. In general, only the pathogenic strains will produce protein toxins, making these proteins, or the genes encoding them, ideal targets for rapid differentiation between potential pathogens and nonpathogens. The following summarizes some of the common pathogens and their toxins (see Table 31.2). Methodologies for detecting the pathogens or their toxins are provided later in the chapter.

CAMPYLOBACTER JEJUNI

There are 12 species of the Gram-negative *Campylobacter* genus known to be of clinical importance. Of these, *C. jejuni* and *C. coli* account for 95% of *Campylobacter* infections in humans (Park 2002). *C. jejuni* was recognized as a foodborne pathogen in the 1970s and is now known to be one of the leading causes of foodborne illness in the industrialized world (Gugnani 1999). The main infective route is through consumption of raw or undercooked poultry products, including eggs (Park 2002). It is thought that *C. jejuni* can spread through a population of chickens via the water system (Tauxe 2002).

Infection with *C. jejuni* causes diarrhea, abdominal pain, and fever (Gugnani 1999). *Campylobacter* species have also been identified as one of the most common infections preceding Guillain-Barré syndrome, a peripheral nervous system condition involving acute flaccid paralysis (Meng and Doyle 2002). *Campylobacter* infections are responsible for about 5% of annual food-related deaths in the United States (Mead et al. 1999).

Campylobacter species are unusual in that they exhibit fastidious growth requirements and display an unusual sensitivity to environmental stress. For instance, members of the genus are microaerophilic and have an optimal growth temperature of 42°C, but are unable to grow at temperatures below 30°C. Desiccation and osmotic stress will kill the bacteria. They cannot grow at salt concentrations above 2%, are killed at pH values less than 4.9, and cannot survive heating and pasteurization processes (Park 2002). Unlike *Campylobacter*, most foodborne pathogens are extremely robust, thereby surviving most treatments used to eradicate pathogens from food. *Campylobacter* is not hardy, and yet it has emerged as the single leading cause of foodborne illness in the Western world (Kopecko et al. 2001, Park 2002).

C. jejuni produces at least one cytotoxin, cytolethal distending toxin (CDT). While little is known about the effects of this toxin, cell culture studies indicate that it causes cell division to stop at the G2 phase (Whitehouse et al. 1998). There is some speculation that this may potentially cause loss of function or erosion of the epithelial layer of the intestine, where the bacteria tend to colonize, leading to diarrhea (Park 2002). Given that the toxin is universal to *C. jejuni*, it makes a good detection target.

SALMONELLA SPP.

Nontyphoidal salmonellosis is one of the major bacterial illnesses caused by food in industrialized countries. There are five species within the *Salmonella* genus, of which *S. enterica*, subspecies *enterica*, is the most relevant to illness in warm-blooded animals. Of the over 1500 serotypes within the subspecies, the two serotypes *S. typhimurium* and *S.*

Table 31.2. Representative Toxins of Common Foodborne Pathogens, and Their Actions in the Pathogenesis of the Bacteria

Organism	Toxin Name	Toxin Type	Size of Toxin	Effect/Mode of Action of Toxin
C. jejuni	Cytolethal distending toxins, CDTA, CDTB and CDTC	Unknown—unrelated to other toxins	30, 29, and 21 kDa	Effects are not fully elucidated, but the toxin seems to stop host cells in the G2 phase of division (Eyigor et al. 1999, Whitehouse et al. 1998).
C. perfringens	*C. perfringens* enterotoxin	Enterotoxin	35 kDa	Type A diarrhea—the enterotoxin binds to a protein receptor in the intestine, forms pores, and results in altered membrane permeability and diarrhea (Granum and Brynstad 1999).
	Perfringolysin O (PLO)	Thiol Activated Cytolysin	53 kDa	Type C human necrotic enteritis—PLO, α-toxin and δ-toxin are all produced during vegetative growth of the organism. The PLO is normally cleaved by trypsin in the intestine, preventing disease. PLO forms cation selective pores in endothelial cells (Brynstad and Granum 2002, Nagahama et al. 2003).
	α-toxin	Phospholipase	43 kDa	α-toxin is the most important of the toxins mediating gas gangrene. It and PLO are involved in inhibiting the migration of inflammatory cells to the sites of infection. The phospholipase increases the ability of endothelial cells to adhere to inflammatory cells (Rossjohn et al. 1999).
	δ-toxin	Haemolysin	42 kDa	δ-toxin specifically lyses cells expressing the ganglioside GM2 (Alouf and Jolivet-Reynaud 1981, Jolivet-Reynaud et al.1993).

(Continues)

Table 31.2. (*Continued*)

Organism	Toxin Name	Toxin Type	Size of Toxin	Effect/Mode of Action of Toxin
E. coli O157:H7	Shiga toxin 1 (Stx1)	Verocytotoxin or Shiga toxin	Subunit A—32 kDa, Subunit B—7 kDa	Toxin exacerbates the severity of intestinal and systemic lesions in human hosts, tending to target the cortex of the kidney. It targets the intestinal villi cells and glomerular endothelial cells because these cells have Gb3 receptors for the toxins (Acheson and Keusch 1999, Melton-Celsa and O'Brien 1998).
	Shiga toxin 2 (Stx2)	Verocytotoxin or Shiga toxin	Subunit A—32 kDa, Subunit B—7 kDa	Toxin exacerbates the severity of intestinal and systemic lesions in human hosts, tending to target the cortex of the kidney. It targets the intestinal villi cells and glomerular endothelial cells because these cells have Gb3 receptors for the toxins. Stx2 is 1000-fold less toxic than Stx1 (Acheson and Keusch 1999, Melton-Celsa and O'Brien 1998).
	Intimin	Adhesin	94–97 kDa	Intimin is an intestinal attachment and effacement factor. The protein is important in intestinal colonization by the pathogen. The intimin receptor in the large intestine has not been elucidated (Kaper et al. 1998).
	α-hemolysin	Cytotoxin (pore forming)	110 kDa	Toxin is Ca^{2+}-dependent. It binds to an unknown receptor in the host endothelial cells, polymorphonuclear leukocytes, monocytes, or T-lymphocytes and undergoes a conformational change to insert in the membrane, thus forming pores and causing osmotic lysis of the cells (Ludwig and Goebel 1999).

	O-antigen	Lipopolysaccharide (LPS)	(nonprotein)	LPS may be directly related to the upregulation of the Gb3 receptor or indirectly related through the upregulation of tumor necrosis factor 1 and other preinflammatory cytokines (Currie et al. 2001, Lingwood et al. 1998, Tesh 1998).
L. monocytogenes	Listeriolysin O (LLO)	Thiol activated cytolysin	58 kDa	Pore-forming protein binds to cholesterol in the host membrane. The protein then oligomerizes, forms a pore, and allows the escape of *L. monocytogenes* from the phagosome of the host into the cytoplasm where it can replicate.
	Phospholipase C-A (PLC-A)	Phospholipase C	33 kDa	Hydrolyses phosphotidylinositol—plays an overlapping role with PLC-B in degrading the phagosome membrane or vacuole, so the bacteria can escape to the cytosol or spread from cell to cell (Bannam and Goldfine 1998, Titball 1999).
	Phospholipase C-B (PLC-B)	Phospholipase C	30 kDa	Broad range phospholipase—plays an overlapping role with PLC-A in degrading the phagosome membrane or vacuole, so the bacteria can escape to the cytosol or spread from cell to cell (Snyder and Marquis 2003, Titball 1999).
	P60 protein (Iap)	Extracellular protein	60 kDa	Extracellular protein associated with the invasion of phagocytic cells. The protein has also been implicated in cell division (Cabanes et al. 2002, Klein and Juneja 1997).
S. aureus	Enterotoxins A-E and G-J	Enterotoxins	22–27 kDa	This group includes haemolysins, nucleases, proteases, leucocidins, collagenases, cell-surface proteins and superantigens, allowing the bacteria to colonize and persist in host organisms (Melton-Celsa and O'Brien 1998).

(Continues)

Table 31.2. (*Continued*)

Organism	Toxin Name	Toxin Type	Size of Toxin	Effect/Mode of Action of Toxin
	α-hemolysin	Thiol-activated cytolysin	33 kDa	These proteins bind to high affinity (unidentified) receptors at low concentrations (< 200 nM), and nonspecifically to cell membranes at higher concentrations (> 200 nM). The hemolysin preferentially attacks endothelial cells and thrombocytes, forming pores and leading to pulmonary edema (Menestrina et al. 2001).
S. typhimurium	TTSS-1	Type III-secretion system	Multiple	A protein transport system involved in translocating a number of effector proteins into host cells. TTSS-1 moves SipA, SopA, SopB, SopD, and SopE2 into host cells. This elicits the infiltration of neutrophils into the intestine through the induction of chemoattractant chemokine secretion into the ileal tissue, resulting in diarrhea (Zhang et al. 2003).
Y. enterocolitica	Y-ST	Enterotoxin	2 kDa	Toxin binds to guanylate cyclase in the brush border of the intestine, and mediates secretion of fluid into the intestine, leading to diarrhea (Takeda et al. 1999, Yoshino et al. 1995).
	Yops proteins	*Yersinia* outer proteins—unconventional toxins	20.8–81.7 kDa (Cornelis et al. 1998)	YopE, YopH, and Yop-O have sequence similarity to *Shigella* α-hemolysins (Dobrindt and Hacker 1999). They are part of a Type III secretion system, which on contact with the host cell interrupts signal transduction pathways in the cells (Carnoy and Simonet 1999). The proteins also interfere with phagocytosis by host cells (Grosdent et al. 2002)

enteriditis caused more than 75% of the human illnesses reported in England and Wales during 2000 (Liebana 2002).

Although it is the most famous of the *S. enterica* serotypes, the occurrence of *S. typhi* has been essentially eradicated in western countries over the past 100 years (Zhang et al. 2003; see Table 31.1 for numbers). In fact, between 1992 and 1997, more than 70% of the *S. typhi* cases in the United States occurred after travelling to the non-Western world (Mead et al. 1999). Because of their prevalence in foodborne illness, discussion in this paper will be restricted to the *S. enterica* serotypes *S. typhimurium* and *S. enteriditis*.

S. enteriditis contamination is most often found in chicken products, but the pathogen may also have a reservoir in the rodent population. New generations of chickens could be infected when rodents contaminate the feeding troughs while sleeping in them at night (Tauxe 2002). In the United States, *S. enteriditis* caused about 6% of *Salmonella* infections in 1980, increasing to 26% of all *Salmonella* infections by 1984. Eighty-two percent of the cases were caused by eating eggs contaminated with the pathogen (Altekruse et al. 1997, Gugnani 1999).

S. typhimurium is most commonly a foodborne pathogen, although there have been a few cases of outbreaks caused by contaminated water sources in the United States. It causes enterocolitis, resulting in diarrheal illness. As of 1998, the U.S. Centers for Disease Control (CDC) claimed that this *S. typhimurium* accounted for 26% of all *Salmonella* isolates causing illness (Zhang et al. 2003). Both *S. enteriditis* and *S. typhimurium* cause nausea, vomiting, abdominal cramps, diarrhea, fever, and headache (Granum and Brynstad 1999).

The type III secretion system from pathogenicity island 1 of *S. typhimurium* (TTSS-1) has been proposed as one of the main virulence determinants required by *S. typhimurium* to cause diarrhea and mortality among cattle. TTSS-1 is thought to work by mediating the translocation of effector proteins into the cytosol of the host cells. The effector proteins, including SopB, SopC, and SopD, form a translocation complex to deliver more effector proteins into the host cytoplasm, then cause fluid accumulation and the influx of neutrophils into the intestine. The end accumulation of fluid into the intestine may be due to a secretory mechanism dependent on the effector proteins, or it may be due to neutrophil-induced tissue damage (Zhang et al. 2003). *S. typhimurium* also produces a number of other toxins. However, most of these do not have equivalent effects between species and are not considered general virulence factors.

E. coli O157:H7

E. coli O157:H7 was responsible for about 3% of the food-related deaths in the United States as of 1999 (Mead et al. 1999). *E. coli* O157:H7 was identified in 1976 and first described as a human pathogen in 1982 during two outbreaks of severe bloody diarrheal syndrome in Oregon and Michigan (Gugnani 1999, Park et al. 1998).

E. coli O157:H7 is transmitted by food or water sources (Tauxe 2002), with secondary person-to-person transmission. The strain produces a number of toxins that can cause bloody diarrhea in infected persons; 4% of those infected develop hemolytic uremic syndrome (HUS), which can cause kidney damage (Mead et al. 1999), and around 1% die (Park et al. 1998). Antimicrobial therapy, such as treatment with ampicillin, tetracycline, erythromycin, or antibiotics from the quinolone family (Neill 1998, Tauxe 2002, Wong et al. 2000, Zhang et al. 2000), are not effective, and only supportive care can be provided (Wong et al. 2000, Zhang et al. 2000). This strain of *E. coli* contaminated the drinking water of Walkerton, Ontario, Canada, in May 2000, causing over 2000 people in a town of 5000 to become ill, 27 to develop HUS, and at least 6 to die (2000). In 1998, a number of illnesses attributed to *E. coli* O157:H7 in Southern Ontario were due to ingestion of contaminated salami (Williams et al. 2000).

E. coli O157:H7 is a robust organism. It is tolerant of cold temperatures, although it does not grow at 4°C. The organism can survive refrigerated storage for more than 2 months (Meng and Doyle 1997), and storage at between −80 and −20°C for 9 months (Park et al. 1998). This means that when a temperature abuse situation (> 6°C) does occur, it is ideally situated to grow and replicate, resulting in bacterial levels high enough to cause infection and illness.

E. coli O157:H7 is also tolerant to pH values as low as 3.6 (Meng and Doyle 1997). This acid tolerance has allowed the occurrence of outbreaks attributed to its contamination of highly acidic foods such as apple cider, mayonnaise, and smoked sausage.

There is speculation that the low number of cells required for infection by *E. coli* O157:H7 is related to the acid resistance of the organism. It is resistant to synthetic gastric fluid (pH 1.5) for longer than the general stomach clearance time of 3 hours, so that a large percentage of the bacteria survive passage to the intestine (Park et al. 1998).

As well, *E. coli* O157:H7 is resistant to drying and fermentation processes used to treat many deli meats such as salamis. These meats are not cooked, but drying and fermentation is considered a valid process to kill any bacteria present. In 1994, however, the United States experienced an outbreak of *E. coli* O157:H7 resulting from contamination of salami, proving that this pathogen is resistant to the fermentation process (Meng and Doyle 1997).

E. coli O157:H7 has been isolated from such diverse sources as ground beef, lettuce, raw cider, raw milk, and untreated water (Altekruse et al. 1997). It is commonly found in the intestinal flora of cattle, and may enter the food chain from this source (O'Brien and Kaper 1998). For example, if fields are fertilized with contaminated manure, then the pathogen can be absorbed into plant tissue (e.g., lettuce), where it is no longer accessible to removal by washing (Solomon et al. 2002).

The pathogenicity of the bacteria is attributed to several virulence factors, including one or more shiga toxins, a hemolysin, the adhesin intimin, and the O-antigen (Osek 2003). Its key virulence factor is the shiga toxin (Stx), which is carried by several different cryptic prophages with wide host ranges (Kim et al. 2001, Perna et al. 2001). These prophages are lysogenic and, because they are cryptic, have lost the ability to enter a lytic cycle of infection. *E. coli* O157:H7 may have evolved as a result of the increased use of antibiotics, leading to an increase in antibiotic resistant phage transfers (Tauxe 2002).

Shiga toxins produced by *E. coli* O157:H7 exacerbate the intestinal and systemic lesions caused by the bacteria in the human host (O'Brien and Kaper 1998). Lesions in the intestine lead to bloody diarrhea, while in the kidney they are associated with HUS. Stx is formed of one A polypeptide and five B polypeptides. The A polypeptide is an N-glycosidase responsible for inhibiting protein synthesis by depurinating an adenine residue in the 28S rRNA of the host cell, while the B pentamer allows the toxin to attach to globotriosylceramide (Gb3), a glycoprotein abundant in the cortex of the human kidney (Park et al. 1998).

Seventy-six percent of *E. coli* O157:H7 strains isolated from infected individuals have been shown to produce a mixture of shiga toxins 1 (Stx1) and 2 (Stx2); 20% produced only Stx 2, and 3% displayed only Stx 1 (Meng and Doyle 1997). Stx1 is highly conserved with the *Shiga disentiriae* shiga toxin, while Stx2 has at least five subgroups (Acheson and Keusch 1999). There is some evidence that Stx2 is more important than Stx1 in causing the progression of hemolytic uremic syndrome (Meng and Doyle 1997).

YERSINIA ENTEROCOLITICA

Yersinia enterocolitica is a Gram-negative microorganism found ubiquitously in the environment. The majority of isolates found from human and environmental samples are nonpathogenic. Little is known about the epidemiology of this organism, although the strongest link to the source of the pathogenic strains is in pigs. Most of the isolates recovered from pigs are the same as those found within humans (Fredriksson-Ahomaa and Korkeala 2003). Outbreaks of *Y. enterocolitica*–mediated illness have also been associated with water and milk consumption (Gugnani 1999).

Y. enterocolitica has been recognized as an emerging foodborne pathogen since the late 1980s. Infection with the pathogen results in diarrhea, fever, headache, rigors, and vomiting and is often associated with infection of the intestinal lymph nodes (Gugnani 1999, Takeda et al. 1999).

One of the major public health challenges in dealing with *Y. enterocolitica* is in detecting the organism, because of its low numbers compared with background flora commonly found in contaminated samples. Direct isolation, even on selective media, is often not successful. *Y. enterocolitica* does have psychrotropic properties, or an ability to grow at temperatures lower than 5°C (Woteki and Kineman 2003). In fact, one of the ways used to isolate this organism has been enrichment on selective media for a number of weeks at 4°C (Fredriksson-Ahomaa and Korkeala 2003). There are also some recent developments in the use of molecular-based methods that may help to alleviate the detection problem (Fredriksson-Ahomaa and Korkeala 2003).

Of the 57 serotypes and six biotypes of *Y. enterocolitica* known (Takeda et al. 1999), only those strains possessing virulence factors are associated with illness. The pathogen produces a heat-stable

enterotoxin, Y-ST, which is involved in causing illness (Gugnani 1999, Takeda et al. 1999). This toxin is a member of a family of heat-stable enterotoxins produced by a number of bacteria such as *E. coli*, *Vibrio* spp., and *Salmonella* spp. These toxins are small peptides with a molecular weight of about 2 kDa. It is thought that Y-ST specifically causes diarrhea by binding to a membrane-bound guanylate cyclase in the brush border membrane of the intestine (Yoshino et al. 1995). This binding results in increased levels of intracellular cGMP (cyclic guanosine monophosphate), which stimulates chloride secretion and/or inhibits absorption. The end result is net intestinal fluid secretion, and diarrhea (Takeda et al. 1999).

The gene encoding Y-ST, *ystA*, has been found by DNA-DNA hybridization in all pathogenic strains of *Y. enterocolitica*, but not in any of the nonpathogenic strains (Takeda et al. 1999). This makes the toxin and its gene a good candidate for detecting specifically pathogenic strains of the bacteria.

LISTERIA MONOCYTOGENES

Listeria monocytogenes is a Gram-positive bacterium responsible for the illness known as listeriosis. The bacterium is widely distributed in the environment and has been isolated from soil, plants, decaying vegetation, silage, water, and sewage. *L. monocytogenes* is also commonly carried in the intestinal tract of humans, as well as cattle, sheep, and goats (Donnelly 2001). Studies as of 1994 showed that 2–6% of people in the United States carry the pathogen (Rocourt 1994). In food, it has been found in raw or processed foods including dairy products, meat, vegetables, and seafood (Gugnani 1999, Meng and Doyle 1997). More specifically, outbreaks have been associated with the consumption of ready-to-eat foods such as coleslaw, milk (contaminated after pasteurization), paté, pork tongue in jelly, and soft cheeses made from raw milk (Altekruse et al. 1997, Donnelly 2001).

In the healthy individual, *L. monocytogenes* may cause meningitis, encephalitis, or septicemia. However, *L. monocytogenes* generally affects only a few segments of society (Donnelly 2001), including the elderly, immunosuppressed individuals, neonates, and pregnant women (Kathariou 2002). In pregnant women, the pathogen can cause bacteremia, which if left untreated can lead to amnionitis and infection of the fetus, resulting in stillbirth or premature birth (Gugnani 1999). Less commonly, infection will result in cutaneous lesions and flu-like symptoms (Meng and Doyle 1997). Although relatively few people become ill with *Listeria* infections (See Table 31.1), there is a 20% mortality rate for those that do. The result is that *L. monocytogenes* is responsible for 28% of food-related deaths in the United States (Mead et al. 1999).

The pathogenic potential of *L. monocytogenes* has been known for decades, but its ability to be transmitted through a food medium was discovered only in 1981 (Meng and Doyle 1997). Listeriosis was first diagnosed in a soldier with meningitis in World War I (Rocourt 1994). It has also been postulated to be the cause of Queen Anne's 17 unsuccessful pregnancies in the 17th century (Saxbe Jr. 1972), due to the pattern of miscarriages, the neonatal deaths experienced by 14 of the children, and the postulated hydroencephaly experienced by her son, the Duke of Gloucester, who survived until the age of 11.

L. monocytogenes is extremely robust and is probably the most important pathogen able to grow at refrigeration temperatures. It is able to grow at temperatures as low as 2°C, and freezing food at −15°C or even repeatedly freezing and thawing food has little effect on the viability of the organism (Woteki and Kineman 2003). It can survive in a pH range of 4.3 to 9.6, making it a contamination factor in a number of cheeses and meats, as well as dairy products (Donnelly 2001). The pathogen is also resistant to desiccation. It can survive in the soil, leaving it able to contaminate crops such as cabbage or lettuce (Kathariou 2002). As well, the organism is able to survive in salt concentrations of up to 25.5%, which would cause the desiccation and death of most bacteria (Donnelly 2001).

Listeriolysin (LLO) is the most important virulence factor produced by *L. monocytogenes* (Jacobs et al. 1999). LLO is a pore-forming protein of the thiol-activated cytolysin family of toxins, which is composed of 529 amino acids, encoded by the *hlyA* gene (Jacobs et al. 1999). The gene is found within a cluster of virulence genes found only in pathogenic varieties of *Listeria,* and its expression is regulated by PrfA, a global virulence transcriptional activator protein (Jacobs et al. 1999).

On infection, a host macrophage will internalize the *Listeria* cell through phagocytosis into its vacuole. The bacterium then secretes LLO, which binds to cholesterol in the host membrane and forms pores in the vacuole wall, allowing the bacterium to escape into the cytoplasm. *L. monocytogenes* cells then replicate and use the host's actin molecules to transport themselves to the host cell's plasma membrane where they force the host cell to extend a pseudopod to an adjacent cell that engulfs the pseudopod. The *L. monocytogenes* again secretes LLO to escape the double-membrane vesicle and starts the cycle over again (Kathariou 2002).

LLO is secreted as a 58-kDa protein (Jacobs et al. 1999). It binds as a monomer to the membrane surface of the target, with subsequent oligomerization into large arc- or ring-shaped structures that puncture the membrane (Alouf 1999). They form pores in the membrane that are about 20 nm in diameter (Jacobs et al. 1999).

The optimal pH for LLO activity is pH 5.6 (Stachowiak and Bielecki 2001). This means that LLO secreted within the acidic phagosome is active, thus allowing pore formation and escape of the cells to the cytoplasm. The cytoplasm, however, has a more neutral pH. This renders the LLO inactive and thus unable to lyse the cell during the bacterium's replicative phase (Beauregard et al. 1997). A number of other external stimuli such as growth temperature and sugar availability also affect the expression and secretion of LLO (Jacobs et al. 1999, Kathariou 2002).

Clostridium perfringens

C. perfringens is an anaerobic Gram-positive, spore-forming bacterium that commonly inhabits soil, water, sewage, and the intestinal tract of a number of warm-blooded animals, including humans. The pathogen is responsible for cases of human gas gangrene and two different foodborne diseases, a mild diarrhea (Type A) and a rare human necrotic enteritis (Type C). Type A diarrhea has become one of the most common foodborne diseases of the industrialized world, often isolated from samples taken from restaurants, hospitals, and old age homes (Brynstad and Granum 2002). *C. perfringens* Type A disease generally occurs 9–17 hours after ingesting the contaminated food, with full recovery occurring within 12–24 hours (Boyd 1988). The type C disease has an incubation time of 5–6 hours followed by an acute sudden onset of severe abdominal pain, diarrhea, vomiting, and necrotic inflammation of the intestine. Death results in 15–25% of those infected (Brynstad and Granum 2002).

Illness generally occurs after eating improperly canned foods. *C. perfringens* is able to grow at any temperature between 15 and 50°C, with an optimum of 45°C (Brynstad and Granum 2002). Unless food is fully sterilized, only the vegetative cells are killed by heat treatment, leaving live *C. perfringens* spores (Woteki and Kineman 2003). The spores survive in an inactive state until conditions are favorable for vegetative growth to resume (Brynstad and Granum 2002).

C. perfringens produces a number of toxins, including α-toxin, phospholipase C, and perfringolysin O (PLO) (Bryant 2003). In fact, members of the species are able to produce a total of 13 different toxins, although each individual cell does not normally produce all of them (Brynstad and Granum 2002).

In the Western world, the *C. perfringens* enterotoxin (CPE) is the most important disease-causing toxin. CPE is produced in sporulating *C. perfringens*, causing the Type A disease, and is thought to be one of the main causes of diarrhea. CPE is a two-domain protein that binds to claudin proteins found in the tight junctions between intestinal cells. It then causes pore formation in the cells, leading to the loss of cations into the host's intestinal space, resulting in water loss and diarrhea (Brynstad and Granum 2002). Since many healthy people have high levels of *C. perfringens* in their feces, it is essential that any test detecting the bacteria during an outbreak situation detect the CPE toxin in both the food and the infected individual to prove the causality of the infection (Brynstad and Granum 2002).

Perfringolysin O (PLO) is found in the Type C human necrotic enteritis–causing *C. perfringens*. It is a thiol-activated cytolysin, similar to LLO from *L. monocytogenes*. Like LLO, PLO binds to cholesterol in membranes, oligomerizes, and forms pores in the membrane (Sekino-Suzuki et al. 1996). As a virulence factor specific to this organism (Feil et al. 1996), the toxin is a suitable target for use in detecting the presence of this foodborne pathogen.

Staphylococcus aureus

Staphylococcus aureus is a Gram-positive microorganism found ubiquitously in the environment. More than 30% of healthy humans carry *S. aureus* (Menestrina et al. 2001). The most commonly contaminated foods are poultry and meat or products produced from them, milk, and seafood (Balaban and Rasooly 2000). Most outbreaks of illness from this organism occur during the late summer when temperatures are warm, and after the holiday season in the winter. In both cases, illness is associated with improper storage of foods (Monday and Bohach 1999). Outbreaks of *Staphylococcus aureus* have been decreasing since 1973 (Mead et al. 1999).

It is the enterotoxins secreted by the *S. aureus*, rather than the organism itself, that are responsible for causing the symptoms of infection. These symptoms include vomiting, nausea, diarrhea, and abdominal pain. The incubation period between ingesting contaminated food and the onset of symptoms is short, that is, between 2 and 6 hours. Because the illness is acute and short-lived, it tends to be underreported (Balaban and Rasooly 2000, Monday and Bohach 1999).

S. aureus secretes a number of toxins. Of these, the nine known enterotoxins (A–E and G–J) are probably the best researched (Balaban and Rasooly 2000, Monday and Bohach 1999). These toxins facilitate the ability of the bacteria to colonize and persist within a broad range of hosts. They include membrane-active hemolysins, nucleases, proteases, leucocidins, collagenases, cell-surface proteins, and superantigens (Monday and Bohach 1999). *Staphylococcus* enterotoxin A is the most commonly found of the enterotoxins in foodborne illness at 77.8% (Balaban and Rasooly 2000).

Another well-researched toxin of *S. aureus* is α-hemolysin, which is a member of the thiol-activated cytolysin family and has a sequence and structure related to LLO and PLO (Menestrina et al. 2001). Like LLO, *S. aureus* α-hemolysin is a pore-forming protein that inserts itself into membranes, oligomerizes, and forms channels in the membrane. The toxin causes the death of phagocytic cells and necrotic lesions in hosts when the bacteria are injected subcutaneously; it also allows replication of *S. aureus* cells inside macrophages. In mouse, α-hemolysin preferentially attacks endothelial cells and thrombocytes, leading to pulmonary edema and eventual death (Menestrina et al. 2001). Mutation of the gene to inactivate α-hemolysin reduces the ability of *S. aureus* to kill mice (Menestrina et al. 2001).

METHODS TO DETECT FOODBORNE PATHOGENS

The ideal pathogen detection test for the food industry should have a number of characteristics. The optimal test for pathogen detection should be simple to perform. It should be sensitive enough to detect pathogens at levels as low as 1 cell/g of food material; some pathogens such as *E. coli* O157:H7 can infect at doses of less than 10 cells. The test should be specific to the pathogenic species within the genus; in *Listeria,* for example, there are six species, of which only *L. monocytogenes* is an important human pathogen. The test should also be rapid, giving results in less than a day. Finally, the test should be amenable to automation and inexpensive (Ingianni et al. 2001).

In this section are described some of the techniques commonly used to detect foodborne pathogens. It should be noted that none of the tests alone satisfies all of the criteria listed above. Where needed, the detection of *L. monocytogenes* is used to illustrate how the methods can be applied, but the methods described below can be used or adapted for use to detect all the foodborne pathogens described above.

CULTURE-DEPENDENT ENRICHMENT METHODS FOR PATHOGEN DETECTION

Most conventional methods for detecting foodborne bacterial pathogens in food and other substrates rely on the use of microbiological media to selectively grow and enumerate bacterial species. The methods are sensitive and inexpensive, and provide qualitative as well as quantitative results. Unfortunately for the food industry, where time and cost are issues, the preparation of media and plates, colony counting, and biochemical characterization of the isolated colonies make for a time-consuming and labor-intensive process (de Boer and Beumer 1999). The success of culturing methods depends on the number and state of the bacteria in the sample, the selectivity of the media (balance between inhibition of competitors and inhibition of the target organism), the conditions of the incubation (time, temperature, O_2), and the selectivity of the isolation medium (distinction between the target organism and competitive microflora) (Beumer and Hazeleger 2003).

The most common methods used to detect *L. monocytogenes* are those developed by the U.S. Department of Agriculture—Food Safety and Inspection Service (USDA-FSIS) for the detection of *Listeria* in meat and poultry products, the Federal Drug Administration (FDA) protocol for detecting the organism in dairy, fruit, vegetable, and seafood products, and the Netherlands Government Food Inspection Service (NGFIS) method, used for all foods (Donnelly 2001). These methods vary only in the type of selective media that are used (Hayes et al. 1992). They are species specific, but because of the strong selective media used, which may reduce growth of *L. monocytogenes*, the methods tend to lose some of their sensitivity. For example, Hayes et al. (Hayes et al. 1992) showed that the USDA-FSIS method was only able to detect *Listeria* in 65% of foods known to be contaminated with the bacteria, while the NGFIS method only gave positive results for 74% of the same cases. If any two of the above three methods were used together, the success in detecting *Listeria* contamination jumped to between 87 and 91%. This is still not an acceptable rate of detection, when deaths may be caused if the pathogen is present but not detected.

Some enrichment-based methods have been modified or automated to increase throughput. For example, automated dilution of the samples before homogenization, and automated homogenization in sterile bags is faster than sample preparation by hand (de Boer and Beumer 1999). Flow cytometry has also been used to speed up counting times, but it cannot be used to distinguish between living and dead cells (de Boer and Beumer 1999).

There are a few problems with enrichment-based methods. First, they are time-consuming. For example, International Organization for Standardization method 11290-1 (Wan et al. 2003), the official method used by Australian food inspectors, requires 5 days for determination of a negative result for *Listeria* contamination. If a positive test result occurs, additional days are required for biochemical tests to identify the strain (Wan et al. 2003). Second, enrichment methods do not account for the recovery of sublethally injured bacteria that may be present as a result of heating, freezing, or acidification of the foods. These bacteria may not be able to form colonies under selective pressure during recovery, but are still virulent and able to grow in food. For example, sublethally injured *L. monocytogenes* cells are capable of repair and growth at storage temperatures of 4°C or higher (Donnelly 2001). Third, after enrichment, selection is often possible only to the genus level. In *Listeria*, to distinguish between the pathogenic *L. monocytogenes* and the other five nonpathogenic species, additional biochemical characterization must still be done. This adds more time to the detection procedure. Finally, there is some evidence (Kathariou 2002, Loncarevic et al. 1996, Ryser et al. 1996) that some *L. monocytogenes* strains are more sensitive to enrichment and selection procedures than others. For instance, serotype 4b is prevalent in 50–70% of listeriosis outbreaks but is rarely found in contaminated food, thereby inviting speculation that its occurrence is underestimated using this method.

To avoid some of these problems, many newer methods rely on the initial use of selective, chromogenic substrates that change color in the presence of some enzymes present only in certain species, which makes these species easier to identify. For instance, *L. monocytogenes* and *L. ivanovii* contain the *plcA* gene, encoding a phosphatidylinositol-specific phospholipase C, which is not found in any of the other *Listeria* strains. When plated on selective BCM plates, positive colonies are turquoise, while all others are white (Jinneman et al. 2003). The two strains can then be separated based on sugar utilization or other plating techniques, thus reducing the number of tests that must be performed to detect *L. monocytogenes* in food samples (Jinneman et al. 2003).

Automated Methods for Detection of Pathogens (Non-Nucleic Acid-Based)

There are a number of commercially available, semiautomated systems for the detection of various foodborne pathogens, including the MicroLog System (Biolog, Inc., Hayward, California), the Microbial Identification System (MIS) (MIDI, Inc., Newark, Delaware), the VITEK System (bioMerieux Vitek, Hazelwood, Missouri), and the Replianalyzer system (Oxoid Inc., Nepean, Ontario) (Odumeru et al. 1999). The Microlog System is based on carbon oxidation profiles, MIS on lipid analysis, and the VITEK and Replianalyzer systems on biochemical selection. According to Odumeru et al. (1999), these systems are able to detect *Listeria* reliably at the genus level 90–100% of the time, depending on the system being used, but biochemical tests are still required for detection at the species level.

Nucleic Acid-Based Methods of Pathogen Detection

One of the biggest problems associated with detection of *Listeria* is the low numbers at which the bacteria are normally found in contaminated food sam-

ples (Hoffman and Wedman 2001). In one outbreak involving hot dogs in the United States, illness was attributed to contamination levels < 0.3 cfu/g of hot dog meat (Donnelly 2001). DNA-based methods of detection employ ways of amplifying the signal of a few cells more quickly than waiting for cell replication. The following methods represent some of the common ways in which nucleic acids have been used as detection targets for foodborne pathogens, and some comparison of the techniques is presented in Table 31.3.

Polymerase Chain Reaction (PCR)

PCR is the basis of many nucleic acid–based detection systems. With this method, total DNA is extracted from the food sample. Next, two oligonucleotide primers are selected that bind to a pathogen-specific target gene at opposite ends of opposing strands of DNA. A DNA polymerase is added along with the four types of deoxynucleotides (dNTPs) and appropriate buffers, and the mixture is inserted into a thermocycler. This machine cycles through a temperature regime that usually involves 94°C to denature the DNA template, an annealing temperature of 50–65°C to allow the oligonucleotide primers to bind to the gene, and 72°C to allow extension by the DNA polymerase. The thermocycler runs through these temperatures a number of times to amplify the gene fragment exponentially. The resulting fragment is stained with ethidium bromide and visualized on an agarose gel under UV light. The presence of a pathogen can be detected by simply determining if a band representing the pathogen gene of interest is present.

Performing PCR takes only a few hours, shortening the time required for pathogen detection dramatically, and is conducive to automation and high throughput processing using 96-well plates. Aznar and Alarcón (2002) and Shearer et al. (2001) claim that PCR-based detection of *L. monocytogenes* is more sensitive than culture-based methods for detecting the pathogen in contaminated food samples because more samples turned up positive for the *L. monocytogenes* using the PCR method. The authors attribute the increase in sensitivity of the PCR method over culturing methods to the fact that the former does not have an initial selection step. The PCR method circumvents the problem that some cells do not grow on the selective media. The authors also claim that some of the false negatives reported for culturing methods are due to methods that use color changes to differentiate between species, and that these methods are influenced by the subjectivity of the observer (Aznar and Alarcón 2002, Shearer et al. 2001).

Among the target genes for PCR detection of *L. monocytogenes* are the *hlyA* gene (encoding LLO; Blais et al. 1997, Hough et al. 2002, Hudson et al. 2001, Lehner et al. 1999, Lunge et al. 2002, Ryser et al. 1996, Weidmann et al. 1997), the *iap* gene (encoding an invasion-associated protein; Cocolin et al. 2002, Schmid et al. 2003), *inl*B (encoding internalin B; Ingianni et al. 2001, Jung et al. 2003, Lunge et al. 2002, Pangallo et al. 2001) and 16S rRNA (Call et al. 2003, Schmid et al. 2003). Among these genes, the most commonly used has been *hlyA* (Aznar and Alarcón 2002).

Wan et al. (2003) compared the use of PCR to detect *L. monocytogenes* in salmon with International Organization for Standardization culturing method 11290-1, and found the two methods gave comparable results in spiked samples if enrichment culturing is used prior to PCR to lower the detection limit for the *L. monocytogenes*. The only difference is that the PCR method requires 58–60 hours to perform rather than 5 days (Wan et al. 2003).

Jung et al. (Jung et al. 2003) also used PCR to detect *L. monocytogenes* using the internalin AB (*inl*AB) gene. PCR allowed the specific detection of a number of serotypes of *L. monocytogenes,* while no bands from other nonpathogenic species of *Listeria* were detected. The limit of detection of the pathogen in pure cultures was 10^5 cells/mL culture. When frankfurters were spiked with the pathogen, the limit of detection was improved to 10 cells in a 25 g sample, provided the sample was first enriched in a modified enrichment broth for at least 16 hours (Jung et al. 2003). This method allowed detection of the pathogen within 24 hours.

PCR methods are able to detect the presence of the pathogen, but they are not able to quantify the level of pathogen contamination. One way to approach this problem is the use of competitive PCR (see Fig. 31.1). In this case, a competitor fragment of DNA that matches the gene to be amplified is introduced into the sample. In general, the competitor fragment is synthesized as a deletion mutant that can be distinguished from the pathogen gene by its smaller size. In order to determine the level of

Table 31.3. Comparison of Nucleic Acid–Based Detection Methods for *L. monocytogenes*

Method of Detection	Gene Detected	Enrichment Culture Time (hrs)	Total Detection Time (hrs)	Number of Cells Detected	Medium in which Pathogens Were Detected	References
PCR	Commercial kit[a]	48	58–60	2–8 cfu/25 g	Salmon	(Wan et al. 2003)
	inlAB	None	Not available	10^5 cfu/mL	Pure culture	(Jung et al. 2003)
	inlAB	16	24	10 cfu/25 g	Frankfurter	(Jung et al. 2003)
Competitive PCR	*hlyA*	None	5	10^3 cfu/0.5 mL	Milk	(Choi and Hong 2003)
	hlyA	15 h	20	1 cfu/0.5 mL	Milk	(Choi and Hong 2003)
PCR with DNA probe	Commercial probe	16 h	24	2-10 cells/g	Milk and various meat products	(Ingianni et al. 2001)
FRET-PCR	*hlyA*	None	2.5	500 cfu/mL	Pure culture	(Koo and Jaykus 2003)
	hlyA	None	2.5	10^3–10^4 cfu/mL	Skim milk	(Koo and Jaykus 2003)
Real-time PCR	Commercial kit	24	30	10^2–10^3 cfu/mL	Pure culture	(Bhagwat 2003)
	Commercial kit	None	8	10^8–10^{10} cfu/25 g	Cabbage	(Hough et al. 2002)
Multiplex PCR with microsphere sorting	23S rRNA	None	4	10^3–10^5 genome copies	Pure culture	(Dunbar et al. 2003)
DNA microarray	*iap, hly, inlB, plcA, plcB, clpE*	None	N/A[b]	N/A	Pure cultures	(Volokhov et al. 2002)
RT-PCR	*iap*	1	54	10–15 cfu/mL	Pure culture	(Klein and Juneja 1997)
	iap	2	55	3 cfu/g	Ground beef	(Klein and Juneja 1997)

[a]The gene amplified is not mentioned for the kit.
[b]N/A, not addressed in paper.

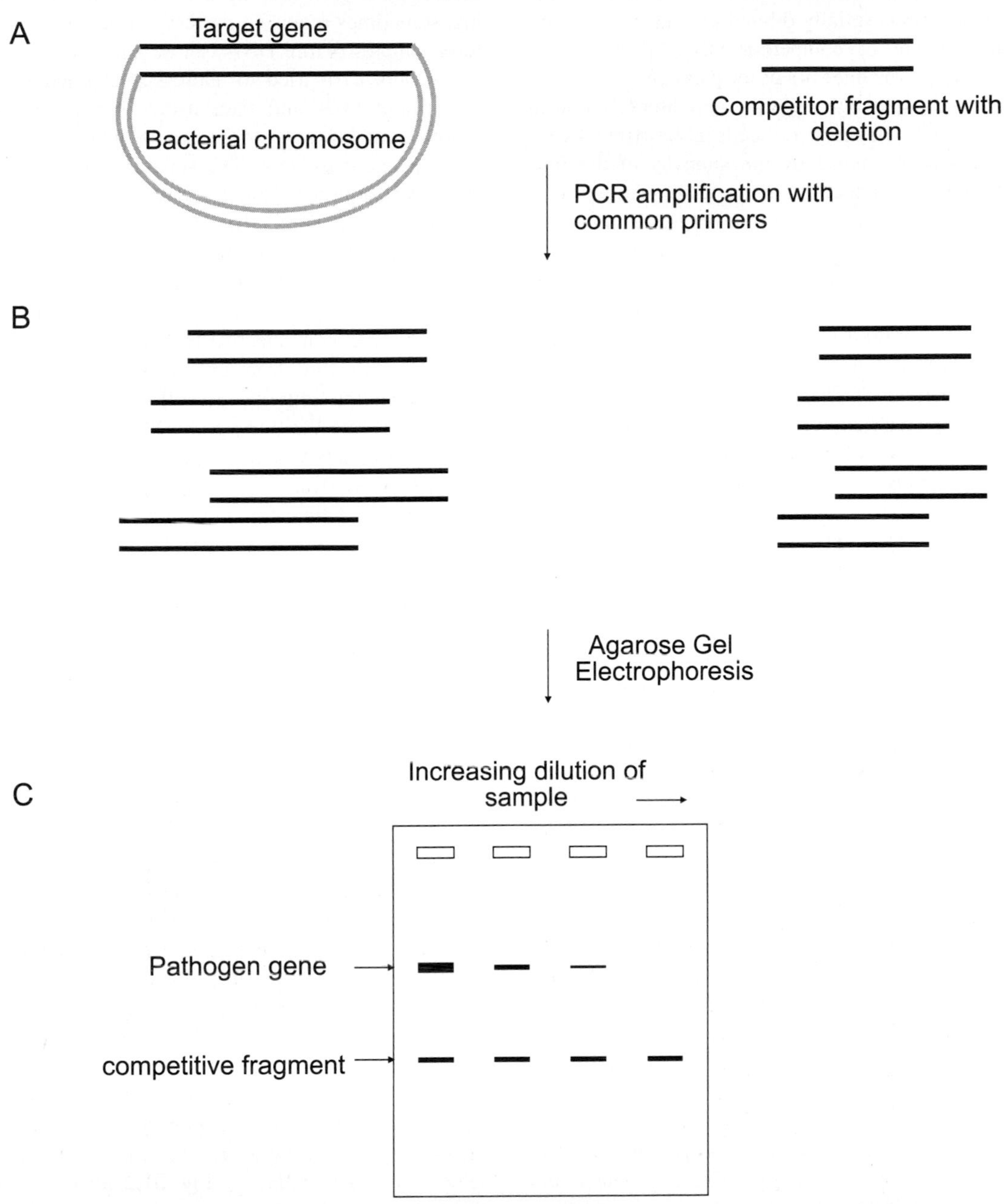

Figure 31.1. Use of competitive PCR to quantify pathogens. **A.** A PCR mixture is made up that contains serially diluted DNA from the sample and a constant amount of a competitive PCR fragment. The competitive fragment has the same sequence at its ends, but an internal deletion makes it shorter than the native pathogen gene for ease of distinguishing between the PCR products. **B.** During PCR amplification, primers complementary to the target gene in the pathogen amplify both the target gene and the competitive gene. **C.** The products of the PCR are analyzed by agarose gel electrophoresis. As the dilution of the sample DNA increases, the resulting PCR product decreases. The intensities of the bands can be compared in order to quantify the number of target genes from the pathogen in the original sample.

pathogen contamination, DNA purified from the food sample is serially diluted and added to a constant amount of competitor DNA. PCR is performed, and the intensity of the pathogen gene's signal is compared to that of the competitor DNA on an agarose gel. The number of cells in the original sample can be estimated by the intensity of the full-length PCR product (from the pathogen) as compared with the intensity of the smaller, competitive PCR product (Schleiss et al. 2003). The advantage to this method over a number of other PCR methods is that no expensive fluorophores or radioactive labels are required to visualize the results (Choi and Hong 2003). Choi and Hong (2003) used a variation of competitive PCR in which the competitor fragment had a restriction endonuclease site removed so that after PCR and digestion with the endonuclease, the competitor was visualized as a slightly larger molecule. The authors also varied the amount of competitor DNA added and kept the sample inoculation constant in order to show that the could quantitatively determine the number of *L. monocytogenes* in artificially inoculated milk. According to Choi and Hong (2003), the method took around 5 hours to complete without enrichment and was able to detect 10^3 cfu/0.5 mL milk using the *hlyA* gene. The detection limit could be reduced to 1 cfu if culture enrichment for 15 hours was conducted first.

PCR-based methods have several limitations. First, the food matrix, which includes complex polysaccharides in feces and fat and proteins in food samples, often interfere with PCR by inhibiting DNA polymerase directly or by binding Mg^{2+} (inhibiting PCR enzymes) (Fratamico and Strobaugh 1998). The matrix-based interference necessitates the use of DNA purification steps that add to both the cost and the completion time of the method. Second, culture enrichment may be required to concentrate the pathogen so that gene of interest can be detected. Third, analysis on an agarose gel is labor intensive when done on the large scale required by the food processing industry. Finally, PCR only detects the presence of DNA. This does not indicate whether the pathogens are dead or alive. The food industry must know whether the food represents a health hazard, not whether the pathogen was present at one point but was killed by the food processing method.

Several groups have developed alternatives to the use of classical PCR for pathogen detection to reduce some of the limitations of PCR. As mentioned before, PCR protocols are generally time consuming, sometimes requiring bacterial enrichment from food samples before DNA can be amplified. Some researchers have tried to address this problem by performing PCR and then using oligonucleotide probes complementary to the gene of interest to capture the amplified DNA. This separates and concentrates the amplified DNA to allow detection of the amplified signal. While Ingianni et al. (2001) were able to detect *L. monocytogenes* in food samples using complementary probes, they admitted that their results were much more reliable after overnight culture enrichment in selective media. This increased their detection time from one working day to two, with a detection limit of 2–10 cells/g sample.

A second rate-limiting step for PCR methods is in the need to analyze and detect the amplified DNA product. When dealing with hundreds or thousands of samples, the time for running agarose gels with the DNA samples or performing DNA hybridization assays becomes significant (Koo and Jaykus 2003). Some groups have improved PCR sensitivity by using fluorescent resonance energy transfer(FRET)-based PCR (Koo and Jaykus 2003). In this method, DNA is analyzed directly after PCR by measuring the fluorescence signal (See Fig. 31.2). This system works by having two DNA probes for the gene of interest, one with a fluorescein label and the other with a quencher label. During the annealing and primer extension steps of the PCR, the fluorescein-labeled oligonucleotide hybridizes to the gene of interest. The hybridized probe is digested by the exonuclease activity of the DNA polymerase as the polymerase amplifies the gene. This digestion releases the fluorophore from the probe. The probe with the quencher label is short and will not anneal to the fluorescein-labeled probe until after the PCR process is completed and the mixture is cooled to room temperature. At this point, any unused fluorescein-labeled probe is quenched due to its hybridization with the quencher probe, leaving only the free fluorophore (Koo and Jaykus 2003; see Fig. 31.2 for more details). The resulting fluorescence is proportional to the number of pathogens in the original sample and obviates the need to run agarose gels for PCR product detection. A single probe containing both the fluorophore and quencher can also be used (Cox et al. 1998), but the double label significantly increases the cost of probe synthesis. This method, targeting *hlyA*, provides a detection limit for *L. monocytogenes* of

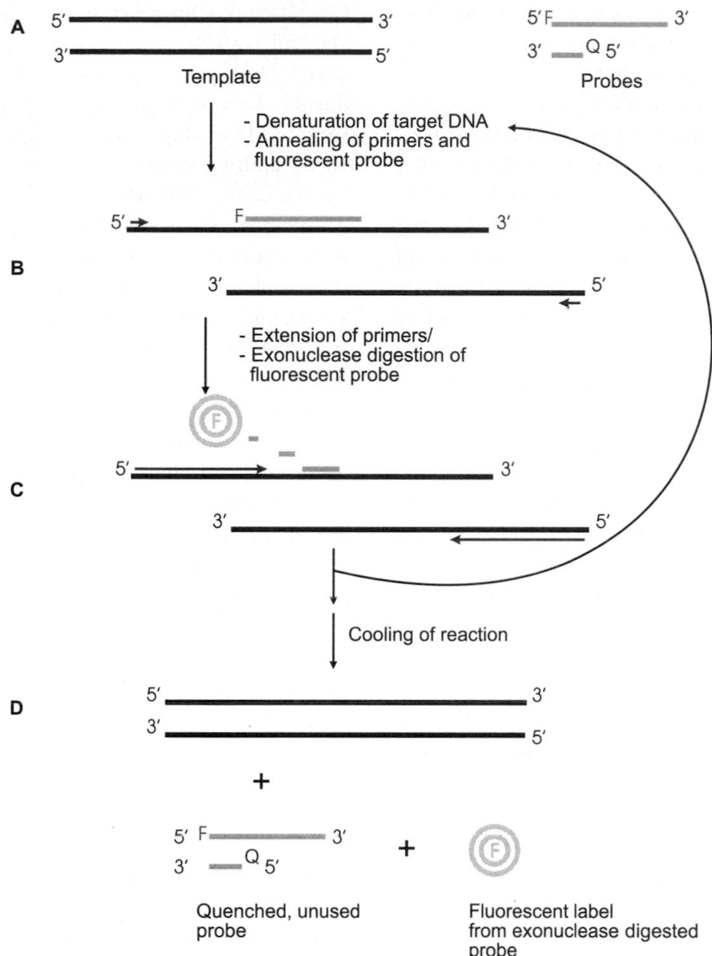

Figure 31.2. Use of fluorescence resonance energy transfer–based PCR to detect pathogens. **A.** This PCR-based method has a number of components. Template DNA from the sample, containing a pathogen-specific target gene, is isolated, and oligonucleotide primers specific to the target gene and used to amplify the target gene are added. A probe specific for a central sequence in the target gene is labeled with a fluorogen and added to the mixture. A short oligonucleotide probe, specific for the fluorescent probe, labeled with a quenching molecule, is also present. This probe has a low melting temperature so that it will not bind to the fluorescent probe during PCR. Finally, the PCR reaction must contain DNA polymerase that retains its 5'-exonuclease activity. **B.** In the first two steps of the PCR, template DNA from the sample is heated to a temperature that will denature all its DNA. The temperature is then lowered to allow annealing of the primers for amplification and the fluorescently labeled probe. **C.** The pathogen-specific target gene is amplified by Taq polymerase. Any fluorescent probe that has annealed to the DNA is digested by the exonuclease activity of the Taq polymerase to release free fluorogen. **D.** After a number of rounds of amplification, PCR products are cooled. At this point, the quencher-labeled probe will bind to any remaining fluorescent probe, preventing fluorescence of that probe. The more pathogens are initially present in the sample, the more copies of the amplified gene will be present. This leads to more of the fluorescent probe binding to the gene and being digested by the DNA polymerase. Only the free fluorogen will fluoresce due to the presence of the quencher-labeled oligonucleotide. At the end of the reaction, the number of pathogen cells in the original sample is estimated from the fluorescence signal; the more pathogen DNA present, the higher the fluorescence. F, fluorescent label; Q, quencher.

about 500 cfu/mL in pure culture, and 10^3–10^4 cfu/25 g of artificially inoculated skim milk (Koo and Jaykus 2003). The total time of end point detection is about 2.5 hours (Koo and Jaykus 2003).

Another way to eliminate the need for agarose gel electrophoresis is use of real-time PCR in a 96-well PCR format. In this method a fluorescent dye, such as SYBR Green I, is used to follow the PCR amplification in real time and can be used to detect the amplified products from a number of genes at the same time (Bhagwat 2003). The genes/pathogens present can be assigned from the melting curves of the PCR products. This protocol also allows multiple pathogens to be detected from a universal enrichment media. For instance, Bhagwat (2003) was able to simultaneously detect *L. monocytogenes*, *S. typhimurium*, and *E. coli* O157:H7 from produce. Including time for culture enrichment, the entire process took around 30 hours, and allowed detection limits of 1 cell/mL for *E. coli* O157:H7 and *Salmonella* spp. and 10^2–10^3 cells/mL for *L. monocytogenes*. Hough et al. (2002) were able to detect between 10^3 and 10^{10} cfu for *Listeria* in 25 g samples of cabbage without any culture enrichment. Because of the lack of enrichment, the entire process from extraction to real-time PCR results was around 8 hours (Hough et al. 2002).

Multiplex PCR is a variation of the traditional PCR. This method makes use of multiple sets of primers to amplify a number of genes or gene fragments simultaneously. For instance, Bhagwat (2003) artificially contaminated produce, cultured the pathogens, then used multiplex PCR to simultaneously detect *L. monocytogenes*, *S. typhimurium*, and *E. coli* O157:H7. Fratamico and Strobaugh (1998) also used the method to detect *Salmonella* and *E. coli* O157:H7 in bovine feces, carcass wash water, apple cider, and ground beef that had been inoculated with the two organisms. After culture enrichment, the authors were able to detect the pathogens at a level of 1 cfu/mL (or g), with a total processing time of 30 hours (Fratamico and Strobaugh 1998). This identification procedure is effective, but not conducive to high-throughput screening because of the need to analyze the PCR products by agarose gel electrophoresis.

In a more elaborate variation for multiplex PCR, Dunbar et al. (2003) amplified portions of the prokaryotic 23S rRNA using universal primers labeled with biotin. After amplification, the PCR product was mixed with fluorescently labeled microspheres displaying pathogen-specific oligonucelotides (see Fig. 31.3). Each type of microsphere has a different spectral pattern, and this allowed each type to display a DNA probe specific for a different DNA sequence derived from a different pathogen (derived from the PCR step). The pathogenic DNA is detected by adding microspheres specific for each pathogen's DNA. The bound microspheres are labeled with another fluorescent label by adding streptavidin-R-phycoerythrin, which binds to the biotin label on the amplified DNA and fluoresces at a different wavelength than the microspheres. Because the spectral label pattern of each type of microsphere is different, each can be sorted according to its spectral signature (Dunbar et al. 2003). The amount of fluorescence from the phycoerythrin can be measured after sorting to determine the quantity of pathogen DNA present in each species. Theoretically, this system could be used to simultaneously detect hundreds of different pathogens. Using the Luminex LabMAP (Luminex, Austin, Texas) system, the authors could detect *L. monocytogenes*, *E. coli*, *Salmonella,* and *C. jejuni* simultaneously at levels of 10^3 to 10^5 organisms (Dunbar et al. 2003). The advantage to this system is that after enrichment and DNA amplification, enumeration of the original number of pathogens takes only 30–40 minutes (Dunbar et al. 2003), and multiple pathogens are detected simultaneously. However, this system is prohibitively expensive, still requires time for culturing, and may experience difficulties with matrix effects.

To address the need of detecting only living pathogens, one can amplify the pathogen RNA rather than DNA. The presence of specific RNA sequences is an indication of live cells. This process works because RNA is much less stable and tends to degrade fairly quickly outside living cells as compared with DNA. Thus, when an organism dies, its RNA is quickly eliminated, whereas the DNA can last for years, depending on storage conditions. Reverse transcription-PCR (RT-PCR) makes use of a reverse transcriptase that is able, in the presence of a complementary primer, to create complementary DNA (cDNA) from an RNA strand corresponding to a transcribed gene. The cDNA is then amplified using PCR primers and DNA polymerase under normal PCR conditions. Analysis of the results is done in the same way as for PCR.

Klein and Juneja (Klein and Juneja 1997) used RT-PCR to detect *L. monocytogenes* in pure culture

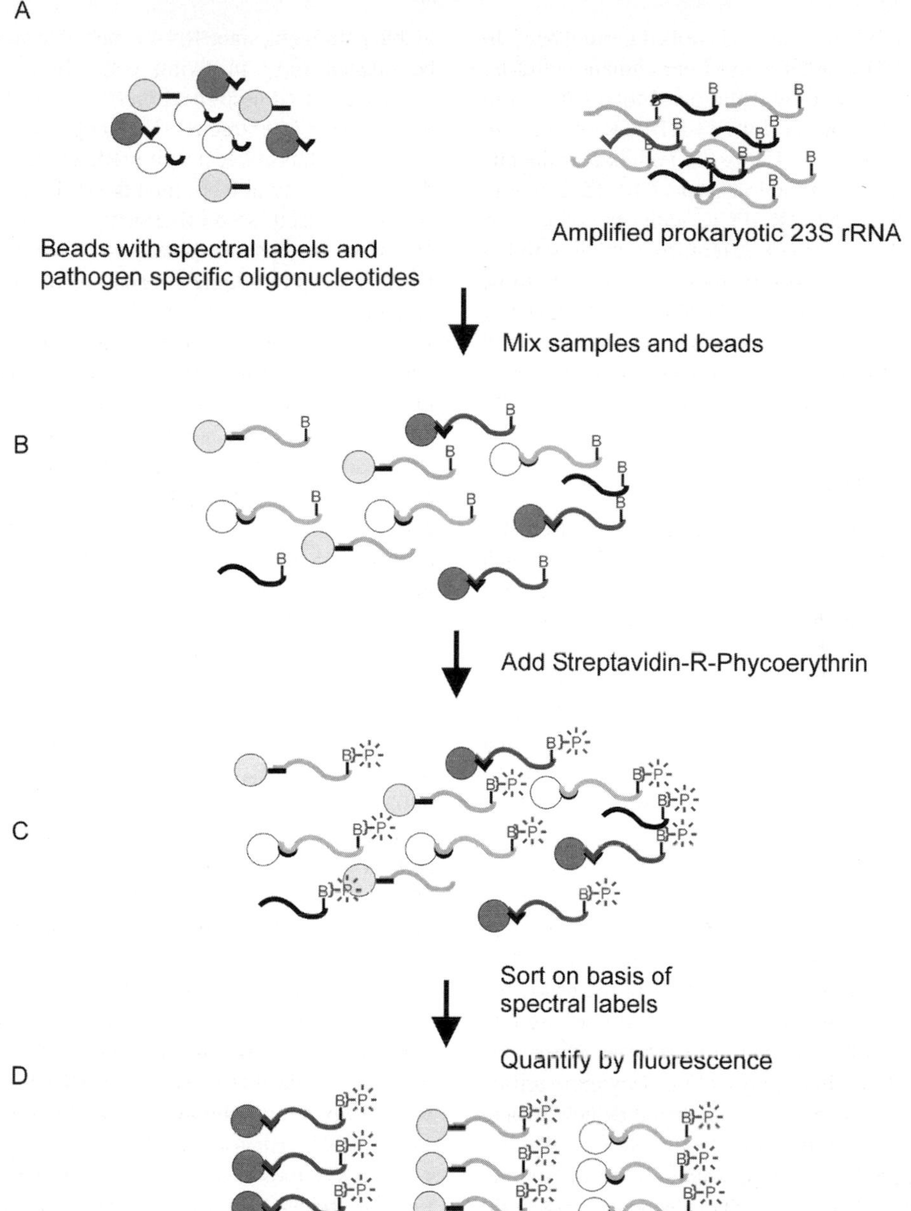

Figure 31.3. Multiplex PCR and the use of fluorescently labeled microspheres to detect pathogens. **A.** DNA is isolated from the food sample and all prokaryotic 23S rRNA gene sequences are amplified using universal primers, which also add a biotin label to the amplified DNA. Fluorescently labeled beads containing a different spectral pattern for each pathogen-specific probe are added to the amplified PCR mix, and the DNA is denatured to create single-stranded fragments. **B.** The amplified pathogenic DNA fragments bind to the beads, displaying their complementary probes. **C.** The bound microspheres are labeled with fluorescent label by adding streptavidin-R-phycoerythrin, which binds to the biotin label on the amplified DNA and fluoresces at a different wavelength than the microspheres. **D.** The beads bound to the pathogenic DNA are sorted based on their spectral patterns, and the amount of fluorescence is measured to quantify the amount of each pathogen present. **B.** biotin; **P,** phycoerythrin.

and artificially contaminated cooked ground beef. In pure culture, the authors used enrichment culturing for 1 hour, followed by isolation of total RNA from the pathogen. They performed RT-PCR on the *iap*, *hlyA*, and *prfA* genes. Using the *iap* gene as the target, the authors were able to detect 10–15 *L. monocytogenes* cells/mL. RT-PCR with the other two genes was not as sensitive, requiring either a higher number of organisms or more enrichment time. When the authors used RT-PCR on the *iap* gene to detect the pathogen in ground beef, they detected organisms at a level of 3 cfu/g of inoculated meat after 2 hours of enrichment culturing.

DNA Microarrays

Microarrays are composed of a number of discreetly located DNA probes fixed on a solid substrate such as glass (Call et al. 2003). Each probe corresponds to an oligonucleotide specific to a target DNA sequence. Call et al. (2003) used probes specific for unique portions of the 16S rRNA gene in *Listeria* to demonstrate how *Listeria* species could be differentiated by this method. In this procedure, PCR is first performed using universal primers to amplify all the 16S rRNA genes present in a sample. The various amplified DNA fragments bind only to the probes for which they have a complementary sequence. Because one of the oligonucleotides used in the PCR contains a fluorescent label, the probes bound to DNA sequences fluoresce. Pathogens are identified by the pattern of fluorescing spots in the array. Alternatively, multiple primer sets can be used to amplify a number of pathogen-specific genes in a multiplex PCR. Microarrays allow many more primer sets to be used in the PCR step than if detection were to be performed by gel electrophoresis. This is because only those products binding to a microarray probe will be detected, so the researcher is not limited by the number of resolvable bands on a gel (Call et al. 2003). Microarrays are relatively cheap and are able to identify a number of pathogens or serotypes at once, but they still require culture enrichment and PCR steps before the results can be determined.

Nucleic Acid Sequence–Based Amplification (NASBA)

Beumer and Hazeleger (Beumer and Hazeleger 2003) used in vitro RNA amplification to detect viable pathogens since RNA is unstable and likely to be present only in living cells. In nucleic acid sequence–based amplification (NASBA), total RNA in a sample is extracted. Messenger RNA is then used as the target because it predicts viability better than rRNA or total RNA, and the PCR amplification step is not used, so no thermocycler is required. In NASBA, all steps take place at 41°C so that genomic DNA stays in double stranded form and does not affect the reaction.

During NASBA, three enzymes are required: reverse transcriptase, RNaseH, and T7 RNA polymerase. Two oligonucleotide primers specific to the gene of interest are used, and a mixture of both NTPs and dNTPs are added. The oligonucleotide complementary to the mRNA binds to its template, and the reverse transcriptase produces a cDNA molecule. The RNaseH is then added to digest the RNA present, at which point the second primer binds to the cDNA, allowing the reverse transcriptase to make the complementary strand. This creates a "minigene" (Cook 2003), which is transcribed into thousands of RNA transcripts by the T7 RNA polymerase. The transcripts can be detected by agarose gel electrophoresis or by probe hybridization (Cook 2003).

NASBA has been used to detect viable *C. jejuni* cells at between 10 and 30 cfu/g of inoculated food after 48 hours of enrichment (Cook 2003; Uyttendaele et al. 1999, 1995a). The technique has also been used to detect viable *L. monocytogenes* at 10 cfu/60 g meat or seafood product (Blais et al. 1997; Uyttendaele et al. 1995b), and *E. coli* at 40 cfu/mL drinking water (Min and Baeumner 2002). All pathogens were detected using 16S rRNA sequences, and *L. monocytogenes* was also detected using the *hlyA* mRNA (Blais et al. 1997). The *L. monocytogenes* assay took 3 days to perform, including culture enrichment (Uyttendaele et al. 1995b).

Aside from ensuring detection of only viable cells, NASBA removes the need for a costly thermocycler. NASBA still requires that nucleic acids be extracted from the food samples, so there may be matrix effects on the enzymes (Cook 2003).

Pathogen Subtyping and Verification Methods

RFLP, RAPD, AFLP, ribotyping, and pulsed-field gel electrophoresis are molecular-based methods that are used for further verification after a food sample is suspected of harboring a pathogen. These methods are not used as routine screening tools. Each allows further characterization of the pathogen in order to know which of the many possible subtypes within a serotype is causing the disease or is the contaminating factor. The information is important from an epidemiological point of view and is needed to identify the precise strains responsible for causing disease, or that contaminate a certain type of food. The main drawback with these methods is that individual cultures have to be isolated and grown before being analyzed independently. The methods are compared in Table 31.4.

RFLP—Restriction Fragment Length Polymorphism

Restriction fragment length polymorphism (RFLP) analysis uses the restriction endonuclease patterns in DNA to determine differences in genetic profiles. Even among closely related individuals, the mutation of a restriction site occurs often enough to allow a pedigree to be formed (Smith and Nelson 1999). Digesting genomic DNA using restriction endonucleases creates fragments of different lengths. The DNA is then run on a capillary electrophoresis and the pattern of bands analyzed. Different groups of organisms, or serotypes within the same species, can be elucidated from the differences in the electrophoresis patterns of DNA fragments. However, digesting genomic DNA often results in too many bands, leading to difficulty in interpretation of results. To minimize interference, RFLP can be done on individual genes. Weidmann et al. (1997) used PCR-RFLP on the *hlyA*, *actA,* and *inlA* genes of *L. monocytogenes* to determine how strains with different pathogenic potentials were related. They found that *L. monocytogenes* could be divided into three lineages, and that all epidemic outbreaks could be traced back to lineage I.

AFLP—Amplified Fragment Length Polymorphism

Amplified fragment length polymorphism (AFLP) analysis is a DNA fingerprinting technique based on the selective amplification of genomic restriction fragments to generate a restriction pattern formed from a large number of fragment bands on gels (Aarts et al. 1999). It essentially combines the reliability of RFLP and the power of PCR. Genomic DNA is digested with two restriction endonucleases, one of which cuts infrequently, and the other on a frequent basis (Blears et al. 1998). Selective pressure for amplification is then applied in two different ways. The first is that adapter oligonucleotides are ligated to the digested DNA. The adapters add a few selective nucleotides after the restriction site. Only those fragments with the complementary sequence between the restriction site and the adaptor will be bound and amplified by PCR. This step results in about 1/16 of the bands being amplified during the PCR (Aarts et al. 1999). After the PCR process, the amplified fragments are separated from the mixture by the presence of a biotin label on the adapter on the infrequently cut end of the fragment. Those fragments are separated from the bulk solution by the use of streptavidin-coated magnetic beads (Blears et al. 1998). Selective amplification results in 50–100 different fragments being amplified (Blears et al. 1998). Analysis is performed and the amplified fragments detected by the presence of a fluorophore on one of the primers. Detection of the fingerprinting pattern can be done either on gels or by automated laser fluorescence analysis. The latter allows comparison of data from other laboratories (Aarts et al. 1999). Genetic polymorphisms are then identified by the presence or absence of different fragments (Blears et al. 1998).

Aarts et al. (1999) used AFLP to differentiate different strains of *L. monocytogenes* on a more discriminating basis than serotyping. In particular, the results match those obtained by other methods such as pulsed-field gel electrophoresis, multilocus PCR-restriction enzyme digestion of the *iap* gene locus, and ribotyping (Aarts et al. 1999).

RAPD—Randomly Amplified Polymorphic DNA

Randomly amplified polymorphic DNA (RAPD) analysis makes use of a short arbitrary primer (e.g., 10 bp) that anneals randomly along genomic DNA to amplify a number of fragments within the genome. As long as the same primer is used for all the test samples, the comparison of the number and sizes of fragments generated allows for discrimination between strains of a pathogen. It does require having a pure culture so that there are no contaminating bands from other organisms or from the DNA in the food (Lawrence and Gilmour 1995). With regard to *L. monocytogenes*, the technique was used

Table 31.4. Comparison of Pathogen Subtyping and Verification Methods

Method	Advantages	Disadvantages	Amount of DNA required
RFLP	• Simplicity • Can be done with restriction endonucleases for serotype level discrimination or combined with PCR for lineage comparisons	• Requires pure cultures of pathogen • Many bands to compare; confusing, may be hard to see differences • Cannot compare organisms at the genus or family level • Partial digestion and faint bands may be a problem	3–5 μg
AFLP	• PCR to amplify a few genes and simplify result analysis • Can compare differences between bands on a gel • Fewer bands to compare than with RFLP • Can compare among labs	• Requires pure cultures of pathogen • Cannot compare organisms at the genus or family level	10–100 ng
RAPD	• Universal primers that are not gene specific	• Requires pure cultures of pathogen • Is not very reproducible between gels or labs • Cannot compare organisms at the genus or family level	10 ng
Ribotyping	• Similar to RFLP, but visualizes only bands correlating to the *rrn* portion of ribosomes so fewer bands to compare • Can compare organisms at the genus and family level	• Requires pure cultures of pathogen	1 μg
Pulsed-field gel Electrophoresis	• PulseNet database available to coordinate and compare results—can track outbreaks • Can compare differences between bands on a gel • Results are comparable among labs	• Time consuming—3 days for results • Requires pure cultures of pathogen • Patterns may change after intestinal passage; differences in patterns may not indicate actual strain differences • Cannot compare organisms at the genus or family level	Measured by turbidity of culture–lysis and DNA digestion performed in agarose plugs
FISH	• May detect one pathogen in a mixed community of microbes	• Cells must be fixed and permeabilized before visualization	Single cell level

Source: Adapted from Gurtler and Mayall 2001, Savelkuoh et al. 1999.

to trace the source of a *L. monocytogenes* contamination in a food plant over a 6-month period. By taking samples over this time, culturing the isolates, and performing RAPD analysis on individual isolates, Lawrence et al. (Lawrence and Gilmour 1995) showed that two strains were present throughout the yearlong study. They showed that the strains were persistent and might be responsible for cross-contamination of the preparation areas within that environment, and they were able to pinpoint the source of the contamination of these strains. These authors also demonstrated that other strains isolated over time were transient in nature, and probably came from various sources of contamination (Lawrence and Gilmour 1995). This method is good for microbial source tracking and determining critical control points for preventing pathogen contamination within the food processing industry.

Ribotyping

Ribotyping is similar to RFLP in that it uses restriction endonuclease digestion of DNA to set a pattern that can be analyzed. This technique relies specifically on the ribosome-encoding genes that are relatively conserved across the bacterial kingdom, and allows lineages to be traced through the appearance of mutations over time. To perform the procedure, genomic DNA from individual strains is first digested with an endonuclease such as *Eco*RI, followed by Southern hybridization with a probe targeted at specific conserved regions of rRNA coding sequence. The probe allows only the DNA fragments encoding rRNA to be detected on the blot (Ryser et al. 1996, Weidmann et al. 1997). Different strains will give rise to different banding patterns that are used to determine lineages.

Weidmann et al. (1997) used ribotyping on *L. monocytogenes* from a number of different isolates to determine the different lineages related to the pathogenicity of the organism. Using ribotyping, it was possible to show that of three distinct lineages, only one was responsible for all the human illnesses (Weidmann et al. 1997).

Pulsed-Field Gel Electrophoresis

Pulsed-field gel electrophoresis allows for the separation of large fragments of DNA, from 10–2000 kilobases (kb; Finney 2000). DNA fragments have an overall negative charge proportional to their sizes, due to the phosphate moiety on each nucleotide. When an electric current is applied to a DNA sample, the DNA moves towards the positive electrode (anode). Because the matrix of the gel acts as a sieve, larger pieces of DNA are retarded in their movement, while smaller fragments travel through the spaces and migrate further on the gel. In a normal agarose gel, the largest size of DNA that can be effectively separated is between 20 and 40 kb. Above this size, limiting mobility occurs because DNA fragments, once molded into a shape that can squeeze through the sieve, migrate at around the same rate.

Pulsed-field gel electrophoresis takes advantage of the time it takes for the large DNA fragments to squeeze into their elongated shapes for movement as a basis for further separation. Larger DNA fragments take longer to be forced into these shapes. Therefore, if the direction of the electric field is changed, then the smaller pieces of DNA will alter their shape faster and start to migrate at the limiting mobility rate. By optimizing the times and angles of this alternating electrophoretic field, the larger pieces of DNA can be resolved on the gel (Finney 2000, Moore and Datta 1994).

Pulsed-field gel electrophoresis can be used for genetic fingerprinting of any pathogen isolate, allowing for the determination of the relatedness between cases of illness such as listeriosis (Donnelly 2001, Moore and Datta 1994). The U.S. CDC initiated a collaborative effort to create an electronic database known as PulseNet that encourages researchers to enter their data to help track outbreaks of *E. coli* O157:H7, nontyphoidal *Salmonella* serotypes, *L. monocytogenes,* and *Shigella* (Swaminthan et al. 2001). As of 2002, 46 state labs, two public labs, the FDA, the USDA, and several Canadian labs have entered data into PulseNet. This allowed for better tracking and earlier detection of possible common-source outbreaks (Kathariou 2002, Swaminthan et al. 2001). Although this method is finding increased use and provides a great deal of information, it is time consuming, requiring about 3 days to obtain results (Finney 2000).

FISH—Fluorescence In Situ Hybridization

Fluorescence in situ hybridization (FISH) is often used to study, in a cultivation-independent way, the structural dynamics of microbial communities. FISH can be used in phylogenetic studies and in assessing the spatial distribution of target microbes in communities such as biofilms (Wagner et al. 1998). The method has only recently been applied to detecting and analyzing microbes in food. To perform FISH, a sample is fixed to a slide, membrane filter, or well using ethanol. In the case of Gram-positive organisms such as *Listeria*, the fixed cells are permeabilized with proteinase K so that fluorescent probes are able to enter the cell. Specific probes are created that are complementary to DNA or rRNA inside the cell, and labeled with a fluorescent tag such as Cy3, Cy5, or CFLUOS. The fixed cells are incubated with the labeled probes under conditions that allow labeling of the microbes of interest but not the background microflora. The labeled cells can then be visualized with a fluorescence microscope (Ootsubo et al. 2003, Schmid et al. 2003, Wagner et al. 1998). For working with *Listeria* in milk samples, Wagner et al. (1998) enriched the bacteria in *Listeria* enrichment broth for 1, 2, or 7 days before analysis. Using primers specific for the 16S rRNA genes, the group was able to detect *Listeria* reliably on the genus level, allowing for further analysis by competitive PCR.

Immunoassay-Based Methods

Immunoassays are based on the natural affinity of antibodies for their antigens, regardless of whether the antibody is against a hapten, a protein, or a carbohydrate on the surface of a cell. These assays are fast, and relatively inexpensive. They allow accurate detection of contaminants after very little sample purification (Hall et al. 1989b). Immunoassays are not as susceptible to matrix effects as PCR assays; samples such as river water can be analyzed directly by ELISA (enzyme-linked immunosorbent assay; Hall et al. 1989a). In addition, immunoassays can be used to provide information in a real-time manner and allow for a timely response if quantities of the pathogen are high enough (Meng and Doyle 2002). The biggest problems with immunoassay-based methods are the low sensitivity of the assays, the low affinity of the antibody to the pathogen or other analytes being measured, and potential interference from contaminants. Improvements in these areas will likely expand the use of immunoassays in a variety of fields, including the food industry. A comparison of various immunoassay methods used to detect foodborne pathogens is presented in Table 31.5.

Types of Antibodies Used in Detection of Pathogens

A number of antibody types and formats are available for immunodetection. These include conventional and heavy-chain antibodies, as well as polyclonal, monoclonal, or recombinant antibodies (see Fig. 31.4).

Polyclonal antibodies have been used as detection vehicles for several decades (Breitling and Dübel 1999). These antibodies are raised by immunizing an animal host with the antigen several times, then harvesting serum from the animal. The serum obtained contains a mixture of antibodies, most of which do not bind to the antigen and were present before immunization. However, if the antigen is immunogenic, containing a number of epitopes to which antibodies can bind, this is a simple way of obtaining a detection reagent for performing immunoassays.

In 1975, Köhler and Milstein were the first to isolate monoclonal antibodies from hybridoma cells (Köhler and Milstein 1975). Monoclonal antibodies are more useful for the specific detection of a molecule than polyclonal antibodies. Raising a monoclonal antibody also begins with the immunization of an animal. Once the immune response has been maximized, the B-cells producing the antibodies are isolated from the spleen of the immunized animal and fused with myeloma cancer cells, to produce a hybridoma cell that has the immortality of a cancer cell and the antibody producing capability of the B-cell.

Monoclonal antibodies have a number of advantages over polyclonal antibodies. First, hybridoma cell lines expressing monoclonal antibodies can be cultured indefinitely in vitro to provide a continuous supply of homogeneous, well-characterized antibodies. Clones can be selected with different specificities and affinities for a molecule, or a family of molecules. Nonspecific antibodies are removed during the selection process, thereby limiting interference in the assay (Deschamps and Hall 1990, Deschamps et al. 1990). However, monoclonal antibodies are expensive to produce, requiring a skillful technician and specialized growth apparatus for tissue culturing,

Table 31.5. Comparison between Immunoassay Methods to Detect Foodborne Pathogens

Method	Advantages	Disadvantages	Detection Limit[a]
ELISA	• Simplicity • Detection within a few hours	• Large quantities of sample required • Background levels can be high	10^3–10^5 cfu/mL
Sandwich ELISA	• Concentrates the antigen in the well • Gives a stronger signal	• Must have a second primary antibody for a different portion of the antigen • More time required than basic ELISA	
Competitive ELISA	• Less false positives • Allows for quantitative results	• Must have pure antigen available	
Fluorescently labeled immunoassay	• Less time required to measure signal	• More expensive than ELISA • Specialized equipment required to analyze results	
Fluorescent immunoassay with microsphere sorting	• 30–40 minutes for detection after enrichment • Sensitive • Multiple pathogens can be detected simultaneously	• Expensive chemicals required • Trained personnel required • Computer sorter is specialized and expensive	500 cfu/mL (Dunbar et al. 2003)
Immunomagnetic assay	• Time—no enrichment of pathogen required	• Expense • Trained personnel required for assay and conjugation of antibody to beads	1–2 cfu/g (Hudson et al. 2001)
Latex agglutination assay	• No skilled help required for assay • Inexpensive • No specialized equipment	• Qualitative results only • Enrichment culturing required	0.3–222 cfu/g (Matar et al. 1997)

[a]Detection limits are for *L. monocytogenes*

and production levels are generally low. Within the food industry, what is needed is a cheap supply of large amounts of antibodies to perform high-throughput screening for pathogens.

Recombinant antibodies fill this void since they can be produced in reasonable quantities in short periods of time from bacterial expression systems. Handling, growth, and storage of the bacteria are simpler and cheaper than the corresponding work with hybridoma cells (Breitling and Dübel 1999). Recombinant antibodies also have the advantage of being clonal—one antibody fragment is produced per bacterial colony. The use of recombinant antibodies offers the opportunity to select binding antibodies from naïve rather than immunized antibody libraries, removing any ethical dilemmas related to immunizing animal hosts with known pathogenic organisms (Churchill et al. 2002) or sacrificing the animals for their spleens. Finally, recombinant antibodies can easily allow one to improve the binding properties and affinities of an antibody through mutagenesis and genetic engineering of the expression vectors and genes (Yau et al. 2003).

Recombinant antibodies come in a number of forms (see Fig. 31.4). The Fab fragment is made up of the first constant domain and the variable domain of each of the heavy and the light chains of the antibody. A disulfide bond between the constant domains joins the two chains. Fab fragments have comparable affinities to the monoclonal antibodies from which they are derived. However, their expression by bacteria may be difficult since the fragment is too large to be expressed intact, and bacteria do not have the machinery to form disulfide bonds correctly in the cytoplasm. Therefore, each chain must be expressed separately and recombined.

To further reduce the size of the expressed fragment, scientists removed the constant domains from the Fab fragments, leaving the heavy and light variable fragment (Fv). These fragments have higher levels of expression in bacterial cells, due to their smaller size, but tend not to combine because the heavy and light chains of the Fv are not attached by a disulfide bond and must be stabilized. The heavy and light chains are joined by creating a short synthetic peptide linker, resulting in the formation of a single-chain variable fragment (scFv) (Fig. 31.4). ScFvs have a number of advantages over other constructs. They are well expressed in bacterial systems and can be expressed as a single gene construct. This eliminates the need for reassociation of the heavy and light chains. They have affinities that are generally comparable to the original antibody, but are somewhat lowered because they contain a single binding domain rather than two. These scFv antibodies have been used to detect pathogens such as *Streptococcus suis* (De Greeff et al. 2000), *Brucella melitensis* (Hayhurst et al. 2003), *Bacillus* (Turnbough Jr. 2003, Zhou et al. 2002), *S. aureus* (Bjerketorp et al. 2002), and *L. monocytogenes* (Churchill, unpublished data).

More recently, a different type of recombinant fragment has been constructed, called the VHH (the designation VHH distinguishes this heavy-chain variable region from the corresponding VH domain of conventional antibodies). Llamas and camels, members of the camelid family, have antibodies that lack a light chain but are stable under biological conditions and have good binding affinities for antigens (Churchill et al. 2002). These antibodies consist of only a heavy-chain dimer (Fig. 31.4) (Hamers-Casterman et al. 1993, Woolven et al. 1999). Within the variable (VHH) domain are several amino acid substitutions in the framework regions. These substitutions stabilize the antibody fragment, increase its binding affinity to antigens, and increase its expression levels, while simultaneously inhibiting any possible interactions with a VL domain (Vu et al. 1997). This allows the easy expression of VHHs as recombinant fragments from bacterial systems. VHHs are smaller than scFvs, and thus have higher expression in bacterial systems. They do not contain a linker that sometimes causes aggregation in scFv constructs, and they are stable, remaining active in some cases for more than a year at 4°C (unpublished observations).

Enzyme-Linked Immunosorbent Assay (ELISA)

Enzyme-linked immunosorbent assay (ELISA) is the most common format used for immunodetection of pathogens. By this method, most pathogens have a detection limit of between 10^3 and 10^5 cfu/mL (de Boer and Beumer 1999). To achieve this detection limit often requires enrichment of the pathogens for at least 16–24 hours before the concentration of the pathogen is adequate for detection by ELISA (de Boer and Beumer 1999).

There are two main ELISA formats used to detect pathogens. In direct ELISA (Fig. 31.5A), the test sample is coated onto a well in a microtiter plate.

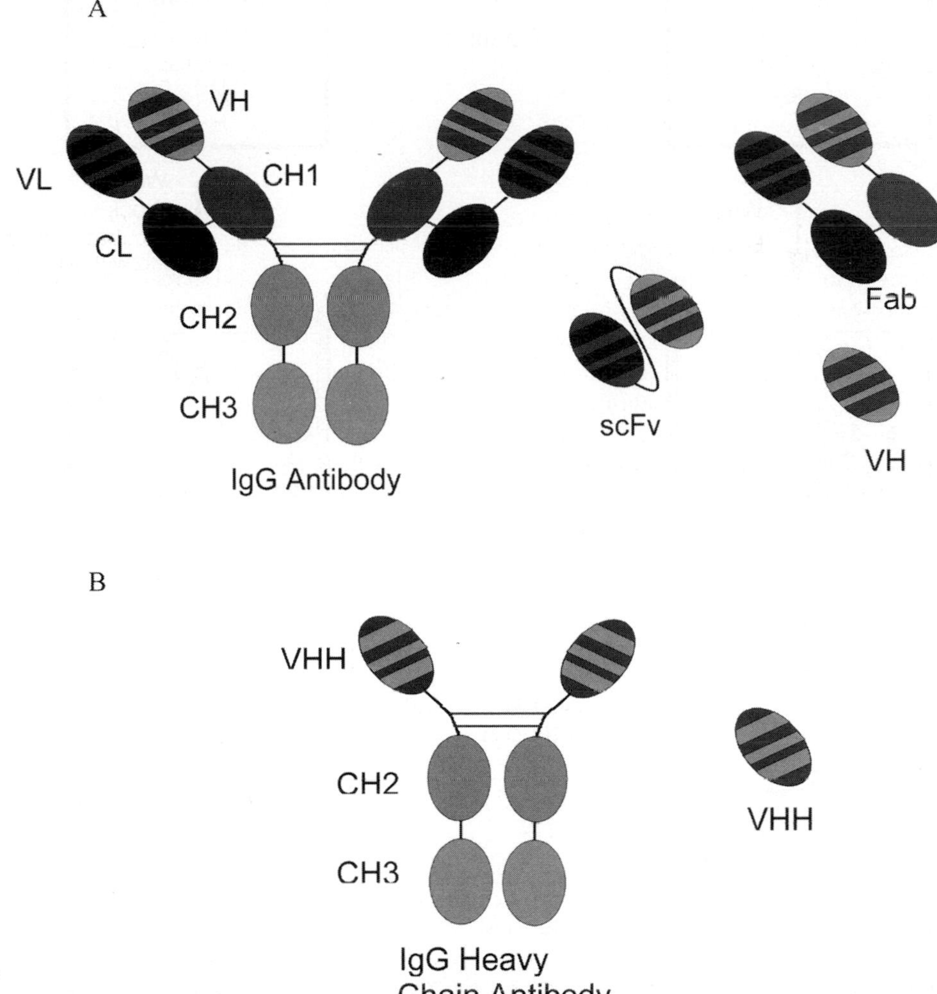

Figure 31.4. A. Conventional IgG antibody and various possible recombinant fragments derived from the antibody. Fab is composed of the first constant domain and the variable domains of both the heavy and light chains. scFv is composed of the variable light and heavy domains connected and stabilized by a flexible peptide linker. VH is derived from only the heavy-chain variable region. **B.** Heavy-chain antibody (HCAb) and its possible recombinant derivatives. The heavy-chain antibody is found only in the camelid family. VHH is derived from the variable region of the HCAb.

The plate is blocked and washed, and then an antibody (primary antibody) specific for the pathogen is added to the well. After a suitable incubation period, the plate is washed to remove unbound antibodies. A secondary antibody conjugated to an enzyme that converts a colorless substrate to a visible product is then added. The secondary antibody has specificity for the primary antibody, so if the primary antibody is bound to any antigen, the secondary antibody will bind to the primary antibody. Finally, the plate is washed again to remove any excess secondary antibody, and a substrate is added that turns color in the presence of the enzyme. The intensity of the color is proportional to the amount of pathogen present in the original sample (Hess et al. 1998). In one variation of this assay, the plate can first be coated with an antibody specific for the pathogen. This allows the pathogen in the sample to be concentrated on the

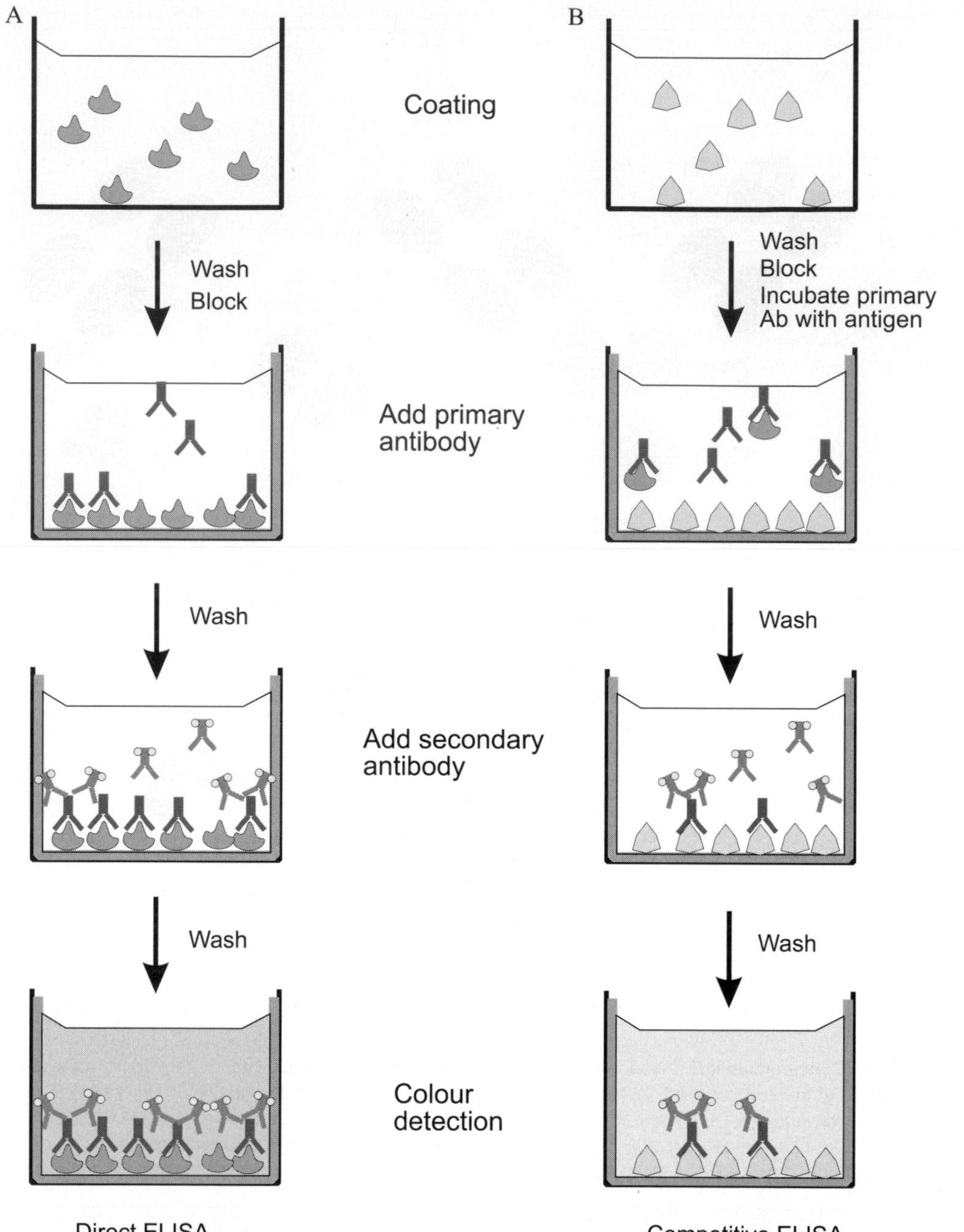

Figure 31.5. A. A schematic of direct ELISA. The antigen is used to coat a well of a microtiter plate. Excess antigen is washed away and the wells blocked to prevent nonspecific binding by other proteins and antibodies. Primary antibody is added to bind the antigen in the well. Excess unbound antibody is washed away, and the secondary antibody, conjugated to a colorimetric enzyme is added. After a final wash, the enzyme substrate is added, and the intensity of color produced is proportional to the quantity of antigen present. **B.** A schematic of competitive ELISA. This format is similar to the direct ELISA except that the well is coated with a known amount of standard antigen and the primary antibody is preincubated with the sample. The more antigen is present in the sample, the less primary antibody will be available to bind to the standard antigen. After adding the secondary antibody and substrate, the final absorbance can be compared to a standard curve to determine the amount of antigen present in the sample.

plate, allowing a more intense signal. This variation is known as a sandwich ELISA (Hess et al. 1998).

More commonly, a competitive ELISA (Fig. 31.5B) is used to detect the presence of pathogens, since this method is less susceptible to nonspecific background binding. In this format, a known amount of antigen is coated onto the plate. The sample to be tested is incubated with the primary antibody and then added to the blocked well. After washing, the secondary antibody is added to the well, and detection is performed as mentioned above. In this case, the more free antigen present in the test solution, the less signal will be detected after treatment with the secondary antibody since the majority of the primary antibody will have bound to the soluble antigen, and is consequently washed away. The amount of inhibition can be compared to an inhibition curve created with known amounts of antigen in the samples, thus allowing a quantitative determination of the amount of pathogen present in the test sample via interpolation using the standard curve (Hess et al. 1998).

Fluorescently Labeled Immunoassays

The ELISA assay can be made more sensitive using fluorescent labels on the antibodies. This type of labeling also decreases the time required for detection by eliminating the need for a colorimetric reaction as the final step of the ELISA. The labeling does, however, increase the cost of the assay.

Going back to the example of detecting *L. monocytogenes*, Sewell et al. (Sewell et al. 2003) used a modified version of an ELISA system with a fluorescent reporter dye. After 52 hours of culture enrichment, the authors carried out the ELFA (enzyme-linked fluorescent assay) procedure. The accuracy of the positive results was 97%, with positive samples requiring confirmation by plating on *Listeria*-selective media (Sewell et al. 2003). This final step, unfortunately, also adds more time to the procedure.

Similar to this approach, Dunbar et al. (2003) used microspheres with different spectral labels, coated with antibodies to each of *E. coli* O157:H7, *Salmonella typhimurium*, *C. jejuni*, and *L. monocytogenes,* respectively, to detect the pathogens after enrichment and amplification. The spheres were sorted by their spectral labels using a fluorescent bead sorter (Luminex Labmap system, Austin, Texas), and the pathogens were detected using a secondary antibody labeled with a fluorophore. The assay is able to detect between 2.5 and 500 organisms/mL depending on the species (Dunbar et al. 2003). This system, although expensive, allows multiple pathogens to be detected simultaneously.

Immunomagnetic Assays

Immunomagnetic assays make use of antibody specificity for a pathogen to concentrate the pathogen before other methods are used to amplify and identify the bacteria. Hudson et al. (2001) used the procedure to isolate *L. monocytogenes* directly from ham. In this procedure, the food was homogenized with some growth medium, the particulate matter removed, and after a number of washes, particles of bacterial size were pelletted and resuspended in a low volume. Commercial immunomagnetic beads coated with an anti-*Listeria* antibody were added to the solution and incubated to allow binding of the *L. monocytogenes* to the beads. The beads were trapped on a magnet, washed, and the DNA extracted for amplification of *L. monocytogenes*–specific genes by PCR. The immunomagnetic separation and concentration cut the detection time to about 1 day, at least for ham, but it is limited in terms of sensitivity, since the recovery of cells on the beads was only about 20% of those initially added (Hudson et al. 2001). Immunomagnetic separation on average allowed detection of 1–2 cfu/g food sample, but the results were somewhat variable in terms of sensitivity (detection limit from 0.1 cfu/g to greater than 5.7 cfu/g) (Hudson et al. 2001), making this method promising, but not ready to be used by the food industry until the efficiency of immunomagnetic isolation is improved.

Surface Plasmon Resonance (SPR)

Surface plasmon resonance (SPR) relies on an optical phenomenon to measure the binding kinetics between molecules (Fig. 31.6A). SPR can detect interactions between unmodified proteins and measure the kinetics in real time (Hashimoto 2000). The most popular equipment for this method is the BIAcore machine (biomolecular interaction assay), in which one of the proteins is coated on a disposable chip that is placed in the machine. Sample buffer is passed over the chip, and its flow is interrupted by the passage of discreet amounts of sample. The sensing mechanism of the machine measures

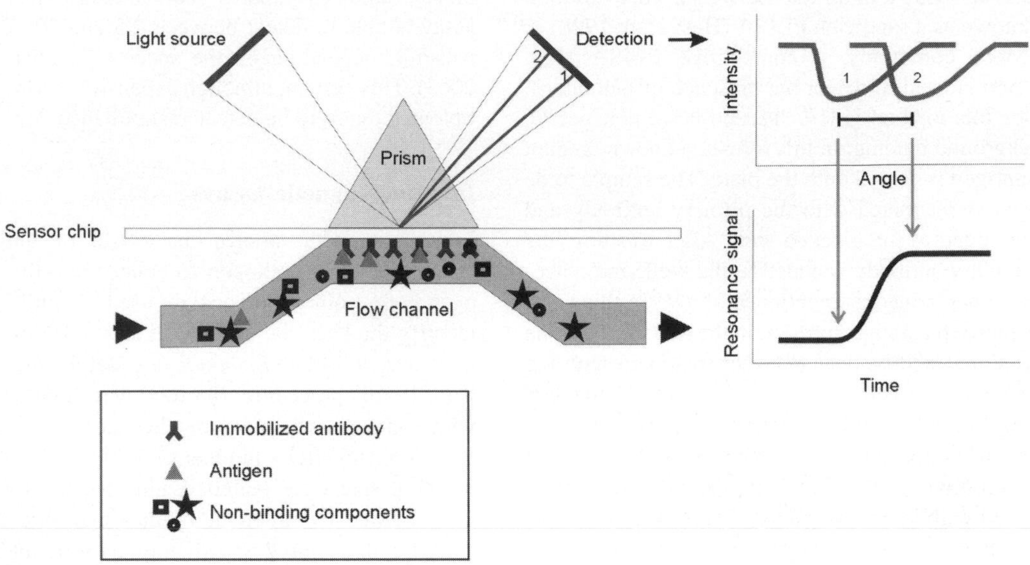

B

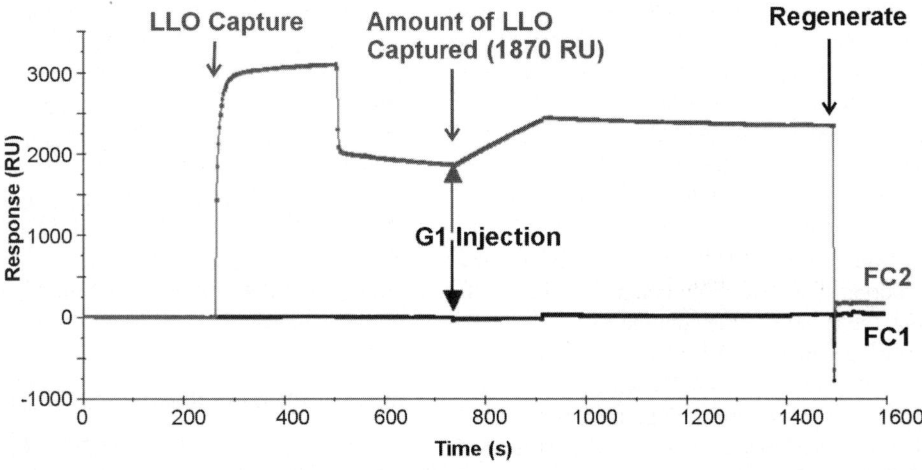

Figure 31.6. A. A description of surface plasmon resonance (SPR). Polarized light is focused on a gold-coated sensor chip. Energy transfer occurs at a critical angle, causing a dip in reflectance. When antigen binds to the antibody, it causes a change in mass, and thus a change in the angle of refraction. This change in angle is measured over time and expressed as resonance units on a sensogram. (Adapted from Hashimoto 2000, Nice and Catimel 1999.) **B.** A typical sensogram of an antigen-antibody interaction. The first peak represents the capture of the antigen, recombinant listeriolysin O (LLO) on the chip. After the flow of LLO stops, the resonance drops to a level representing only the LLO bound to the chip. Fractionated serum from a llama immunized with LLO is passed across the chip, starting at the G1 injection site. This fraction contains only heavy-chain antibodies from the llama. The increased number of resonance units in the sensogram after the injection is over represents the binding of the antibodies to the LLO. The drop in resonance units at the end of the sensogram represents the stripping of the LLO from the chip and the regeneration of the chip. (Sensogram courtesy of T. Hirama, National Research Council Canada.)

the angle of refraction of a beam of light over the coated chip. When the protein of interest binds to the antibody coated on the chip, the mass changes, resulting in a change of the angle of light refraction. The results are displayed in real time as the optical response, measured in resonance units (RU), plotted against time (Bamdad 1997, Nice and Catimel 1999). The number of RUs is proportional to the amount of protein binding to the coated chip (Bamdad 1997). The data generated can be used to calculate the antigen concentration and equilibrium dissociation binding constants.

Using SPR, each binding experiment takes only 5 to 10 minutes (Bamdad 1997). Once the conditions of the experiment are established, this offers considerable time saving over traditional methods of determining kinetic rates. However, determining the right conditions for running SPR experiments as well as regenerating the chip can take from days to months, and this represents a major limitation of the system if many different proteins are being analyzed.

SPR is good for measuring antibody affinity for toxins secreted from pathogens. The conditions for different antibodies against toxins are about the same, making optimization of the conditions less time consuming than may otherwise be the case. Samples containing the antigen can be passed over the antibody-coated chip and the change in resonance units compared to a standard curve in order to quantify the number of pathogens present. Figure 31.6B illustrates a typical SPR profile of the binding of llama heavy-chain antibodies to the *L. monocytogenes* toxin LLO.

Given the correct antigen, such as a secreted toxin, SPR can be a useful method for screening samples for the presence of specific pathogens. However, the cost of the instrument and the time required for optimizing sample purification severely limit the application of this technology as a routine screening method.

Latex Agglutination Assays

One simple assay that has been developed for the detection of foodborne pathogens is the latex agglutination assay. The assay makes use of latex bead–bound antibodies specific to the antigen of choice. Antibodies on each bead bind to the antigen. Each antigen can bind to more than one antibody bead, causing the beads to agglutinate.

Matar et al. (1997) used antibodies specific to the *L. monocytogenes* LLO toxin to detect the pathogen in foods and pure cultures. The toxin was used instead of cells because antibodies to cell surface proteins tend to be more genus specific. Additionally, LLO is a secreted protein, so that the supernatants of enrichment cultures can be used directly without further extraction. After culture enrichment following the USDA method, the latex agglutination assay was able to detect LLO at concentrations indicating contamination of the original food sample with between 0.3 and 220 cfu/g food sample (Matar et al. 1997). The agglutination test itself gives qualitative (such as positive or negative) results (Matar et al. 1997) rather than quantitative results.

Although this test still requires culture enrichment and then 48 hours to give results, it does have one major advantage over most of the molecular methods available, that is, no specialized equipment is required. Only the bare eye and a light source are needed to see the agglutination result, and there is no need for any expensive reagents, making the test suitable for use in the field or for use in third world countries (Matar et al. 1997).

CONCLUSIONS

Despite considerable efforts made to prevent contamination of food, new pathogens often emerge as a result of changing demography, food consumption habits, food technology, commerce, changes in water sources and environmental factors (Gugnani 1999), and microbial adaptation (Altekruse et al. 1997). Because of this, the food industry must have the ability to detect a number of different pathogens, at a cost that is affordable. They must also have the ability to easily incorporate new pathogen tests into regimes to ensure that any emerging pathogen is not overlooked.

For the food industry, the ideal pathogen detection test should have at least some of the following characteristics: the optimal test for pathogen detection should be simple to perform by nonspecialized personnel. It should be sensitive enough to detect low levels of pathogens, and yet specific for detection of the pathogenic species of interest. The test should be rapid, suitable for automation, and inexpensive (Ingianni et al. 2001). Technologies are becoming available that satisfy a number of these criteria, but none so far has been able to satisfy all, especially for determination of multiple pathogens.

As this article has illustrated, a number of technologies are available or being developed to detect foodborne pathogens. They promise to improve the efficiency, sensitivity and specificity of pathogen detection. However one cannot overstate the importance of preventing foodborne pathogens from entering the food chain in the first place. Innovations in food production such as hazard analysis critical control point (HACCP) programs play a great role in preventing the spread of the pathogens, as do the efforts of inspectors and consumers in reporting unsafe food handling practices. Also, technologies such as pasteurization and new technologies like postprocessing irradiation treatment also play or will play roles in protecting the safety of consumers.

GLOSSARY

AFLP—Amplified fragment length polymorphism.

cDNA—Copy DNA; created when reverse transcriptase is used to create a DNA copy of a transcribed mRNA.

cfu—Colony forming unit.

DNTP—Deoxynucleotide triphosphate; four different dNTPs form the building blocks of DNA.

ELISA—Enzyme-linked immunosorbent assay.

nterotoxin—Protein toxins active in the intestine or on intestinal cells that causes diarrhea.

Gram-negative bacteria—Gram staining is the first step in the classification of bacteria. Gram-negative bacteria stain pink because an outer membrane prevents the entrance of the crystal violet dye. The pink color results from secondary staining with safranin dye.

Gram-positive bacteria—Gram staining is the first step in the classification of bacteria. Gram-positive bacteria stain purple in the test because they lack an outer membrane and are colored by the crystal violet dye.

NASBA—Nucleic acid sequence–based amplification.

NTP—Nucleotide triphosphate; four different NTPs form the building blocks of RNA.

Oligonucleotide—A short DNA sequence, generally complementary to a gene of interest.

PCR—Polymerase chain reaction.

Primary antibody—An antibody with specificity for the antigen.

Primer—DNA oligonucleotide complementary to a target DNA; used to start the next strand in DNA or RNA synthesis in amplification or in transcription of RNA.

Probe—DNA oligonucleotide labeled with a tag (e.g., fluorescein) for detection of an amplified gene.

Psychrotropic—Able to grow and replicate at low temperatures.

RAPD—Randomly amplified polymorphic DNA.

RFLP—Random fragment length polymorphism.

RT-PCR—Reverse transcription PCR.

Secondary antibody—An antibody with species-level specificity for the primary antibody. Secondary antibodies are often conjugated to a colorimetric substrate, allowing detection of the antibody-antigen interaction.

REFERENCES

Anonymous. 1992. *Clostridium botulinum.* U.S. Food and Drug Administration.

Anonymous. 2000. Waterborne outbreak of gastroenteritis associated with a contaminated municipal water supply, Walkerton, Ontario, May–June 2000. Can Comm Dis Rep 26(20): 170–173.

Aarts HJM, Hakemulder LE, Van Hoef AMA. 1999. Genomic typing of *Listeria monocytogenes* strains by automated laser fluorescence analysis of amplified fragment length polymorphism fingerprint patterns. Int J Food Microbiol 49:95–102.

Acheson DWK, Keusch GT. 1999. The family of Shiga toxins. In: JE Alouf, JH Freer, editors. The Comprehensive Sourcebook of Bacterial Protein Toxins, 2nd edition. Toronto: Academic Press. Pp. 229–242.

Alouf JE. 1999. Introduction to the family of the structurally related cholesterol-binding cytolysins ("sulfydryl-activated" toxins). In: JE Alouf, JH Freer, editors. The Comprehensive Sourcebook of Bacterial Protein Toxins. Toronto: Academic Press. Pp. 443–456.

Alouf JE, Jolivet-Reynaud C. 1981. Purification and characterization of *Clostridium perfringens* delta-toxin. Infect Immun 31(2): 537–546.

Altekruse SF, Cohen ML, Swerdlow DL. 1997. Emerging foodborne diseases. Emerg Infect Dis 3(3): 285–293.

Aznar R, Alarcón B. 2002. On the specificity of PCR detection of *Listeria monocytogenes* in food: A comparison of published primers. Sys Appl Microbiol 25:109–119.

Balaban N, Rasooly A. 2000. Staphylococcal enterotoxins. Int J Food Microbiol 61:1–10.

Bamdad C. 1997. Surface plasmon resonance for measurements of biological interest. In: FM Ausubel, R Brent, RE Kingston, DD Moore, JG Seidman, JA Smith, K Struhl, editors. Current Protocols in Molecular Biology. Canada: John Wiley and Sons Inc. Pp. 20.4.1–20.4.12.

Bannam T, Goldfine H. 1998. Mutagenesis of active-site histidines of *Listeria monocytogenes* phosphatidylinositol-specific phospholipase C: Effects on enzyme activity and biological function. Infect Immun 67(1): 182–186.

Beauregard KE, Lee K-D, Collier RJ, Swanson JA. 1997. pH-dependent perforation of macrophage phagosomes by listeriolysin O from *Listeria monocytogenes*. J Exp Med 186(7): 1159–1163.

Beumer RR, Hazeleger WC. 2003. *Listeria monocytogenes*: Diagnostic problems. FEMS Immunol Med Microbiol 35:191–197.

Bhagwat AA. 2003. Simultaneous detection of *Escherichia coli* O157:H7, *Listeria monocytogenes* and *Salmonella* strains by real-time PCR. Int J Food Microbiol 84:217–224.

Bjerketorp J, Nilsson M, Ljungh Å, Flock J-I, Jacobsson K, Frykberg L. 2002. A novel von Willembrand factor binding protein expressed by *Staphylococcus aureus*. Microbiol 148:2037–2044.

Blais BW, Turner G, Sooknanan R, Malek LT. 1997. A nucleic acid sequence-based amplification system for detection of *Listeria monocytogenes hlyA* sequences. Appl Envir Biol 63(1): 310–313.

Blears MJ, de Grandis SA, Lee H, Trevors JT. 1998. Amplified fragment length polymorphism (AFLP): A review of the procedure and its applications. J Ind Microbiol 21:99–114.

Boyd RF. 1988. General Microbiology. St. Louis: Times Mirror/Mosby College Publishing.

Breitling F, Dübel S. 1999. Recombinant Antibodies. New York: John Wiley and Sons, Inc. 154 pp.

Bryant AE. 2003. Biology and pathogenesis of thrombosis and procoagulant activity in invasive infections caused by group A Streptocicci and *Clostridium perfringens*. Clin Microbiol Rev 16(3): 451–462.

Brynstad S, Granum PE. 2002. *Clostridium perfringens* and foodborne infections. Int J Food Microbiol 74:195–202.

Cabanes D, Dehoux P, Dussurget O, Frangeul L, Cossart P. 2002. Surface proteins and the pathogenic potential of *Listeria monocytogenes*. Trends Microbiol 10(5): 238–265.

Call DR, Borucki MK, Loge FJ. 2003. Detection of bacterial pathogens in environmental samples using DNA microarrays. J Micriobiol Meth 53:235–243.

Carnoy C, Simonet M. 1999. *Yesinia pseudotuberculosis* superantigenic toxins. In: JE Alouf, JH Freer, editors. The Comprehensive Sourcebook of Bacterial Protein Toxins, 2nd edition. Toronto: Academic Press. Pp. 611–622.

Choi WS, Hong C-H. 2003. Rapid enumeration of *Listeria monocytogenes* in milk using competitive PCR. Int J Food Microbiol 84:79–85.

Churchill RLT, Sheedy C, Yau KYF, Hall JC. 2002. Evolution of antibodies for environmental monitoring: From mice to plants. Anal Chim Acta 468:185–197.

Cocolin L, Rantsiou K, Iacumin L, Cantoni C, Comi G. 2002. Direct identification in food samples of *Listeria* spp. and *Listeria monocytogenes* by molecular methods. Appl Envir Biol 68(12): 6273–6282.

Cook N. 2003. The use of NASBA for the detection of microbial pathogens in food and environmental samples. J Micriobiol Meth 53:165–174.

Cornelis GR, Boland A, Boyd AP, Geuijen C, Iriarte M, Neyt C, Sory M-P, Stanier I. 1998. The virulence plasmid of *Yesinia*, an antihost genome. Microbiol Mol Biol Rev 62(4): 1315–1352.

Cox T, Frazier C, Tuttle J, Flood S, Yago L, Yamashiro CT, Behari R, Paszko C, Cano RJ. 1998. Rapid detection of *Listeria monocytogenes* in dairy samples utilizing a PCR-based fluorogenic 5' nuclease assay. J Ind Microbiol 21:167–174.

Currie CG, McCallum K, Poxton IR. 2001. Mucosal and systemic antibody responses to the lipopolysaccharide of *Escherichia coli* O157:H7 in health and disease. J Med Microbiol 50:345–354.

de Boer E, Beumer RR. 1999. Methodology for detection and typing of foodborne microorganisms. Int J Food Microbiol 50:119–130.

De Greeff A, van Alphen L, Smith HE. 2000. Selection of recombinant antibodies specific for pathogenic *Streptococcus suis* by subtractive phage display. Infect Immun 68(7): 3949–3955.

Deschamps RJA, Hall JC. 1990. Polyclonal and monoclonal immunoassays for picloram detection. In: JM Van Emon, RO Mumma, editors. Immunochemical Methods for Environmental Analysis. Miami Beach, Florida: American Chemical Society. Pp. 66–78.

Deschamps RJA, Hall JC, McDermott MR. 1990. Polyclonal and monoclonal enzyme immunoassays for picloram detection in water, soil, plants and urine. J Agric Food Chem 38:1881–1886.

Dobrindt U, Hacker J. 1999. Plasmids, phages and the pathogenicity islands: Lessons on the evolution of bacterial toxins. In: JE Alouf, JH Freer, editors. The Comprehensive Sourcebook of Bacterial Protein Toxins, 2nd edition. Toronto: Academic Press. Pp. 3–23.

Donnelly CW. 2001. *Listeria monocytogenes*: A continuing challenge. Nutr Rev 59(6): 183–194.

Dunbar SA, Vander Zee CA, Oliver KG, Karem KL, Jacobson JW. 2003. Quantitative, multiplexed detection of bacterial pathogens: DNA and protein applications of the Luminex LabMAP™ system. J Micriobiol Meth 53:245–252.

Eyigor A, Dawson KA, Langlois BE, Pickett CL. 1999. Detection of cytolethal distending toxin activity and *cdt* genes in *Campylobacter* spp. isolated from chicken carcasses. Appl Envir Biol 65(4): 1501–1505.

Feil SC, Rossjohn J, Rohde K, Tweten RK, Parker MW. 1996. Crystalization and preliminary X-ray analysis of a thiol-activated cytolysin. FEBS Lett 397:290–292.

Finney M. 2000. Pulsed-field gel electrophoresis. In: FM Ausubel, R Brent, RE Kingston, DD Moore, JG Seidman, JA Smith, K Struhl, editors. Current Protocols in Molecular Biology. Canada: John Wiley and Sons, Inc. Pp. 2.5B.1–2.5B.12.

Fratamico PM, Strobaugh TP. 1998. Simultaneous detection of *Salmonella* spp. and *Escherichia coli* O157:H7 by multiplex PCR. J Ind Microbiol 21:92–98.

Fredriksson-Ahomaa M, Korkeala H. 2003. Low occurrence of pathogenic *Yersinia enterocolitca* in clinical, food, and environmental samples: A methodological problem. Clin Microbiol Rev 16(2): 220–229.

Granum PE, Brynstad S. 1999. Bacterial toxins as food poisons. In: JE Alouf, JH Freer, editors. The Comprehensive Sourcebook of Bacterial Protein Toxins, 2nd edition. Toronto: Academic Press. Pp. 669–690.

Grosdent N, Maridonneau-Parini I, Sory P-P, Cornelis GR. 2002. Role of Yops and adhesins in resistance of *Yersinia enterocolitica* to phagocytosis. Infect Immun 70(8): 4165–4176.

Gugnani HC. 1999. Some emerging food and water borne pathogens. J Commun Dis 31(2): 65–72.

Gurtler V, Mayall BC. 2001. Genomic approaches to typing, taxonomy and evolution of bacterial isolates. Int J Syst Evol Microbiol 51:3–16.

Hall JC, Deschamps RJA, Krieg KK. 1989a. Immunoassays for the detection of 2,4-D and picloram in river water and urine. J Agric Food Chem 37:981–984.

Hall JC, Deschamps RJA, McDermott MR. 1989b. Immunoassays for detecting pesticides in the environment. Highlights 12(2): 19–23.

Hamers-Casterman C, Atarhouch T, Muyldermans S, Robinson G, Hamers C, Bjyana Songa E, Bendahman N, Hamers R. 1993. Naturally occuring antibodies devoid of light chains. Nature 363:446–448.

Hashimoto S. 2000. Principles of BIACORE. In: K Nagata, H Handa, editors. Real-Time Analysis of Biomolecular Intereactions: Applications of BIACORE. Tokyo: Springer-Verlag. Pp. 23–32.

Hayes PS, Graves LS, Swaminthan B, Ajello GW, Malcolm GB, Weaver RE, Ransom R, Deaver K, Plikaytis BD, Schuchat A et al. 1992. Comparison of three selective enrichment methods for the isolation of *Listeria monocytogenes* from naturally contaminated foods. J Food Prot 55(12): 952–959.

Hayhurst A, Happe S, Mabry R, Koch Z, Iverson BL, Georgiou G. 2003. Isolation and expression of recombinant antibody fragments to the biological warfare pathogen *Brucella melitensis*. J Immun Meth 275:185–196.

Hess J, Miko D, Catic A, Lehmensiek V, Russell DG, Kaufmann SHE. 1998. *Mycobacterium bovis* bacille Calmette-Guérin strains secreting listeriolysin of *Listeria monocytogenes*. Proc Natl Acad Sci USA 95:5299–5304.

Hoffman AD, Wedman M. 2001. Comparative evaluation of culture- and BAX polymerase chain reaction-based detection methods for *Listeria* spp. and *Listeria monocytogenes* in environmental and raw fish samples. J Food Prot 64(10): 1521–1526.

Hough AJ, Harbison S-A, Savill MG, Melton LD, Fletcher G. 2002. Rapid enumeration of *Listeria monocytogenes* in artificially contaminated cabbage using real-time polymerase chain reaction. J Food Prot 65(8): 1329–1332.

Hudson JA, Lake RJ, Savill MG, Scholes P, McCormick RE. 2001. Rapid detection of *Listeria monocytogenes* in ham samples using immunomagnetic separation followed by polymerase chain reaction. J Appl Microbiol 90:614–621.

Ingianni A, Floris M, Palomba P, Madeddu MA, Quartuccio M, Pompei R. 2001. Rapid detection of *Listeria monocytogenes* in foods, by a combination of PCR and DNA probe. Mol Cell Probes 15:275–280.

Jacobs T, Darji A, Weiss S, Chakraborty T. 1999. Listeriolysin, the thiol-activated haemolysin of *Listeria monocytogenes*. In: JE Alouf, editor. The Comprehensive Sourcebook of Bacterial Protein Toxins. Toronto: Academic Press. Pp. 511–521.

Jinneman KC, Hunt JM, Eklund CA, Wernberg JS, Sado P, Johnson JM, Richter RS, Torres ST, Ayotte E, Eliasberg SJ, et al. 2003. Evaluation and interlaboratory validation of a selective agar for phosphatidylinositol-specific phospholipase C activity using a chromogenic substrate to detect *Listeria monocytogenes* from foods. J Food Prot 66(3): 441–445.

Jolivet-Reynaud C, Hauttecouer B, Alouf JE. 1993. Interaction of *Clostridium perfringens* delta toxin with erythrocyte and liposome membranes and relation with the specific binding to the ganglioside GM2. Toxicon 27(10): 1113–1116.

Jung YS, Frank JF, Brackett RE, Chen J. 2003. Polymerase chain reaction detection of *Listeria monocytogenes* of frankfurters using oligonucleotide primers targetting the genes encoding internalin AB. J Food Prot 66(2): 237–241.

Kaper JB, Gansheroff LJ, Wachtel MR, O'Brien AD. 1998. Intimin-mediated adherence of shiga toxin-producing *Escherichia coli* and attaching-and-effacing pathogens. In: JB Kaper, AD O'Brien, editors. *Escherichia coli* O157:H7 and Other Shiga Toxin-Producing *E coli* strains. Washington, D.C.: ASM Press. Pp. 148–156.

Kathariou S. 2002. *Listeria monocytogenes* virulence and pathogenicity, a food safety perspective. J Food Prot 65(11): 1811–1829.

Kim J, Nietfield J, Ju J, Wise J, Fegan N, Desmarchelier P, Benson AK. 2001. Ancestral divergence, genome diversification, and phylogeographic variation in subpopulations of sorbitol-negative, β-glucuronidase-negative enterohemorrhagic *Escherichia coli* O157. J Bacteriol 183(23): 6885–6897.

Klein PG, Juneja VK. 1997. Sensitive detection of viable *Listeria monocytogenes* by reverse transcription-PCR. Appl Envir Biol 63(11): 4441–4448.

Köhler G, Milstein C. 1975. Continuous cultures of fused cells secreting antibody of predefined specificity. Nature 256:495–497.

Koo K, Jaykus L-A. 2003. Detection of *Listeria monocytogenes* from a model food by fluorescence resonance energy transfer-based PCR with an asymetric fluorogenic probe set. Appl Envir Biol 69(2): 1082–1088.

Kopecko DJ, Hu L, Zaal JM. 2001. *Campylobacter jejuni*—Microtubule-dependent invasion. Trends Microbiol 9(8): 389–396.

Lawrence LM, Gilmour A. 1995. Characterization of *Listeria monocytogenes* isolated from poultry products and from the poultry-processing environment by random amplification of polymorphic DNA and multilocus enzyme electgrophoresis. Appl Envir Biol 61(6): 2139–2144.

Lehner A, Loncarevic S, Wagner M, Kreike J, Brandl E. 1999. A rapid differentiation of *Listeria monocytogenes* by use of PCR-SSCP in the listeriolysin O (*hlyA*) locus. J Micriobiol Meth 34:165–171.

Liebana E. 2002. Molecular tools for epidemiological investigations of *S. enterica* subspecies *enterica* infections. Res Vet Med 72:169–175.

Lingwood CA, Mylvaganam M, Arab S, Khine AA, Magnusson G, Grinstein S, Niholm P-G. 1998. Shiga toxin (verocytoxin) binding to its receptor glycolipid. In: JB Kaper, AD O'Brien, editors. *Escherichia coli* O157:H7 and Other Shiga Toxin-Producing *E coli* strains. Washington, D.C.: ASM Press. Pp. 129–139.

Loncarevic S, Tham W, Danielson-Tham ML. 1996. The clones of *Listeria monocytogenes* detected in food depend on the method used. Lett Appl Microbiol 22(5): 381–384.

Ludwig A, Goebel W. 1999. The family of the multigenic encoded RTX toxins. In: JE Alouf, JH Freer, editors. The Comprehensive Sourcebook of Bacterial Protein Toxins, 2nd edition. Toronto: Academic Press. Pp. 330–348.

Lunge VR, Miller BJ, Livak KJ, Batt CA. 2002. Factors affecting the performance of 5' nuclease PCR assays for *Listeria monocytogenes* detection. J Micriobiol Meth 51:361–368.

MacKay B. 2002. Walkerton, 2 years later: "Memory fades very quickly". Can Med Assoc J 166(10): 1326.

Matar GM, Hayes PS, Ribb WF, Swaminthan B. 1997. Listeriolysin O-based latex agglutination test for the rapid detection of *Listeria monocytogenes* in foods. J Food Prot 60(9): 1038–1040.

Mead PS, Slusker L, Dietz V, McCaig LF, Bresee JS, Shapiro C, Griffin PM, Tauxe RV. 1999. Food-related illness and death in the United States. Emerg Infect Dis 5(5): 607–624.

Melton-Celsa AR, O'Brien AD. 1998. Structure, biology, and relative toxicity of shiga toxin family mambers for cells and animals. In: JB Kaper, AD O'Brien, editors. *Escherichia coli* O157:H7 and Other Shiga Toxin-Producing *E coli* strains. Washington, D.C.: ASM Press. Pp. 121–128.

Menestrina G, Serra MD, Prévost G. 2001. Mode of action of β-barrel pore-forming toxins of the staphylococcal α-hemolysin family. Toxicon 39:1661–1672.

Meng J, Doyle MP. 1997. Emerging issues in microbiological food safety. Annu Rev Nutr 17:255–275.

———. 2002. Introduction: microbiological food safety. Microb Infect 4:395-397.

Min J, Baeumner AJ. 2002. Highly sensitive and specific detection of viable *Escherichia coli* in drinking water. Anal Biochem 303:186–193.

Monday SR, Bohach GA. 1999. Properties of *Staphylococcus aureus* enterotoxins and toxic shock syndrome toxin-1. In: JE Alouf, JH Freer, editors. The

Comprehensive Sourcebook of Bacterial Protein Toxins, 2nd edition. Toronto: Academic Press. Pp. 589–610.

Moore MA, Datta AR. 1994. DNA fingerprinting of *Listeria monocytogenes* strains by pulsed-field gel electrophoresis. Food Microbiol 11:31–38.

Nagahama M, Hayashi S, Morimitsu S, Sakurai J. 2003. Biological activities and pore formation of *Clostridium perfringens* beta toxin in HL 60 cells. J Biol Chem 278(38): 36934–36941.

Neill MA. 1998. Treatment of disease due to shiga toxin-associated *Escherichia coli*: Infectious disease management. In: JB Kaper, AD O'Brien, editors. *Escherichia coli* O157:H7 and Other Shiga Toxin-Producing *E coli* Strains. Washington, D.C.: ASM Press. Pp. 357–363.

Neyt C, Iriarte M, Thi VH, Cornelis GR. 1997. Virulence and arsenic resistance in Yersiniae. J Bacteriol 179(3): 612–619.

Nice EC, Catimel B. 1999. Instrumental biosensors: New perspectives for analysis of biomolecular interactions. BioEssays 21:339–352.

O'Brien AD, Kaper JB. 1998. Shiga toxin-producing *Escherichia coli*: yesterday, today and tomorrow. In: AD O'Brien, JB Kaper, editors. *Escherichia coli* O157:H7 and Other Shiga Toxin-Producing *E coli* Strains. Washington, D.C.: ASM Press. Pp. 1–12.

Odumeru JA, Steele M, Fruhner L, Larkin C, Jiang J, Mann E, McNab WB. 1999. Evaluation of the accuracy and repeatability of identification of food-borne pathogens by automated bacterial identification systems. J Clin Microbiol 37(4): 944–949.

Ootsubo M, Shimizu T, Tanaka R, Sawabe T, Tajima K, Ezura Y. 2003. Seven-hour fluorescence *in situ* hybridization technique for enumeration of Enterobacteriaceae in food and environmental water sample. J Appl Microbiol 95: 1182–1190.

Osek J. 2003. Development of a multiplex PCR approach for the identification of Shiga toxin-producing *Escherichia coli* strains and their major virulence factors. J Appl Microbiol 95:1217–1225.

Pangallo D, Kaclíková E, Kuchta T, Drahovská H. 2001. Detection of *Listeria monocytogenes* by polymerase chain reaction oriented to the *inlB* gene. Microbiologica 24:333–339.

Park CH, Martin EA, White EL. 1998. Isolation of a nonpathogenic strain of *Citrobacter sedlakii* which expresses *Escherichia* O157 antigen. J Clin Microbiol 36(5): 1408–1409.

Park SF. 2002. The physiology of *Campylobacter* species and its relevance to their role as foodborne pathogens. Int J Food Microbiol 74:177–188.

Perna NT, Plunkett III G, Burland V, Mau B, Glasner JD, Rose DJ, Mayhew GF, Evans PS, Gregor J, Kirkpatrick HA, et al. 2001. Genome sequence of enterohemorrhagic *Escherichia coli* O157:H7. Nature 409:529–533.

Rocourt J. 1994. *Listeria monocytogenes*: The state of the science. Dairy Food Envir Sanit 14(2): 70–82.

Rossjohn J, Tweten RK, Rood JI, Parker MW. 1999. Perfringolysin O. In: JE Alouf, editor. The Comprehensive Sourcebook of Bacterial Protein Toxins. Toronto: Academic Press. Pp. 496–510.

Ryser ET, Arimi SM, Bunduki MM-C, Donnelly CW. 1996. Recovery of different *Listeria* ribotypes from naturally contaminated, raw refrigerated meat and poultry products with two primary enrichment media. Appl Envir Biol 62(5): 1781–1787.

Savelkuoh PHM, Aarts HJM, de Haas J, Dijkshoorn L, Duim B, Osten M, Rademaker JLW, Schouls L, Lenstra JA. 1999. Amplified-fragment length polymorphism analysis: The state of the art. J Clin Microbiol 37(10): 3083–3091.

Saxbe Jr. WB. 1972. *Listeria monocytogenes* and Queen Anne. Pediatrics 49(1): 97–101.

Schleiss MR, Bourne N, Bravo FJ, Jensen NJ, Bernstein DI. 2003. Quantitative-competitive PCR monitoring of viral load following experimental guinea pig cytomegalovirus infection. J Virol Meth 108:103–110.

Schlundt J. 2001. Emerging food-borne pathogens. Biomed Envir Sci 14:44–52.

Schmid M, Walchor M, Bubert A, Wagner M, Wagner M, Schleifer K-H. 2003. Nucleic acid-based, cultivation-independent detection of *Listeria* spp. and genotypes of *Listeria monocytogenes*. FEMS Immunol Med Microbiol 35:215–225.

Sekino-Suzuki N, Nakamura M, Mitsui K-i, Ohno-Iwashita Y. 1996. Contribution of individual tryptophan residues to the structure and activity of q-toxin (perfringolysin O), a cholesterol binding protein. Eur J Biochem 241:941–947.

Sewell AM, Warburton DW, Boville A, Daley EF, Mullen K. 2003. The development of an efficient and rapid enzyme linked fluorescent assay method for the detection of *Listeria* spp. from foods. Int J Food Microbiol 81:123–129.

Shearer AEH, Strapp CM, Joerger RD. 2001. Evaluation of a polymerase chain reaction-based system for detection of *Salmonella enteritidis*, *Escherichia coli* O157:H7, *Listeria* spp. and *Listeria monocytogenes* on fresh fruits and vegetables. J Food Prot 64(6): 788–795.

Smith A, Nelson RJ. 1999. Capillary electrophoresis of DNA. In: FM Ausubel, R Brent, RE Kingston, DD

Moore, JG Seidman, JA Smith, K Struhl, editors. Current Protocols in Molecular Biology. Canada: John Wiley and Sons, Inc. Pp. 2.8.1–2.8.17.

Snyder A, Marquis H. 2003. Restricted translocation across the cell wall regulates secretion of the broad-range phospholipase C of *Listeria monocytogenes*. J Bacteriol 185(20): 5953–5958.

Solomon EB, Yaron S, Matthews KR. 2002. Transmission of *Escherichia coli* O157:H7 from contaminated manure and irrigation water to lettuce plant tissue and its subsequent internalization. Appl Envir Biol 68(1): 397–400.

Stachowiak R, Bielecki J. 2001. Contribution of hemolysin and phospholipase activity to cytolytic properties and viability of *Listeria monocytogenes*. Acta Microbiol Pol 50(3–4): 243–250.

Swaminthan B, Barrett TJ, Hunter SB, Tauxe RV, Force CPT. 2001. PulseNet: The molecular subtyping network for foodborne bacterial disease surveillance, United States. Emerg Infect Dis 7(3): 382–389.

Takeda T, Yoshino K-I, Ramamurthy T, Nair GB. 1999. Heat-stable enterotoxins for *Vibrio* and *Yersinia* species. In: JE Alouf, JH Freer, editors. The Comprehensive Sourcebook of Bacterial Protein Toxins, 2nd edition. Toronto: Academic Press. Pp. 545–556.

Tauxe RV. 1998. Public health perspective on immunoprophylactic strategies for *Escherichia coli* O157:H7: who or what would we immunize? In: JB Kaper, AD O'Brien, editors. *Escherichia coli* O157:H7 and Other Shiga Toxin-Producing *E coli* Strains. Washington, D.C.: ASM Press. Pp. 445–452.

———. 2002. Emerging foodborne pathogens. Int J Food Microbiol 78:31–41.

Tesh VL. 1998. Cytokine response to shiga toxin. In: JB Kaper, AD O'Brien, editors. *Escherichia coli* O157:H7 and Other Shiga Toxin-Producing *E coli* Strains. Washington, D.C.: ASM Press. Pp. 226–235.

Titball RW. 1999. Membrane-damaging and cytotoxic phospholipases. In: JE Alouf, JH Freer, editors. The Comprehensive Sourcebook of Bacterial Protein Toxins, 2nd edition. Toronto: Academic Press. Pp. 310–329.

Turnbough Jr. CL. 2003. Discovery of phage display peptide ligands for species-specific detection of *Bacillus* spores. J Micriobiol Meth 53:263–271.

Uyttendaele M, Debvrere J, Lindqvist R. 1999. Evaluation of buoyant density centrifugation as a sample preparation method for NASBA-ELGA detection of *Campylobacter jejuni* in foods. Food Microbiol 16:575–582.

Uyttendaele M, Schukkink R, Debvrere J. 1995a. Detection of *Campylobacter jejuni* added to foods by using a combined selective enrichment and nucleic acid sequence-based amplification (NASBA). Appl Envir Biol 61(4): 1341–1347.

Uyttendaele M, Schukkink R, van Gemen B, Debvrere J. 1995b. Development of NASBA, a nucleic acid amplification system, for identification of *Listeria monocytogenes* and comparison to ELISA and a modified FDA method. Int J Food Microbiol 27:77–89.

Volokhov D, Rasooly A, Chumankov K, Chizhikov V. 2002. Identification of *Listeria* species by microarray-based assay. J Clin Microbiol 40(12): 4720–2728.

Vu KB, Ghahroudi MA, Wyns L, Muyldermans S. 1997. Comparison of llama VH sequences from conventional and heavy chain antibodies. Mol Immunol 34(16–17): 1121–1131.

Wagner M, Schmid M, Juretschko S, Trebesius K-H, Bubert A, Goebel W, Schleifer K-H. 1998. In situ detection of a virulence factor mRNA and 16S rRNA in *Listeria monocytogenes*. FEMS Microbiol Lett 160:159–168.

Wan J, King K, Forsyth S, Coventry MJ. 2003. Detection of *Listeria monocytogenes* in salmon using Probelia polymerase chain reaction system. J Food Prot 66(3): 436–440.

Weidmann M, Bruce JL, Keating C, Johnson AE, McDonough PL, Batt CA. 1997. Ribotypes and virulence gene polymorphisms suggest three distinct *Listeria monocytogenes* lineages with differences in pathogenic potential. Infect Immun 65(7): 2707–2716.

Whitehouse CA, Balbo PB, Pesci EC, Cottle DL, Mirabito PM, Pickett CL. 1998. *Campulobacter jejuni* cytolethal distending toxin causes a G_2-phase cell cycle block. Infect Immun 66(5): 1934–1940.

Williams RC, Isaacs S, Decou ML, Richardson EA, Buffett MC, Slinger RW, Brodski MH, Ciebin BW, Ellis A, Hocken J et al. 2000. Illness outbreak associated with *Escherichia coli* O157:H7 in Genoa salami. Can Med Assoc J 162(10): 1409–1413.

Wong CS, Jelacic S, Habeeb RL, Watkins SL, Tarr PI. 2000. The risk of the hemolytic-uremic syndrome after antibiotic treatment of *Escherichia coli* O157:H7 infections. N Eng J Med 342:1930–1936.

Woolven BP, Frenkin LGJ, van der Logt P, Nicholls PJ. 1999. The structure of the llama heavy chain constant genes reveals a mechanism for heavy-chain antibody formation. Immunogenet 50:98–101.

Woteki CE, Kineman BD. 2003. Challenges and approaches to reducing foodborne illness. Annu Rev Nutr 23:314–344.

Yau KYF, Lee H, Hall JC. 2003. Emerging trends in the synthesis and improvement of hapten-specific recombinant antibodies. Biotech Adv 21:599–637.

Yoshino K-I, Takao T, Huang X, Murata H, Nakao H, Tae T, Shimonishi Y. 1995. Characterization of a highly toxic, large molecular size heat-stable enterotoxin produced by a clinical isolate of *Yersinia enterocolitica*. FEBS Lett 362:319–322.

Zhang S, Kingsley RA, Santos RL, Andrews-Polymenis H, Raffatellu M, Figueiredo J, Nunes J, Ysolis RM, Adams LG, Bäumler AJ. 2003. Molecular pathogenesis of *Salmonella enterica* serotype typhimurium-induced diarrhea. Infect Immun 71(1): 1–12.

Zhang X, McDaniel AD, Wolf LE, Keusch GT, Waldor MK, Acheson DWK. 2000. Quinolone antibiotics induce shiga toxin-encoding bacteriophages, toxin production and death in mice. J Infect Dis 181:664–670.

Zhou B, Wirsching P, Janda KD. 2002. Human antibodies against spores of the genus *Bacillus*: A model study for the detection and protection against anthrax and the bioterrorist threat. Proc Natl Acad Sci USA 99(8): 5241–5246.

Index